3rd Edition

Introduction to ORGANIC CHEMISTRY

William H. Brown
Beloit College

Thomas Poon
Claremont McKenna College
Scripps College
Pitzer College

WILEY

John Wiley & Sons, Inc.

Project Editor: *Jennifer Yee*
Acquistions Editor: *Kevin Molloy*
Senior Media Editor: *Martin Batey*
Marketing Manager: *Amanda Wygal*
Production Editor: *Sandra Dumas*
Senior Designer: *Kevin Murphy*
Cover and Interior Design: *Nancy Field*
Cover Photo: *Ray Coleman/Photo Researchers, Inc.*
Senior Photo Editor: *Lisa Gee*
Illustration Editor: *Sandra Rigby*
Editorial Program Assistant: *Catherine Donovan*
Production Management Services: *Preparé*

This book was typeset in 10/12 New Baskerville Roman by Preparé and printed and bound
by Von Hoffmann Corporation. The cover was printed by Von Hoffmann Corporation.

The paper in this book was manufactured by a mill whose forest management programs include
sustained yield harvesting of its timberlands. Sustained yield harvesting principles ensure that
the number of trees cut each year does not exceed the amount of new growth.

This book is printed on acid-free paper.   ∞

Brown, William, H., Poon, Thomas
Introduction To Organic Chemistry—Third Edition.

ISBN 0-471-44451-0
Wiley International Edition   0-471-45161-4

Printed in the United States of America.

10   9   8   7   6   5   4   3

To Carolyn,
with whom life is a joy

Bill Brown

To My mother and father
for their continuing love
and support.

Thomas Poon

About the Authors

William H. Brown is Professor Emeritus at Beloit College, where he was twice named Teacher of the Year. He is also the author of two other college textbooks; *Organic Chemistry* 3/e published in 2002, and *General, Organic, and Biochemistry* 7/e coauthored with Fred Bettelheim and Jerry March, published in 2004. He received his PhD from Columbia University under the direction of Gilbert Stork and did postdoctoral work at California Institute of Technology and the University of Arizona. Twice he was Director of a Beloit College World Affairs Center seminar at the University of Glasgow, Scotland. In 1999, he retired from Beloit College to devote more time to writing and development of educational materials. Although officially retired, he continues to teach Special Topics in Organic Synthesis on a yearly basis.

Bill and his wife Carolyn enjoy hiking in the canyon country of the Southwest. In addition, they both enjoy quilting and quilts.

Thomas Poon is Associate Professor of Chemistry in the Joint Science Department of Claremont McKenna, Pitzer, and Scripps Colleges, three of the five undergraduate institutions that make up the Claremont Colleges in Claremont, California. He received his B.S. degree from Fairfield University (CT) and his Ph.D. from the University of California, Los Angeles under the direction of Christopher S. Foote. Poon was a Camille and Henry Dreyfus Postdoctoral Fellow under Bradford P. Mundy at Colby College (ME) before joining the faculty at Randolph-Macon College (VA) where he received the Thomas Branch Award for Excellence in Teaching in 1999. He was a visiting scholar at Columbia University (NY) in 2002 (and again in 2004) where he worked on projects in both research and education with his friend and mentor, Nicholas J. Turro. His teaching duties include organic chemistry, forensic chemistry, and upper level courses in photochemistry and advanced laboratory techniques. His favorite activity is working alongside undergraduates in the laboratory on research problems involving the investigation of synthetic methodology in zeolites, zeolite photochemistry, and reactions of singlet oxygen. When not in the lab, he enjoys running, rollerhockey, hiking, and playing guitar, all in the company of his faithful dog Sami.

Preface

GOALS OF THIS TEXT

Our goals with this text are two-fold. The first is to help you understand the logic of organic chemistry and the connectivity of ideas within organic chemistry. To this end, the organization and the order of presentation of topics within the text stress the connections between what you have learned already and the topics you are currently studying. Our second goal is to enable you to see the connections between organic chemistry and the world around you. To this end, we provide a myriad of examples throughout the text to highlight the applications of organic chemistry to the health and biological sciences, and, even more broadly, to the world around us. We hope that this logical development of topics and connectivity to the world around you will help you understand organic chemistry and to appreciate the contributions organic chemistry has made and will continue to make to your lives.

NEW TO THIS EDITION

Attention to Visual Learning: Research in knowledge and cognition has shown that visualization and organization can greatly enhance learning. We strive to incorporate such learning devices by using callouts (short text boxes) to highlight important features of many of the illustrations throughout the text. This places most of the important information in one location. When you try to recall a concept or attempt to solve a problem, try to visualize the relevant illustration from the text. You may be pleasantly surprised to find that the visual cues provided by the callouts help you to remember the content as well as the context of the illustration.

Looking Ahead Problems: To the end-of-chapter problems, we have added a new section called Looking Ahead. Within this section, we present problems that apply concepts from current or past chapters to future chapters. The purpose of these problems is to provide a constructivist approach to learning organic chemistry, and to show you that concepts constantly build on each other throughout the course.

Infrared and Nuclear Magnetic Resonance Spectroscopy We have moved these two chapters forward from being the last two chapters in the text to now fall in approximately the middle of the text. We find that more and more instructors now include the treatment of spectroscopy as an integral part of their courses.

Chirality and Molecular Handedness We have moved this chapter back in the text so that it now follows Chapter 5: Reactions of Alkenes. We find that students are better able to deal with the study of organic molecules as three-dimensional objects after they have an appreciation of the stereochemistry of alkene addition and oxidation reactions.

The Organic Chemistry of Metabolism This chapter has been expanded to include an overview of the citric acid cycle and to show that the major reactions of the cycle have counterparts in the organic reactions studied in previous chapters. This addition is but one more example of our efforts to make connections between organic chemistry and the health and biological sciences.

THE AUDIENCE

This book provides an introduction to organic chemistry for students who are aiming toward careers in the sciences and who require a grounding in organic chemistry. For this reason, we make a special effort throughout to show the interrelation between organic chemistry and other areas of science, particularly the biological and health sciences. While studying with this book, we hope that you will see that organic chemistry is a tool for these many disciplines, and that organic compounds, both natural and synthetic, are all round you, in pharmaceuticals, plastics, fibers, agrochemicals, surface coatings, toiletry preparations and cosmetics, food additives, adhesives, and elastomers. Furthermore, we hope that you will experience that organic chemistry is a dynamic and ever-expanding area of science waiting openly for those who are prepared, both by training and an inquisitive nature, to ask questions and explore.

ORGANIZATION: AN OVERVIEW

Chapters 1–10 begin study of organic compounds by first reviewing the fundamentals of covalent bonding, the shapes of molecules, and of acid/base chemistry. The structures and typical reactions of the important classes of organic compounds are then discussed; alkanes, alkenes and alkynes, haloalkanes, alcohols and ethers, benzene and its derivatives, and amines.

Chapters 11 and 12 introduce IR, ^{1}H-NMR and ^{13}C-NMR spectroscopy. Discussion of spectroscopy requires no more background than what students receive in general chemistry. These two chapters are freestanding, and can be taken up in any order appropriate to a particular course.

Chapters 13–17 continue the study of organic compounds, including aldehydes and ketones, carboxylic acids, and finally carboxylic acids and their derivatives. Chapter 16 concludes with an introduction to the aldol, Claisen, and Michael reactions, all three important means for the formation of new carbon-carbon bonds. Chapter 17 provides a brief introduction to organic polymer chemistry.

Chapters 18–21 present an introduction to the organic chemistry of carbohydrates, amino acids and proteins, nucleic acids, and lipids. Chapter 22, The Organic Chemistry of Metabolism, demonstrates how the chemistry developed to this point can be applied to an understanding of three major metabolic pathways—glycolysis, the β-oxidation of fatty acids, and the citric acid cycle.

Chapter-by-Chapter

Chapter 1 begins with a review of the electronic structure of atoms and molecules, and uses the VSEPR model to predict shapes of molecules and polyatomic ions. The theory of resonance is introduced midway through Chapter 1 and, with it, the use of curved arrows and electron pushing. The discussion of resonance is followed by an introduction to the orbital overlap description of covalent bonding. Chapter 1 concludes with an introduction to the concept of a functional group, and in particular to hydroxyl, amino, carbonyl, and carboxyl groups.

Chapter 2 provides a general introduction to acid-base chemistry, and concentrates on two major themes: the quantitative determination of the position of equilibrium in an acid-base reaction, and the qualitative relationship between structure and acidity.

Chapter 3 opens with a description of the structure, nomenclature, and conformations of alkanes and cycloalkanes. The IUPAC system is introduced in Section 3.4 through the naming of alkanes and, in Section 3.5, the IUPAC system is presented as a general system of nomenclature. The concept of stereoisomerism is introduced in this chapter with a discussion of cis-trans isomerism in cycloalkanes.

Chapter 4 presents the structure, nomenclature, and physical properties of alkenes and alkynes.

Chapter 5 opens with an introduction to chemical energetics and the concept of a reaction mechanism. The focus of the chapter is then on the reactions of alkenes, organized in the order of electrophilic addition, oxidation, and reduction. The twin concepts of regioselectivity and stereoselectivity are introduced in the context of electrophilic additions to alkenes.

Chapter 6 introduces the concepts of chirality, enantiomers, and diastereomers. This material is coupled with the liberal use of molecular models and encourages students to think of organic molecules as three-dimensional objects and to treat them as such in order to gain a deeper understanding of organic and biochemical reactions.

Chapter 7 introduces haloalkanes and uses them as a vehicle for the discussion of nucleophilic substitution and β-elimination. The concepts of one-step and two-step nucleophilic substitutions along with the S_N1 and S_N2 terminology are introduced first, followed by the concepts of E1 and E2 reactions of haloalkanes. The chapter concludes with a discussion of nucleophilic substitution versus β-elimination.

Chapter 8 concentrates on the stucture and characteristic reactions of alcohols, including their conversion to alkyl halides, acid-catalyzed dehydration, and oxidation of primary and secondary alcohols. There then follows a brief introduction to the structure, preparation, and acid-catalyzed ring opening of epoxides. This chapter concludes with a discussion of the acidity of thiols and their oxidation to disulfides. This chapter is newly structured such that each functional group is given its own section in the chapter.

Chapter 9 opens with the structure and nomenclature of benzene and its derivatives, and then presents several important aromatic heterocyclic compounds. The mechanism of electrophilic aromatic substitution, including the theory of directing effects, is presented in detail. The chapter concludes with a discussion of the structure and acidity of phenols.

Chapter 10 presents the structure and nomenclature of amines and then concentrates on the basicity of amines. New to this edition is a discussion of the reaction of primary aromatic amines with nitrous acid and the ways in which the resulting arenediazonium ions are useful in the synthesis of more complex molecules.

Chapters 11–12 present the fundamentals of IR, ^{1}H-NMR, and ^{13}C-NMR spectroscopy, and how these tools are used in the elucidation of molecular structure.

Chapters 13–16 develop the chemistry of carbonyl-containing compounds. First is the chemistry of aldehydes and ketones in Chapter 13, followed by the chemistry of carboxylic acids in Chapter 14, and then the chemistry of carboxylic acid derivatives in Chapter 15. A major theme in these chapters is the addition of nucleophiles to a carbonyl carbon to form tetrahedral carbonyl addition products. Chapter 16 introduces the concept of an enolate anion and its involvement in aldol, Claisen, and Dieckmann reactions to form new carbon-carbon bonds. New to this edition is the Michael reaction, an important reaction to create new carbon-carbon bonds.

Chapter 17 provides a brief introduction to the organic chemistry of polymers and emphasizes their prevalence and importance in our world.

Chapters 18–22 present the chemistry of carbohydrates (Chapter 18), amino acids and proteins (Chapter 19), nucleic acids (Chapter 20), and lipids (Chapter 21). Chapter 22 presents a discussion of three key metabolic pathways, namely glycolysis, β-oxidation of fatty acids, and the citric acid cycle. The purpose of this material is to show that the reactions of these pathways are the biochemical equivalents of organic functional group reactions already studied in previous chapters.

SPECIAL FEATURES

Applications of organic chemistry to the world around us, particularly to the health and biological sciences

Chapter 1 Covalent Bonding and Shapes of Molecules

Tetrafluoroethylene   Propylene glycol
Freon-11 and Freon-12   Ephedrine
Dihydroxyacetone

Chapter 2 Acids and Bases

Chapter 3 Alkanes and Cycloalkanes

Tetrodotoxin   Petroleum
Paraffin wax   Gasoline
Petrolatum   Asphalt
Vaseline   Gasohol
Natural gas

Chapter 4 Alkenes and Alkynes

Vitamin A and vision β-Carotene
Myrcene   Santonin
Geraniol   Periplanone
Limonene   Gossypol
Menthol   Zoapatanol
Farnesol   Pyrethrins
Lycopene

Chapter 5 Reactions of Alkenes

Limonene
Terpin hydrate

Chapter 6 Chirality and the Handedness of Molecules

Ibuprofen   Captopril
Tartaric acid   L-DOPA
Naproxen   Dopamine
Carvone   Fluoxetine (Prozac)
Ephedrine   Sertraline (Zoloft)
Loratidine (Claritin)   Paroxetine (Paxil)
Fexophenadine (Allegra) Triamcinolone acetonide (Azmacort)

Chapter 14 Carboxylic Acids

Oxalic acid
Salicylic acid
Glyceric acid
Mevalonic acid
Salicin
Aspirin
Ibuprofen
Naproxen
Ketoprofen
Ethyl formate
Isopentyl acetate
Octyl acetate
Methyl butanoate
Ethyl butanoate
Methyl 2-aminobenzoate
Pyrethrins

Terephthalic acid
4-Aminobutanoicacid (GABA)
3-Oxobutanoic acid
3-Hydroxybutanoic acid
Megatomoic acid
Monopotassium oxalate
Potassium sorbate
Zinc 10-undecenoate
Formic acid
Methyl salicylate
Benzocaine
Methylparaben
Propylparaben
Procaine
Meclizine (Bonine)
Albuterol (Proventil)

Chapter 15 Functional Derivatives of Carboxylic Acids

Octyl *p*-methoxycinnamate
Homosalate
Padimate A
Coumarin
Dicoumarol
Warfarin
Glucose 6-phosphate
Pyridoxal phosphate
Amoxicillin
Keflex
Aspirin
Methyl salicylate
Spermaceti
Phenacetin

Acetaminophen
Niacin
Barbiturates
Barbital
Meprobamate
Phenobarbital
N,N-Diethyl-*m*-toluamide (Deet)
Procaine
Propanil
Lidocaine (Xylocaine)
Etidocaine (Duranest)
Mepivacaine (Carbocaine)
Diethylcarbamazine

Chapter 16 Enolate Anions

Mevastatin
Lovastatin (Mevacor)
Simvastatin (Zocor)
Oxanamide

Pindone
Fentanyl
Meclizine

Chapter 17 Organic Polymer Chemistry

Polyamides
Nylon-66
Nylon-6
Kevlar
Polyesters
Poly(ethylene terephthalate)
Dacron polyester
Mylar
Polycarbonates

Polyethylene
Polypropylene
Poly(vinyl chloride) (PVC)
Polyacrylonitrile (Orlon)
Polytetrafluoroethylene (Teflon)
Polystyrene
Poly(ethyl acrylate)
Poly(methyl methacrylate) (Plexiglas)
Low-density polyethylene (LDPE)

Lexan   High-density polyethylene (HDPE)
Polyurethanes   Nylon 6,10
Spandex   Nomex
Lycra   Poly(vinyl acetate)
Epoxy resins

End-of-Chapter Problems

There are plentiful end-of-chapter problems. All problems are categorized by topic. A problem number set in red indicates an applied (real-world) problem.

Organic Synthesis

In this text, we treat organic synthesis and all of the challenges it presents as a teaching tool. We recognize that the majority of students taking this course are intending toward careers in the health and biological sciences, and that very few intend to become synthetic organic chemists. We recognize also that what organic chemists do best is to synthesize new compounds; that is, they make things. Furthermore, we recognize that one of the keys to mastering organic chemistry is extensive problem solving. To this end, we have developed a large number of synthetic problems in which the target molecule is one with an applied, real-world use. Our purpose in this regard is to provide drill in recognizing and using particular reactions within the context of real syntheses. It is not our intent, for example, that you be able to propose a synthesis for procaine (Novocaine), but rather that when you are given an outline of the steps by which it can be made, you can supply necessary reagents.

Molecular Models

The authors have developed a set of more than 275 molecular models for incorporation into this text. Models have been prepared using CambridgeSoft Corporation's Chem3D. We have given further emphasis by adding electron density models, which show calculated values for regions of positive and negative charge within molecules and ions. The purpose of these two types of models is to assist you to visualize organic molecules as three-dimensional objects and to appreciate that the attraction between unlike charges is a major factor in directing molecular reactivity.

In-Chapter Examples

There is an abundance of in-chapter examples, each with a detailed solution. Following each in-chapter example is a comparable practice problem designed to give you the opportunity to solve a related problem.

Full-Color Art Program

One of the distinctive features of this text is its visual impact. The text's extensive full-color art program includes over 200 pieces of art by professional artists John and Bette Woolsey.

Margin Definitions and Glossary of Key Terms

Each bold-faced term has a freestanding definition placed in the margin adjacent to where the term is first introduced. All margin definitions are then collected in a Glossary, with a notation to show the section of the text in which the term is first introduced.

End-of-Chapter Summaries and Summaries of Key Reactions

End-of-chapter summaries highlight all important new terms and concepts found in the chapter. In addition, each new reaction is annotated and keyed to the section where it is discussed.

SUPPORT PACKAGE FOR STUDENTS

Multimedia Tutors: Several Quicktime™ movies will be available from the course web site. Some movies will contain complex animations and text annotations that help to explain the more difficult concepts encountered in organic chemistry. Other movies will be short online lectures given by one of the coauthors. Students may find these lectures useful as an alternative presentation of the materials in the text.

 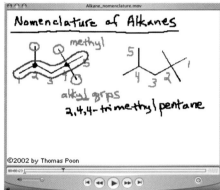

Student Flash Cards: by Thomas Poon. A set of multimedia flash cards will be available through the text web site. These virtual flash cards will organize the key reactions and concepts from the course in a useful manner for learning the material. Mechanisms of each reaction will be provided as well as examples of their use.

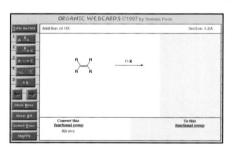 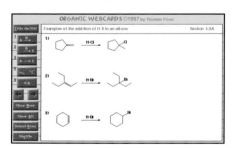

Student Study Guide: by Mark Erickson, Hartwick College, contains detailed solutions to all problems.

Student On-Line Quizzing: The text web site features quizzing, allowing you to practice and review at your own pace.

SUPPORT PACKAGE FOR INSTRUCTORS

PowerPoint Presentation™: by William Brown, contains a pre-built set of approximately 700 PowerPoint™ slides corresponding to every chapter in the text. These slides can be used in conjunction with the freely distributible PowerPoint Viewer™, or edited and customized with the PowerPoint™ application program.

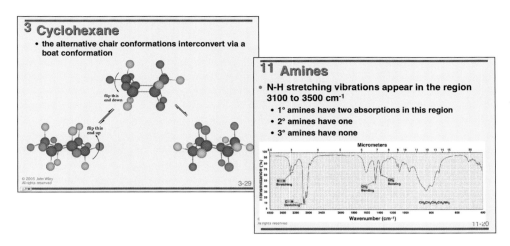

Egrade Homework Management (optional): an on-line quizzing and homework management system that allows students to take practice tests and email homework assignments directly to their instructors.

Digital Image Archive: A text web site that includes downloadable files of all text images in JPEG format.

Test Bank: by Jeffrey Elbert of the University of Northern Iowa contains 45 fill-in-the-blank, true/false, and multiple choice questions per chapter for instructors to use as tests, quizzes, or homework assignments.

Computerized Test Bank CD: IBM and Macintosh versions of the entire Test Bank, with full editing features to allow instructors to easily customize tests.

Overhead Transparency Acetates: A selection of 100 full-color figures from the text.

ACKNOWLEDGMENTS

While one or a few persons are listed as "authors" of any textbook, the book is in fact the product of collaboration of many individuals, some obvious and some not so obvious. It is with gratitude that we acknowledge the contributions of the many. We begin with David Harris through whose efforts this text was launched into production, and then put in the capable hands of Jennifer Yee, Project Editor. We also thank Kevin Molloy, Acquisitions Editor; Amanda Wygal, Marketing Manager; Sandra Dumas, Production Editor; Sandra Rigby, Illustrations Editor; Kevin Murphy, Senior Designer; Lisa Gee, Photo Editor; Martin Batey, Senior New Media Editor; and Catherine Donovan, Editorial Program Assistant.

List of Reviewers

The authors gratefully acknowledge the following reviewers for their valuable critiques of this book in its many stages:

Jennifer Batten, *Grand Rapids Community College*

Dana S. Chatellier, *University of Delaware*

Peter Hamlet, *Pittsburg State University*

John F. Helling, *University of Florida, Gainesville*

Klaus Himmeldirk, *Ohio University, Athens*

Dennis Neil Kevill, *Northern Illinois University*

Dalila G. Kovacs, *Michigan State University, East Lansing*

Tom Munson, *Concordia University*

Robert H. Paine, *Rochester Institute of Technology*

Jeff Piquette, *University of Southern Colorado, Pueblo*

Joe Saunders, *Pennsylvania State University*

K. Barbara Schowen, *University of Kansas, Lawrence*

Robert P. Smart, *Grand Valley State University*

Joshua R. Smith, *Humboldt State University*

Richard T. Taylor, *Miami University, Oxford*

Kjirsten Wayman, *Humboldt State University*

Contents Overview

Contents

9 **Benzene and Its Derivatives**   235

10 **Amines**   277

14 Carboxylic Acids   402

15 Functional Derivatives of Carboxylic Acids   430

22 The Organic Chemistry of Metabolism   619

Chemical Connections Boxes

Each of the 34 boxes illustrates an application of organic chemistry to an everyday setting. Following is a complete list of these boxes by chapter.

1 Covalent Bonding and Shapes of Molecules

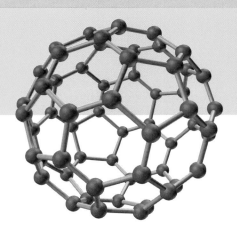

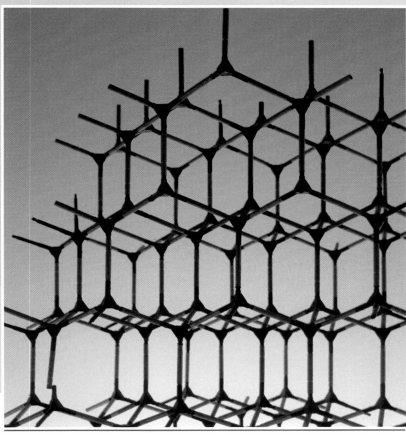

A model of the structure of diamond, one form of pure carbon. Each carbon atom in diamond is bonded to four other carbon atoms at the corners of a tetrahedron. Inset: A model of buckyball, a form of carbon with a molecular formula of C_{60}. *(Charles D. Winters)*

1.1 INTRODUCTION

According to the simplest definition, **organic chemistry** is the study of the compounds of carbon. As you study this text, you will realize that organic compounds are everywhere around us—in our foods, flavors, and fragrances; in our medicines, toiletries, and cosmetics; in our plastics, films, fibers, and resins; in our paints and varnishes; in our glues and adhesives; and, of course, in our bodies and those of all living things.

Perhaps the most remarkable feature of organic chemistry is that it is the chemistry of carbon and only a few other elements—chiefly hydrogen, oxygen, and nitrogen. Chemists have discovered or made well over 10 million organic compounds.

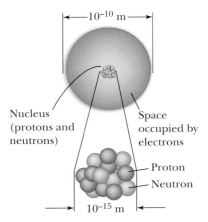

Figure 1.1
A schematic view of an atom. Most of the mass of an atom is concentrated in its small, dense nucleus.

Shell A region of space around a nucleus where electrons are found.

Orbital A region of space where an electron or pair of electrons spends 90 to 95% of its time.

While the majority of them contain carbon and just those three elements, many also contain sulfur, phosphorus, and a halogen (fluorine, chlorine, bromine, or iodine).

Let us begin our study of organic chemistry with a review of how carbon, hydrogen, oxygen, and nitrogen combine by sharing electron pairs to form molecules.

1.2 ELECTRONIC STRUCTURE OF ATOMS

You are already familiar with the fundamentals of the electronic structure of atoms from a previous study of chemistry. Briefly, an atom contains a small, dense nucleus made of neutrons and positively charged protons. Most of the mass of an atom is contained in its nucleus, which is surrounded by a much larger extranuclear space that contains the negatively charged electrons. The nucleus of an atom has a diameter of 10^{-14} to 10^{-15} meter (m). The extranuclear space, where the electrons are found, is a much larger area, with a diameter of approximately 10^{-10} m (Figure 1.1).

Electrons do not move freely in the space around a nucleus, but rather are confined to regions of space called **principal energy levels** or, more simply, **shells**. We number these shells 1, 2, 3, and so forth from the inside out. Each shell can contain up to $2n^2$ electrons, where n is the number of the shell. Thus, the first shell can hold 2 electrons, the second 8 electrons, the third 18, the fourth 32, and so on (Table 1.1). Electrons in the first shell are nearest to the positively charged nucleus and are held most strongly by it; these electrons are said to be the lowest in energy. Electrons in higher numbered shells are farther from the nucleus and are held less strongly to it; these electrons are said to be higher in energy.

Shells are divided into subshells designated by the letters s, p, d, and f, and within these subshells, electrons are grouped in orbitals (Table 1.2). An **orbital** is a region of space that can hold 2 electrons. The first shell contains a single orbital called a $1s$ orbital. The second shell contains one $2s$ orbital and three $2p$ orbitals.

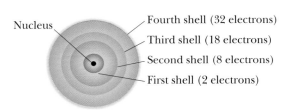

TABLE 1.1 Distribution of Electrons in Shells

Shell	Number of Electrons Shell Can Hold	Relative Energies of Electrons in Each Shell
4	32	Higher
3	18	
2	8	⬆
1	2	Lower

TABLE 1.2 Distribution of Orbitals within Shells

Shell	Orbitals Contained in Each Shell	Maximum Number of Electrons Shell Can Hold
4	One $4s$, three $4p$, five $4d$, and seven $4f$ orbitals	$2 + 6 + 10 + 14 = 32$
3	One $3s$, three $3p$, and five $3d$ orbitals	$2 + 6 + 10 = 18$
2	One $2s$ and three $2p$ orbitals	$2 + 6 = 8$
1	One $1s$ orbital	2

All *p* orbitals come in sets of three and can hold up to 6 electrons. The third shell contains one 3*s* orbital, three 3*p* orbitals, and five 3*d* orbitals. All *d* orbitals come in sets of five and can hold up to 10 electrons. All *f* orbitals come in sets of seven and can hold up to 14 electrons. In this course, we focus on compounds of carbon with hydrogen, oxygen, and nitrogen, all of which use only electrons in *s* and *p* orbitals for covalent bonding. Therefore, we are concerned primarily with *s* and *p* orbitals.

A. Electron Configuration of Atoms

The electron configuration of an atom is a description of the orbitals the electrons in the atom occupy. Every atom has an infinite number of possible electron configurations. At this stage, we are concerned only with the **ground-state electron configuration**—the electron configuration of lowest energy. Table 1.3 shows ground-state electron configurations for the first 18 elements of the Periodic Table. We determine the ground-state electron configuration of an atom with the use of the following three rules:

Ground-state electron configuration The electron configuration of lowest energy for an atom, molecule, or ion.

Rule 1. *Orbitals fill in order of increasing energy from lowest to highest.*
Example: In this course, we are concerned primarily with elements of the first, second, and third periods of the Periodic Table. Orbitals in these elements fill in the order 1*s*, 2*s*, 2*p*, 3*s*, and 3*p* (Figure 1.2).

Rule 2. *Each orbital can hold up to two electrons with their spins paired.*
Example: With four electrons, the 1*s* and 2*s* orbitals are filled, and we write them as $1s^2 2s^2$. With an additional six electrons, a set of three 2*p* orbitals is filled. We can write them in an expanded form as $2p_x^2 2p_y^2 2p_z^2$. Alternatively, we can group the three filled 2*p* orbitals and write them in a condensed form as $2p^6$.

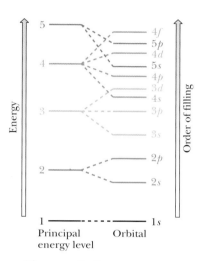

Figure 1.2
Relative energies and order of filling of orbitals through the 5*p* orbitals.

TABLE 1.3	**Ground-State Electron Configurations for Elements 1–18. In this table, we use the symbol of the noble gas immediately preceding the particular element to indicate the electron configuration of all filled shells. The valence shell of helium, for example, is $1s^2$, and the valence shell of neon is $2s^2 2p^6$. We show neon's configuration with the shorthand notation $[He]2s^2 2p^6$.**

First Period*	**Second Period**	**Third Period**
H 1 $1s^1$	Li 3 [He] $2s^1$	Na 11 [Ne] $3s^1$
He 2 $1s^2$	Be 4 [He] $2s^2$	Mg 12 [Ne] $3s^2$
	B 5 [He] $2s^2 2p_x^1$	Al 13 [Ne] $3s^2 3p_x^1$
	C 6 [He] $2s^2 2p_x^1 2p_y^1$	Si 14 [Ne] $3s^2 3p_x^1 3p_y^1$
	N 7 [He] $2s^2 2p_x^1 2p_y^1 2p_z^1$	P 15 [Ne] $3s^2 3p_x^1 3p_y^1 3p_z^1$
	O 8 [He] $2s^2 2p_x^2 2p_y^1 2p_z^1$	S 16 [Ne] $3s^2 3p_x^2 3p_y^1 3p_z^1$
	F 9 [He] $2s^2 2p_x^2 2p_y^2 2p_z^1$	Cl 17 [Ne] $3s^2 3p_x^2 3p_y^2 3p_z^1$
	Ne 10 [He] $2s^2 2p_x^2 2p_y^2 2p_z^2$	Ar 18 [Ne] $3s^2 3p_x^2 3p_y^2 3p_z^2$

*Elements are listed by symbol, atomic number, and ground-state electron configuration, in that order.

Figure 1.3
The pairing of
electron spins.

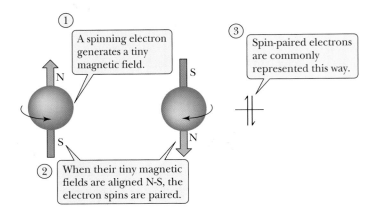

① A spinning electron generates a tiny magnetic field.

② When their tiny magnetic fields are aligned N-S, the electron spins are paired.

③ Spin-paired electrons are commonly represented this way.

Spin pairing means that each electron spins in a direction opposite that of its partner (Figure 1.3). We show this pairing by writing two arrows, one with its head up and the other with its head down.

Rule 3.   *When orbitals of equivalent energy are available, but there are not enough electrons to fill them completely, then we add one electron to each equivalent orbital before we add a second electron to any one of them.*

Example: After the $1s$ and $2s$ orbitals are filled with four electrons, we add a fifth electron to the $2p_x$ orbital, a sixth to the $2p_y$ orbital, and a seventh to the $2p_z$ orbital. Only after each $2p$ orbital contains one electron do we add a second electron to the $2p_x$ orbital. For eight electrons, we write the ground-state electron configuration as $1s^2 2s^2 2p_x^2 2p_y^1 2p_z^1$ or, alternatively, $1s^2 2s^2 2p^4$.

EXAMPLE 1.1

Write ground-state electron configurations for these elements:

(a) Lithium   (b) Oxygen   (c) Chlorine

SOLUTION

(a) Lithium (atomic number 3): $1s^2 2s^1$. Alternatively, we can write the ground-state electron configuration as [He] $2s^1$.
(b) Oxygen (atomic number 8): $1s^2 2s^2 2p_x^2 2p_y^1 2p_z^1$. Alternatively, we can group the four electrons of the $2p$ orbitals together and write the ground-state electron configuration as $1s^2 2s^2 2p^4$. We can also write it as [He] $2s^2 2p^4$.
(c) Chlorine (atomic number 17): $1s^2 2s^2 2p^6 3s^2 3p^5$. Alternatively, we can write it as [Ne] $3s^2 3p^5$.

Practice Problem 1.1

Write and compare the ground-state electron configurations for the elements in each set.

(a) Carbon and silicon   (b) Oxygen and sulfur
(c) Nitrogen and phosphorus

B. Lewis Structures

In discussing the physical and chemical properties of an element, chemists often focus on the outermost shell of its atoms, because electrons in this shell are the ones involved in the formation of chemical bonds and in chemical reactions. We call outer-shell electrons **valence electrons**, and we call the energy level in which they are found

Valence electrons Electrons in the valence (outermost) shell of an atom.

TABLE 1.4 Lewis Structures for Elements 1–18 of the Periodic Table

1A	2A	3A	4A	5A	6A	7A	8A
H·							He:
Li·	Be:	B:	·C:	·N:	:O:	:F:	:Ne:
Na·	Mg:	Al:	·Si:	·P:	:S:	:Cl:	:Ar:

the **valence shell**. Carbon, for example, with a ground-state electron configuration of $1s^2 2s^2 2p^2$, has four valence (outer-shell) electrons.

To show the outermost electrons of an atom, we commonly use a representation called a **Lewis structure**, after the American chemist Gilbert N. Lewis (1875–1946), who devised this notation. A Lewis structure shows the symbol of the element, surrounded by a number of dots equal to the number of electrons in the outer shell of an atom of that element. In Lewis structures, the atomic symbol represents the nucleus and all filled inner shells. Table 1.4 shows Lewis structures for the first 18 elements of the Periodic Table. As you study the entries in the table, note that, with the exception of helium, the number of valence electrons of the element corresponds to the group number of the element in the Periodic Table; for example, oxygen, with six valence electrons, is in Group 6A.

The noble gases helium and neon have filled valence shells. The valence shell of helium is filled with two electrons; that of neon is filled with eight electrons. Neon and argon have in common an electron configuration in which the s and p orbitals of their valence shells are filled with eight electrons. The valence shells of all other elements shown in Table 1.4 contain fewer than eight electrons.

Compare the Lewis structures given in Table 1.4 with the ground-state electron configurations listed in Table 1.3. Table 1.4 shows the Lewis structure of boron (B), for example, with three valence electrons: the paired $2s$ electrons and the single $2p_x$ electron listed in Table 1.3. Table 1.4 shows the Lewis structure of carbon (C) with four valence electrons: the two paired $2s$ electrons and the single $2p_x$ and $2p_y$ electrons listed in Table 1.3.

Notice also from Table 1.4 that, for C, N, O, and F in period 2 of the Periodic Table, the valence electrons belong to the second shell. It requires 8 electrons to fill this shell. For Si, P, S, and Cl in period 3 of the Periodic Table, the valence electrons belong to the third shell. With 8 electrons, this shell is only partially filled: the $3s$ and $3p$ orbitals are fully occupied, but the five $3d$ orbitals can accommodate an additional 10 valence electrons. Because of the differences in number and kind of valence shell orbitals available to elements of the second and third periods, significant differences exist in the covalent bonding of oxygen and sulfur and of nitrogen and phosphorus. For example, although oxygen and nitrogen can accommodate no more than 8 electrons in their valence shells, many phosphorus-containing compounds have 10 electrons in the valence shell of phosphorus, and many sulfur-containing compounds have 10 and even 12 electrons in the valence shell of sulfur.

Valence shell The outermost electron shell of an atom.

Lewis structure of an atom The symbol of an element surrounded by a number of dots equal to the number of electrons in the valence shell of the atom.

Gilbert N. Lewis (1875–1946) introduced the theory of the electron pair that extended our understanding of covalent bonding and of the concept of acids and bases. It is in his honor that we often refer to an "electron dot" structure as a Lewis structure. *(UPI/Bettmann)*

1.3 LEWIS MODEL OF BONDING

A. Formation of Ions

In 1916, Lewis devised a beautifully simple model that unified many of the observations about chemical bonding and reactions of the elements. He pointed out that the chemical inertness of the noble gases (Group 8A) indicates a high degree

Noble gas	Noble gas notation
He	$1s^2$
Ne	[He] $2s^2 2p^6$
Ar	[Ne] $3s^2 3p^6$
Kr	[Ar] $4s^2 4p^6$
Xe	[Kr] $5s^2 5p^6$

of stability of the electron configurations of these elements: helium with a valence shell of two electrons ($1s^2$), neon with a valence shell of eight electrons ($2s^22p^6$), argon with a valence shell of eight electrons ($3s^23p^6$), and so forth.

The tendency of atoms to react in ways that achieve an outer shell of eight valence electrons is particularly common among elements of Groups 1A–7A (the main-group elements). We give this tendency the special name, the **octet rule**. An atom with almost eight valence electrons tends to gain the needed electrons to have eight electrons in its valence shell and an electron configuration like that of the noble gas nearest it in atomic number. In gaining electrons, the atom becomes a negatively charged ion called an **anion**. An atom with only one or two valence electrons tends to lose the number of electrons required to have the same electron configuration as the noble gas nearest it in atomic number. In losing one or more electrons, the atom becomes a positively charged ion called a **cation**.

Octet rule The tendency among atoms of Group 1A–7A elements to react in ways that achieve an outer shell of eight valence electrons.

Anion An atom or group of atoms bearing a negative charge.

Cation An atom or group of atoms bearing a positive charge.

Example 1.2

Show how the loss of one electron from a sodium atom to form a sodium ion leads to a stable octet:

$$Na \longrightarrow Na^+ + e^-$$

A sodium atom   A sodium ion   An electron

SOLUTION

To see how this chemical change leads to a stable octet, write the condensed ground-state electron configuration for a sodium atom and for a sodium ion, and then compare the two:

Na (11 electrons): $1s^22s^22p^63s^1$

Na$^+$ (10 electrons): $1s^22s^22p^6$

A sodium atom has one electron in its valence shell. The loss of this one valence electron changes the sodium atom to a sodium ion, Na$^+$, which has a complete octet of electrons in its valence shell and the same electron configuration as neon, the noble gas nearest to it in atomic number.

Practice Problem 1.2

Show how the gain of two electrons by a sulfur atom to form a sulfide ion leads to a stable octet:

$$S + 2e^- \longrightarrow S^{2-}$$

B. Formation of Chemical Bonds

According to the Lewis model of bonding, atoms interact with each other in such a way that each atom participating in a chemical bond acquires a valence-shell electron configuration the same as that of the noble gas closest to it in atomic number. Atoms acquire completed valence shells in two ways:

1. An atom may lose or gain enough electrons to acquire a filled valence shell. An atom that gains electrons becomes an anion, and an atom that loses electrons becomes a cation. A chemical bond between an anion and a cation is called an **ionic bond**.
2. An atom may share electrons with one or more other atoms to acquire a filled valence shell. A chemical bond formed by sharing electrons is called a **covalent bond**.

Ionic bond A chemical bond resulting from the electrostatic attraction of an anion and a cation.

Covalent bond A chemical bond resulting from the sharing of one or more pairs of electrons.

We now ask how we can find out whether two atoms in a compound are joined by an ionic bond or a covalent bond? One way to answer this question is to consider the relative positions of the two atoms in the Periodic Table. Ionic bonds usually form between a metal and a nonmetal. An example of an ionic bond is that formed between the metal sodium and the nonmetal chlorine in the compound sodium chloride, Na^+Cl^-. By contrast, when two nonmetals or a metalloid and a nonmetal combine, the bond between them is usually covalent. Examples of compounds containing covalent bonds between nonmetals include Cl_2, H_2O, CH_4, and NH_3. Examples of compounds containing covalent bonds between a metalloid and a nonmetal include BF_3, $SiCl_4$, and AsH_4.

Another way to identify the type of bond is to compare the electronegativities of the atoms involved, which is the subject of the next subsection.

C. Electronegativity and Chemical Bonds

Electronegativity is a measure of the force of an atom's attraction for electrons that it shares in a chemical bond with another atom. The most widely used scale of electronegativities (Table 1.5) was devised by Linus Pauling in the 1930s. On the Pauling scale, fluorine, the most electronegative element, is assigned an electronegativity of 4.0, and all other elements are assigned values in relation to fluorine.

As you study the electronegativity values in this table, note that they generally increase from left to right within a period of the Periodic Table and generally increase from bottom to top within a group. Values increase from left to right because of the increasing positive charge on the nucleus, which leads to a stronger attraction for electrons in the valence shell. Values increase going up a column because of the decreasing distance of the valence electrons from the nucleus, which leads to stronger attraction between a nucleus and its valence electrons.

Note that the values given in Table 1.5 are only approximate. The electronegativity of a particular element depends not only on its position in the Periodic Table, but also on its oxidation state. The electronegativity of Cu(I) in Cu_2O, for example, is 1.8, whereas the electronegativity of Cu(II) in CuO is 2.0. In spite of these variations, electronegativity is still a useful guide to the distribution of electrons in a chemical bond.

Electronegativity A measure of the force of an atom's attraction for electrons it shares in a chemical bond with another atom.

Linus Pauling (1901–1994) was the first person ever to receive two unshared Nobel Prizes. He received the Nobel Prize for Chemistry in 1954 for his contributions to the nature of chemical bonding. He received the Nobel Prize for Peace in 1962 for his efforts on behalf of international control of nuclear weapons and against nuclear testing. *(UPI/Bettmann)*

TABLE 1.5 Electronegativity Values for Some Atoms (Pauling Scale)

1A	2A	3B	4B	5B	6B	7B	8B			1B	2B	3A	4A	5A	6A	7A
															H 2.1	
Li 1.0	Be 1.5											B 2.0	C 2.5	N 3.0	O 3.5	F 4.0
Na 0.9	Mg 1.2											Al 1.5	Si 1.8	P 2.1	S 2.5	Cl 3.0
K 0.8	Ca 1.0	Sc 1.3	Ti 1.5	V 1.6	Cr 1.6	Mn 1.5	Fe 1.8	Co 1.8	Ni 1.8	Cu 1.9	Zn 1.6	Ga 1.6	Ge 1.8	As 2.0	Se 2.4	Br 2.8
Rb 0.8	Sr 1.0	Y 1.2	Zr 1.4	Nb 1.6	Mo 1.8	Tc 1.9	Ru 2.2	Rh 2.2	Pd 2.2	Ag 1.9	Cd 1.7	In 1.7	Sn 1.8	Sb 1.9	Te 2.1	I 2.5
Cs 0.7	Ba 0.9	La 1.1	Hf 1.3	Ta 1.5	W 1.7	Re 1.9	Os 2.2	Ir 2.2	Pt 2.2	Au 2.4	Hg 1.9	Tl 1.8	Pb 1.8	Bi 1.9	Po 2.0	At 2.2

<1.0
1.0 – 1.4
1.5 – 1.9
2.0 – 2.4
2.5 – 2.9
3.0 – 4.0

EXAMPLE 1.3

Judging from their relative positions in the Periodic Table, which element in each pair has the larger electronegativity?

(a) Lithium or carbon   (b) Nitrogen or oxygen
(c) Carbon or oxygen

SOLUTION

The elements in these pairs are all in the second period of the Periodic Table. Electronegativity in this period increases from left to right.

(a) C > Li   (b) O > N   (c) O > C

Practice Problem 1.3

Judging from their relative positions in the Periodic Table, which element in each pair has the larger electronegativity?

(a) Lithium or potassium   (b) Nitrogen or phosphorus
(c) Carbon or silicon

Ionic Bonds

An ionic bond forms by the transfer of electrons from the valence shell of an atom of lower electronegativity to the valence shell of an atom of higher electronegativity. The more electronegative atom gains one or more valence electrons and becomes an anion; the less electronegative atom loses one or more valence electrons and becomes a cation.

As a guideline, we say that this type of electron transfer to form an ionic compound is most likely to occur if the difference in electronegativity between two atoms is approximately 1.9 or greater. A bond is more likely to be covalent if this difference is less than 1.9. Note that the value 1.9 is somewhat arbitrary: Some chemists prefer a slightly larger value, others a slightly smaller value. The essential point is that the value 1.9 gives us a guidepost against which to decide whether a bond is more likely to be ionic or more likely to be covalent.

An example of an ionic bond is that formed between sodium (electronegativity 0.9) and fluorine (electronegativity 4.0). The difference in electronegativity between these two elements is 3.1. In forming Na^+F^-, the single $3s$ valence electron of sodium is transferred to the partially filled valence shell of fluorine:

$$Na(1s^2 2s^2 2p^6 3s^1) + F(1s^2 2s^2 2p^5) \longrightarrow Na^+(1s^2 2s^2 2p^6) + F^-(1s^2 2s^2 2p^6)$$

As a result of this transfer of one electron, both sodium and fluorine form ions that have the same electron configuration as neon, the noble gas closest to each in atomic number. In the following equation, we use a single-headed curved arrow to show the transfer of one electron from sodium to fluorine:

$$Na \cdot + \cdot \ddot{F} : \longrightarrow Na^+ : \ddot{F} :^-$$

Covalent Bonds

A covalent bond forms when electron pairs are shared between two atoms whose difference in electronegativity is 1.9 or less. According to the Lewis model, an electron pair in a covalent bond functions in two ways simultaneously: It is shared by two atoms, and, at the same time, it fills the valence shell of each atom.

The simplest example ... valent bond is that in a hydrogen molecule, H_2. When two hydrogen atoms by ... single electrons from each atom combine to form an electron pair. A bond ... sharing a pair of electrons is called a *single bond* and is represented by a ... completes the valence shell of each hydrogen. Thus, in the two hydroge ... ctrons in its valence shell and an electron configuration H_2, each hydr ... gas nearest to it in atomic number: like that of

$$ \text{H} \!-\! \text{H} \qquad \Delta H^0 = -104 \text{ kcal/mol } (-435 \text{ kJ/mol}) $$

... ounts for the stability of covalently bonded atoms in the ... a covalent bond, an electron pair occupies the region ... serves to shield one positively charged nucleus from the ... her positively charged nucleus. At the same time, an elec- ... uclei. In other words, an electron pair in the space between ... together and fixes the internuclear distance to within very ... tance between nuclei participating in a chemical bond is ... very covalent bond has a definite bond length. In H—H, it ... $= 10^{-12}$ m.

... lent bonds involve the sharing of electrons, they differ widely ... ing. We classify covalent bonds into two categories—nonpolar ... covalent—depending on the difference in electronegativity ... ded atoms. In a **nonpolar covalent bond**, electrons are shared ... **polar covalent bond**, they are shared unequally. It is important to real- ... sharp line divides these two categories, nor, for that matter, does a sharp ... de polar covalent bonds and ionic bonds. Nonetheless, the rule-of-thumb ... elines in Table 1.6 will help you decide whether a given bond is more likely to ... e nonpolar covalent, polar covalent, or ionic.

A covalent bond between carbon and hydrogen, for example, is classified as nonpolar covalent because the difference in electronegativity between these two atoms is $2.5 - 2.1 = 0.4$ unit. An example of a polar covalent bond is that of H—Cl. The difference in electronegativity between chlorine and hydrogen is $3.0 - 2.1 = 0.9$ unit.

Nonpolar covalent bond A covalent bond between atoms whose difference in electronegativity is less than approximately 0.5.

Polar covalent bond A covalent bond between atoms whose difference in electronegativity is between approximately 0.5 and 1.9.

TABLE 1.6 Classification of Chemical Bonds

Difference in Electronegativity between Bonded Atoms	Type of Bond	Most Likely Formed Between
Less than 0.5	Nonpolar covalent ⎱	Two nonmetals or a
0.5 to 1.9	Polar covalent ⎰	nonmetal and a metalloid
Greater than 1.9	Ionic	A metal and a nonmetal

EXAMPLE 1.4

Classify each bond as nonpolar covalent, polar covalent, or ionic:

(a) O—H (b) N—H (c) Na—F (d) C—Mg

SOLUTION

On the basis of differences in electronegativity between
of these bonds are polar covalent and one is ionic:

Bond	Difference in Electronegativity	Type of B
(a) O—H	$3.5 - 2.1 = 1.4$	polar covale.
(b) N—H	$3.0 - 2.1 = 0.9$	polar covalent
(c) Na—F	$4.0 - 0.9 = 3.1$	ionic
(d) C—Mg	$2.5 - 1.2 = 1.3$	polar covalent

Practice Problem 1.4

Classify each bond as nonpolar covalent, polar covalent, or ionic:

(a) S—H   (b) P—H   (c) C—F   (d) C—Cl

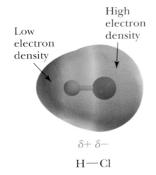

Low
electron
density

High
electron
density

$\delta+ \; \delta-$

H—Cl

Figure 1.4

An electron density model of
HCl. Red indicates a region
of high electron density, and
blue indicates a region of low
electron density.

An important consequence of the unequal sharing of electrons in a polar co
lent bond is that the more electronegative atom gains a greater fraction of th
shared electrons and acquires a partial negative charge, which we indicate by the
symbol $\delta-$ (read "delta minus"). The less electronegative atom has a lesser fraction
of the shared electrons and acquires a partial positive charge, which we indicate by
the symbol $\delta+$ (read "delta plus"). This separation of charge produces a **dipole**
(two poles). We can also show the presence of a bond dipole by an arrow, with the
head of the arrow near the negative end of the dipole and a cross on the tail of the
arrow near the positive end (Figure 1.4).

We can display the polarity of a covalent bond by a type of molecular model
called an *electron density model*. In this type of model, a blue color shows the presence
of a $\delta+$ charge and a red color shows the presence of a $\delta-$ charge. Figure 1.4 shows
an electron density model of HCl. The ball-and-stick model in the center shows the
orientation of the two atoms in space. The transparent surface surrounding the
ball-and-stick model shows the relative sizes of the atoms (equivalent to the size
shown by a space-filling model). Colors on the surface show the distribution of electron density. We see by the blue color that hydrogen bears a $\delta+$ charge and by the
red color that chlorine bears a $\delta-$ charge.

EXAMPLE 1.5

Using the symbols $\delta-$ and $\delta+$, indicate the direction of polarity in these polar
covalent bonds:

(a) C—O   (b) N—H   (c) C—Mg

SOLUTION

For (a), carbon and oxygen are both in period 2 of the Periodic Table. Because
oxygen is farther to the right than carbon, it is more electronegative. For (b),
nitrogen is more electronegative than hydrogen. For (c), magnesium is a metal

located at the far left of the Periodic Table, and carbon is a nonmetal located at the right. All nonmetals, including hydrogen, have a greater electronegativity than do the metals in columns 1A and 2A. The electronegativity of each element is given below the symbol of the element:

(a) $\overset{\delta+}{C}—\overset{\delta-}{O}$

2.5 3.5

(b) $\overset{\delta-}{N}—\overset{\delta+}{H}$

3.0 2.1

(c) $\overset{\delta-}{C}—\overset{\delta+}{Mg}$

2.5 1.2

Practice Problem 1.5

Using the symbols δ− and δ+, indicate the direction of polarity in these polar covalent bonds:

(a) C—N

(b) N—O

(c) C—Cl

D. Drawing Lewis Structures of Molecules and Ions

The ability to draw Lewis structures for molecules and ions is a fundamental skill in the study of organic chemistry. The following guidelines will help you to do this (as you study these guidelines, look at the examples in Table 1.7):

Step 1: *Determine the number of valence electrons in the molecule or ion.*
To do so, add the number of valence electrons contributed by each atom. For ions, add one electron for each negative charge on the ion, and subtract one electron for each positive charge on the ion. For example, the Lewis structure of the water molecule, H_2O, must show eight valence electrons: one from each hydrogen and six from oxygen. The Lewis structure for the hydroxide ion, OH^-, must also show eight valence electrons: one from hydrogen, six from oxygen, plus one for the negative charge on the ion.

Step 2: *Determine the arrangement of atoms in the molecule or ion.*
Except for the simplest molecules and ions, this arrangement must be determined experimentally. For some molecules and ions we give as examples, we

TABLE 1.7 Lewis Structures for Several Compounds. The number of valence electrons in each molecule is given in parentheses after the molecule's molecular formula.

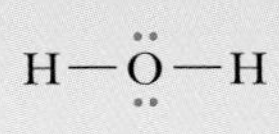 H—Ö—H	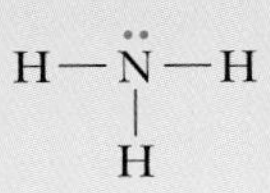 H—N̈—H with H below	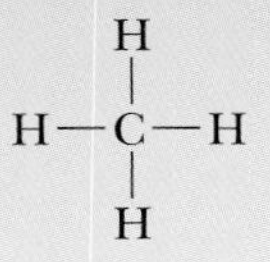 H—C—H with H above and below	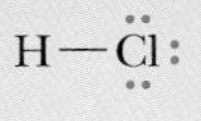 H—C̈l:
H_2O (8)	NH_3 (8)	CH_4 (8)	HCl (8)
Water	Ammonia	Methane	Hydrogen chloride
C=C (H,H,H,H)	H—C≡C—H	C=Ö (H,H)	O=C(O)(O) (H,H)
C_2H_4 (12)	C_2H_2 (10)	CH_2O (12)	H_2CO_3 (24)
Ethylene	Acetylene	Formaldehyde	Carbonic acid

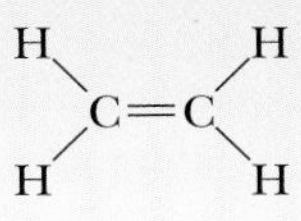

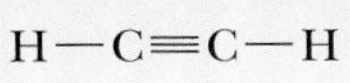

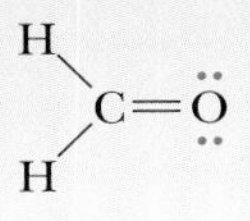

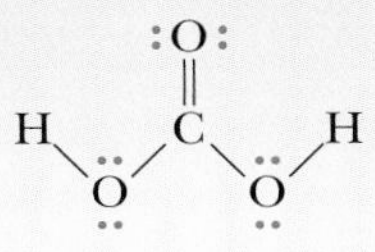

ask you to propose an arrangement of atoms. For most, however, we give you the experimentally determined arrangement.

Step 3: *Arrange the remaining electrons in pairs so that each atom in the molecule or ion has a complete outer shell.*

To accomplish this, connect the atoms with single bonds. Then arrange the remaining electrons in pairs so that each atom in the molecule or ion has a complete outer shell. Each hydrogen atom must be surrounded by two electrons. Each atom of carbon, oxygen, and nitrogen, as well as each atom of a halogen, must be surrounded by eight electrons (per the octet rule).

> **Bonding electrons** Valence electrons shared in a covalent bond.

> **Nonbonding electrons** Valence electrons not involved in forming covalent bonds; that is, unshared electrons.

Step 4: *Show a pair of **bonding electrons** as a single line; show a pair of **nonbonding electrons** as a pair of Lewis dots.*

Step 5: *Use multiple bonds where necessary.*

In a **single bond**, two atoms share one pair of electrons. It is sometimes necessary for atoms to share more than one pair of electrons. In a **double bond**, they share two pairs of electrons; we show a double bond by two lines between the bonded atoms. In a **triple bond**, two atoms share three pairs of electrons; we show a triple bond by three lines between the bonded atoms.

From the study of the compounds in Table 1.7 and other organic compounds, we can make the following generalizations:
In neutral (uncharged) organic compounds,

- H has one bond
- C has four bonds
- N has three bonds and one unshared pair of electrons
- O has two bonds and two unshared pair of electrons
- F, Cl, Br, and I have one bond and three unshared pairs of electrons

EXAMPLE 1.6

Draw Lewis structures, showing all valence electrons, for these molecules:

(a) H_2O_2 (b) CH_3OH (c) CH_3Cl

SOLUTION

(a) A Lewis structure for hydrogen peroxide, H_2O_2, must show 6 valence electrons from each oxygen and 1 from each hydrogen, for a total of $12 + 2 = 14$ valence electrons. We know that hydrogen forms only one covalent bond, so the order of attachment of atoms must be as follows:

$$H—O—O—H$$

The three single bonds account for 6 valence electrons. We place the remaining 8 valence electrons on the oxygen atoms to give each a complete octet:

$$H—\ddot{\underset{..}{O}}—\ddot{\underset{..}{O}}—H$$

Lewis structure

Ball-and-stick models show only nuclei and covalent bonds; they do not show unshared pairs of electrons

(b) A Lewis structure for methanol, CH_3OH, must show 4 valence electrons from carbon, 1 from each hydrogen, and 6 from oxygen, for a total of $4 + 4 + 6 = 14$ valence electrons. The order of attachment of atoms in methanol is given on the left. The five single bonds in this partial structure account for 10 valence electrons. We place the remaining 4 valence electrons on oxygen as two Lewis dot pairs to give it a complete octet.

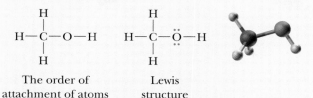

The order of attachment of atoms

Lewis structure

(c) A Lewis structure for chloromethane, CH_3Cl, must show 4 valence electrons from carbon, 1 from each hydrogen, and 7 from chlorine, for a total of $4 + 3 + 7 = 14$. Carbon has four bonds, one to each of the hydrogens and one to chlorine. We place the remaining 6 valence electrons on chlorine as three Lewis dot pairs to complete its octet.

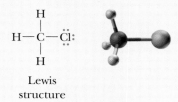

Lewis structure

Practice Problem 1.6

Draw Lewis structures, showing all valence electrons, for these molecules:

(a) C_2H_6 (b) CS_2 (c) HCN

E. Formal Charge

Throughout this course, we deal not only with molecules, but also with polyatomic cations and polyatomic anions. Examples of polyatomic cations are the hydronium ion, H_3O^+, and the ammonium ion, NH_4^+. An example of a polyatomic anion is the bicarbonate ion, HCO_3^-. It is important that you be able to determine which atom or atoms in a molecule or polyatomic ion bear the positive or negative charge. The charge on an atom in a molecule or polyatomic ion is called its **formal charge**. To derive a formal charge,

Formal charge The charge on an atom in a molecule or polyatomic ion.

Step 1: *Write a correct Lewis structure for the molecule or ion.*

Step 2: *Assign to each atom all its unshared (nonbonding) electrons and one-half its shared (bonding) electrons.*

Step 3: *Compare the number arrived at in Step 2 with the number of valence electrons in the neutral, unbonded atom.* If the number of electrons assigned to a bonded atom is less than that assigned to the unbonded atom, then more positive charges are in the nucleus than counterbalancing negative charges, and the atom has a positive formal charge. Conversely, if the number of electrons assigned to a bonded atom is greater than that assigned to the unbonded atom, then the atom has a negative formal charge.

$$\text{Formal charge} = \begin{matrix} \text{Number of valence} \\ \text{electrons in neutral} \\ \text{unbonded atom} \end{matrix} - \left(\begin{matrix} \text{All unshared} \\ \text{electrons} \end{matrix} + \begin{matrix} \text{One-half of all} \\ \text{shared electrons} \end{matrix} \right)$$

EXAMPLE 1.7

Draw Lewis structures for these ions, and show which atom in each bears the formal charge:

(a) H_3O^+ (b) CH_3O^-

SOLUTION

(a) The Lewis structure for the hydronium ion must show 8 valence electrons: 3 from the three hydrogens, 6 from oxygen, minus 1 for the single positive charge. A neutral, unbonded oxygen atom has 6 valence electrons. To the oxygen atom in H_3O^+, we assign two unshared electrons and one from each shared pair of electrons, giving it a formal charge of $6 - (2 + 3) = +1$.

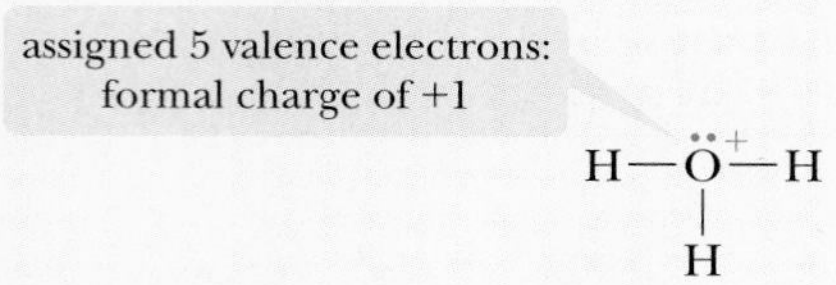

assigned 5 valence electrons:
formal charge of +1

$$H-\overset{..}{\underset{|}{O}}{}^{+}-H$$
$$|$$
$$H$$

(b) The Lewis structure for the methoxide ion, CH_3O^-, must show 12 valence electrons: 4 from carbon, 6 from oxygen, 3 from the hydrogens, plus 1 for the single negative charge. To carbon, we assign 1 electron from each shared pair, giving it a formal charge of $4 - 4 = 0$. To oxygen, we assign 7 valence electrons, giving it a formal charge of $6 - 7 = -1$.

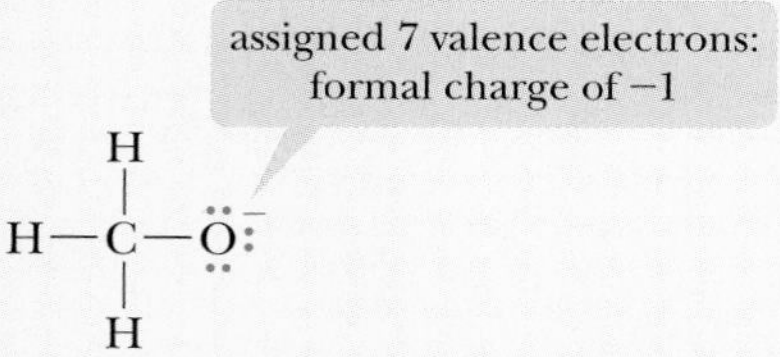

assigned 7 valence electrons:
formal charge of −1

$$\overset{H}{\underset{H}{\overset{|}{H-C}}}-\overset{..}{\underset{..}{O}}{}^{:-}$$

Practice Problem 1.7

Draw Lewis structures for these ions, and show which atom in each bears the formal charge(s):

(a) $CH_3NH_3^+$ (b) CH_3^+

In writing Lewis structures for molecules and ions, you must remember that elements of the second period, including carbon, nitrogen, and oxygen, can accommodate no more than eight electrons in the four orbitals ($2s$, $2p_x$, $2p_y$, and $2p_z$) of their valence shells. Following are two Lewis structures for nitric acid, HNO_3, each with the correct number of valence electrons, namely, 24; one structure is acceptable and the other is not:

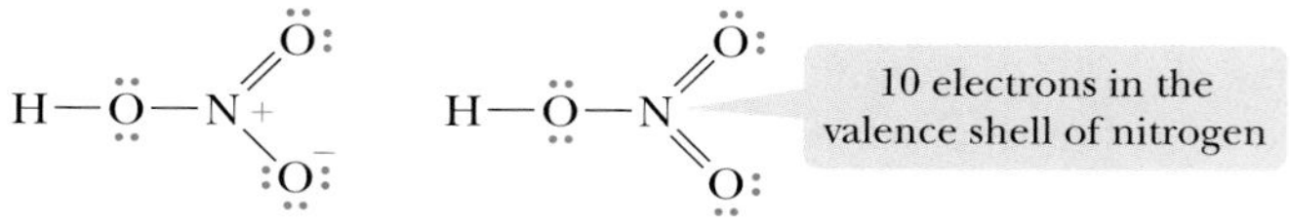

$$H-\overset{..}{\underset{..}{O}}-N^+\diagup\overset{..}{O}:\diagdown:\overset{..}{\underset{..}{O}}:^-$$ $$H-\overset{..}{\underset{..}{O}}-N\diagup\overset{..}{O}:\diagdown\overset{..}{\underset{..}{O}}:$$ 10 electrons in the valence shell of nitrogen

An acceptable   Not an acceptable
Lewis structure   Lewis structure

The structure on the left is an acceptable Lewis structure. It shows the required 24 valence electrons, and each oxygen and nitrogen has a completed valence shell of

8 electrons. Further, the structure on the left shows a positive formal charge on nitrogen and a negative formal charge on one of the oxygens. An acceptable Lewis structure must show these formal charges. The structure on the right is *not* an acceptable Lewis structure. Although it shows the correct number of valence electrons, it places 10 electrons in the valence shell of nitrogen, yet the four orbitals of the second shell ($2s$, $2p_x$, $2p_y$, and $2p_z$) can hold no more than 8 valence electrons!

1.4 BOND ANGLES AND THE SHAPES OF MOLECULES

In Section 1.3, we used a shared pair of electrons as the fundamental unit of a covalent bond and drew Lewis structures for several small molecules containing various combinations of single, double, and triple bonds. (See, for example, Table 1.7.) We can predict bond angles in these and other molecules in a very straightforward way by using the **valence-shell electron-pair repulsion (VSEPR) model**. According to this model, the valence electrons of an atom may be involved in the formation of single, double, or triple bonds, or they may be unshared. Each combination creates a negatively charged region of space. Because like charges repel each other, the various regions of electron density around an atom spread so that each is as far away from the others as possible.

You can demonstrate the bond angles predicted by this model in a very simple way. Imagine that a balloon represents a region of electron density. If you tie two balloons together by their ends, they assume the shapes shown in Figure 1.5. The point where they are tied together represents the atom about which you want to predict a bond angle, and the balloons represent regions of electron density about that atom.

We use the VSEPR model in the following way to predict the shape of methane, CH_4. The Lewis structure for CH_4 shows a carbon atom surrounded by four regions of electron density. Each region contains a pair of electrons forming a bond to a hydrogen atom. According to the VSEPR model, the four regions radiate from carbon so that they are as far away from each other as possible. The maximum separation occurs when the angle between any two pairs of electrons is 109.5°. Therefore, we predict that the H—C—H bond angles are 109.5° and that the shape of the molecule is **tetrahedral** (Figure 1.6). The H—C—H bond angles in methane have been measured experimentally and found to be 109.5°. Thus, the bond angles and shape of methane predicted by the VSEPR model are identical to those observed.

We can predict the shape of an ammonia molecule, NH_3, in the same manner. The Lewis structure of NH_3 shows nitrogen surrounded by four regions of electron density. Three regions contain single pairs of electrons forming covalent bonds

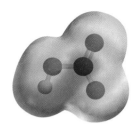

HNO₃

The Lewis structure of HNO_3 shows the negative formal charge localized on one of the oxygen atoms. The electron density model, on the other hand, shows that the negative charge is distributed equally over the two oxygen atoms on the right. The concept of resonance can explain this phenomenon and will be discussed in Section 1.6. Notice also the intense blue color on nitrogen, which is due to its positive formal charge.

(a)

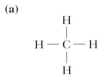

(b)

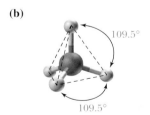

Figure 1.6
The shape of a methane molecule, CH_4. (a) Lewis structure and (b) shape. The hydrogens occupy the four corners of a regular tetrahedron, and all H—C—H bond angles are 109.5°.

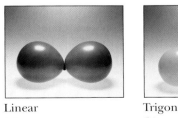  

Linear	Trigonal planar	Tetrahedral
(a)	**(b)**	**(c)**

Figure 1.5
Balloon models used to predict bond angles. (a) Two balloons assume a linear shape with a bond angle of 180° about the tie point. (b) Three balloons assume a trigonal planar shape with bond angles of 120° about the tie point. (c) Four balloons assume a tetrahedral shape with bond angles of 109.5° about the tie point. *(Charles D. Winters)*

(a)

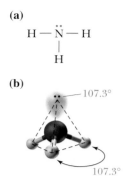

Figure 1.7
The shape of an ammonia molecule, NH₃. (a) Lewis structure and (b) ball-and-stick model. We describe the geometry of an ammonia molecule as **pyramidal**; that is, the molecule has a shape like a triangular pyramid with the three hydrogens at the base and nitrogen at the apex.

(a)

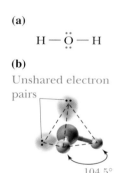

Figure 1.8
The shape of a water molecule, H₂O.
(a) A Lewis structure and
(b) a ball-and-stick model.

with hydrogen atoms. The fourth region contains an unshared pair of electrons (Figure 1.7). Using the VSEPR model, we predict that the four regions of electron density around nitrogen are arranged in a tetrahedral manner and that all H—N—H bond angles are 109.5°. The observed bond angles are 107.3°. We account for this small difference between the predicted and observed angles by proposing that the unshared pair of electrons on nitrogen repels adjacent electron pairs more strongly than bonding pairs repel each other.

Figure 1.8 shows a Lewis structure and a ball-and-stick model of a water molecule. In H₂O, oxygen is surrounded by four regions of electron density. Two of these regions contain pairs of electrons forming covalent bonds to two hydrogens; the remaining two contain unshared electron pairs. Using the VSEPR model, we predict that the four regions of electron density around oxygen are arranged in a tetrahedral manner and that the H—O—H bond angle is 109.5°. Experimental measurements show that the actual H—O—H bond angle is 104.5°, a value smaller than that predicted. We explain this difference between the predicted and observed bond angle by proposing, as we did for NH₃, that unshared pairs of electrons repel adjacent pairs more strongly than do bonding pairs. Note that the distortion from 109.5° is greater in H₂O, which has two unshared pairs of electrons, than it is in NH₃, which has only one unshared pair.

A general prediction emerges from this discussion of the shapes of CH₄, NH₃, and H₂O molecules. If a Lewis structure shows four regions of electron density around an atom, the VSEPR model predicts a tetrahedral distribution of electron density and bond angles of approximately 109.5°.

In many of the molecules we encounter, an atom is surrounded by three regions of electron density. Figure 1.9 shows Lewis structures for formaldehyde (CH₂O) and ethylene (C₂H₄).

In the VSEPR model, we treat a double bond as a single region of electron density. In formaldehyde, carbon is surrounded by three regions of electron density: Two regions contain single pairs of electrons, which form single bonds to hydrogen atoms; the third region contains two pairs of electrons, which form a double bond to oxygen. In ethylene, each carbon atom is also surrounded by three regions of electron density; two contain single pairs of electrons, and the third contains two pairs of electrons.

Three regions of electron density about an atom are farthest apart when they lie in a plane and make angles of 120° with each other. Thus, we predict that the H—C—H and H—C—O bond angles in formaldehyde and the H—C—H and H—C—C bond angles in ethylene are all 120°.

In still other types of molecules, a central atom is surrounded by only two regions of electron density. Figure 1.10 shows Lewis structures and ball-and-stick models of carbon dioxide (CO₂) and acetylene (C₂H₂).

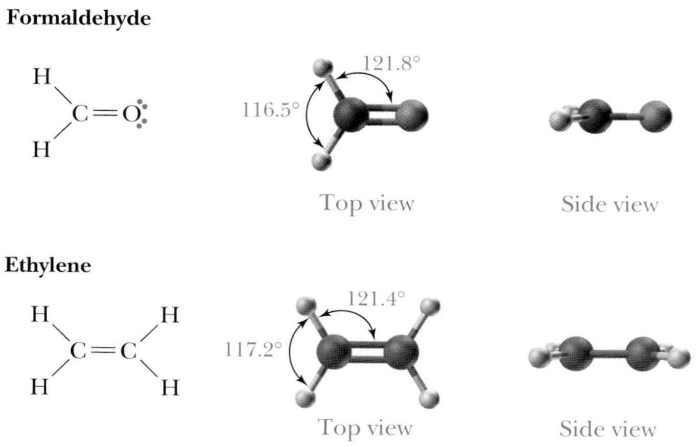

Formaldehyde

Ethylene

Figure 1.9
Shapes of formaldehyde (CH₂O) and ethylene (C₂H₄).

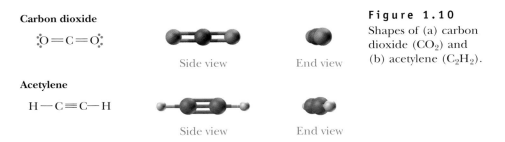

Figure 1.10
Shapes of (a) carbon dioxide (CO_2) and (b) acetylene (C_2H_2).

In carbon dioxide, carbon is surrounded by two regions of electron density, each containing two pairs of electrons and forming a double bond to an oxygen atom. In acetylene, each carbon is also surrounded by two regions of electron density, one containing a single pair of electrons and forming a single bond to a hydrogen atom and the other containing three pairs of electrons and forming a triple bond to a carbon atom. In each case, the two regions of electron density are farthest apart if they form a straight line through the central atom and create an angle of 180°. Both carbon dioxide and acetylene are linear molecules. Table 1.8 summarizes the predictions of the VSEPR model.

TABLE 1.8 Predicted Molecular Shapes (VSEPR Model)

Regions of Electron Density around Central Atom	Predicted Distribution of Electron Density	Predicted Bond Angles	Examples (Shape of the Molecule)
4	Tetrahedral	109.5°	Methane (tetrahedral) Ammonia (pyramidal) Water (bent)
3	Trigonal planar	120°	Ethylene (planar) Formaldehyde (planar)
2	Linear	180°	Carbon dioxide (linear) Acetylene (linear)

EXAMPLE 1.8

Predict all bond angles in these molecules:

(a) CH_3Cl

(b) $CH_2 \!=\! CHCl$

SOLUTION

(a) The Lewis structure for CH_3Cl shows the carbon surrounded by four regions of electron density. Therefore, we predict that the distribution of electron pairs about carbon is tetrahedral, that all bond angles are 109.5°, and that the shape of CH_3Cl is tetrahedral:

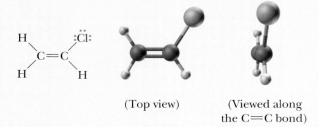

(b) The Lewis structure for $CH_2\!=\!CHCl$ shows each carbon surrounded by three regions of electron density. Therefore, we predict that all bond angles are 120°.

(Top view)   (Viewed along the C═C bond)

Practice Problem 1.8

Predict all bond angles for these molecules:

(a) CH_3OH (b) CH_2Cl_2 (c) H_2CO_2 (carbonic acid)

CHEMICAL CONNECTIONS

Buckyball: A New Form of Carbon

A favorite chemistry examination question is What are the elemental forms of carbon? The usual answer is that pure carbon is found in two forms: graphite and diamond. These forms have been known for centuries, and it was generally believed that they were the only forms of carbon having extended networks of C atoms in well-defined structures.

But that is not so! The scientific world was startled in 1985 when Richard Smalley of Rice University and Harry W. Kroto of the University of Sussex, England, and their coworkers announced that they had detected a new form of carbon with a molecular formula C_{60}. They suggested that the molecule has a structure resembling a soccer ball: 12 five-membered rings and 20 six-membered rings arranged such that each five-membered ring is surrounded by six-membered rings. This structure reminded its discoverers of a geodesic dome, a structure invented by the innovative American engineer and philosopher R. Buckminster Fuller. Therefore, the official name of the new allotrope of

carbon has become fullerene. Kroto, Smalley, and Robert F. Curl were awarded the Nobel prize for chemistry in 1996 for their work with fullerenes. Many higher fullerenes, such as C_{70} and C_{84}, have also been isolated and studied.

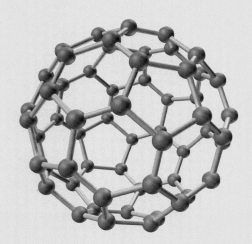

1.5 POLAR AND NONPOLAR MOLECULES

In Section 1.3C, we used the terms "polar" and "dipole" to describe a covalent bond in which one atom bears a partial positive charge and the other bears a partial negative charge. We also saw that we can use the difference in electronegativity between bonded atoms to determine the polarity of a covalent bond. We can now combine our understanding of bond polarity and molecular geometry (Section 1.4) to predict the polarity of molecules.

A molecule will be polar if (1) it has polar bonds and (2) its centers of partial positive and partial negative charge lie at different places within the molecule. Consider first carbon dioxide, CO_2, a molecule with two polar carbon–oxygen double bonds. Because carbon dioxide is a linear molecule, the centers of negative and positive partial charge coincide; therefore, this molecule is nonpolar.

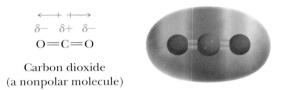

Carbon dioxide
(a nonpolar molecule)

In a water molecule, each O—H bond is polar, with oxygen, the more electronegative atom, bearing a partial negative charge and each hydrogen bearing a partial positive charge. Because water is a bent molecule, the center of its partial positive charge is between the two hydrogen atoms, and the center of its partial negative charge is on oxygen. Thus, water has polar bonds and, because of its geometry, is a polar molecule.

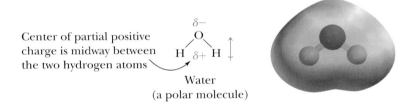

Center of partial positive charge is midway between the two hydrogen atoms

Water
(a polar molecule)

Ammonia has three polar N—H bonds, and because of its geometry, the centers of partial positive and negative charges are at different places within the molecule. Thus, ammonia has polar bonds and, because of its geometry, is a polar molecule.

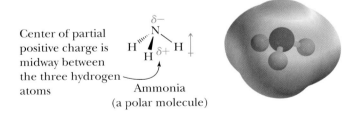

Center of partial positive charge is midway between the three hydrogen atoms

Ammonia
(a polar molecule)

EXAMPLE 1.9

Which of these molecules are polar? For each that is, specify the direction of its polarity.

(a) CH_3Cl (b) CH_2O (c) C_2H_2

SOLUTION

Both chloromethane (CH_3Cl) and formaldehyde (CH_2O) have polar bonds and, because of their geometry, are polar molecules. Because acetylene (C_2H_2) is linear, it is a nonpolar molecule.

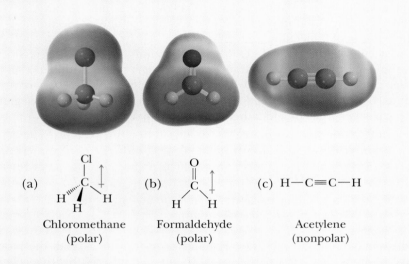

(a) Chloromethane (polar)   (b) Formaldehyde (polar)   (c) H—C≡C—H Acetylene (nonpolar)

Practice Problem 1.9

Both carbon dioxide (CO_2) and sulfur dioxide (SO_2) are triatomic molecules. Account for the fact that carbon dioxide is a nonpolar molecule, whereas sulfur dioxide is a polar molecule.

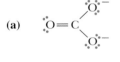

Figure 1.11
Three Lewis structures for the carbonate ion.

Resonance contributing structures Representations of a molecule or ion that differ only in the distribution of valence electrons.

Resonance hybrid A molecule or ion that is best described as a composite of a number of contributing structures.

Double-headed arrow A symbol used to connect contributing structures.

1.6 RESONANCE

As chemists developed a better understanding of covalent bonding in organic compounds, it became obvious that, for a great many molecules and ions, no single Lewis structure provides a truly accurate representation. For example, Figure 1.11 shows three Lewis structures for the carbonate ion, CO_3^{2-}, each of which shows carbon bonded to three oxygen atoms by a combination of one double bond and two single bonds. Each Lewis structure implies that one carbon–oxygen bond is different from the other two. This, however, is not the case; it has been shown that all three carbon–oxygen bonds are identical.

To describe the carbonate ion, as well as other molecules and ions for which no single Lewis structure is adequate, we turn to the theory of resonance.

A. The Theory of Resonance

The theory of resonance was developed by Linus Pauling in the 1930s. According to this theory, many molecules and ions are best described by writing two or more Lewis structures and considering the real molecule or ion to be a composite of these structures. We call individual Lewis structures **resonance contributing structures**. We show that the real molecule or ion is a **resonance hybrid** of the various contributing structures by interconnecting them with **double-headed arrows**.

Figure 1.12 shows three resonance contributing structures for the carbonate ion. The three are equivalent, meaning that they have identical patterns of covalent bonding and are of equal energy.

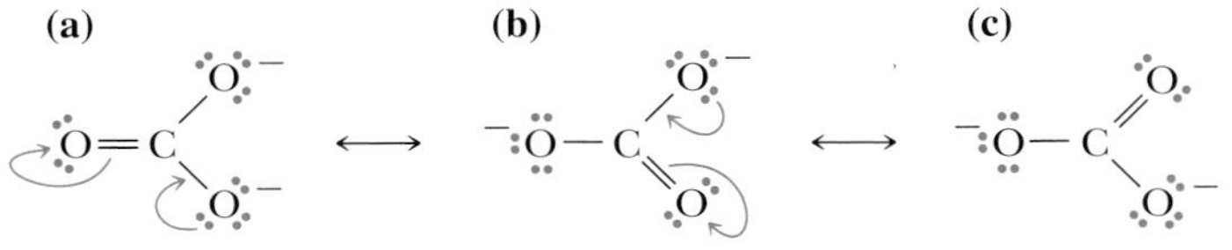

(a)   **(b)**   **(c)**

Figure 1.12
The carbonate ion represented as a hybrid of three equivalent resonance contributing
structures. Curved arrows show the redistribution of valence electrons between one
contributing structure and the next.

The use of the term "resonance" for this theory of covalent bonding might suggest to you that bonds and electron pairs constantly change back and forth from
one position to another over time. This notion is not at all correct. The carbonate
ion, for example, has one and only one real structure. The problem is ours: How
do we draw that one real structure? The resonance method is a way to describe the
real structure and at the same time retain Lewis structures with electron-pair bonds.
Thus, although we realize that the carbonate ion is not accurately represented by
any one resonance contributing structure shown in Figure 1.12, we continue to
represent it as one of these for convenience. We understand, of course, that what is
intended is the resonance hybrid.

B. Curved Arrows and Electron Pushing

Notice in Figure 1.12 that the only change from resonance contributing structure
(a) to (b) and then from (b) to (c) is a redistribution of valence electrons. To show
how this redistribution of valence electrons occurs, chemists use a symbol called a
curved arrow, which shows the repositioning of an electron pair from its origin (the
tail of the arrow) to its destination (the head of the arrow). The repositioning may
be from an atom to an adjacent bond or from a bond to an adjacent atom.

Curved arrow A symbol used to
show the redistribution of valence
electrons.

A curved arrow is nothing more than a bookkeeping symbol for keeping track
of electron pairs or, as some call it, **electron pushing**. Do not be misled by its sim-
plicity. Electron pushing will help you see the relationship between contributing
structures. Furthermore, it will help you follow bond-breaking and bond-forming
steps in organic reactions. Understanding this type of electron pushing is a survival
skill in organic chemistry.

C. Rules for Writing Acceptable Resonance Contributing Structures

You must follow these four rules in writing acceptable resonance contributing
structures:

1. All resonance contributing structures must have the same number of valence
   electrons.
2. All resonance contributing structures must obey the rules of covalent bonding;
   thus, no contributing structure may have more than 2 electrons in the valence
   shell of hydrogen or more than 8 electrons in the valence shell of a second-
   period element. Third-period elements, such as sulfur and phosphorus, may
   have up to 12 electrons in their valence shells.
3. The positions of all nuclei must be the same; that is, resonance contributing
   structures differ only in the distribution of valence electrons.
4. All resonance contributing structures must have the same total number of
   paired and unpaired electrons.

EXAMPLE 1.10

Which sets are pairs of resonance contributing structures?

(a) 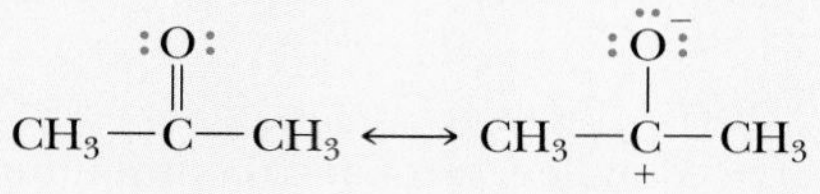 CH₃—C(=O)—CH₃ ⟷ CH₃—C(—O⁻)—CH₃⁺

(b) 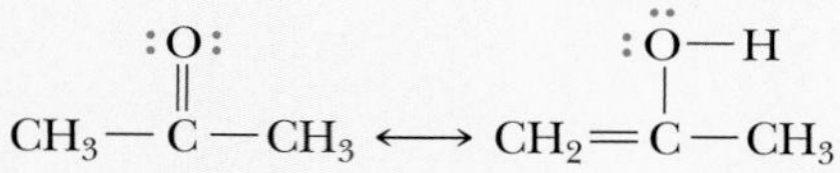 CH₃—C(=O)—CH₃ ⟷ CH₂=C(—O—H)—CH₃

SOLUTION

(a) A pair of resonance contributing structures. They differ only in the distribution of valence electrons.
(b) Not a pair of resonance contributing structures. They differ in the arrangement of their atoms.

Practice Problem 1.10

Which sets are pairs of resonance contributing structures?

(a) 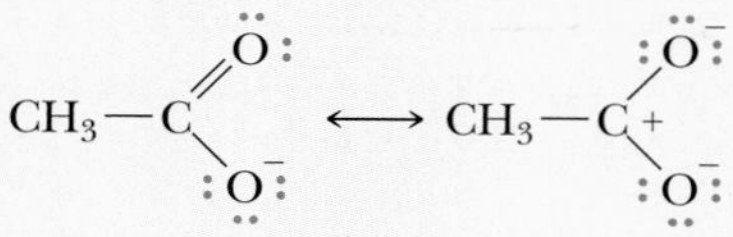 CH₃—C(=O)(O⁻) ⟷ CH₃—C⁺(O⁻)(O⁻)

(b) 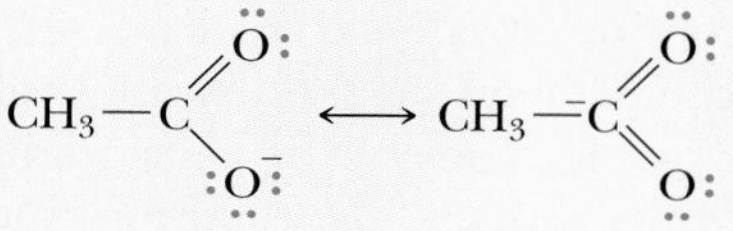 CH₃—C(=O)(O⁻) ⟷ CH₃=C(O⁻)(O:)

EXAMPLE 1.11

Draw the resonance contributing structure indicated by the curved arrows. Be certain to show all valence electrons and all formal charges.

(a) 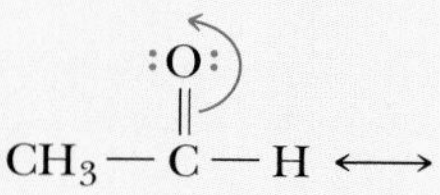 CH₃—C(=O)—H ⟷

(b) H—C⁻(H)—C(=O)—H ⟷

(c) CH₃—O—C⁺(H)—H ⟷

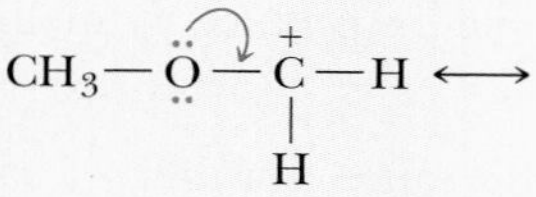

SOLUTION

(a) 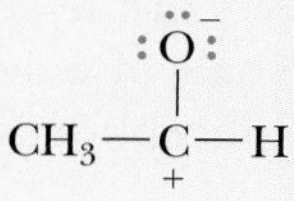 CH₃—C⁺(—O⁻)—H

(b) H—C(=C)(H)—H with O⁻

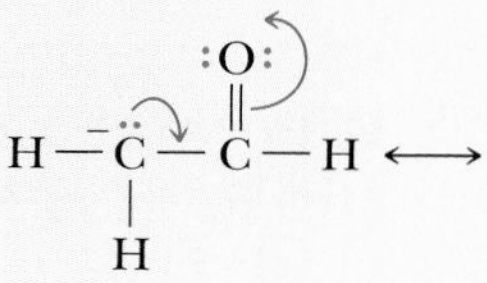

(c) CH₃—O⁺=C(H)—H

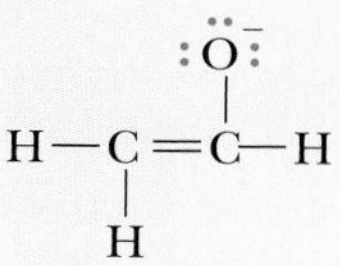

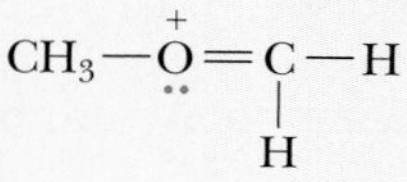

Practice Problem 1.11

Use curved arrows to show the redistribution of valence electrons in converting resonance contributing structure (a) to (b) and then (b) to (c). Also show, using curved arrows, how (a) can be converted to (c) without going through (b).

$$\underset{(a)}{CH_3-C\begin{smallmatrix}\ddot{O}:\\\\:\ddot{O}:^-\end{smallmatrix}} \longleftrightarrow \underset{(b)}{CH_3-\overset{+}{C}\begin{smallmatrix}:\ddot{O}:^-\\\\:\ddot{O}:^-\end{smallmatrix}} \longleftrightarrow \underset{(c)}{CH_3-C\begin{smallmatrix}:\ddot{O}:^-\\\\O:\end{smallmatrix}}$$

1.7 ORBITAL OVERLAP MODEL OF COVALENT BONDING

As much as the Lewis and VSEPR models help us to understand covalent bonding and the geometry of molecules, they leave many questions unanswered. The most important of these questions is the relation between molecular structure and chemical reactivity. For example, carbon–carbon double bonds are different in chemical reactivity from carbon–carbon single bonds. Most carbon–carbon single bonds are quite unreactive, but carbon–carbon double bonds react with a wide variety of reagents. The Lewis model gives us no way to account for these differences. Therefore, let us turn to a newer model of covalent bonding, namely, the formation of covalent bonds by the overlap of atomic orbitals.

A. Shapes of Atomic Orbitals

One way to visualize the electron density associated with a particular orbital is to draw a boundary surface around the region of space that encompasses some arbitrary percentage of the negative charge associated with that orbital. Most commonly, we draw the boundary surface at 95%. Drawn in this manner, all *s* orbitals have the shape of a sphere with its center at the nucleus (Figure 1.13). Of the various *s* orbitals, the sphere representing the 1*s* orbital is the smallest. A 2*s* orbital is a larger sphere, and a 3*s* orbital is an even larger sphere.

Figure 1.14 shows the three-dimensional shapes of the three 2*p* orbitals, combined in one diagram to illustrate their relative orientations in space. Each 2*p* orbital consists of two lobes arranged in a straight line with the nucleus in the middle. The three 2*p* orbitals are mutually perpendicular and are designated $2p_x$, $2p_y$, and $2p_z$.

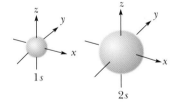

Figure 1.13
Shapes of 1*s* and 2*s* atomic orbitals.

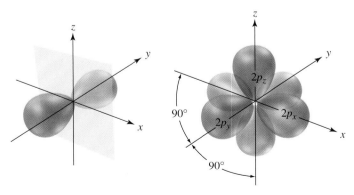

Figure 1.14
Shapes of $2p_x$, $2p_y$, and $2p_z$ atomic orbitals. The three 2*p* orbitals are mutually perpendicular. One lobe of each orbital is shown in red, the other in blue.

B. Formation of a Covalent Bond by the Overlap of Atomic Orbitals

According to the orbital overlap model, a covalent bond is formed when a portion of an atomic orbital of one atom overlaps a portion of an atomic orbital of another atom. In forming the covalent bond in H_2, for example, two hydrogens approach each other so that their $1s$ atomic orbitals overlap to form a sigma covalent bond (Figure 1.15). A **sigma (σ) bond** is a covalent bond in which orbitals overlap along the axis joining the two nuclei.

Sigma (σ) bond A covalent bond in which the overlap of atomic orbitals is concentrated along the bond axis.

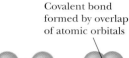

Covalent bond formed by overlap of atomic orbitals

Figure 1.15
Formation of the covalent bond in H_2 by the overlap of the $1s$ atomic orbitals of each hydrogen.

C. Hybridization of Atomic Orbitals

The formation of a covalent bond between two hydrogen atoms is straightforward. The formation of covalent bonds with second-period elements, however, presents the following problem: In forming covalent bonds, atoms of carbon, nitrogen, and oxygen (all second-period elements), use $2s$ and $2p$ atomic orbitals. The three $2p$ atomic orbitals are at angles of 90° to one another (Figure 1.14), and if atoms of second-period elements used these orbitals to form covalent bonds, the bond angles around each would be approximately 90°. Bond angles of 90°, however, are rarely observed in organic molecules. What we find, instead, are bond angles of approximately 109.5° in molecules with only single bonds, 120° in molecules with double bonds, and 180° in molecules with triple bonds:

H
| 109.5°
C
H   H
H

120° (approx)
H   H
C=C
H   H

180°
H—C≡C—H

Hybrid orbital An orbital produced from the combination of two or more atomic orbitals.

To account for these observed bond angles, Pauling proposed that atomic orbitals combine to form new orbitals, called **hybrid orbitals**. The number of hybrid orbitals formed is equal to the number of atomic orbitals combined. Elements of the second period form three types of hybrid orbitals, designated sp^3, sp^2, and sp, each of which can contain up to two electrons.

D. sp^3 Hybrid Orbitals: Bond Angles of Approximately 109.5°

sp^3 Hybrid orbital An orbital produced by the combination of one s atomic orbital and three p atomic orbitals.

The combination of one $2s$ atomic orbital and three $2p$ atomic orbitals forms four equivalent **sp^3 hybrid orbitals**. Because they are derived from four atomic orbitals, sp^3 hybrid orbitals always occur in sets of four. Each sp^3 hybrid orbital consists of a larger lobe pointing in one direction and a smaller lobe pointing in the opposite direction. The axes of the four sp^3 hybrid orbitals point toward the corners of a regular tetrahedron, and sp^3 hybridization results in bond angles of approximately 109.5° (Figure 1.16).

Figure 1.16
sp^3 Hybrid orbitals.
(a) Representation of a single sp^3 hybrid orbital showing two lobes of unequal size.
(b) Three-dimensional representation of four sp^3 hybrid orbitals, which point toward the corners of a regular tetrahedron. The smaller lobes of each sp^3 hybrid orbital are hidden behind the larger lobes.

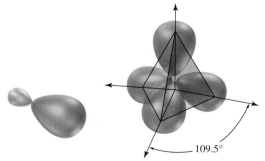

(a) An sp^3 orbital   **(b)** Four tetrahedral sp^3 orbitals

109.5°

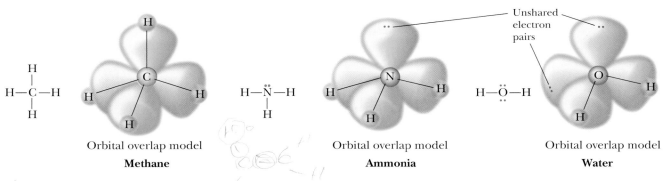

Orbital overlap model
Methane

Orbital overlap model
Ammonia

Orbital overlap model
Water

Figure 1.17
Orbital overlap pictures of methane, ammonia, and water.

Keep in mind that superscripts in the designation of hybrid orbitals tell you how many atomic orbitals have been combined to form the hybrid orbitals. The designation sp^3, for example, tells you that *one s* atomic orbital and *three p* atomic orbitals are combined in forming the hybrid orbital. Do not confuse this use of superscripts with how we use superscripts in writing a ground-state electron configuration—for example, $1s^2 2s^2 2p^5$ for fluorine. In the case of a ground-state electron configuration, superscripts tell you the number of electrons in each orbital or set of orbitals.

In Section 1.3, we described the covalent bonding in CH_4, NH_3, and H_2O in terms of the Lewis model, and in Section 1.4 we used the VSEPR model to predict bond angles of approximately 109.5° in each molecule. Now let us consider the bonding in these molecules in terms of the overlap of atomic orbitals. To bond with four other atoms with bond angles of 109.5°, carbon uses a set of four sp^3 hybrid orbitals. Carbon has four valence electrons, and one electron is placed in each sp^3 hybrid orbital. Each partially filled sp^3 hybrid orbital then overlaps with a partially filled $1s$ atomic orbital of hydrogen to form a sigma (σ) bond, and hydrogen atoms occupy the corners of a regular tetrahedron (Figure 1.17).

In bonding with three other atoms, the five valence electrons of nitrogen are distributed so that one sp^3 hybrid orbital is filled with a pair of electrons and the remaining three sp^3 hybrid orbitals have one electron each. Overlapping of these partially filled sp^3 hybrid orbitals with $1s$ atomic orbitals of hydrogen atoms produces the NH_3 molecule (Figure 1.17).

In bonding with two other atoms, the six valence electrons of oxygen are distributed so that two sp^3 hybrid orbitals are filled and the remaining two have one electron each. Each partially filled sp^3 hybrid orbital overlaps with a $1s$ atomic orbital of hydrogen, and hydrogen atoms occupy two corners of a regular tetrahedron. The remaining two corners of the tetrahedron are occupied by unshared pairs of electrons (Figure 1.17).

E. sp^2 Hybrid Orbitals: Bond Angles of Approximately 120°

The combination of one $2s$ atomic orbital and two $2p$ atomic orbitals forms three equivalent **sp^2 hybrid orbitals**. Because they are derived from three atomic orbitals, sp^2 hybrid orbitals always occur in sets of three. Each sp^2 hybrid orbital consists of two lobes, one larger than the other. The three sp^2 hybrid orbitals lie in a plane and are directed toward the corners of an equilateral triangle; the angle between sp^2 hybrid orbitals is 120°. The third $2p$ atomic orbital (remember, $2p_x$, $2p_y$, and $2p_z$!) is not involved in hybridization and consists of two lobes lying perpendicular to the plane

sp^2 Hybrid orbital An orbital produced by the combination of one s atomic orbital and two p atomic orbitals.

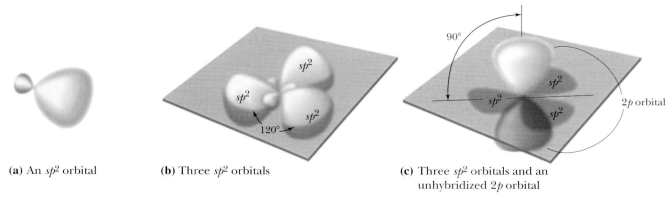

(a) An sp^2 orbital   **(b)** Three sp^2 orbitals   **(c)** Three sp^2 orbitals and an
unhybridized $2p$ orbital

Figure 1.18
sp^2 Hybrid orbitals. (a) A single sp^2 hybrid orbital showing two lobes of unequal size. (b) The
three sp^2 hybrid orbitals with their axes in a plane at angles of 120°. (c) The unhybridized $2p$
atomic orbital perpendicular to the plane created by the three sp^2 hybrid orbitals.

of the sp^2 hybrid orbitals. Figure 1.18 shows three equivalent sp^2 hybrid orbitals, along
with the remaining unhybridized $2p$ atomic orbital.

Second-period elements use sp^2 hybrid orbitals to form double bonds.
Figure 1.19(a) shows a Lewis structure for ethylene, C_2H_4. A sigma bond between
the carbons in ethylene forms by the overlap of sp^2 hybrid orbitals along a common
axis [Figure 1.19(b)]. Each carbon also forms sigma bonds to two hydrogens. The
remaining $2p$ orbitals on adjacent carbon atoms lie parallel to each other and overlap to form a pi bond [Figure 1.19(c)]. A **pi (π) bond** is a covalent bond formed
by the overlap of parallel p orbitals. Because of the lesser degree of overlap of
orbitals forming pi bonds compared with those forming sigma bonds, pi bonds are
generally weaker than sigma bonds.

The orbital overlap model describes all double bonds in the same way that we
have described carbon–carbon double bonds. In formaldehyde, CH_2O, the simplest organic molecule containing a carbon–oxygen double bond, carbon forms
sigma bonds to two hydrogens by the overlap of an sp^2 hybrid orbital of carbon and
the $1s$ atomic orbital of each hydrogen. Carbon and oxygen are joined by a sigma

Pi (π) bond A covalent bond
formed by the overlap of parallel p
orbitals.

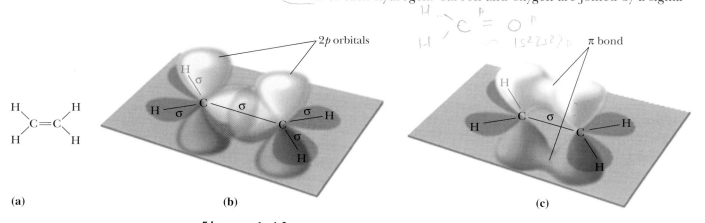

(a)   **(b)**   **(c)**

Figure 1.19
Covalent bond formation in ethylene. (a) Lewis structure, (b) overlap of sp^2 hybrid orbitals
forms a sigma (σ) bond between the carbon atoms, and (c) overlap of parallel $2p$ orbitals
forms a pi (π) bond. Ethylene is a planar molecule; that is, the two carbons of the double
bond and the four atoms bonded to them all lie in the same plane.

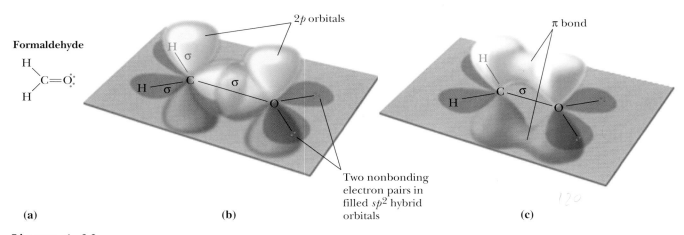

Formaldehyde

$$\underset{H}{\overset{H}{>}}C=\ddot{O}:$$

Figure 1.20
A carbon–oxygen double bond. (a) Lewis structure of formaldehyde, CH_2O, (b) the sigma (σ) bond framework and nonoverlapping parallel $2p$ atomic orbitals, and (c) overlap of parallel $2p$ atomic orbitals to form a pi (π) bond.

bond formed by the overlap of sp^2 hybrid orbitals and a pi bond formed by the overlap of unhybridized $2p$ atomic orbitals (Figure 1.20).

F. *sp* Hybrid Orbitals: Bond Angles of Approximately 180°

The combination of one $2s$ atomic orbital and one $2p$ atomic orbital forms two equivalent ***sp* hybrid orbitals**. Because they are derived from two atomic orbitals, sp hybrid orbitals always occur in sets of two. The two sp hybrid orbitals lie at an angle of 180°. The axes of the unhybridized $2p$ atomic orbitals are perpendicular to each other and to the axis of the two sp hybrid orbitals. Figure 1.21 shows the two sp hybrid orbitals on the *x*-axis and the unhybridized $2p$ orbitals on the *y*-axis and *z*-axis.

Figure 1.22 shows a Lewis structure and an orbital overlap diagram for acetylene, C_2H_2. A carbon–carbon triple bond consists of one sigma bond and two pi bonds. The sigma bond is formed by the overlap of sp hybrid orbitals. One pi bond is formed by the overlap of a pair of parallel $2p$ atomic orbitals. The second pi bond is formed by the overlap of a second pair of parallel $2p$ atomic orbitals.

sp **Hybrid orbital** A hybrid atomic orbital produced by the combination of one *s* atomic orbital and one *p* atomic orbital.

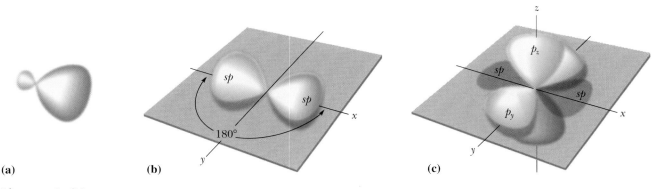

Figure 1.21
sp Hybrid orbitals. (a) A single *sp* hybrid orbital consisting of two lobes of unequal size. (b) Two *sp* hybrid orbitals in a linear arrangement. (c) Unhybridized $2p$ atomic orbitals are perpendicular to the line created by the axes of the two *sp* hybrid orbitals.

Acetylene

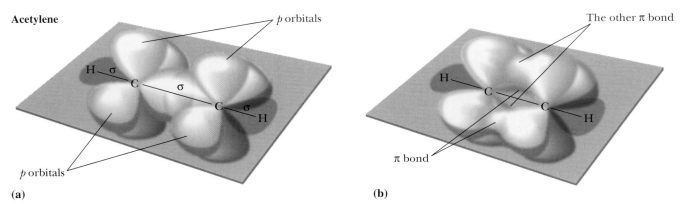

Figure 1.22
Covalent bonding in acetylene. (a) The sigma bond framework shown along with nonoverlapping $2p$ atomic orbitals. (b) Formation of two pi bonds by overlap of two sets of parallel $2p$ atomic orbitals.

Table 1.9 summarizes the relationship among the number of groups bonded to carbon, orbital hybridization, and the types of bonds involved.

TABLE 1.9   Covalent Bonding of Carbon

Groups Bonded to Carbon	Orbital Hybridization	Predicted Bond Angles	Types of Bonds to Carbon	Example	Name
4	sp^3	109.5°	four sigma bonds	H—C—C—H (ethane structure)	ethane
3	sp^2	120°	three sigma bonds and one pi bond	C=C (ethylene structure)	ethylene
2	sp	180°	two sigma bonds and two pi bonds	H—C≡C—H	acetylene

EXAMPLE 1.12

Describe the bonding in acetic acid, CH_3COOH, in terms of the orbitals involved, and predict all bond angles.

SOLUTION

The following are three identical Lewis structures: Labels on the first structure point to atoms and show hybridization. Labels on the second structure point to bonds and show the type of bond, either sigma or pi. Labels on the third structure point to atoms and show bond angles about each atom as predicted by the valence-shell electron-pair repulsion model.

Practice Problem 1.12

Describe the bonding in these molecules in terms of the atomic orbitals involved, and predict all bond angles:

(a) $CH_3CH{=}CH_2$ (b) CH_3NH_2

1.8 FUNCTIONAL GROUPS

There are over 10 million organic compounds that have been discovered or made by organic chemists! Surely it would seem to be an almost impossible task to learn the physical and chemical properties of this many compounds. Fortunately, the study of organic compounds is not as formidable a task as you might think. While organic compounds can undergo a wide variety of chemical reactions, only certain portions of their structure are changed in any particular reaction. The part of an organic molecule that undergoes chemical reactions is called a **functional group**, and, as we will see, the same functional group, in whatever organic molecule we find it, undergoes the same types of chemical reactions. Therefore, you do not have to study the chemical reactions of even a fraction of the 10 million known organic compounds. Instead you need only to identify a few characteristic types of functional groups and then study the chemical reactions that each undergoes.

Functional groups are also important because they are the units by which we divide organic compounds into families of compounds. For example, we group those compounds which contain an —OH (hydroxyl) group bonded to a tetrahedral carbon into a family called alcohols, and compounds containing a —COOH (carboxyl) group into a family called carboxylic acids. In Table 1.10, we introduce five of the most common functional groups. A complete list of all functional groups we will study is on the inside back cover of the text.

Functional group An atom or a group of atoms within a molecule that shows a characteristic set of physical and chemical properties.

TABLE 1.10   Five Common Functional Groups

Functional group	Name of group	Present in	Example	Name of example
—OH	hydroxyl	alcohols	CH_3CH_2OH	Ethanol
—NH$_2$	amino	amines	$CH_3CH_2NH_2$	Ethanamine
$-\overset{\text{O}}{\overset{\|}{C}}-H$	carbonyl	aldehydes	$CH_3\overset{\text{O}}{\overset{\|}{C}}H$	Ethanal
$-\overset{\text{O}}{\overset{\|}{C}}-$	carbonyl	ketones	$CH_3\overset{\text{O}}{\overset{\|}{C}}CH_3$	Acetone
$-\overset{\text{O}}{\overset{\|}{C}}-OH$	carboxyl	carboxylic acids	$CH_3\overset{\text{O}}{\overset{\|}{C}}OH$	Acetic acid

At this point, our concern is only pattern recognition—that is, how to recognize these five functional group when you see them and how to draw structural formulas of molecules containing them.

Finally, functional groups serve as the basis for naming organic compounds. Ideally, each of the 10 million or more organic compounds must have a name that is different from every other compound.

To summarize, functional groups

- are sites of chemical reaction; a particular functional group, in whatever compound we find it, undergoes the same types of chemical reactions.
- determine, in large measure, the physical properties of a compound.
- are the units by which we divide organic compounds into families.
- serve as a basis for naming organic compounds.

A. Alcohols

Hydroxyl group An —OH group.

The functional group of an **alcohol** is an **—OH (hydroxyl)** group bonded to a tetrahedral (sp^3 hybridized) carbon atom. In the general formula on the left, we use the symbol R to indicate either a hydrogen or another carbon group. The important point in the general structure is that the —OH group is bonded to a tetrahedral carbon atom:

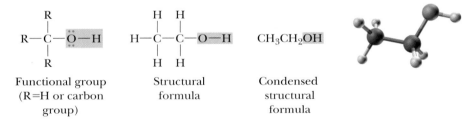

Functional group   Structural   Condensed
(R=H or carbon   formula   structural
group)   formula

The rightmost representation of this alcohol is a **condensed structural formula**, CH_3CH_2OH. In a condensed structural formula, CH_3 indicates a carbon bonded to three hydrogens, CH_2 indicates a carbon bonded to two hydrogens, and CH indicates a carbon bonded to one hydrogen. We generally do not show unshared pairs of electrons in a condensed structural formula.

Alcohols are classified as **primary (1°)**, **secondary (2°)**, or **tertiary (3°)**, depending on the number of carbon atoms bonded to the carbon bearing the —OH group:

A 1° alcohol   A 2° alcohol   A 3° alcohol

EXAMPLE 1.13

Write condensed structural formulas for the two alcohols with molecular formula C_3H_8O. Classify each as primary, secondary, or tertiary.

SOLUTION

First, bond the three carbon atoms in a chain with the —OH (hydroxyl) group bonded to the end carbon or the middle carbon of the chain. Then, to complete each structural formula, add seven hydrogens so that each carbon has four bonds to it:

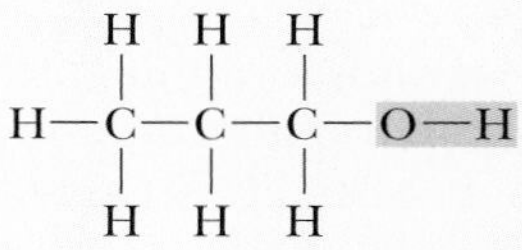

H—C—C—C—O—H   or $CH_3CH_2CH_2OH$

A 1° alcohol

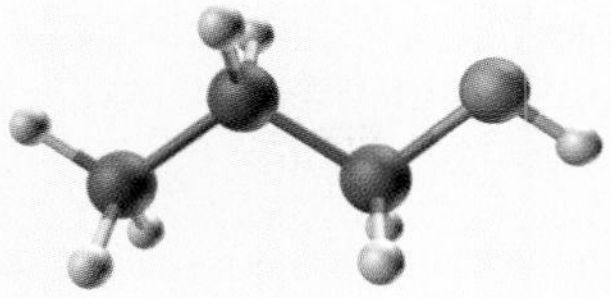

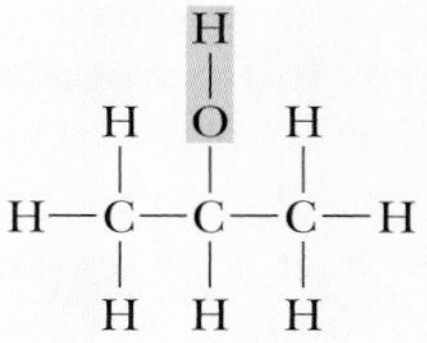

H—C—C—C—H   or CH_3CHCH_3

A 2° alcohol

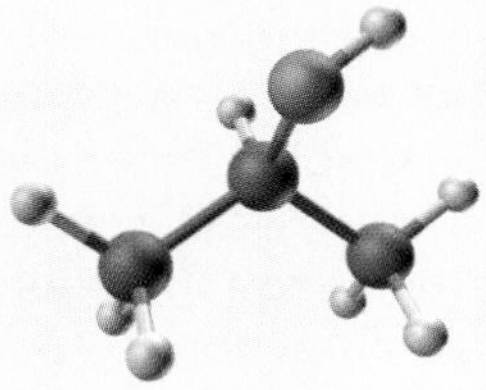

Practice Problem 1.13

Write condensed structural formulas for the four alcohols with molecular formula $C_4H_{10}O$. Classify each as primary, secondary, or tertiary.

B. Amines

The functional group of an amine is an **amino group**—a nitrogen atom bonded to one, two, or three carbon atoms. In a **primary (1°) amine**, nitrogen is bonded to one carbon atom. In a **secondary (2°) amine**, it is bonded to two carbon atoms, and in a **tertiary (3°) amine,** it is bonded to three carbon atoms. The second and third structural formulas can be written in a more abbreviated form by collecting the CH_3 groups and writing them as $(CH_3)_2NH$ and $(CH_3)_3N$, respectively.

Amino group An sp^3 hybridized nitrogen atom bonded to one, two, or three carbon groups.

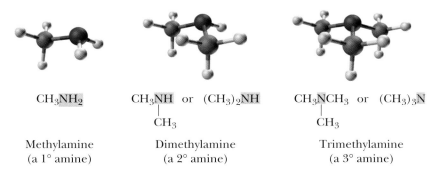

CH_3NH_2

Methylamine
(a 1° amine)

CH_3NH or $(CH_3)_2NH$
|
CH_3

Dimethylamine
(a 2° amine)

CH_3NCH_3 or $(CH_3)_3N$
|
CH_3

Trimethylamine
(a 3° amine)

EXAMPLE 1.14

Write condensed structural formulas for the two primary (1°) amines with molecular formula C_3H_9N.

SOLUTION

For a primary amine, draw a nitrogen atom bonded to two hydrogens and one carbon. The nitrogen may be bonded to the three-carbon chain in two different ways. Then add the seven hydrogens to give each carbon four bonds and give the correct molecular formula:

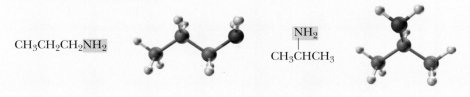

$CH_3CH_2CH_2NH_2$

NH_2
|
CH_3CHCH_3

Practice Problem 1.14

Write condensed structural formulas for the three secondary amines with molecular formula $C_4H_{11}N$.

C. Aldehydes and Ketones

Carbonyl group A C=O group.

Both aldehydes and ketones contain a **C=O (carbonyl)** group. The **aldehyde** functional group contains a carbonyl group bonded to a hydrogen. In formaldehyde, CH_2O, the simplest aldehyde, the carbonyl carbon is bonded to two hydrogen atoms. In a condensed structural formula, the aldehyde group may be written showing the carbon–oxygen double bond as CH=O, or, alternatively, it may be written —CHO. The functional group of a **ketone** is a carbonyl group bonded to two carbon atoms.

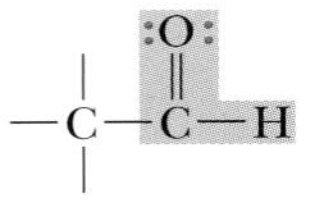

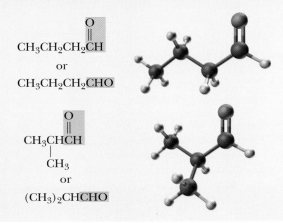

Functional group

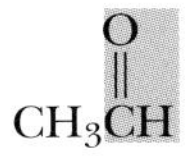

An aldehyde

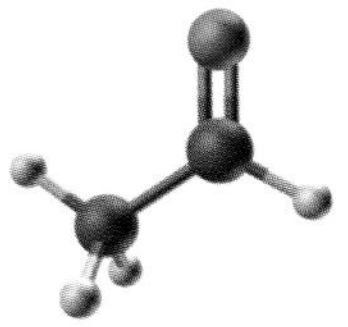

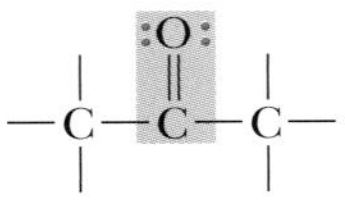

Functional group

A ketone

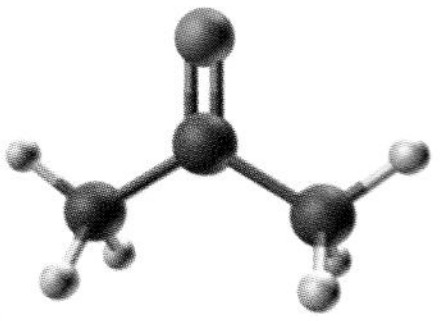

Example 1.15

Write condensed structural formulas for the two aldehydes with molecular formula C_4H_8O.

SOLUTION

First, draw the functional group of an aldehyde and add the remaining carbons, which may be bonded in two different ways. Then, add seven hydrogens to complete the four bonds of each carbon and give the correct molecular formula: Note that the aldehyde group may be written showing the carbon–oxygen double bond as C=O, or, alternatively, it may be written —CHO.

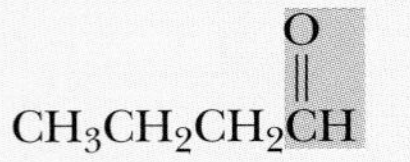

or

$CH_3CH_2CH_2CHO$

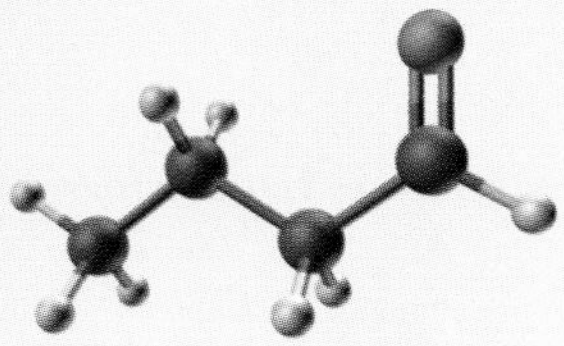

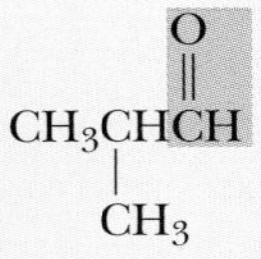

or

$(CH_3)_2CHCHO$

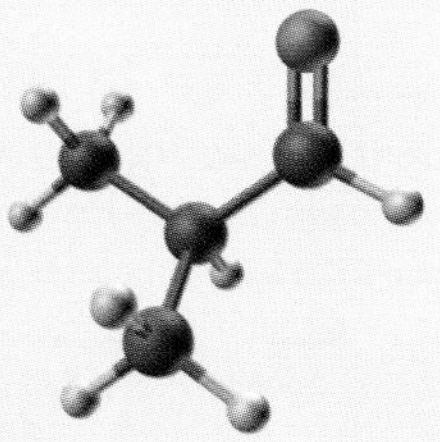

Practice Problem 1.15

Write condensed structural formulas for the three ketones with molecular formula $C_5H_{10}O$.

D. Carboxylic Acids

The functional group of a carboxylic acid is a —**COOH** (**carboxyl**: *carb*onyl + hydr*oxyl*) group:

Carboxyl group A —COOH group.

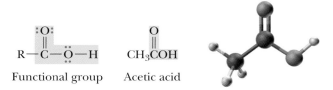

Functional group   Acetic acid

EXAMPLE 1.16

Write a condensed structural formula for the single carboxylic acid with molecular formula $C_3H_6O_2$.

SOLUTION

$$CH_3CH_2COH \quad \text{or} \quad CH_3CH_2COOH$$

Practice Problem 1.16

Write condensed structural formulas for the two carboxylic acids with molecular formula $C_4H_8O_2$.

SUMMARY

An atom consists of a small, dense nucleus and electrons concentrated about the nucleus in regions of space called **shells** (Section 1.2A). Each shell can contain as many as $2n^2$ electrons, where n is the number of the shell. Each shell is subdivided into regions of space called **orbitals**. The first shell ($n = 1$) has a single s orbital and can hold $2 \times 1^2 = 2$ electrons. The second shell ($n = 2$) has one s orbital and three p orbitals and can hold $2 \times 2^2 = 8$ electrons. The **Lewis structure** (Section 1.2B) of an element shows the symbol of the element surrounded by a number of dots equal to the number of electrons in its **valence shell**. According to the **Lewis model of bonding** (Section 1.3), atoms bond together in such a way that each atom participating in a chemical bond acquires a completed valence-shell electron configuration resembling that of the noble gas nearest it in atomic number. Atoms that lose sufficient electrons to acquire a completed valence shell become **cations**; atoms that gain sufficient electrons to acquire a completed valence shell become **anions**. An **ionic bond** is a chemical bond formed by the attractive force between an anion and a cation. A **covalent bond** is a chemical bond formed by the sharing of electron pairs between atoms. The tendency of main-group elements (those of Groups 1A–7A) to achieve an outer shell of eight valence electrons is called the **octet rule**.

Electronegativity (Section 1.3C) is a measure of the force of attraction by an atom for electrons it shares in a chemical bond with another atom. Electronegativity increases from left to right and from bottom to top in the Periodic Table. As a rough guideline, we say that a **nonpolar covalent bond** (Section 1.3C) is a covalent bond in which the difference in electronegativity between the bonded atoms is less than 0.5 unit. A **polar covalent bond** is a covalent bond in which the difference in electronegativity between the bonded atoms is between 0.5 and 1.9 units.

In a polar covalent bond, the more electronegative atom bears a partial negative charge ($\delta-$) and the less electronegative atom bears a partial positive charge ($\delta+$).

A **Lewis structure** (Section 1.3D) for a molecule or an ion must show (1) the correct arrangement of atoms, (2) the correct number of valence electrons, (3) no more than two electrons in the outer shell of hydrogen, (4) no more than eight electrons in the outer shell of any second-period element, and (5) all formal charges. **Formal charge** is the charge on an atom in a molecule or polyatomic ion (Section 1.3E).

We can predict bond angles of molecules and polyatomic ions using Lewis structures and the **valence-shell electron-pair repulsion (VSEPR) model** (Section 1.4). For atoms surrounded by four regions of electron density, we predict bond angles of 109.5°; by three regions of electron density, bond angles of 120°; and by two regions of electron density, bond angles of 180°.

A molecule is polar if it has one or more polar bonds and its centers of partial positive and negative charge are at different places within the molecule (Section 1.5).

According to the **theory of resonance** (Section 1.6A), a molecule or ion for which no single Lewis structure is adequate is best described by writing two or more **resonance contributing structures** and considering the real molecule or ion to be a **hybrid** of the various contributing structures. Resonance contributing structures are interconnected by **double-headed arrows**. We show how valence electrons are redistributed from one contributing structure to the next by **curved arrows** (Section 1.6B). A curved arrow extends from where the electrons are initially shown (on an atom or in a covalent bond) to their new location (an adjacent atom or an adjacent covalent bond). The use of curved arrows in this way is commonly referred to as **electron pushing**.

According to the **orbital overlap model**, the formation of a covalent bond results from the overlap of atomic orbitals (Section 1.7B). The greater the overlap, the stronger is the resulting covalent bond. The combination of atomic orbitals is called **hybridization** (Section 1.7C), and the resulting orbitals are called **hybrid orbitals.** The combination of one $2s$ atomic orbital and three $2p$ atomic orbitals produces four equivalent sp^3 **hybrid orbitals**, each pointing toward a corner of a regular tetrahedron at angles of 109.5°.

The combination of one $2s$ atomic orbital and two $2p$ atomic orbitals produces three equivalent sp^2 **hybrid orbitals**, the axes of which lie in a plane at angles of 120°. Most C=C and C=O double bonds are a combination of one **sigma (σ) bond** formed by the overlap of sp^2 hybrid orbitals and one **pi (π) bond** formed by the overlap of parallel $2p$ atomic orbitals.

The combination of one $2s$ atomic orbital and one $2p$ atomic orbital produces two equivalent sp **hybrid orbitals**, the axes of which lie in a plane at an angle of 180°. All C≡C triple bonds are a combination of one sigma bond formed by the overlap of sp hybrid orbitals and two pi bonds formed by the overlap of two pairs of parallel $2p$ atomic orbitals.

Functional groups (Section 1.8) are characteristic structural units by which we divide organic compounds into classes and that serve as a basis for nomenclature. They are also sites of chemical reactivity; a particular functional group, in whatever compound we find it, undergoes the same types of reactions. Important functional groups for us at this stage in the course are the **hydroxyl group** of 1°, 2°, and 3° alcohols, the **amino group** of 1°, 2°, and 3° amines, the **carbonyl group** of aldehydes and ketones, and the **carboxyl group** of carboxylic acids.

Problems

A problem number set in red indicates an applied "real-world" problem.

Electronic Structure of Atoms

1.17 Write the ground-state electron configuration for each element. The atomic number for each is provided in parentheses.

(a) Sodium (11) (b) Magnesium (12) (c) Oxygen (8) (d) Nitrogen (7)

1.18 Write the ground-state electron configuration for each element:

(a) Potassium (b) Aluminum

(c) Phosphorus (d) Argon

1.19 Which element has the ground-state electron configuration

(a) $1s^22s^22p^63s^23p^4$ (b) $1s^22s^22p^4$

1.20 Which element or ion does not have the ground-state electron configuration $1s^22s^22p^63s^23p^6$?

(a) S^{2-} (b) Cl^- (c) Ar (d) Ca^{2+} (e) K

1.21 Define *valence shell* and *valence electron*.

1.22 How many electrons are in the valence shell of each element?

(a) Carbon (b) Nitrogen (c) Chlorine (d) Aluminum (e) Oxygen

1.23 How many electrons are in the valence shell of each ion?

(a) H^+ (b) H^-

Lewis Structures

1.24 Judging from their relative positions in the Periodic Table, which element in each set is more electronegative?

(a) Carbon or nitrogen   **(b)** Chlorine or bromine

(c) Oxygen or sulfur

1.25 Which compounds have nonpolar covalent bonds, which have polar covalent bonds, and which have ionic bonds?

(a) LiF   **(b)** CH_3F **(c)** $MgCl_2$ **(d)** HCl

1.26 Using the symbols $\delta-$ and $\delta+$, indicate the direction of polarity, if any, in each covalent bond:

(a) C—Cl   **(b)** S—H   **(c)** C—S   **(d)** P—H

1.27 Write Lewis structures for each of the following compounds, showing all valence electrons (none of the compounds contains a ring of atoms):

(a) Hydrogen peroxide, H_2O_2 **(b)** Hydrazine, N_2H_4

(c) Methanol, CH_3OH **(d)** Methanethiol, CH_3SH

(e) Methanamine, CH_3NH_2 **(f)** Chloromethane, CH_3Cl

(g) Dimethyl ether, CH_3OCH_3 **(h)** Ethane, C_2H_6

(i) Ethylene, C_2H_4 **(j)** Acetylene, C_2H_2

(k) Carbon dioxide, CO_2 **(l)** Formaldehyde, CH_2O

(m) Acetone, CH_3COCH_3 **(n)** Carbonic acid, H_2CO_3

(o) Acetic acid, CH_3COOH

1.28 Write Lewis structures for these ions:

(a) Bicarbonate ion, HCO_3^- **(b)** Carbonate ion, CO_3^{2-}

(c) Acetate ion, CH_3COO^- **(d)** Chloride ion, Cl^-

1.29 Why are the following molecular formulas impossible?

(a) CH_5 **(b)** C_2H_7

1.30 Following the rule that each atom of carbon, oxygen, and nitrogen reacts to achieve a complete outer shell of eight valence electrons, add unshared pairs of electrons as necessary to complete the valence shell of each atom in the following ions. Then, assign formal charges as appropriate:

$$\begin{array}{ll}
\textbf{(a)} \quad \begin{array}{c} \quad\;\; O \\ \quad\;\; \| \\ H{-}O{-}C{-}O \end{array}
& \qquad
\textbf{(b)} \quad \begin{array}{c} H \;\; H \\ | \quad | \\ H{-}C{-}C{-}O \\ | \quad | \\ H \;\; H \end{array}
\end{array}$$

$$\begin{array}{ll}
\textbf{(c)} \quad \begin{array}{c} H \;\; H \\ | \quad | \\ H{-}C{-}C \\ | \quad | \\ H \;\; H \end{array}
& \qquad
\textbf{(d)} \quad \begin{array}{c} H \;\; H \;\; O \\ | \quad | \quad \| \\ H{-}N{-}C{-}C{-}O \\ \quad\;\; | \quad | \\ \quad\;\; H \;\; H \end{array}
\end{array}$$

1.31 The following Lewis structures show all valence electrons. Assign formal charges in each structure as appropriate.

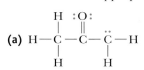

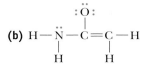

1.32 Each compound contains both ionic and covalent bonds. Draw a Lewis structure for each, and show by charges which bonds are ionic and by dashes which bonds are covalent.

(a) NaOH   (b) NaHCO$_3$ (c) NH$_4$Cl
(d) CH$_3$COONa   (e) CH$_3$ONa

1.33 Silver and oxygen can form a stable compound. Predict the formula of this compound, and state whether the compound consists of ionic or covalent bonds.

Polarity of Covalent Bonds

1.34 Which statement is true about electronegativity?
(a) Electronegativity increases from left to right in a period of the Periodic Table.
(b) Electronegativity increases from top to bottom in a column of the Periodic Table.
(c) Hydrogen, the element with the lowest atomic number, has the smallest electronegativity.
(d) The higher the atomic number of an element, the greater is its electronegativity.

1.35 Why does fluorine, the element in the upper right corner of the Periodic Table, have the largest electronegativity of any element?

1.36 Arrange the single covalent bonds within each set in order of increasing polarity:
(a) C—H, O—H, N—H   (b) C—H, C—Cl, C—I
(c) C—C, C—O, C—N   (d) C—Li, C—Hg, C—Mg

1.37 Using the values of electronegativity given in Table 1.5, predict which indicated bond in each set is more polar and, using the symbols $\delta+$ and $\delta-$, show the direction of its polarity:
(a) CH$_3$—OH or CH$_3$O—H   (b) H—NH$_2$ or CH$_3$—NH$_2$
(c) CH$_3$—SH or CH$_3$S—H   (d) CH$_3$—F or H—F

1.38 Identify the most polar bond in each molecule:
(a) HSCH$_2$CH$_2$OH   (b) CHCl$_2$F   (c) HOCH$_2$CH$_2$NH$_2$

1.39 Predict whether the carbon–metal bond in each of these organometallic compounds is nonpolar covalent, polar covalent, or ionic. For each polar covalent bond, show its direction of polarity using the symbols $\delta+$ and $\delta-$.

$$\text{(a)   CH}_3\text{CH}_2\text{—Pb—CH}_2\text{CH}_3$$
with CH$_2$CH$_3$ above and CH$_2$CH$_3$ below the Pb

(b) CH$_3$—Mg—Cl   (c) CH$_3$—Hg—CH$_3$

Tetraethyllead   Methylmagnesium chloride   Dimethylmercury

Bond Angles and Shapes of Molecules

1.40 Use the VSEPR model to predict bond angles about each highlighted atom:

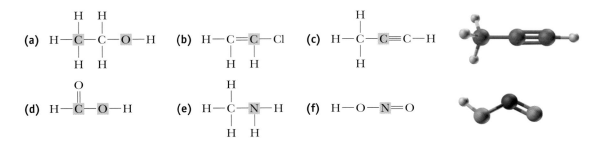

1.41 Use the VSEPR model to predict bond angles about each atom of carbon, nitrogen, and oxygen in these molecules. (*Hint:* First add unshared pairs of electrons as necessary to complete the valence shell of each atom, and then make your predictions of bond angles.)

(a) CH_3—CH_2—CH_2—OH

(b) CH_3—CH_2—$\overset{\overset{\displaystyle O}{\|}}{C}$—H

(c) CH_3—CH=CH_2

(d) CH_3—C≡C—CH_3

(e) CH_3—$\overset{\overset{\displaystyle O}{\|}}{C}$—O—$CH_3$

(f) CH_3—$\overset{\overset{\displaystyle CH_3}{|}}{N}$—$CH_3$

1.42 Silicon is immediately below carbon in the Periodic Table. Predict the C—Si—C bond angle in tetramethylsilane, $(CH_3)_4Si$.

Polar and Nonpolar Molecules

1.43 Draw a three-dimensional representation for each molecule. Indicate which molecules are polar and the direction of their polarity:

(a) CH_3F (b) CH_2Cl_2 (c) $CHCl_3$ (d) CCl_4
(e) CH_2=CCl_2 (f) CH_2=CHCl (g) CH_3C≡N (h) $(CH_3)_2C$=O

1.44 Tetrafluoroethylene, C_2F_4, is the starting material for the synthesis of the polymer poly(tetrafluoroethylene), commonly known as Teflon. Molecules of tetrafluoroethylene are nonpolar. Propose a structural formula for this compound.

1.45 Until several years ago, the two chlorofluorocarbons (CFCs) most widely used as heat transfer media for refrigeration systems were Freon-11 (trichlorofluoromethane, CCl_3F) and Freon-12 (dichlorodifluoromethane, CCl_2F_2). Draw a three-dimensional representation of each molecule, and indicate the direction of its polarity.

Resonance Contributing Structures

1.46 Which of these statements are true about resonance contributing structures?
  (a) All resonance contributing structures must have the same number of valence electrons.
  (b) All resonance contributing structures must have the same arrangement of atoms.
  (c) All atoms in a resonance contributing structure must have complete valence shells.
  (d) All bond angles in sets of resonance contributing structures must be the same.

1.47 Draw the resonance contributing structure indicated by the curved arrow(s), and assign formal charges as appropriate:

(a) H—$\ddot{\ddot{O}}$—$C\overset{\overset{\displaystyle :\ddot{O}:}{}}{\underset{:\ddot{O}:^-}{}}$ ⟷ (b) $\overset{\displaystyle H}{\underset{\displaystyle H}{}}C$=$\ddot{O}$ ⟷ (c) CH_3—$\ddot{\ddot{O}}$—$C\overset{\overset{\displaystyle :O:}{}}{\underset{:\ddot{O}:^-}{}}$ ⟷

1.48 Using the VSEPR model, predict the bond angles about the carbon atom in each pair of resonance contributing structures in Problem 1.47. In what way do the bond angles change from one contributing structure to the other?

Hybridization of Atomic Orbitals

1.49 State the hybridization of each highlighted atom:

(a) H—$\overset{\overset{\displaystyle H}{|}}{\underset{\underset{\displaystyle H}{|}}{C}}$—$\overset{\overset{\displaystyle H}{|}}{\underset{\underset{\displaystyle H}{|}}{C}}$—H

(b) $\overset{\displaystyle H}{\underset{\displaystyle H}{}}C$=$C\overset{\displaystyle H}{\underset{\displaystyle H}{}}$

(c) H—C≡C—H

(d) H—$\overset{\overset{\displaystyle H}{|}}{\underset{\underset{\displaystyle H}{|}}{C}}$—$\overset{\overset{\displaystyle H}{|}}{\underset{\underset{\displaystyle H}{|}}{N}}$—H

(e) H—$\overset{\overset{\displaystyle O}{\|}}{C}$—O—H

(f) $\overset{\displaystyle H}{\underset{\displaystyle H}{}}C$=O

1.50 Describe each highlighted bond in terms of the overlap of hybrid orbitals:

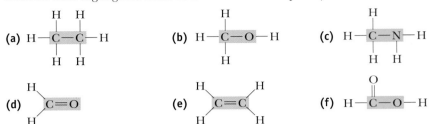

Functional Groups

1.51 Draw Lewis structures for these functional groups. Be certain to show all valence electrons on each:
  (a) Carbonyl group
  (b) Carboxyl group
  (c) Hydroxyl group
  (d) Primary amino group

1.52 Draw the structure for a compound with the molecular formula
  (a) C_2H_6O that is an alcohol.
  (b) C_3H_6O that is an aldehyde.
  (c) C_3H_6O that is a ketone.
  (d) $C_3H_6O_2$ that is a carboxylic acid.
  (e) $C_4H_{11}N$ that is a tertiary amine.

1.53 Draw condensed structural formulas for all compounds of molecular formula C_4H_8O that contain
  (a) A carbonyl group. (There are two aldehydes and one ketone.)
  (b) A carbon–carbon double bond and a hydroxyl group. (There are eight.)

1.54 Draw structural formulas for
  (a) The eight alcohols with molecular formula $C_5H_{12}O$.
  (b) The eight aldehydes with molecular formula $C_6H_{12}O$.
  (c) The six ketones with molecular formula $C_6H_{12}O$.
  (d) The eight carboxylic acids with molecular formula $C_6H_{12}O_2$.
  (e) The three tertiary amines with molecular formula $C_5H_{13}N$.

1.55 Identify the functional groups in each compound (we study each compound in more detail in the indicated section):

(a) 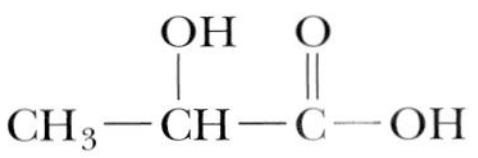 $CH_3-\overset{\overset{\displaystyle OH}{|}}{CH}-\overset{\overset{\displaystyle O}{\|}}{C}-OH$
Lactic acid
(Section 22.5A)

(b) $HO-CH_2-CH_2-OH$
Ethylene glycol
(Section 8.2B)

(c) 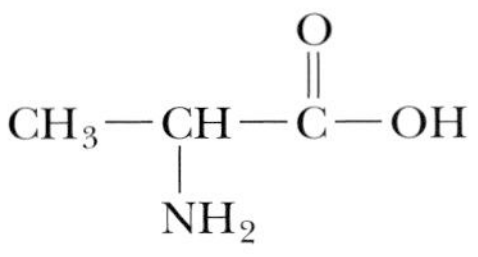 $CH_3-\underset{\underset{\displaystyle NH_2}{|}}{CH}-\overset{\overset{\displaystyle O}{\|}}{C}-OH$
Alanine
(Section 19.2)

(d) 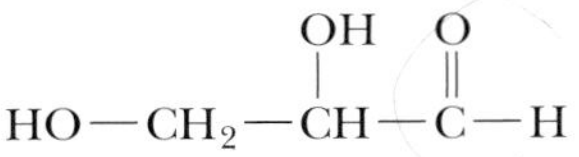 $HO-CH_2-\overset{\overset{\displaystyle OH}{|}}{CH}-\overset{\overset{\displaystyle O}{\|}}{C}-H$
Glyceraldehyde
(Section 18.2A)

(e) 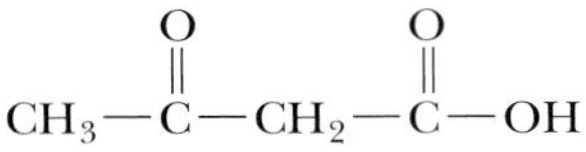 $CH_3-\overset{\overset{\displaystyle O}{\|}}{C}-CH_2-\overset{\overset{\displaystyle O}{\|}}{C}-OH$
Acetoacetic acid
(Section 14.3B)

(f) $H_2NCH_2CH_2CH_2CH_2CH_2CH_2NH_2$
1,6-Hexanediamine
(Section 17.5A)

1.56 Dihydroxyacetone, $C_3H_6O_3$, the active ingredient in many sunless tanning lotions, contains two 1° hydroxyl groups, each on a different carbon, and one ketone group. Draw a structural formula for dihydroxyacetone.

1.57 Propylene glycol, $C_3H_8O_2$, commonly used in airplane deicers, contains a 1° alcohol and a 2° alcohol. Draw a structural formula for propylene glycol.

1.58 Ephedrine is a molecule found in the dietary supplement ephedra, which has been linked to adverse health reactions such as heart attacks, strokes, and heart palpitations. The use of ephedra in dietary supplements is now banned by the FDA.
   (a) Identify at least two functional groups in ephedrine.
   (b) Would you predict ephedrine to be polar or nonpolar?

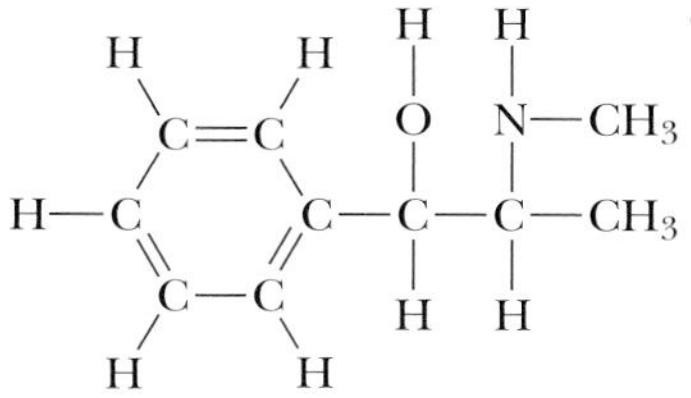

1.59 Ozone (O_3) and carbon dioxide (CO_2) are both known as greenhouse gases. Compare and contrast their shapes, and indicate the hybridization of each atom in the two molecules.

Looking Ahead

1.60 Allene, C_3H_4, has the structural formula $H_2C{=}C{=}CH_2$. Determine the hybridization of each carbon in allene and predict the shape of the molecule.

1.61 Dimethylsulfoxide, $(CH_3)_2SO$, is a common solvent used in organic chemistry.
   (a) Write a Lewis structure for dimethylsulfoxide.
   (b) Predict the hybridization of the sulfur atom in the molecule.
   (c) Predict the geometry of dimethylsulfoxide.
   (d) Is dimethylsulfoxide a polar or a nonpolar molecule?

1.62 In Chapter 5, we study a group of organic cations called carbocations. Following is the structure of one such carbocation, the *tert*-butyl cation:

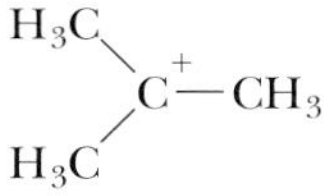

   (a) How many electrons are in the valence shell of the carbon bearing the positive charge?
   (b) Predict the bond angles about this carbon.
   (c) Given the bond angles you predicted in **(b)**, what hybridization do you predict for this carbon?

1.63 We also study the isopropyl cation, $(CH_3)_2CH^+$, in Chapter 5.
   (a) Write a Lewis structure for this cation. Use a plus sign to show the location of the positive charge.
   (b) How many electrons are in the valence shell of the carbon bearing the positive charge?
   (c) Use the VSEPR model to predict all bond angles about the carbon bearing the positive charge.
   (d) Describe the hybridization of each carbon in this cation.

1.64 In Chapter 9, we study benzene, C_6H_6, and its derivatives.

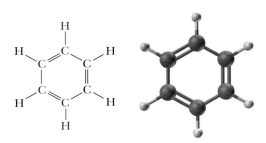

  (a) Predict each H—C—C and each C—C—C bond angle on benzene.
  (b) State the hybridization of each carbon in benzene.
  (c) Predict the shape of a benzene molecule.

1.65 Explain why *all* the carbon–carbon bonds in benzene are equal in length.

1.39×10^{-10} m

2 Acids and Bases

Citrus fruits are sources of citric acid. Lemon juice, for example, contains 5–8% citric acid. Inset: A model of citric acid. (*Corbis Digital Stock*)

2.1 INTRODUCTION

A great many organic reactions are acid–base reactions. In this and later chapters, we will study the acid–base properties of the major classes of organic compounds, including alcohols, phenols, carboxylic acids, carbonyl compounds containing α-hydrogens, amines, amino acids and proteins, and, finally, nucleic acids. Furthermore, many organic reactions are catalyzed by proton-donating acids, such as H_3O^+ and $CH_3OH_2^+$. Others are catalyzed by Lewis acids, such as $AlCl_3$. It is essential, therefore, that you have a good grasp of the fundamentals of acid–base chemistry.

2.2   ARRHENIUS ACIDS AND BASES

The first useful definitions of an acid and a base were put forward by Svante Arrhenius in 1884. According to the original Arrhenius definitions, an acid is a substance that dissolves in water to produce H^+ ions and a base is a substance that dissolves in water to produce OH^- ions. Today we know that an H^+ ion does not exist in water because it reacts immediately with an H_2O molecule to give a hydronium ion, H_3O^+:

$$H^+(aq) + H_2O(l) \longrightarrow H_3O^+(aq)$$

$$\text{Hydronium ion}$$

Apart from this modification, the Arrhenius definitions of acid and base are still valid and useful today, as long as we are talking about aqueous solutions.

When an acid dissolves in water, it reacts with the water to produce H_3O^+. For example, when HCl dissolves in water, it reacts with a water molecule to give hydronium ion and chloride ion:

$$H_2O(l) + HCl(aq) \longrightarrow H_3O^+(aq) + Cl^-(aq)$$

We can show the transfer of a proton from an acid to a base by using a symbol called a **curved arrow**. First we write the Lewis structure of each reactant and product, showing all valence electrons on reacting atoms. Then we use curved arrows to show the change in position of electron pairs during the reaction. The tail of the curved arrow is located at an electron pair. The head of the curved arrow shows the new position of the electron pair. Whenever we use curved arrows to show changes in electron position, the arrows will signify that an electron pair originating from an atom will form a new bond, while an electron pair originating from a bond will break that bond.

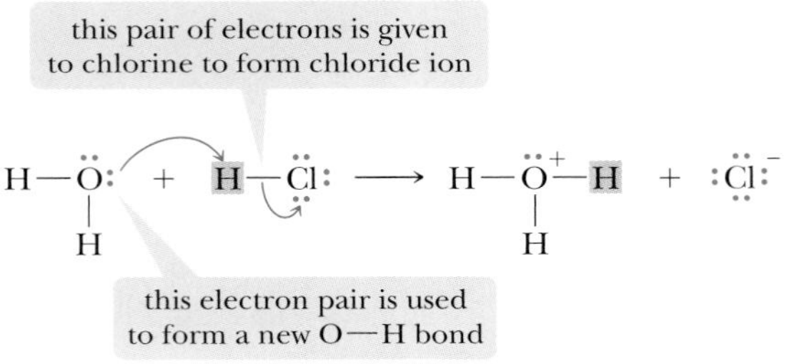

In this equation, the leftmost curved arrow shows that an unshared pair of electrons on oxygen changes position to form a new covalent bond with hydrogen. The rightmost curved arrow shows that the pair of electrons of the H—Cl bond is given entirely to chlorine to form a chloride ion. Thus, in the reaction of HCl with H_2O, a proton is transferred from HCl to H_2O, and in the process, an O—H bond forms and an H—Cl bond breaks.

With bases, the situation is slightly different. Many bases are metal hydroxides, such as KOH, NaOH, $Mg(OH)_2$, and $Ca(OH)_2$. These compounds are ionic solids, and when they dissolve in water, their ions merely separate, with each ion solvated by water molecules, as illustrated in the following reaction:

$$NaOH(s) \xrightarrow{H_2O} Na^+(aq) + OH^-(aq)$$

Other bases are not hydroxides. Instead, they produce OH^- ions in water by reacting with water molecules. The most important examples of this kind of base are ammonia, NH_3, and amines (Section 1.8B). When ammonia dissolves in water, it reacts with the water to produce ammonium ions and hydroxide ions:

$$NH_3(aq) + H_2O(l) \rightleftharpoons NH_4^+(aq) + OH^-(aq)$$

As we will see in Section 2.5, ammonia is a weak base, and the position of the equilibrium for its reaction with water lies considerably toward the left. In a 1.0-M solution of NH_3 in water, for example, only about four molecules of NH_3 out of every thousand react with the water to form NH_4^+ and OH^-. Thus, when ammonia dissolves in water, it exists primarily as NH_3 molecules. Nevertheless, some OH^- ions are produced; therefore, NH_3 is a base.

We indicate how the reaction of ammonia with water takes place by using curved arrows to show the transfer of a proton from a water molecule to an ammonia molecule:

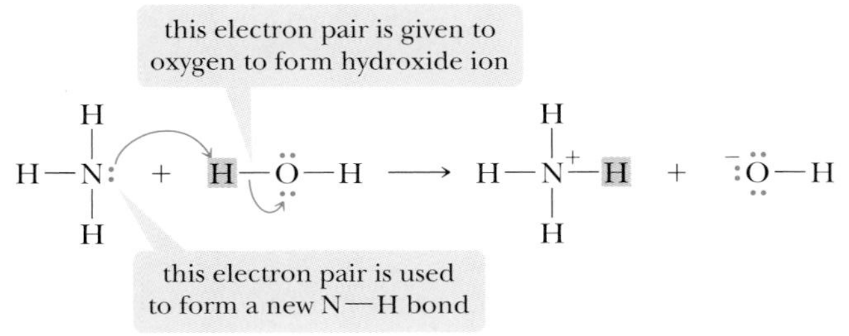

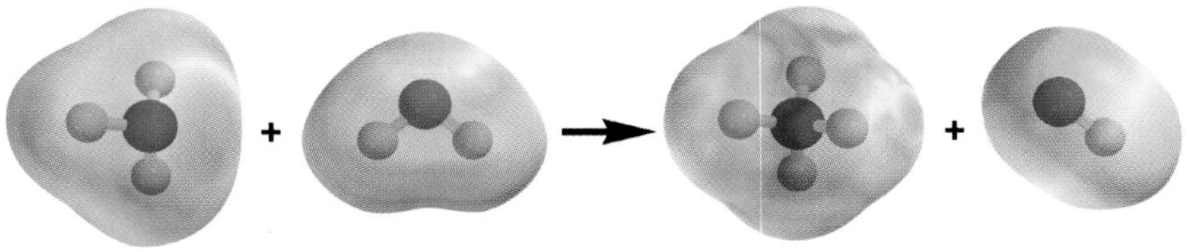

Notice the increase in charge distribution upon protonation of NH_3 and deprotonation of H_2O. The nitrogen of NH_4^+ shows a more intense blue than the nitrogen of NH_3, and the oxygen of OH^- shows a more intense red than the oxygen of H_2O.

Here, the leftmost curved arrow shows that the unshared pair of electrons on nitrogen changes position to form a new covalent bond with a hydrogen of a water molecule. At the same time as the new N—H bond forms, an O—H bond of a water molecule breaks, and, as the rightmost arrow indicates, the pair of electrons forming the H—O bond moves entirely to oxygen, forming OH^-. Thus, ammonia produces an OH^- ion by a proton-transfer reaction from a water molecule and leaves OH^- behind.

The Arrhenius concept of acids and bases is so intimately tied to reactions that take place in water that it has no good way to deal with acid–base reactions in nonaqueous solutions. For this reason, we concentrate in this chapter on the Brønsted–Lowry definitions of acids and bases, which are more useful to us in our discussion of reactions of organic compounds.

2.3 BRØNSTED–LOWRY ACIDS AND BASES

In 1923, the Danish chemist Johannes Brønsted and the English chemist Thomas Lowry independently proposed the following definitions: An **acid** is a **proton donor**, a **base** is a **proton acceptor**, and an acid–base reaction is a **proton-transfer reaction**. Furthermore, according to the Brønsted–Lowry definitions, any pair of molecules or ions that can be interconverted by the transfer of a proton is called a **conjugate acid–base pair**. When an acid transfers a proton to a base, the acid is converted to its **conjugate base**. When a base accepts a proton, the base is converted to its **conjugate acid**.

Brønsted–Lowry acid A proton donor.

Brønsted–Lowry base A proton acceptor.

Conjugate base The species formed when an acid donates a proton.

Conjugate acid The species formed when a base accepts a proton.

We can illustrate these relationships by examining the reaction of hydrogen chloride with water to form chloride ion and hydronium ion:

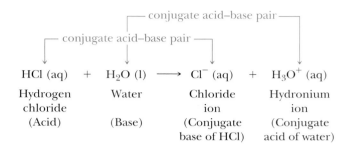

$$HCl\ (aq)\ +\ H_2O\ (l)\ \longrightarrow\ Cl^-\ (aq)\ +\ H_3O^+\ (aq)$$

| Hydrogen chloride (Acid) | Water (Base) | Chloride ion (Conjugate base of HCl) | Hydronium ion (Conjugate acid of water) |

In this reaction, the acid HCl donates a proton and is converted to its conjugate base, Cl^-. The base H_2O accepts a proton and is converted to its conjugate acid, H_3O^+.

We have illustrated the application of the Brønsted–Lowry definitions with water as a reactant. These definitions, however, do not require water as a reactant. Consider the following reaction between acetic acid and ammonia:

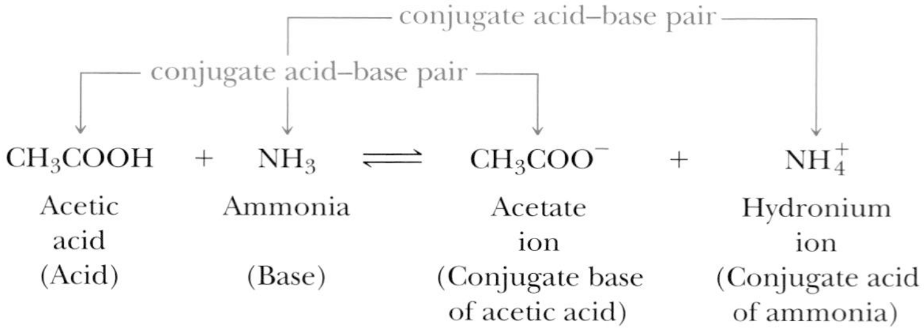

$$CH_3COOH\ +\ NH_3\ \rightleftharpoons\ CH_3COO^-\ +\ NH_4^+$$

| Acetic acid (Acid) | Ammonia (Base) | Acetate ion (Conjugate base of acetic acid) | Hydronium ion (Conjugate acid of ammonia) |

We can use curved arrows to show how this reaction takes place:

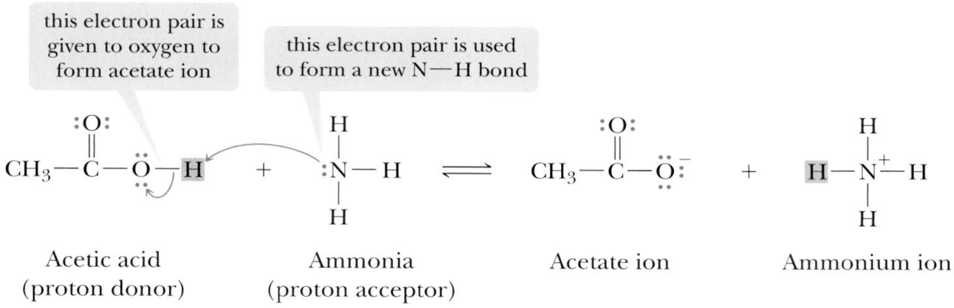

this electron pair is given to oxygen to form acetate ion

this electron pair is used to form a new N—H bond

$$CH_3-\overset{\overset{:O:}{\|}}{C}-\overset{..}{\underset{..}{O}}-H\ +\ \overset{H}{\underset{H}{:N-H}}\ \rightleftharpoons\ CH_3-\overset{\overset{:O:}{\|}}{C}-\overset{..}{\underset{..}{O}}:^-\ +\ \overset{H}{\underset{H}{H-\overset{+}{N}-H}}$$

| Acetic acid (proton donor) | Ammonia (proton acceptor) | Acetate ion | Ammonium ion |

The rightmost curved arrow shows that the unshared pair of electrons on nitrogen becomes shared between N and H to form a new H—N bond. At the same time that the H—N bond forms, the O—H bond breaks, and the electron pair of the O—H bond moves entirely to oxygen to form the —O^- of the acetate ion. The result of these two electron-pair shifts is the transfer of a proton from an acetic acid molecule to an ammonia molecule. Table 2.1 gives examples of common acids and their conjugate bases. As you study the examples of conjugate acid–base pairs in the table, note the following points:

1. An acid can be positively charged, neutral, or negatively charged. Examples of these charge types are H_3O^+, H_2CO_3, and $H_2PO_4^-$.

TABLE 2.1 Some Acids and Their Conjugate Bases

	Acid	Name	Conjugate Base	Name	
Strong Acids	HI	Hydroiodic acid	I^-	Iodide ion	Weak Bases
	HCl	Hydrochloric acid	Cl^-	Chloride ion	
	H_2SO_4	Sulfuric acid	HSO_4^-	Hydrogen sulfate ion	
	HNO_3	Nitric acid	NO_3^-	Nitrate ion	
	H_3O^+	Hydronium ion	H_2O	Water	
	HSO_4^-	Hydrogen sulfate ion	SO_4^{2-}	Sulfate ion	
	H_3PO_4	Phosphoric acid	$H_2PO_4^-$	Dihydrogen phosphate ion	
	CH_3COOH	Acetic acid	CH_3COO^-	Acetate ion	
	H_2CO_3	Carbonic acid	HCO_3^-	Bicarbonate ion	
	H_2S	Hydrogen sulfide	HS^-	Hydrogen sulfide ion	
	$H_2PO_4^-$	Dihydrogen phosphate ion	HPO_4^{2-}	Hydrogen phosphate ion	
	NH_4^+	Ammonium ion	NH_3	Ammonia	
	HCN	Hydrocyanic acid	CN^-	Cyanide ion	
	C_6H_5OH	Phenol	$C_6H_5O^-$	Phenoxide ion	
	HCO_3^-	Bicarbonate ion	CO_3^{2-}	Carbonate ion	
	HPO_4^{2-}	Hydrogen phosphate ion	PO_4^{3-}	Phosphate ion	
Weak Acids	H_2O	Water	OH^-	Hydroxide ion	Strong Bases
	C_2H_5OH	Ethanol	$C_2H_5O^-$	Ethoxide ion	

2. A base can be negatively charged or neutral. Examples of these charge types are Cl^- and NH_3.

3. Acids are classified as monoprotic, diprotic, or triprotic, depending on the number of protons each may give up. Examples of **monoprotic acids** include HCl, HNO_3, and CH_3COOH. Examples of **diprotic acids** include H_2SO_4 and H_2CO_3. An example of a **triprotic acid** is H_3PO_4. Carbonic acid, for example, loses one proton to become bicarbonate ion and then a second proton to become carbonate ion:

$$H_2CO_3 + H_2O \rightleftharpoons HCO_3^- + H_3O^+$$

Carbonic   Bicarbonate
acid   ion

$$HCO_3^- + H_2O \rightleftharpoons CO_3^{2-} + H_3O^+$$

Bicarbonate   Carbonate
ion   ion

4. Several molecules and ions appear in both the acid and conjugate base columns; that is, each can function as either an acid or a base. The bicarbonate ion, HCO_3^-, for example, can give up a proton to become CO_3^{2-} (in which case it is an acid) or can accept a proton to become H_2CO_3 (in which case it is a base).

5. There is an inverse relationship between the strength of an acid and the strength of its conjugate base: The stronger the acid, the weaker is its conjugate base. HI, for example, is the strongest acid listed in Table 2.1, and I^-, its conjugate base, is the weakest base. As another example, CH_3COOH (acetic acid) is a stronger acid than H_2CO_3 (carbonic acid); conversely, CH_3COO^- (acetate ion) is a weaker base than HCO_3^- (bicarbonate ion).

EXAMPLE 2.1

Write the following acid–base reaction as a proton-transfer reaction. Label which reactant is the acid and which the base, as well as which product is the conjugate base of the original acid and which is the conjugate acid of the original base. Use curved arrows to show the flow of electrons in the reaction.

$$CH_3\overset{\overset{\displaystyle O}{\|}}{C}OH + HCO_3^- \longrightarrow CH_3\overset{\overset{\displaystyle O}{\|}}{C}O^- + H_2CO_3$$

| Acetic acid | Bicarbonate ion | Acetate ion | Carbonic acid |

SOLUTION

First, write a Lewis structure for each reactant by showing all valence electrons on the reacting atoms: Acetic acid is the acid (proton donor), and bicarbonate ion is the base (proton acceptor). Although bicarbonate can act as an acid (as shown in Table 2.1), acetic acid is a stronger acid and, therefore, will donate the proton in this reaction.

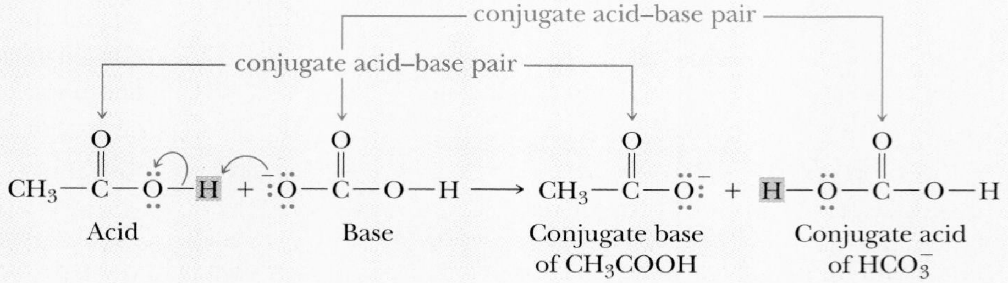

Practice Problem 2.1

Write each acid–base reaction as a proton-transfer reaction. Label which reactant is the acid and which the base, as well as which product is the conjugate base of the original acid and which is the conjugate acid of the original base. Use curved arrows to show the flow of electrons in each reaction.

(a) $CH_3SH + OH^- \longrightarrow CH_3S^- + H_2O$

(b) $CH_3OH + NH_2^- \longrightarrow CH_3O^- + NH_3$

2.4 QUANTITATIVE MEASURE OF ACID AND BASE STRENGTH

Strong acid An acid that is completely ionized in aqueous solution.

Strong base A base that is completely ionized in aqueous solution.

A **strong acid** or **strong base** is one that ionizes completely in aqueous solution. When HCl is dissolved in water, a proton is transferred completely from HCl to H_2O to form Cl^- and H_3O^+. There is no tendency for the reverse reaction to occur—that is, for the transfer of a proton from H_3O^+ to Cl^- to form HCl and H_2O. Therefore, when we compare the relative acidities of HCl and H_3O^+, we conclude that HCl is the stronger acid and H_3O^+ is the weaker acid. Similarly, H_2O is the stronger base and Cl^- is the weaker base.

$$HCl + H_2O \longrightarrow Cl^- + H_3O^+$$

| Acid (stronger acid) | Base (stronger base) | Conjugate base of HCl (weaker base) | Conjugate acid of H_2O (weaker acid) |

Examples of strong acids in aqueous solution are HCl, HBr, HI, HNO_3, $HClO_4$, and H_2SO_4. Examples of strong bases in aqueous solution are LiOH, NaOH, KOH, $Ca(OH)_2$, and $Ba(OH)_2$.

A **weak acid** or **weak base** is one that only partially ionizes in aqueous solution. Most organic acids and bases are weak. Among the most common organic acids we deal with are the carboxylic acids, which contain a carboxyl group, —COOH (Section 1.8D), as shown in the following reaction:

Weak acid An acid that only partially ionizes in aqueous solution.

Weak base A base that only partially ionizes in aqueous solution.

$$CH_3\overset{\overset{\displaystyle O}{\|}}{C}OH \;+\; H_2O \;\rightleftharpoons\; CH_3\overset{\overset{\displaystyle O}{\|}}{C}O^- \;+\; H_3O^+$$

Acid	Base	Conjugate base	Conjugate acid
(weaker acid)	(weaker base)	of CH_3COOH (stronger base)	of H_2O (stronger acid)

The equation for the ionization of a weak acid, HA, in water and the acid ionization constant K_a for this equilibrium are, respectively,

$$HA + H_2O \rightleftharpoons A^- + H_3O^+$$

and

$$K_a = K_{eq}[H_2O] = \frac{[H_3O^+][A^-]}{[HA]}$$

Because acid ionization constants for weak acids are numbers with negative exponents, they are often expressed as $\mathbf{pK_a} = -\log_{10} K_a$. Table 2.2 gives names, molecular formulas, and values of pK_a for some organic and inorganic acids. Note that the larger the value of pK_a, the weaker is the acid. Note also the inverse relationship between the strengths of the conjugate acid–conjugate base pairs; the stronger the acid, the weaker is its conjugate base.

The pH of this soft drink is 3.12. Soft drinks are often quite acidic. (*Charles D. Winters*).

TABLE 2.2 pK_a Values for Some Organic and Inorganic Acids

	Acid	Formula	pK_a	Conjugate Base	
Weaker acid	ethane	CH_3CH_3	51	$CH_3CH_2^-$	Stronger base
	ammonia	NH_3	38	NH_2^-	
	ethanol	CH_3CH_2OH	15.9	$CH_3CH_2O^-$	
	water	H_2O	15.7	HO^-	
	methylammonium ion	$CH_3NH_3^+$	10.64	CH_3NH_2	
	bicarbonate ion	HCO_3^-	10.33	CO_3^{2-}	
	phenol	C_6H_5OH	9.95	$C_6H_5O^-$	
	ammonium ion	NH_4^+	9.24	NH_3	
	carbonic acid	H_2CO_3	6.36	HCO_3^-	
	acetic acid	CH_3COOH	4.76	CH_3COO^-	
	benzoic acid	C_6H_5COOH	4.19	$C_6H_5COO^-$	
	phosphoric acid	H_3PO_4	2.1	$H_2PO_4^-$	
	hydronium ion	H_3O^+	−1.74	H_2O	
	sulfuric acid	H_2SO_4	−5.2	HSO_4^-	
	hydrogen chloride	HCl	−7	Cl^-	
Stronger acid	hydrogen bromide	HBr	−8	Br^-	Weaker base
	hydrogen iodide	HI	−9	I^-	

EXAMPLE 2.2

For each value of pK_a, calculate the corresponding value of K_a. Which compound is the stronger acid?

(a) Ethanol, pK_a = 15.9   (b) Carbonic acid, pK_a = 6.36

SOLUTION

(a) For ethanol, K_a = 1.3 × 10^{-16} (b) For carbonic acid, K_a = 4.4 × 10^{-7}

Because the value of pK_a for carbonic acid is smaller than that for ethanol, carbonic acid is the stronger acid and ethanol is the weaker acid.

Practice Problem 2.2

For each value of K_a, calculate the corresponding value of pK_a. Which compound is the stronger acid?

(a) Acetic acid, K_a = 1.74 × 10^{-5} (b) Water, K_a = 2.00 × 10^{-16}

Caution: In exercises such as Example 2.2 and Problem 2.2, we ask you to select the stronger acid. You must remember that these and all other acids with ionization constants considerably less than 1.00 are *weak* acids. Thus, although acetic acid is a considerably stronger acid than water, it still is only slightly ionized in water. The ionization of acetic acid in a 0.1-M solution, for example, is only about 1.3%; the major form of this weak acid that is present in a 0.1-M solution is the un-ionized acid!

$$\underset{\substack{\text{Forms present in}\\\text{0.1 M acetic acid}}}{} \quad \underset{98.7\%}{CH_3\overset{\overset{\displaystyle O}{\|}}{C}OH} + H_2O \rightleftharpoons \underset{1.3\%}{CH_3\overset{\overset{\displaystyle O}{\|}}{C}O^-} + H_3O^+$$

2.5   THE POSITION OF EQUILIBRIUM IN AN ACID–BASE REACTION

We know that HCl reacts with H_2O according to the following equilibrium:

$$HCl + H_2O \longrightarrow Cl^- + H_3O^+$$

We also know that HCl is a strong acid, which means that the position of this equilibrium lies very far to the right.

As we have seen, acetic acid reacts with H_2O according to the following equilibrium:

$$\underset{\text{Acetic acid}}{CH_3COOH} + H_2O \rightleftharpoons \underset{\text{Acetate ion}}{CH_3COO^-} + H_3O^+$$

Acetic acid is a weak acid. Only a few acetic acid molecules react with water to give acetate ions and hydronium ions, and the major species present in equilibrium in aqueous solution is CH_3COOH. The position of this equilibrium, therefore, lies very far to the left.

In the preceding two acid–base reactions, water was the base (proton acceptor). But what if we have a base other than water as the proton acceptor? How can we determine which are the major species present at equilibrium? That is, how can we determine whether the position of equilibrium lies toward the left or toward the right?

As an example, let us examine the acid–base reaction between acetic acid and ammonia to form acetate ion and ammonium ion:

$$CH_3COOH \quad + \quad NH_3 \quad \overset{?}{\rightleftharpoons} \quad CH_3COO^- \quad + \quad NH_4^+$$

Acetic acid	Ammonia	Acetate ion	Ammonium ion
(Acid)	(Base)	(Conjugate base of CH_3COOH)	(Conjugate acid of NH_3)

Vinegar (which contains acetic acid) and baking soda (sodium bicarbonate) react to produce sodium acetate, carbon dioxide, and water. The carbon dioxide inflates the balloon. *(Charles D. Winters)*

As indicated by the question mark over the equilibrium arrow, we want to determine whether the position of this equilibrium lies toward the left or toward the right. There are two acids present: acetic acid and ammonium ion. There are also two bases present: ammonia and acetate ion. From Table 2.2, we see that CH_3COOH (pK_a 4.76) is the stronger acid, which means that CH_3COO^- is the weaker conjugate base. Conversely, NH_4^+ (pK_a 9.24) is the weaker acid, which means that NH_3 is the stronger conjugate base. We can now label the relative strengths of each acid and base in the equilibrium:

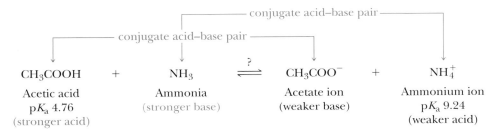

In an acid–base reaction, the position of equilibrium always favors reaction of the stronger acid and stronger base to form the weaker acid and weaker base. Thus, at equilibrium, the major species present are the weaker acid and weaker base. In the reaction between acetic acid and ammonia, therefore, the equilibrium lies to the right, and the major species present are acetate ion and ammonium ion:

$$CH_3COOH \quad + \quad NH_3 \quad \rightleftharpoons \quad CH_3COO^- \quad + \quad NH_4^+$$

Acetic acid	Ammonia	Acetate ion	Ammonium ion
(stronger acid)	(stronger base)	(weaker base)	(weaker acid)

To summarize, the four steps we use to determine the position of an acid–base equilibrium are as follows:

1. Identify the two acids in the equilibrium; one is on the left side of the equilibrium, the other on the right side.
2. Using the information in Table 2.2, determine which acid is the stronger and which the weaker.
3. Identify the stronger base and weaker base in each equilibrium. Remember that the stronger acid gives the weaker conjugate base and the weaker acid gives the stronger conjugate base.
4. The stronger acid and stronger base react to give the weaker acid and weaker base, and the position of equilibrium lies on the side of the weaker acid and weaker base.

EXAMPLE 2.3

For each acid–base equilibrium, label the stronger acid, the stronger base, the weaker acid, and the weaker base. Then predict whether the position of equilibrium lies toward the right or toward the left.

(a) $H_2CO_3 + OH^- \rightleftharpoons HCO_3^- + H_2O$

    Carbonic   Bicarbonate
      acid   ion

(b) $C_6H_5OH + HCO_3^- \rightleftharpoons C_6H_5O^- + H_2CO_3$

     Phenol   Bicarbonate   Phenoxide   Carbonic
                   ion   ion   acid

SOLUTION

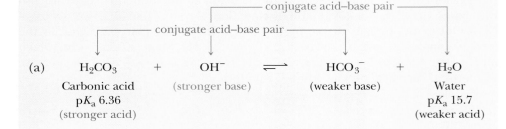

(a) H_2CO_3 + OH^- $\rightleftharpoons$ HCO_3^- + H_2O

      Carbonic acid   (stronger base)   (weaker base)   Water
      pK_a 6.36 pK_a 15.7
      (stronger acid)   (weaker acid)

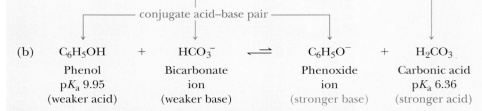

(b) C_6H_5OH + HCO_3^- $\rightleftharpoons$ $C_6H_5O^-$ + H_2CO_3

      Phenol   Bicarbonate   Phenoxide   Carbonic acid
      pK_a 9.95   ion   ion pK_a 6.36
      (weaker acid)   (weaker base)   (stronger base)   (stronger acid)

Arrows over each equilibrium show the conjugate acid–base pairs. The position of equilibrium in (a) lies toward the right. In (b) it lies toward the left.

Practice Problem 2.3

For each acid–base equilibrium, label the stronger acid, the stronger base, the weaker acid, and the weaker base. Then predict whether the position of equilibrium lies toward the right or the left.

(a) $CH_3NH_2 + CH_3COOH \rightleftharpoons CH_3NH_3^+ + CH_3COO^-$

     Methylamine   Acetic acid   Methylammonium   Acetate ion
                                     ion

(b) $CH_3CH_2O^- + NH_3 \rightleftharpoons CH_3CH_2OH + NH_2^-$

     Ethoxide   Ammonia   Ethanol   Amide
      ion   ion

2.6 MOLECULAR STRUCTURE AND ACIDITY

Now let us examine the relationship between the acidity of organic compounds and their molecular structure. The most important factor in determining the relative acidities of organic acids is the relative stability of the anion, A^-, formed when the acid, HA, transfers a proton to a base. We can understand the relationship involved by considering (A) the electronegativity of the atom bonded to H, (B) resonance, and (C) the inductive effect. We will look at each of these factors briefly in this chapter. We will study them more fully in later chapters when we deal with particular functional groups.

A. Electronegativity: Acidity of HA within a Period of the Periodic Table

The relative acidity of the hydrogen acids within a period of the Periodic Table is determined by the stability of A^-—that is, by the stability of the anion that forms when a proton is transferred from HA to a base. The greater the electronegativity of A, the greater is the stability of the anion, A^-, and the stronger is the acid, HA:

	H_3C-H	H_2N-H	$HO-H$	$F-H$
pK_a	51	38	15.7	3.5
Electronegativity of A in A–H	2.5	3.0	3.5	4.0

Increasing acid strength ⟶

B. Resonance Effect: Delocalization of Charge in A^-

Carboxylic acids are weak acids: Values of pK_a for most unsubstituted carboxylic acids fall within the range from 4 to 5. The value of pK_a for acetic acid, for example, is 4.76:

$$CH_3COOH + H_2O \rightleftharpoons CH_3COO^- + H_3O^+ \qquad pK_a = 4.76$$

Values of pK_a for most alcohols—compounds that also contain an $-OH$ group—fall within the range from 15 to 18; the value of pK_a for ethanol, for example, is 15.9:

$$CH_3CH_2O-H + H_2O \rightleftharpoons CH_3CH_2O^- + H_3O^+ \qquad pK_a = 15.9$$

An alcohol · · · · · · · · · · · · · · · An alkoxide ion

Thus, alcohols are slightly weaker acids than water ($pK_a = 15.7$) and are much weaker acids than carboxylic acids.

We account for the greater acidity of carboxylic acids compared with alcohols in part by using the resonance model and looking at the relative stabilities of the alkoxide ion and the carboxylate ion. Our guideline is this: The more stable the anion, the farther the position of equilibrium is shifted toward the right and the more acidic is the compound.

There is no resonance stabilization in an alkoxide anion. The ionization of a carboxylic acid, however, gives an anion for which we can write two equivalent contributing structures in which the negative charge of the anion is delocalized; that is, it is spread evenly over the two oxygen atoms:

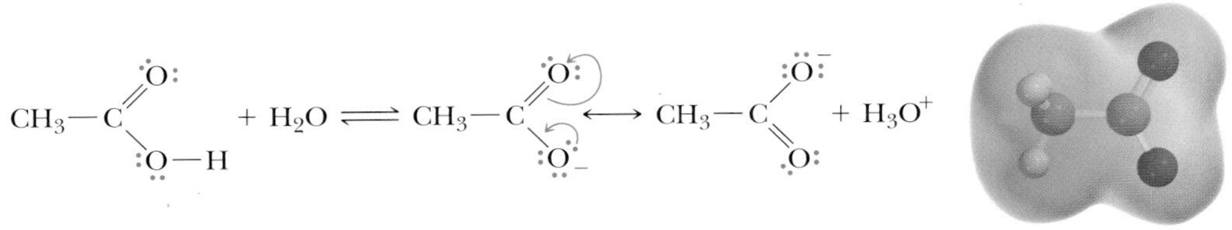

These contributing structures are equivalent; the carboxylate anion is stabilized by delocalization of the negative charge. The electron density map shows the negative charge equally delocalized on the two carboxylate oxygens.

Because of the delocalization of its charge, a carboxylate anion is significantly more stable than an alkoxide anion. Therefore, the equilibrium for the ionization of a carboxylic acid is shifted to the right relative to that for the ionization of an alcohol, and a carboxylic acid is a stronger acid than an alcohol.

C. The Inductive Effect: Withdrawal of Electron Density from the HA Bond

Inductive effect The polarization of electron density transmitted through covalent bonds caused by a nearby atom of higher electronegativity.

The **inductive effect** is the polarization of electron density transmitted through covalent bonds by a nearby atom of higher electronegativity. We see the operation of the inductive effect when we compare acetic acid to trifluoroacetic acid in the following way: Fluorine is more electronegative than carbon and polarizes the electrons of the C—F bond, creating a partial positive charge on the carbon. The partial positive charge, in turn, withdraws electron density from the negatively charged carboxylate group. The withdrawal of electron density delocalizes the negative charge and makes the conjugate base of trifluoroacetic acid more stable than the conjugate base of acetic acid. The delocalizing effect is apparent when the electron density map of each conjugate base is compared:

Acetic acid
$pK_a = 4.76$

Trifluoroacetic acid
$pK_a = 0.23$

Notice that the oxygen atoms on the trifluoroacetate ion are less negative (represented by a lighter shade of red). As we have seen before, this causes the equilibrium for ionization of trifluoroacetic acid to shift to the right relative to the ionization of acetic acid, making trifluoroacetic acid more acidic than acetic acid.

2.7 LEWIS ACIDS AND BASES

Lewis acid Any molecule or ion that can form a new covalent bond by accepting a pair of electrons.

Lewis base Any molecule or ion that can form a new covalent bond by donating a pair of electrons.

Gilbert Lewis, who proposed that covalent bonds are formed by the sharing of one or more pairs of electrons (Section 1.3), further expanded the theory of acids and bases to include a group of substances not included in the Brønsted–Lowry concept. According to the Lewis definition, an **acid** is a species that can form a new covalent bond by accepting a pair of electrons; a **base** is a species that can form a new covalent bond by donating a pair of electrons. In the following general equation, the Lewis acid, A, accepts a pair of electrons in forming the new covalent bond and acquires a negative formal charge, while the Lewis base, :B, donates the pair of electrons in forming the new covalent bond and acquires a positive formal charge:

new covalent bond formed in this Lewis acid–base reaction

$$A + :B \rightleftharpoons \bar{A}—\overset{+}{B}$$

Lewis   Lewis
acid   base

Note that, although we speak of a Lewis base as "donating" a pair of electrons, the term is not fully accurate. "Donating" in this case does not imply that the electron pair under consideration is removed completely from the valence shell of the base. Rather, "donating" means that the electron pair is shared with another atom to form a covalent bond.

As we will see in the chapters that follow, a great many organic reactions can be interpreted as Lewis acid–base reactions. Perhaps the most important (but not the only) Lewis acid is the proton. Isolated protons, of course, do not exist in solution; rather, a proton attaches itself to the strongest available Lewis base. When HCl is dissolved in water, for example, the strongest available Lewis base is an H_2O molecule, and the following proton-transfer reaction takes place:

$$
\underset{\underset{H}{|}}{H-\overset{\cdot\cdot}{O}\colon} \;+\; H-\overset{\cdot\cdot}{\underset{\cdot\cdot}{Cl}}\colon \longrightarrow \underset{\underset{H}{|}}{H-\overset{\cdot\cdot+}{O}-H} \;+\; \colon\overset{\cdot\cdot}{\underset{\cdot\cdot}{Cl}}\colon^{-}
$$

Hydronium
ion

When HCl is dissolved in methanol, the strongest available Lewis base is a CH_3OH molecule, and this proton-transfer reaction takes place:

$$
\underset{\underset{H}{|}}{CH_3-\overset{\cdot\cdot}{O}\colon} \;+\; H-\overset{\cdot\cdot}{\underset{\cdot\cdot}{Cl}}\colon \longrightarrow \underset{\underset{H}{|}}{CH_3-\overset{\cdot\cdot+}{O}-H} \;+\; \colon\overset{\cdot\cdot}{\underset{\cdot\cdot}{Cl}}\colon^{-}
$$

Methanol   An oxonium
                                        ion

An **oxonium** ion is an ion that contains an oxygen atom with three bonds and bears a positive charge.

Table 2.3 gives examples of the most important types of Lewis bases we will encounter in this text, arranged in order of their increasing strength in proton-transfer reactions. Ethers are organic derivatives of water in which both hydrogens of water are replaced by carbon groups. We study the properties of ethers along with those of alcohols in Chapter 8. We study the properties of amines in Chapter 10.

TABLE 2.3 Some Organic Lewis Bases and Their Relative Strengths in Proton-Transfer Reactions

Halide ions	Water, alcohols, and ethers	Ammonia and Amines	Hydroxide ion and alkoxide ions	Amide ions
$\colon\overset{\cdot\cdot}{\underset{\cdot\cdot}{Cl}}\colon^{-}$	$H-\overset{\cdot\cdot}{\underset{\cdot\cdot}{O}}-H$	$H-\underset{\underset{H}{\mid}}{\overset{\cdot\cdot}{N}}-H$	$H-\overset{\cdot\cdot}{\underset{\cdot\cdot}{O}}\colon^{-}$	$H-\underset{\underset{H}{\mid}}{\overset{\cdot\cdot}{N}}\colon^{-}$
$\colon\overset{\cdot\cdot}{\underset{\cdot\cdot}{Br}}\colon^{-}$	$CH_3-\overset{\cdot\cdot}{\underset{\cdot\cdot}{O}}-H$	$CH_3-\underset{\underset{H}{\mid}}{\overset{\cdot\cdot}{N}}-H$	$CH_3-\overset{\cdot\cdot}{\underset{\cdot\cdot}{O}}\colon^{-}$	$CH_3-\underset{\underset{H}{\mid}}{\overset{\cdot\cdot}{N}}\colon^{-}$
$\colon\overset{\cdot\cdot}{\underset{\cdot\cdot}{I}}\colon^{-}$	$CH_3-\overset{\cdot\cdot}{\underset{\cdot\cdot}{O}}-CH_3$	$CH_3-\underset{\underset{CH_3}{\mid}}{\overset{\cdot\cdot}{N}}-H$		$CH_3-\underset{\underset{CH_3}{\mid}}{\overset{\cdot\cdot}{N}}\colon^{-}$
		$CH_3-\underset{\underset{CH_3}{\mid}}{N}-CH_3$		
Very weak	Weak	Strong	Stronger	Very strong

EXAMPLE 2.4

Complete this acid–base reaction: First add unshared pairs of electrons on the reacting atoms to give each atom a complete octet. Use curved arrows to show the redistribution of electrons in the reaction. In addition, predict whether the position of this equilibrium lies toward the left or the right.

$$CH_3-O^+-H + CH_3-N-H \rightleftharpoons$$
$$\quad\quad\quad | \quad\quad\quad\quad\quad\quad |$$
$$\quad\quad\quad H \quad\quad\quad\quad\quad\quad H$$

SOLUTION

Proton transfer takes place to form an alcohol and an ammonium ion. We know from Table 2.3 that amines are stronger bases than alcohols. We also know that the weaker the base, the stronger its conjugate acid, and vice versa. From this analysis, we conclude that the position of this equilibrium lies to the right, on the side of the weaker acid and the weaker base.

$$CH_3-\overset{..}{O}\overset{+}{-}H + CH_3-\overset{..}{N}-H \rightleftharpoons CH_3-\overset{..}{O}: + CH_3-\overset{\overset{H}{|}}{N}{}^+-H$$
$$\quad\quad | \quad\quad\quad\quad | \quad\quad\quad\quad\quad | \quad\quad\quad\quad |$$
$$\quad\quad H \quad\quad\quad\quad H \quad\quad\quad\quad H \quad\quad\quad\quad H$$

Stronger   Stronger   Weaker   Weaker
acid   base   base   acid

Practice Problem 2.4

Complete this acid–base reaction. First add unshared pairs of electrons on the reacting atoms to give each atom a complete octet. Use curved arrows to show the redistribution of electrons in the reaction. In addition, predict whether the position of the equilibrium lies toward the left or the right.

$$\quad\quad\quad\quad\quad\quad\quad\quad\quad H$$
$$\quad\quad\quad\quad\quad\quad\quad\quad\quad |$$
$$CH_3-O^- + CH_3-N^+-CH_3 \rightleftharpoons$$
$$\quad\quad\quad\quad\quad\quad\quad\quad\quad |$$
$$\quad\quad\quad\quad\quad\quad\quad\quad\quad CH_3$$

Another type of Lewis acid we will encounter in later chapters is an organic cation in which a carbon is bonded to only three atoms and bears a positive formal charge. Such carbon cations are called carbocations. Consider the reaction that occurs when the following organic cation reacts with a bromide ion:

$$\quad\quad\quad\quad\quad\quad\quad\quad\quad\quad\quad\quad\quad\quad :\overset{..}{\underset{..}{Br}}:$$
$$\quad\quad\quad\quad\quad\quad\quad\quad\quad\quad\quad\quad\quad\quad |$$
$$CH_3-\overset{+}{C}H-CH_3 + :\overset{..}{\underset{..}{Br}}: \longrightarrow CH_3-CH-CH_3$$

An organic   Bromide ion   2-Bromopropane
cation   (a Lewis base)
(a Lewis acid)

In this reaction, the organic cation is the electron pair acceptor (the Lewis acid), and bromide ion is the electron pair donor (the Lewis base).

EXAMPLE 2.5

Complete the following Lewis acid–base reaction. Show all electron pairs on the reacting atoms and use curved arrows to show the flow of electrons in the reaction:

$$CH_3 \overset{+}{-} \overset{+}{CH} - CH_3 + H_2O \longrightarrow$$

SOLUTION

The trivalent carbon atom in the organic cation has an empty orbital in its valence shell and, therefore, is the Lewis acid. Water is the Lewis base.

| Lewis acid | Lewis base | An oxonium ion |

Practice Problem 2.5

Write an equation for the reaction between each Lewis acid–base pair, showing electron flow by means of curved arrows. (*Hint:* Aluminum is in Group 3A of the Periodic Table, just under boron. Aluminum in $AlCl_3$ has only six electrons in its valence shell and thus has an incomplete octet.)

(a) $Cl^- + AlCl_3 \longrightarrow$ (b) $CH_3Cl + AlCl_3 \longrightarrow$

SUMMARY

According to the **Arrhenius definitions**, an acid is a substance that dissolves in aqueous solution to produce H_3O^+ ions; a base is a substance that dissolves in aqueous solution to produce OH^- ions (Section 2.2). According to the **Brønsted–Lowry** definitions (Section 2.3), an acid is a proton donor and a base is a proton acceptor. Neutralization of an acid by a base is a **proton-transfer reaction** in which the acid is transformed into its **conjugate base**, and the base is transformed into its **conjugate acid**.

A **strong acid** or **strong base** is one that is completely ionized in water. A weak acid or weak base is one that is only partially ionized in water (Section 2.4). The strength of a weak acid is expressed by its **ionization constant, K_a**

(Section 2.4). The larger the value of K_a, the stronger the acid. $pK_a = -\log K_a$.

In an acid–base reaction, the position of equilibrium favors the reaction of the stronger acid and the stronger base to form the weaker acid and the weaker base (Section 2.5).

The relative acidities of the organic acids, HA, are determined by (1) the electronegativity of A, (2) resonance stabilization of the conjugate base, A^-, and (3) the electron-withdrawing inductive effect, which also stabilizes the conjugate base (Section 2.6).

According to the **Lewis definitions** (Section 2.7), an **acid** is a species that forms a new covalent bond by accepting a pair of electrons; a **base** is a species that forms a new covalent bond by donating a pair of electrons.

KEY REACTIONS

1. Proton-Transfer Reaction (Section 2.3)

This reaction involves the transfer of a proton from a proton donor (a Brønsted–Lowry acid) to a proton acceptor (a Brønsted–Lowry base):

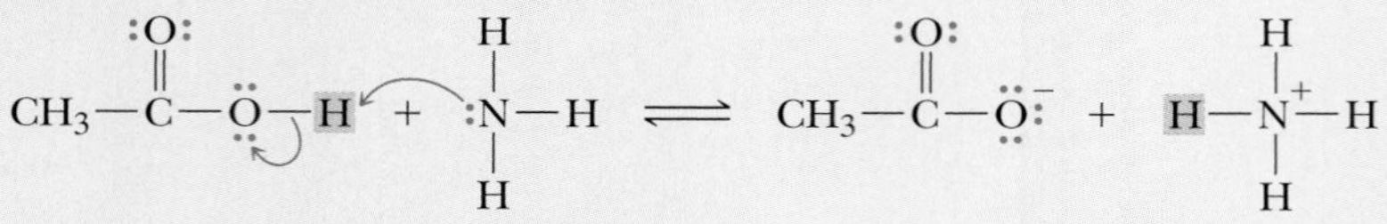

Acetic acid   Ammonia   Acetate   Ammonium
(proton donor) (proton acceptor)   ion   ion

2. Position of Equilibrium in an Acid–Base Reaction (Section 2.5)

Equilibrium favors reaction of the stronger acid with the stronger base to give the weaker acid and the weaker base:

$$CH_3COOH + NH_3 \rightleftharpoons CH_3COO^- + NH_4^+$$

Acetic acid	Ammonium ion
pK_a 4.76	pK_a 9.24
(stronger acid)	(weaker acid)

3. Lewis Acid–Base Reaction (Section 2.7)

A Lewis acid–base reaction involves sharing an electron pair between an electron pair donor (a Lewis base) and an electron pair acceptor (a Lewis acid):

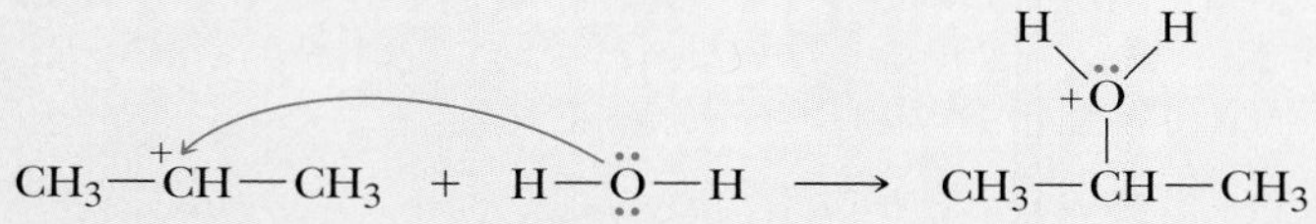

A Lewis acid   A Lewis base   An oxonium ion

PROBLEMS

A problem number set in red indicates an applied "real-world" problem.

Arrhenius Acids and Bases

2.6 Complete the net ionic equation for each acid placed in water. Use curved arrows to show the flow of electron pairs in each reaction. Also, for each reaction, determine the direction of equilibrium, using Table 2.2 as a reference for the pK_a values of proton acids.

(a) $NH_4^+ + H_2O \rightleftharpoons$ (b) $HCO_3^- + H_2O \rightleftharpoons$

(c) $CH_3-\overset{\overset{\displaystyle O}{\|}}{C}-OH + H_2O \rightleftharpoons$

2.7 Complete the net ionic equation for each base placed in water. Use curved arrows to show the flow of electron pairs in each reaction. Also, for each reaction, determine the direction of equilibrium, using Table 2.2 as a reference for the pK_a values of proton acids formed.

(a) $CH_3NH_2 + H_2O \rightleftharpoons$ (b) $HSO_4^- + H_2O \rightleftharpoons$

(c) $Br^- + H_2O \rightleftharpoons$ (d) $CO_3^{2-} + H_2O \rightleftharpoons$

Brønsted–Lowry Acids and Bases

2.8 Complete a net ionic equation for each proton-transfer reaction, using curved arrows to show the flow of electron pairs in each reaction. In addition, write Lewis structures for all starting materials and products. Label the original acid and its conjugate base;

label the original base and its conjugate acid. If you are uncertain about which substance in each equation is the proton donor, refer to Table 2.2 for the pK_a values of proton acids.

(a) $NH_3 + HCl \longrightarrow$ (b) $CH_3CH_2O^- + HCl \longrightarrow$

(c) $HCO_3^- + OH^- \longrightarrow$ (d) $CH_3COO^- + NH_4^+ \longrightarrow$

(e) $NH_4^+ + OH^- \longrightarrow$ (f) $CH_3COO^- + CH_3NH_3^+ \longrightarrow$

(g) $CH_3CH_2O^- + NH_4^+ \longrightarrow$ (h) $CH_3NH_3^+ + OH^- \longrightarrow$

2.9 Each of these molecules and ions can function as a base. Complete the Lewis structure of each base, and write the structural formula of the conjugate acid formed by its reaction with HCl.

(a) CH_3CH_2OH (b) $\overset{\displaystyle O}{\overset{\|}{H}}CH$ (c) $(CH_3)_2NH$ (d) HCO_3^-

2.10 Offer an explanation for the following observations:

(a) H_3O^+ is a stronger acid than NH_4^+.

(b) Nitric acid, HNO_3, is a stronger acid than nitrous acid, HNO_2 (pK_a 3.7).

(c) Ethanol, CH_3CH_2OH, and water have approximately the same acidity.

(d) Trichloroacetic acid, CCl_3COOH (pK_a 0.64), is a stronger acid than acetic acid, CH_3COOH (pK_a 4.74).

(e) Trifluoroacetic acid, CF_3COOH (pK_a 0.23), is a stronger acid than trichloroacetic acid, CCl_3COOH (pK_a 0.64).

2.11 Select the most acidic proton in the following compounds:

(a) $H_3C \overset{\displaystyle O}{\overset{\|}{-}C} - CH_2 - \overset{\displaystyle O}{\overset{\|}{C}} - CH_3$ (b) $H_2N - \overset{\displaystyle NH_2^+}{\overset{\|}{C}} - NH_2$

Quantitative Measure of Acid Strength

2.12 Which has the larger numerical value?

(a) The pK_a of a strong acid or the pK_a of a weak acid?

(b) The K_a of a strong acid or the K_a of a weak acid?

2.13 In each pair, select the stronger acid:

(a) Pyruvic acid (pK_a 2.49) or lactic acid (pK_a 3.85)

(b) Citric acid (pK_{a1} 3.08) or phosphoric acid (pK_{a1} 2.10)

(c) Nicotinic acid (niacin, K_a 1.4×10^{-5}) or acetylsalicylic acid (aspirin, K_a 3.3×10^{-4})

(d) Phenol (K_a 1.12×10^{-10}) or acetic acid (K_a 1.74×10^{-5})

2.14 Arrange the compounds in each set in order of increasing acid strength. Consult Table 2.2 for pK_a values of each acid.

(a) CH_3CH_2OH $HO\overset{\displaystyle O}{\overset{\|}{C}}O^-$ $C_6H_5\overset{\displaystyle O}{\overset{\|}{C}}OH$

       Ethanol   Bicarbonate ion   Benzoic acid

(b) $HO\overset{\displaystyle O}{\overset{\|}{C}}OH$ $CH_3\overset{\displaystyle O}{\overset{\|}{C}}OH$ HCl

     Carbonic acid   Acetic acid   Hydrogen chloride

2.15 Arrange the compounds in each set in order of increasing base strength. Consult Table 2.2 for pK_a values of the conjugate acid of each base. (*Hint:* The stronger the acid, the weaker is its conjugate base, and vice versa.)

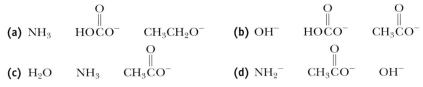

(a) NH_3 $HO\overset{\displaystyle O}{\overset{\|}{C}}O^-$ $CH_3CH_2O^-$ (b) OH^- $HO\overset{\displaystyle O}{\overset{\|}{C}}O^-$ $CH_3\overset{\displaystyle O}{\overset{\|}{C}}O^-$

(c) H_2O NH_3 $CH_3\overset{\displaystyle O}{\overset{\|}{C}}O^-$ (d) NH_2^- $CH_3\overset{\displaystyle O}{\overset{\|}{C}}O^-$ OH^-

Position of Equilibrium in Acid–Base Reactions

2.16 Unless under pressure, carbonic acid in aqueous solution breaks down into carbon dioxide and water, and carbon dioxide is evolved as bubbles of gas. Write an equation for the conversion of carbonic acid to carbon dioxide and water.

2.17 For each of the following compounds, will carbon dioxide be evolved when sodium bicarbonate is added to an aqueous solution of the compound?
(a) H_2SO_4 **(b)** CH_3CH_2OH **(c)** NH_4Cl

2.18 Acetic acid, CH_3COOH, is a weak organic acid, pK_a 4.76. Write equations for the equilibrium reactions of acetic acid with each base. Which equilibria lie considerably toward the left? Which lie considerably toward the right?
(a) $NaHCO_3$ **(b)** NH_3 **(c)** H_2O **(d)** $NaOH$

2.19 For an acid–base reaction, one way to indicate the predominant species at equilibrium is to say that the reaction arrow points to the acid with the higher value of pK_a. For example

$$NH_4^+ + H_2O \longleftarrow NH_3 + H_3O^+$$
$$pK_a\ 9.24 \qquad\qquad pK_a\ -1.74$$

$$NH_4^+ + OH^- \longrightarrow NH_3 + H_2O$$
$$pK_a\ 9.24 \qquad\qquad pK_a\ 15.7$$

Explain why this rule works.

Lewis Acids and Bases

2.20 Complete the following acid–base reactions, using curved arrow notation to show the flow of electron pairs.

(a) BF_3 + (oxetane structure) ⟶   **(b)** CH_3—C(—CH$_3$)(—CH$_3$)—Cl + Al(—Cl)(—Cl)—Cl ⟶

2.21 Complete equations for these reactions between Lewis acid–Lewis base pairs. Label which starting material is the Lewis acid and which is the Lewis base, and use a curved arrow to show the flow of the electron pair in each reaction. In solving these problems, it is essential that you show all valence electrons for the atoms participating directly in each reaction.

(a) CH_3—$\overset{+}{C}H$—CH_3 + CH_3—O—H ⟶

(b) CH_3—$\overset{+}{C}H$—CH_3 + Br^- ⟶

(c) CH_3—$\overset{+}{C}$(CH$_3$)(CH$_3$) + H—O—H ⟶

2.22 Use curved arrow notation to show the flow of electron pairs in each Lewis acid–base reaction. Be certain to show all valence electron pairs on each atom participating in the reaction.

(a) CH_3—C(=O)—CH_3 + :CH_3 ⟶ CH_3—C(O$^-$)(CH$_3$)—CH$_3$

(b) CH_3—C(=$\overset{+}{O}$H)—CH_3 + :$\overset{-}{C}N$ ⟶ CH_3—C(OH)(CN)—CH$_3$

(c) CH_3O^- + CH_3—Br ⟶ CH_3—O—CH_3 + Br^-

Looking Ahead

2.23 Alcohols (Chapter 8) are weak organic acids, pK_a 15–18. The pK_a of ethanol, CH_3CH_2OH, is 15.9. Write equations for the equilibrium reactions of ethanol with each base. Which equilibria lie considerably toward the right? Which lie considerably toward the left?

(a) $NaHCO_3$ **(b)** NaOH **(c)** $NaNH_2$ **(d)** NH_3

2.24 Phenols (Chapter 9) are weak acids, and most are insoluble in water. Phenol, C_6H_5OH (pK_a 9.95), for example, is only slightly soluble in water, but its sodium salt, $C_6H_5O^-Na^+$, is quite soluble in water. Will phenol dissolve in

(a) Aqueous NaOH? **(b)** Aqueous $NaHCO_3$? **(c)** Aqueous Na_2CO_3?

2.25 Carboxylic acids (Chapter 14) of six or more carbons are insoluble in water, but their sodium salts are very soluble in water. Benzoic acid, C_6H_5COOH (pK_a 4.19), for example, is insoluble in water, but its sodium salt, $C_6H_5COO^-Na^+$, is quite soluble in water. Will benzoic acid dissolve in

(a) Aqueous NaOH? **(b)** Aqueous $NaHCO_3$? **(c)** Aqueous Na_2CO_3?

2.26 As we shall see in Chapter 16, hydrogens on a carbon adjacent to a carbonyl group are far more acidic than those not adjacent to a carbonyl group. The highlighted H in propanone, for example, is more acidic than the highlighted H in ethane:

$$CH_3\overset{\overset{\displaystyle O}{\|}}{C}CH_2—\boxed{H} \qquad CH_3CH_2—\boxed{H}$$

Propanone   Ethane
$pK_a = 22$ $pK_a = 51$

Account for the greater acidity of propanone in terms of (a) the inductive effect and (b) the resonance effect.

2.27 Explain why the protons in dimethyl ether, $CH_3—O—CH_3$, are not very acidic.

2.28 Predict whether sodium hydride, NaH, will act as a base or an acid, and provide a rationale for your decision.

2.29 Alanine is one of the 20 amino acids (it contains both an amino and a carboxyl group) found in proteins (Chapter 19). Is alanine better represented by the structural formula A or B? Explain.

$$CH_3—CH—\overset{\overset{\displaystyle O}{\|}}{C}—OH \qquad CH_3—CH—\overset{\overset{\displaystyle O}{\|}}{C}—O^-$$
$$\quad\quad\;\; | \qquad\qquad\qquad\qquad |$$
$$\quad\quad NH_2 \qquad\qquad\qquad\quad NH_3^+$$
$$\quad\quad (A) \qquad\qquad\qquad\qquad (B)$$

2.30 Glutamic acid is another of the amino acids found in proteins (Chapter 19):

$$\text{Glutamic acid}\qquad HO—\overset{\overset{\displaystyle O}{\|}}{C}—CH_2—CH_2—CH—\overset{\overset{\displaystyle O}{\|}}{C}—OH$$
$$\qquad\qquad\qquad\qquad\qquad\qquad\qquad\qquad\;\; |$$
$$\qquad\qquad\qquad\qquad\qquad\qquad\qquad\quad NH_3^+$$

Glutamic acid has two carboxyl groups, one with pK_a 2.10, the other with pK_a 4.07.

(a) Which carboxyl group has which pK_a?

(b) Account for the fact that one carboxyl group is a considerably stronger acid than the other.

3

Alkanes and Cycloalkanes

Bunsen burners burn natural gas, which is primarily methane with small amounts of ethane, propane, butane, and 2-methylbutane. Inset: A model of methane. *(Charles D. Winters)*

Hydrocarbon A compound that contains only carbon atoms and hydrogen atoms.

Alkane A saturated hydrocarbon whose carbon atoms are arranged in an open chain.

Saturated hydrocarbon A hydrocarbon containing only carbon–carbon single bonds.

Aliphatic hydrocarbon An alternative word to describe an alkane.

3.1 INTRODUCTION

In this chapter, we begin our study of organic compounds with the physical and chemical properties of alkanes, the simplest types of organic compounds. Actually, alkanes are members of a larger class of organic compounds called **hydrocarbons**. A hydrocarbon is a compound composed of only carbon and hydrogen. Figure 3.1 shows the four classes of hydrocarbons, along with the characteristic type of bonding between carbon atoms in each.

Alkanes are **saturated hydrocarbons**; that is, they contain only carbon–carbon single bonds. In this context, "saturated" means that each carbon has the maximum number of hydrogens bonded to it. We often refer to alkanes as **aliphatic hydro-**

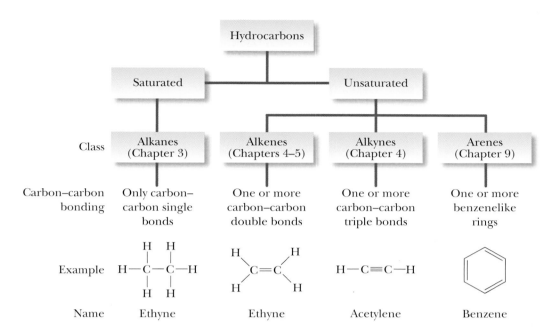

Figure 3.1
The four classes of hydrocarbons.

carbons, because the physical properties of the higher members of this class resemble those of the long carbon-chain molecules we find in animal fats and plant oils (Greek: *aleiphar*, fat or oil).

A hydrocarbon that contains one or more carbon–carbon double bonds, triple bonds, or benzene rings is classified as an **unsaturated hydrocarbon**. We study alkanes (saturated hydrocarbons) in this chapter. We study alkenes and alkynes (both unsaturated hydrocarbons) in Chapters 4 and 5, and arenes (also unsaturated hydrocarbons) in Chapter 9.

3.2 STRUCTURE OF ALKANES

Methane (CH_4) and ethane (C_2H_6) are the first two members of the alkane family. Figure 3.2 shows Lewis structures and ball-and-stick models for these molecules. The shape of methane is tetrahedral, and all H—C—H bond angles are 109.5°. Each carbon atom in ethane is also tetrahedral, and all bond angles are approximately 109.5°.

Although the three-dimensional shapes of larger alkanes are more complex than those of methane and ethane, the four bonds about each carbon atom are still arranged in a tetrahedral manner, and all bond angles are still approximately 109.5°.

The next members of the alkane family are propane, butane, and pentane. In the representations that follow, these hydrocarbons are drawn first as condensed structural formulas that show all carbons and hydrogens. They are also drawn in an even more abbreviated form called a **line-angle formula**. In this type of representation, a line represents a carbon—carbon bond and an angle represents a carbon

Butane is the fuel in this lighter. Butane molecules are present in the liquid and gaseous states in the lighter. (*Charles D. Winters*)

Line-angle formula An abbreviated way to draw structural formulas in which each line ending represents a carbon atom and a line represents a bond.

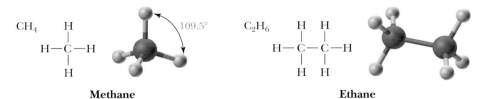

Methane   **Ethane**

Figure 3.2
Methane and ethane.

A tank for propane fuel.
(Charles D. Winters)

TABLE 3.1   Names, Molecular Formulas, and Condensed Structural Formulas for the First 20 Alkanes with Unbranched Chains

Name	Molecular Formula	Condensed Structural Formula	Name	Molecular Formula	Condensed Structural Formula
methane	CH_4	CH_4	undecane	$C_{11}H_{24}$	$CH_3(CH_2)_9CH_3$
ethane	C_2H_6	CH_3CH_3	dodecane	$C_{12}H_{26}$	$CH_3(CH_2)_{10}CH_3$
propane	C_3H_8	$CH_3CH_2CH_3$	tridecane	$C_{13}H_{28}$	$CH_3(CH_2)_{11}CH_3$
butane	C_4H_{10}	$CH_3(CH_2)_2CH_3$	tetradecane	$C_{14}H_{30}$	$CH_3(CH_2)_{12}CH_3$
pentane	C_5H_{12}	$CH_3(CH_2)_3CH_3$	pentadecane	$C_{15}H_{32}$	$CH_3(CH_2)_{13}CH_3$
hexane	C_6H_{14}	$CH_3(CH_2)_4CH_3$	hexadecane	$C_{16}H_{34}$	$CH_3(CH_2)_{14}CH_3$
heptane	C_7H_{16}	$CH_3(CH_2)_5CH_3$	heptadecane	$C_{17}H_{36}$	$CH_3(CH_2)_{15}CH_3$
octane	C_8H_{18}	$CH_3(CH_2)_6CH_3$	octadecane	$C_{18}H_{38}$	$CH_3(CH_2)_{16}CH_3$
nonane	C_9H_{20}	$CH_3(CH_2)_7CH_3$	nonadecane	$C_{19}H_{40}$	$CH_3(CH_2)_{17}CH_3$
decane	$C_{10}H_{22}$	$CH_3(CH_2)_8CH_3$	eicosane	$C_{20}H_{42}$	$CH_3(CH_2)_{18}CH_3$

atom. A line ending represents a —CH_3 group. Although hydrogen atoms are not shown in line-angle formulas, they are assumed to be there in sufficient numbers to give each carbon four bonds.

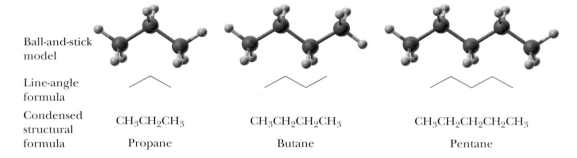

Ball-and-stick model			
Line-angle formula			
Condensed structural formula	$CH_3CH_2CH_3$	$CH_3CH_2CH_2CH_3$	$CH_3CH_2CH_2CH_2CH_3$
	Propane	Butane	Pentane

We can write structural formulas for alkanes in still another abbreviated form. The structural formula of pentane, for example, contains three CH_2 (methylene) groups in the middle of the chain. We can collect these groups together and write the structural formula as $CH_3(CH_2)_3CH_3$. Table 3.1 gives the names and molecular formulas of the first 20 alkanes. Note that the names of all these alkanes end in -ane. We will have more to say about naming alkanes in Section 3.4.

Alkanes have the general molecular formula C_nH_{2n+2}. Thus, given the number of carbon atoms in an alkane, it is easy to determine the number of hydrogens in the molecule and also its molecular formula. For example, decane, with 10 carbon atoms, must have $(2 \times 10) + 2 = 22$ hydrogens and a molecular formula of $C_{10}H_{22}$.

3.3   CONSTITUTIONAL ISOMERISM IN ALKANES

Constitutional isomers Compounds with the same molecular formula, but a different order of attachment of their atoms.

Constitutional isomers are compounds that have the same molecular formula, but different structural formulas. By "different structural formulas," we mean that these compounds differ in the kinds of bonds they have (single, double, or triple) or in their connectivity (the order of attachment among their atoms).

For the molecular formulas CH_4, C_2H_6, and C_3H_8, only one order of attachment of atoms is possible. For the molecular formula C_4H_{10}, two orders of attachment of atoms are possible. In one of these, named butane, the four carbons are bonded in a chain; in the other, named 2-methylpropane, three carbons are bonded in a chain, with the fourth carbon as a branch on the chain.

$CH_3CH_2CH_2CH_3$

Butane
(boiling point
= −0.5°C)

$\underset{|}{CH_3}$
CH_3CHCH_3

2-Methylpropane
(boiling point
= −11.6°C)

Butane and 2-methylpropane are constitutional isomers; they are different compounds and have different physical and chemical properties. Their boiling points, for example, differ by approximately 11°C.

In Section 1.8, we encountered several examples of constitutional isomers, although we did not call them that at the time. We saw that there are two alcohols with molecular formula C_3H_8O, two aldehydes with molecular formula C_4H_8O, and two carboxylic acids with molecular formula $C_4H_8O_2$.

To find out whether two or more structural formulas represent constitutional isomers, write the molecular formula of each and then compare them. All compounds that have the same molecular formula, but different structural formulas, are constitutional isomers.

EXAMPLE 3.1

Do the structural formulas in each pair represent the same compound or constitutional isomers?

(a) $CH_3CH_2CH_2CH_2CH_2CH_3$ and $CH_3CH_2CH_2$
$\phantom{CH_3CH_2CH_2CH_2CH_2CH_3 \quad \text{and} \quad CH_3CH_2CH_2}$ |
$\phantom{CH_3CH_2CH_2CH_2CH_2CH_3 \quad \text{and} \quad CH_3CH}$ $CH_2CH_2CH_3$ (each is C_6H_{14})

(b) $\underset{|}{CH_3}$ $\underset{|}{CH_3}$
CH_3CHCH_2CH and $CH_3CH_2CHCHCH_3$ (each is C_7H_{16})
CH_3 CH_3

SOLUTION

To determine whether these structural formulas represent the same compound or constitutional isomers, first find the longest chain of carbon atoms in each. Note that it makes no difference whether the chain is drawn straight or bent. Second, number the longest chain from the end nearest the first branch. Third, compare the lengths of each chain and the sizes and locations of any branches. Structural formulas that have the same order of attachment of atoms represent the same compound; those which have different orders of attachment of atoms represent constitutional isomers.

(a) Each structural formula has an unbranched chain of six carbons. The two structures are identical and represent the same compound:

$$\underset{1}{CH_3}\underset{2}{CH_2}\underset{3}{CH_2}\underset{4}{CH_2}\underset{5}{CH_2}\underset{6}{CH_3} \quad \text{and} \quad \underset{1}{CH_3}\underset{2}{CH_2}\underset{3}{CH_2}$$
$$\underset{4}{CH_2}\underset{5}{CH_2}\underset{6}{CH_3}$$

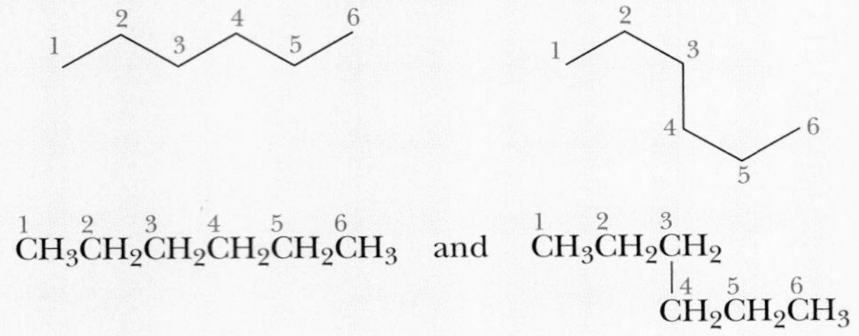

(b) Each structural formula has a chain of five carbons with two CH_3 branches. Although the branches are identical, they are at different locations on the chains. Therefore, these structural formulas represent constitutional isomers:

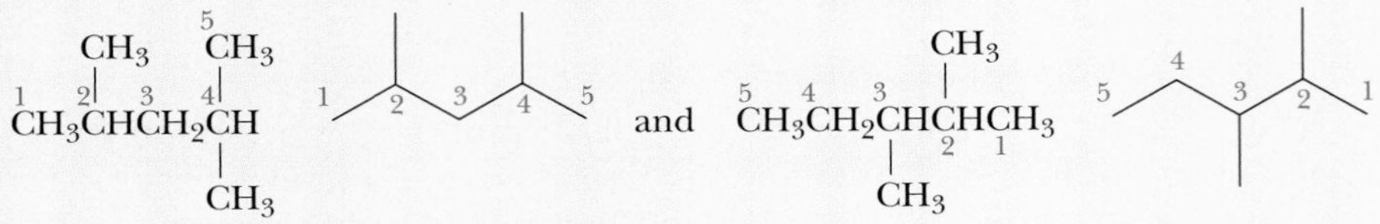

Practice Problem 3.1

Do the structural formulas in each pair represent the same compound or constitutional isomers?

(a) 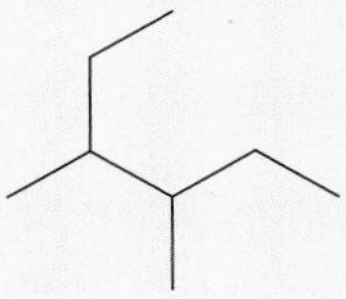 and

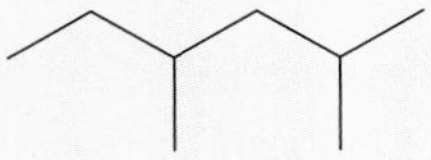

(b) 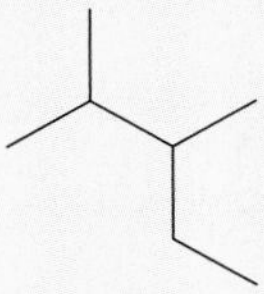 and

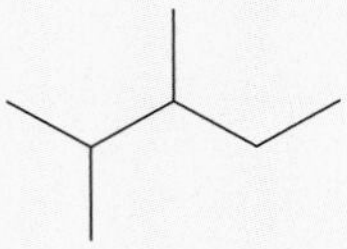

EXAMPLE 3.2

Draw structural formulas for the five constitutional isomers with molecular formula C_6H_{14}.

SOLUTION

In solving problems of this type, you should devise a strategy and then follow it. Here is one such strategy: First, draw a line-angle formula for the constitutional isomer with all six carbons in an unbranched chain. Then, draw line-angle formulas for all constitutional isomers with five carbons in a chain and one carbon as a branch on the chain. Finally, draw line-angle formulas for all constitutional isomers with four carbons in a chain and two carbons as branches.

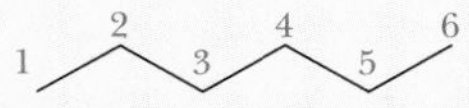

Six carbons in an
unbranched chain

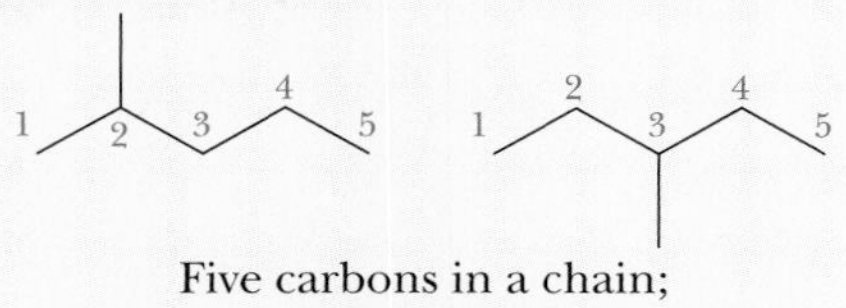

Five carbons in a chain;
one carbon as a branch

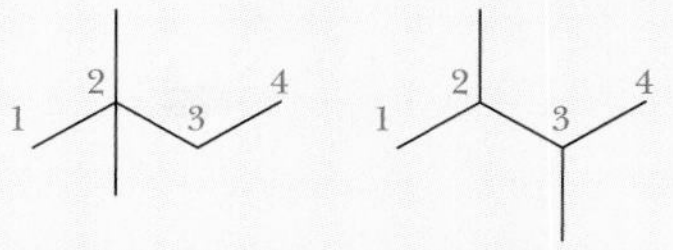

Four carbons in a chain;
two carbons as branches

No constitutional isomers with only three carbons in the longest chain are possible for C_6H_{14}.

Practice Problem 3.2

Draw structural formulas for the three constitutional isomers with molecular formula C_5H_{12}.

The ability of carbon atoms to form strong, stable bonds with other carbon atoms results in a staggering number of constitutional isomers. As the following table shows, there are 3 constitutional isomers with molecular formula C_5H_{12}, 75 constitutional isomers for molecular formula $C_{10}H_{22}$, and almost 37 million constitutional isomers with molecular formula $C_{25}H_{52}$:

Carbon Atoms	Constitutional Isomers
1	0
5	3
10	75
15	4,347
25	36,797,588

Thus, for even a small number of carbon and hydrogen atoms, a very large number of constitutional isomers is possible. In fact, the potential for structural and functional group individuality among organic molecules made from just the basic building blocks of carbon, hydrogen, nitrogen, and oxygen is practically limitless.

3.4 NOMENCLATURE OF ALKANES

A. The IUPAC System

Ideally, every organic compound should have a name from which its structural formula can be drawn. For this purpose, chemists have adopted a set of rules established by an organization called the International Union of Pure and Applied Chemistry (IUPAC).

TABLE 3.2	Prefixes Used in the IUPAC System to Show the Presence of 1 to 20 Carbons in an Unbranched Chain		
Prefix	**Number of Carbon Atoms**	**Prefix**	**Number of Carbon Atoms**
meth-	1	undec-	11
eth-	2	dodec-	12
prop-	3	tridec-	13
but-	4	tetradec-	14
pent-	5	pentadec-	15
hex-	6	hexadec-	16
hept-	7	heptadec-	17
oct-	8	octadec-	18
non-	9	nonadec-	19
dec-	10	eicos-	20

The IUPAC name of an alkane with an unbranched chain of carbon atoms consists of two parts: (1) a prefix that indicates the number of carbon atoms in the chain and (2) the ending **-ane** to show that the compound is a saturated hydrocarbon. Table 3.2 gives the prefixes used to show the presence of 1 to 20 carbon atoms.

The first four prefixes listed in Table 3.2 were chosen by the IUPAC because they were well established in the language of organic chemistry. In fact, they were well established even before there were hints of the structural theory underlying the discipline. For example, the prefix *but-* appears in the name *butyric acid*, a compound of four carbon atoms formed by the air oxidation of butter fat (Latin: *butyrum*, butter). Prefixes to show five or more carbons are derived from Greek or Latin numbers. (See Table 3.1 for the names, molecular formulas, and condensed structural formulas for the first 20 alkanes with unbranched chains.)

The IUPAC name of an alkane with a branched chain consists of a parent name that indicates the longest chain of carbon atoms in the compound and substituent names that indicate the groups bonded to the parent chain.

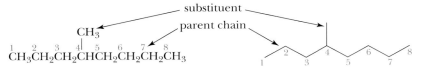

4-Methyloctane

A substituent group derived from an alkane by the removal of a hydrogen atom is called an **alkyl group** and is commonly represented by the symbol **R-**. We name alkyl groups by dropping the *-ane* from the name of the parent alkane and adding the suffix *-yl*. Table 3.3 gives names and structural formulas for eight of the most common alkyl groups. The prefix *sec-* is an abbreviation for secondary, meaning a carbon bonded to two other carbons. The prefix *tert-* is an abbreviation for tertiary, meaning a carbon bonded to three other carbons. Note that when these two prefixes are part of a name, they are always italicized.

The rules of the IUPAC system for naming alkanes are as follows:

1. The name for an alkane with an unbranched chain of carbon atoms consists of a prefix showing the number of carbon atoms in the chain and the ending *-ane*.
2. For branched-chain alkanes, take the longest chain of carbon atoms as the parent chain, and its name becomes the root name.

Alkyl group A group derived by removing a hydrogen from an alkane; given the symbol R-.

R- A symbol used to represent an alkyl group.

TABLE 3.3 Names of the Most Common Alkyl Groups

Name	Condensed Structural Formula	Name	Condensed Structural Formula
methyl	$-CH_3$	isobutyl	$-CH_2CHCH_3$
ethyl	$-CH_2CH_3$		$\quad\quad\ \|$
propyl	$-CH_2CH_2CH_3$		$\quad\quad\ CH_3$
isopropyl	$-CHCH_3$	sec-butyl	$-CHCH_2CH_3$
	$\quad\ \|$		$\quad\ \|$
	$\quad\ CH_3$		$\quad\ CH_3$
			$\quad\ CH_3$
butyl	$-CH_2CH_2CH_2CH_3$	tert-butyl	$\quad\ \|$
			$-CCH_3$
			$\quad\ \|$
			$\quad\ CH_3$

3. Give each substituent on the parent chain a name and a number. The number shows the carbon atom of the parent chain to which the substituent is bonded. Use a hyphen to connect the number to the name:

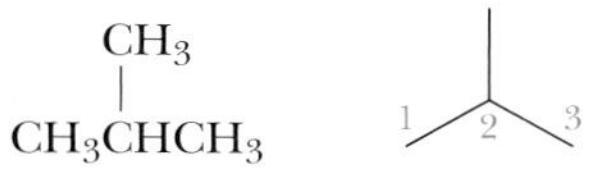

$$CH_3$$
$$|$$
$$CH_3CHCH_3$$

2-Methylpropane

4. If there is one substituent, number the parent chain from the end that gives it the lower number:

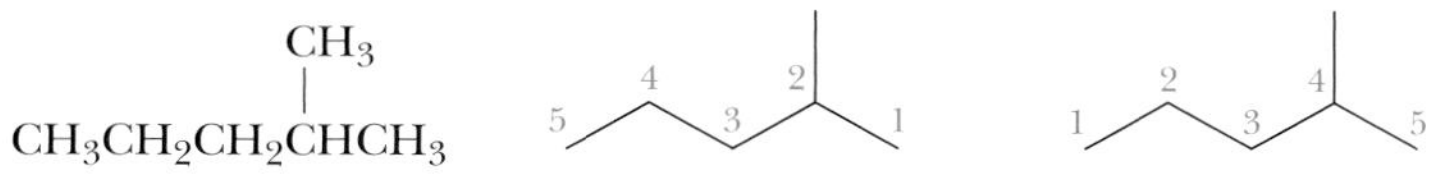

$$CH_3$$
$$|$$
$$CH_3CH_2CH_2CHCH_3$$

2-Methylpentane   (not 4-methylpentane)

5. If there are two or more identical substituents, number the parent chain from the end that gives the lower number to the substituent encountered first. The number of times the substituent occurs is indicated by the prefix *di-*, *tri-*, *tetra-*, *penta-*, *hexa-*, and so on. A comma is used to separate position numbers:

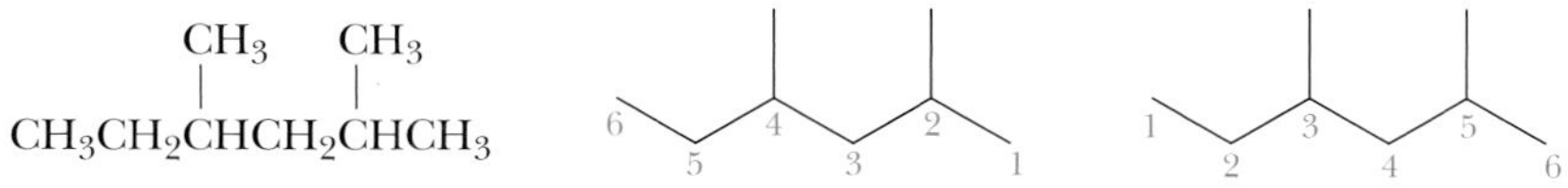

$$CH_3 \quad\ CH_3$$
$$| \quad\quad\ |$$
$$CH_3CH_2CHCH_2CHCH_3$$

2,4-Dimethylhexane   (not 3,5-dimethylhexane)

6. If there are two or more different substituents, list them in alphabetical order, and number the chain from the end that gives the lower number to the substituent encountered first. If there are different substituents in equivalent positions on opposite ends of the parent chain, the substituent of lower alphabetical order is given the lower number:

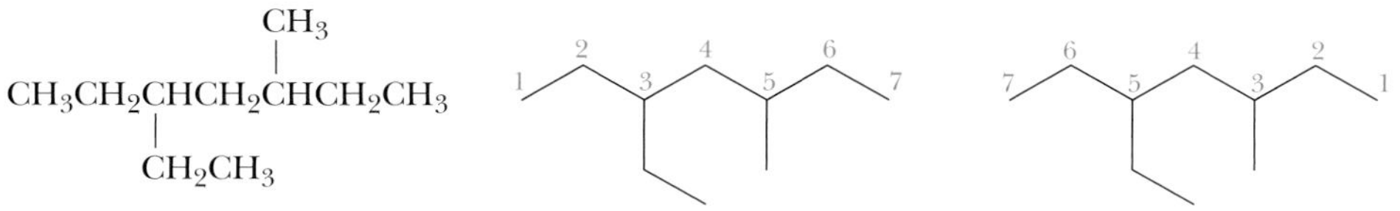

$$CH_3$$
$$|$$
$$CH_3CH_2CHCH_2CHCH_2CH_3$$
$$|$$
$$CH_2CH_3$$

3-Ethyl-5-methylheptane   (not 3-methyl-5-ethylheptane)

7. The prefixes di-, tri-, tetra-, and so on are not included in alphabetizing. Neither are the hyphenated prefixes *sec-* and *tert-*. Alphabetize the names of the substituents first, and then insert the prefix. In the following example, the alphabetizing parts are ethyl and methyl, not ethyl and dimethyl:

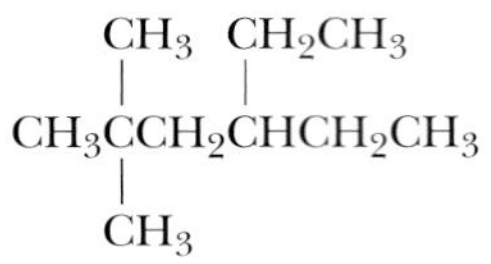

$$CH_3 \quad CH_2CH_3$$
$$CH_3CCH_2CHCH_2CH_3$$
$$CH_3$$

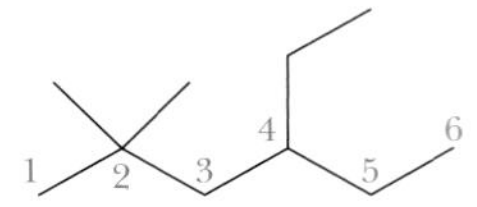

4-Ethyl-2,2-dimethylhexane
(not 2,2-dimethyl-4-ethylhexane)

EXAMPLE 3.3

Write IUPAC names for these alkanes:

(a)

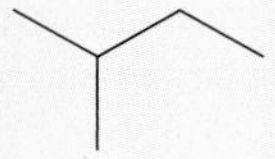

(b)

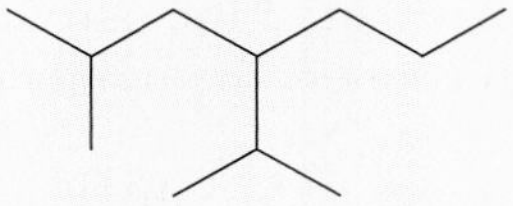

SOLUTION

Number the longest chain in each compound from the end of the chain toward the substituent encountered first (rule 4). List the substituents in (b) in alphabetical order (rule 6).

(a) 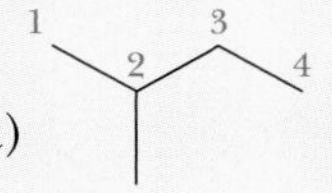   2-Methylbutane

(b) 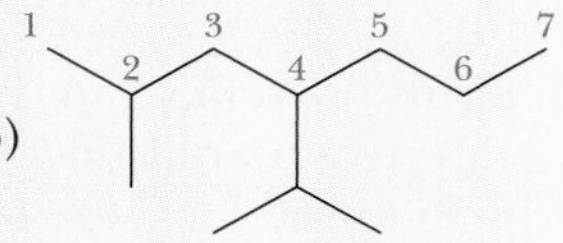   4-Isopropyl-2-methylheptane

Practice Problem 3.3

Write IUPAC names for these alkanes:

(a) 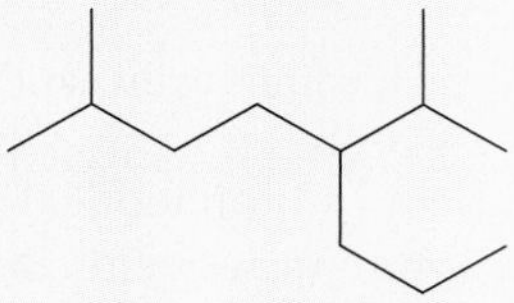   (b)

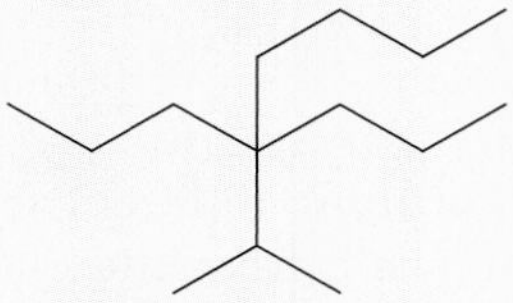

B. Common Names

In the older system of common nomenclature, the total number of carbon atoms in an alkane, regardless of their arrangement, determines the name. The first three alkanes are methane, ethane, and propane. All alkanes of formula C_4H_{10} are called butanes, all those of formula C_5H_{12} are called pentanes, and all those of formula C_6H_{14} are called hexanes. For alkanes beyond propane, *iso* indicates that one end of an otherwise unbranched chain terminates in a $(CH_3)_2CH$—group. Following are examples of common names:

$$CH_3CH_2CH_2CH_3 \qquad CH_3\overset{\overset{\displaystyle CH_3}{|}}{CH}CH_3 \qquad CH_3CH_2CH_2CH_2CH_3 \qquad CH_3CH_2\overset{\overset{\displaystyle CH_3}{|}}{CH}CH_3$$

Butane   Isobutane   Pentane   Isopentane

This system of common names has no good way of handling other branching patterns, so, for more complex alkanes, it is necessary to use the more flexible IUPAC system of nomenclature.

In this text, we concentrate on IUPAC names. However, we also use common names, especially when the common name is used almost exclusively in the everyday discussions of chemists and biochemists. When both IUPAC and common names are given in the text, we always give the IUPAC name first, followed by the common name in parentheses. In this way, you should have no doubt about which name is which.

C. Classification of Carbon and Hydrogen Atoms

We classify a carbon atom as primary (1°), secondary (2°), tertiary (3°), or quaternary (4°), depending on the number of carbon atoms bonded to it. A carbon bonded to one carbon atom is a primary carbon; a carbon bonded to two carbon atoms is a secondary carbon, and so forth. For example, propane contains two primary carbons and one secondary carbon, 2-methylpropane contains three primary carbons and one tertiary carbon, and 2,2,4-trimethylpentane contains five primary carbons, one secondary carbon, one tertiary carbon, and one quaternary carbon:

two 1° carbons   a 3° carbon   a 4° carbon

$$CH_3-CH_2-CH_3 \qquad CH_3-\underset{\underset{\displaystyle CH_3}{|}}{CH}-CH_3 \qquad CH_3-\underset{\underset{\displaystyle CH_3}{|}}{\overset{\overset{\displaystyle CH_3}{|}}{C}}-CH_2-\underset{\underset{\displaystyle CH_3}{|}}{CH}-CH_3$$

a 2° carbon

Propane   2-Methylpropane   2,2,4-Trimethylpentane

Similarly, hydrogens are also classified as primary, secondary, or tertiary, depending on the type of carbon to which each is bonded. Those bonded to a primary carbon are classified as primary hydrogens, those on a secondary carbon are secondary hydrogens, and those on a tertiary carbon are tertiary hydrogens.

3.5 CYCLOALKANES

A hydrocarbon that contains carbon atoms joined to form a ring is called a *cyclic hydrocarbon*. When all carbons of the ring are saturated, we call the hydrocarbon a **cycloalkane**. Cycloalkanes of ring sizes ranging from 3 to over 30 abound in nature, and, in principle, there is no limit to ring size. Five-membered (cyclopentane) and six-membered (cyclohexane) rings are especially abundant in nature and have received special attention.

Figure 3.3 shows the structural formulas of cyclobutane, cyclopentane, and cyclohexane. When writing structural formulas for cycloalkanes, chemists rarely show all carbons and hydrogens. Rather, they use line-angle formulas to represent cycloalkane rings. Each ring is represented by a regular polygon having the same number of sides as there are carbon atoms in the ring. For example, chemists represent cyclobutane by a square, cyclopentane by a pentagon, and cyclohexane by a hexagon.

Cycloalkane A saturated hydrocarbon that contains carbon atoms joined to form a ring.

Figure 3.3
Examples of cycloalkanes.

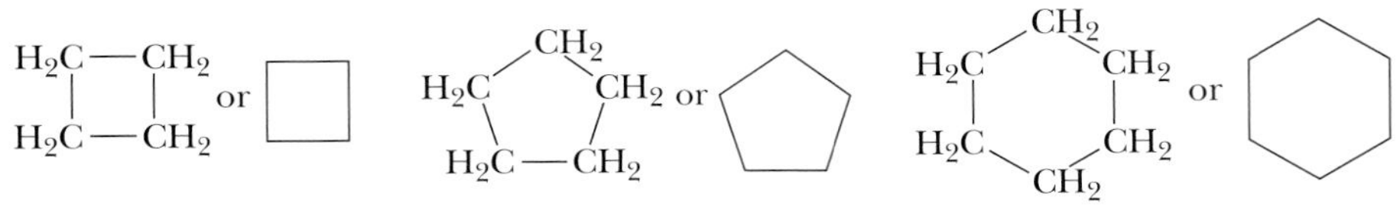

H_2C—CH_2 | H_2C—CH_2 or ▢

Cyclobutane

CH_2 | H_2C CH_2 | H_2C—CH_2 or ⬠

Cyclopentane

CH_2 | H_2C CH_2 | H_2C CH_2 | CH_2 or ⬡

Cyclohexane

Cycloalkanes contain two fewer hydrogen atoms than an alkane with the same number of carbon atoms. For instance, compare the molecular formulas of cyclohexane (C_6H_{12}) and hexane (C_6H_{14}). The general formula of a cycloalkane is C_nH_{2n}.

To name a cycloalkane, prefix the name of the corresponding open-chain hydrocarbon with *cyclo-*, and name each substituent on the ring. If there is only one substituent, there is no need to give it a number. If there are two substituents, number the ring by beginning with the substituent of lower alphabetical order. If there are three or more substituents, number the ring so as to give them the lowest set of numbers, and then list the substituents in alphabetical order.

EXAMPLE 3.4

Write the molecular formula and IUPAC name for each cycloalkane:

(a) 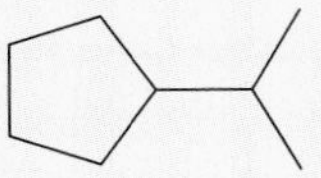   (b)

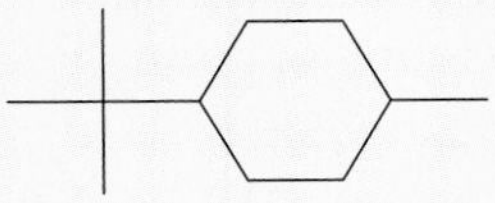

SOLUTION

(a) The molecular formula of this cycloalkane is C_8H_{16}. Because there is only one substituent on the ring, there is no need to number the atoms of the ring. The IUPAC name of this compound is isopropylcyclopentane.
(b) Number the atoms of the cyclohexane ring by beginning with *tert*-butyl, the substituent of lower alphabetical order. The compound's name is 1-*tert*-butyl-4-methylcyclohexane, and its molecular formula is $C_{11}H_{22}$.

Practice Problem 3.4

Write the molecular formula and IUPAC name for each cycloalkane:

(a) 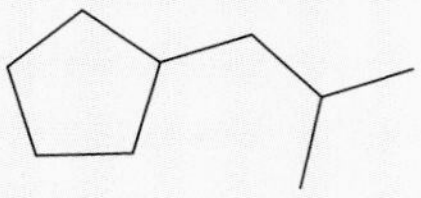   (b) 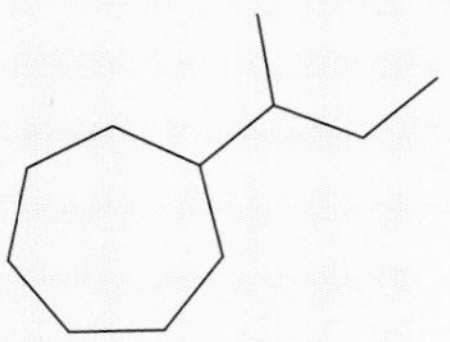   (c)

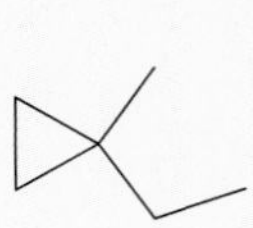

3.6   THE IUPAC SYSTEM: A GENERAL SYSTEM OF NOMENCLATURE

The naming of alkanes and cycloalkanes in Sections 3.4 and 3.5 illustrates the application of the IUPAC system of nomenclature to two specific classes of organic compounds. Now let us describe the general approach of the IUPAC system. The name we give to any compound with a chain of carbon atoms consists of three parts: a prefix, an infix (a modifying element inserted into a word), and a suffix. Each part provides specific information about the structural formula of the compound.

1. The prefix shows the number of carbon atoms in the parent chain. Prefixes that show the presence of 1 to 20 carbon atoms in a chain were given in Table 3.2.

2. The infix shows the nature of the carbon–carbon bonds in the parent chain;

Infix	Nature of Carbon–Carbon Bonds in the Parent Chain
-an-	all single bonds
-en-	one or more double bonds
-yn-	one or more triple bonds

3. The suffix shows the class of compound to which the substance belongs;

Suffix	Class of Compound
-e	hydrocarbon
-ol	alcohol
-al	aldehyde
-one	ketone
-oic acid	carboxylic acid

EXAMPLE 3.5

Following are IUPAC names and structural formulas for four compounds:

$$\text{O}$$
$$\parallel$$
(a) $CH_2=CHCH_3$　(b) CH_3CH_2OH　(c) $CH_3CH_2CH_2CH_2COH$　(d) $HC\equiv CH$

　　Propene　　　　　Ethanol　　　　Pentanoic acid　　　　Ethyne

Divide each name into a prefix, an infix, and a suffix, and specify the information about the structural formula that is contained in each part of the name.

SOLUTION

(a) propene
a carbon–carbon double bond
a hydrocarbon
three carbon atoms

(b) ethanol
only carbon–carbon single bonds
an —OH (hydroxyl) group
two carbon atoms

(c) pentanoic acid
only carbon–carbon single bonds
a —COOH (carboxyl) group
five carbon atoms

(d) ethyne
a carbon–carbon triple bond
a hydrocarbon
two carbon atoms

Practice Problem 3.5

Combine the proper prefix, infix, and suffix, and write the IUPAC name for each compound:

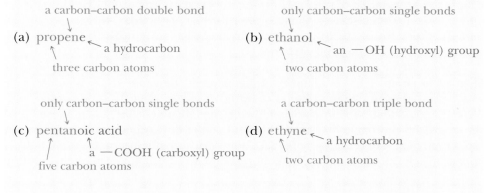

(a) CH_3CCH_3 (with O double-bonded above second C)　(b) $CH_3CH_2CH_2CH_2CH$ (with O double-bonded above last C)　(c) 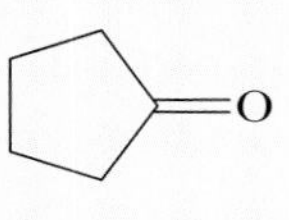 cyclopentanone (ring =O)　(d) 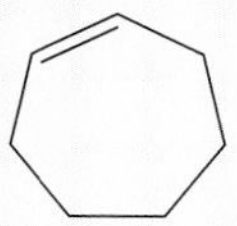 cycloheptene ring

3.7 CONFORMATIONS OF ALKANES AND CYCLOALKANES

Even though structural formulas are useful for showing the order of attachment of atoms, they do not show three-dimensional shapes. As chemists try to understand more and more about the relationships between structure and the chemical and physical properties of molecules, it becomes increasingly important to know more about the three-dimensional shapes of molecules.

In this section, we concentrate on ways to visualize molecules as three-dimensional objects and to visualize not only bond angles within molecules, but also distances between various atoms and groups of atoms not bonded to each other. We also describe strain, which we divide into three types: torsional strain, angle strain, and steric strain. We urge you to build models and to study and manipulate them. Organic molecules are three-dimensional objects, and it is essential that you become comfortable in dealing with them as such.

A. Alkanes

Alkanes of two or more carbons can be twisted into a number of different three-dimensional arrangements of their atoms by rotating about one or more carbon–carbon bonds. Any three-dimensional arrangement of atoms that results from rotation about a single bond is called a **conformation**. Figure 3.4(a) shows a ball-and-stick model of a **staggered conformation** of ethane. In this conformation, the three C—H bonds on one carbon are as far apart as possible from the three C—H bonds on the adjacent carbon. Figure 3.4(b), called a Newman projection, is a shorthand way of representing the staggered conformation of ethane. In a **Newman projection**, we view a molecule along the axis of a C—C bond. The three atoms or groups of atoms nearer your eye appear on lines extending from the center of the circle at angles of 120°. The three atoms or groups of atoms on the carbon farther from your eye appear on lines extending from the circumference of the circle at angles of 120°. Remember that bond angles about each carbon in ethane are approximately 109.5° and not 120°, as this Newman projection might suggest.

Figure 3.5 shows a ball-and-stick model and a Newman projection of an **eclipsed conformation** of ethane. In this conformation, the three C—H bonds on one carbon are as close as possible to the three C—H bonds on the adjacent carbon. In other words, hydrogen atoms on the back carbon are eclipsed by the hydrogen atoms on the front carbon.

For a long time, chemists believed that rotation about the C—C single bond in ethane was completely free. Studies of ethane and other molecules, however, have shown that a potential energy difference exists between its staggered and eclipsed conformations and that rotation is not completely free. In ethane, the potential energy of the eclipsed conformation is a maximum and that of the staggered conformation is a minimum. The difference in potential energy between these two conformations is approximately 3.0 kcal/mol (12.6 kJ/mol), which means that, at room temperature,

Conformation Any three-dimensional arrangement of atoms in a molecule that results by rotation about a single bond.

Staggered conformation A conformation about a carbon–carbon single bond in which the atoms on one carbon are as far apart as possible from the atoms on the adjacent carbon.

Newman projection A way to view a molecule by looking along a carbon–carbon bond.

Eclipsed conformation A conformation about a carbon–carbon single bond in which the atoms on one carbon are as close as possible to the atoms on the adjacent carbon.

Figure 3.4
A staggered conformation of ethane. (a) Ball-and-stick model and (b) Newman projection.

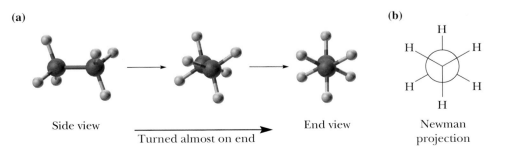

(a)

Side view

Turned almost on end

End view

(b)

Newman projection

(a) **(b)** **(c)**

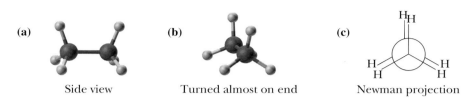

Side view   Turned almost on end   Newman projection

Figure 3.5
An eclipsed conformation of ethane. (a, b) Ball-and-stick models and (c) Newman projection.

the ratio of ethane molecules in a staggered conformation to those in an eclipsed conformation is approximately 100 to 1.

The strain induced in the eclipsed conformation of ethane is an example of torsional strain. **Torsional strain** (also called eclipsed interaction strain) is strain that arises when nonbonded atoms separated by three bonds are forced from a staggered conformation to an eclipsed conformation.

Torsional strain (also called eclipsed interaction strain) Strain that arises when atoms separated by 3 bonds are forced from a staggered conformation to an eclipsed conformation.

EXAMPLE 3.6

Draw Newman projections for one staggered conformation and one eclipsed conformation of propane.

SOLUTION

Following are Newman projections and ball-and-stick models of these conformations:

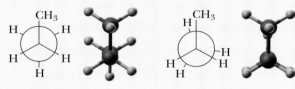

Staggered conformation   Eclipsed conformation

Practice Problem 3.6

Draw Newman projections for two staggered and two eclipsed conformations of 1,2-dichloroethane.

B. Cycloalkanes

We limit our discussion to the conformations of cyclopentanes and cyclohexanes, because these are the most common carbon rings in the molecules of nature.

Cyclopentane

We can draw cyclopentane [Figure 3.6(a)] as a planar conformation with all C—C—C bond angles equal to 108° [Figure 3.6(b)]. This angle differs only slightly from the tetrahedral angle of 109.5°; consequently, there is little angle strain in the planar conformation of cyclopentane. **Angle strain** results when a bond

Angle strain The strain that arises when a bond angle is either compressed or expanded compared with its optimal value.

(a) **(b)** **(c)**

puckering relieves some of the torsional strain

Planar conformation   Puckered envelope conformation

Figure 3.6
Cyclopentane. (a) Structural formula. (b) In the planar conformation, there are 10 pairs of eclipsed C—H interactions. (c) The most stable conformation is a puckered "envelope" conformation.

Figure 3.7
Cyclohexane. The most stable conformation is the puckered "chair" conformation.

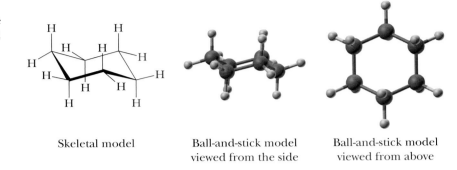

Skeletal model   Ball-and-stick model   Ball-and-stick model
                        viewed from the side   viewed from above

angle in a molecule is either expanded or compressed compared with its optimal values. There are 10 fully eclipsed C—H bonds creating a torsional strain of approximately 10 kcal/mol (42 kJ/mol). To relieve at least a part of this strain, the atoms of the ring twist into the "envelope" conformation [Figure 3.6(c)]. In this conformation, four carbon atoms are in a plane, and the fifth is bent out of the plane, rather like an envelope with its flap bent upward.

In the envelope conformation, the number of eclipsed hydrogen interactions is reduced, thereby decreasing torsional strain. The C—C—C bond angles, however, are also reduced, which increases angle strain. The observed C—C—C bond angles in cyclopentane are 105°, indicating that, in its conformation of lowest energy, cyclopentane is slightly puckered. The strain energy in cyclopentane is approximately 5.6 kcal/mol (23.4 kJ/mol).

Cyclohexane

Cyclohexane adopts a number of puckered conformations, the most stable of which is a **chair conformation**. In this conformation (Figure 3.7), all C—C—C bond angles are 109.5° (minimizing angle strain), and hydrogens on adjacent carbons are staggered with respect to one another (minimizing torsional strain). Thus, there is very little strain in a chair conformation of cyclohexane.

In a chair conformation, the C—H bonds are arranged in two different orientations. Six C—H bonds are called **equatorial bonds**, and the other six are called **axial bonds**. One way to visualize the difference between these two types of bonds is to imagine an axis through the center of the chair, perpendicular to the floor [Figure 3.8(a)]. Equatorial bonds are approximately perpendicular to our imaginary axis and alternate first slightly up and then slightly down as you move from one carbon of the ring to the next. Axial bonds are parallel to the imaginary axis. Three axial bonds point up; the other three point down. Notice that axial bonds alternate also, first up and then down as you move from one carbon of the ring to the next. Notice further that if the axial

Chair conformation The most stable puckered conformation of a cyclohexane ring; all bond angles are approximately 109.5°, and bonds to all adjacent carbons are staggered.

Equatorial bond A bond on a chair conformation of a cyclohexane ring that extends from the ring roughly perpendicular to the imaginary axis of the ring.

Axial bond A bond on a chair conformation of a cyclohexane ring that extends from the ring parallel to the imaginary axis of the ring.

Axis through the center of the ring

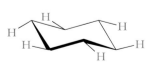

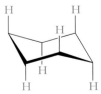

Figure 3.8
Chair conformation of cyclohexane, showing axial and equatorial C—H bonds.

(a) Ball-and-stick model showing all 12 hydrogens

(b) The six equatorial C—H bonds shown in red

(c) The six axial C—H bonds shown in blue

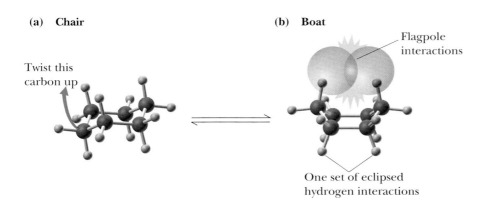

(a) Chair

Twist this carbon up

(b) Boat

Flagpole interactions

One set of eclipsed hydrogen interactions

Figure 3.9
Conversion of (a) a chair conformation to (b) a boat conformation. In the boat conformation, there is torsional strain due to the four sets of eclipsed hydrogen interactions and steric strain due to the one set of flagpole interactions. A chair conformation is more stable than a boat conformation.

bond on a carbon points upward, then the equatorial bond on that carbon points slightly downward. Conversely, if the axial bond on a particular carbon points downward, then the equatorial bond on that carbon points slightly upward.

There are many other nonplanar conformations of cyclohexane, one of which is the **boat conformation**. You can visualize the interconversion of a chair conformation to a boat conformation by twisting the ring as illustrated in Figure 3.9. A boat conformation is considerably less stable than a chair conformation. In a boat conformation, torsional strain is created by four sets of eclipsed hydrogen interactions, and steric strain is created by the one set of flagpole interactions. **Steric strain** (also called nonbonded interaction strain) results when nonbonded atoms separated by four or more bonds are forced abnormally close to each other—that is, when they are forced closer than their atomic (contact) radii allow. The difference in potential energy between chair and boat conformations is approximately 6.5 kcal/mol (27 kJ/mol), which means that, at room temperature, approximately 99.99% of all cyclohexane molecules are in the chair conformation.

For cyclohexane, the two equivalent chair conformations can interconvert by one chair twisting first into a boat and then into the other chair. When one chair is converted to the other, a change occurs in the relative orientations in space of the hydrogen atoms bonded to each carbon: All hydrogen atoms equatorial in one chair become axial in the other, and vice versa (Figure 3.10). The interconversion of one chair conformation of cyclohexane to the other occurs rapidly at room temperature.

Boat conformation A puckered conformation of a cyclohexane ring in which carbons 1 and 4 of the ring are bent toward each other.

Steric strain The strain that arises when atoms separated by four or more bonds are forced abnormally close to one another.

Hydrogen switches from equatorial to axial upon chair-to-chair interconversion, but remains pointing upward

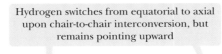

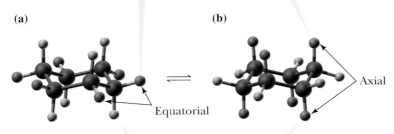

(a)

(b)

Axial

Equatorial

Hydrogen switches from axial to equatorial upon chair-to-chair interconversion, but remains pointing downward

Figure 3.10
Interconversion of chair cyclohexanes. All C—H bonds that are equatorial in one chair are axial in the alternative chair, and vice versa.

EXAMPLE 3.7

Following is a chair conformation of cyclohexane showing a methyl group and one hydrogen:

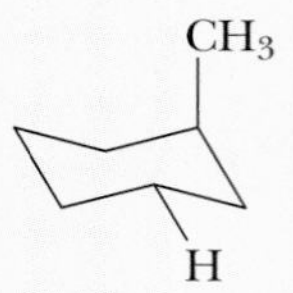

(a) Indicate by a label whether each group is equatorial or axial.
(b) Draw the other chair conformation, and again label each group as equatorial or axial.

SOLUTION

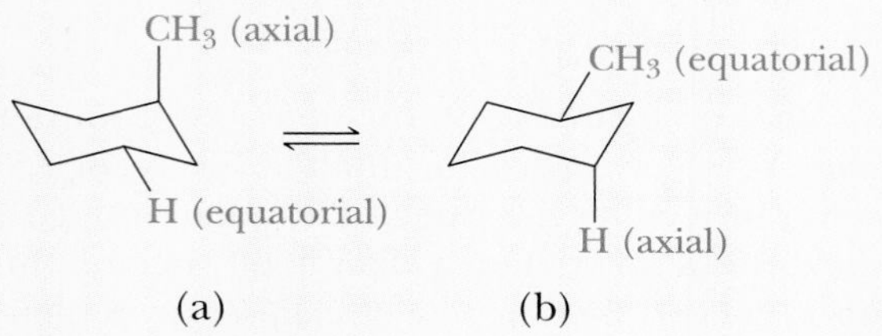

CH₃ (axial)   CH₃ (equatorial)

H (equatorial)   H (axial)

(a)   (b)

Practice Problem 3.7

Following is a chair conformation of cyclohexane with carbon atoms numbered 1 through 6:

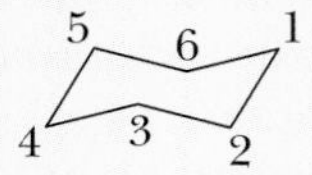

(a) Draw hydrogen atoms that are above the plane of the ring on carbons 1 and 2 and below the plane of the ring on carbon 4.
(b) Which of these hydrogens are equatorial? Which are axial?
(c) Draw the other chair conformation. Now which hydrogens are equatorial? Which are axial? Which are above the plane of the ring, and which are below it?

If we replace a hydrogen atom of cyclohexane by an alkyl group, the group occupies an equatorial position in one chair and an axial position in the other chair. This means that the two chairs are no longer equivalent and no longer of equal stability.

A convenient way to describe the relative stabilities of chair conformations with equatorial or axial substituents is in terms of a type of steric strain called **axial–axial (diaxial) interaction**. *Axial–axial interaction* refers to the steric strain existing between an axial substituent and an axial hydrogen (or other group) on the same side of the ring. Consider methylcyclohexane (Figure 3.11). When the —CH₃ is equatorial, it is staggered with respect to all other groups on its adjacent carbon atoms. When the —CH₃ is axial, it is parallel to the axial C—H bonds on carbons 3 and 5. Thus, for axial methylcyclohexane, there are two unfavorable methyl–hydrogen axial–axial interactions. For methylcyclohexane, the equatorial methyl conformation is favored over the axial methyl conformation by approximately 1.74 kcal/mol (7.28 kJ/mol). At equilibrium at room temperature, approximately 95% of all methylcyclohexane molecules have their methyl group equatorial, and less than 5% have their methyl group axial.

Diaxial interactions Interactions between groups in parallel axial positions on the same side of a chair conformation of a cyclohexane ring.

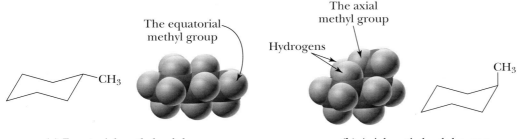

(a) Equatorial methylcyclohexane   (b) Axial methylcyclohexane

The equatorial methyl group

The axial methyl group

Hydrogens

CH₃

Figure 3.11
Two chair conformations of methylcyclohexane. The two axial–axial interactions (steric strain) make conformation (b) less stable than conformation (a) by approximately 1.74 kcal/mol (7.28 kJ/mol).

As the size of the substituent increases, the conformations with the group equatorial increases. When the group is as large as *tert*-butyl, the equatorial conformation is approximately 4,000 times more abundant at room temperature than the axial conformation, and, in effect, the ring is "locked" into a chair conformation with the *tert*-butyl group equatorial.

EXAMPLE 3.8

Label all axial–axial interactions in the following chair conformation:

H₃C, CH₃, CH₃, H, and H labels shown on chair conformation structure

SOLUTION

There are four axial–axial interactions: Each axial methyl group has two sets of axial–axial interactions with parallel hydrogen atoms on the same side of the ring. The equatorial methyl group has no axial–axial interactions.

Practice Problem 3.8

The conformational equilibria for methyl-, ethyl-, and isopropylcyclohexane are all about 95% in favor of the equatorial conformation, but the conformational equilibrium for *tert*-butylcyclohexane is almost completely on the equatorial side. Explain why the conformational equilibria for the first three compounds are comparable, but that for *tert*-butylcyclohexane lies considerably farther toward the equatorial conformation.

CHEMICAL CONNECTIONS   3A

The Poisonous Puffer Fish

Nature is by no means limited to carbon in six-membered rings. Tetrodotoxin, one of the most potent toxins known, is composed of a set of interconnected six-membered rings, each in a chair conformation. All but one of these rings have atoms other than carbon in them. Tetrodotoxin is produced in the liver and ovaries of many species of *Tetraodontidae*, especially the puffer fish, so called because it inflates itself to an almost spherical spiny ball when alarmed. The puffer is evidently a species that is highly preoccupied with defense, but the Japanese are not put off. They regard the puffer, called *fugu* in Japanese, as a delicacy. To serve it in a public restaurant, a chef must be registered as sufficiently skilled in removing the toxic organs so as to make the flesh safe to eat.

Symptoms of tetrodotoxin poisoning begin with attacks of severe weakness, progressing to complete paralysis and eventual death. Tetrodotoxin exerts its severe poisonous effect by blocking Na^+ ion channels in excitable membranes. The $=NH_2^+$ end of tetrodotoxin lodges in the mouth of a Na^+ ion channel, thus blocking further transport of Na^+ ions through the channel.

A puffer fish inflated. *(Tim Rock/Animals Animals)*

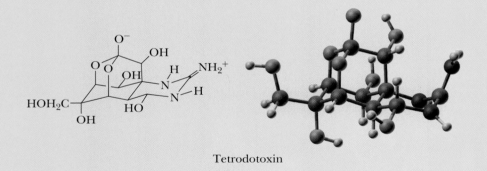

Tetrodotoxin

3.8   CIS–TRANS ISOMERISM IN CYCLOALKANES

Cis–trans isomers Isomers that have the same order of attachment of their atoms, but a different arrangement of their atoms in space, due to the presence of either a ring or a carbon–carbon double bond.

Cis A prefix meaning "on the same side."

Trans A prefix meaning "across from."

Cycloalkanes with substituents on two or more carbons of the ring show a type of isomerism called **cis–trans isomerism**. Cis-trans isomers have (1) the same molecular formula, (2) the same order of attachment of atoms, and (3) an arrangement of atoms that cannot be interchanged by rotation about sigma bonds under ordinary conditions. By way of comparison, the potential energy difference between conformations is so small that they can be interconverted easily at or near room temperature by rotation about single bonds.

We can illustrate cis–trans isomerism in cycloalkanes using 1,2-dimethylcyclo-pentane as an example. In the following structural formula, the cyclopentane ring is drawn as a planar pentagon viewed edge on (in determining the number of cis–trans isomers in a substituted cycloalkane, it is adequate to draw the cycloalkane ring as a planar polygon):

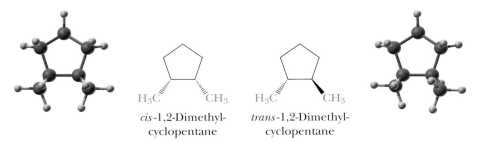

cis-1,2-Dimethyl-
cyclopentane

trans-1,2-Dimethyl-
cyclopentane

Carbon–carbon bonds of the ring that project forward are shown as heavy lines. When viewed from this perspective, substituents bonded to the cyclopentane ring project above and below the plane of the ring. In one isomer of 1,2-dimethylcyclopentane, the methyl groups are on the same side of the ring (either both above or both below the plane of the ring); in the other isomer, they are on opposite sides of the ring (one above and one below the plane of the ring).

Alternatively, the cyclopentane ring can be viewed from above, with the ring in the plane of the paper. Substituents on the ring then either project toward you (that is, they project up above the page) and are shown by solid wedges, or they project away from you (they project down below the page) and are shown by broken wedges. In the following structural formulas, only the two methyl groups are shown (hydrogen atoms of the ring are not shown):

cis-1,2-Dimethyl-
cyclopentane

trans-1,2-Dimethyl-
cyclopentane

EXAMPLE 3.9

Which cycloalkanes show cis–trans isomerism? For each that does, draw both isomers.

(a) Methylcyclopentane   (b) 1,1-Dimethylcyclobutane
(c) 1,3-Dimethylcyclobutane

SOLUTION

(a) Methylcyclopentane does not show cis–trans isomerism: It has only one sub-stituent on the ring.
(b) 1,1-Dimethylcyclobutane does not show cis–trans isomerism: Only one arrangement is possible for the two methyl groups on the ring, and they must be trans.

(c) 1,3-Dimethylcyclobutane shows cis–trans isomerism. Note that, in these structural formulas, we show only the hydrogens on carbons bearing the methyl groups.

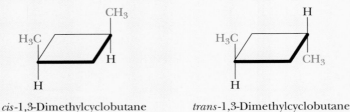

cis-1,3-Dimethylcyclobutane   trans-1,3-Dimethylcyclobutane

Practice Problem 3.9

Which cycloalkanes show cis–trans isomerism? For each that does, draw both isomers.

(a) 1,3-Dimethylcyclopentane   (b) Ethylcyclopentane
(c) 1-Ethyl-2-methylcyclobutane

Two cis–trans isomers exist for 1,4-dimethylcyclohexane. For the purposes of determining the number of cis–trans isomers in substituted cycloalkanes, it is adequate to draw the cycloalkane ring as a planar polygon, as is done in the following disubstituted cyclohexanes:

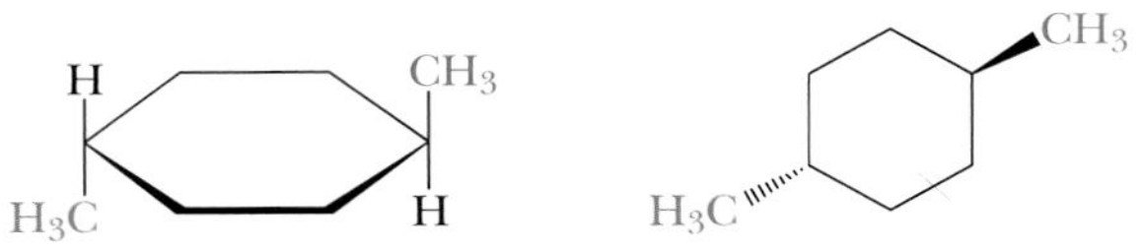

trans-1,4-Dimethylcyclohexane 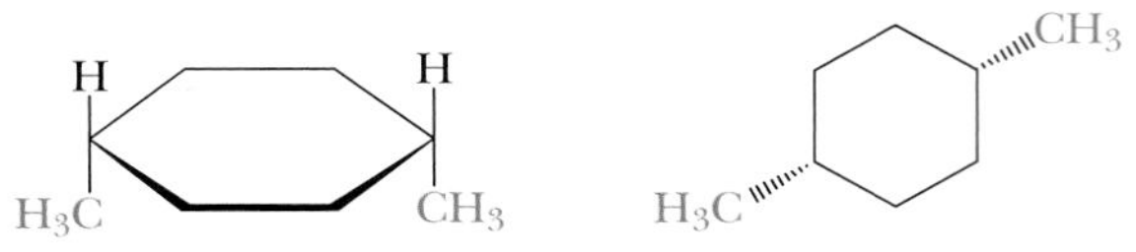   cis-1,4-Dimethylcyclohexane

We can also draw the cis and trans isomers of 1,4-dimethylcyclohexane as nonplanar chair conformations. In working with alternative chair conformations, it is helpful to remember that all groups axial on one chair are equatorial in the alternative chair, and vice versa. In one chair conformation of trans-1,4-dimethylcyclohexane, the two methyl groups are axial; in the alternative chair conformation, they are equatorial. Of these chair conformations, the one with both methyls equatorial is considerably more stable.

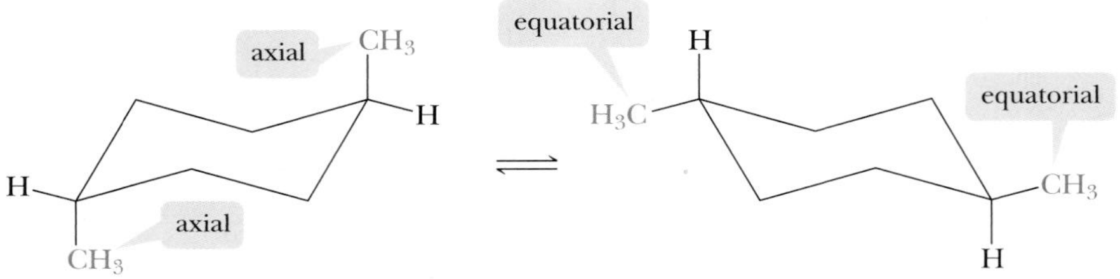

axial   CH3   equatorial   H   equatorial
H   axial   CH3   H3C   CH3
CH3   H
(less stable)   (more stable)

trans-1,4-Dimethylcyclohexane

The alternative chair conformations of cis-1,4-dimethylcyclohexane are of equal energy. In each chair conformation, one methyl group is equatorial and the other is axial.

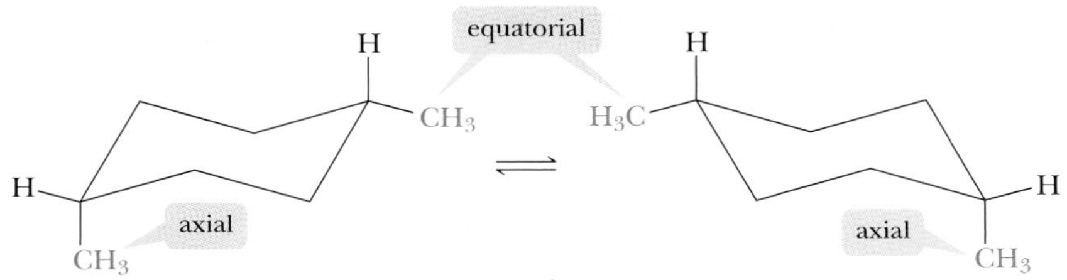

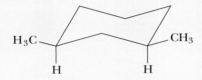

cis-1,4-Dimethylcyclohexane
(these conformations are of equal stability)

EXAMPLE 3.10

Following is a chair conformation of 1,3-dimethylcyclohexane:

(a) Is this a chair conformation of *cis*-1,3-dimethylcyclohexane or of *trans*-1,3-dimethylcyclohexane?
(b) Draw the alternative chair conformation. Of the two chair conformations, which is the more stable?
(c) Draw a planar hexagon representation of the isomer shown in this example.

SOLUTION

(a) The isomer shown is *cis*-1,3-dimethylcyclohexane; the two methyl groups are on the same side of the ring.

(b)

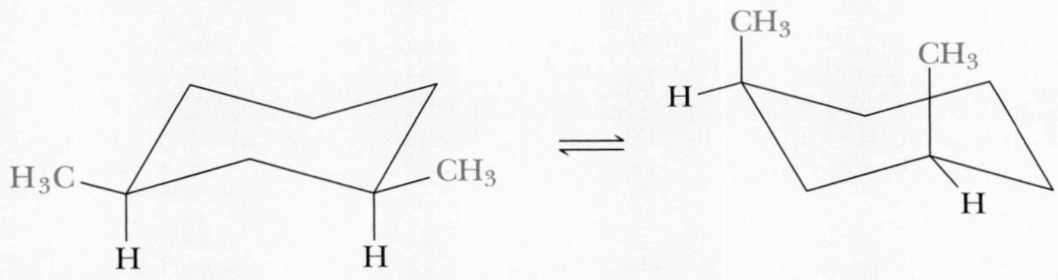

(more stable)   (less stable)

(c)   or

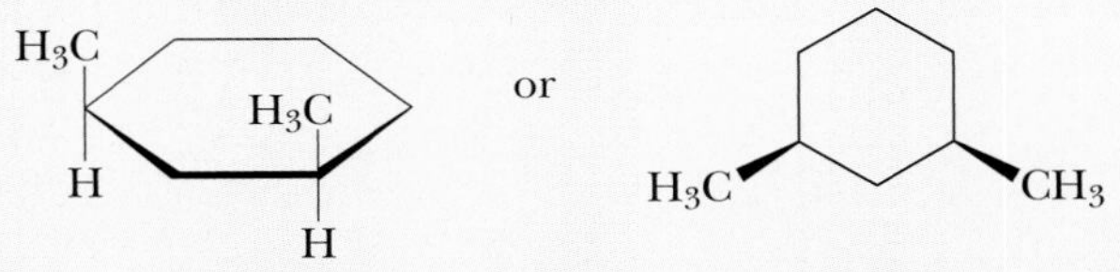

Practice Problem 3.10

Following is a planar hexagon representation of one isomer of 1,2,4-trimethylcyclohexane:

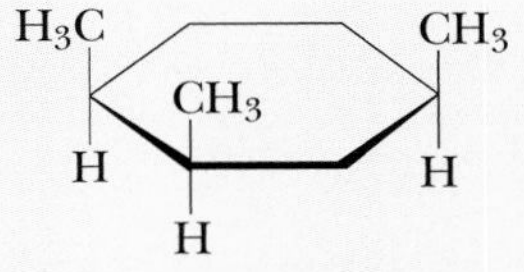

Draw the alternative chair conformations of this compound, and state which is the more stable.

3.9 PHYSICAL PROPERTIES OF ALKANES AND CYCLOALKANES

The most important property of alkanes and cycloalkanes is their almost complete lack of polarity. As we saw in Section 1.3C, the difference in electronegativity between carbon and hydrogen is $2.5 - 2.1 = 0.4$ on the Pauling scale, and given this small difference, we classify a C—H bond as nonpolar covalent. Therefore, alkanes are nonpolar compounds, and there is only weak interaction between their molecules.

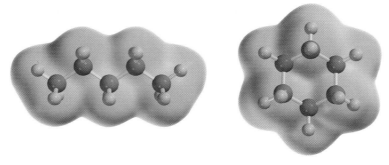

Pentane and cyclohexane. The electron density models show no evidence of any polarity in alkanes and cycloalkanes.

A. Boiling Points

The boiling points of alkanes are lower than those of almost any other type of compound of the same molecular weight. In general, both boiling and melting points of alkanes increase with increasing molecular weight (Table 3.4).

Alkanes containing 1 to 4 carbons are gases at room temperature, and those containing 5 to 17 carbons are colorless liquids. High-molecular-weight alkanes (those with 18 or more carbons) are white, waxy solids. Several plant waxes are high-molecular-weight alkanes. The wax found in apple skins, for example, is an unbranched alkane with molecular formula $C_{27}H_{56}$. Paraffin wax, a mixture of

TABLE 3.4   Physical Properties of Some Unbranched Alkanes

Name	Condensed Structrual Formula	Melting Point (°C)	Boiling Point (°C)	*Density of Liquid (g/mL at 0°C)
methane	CH_4	−182	−164	(a gas)
ethane	CH_3CH_3	−183	−88	(a gas)
propane	$CH_3CH_2CH_3$	−190	−42	(a gas)
butane	$CH_3(CH_2)_2CH_3$	−138	0	(a gas)
pentane	$CH_3(CH_2)_3CH_3$	−130	36	0.626
hexane	$CH_3(CH_2)_4CH_3$	−95	69	0.659
heptane	$CH_3(CH_2)_5CH_3$	−90	98	0.684
octane	$CH_3(CH_2)_6CH_3$	−57	126	0.703
nonane	$CH_3(CH_2)_7CH_3$	−51	151	0.718
decane	$CH_3(CH_2)_8CH_3$	−30	174	0.730

*For comparison, the density of H_2O is 1 g/mL at 4°C.

high-molecular-weight alkanes, is used for wax candles, in lubricants, and to seal home canned jams, jellies, and other preserves. Petrolatum, so named because it is derived from petroleum refining, is a liquid mixture of high-molecular-weight alkanes. Sold as mineral oil and Vaseline, petrolatum is used as an ointment base in pharmaceuticals and cosmetics and as a lubricant and rust preventative.

B. Dispersion Forces and Interactions between Alkane Molecules

Methane is a gas at room temperature and atmospheric pressure. It can be converted to a liquid if cooled to $-164°C$ and to a solid if further cooled to $-182°C$. The fact that methane (or any other compound, for that matter) can exist as a liquid or a solid depends on the existence of forces of attraction between particles of the pure compound. Although the forces of attraction between particles are all electrostatic in nature, they vary widely in their relative strengths. The strongest attractive forces are between ions—for example, between Na^+ and Cl^- in NaCl (188 kcal/mol, 787 kJ/mol). Hydrogen bonding is a weaker attractive force (2–10 kcal/mol, 8–42 kJ/mol). We will have more to say about hydrogen bonding in Chapter 8 when we discuss the physical properties of alcohols—compounds containing polar $O{-}H$ groups.

Dispersion forces (0.02–2 kcal/mol, 0.08–8 kJ/mol) are the weakest intermolecular attractive forces. It is the existence of dispersion forces that accounts for the fact that low-molecular-weight, nonpolar substances such as methane can be liquefied. When we convert methane from a liquid to a gas at $-164°C$, for example, the process of separating its molecules requires only enough energy to overcome the very weak dispersion forces.

To visualize the origin of these forces, it is necessary to think in terms of instantaneous distributions of electron density rather than average distributions. Over time, the distribution of electron density in a methane molecule is symmetrical [Figure 3.12(a)], and there is no separation of charge. However, at any instant, there is a nonzero probability that the electron density is polarized (shifted) more toward one part of a methane molecule than toward another. This temporary polarization creates temporary partial positive and partial negative charges, which in turn induce temporary partial positive and negative charges in adjacent methane molecules [Figure 3.12(b)]. **Dispersion forces** are weak electrostatic attractive forces that occur between temporary partial positive and partial negative charges in adjacent atoms or molecules.

Because interactions between alkane molecules consist only of these very weak dispersion forces, the boiling points of alkanes are lower than those of almost any other type of compound of the same molecular weight. As the number of atoms and the molecular weight of an alkane increase, the strength of the dispersion forces among alkane molecules increases, and consequently, boiling points increase.

Dispersion forces Very weak intermolecular forces of attraction resulting from the interaction of temporary induced dipoles.

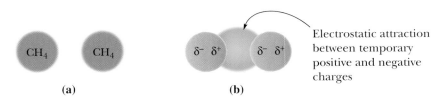

(a)   (b)

Electrostatic attraction between temporary positive and negative charges

Figure 3.12

Dispersion forces. (a) The average distribution of electron density in a methane molecule is symmetrical, and there is no polarity. (b) Temporary polarization of one molecule induces temporary polarization in an adjacent molecule. Electrostatic attractions between temporary partial positive and partial negative charges are called *dispersion forces*.

C. Melting Point and Density

The melting points of alkanes increase with increasing molecular weight. The increase, however, is not as regular as that observed for boiling points, because the ability of molecules to pack into ordered patterns of solids changes as the molecular size and shape change.

The average density of the alkanes listed in Table 3.4 is about 0.7 g/mL; that of higher-molecular-weight alkanes is about 0.8 g/mL. All liquid and solid alkanes are less dense than water (1.0 g/mL); therefore, they float on water.

D. Constitutional Isomers Have Different Physical Properties

Alkanes that are constitutional isomers are different compounds and have different physical properties. Table 3.5 lists the boiling points, melting points, and densities of the five constitutional isomers with molecular formula C_6H_{14}. The boiling point of each of its branched-chain isomers is lower than that of hexane itself, and the more branching there is, the lower is the boiling point. These differences in boiling points are related to molecular shape in the following way: The only forces of attraction between alkane molecules are dispersion forces. As branching increases, the shape of an alkane molecule becomes more compact, and its surface area decreases. As the surface area decreases, the strength of the dispersion forces decreases, and boiling points also decrease. Thus, for any group of alkane constitutional isomers, it is usually observed that the least-branched isomer has the highest boiling point and the most-branched isomer has the lowest boiling point. The trend in melting points is less obvious, but, as previously mentioned, it correlates with a molecule's ability to pack into ordered patterns of solids.

TABLE 3.5	Physical Properties of the Isomeric Alkanes with Molecular Formula C_6H_{14}		
Name	**Melting Point (°C)**	**Boiling Point (°C)**	**Density (g/mL)**
hexane	−95	69	0.659
3-methylpentane	−6	64	0.664
2-methylpentane	−23	62	0.653
2,3-dimethylbutane	−129	58	0.662
2,2-dimethylbutane	−100	50	0.649

more surface area, an increase in dispersion forces, and a higher boiling point

Hexane

smaller surface area, a decrease in dispersion forces, and a lower boiling point

2,2-Dimethylbutane

EXAMPLE 3.11

Arrange the alkanes in each set in order of increasing boiling point:

(a) Butane, decane, and hexane
(b) 2-Methylheptane, octane, and 2,2,4-trimethylpentane

SOLUTION

(a) All of the compounds are unbranched alkanes. As the number of carbon atoms in the chain increases, the dispersion forces among molecules increase, and the boiling points increase. Decane has the highest boiling point, butane the lowest:

| Butane | Hexane | Decane |
| (bp −0.5°C) | (bp 69°C) | (bp 174°C) |

(b) These three alkanes are constitutional isomers with molecular formula C_8H_{18}. Their relative boiling points depend on the degree of branching. 2,2,4-Trimethylpentane, the most highly branched isomer, has the smallest surface area and the lowest boiling point. Octane, the unbranched isomer, has the largest surface area and the highest boiling point.

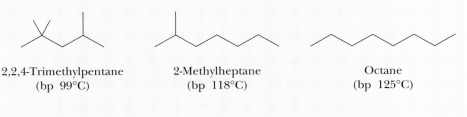

| 2,2,4-Trimethylpentane | 2-Methylheptane | Octane |
| (bp 99°C) | (bp 118°C) | (bp 125°C) |

Practice Problem 3.11

Arrange the alkanes in each set in order of increasing boiling point:

(a) 2-Methylbutane, 2,2-dimethylpropane, and pentane
(b) 3,3-Dimethylheptane, 2,2,4-trimethylhexane, and nonane

3.10 REACTIONS OF ALKANES

The most important chemical property of alkanes and cycloalkanes is their inertness. They are quite unreactive toward most reagents, a behavior consistent with the fact that they are nonpolar compounds containing only strong sigma bonds. Under certain conditions, however, alkanes and cycloalkanes do react, with oxygen, O_2. By far their most important reaction with oxygen is oxidation (combustion) to form carbon dioxide and water. The oxidation of saturated hydrocarbons is the basis for their use as energy sources for heat [natural gas, liquefied petroleum gas (LPG), and fuel oil] and power (gasoline, diesel fuel, and aviation fuel). Following are balanced equations for the complete combustion of methane, the major component of natural gas, and for propane, the major component of LPG:

$$CH_4 + 2O_2 \longrightarrow CO_2 + 2H_2O \qquad \Delta H° = -212 \text{ kcal/mol } (-886 \text{ kJ/mol})$$
Methane

$$CH_3CH_2CH_3 + 5O_2 \longrightarrow 3CO_2 + 4H_2O \qquad \Delta H° = -530 \text{ kcal/mol } (-2,220 \text{ kJ/mol})$$
Propane

3.11 SOURCES OF ALKANES

The three major sources of alkanes throughout the world are the fossil fuels: natural gas, petroleum, and coal. Fossil fuels account for approximately 90% of the total energy consumed in the United States. Nuclear electric power and hydroelectric power make up most of the remaining 10%. In addition, fossil fuels provide the bulk of the raw material for the organic chemicals consumed worldwide.

A. Natural Gas

Natural gas consists of approximately 90–95% methane, 5–10% ethane, and a mixture of other relatively low-boiling alkanes—chiefly propane, butane, and 2-methylpropane. The current widespread use of ethylene as the organic chemical industry's most important building block is largely the result of the ease with which ethane can be separated from natural gas and cracked into ethylene. Cracking is a process whereby a saturated hydrocarbon is converted into an unsaturated hydrocarbon plus H_2. Ethane is cracked by heating it in a furnace at 800 to 900°C for a fraction of a second. The production of ethylene in the United States in 1997 was 51.1 billion pounds, making it the number-one organic compound produced by the U.S. chemical industry, on a weight basis. The bulk of the ethylene produced is used to create organic polymers, as described in Chapter 17.

$$CH_3CH_3 \xrightarrow[\text{(thermal cracking)}]{800-900°C} CH_2{=}CH_2 + H_2$$
$$\text{Ethane} \qquad\qquad\qquad \text{Ethylene}$$

B. Petroleum

A petroleum refinery.
(K. Straiton/Photo Researchers, Inc.)

Petroleum is a thick, viscous liquid mixture of literally thousands of compounds, most of them hydrocarbons, formed from the decomposition of marine plants and animals. Petroleum and petroleum-derived products fuel automobiles, aircraft, and trains. They provide most of the greases and lubricants required for the machinery of our highly industrialized society. Furthermore, petroleum, along with natural gas, provides close to 90% of the organic raw materials used in the synthesis and manufacture of synthetic fibers, plastics, detergents, drugs, dyes, and a multitude of other products.

It is the task of a petroleum refinery to produce usable products, with a minimum of waste, from the thousands of different hydrocarbons in this liquid mixture. The various physical and chemical processes for this purpose fall into two broad categories: separation processes, which separate the complex mixture into various fractions, and re-forming processes, which alter the molecular structure of the hydrocarbon components themselves.

The fundamental separation process utilized in refining petroleum is fractional distillation (Figure 3.13). Practically all crude oil that enters a refinery goes to distillation units, where it is heated to temperatures as high as 370 to 425°C and separated into fractions. Each fraction contains a mixture of hydrocarbons that boils within a particular range:

1. Gases boiling below 20°C are taken off at the top of the distillation column. This fraction is a mixture of low-molecular-weight hydrocarbons, predominantly propane, butane, and 2-methylpropane, substances that can be liquefied under pressure at room temperature. The liquefied mixture, known as liquefied petroleum gas (LPG), can be stored and shipped in metal tanks and is a convenient source of gaseous fuel for home heating and cooking.

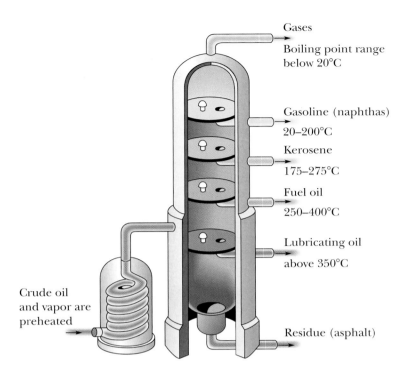

Figure 3.13
Fractional distillation of petroleum. The lighter, more volatile fractions are removed from higher up the column and the heavier, less volatile fractions from lower down.

2. Naphthas, bp 20 to 200°C, are a mixture of C_5 to C_{12} alkanes and cycloalkanes. Naphthas also contain small amounts of benzene, toluene, xylene, and other aromatic hydrocarbons (Chapter 9). The light naphtha fraction, bp 20 to 150°C, is the source of straight-run gasoline and averages approximately 25% of crude petroleum. In a sense, naphthas are the most valuable distillation fractions, because they are useful not only as fuel, but also as sources of raw materials for the organic chemical industry.

3. Kerosene, bp 175 to 275°C, is a mixture of C_9 to C_{15} hydrocarbons.

4. Fuel oil, bp 250 to 400°C, is a mixture of C_{15} to C_{18} hydrocarbons. Diesel fuel is obtained from this fraction.

5. Lubricating oil and heavy fuel oil distill from the column at temperatures above 350°C.

6. Asphalt is the black, tarry residue remaining after the removal of the other volatile fractions.

The two most common re-forming processes are cracking, illustrated by the thermal conversion of ethane to ethylene (Section 3.11A), and catalytic re-forming, illustrated by the conversion of hexane first to cyclohexane and then to benzene:

$$CH_3CH_2CH_2CH_2CH_2CH_3 \quad \xrightarrow[-H_2]{\text{catalyst}} \quad \bigcirc \quad \xrightarrow[-3H_2]{\text{catalyst}} \quad \bigcirc$$

Hexane   Cyclohexane   Benzene

C. Coal

To understand how coal can be used as a raw material for the production of organic compounds, it is necessary to discuss synthesis gas. Synthesis gas is a mixture of carbon monoxide and hydrogen in varying proportions, depending on the means

CHEMICAL CONNECTIONS 3B

Octane Rating: What Those Numbers at the Pump Mean

Gasoline is a complex mixture of C_6 to C_{12} hydrocarbons. The quality of gasoline as a fuel for internal combustion engines is expressed in terms of an *octane rating*. Engine knocking occurs when a portion of the air–fuel mixture explodes prematurely (usually as a result of heat developed during compression) and independently of ignition by the spark plug. Two compounds were selected as reference fuels. One of these, 2,2,4-trimethylpentane (isooctane), has very good antiknock properties (the fuel–air mixture burns smoothly in the combustion chamber) and was assigned an octane rating of 100. (The name *isooctane* is a trivial name; its only relation to the name 2,2,4-trimethylpentane is that both compounds show eight carbon atoms.) Heptane, the other reference compound, has poor antiknock properties and was assigned an octane rating of 0.

heptane that has antiknock properties equivalent to those of the gasoline. For example, the antiknock properties of 2-methylhexane are the same as those of a mixture of 42% isooctane and 58% heptane; therefore, the octane rating of 2-methylhexane is 42. Octane itself has an octane rating of -20, which means that it produces even more engine knocking than heptane. Ethanol, the additive to gasohol, has an octane rating of 105. Benzene and toluene have octane ratings of 106 and 120, respectively.

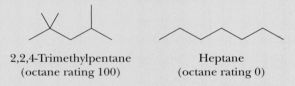

2,2,4-Trimethylpentane
(octane rating 100)

Heptane
(octane rating 0)

The octane rating of a particular gasoline is that percentage of isooctane in a mixture of isooctane and

Typical octane ratings of commonly available gasolines. *(Charles D. Winters)*

by which it is manufactured. Synthesis gas is prepared by passing steam over coal. It is also prepared by the partial oxidation of methane with oxygen.

$$C + H_2O \xrightarrow{heat} CO + H_2$$
Coal

$$CH_4 + \frac{1}{2}O_2 \xrightarrow{catalyst} CO + 2H_2$$
Methane

Two important organic compounds produced today almost exclusively from carbon monoxide and hydrogen are methanol and acetic acid. In the production of methanol, the ratio of carbon monoxide to hydrogen is adjusted to 1:2, and the mixture is passed over a catalyst at elevated temperature and pressure:

$$CO + 2H_2 \xrightarrow{catalyst} CH_3OH$$
methanol

The treatment of methanol, in turn, with carbon monoxide over a different catalyst gives acetic acid:

$$CH_3OH + CO \xrightarrow{\text{catalyst}} CH_3\overset{\displaystyle O}{\overset{\displaystyle \|}{C}}OH$$

Methanol   Acetic acid

Because the processes for making methanol and acetic acid directly from carbon monoxide are commercially proven, it is likely that the decades ahead will see the development of routes to other organic chemicals from coal via methanol.

SUMMARY

A **hydrocarbon** is a compound that contains only carbon and hydrogen. A **saturated hydrocarbon** contains only single bonds. Alkanes have the general formula C_nH_{2n+2}. **Constitutional isomers** (Section 3.3) have the same molecular formula, but a different connectivity (a different order of attachment of their atoms). Alkanes are named according to a set of rules developed by the **International Union of Pure and Applied Chemistry** (**IUPAC**; Section 3.4A). A carbon atom is classified as **primary (1°), secondary (2°), tertiary (3°),** or **quaternary (4°)**, depending on the number of carbon atoms bonded to it (Section 3.4C). A hydrogen atom is classified as primary (1°), secondary (2°), or tertiary (3°), depending on the type of carbon to which it is bonded.

An alkane that contains carbon atoms bonded to form a ring is called a **cycloalkane** (Section 3.5). To name a cycloalkane, prefix the name of the open-chain hydrocarbon with "*cyclo-*." Five-membered rings (cyclopentanes) and six-membered rings (cyclohexanes) are especially abundant in the biological world.

The IUPAC system is a general system of nomenclature (Section 3.6). The IUPAC name of a compound consists of three parts: (1) a **prefix** that indicates the number of carbon atoms in the parent chain, (2) an **infix** that indicates the nature of the carbon–carbon bonds in the parent chain, and (3) a **suffix** that indicates the class to which the compound belongs. Substituents derived from alkanes by the removal of a hydrogen atom are called **alkyl groups** and are given the symbol **R**. The name of an alkyl group is formed by dropping the suffix *-ane* from the name of the parent alkane and adding *-yl* in its place.

A **conformation** is any three-dimensional arrangement of the atoms of a molecule that results from rotation about a single bond (Section 3.7). One convention for showing conformations is the **Newman projection**. Staggered conformations are lower in energy (more stable) than eclipsed conformations.

Molecular strain (Section 3.7) is of three types:

1. **torsional strain** (also called eclipsed interaction strain) results when nonbonded atoms separated by three bonds are forced from a staggered conformation to an eclipsed conformation,

2. **angle strain** results when a bond angle in a molecule is either expanded or compressed compared with its optimal values, and

3. **steric strain** (also called **nonbonded interaction strain**) results when nonbonded atoms separated by four or more bonds are forced abnormally close to each other—that is, when they are forced closer than their atomic (contact) radii would otherwise allow.

Cyclopentanes, cyclohexanes, and all larger cycloalkanes exist in dynamic equilibrium between a set of puckered conformations. The lowest energy conformation of cyclopentane is an envelope conformation. The lowest energy conformations of cyclohexane are two interconvertible **chair conformations** (Section 3.7B). In a chair conformation, six bonds are axial and six are **equatorial**. Bonds axial in one chair are equatorial in the alternative chair, and vice versa. A **boat conformation** is higher in energy than chair conformations. The more stable conformation of a substituted cyclohexane is the one that minimizes **axial–axial interactions.**

Cis–trans isomers (Section 3.8) have the same molecular formula and the same order of attachment of atoms, but arrangements of atoms in space that cannot be interconverted by rotation about single bonds. **Cis** means that substituents are on the same side of the ring; **trans** means that they are on opposite sides of the ring. Most cycloalkanes with substituents on two or more carbons of the ring show cis–trans isomerism.

Alkanes are nonpolar compounds, and the only forces of attraction between their molecules are **dispersion forces** (Section 3.9), which are weak electrostatic interactions between temporary partial positive and negative charges of atoms or molecules. Low-molecular-weight alkanes, such as methane, ethane, and propane, are gases at room temperature and atmospheric pressure. Higher-molecular-weight alkanes, such as those in gasoline and kerosene, are liquids. Very high-molecular-weight alkanes, such as those in paraffin wax, are solids. Among a set of alkane constitutional isomers, the least branched isomer generally has the highest boiling point; the most branched isomer generally has the lowest boiling point.

Natural gas (Section 3.11A) consists of 90–95% methane, with lesser amounts of ethane and other lower-molecular-weight hydrocarbons. **Petroleum** (Section 3.11B) is a liquid mixture of literally thousands of different hydrocarbons. **Synthesis gas**, a mixture of carbon monoxide and hydrogen, can be derived from natural gas and coal (Section 3.11C).

KEY REACTIONS

1. Oxidation of Alkanes (Section 3.10)

The oxidation of alkanes to carbon dioxide and water is the basis for their use as energy sources of heat and power:

$$CH_3CH_2CH_3 + 5O_2 \longrightarrow 3CO_2 + 4H_2O + \text{energy}$$

PROBLEMS

A problem number set in red indicates an applied "real-world" problem.

Structure of Alkanes

3.12 For each condensed structural formula, write a line-angle formula:

$$\overset{\overset{\displaystyle CH_2CH_3}{|}}{}\overset{\overset{\displaystyle CH_3}{|}}{}$$

(a) 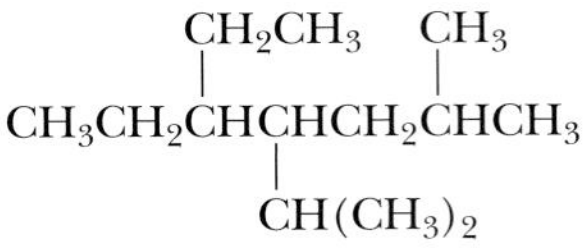 $CH_3CH_2\underset{\underset{\displaystyle CH(CH_3)_2}{|}}{C}HCHCH_2CHCH_3$

(b) 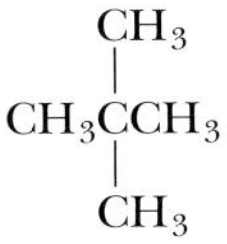 $CH_3\underset{\underset{\displaystyle CH_3}{|}}{\overset{\overset{\displaystyle CH_3}{|}}{C}}CH_3$

(c) $(CH_3)_2CHCH(CH_3)_2$

(d) 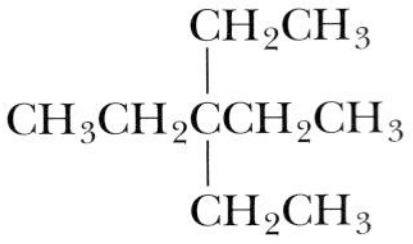 $CH_3CH_2\underset{\underset{\displaystyle CH_2CH_3}{|}}{\overset{\overset{\displaystyle CH_2CH_3}{|}}{C}}CH_2CH_3$

(e) $(CH_3)_3CH$

(f) $CH_3(CH_2)_3CH(CH_3)_2$

3.13 Write a condensed structural formula and the molecular formula of each alkane:

(a)   (b)   (c)

3.14 For each of the following condensed structural formulas, provide an even more abbreviated formula, using parentheses and subscripts:

(a) 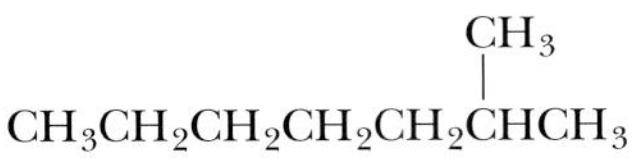 $CH_3CH_2CH_2CH_2CH_2\underset{\underset{\displaystyle }{}}{\overset{\overset{\displaystyle CH_3}{|}}{C}}HCH_3$

(b) $H\underset{\underset{\displaystyle CH_2CH_2CH_3}{|}}{\overset{\overset{\displaystyle CH_2CH_2CH_3}{|}}{C}}CH_2CH_2CH_3$

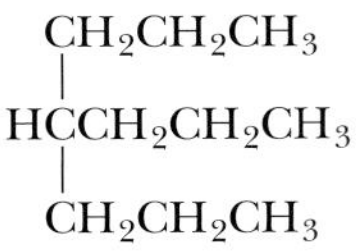

(c) $CH_3\underset{\underset{\displaystyle CH_2CH_2CH_3}{|}}{\overset{\overset{\displaystyle CH_2CH_2CH_3}{|}}{C}}CH_2CH_2CH_2CH_2CH_3$

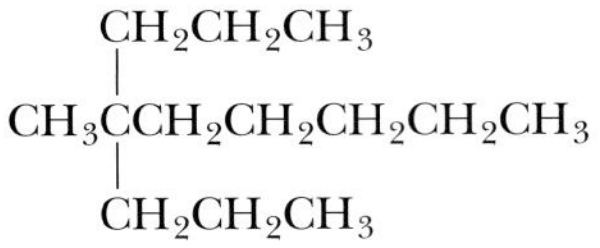

Constitutional Isomerism

3.15 Which statements are true about constitutional isomers?
  (a) They have the same molecular formula.
  (b) They have the same molecular weight.
  (c) They have the same order of attachment of atoms.
  (d) They have the same physical properties.

3.16 Each member of the following set of compounds is an alcohol; that is, each contains an —OH (hydroxyl group, Section 1.8A):

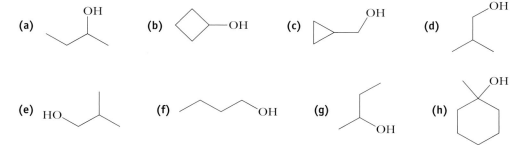

Which structural formulas represent (1) the same compound, (2) different compounds that are constitutional isomers, or (3) different compounds that are not constitutional isomers?

3.17 Each member of the following set of compounds is an amine; that is, each contains a nitrogen bonded to one, two, or three carbon groups (Section 1.8B):

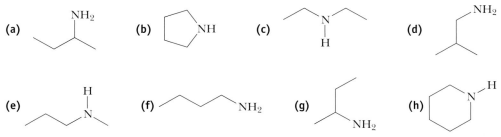

Which structural formulas represent (1) the same compound, (2) different compounds that are constitutional isomers, or (3) different compounds that are not constitutional isomers?

3.18 Each member of the following set of compounds is either an aldehyde or a ketone (Section 1.8C):

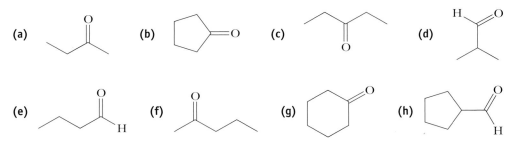

Which structural formulas represent (1) the same compound, (2) different compounds that are constitutional isomers, or (3) different compounds that are not constitutional isomers?

3.19 For each pair of compounds, tell whether the structural formulas shown represent
(**1**) the same compound,
(**2**) different compounds that are constitutional isomers, or
(**3**) different compounds that are not constitutional isomers:

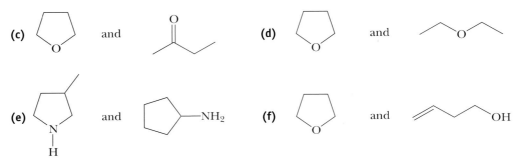

(c) [structure] and [structure]   **(d)** [structure] and [structure]

(e) [structure] and [structure]—NH₂   **(f)** [structure] and [structure]OH

3.20 Name and draw line-angle formulas for the nine constitutional isomers with molecular formula C_7H_{16}.

3.21 Tell whether the compounds in each set are constitutional isomers:

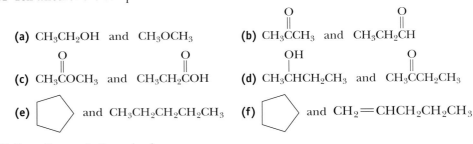

(a) CH_3CH_2OH and CH_3OCH_3

(b) $CH_3\overset{O}{\overset{\|}{C}}CH_3$ and $CH_3CH_2\overset{O}{\overset{\|}{C}}H$

(c) $CH_3\overset{O}{\overset{\|}{C}}OCH_3$ and $CH_3CH_2\overset{O}{\overset{\|}{C}}OH$

(d) $CH_3\overset{OH}{\overset{|}{C}H}CH_2CH_3$ and $CH_3\overset{O}{\overset{\|}{C}}CH_2CH_3$

(e) [pentagon] and $CH_3CH_2CH_2CH_2CH_3$

(f) [pentagon] and $CH_2{=}CHCH_2CH_2CH_3$

3.22 Draw line-angle formulas for
  (a) The four alcohols with molecular formula $C_4H_{10}O$.
  (b) The two aldehydes with molecular formula C_4H_8O.
  (c) The one ketone with molecular formula C_4H_8O.
  (d) The three ketones with molecular formula $C_5H_{10}O$.
  (e) The four carboxylic acids with molecular formula $C_5H_{10}O_2$.

Nomenclature of Alkanes and Cycloalkanes

3.23 Write IUPAC names for these alkanes and cycloalkanes:

(a) $CH_3\overset{\underset{|}{CH_3}}{C}HCH_2CH_2CH_3$ **(b)** $CH_3\overset{\underset{|}{CH_3}}{C}HCH_2CH_2\overset{\underset{|}{CH_3}}{C}HCH_3$ **(c)** $CH_3(CH_2)_4\overset{\underset{|}{CH_2CH_3}}{C}HCH_2CH_3$

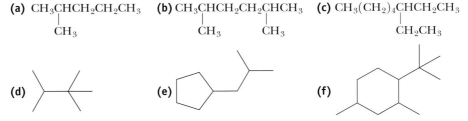

(d) [structure]   **(e)** [structure]   **(f)** [structure]

3.24 Write line-angle formulas for these alkanes:
  (a) 2,2,4-Trimethylhexane   **(b)** 2,2-Dimethylpropane
  (c) 3-Ethyl-2,4,5-trimethyloctane   **(d)** 5-Butyl-2,2-dimethylnonane
  (e) 4-Isopropyloctane   **(f)** 3,3-Dimethylpentane
  (g) *trans*-1,3-Dimethylcyclopentane **(h)** *cis*-1,2-Diethylcyclobutane

3.25 Explain why each of the following names is an incorrect IUPAC name and write the correct IUPAC name for the intended compound:
  (a) 1,3-Dimethylbutane   **(b)** 4-Methylpentane
  (c) 2,2-Diethylbutane   **(d)** 2-Ethyl-3-methylpentane
  (e) 2-Propylpentane   **(f)** 2,2-Diethylheptane
  (g) 2,2-Dimethylcyclopropane   **(h)** 1-Ethyl-5-methylcyclohexane

3.26 Draw a structural formula for each compound:
  (a) Ethanol   **(b)** Ethanal   **(c)** Ethanoic acid
  (d) Butanone   **(e)** Butanal   **(f)** Butanoic acid

(g) Propanal   (h) Cyclopropanol   (i) Cyclopentanol

(j) Cyclopentene   (k) Cyclopentanone

3.27 Write the IUPAC name for each compound:

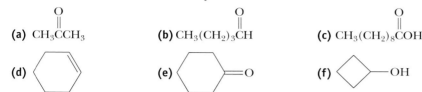

$$\text{(a) } CH_3\overset{\overset{\displaystyle O}{\|}}{C}CH_3 \qquad \text{(b) } CH_3(CH_2)_3\overset{\overset{\displaystyle O}{\|}}{C}H \qquad \text{(c) } CH_3(CH_2)_8\overset{\overset{\displaystyle O}{\|}}{C}OH$$

Conformations of Alkanes and Cycloalkanes

3.28 How many different staggered conformations are there for 2-methylpropane? How many different eclipsed conformations are there?

3.29 Looking along the bond between carbons 2 and 3 of butane, there are two different staggered conformations and two different eclipsed conformations. Draw Newman projections of each, and arrange them in order from the most stable conformation to the least stable conformation.

3.30 Explain why each of the following Newman projections might not represent the most stable conformation of that molecule:

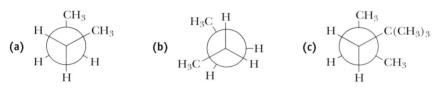

3.31 Explain why the following are not different conformations of 3-hexene:

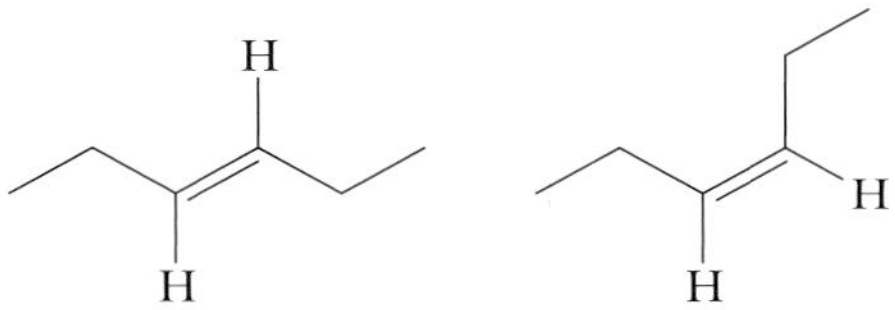

3.32 Which of the following two conformations is the more stable? (*Hint:* Use molecular models to compare structures):

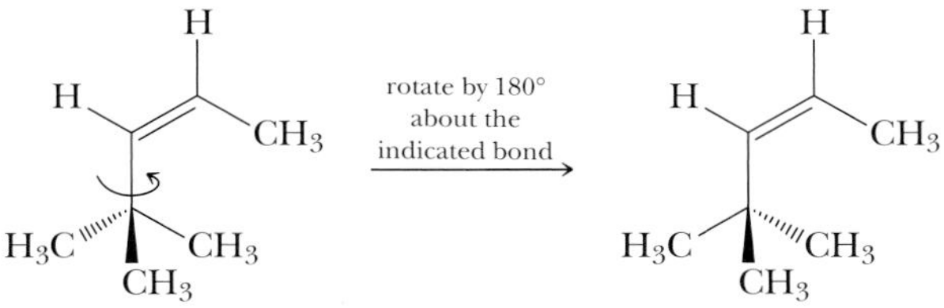

3.33 Determine whether the following pairs of structures in each set represent the same molecule or constitutional isomers, and if they are the same molecule, determine whether they are in the same or different conformations:

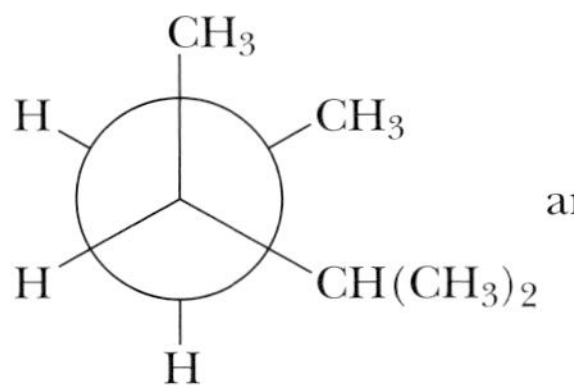

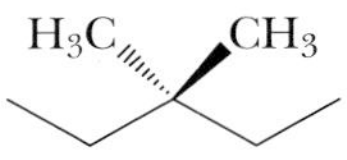

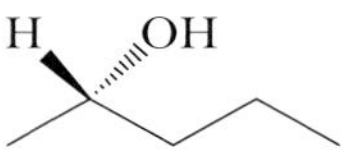

(c) [structure: H₃C and CH₃ substituted cyclohexane] and [structure with H₃C wedge and CH₃ hash cyclohexane]

(d) [Newman projection with H₃C, C(CH₃)₃] and [branched structure]

Cis–trans Isomerism in Cycloalkanes

3.34 What structural feature of cycloalkanes makes cis–trans isomerism in them possible?

3.35 Is cis–trans isomerism possible in alkanes?

3.36 Name and draw structural formulas for the cis and trans isomers of 1,2-dimethylcyclopropane.

3.37 Name and draw structural formulas for all cycloalkanes with molecular formula C_5H_{10}. Be certain to include cis–trans isomers, as well as constitutional isomers.

3.38 Using a planar pentagon representation for the cyclopentane ring, draw structural formulas for the cis and trans isomers of
(a) 1,2-Dimethylcyclopentane   **(b)** 1,3-Dimethylcyclopentane

3.39 Draw the alternative chair conformations for the cis and trans isomers of 1,2-dimethylcyclohexane, 1,3-dimethylcyclohexane, and 1,4-dimethylcyclohexane.
(a) Indicate by a label whether each methyl group is axial or equatorial.
(b) For which isomer(s) are the alternative chair conformations of equal stability?
(c) For which isomer(s) is one chair conformation more stable than the other?

3.40 Use your answers from Problem 3.39 to complete the following table, showing correlations between cis, trans isomers and axial, equatorial positions for disubstituted derivatives of cyclohexane:

Position of Substitution	cis		trans	
1,4-	a,e or	e,a	e,e or	a,a
1,3-	___ or	___	___ or	___
1,2-	___ or	___	___ or	___

3.41 There are four cis–trans isomers of 2-isopropyl-5-methylcyclohexanol:

[structure of 2-isopropyl-5-methylcyclohexanol with positions 1, 2, 5 labeled, OH and isopropyl groups]

2-Isopropyl-5-methylcyclohexanol

(a) Using a planar hexagon representation for the cyclohexane ring, draw structural formulas for these four isomers.
(b) Draw the more stable chair conformation for each of your answers in Part (a).
(c) Of the four cis–trans isomers, which is the most stable? If you answered this part correctly, you picked the isomer found in nature and given the name menthol.

3.42 Draw alternative chair conformations for each substituted cyclohexane, and state which chair is the more stable:

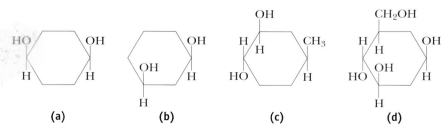

(a)   **(b)**   **(c)**   **(d)**

Peppermint plant *(Mentha piperita)*, a source of menthol, is a perennial herb with aromatic qualities used in candies, gums, hot and cold beverages, and garnish for punch and fruit. *(John Kaprielian/ Photo Researchers, Inc.)*

3.43 What kinds of conformations do the six-membered rings exhibit in adamantane?

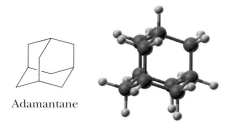

Adamantane

Physical Properties of Alkanes and Cycloalkanes

3.44 In Problem 3.20, you drew structural formulas for all constitutional isomers with molecular formula C_7H_{16}. Predict which isomer has the lowest boiling point and which has the highest.

3.45 What generalizations can you make about the densities of alkanes relative to that of water?

3.46 What unbranched alkane has about the same boiling point as water? (See Table 3.4.) Calculate the molecular weight of this alkane, and compare it with that of water.

3.47 As you can see from Table 3.4, each CH_2 group added to the carbon chain of an alkane increases the boiling point of the alkane. The increase is greater going from CH_4 to C_2H_6 and from C_2H_6 to C_3H_8 than it is from C_8H_{18} to C_9H_{20} or from C_9H_{20} to $C_{10}H_{22}$. What do you think is the reason for this trend?

3.48 Dodecane, $C_{12}H_{26}$, is an unbranched alkane. Predict the following:
  (a) Will it dissolve in water?
  (b) Will it dissolve in hexane?
  (c) Will it burn when ignited?
  (d) Is it a liquid, solid, or gas at room temperature and atmospheric pressure?
  (e) Is it more or less dense than water?

3.49 As stated in Section 3.9A, the wax found in apple skins is an unbranched alkane with molecular formula $C_{27}H_{56}$. Explain how the presence of this alkane prevents the loss of moisture from within an apple.

Reactions of Alkanes

3.50 Write balanced equations for the combustion of each hydrocarbon. Assume that each is converted completely to carbon dioxide and water.
  (a) Hexane   **(b)** Cyclohexane   **(c)** 2-Methylpentane

3.51 Following are heats of combustion of methane and propane:

Hydrocarbon	Component of	$\Delta H°$ [kcal/mol (kJ/mol)]
CH_4	natural gas	−212 (−886)
$CH_3CH_2CH_3$	LPG	−530 (−2220)

On a gram-for-gram basis, which of these hydrocarbons is the better source of heat energy?

3.52 When ethanol is added to gasoline to produce gasohol, the ethanol promotes more complete combustion of the gasoline and is an octane booster (Section 3.11B). Compare the heats of combustion of 2,2,4-trimethylpentane (1304 kcal/mol) and ethanol (327 kcal/mol). Which has the higher heat of combustion in kcal/mol? in kcal/g?

Looking Ahead

3.53 Explain why 1,2-dimethylcyclohexane can exist as cis–trans isomers, while 1,2-dimethylcyclododecane cannot.

3.54 On the left is a representation of the glucose molecule (we discuss the structure and chemistry of glucose in Chapter 18):

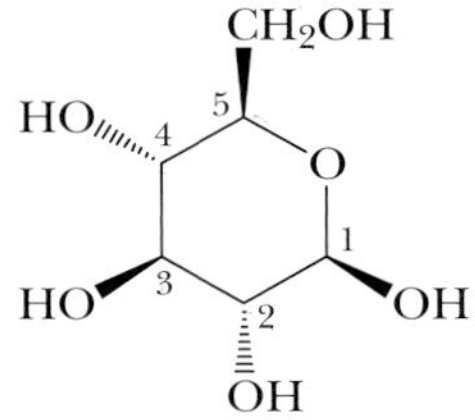

Glucose 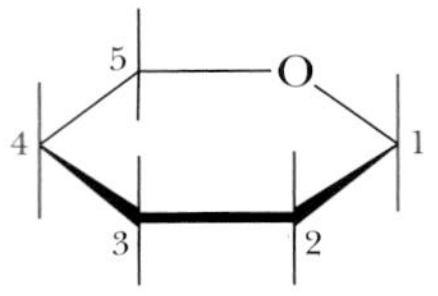   **(a)** 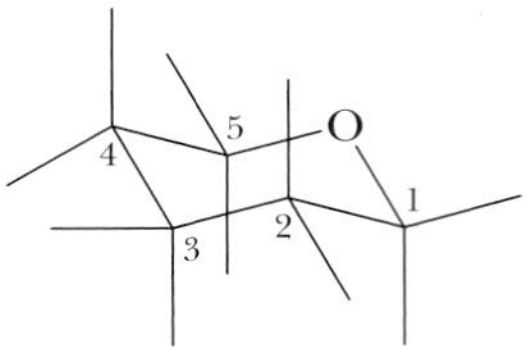   **(b)**

(a) Convert this representation to a planar hexagon representation.

(b) Convert this representation to a chair conformation. Which substituent groups in the chair conformation are equatorial? Which are axial?

3.55 Following is the structural formula of cholic acid (Section 21.5A), a component of human bile whose function is to aid in the absorption and digestion of dietary fats:

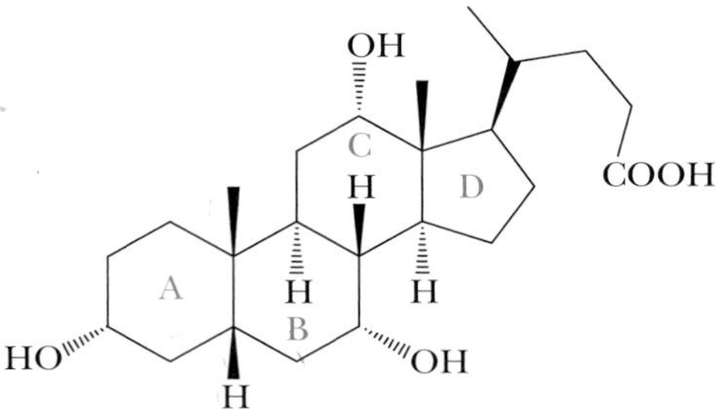

Cholic acid

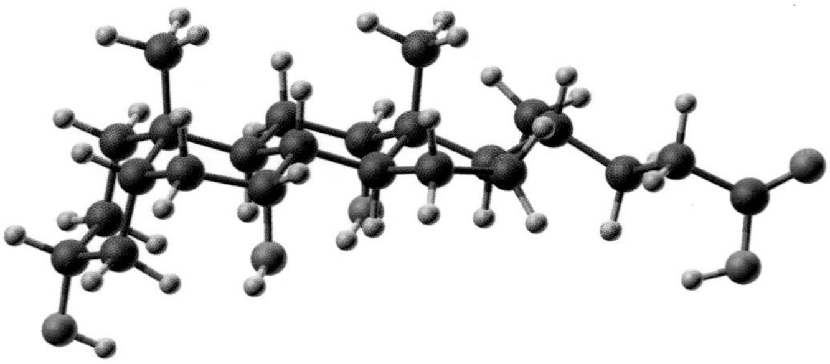

(a) What is the conformation of ring A? of ring B? of ring C? of ring D?

(b) There are hydroxyl groups on rings A, B, and C. Tell whether each is axial or equatorial.

(c) Is the methyl group at the junction of rings A and B axial or equatorial to ring A? Is it axial or equatorial to ring B?

(d) Is the methyl group at the junction of rings C and D axial or equatorial to ring C?

3.56 Following is the structural formula and ball-and-stick model of cholestanol:

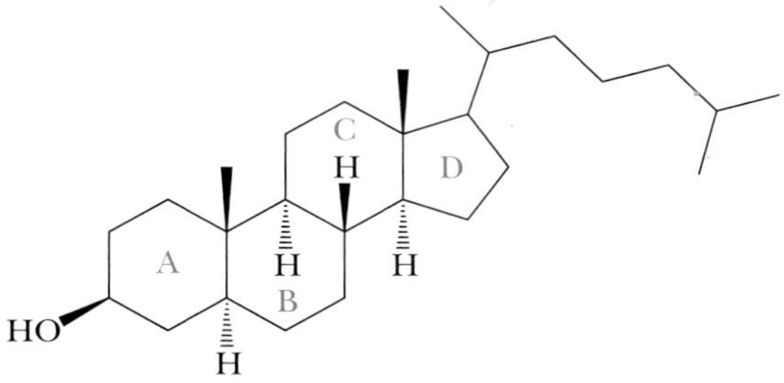

Cholestanol

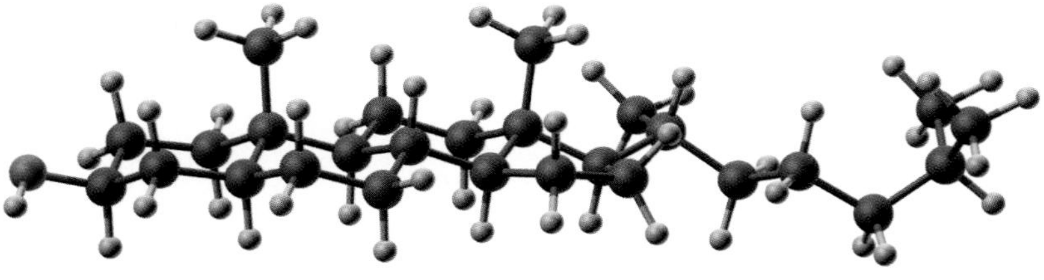

The only difference between this compound and cholesterol (Section 21.5A) is that cholesterol has a carbon–carbon double bond in ring B.

(a) Describe the conformation of rings A, B, C, and D in cholestanol.

(b) Is the hydroxyl group on ring A axial or equatorial?

(c) Consider the methyl group at the junction of rings A and B. Is it axial or equatorial to ring A? Is it axial or equatorial to ring B?

(d) Is the methyl group at the junction of rings C and D axial or equatorial to ring C?

3.57 As we have seen in Section 3.5, the IUPAC system divides the name of a compound into a prefix (showing the number of carbon atoms), an infix (showing the presence of carbon–carbon single, double, or triple bonds), and a suffix (showing the presence of an alcohol, amine, aldehyde, ketone, or carboxylic acid). Assume for the purposes of this problem that, to be alcohol (-ol) or amine (-amine), the hydroxyl or amino group must be bonded to a tetrahedral (sp^3 hybridized) carbon atom.

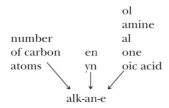

Given this information, write the structural formula of a compound with an unbranched chain of four carbon atoms that is an:

(a) alkane	**(b)** alkene	**(c)** alkyne
(d) alkanol	**(e)** alkenol	**(f)** alkynol
(g) alkanamine	**(h)** alkenamine	**(i)** alkynamine
(j) alkanal	**(k)** alkenal	**(l)** alkynal
(m) alkanone	**(n)** alkenone	**(o)** alkynone
(p) alkanoic acid	**(q)** alkenoic acid	**(r)** alkynoic acid

(Note: There is only one structural formula possible for some parts of this problem. For other parts, two or more structural formulas are possible. Where two are more are possible, we will deal with how the IUPAC system distinguishes among them when we come to the chapters on those particular functional groups.)

4 Alkenes and Alkynes

Carotene and carotene-like molecules are partnered with chlorophyll in nature to assist in the harvest of sunlight. In autumn, green chlorophyll molecules are destroyed and the yellows and reds of carotene and related molecules become visible. The red color of tomatoes comes from lycopene, a molecule closely related to carotene. See Problems 4.33 and 4.34. Inset: A model of β-carotene. *(Charles D. Winters)*

4.1 INTRODUCTION

In this chapter, we begin our study of unsaturated hydrocarbons. A hydrocarbon is unsaturated when it has fewer hydrogens bonded to carbon than an alkane has. There are three classes of unsaturated hydrocarbons: alkenes, alkynes, and arenes. **Alkenes** contain one or more carbon–carbon double bonds, and **alkynes** contain one or more carbon–carbon triple bonds. Ethene (ethylene) is the simplest alkene, and ethyne (acetylene) is the simplest alkyne:

Alkene An unsaturated hydrocarbon that contains a carbon–carbon double bond.

Alkyne An unsaturated hydrocarbon that contains a carbon–carbon triple bond.

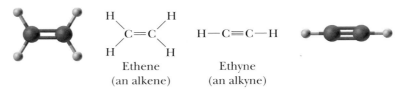

Ethene
(an alkene)

Ethyne
(an alkyne)

CHEMICAL CONNECTIONS 4A

Ethylene, a Plant Growth Regulator

As we have noted, ethylene occurs only in trace amounts in nature. Still, scientists have discovered that this small molecule is a natural ripening agent for fruits. Thanks to this knowledge, fruit growers can pick fruit while it is green and less susceptible to bruising. Then, when they are ready to pack the fruit for shipment, the growers can treat it with ethylene gas to induce ripening. Alternatively, the fruit can be treated with ethephon (Ethrel), which slowly releases ethylene and initiates ripening.

$$\text{Ethephon} \qquad \underset{\underset{\displaystyle \text{OH}}{|}}{\overset{\overset{\displaystyle O}{\|}}{Cl-CH_2-CH_2-P}}-OH$$

The next time you see ripe bananas in the market, you might wonder when they were picked and whether their ripening was artificially induced.

Arenes are the third class of unsaturated hydrocarbons. The simplest arene is benzene:

Arene A compound containing one or more benzene rings.

Benzene
(an arene)

The chemistry of benzene and its derivatives is quite different from that of alkenes and alkynes. Even though we do not study the chemistry of arenes until Chapter 9, we will show structural formulas of compounds containing benzene rings in earlier chapters. What you need to remember at this point is that a benzene ring is not chemically reactive under any of the conditions we describe in Chapters 4–8.

Compounds containing carbon–carbon double bonds are especially widespread in nature. Furthermore, several low-molecular-weight alkenes, including ethylene and propene, have enormous commercial importance in our modern, industrialized society. The organic chemical industry produces more pounds of ethylene worldwide than any other chemical. Annual production in the United States alone exceeds 55 billion pounds.

What is unusual about ethylene is that it occurs only in trace amounts in nature. The enormous amounts of it required to meet the needs of the chemical industry are derived the world over by thermal cracking of hydrocarbons. In the United States and other areas of the world with vast reserves of natural gas, the major process for the production of ethylene is thermal cracking of the small quantities of ethane extracted from natural gas (Section 3.11A):

$$\underset{\text{Ethane}}{CH_3CH_3} \xrightarrow[\text{(thermal cracking)}]{800-900°C} \underset{\text{Ethylene}}{CH_2{=}CH_2} + H_2$$

Europe, Japan, and other areas of the world with limited supplies of natural gas depend almost entirely on thermal cracking of petroleum for their ethylene.

The crucial point to recognize is that ethylene and all of the commercial and industrial products made from it are derived from either natural gas or petroleum—both nonrenewable natural resources!

4.2   STRUCTURE

A. Shapes of Alkenes

Using the valence-shell electron-pair repulsion model (Section 1.4), we predict a value of 120° for the bond angles about each carbon in a double bond. The observed H—C—C bond angle in ethylene is 121.7°, a value close to that predicted by this model. In other alkenes, deviations from the predicted angle of 120° may be somewhat larger as a result of strain between groups bonded to one or both carbons of the double bond. The C—C—C bond angle in propene, for example, is 124.7°:

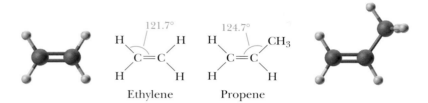

Ethylene   Propene

B. Orbital Overlap Model of a Carbon–Carbon Double Bond

In Section 1.7D, we described the formation of a carbon–carbon double bond in terms of the overlap of atomic orbitals. A carbon–carbon double bond consists of one sigma bond and one pi bond. Each carbon of the double bond uses its three sp^2 hybrid orbitals to form sigma bonds to three atoms. The unhybridized $2p$ atomic orbitals, which lie perpendicular to the plane created by the axes of the three sp^2 hybrid orbitals, combine to form the pi bond of the carbon–carbon double bond.

It takes approximately 63 kcal/mol (264 kJ/mol) to break the pi bond in ethylene; that is, to rotate one carbon by 90° with respect to the other so that no overlap occurs between $2p$ orbitals on adjacent carbons (Figure 4.1). This energy is considerably greater than the thermal energy available at room temperature, and, as a consequence, rotation about a carbon–carbon double bond is severely restricted. You might compare rotation about a carbon–carbon double bond, such as the bond in

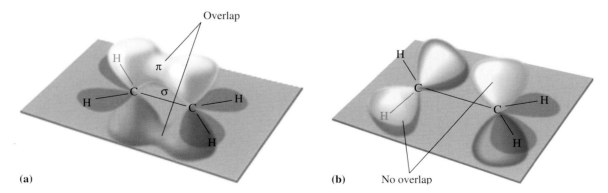

(a)   **(b)**   No overlap

Figure 4.1
Restricted rotation about the carbon–carbon double bond in ethylene. (a) Orbital overlap model showing the pi bond. (b) The pi bond is broken by rotating the plane of one H—C—H group by 90° with respect to the plane of the other H—C—H group.

ethylene, with that about a carbon–carbon single bond, such as the bond in ethane (Section 3.7A), as follows: Whereas rotation about the carbon–carbon single bond in ethane is relatively free (the energy barrier is approximately 3 kcal/mol), rotation about the carbon–carbon double bond in ethylene is restricted (the energy barrier is approximately 63 kcal/mol).

C. Cis–Trans Isomerism in Alkenes

Because of restricted rotation about a carbon–carbon double bond, an alkene in which each carbon of the double bond has two different groups bonded to it shows cis–trans isomerism. Consider, for example, 2-butene: In *cis*-2-butene, the two methyl groups are on the same side of the double bond; in *trans*-2-butene, the two methyl groups are on opposite sides of the double bond:

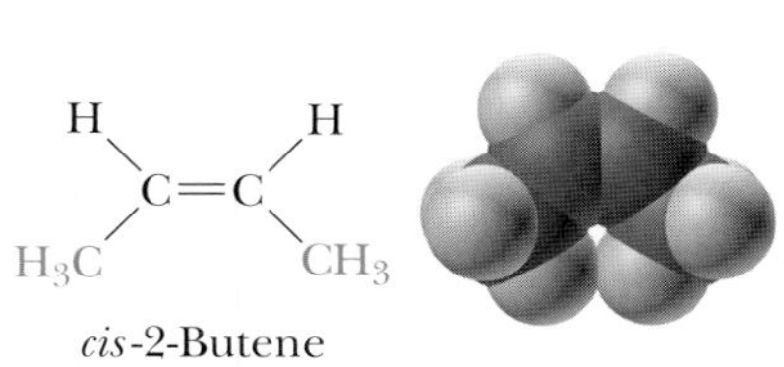

Cis–trans isomerism Isomers that have the same order of attachment of their atoms, but a different arrangement of their atoms in space due to the presence of either a ring (Chapter 3) or a carbon–carbon double bond (Chapter 4).

cis-2-Butene
mp −139°C, bp 4°C

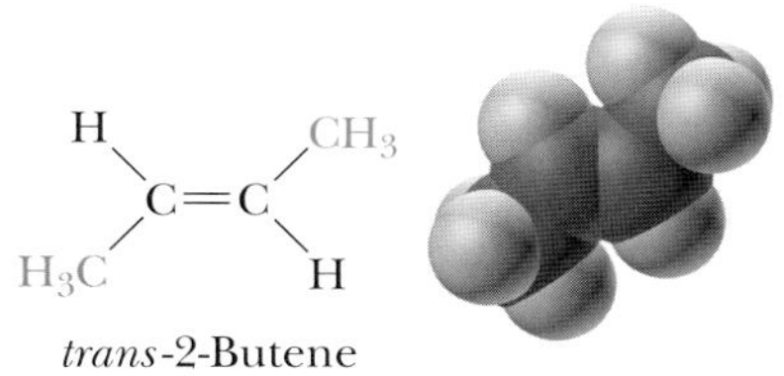

trans-2-Butene
mp −106°C, bp 1°C

These two compounds cannot be converted into one another at room temperature because of the restricted rotation about the double bond; they are different compounds, with different physical and chemical properties.

   Cis alkenes are less stable than their trans isomers because of nonbonded interaction strain between alkyl substituents on the same side of the double bond in the cis isomer, as can be seen in space-filling models of the cis and trans isomers of 2-butene. This is the same type of steric strain that results in the preference for equatorial methylcyclohexane over axial methylcyclohexane (Section 3.7B).

D. Shapes of Alkynes

The functional group of an alkyne is a **carbon–carbon triple bond**. The simplest alkyne is ethyne, C_2H_2. Ethyne is a linear molecule; all of its bond angles are 180° (Figure 1.12).

   According to the valence bond model (Section 1.7E), a triple bond is described in terms of the overlap of *sp* hybrid orbitals of adjacent carbons to form a sigma bond, the overlap of parallel $2p_y$ orbitals to form one pi bond, and the overlap of parallel $2p_z$ orbitals to form the second pi bond. In ethyne, each carbon forms a bond to a hydrogen by the overlap of an *sp* hybrid orbital of carbon with a 1*s* atomic orbital of hydrogen.

The combustion of acetylene yields energy that produces the very hot temperatures of an oxyacetylene torch.
(Charles D. Winters)

4.3 NOMENCLATURE

Alkenes are named using the IUPAC system, but, as we shall see, some are still referred to by their common names.

A. IUPAC Names

We form IUPAC names of alkenes by changing the **-an-** infix of the parent alkane to **-en-** (Section 3.6). Hence, $CH_2{=}CH_2$ is named ethene, and $CH_3CH{=}CH_2$ is named propene. In higher alkenes, where isomers exist that differ in the location of the double bond, we use a numbering system. We number the longest carbon chain that contains the double bond in the direction that gives the carbon atoms of the double bond the lower set of numbers. We then use the number of the first carbon of the double bond to show its location. We name branched or substituted alkenes in a manner similar to the way we name alkanes. We number the carbon atoms, locate the double bond, locate and name substituent groups, and name the main chain.

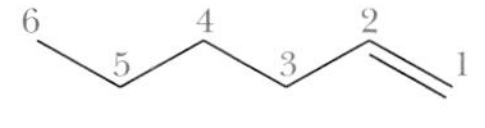

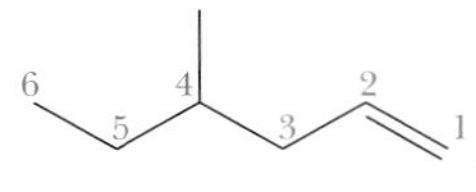

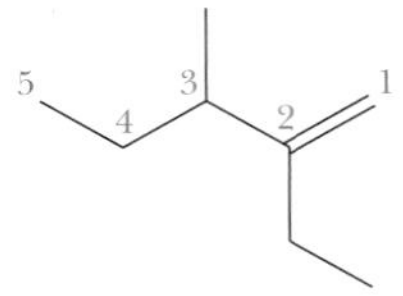

$$CH_3CH_2CH_2CH_2CH{=}CH_2$$
1-Hexene

$$CH_3CH_2CHCH_2CH{=}CH_2$$
   |
  CH_3
4-Methyl-1-hexene

$$CH_3CH_2CHC{=}CH_2$$
   |
  CH_2CH_3
2-Ethyl-3-methyl-1-pentene

Note that there is a six-carbon chain in 2-ethyl-3-methyl-1-pentene. However, because the longest chain that contains the double bond has only five carbons, the parent hydrocarbon is pentane, and we name the molecule as a disubstituted 1-pentene.

We form IUPAC names of alkynes by changing the **-an-** infix of the parent alkane to **-yn-** (Section 3.6). Thus, HC≡CH is named ethyne, and $CH_3C{≡}CH$ is named propyne. The IUPAC system retains the name *acetylene*; therefore, there are two acceptable names for HC≡CH: *ethyne* and *acetylene*. Of these two names, *acetylene* is used much more frequently. For larger molecules, we number the longest carbon chain that contains the triple bond from the end that gives the triply bonded carbons the lower set of numbers. We indicate the location of the triple bond by the number of the first carbon of the triple bond.

$$CH_3CHC{≡}CH$$
   |
  CH_3
3-Methyl-1-butyne

$$CH_3CH_2C{≡}CCH_2CCH_3$$
              |
             CH_3 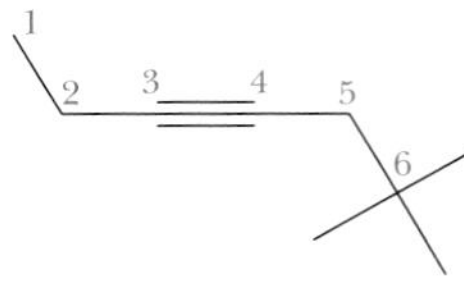
6,6-Dimethyl-3-heptyne

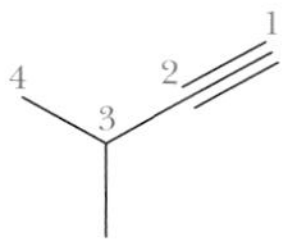

EXAMPLE 4.1

Write the IUPAC name of each unsaturated hydrocarbon:

(a) $CH_2{=}CH(CH_2)_5CH_3$

(b)
$$\begin{array}{c} H_3C \\ \\ H_3C \end{array}C{=}C\begin{array}{c} CH_3 \\ \\ H \end{array}$$

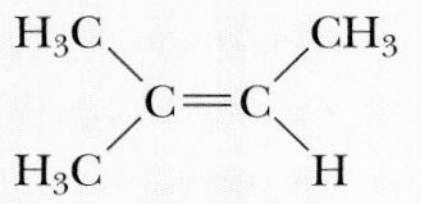

(c) $CH_3(CH_2)_2C{≡}CCH_3$

SOLUTION

(a) 1-Octene   (b) 2-Methyl-2-butene   (c) 2-Hexyne

Practice Problem 4.1 ────────────────────────────

Write the IUPAC name of each unsaturated hydrocarbon:

(a) 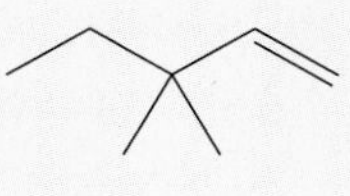   (b) 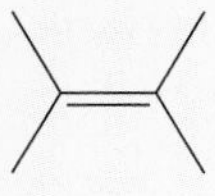   (c)

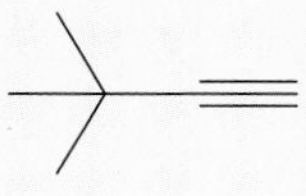

B. Common Names

Despite the precision and universal acceptance of IUPAC nomenclature, some alkenes—particularly those of low molecular weight—are known almost exclusively by their common names, as illustrated by the common names of these alkenes:

			CH_3
	$CH_2{=}CH_2$	$CH_3CH{=}CH_2$	$CH_3C{=}CH_2$
IUPAC name:	Ethene	Propene	2-Methylpropene
Common name:	Ethylene	Propylene	Isobutylene

C. Systems for Designating Configuration in Alkenes

The Cis–Trans System

The most common method for specifying the configuration of a disubstituted alkene uses the prefixes *cis* and *trans*. In this system, the orientation of the atoms of the parent chain determines whether the alkene is cis or trans. Following is a structural formula for the cis isomer of 4-methyl-2-pentene:

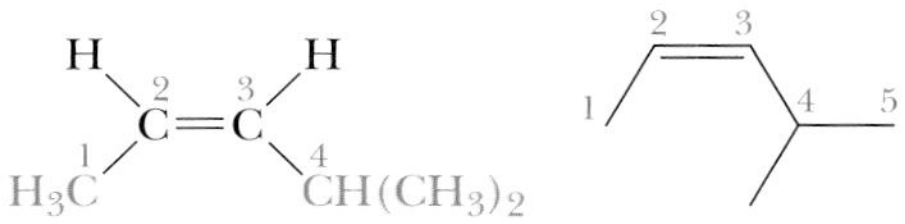

cis-4-Methyl-2-pentene

In this example, carbon atoms of the main chain (carbons 1 and 4) are on the same side of the double bond; therefore, the configuration of this alkene is cis.

EXAMPLE 4.2

Name each alkene, and, using the cis–trans system, show the configuration about each double bond:

(a) 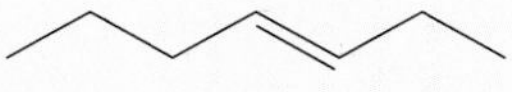 [structure]

(b) 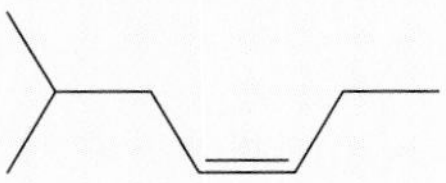 [structure]

SOLUTION

(a) The chain contains seven carbon atoms and is numbered from the end that gives the lower number to the first carbon of the double bond. The carbon atoms of the parent chain are on opposite sides of the double bond. The compound's name is *trans*-3-heptene.

(b) The longest chain contains seven carbon atoms and is numbered from the right, so that the first carbon of the double bond is carbon 3 of the chain. The carbon atoms of the parent chain are on the same side of the double bond. The compound's name is *cis*-6-methyl-3-heptene.

Practice Problem 4.2

Name each alkene, and, using the cis–trans system, specify its configuration:

(a)

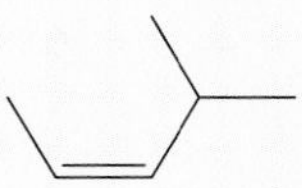

(b)

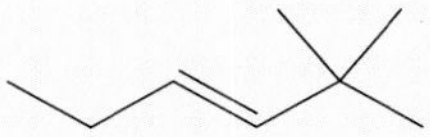

The **E,Z system** must be used for tri- and tetrasubstituted alkenes. This system uses
a set of rules to assign priorities to the substituents on each carbon of a double
bond. If the groups of higher priority are on the same side of the double bond, the
configuration of the alkene is **Z** (German: *zusammen*, together). If the groups of
higher priority are on opposite sides of the double bond, the configuration is **E**
(German: *entgegen*, opposite).

The E,Z System

E,Z system A system used to specify
the configuration of groups about a
carbon–carbon double bond.

Z From the German *zusammen,*
together; specifies that groups of
higher priority on the carbons of a
double bond are on the same side.

E From the German *entgegen,*
opposite; specifies that groups of
higher priority on the carbons of a
double bond are on opposite sides.

higher   higher   higher   lower

C=C   C=C

lower   lower   lower   higher

Z (zusammen)   **E** (entgegen)

The first step in assigning an E or a Z configuration to a double bond is to label the
two groups bonded to each carbon in order of priority.

Priority Rules

1. Priority is based on atomic number: The higher the atomic number, the higher
   is the priority. Following are several substituents arranged in order of increasing priority (the atomic number of the atom determining priority is shown in
   parentheses):

   (1)   (6)   (7)   (8)   (16)   (17)   (35)   (53)

   $-H, -CH_3, -NH_2, -OH, -SH, -Cl, -Br, -I$

   → Increasing priority

2. If priority cannot be assigned on the basis of the atoms that are bonded directly
   to the double bond, look at the next set of atoms, and continue until a priority
   can be assigned. Priority is assigned at the first point of difference. Following is
   a series of groups, arranged in order of increasing priority (again, numbers in
   parentheses give the atomic number of the atom on which the assignment of priority is based):

   (1)   (6)   (7)   (8)   (17)

   $-CH_2-H$ $-CH_2-CH_3$ $-CH_2-NH_2$ $-CH_2-OH$ $-CH_2-Cl$

   → Increasing priority

3. In order to compare carbons that are not sp^3 hybridized, the carbons must be
   manipulated in a way that allows us to maximize the number of groups bonded
   to them. Thus, we treat atoms participating in a double or triple bond as if they
   are bonded to an equivalent number of similar atoms by single bonds; that is,
   atoms of a double bond are replicated. Accordingly,

$-CH=CH_2$ is treated as $-\overset{\text{C}}{\underset{}{C}}H-\overset{\text{C}}{\underset{}{C}}H_2$ and $-\overset{\text{O}}{\underset{}{C}}H$ is treated as $-\overset{\text{O}-C}{\underset{H}{C}}-O$

EXAMPLE 4.3

Assign priorities to the groups in each set:

(a) 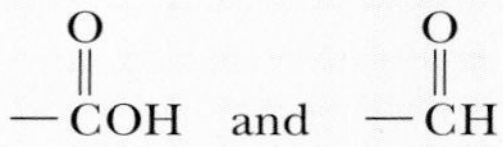 $-\overset{\overset{\text{O}}{\|}}{\text{C}}\text{OH}$ and $-\overset{\overset{\text{O}}{\|}}{\text{C}}\text{H}$

(b) 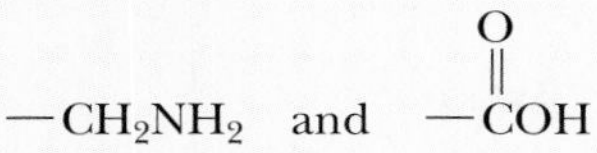 $-\text{CH}_2\text{NH}_2$ and $-\overset{\overset{\text{O}}{\|}}{\text{C}}\text{OH}$

SOLUTION

(a) The first point of difference is the O of the —OH in the carboxyl group, compared with the —H in the aldehyde group. The carboxyl group is higher in priority:

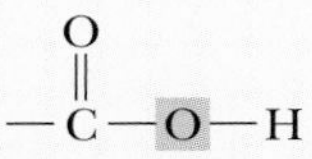

$-\overset{\overset{\text{O}}{\|}}{\text{C}}-\text{O}-\text{H}$ 　　　　 $-\overset{\overset{\text{O}}{\|}}{\text{C}}-\text{H}$

Carboxyl group 　　　　 Aldehyde group
(higher priority) 　　　　 (lower priority)

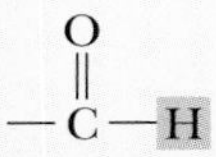

(b) Oxygen has a higher priority (higher atomic number) than nitrogen. Therefore, the carboxyl group has a higher priority than the primary amino group:

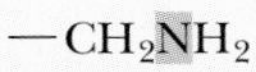

$-\text{CH}_2\text{NH}_2$ 　　　　 $-\overset{\overset{\text{O}}{\|}}{\text{C}}\text{OH}$
lower 　　　　 higher
priority 　　　　 priority

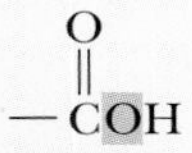

EXAMPLE 4.4

Name each alkene and specify its configuration by the E,Z system:

(a) 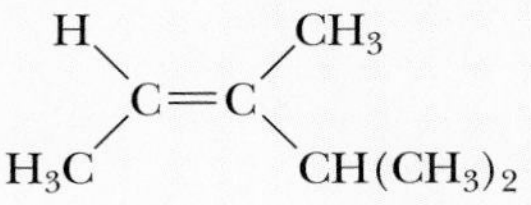 $\overset{\text{H}}{\underset{\text{H}_3\text{C}}{}}\text{C}=\text{C}\overset{\text{CH}_3}{\underset{\text{CH(CH}_3)_2}{}}$

(b) 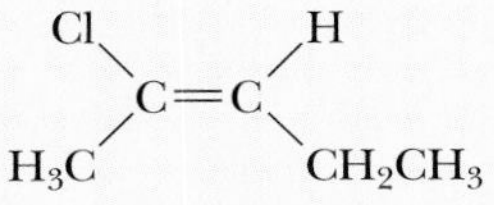 $\overset{\text{Cl}}{\underset{\text{H}_3\text{C}}{}}\text{C}=\text{C}\overset{\text{H}}{\underset{\text{CH}_2\text{CH}_3}{}}$

SOLUTION

(a) The group of higher priority on carbon 2 is methyl; that of higher priority on carbon 3 is isopropyl. Because the groups of higher priority are on the same side of the carbon–carbon double bond, the alkene has the Z configuration. Its name is (Z)-3,4-dimethyl-2-pentene.

(b) Groups of higher priority on carbons 2 and 3 are —Cl and —CH$_2$CH$_3$. Because these groups are on opposite sides of the double bond, the configuration of this alkene is E, and its name is (E)-2-chloro-2-pentene.

Practice Problem 4.3

Name each alkene and specify its configuration by the E,Z system:

(a) 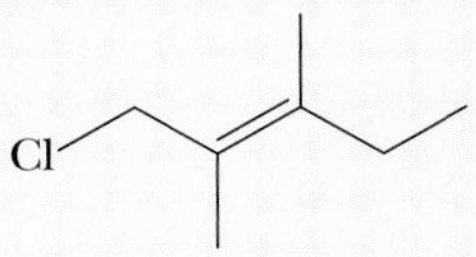 　　　　 (b) 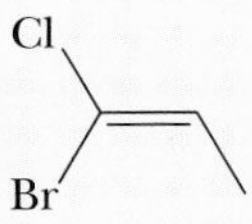 　　　　 (c)

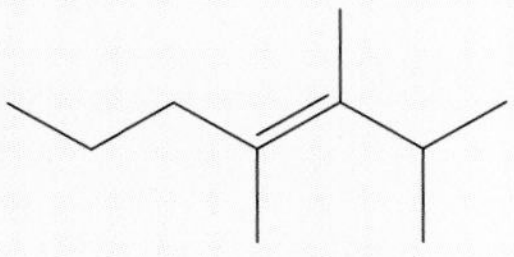

D. Cycloalkenes

In naming **cycloalkenes**, we number the carbon atoms of the ring double bond 1 and 2 in the direction that gives the substituent encountered first the smaller number. We name and locate substituents and list them in alphabetical order, as in the following compounds:

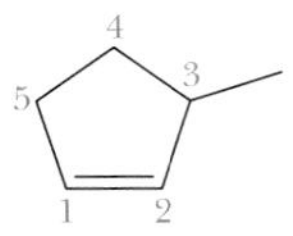

3-Methylcyclopentene
(not 5-methylcyclopentene)

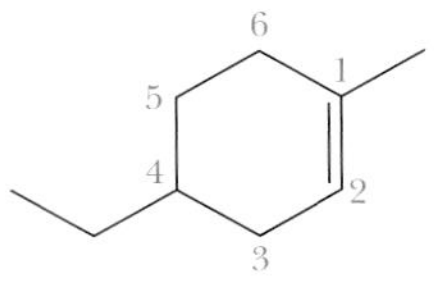

4-Ethyl-1-methylcyclohexene
(not 5-ethyl-2-methylcyclohexene)

EXAMPLE 4.5

Write the IUPAC name for each cycloalkene:

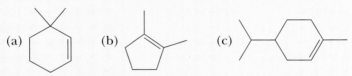

(a)   (b)   (c)

SOLUTION

(a) 3,3-Dimethylcyclohexene
(b) 1,2-Dimethylcyclopentene
(c) 4-Isopropyl-1-methylcyclohexene

Practice Problem 4.4

Write the IUPAC name for each cycloalkene:

(a) 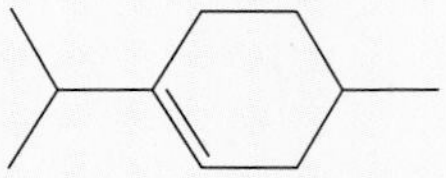   (b) 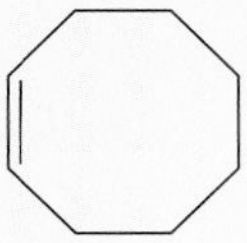   (c)

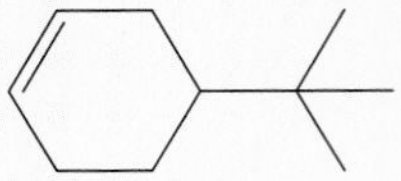

E. Cis–Trans Isomerism in Cycloalkenes

Following are structural formulas for four cycloalkenes:

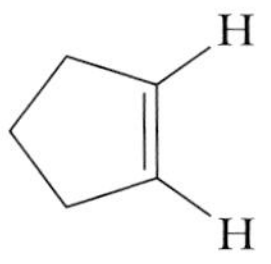

Cyclopentene 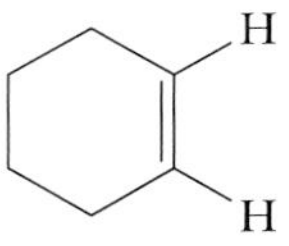   Cyclohexene 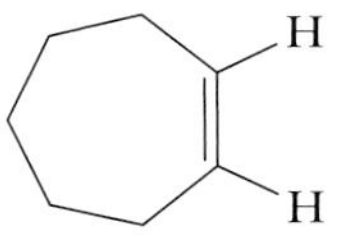   Cycloheptene 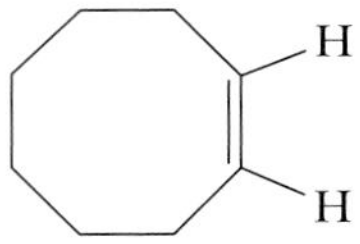   Cyclooctene

In these representations, the configuration about each double bond is cis. Because of angle strain, it is not possible to have a trans configuration in cycloalkenes of seven or fewer carbons. To date, *trans*-cyclooctene is the smallest *trans*-cycloalkene that has been prepared in pure form and is stable at room temperature. Yet, even in this *trans*-

cycloalkene, there is considerable intramolecular strain. *cis*-Cyclooctene is more stable than its trans isomer by 9.1 kcal/mol (38 kJ/mol).

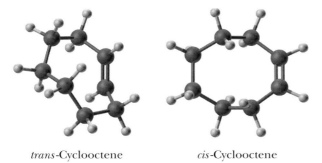

trans-Cyclooctene   *cis*-Cyclooctene

F. Dienes, Trienes, and Polyenes

We name alkenes that contain more than one double bond as alkadienes, alkatrienes, and so on. We refer to those that contain several double bonds more generally as polyenes (Greek: *poly*, many). Following are three examples of dienes:

$$CH_2{=}CHCH_2CH{=}CH_2$$

1,4-Pentadiene

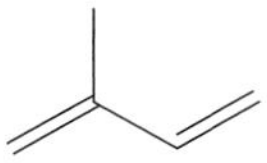

$$\underset{\text{CH}_3}{\overset{\displaystyle|}{CH_2{=}CCH{=}CH_2}}$$

2-Methyl-1,3-butadiene
(Isoprene)

1,3-Cyclopentadiene

G. Cis–Trans Isomerism in Dienes, Trienes, and Polyenes

Thus far, we have considered cis–trans isomerism in alkenes containing only one carbon–carbon double bond. For an alkene with one carbon–carbon double bond that can show cis–trans isomerism, two cis–trans isomers are possible. For an alkene with **n** carbon–carbon double bonds, each of which can show cis–trans isomerism, 2^n cis–trans isomers are possible.

EXAMPLE 4.6

How many cis–trans isomers are possible for 2,4-heptadiene?

SOLUTION

This molecule has two carbon–carbon double bonds, each of which exhibits cis–trans isomerism. As the following table shows, $2^2 = 4$ cis–trans isomers are possible (to the right of the table are line angle-formulas for two of these isomers):

Double bond	
$C_2{-}C_3$	$C_4{-}C_5$
trans	trans
trans	cis
cis	trans
cis	cis

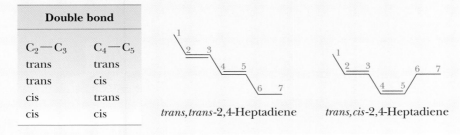

trans,trans-2,4-Heptadiene   *trans,cis*-2,4-Heptadiene

Practice Problem 4.5

Draw structural formulas for the other two cis–trans isomers of 2,4-heptadiene.

EXAMPLE 4.7

Draw all possible cis–trans isomers for the following unsaturated alcohol:

$$CH_3C{=}CHCH_2CH_2C{=}CHCH_2OH$$

with CH_3 groups above the two central carbons.

SOLUTION

Cis–trans isomerism is possible only about the double bond between carbons 2 and 3 of the chain. It is not possible for the other double bond, because carbon 7 has two identical groups on it. Thus, $2^1 = 2$ cis–trans isomers are possible. The trans isomer of this alcohol, named geraniol, is a major component of the oils of rose, citronella, and lemongrass.

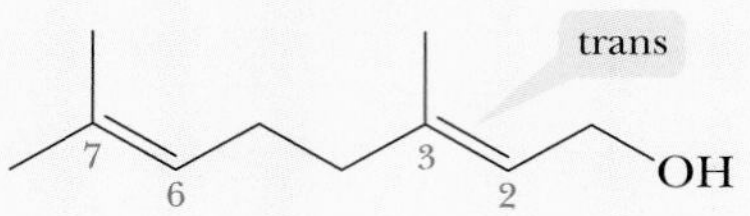

The trans isomer 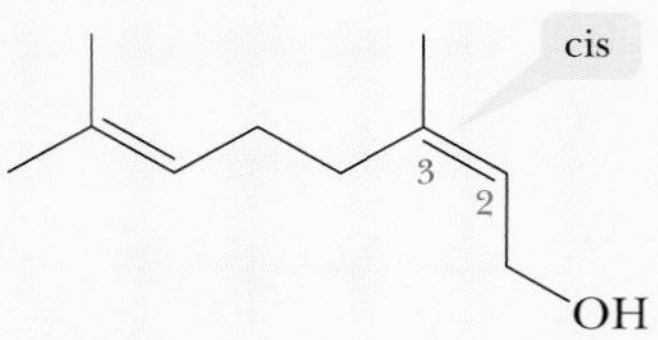   The cis isomer

---

Practice Problem 4.6

How many cis–trans isomers are possible for the following unsaturated alcohol?

$$CH_3C{=}CHCH_2CH_2C{=}CHCH_2CH_2C{=}CHCH_2OH$$

with CH_3 groups above three carbons.

Vitamin A is an example of a biologically important compound for which a number of cis–trans isomers are possible. There are four carbon–carbon double bonds in the chain of carbon atoms bonded to the substituted cyclohexene ring, and each has the potential for cis–trans isomerism. Thus, $2^4 = 16$ cis–trans isomers are possible for this structural formula. Vitamin A is the all-trans isomer. The enzyme-catalyzed oxidation of vitamin A converts the primary hydroxyl group to an aldehyde group, to give retinal, the biologically active form of the vitamin:

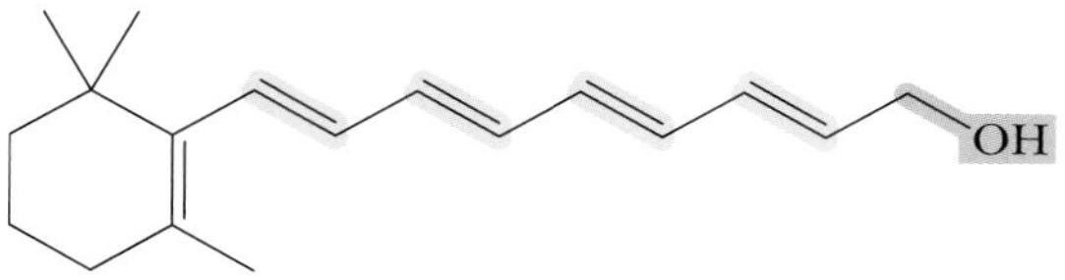

Vitamin A (retinol)

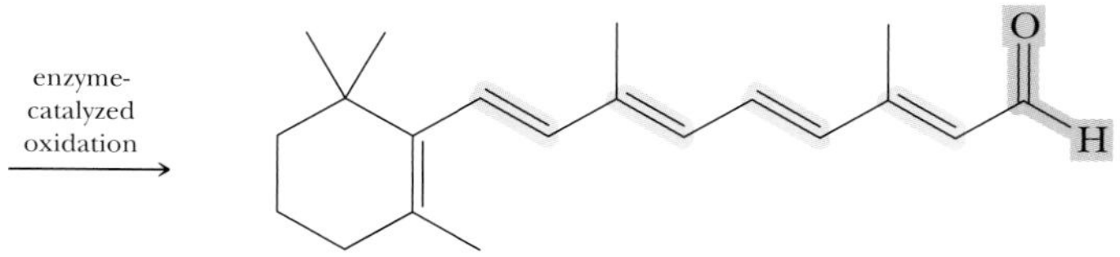

enzyme-
catalyzed
oxidation

Vitamin A aldehyde (retinal)

CHEMICAL CONNECTIONS 4B

Cis–Trans Isomerism in Vision

The retina—the light-detecting layer in the back of our eyes—contains reddish compounds called *visual pigments*. Their name, *rhodopsin*, is derived from the Greek word meaning "rose colored." Each rhodopsin molecule is a combination of one molecule of a protein called opsin and one molecule of 11-*cis*-retinal, a derivative of vitamin A in which the CH_2OH group of the vitamin is converted to an aldehyde group, —CHO, and the double bond between carbons 11 and 12 of the side chain is in the less stable cis configuration. When rhodopsin absorbs light energy, the less stable 11-cis double bond is converted to the more stable 11-trans double bond. This isomerization changes the shape of the rhodopsin molecule, which in turn causes the neurons of the optic nerve to fire and produce a visual image:

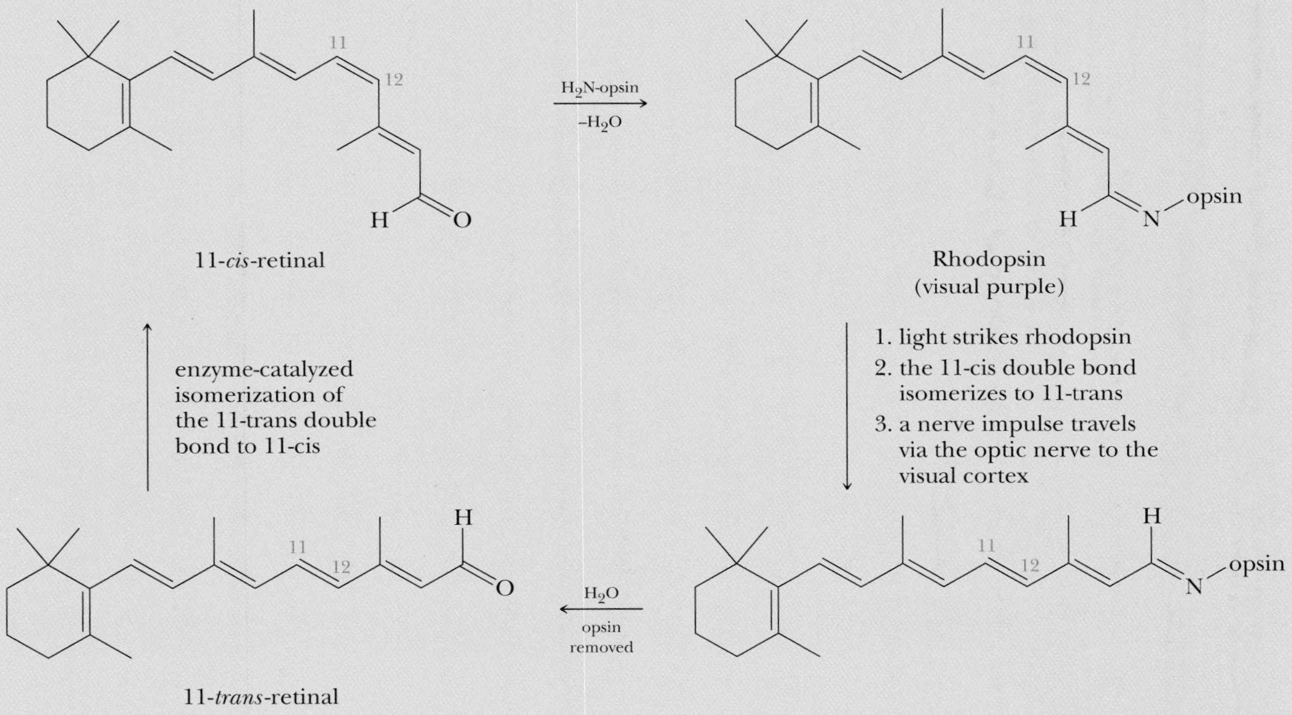

11-*cis*-retinal

H₂N-opsin
−H₂O

Rhodopsin
(visual purple)

enzyme-catalyzed
isomerization of
the 11-trans double
bond to 11-cis

1. light strikes rhodopsin
2. the 11-cis double bond isomerizes to 11-trans
3. a nerve impulse travels via the optic nerve to the visual cortex

11-*trans*-retinal

H₂O
opsin
removed

The retina of vertebrates contains two kinds of cells that contain rhodopsin: rods and cones. Cones function in bright light and are used for color vision; they are concentrated in the central portion of the retina, called the *macula*, and are responsible for the greatest visual acuity. The remaining area of the retina consists mostly of rods, which are used for peripheral and night vision. 11-*cis*-Retinal is present in both cones and rods. Rods have one kind of opsin, whereas cones have three kinds—one for blue, one for green, and one for red color vision.

4.4 PHYSICAL PROPERTIES

Alkenes and alkynes are nonpolar compounds, and the only attractive forces between their molecules are dispersion forces (Section 3.9B). Therefore, their physical properties are similar to those of alkanes (Section 3.9) with the same carbon skeletons. Alkenes and alkynes that are liquid at room temperature have densities less than 1.0 g/mL. Thus, they are less dense than water. Like alkanes, alkenes and alkynes are nonpolar and are soluble in each other. Because of their contrasting polarity with water, they do not dissolve in water. Instead, they form two layers when mixed with water or another polar organic liquid such as ethanol.

Tetramethylethylene and dimethylacetylene. Both a carbon double bond and a carbon-carbon triple bond are sites of high electron density and, therefore, sites of chemical reactivity.

EXAMPLE 4.8

Describe what will happen when 1-nonene is added to the following compounds:

(a) Water   (b) 8-Methyl-1-nonyne

SOLUTION

(a) 1-Nonene is an alkene and, therefore, nonpolar. It will not dissolve in a polar solvent such as water. Water and 1-nonene will form two layers; water, which has the higher density will be the lower layer, and 1-nonene will be the upper layer.
(b) Because alkenes and alkynes are both nonpolar, they will dissolve in one another.

4.5   NATURALLY OCCURRING ALKENES: THE TERPENES

Terpene A compound whose carbon skeleton can be divided into two or more units identical with the carbon skeleton of isoprene.

Among the compounds found in the essential oils of plants is a group of substances called **terpenes**, all of which have in common the property that their carbon skeletons can be divided into two or more carbon units that are identical with the carbon skeleton of isoprene. Carbon 1 of an **isoprene unit** is called the head, and carbon 4 is called the tail. A terpene is a compound in which the tail of one isoprene unit becomes bonded to the head of another isoprene unit.

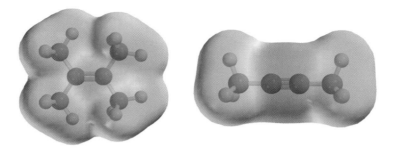

2-Methyl-1,3-butadiene (Isoprene)   Isoprene unit

Terpenes are among the most widely distributed compounds in the biological world, and a study of their structure provides a glimpse of the wondrous diversity that nature can generate from a simple carbon skeleton. Terpenes also illustrate an impor-

CHEMICAL CONNECTIONS 4C

Why Plants Emit Isoprene

Names like Virginia's *Blue Ridge*, Jamaica's *Blue Mountain Peak*, and Australia's *Blue Mountains* remind us of the bluish haze that hangs over wooded hills in the summertime. In the 1950s, it was discovered that this haze is rich in isoprene, which means that isoprene is far more abundant in the atmosphere than anyone thought. The haze is caused by the scattering of light from an aerosol produced by the photooxidation of isoprene and other hydrocarbons. Scientists now estimate that the global emission of isoprene by plants is 3×10^{11} kg/yr (3.3×10^8 ton/yr), which represents approximately 2% of all carbon fixed by photosynthesis. A recent study of hydrocarbon emissions in the Atlanta area revealed that plants are by far the largest emitters of hydrocarbons, with plant-derived isoprene accounting for almost 60% of the total.

Why do plants emit so much isoprene into the atmosphere rather than use it for the synthesis of terpenes and other natural products? Tom Sharkey, a University of Wisconsin plant physiologist, found that the emission of isoprene is extremely sensitive to temperature. Plants grown at 20°C do not emit isoprene, but they begin to emit it when the temperature of their leaves increases to 30°C. In certain plants, isoprene emission can increase as much as

tenfold for a 10°C increase in leaf temperature. Sharkey studied the relationship between temperature-induced leaf damage and isoprene concentration in leaves of the kudzu plant, a nonnative invasive vine. He discovered that leaf damage, as measured by the destruction of chlorophyll, begins to occur at 37.5°C in the absence of isoprene, but not until 45°C in its presence. Sharkey speculates that isoprene dissolves in leaf membranes and in some way increases their tolerance to heat stress. Because isoprene is made rapidly and also lost rapidly, its concentration correlates with temperature throughout the day.

The haze of the Smoky Mountains is caused by light-scattering from the aerosol produced by the photooxidation of isoprene and other hydrocarbons. See the Chemistry in Action box "Why Plants Emit Isoprene". Inset: A model of isoprene *(Digital Vision)*

tant principle of the molecular logic of living systems, namely, that, in building large molecules, small subunits are strung together enzymatically by an iterative process and are then chemically modified by precise enzyme-catalyzed reactions. Chemists use the same principles in the laboratory, but our methods do not have the precision and selectivity of the enzyme-catalyzed reactions of cellular systems.

Probably the terpenes most familiar to you, at least by odor, are components of the so-called essential oils obtained by steam distillation or ether extraction of various parts of plants. Essential oils contain the relatively low-molecular-weight substances that are in large part responsible for characteristic plant fragrances. Many essential oils, particularly those from flowers, are used in perfumes.

Figure 4.2
Four terpenes, each derived
from two isoprene units
(highlighted) bonded from
the tail of the first unit to the
head of the second unit. In
limonene and menthol, he
formation of an additional
carbon–carbon bond creates
a six-membered ring.

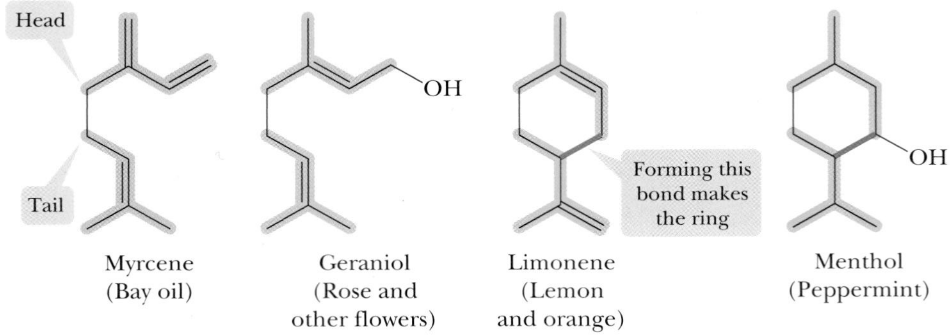

Myrcene
(Bay oil)

Geraniol
(Rose and
other flowers)

Limonene
(Lemon
and orange)

Menthol
(Peppermint)

One example of a terpene obtained from an essential oil is myrcene, $C_{10}H_{16}$, a component of bayberry wax and oils of bay and verbena. Myrcene is a triene with a parent chain of eight carbon atoms and two one-carbon branches (Figure 4.2).

Farnesol, a terpene with molecular formula $C_{15}H_{26}O$, includes three isoprene units:

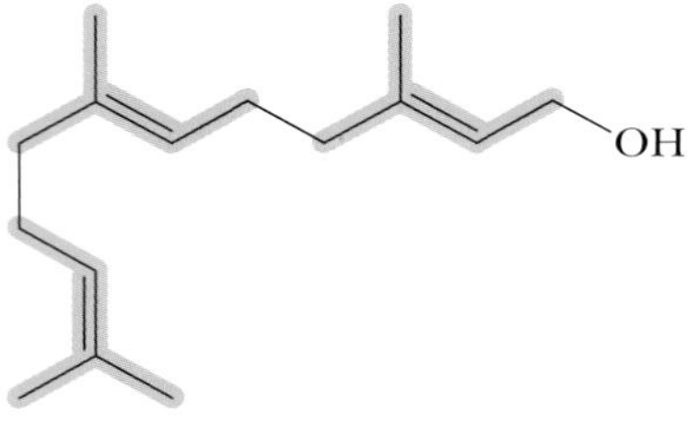

Farnesol
(Lily-of-the-valley)

Derivatives of both farnesol and geraniol are intermediates in the biosynthesis of cholesterol (Section 21.5B).

Vitamin A (Section 4.3G), a terpene with molecular formula $C_{20}H_{30}O$, consists of four isoprene units linked head-to-tail and cross-linked at one point to form a six-membered ring.

SUMMARY

An **alkene** is an unsaturated hydrocarbon that contains a carbon–carbon double bond. Alkenes have the general formula C_nH_{2n}. An **alkyne** is an unsaturated hydrocarbon that contains a carbon–carbon triple bond. Alkynes have the general formula C_nH_{2n-2}. According to the **orbital overlap model** (Section 4.2B), a carbon–carbon double bond consists of one sigma bond formed by the overlap of sp^2 hybrid orbitals and one pi bond formed by the overlap of parallel $2p$ atomic orbitals. It takes approximately 63 kcal/mol (264 kJ/mol) to break the pi bond in ethylene. A carbon–carbon triple bond consists of one sigma bond formed by the overlap of sp hybrid orbitals and two pi bonds formed by the overlap of pairs of parallel $2p$ orbitals.

The structural feature that makes **cis–trans isomerism** possible in alkenes is restricted rotation about the two carbons of the double bond (Section 4.2C). To date, *trans*-cyclooctene is the smallest trans cycloalkene that has been prepared in pure form and is stable at room temperature.

According to the IUPAC system (Section 4.3A), we show the presence of a **carbon–carbon double bond** by changing the infix of the parent hydrocarbon from -an- to -en-. We show the presence of a **carbon–carbon triple bond** by changing the infix of the parent alkane from -an- to -yn-.

The orientation of the carbon atoms of the parent chain about the double bond determines whether an alkene is cis or trans (Section 4.3C). If atoms of the parent are on the same side of the double bond, the configuration of the alkene is cis; if they are on opposite sides, the configuration is trans. Using a set of priority rules, we can also specify the configuration of a carbon–carbon double bond by the **E,Z system** (Section 4.3C). If the two groups of higher priority are on the same side of the double bond, the configuration of the alkene is **Z** (German: *zusammen*, together); if they are on opposite sides, it is **E** (German: *entgegen*, opposite).

To name an alkene containing two or more double bonds, we change the infix to *-adien-*, *-atrien-*, and so on (Section 4.3F). Compounds containing several double bonds are called polyenes.

Alkenes and alkynes are nonpolar compounds, and the only interactions between their molecules are **dispersion forces**. The physical properties of alkenes and alkynes are similar to those of alkanes (Section 4.4).

The characteristic structural feature of a **terpene** (Section 4.5) is a carbon skeleton that can be divided into two or more **isoprene units**. Terpenes illustrate an important principle of the molecular logic of living systems, namely, that, in building large molecules, small subunits are strung together by an iterative process and are then chemically modified by precise enzyme-catalyzed reactions.

PROBLEMS

A problem number set in red indicates an applied real-world" problem.

Structure of Alkenes and Alkynes

4.7 Each carbon atom in ethane and in ethylene is surrounded by eight valence electrons and has four bonds to it. Explain how the VSEPR model (Section 1.4) predicts a bond angle of 109.5° about each carbon in ethane, but an angle of 120° about each carbon in ethylene.

4.8 Use the valence-shell electron-pair repulsion (VSEPR) model to predict all bond angles about each of the following highlighted carbon atoms:

(a) 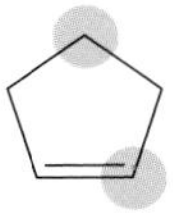   **(b)** 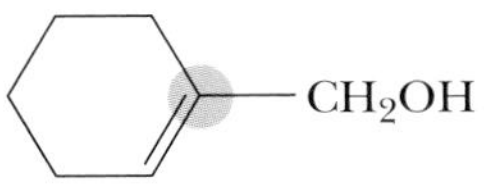 ⬡—CH_2OH **(c)** $HC{\equiv}C{-}CH{=}CH_2$ **(d)**

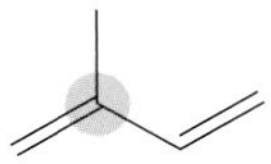

4.9 For each highlighted carbon atom in Problem 4.8, identify which orbitals are used to form each sigma bond and which are used to form each pi bond.

4.10 Predict all bond angles about each highlighted carbon atom:

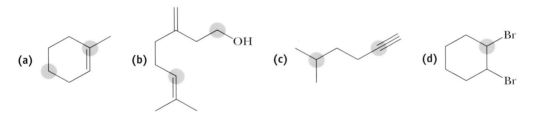

(a)   **(b)**   **(c)**   **(d)**

4.11 For each highlighted carbon atom in Problem 4.10, identify which orbitals are used to form each sigma bond and which are used to form each pi bond.

4.12 Following is the structure of 1,2-propadiene (allene):

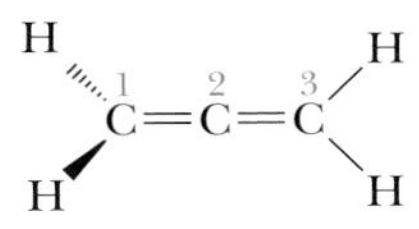

$$\underset{H}{\overset{H}{\diagdown}}\overset{1}{C}{=}\overset{2}{C}{=}\overset{3}{C}\underset{H}{\overset{H}{\diagup}}$$

1,2-Propadiene   Ball-and-stick model
(Allene)

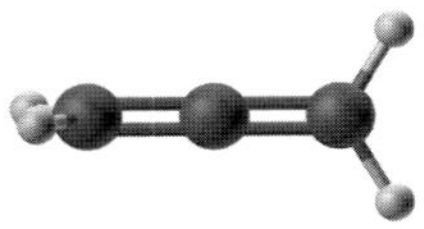

The plane created by $H{-}C{-}H$ of carbon 1 is perpendicular to that created by $H{-}C{-}H$ of carbon 3.
(a) State the orbital hybridization of each carbon in allene.
(b) Account for the molecular geometry of allene in terms of the orbital overlap model. Specifically, explain why all four hydrogen atoms are not in the same plane.

Nomenclature of Alkenes and Alkynes

4.13 Draw a structural formula for each compound:
- **(a)** *trans*-2-Methyl-3-hexene
- **(b)** 2-Methyl-3-hexyne
- **(c)** 2-Methyl-1-butene
- **(d)** 3-Ethyl-3-methyl-1-pentyne
- **(e)** 2,3-Dimethyl-2-butene
- **(f)** *cis*-2-Pentene
- **(g)** (*Z*)-1-Chloropropene
- **(h)** 3-Methylcyclohexene

4.14 Draw a structural formula for each compound:
- **(a)** 1-Isopropyl-4-methylcyclohexene
- **(b)** (6*E*)-2,6-Dimethyl-2,6-octadiene
- **(c)** *trans*-1,2-Diisopropylcyclopropane
- **(d)** 2-Methyl-3-hexyne
- **(e)** 2-Chloropropene
- **(f)** Tetrachloroethylene

4.15 Write the IUPAC name for each compound:

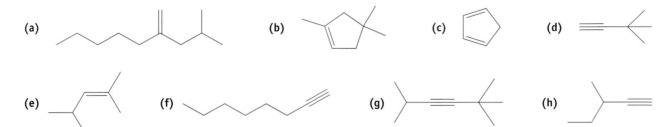

4.16 Explain why each name is incorrect, and then write a correct name for the intended compound:
- **(a)** 1-Methylpropene
- **(b)** 3-Pentene
- **(c)** 2-Methylcyclohexene
- **(d)** 3,3-Dimethylpentene
- **(e)** 4-Hexyne
- **(f)** 2-Isopropyl-2-butene

4.17 Explain why each name is incorrect, and then write a correct name for the intended compound:
- **(a)** 2-Ethyl-1-propene
- **(b)** 5-Isopropylcyclohexene
- **(c)** 4-Methyl-4-hexene
- **(d)** 2-*sec*-Butyl-1-butene
- **(e)** 6,6-Dimethylcyclohexene
- **(f)** 2-Ethyl-2-hexene

Cis–Trans Isomerism in Alkenes and Cycloalkenes

4.18 Which of these alkenes show cis–trans isomerism? For each that does, draw structural formulas for both isomers.
- **(a)** 1-Hexene
- **(b)** 2-Hexene
- **(c)** 3-Hexene
- **(d)** 2-Methyl-2-hexene
- **(e)** 3-Methyl-2-hexene
- **(f)** 2,3-Dimethyl-2-hexene

4.19 Which of these alkenes show cis–trans isomerism? For each that does, draw structural formulas for both isomers.
- **(a)** 1-Pentene
- **(b)** 2-Pentene
- **(c)** 3-Ethyl-2-pentene
- **(d)** 2,3-Dimethyl-2-pentene
- **(e)** 2-Methyl-2-pentene
- **(f)** 2,4-Dimethyl-2-pentene

4.20 Which alkenes can exist as pairs of cis-trans isomers? For each alkene that does, draw the trans isomer.
- **(a)** $CH_2\text{=}CHBr$
- **(b)** $CH_3CH\text{=}CHBr$
- **(c)** $(CH_3)_2C\text{=}CHCH_3$
- **(d)** $(CH_3)_2CHCH\text{=}CHCH_3$

4.21 There are three compounds with molecular formula $C_2H_2Br_2$. Two of these compounds have a dipole greater than zero, and one has no dipole. Draw structural formulas for the three compounds, and explain why two have dipole moments but the third one has none.

4.22 Name and draw structural formulas for all alkenes with molecular formula C_5H_{10}. As you draw these alkenes, remember that cis and trans isomers are different compounds and must be counted separately.

4.23 Name and draw structural formulas for all alkenes with molecular formula C_6H_{12} that have the following carbon skeletons (remember cis and trans isomers):

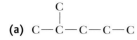

(a)

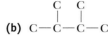

(b) (c)

4.24 Arrange the groups in each set in order of increasing priority:
(a) $-CH_3$, $-Br$, $-CH_2CH_3$
(b) $-OCH_3$, $-CH(CH_3)_2$, $-CH_2CH_2NH_2$
(c) $-CH_2OH$, $-COOH$, $-OH$
(d) $-CH=CH_2$, $-CH=O$, $-CH(CH_3)_2$

4.25 Draw the structural formula for at least one bromoalkene with molecular formula C_5H_9Br that (a) shows E,Z isomerism and (b) does not show E,Z isomerism.

4.26 For each molecule that shows cis–trans isomerism, draw the cis isomer:

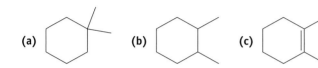

(a) (b) (c) (d)

4.27 Explain why each name is incorrect or incomplete, and then write a correct name:
(a) (Z)-2-Methyl-1-pentene (b) (E)-3,4-Diethyl-3-hexene
(c) trans-2,3-Dimethyl-2-hexene (d) (1Z,3Z)-2,3-Dimethyl-1,3-butadiene

4.28 Draw structural formulas for all compounds with molecular formula C_5H_{10} that are
(a) Alkenes that do not show cis–trans isomerism.
(b) Alkenes that do show cis–trans isomerism.
(c) Cycloalkanes that do not show cis–trans isomerism.
(d) Cycloalkanes that do show cis–trans isomerism.

4.29 β-Ocimene, a triene found in the fragrance of cotton blossoms and several essential oils, has the IUPAC name (3Z)-3,7-dimethyl-1,3,6-octatriene. Draw a structural formula for β-ocimene.

4.30 Oleic acid and elaidic acid are, respectively, the cis and trans isomers of 9-octadecenoic acid. One of these fatty acids, a colorless liquid that solidifies at 4°C, is a major component of butterfat. The other, a white solid with a melting point of 44–45°C, is a major component of partially hydrogenated vegetable oils. Which of these two fatty acids is the cis isomer and which is the trans isomer?

4.31 Determine whether the structures in each set represent the same molecule, cis–trans isomers, or constitutional isomers. If they are the same molecule, determine whether they are in the same or different conformations.

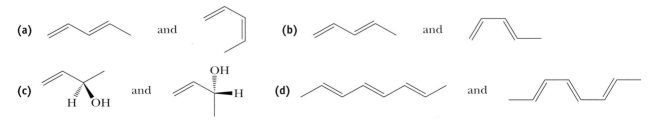

(a) and (b) and
(c) and (d) and

Terpenes

4.32 Show that the structural formula of vitamin A (Section 4.3G) can be divided into four isoprene units joined by head-to-tail linkages and cross-linked at one point to form the six-membered ring.

4.33 Following is the structural formula of lycopene, a deep-red compound that is partially responsible for the red color of ripe fruits, especially tomatoes:

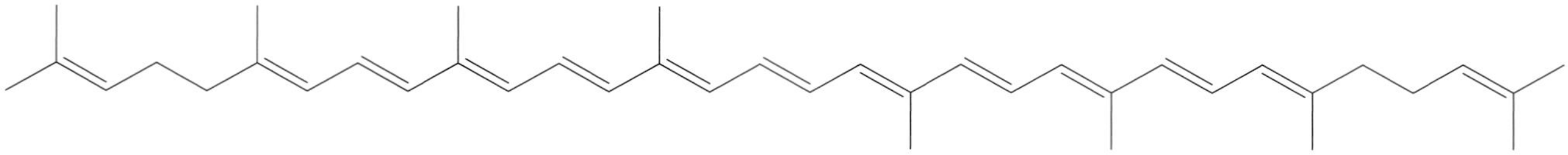

Lycopene

Approximately 20 mg of lycopene can be isolated from 1 kg of fresh, ripe tomatoes.

(a) Show that lycopene is a terpene; that is, show that lycopene's carbon skeleton can be divided into two sets of four isoprene units with the units in each set joined head-to-tail.

(b) How many of the carbon–carbon double bonds in lycopene have the possibility for cis–trans isomerism? Lycopene is the all-trans isomer.

4.34 As you might suspect, β-carotene, a precursor of vitamin A, was first isolated from carrots. Dilute solutions of β-carotene are yellow—hence its use as a food coloring. In plants, it is almost always present in combination with chlorophyll to assist in the harvesting of the energy of sunlight. As tree leaves die in the fall, the green of their chlorophyll molecules is replaced by the yellows and reds of carotene and carotene-related molecules.

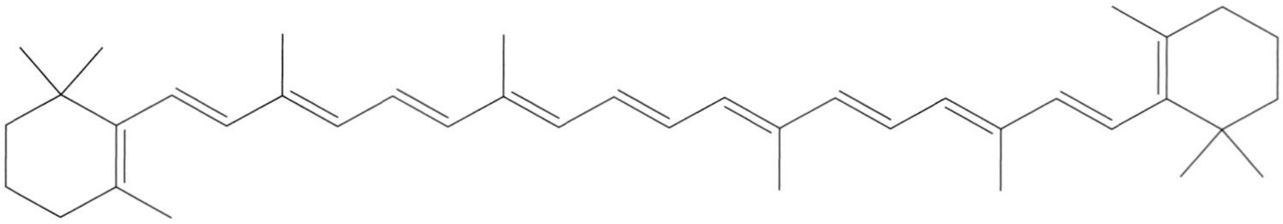

β-Carotene

(a) Compare the carbon skeletons of β-carotene and lycopene. What are the similarities? What are the differences?

(b) Show that β-carotene is a terpene.

4.35 α-Santonin, isolated from the flower heads of certain species of artemisia, is an anthelmintic—that is, a drug used to rid the body of worms (helminths). It has been estimated that over one-third of the world's population is infested with these parasites. Farnesol is an alcohol with a florid odor:

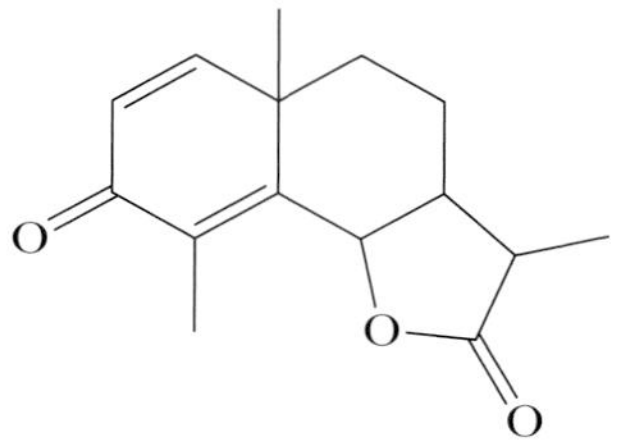

Santonin 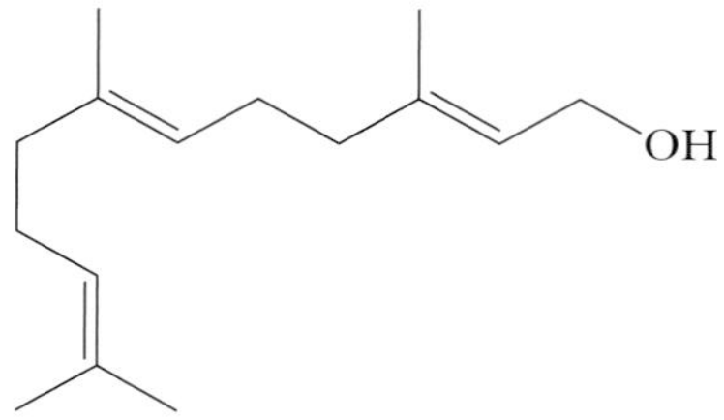   Farnesol

Locate the three isoprene units in santonin, and show how the carbon skeleton of farnesol might be coiled and then cross-linked to give santonin. Two different coiling patterns of the carbon skeleton of farnesol can lead to santonin. Try to find them both.

4.36 Periplanone is a pheromone (a chemical sex attractant) isolated from a species of cockroach. Show that the carbon skeleton of periplanone classifies it as a terpene:

Periplanone

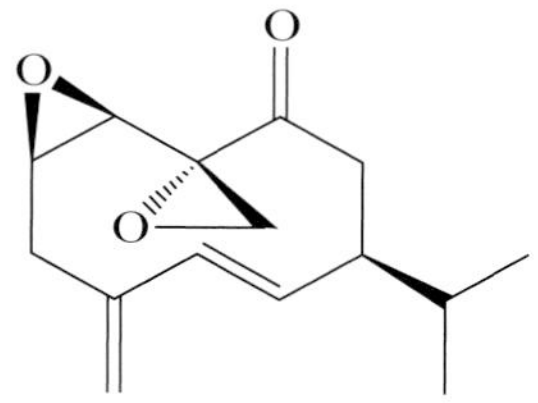

4.37 Gossypol, a compound found in the seeds of cotton plants, has been used as a male contraceptive in overpopulated countries such as China. Show that gossypol is a terpene:

Gossypol

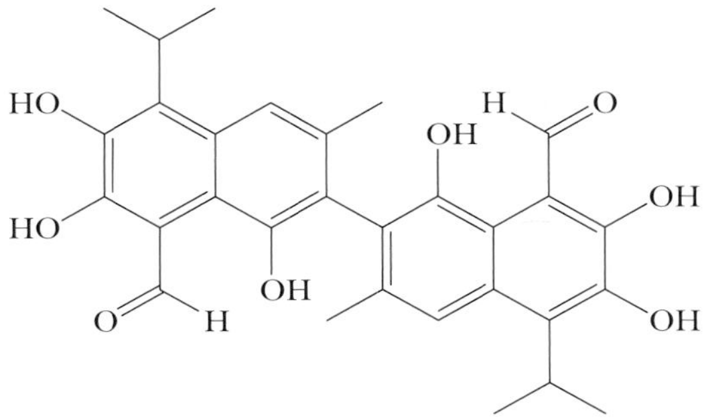

4.38 In many parts of South America, extracts of the leaves and twigs of *Montanoa tomentosa* are used as a contraceptive, to stimulate menstruation, to facilitate labor, and as an abortifacient. The compound responsible for these effects is zoapatanol:

Zoapatanol

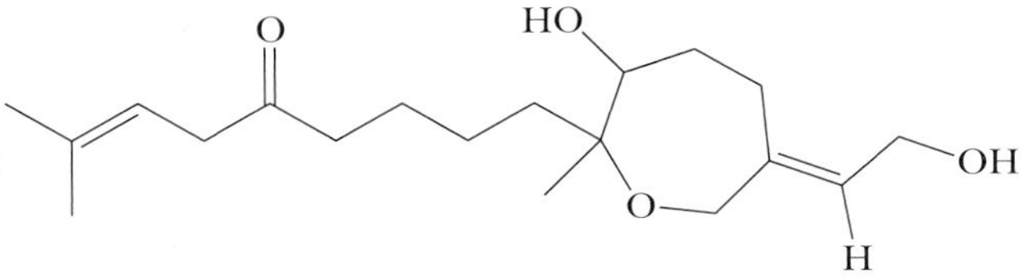

(a) Show that the carbon skeleton of zoapatanol can be divided into four isoprene units bonded head-to-tail and then cross-linked in one point along the chain.

(b) Specify the configuration about the carbon–carbon double bond to the seven-membered ring, according to the E,Z system.

(c) How many cis–trans isomers are possible for zoapatanol? Consider the possibilities for cis–trans isomerism in cyclic compounds and about carbon–carbon double bonds.

4.39 Pyrethrin II and pyrethrosin are natural products isolated from plants of the chrysanthemum family:

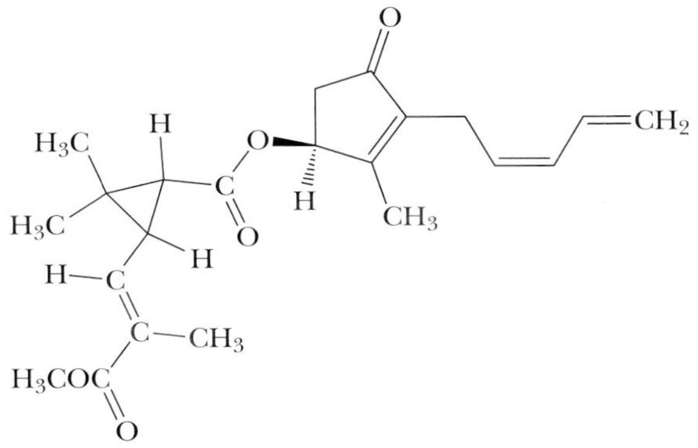

Pyrethrin II 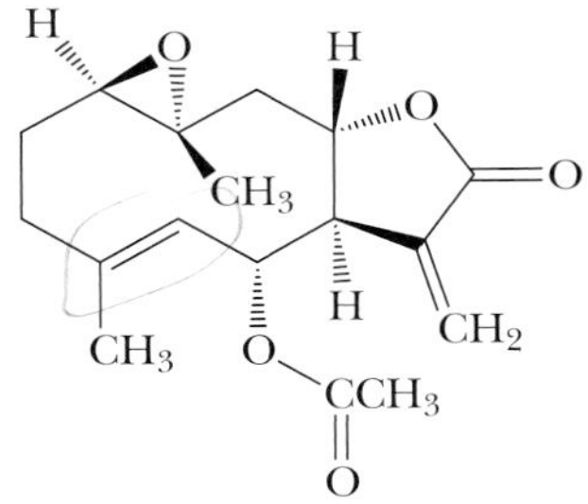   Pyrethrosin

Chrysanthemum blossoms.
(Scott Camazine/Photo Researchers, Inc.)

Pyrethrin II is a natural insecticide and is marketed as such.

(a) Label all carbon–carbon double bonds in each about which cis–trans isomerism is possible.

(b) Why are cis–trans isomers possible about the three-membered ring in pyrethrin II, but not about its five-membered ring?

(c) Show that the ring system of pyrethrosin is composed of three isoprene units.

4.40 Cuparene and herbertene are naturally occurring compounds isolated from various species of lichen:

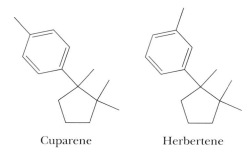

Cuparene   Herbertene

Determine whether one or both of these compounds can be classified as terpenes.

Looking Ahead

4.41 Explain why the $=C{-}C$ single bond in 1,3-butadiene is slightly shorter than the $=C{-}C$ single bond in 1-butene:

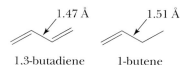

1,3-butadiene   1-butene

4.42 What effect might the ring size in the following cycloalkenes have on the reactivity of the $C{=}C$ double bond in each?

4.43 What effect might each substituent have on the electron density surrounding the alkene $C{=}C$ bond?

(a)   OCH_3 **(b)** 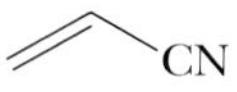 CN **(c)** 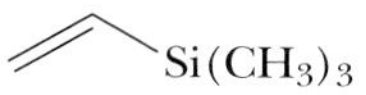 $Si(CH_3)_3$

4.44 In Section 21.1 on the biochemistry of fatty acids, we will study the following three long-chain unsaturated carboxylic acids:

Oleic acid $CH_3(CH_2)_7CH{=}CH(CH_2)_7COOH$

Linoleic acid $CH_3(CH_2)_4CH{=}CHCH_2CH{=}CH(CH_2)_7COOH$

Linolenic acid $CH_3CH_2CH{=}CHCH_2CH{=}CHCH_2CH{=}CH(CH_2)_7COOH$

Each has 18 carbons and is a component of animal fats, vegetable oils, and biological membranes. Because of their presence in animal fats, they are called fatty acids.

(a) How many cis–trans isomers are possible for each fatty acid?

(b) These three fatty acids occur in biological membranes almost exclusively in the cis configuration. Draw line-angle formulas for each fatty acid, showing the cis configuration about each carbon–carbon double bond.

4.45 Assign an E or a Z configuration and a cis or a trans configuration to these carboxylic acids, each of which is an intermediate in the citric acid cycle (Section 22.7; under each is given its common name):

(a)

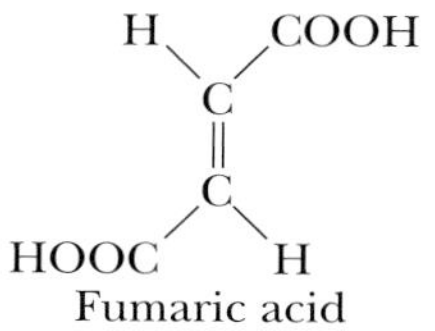

```
H        COOH
 \      /
   C
   ‖
   C
 /      \
HOOC      H
```
Fumaric acid

(b)

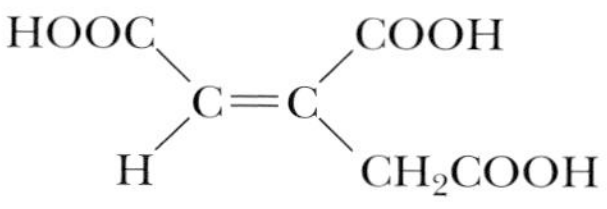

```
HOOC        COOH
    \      /
     C = C
    /      \
   H        CH2COOH
```
Aconitic acid

5 Reactions of Alkenes

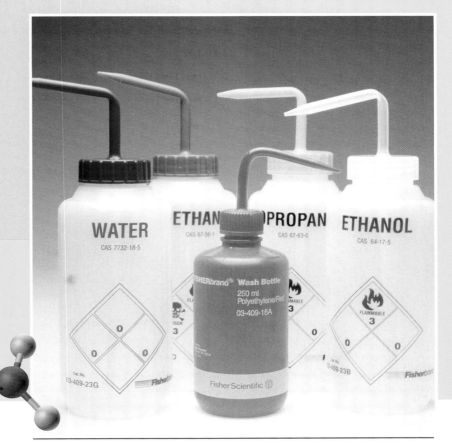

These wash bottles are made of
polyethylene. Inset: A model
of ethylene. *(Charles D. Winters)*

I n this chapter, we begin our systematic study of reaction mechanisms, one
of the most important unifying concepts in organic chemistry. We use the
reactions of alkenes as the vehicle to introduce this concept.

5.1 INTRODUCTION

The most characteristic reaction of alkenes is **addition to the carbon–carbon double bond** in such a way that the pi bond is broken and, in its place, sigma bonds are
formed to two new atoms or groups of atoms. Several examples of reactions at the
carbon–carbon double bond are shown in Table 5.1, along with the descriptive
name(s) associated with each.

TABLE 5.1 Characteristic Addition Reactions of Alkenes

Reaction	Descriptive Name(s)
$\overset{\displaystyle\diagdown}{\diagup}C=C\overset{\displaystyle\diagup}{\diagdown}$ + HCl $\longrightarrow$ $-\overset{\mid}{\underset{\mid}{C}}-\overset{\mid}{\underset{\mid}{C}}-$ H Cl	hydrochlorination (hydrohalogenation)
$C=C$ + H_2O $\longrightarrow$ $-\overset{\mid}{\underset{\mid}{C}}-\overset{\mid}{\underset{\mid}{C}}-$ H OH	hydration
$C=C$ + Br_2 $\longrightarrow$ $-\overset{\mid}{\underset{\mid}{C}}-\overset{\mid}{\underset{\mid}{C}}-$ Br Br	bromination (halogenation)
$C=C$ + OsO_4 $\longrightarrow$ $-\overset{\mid}{\underset{\mid}{C}}-\overset{\mid}{\underset{\mid}{C}}-$ HO OH	hydroxylation (oxidation)
$C=C$ + H_2 $\longrightarrow$ $-\overset{\mid}{\underset{\mid}{C}}-\overset{\mid}{\underset{\mid}{C}}-$ H H	hydrogenation (reduction)

From the perspective of the chemical industry, the single most important reaction of ethylene and other low-molecular-weight alkenes is the production of **chain-growth polymers** (Greek: *poly*, many, and *meros*, part). In the presence of certain catalysts called *initiators*, many alkenes form polymers by the addition of **monomers** (Greek: *mono*, one, and *meros*, part) to a growing polymer chain, as illustrated by the formation of polyethylene from ethylene:

$$n\text{CH}_2 = \text{CH}_2 \xrightarrow{\text{initiator}} \text{--}(\text{CH}_2\text{CH}_2\text{)}_n$$

In alkene polymers of industrial and commercial importance, n is a large number, typically several thousand. We discuss this alkene reaction in Chapter 17.

5.2 REACTION MECHANISMS

A **reaction mechanism** describes in detail how a chemical reaction occurs. It describes which bonds break and which new ones form, as well as the order and relative rates of the various bond-breaking and bond-forming steps. If the reaction takes place in solution, the reaction mechanism describes the role of the solvent; if the reaction involves a catalyst, the reaction mechanism describes the role of the catalyst.

A. Energy Diagrams and Transition States

To understand the relationship between a chemical reaction and energy, think of a chemical bond as a spring. As a spring is stretched from its resting position, its energy increases. As it returns to its resting position, its energy decreases. Similarly, during a chemical reaction, bond breaking corresponds to an increase in energy, and bond forming corresponds to a decrease in energy. We use an **energy diagram** to show the changes in energy that occur in going from reactants to products. Energy is measured along the vertical axis, and the change in position of the atoms during a reaction is measured on the horizontal axis, called the **reaction coordinate**. The reaction coordinate indicates how far the reaction has progressed, from no reaction to a completed reaction.

Reaction mechanism A step-by-step description of how a chemical reaction occurs.

Energy diagram A graph showing the changes in energy that occur during a chemical reaction; energy is plotted on the y-axis, and the progress of the reaction is plotted on the x-axis.

Reaction coordinate A measure of the progress of a reaction, plotted on the x-axis in an energy diagram.

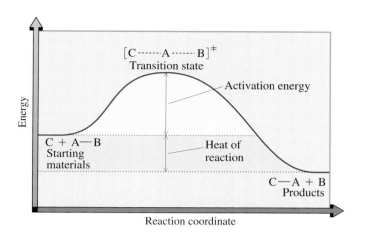

Figure 5.1
An energy diagram for a one-step reaction between C and A—B. The dashed lines indicate that, in the transition state, the new C—A bond is partially formed and the A—B bond is partially broken. The energy of the reactants is higher than that of the products.

Figure 5.1 shows an energy diagram for the reaction of C + A—B to form C—A + B. This reaction occurs in one step, meaning that bond breaking in reactants and bond forming in products occur simultaneously.

The difference in energy between the reactants and products is called the **heat of reaction**, **ΔH**. If the energy of the products is lower than that of the reactants, heat is released and the reaction is called **exothermic**. If the energy of the products is higher than that of the reactants, heat is absorbed and the reaction is called **endothermic**. The one-step reaction shown in Figure 5.1 is exothermic.

A **transition state** is the point on the reaction coordinate at which the energy is at a maximum. At the transition state, sufficient energy has become concentrated in the proper bonds so that bonds in the reactants break. As they break, energy is redistributed and new bonds form, giving products. Once the transition state is reached, the reaction proceeds to give products, with the release of energy.

A transition state has a definite geometry, a definite arrangement of bonding and nonbonding electrons, and a definite distribution of electron density and charge. Because a transition state is at an energy maximum on an energy diagram, we cannot isolate it and we cannot determine its structure experimentally. Its lifetime is on the order of a picosecond (the duration of a single bond vibration). As we will see, however, even though we cannot observe a transition state directly by any experimental means, we can often infer a great deal about its probable structure from other experimental observations.

For the reaction shown in Figure 5.1, we use dashed lines to show the partial bonding in the transition state. As C begins to form a new covalent bond with A (as shown by the dashed line), the covalent bond between A and B begins to break (also shown by a dashed line). Upon completion of the reaction, the A—B bond is fully broken and the C—A bond is fully formed.

The difference in energy between the reactants and the transition state is called the **activation energy**. The activation energy is the minimum energy required for a reaction to occur; it can be considered an energy barrier for the reaction. The activation energy determines the rate of a reaction—that is, how fast the reaction occurs. If the activation energy is large, only a very few molecular collisions occur with sufficient energy to reach the transition state, and the reaction is slow. If the activation energy is small, many collisions generate sufficient energy to reach the transition state, and the reaction is fast.

In a reaction that occurs in two or more steps, each step has its own transition state and activation energy. Shown in Figure 5.2 is an energy diagram for the conversion of reactants to products in two steps. A **reaction intermediate** corresponds to an energy minimum between two transition states, in this case transition States 1 and 2. Note that because the energies of the reaction intermediates we describe are higher than the energies of either the reactants or the products, these intermediates are highly reactive, and rarely, if ever, can one be isolated.

Heat of reaction The difference in energy between reactants and products.

Exothermic reaction A reaction in which the energy of the products is lower than the energy of the reactants; a reaction in which heat is liberated.

Endothermic reaction A reaction in which the energy of the products is higher than the energy of the reactants; a reaction in which heat is absorbed.

Transition state An unstable species of maximum energy formed during the course of a reaction; a maximum on an energy diagram.

Activation energy The difference in energy between reactants and the transition state.

Reaction intermediate An unstable species that lies in an energy minimum between two transition states.

Figure 5.2
Energy diagram for a two-step reaction involving the formation of an intermediate. The energy of the reactants is higher than that of the products, and energy is released in the conversion of A + B to C + D.

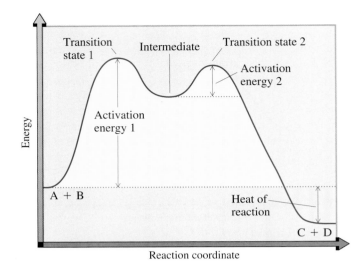

Rate-determining step The step in a reaction sequence that crosses the highest energy barrier; the slowest step in a multistep reaction.

The slowest step in a multistep reaction, called the **rate-determining step**, is the step that crosses the highest energy barrier. In the two-step reaction shown in Figure 5.2, Step 1 crosses the higher energy barrier and is, therefore, the rate-determining step.

EXAMPLE 5.1

Draw an energy diagram for a two-step exothermic reaction in which the second step is rate determining.

SOLUTION

A two-step reaction involves the formation of an intermediate. In order for the reaction to be exothermic, the products must be lower in energy than the reactants. In order for the second step to be rate determining, it must cross the higher energy barrier.

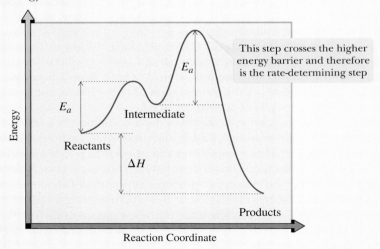

Practice Problem 5.1

In what way would the energy diagram drawn in Example 5.1 change if the reaction were endothermic?

B. Developing a Reaction Mechanism

To develop a reaction mechanism, chemists begin by designing experiments that will reveal details of a particular chemical reaction. Next, through a combination of experience and intuition, they propose several sets of steps or mechanisms, each of which might account for the overall chemical transformation. Finally, they test each proposed mechanism against the experimental observations to exclude those mechanisms which are not consistent with the facts.

A mechanism becomes generally established by excluding reasonable alternatives and by showing that it is consistent with every test that can be devised. This, of course, does not mean that a generally accepted mechanism is a completely accurate description of the chemical events, but only that it is the best chemists have been able to devise. It is important to keep in mind that, as new experimental evidence is obtained, it may be necessary to modify a generally accepted mechanism or possibly even discard it and start all over again.

Before we go on to consider reactions and reaction mechanisms, we might ask why it is worth the trouble to establish them and your time to learn about them. One reason is very practical: Mechanisms provide a theoretical framework within which to organize a great deal of descriptive chemistry. For example, with insight into how reagents add to particular alkenes, it is possible to make generalizations and then predict how the same reagents might add to other alkenes. A second reason lies in the intellectual satisfaction derived from constructing models that accurately reflect the behavior of chemical systems. Finally, to a creative scientist, a mechanism is a tool to be used in the search for new knowledge and new understanding. A mechanism consistent with all that is known about a reaction can be used to make predictions about chemical interactions as yet unexplored, and experiments can be designed to test these predictions. Thus, reaction mechanisms provide a way not only to organize knowledge, but also to extend it.

5.3 ELECTROPHILIC ADDITION REACTIONS

We begin our introduction to the chemistry of alkenes with an examination of three types of addition reactions: the addition of hydrogen halides (HCl, HBr, and HI), water (H_2O), and halogens (Br_2, Cl_2). We first study some of the experimental observations about each addition reaction and then its mechanism. By examining these particular reactions, we develop a general understanding of how alkenes undergo addition reactions.

A. Addition of Hydrogen Halides

The hydrogen halides HCl, HBr, and HI add to alkenes to give haloalkanes (alkyl halides). These additions may be carried out either with the pure reagents or in the presence of a polar solvent such as acetic acid. The addition of HCl to ethylene gives chloroethane (ethyl chloride):

$$CH_2{=}CH_2 + HCl \longrightarrow \overset{\displaystyle H}{\underset{\displaystyle |}{C}}H_2{-}\overset{\displaystyle Cl}{\underset{\displaystyle |}{C}}H_2$$

Ethylene   Chloroethane

The addition of HCl to propene gives 2-chloropropane (isopropyl chloride); hydrogen adds to carbon 1 of propene and chlorine adds to carbon 2. If the orientation of addition were reversed, 1-chloropropane (propyl chloride) would be formed.

The observed result is that 2-chloropropane is formed to the virtual exclusion of 1-chloropropane:

$$CH_3CH=CH_2 + HCl \longrightarrow \underset{\text{2-Chloropropane}}{CH_3\overset{\overset{\displaystyle Cl}{|}}{C}H-\overset{\overset{\displaystyle H}{|}}{C}H_2} + \underset{\substack{\text{1-Chloropropane}\\\text{(not observed)}}}{CH_3\overset{\overset{\displaystyle H}{|}}{C}H-\overset{\overset{\displaystyle Cl}{|}}{C}H_2}$$

Propene

We say that the addition of HCl to propene is highly regioselective and that 2-chloropropane is the major product of the reaction. A **regioselective reaction** is a reaction in which one direction of bond forming or breaking occurs in preference to all other directions.

Vladimir Markovnikov observed this regioselectivity and made the generalization, known as **Markovnikov's rule**, that, in the addition of HX to an alkene, hydrogen adds to the doubly bonded carbon that has the greater number of hydrogens already bonded to it. Although Markovnikov's rule provides a way to predict the product of many alkene addition reactions, it does not explain why one product predominates over other possible products.

Regioselective reaction A reaction in which one direction of bond forming or bond breaking occurs in preference to all other directions.

Markovnikov's rule In the addition of HX or H$_2$O to an alkene, hydrogen adds to the carbon of the double bond having the greater number of hydrogens.

EXAMPLE 5.2

Name and draw a structural formula for the major product of each alkene addition reaction:

$$(a)\ CH_3\overset{\overset{\displaystyle CH_3}{|}}{C}=CH_2 + HI \longrightarrow \qquad (b)\ \text{[cyclopentene with } CH_3] + HCl \longrightarrow$$

SOLUTION

Using Markovnikov's rule, we predict that 2-iodo-2-methylpropane is the product in (a) and 1-chloro-1-methylcyclopentane is the product in (b):

(a) $CH_3\overset{\overset{\displaystyle CH_3}{|}}{\underset{\underset{\displaystyle I}{|}}{C}}CH_3$

2-Iodo-2-methylpropane

(b) [cyclopentane with Cl and CH$_3$]

1-Chloro-1-methylcyclopentane

Practice Problem 5.2

Name and draw a structural formula for the major product of each alkene addition reaction:

(a) $CH_3CH=CH_2 + HI \longrightarrow$ (b) [cyclohexane]$=CH_2 + HI \longrightarrow$

Chemists account for the addition of HX to an alkene by a two-step mechanism, which we illustrate by the reaction of 2-butene with hydrogen chloride to give 2-chlorobutane. Let us first look at this two-step mechanism in general and then go back and study each step in detail.

Mechanism: Electrophilic Addition of HCl to 2-Butene

Step 1: The reaction begins with the transfer of a proton from HCl to 2-butene, as shown by the two curved arrows on the left side of Step 1:

$$CH_3CH = CHCH_3 \; + \; H-Cl: \xrightleftharpoons[\text{determining}]{\text{slow, rate}} \; CH_3CH-\overset{\displaystyle H}{\underset{+}{C}}HCH_3 \; + \; :\overset{..}{\underset{..}{Cl}}:^-$$

sec-Butyl cation
(a 2° carbocation
intermediate)

The first curved arrow shows the breaking of the pi bond of the alkene and its electron pair now forming a new covalent bond with the hydrogen atom of HCl. The second curved arrow shows the breaking of the polar covalent bond in HCl and this electron pair being given entirely to chlorine, forming chloride ion. Step 1 in this mechanism results in the formation of an organic cation and chloride ion.

Step 2: The reaction of the *sec*-butyl cation (a Lewis acid) with chloride ion (a Lewis base) completes the valence shell of carbon and gives 2-chlorobutane:

$$:\overset{..}{\underset{..}{Cl}}:^- + \; CH_3\overset{+}{C}HCH_2CH_3 \xrightarrow{\text{fast}} CH_3\overset{\displaystyle :\overset{..}{\underset{..}{Cl}}:}{\underset{|}{C}}HCH_2CH_3$$

Chloride ion   *sec*-Butyl cation   2-Chlorobutane
(a Lewis base)   (a Lewis acid)

Now let us go back and look at the individual steps in more detail. There is a great deal of important organic chemistry embedded in these two steps, and it is crucial that you understand it now.

Step 1 results in the formation of an organic cation. One carbon atom in this cation has only six electrons in its valence shell and carries a charge of +1. A species containing a positively charged carbon atom is called a **carbocation** (*carbo*n + *cation*). Carbocations are classified as primary (1°), secondary (2°), or tertiary (3°), depending on the number of carbon atoms bonded to the carbon bearing the positive charge. All carbocations are Lewis acids (Section 2.7). They are also electrophiles. The term **electrophile** quite literally means "electron lover."

In a carbocation, the carbon bearing the positive charge is bonded to three other atoms, and, as predicted by the valence-shell electron-pair repulsion (VSEPR) model, the three bonds about that carbon are coplanar and form bond angles of approximately 120°. According to the orbital overlap model, the electron-deficient carbon of a carbocation uses its sp^2 hybrid orbitals to form sigma bonds to the three attached groups. The unhybridized $2p$ orbital lies perpendicular to the sigma bond framework and contains no electrons. A Lewis structure and an orbital overlap diagram for the *tert*-butyl cation are shown in Figure 5.3.

Carbocation A species containing a carbon atom with only three bonds to it and bearing a positive charge.

Electrophile Any molecule or ion that can accept a pair of electrons to form a new covalent bond; a Lewis acid.

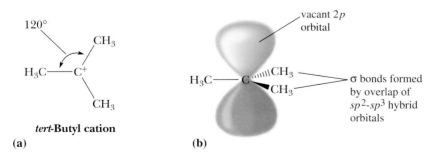

(a)   (b)

Figure 5.3
The structure of the *tert*-butyl cation. (a) Lewis structure and (b) an orbital picture.

Figure 5.4
Energy diagram for the
two-step addition of HCl
to 2-butene. The
reaction is exothermic.

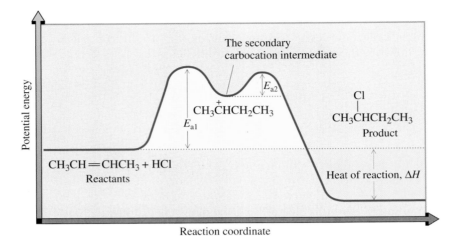

Figure 5.4 shows an energy diagram for the two-step reaction of 2-butene with HCl. The slower, rate-determining step (the one that crosses the higher energy barrier) is Step 1, which leads to the formation of the 2° carbocation intermediate. This intermediate lies in an energy minimum between the transition states for Steps 1 and 2. As soon as the carbocation intermediate (a Lewis acid) forms, it reacts with chloride ion (a Lewis base) in a Lewis acid–base reaction to give 2-chlorobutane. Note that the energy level for 2-chlorobutane (the product) is lower than the energy level for 2-butene and HCl (the reactants). Thus, in this alkene addition reaction, heat is released; the reaction is, accordingly, exothermic.

Relative Stabilities of Carbocations: Regioselectivity and Markovnikov's Rule

The reaction of HX and an alkene can, at least in principle, give two different carbocation intermediates, depending on which of the doubly bonded carbon atoms forms a bond with H^+, as illustrated by the reaction of HCl with propene:

$$CH_3CH \!=\! CH_2 \; + \; H \!-\! \overset{\cdot\cdot}{\underset{\cdot\cdot}{Cl}} \!: \; \longrightarrow \; CH_3CH_2\overset{+}{C}H_2 \; \xrightarrow{\; :\overset{\cdot\cdot}{\underset{\cdot\cdot}{Cl}}:^- \;} \; CH_3CH_2CH_2\overset{\cdot\cdot}{\underset{\cdot\cdot}{Cl}}:$$

Propene   Propyl cation   1-Chloropropane
                                        (a 1° carbocation)   (not formed)

$$CH_3CH \!=\! CH_2 \; + \; H \!-\! \overset{\cdot\cdot}{\underset{\cdot\cdot}{Cl}} \!: \; \longrightarrow \; CH_3\overset{+}{C}HCH_3 \; \xrightarrow{\; :\overset{\cdot\cdot}{\underset{\cdot\cdot}{Cl}}:^- \;} \; CH_3\overset{\overset{\displaystyle :\overset{\cdot\cdot}{\underset{\cdot\cdot}{Cl}}:}{|}}{C}HCH_3$$

Propene   Isopropyl cation   2-Chloropropane
                                      (a 2° carbocation) (product formed)

The observed product is 2-chloropropane. Because carbocations react very quickly with chloride ions, the absence of 1-chloropropane as a product tells us that the 2° carbocation is formed in preference to the 1° carbocation.

Similarly, in the reaction of HCl with 2-methylpropene, the transfer of a proton to the carbon–carbon double bond might form either the isobutyl cation (a 1° carbocation) or the *tert*-butyl cation (a 3° carbocation):

$$\underset{\text{2-Methylpropene}}{\overset{\displaystyle \overset{CH_3}{\underset{|}{}}}{CH_3C \!=\! CH_2}} \; + \; H \!-\! \overset{\cdot\cdot}{\underset{\cdot\cdot}{Cl}} \!: \; \longrightarrow \; \underset{\substack{\text{Isobutyl cation} \\ \text{(a 1° carbocation)}}}{\overset{\displaystyle \overset{CH_3}{\underset{|}{}}}{CH_3\overset{+}{C}HCH_2}} \; \xrightarrow{\; :\overset{\cdot\cdot}{\underset{\cdot\cdot}{Cl}}:^- \;} \; \underset{\substack{\text{1-Chloro-} \\ \text{2-methylpropane} \\ \text{(not formed)}}}{\overset{\displaystyle \overset{CH_3}{\underset{|}{}}}{CH_3CHCH_2\overset{\cdot\cdot}{\underset{\cdot\cdot}{Cl}}:}}$$

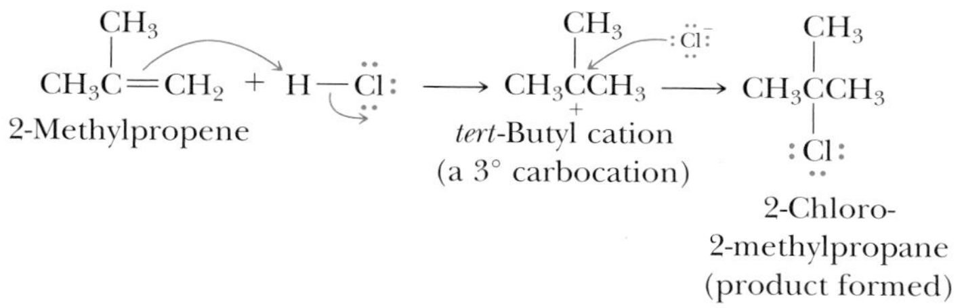

$$CH_3C{=}CH_2 \ + \ H{-}\ddot{\underset{\cdot\cdot}{Cl}}: \ \longrightarrow \ CH_3\overset{CH_3}{\underset{+}{C}}CH_3 \ \longrightarrow \ CH_3\overset{CH_3}{\underset{\underset{\cdot\cdot}{:Cl:}}{C}}CH_3$$

2-Methylpropene   *tert*-Butyl cation   2-Chloro-
                           (a 3° carbocation)   2-methylpropane
                                               (product formed)

In this reaction, the observed product is 2-chloro-2-methylpropane, indicating that the 3° carbocation forms in preference to the 1° carbocation.

From such experiments and a great amount of other experimental evidence, we learn that a 3° carbocation is more stable and requires a lower activation energy for its formation than a 2° carbocation. A 2° carbocation, in turn, is more stable and requires a lower activation energy for its formation than a 1° carbocation. It follows that a more stable carbocation intermediate forms faster than a less stable carbocation intermediate. Following is the order of stability of four types of alkyl carbocations:

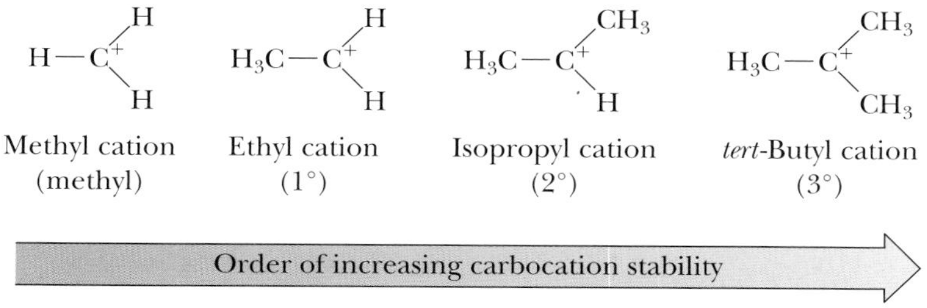

Methyl cation   Ethyl cation   Isopropyl cation   *tert*-Butyl cation
(methyl)   (1°)   (2°)   (3°)

Order of increasing carbocation stability →

Although the concept of the relative stabilities of carbocations had not been developed in Markovnikov's time, their relative stabilities is the underlying basis for his rule; that is, the proton of H—X adds to the less substituted carbon of a double bond because this mode of addition produces the more stable carbocation intermediate.

Now that we know the order of stability of carbocations, how do we account for it? The principles of physics teach us that a system bearing a charge (either positive or negative) is more stable if the charge is delocalized. Using this principle, we can explain the order of stability of carbocations if we assume that alkyl groups bonded to a positively charged carbon release electrons toward the cationic carbon and thereby help delocalize the charge on the cation. The electron-releasing ability of alkyl groups bonded to a cationic carbon is accounted for by the **inductive effect** (Section 2.6C).

The inductive effect operates in the following way: The electron deficiency of the carbon atom bearing a positive charge exerts an electron-withdrawing inductive effect that polarizes electrons from adjacent sigma bonds toward it. Thus, the positive charge of the cation is not localized on the trivalent carbon, but rather is delocalized over nearby atoms. The larger the volume over which the positive charge is delocalized, the greater is the stability of the cation. Thus, as the number of alkyl groups bonded to the cationic carbon increases, the stability of the cation increases as well. Figure 5.5 illustrates the electron-withdrawing inductive effect of the positively charged carbon and the resulting delocalization of charge. According to quantum mechanical calculations, the charge on carbon in the methyl cation is approximately +0.645, and the charge on each of the hydrogen atoms is +0.118. Thus, even in the methyl cation, the positive charge is not localized on carbon. Rather, it is delocalized over the volume of space occupied by the entire ion. The polarization of electron density and the delocalization of charge are even more extensive in the *tert*-butyl cation.

Figure 5.5
Methyl and *tert*-butyl cations.
Delocalization of positive
charge by the electron-withdrawing inductive effect of
the trivalent, positively
charged carbon according to
molecular orbital calculations.

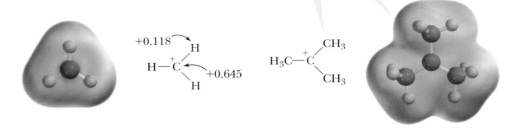

The methyl groups donate electron density
toward the carbocation carbon,
thus delocalizing the positive charge.

EXAMPLE 5.3

Arrange these carbocations in order of increasing stability:

(a)   (b)   (c)

SOLUTION

Carbocation (a) is secondary, (b) is tertiary, and (c) is primary. In order of increasing stability, they are $c < a < b$.

Practice Problem 5.3

Arrange these carbocations in order of increasing stability:

(a) $-CH_3$ (b) $-CH_3$

(c) $-CH_2$

EXAMPLE 5.4

Propose a mechanism for the addition of HI to methylenecyclohexane to give 1-iodo-1-methylcyclohexane:

$=CH_2 + HI \longrightarrow$

Methylenecyclohexane   1-Iodo-1-methylcyclohexane

Which step in your mechanism is rate determining?

SOLUTION

Propose a two-step mechanism similar to that proposed for the addition of HCl to propene.

Step 1: A rate-determining proton transfer from HI to the carbon–carbon double bond gives a 3° carbocation intermediate:

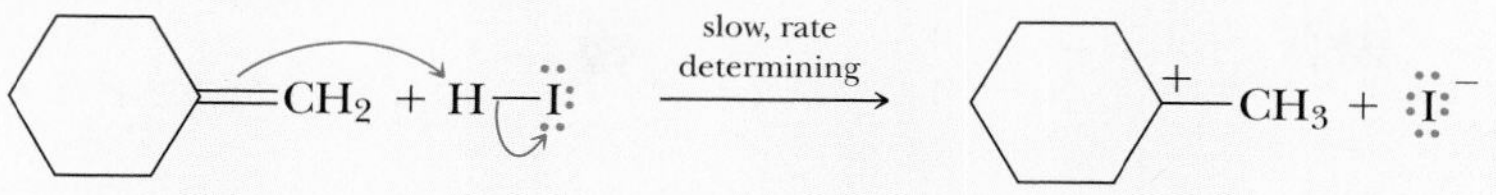

Methylenecyclohexane $\xrightarrow[\text{determining}]{\text{slow, rate}}$ A 3° carbocation intermediate

Step 2: Reaction of the 3° carbocation intermediate (a Lewis acid) with iodide ion (a Lewis base) completes the valence shell of carbon and gives the product:

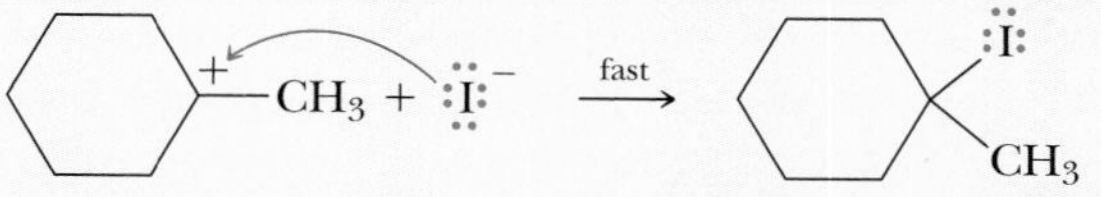

$\xrightarrow{\text{fast}}$

1-Iodo-1-methylcyclohexane

Practice Problem 5.4

Propose a mechanism for the addition of HI to 1-methylcyclohexene to give 1-iodo-1-methylcyclohexane. Which step in your mechanism is rate determining?

B. Addition of Water: Acid-Catalyzed Hydration

In the presence of an acid catalyst—most commonly, concentrated sulfuric acid—water adds to the carbon–carbon double bond of an alkene to give an alcohol. The addition of water is called **hydration**. In the case of simple alkenes, H adds to the carbon of the double bond with the greater number of hydrogens and OH adds to the carbon with the lesser number of hydrogens. Thus, H—OH adds to alkenes in accordance with Markovnikov's rule:

Hydration Addition of water.

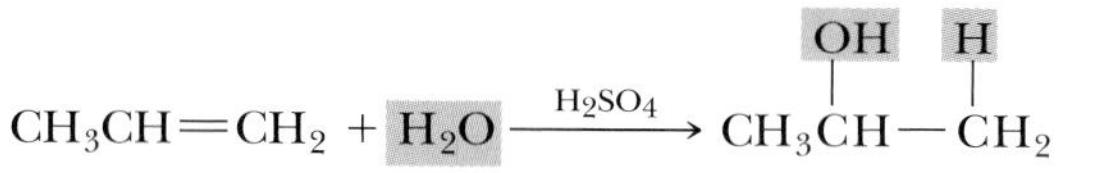

$$CH_3CH=CH_2 + H_2O \xrightarrow{H_2SO_4} CH_3CH-CH_2$$
$$\text{Propene} \qquad\qquad\qquad \text{2-Propanol}$$ 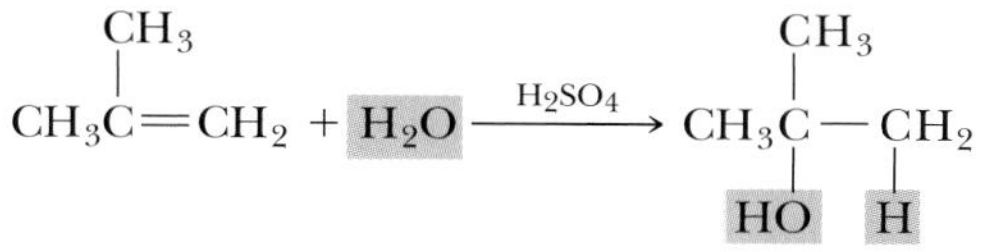
(OH H)

$$CH_3\underset{|}{\overset{CH_3}{C}}=CH_2 + H_2O \xrightarrow{H_2SO_4} CH_3\underset{|}{\overset{CH_3}{C}}-CH_2$$
$$\text{2-Methylpropene} \qquad\qquad \text{2-Methyl-2-propanol}$$
(HO H)

EXAMPLE 5.5

Draw a structural formula for the product of the acid-catalyzed hydration of 1-methylcyclohexene.

SOLUTION

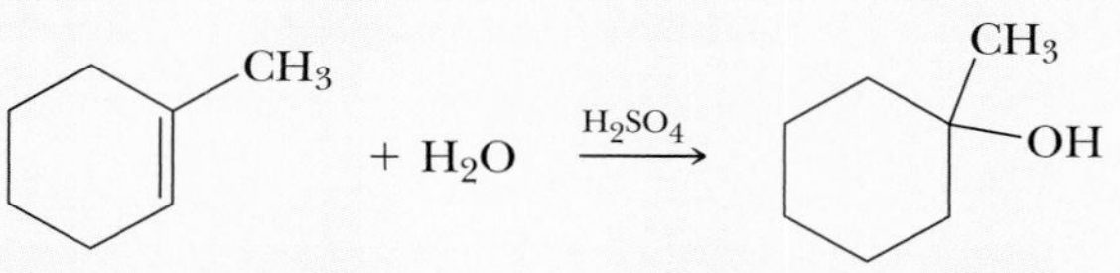

1-Methylcyclohexene $+ H_2O \xrightarrow{H_2SO_4}$ 1-Methylcyclohexanol

Practice Problem 5.5

Draw a structural formula for the product of each alkene hydration reaction:

(a)   [structure] + H_2O $\xrightarrow{H_2SO_4}$ (b)   [structure] + H_2O $\xrightarrow{H_2SO_4}$

     The mechanism for the acid-catalyzed hydration of alkenes is quite similar to what we have already proposed for the addition of HCl, HBr, and HI to alkenes and is illustrated by the hydration of propene to 2-propanol. This mechanism is consistent with the fact that acid is a catalyst. An H_3O^+ is consumed in Step 1, but another is generated in Step 3.

Mechanism: Acid-Catalyzed Hydration of Propene

Step 1: Proton transfer from the acid catalyst to propene gives a 2° carbocation intermediate (a Lewis acid):

$$CH_3CH{=}CH_2 + H{-}\overset{+}{\underset{H}{\ddot{O}}}{-}H \underset{\text{slow, rate determining}}{\rightleftharpoons} CH_3\overset{+}{C}HCH_3 + \underset{H}{\ddot{O}}{-}H$$

A 2° carbocation intermediate

Step 2: Reaction of the carbocation intermediate (a Lewis acid) with water (a Lewis base) completes the valence shell of carbon and gives an **oxonium ion**:

$$CH_3\overset{+}{C}HCH_3 + \underset{H}{\ddot{O}}{-}H \xrightarrow{\text{fast}} CH_3CHCH_3$$

$$\overset{\overset{+}{O}}{\underset{H\quad H}{}}$$

An oxonium ion

Oxonium ion An ion in which oxygen is bonded to three other atoms and bears a positive charge.

Step 3: Proton transfer from the oxonium ion to water gives the alcohol and generates a new molecule of the catalyst:

$$CH_3CHCH_3 + H{-}\ddot{O}{-}H \xrightarrow{\text{fast}} CH_3CHCH_3 + H{-}\overset{+}{\ddot{O}}{-}H$$

$$\overset{+}{O} \qquad\qquad\qquad \ddot{O}{:} \qquad\qquad H$$
$$H\quad H \qquad\qquad\qquad H$$

EXAMPLE 5.6

Propose a mechanism for the acid-catalyzed hydration of methylenecyclohexane to give 1-methylcyclohexanol. Which step in your mechanism is rate determining?

SOLUTION

Propose a three-step mechanism similar to that for the acid-catalyzed hydration of propene. The formation of the 3° carbocation intermediate in Step 1 is rate determining.

Step 1: Proton transfer from the acid catalyst to the alkene gives a 3° carbocation intermediate (a Lewis acid):

$$\text{[cyclohexane ring]}{=}CH_2 + H{-}\overset{+}{\ddot{O}}{-}H \underset{\text{slow, rate determining}}{\rightleftharpoons} \text{[cyclohexane ring]}\overset{+}{}{-}CH_3 + \ddot{O}{-}H$$

A 3° carbocation intermediate

Step 2: Reaction of the carbocation intermediate (a Lewis acid) with water (a Lewis base) completes the valence shell of carbon and gives an oxonium ion:

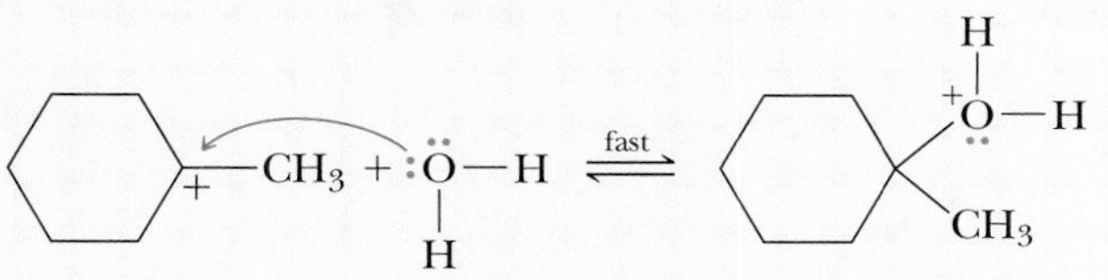

An oxonium ion

Step 3: Proton transfer from the oxonium ion to water gives the alcohol and generates a new molecule of the catalyst:

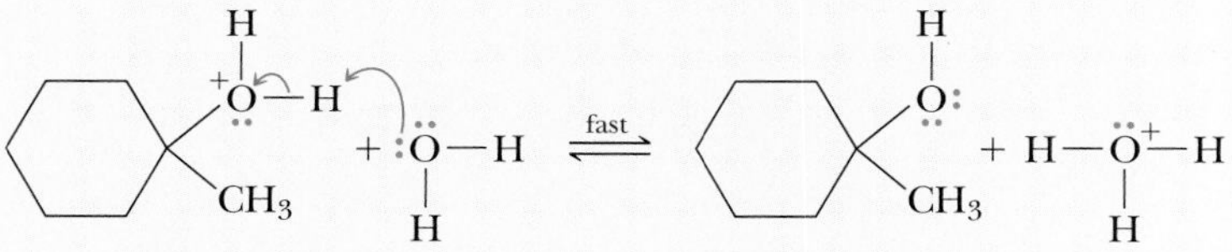

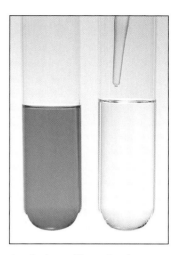

A solution of bromine in dichloromethane is red. Add a few drops of an alkene and the red color disappears. *(Charles D. Winters)*

Practice Problem 5.6

Propose a mechanism for the acid-catalyzed hydration of 1-methylcyclohexene to give 1-methylcyclohexanol. Which step in your mechanism is rate determining?

C. Addition of Bromine and Chlorine

Chlorine (Cl_2) and bromine (Br_2) react with alkenes at room temperature by the addition of halogen atoms to the two carbon atoms of the double bond, forming two new carbon–halogen bonds:

$$CH_3CH = CHCH_3 + Br_2 \xrightarrow{CH_2Cl_2} CH_3CH - CHCH_3$$

2-Butene   2,3-Dibromobutane

Fluorine, F_2, also adds to alkenes, but because its reactions are very fast and difficult to control, addition of fluorine is not a useful laboratory reaction. Iodine, I_2, also adds, but the reaction is not preparatively useful.

The addition of bromine and chlorine to a cycloalkene gives a trans dihalocycloalkane. For example, the addition of bromine to cyclohexene gives *trans*-1,2-dibromocyclohexane; the cis isomer is not formed. Thus, the addition of a halogen to a cycloalkene is stereoselective. A **stereoselective reaction** is a reaction in which one stereoisomer is formed or destroyed in preference to all others that might be formed or destroyed. Addition of bromine to a cycloalkene is highly stereoselective; the halogen atoms always add trans to each other:

Stereoselective reaction A reaction in which one stereoisomer is formed or destroyed in preference to all others that might be formed or destroyed.

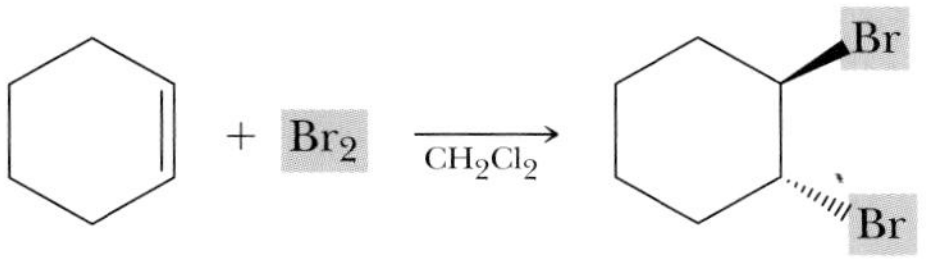

Cyclohexene   *trans*-1,2-Dibromocyclohexane

The reaction of bromine with an alkene is a particularly useful qualitative test for the presence of a carbon–carbon double bond. If we dissolve bromine in dichloromethane, the solution turns red. Both alkenes and dibromoalkanes are colorless. If we now mix a few drops of the bromine solution with an alkene, a dibromoalkane is formed, and the solution becomes colorless.

EXAMPLE 5.7

Complete these reactions, showing the relative orientation of the substituents in the product:

(a) [cyclopentene structure] + Br$_2$ $\xrightarrow{\text{CH}_2\text{Cl}_2}$

(b) [1-methylcyclohexene structure with CH$_3$] + Cl$_2$ $\xrightarrow{\text{CH}_2\text{Cl}_2}$

SOLUTION

The halogen atoms are trans to each other in each product:

(a) [cyclopentene structure] + Br$_2$ $\xrightarrow{\text{CH}_2\text{Cl}_2}$ [trans-1,2-dibromocyclopentane structure with Br and Br]

(b) [1-methylcyclohexene structure with CH$_3$] + Cl$_2$ $\xrightarrow{\text{CH}_2\text{Cl}_2}$ [cyclohexane product with CH$_3$, Cl, and Cl]

Practice Problem 5.7

Complete these reactions:

(a) $\underset{\underset{\text{CH}_3}{|}}{\overset{\overset{\text{CH}_3}{|}}{\text{CH}_3\text{CCH}}}{=}\text{CH}_2$ + Br$_2$ $\xrightarrow{\text{CH}_2\text{Cl}_2}$

(b) [methylenecyclohexane structure with CH$_2$] + Cl$_2$ $\xrightarrow{\text{CH}_2\text{Cl}_2}$

Anti Selectivity and Bridged Halonium Ion Intermediates

We explain the addition of bromine and chlorine to cycloalkenes, as well as their selectivity (they always add trans to each other), by a two-step mechanism that involves a halogen atom bearing a positive charge, called a **halonium ion**. The cyclic structure of which this ion is a part is called a **bridged halonium ion.** The bridged bromonium ion shown in the mechanism may look odd to you, but it is an acceptable Lewis structure. A calculation of formal charge places a positive charge on bromine. Then, in Step 2, a bromide ion reacts with the bridged intermediate from the side opposite that occupied by the bromine atom, giving the dibromoalkane. Thus, bromine atoms add from opposite faces of the carbon–carbon double bond. We say that this addition occurs with **anti selectivity**. Alternatively, we say that the addition of halogens is stereoselective involving anti addition of the halogen atoms.

Halonium ion An ion in which a halogen atom bears a positive charge.

Anti stereoselectivity Addition of atoms or groups of atoms from opposite sides or faces of a carbon–carbon double bond.

Mechanism: Addition of Bromine with Anti Selectivity

Step 1: Reaction of the pi electrons of the carbon–carbon double bond with bromine forms a bridged bromonium ion intermediate in which bromine bears a positive formal charge:

A bridged bromonium ion intermediate

Step 2: A bromide ion (a Lewis base) attacks carbon (a Lewis acid) from the side opposite the bridged bromonium ion, opening the three-membered ring:

Anti (coplanar) orientation of added bromine atoms   A Newman projection of the product

The addition of chlorine or bromine to cyclohexene and its derivatives gives a trans diaxial product because only axial positions on adjacent atoms of a cyclohexane ring are anti and coplanar. The initial trans diaxial conformation of the product is in equilibrium with the trans diequatorial conformation, and, in simple derivatives of cyclohexane, the latter is more stable and predominates.

trans Diaxial   trans Diequatorial (more stable)

5.4 OXIDATION OF ALKENES: FORMATION OF GLYCOLS

Recall from your course in general chemistry that oxidation and reduction can be defined in terms of the loss or gain of oxygens or hydrogens by a compound. For organic compounds, we define oxidation and reduction as follows:

oxidation: addition of O to, or removal of H from, a carbon atom

reduction: removal of O from, or addition of H to, a carbon atom

Osmium tetroxide, OsO_4, and certain other transition metal oxides are effective oxidizing agents for the conversion of an alkene to a **glycol**—a compound with two hydroxyl groups on adjacent carbons. The oxidation of an alkene by osmium tetroxide is stereoselective, involving **syn addition** (addition from the same side) of —OH groups to the carbons of the double bond. For example, the oxidation of cyclopentene by OsO_4 gives *cis*-1,2-cyclopentanediol, a cis glycol:

Glycol A compound with two hydroxyl (—OH) groups on adjacent carbons.

Syn addition Addition of atoms or groups of atoms from the same side or face of a carbon–carbon double bond.

A cyclic osmate   *cis*-1,2-Cyclopentanediol (a cis glycol)

Note that both cis and trans isomers are possible for this glycol, but only the cis isomer is formed.

The syn selectivity of the osmium tetroxide oxidation of alkenes is accounted for by the formation of a cyclic osmate in which oxygen atoms of OsO_4 form new covalent bonds with each carbon of the double bond in such a way that the five-membered osmium-containing ring fuses cis to the original alkene. Osmates can be isolated and characterized. Usually, however, they are treated directly with a reducing agent, such as $NaHSO_3$, which cleaves osmium–oxygen bonds to give a cis glycol and reduced forms of osmium.

The drawbacks of OsO_4 are that it is both expensive and highly toxic. One strategy to circumvent the high cost of OsO_4 is to use it in catalytic amounts along with stoichiometric amounts of another oxidizing agent, which reoxidizes the reduced forms of osmium and thus recycles the osmium reagent. Oxidizing agents commonly used for this purpose are hydrogen peroxide (H_2O_2) and *tert*-butyl hydroperoxide ((CH_3)$_3$COOH). When this procedure is used, there is no need for a reducing step using $NaHSO_3$.

5.5 REDUCTION OF ALKENES: FORMATION OF ALKANES

A. Catalytic Reduction

Most alkenes react quantitatively with molecular hydrogen, H_2, in the presence of a transition metal catalyst to give alkanes. Commonly used transition metal catalysts include platinum, palladium, ruthenium, and nickel. Yields are usually quantitative or nearly so. Because the conversion of an alkene to an alkane involves reduction by hydrogen in the presence of a catalyst, the process is called **catalytic reduction** or, alternatively, **catalytic hydrogenation**.

Cyclohexene + H_2 $\xrightarrow[\text{25°C, 3 atm}]{\text{Pd}}$ Cyclohexane

The metal catalyst is used as a finely powdered solid, which may be supported on some inert material such as powdered charcoal or alumina. The reaction is carried out by dissolving the alkene in ethanol or another nonreacting organic solvent, adding the solid catalyst, and exposing the mixture to hydrogen gas at pressures from 1 to 100 atm. Alternatively, the metal may be chelated with certain organic molecules and used in the form of a soluble complex.

Catalytic reduction is stereoselective, the most common pattern being the **syn addition** of hydrogens to the carbon–carbon double bond. The catalytic reduction of 1,2-dimethylcyclohexene, for example, yields *cis*-1,2-dimethylcyclohexane:

1,2-Dimethylcyclohexene + H_2 $\xrightarrow{\text{Pt}}$ *cis*-1,2-Dimethylcyclohexane

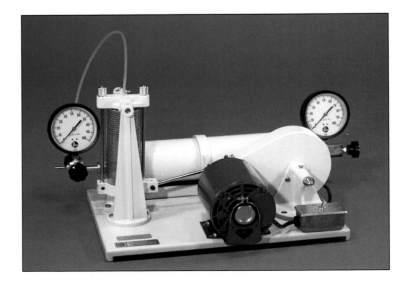

A Paar shaker-type hydrogenation apparatus. *(Paar Instrument Co., Moline, IL.)*

(a) metal surface

Figure 5.6
Syn addition of hydrogen to an alkene involving a transition metal catalyst. (a) Hydrogen and the alkene are adsorbed on the metal surface, and (b) one hydrogen atom is transferred to the alkene, forming a new C—H bond. The other carbon remains adsorbed on the metal surface. (c) A second C—H bond forms and the alkane is desorbed.

The transition metals used in catalytic reduction are able to adsorb large quantities of hydrogen onto their surfaces, probably by forming metal–hydrogen sigma bonds. Similarly, these transition metals adsorb alkenes on their surfaces, with the formation of carbon–metal bonds [Figure 5.6(a)]. Hydrogen atoms are added to the alkene in two steps.

B. Heats of Hydrogenation and the Relative Stabilities of Alkenes

The **heat of hydrogenation** of an alkene is defined as its heat of reaction, $\Delta H°$, with hydrogen, to form an alkane. Table 5.2 lists the heats of hydrogenation of several alkenes.

Three important points follow from the information given in the table:

TABLE 5.2 Heats of Hydrogenation of Several Alkenes

Name	Structural Formula	$\Delta H°$ (kcal/mol)	
Ethylene	$CH_2{=}CH_2$	$-137\ (-32.8)$	
Propene	$CH_3CH{=}CH_2$	$-126\ (-30.1)$	Ethylene
1-Butene	$CH_3CH_2CH{=}CH_2$	$-127\ (-30.3)$	
cis-2-Butene	H₃C, CH₃ C=C H, H	$-120\ (-28.6)$	
trans-2-Butene	H₃C, H C=C H, CH₃	$-115\ (-27.6)$	trans-2-Butene
2-Methyl-2-butene	H₃C, CH₃ C=C H₃C, H	$-113\ (-26.9)$	
2,3-Dimethyl-2-butene	H₃C, CH₃ C=C H₃C, CH₃	$-111\ (-26.6)$	2,3-Dimethyl-2-butene

Figure 5.7

Heats of hydrogenation of *cis*-2-butene and *trans*-2-butene. *Trans*-2-Butene is more stable than *cis*-2-butene by 1.0 kcal/mol (4.2 kJ/mol).

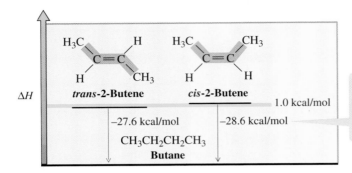

A higher heat of hydrogenation means that more heat is released and indicates that the cis alkene starts at a higher (less stable) energy level.

1. The reduction of an alkene to an alkane is an exothermic process. This observation is consistent with the fact that, during hydrogenation, there is net conversion of a weaker pi bond to a stronger sigma bond; that is, one sigma bond (H—H) and one pi bond (C=C) are broken, and two new sigma bonds (C—H) are formed.

2. The heats of hydrogenation depend on the degree of substitution of the carbon–carbon double bond: The greater the substitution, the lower is the heat of hydrogenation. Compare, for example, the heats of hydrogenation of ethylene (no substituents), propene (one substituent), 1-butene (one substituent), and the cis and trans isomers of 2-butene (two substituents each).

3. The heat of hydrogenation of a trans alkene is lower than that of the isomeric cis alkene. Compare, for example, the heats of hydrogenation of *cis*-2-butene and *trans*-2-butene. Because the reduction of each alkene gives butane, any difference in their heats of hydrogenation must be due to a difference in relative energy between the two alkenes (Figure 5.7). The alkene with the lower (less negative) value of $\Delta H°$ is the more stable alkene.

We explain the greater stability of trans alkenes relative to cis alkenes in terms of nonbonded interaction strain. In *cis*-2-butene, the two —CH$_3$ groups are sufficiently close to each other that there is repulsion between their electron clouds. This repulsion is reflected in the larger heat of hydrogenation (decreased stability) of *cis*-2-butene compared with that of *trans*-2-butene (approximately 1.0 kcal/mol).

Summary

A **reaction mechanism** (Section 5.2) is a description of (1) how and why a chemical reaction occurs, (2) which bonds break and which new ones form, (3) the order and relative rates with which the various bond-breaking and bond-forming steps take place, and (4) the role of the catalyst if the reaction involves one. **Transition state theory** (Section 5.2A) provides a model for understanding the relationships among reaction rates, molecular structure, and energetics. A key postulate of transition state theory is that a **transition state** is formed. The difference in energy between the reactants and the transition state is called the **activation energy**. An **intermediate** is an energy minimum between two transition states. The slowest step in a multistep reaction, called the **rate-determining step**, is the one that crosses the highest energy barrier.

A characteristic reaction of alkenes is **addition**, during which a pi bond is broken and sigma bonds to two new atoms or groups of atoms are formed.

An **electrophile** (Section 5.3A) is any molecule or ion that can accept a pair of electrons to form a new covalent bond. All electrophiles are Lewis acids. The rate-determining step in **electrophilic addition** to an alkene is the reaction of an electrophile with a carbon–carbon double bond to form a **carbocation**—an ion that contains a carbon with only six electrons in its valence shell and has a positive charge. Carbocations are planar, with bond angles of 120° about the positive carbon. The order of stability of carbocations is $3° > 2° > 1° > $ methyl (Section 5.3A).

KEY REACTIONS

1. Addition of H—X (Section 5.3A)
Addition of H—X is regioselective and follows Markovnikov's rule. Reaction occurs in two steps and involves the formation of a carbocation intermediate:

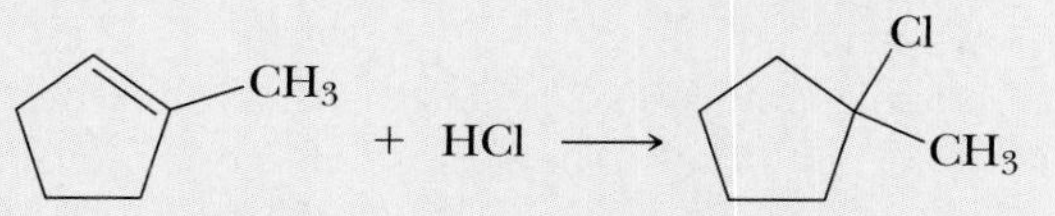

2. Acid-Catalyzed Hydration (Section 5.3B)
Hydration is regioselective and follows Markovnikov's rule. Reaction occurs in two steps and involves the formation of a carbocation intermediate:

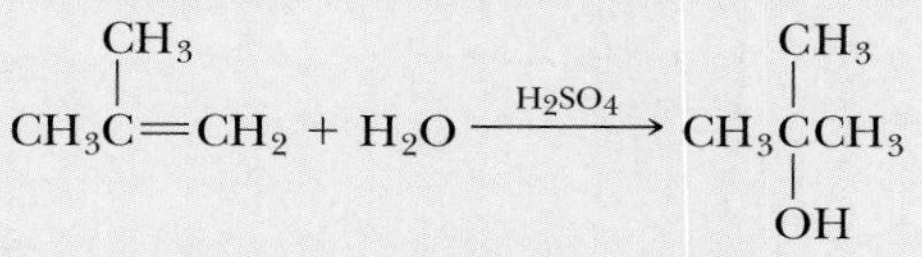

3. Addition of Bromine and Chlorine (Section 5.3C)
Addition occurs in two steps, and involves anti addition by way of a bridged bromonium or chloronium ion intermediate:

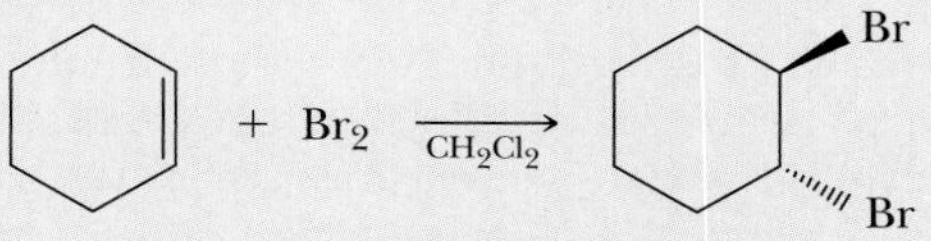

4. Oxidation: Formation of Glycols (Section 5.4B)
Oxidation occurs by the syn addition of —OH groups to the double bond via a cyclic osmate:

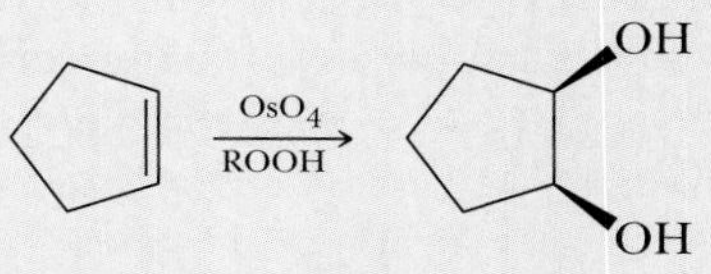

5. Reduction: Formation of Alkanes (Section 5.5)
Catalytic reduction involves predominantly the syn addition of hydrogen:

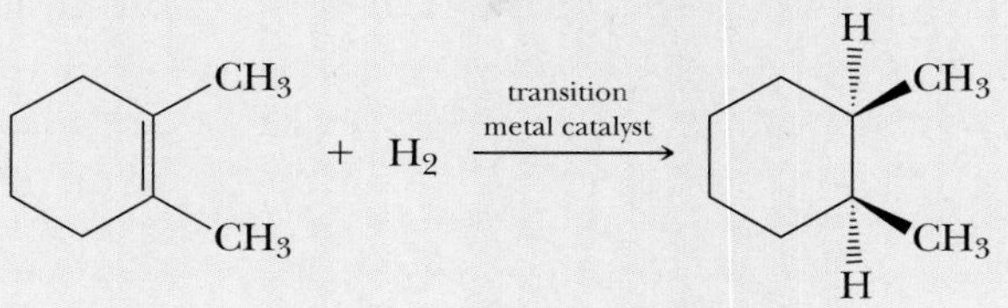

PROBLEMS

A problem number set in red indicates an applied "real-world" problem.

Energy Diagrams

5.8 Describe the differences between a transition state and a reaction intermediate.

5.9 Sketch an energy diagram for a one-step reaction that is very slow and only slightly exothermic. How many transition states are present in this reaction? How many intermediates are present?

5.10 Sketch an energy diagram for a two-step reaction that is endothermic in the first step, exothermic in the second step, and exothermic overall. How many transition states are present in this two-step reaction? How many intermediates are present?

5.11 Determine whether each of the following statements is true or false, and provide a rationale for your decision:

  (a) A transition state can never be lower in energy than the reactants from which it was formed.

  (b) An endothermic reaction cannot have more than one intermediate.

  (c) An exothermic reaction cannot have more than one intermediate.

Electrophilic Additions

5.12 From each pair, select the more stable carbocation:

  (a) $CH_3CH_2CH_2^+$ or $CH_3\overset{+}{C}HCH_3$ **(b)** $CH_3\overset{\underset{CH_3}{|}}{C}H\overset{+}{C}HCH_3$ or $CH_3\overset{\underset{CH_3}{|}}{\overset{+}{C}}CH_2CH_3$

5.13 From each pair, select the more stable carbocation:

(a) 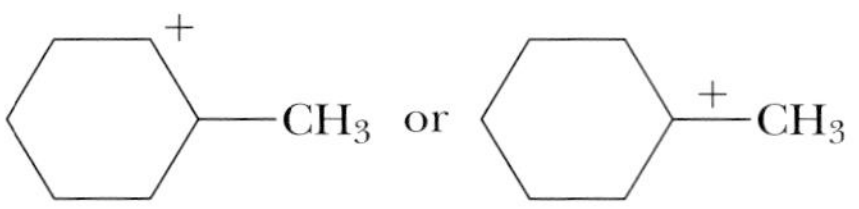 or   **(b)** 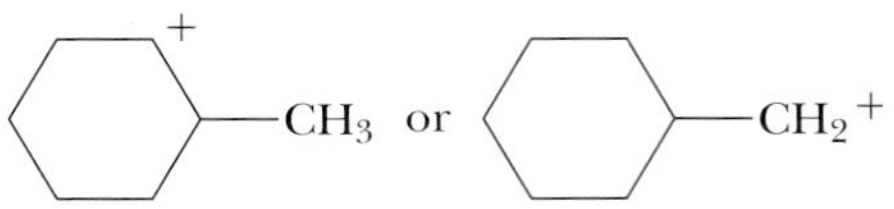 or

5.14 Draw structural formulas for the isomeric carbocation intermediates formed by the reaction of each alkene with HCl. Label each carbocation as primary, secondary, or tertiary, and state which, if either, of the isomeric carbocations is formed more readily.

(a) 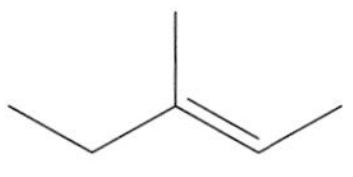   **(b)** 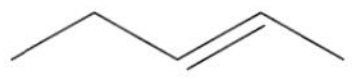   **(c)** 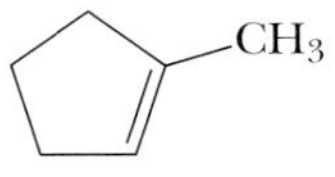   **(d)**

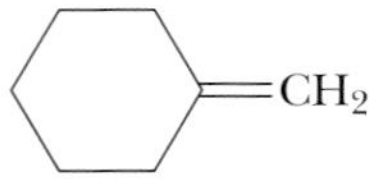

5.15 From each pair of compounds, select the one that reacts more rapidly with HI, draw the structural formula of the major product formed in each case, and explain the basis for your ranking:

(a) 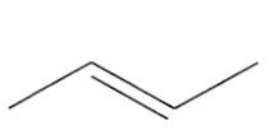 and 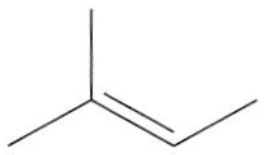   **(b)** 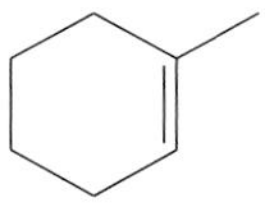 and

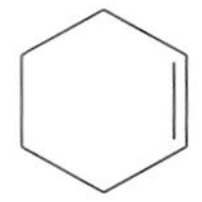

5.16 Complete these equations by predicting the major product formed in each reaction:

(a) 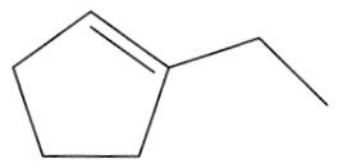 + HCl $\longrightarrow$ **(b)** 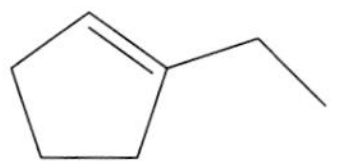 + H_2O $\xrightarrow{H_2SO_4}$

(c)  + HI $\longrightarrow$ **(d)** 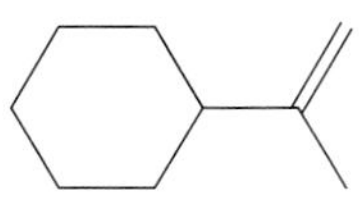 + HCl $\longrightarrow$

(e) 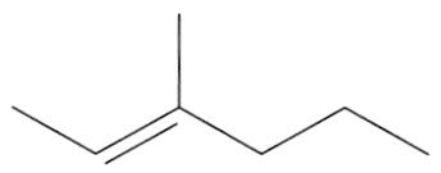 + H_2O $\xrightarrow{H_2SO_4}$ **(f)**  + H_2O $\xrightarrow{H_2SO_4}$

5.17 The reaction of 2-methyl-2-pentene with each reagent is regioselective. Draw a structural formula for the product of each reaction, and account for the observed regioselectivity.

  (a) HI   **(b)** H_2O in the presence of H_2SO_4

5.18 The addition of bromine and chlorine to cycloalkenes is stereoselective. Predict the stereochemistry of the product formed in each reaction:

  (a) 1-Methylcyclohexene + Br_2 **(b)** 1,2-Dimethylcyclopentene + Cl_2

5.19 Draw a structural formula for an alkene with the indicated molecular formula that gives the compound shown as the major product. Note that more than one alkene may give the same compound as the major product.

(a) $C_5H_{10} + H_2O \xrightarrow{H_2SO_4}$ 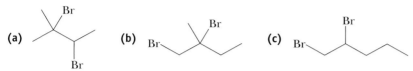   **(b)** $C_5H_{10} + Br_2 \longrightarrow$

(c) $C_7H_{12} + HCl \longrightarrow$

5.20 Draw the structural formula for an alkene with molecular formula C_5H_{10} that reacts with Br_2 to give each product:

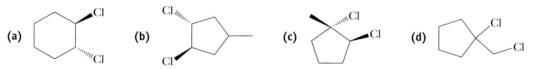

(a)   **(b)**   **(c)**

5.21 Draw the structural formula for a cycloalkene with molecular formula C_6H_{10} that reacts with Cl_2 to give each compound:

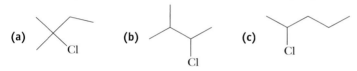

(a)   **(b)**   **(c)**   **(d)**

5.22 Draw the structural formula for an alkene with molecular formula C_5H_{10} that reacts with HCl to give the indicated chloroalkane as the major product:

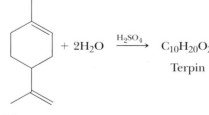

(a)   **(b)**   **(c)**

5.23 Draw the structural formula of an alkene that undergoes acid-catalyzed hydration to give the indicated alcohol as the major product. More than one alkene may give each compound as the major product.
(a) 3-Hexanol   **(b)** 1-Methylcyclobutanol
(c) 2-Methyl-2-butanol   **(d)** 2-Propanol

5.24 Draw the structural formula of an alkene that undergoes acid-catalyzed hydration to give each alcohol as the major product. More than one alkene may give each compound as the major product.
(a) Cyclohexanol   **(b)** 1,2-Dimethylcyclopentanol
(c) 1-Methylcyclohexanol   **(d)** 1-Isopropyl-4-methylcyclohexanol

5.25 Terpin is prepared commercially by the acid-catalyzed hydration of limonene:

$+ 2H_2O \xrightarrow{H_2SO_4} C_{10}H_{20}O_2$

Terpin

Limonene

(a) Propose a structural formula for terpin and a mechanism for its formation.
(b) How many cis–trans isomers are possible for the structural formula you propose?
(c) Terpin hydrate, the isomer in terpin in which the one-carbon and three-carbon substituents are cis to each other, is used as an expectorant in cough medicines. Draw the alternative chair conformations for terpin hydrate, and state which of the two is the more stable.

5.26 The treatment of 2-methylpropene with methanol in the presence of a sulfuric acid catalyst gives *tert*-butyl methyl ether:

$$CH_3C{=}CH_2 \;+\; CH_3OH \xrightarrow{H_2SO_4} CH_3C{-}OCH_3$$

$$\underset{\text{2-Methylpropene}}{\overset{CH_3}{|}} \quad \underset{\text{Methanol}}{} \quad \underset{\textit{tert}\text{-Butyl methyl ether}}{\overset{CH_3}{|}\,\underset{CH_3}{|}}$$

Propose a mechanism for the formation of this ether.

5.27 The treatment of 1-methylcyclohexene with methanol in the presence of a sulfuric acid catalyst gives a compound with molecular formula $C_8H_{16}O$:

$$\text{1-Methylcyclohexene} \;+\; CH_3OH \xrightarrow{H_2SO_4} C_8H_{16}O$$

Propose a structural formula for this compound and a mechanism for its formation.

5.28 *cis*-3-Hexene and *trans*-3-hexene are different compounds and have different physical and chemical properties. Yet, when treated with H_2O/H_2SO_4, each gives the same alcohol. What is the alcohol, and how do you account for the fact that each alkene gives the same one?

Oxidation–Reduction

5.29 Which of these transformations involve oxidation, which involve reduction, and which involve neither oxidation nor reduction?

(a) $CH_3\overset{OH}{\overset{|}{C}}HCH_3 \longrightarrow CH_3\overset{O}{\overset{\|}{C}}CH_3$ **(b)** $CH_3\overset{OH}{\overset{|}{C}}HCH_3 \longrightarrow CH_3CH{=}CH_2$

(c) $CH_3CH{=}CH_2 \longrightarrow CH_3CH_2CH_3$

5.30 Write a balanced equation for the combustion of 2-methylpropene in air to give carbon dioxide and water. The oxidizing agent is O_2, which makes up approximately 20% of air.

5.31 Draw the product formed by treating each alkene with aqueous $OsO_4/ROOH$:
(a) 1-Methylcyclopentene   **(b)** 1-Cyclohexylethylene   **(c)** *cis*-2-Pentene

5.32 What alkene, when treated with $OsO_4/ROOH$, gives each glycol?

(a)   **(b)**   **(c)**

5.33 Draw the product formed by treating each alkene with H_2/Ni:

(a)   **(b)**   **(c)**   **(d)**

5.34 Hydrocarbon A, C_5H_8, reacts with 2 moles of Br_2 to give 1,2,3,4-tetrabromo-2-methylbutane. What is the structure of hydrocarbon A?

5.35 Two alkenes, A and B, each have the formula C_5H_{10}. Both react with H_2/Pt and with HBr to give identical products. What are the structures of A and B?

Synthesis

5.36 Show how to convert ethylene into these compounds:

    (a) Ethane   **(b)** Ethanol   **(c)** Bromoethane

    (d) 1,2-Dibromoethane   **(e)** 1,2-Ethanediol   **(f)** Chloroethane

5.37 Show how to convert cyclopentene into these compounds:

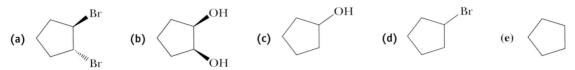

5.38 Show how to convert 1-butene into these compounds:

    (a) Butane   **(b)** 2-Butanol

    (c) 2-Bromobutane   **(d)** 1,2-Dibromobutane

5.39 Show how the following compounds can be synthesized in good yields from an alkene:

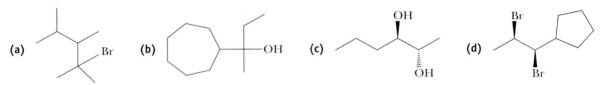

Looking Ahead

5.40 Each of the following 2° carbocations is more stable than the tertiary carbocation shown:

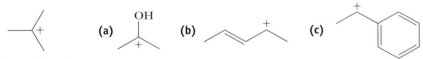

a tertiary carbocation

Provide an explanation for each cation's enhanced stability.

5.41 Recall that an alkene possesses a π cloud of electrons above and below the plane of the C=C bond. Any reagent can therefore react with either face of the double bond. Determine whether the reaction of each of the given reagents with the top face of *cis*-2-butene will produce the same product as the reaction of the same reagent with the bottom face. (*Hint:* Build molecular models of the products and compare them.)

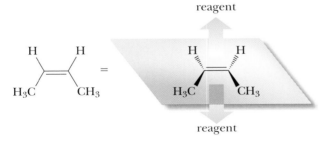

    (a) H_2/Pt **(b)** $OsO_4/ROOH$ **(c)** Br_2/CH_2Cl_2

5.42 Each of the following reactions yields two products in differing amounts:

Draw the products of each reaction and determine which product is favored.

6 Chirality: The Handedness of Molecules

Tartaric acid is found in grapes and other fruits, both free and as its salts (see Section 6.5B). Inset: A model of tartaric acid. *(Pierre-Louis Martin/Photo Researchers, Inc.)*

6.1 INTRODUCTION

Mirror image The reflection of an object in a mirror.

In this chapter, we will explore the relationships between three-dimensional objects and their mirror images. When you look in a mirror, you see a reflection, or **mirror image**, of yourself. Now, suppose your mirror image becomes a three-dimensional object. We could then ask, "What is the relationship between you and your mirror image?" By relationship, we mean "Can your reflection be superposed on the original 'you' in such a way that every detail of the reflection corresponds exactly to the original?" The answer is that you and your mirror image are not superposable. If you have a ring on the little finger of your right hand, for example, your mirror image has the ring on the little finger of its left hand. If you part your hair on your right side, the part will be on the left side in your mirror image. Simply stated, you and your reflection are different objects. You cannot superpose one on the other.

An understanding of relationships of this type is fundamental to an understanding of organic chemistry and biochemistry. In fact, the ability to deal with molecules as three-dimensional objects is a survival skill in organic chemistry and biochemistry.

6.2 STEREOISOMERS

Stereoisomers have the same molecular formula and the same order of attachment of atoms in their molecules, but different three-dimensional orientations of their atoms in space. The one example of stereoisomers we have seen thus far is that of cis–trans isomers in cycloalkanes (Section 3.8) and alkenes (Section 4.2C):

The horns of this African gazelle show chirality and are mirror images of each other. *(William H. Brown)*

cis-1,2-Dimethyl-
cyclohexane   and   trans-1,2-Dimethyl-
cyclohexane   cis-2-Butene   trans-2-Butene

Stereoisomers Isomers that have the same molecular formula and the same connectivity, but different orientations of their atoms in space.

In this chapter, we study enantiomers and diastereomers (Figure 6.1).

6.3 ENANTIOMERISM

A. What Are Enantiomers?

Enantiomers are stereoisomers that are nonsuperposable mirror images of each other. The significance of enantiomerism is that, except for inorganic and a few simple organic compounds, the vast majority of molecules in the biological world show this type of isomerism, including carbohydrates (Chapter 18), lipids (Chapter 21), amino acids and proteins (Chapter 19), and nucleic acids (DNA and RNA, Chapter 20). Further, approximately one-half of the medications used in human medicine also show this type of isomerism.

As an example of a molecule that exhibits enantiomerism, let us consider 2-butanol. As we go through the discussion of this molecule, we focus on carbon 2,

Enantiomers Stereoisomers that are nonsuperposable mirror images; the term refers to a relationship between pairs of objects.

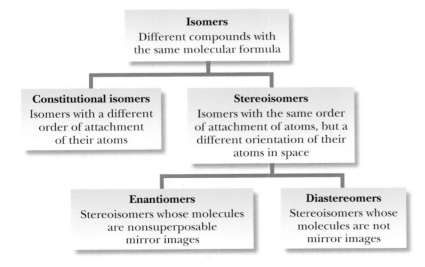

Figure 6.1
Relationships among isomers.

the carbon bearing the —OH group. What makes this carbon of interest is that it has four different groups bonded to it. The most common cause of enantiomerism among organic molecules is a carbon with four different groups bonded to it.

$$\underset{\text{2-Butanol}}{\overset{\overset{\displaystyle OH}{|}}{CH_3CHCH_2CH_3}}$$

The structural formula we have just drawn does not show the shape of 2-butanol or the orientation of its atoms in space. To do this, we must consider the molecule as a three-dimensional object. On the left is what we will call the "original molecule" and a ball-and-stick model of it. In this drawing, the —OH and —CH₃ groups on carbon-2 are in the plane of the paper, the —H is behind the plane and the —CH₂CH₃ group is in front of it:

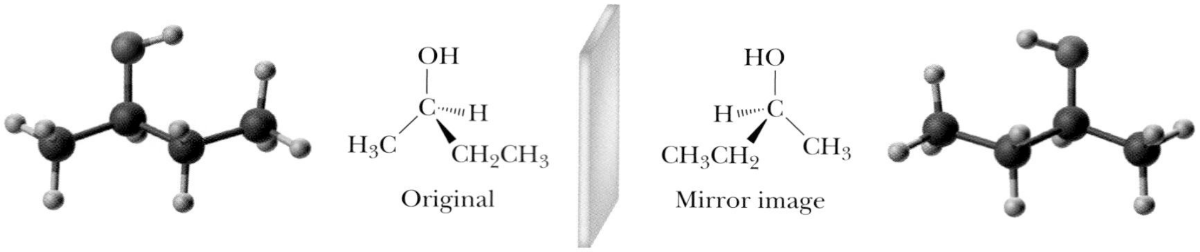

Original   Mirror image

To the right in the preceding diagram is the mirror image of the original molecule. Every molecule and, in fact, every object in the world around us, has a mirror image. The question we now need to ask is "What is the relationship between the original of 2-butanol and its mirror image?" To answer this question, you need to imagine that you can pick up the mirror image and move it in space in any way you wish. If you can move the mirror image in space and find that it fits over the original so that every bond, atom, and detail of the mirror image exactly matches the bonds, atoms, and details of the original, then the two are **superposable**. In this case, the mirror image and the original represent the same molecule; they are only oriented differently in space. If, however, no matter how you turn the mirror image in space, it will not fit exactly on the original with every detail matching, then the two are **nonsuperposable**; they are different molecules.

The key point here is that, either an object is superposable on its mirror image or it isn't. Now let us look at 2-butanol and its mirror image; are they or are they not superposable?

The following drawings illustrate one way to see that the mirror image of 2-butanol is not superposable on the original molecule:

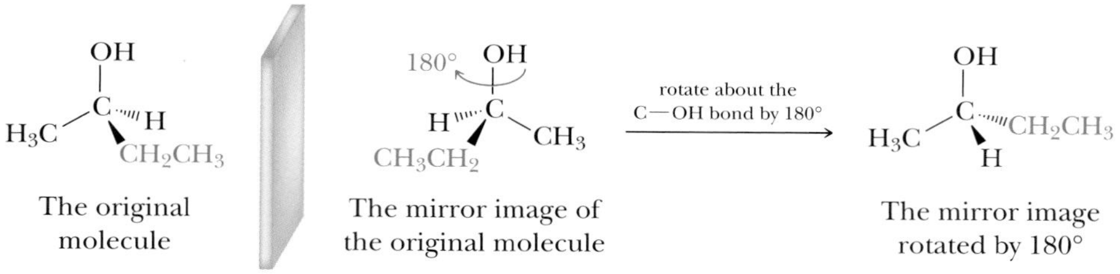

The original   The mirror image of   The mirror image
molecule   the original molecule   rotated by 180°

Imagine that you hold the mirror image by the C—OH bond and rotate the bottom part of the molecule by 180° about this bond. The —OH group retains its position in space, but the —CH₃ group, which was to the right and in the plane of the paper, is still in the plane of the paper, but now to the left. Similarly, the —CH₂CH₃ group, which was in front of the plane of the paper and to the left, is now behind the plane and to the right.

Now move the rotated mirror image in space, and try to fit it on the original so that all bonds and atoms match:

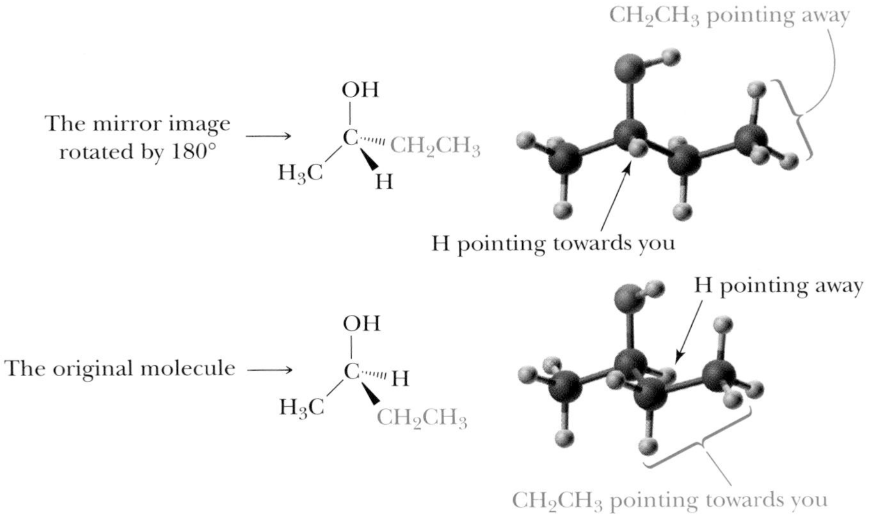

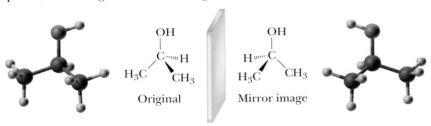

By rotating the mirror image as we did, its —OH and —CH₃ groups now fit exactly on top of the —OH and —CH₃ groups of the original. But the —H and —CH₂CH₃ groups of the two do not match: The —H is away from you in the original, but toward you in the mirror image; the —CH₂CH₃ group is toward you in the original, but away from you in the mirror image. We conclude that the original of 2-butanol and its mirror image are nonsuperposable and, therefore, are different compounds.

To summarize, we can rotate the mirror image of 2-butanol in space in any way we want, but as long as no bonds are broken or rearranged, only two of the four groups bonded to carbon-2 of the mirror image can be made to coincide with those on the original. Because 2-butanol and its mirror image are not superposable, they are enantiomers. Like gloves, enantiomers always occur in pairs.

Objects that are not superposable on their mirror images are said to be **chiral** (pronounced ki′-ral, rhymes with spiral; from the Greek: *cheir*, hand); that is, they show handedness. Chirality is encountered in three-dimensional objects of all sorts. Your left hand is chiral, and so is your right hand. A spiral binding on a notebook is chiral. A machine screw with a right-handed twist is chiral. A ship's propeller is chiral. As you examine the objects in the world around you, you will undoubtedly conclude that the vast majority of them are chiral.

As we said before we examined the original and the mirror image of 2-butanol, the most common cause of enantiomerism in organic molecules is the presence of a carbon with four different groups bonded to it. Let us examine this statement further by considering a molecule such as 2-propanol, which has no such carbon. In this molecule, carbon-2 is bonded to three different groups, but no carbon is bonded to four different groups. The question we ask is "Is the mirror image of 2-propanol superposable on the original, or isn't it?"

In the following diagram, on the left is a three-dimensional representation of 2-propanol, on the right its mirror image:

Left- and right-handed sea shells. If you cup a right-handed shell in your right hand with your thumb pointing from the narrow end to the wide end, the opening will be on your right.
(*Charles D. Winters*)

Chiral From the Greek *cheir*, meaning hand; objects that are not superposable on their mirror images.

Figure 6.2
Planes of symmetry in (a) a beaker, (b) a cube, and (c) 2-propanol. The beaker and 2-propanol each have one plane of symmetry; the cube has several planes of symmetry, only three of which are shown in the figure.

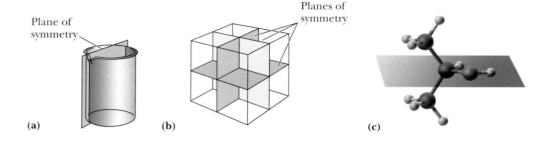

The question we now ask is "What is the relationship of the mirror image to the original?" This time, let us rotate the mirror image by 120° about the C—OH bond and then compare it with the original. When we do this rotation, we see that all atoms and bonds of the mirror image fit exactly on the original. This means that the structures we first drew for the original and its mirror image are, in fact, the same molecule viewed from different perspectives:

If an object and its mirror image are superposable, then the object and its mirror image are identical, and there is no possibility of enantiomerism. We say that such an object is **achiral** (without chirality).

An achiral object has at least one plane of symmetry. A **plane of symmetry** (also called a *mirror plane*) is an imaginary plane passing through an object and dividing it so that one-half of the object is the reflection of the other half. The beaker shown in Figure 6.2 has a single plane of symmetry, whereas a cube has several planes of symmetry. 2-Propanol also has a single plane of symmetry.

To repeat, the most common cause of chirality in organic molecules is a tetrahedral carbon atom with four different groups bonded to it. We call such a carbon atom a **stereocenter**. 2-Butanol has one stereocenter; 2-propanol has none.

As another example of a molecule with a stereocenter, consider 2-hydroxypropanoic acid, more commonly named lactic acid. Lactic acid is a product of anaerobic glycolysis and is what gives sour cream its sour taste. Figure 6.3 shows three-dimensional representations of lactic acid and its mirror image. In these representations, all bond angles about the central carbon atom are approximately 109.5°, and the four bonds projecting from it are directed toward the corners of a regular tetrahedron. Lactic acid shows enantiomerism; that is, it and its mirror image are not superposable, but rather are different molecules.

Achiral An object that lacks chirality; an object that has no handedness.

Plane of symmetry An imaginary plane passing through an object and dividing it such that one half is the mirror image of the other half.

Stereocenter An atom that has four different groups bonded to it.

Figure 6.3
Three-dimensional representations of lactic acid and its mirror image.

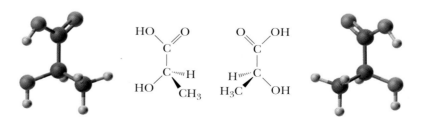

B. Drawing Enantiomers

Now that we know what enantiomers are, we can think about how to represent their three-dimensional structures on a two-dimensional page. Let us take one of the enantiomers of 2-butanol as an example. Following are four different representations of this enantiomer:

(1)   (2)   (3)   (4)

In our initial discussions of 2-butanol, we used (1) to show the tetrahedral geometry of the stereocenter; in it, two groups are in the plane of the paper, one is coming out of the plane towards us, and one is behind the plane, away from us. We can turn (1) slightly in space and tip it a bit to place the carbon framework in the plane of the paper. Doing so gives us representation (2), in which we still have two groups in the plane of the paper, one coming towards us and one going away from us. For an even more abbreviated representation of this enantiomer of 2-butanol, we can turn (2) into the line-angle formula (3). Although we don't normally show hydrogens in a line-angle formula, we do in (3) just to remind ourselves that the fourth group on this stereocenter is really there and that it is H. Finally, we can carry the abbreviation a step further and write 2-butanol as (4). Here, we omit the H on the stereocenter, but we know that it must be there (carbon needs four bonds), and we know that it must be behind the plane of the paper. Clearly, the abbreviated formulas (3) and (4) are the easiest to write, and we will rely on these representations throughout the remainder of the text. When you have to write three-dimensional representations of stereocenters, try to keep the carbon framework in the plane of the paper and the other two atoms or groups of atoms on the stereocenter toward and away from you, respectively. Using representation (4) as a model, we get the following two different representations of its enantiomer:

One enantiomer   Alternative representations
of 2-butanol   for its mirror image

Notice that in the second alternative, the carbon skeleton has been reversed.

EXAMPLE 6.1

Each of the following molecules has one stereocenter:

(a) CH₃CHCH₂CH₃   (b)

Draw stereorepresentations of the enantiomers of each.

SOLUTION

You will find it helpful to study the models of each pair of enantiomers and to view them from different perspectives. As you work with these models, notice that each enantiomer has a carbon atom bonded to four different groups, which makes the molecule chiral. The hydrogen at the stereocenter is shown in (a), but not in (b).

(a) 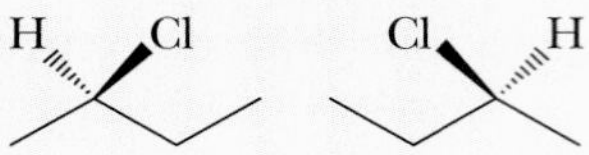   (b)

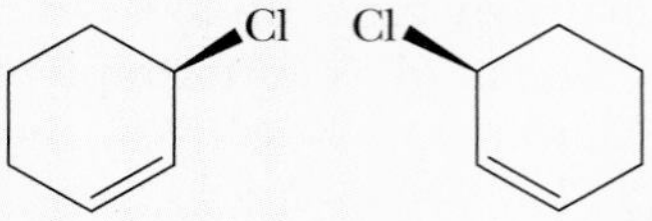

Practice Problem 6.1

Each of the following molecules has one stereocenter:

(a) 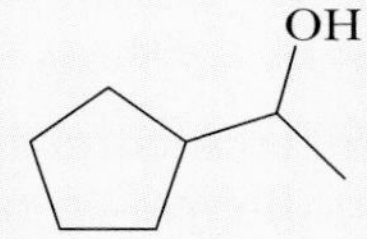   (b)

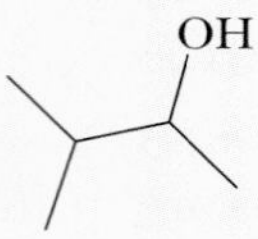

Draw stereorepresentations of the enantiomers of each.

6.4 NAMING STEREOCENTERS: THE (R,S) SYSTEM

R,S system A set of rules for specifying the configuration about a stereocenter.

Because enantiomers are different compounds, each must have a different name. The over-the-counter drug ibuprofen, for example, shows enantiomerism and can exist as the pair of enantiomers shown here:

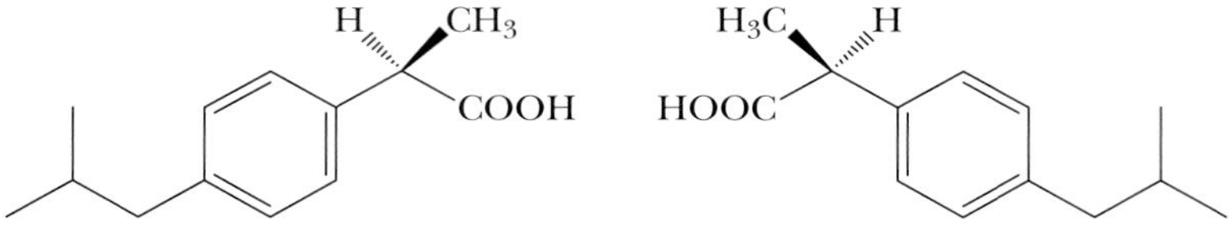

The inactive enantiomer of ibuprofen   The active enantiomer

Only one enantiomer of ibuprofen is biologically active. This enantiomer reaches therapeutic concentrations in the human body in approximately 12 minutes. However, in this case, the inactive enantiomer is not wasted. The body converts it to the active enantiomer, but that takes time.

What we need is a way to name each enantiomer of ibuprofen (or any other pair of enantiomers for that matter) so that we can refer to them in conversation or in writing. To do so, chemists have developed the **R,S system**. The first step in assigning an R or S configuration to a stereocenter is to arrange the groups bonded to it in order of priority. For this, we use the same set of priority rules we used in Section 4.3C to assign an E,Z configuration to an alkene.

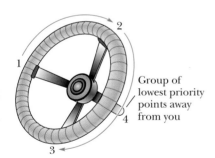

Group of lowest priority points away from you

To assign an R or S configuration to a stereocenter,

1. Locate the stereocenter, identify its four substituents, and assign a priority from 1 (highest) to 4 (lowest) to each substituent.
2. Orient the molecule in space so that the group of lowest priority (4) is directed away from you, as would be, for instance, the steering column of a car. The three groups of higher priority (1–3) then project towards you, as would the spokes of a steering wheel.
3. Read the three groups projecting towards you in order, from highest priority (1) to lowest priority (3):
4. If reading the groups proceeds in a clockwise direction, the configuration is designated **R** (Latin: *rectus*, straight, correct); if reading proceeds in a counterclockwise direction, the configuration is **S** (Latin: *sinister*, left). You can also visualize this situation as follows: Turning the steering wheel to the right equals R, and turning it to the left equals S.

R From the Latin *rectus*, meaning right; used in the R,S system to show that the order of priority of groups on a stereocenter is clockwise.

S From the Latin *sinister*, meaning left; used in the R,S system to show that the order of priority of groups on a stereocenter is counterclockwise.

EXAMPLE 6.2

Assign an R or S configuration to each stereocenter:

(a)

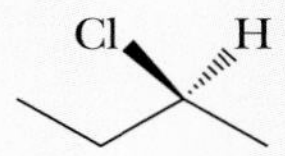

(b)

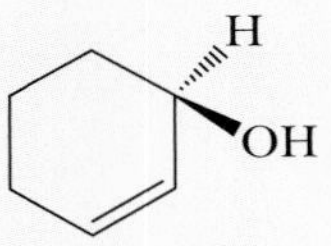

SOLUTION

View each molecule through the stereocenter and along the bond from the stereocenter toward the group of lowest priority.

(a) The order of priority is $-Cl > -CH_2CH_3 > -CH_3 > -H$. The group of lowest priority, H, points away from you. Reading the groups in the order 1, 2, 3 occurs in the counterclockwise direction, so the configuration is S.

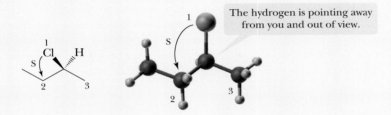

The hydrogen is pointing away from you and out of view.

(b) The order of priority is $-OH > -CH=CH > -CH_2-CH_2 > -H$. With hydrogen, the group of lowest priority, pointing away from you, reading the groups in the order 1, 2, 3 occurs in the clockwise direction, so the configuration is R.

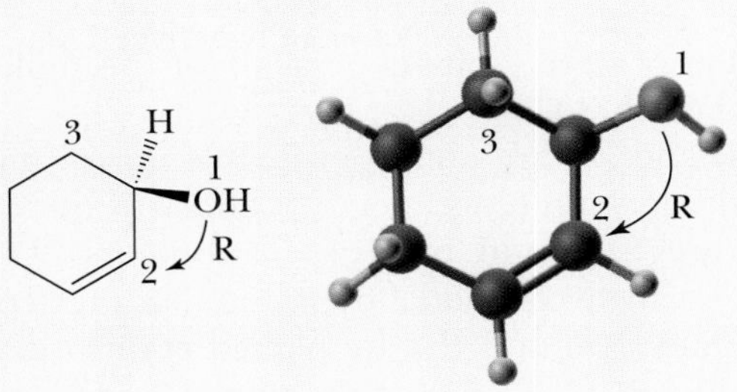

Practice Problem 6.2

Assign an R or S configuration to each stereocenter:

(a)

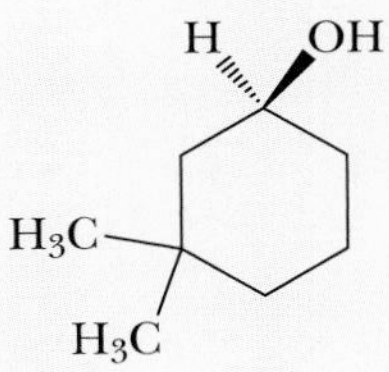

(b)

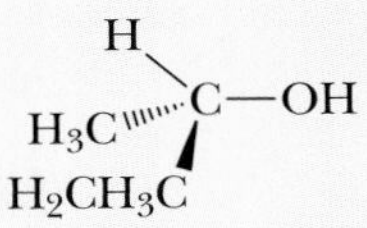

(c)

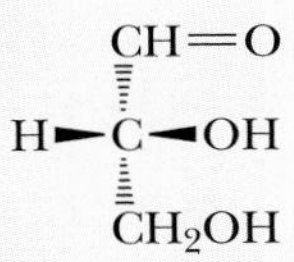

Now let us return to our three-dimensional drawing of the enantiomers of ibuprofen and assign each an R or S configuration. In order of decreasing priority, the groups bonded to the stereocenter are $-COOH > -C_6H_4 > -CH_3 > H$. In the enantiomer on the left, reading the groups on the stereocenter in order of priority occurs clockwise. Therefore, this enantiomer is (*R*)-ibuprofen, and its mirror image is (*S*)-ibuprofen:

(*R*)-Ibuprofen
(the inactive enantiomer)

(*S*)-Ibuprofen
(the active enantiomer)

6.5 ACYCLIC MOLECULES WITH TWO STEREOCENTERS

We have now seen several examples of molecules with one stereocenter and verified that, for each, two stereoisomers (one pair of enantiomers) are possible. Now let us consider molecules with two stereocenters. To generalize, for a molecule with n stereocenters, the maximum number of stereoisomers possible is 2^n. We have already verified that, for a molecule with one stereocenter, $2^1 = 2$ stereoisomers are possible. For a molecule with two stereocenters, $2^2 = 4$ stereoisomers are possible; for a molecule with three stereocenters, $2^3 = 8$ stereoisomers are possible, and so forth.

A. Enantiomers and Diastereomers

We begin our study of molecules with two stereocenters by considering 2,3,4-trihydroxybutanal. Its two stereocenters are marked with asterisks:

$$HOCH_2 - \overset{*}{C}H - \overset{*}{C}H - CH = O$$
$$\qquad\quad | \qquad\; |$$
$$\qquad\quad OH \quad\; OH$$

2,3,4-Trihydroxybutanal

The maximum number of stereoisomers possible for this molecule is $2^2 = 4$, each of which is drawn in Figure 6.4.

Stereoisomers (a) and (b) are nonsuperposable mirror images and are, therefore, a pair of enantiomers. Stereoisomers (c) and (d) are also nonsuperposable mirror images and are a second pair of enantiomers. We describe the four stereoisomers of 2,3,4-trihydroxybutanal by saying that they consist of two pairs of enantiomers. Enantiomers (a) and (b) are named **erythrose**, which is synthesized in erythrocytes (red blood cells)—hence the name. Enantiomers (c) and (d) are named **threose**. Erythrose and threose belong to the class of compounds called carbohydrates, which we discuss in Chapter 18.

Figure 6.4

The four stereoisomers of 2,3,4-trihydroxybutanal, a compound with two stereocenters. Configurations (a) and (b) are (2*R*,3*R*) and (2*S*,3*S*), respectively. Configurations (c) and (d) are (2*R*,3*S*) and (2*S*,3*R*), respectively.

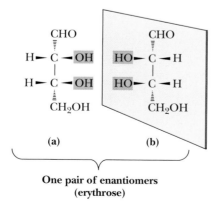

(a)   (b)

One pair of enantiomers
(erythrose)

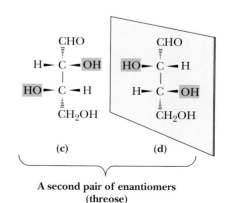

(c)   (d)

A second pair of enantiomers
(threose)

We have specified the relationship between (a) and (b) and between (c) and (d). What is the relationship between (a) and (c), between (a) and (d), between (b) and (c), and between (b) and (d)? The answer is that they are diastereomers. **Diastereomers** are stereoisomers that are not enantiomers; that is, they are stereoisomers that are not mirror images of each other.

Diastereomers Stereoisomers that are not mirror images of each other; the term refers to relationships among objects.

EXAMPLE 6.3

Following are stereorepresentations of the four stereoisomers of 1,2,3-butanetriol:

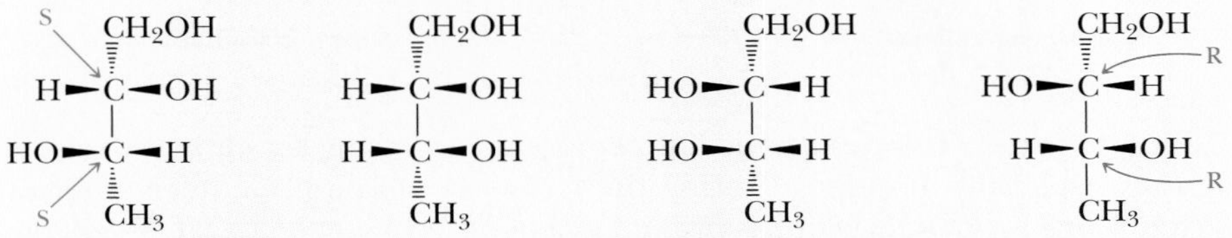

Configurations are given for the stereocenters in (1) and (4).

(a) Which compounds are enantiomers?
(b) Which compounds are diastereomers?

SOLUTION

(a) Enantiomers are stereoisomers that are nonsuperposable mirror images. Compounds (1) and (4) are one pair of enantiomers, and compounds (2) and (3) are a second pair of enantiomers. Note that the configurations of the stereocenters in (1) are the opposite of those in (4), its enantiomer.
(b) Diastereomers are stereoisomers that are not mirror images. Compounds (1) and (2), (1) and (3), (2) and (4), and (3) and (4) are diastereomers.

Practice Problem 6.3

Following are stereorepresentations of the four stereoisomers of 3-chloro-2-butanol:

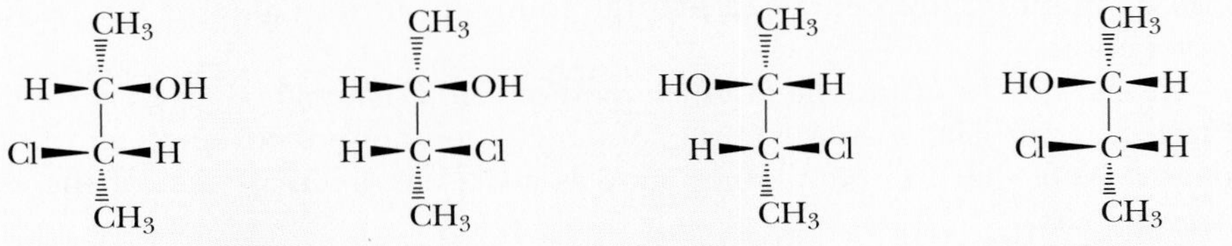

(a) Which compounds are enantiomers?
(b) Which compounds are diastereomers?

B. Meso Compounds

Certain molecules containing two or more stereocenters have special symmetry properties that reduce the number of stereoisomers to fewer than the maximum number predicted by the 2^n rule. One such molecule is 2,3-dihydroxybutanedioic acid, more commonly named tartaric acid:

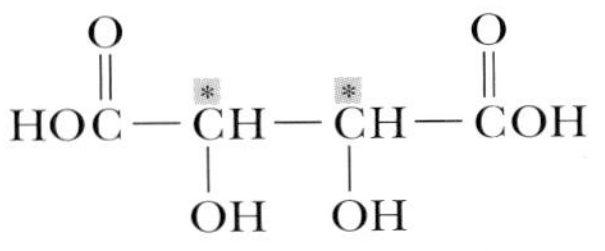

Meso compound An achiral compound possessing two or more stereocenters.

2,3-Dihydroxybutanedioic acid
(Tartaric acid)

Figure 6.5
Stereoisomers of tartaric acid. One pair of enantiomers and one meso compound. The presence of an internal plane of simmetry indicates that the molecule is achiral.

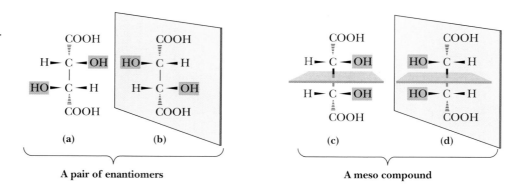

A pair of enantiomers   A meso compound

Tartaric acid is a colorless, crystalline compound occurring largely in the vegetable kingdom, especially in grapes. During the fermentation of grape juice, potassium bitartrate (one —COOH group is present as a potassium salt, —$COO^- K^+$) deposits as a crust on the sides of wine casks. Then, collected and purified, it is sold commercially as cream of tartar.

Carbons 2 and 3 of tartaric acid are stereocenters, and, from the 2^n rule, the maximum number of stereoisomers possible is $2^2 = 4$. Figure 6.5 shows the two pairs of mirror images of this compound. Structures (a) and (b) are nonsuperposable mirror images and, therefore, are a pair of enantiomers. Structures (c) and (d) are also mirror images, but they are superposable. To see this, imagine that you rotate (d) by 180° in the plane of the paper, lift it out of the plane of the paper, and place it on top of (c). If you do this mental manipulation correctly, you will find that (d) is superposable on (c). Therefore, (c) and (d) are *not* different molecules; they are the same molecule, just oriented differently. Because (c) and its mirror image are superposable, (c) is achiral.

Another way to verify that (c) is achiral is to see that it has a plane of symmetry which bisects the molecule in such a way that the top half is the reflection of the bottom half. Thus, even though (c) has two stereocenters, it is achiral. The stereoisomer of tartaric acid represented by (c) or (d) is called a **meso compound**, defined as an achiral compound that contains two or more stereocenters.

We can now return to the original question: How many stereoisomers are there of tartaric acid? The answer is three: one meso compound and one pair of enantiomers. Note that the meso compound is a diastereomer of each of the other stereoisomers.

EXAMPLE 6.4

Following are stereorepresentations of the three stereoisomers of 2,3-butanediol:

CH₃   CH₃   CH₃

H—C—OH   H—C—OH   HO—C—H

HO—C—H   H—C—OH   H—C—OH

CH₃   CH₃   CH₃

(1)   (2)   (3)

(a) Which are enantiomers?
(b) Which is the meso compound?

SOLUTION

(a) Compounds (1) and (3) are enantiomers.

(b) Compound (2) is a meso compound.

Practice Problem 6.4 ———————————————————————————————

Following are four Newman projection formulas for tartaric acid:

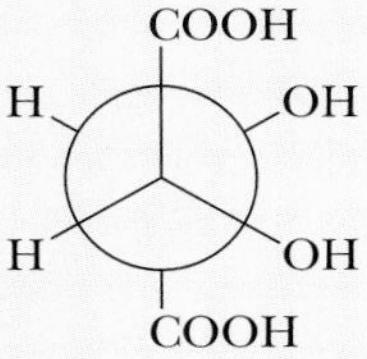

(1) 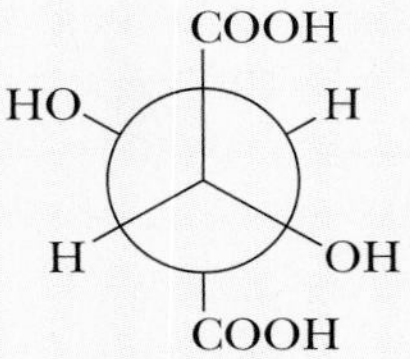   (2)

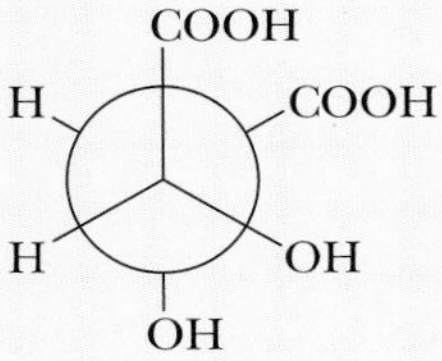

(3) 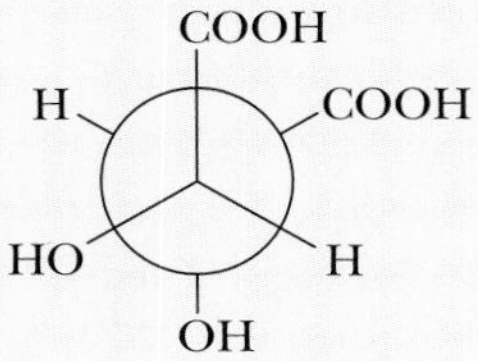   (4)

(a) Which represent the same compound?

(b) Which represent enantiomers?

(c) Which represent(s) meso tartaric acid?

6.6 CYCLIC MOLECULES WITH TWO STEREOCENTERS

In this section, we concentrate on derivatives of cyclopentane and cyclohexane containing two stereocenters. We can analyze chirality in these cyclic compounds in the same way we analyzed it in acyclic compounds.

A. Disubstituted Derivatives of Cyclopentane

Let us start with 2-methylcyclopentanol, a compound with two stereocenters. Using the 2^n rule, we predict a maximum of $2^2 = 4$ stereoisomers. Both the cis isomer and the trans isomer are chiral, with the cis isomer existing as one pair of enantiomers and the trans isomer existing as a second pair:

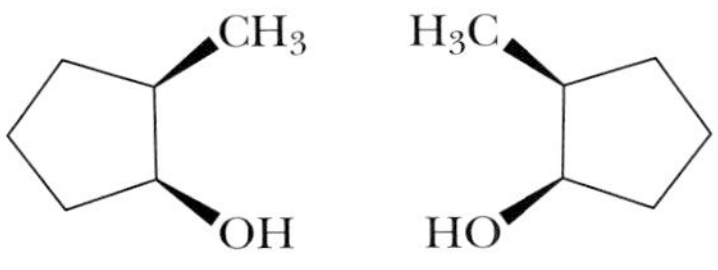

cis-2-Methylcyclopentanol   *trans*-2-Methylcyclopentanol
(a pair of enantiomers)   (a pair of enantiomers)

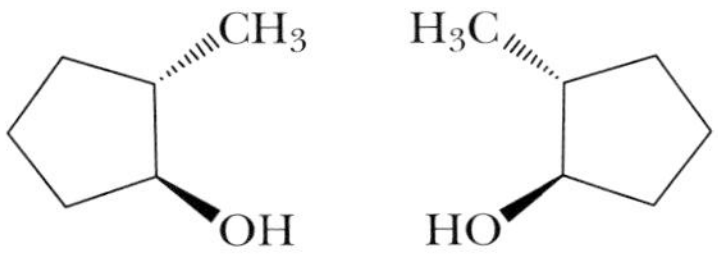

1,2-Cyclopentanediol also has two stereocenters; therefore, the 2^n rule predicts a maximum of $2^2 = 4$ stereoisomers. As seen in the following stereodrawings, only three stereoisomers exist for this compound:

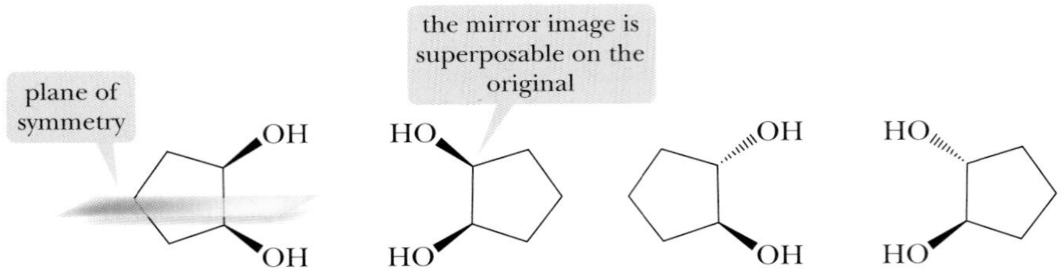

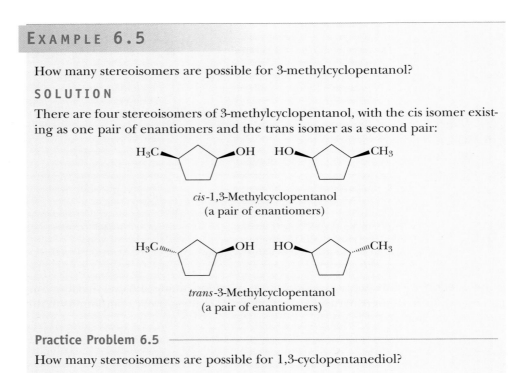

cis-1,2-Cyclopentanediol
(a meso compound)

trans-1,2-Cyclopentanediol
(a pair of enantiomers)

The cis isomer is achiral (meso) because it and its mirror image are superposable. Alternatively, the cis isomer is achiral because it possesses a plane of symmetry that bisects the molecule into two mirror-image halves. The trans isomer is chiral and exists as a pair of enantiomers.

EXAMPLE 6.5

How many stereoisomers are possible for 3-methylcyclopentanol?

SOLUTION

There are four stereoisomers of 3-methylcyclopentanol, with the cis isomer existing as one pair of enantiomers and the trans isomer as a second pair:

cis-1,3-Methylcyclopentanol
(a pair of enantiomers)

trans-3-Methylcyclopentanol
(a pair of enantiomers)

Practice Problem 6.5

How many stereoisomers are possible for 1,3-cyclopentanediol?

B. Disubstituted Derivatives of Cyclohexane

As an example of a disubstituted cyclohexane, let us consider the methylcyclohexanols. 4-Methylcyclohexanol can exist as two stereoisomers—a pair of cis–trans isomers:

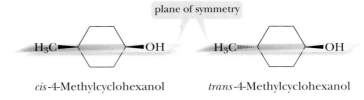

cis-4-Methylcyclohexanol   *trans*-4-Methylcyclohexanol

Both the cis and the trans isomer are achiral. In each, a plane of symmetry runs through the CH_3 and OH groups and the two attached carbons.

3-Methylcyclohexanol has two stereocenters and exists as $2^2 = 4$ stereoisomers, with the cis isomer existing as one pair of enantiomers and the trans isomer as a second pair:

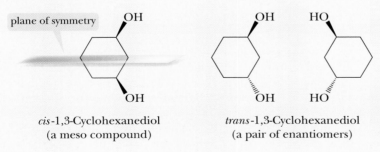

cis-3-Methylcyclohexanol
(a pair of enantiomers)

trans-3-Methylcyclohexanol
(a pair of enantiomers)

Similarly, 2-methylcyclohexanol has two stereocenters and exists as $2^2 = 4$ stereoisomers, with the cis isomer existing as one pair of enantiomers and the trans isomer as a second pair:

cis-2-Methylcyclohexanol
(a pair of enantiomers)

trans-2-Methylcyclohexanol
(a pair of enantiomers)

EXAMPLE 6.6

How many stereoisomers exist for 1,3-cyclohexanediol?

SOLUTION

1,3-Cyclohexanediol has two stereocenters, and, according to the 2^n rule, a maximum of $2^2 = 4$ stereoisomers is possible. The trans isomer of this compound exists as a pair of enantiomers. The cis isomer has a plane of symmetry and is a meso compound. Therefore, although the 2^n rule predicts a maximum of four stereoisomers for 1,3-cyclohexanediol, only three exist—one pair of enantiomers and one meso compound:

plane of symmetry

cis-1,3-Cyclohexanediol
(a meso compound)

trans-1,3-Cyclohexanediol
(a pair of enantiomers)

Practice Problem 6.6

How many stereoisomers exist for 1,4-cyclohexanediol?

6.7   MOLECULES WITH THREE OR MORE STEREOCENTERS

The 2^n rule applies equally well to molecules with three or more stereocenters. Here is a disubstituted cyclohexanol with three stereocenters, each marked with an asterisk:

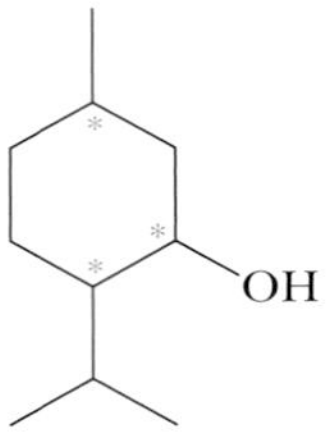

2-Isopropyl-5-methyl-cyclohexanol

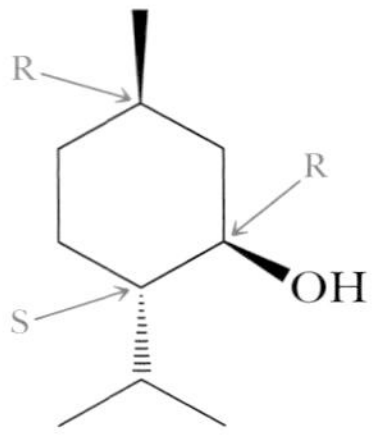

Menthol

There is a maximum of $2^3 = 8$ stereoisomers possible for this molecule. Menthol, one of the eight, has the configuration shown on the right. The configuration at each stereocenter is indicated. Menthol is present in peppermint and other mint oils.

Cholesterol, a more complicated molecule, has eight stereocenters:

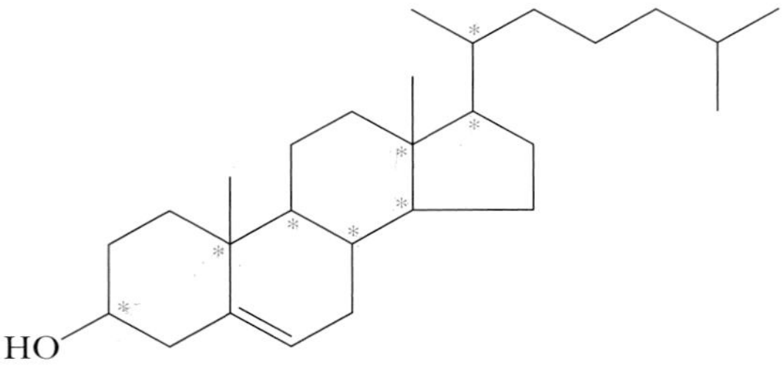

Cholesterol has 8 stereocenters; 256 stereoisomers are possible

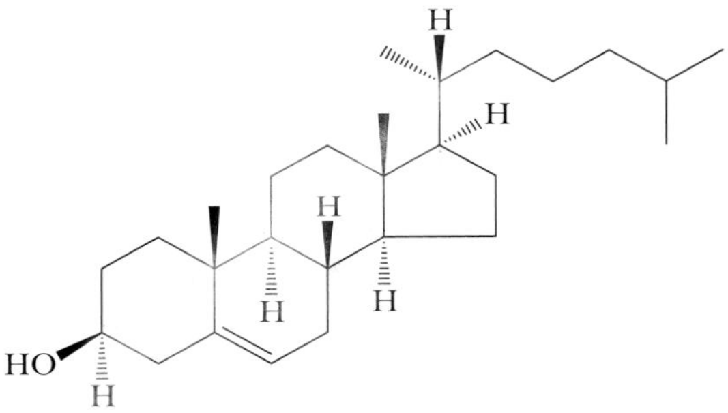

This is the stereoisomer found in human metabolism

To identify the stereocenters, remember to add an appropriate number of hydrogens to complete the tetravalence of each carbon you think might be a stereocenter.

6.8   PROPERTIES OF STEREOISOMERS

Enantiomers have identical physical and chemical properties in achiral environments. The enantiomers of tartaric acid (Table 6.1), for example, have the same melting point, the same boiling point, the same solubilities in water and other common solvents, and the same values of pK_a (the acid ionization constant), and they all undergo the same acid–base reactions. The enantiomers of tartaric acid do, however, differ in optical activity (the ability to rotate the plane of polarized light), a property that is discussed in the next section.

Diastereomers have different physical and chemical properties, even in achiral environments. Meso-tartaric acid has different physical properties from those of the enantiomers.

TABLE 6.1 Some Physical Properties of the Stereoisomers of Tartaric Acid

	(R,R)-Tartaric acid	(S,S)-Tartaric acid	Meso-tartaric acid
Specific rotation*	+12.7	−12.7	0
Melting point (°C)	171–174	171–174	146–148
Density at 20°C (g/cm³)	1.7598	1.7598	1.660
Solubility in water at 20°C (g/100 mL)	139	139	125
pK_1 (25°C)	2.98	2.98	3.23
pK_2 (25°C)	4.34	4.34	4.82

$$
\begin{array}{ccc}
\text{COOH} & \text{COOH} & \text{COOH} \\
\text{H}-\text{C}-\text{OH} & \text{HO}-\text{C}-\text{H} & \text{H}-\text{C}-\text{OH} \\
\text{HO}-\text{C}-\text{H} & \text{H}-\text{C}-\text{OH} & \text{H}-\text{C}-\text{OH} \\
\text{COOH} & \text{COOH} & \text{COOH}
\end{array}
$$

*Specific rotation is discussed in the next section.

6.9 OPTICAL ACTIVITY: HOW CHIRALITY IS DETECTED IN THE LABORATORY

As we have already established, enantiomers are different compounds, and we must expect, therefore, that they differ in some properties. One property that differs between enantiomers is their effect on the plane of polarized light. Each member of a pair of enantiomers rotates the plane of polarized light, and for this reason, enantiomers are said to be **optically active**. To understand how optical activity is detected in the laboratory, we must first understand plane-polarized light and a polarimeter, the instrument used to detect optical activity.

Optically active Showing that a compound rotates the plane of polarized light.

A. Plane-Polarized Light

Ordinary light consists of waves vibrating in all planes perpendicular to its direction of propagation (Figure 6.6). Certain materials, such as calcite and Polaroid™ sheet (a plastic film containing properly oriented crystals of an organic substance embedded in it), selectively transmit light waves vibrating in parallel planes. Electromagnetic radiation vibrating in only parallel planes is said to be **plane polarized**.

Plane-polarized light Light vibrating only in parallel planes.

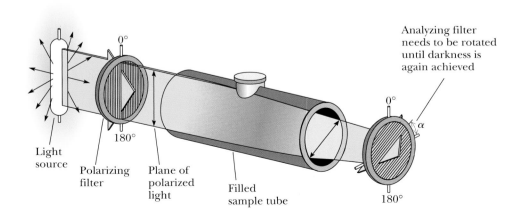

Light source
Polarizing filter
Plane of polarized light
Filled sample tube
0°
180°
0°
180°
α
Analyzing filter needs to be rotated until darkness is again achieved

Figure 6.6
Schematic diagram of a polarimeter with its sample tube containing a solution of an optically active compound. The analyzing filter has been turned clockwise by α degrees to restore the dark field.

A polarimeter is used to measure the rotation of plane-polarized light as it passes through a sample. *(Richard Magna, 1992, Fundamental Photographs)*

Polarimeter An instrument for measuring the ability of a compound to rotate the plane of polarized light.

Observed rotation The number of degrees through which a compound rotates the plane of polarized light.

Dextrorotatory Rotating the plane of polarized light in a polarimeter to the right.

Levorotatory Rotating the plane of polarized light in a polarimeter to the left.

Specific rotation Observed rotation of the plane of polarized light when a sample is placed in a tube 1.0 dm long at a concentration of 1.0 g/100 mL.

B. A Polarimeter

A **polarimeter** consists of a light source, a polarizing filter and an analyzing filter (each made of calcite or Polaroid™ film), and a sample tube (Figure 6.6). If the sample tube is empty, the intensity of light reaching the detector (in this case, your eye) is at its maximum when the polarizing axes of the two filters are parallel. If the analyzing filter is turned either clockwise or counterclockwise, less light is transmitted. When the axis of the analyzing filter is at right angles to the axis of the polarizing filter, the field of view is dark. This position of the analyzing filter is taken to be 0° on the optical scale.

The ability of molecules to **rotate the plane of polarized light** can be observed with the use of a polarimeter in the following way: First, a sample tube filled with solvent is placed in the polarimeter, and the analyzing filter is adjusted so that no light passes through to the observer; that is, the filter is set to 0°. Then we place a solution of an optically active compound in the sample tube. When we do so, we find that a certain amount of light now passes through the analyzing filter. We also find that the plane of polarized light from the polarizing filter has been rotated so that it is no longer at an angle of 90° to the analyzing filter. Consequently, we rotate the analyzing filter to restore darkness in the field of view. The number of degrees, α, through which we must rotate the analyzing filter to restore darkness to the field of view is called the **observed rotation**. If we must turn the analyzing filter to the right (clockwise) to restore the dark field, we say that the compound is **dextrorotatory** (Latin: *dexter*, on the right side); if we must turn it to the left (counterclockwise), we say that the compound is **levorotatory** (Latin: *laevus*, on the left side).

The magnitude of the observed rotation for a particular compound depends on its concentration, the length of the sample tube, the temperature, the solvent, and the wavelength of the light used. The **specific rotation**, $[\alpha]$, is defined as the observed rotation at a specific cell length and sample concentration:

$$\text{Specific rotation} = [\alpha]\frac{T}{\lambda} = \frac{\text{Observed rotation (degrees)}}{\text{Length (dm)} \times \text{Concentration}}$$

The standard cell length is 1 decimeter (1 dm = 0.1 m). For a pure liquid sample, the concentration is expressed in grams per milliliter (g/mL; density). The concentration of a sample dissolved in a solvent is also usually expressed as grams per 100 milliliters of solution. The temperature (T, in degrees centigrade) and wavelength (λ, in nanometers) of light are designated, respectively, as superscripts and subscripts. The light source most commonly used in polarimetry is the sodium D line ($\lambda = 589$ nm), the same line responsible for the yellow color of sodium-vapor lamps.

In reporting either observed or specific rotation, it is common to indicate a dextrorotatory compound with a plus sign in parentheses, (+), and a levorotatory compound with a minus sign in parentheses, (−). For any pair of enantiomers, one enantiomer is dextrorotatory and the other is levorotatory. For each member, the value of the specific rotation is exactly the same, but the sign is opposite. Following are the specific rotations of the enantiomers of 2-butanol at 25°C, observed with the D line of sodium:

(*S*)-(+)-2-Butanol   (*R*)-(−)-2-Butanol
$[\alpha]_D^{25}$ +13.52° $[\alpha]_D^{25}$ −13.52°

EXAMPLE 6.7

A solution is prepared by dissolving 4.00 mg of testosterone, a male sex hormone (Table 17.3), in 10.0 mL of ethanol and placing it in a sample tube 1.00 dm in length. The observed rotation of this sample at 25°C, using the D line of sodium, is +4.36°. Calculate the specific rotation of testosterone.

SOLUTION

The concentration of testosterone is 4.00 mg/100 mL = 0.0400 g/100 mL. The length of the sample tube is 1.00 dm. Inserting these values into the equation for calculating specific rotation gives

$$\text{Specific rotation} = \frac{\text{Observed rotation}}{\text{Length} \times \text{Concentration}} = \frac{+4.36°}{1.00 \times 0.0400} = +109°$$

Practice Problem 6.7

The specific rotation of progesterone, a female sex hormone (Table 17.3), is +172°, measured at 20°C. Calculate the observed rotation for a solution prepared by dissolving 40 mg of progesterone in 100 mL of dioxane and placing it in a sample tube 1.00 dm long.

C. Racemic Mixtures

An equimolar mixture of two enantiomers is called a **racemic mixture**, a term derived from the name "racemic acid" (Latin: *racemus*, a cluster of grapes), originally given to an equimolar mixture of the enantiomers of tartaric acid (Table 6.1). Because a racemic mixture contains equal numbers of the dextrorotatory and the levorotatory molecules, its specific rotation is zero. Alternatively, we say that a racemic mixture is optically inactive. A racemic mixture is indicated by adding the prefix (±) to the name of the compound.

Racemic mixture A mixture of equal amounts of two enantiomers.

6.10 THE SIGNIFICANCE OF CHIRALITY IN THE BIOLOGICAL WORLD

Except for inorganic salts and a relatively few low-molecular-weight organic substances, the molecules in living systems, both plant and animal, are chiral. Although these molecules can exist as a number of stereoisomers, almost invariably only one stereoisomer is found in nature. Of course, instances do occur in which more than one stereoisomer is found, but these rarely exist together in the same biological system.

A. Chirality in Biomolecules

Perhaps the most conspicuous examples of chirality among biological molecules are the enzymes, all of which have many stereocenters. An example is chymotrypsin, an enzyme found in the intestines of animals that catalyzes the digestion of proteins (Section 19.4). Chymotrypsin has 251 stereocenters. The maximum number of stereoisomers possible is thus 2^{251}, a staggeringly large number, almost beyond comprehension. Fortunately, nature does not squander its precious energy and resources unnecessarily: Only one of these stereoisomers is produced and used by any given organism. Because enzymes are chiral substances, most either produce or react with only substances that match their stereochemical requirements.

Figure 6.7
A schematic diagram of an enzyme surface capable of interacting with (R)-$(+)$-glyceraldehyde at three binding sites, but with (S)-$(-)$-glyceraldehyde at only two of these sites.

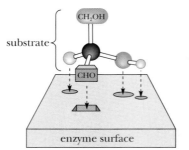

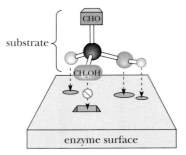

This enantiomer of glyceraldehyde fits the three specific binding sites on the enzyme surface

This enantiomer of glyceraldehyde does not fit the same binding sites

B. How an Enzyme Distinguishes between a Molecule and Its Enantiomer

An enzyme catalyzes a biological reaction of a molecule by first positioning it at a **binding site** on the enzyme's surface. An enzyme with binding sites specific for three of the four groups on a stereocenter can distinguish between a molecule and its enantiomer or one of its diastereomers. Assume, for example, that an enzyme involved in catalyzing a reaction of glyceraldehyde has on its surface a binding site specific for —H, a second specific for —OH, and a third specific for —CHO. Assume further that the three sites are arranged on the enzyme surface as shown in Figure 6.7. The enzyme can distinguish (R)-$(+)$-glyceraldehyde (the natural, or biologically active, form) from its enantiomer because the natural enantiomer can be absorbed, with three groups interacting with their appropriate binding sites; for the S enantiomer, at best only two groups can interact with these binding sites.

Because interactions between molecules in living systems take place in a chiral environment, it should come as no surprise that a molecule and its enantiomer or one of its diastereomers elicit different physiological responses. As we have already seen, (S)-ibuprofen is active as a pain and fever reliever, whereas its R enantiomer is inactive. The S enantiomer of the closely related analgesic naproxen is also the active pain reliever of this compound, but its R enantiomer is a liver toxin!

(S)-Ibuprofen

(S)-Naproxen

6.11   SEPARATION OF ENANTIOMERS: RESOLUTION

Resolution Separation of a racemic mixture into its enantiomers.

Resolution is the separation of a racemic mixture into its enantiomers. Separating enantiomers is, in general, difficult, but scientists have developed a number of ways to do it. In this section, we illustrate just one of the several laboratory methods for resolution, namely, the use of enzymes as chiral catalysts. The principle involved in the method we describe is that a particular enzyme will catalyze a reaction of a chiral molecule, but not of its enantiomer.

One class of enzymes that has received particular attention in this regard is the esterases, which catalyze the hydrolysis of esters to give an alcohol and a carboxylic acid. We illustrate this method by describing the resolution of (R,S)-naproxen. The

Chiral Drugs

Some of the common drugs used in human medicine (for example, aspirin, Section 14.4B) are achiral. Others are chiral and are sold as single enantiomers. The penicillin and erythromycin classes of antibiotics and the drug captopril are all chiral drugs. Captopril, which is highly effective for the treatment of high blood pressure and congestive heart failure, was developed in a research program designed to discover effective inhibitors of angiotensin-converting enzyme (ACE). Captopril is manufactured and sold as the (S,S)-stereoisomer. A large number of chiral drugs, however, are sold as racemic mixtures. The popular analgesic ibuprofen (the active ingredient in Motrin®, Advil®, and many other nonaspirin analgesics) is an example. Only the S enantiomer of the pain reliever ibuprofen is biologically active. The body, however, converts the inactive R enantiomer to the active S enantiomer.

Captopril

(S)-Ibuprofen

For racemic drugs, most often only one enantiomer exerts the beneficial effect, whereas the other enantiomer either has no effect or exerts a detrimental effect. Thus, enantiomerically pure drugs should, more often than not, be more effective than their racemic counterparts. A case in point is 3,4-dihydroxyphenylalanine, which is used in the treatment of Parkinson's disease. The active drug is dopamine.

Unfortunately, this compound does not cross the blood–brain barrier to the required site of action in the brain. Consequently, what is administered instead is the prodrug, a compound that is not active by itself, but is converted in the body to an active drug. 3,4-Dihydroxyphenylalanine is such a prodrug; it crosses the blood–brain barrier and then undergoes decarboxylation, catalyzed by the enzyme dopamine decarboxylase, to give dopamine:

(S)-(−)-3,4-Dihydroxyphenylalanine
(L-DOPA)
$[\alpha]_D^{13}$ −13.1°

Dopamine

Dopamine decarboxylase is specific for the S enantiomer, which is commonly known as L-DOPA. It is essential, therefore, to administer the enantiomerically pure prodrug. Were the prodrug to be administered in a racemic form, there could be a dangerous buildup of the R enantiomer, which cannot be metabolized by the enzymes present in the brain.

Recently, the U.S. Food and Drug Administration established new guidelines for the testing and marketing of chiral drugs. After reviewing these guidelines, many drug companies have decided to develop only single enantiomers of new chiral drugs. In addition to regulatory pressure, there are patent considerations: If a company has patents on a racemic drug, a new patent can often be taken out on one of its enantiomers.

ethyl esters of both (R)- and (S)-naproxen are solids with very low solubilities in water. Chemists then use an esterase in alkaline solution to hydrolyze selectively the (S)-ester, which goes into the aqueous solution as the sodium salt of the (S)-carboxylic acid. The (R)-ester is unaffected by these conditions. Filtering the alkaline solution recovers the crystals of the (R)-ester. After the crystals are removed, the alkaline solution is acidified to precipitate pure (S)-naproxen. The recovered (R)-ester can be racemized (converted to an R,S-mixture) and again treated with the esterase. Thus, by recycling the (R)-ester, all the racemic ester is converted to (S)-naproxen.

Ethyl ester of (S)-naproxen

Ethyl ester of (R)-naproxen
(not affected by the esterase)

1. esterase | NaOH, H_2O
2. HCl, H_2O

(S)-Naproxen

The sodium salt of (S)-naproxen is the active ingredient in Aleve® and a score of other over-the-counter nonsteroidal anti-inflammatory preparations.

Summary

Stereoisomers (Section 6.2) have the same order of attachment of atoms, but a different three-dimensional orientation of their atoms in space. A **mirror image** is the reflection of an object in a mirror (Section 6.3). **Enantiomers** are a pair of stereoisomers that are nonsuperposable mirror images (Section 6.3). A molecule that is not superposable on its mirror image is said to be **chiral**. Chirality is a property of an object as a whole, not of a particular atom. An object is **achiral** if it possesses a **plane of symmetry**—an imaginary plane passing through the object and dividing it such that one half is the reflection of the other half.

A **stereocenter** (Section 6.3) is an atom at which the interchange of two atoms or groups of atoms bonded to it produces a different stereoisomer. The most common type of stereocenter among organic compounds is a tetrahedral carbon atom with four different groups bonded to it.

The **configuration** at a stereocenter can be specified by the **R,S convention** (Section 6.4). To apply this convention, (1) each atom or group of atoms bonded to the stereocenter is assigned a priority and numbered from highest priority to

lowest priority, (2) the molecule is oriented in space so that the group of lowest priority is directed away from the observer, and (3) the remaining three groups are read in order, from highest priority to lowest priority. If the reading of groups is clockwise, the configuration is **R** (Latin: *rectus*, right). If the reading is counterclockwise, the configuration is **S** (Latin: *sinister*, left).

For a molecule with *n* stereocenters, the maximum number of stereoisomers possible is 2^n (Section 6.5). **Diastereomers** are stereoisomers that are not mirror images (Section 6.5A). Certain molecules have special symmetry properties that reduce the number of stereoisomers to fewer than that predicted by the 2^n rule. A **meso compound** (Section 6.5B) contains two or more stereocenters assembled in such a way that its molecules are achiral.

Enantiomers have identical physical and chemical properties in achiral environments (Section 6.8). Diastereomers have different physical and chemical properties.

Light that vibrates in only parallel planes is said to be **plane polarized** (Section 6.9A). A **polarimeter** (Section 6.9B) is an instrument used to detect and measure the magnitude of optical activity. **Observed rotation** is the number of degrees the plane of polarized light is rotated. **Specific rotation** is the observed rotation measured with a cell 1 dm long. The concentration of a pure liquid sample is expressed in grams per milliliter (density) whereas the concentration of a solution is expressed in grams of sample per 100 mL of solution. If the analyzing filter must be turned clockwise to restore the zero point, the compound is **dextrorotatory**. If the analyzing filter must be turned counterclockwise to restore the zero point, the compound is **levorotatory**. A compound is said to be **optically active** if it rotates the plane of polarized light. Each member of a pair of enantiomers rotates the plane of polarized light an equal number of degrees, but opposite in direction (Section 6.9B). A **racemic mixture** (Section 6.9C) is a mixture of equal amounts of two enantiomers and has a specific rotation of zero.

An enzyme catalyzes biological reactions of molecules by first positioning them at a binding site on its surface (Section 6.10). An enzyme with a binding site specific for three of the four groups on a stereocenter can distinguish between a molecule and its enantiomer.

Resolution (Section 6.11) is the experimental process of separating a mixture of enantiomers into two pure enantiomers. One means of resolution is to treat the racemic mixture with an enzyme that catalyzes a specific reaction of one enantiomer, but not the other.

PROBLEMS

A problem number set in red indicates an applied "real-world" problem.

Chirality

6.8 Define the term *stereoisomer*. Name four types of stereoisomers.

6.9 In what way are constitutional isomers different from stereoisomers? In what way are they the same?

6.10 Which of these objects are chiral (assume that there is no label or other identifying mark)?
- **(a)** A pair of scissors
- **(b)** A tennis ball
- **(c)** A paper clip
- **(d)** A beaker
- **(e)** The swirl created in water as it drains out of a sink or bathtub

6.11 Think about the helical coil of a telephone cord or the spiral binding on a notebook, and suppose that you view the spiral from one end and find that it has a left-handed twist. If you view the same spiral from the other end, does it have a right-handed twist or a left-handed twist from that end as well?

6.12 Next time you have the opportunity to view a collection of augers or other seashells that have a helical twist, study the chirality of their twists. Do you find an equal number of left-handed and right-handed augers, or, for example, do they all have the same handedness? What about the handedness of augers compared with that of other spiral shells?

6.13 Next time you have an opportunity to examine any of the seemingly endless varieties of spiral pasta (rotini, fusilli, radiatori, tortiglioni), examine their twist. Do the twists of any one kind all have a right-handed twist, do they all have a left-handed twist, or are they a racemic mixture?

6.14 One reason we can be sure that sp^3-hybridized carbon atoms are tetrahedral is the number of stereoisomers that can exist for different organic compounds.
- **(a)** How many stereoisomers are possible for $CHCl_3$, CH_2Cl_2, and $CHBrClF$ if the four bonds to carbon have a tetrahedral geometry?
- **(b)** How many stereoisomers are possible for each of the compounds if the four bonds to the carbon have a square planar geometry?

6.15 Which of the following statements is true?
- **(a)** Enantiomers are always chiral.
- **(b)** A diastereomer of a chiral molecule must also be chiral.
- **(c)** A molecule that possesses an internal plane of symmetry can never be chiral.
- **(d)** An achiral molecule will always have an enantiomer.
- **(e)** An achiral molecule will always have a diastereomer.
- **(f)** A chiral molecule will always have an enantiomer.
- **(g)** A chiral molecule will always have a diastereomer.

Median cross section through the shell of a chambered nautilus found in the deep waters of the Pacific Ocean. The shell shows handedness; this cross section is right-handed spiral.
(Photo Disc. Inc./Getty Images)

Enantiomers

6.16 Which compounds contain stereocenters?
- **(a)** 2-Chloropentane
- **(b)** 3-Chloropentane
- **(c)** 3-Chloro-1-pentene
- **(d)** 1,2-Dichloropropane

6.17 Using only C, H, and O, write structural formulas for the lowest-molecular-weight chiral molecule of each of the following compounds:
- **(a)** Alkane
- **(b)** Alcohol
- **(c)** Aldehyde
- **(d)** Ketone
- **(e)** Carboxylic acid

6.18 Which alcohols with molecular formula $C_5H_{12}O$ are chiral?

6.19 Which carboxylic acids with molecular formula $C_6H_{12}O_2$ are chiral?

6.20 Draw the enantiomer for each molecule:

6.21 Mark each stereocenter in these molecules with an asterisk (note that not all contain stereocenters):

6.22 Mark each stereocenter in these molecules with an asterisk (note that not all contain stereocenters):

6.23 Mark each stereocenter in these molecules with an asterisk (note that not all contain stereocenters):

(a) $\underset{\overset{|}{OH}}{\overset{\overset{CH_3}{|}}{CH_3CCH=CH_2}}$

(b) $\underset{\overset{|}{CH_3}}{\overset{\overset{COOH}{|}}{HCOH}}$

(c) $\underset{\overset{|}{NH_2}}{\overset{\overset{CH_3}{|}}{CH_3CHCHCOOH}}$

(d) $\overset{\overset{O}{\parallel}}{CH_3CCH_2CH_3}$

(e) $\underset{\overset{|}{CH_2OH}}{\overset{\overset{CH_2OH}{|}}{HCOH}}$

(f) $\underset{\overset{|}{}}{\overset{\overset{OH}{|}}{CH_3CH_2CHCH=CH_2}}$

(g) $\underset{\overset{|}{CH_2COOH}}{\overset{\overset{CH_2COOH}{|}}{HOCCOOH}}$

6.24 Following are eight stereorepresentations of lactic acid:

(a) $\underset{H_3C}{\overset{COOH}{\underset{H}{\diagup C \diagdown}}OH}$

(b) $\underset{HOOC}{\overset{CH_3}{\underset{HO}{\diagup C \diagdown}}H}$

(c) $\underset{H}{\overset{COOH}{\underset{HO}{\diagup C \diagdown}}CH_3}$

(d) $\underset{HO}{\overset{CH_3}{\underset{H}{\diagup C \diagdown}}COOH}$

(e) $\underset{CH_3}{\overset{COOH}{H\!\!-\!\!C\!\!-\!\!OH}}$

(f) $\underset{COOH}{\overset{CH_3}{H\!\!-\!\!C\!\!-\!\!OH}}$

(g) $\underset{H}{\overset{OH}{H_3C\!\!-\!\!C\!\!-\!\!COOH}}$

(h) $\underset{OH}{\overset{CH_3}{H\!\!-\!\!C\!\!-\!\!COOH}}$

Take (a) as a reference structure. Which stereorepresentations are identical with (a) and which are mirror images of (a)?

Designation of Configuration: The R,S Convention

6.25 Assign priorities to the groups in each set:
   (a) —H   —CH$_3$ —OH   —CH$_2$OH
   (b) —CH$_2$CH=CH$_2$ —CH=CH$_2$ —CH$_3$ —CH$_2$COOH
   (c) —CH$_3$ —H   —COO$^-$ —NH$_3$$^+$
   (d) —CH$_3$ —CH$_2$SH   —NH$_3$$^+$ —COO$^-$

6.26 Which molecules have R configurations?

(a) $\underset{CH_2OH}{\overset{CH_3}{\underset{Br}{C}\diagdown H}}$

(b) $\underset{Br}{\overset{H}{HOCH_2\diagup C\diagdown CH_3}}$

(c) $\underset{H}{\overset{CH_2OH}{H_3C\diagup C\diagdown Br}}$

(d) $\underset{CH_3}{\overset{Br}{H\diagup C\diagdown CH_2OH}}$

6.27 Following are structural formulas for the enantiomers of carvone:

(−)-Carvone
(Spearmint
oil)

(+)-Carvone
(Caraway and
dillseed oil)

Each enantiomer has a distinctive odor characteristic of the source from which it can be isolated. Assign an R or S configuration to the stereocenter in each. How can they have such different properties when they are so similar in structure?

6.28 Following is a staggered conformation of one of the stereoisomers of 2-butanol:

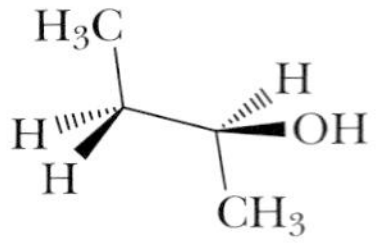

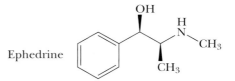

(a) Is this (*R*)-2-butanol or (*S*)-2-butanol?

(b) Draw a Newman projection for this staggered conformation, viewed along the bond between carbons 2 and 3.

(c) Draw a Newman projection for one more staggered conformation of this molecule. Which of your conformations is the more stable? Assume that —OH and —CH₃ are comparable in size.

Molecules with Two or More Stereocenters

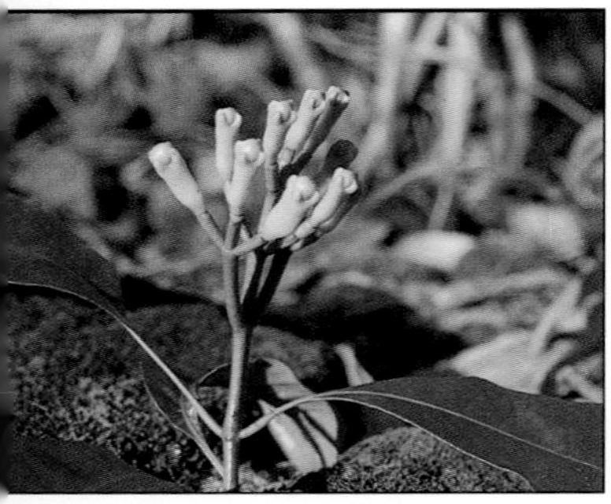

6.29 Write the structural formula of an alcohol with molecular formula $C_6H_{14}O$ that contains two stereocenters.

6.30 For centuries, Chinese herbal medicine has used extracts of *Ephedra sinica* to treat asthma. Investigation of this plant resulted in the isolation of ephedrine, a potent dilator of the air passages of the lungs. The naturally occurring stereoisomer is levorotatory and has the following structure:

Ephedrine

Assign an R or S configuration to each stereocenter.

6.31 The specific rotation of naturally occurring ephedrine, shown in Problem 6.30, is −41°. What is the specific rotation of its enantiomer?

6.32 Label each stereocenter in these molecules with an asterisk:

(a) CH₃CHCHCOOH
       | |
      HO OH

(b)
CH₂—COOH
|
CH—COOH
|
HO—CH—COOH

(c)

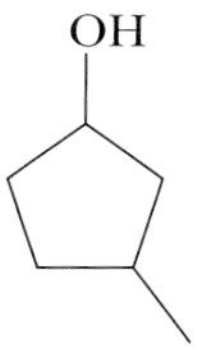

(d)

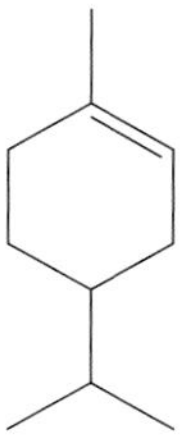

(e)

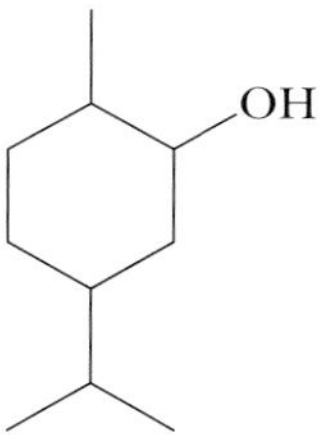

(f)

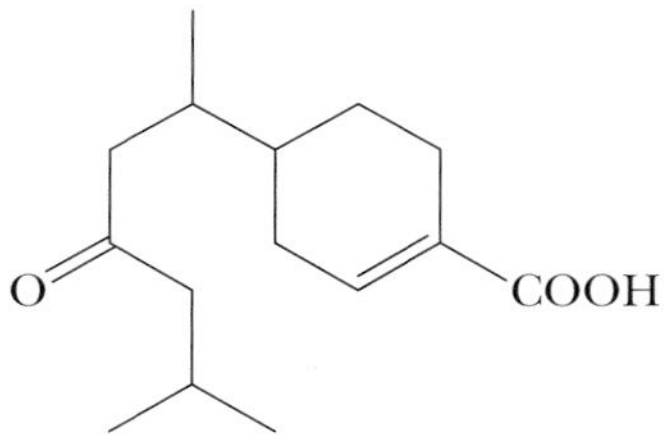

Ephedra sinica, a source of ephedrine, a potent bronchodilator. *(Paolo Koch/Photo Researchers, Inc.)*

(g)

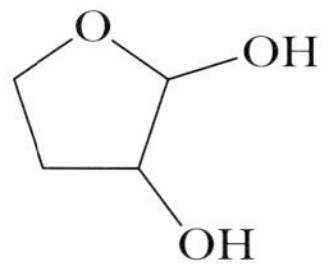

(h)

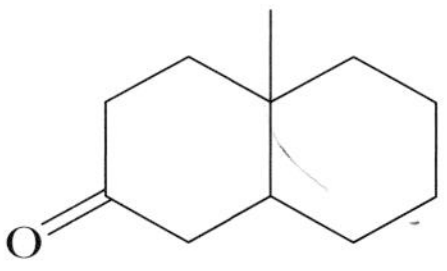

How many stereoisomers are possible for each molecule?

6.33 Label the four stereocenters in amoxicillin, which belongs to the family of semisynthetic penicillins:

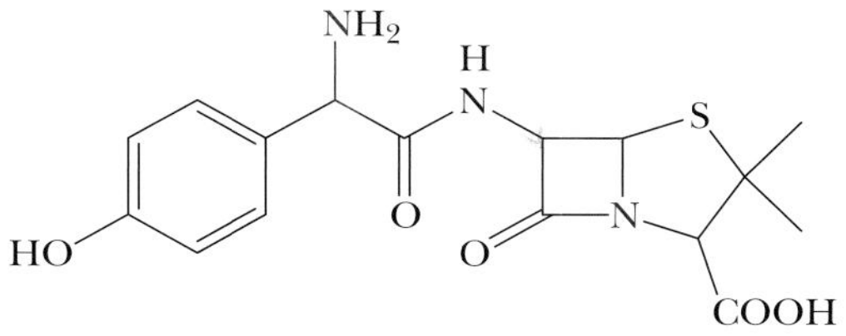

Amoxicillin

6.34 Label all stereocenters in loratadine (Claritin®) and fexofenadine (Allegra®), now the top-selling antihistamines in the United States:

(a)

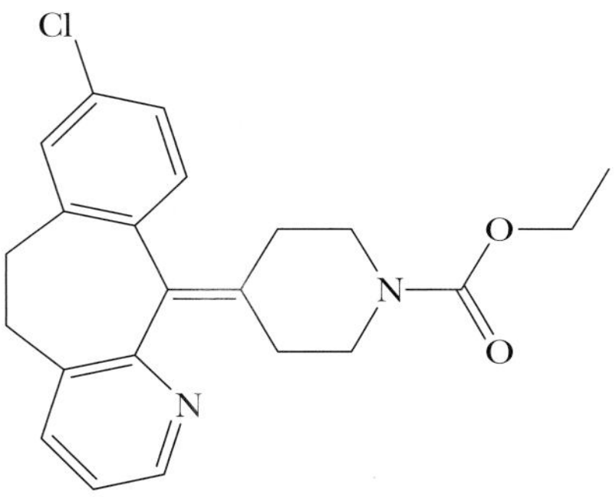

Loratadine
(Claritin)

(b)

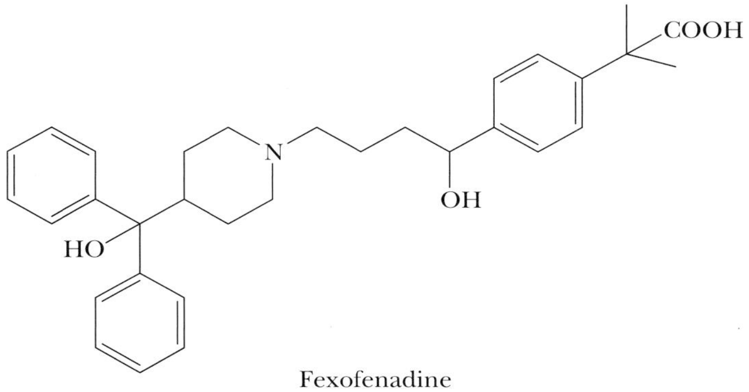

Fexofenadine
(Allegra)

How many stereoisomers are possible for each compound?

6.35 Following are structural formulas for three of the most widely prescribed drugs used to treat depression:

(a) 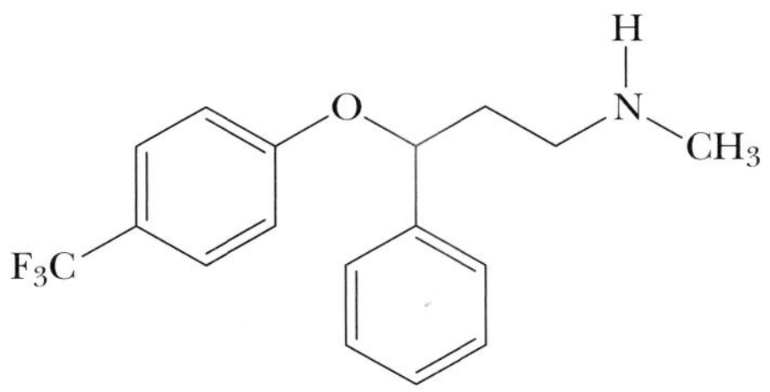 Fluoxetine
(Prozac®)

(b) 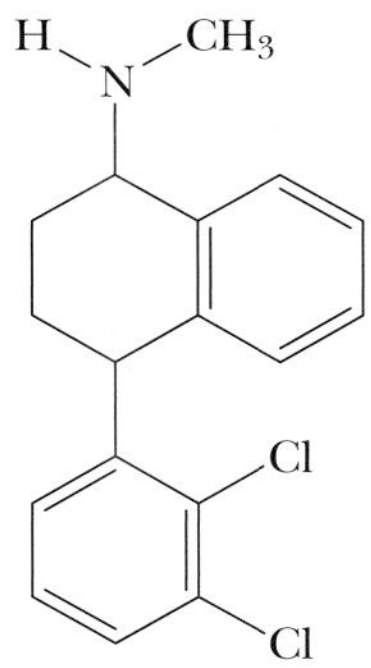 Sertraline
(Zoloft®)

(c) 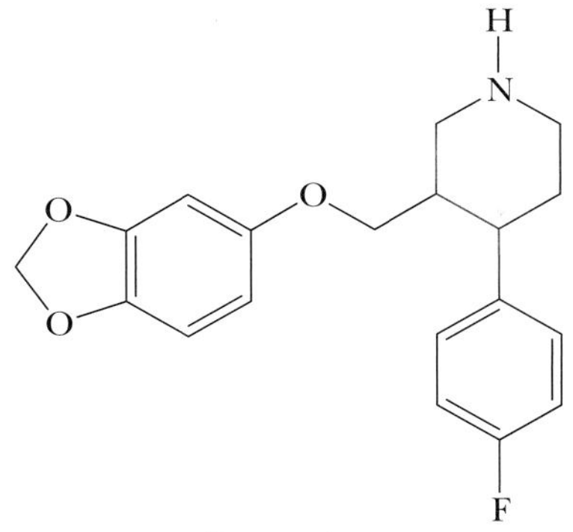 Paroxetine
(Paxil®)

Label all stereocenters in each compound, and state the number of stereoisomers possible for each.

168   Chapter 6   CHIRALITY: THE HANDEDNESS OF MOLECULES

6.36 Triamcinolone acetonide, the active ingredient in Azmacort® Inhalation Aerosol, is a steroid used to treat bronchial asthma:

Triamcinolone acetonide

(a) Label the eight stereocenters in this molecule.

(b) How many stereoisomers are possible for the molecule? (Of this number, only one is the active ingredient in Azmacort.)

6.37 Which of these structural formulas represent meso compounds?

6.38 Draw a Newman projection, viewed along the bond between carbons 2 and 3, for both the most stable and the least stable conformations of meso-tartaric acid:

6.39 How many stereoisomers are possible for 1,3-dimethylcyclopentane? Which are pairs of enantiomers? Which are meso compounds?

6.40 In Problem 3.39, you were asked to draw the more stable chair conformation of glucose, a molecule in which all groups on the six-membered ring are equatorial:

(a) Identify all stereocenters in this molecule.

(b) How many stereoisomers are possible?

(c) How many pairs of enantiomers are possible?

(d) What is the configuration (*R* or *S*) at carbons 1 and 5 in the stereoisomer shown?

6.41 What is a racemic mixture? Is a racemic mixture optically active? That is, will it rotate the plane of polarized light?

Looking Ahead

6.42 Predict the product(s) of the following reactions (in cases where more than one stereoisomer is possible, show each stereoisomer):

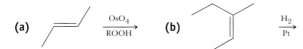

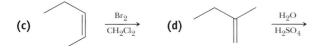

6.43 What alkene, when treated with H₂/Pd, will ensure a high yield of the stereoisomer shown:

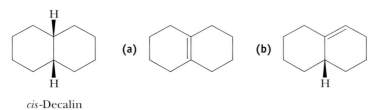

cis-Decalin

6.44 Which of the following reactions will yield a racemic mixture of products?

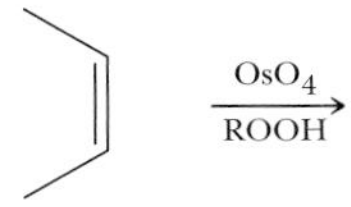

6.45 Draw all the stereoisomers that can be formed in the following reaction:

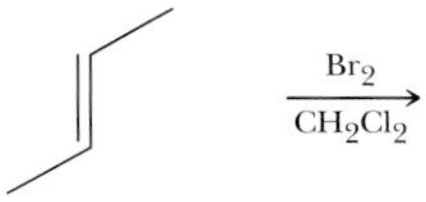

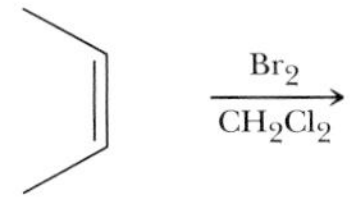

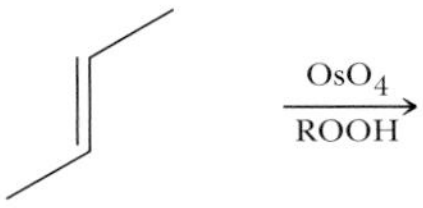

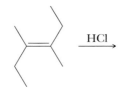

Comment on the utility of this particular reaction as a synthetic method.

6.46 Explain why the products of the following reaction do not rotate plane-polarized light:

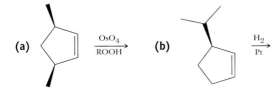

7

Haloalkanes

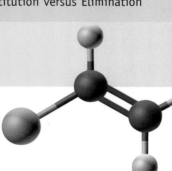

Compact disks are made from poly(vinyl chloride), which is in turn made from 1,2-dichloroethane. Inset: A molecule of vinyl chloride. *(Charles D. Winter)*

7.1   INTRODUCTION

Alkyl halide A compound containing a halogen atom covalently bonded to an alkyl group; given the symbol RX.

Compounds containing a halogen atom covalently bonded to an sp^3 hybridized carbon atom are named *haloalkanes* or, in the common system of nomenclature, *alkyl halides*. The general symbol for an **alkyl halide** is R—X, where X may be F, Cl, Br, or I:

$$R - \overset{..}{\underset{..}{X}} :$$

A haloalkane (An alkyl halide)

In this chapter, we study two characteristic reactions of haloalkanes: nucleophilic substitution and β-elimination. By these reactions, haloalkanes can be converted to alcohols, ethers, thiols, amines, and alkenes and are thus versatile molecules. Indeed, haloalkanes are often used as starting materials for the synthesis of many useful compounds encountered in all walks of life, such as medicine, food chemistry, and agriculture (to name a few).

7.2 NOMENCLATURE

A. IUPAC Names

IUPAC names for haloalkanes are derived by naming the parent alkane according to the rules given in Section 3.4A:

- Locate and number the parent chain from the direction that gives the substituent encountered first the lower number.
- Show halogen substituents by the prefixes *fluoro-*, *chloro-*, *bromo-*, and *iodo-*, and list them in alphabetical order along with other substituents.
- Use a number preceding the name of the halogen to locate each halogen on the parent chain.
- In haloalkenes, the location of the double bond determines the numbering of the parent hydrocarbon. In molecules containing functional groups designated by a suffix (for example, *-ol, -al, -one, -oic acid*), the location of the functional group indicated by the suffix determines the numbering:

3-Bromo-2-methylpentane   4-Bromocyclohexene   *trans*-2-Chlorocyclohexanol

B. Common Names

Common names of haloalkanes consist of the common name of the alkyl group, followed by the name of the halide as a separate word. Hence, the name **alkyl halide** is a common name for this class of compounds. In the following examples, the IUPAC name of the compound is given first, followed by its common name, in parentheses:

$$CH_3CHCH_2CH_3 \qquad CH_2{=}CHCl$$

(with F on the second carbon)

2-Fluorobutane   Chloroethene
(*sec*-Butyl fluoride)   (Vinyl chloride)

Several of the polyhalomethanes are common solvents and are generally referred to by their common, or trivial, names. Dichloromethane (methylene chloride) is the most widely used haloalkane solvent. Compounds of the type CHX_3 are called **haloforms**. The common name for $CHCl_3$, for example, is *chloroform*. The common name for CH_3CCl_3 is *methyl chloroform*. Methyl chloroform and trichloroethylene are solvents for commercial dry cleaning.

CH_2Cl_2	$CHCl_3$	CH_3CCl_3	$CCl_2{=}CHCl$
Dichloromethane	Trichloromethane	1,1,1-Trichloroethane	Trichloroethylene
(Methylene chloride)	(Chloroform)	(Methyl chloroform)	(Trichlor)

EXAMPLE 7.1

Write the IUPAC name for each compound:

(a) 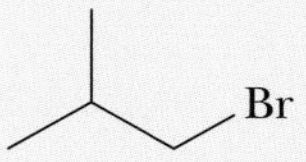 [structure: isobutyl bromide]   (b) 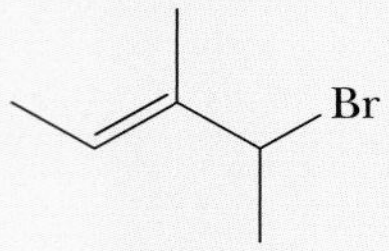 [structure]   (c) 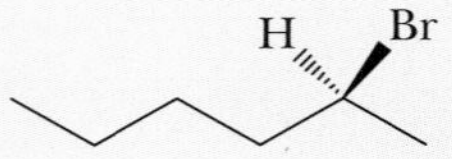 [structure]

SOLUTION

(a) 1-Bromo-2-methylpropane. Its common name is isobutyl bromide.
(b) (*E*)-4-Bromo-3-methyl-2-pentene or *trans*-4-bromo-3-methyl-2-pentene.
(c) (*S*)-2-Bromohexane.

Practice Problem 7.1

Write the IUPAC name for each compound:

(a) 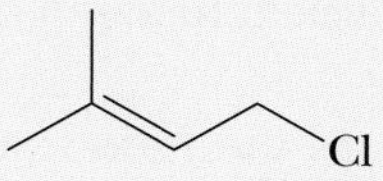 [structure]   (b) 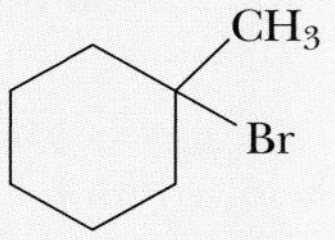 [structure with CH_3 and Br]

(c) $CH_3\overset{\underset{|}{Cl}}{CH}CH_2Cl$ (d) 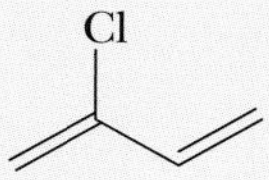 [structure with Cl]

Of all the haloalkanes, the **chlorofluorocarbons (CFCs)** manufactured under the trade name Freon® are the most widely known. CFCs are nontoxic, nonflammable, odorless, and noncorrosive. Originally, they seemed to be ideal replacements for the hazardous compounds such as ammonia and sulfur dioxide formerly used as heat-transfer agents in refrigeration systems. Among the CFCs most widely used for this purpose were trichlorofluoromethane (CCl_3F, Freon-11) and dichlorodifluoromethane (CCl_2F_2, Freon-12). The CFCs also found wide use as industrial cleaning solvents to prepare surfaces for coatings, to remove cutting oils and waxes from millings, and to remove protective coatings. In addition, they were employed as propellants in aerosol sprays.

7.3 NUCLEOPHILIC ALIPHATIC SUBSTITUTION AND β-ELIMINATION

Nucleophile An atom or a group of atoms that donates a pair of electrons to another atom or group of atoms to form a new covalent bond.

Nucleophilic substitution A reaction in which one nucleophile is substituted for another.

A **nucleophile** (nucleus-loving reagent) is any reagent that donates an unshared pair of electrons to form a new covalent bond. **Nucleophilic substitution** is any reaction in which one nucleophile is substituted for another. In the following general equations, $Nu:^-$ is the nucleophile, X is the leaving group, and substitution takes place on an sp^3 hybridized carbon atom:

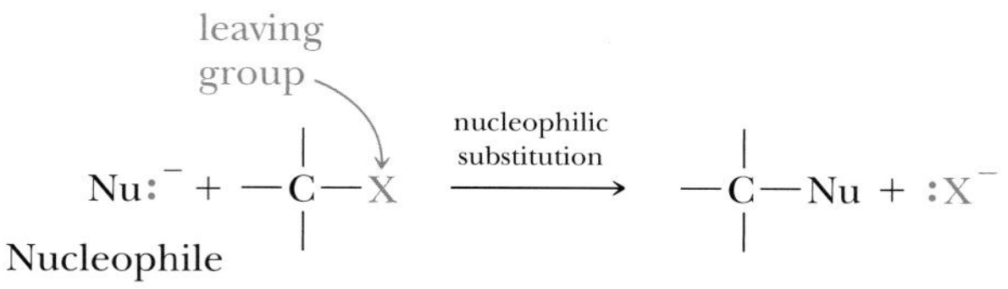

$$Nu:^- + -\underset{|}{\overset{|}{C}}-X \xrightarrow[\text{substitution}]{\text{nucleophilic}} -\underset{|}{\overset{|}{C}}-Nu + :X^-$$

Halide ions are among the best and most important leaving groups.

CHEMICAL CONNECTIONS

The Environmental Impact of Chlorofluorocarbons

Concern about the environmental impact of CFCs arose in the 1970s when researchers found that more than 4.5×10^5 kg/yr of these compounds were being emitted into the atmosphere. In 1974, Sherwood Rowland and Mario Molina announced their theory, which has since been amply confirmed, that CFCs catalyze the destruction of the stratospheric ozone layer. When released into the air, CFCs escape to the lower atmosphere. Because of their inertness, however, they do not decompose there. Slowly, they find their way to the stratosphere, where they absorb ultraviolet radiation from the sun and then decompose. As they do so, they set up a chemical reaction that leads to the destruction of the stratospheric ozone layer, which shields the earth against short-wavelength ultraviolet radiation from the sun. An increase in short-wavelength ultraviolet radiation reaching the earth is believed to promote the destruction of certain crops and agricultural species and even to increase the incidence of skin cancer in light-skinned individuals.

The concern about CFCs prompted two conventions, one in Vienna in 1985 and one in Montreal in 1987, held by the United Nations Environmental Program. The 1987 meeting produced the Montreal Protocol, which set limits on the production and use of ozone-depleting CFCs and urged the complete phaseout of their production by the year 1996. This phaseout resulted in enormous costs for manufacturers and is not yet complete in developing countries.

Rowland, Molina, and Paul Crutzen (a Dutch chemist at the Max Planck Institute for Chemistry in Germany) were awarded the 1995 Nobel prize for chemistry. As the Royal Swedish Academy of Sciences noted in awarding the prize, "By explaining the chemical mechanisms that affect the thickness of the ozone layer, these three researchers have contributed to our salvation from a global environmental problem that could have catastrophic consequences."

The chemical industry responded to the crisis by developing replacement refrigerants that have a much lower ozone-depleting potential. The most prominent replacements are the hydrofluorocarbons (HFCs) and hydrochlorofluorocarbons (HCFCs), such as the following:

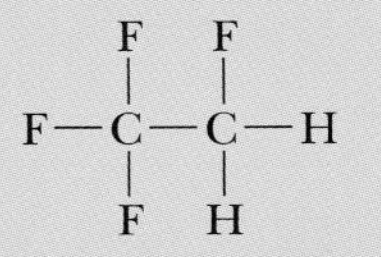

HFC-134a 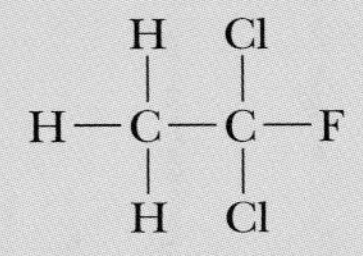   HCFC-141b

These compounds are much more chemically reactive in the atmosphere than the Freons are and are destroyed before they reach the stratosphere. However, they cannot be used in air conditioners in 1994 and earlier model cars.

Because all nucleophiles are also bases, nucleophilic substitution and base-promoted **β-elimination** are competing reactions. The ethoxide ion, for example, is both a nucleophile and a base. With bromocyclohexane, it reacts as a nucleophile (pathway shown in red) to give ethoxycyclohexane (cyclohexyl ethyl ether) and as a base (pathway shown in blue) to give cyclohexene and ethanol:

β-Elimination reaction The removal of atoms or groups of atoms from two adjacent carbon atoms, as for example, the removal of H and X from an alkyl halide or H and OH from an alcohol to form a carbon–carbon double bond.

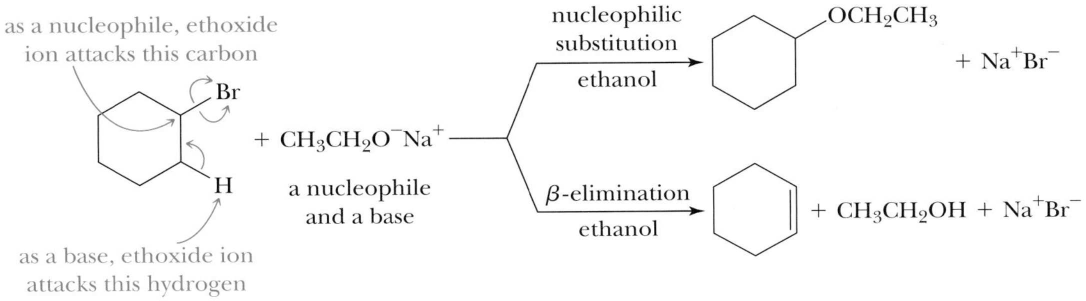

as a nucleophile, ethoxide ion attacks this carbon

as a base, ethoxide ion attacks this hydrogen

$+ \; CH_3CH_2O^- Na^+$

a nucleophile and a base

nucleophilic substitution
ethanol

OCH_2CH_3

$+ \; Na^+Br^-$

β-elimination
ethanol

$+ \; CH_3CH_2OH \; + \; Na^+Br^-$

In this chapter, we study both of these organic reactions. Using them, we can convert haloalkanes to compounds with other functional groups including alcohols, ethers, thiols, sulfides, amines, nitriles, alkenes, and alkynes. Thus, an understanding of nucleophilic substitution and β-elimination opens entirely new areas of organic chemistry.

7.4   NUCLEOPHILIC ALIPHATIC SUBSTITUTION

Nucleophilic substitution is one of the most important reactions of haloalkanes and can lead to a wide variety of new functional groups, several of which are illustrated in Table 7.1. As you study the entries in this table, note the following points:

1. If the nucleophile is negatively charged, as, for example, OH^- and RS^-, then the atom donating the pair of electrons in the substitution reaction becomes neutral in the product.
2. If the nucleophile is uncharged, as, for example, NH_3 and CH_3OH, then the atom donating the pair of electrons in the substitution reaction becomes positively charged in the product. The products then often undergo a second step involving proton transfer to yield a neutral substitution product.

TABLE 7.1   Some Nucleophilic Substitution Reactions

Reaction: $Nu^- + CH_3X \longrightarrow CH_3Nu + :X^-$

Nucleophile	Product	Class of Compound Formed
HO^-	CH_3OH	An alcohol
RO^-	CH_3OR	An ether
HS^-	CH_3SH	A thiol (a mercaptan)
RS^-	CH_3SR	A sulfide (a thioether)
$:I^-$	CH_3I	An alkyl iodide
$:NH_3$	$CH_3NH_3^+$	An alkylammonium ion
HOH	$CH_3\overset{+}{O}-H$, H	An alcohol (after proton transfer)
CH_3OH	$CH_3\overset{+}{O}-CH_3$, H	An ether (after proton transfer)

EXAMPLE 7.2

Complete these nucleophilic substitution reactions:

(a) 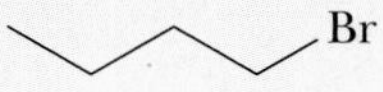 ∿∿Br + Na^+OH^- $\longrightarrow$ (b) 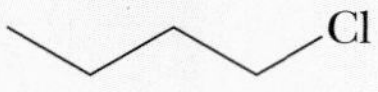 ∿∿Cl + NH_3 $\longrightarrow$

SOLUTION

(a) Hydroxide ion is the nucleophile and bromine is the leaving group:

$$\text{1-Bromobutane} + Na^+OH^- \longrightarrow \text{1-Butanol} + Na^+Br^-$$

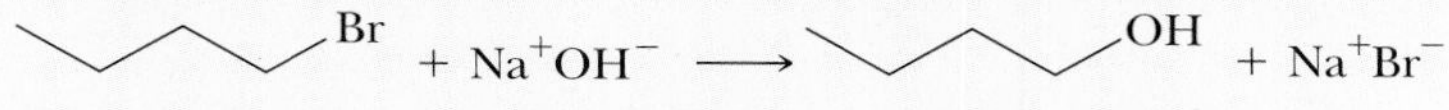

1-Bromobutane	Sodium hydroxide		1-Butanol	Sodium bromide

(b) Ammonia is the nucleophile and chlorine is the leaving group:

$$\text{1-Chlorobutane} + NH_3 \longrightarrow \text{Butylammonium chloride}$$

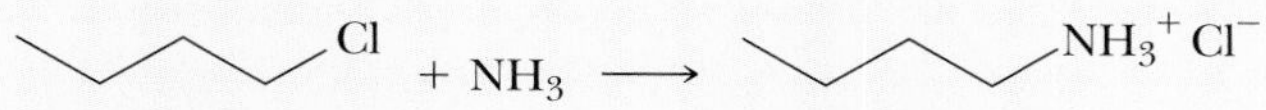

1-Chlorobutane	Ammonia		Butylammonium chloride

Practice Problem 7.2

Complete these nucleophilic substitution reactions:

(a) 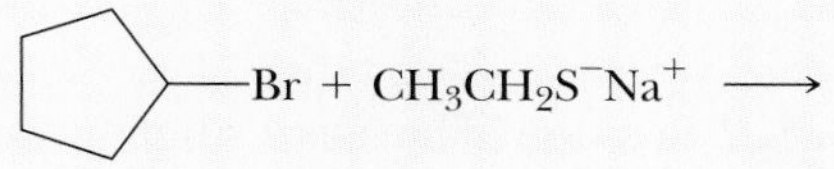 ⬠—Br + $CH_3CH_2S^-Na^+$ ⟶

(b) 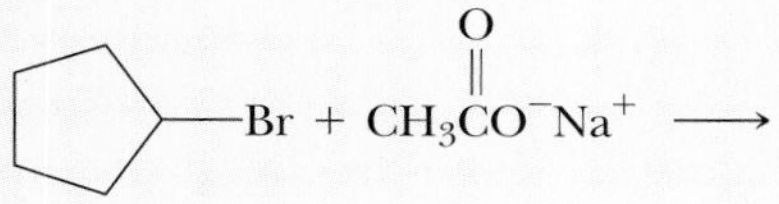 ⬠—Br + $CH_3\overset{O}{\overset{\|}{C}}O^-Na^+$ ⟶

7.5 MECHANISMS OF NUCLEOPHILIC ALIPHATIC SUBSTITUTION

On the basis of a wealth of experimental observations developed over a 70-year period, chemists have proposed two limiting mechanisms for nucleophilic substitutions. A fundamental difference between them is the timing of bond breaking between carbon and the leaving group and of bond forming between carbon and the nucleophile.

A. S$_N$2 Mechanism

At one extreme, the two processes are *concerted*, meaning that bond breaking and bond forming occur simultaneously. Thus, the departure of the leaving group is assisted by the incoming nucleophile. This mechanism is designated **S$_N$2**, where *S* stands for *Substitution*, *N* for *Nucleophilic*, and 2 for a **bimolecular reaction**. This type of substitution reaction is classified as bimolecular because both the haloalkane and the nucleophile are involved in the rate-determining step. That is, both species contribute to the rate law of the reaction:

$$\text{Rate} = k[\text{haloalkane}][\text{nucleophile}]$$

Bimolecular reaction A reaction in which two species are involved in the reaction leading to the transition state of the rate-determining step.

Following is an S$_N$2 mechanism for the reaction of hydroxide ion and bromomethane to form methanol and bromide ion:

Mechanism: An S$_N$2 Reaction

The nucleophile attacks the reactive center from the side opposite the leaving group; that is, an S$_N$2 reaction involves a backside attack by the nucleophile.

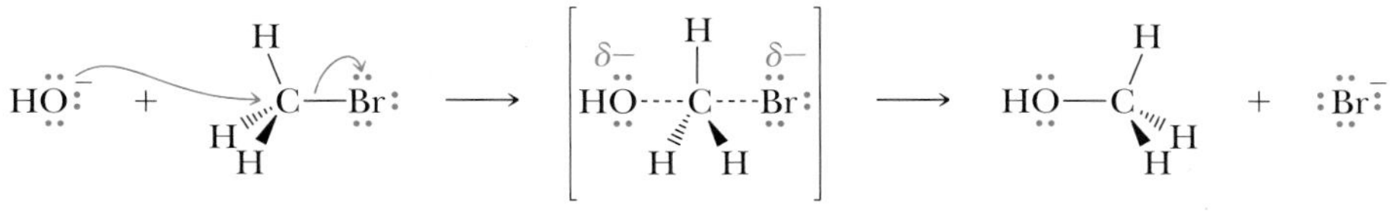

Reactants	Transition state with simultaneous bond breaking and bond forming	Products

Figure 7.1
An energy diagram for
an S_N2 reaction. There is
one transition state and
no reactive intermediate.

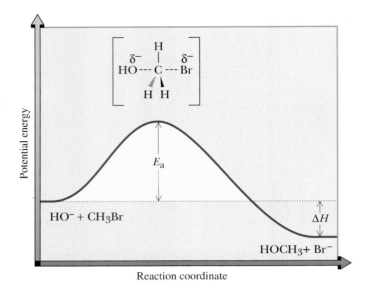

Figure 7.1 shows an energy diagram for an S_N2 reaction. There is a single transition state and no reactive intermediate.

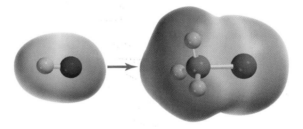

Nucleophilic attack from the side
opposite the leaving group.

An S_N2 reaction is driven by the attraction between the negative charge of the nucleophile (in this case the negatively charged oxygen of the hydroxide ion) and the center of positive charge of the electrophile (in this case the partial positive charge on the carbon bearing the chlorine leaving group).

B. S_N1 Mechanism

In the other limiting mechanism, called S_N1, bond breaking between carbon and the leaving group is completed before bond forming with the nucleophile begins. In the designation $\mathbf{S_N1}$, *S* stands for *Substitution*, *N* stands for *Nucleophilic*, and *1* stands for a ***uni*molecular reaction**. This type of substitution is classified as unimolecular because only the haloalkane is involved in the rate-determining step; that is, only the haloalkane contributes to the rate law governing the rate-determining step:

$$\text{Rate} = k[\text{haloalkane}]$$

An S_N1 reaction is illustrated by the **solvolysis** reaction of 2-bromo-2-methyl-propane (*tert*-butyl bromide) in methanol to form 2-methoxy-2-methylpropane (*tert*-butyl methyl ether).

Unimolecular reaction A reaction in which only one species is involved in the reaction leading to the transition state of the rate-determining step.

Solvolysis A nucleophilic substitution reaction in which the solvent is the nucleophile.

Mechanism: An S_N1 Reaction

Step 1: The ionization of a C—X bond forms a 3° carbocation intermediate:

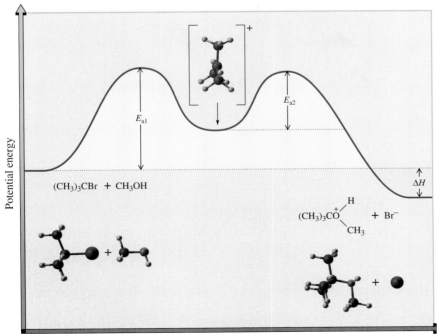

Figure 7.2 shows an energy diagram for the S_N1 reaction of 2-bromo-2-methyl-propane and methanol. There is one transition state leading to formation of the carbocation intermediate in Step 1 and a second transition state for reaction of the carbocation intermediate with methanol in Step 2 to give the oxonium ion. The reaction leading to formation of the carbocation intermediate crosses the higher energy barrier and is, therefore, the rate-determining step.

Figure 7.2
An energy diagram for the S_N1 reaction of 2-bromo-2-methylpropane and methanol. There is one transition state leading to formation of the carbocation intermediate in Step 1 and a second transition state for the reaction of the carbocation intermediate with methanol in Step 2. Step 1 crosses the higher energy barrier and is, therefore, rate determining.

If an S_N1 reaction is carried out at a tetrahedral stereocenter, the major product is a racemic mixture. We can illustrate this result with the following example: Upon ionization, the R enantiomer forms an achiral carbocation intermediate. Attack by the nucleophile from the left face of the carbocation intermediate gives the S enantiomer; attack from the right face gives the R enantiomer. Because attack by the nucleophile occurs with equal probability from either face of the planar carbocation intermediate, the R and S enantiomers are formed in equal amounts, and the product is a racemic mixture.

R enantiomer   Planar carbocation (achiral)   S enantiomer   R enantiomer
                                                          A racemic mixture

7.6 EXPERIMENTAL EVIDENCE FOR S_N1 AND S_N2 MECHANISMS

Let us now examine some of the experimental evidence on which these two contrasting mechanisms are based. As we do, we consider the following questions:

1. What effect does the structure of the nucleophile have on the rate of reaction?
2. What effect does the structure of the haloalkane have on the rate of reaction?
3. What effect does the structure of the leaving group have on the rate of reaction?
4. What is the role of the solvent?

A. Structure of the Nucleophile

Nucleophilicity is a kinetic property, which we measure by relative rates of reaction. We can establish the relative nucleophilicities for a series of nucleophiles by measuring the rate at which each displaces a leaving group from a haloalkane—for example, the rate at which each displaces bromide ion from bromoethane in ethanol at 25°C:

$$CH_3CH_2Br + NH_3 \longrightarrow CH_3CH_2NH_3^+ + Br^-$$

Relative nucleophilicity The relative rates at which a nucleophile reacts in a reference nucleophilic substitution reaction.

From these studies, we can then make correlations between the structure of the nucleophile and its **relative nucleophilicity**. Table 7.2 lists the types of nucleophiles we deal with most commonly in this text.

Because the nucleophile participates in the rate-determining step in an S_N2 reaction, the better the nucleophile, the more likely it is that the reaction will occur by that mechanism. The nucleophile does not participate in the rate-determining step for an S_N1 reaction. Thus, an S_N1 reaction can, in principle, occur at approximately the same rate with any of the common nucleophiles, regardless of their relative nucleophilicities.

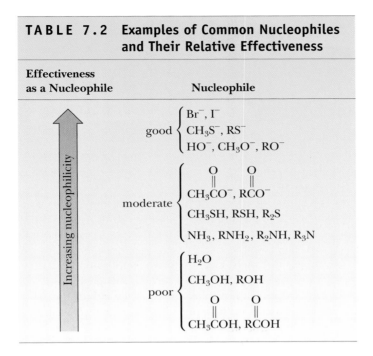

TABLE 7.2 Examples of Common Nucleophiles and Their Relative Effectiveness

Effectiveness as a Nucleophile		Nucleophile
(Increasing nucleophilicity ↑)	good	Br^-, I^- CH_3S^-, RS^- HO^-, CH_3O^-, RO^-
	moderate	$CH_3\overset{\displaystyle O}{\overset{\|}{C}}O^-$, $R\overset{\displaystyle O}{\overset{\|}{C}}O^-$ CH_3SH, RSH, R_2S NH_3, RNH_2, R_2NH, R_3N
	poor	H_2O CH_3OH, ROH $CH_3\overset{\displaystyle O}{\overset{\|}{C}}OH$, $R\overset{\displaystyle O}{\overset{\|}{C}}OH$

B. Structure of the Haloalkane

S_N1 reactions are governed mainly by **electronic factors**, namely, the relative stabilities of carbocation intermediates. S_N2 reactions, by contrast, are governed mainly by **steric factors**, and their transition states are particularly sensitive to crowding about the site of reaction. The distinction is as follows:

Steric hindrance The ability of groups, because of their size, to hinder access to a reaction site within a molecule.

1. *Relative stabilities of carbocations.* As we learned in Section 5.3A, 3° carbocations are the most stable carbocations, requiring the lowest activation energy for their formation, whereas 1° carbocations are the least stable requiring the highest activation energy for their formation. In fact, 1° carbocations are so unstable that they have never been observed in solution. Therefore, 3° haloalkanes are most likely to react by carbocation formation; 2° haloalkanes are less likely to react in this manner, and methyl and 1° haloalkanes never react in that manner.

2. *Steric hindrance.* To complete a substitution reaction, the nucleophile must approach the substitution center and begin to form a new covalent bond to it. If we compare the ease of approach by the nucleophile to the substitution center of a 1° haloalkane with that of a 3° haloalkane, we see that the approach is considerably easier in the case of the 1° haloalkane. Two hydrogen atoms and one alkyl group screen the backside of the substitution center of a 1° haloalkane. In contrast, three alkyl groups screen the backside of the substitution center of a 3° haloalkane. This center in bromoethane is easily accessible to a nucleophile, while there is extreme crowding around it in 2-bromo-2-methylpropane:

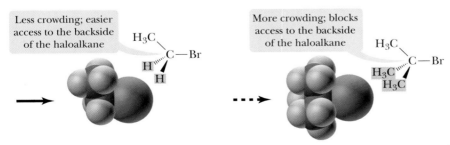

Less crowding; easier access to the backside of the haloalkane

More crowding; blocks access to the backside of the haloalkane

Bromoethane (Ethyl bromide)   2-Bromo-2-methylpropane (*tert*-Butyl bromide)

Figure 7.3
Effect of electronic and steric factors in competition between S_N1 and S_N2 reactions of haloalkanes.

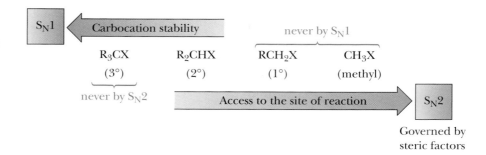

Given the competition between electronic and steric factors, we find that 3° haloalkanes react by an S_N1 mechanism because 3° carbocation intermediates are particularly stable and because the backside approach of a nucleophile to the substitution center in a 3° haloalkane is hindered by the three groups surrounding it; 3° haloalkanes never react by an S_N2 mechanism. Halomethanes and 1° haloalkanes have little crowding around the substitution center and react by an S_N2 mechanism; they never react by an S_N1 mechanism, because methyl and primary carbocations are so unstable. Secondary haloalkanes may react by either an S_N1 or an S_N2 mechanism, depending on the nucleophile and solvent. The competition between electronic and steric factors and their effects on relative rates of nucleophilic substitution reactions of haloalkanes are summarized in Figure 7.3.

C. The Leaving Group

In the transition state for nucleophilic substitution on a haloalkane, the leaving group develops a partial negative charge in both S_N1 and S_N2 reactions; therefore, the ability of a group to function as a leaving group is related to how stable it is as an anion. The most stable anions and the best leaving groups are the conjugate bases of strong acids. Thus, we can use the information on the relative strengths of organic and inorganic acids in Table 2.1 to determine which anions are the best leaving groups:

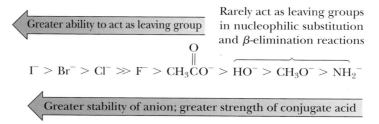

The best leaving groups in this series are the halogens I^-, Br^-, and Cl^-. Hydroxide ion (OH^-), methoxide ion (CH_3O^-), and amide ion (NH_2^-) are such poor leaving groups that they rarely, if ever, are displaced in nucleophilic substitution reactions.

D. The Solvent

Solvents provide the medium in which reactants are dissolved and in which nucleophilic substitution reactions take place. Common solvents for these reactions are divided into two groups: **protic** and **aprotic**.

    Protic solvents contain —OH groups and are hydrogen-bond donors. Common protic solvents for nucleophilic substitution reactions are water, low-molecular-weight alcohols, and low-molecular-weight carboxylic acids (Table 7.3). Each is able to solvate both the anionic and cationic components of ionic compounds

Protic solvent A hydrogen bond donor solvent, as for example water, ethanol, and acetic acid.

TABLE 7.3 Common Protic Solvents

Protic Solvent	Structure	Polarity of Solvent	Notes
Water	H_2O		These solvents favor S_N1 reactions. The greater the polarity of the solvent, the easier it is to form carbocations in it, because both the carbocation and the negatively charged leaving group can be solvated.
Formic acid	HCOOH	Increasing ↑	
Methanol	CH_3OH		
Ethanol	CH_3CH_2OH		
Acetic acid	CH_3COOH		

by electrostatic interaction between its partially negatively charged oxygen(s) and the cation and between its partially positively charged hydrogen and the anion. These same properties aid in the ionization of C—X bonds to give an X^- anion and a carbocation; thus, protic solvents are good solvents in which to carry out S_N1 reactions.

Aprotic solvents do not contain —OH groups and cannot function as hydrogen-bond donors. Table 7.4 lists the aprotic solvents most commonly used for nucleophilic substitution reactions. Dimethyl sulfoxide and acetone are polar aprotic solvents; dichloromethane and diethyl ether are nonpolar aprotic solvents. The aprotic solvents listed in the table are particularly good ones in which to carry out S_N2 reactions. Polar aprotic solvents are able to solvate only cations; they are not able to solvate anions; therefore, they allow for "naked" and highly reactive anions as nucleophiles.

Aprotic solvent A solvent that cannot serve as a hydrogen bond donor, as for example, acetone, diethyl ether, and dichloromethane.

TABLE 7.4 Common Aprotic Solvents

Aprotic Solvent	Structure	Polarity of Solvent	Notes
Dimethyl sulfoxide (DMSO)	$\overset{\text{O}}{\overset{\|}{CH_3SCH_3}}$		These solvents favor S_N2 reactions. Although solvents at the top of this list are polar, the formation of carbocations in them is far more difficult than in protic solvents, because the anionic leaving group cannot be solvated by these solvents.
Acetone	$\overset{\text{O}}{\overset{\|}{CH_3CCH_3}}$	Increasing ↑	
Dichloromethane	CH_2Cl_2		
Diethyl ether	$(CH_3CH_2)_2O$		

Table 7.5 summarizes the factors favoring S_N1 or S_N2 reactions; it also shows the change in configuration when nucleophilic substitution takes place at a stereocenter.

7.7 ANALYSIS OF SEVERAL NUCLEOPHILIC SUBSTITUTION REACTIONS

Predictions about the mechanism for a particular nucleophilic substitution reaction must be based on considerations of the structure of the haloalkane, the nucleophile, the leaving group, and the solvent. Following are analyses of three such reactions:

TABLE 7.5   Summary of S_N1 versus S_N2 Reactions of Haloalkanes

Type of Haloalkane	S_N2	S_N1
Methyl CH_3X	S_N2 is favored.	S_N1 does not occur. The methyl cation is so unstable that it is never observed in solution.
Primary RCH_2X	S_N2 is favored.	S_N1 does not occur. Primary carbocations are so unstable that they are not observed in solution.
Secondary R_2CHX	S_N2 is favored in aprotic solvents with good nucleophiles.	S_N1 is favored in protic solvents with poor nucleophiles.
Tertiary R_3CX	S_N2 does not occur, because of steric hindrance around the substitution center.	S_N1 is favored because of the ease of formation of tertiary carbocations.
Substitution at a stereocenter	Inversion of configuration. The nucleophile attacks the stereocenter from the side opposite the leaving group.	Racemization. The carbocation intermediate is planar, and an attack by the nucleophile occurs with equal probability from either side.

Nucleophilic Substitution 1

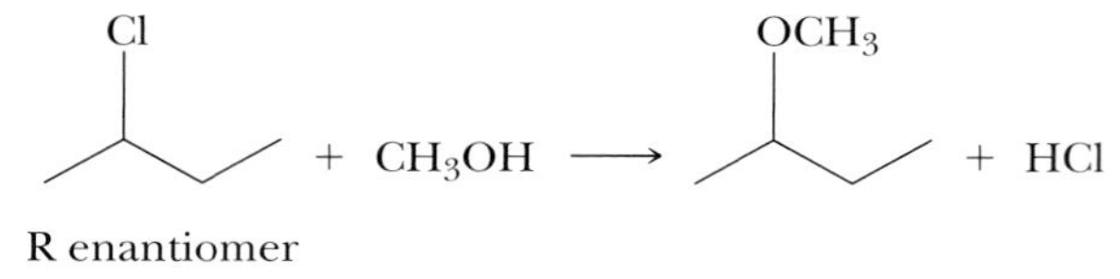

Methanol is a polar protic solvent and a good one in which to form carbocations. 2-Chlorobutane ionizes in methanol to form a 2° carbocation intermediate. Methanol is a weak nucleophile. From this analysis, we predict that reaction is by an S_N1 mechanism. The 2° carbocation intermediate (an electrophile) then reacts with methanol (a nucleophile) followed by proton transfer to give the observed product. The product is formed as a 50:50 mixture of R and S configurations; that is, it is formed as a racemic mixture.

Nucleophilic Substitution 2

This is a 1° bromoalkane in the presence of iodide ion, a good nucleophile. Because 1° carbocations are so unstable, they never form in solution, and an S_N1 reaction is not possible. Dimethyl sulfoxide (DMSO), a polar aprotic solvent, is a good solvent in which to carry out S_N2 reactions. From this analysis, we predict that reaction is by an S_N2 mechanism.

Nucleophilic Substitution 3

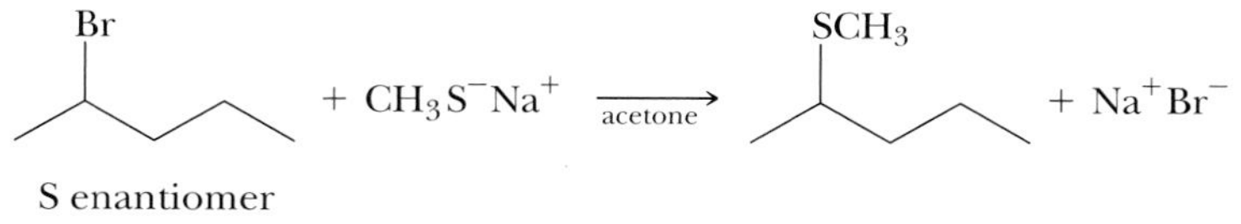

Bromine ion is a good leaving group on a 2° carbon. The methylsulfide ion is a good nucleophile. Acetone, a polar aprotic solvent, is a good medium in which to carry out S_N2 reactions, but a poor medium in which to carry out S_N1 reactions. We predict that reaction is by an S_N2 mechanism and that the product formed has the R configuration.

EXAMPLE 7.3

Write the expected product for each nucleophilic substitution reaction, and predict the mechanism by which the product is formed:

(a) 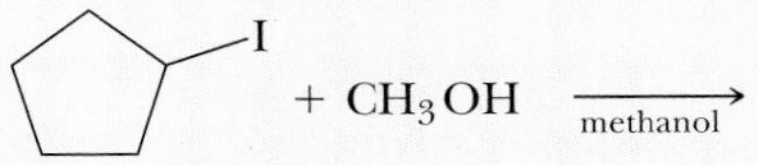 $+ CH_3OH$ $\xrightarrow{\text{methanol}}$

(b) 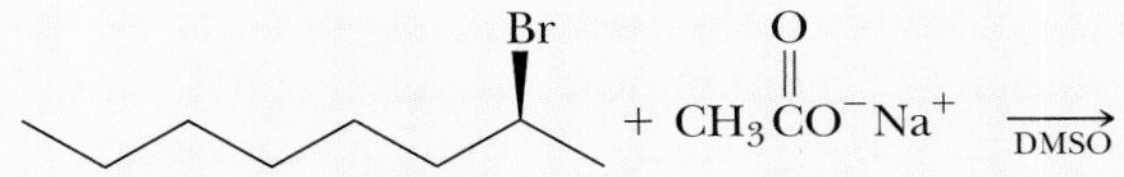 $+ CH_3\overset{\displaystyle O}{\overset{\|}{C}}O^- Na^+$ $\xrightarrow{\text{DMSO}}$

SOLUTION

(a) Methanol is a poor nucleophile. It is also a polar protic solvent that is able to solvate carbocations. Ionization of the carbon–iodine bond forms a 2° carbocation intermediate. We predict an S_N1 mechanism:

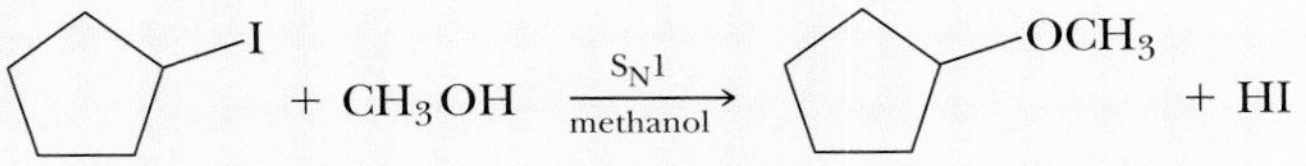

$+ CH_3OH$ $\xrightarrow[\text{methanol}]{S_N1}$ $+ HI$

(b) Bromide is a good leaving group on a 2° carbon. Acetate ion is a moderate nucleophile. DMSO is a particularly good solvent for S_N2 reactions. We predict substitution by an S_N2 mechanism with inversion of configuration at the stereocenter:

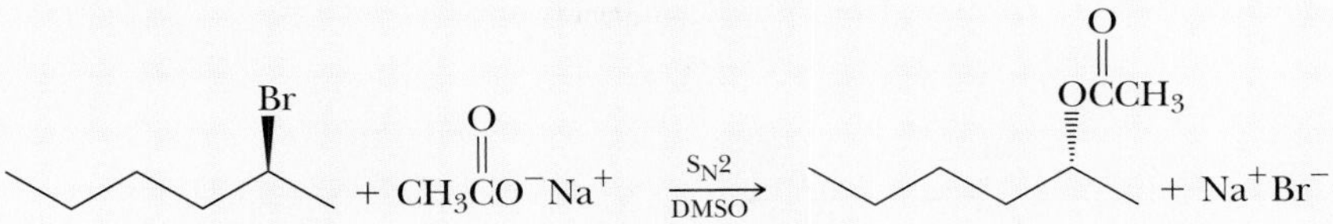

$+ CH_3\overset{\displaystyle O}{\overset{\|}{C}}O^- Na^+$ $\xrightarrow[\text{DMSO}]{S_N2}$ $+ Na^+Br^-$

Practice Problem 7.3

Write the expected product for each nucleophilic substitution reaction, and predict the mechanism by which the product is formed:

(a) 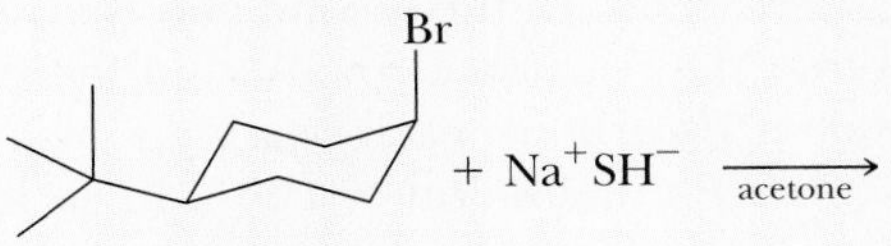 $+ Na^+SH^-$ $\xrightarrow{\text{acetone}}$

(b) 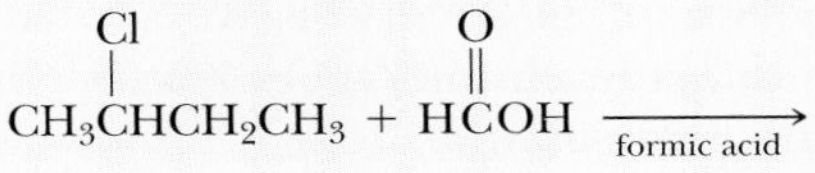 $CH_3\overset{\displaystyle Cl}{\overset{|}{C}}HCH_2CH_3 + H\overset{\displaystyle O}{\overset{\|}{C}}OH$ $\xrightarrow{\text{formic acid}}$

7.8 β-ELIMINATION

Dehydrohalogenation Removal of
—H and —X from adjacent
carbons; a type of β-elimination.

In this section, we study a type of β-elimination called **dehydrohalogenation**. In the presence of a strong base, such as hydroxide ion or ethoxide ion, halogen can be removed from one carbon of a haloalkane and hydrogen from an adjacent carbon to form a carbon–carbon double bond:

$$-\overset{\underset{|}{\beta}}{\underset{\underset{H}{|}}{C}}-\overset{\underset{|}{\alpha}}{\underset{\underset{X}{|}}{C}}- \; + \; CH_3CH_2O^-Na^+ \; \xrightarrow[CH_3CH_2OH]{} \; \overset{}{\underset{}{C}}{=}\overset{}{\underset{}{C}} \; + \; CH_3CH_2OH \; + \; Na^+X^-$$

A haloalkane   Base   An alkene

As the equation shows, we call the carbon bearing the halogen the *α-carbon* and the adjacent carbon the *β-carbon*.

Because most nucleophiles can also act as bases and vice versa, it is important to keep in mind that β-elimination and nucleophilic substitution are competing reactions. In this section, we concentrate on β-elimination. In Section 7.10, we examine the results of competition between the two.

Common strong bases used for β-elimination are OH^-, OR^-, and NH_2^-. Following are three examples of base-promoted β-elimination reactions:

1-Bromooctane   +   *t*-BuO⁻K⁺   ⟶
                      Potassium
                      tert-butoxide

                                        +   *t*-BuOH + Na⁺Br⁻
1-Octene

2-Bromo-2- $\xrightarrow[CH_3CH_2OH]{CH_3CH_2O^-Na^+}$ 2-Methyl-   2-Methyl-1-butene
methylbutane   2-butene
                                                   (major product)

1-Bromo-1-methyl- $\xrightarrow[CH_3OH]{CH_3O^-Na^+}$ 1-Methyl-   +   Methylene-
cyclopentane   cyclopentene   cyclopentane
                                              (major product)

Zaitsev's rule A rule stating that
the major product from a
β-elimination reaction is the most
stable alkene; that is, the major
product is the alkene with the
greatest number of substituents on
the carbon–carbon double bond.

In the first example, the base is shown as a reactant. In the second and third examples, the base is a reactant, but is shown over the reaction arrow. Also in the second and third examples, there are nonequivalent β-carbons, each bearing a hydrogen; therefore, two alkenes are possible from each β-elimination reaction. In each case, the major product of these and most other β-elimination reactions is the more substituted (and therefore the more stable—see Section 5.5B) alkene. We say that each reaction follows **Zaitsev's rule** or, alternatively, that each undergoes Zaitsev elimination, to honor the chemist who first made this generalization.

EXAMPLE 7.4

Predict the β-elimination product(s) formed when each bromoalkane is treated with sodium ethoxide in ethanol (if two might be formed, predict which is the major product):

(a)

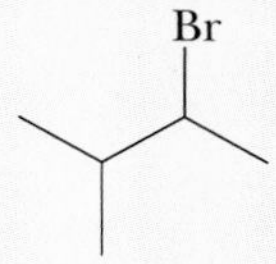

Br

(b)

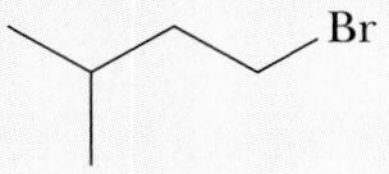

Br

SOLUTION

(a) There are two nonequivalent β-carbons in this bromoalkane, and two alkenes are possible. 2-Methyl-2-butene, the more substituted alkene, is the major product:

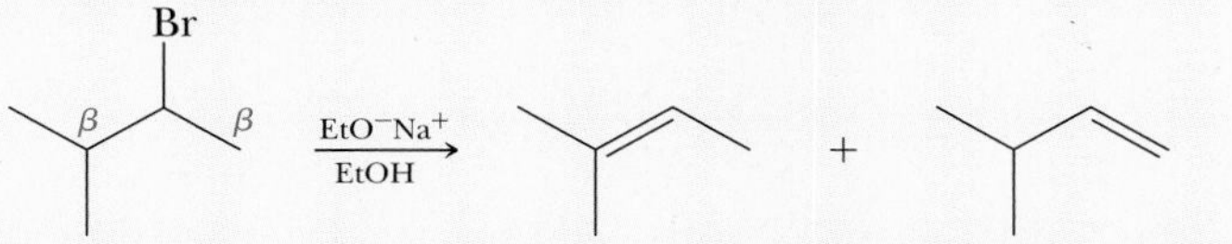

Br

β   β

$\xrightarrow[\text{EtOH}]{\text{EtO}^-\text{Na}^+}$

+

2-Methyl-2-butene   3-Methyl-1-butene
(major product)

(b) There is only one β-carbon in this bromoalkane, and only one alkene is possible:

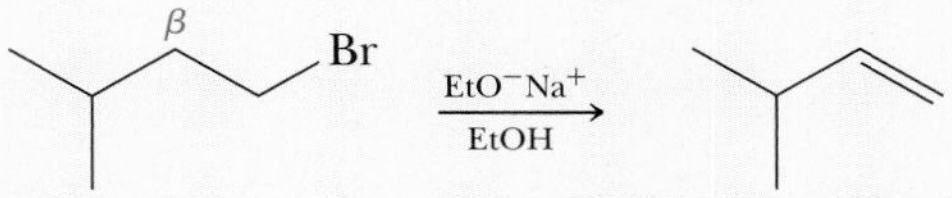

β   Br

$\xrightarrow[\text{EtOH}]{\text{EtO}^-\text{Na}^+}$

3-Methyl-1-butene

Practice Problem 7.4

Predict the β-elimination products formed when each chloroalkane is treated with sodium ethoxide in ethanol (if two products might be formed, predict which is the major product):

(a)

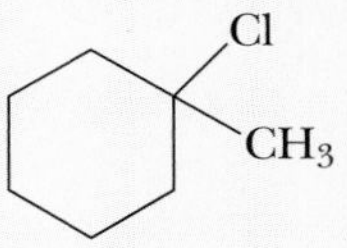

Cl

CH₃

(b)

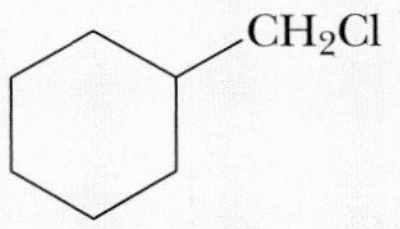

CH₂Cl

(c)

Cl   CH₃

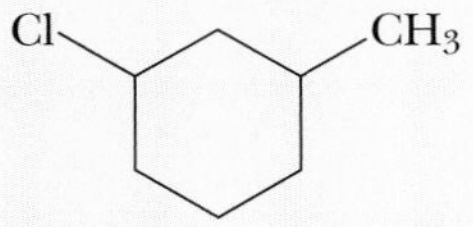

7.9 MECHANISMS OF β-ELIMINATION

There are two limiting mechanisms of β-elimination reactions. A fundamental difference between them is the timing of the bond-breaking and bond-forming steps. Recall that we made this same statement about the two limiting mechanisms for nucleophilic substitution reactions in Section 7.5.

A. E1 Mechanism

At one extreme, breaking of the C—X bond is complete before any reaction occurs with base to lose a hydrogen and before the carbon–carbon double bond is formed. This mechanism is designated **E1**, where *E* stands for *e*limination and *1* stands for a *uni*molecular reaction; only *one* species, in this case the haloalkane, is involved in the rate-determining step. The rate law for an E1 reaction has the same form as that for an S_N1 reaction:

$$\text{Rate} = k[\text{haloalkane}]$$

The mechanism for an E1 reaction is illustrated by the reaction of 2-bromo-2-methylpropane to form 2-methylpropene. In this two-step mechanism, the rate-determining step is the ionization of the carbon–halogen bond to form a carbocation intermediate (just as it is in an S_N1 mechanism).

Mechanism: E1 Reaction of 2-Bromo-2-methylpropane

Step 1: Rate-determining ionization of the C—Br bond gives a carbocation intermediate:

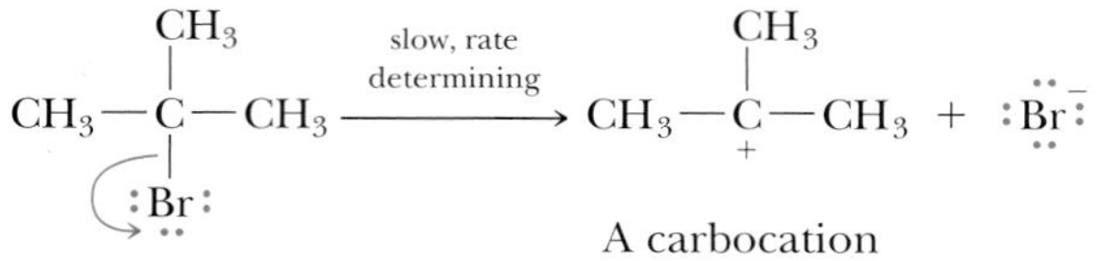

A carbocation intermediate

Step 2: Proton transfer from the carbocation intermediate to methanol (which in this instance is both the solvent and a reactant) gives the alkene:

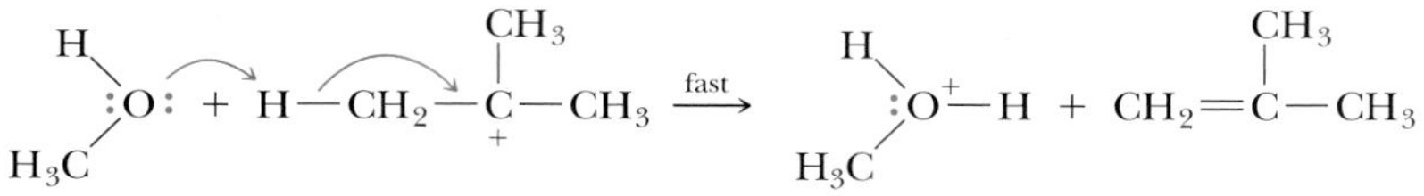

B. E2 Mechanism

At the other extreme is a concerted process. In an **E2** reaction, *E* stands for *e*limination, and *2* stands for *bi*molecular. Because the base removes a β-hydrogen at the same time the C—X bond is broken to form a halide ion, the rate law for the rate-determining step is dependent on both the haloalkane and the base:

$$\text{Rate} = k[\text{haloalkane}][\text{base}]$$

The stronger the base, the more likely it is that the E2 mechanism will be in operation. We illustrate an E2 mechanism by the reaction of 1-bromopropane with sodium ethoxide.

Mechanism: E2 Reaction of 1-Bromopropane

In this mechanism, proton transfer to the base, formation of the carbon–carbon double bond, and the ejection of bromide ion occur simultaneously; that is, all bond-forming and bond-breaking steps occur at the same time.

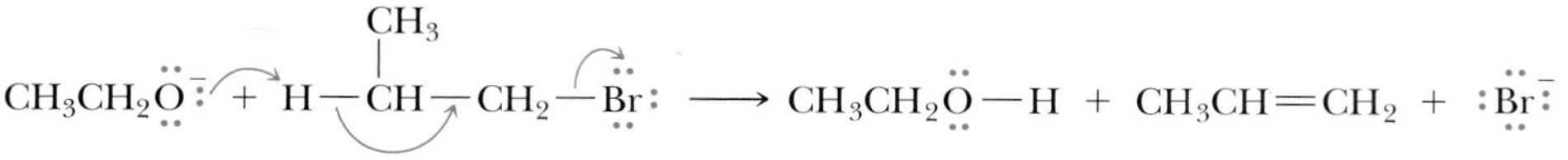

For both E1 and E2 reactions, the major product is that formed in accordance with Zaitsev's rule as illustrated by this E2 reaction:

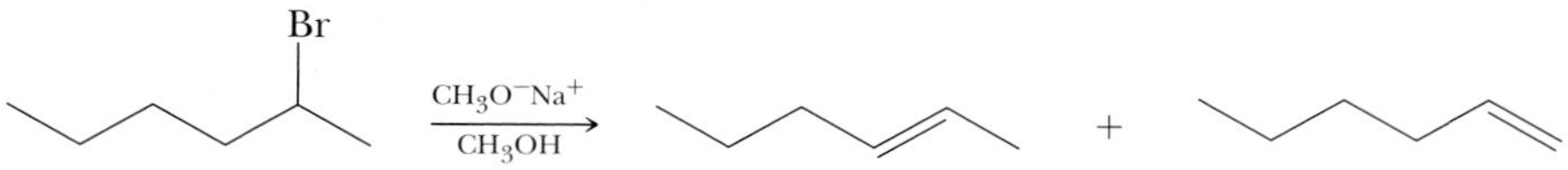

2-Bromohexane — 2-Hexene (74%) + 1-Hexene (26%)

Table 7.6 summarizes these generalizations about β-elimination reactions of haloalkanes.

TABLE 7.6 Summary of E1 versus E2 Reactions of Haloalkanes

Haloalkane	E1	E2
Primary RCH_2X	E1 does not occur. Primary carbocations are so unstable that they are never observed in solution.	E2 is favored.
Secondary R_2CHX	Main reaction with weak bases such as H_2O and ROH.	Main reaction with strong bases such as OH^- and OR^-.
Tertiary R_3CX	Main reaction with weak bases such as H_2O and ROH.	Main reaction with strong bases such as OH^- and OR^-.

EXAMPLE 7.5

Predict whether each β-elimination reaction proceeds predominantly by an E1 or E2 mechanism, and write a structural formula for the major organic product:

(a)
$$\underset{\underset{Cl}{|}}{\overset{\overset{CH_3}{|}}{CH_3CCH_2CH_3}} + NaOH \xrightarrow[H_2O]{80°C}$$

(b)
$$\underset{\underset{Cl}{|}}{\overset{\overset{CH_3}{|}}{CH_3CCH_2CH_3}} \xrightarrow{CH_3COOH}$$

SOLUTION

(a) A 3° chloroalkane is heated with a strong base. Elimination by an E2 reaction predominates, giving 2-methyl-2-butene as the major product:

$$\underset{\underset{Cl}{|}}{\overset{\overset{CH_3}{|}}{CH_3CCH_2CH_3}} + NaOH \xrightarrow[H_2O]{80°C} \overset{\overset{CH_3}{|}}{CH_3C}=CHCH_3 + NaCl + H_2O$$

(b) A 3° chloroalkane dissolved in acetic acid, a solvent that promotes the formation of carbocations, forms a 3° carbocation that then loses a proton to give 2-methyl-2-butene as the major product. The reaction is by an E1 mechanism:

$$\underset{\underset{Cl}{|}}{\overset{\overset{CH_3}{|}}{CH_3CCH_2CH_3}} \xrightarrow{CH_3COOH} \overset{\overset{CH_3}{|}}{CH_3C}=CHCH_3 + HCl$$

Practice Problem 7.5

Predict whether each elimination reaction proceeds predominantly by an E1 or E2 mechanism, and write a structural formula for the major organic product:

(a) 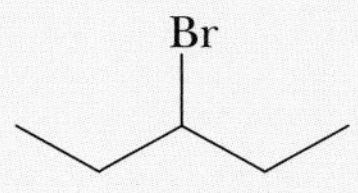
$+ CH_3O^-Na^+ \xrightarrow[methanol]{}$

(b) 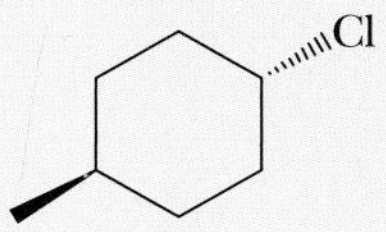
$+ CH_3CH_2O^-Na^+ \xrightarrow[ethanol]{}$

7.10 SUBSTITUTION VERSUS ELIMINATION

Thus far, we have considered two types of reactions of haloalkanes: nucleophilic substitution and β-elimination. Many of the nucleophiles we have examined—for example, hydroxide ion and alkoxide ions—are also strong bases. Accordingly, nucleophilic substitution and β-elimination often compete with each other, and the ratio of products formed by these reactions depends on the relative rates of the two reactions:

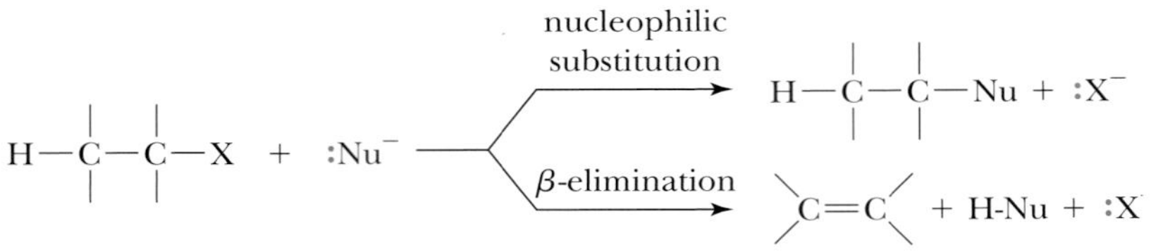

A. S$_N$1-versus-E1 Reactions

Reactions of secondary and tertiary haloalkanes in polar protic solvents give mixtures of substitution and elimination products. In both reactions, Step 1 is the formation of a carbocation intermediate. This step is then followed by either (1) the loss of a hydrogen to give an alkene (E1) or (2) reaction with solvent to give a substitution product (S$_N$1). In polar protic solvents, the products formed depend only on the structure of the particular carbocation. For example, *tert*-butyl chloride and *tert*-butyl iodide in 80% aqueous ethanol both react with solvent, giving the same mixture of substitution and elimination products:

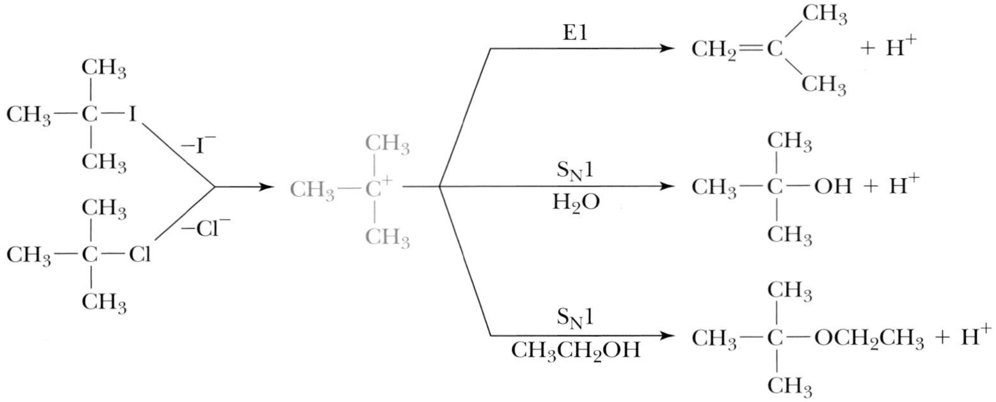

Because iodide ion is a better leaving group than chloride ion, *tert*-butyl iodide reacts over 100 times faster than *tert*-butyl chloride. Yet the ratio of products is the same.

B. S$_N$2-versus-E2 Reactions

It is considerably easier to predict the ratio of substitution to elimination products for reactions of haloalkanes with reagents that act as both nucleophiles and bases. The guiding principles are as follows:

1. Branching at the α-carbon or β-carbon(s) increases steric hindrance about the α-carbon and significantly retards S$_N$2 reactions. By contrast, branching at the α-carbon or β-carbon(s) increases the rate of E2 reactions because of the increased stability of the alkene product.

2. The greater the nucleophilicity of the attacking reagent, the greater is the S_N2-to-E2 ratio. Conversely, the greater the basicity of the attacking reagent, the greater is the E2-to-S_N2 ratio.

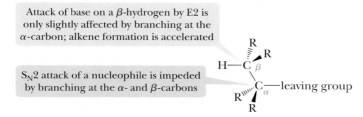

Attack of base on a β-hydrogen by E2 is only slightly affected by branching at the α-carbon; alkene formation is accelerated

S_N2 attack of a nucleophile is impeded by branching at the α- and β-carbons

Primary halides react with bases/nucleophiles to give predominantly substitution products. With strong bases, such as hydroxide ion and ethoxide ion, a percentage of the product is formed by an E2 reaction, but it is generally small compared with that formed by an S_N2 reaction. With strong, bulky bases, such as *tert*-butoxide ion, the E2 product becomes the major product. Tertiary halides react with all strong bases/good nucleophiles to give only elimination products.

Secondary halides are borderline, and substitution or elimination may be favored, depending on the particular base/nucleophile, solvent, and temperature at which the reaction is carried out. Elimination is favored with strong bases/good nucleophiles—for example, hydroxide ion and ethoxide ion. Substitution is favored with weak bases/poor nucleophiles—for example, acetate ion. Table 7.7 summarizes these generalizations about substitution versus elimination reactions of haloalkanes.

TABLE 7.7 Summary of Substitution versus Elimination Reactions of Haloalkanes

Halide	Reaction	Comments
Methyl CH_3X	S_N2	The only substitution reactions observed.
	S_N1	S_N1 reactions of methyl halides are never observed. The methyl cation is so unstable that it is never formed in solution.
Primary RCH_2X	S_N2	The main reaction with strong bases such as OH^- and EtO^-. Also, the main reaction with good nucleophiles/weak bases, such as I^- and CH_3COO^-.
	E2	The main reaction with strong, bulky bases, such as potassium *tert*-butoxide.
	$S_N1/E1$	Primary cations are never formed in solution; therefore, S_N1 and E1 reactions of primary halides are never observed.
Secondary R_2CHX	S_N2	The main reaction with weak bases/good nucleophiles, such as I^- and CH_3COO^-.
	E2	The main reaction with strong bases/good nucleophiles, such as OH^- and $CH_3CH_2O^-$.
	$S_N1/E1$	Common in reactions with weak nucleophiles in polar protic solvents, such as water, methanol, and ethanol.
Tertiary R_3CX	S_N2	S_N2 reactions of tertiary halides are never observed because of the extreme crowding around the 3° carbon.
	E2	Main reaction with strong bases, such as HO^- and RO^-.
	$S_N1/E1$	Main reactions with poor nucleophiles/weak bases.

EXAMPLE 7.6

Predict whether each reaction proceeds predominantly by substitution (S_N1 or S_N2) or elimination (E1 or E2) or whether the two compete, and write structural formulas for the major organic product(s):

(a) 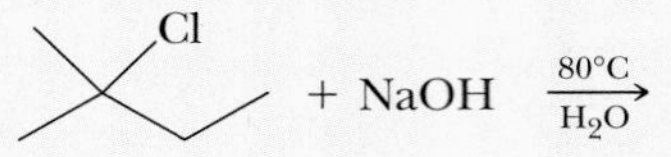 [structure: 2-chloro-2-methylbutane] + NaOH $\xrightarrow[\text{H}_2\text{O}]{80°C}$ (b) 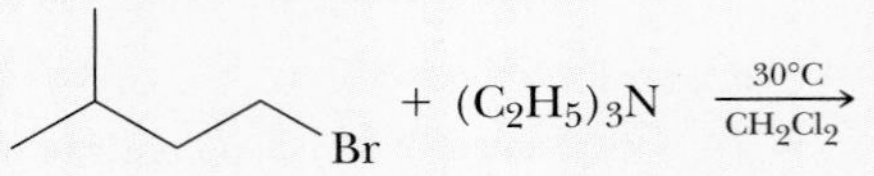 [structure: 1-bromo-3-methylbutane] + $(C_2H_5)_3N$ $\xrightarrow[\text{CH}_2\text{Cl}_2]{30°C}$

SOLUTION

(a) A 3° halide is heated with a strong base/good nucleophile. Elimination by an E2 reaction predominates to give 2-methyl-2-butene as the major product:

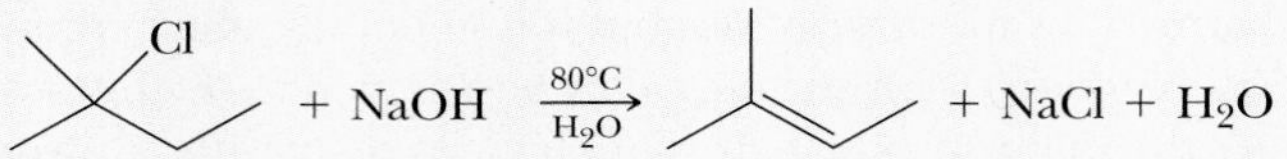

[structure] + NaOH $\xrightarrow[\text{H}_2\text{O}]{80°C}$ [structure 2-methyl-2-butene] + NaCl + H_2O

(b) Reaction of a 1° halide with triethylamine, a moderate nucleophile/weak base, gives substitution by an S_N2 reaction:

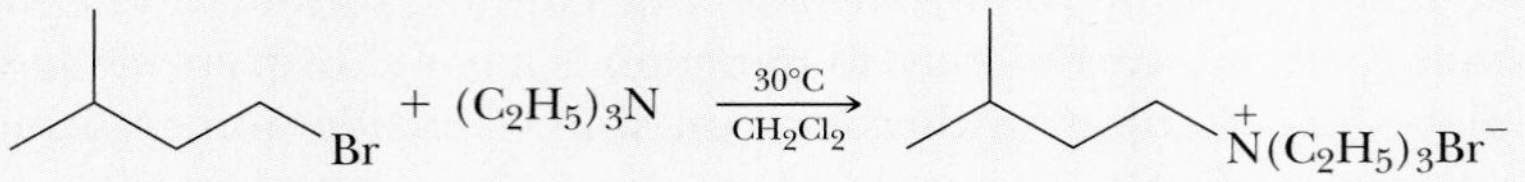

[structure] + $(C_2H_5)_3N$ $\xrightarrow[\text{CH}_2\text{Cl}_2]{30°C}$ [structure] $\overset{+}{N}(C_2H_5)_3Br^-$

Practice Problem 7.6

Predict whether each reaction proceeds predominantly by substitution (S_N1 or S_N2) or elimination (E1 or E2) or whether the two compete, and write structural formulas for the major organic product(s):

(a) 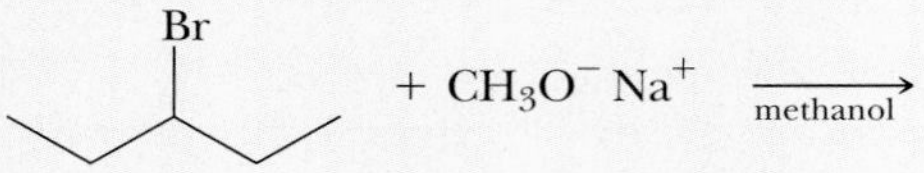 [structure: 2-bromopentane] + $CH_3O^- Na^+$ $\xrightarrow{\text{methanol}}$

(b) 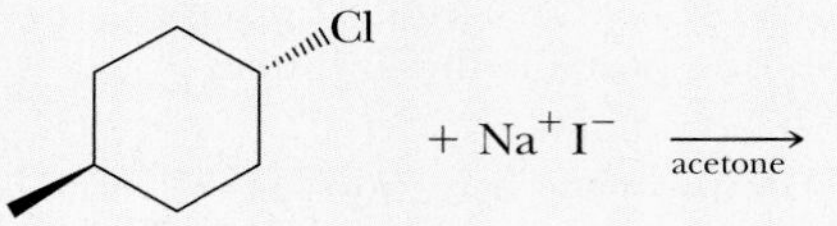 [structure: cis-chloro-methylcyclohexane] + $Na^+ I^-$ $\xrightarrow{\text{acetone}}$

SUMMARY

Haloalkanes contain a halogen covalently bonded to an sp^3-hybridized carbon (Section 7.1). In the IUPAC system, halogen atoms are named as fluoro-, chloro-, bromo-, or iodo-substituents and are listed in alphabetical order with other substituents (Section 7.2A). In the common system, haloalkanes are named **alkyl halides**. Common names are derived by naming the alkyl group, followed by the name of the halide as a separate word (Section 7.2B). Compounds of the type CHX_3 are called **haloforms**.

A **nucleophile** (Section 7.3) is any molecule or ion with an unshared pair of electrons that can be donated to another atom or ion to form a new covalent bond; alternatively, a nucleophile is a Lewis base. An **S_N2 reaction** (Section 7.5A) occurs in one step. The departure of the leaving group is assisted by the incoming nucleophile, and both nucleophile and leaving group are involved in the transition state. S_N2 reactions are stereoselective; reaction at a stereocenter proceeds with inversion of configuration.

An **S_N1 reaction** occurs in two steps (Section 7.5B). Step 1 is a slow, rate-determining ionization of the C–X bond to form a carbocation intermediate, followed in Step 2 by its rapid reaction with a nucleophile to complete the substitution. For S_N1 reactions taking place at a stereocenter, the major reaction occurs with racemization.

The **nucleophilicity** of a reagent is measured by the rate of its reaction in a reference nucleophilic substitution (Section 7.6A). S_N1 reactions are governed by **electronic factors**, namely, the relative stabilities of carbocation intermediates. S_N2 reactions are governed by **steric factors**, namely, the degree of crowding around the site of substitution.

The ability of a group to function as a leaving group is related to its stability as an anion (Section 7.6C). The most stable anions and the best leaving groups are the conjugate bases of strong acids.

Protic solvents contain —OH groups (Section 7.6D). Protic solvents interact strongly with polar molecules and ions and are good solvents in which to form carbocations. Protic solvents favor S_N1 reactions. **Aprotic solvents** do not contain —OH groups. Common aprotic solvents are dimethyl sulfoxide, acetone, diethyl ether, and dichloromethane. Aprotic solvents do not interact as strongly with polar molecules and ions, and carbocations are less likely to form in them. Aprotic solvents favor S_N2 reactions.

Dehydrohalogenation, a type of **β-elimination reaction**, is the removal of H and X from adjacent carbon atoms (Section 7.8). A β-elimination that gives the most highly substituted alkene is called **Zaitsev elimination**. An **E1 reaction** occurs in two steps: breaking the C–X bond to form a carbocation intermediate, followed by the loss of an H^+ to form an alkene. An **E2 reaction** occurs in one step: reaction with base to remove an H^+, formation of the alkene, and departure of the leaving group, all occurring simultaneously.

KEY REACTIONS

1. Nucleophilic Aliphatic Substitution: S_N2 (Section 7.5A)

S_N2 reactions occur in one step, and both the nucleophile and the leaving group are involved in the transition state of the rate-determining step. The nucleophile may be negatively charged or neutral. S_N2 reactions result in an inversion of configuration at the reaction center. They are accelerated in polar aprotic solvents, compared with polar protic solvents. S_N2 reactions are governed by steric factors, namely, the degree of crowding around the site of reaction.

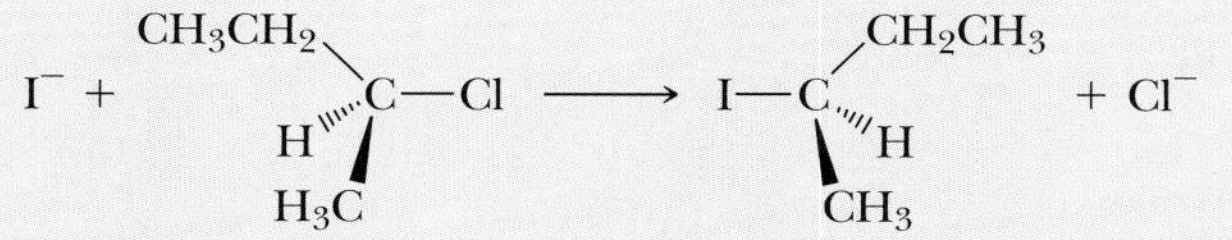

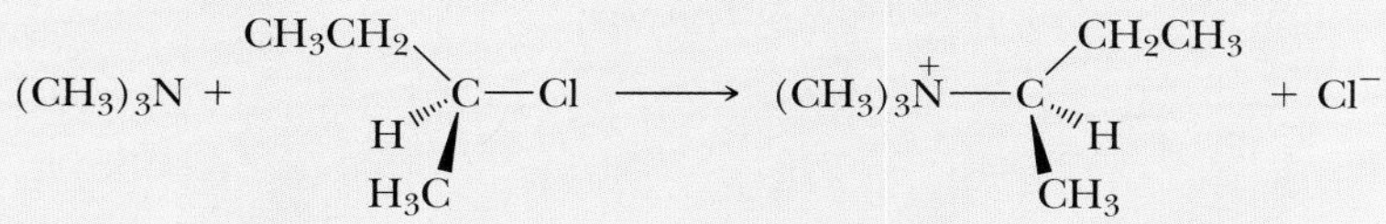

2. Nucleophilic Aliphatic Substitution: S_N1 (Section 7.5B)

An S_N1 reaction occurs in two steps. Step 1 is a slow, rate-determining ionization of the C–X bond to form a carbocation intermediate, followed in Step 2 by its rapid reaction with a nucleophile to complete the substitution. Reaction at a stereocenter gives a racemic product. S_N1 reactions are governed by electronic factors, namely, the relative stabilities of carbocation intermediates:

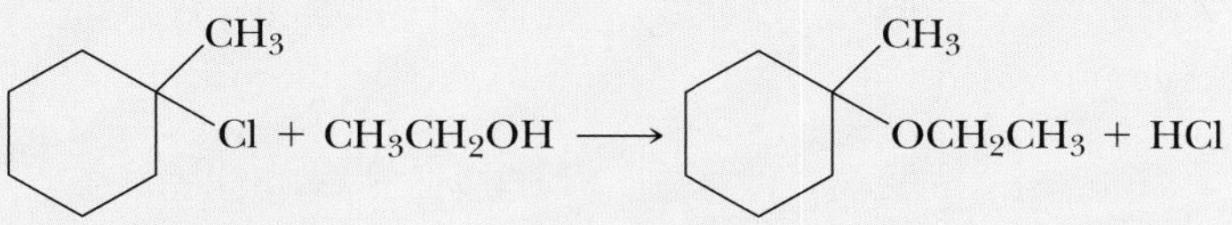

3. β-Elimination: E1 (Section 7.9A)

E1 reactions involve the elimination of atoms or groups of atoms from adjacent carbons. Reaction occurs in two steps and involves the formation of a carbocation intermediate:

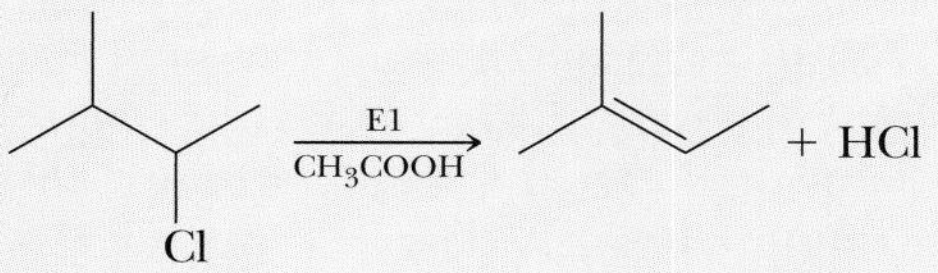

4. β-Elimination: E2 (Section 7.9B)
An E2 reaction occurs in one step: reaction with base to remove a hydrogen, formation of the alkene, and departure of the leaving group, all occurring simultaneously:

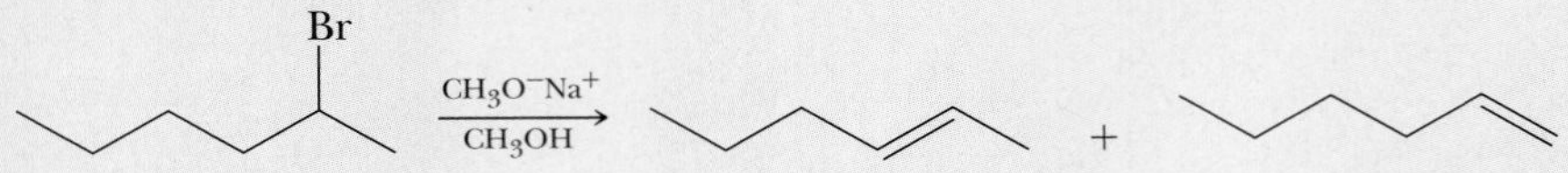

(74%) (26%)

PROBLEMS

A problem number set in red indicates an applied "real-world" problem.

Nomenclature

7.7 Write the IUPAC name for each compound:

(a) $CH_2{=}CF_2$ **(b)** 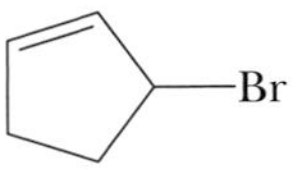

(c) 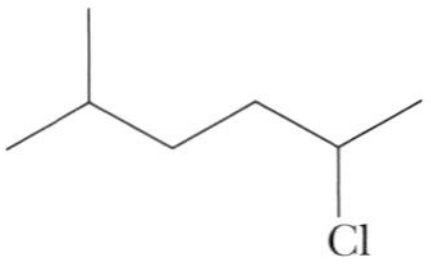

(d) $Cl(CH_2)_6Cl$ **(e)** CF_2Cl_2 **(f)** 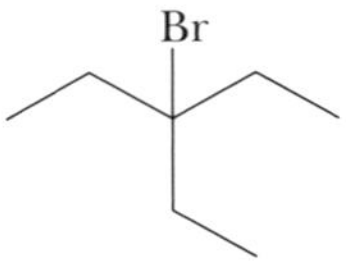

7.8 Write the IUPAC name for each compound (be certain to include a designation of configuration, where appropriate, in your answer):

(a) 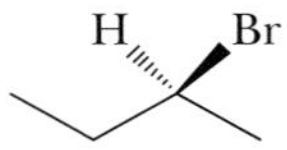

(b) 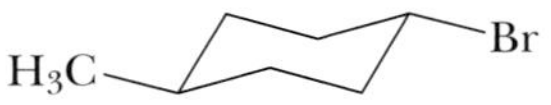

(c) 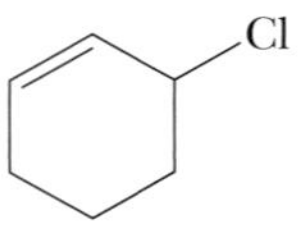

(d) 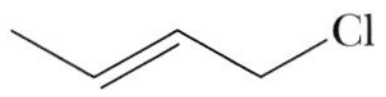

(e) 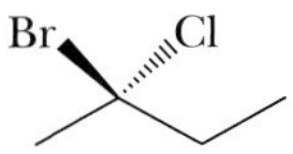

(f) 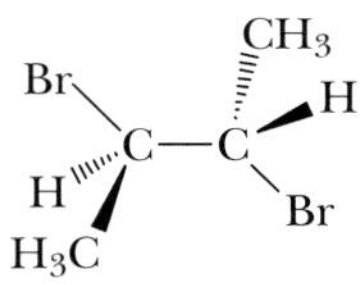

7.9 Draw a structural formula for each compound (given are IUPAC names):
(a) 3-Bromopropene **(b)** (*R*)-2-Chloropentane
(c) meso-3,4-Dibromohexane **(d)** *trans*-1-Bromo-3-isopropylcyclohexane
(e) 1,2-Dichloroethane **(f)** Bromocyclobutane

7.10 Draw a structural formula for each compound (given are common names):
(a) Isopropyl chloride **(b)** *sec*-Butyl bromide **(c)** Allyl iodide
(d) Methylene chloride **(e)** Chloroform **(f)** *tert*-Butyl chloride
(g) Isobutyl chloride

7.11 Which compounds are 2° alkyl halides?
(a) Isobutyl chloride **(b)** 2-Iodooctane **(c)** *trans*-1-Chloro-4-methylcyclohexane

Synthesis of Alkyl Halides

7.12 What alkene or alkenes and reaction conditions give each alkyl halide in good yield? (*Hint:* Review Chapter 5.)

(a) 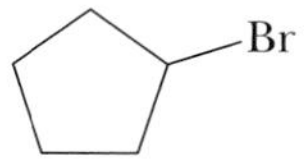

(b) $CH_3\overset{CH_3}{\underset{Br}{C}}CH_2CH_2CH_3$

(c) 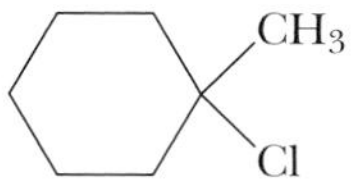

7.13 Show reagents and conditions that bring about these conversions:

(a) 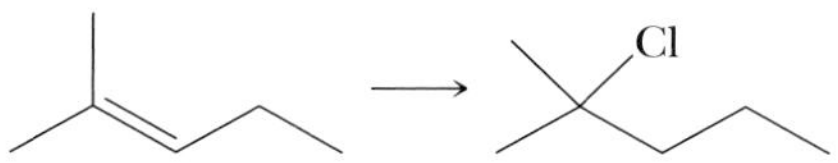

(b) $CH_3CH_2CH{=}CH_2 \longrightarrow CH_3CH_2\overset{I}{C}HCH_3$

(c) $CH_3CH{=}CHCH_3 \longrightarrow CH_3\overset{Cl}{C}HCH_2CH_3$

(d) 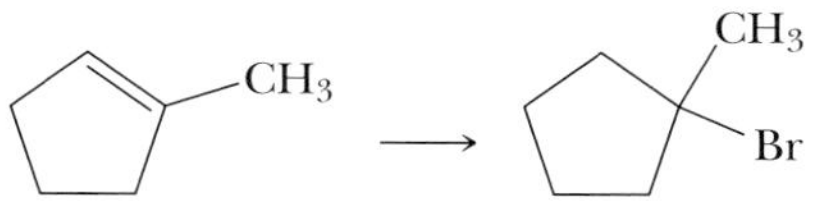

Nucleophilic Aliphatic Substitution

7.14 Write structural formulas for these common organic solvents:
(a) Dichloromethane **(b)** Acetone **(c)** Ethanol
(d) Diethyl ether **(e)** Dimethyl sulfoxide

7.15 Arrange these protic solvents in order of increasing polarity:
(a) H_2O **(b)** CH_3CH_2OH **(c)** CH_3OH

7.16 Arrange these aprotic solvents in order of increasing polarity:
(a) Acetone **(b)** Pentane **(c)** Diethyl ether

7.17 From each pair, select the better nucleophile:
(a) H_2O or OH^- **(b)** CH_3COO^- or OH^- **(c)** CH_3SH or CH_3S^-

7.18 Which statements are true for S_N2 reactions of haloalkanes?
(a) Both the haloalkane and the nucleophile are involved in the transition state.
(b) The reaction proceeds with inversion of configuration at the substitution center.
(c) The reaction proceeds with retention of optical activity.
(d) The order of reactivity is $3° > 2° > 1° >$ methyl.
(e) The nucleophile must have an unshared pair of electrons and bear a negative charge.
(f) The greater the nucleophilicity of the nucleophile, the greater is the rate of reaction.

7.19 Complete these S_N2 reactions:

(a) $Na^+I^- + CH_3CH_2CH_2Cl \xrightarrow[\text{acetone}]{}$

(b) NH_3 + 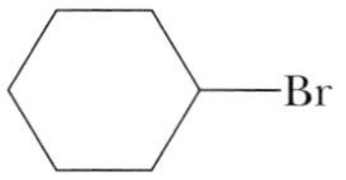 $\xrightarrow[\text{ethanol}]{}$

(c) $CH_3CH_2O^-Na^+ + CH_2{=}CHCH_2Cl \xrightarrow[\text{ethanol}]{}$

7.20 Complete these S_N2 reactions:

(a) 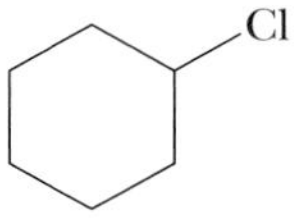 $+ CH_3\overset{O}{\overset{\|}{C}}O^-\,Na^+ \xrightarrow[\text{ethanol}]{}$

(b) $CH_3\overset{I}{C}HCH_2CH_3 + CH_3CH_2S^-Na^+ \xrightarrow[\text{acetone}]{}$

(c) $\underset{\text{acetone}}{\overset{\underset{|}{\text{CH}_3}}{\text{CH}_3\text{CHCH}_2\text{CH}_2\text{Br}}} + \text{Na}^+\text{I}^- \longrightarrow$

(d) $(\text{CH}_3)_3\text{N} + \text{CH}_3\text{I} \xrightarrow[\text{acetone}]{}$

(e) 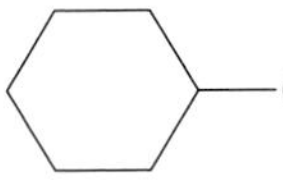 ⬡—$\text{CH}_2\text{Br} + \text{CH}_3\text{O}^-\text{Na}^+ \xrightarrow[\text{methanol}]{}$

(f) 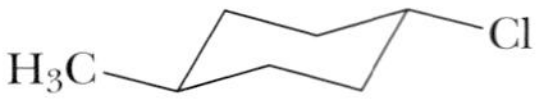 H_3C—⬡—$\text{Cl} + \text{CH}_3\text{S}^-\text{Na}^+ \xrightarrow[\text{ethanol}]{}$

(g) 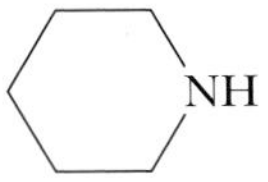 ⬡$\text{NH} + \text{CH}_3(\text{CH}_2)_6\text{CH}_2\text{Cl} \xrightarrow[\text{ethanol}]{}$

(h) 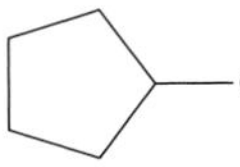 ⬠—$\text{CH}_2\text{Cl} + \text{NH}_3 \xrightarrow[\text{ethanol}]{}$

7.21 You were told that each reaction in Problem 7.20 proceeds by an S_N2 mechanism. Suppose you were not told the mechanism. Describe how you could conclude, from the structure of the haloalkane, the nucleophile, and the solvent, that each reaction is in fact an S_N2 reaction.

7.22 In the following reactions, a haloalkane is treated with a compound that has two nucleophilic sites. Select the more nucleophilic site in each part, and show the product of each S_N2 reaction:

(a) $\text{HOCH}_2\text{CH}_2\text{NH}_2 + \text{CH}_3\text{I} \xrightarrow[\text{ethanol}]{}$

(b) 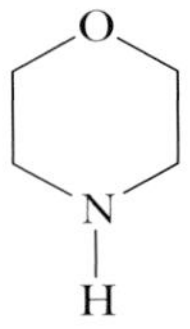 [morpholine with O at top and N—H at bottom] $+ \text{CH}_3\text{I} \xrightarrow[\text{ethanol}]{}$

(c) $\text{HOCH}_2\text{CH}_2\text{SH} + \text{CH}_3\text{I} \xrightarrow[\text{ethanol}]{}$

7.23 Which statements are true for S_N1 reactions of haloalkanes?
(a) Both the haloalkane and the nucleophile are involved in the transition state of the rate-determining step.
(b) The reaction at a stereocenter proceeds with retention of configuration.
(c) The reaction at a stereocenter proceeds with loss of optical activity.
(d) The order of reactivity is $3° > 2° > 1° > $ methyl.
(e) The greater the steric crowding around the reactive center, the lower is the rate of reaction.
(f) The rate of reaction is greater with good nucleophiles compared with poor nucleophiles.

7.24 Draw a structural formula for the product of each S_N1 reaction:

(a) $\underset{\text{S enantiomer}}{\overset{\underset{|}{\text{Cl}}}{\text{CH}_3\text{CHCH}_2\text{CH}_3}} + \text{CH}_3\text{CH}_2\text{OH} \xrightarrow[\text{ethanol}]{}$

(b) [cyclopentane ring with two Cl substituents] $+ \text{CH}_3\text{OH} \xrightarrow[\text{methanol}]{}$

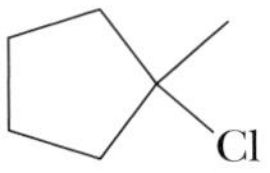

(c) $\underset{\underset{|}{\text{CH}_3}}{\overset{\overset{|}{\text{CH}_3}}{\text{CH}_3\text{CCl}}} + \text{CH}_3\overset{\overset{\text{O}}{\|}}{\text{COH}} \xrightarrow[\text{acetic acid}]{}$

(d) 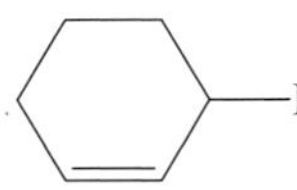 [cyclohexene ring]—$\text{Br} + \text{CH}_3\text{OH} \xrightarrow[\text{methanol}]{}$

7.25 You were told that each substitution reaction in Problem 7.24 proceeds by an S_N1 mechanism. Suppose that you were not told the mechanism. Describe how you could conclude, from the structure of the haloalkane, the nucleophile, and the solvent, that each reaction is in fact an S_N1 reaction.

7.26 Select the member of each pair that undergoes nucleophilic substitution in aqueous ethanol more rapidly:

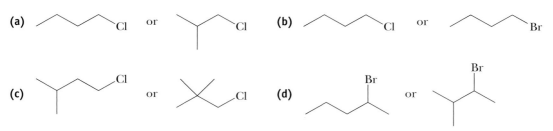

7.27 Propose a mechanism for the formation of the products (but not their relative percentages) in this reaction:

$$\underset{\overset{|}{CH_3}}{\overset{\overset{CH_3}{|}}{CH_3CCl}} \xrightarrow[\overset{80\%CH_3CH_2OH}{25°C}]{20\%H_2O,} \underset{\overset{|}{CH_3}}{\overset{\overset{CH_3}{|}}{CH_3COCH_2CH_3}} + \underset{\overset{|}{CH_3}}{\overset{\overset{CH_3}{|}}{CH_3COH}} + \underset{15\%}{\overset{\overset{CH_3}{|}}{CH_3C{=}CH_2}} + HCl$$

$$\underbrace{}_{85\%}$$

7.28 The rate of reaction in Problem 7.27 increases by 140 times when carried out in 80% water to 20% ethanol, compared with 40% water to 60% ethanol. Account for this difference.

7.29 Select the member of each pair that shows the greater rate of S_N2 reaction with KI in acetone:

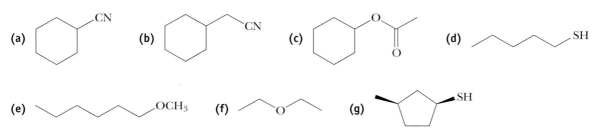

7.30 What hybridization best describes the reacting carbon in the S_N2 transition state?

7.31 Haloalkenes such as vinyl bromide, $CH_2{=}CHBr$, undergo neither S_N1 nor S_N2 reactions. What factors account for this lack of reactivity?

7.32 Show how you might synthesize the following compounds from a haloalkane and a nucleophile:

7.33 Show how you might synthesize each compound from a haloalkane and a nucleophile:

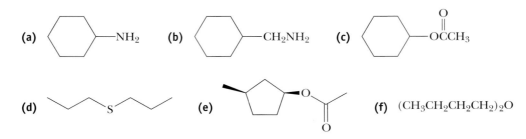

(a) —NH₂   (b) —CH₂NH₂   (c) —OCCH₃

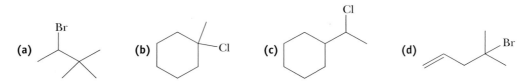

(d)   (e)   (f) $(CH_3CH_2CH_2CH_2)_2O$

β-Eliminations

7.34 Draw structural formulas for the alkene(s) formed by treating each of the following haloalkanes with sodium ethoxide in ethanol. Assume that elimination is by an E2 mechanism. Where two alkenes are possible, use Zaitsev's rule to predict which alkene is the major product:

7.35 Which of the following haloalkanes undergo dehydrohalogenation to give alkenes that do not show cis–trans isomerism?
(a) 2-Chloropentane   (b) 2-Chlorobutane
(c) Chlorocyclohexane   (d) Isobutyl chloride

7.36 How many isomers, including cis–trans isomers, are possible for the major product of dehydrohalogenation of each of the following haloalkanes?
(a) 3-Chloro-3-methylhexane   (b) 3-Bromohexane

7.37 What haloalkane might you use as a starting material to produce each of the following alkenes in high yield and uncontaminated by isomeric alkenes?

(a) =CH₂

CH_3
(b) $CH_3CHCH_2CH{=}CH_2$

7.38 For each of the following alkenes, draw structural formulas of all chloroalkanes that undergo dehydrohalogenation when treated with KOH to give that alkene as the major product (for some parts, only one chloroalkane gives the desired alkene as the major product; for other parts, two chloroalkanes may work):

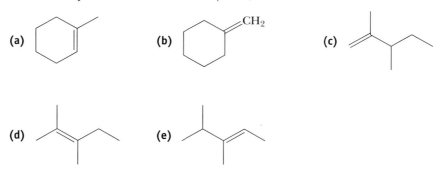

7.39 When *cis*-4-chlorocyclohexanol is treated with sodium hydroxide in ethanol, it gives only the substitution product *trans*-1,4-cyclohexanediol (1). Under the same experimental conditions, *trans*-4-chlorocyclohexanol gives 3-cyclohexenol (2) and the bicyclic ether (3):

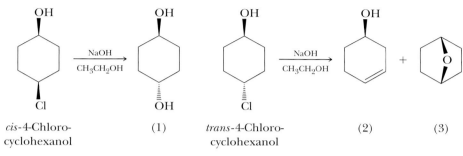

cis-4-Chloro-   (1)   *trans*-4-Chloro-   (2)   (3)
cyclohexanol   cyclohexanol

(a) Propose a mechanism for the formation of product (1), and account for its configuration.

(b) Propose a mechanism for the formation of product (2).

(c) Account for the fact that the bicyclic ether (3) is formed from the trans isomer, but not from the cis isomer.

Synthesis

7.40 Show how to convert the given starting material into the desired product (note that some syntheses require only one step, whereas others require two or more steps):

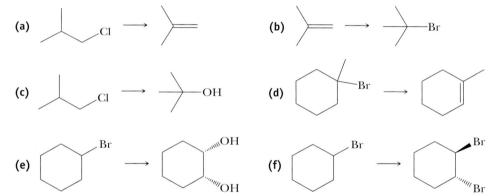

Looking Ahead

7.41 The Williamson ether synthesis involves treating a haloalkane with a metal alkoxide. Following are two reactions intended to give benzyl *tert*-butyl ether. One reaction gives the ether in good yield, the other does not. Which reaction gives the ether? What is the product of the other reaction, and how do you account for its formation?

(a) $CH_3\underset{\underset{CH_3}{|}}{\overset{\overset{CH_3}{|}}{C}}O^- K^+$ + ⟨benzene⟩—CH_2Cl $\xrightarrow{DMSO}$ $CH_3\underset{\underset{CH_3}{|}}{\overset{\overset{CH_3}{|}}{C}}OCH_2$—⟨benzene⟩ + KCl

(b) ⟨benzene⟩—$CH_2O^- K^+$ + $CH_3\underset{\underset{CH_3}{|}}{\overset{\overset{CH_3}{|}}{C}}Cl$ $\xrightarrow{DMSO}$ $CH_3\underset{\underset{CH_3}{|}}{\overset{\overset{CH_3}{|}}{C}}OCH_2$—⟨benzene⟩ + KCl

7.42 The following ethers can, in principle, be synthesized by two different combinations of haloalkane or halocycloalkane and metal alkoxide. Show one combination that forms ether bond (1) and another that forms ether bond (2). Which combination gives the higher yield of ether?

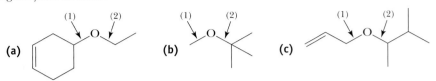

7.43 Propose a mechanism for this reaction:

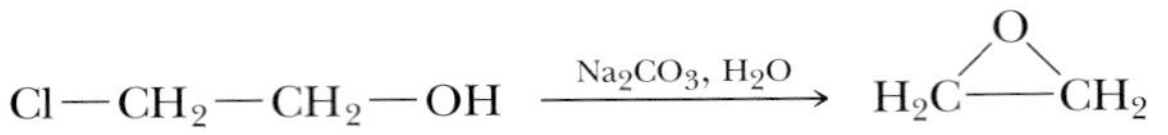

$$Cl-CH_2-CH_2-OH \xrightarrow{Na_2CO_3,\ H_2O} \underset{\text{Ethylene oxide}}{H_2C \overset{O}{\diagup\diagdown} CH_2}$$

2-Chloroethanol

7.44 An OH group is a poor leaving group, and yet substitution occurs readily in the following reaction. Propose a mechanism for this reaction that shows how OH overcomes its limitation of being a poor leaving group.

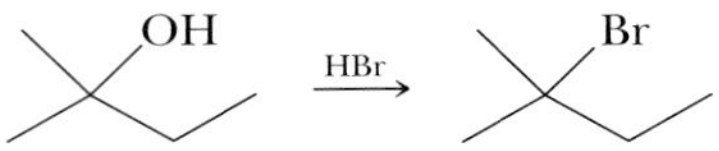

7.45 Explain why (*S*)-2-bromobutane becomes optically inactive when treated with sodium bromide in DMSO:

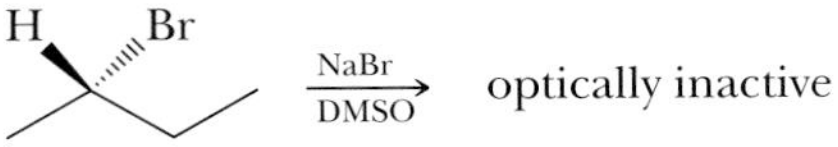

optically active $\xrightarrow[\text{DMSO}]{\text{NaBr}}$ optically inactive

7.46 Explain why phenoxide is a much poorer nucleophile than cyclohexoxide:

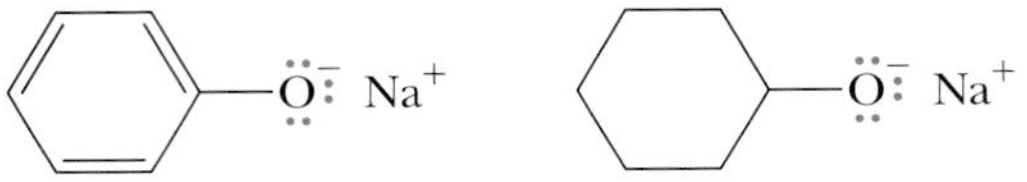

Sodium phenoxide   Sodium cyclohexoxide

7.47 In ethers, each side of the oxygen is essentially an OR group and is thus a poor leaving group. Epoxides are three-membered ring ethers. Explain why an epoxide reacts readily with a nucleophile despite being an ether.

$$R-O-R + :Nu^- \longrightarrow \text{no reaction}$$

An ether

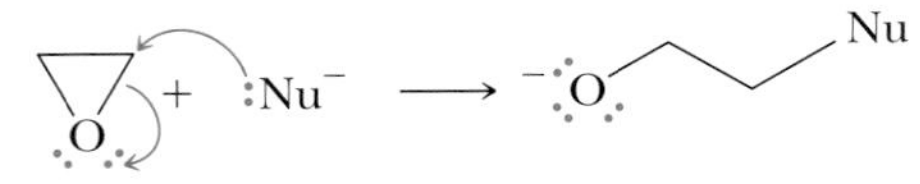

An epoxide

8 Alcohols, Ethers, and Thiols

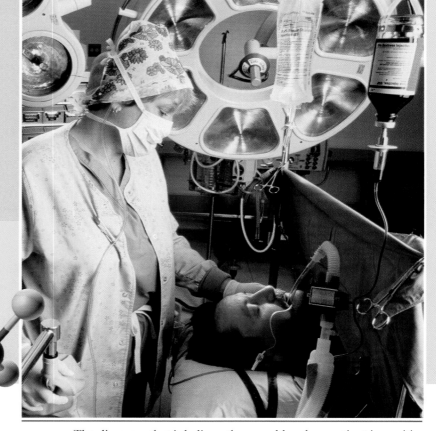

The discovery that inhaling ethers could make a patient insensitive to pain revolutionized the practice of medicine. Inset: A model of isoflurane, $CF_3CHClOCHF_2$, a halogenated ether widely used as an inhalation anesthetic in both human and veterinary medicine. *(Allan Levenson/Stone/Getty Images)*

8.1 INTRODUCTION

In this chapter, we study the physical and chemical properties of alcohols and ethers, two classes of oxygen-containing compounds. We also study thiols, a class of sulfur-containing compounds. Thiols are like alcohols in structure, except that they contain an —SH group rather than an —OH group.

CH_3CH_2OH	$CH_3CH_2OCH_2CH_3$	CH_3CH_2SH
Ethanol	Diethyl ether	Ethanethiol
(an alcohol)	(an ether)	(a thiol)

These three compounds are certainly familiar to you. Ethanol is the fuel additive in gasohol, the alcohol in alcoholic beverages, and an important industrial and laboratory solvent. Diethyl ether was the first inhalation anesthetic used in general surgery. It is also an important industrial and laboratory solvent. Ethanethiol, like other low-molecular-weight thiols, has a stench. Smells such as those from skunks, rotten eggs, and sewage are caused by thiols.

Alcohols are particularly important in both laboratory and biochemical transformations of organic compounds. They can be converted into other types of compounds, such as alkenes, haloalkanes, aldehydes, ketones, carboxylic acids, and esters. Not only can alcohols be converted to these compounds, but they also can be prepared from them. Thus, alcohols play a central role in the interconversion of organic functional groups.

8.2 ALCOHOLS

A. Structure

Alcohol A compound containing an —OH (hydroxyl) group bonded to an sp^3 hybridized carbon.

The functional group of an **alcohol** is an **—OH (hydroxyl) group** bonded to an sp^3 hybridized carbon atom (Section 1.8A). The oxygen atom of an alcohol is also sp^3 hybridized. Two sp^3 hybrid orbitals of oxygen form sigma bonds to atoms of carbon and hydrogen. The other two sp^3 hybrid orbitals of oxygen each contain an unshared pair of electrons. Figure 8.1 shows a Lewis structure and ball-and-stick model of methanol, CH_3OH, the simplest alcohol.

(a)

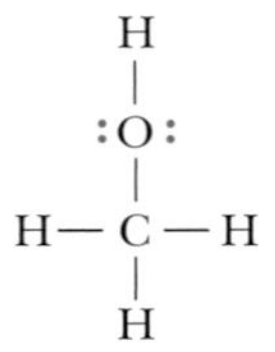

$$H$$
$$|$$
$$:O:$$
$$|$$
$$H—C—H$$
$$|$$
$$H$$

(b)

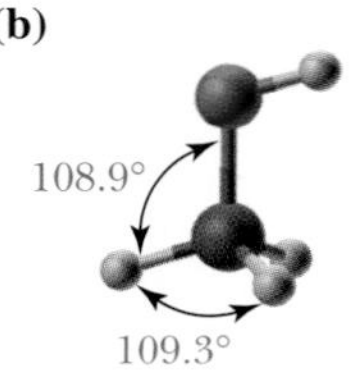

108.9°

109.3°

Figure 8.1
Methanol, CH_3OH.
(a) Lewis structure and (b) ball-and-stick model. The measured H—C—O bond angle in methanol is 108.6°, very close to the tetrahedral angle of 109.5°.

B. Nomenclature

We derive the IUPAC names for alcohols in the same manner as those for alkanes, with the exception that the ending of the parent alkane is changed from *-e* to *-ol*. The ending *-ol* tells us that the compound is an alcohol.

1. Select, as the parent alkane, the longest chain of carbon atoms that contains the —OH, and number that chain from the end closer to the —OH group. In numbering the parent chain, the location of the —OH group takes precedence over alkyl groups and halogens.
2. Change the suffix of the parent alkane from *-e* to *-ol* (Section 3.6), and use a number to show the location of the —OH group. For cyclic alcohols, numbering begins at the carbon bearing the —OH group.
3. Name and number substituents and list them in alphabetical order.

To derive common names for alcohols, we name the alkyl group bonded to —OH and then add the word *alcohol*. Following are the IUPAC names and, in parentheses, the common names of eight low-molecular-weight alcohols:

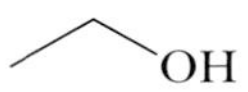

Ethanol
(Ethyl alcohol)

1-Propanol
(Propyl alcohol)

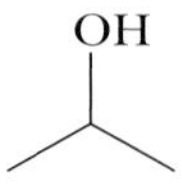

2-Propanol
(Isopropyl alcohol)

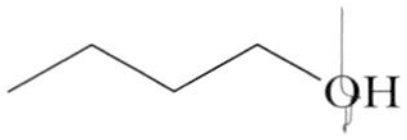

1-Butanol
(Butyl alcohol)

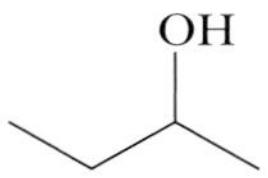

2-Butanol
(*sec*-Butyl alcohol)

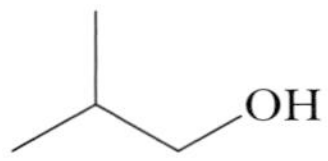

2-Methyl-1-propanol
(Isobutyl alcohol)

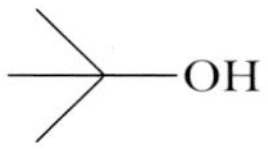

2-Methyl-2-propanol
(*tert*-Butyl alcohol)

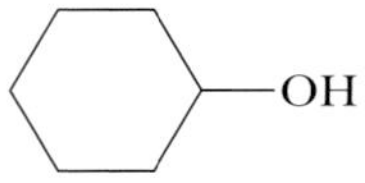

Cyclohexanol
(Cyclohexyl alcohol)

EXAMPLE 8.1

Write the IUPAC name for each alcohol:

(a) $CH_3(CH_2)_6CH_2OH$ (b) 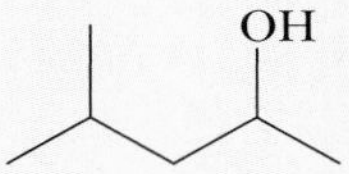   (c)

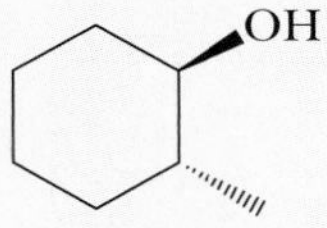

SOLUTION

(a) 1-Octanol   (b) 4-Methyl-2-pentanol
(c) *trans*-2-Methylcyclohexanol or (1R,2R)-2-Methylcyclohexanol

Practice Problem 8.1

Write the IUPAC name for each alcohol:

(a) 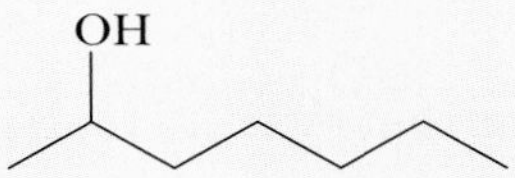   (b) 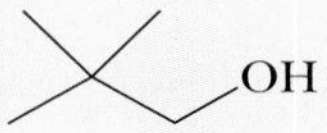   (c)

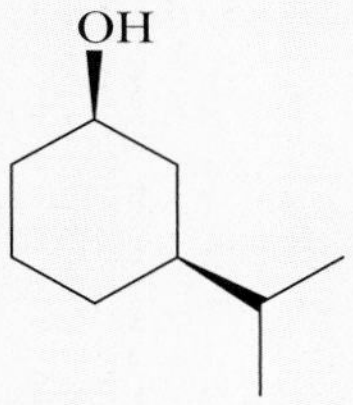

We classify alcohols as **primary (1°)**, **secondary (2°)**, or **tertiary (3°)**, depending on whether the —OH group is on a primary, secondary, or tertiary carbon
(Section 1.8A).

EXAMPLE 8.2

Classify each alcohol as primary, secondary, or tertiary:

(a) 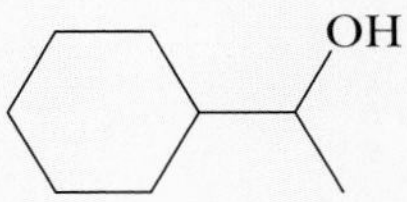   (b) 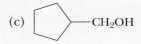   (c)

SOLUTION

(a) Secondary (2°)   (b) Tertiary (3°)   (c) Primary (1°)

Practice Problem 8.2

Classify each alcohol as primary, secondary, or tertiary:

(a) 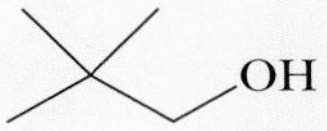   (b)

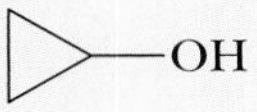

(c) $CH_2=CHCH_2OH$ (d)

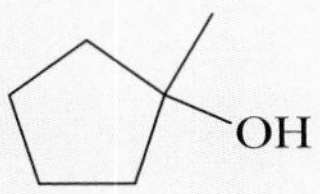

In the IUPAC system, a compound containing two hydroxyl groups is named as
a **diol**, one containing three hydroxyl groups is named as a **triol**, and so on. In
IUPAC names for diols, triols, and so on, the final -*e* of the parent alkane name is
retained, as for example, in 1,2-ethanediol.

Glycol A compound with two hydroxyl (—OH) groups on adjacent carbons.

As with many organic compounds, common names for certain diols and triols have persisted. Compounds containing two hydroxyl groups on adjacent carbons are often referred to as **glycols** (Section 5.4). Ethylene glycol and propylene glycol are synthesized from ethylene and propylene, respectively—hence their common names:

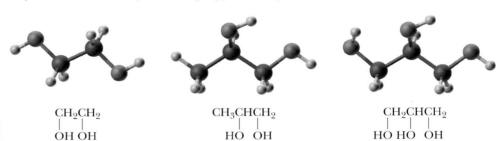

CH₂CH₂	CH₃CHCH₂	CH₂CHCH₂

CH₂CH₂
OH OH
1,2-Ethanediol
(Ethylene glycol)

CH₃CHCH₂
HO OH
1,2-Propanediol
(Propylene glycol)

CH₂CHCH₂
HO HO OH
1,2,3-Propanetriol
(Glycerol, Glycerin)

We often refer to compounds containing —OH and C=C groups as unsaturated alcohols. To name an unsaturated alcohol,

1. Number the parent alkane so as to give the —OH group the lowest possible number.
2. Show the double bond by changing the infix of the parent alkane from -an- to -en- (Section 3.6), and show the alcohol by changing the suffix of the parent alkane from -e to -ol.
3. Use numbers to show the location of both the carbon–carbon double bond and the hydroxyl group.

Ethylene glycol is a polar molecule and dissolves readily in water, a polar solvent. *(Charles D. Winter)*

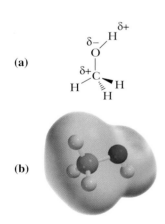

Figure 8.2
Polarity of the C—O—H bond in methanol.
(a) There are partial positive charges on carbon and hydrogen and a partial negative charge on oxygen.
(b) An electron density map showing the partial negative charge (in red) around oxygen and a partial positive charge (in blue) around hydrogen of the —OH group.

EXAMPLE 8.3

Write the IUPAC name for each unsaturated alcohol:

(a) CH₂=CHCH₂OH   (b) [cyclohexene-OH]   (c) HO [cis-hexenol]

SOLUTION

(a) 2-Propen-1-ol. Its common name is allyl alcohol.
(b) 2-Cyclohexenol.
(c) *cis*-3-Hexen-1-ol. This unsaturated alcohol is sometimes called leaf alcohol because of its occurrence in leaves of fragrant plants, including trees and shrubs.

Practice Problem 8.3

Write the IUPAC name for each unsaturated alcohol:

(a) [structure]—OH   (b) [cyclopentene]—OH

C. Physical Properties

The most important physical property of alcohols is the polarity of their —OH groups. Because of the large difference in electronegativity (Table 1.5) between oxygen and carbon (3.5 − 2.5 = 1.0) and between oxygen and hydrogen (3.5 − 2.1 = 1.4), both the C—O and O—H bonds of an alcohol are polar covalent, and alcohols are polar molecules, as illustrated in Figure 8.2 for methanol.

CHEMICAL CONNECTIONS 8A

Nitroglycerin: An Explosive and a Drug

In 1847, Ascanio Sobrero (1812–1888) discovered that 1,2,3-propanetriol, more commonly named glycerin, reacts with nitric acid in the presence of sulfuric acid to give a pale yellow, oily liquid called nitroglycerin:

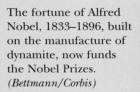

1,2,3-Propanetriol   1,2,3-Propanetriol trinitrate
(Glycerol, Glycerin)   (Nitroglycerin)

The fortune of Alfred Nobel, 1833–1896, built on the manufacture of dynamite, now funds the Nobel Prizes. *(Bettmann/Corbis)*

Sobrero also discovered the explosive properties of the compound: When he heated a small quantity of it, it exploded! Soon, nitroglycerin became widely used for blasting in the construction of canals, tunnels, roads, and mines and, of course, for warfare.

One problem with the use of nitroglycerin was soon recognized: It was difficult to handle safely, and accidental explosions occurred frequently. The Swedish chemist Alfred Nobel (1833–1896) solved the problem: He discovered that a claylike substance called diatomaceous earth absorbs nitroglycerin so that it will not explode without a fuse. He gave the name *dynamite* to this mixture of nitroglycerine, diatomaceous earth, and sodium carbonate.

Surprising as it may seem, nitroglycerin is used in medicine to treat angina pectoris, the symptoms of which are sharp chest pains caused by a reduced flow of blood in the coronary artery. Nitroglycerin, which is available in liquid (diluted with alcohol to render it nonexplosive), tablet, or paste form, relaxes the smooth muscles of blood vessels, causing dilation of the coronary artery. This dilation, in turn, allows more blood to reach the heart.

When Nobel became ill with heart disease, his physicians advised him to take nitroglycerin to relieve his chest pains. He refused, saying he could not understand how the explosive could relieve chest pains. It took science more than 100 years to find the answer. We now know that it is nitric oxide, NO, derived from the nitro groups of nitroglycerin, that relieves the pain.

Table 8.1 lists the boiling points and solubilities in water for five groups of alcohols and alkanes of similar molecular weight. Notice that, of the compounds compared in each group, the alcohol has the higher boiling point and is the more soluble in water.

Alcohols have higher boiling points than alkanes of similar molecular weight, because alcohols are polar molecules and can associate in the liquid state by a type of intermolecular attraction called **hydrogen bonding** (Figure 8.3). The strength of hydrogen bonding between alcohol molecules is approximately 2 to 5 kcal/mol (8.4 to 21 kJ/mol). For comparison, the strength of the O—H covalent bond in an alcohol molecule is approximately 110 kcal/mol (460 kJ/mol). As we see by comparing these numbers, an O----H hydrogen bond is considerably weaker than an O—H covalent bond. Nonetheless, it is sufficient to have a dramatic effect on the physical properties of alcohols.

Because of hydrogen bonding between alcohol molecules in the liquid state, extra energy is required to separate each hydrogen-bonded alcohol molecule from its neighbors—hence the relatively high boiling points of alcohols compared with those of alkanes. The presence of additional hydroxyl groups in a molecule further increases the extent of hydrogen bonding, as can be seen by comparing the boiling points of 1-pentanol (138°C) and 1,4-butanediol (230°C), both of which have approximately the same molecular weight.

Hydrogen bonding The attractive force between a partial positive charge on hydrogen and partial negative charge on a nearby oxygen, nitrogen, or fluorine atom.

TABLE 8.1 Boiling Points and Solubilities in Water of Five Groups of Alcohols and Alkanes of Similar Molecular Weight

Structural Formula	Name	Molecular Weight	Boiling Point (°C)	Solubility in Water
CH_3OH	methanol	32	65	infinite
CH_3CH_3	ethane	30	−89	insoluble
CH_3CH_2OH	ethanol	46	78	infinite
$CH_3CH_2CH_3$	propane	44	−42	insoluble
$CH_3CH_2CH_2OH$	1-propanol	60	97	infinite
$CH_3CH_2CH_2CH_3$	butane	58	0	insoluble
$CH_3CH_2CH_2CH_2OH$	1-butanol	74	117	8 g/100 g
$CH_3CH_2CH_2CH_2CH_3$	pentane	72	36	insoluble
$CH_3CH_2CH_2CH_2CH_2OH$	1-pentanol	88	138	2.3 g/100 g
$HOCH_2CH_2CH_2CH_2OH$	1,4-butanediol	90	230	infinite
$CH_3CH_2CH_2CH_2CH_2CH_3$	hexane	86	69	insoluble

Figure 8.3
The association of ethanol molecules in the liquid state. Each O—H can participate in up to three hydrogen bonds (one through hydrogen and two through oxygen). Only two of these three possible hydrogen bonds per molecule are shown in the figure.

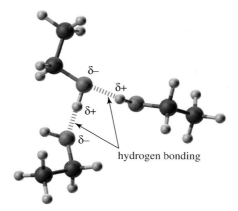

hydrogen bonding

Because of increased dispersion forces (Section 3.9B) between larger molecules, boiling points of all types of compounds, including alcohols, increase with increasing molecular weight. (Compare, for example, the boiling points of ethanol, 1-propanol, 1-butanol, and 1-pentanol.)

Alcohols are much more soluble in water than are alkanes, alkenes, and alkynes of comparable molecular weight. Their increased solubility is due to hydrogen bonding between alcohol molecules and water. Methanol, ethanol, and 1-propanol are soluble in water in all proportions. As molecular weight increases, the physical properties of alcohols become more like those of hydrocarbons of comparable molecular weight. Alcohols of higher molecular weight are much less soluble in water because of the increase in size of the hydrocarbon portion of their molecules.

8.3 REACTIONS OF ALCOHOLS

In this section, we study the acidity and basicity of alcohols, their dehydration to alkenes, their conversion to haloalkanes, and their oxidation to aldehydes, ketones, or carboxylic acids.

A. Acidity of Alcohols

Alcohols have about the same pK_a values as water (15.7), which means that aqueous solutions of alcohols have about the same pH as that of pure water. The pK_a of methanol, for example, is 15.5:

$$CH_3\overset{..}{\underset{..}{O}}-H + :\overset{..}{\underset{\underset{H}{|}}{O}}-H \rightleftharpoons CH_3\overset{..}{\underset{..}{O}}:^- + H-\overset{\overset{+}{..}}{\underset{\underset{H}{|}}{O}}-H$$

$$K_a = \frac{[CH_3O^-][H_3O^+]}{[CH_3OH]} = 3.2 \times 10^{-16}$$

$$pK_a = 15.5$$

Table 8.2 gives the acid ionization constants for several low-molecular-weight alcohols. Methanol and ethanol are about as acidic as water. Higher-molecular-weight, water-soluble alcohols are slightly weaker acids than water. Even though alcohols have some slight acidity, they are not strong enough acids to react with weak bases such as sodium bicarbonate or sodium carbonate. (At this point, it would be worthwhile to review Section 2.5 and the discussion of the position of equilibrium in acid–base reactions.) Note that, although acetic acid is a "weak acid" compared with acids such as HCl, it is still 10^{10} times stronger as an acid than alcohols are.

TABLE 8.2 pK_a Values for Selected Alcohols in Dilute Aqueous Solution*

Compound	Structural Formula	pK_a	
hydrogen chloride	HCl	-7	Stronger acid
acetic acid	CH_3COOH	4.8	
methanol	CH_3OH	15.5	
water	H_2O	15.7	↑
ethanol	CH_3CH_2OH	15.9	
2-propanol	$(CH_3)_2CHOH$	17	Weaker acid
2-methyl-2-propanol	$(CH_3)_3COH$	18	

* Also given for comparison are pK_a values for water, acetic acid, and hydrogen chloride.

B. Basicity of Alcohols

In the presence of strong acids, the oxygen atom of an alcohol is a weak base and reacts with an acid by proton transfer to form an oxonium ion:

$$CH_3CH_2-\overset{..}{\underset{..}{O}}-H + H-\overset{\overset{+}{..}}{\underset{\underset{H}{|}}{O}}-H \xrightarrow{H_2SO_4} CH_3CH_2-\overset{\overset{+}{..}}{\underset{\underset{H}{|}}{O}}-H + :\overset{..}{O}-H$$

Ethanol   Hydronium ion   Ethyloxonium ion
         $(pK_a - 1.7)$ $(pK_a - 2.4)$

Thus, alcohols can function as both weak acids and weak bases.

C. Reaction with Active Metals

Like water, alcohols react with Li, Na, K, Mg, and other active metals to liberate hydrogen and to form metal alkoxides. In the following oxidation–reduction reaction, Na is oxidized to Na^+ and H^+ is reduced to H_2:

$$2\ CH_3OH + 2\ Na \longrightarrow 2\ CH_3O^-Na^+ + H_2$$

Sodium methoxide

To name a metal alkoxide, name the cation first, followed by the name of the anion. The name of an alkoxide ion is derived from a prefix showing the number of carbon atoms and their arrangement (*meth-*, *eth-*, *isoprop-*, tert-*but-*, and so on) followed by the suffix -*oxide*.

Methanol reacts with sodium metal with the evolution of hydrogen gas.
(Charles D. Winters)

Alkoxide ions are somewhat stronger bases than is the hydroxide ion. In addition to sodium methoxide, the following metal salts of alcohols are commonly used in organic reactions requiring a strong base in a nonaqueous solvent; sodium ethoxide in ethanol and potassium *tert*-butoxide in 2-methyl-2-propanol (*tert*-butyl alcohol):

$$CH_3CH_2O^-Na^+ \qquad \begin{array}{c} CH_3 \\ | \\ CH_3CO^-K^+ \\ | \\ CH_3 \end{array}$$

Sodium ethoxide   Potassium *tert*-butoxide

As we saw in Chapter 7, alkoxide ions can also be used as nucleophiles in substitution reactions.

EXAMPLE 8.4

Write a balanced equation for the reaction of cyclohexanol with sodium metal.

SOLUTION

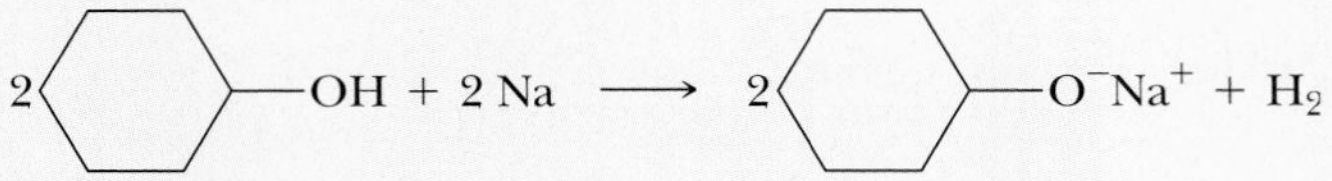

$$2 \langle\hexagon\rangle\text{—OH} + 2\,Na \longrightarrow 2 \langle\hexagon\rangle\text{—O}^-Na^+ + H_2$$

Cyclohexanol   Sodium
                            cyclohexoxide

Practice Problem 8.4

Predict the position of equilibrium for the following acid–base reaction. (*Hint:* Review Section 2.5.)

$$CH_3CH_2O^-Na^+ + CH_3\overset{\overset{\displaystyle O}{\|}}{C}OH \rightleftharpoons CH_3CH_2OH + CH_3\overset{\overset{\displaystyle O}{\|}}{C}O^-Na^+$$

D. Conversion to Haloalkanes

The conversion of an alcohol to an alkyl halide involves substituting halogen for —OH at a saturated carbon. The most common reagents for this conversion are the halogen acids and $SOCl_2$.

Reaction with HCl, HBr, and HI

Water-soluble tertiary alcohols react very rapidly with HCl, HBr, and HI. Mixing a tertiary alcohol with concentrated hydrochloric acid for a few minutes at room temperature converts the alcohol to a water-insoluble chloroalkane that separates from the aqueous layer.

$$\begin{array}{c} CH_3 \\ | \\ CH_3COH \\ | \\ CH_3 \end{array} + HCl \xrightarrow{25^\circ C} \begin{array}{c} CH_3 \\ | \\ CH_3CCl \\ | \\ CH_3 \end{array} + H_2O$$

2-Methyl-   2-Chloro-
2-propanol   2-methylpropane

Low-molecular-weight, water-soluble primary and secondary alcohols do not react under these conditions.

Water-insoluble tertiary alcohols are converted to tertiary halides by bubbling gaseous HX through a solution of the alcohol dissolved in diethyl ether or tetrahydrofuran (THF):

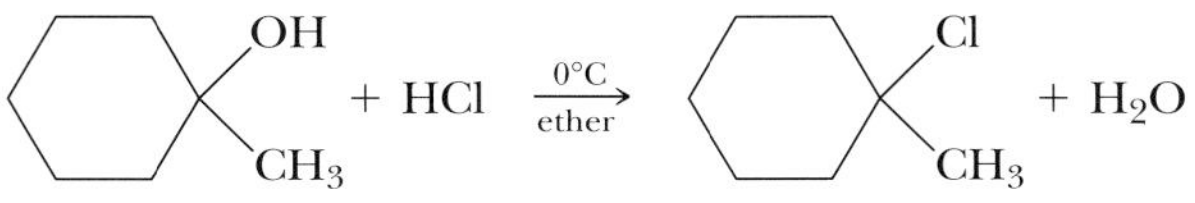

1-Methyl-   1-Chloro-1-methyl
cyclohexanol   cyclohexane

Water-insoluble primary and secondary alcohols react only slowly under these conditions.

Primary and secondary alcohols are converted to bromoalkanes and iodoalkanes by treatment with concentrated hydrobromic and hydroiodic acids. For example, heating 1-butanol with concentrated HBr gives 1-bromobutane:

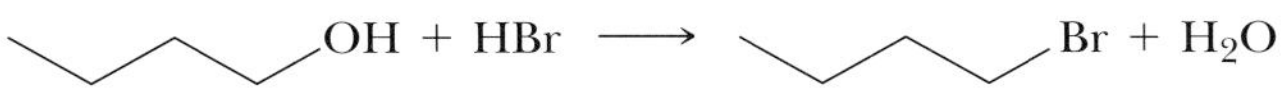

1-Butanol   1-Bromobutane
                   (Butyl bromide)

On the basis of observations of the relative ease of reaction of alcohols with HX ($3° > 2° > 1°$), it has been proposed that the conversion of tertiary and secondary alcohols to haloalkanes by concentrated HX occurs by an S_N1 mechanism and involves the formation of a carbocation intermediate.

Mechanism: Reaction of a Tertiary Alcohol with HCl: An S_N1 Reaction

Step 1: Rapid and reversible proton transfer from the acid to the OH group gives an oxonium ion. The result of this proton transfer is to convert the leaving group from OH⁻, a poor leaving group, to H_2O, a better leaving group:

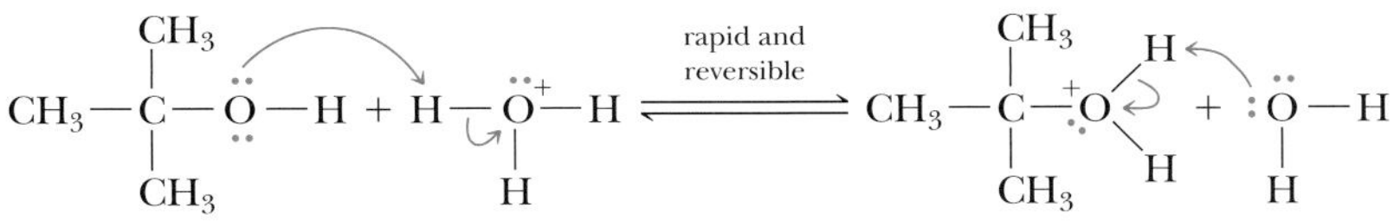

2-Methyl-2-propanol   An oxonium ion
(*tert*-Butyl alcohol)

Step 2: Loss of water from the oxonium ion gives a 3° carbocation intermediate:

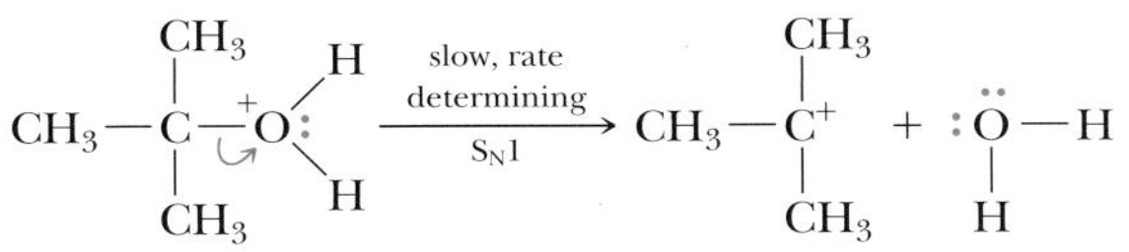

An oxonium ion   A 3° carbocation
                        intermediate

Step 3: Reaction of the 3° carbocation intermediate (an electrophile) with chloride ion (a nucleophile) gives the product:

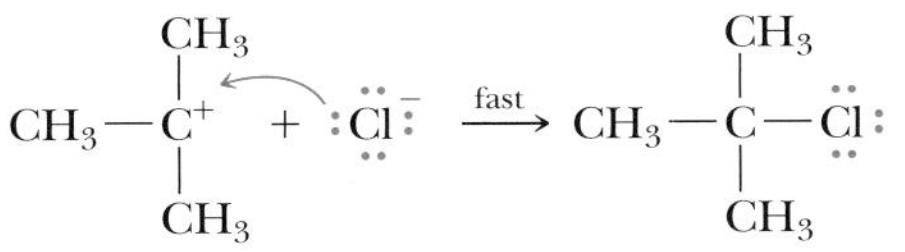

2-Chloro-2-methylpropane
(*tert*-Butyl chloride)

Primary alcohols react with HX by an S_N2 mechanism. In the rate-determining step, the halide ion displaces H_2O from the carbon bearing the oxonium ion. The displacement of H_2O and the formation of the C—X bond are simultaneous.

Mechanism: Reaction of a Primary Alcohol with HBr: An S_N2 Reaction

Step 1: Rapid and reversible proton transfer to the OH group which converts the leaving group from OH^-, a poor leaving group, to H_2O, a better leaving group:

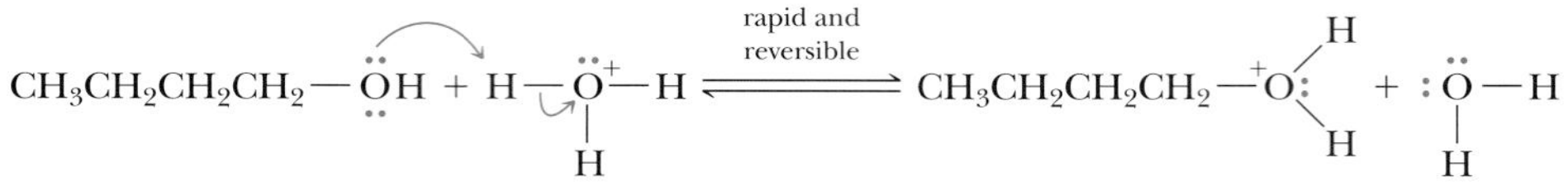

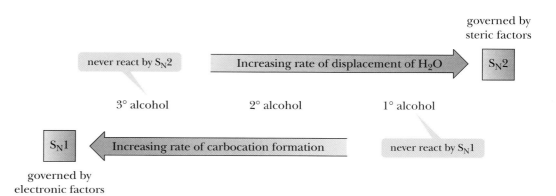

An oxonium ion

Step 2: The nucleophilic displacement of H_2O by Br^- gives the bromoalkane:

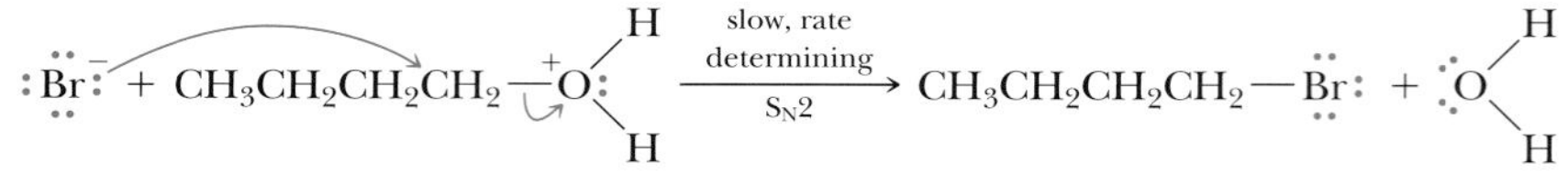

Why do tertiary alcohols react with HX by formation of carbocation intermediates, whereas primary alcohols react by direct displacement of —OH (more accurately, by displacement of —OH_2^+)? The answer is a combination of the same two factors involved in nucleophilic substitution reactions of haloalkanes (Section 7.6B):

1. *Electronic factors*   Tertiary carbocations are the most stable (require the lowest activation energy for their formation), whereas primary carbocations are the least stable (require the highest activation energy for their formation). Therefore, tertiary alcohols are most likely to react by carbocation formation; secondary alcohols are intermediate, and primary alcohols rarely, if ever, react by carbocation formation.

2. *Steric factors*   To form a new carbon–halogen bond, halide ion must approach the substitution center and begin to form a new covalent bond to it. If we compare the ease of approach to the substitution center of a primary oxonium ion with that of a tertiary oxonium ion, we see that approach is considerably easier in the case of a primary oxonium ion. Two hydrogen atoms and one alkyl group screen the back side of the substitution center of a primary oxonium ion, whereas three alkyl groups screen the back side of the substitution center of a tertiary oxonium ion.

Reaction with Thionyl Chloride

The most widely used reagent for the conversion of primary and secondary alcohols to alkyl chlorides is thionyl chloride, $SOCl_2$. The by-products of this nucleophilic substitution reaction are HCl and SO_2, both given off as gases. Often, an organic base such as pyridine (Section 10.2) is added to react with and neutralize the HCl by-product:

$$\text{1-Heptanol} \quad \text{OH} + SOCl_2 \xrightarrow{\text{pyridine}} \text{Cl} + SO_2 + HCl$$

1-Heptanol   Thionyl   1-Chloroheptane
                chloride

E. Acid-Catalyzed Dehydration to Alkenes

An alcohol can be converted to an alkene by **dehydration**—that is, by the elimination of a molecule of water from adjacent carbon atoms. In the laboratory, the dehydration of an alcohol is most often brought about by heating it with either 85% phosphoric acid or concentrated sulfuric acid. Primary alcohols are the most difficult to dehydrate and generally require heating in concentrated sulfuric acid at temperatures as high as 180°C. Secondary alcohols undergo acid-catalyzed dehydration at somewhat lower temperatures. The acid-catalyzed dehydration of tertiary alcohols often requires temperatures only slightly above room temperature:

Dehydration Elimination of a molecule of water from a compound.

$$CH_3CH_2OH \xrightarrow[180°C]{H_2SO_4} CH_2{=}CH_2 + H_2O$$

Cyclohexanol   Cyclohexene

$$\underset{\underset{CH_3}{|}}{\overset{\overset{CH_3}{|}}{CH_3COH}} \xrightarrow[50°C]{H_2SO_4} \overset{\overset{CH_3}{|}}{CH_3C}{=}CH_2 + H_2O$$

2-Methyl-2-propanol   2-Methylpropene
(*tert*-Butyl alcohol)   (Isobutylene)

Thus, the ease of acid-catalyzed dehydration of alcohols occurs in this order:

1° alcohol   <   2° alcohol   <   3° alcohol

Ease of dehydration of alcohols ➔

When isomeric alkenes are obtained in the acid-catalyzed dehydration of an alcohol, the more stable alkene (the one with the greater number of substituents on the double bond; see Section 5.5B) generally predominates; that is, the acid-catalyzed dehydration of alcohols follows Zaitsev's rule (Section 7.8):

$$\underset{\text{2-Butanol}}{\overset{\overset{OH}{|}}{CH_3CH_2CHCH_3}} \xrightarrow[\text{heat}]{85\% \ H_3PO_4} \underset{\substack{\text{2-Butene}\\(80\%)}}{CH_3CH{=}CHCH_3} + \underset{\substack{\text{1-Butene}\\(20\%)}}{CH_3CH_2CH{=}CH_2}$$

EXAMPLE 8.5

For each of the following alcohols, draw structural formulas for the alkenes that form upon acid-catalyzed dehydration of that alcohol, and predict which alkene is the major product:

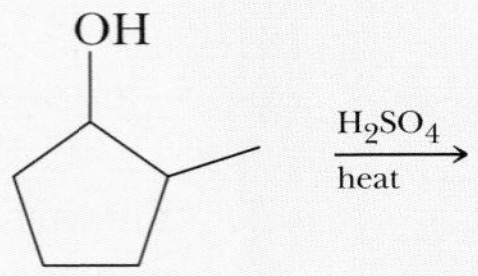

SOLUTION

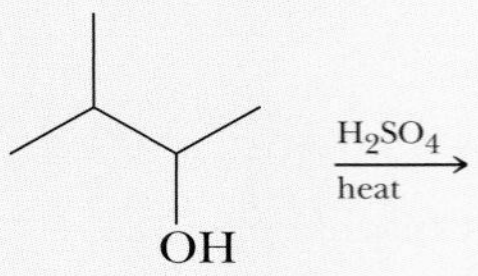

(a) The elimination of H_2O from carbons 2 and 3 gives 2-methyl-2-butene; the elimination of H_2O from carbons 1 and 2 gives 3-methyl-1-butene. 2-Methyl-2-butene, with three alkyl groups (three methyl groups) on the double bond, is the major product. 3-Methyl-1-butene, with only one alkyl group (an isopropyl group) on the double bond, is the minor product:

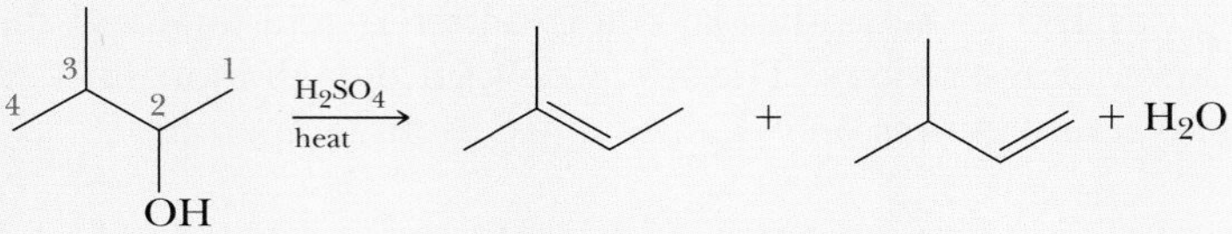

3-Methyl-2-butanol   2-Methyl-2-butene   3-Methyl-1-butene
                          (major product)

(b) The major product, 1-methylcyclopentene, has three alkyl substituents on the double bond. The minor product, 3-methylcyclopentene, only two alkyl substituents on the double bond:

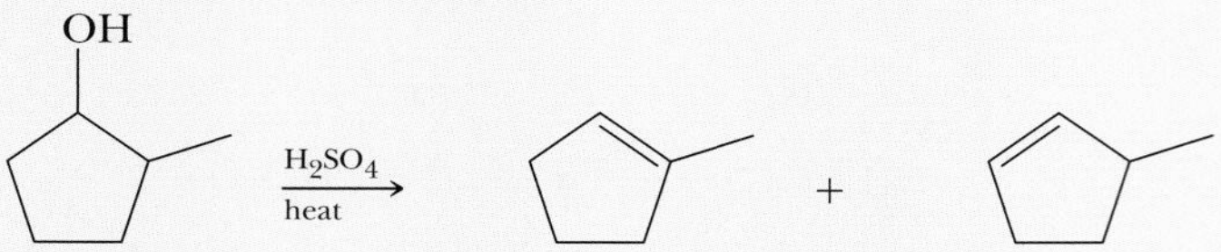

2-Methylcyclopentanol   1-Methylcyclopentene 3-Methylcyclopentene
                          (major product)

Practice Problem 8.5

For each of the following alcohols, draw structural formulas for the alkenes that form upon acid-catalyzed dehydration of that alcohol, and predict which alkene is the major product from each alcohol:

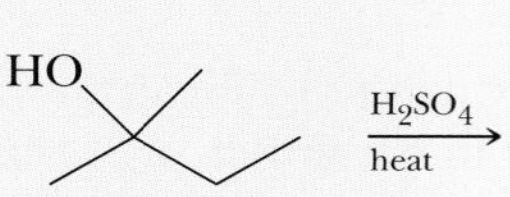

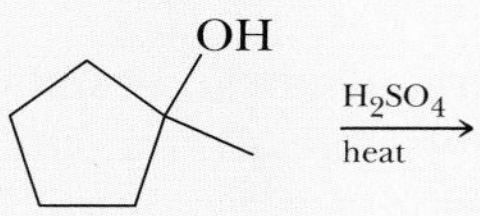

On the basis of the relative ease of dehydration of alcohols ($3° > 2° > 1°$), chemists propose a three-step mechanism for the acid-catalyzed dehydration of secondary and tertiary alcohols. This mechanism involves the formation of a carbocation intermediate in the rate-determining step and therefore is an E1 mechanism.

Mechanism: Acid-Catalyzed Dehydration of 2-Butanol: An E1 Mechanism

Step 1: Proton transfer from H_3O^+ to the OH group of the alcohol gives an oxonium ion. A result of this step is to convert OH^-, a poor leaving group, into H_2O, a better leaving group:

$$CH_3CHCH_2CH_3 + H-\overset{+}{\underset{H}{O}}-H \underset{reversible}{\overset{rapid\ and}{\rightleftharpoons}} CH_3CHCH_2CH_3 + \overset{..}{\underset{H}{O}}-H$$

An oxonium ion

Step 2: Breaking of the C—O bond gives a 2° carbocation intermediate and H_2O:

H_2O is a good leaving group

$$CH_3\overset{+}{C}HCH_2CH_3 \underset{determining}{\overset{slow,\ rate}{\rightleftharpoons}} CH_3\overset{+}{C}HCH_2CH_3 + H_2\overset{..}{O}$$

A 2° carbocation intermediate

Step 3: Proton transfer from the carbon adjacent to the positively charged carbon to H_2O gives the alkene and regenerates the catalyst. The sigma electrons of a C—H bond become the pi electrons of the carbon–carbon double bond:

$$CH_3-\overset{+}{C}H-CH-CH_3 + :\overset{..}{\underset{H}{O}}-H \overset{rapid}{\longrightarrow} CH_3-CH=CH-CH_3 + H-\overset{+}{\underset{H}{O}}-H$$

Because the rate-determining step in the acid-catalyzed dehydration of secondary and tertiary alcohols is the formation of a carbocation intermediate, the relative ease of dehydration of these alcohols parallels the ease of formation of carbocations.

Primary alcohols react by the following two-step mechanism, in which Step 2 is the rate-determining step:

Mechanism: Acid-Catalyzed Dehydration of a Primary Alcohol: An E2 Mechanism

Step 1: Proton transfer from H_3O^+ to the OH group of the alcohol gives an oxonium ion:

$$CH_3CH_2-\overset{..}{\underset{..}{O}}-H + H-\overset{+}{\underset{H}{O}}-H \underset{reversible}{\overset{rapid\ and}{\rightleftharpoons}} CH_3CH_2-\overset{+}{\underset{H}{O}}\overset{H}{<} + :\overset{..}{\underset{H}{O}}-H$$

Step 2: Simultaneous proton transfer to solvent and loss of H_2O gives the alkene:

$$H-\overset{..}{\underset{H}{O}}: + H-\overset{H}{\underset{H}{C}}-CH_2-\overset{+}{\underset{H}{O}}: \underset{E2}{\overset{slow,\ rate\ determining}{\longrightarrow}} H-\overset{+}{\underset{H}{O}}-H + \overset{H}{\underset{H}{>}}C=C\overset{H}{\underset{H}{<}} + :\overset{..}{\underset{H}{O}}-H$$

In Section 5.3B, we discussed the acid-catalyzed hydration of alkenes to give alcohols. In the current section, we discussed the acid-catalyzed dehydration of alcohols to give alkenes. In fact, hydration–dehydration reactions are reversible.

Alkene hydration and alcohol dehydration are competing reactions, and the following equilibrium exists:

$$\underset{\text{An alkene}}{\overset{}{\diagdown}{C}{=}{C}\diagup} + \boxed{H_2O} \underset{}{\overset{\text{acid}\atop\text{catalyst}}{\rightleftharpoons}} \underset{\text{An alcohol}}{-\overset{|}{\underset{|}{C}}-\overset{|}{\underset{|}{C}}-}$$

How, then, do we control which product will predominate? Recall that LeChâtelier's principle states that a system in equilibrium will respond to a stress in the equilibrium by counteracting that stress. This response allows us to control these two reactions to give the desired product. Large amounts of water (achieved with the use of dilute aqueous acid) favor alcohol formation, whereas a scarcity of water (achieved with the use of concentrated acid) or experimental conditions by which water is removed (for example, heating the reaction mixture above 100°C) favor alkene formation. Thus, depending on the experimental conditions, it is possible to use the hydration–dehydration equilibrium to prepare either alcohols or alkenes, each in high yields.

F. Oxidation of Primary and Secondary Alcohols

The oxidation of a primary alcohol gives an aldehyde or a carboxylic acid, depending on the experimental conditions. Secondary alcohols are oxidized to ketones. Tertiary alcohols are not oxidized. Following is a series of transformations in which a primary alcohol is oxidized first to an aldehyde and then to a carboxylic acid. The fact that each transformation involves oxidation is indicated by the symbol O in brackets over the reaction arrow:

$$\underset{\substack{\text{A primary}\\\text{alcohol}}}{CH_3-\overset{OH}{\underset{H}{\overset{|}{\underset{|}{C}}}}-H} \xrightarrow{[O]} \underset{\text{An aldehyde}}{CH_3-\overset{O}{\overset{\|}{C}}-H} \xrightarrow{[O]} \underset{\substack{\text{A carboxylic}\\\text{acid}}}{CH_3-\overset{O}{\overset{\|}{C}}-OH}$$

The reagent most commonly used in the laboratory for the oxidation of a primary alcohol to a carboxylic acid and a secondary alcohol to a ketone is chromic acid, H_2CrO_4. Chromic acid is prepared by dissolving either chromium(VI) oxide or potassium dichromate in aqueous sulfuric acid:

$$\underset{\substack{\text{Chromium(VI)}\\\text{oxide}}}{CrO_3} + H_2O \xrightarrow{H_2SO_4} \underset{\text{Chromic acid}}{H_2CrO_4}$$

$$\underset{\substack{\text{Potassisum}\\\text{dichromate}}}{K_2Cr_2O_7} \xrightarrow{H_2SO_4} H_2Cr_2O_7 \xrightarrow{H_2O} \underset{\text{Chromic acid}}{2\ H_2CrO_4}$$

The oxidation of 1-octanol by chromic acid in aqueous sulfuric acid gives octanoic acid in high yield. These experimental conditions are more than sufficient to oxidize the intermediate aldehyde to a carboxylic acid:

$$\underset{\text{1-Octanol}}{CH_3(CH_2)_6CH_2OH} \xrightarrow[H_2SO_4,\ H_2O]{CrO_3} \left[\underset{\substack{\text{Octanal}\\\text{(not isolated)}}}{CH_3(CH_2)_6\overset{O}{\overset{\|}{C}}H} \right] \longrightarrow \underset{\text{Octanoic acid}}{CH_3(CH_2)_6\overset{O}{\overset{\|}{C}}OH}$$

The form of Cr(VI) commonly used for the oxidation of a primary alcohol to an aldehyde is prepared by dissolving CrO_3 in aqueous HCl and adding pyridine to precipitate **pyridinium chlorochromate (PCC)** as a solid. PCC oxidations are carried out in aprotic solvents, most commonly dichloromethane, CH_2Cl_2:

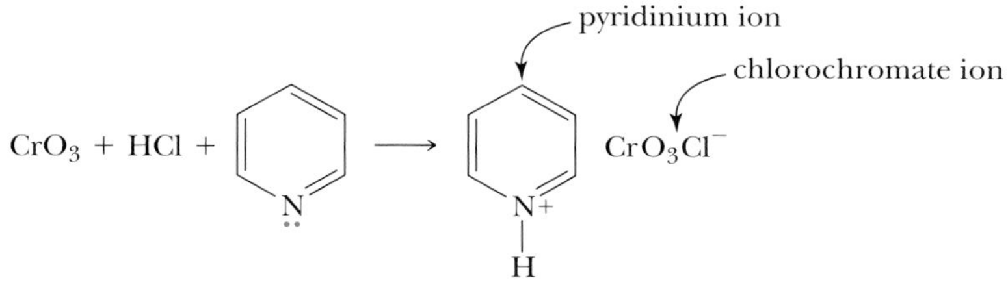

Pyridine   Pyridinium
chlorochromate
(PCC)

This reagent is not only selective for the oxidation of primary alcohols to aldehydes, but also has little effect on carbon–carbon double bonds or other easily oxidized functional groups. In the following example, geraniol is oxidized to geranial without affecting either carbon–carbon double bond:

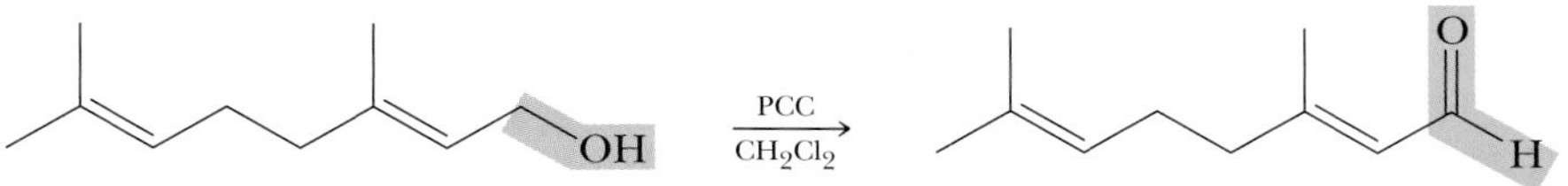

Geraniol   Geranial

Secondary alcohols are oxidized to ketones by both chromic acid and PCC:

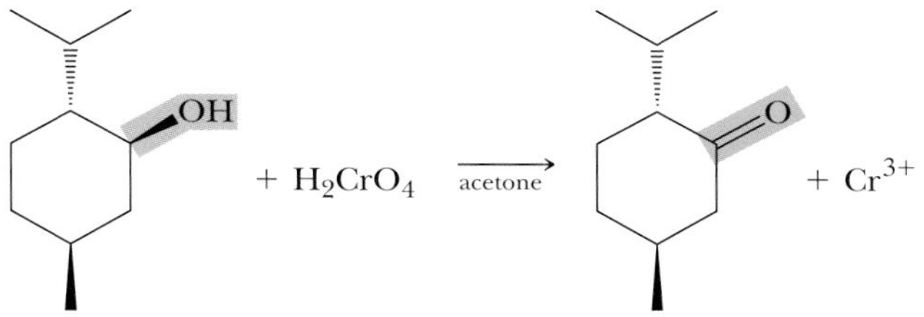

2-Isopropyl-5-methyl-   2-Isopropyl-5-methyl-
cyclohexanol   cyclohexanone
(Menthol)   (Menthone)

Tertiary alcohols are resistant to oxidation, because the carbon bearing the —OH is bonded to three carbon atoms and therefore cannot form a carbon—oxygen double bond:

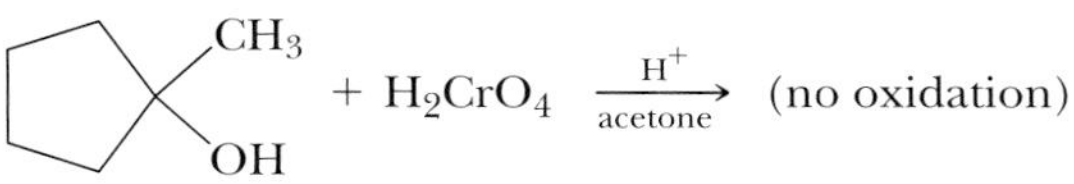

1-Methylcyclopentanol

Note that the essential feature of the oxidation of an alcohol is the presence of at least one hydrogen on the carbon bearing the OH group. Tertiary alcohols lack such a hydrogen; therefore, they are not oxidized.

EXAMPLE 8.6

Draw the product of the treatment of each of the following alcohols with PCC:

(a) 1-Hexanol   (b) 2-Hexanol   (c) Cyclohexanol

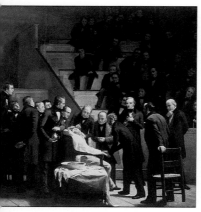

This painting by Robert Hinckley shows the first use of diethyl ether as an anesthetic in 1846. Dr. Robert John Collins was removing a tumor from the patient's neck, and the dentist W. T. G. Morton - who discovered its anesthetic properties - administered the ether. *(Boston Medical Library in the Francis A. Courtney Library of Medicine)*

Ether A compound containing an oxygen atom bonded to two carbon atoms.

SOLUTION

1-Hexanol, a primary alcohol, is oxidized to hexanal. 2-Hexanol, a secondary alcohol, is oxidized to 2-hexanone. Cyclohexanol, a secondary alcohol, is oxidized to cyclohexanone.

(a)   Hexanal   (b)   2-Hexanone   (c)   Cyclohexanone

Practice Problem 8.6

Draw the product of the treatment of each alcohol in Example 8.6 with chromic acid.

8.4   ETHERS

A. Structure

The functional group of an **ether** is an atom of oxygen bonded to two carbon atoms. Figure 8.4 shows a Lewis structure and ball-and-stick model of dimethyl ether, CH_3OCH_3, the simplest ether. In dimethyl ether, two sp^3 hybrid orbitals of oxygen form sigma bonds to carbon atoms. The other two sp^3 hybrid orbitals of oxygen each contain an unshared pair of electrons. The C—O—C bond angle in dimethyl ether is 110.3°, close to the predicted tetrahedral angle of 109.5°.

In ethyl vinyl ether, the ether oxygen is bonded to one sp^3 hybridized carbon and one sp^2 hybridized carbon:

$$CH_3CH_2-O-CH=CH_2$$
Ethyl vinyl ether

Alkoxy group An —OR group, where R is an alkyl group.

B. Nomenclature

In the IUPAC system, ethers are named by selecting the longest carbon chain as the parent alkane and naming the —OR group bonded to it as an **alkoxy** (*alk*yl + *oxy*gen) group. Common names are derived by listing the alkyl groups bonded to oxygen in alphabetical order and adding the word *ether*.

$$CH_3CH_2OCH_2CH_3$$
Ethoxyethane
(Diethyl ether)

$$CH_3OCCH_3$$ (with CH_3 above and CH_3 below)
2-Methoxy-2-methylpropane
(methyl *tert*-butyl ether, MTBE)

trans-2-Ethoxycyclohexanol

(a)

H—C—O—C—H (with H above and below each carbon)

(b)

110.3°

Figure 8.4
Dimethyl ether, CH_3OCH_3.
(a) Lewis structure and
(b) ball-and-stick model.

Chemists almost invariably use common names for low-molecular-weight ethers. For example, although ethoxyethane is the IUPAC name for $CH_3CH_2OCH_2CH_3$, it is rarely called that, but rather is called diethyl ether, ethyl ether, or, even more commonly, simply ether. The abbreviation for *tert*-butyl methyl ether, used at one time as an octane-improving additive to gasolines, is *MTBE*, after the common name of methyl *tert*-butyl ether.

CHEMICAL CONNECTIONS 8B

Blood Alcohol Screening

Potassium dichromate oxidation of ethanol to acetic acid is the basis for the original breath alcohol screening test used by law enforcement agencies to determine a person's blood alcohol content. The test is based on the difference in color between the dichromate ion (reddish orange) in the reagent and the chromium(III) ion (green) in the product. Thus, color change can be used as a measure of the quantity of ethanol present in a breath sample:

$$CH_3CH_2OH \quad + \quad Cr_2O_7{}^{2-} \xrightarrow[H_2O]{H_2SO_4}$$

Ethanol   Dichromate ion
(reddish orange)

$$\underset{\text{Acetic acid}}{CH_3\overset{\displaystyle O}{\overset{\|}{C}}OH} \quad + \quad \underset{\substack{\text{Chromium(III)}\\ \text{ion (green)}}}{Cr^{3+}}$$

In its simplest form, a breath alcohol screening test consists of a sealed glass tube containing a potassium dichromate–sulfuric acid reagent impregnated on silica gel. To administer the test, the ends of the tube are broken off, a mouthpiece is fitted to one end, and the other end is inserted into the neck of a plastic bag. The person being tested then blows into the mouthpiece until the plastic bag is inflated.

As breath containing ethanol vapor passes through the tube, reddish orange dichromate ion is reduced to green chromium(III) ion. The concentration of ethanol in the breath is then estimated by measuring how far the green color extends along the length of the tube. When it extends beyond the halfway point, the person is judged as having a sufficiently high blood alcohol content to warrant further, more precise testing.

The Breathalyzer, a more precise testing device, operates on the same principle as the simplified screening test. In a Breathalyzer test, a measured volume of breath is bubbled through a solution of potassium dichromate in aqueous sulfuric acid, and the color change is measured spectrophotometrically.

Both tests measure alcohol in the breath. The legal definition of being under the influence of alcohol is based on *blood* alcohol content, not breath alcohol content. The chemical correlation between these two measurements is that air deep within the lungs is in equilibrium with blood passing through the pulmonary arteries, and an equilibrium is established between blood alcohol and breath alcohol. It has been determined by tests in persons drinking alcohol that 2100 mL of breath contains the same amount of ethanol as 1.00 mL of blood.

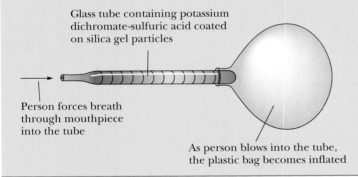

Glass tube containing potassium dichromate-sulfuric acid coated on silica gel particles

Person forces breath through mouthpiece into the tube

As person blows into the tube, the plastic bag becomes inflated

A device for testing the breath for the presence of ethanol. When ethanol is oxidized by potassium dichromate, the reddish-orange color of dichromate ion turns to green as it is reduced to chromium(III) ion. *(Charles D. Winters)*

Cyclic ethers are heterocyclic compounds in which the ether oxygen is one of the atoms in a ring. These ethers are generally known by their common names:

Cyclic ether An ether in which the oxygen is one of the atoms of a ring.

Ethylene oxide   Tetrahydrofuran (THF)   1,4-Dioxane

Figure 8.5
Ethers are polar molecules, but because of steric hindrance, only weak attractive interactions exist between their molecules in the pure liquid.

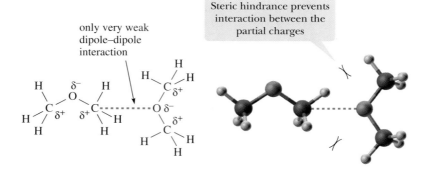

C. Physical Properties

Ethers are polar compounds in which oxygen bears a partial negative charge and each carbon bonded to it bears a partial positive charge (Figure 8.5). Because of steric hindrance, however, only weak forces of attraction exist between ether molecules in the pure liquid. Consequently, boiling points of ethers are much lower than those of alcohols of comparable molecular weight (Table 8.3). Boiling points of ethers are close to those of hydrocarbons of comparable molecular weight (compare Tables 3.4 and 8.3).

TABLE 8.3 Boiling Points and Solubilities in Water of Some Alcohols and Ethers of Comparable Molecular Weight

Structural Formula	Name	Molecular Weight	Boiling Point (°C)	Solubility in Water
CH_3CH_2OH	ethanol	46	78	infinite
CH_3OCH_3	dimethyl ether	46	−24	7.8 g/100 g
$CH_3CH_2CH_2CH_2OH$	1-butanol	74	117	7.4 g/100 g
$CH_3CH_2OCH_2CH_3$	diethyl ether	74	35	8 g/100 g
$CH_3CH_2CH_2CH_2CH_2OH$	1-pentanol	88	138	2.3 g/100 g
$HOCH_2CH_2CH_2CH_2OH$	1,4-butanediol	90	230	infinite
$CH_3CH_2CH_2CH_2OCH_3$	butyl methyl ether	88	71	slight
$CH_3OCH_2CH_2OCH_3$	ethylene glycol dimethyl ether	90	84	infinite

Because the oxygen atom of an ether carries a partial negative charge, ethers form hydrogen bonds with water (Figure 8.6) and are more soluble in water than are hydrocarbons of comparable molecular weight and shape (compare data in Tables 3.4 and 8.3).

The effect of hydrogen bonding is illustrated dramatically by comparing the boiling points of ethanol (78°C) and its constitutional isomer dimethyl ether (−24°C). The difference in boiling points between these two compounds is due to the polar O—H group in the alcohol, which is capable of forming intermolecular hydrogen bonds. This hydrogen bonding increases the attractive force between molecules of ethanol; thus, ethanol has a higher boiling point than dimethyl ether:

$$CH_3CH_2OH \qquad\qquad CH_3OCH_3$$

Ethanol   Dimethyl ether
bp 78°C   bp −24°C

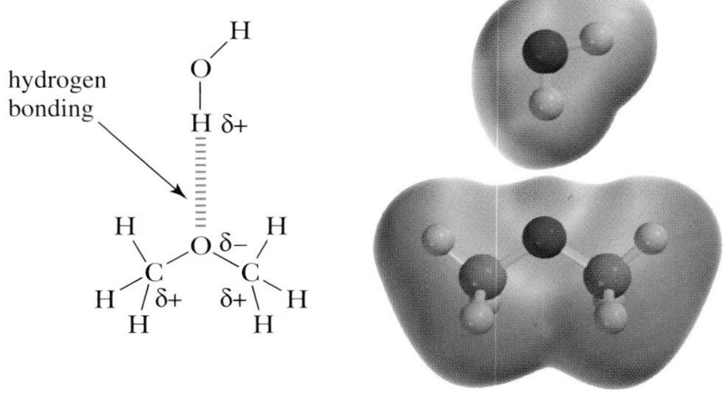

Dimethyl ether in water. The partially negative oxygen of the ether is the hydrogen bond acceptor, and a partially positive hydrogen of a water molecule is the hydrogen bond donor.

Figure 8.6
Ethers are hydrogen-bond acceptors only. They are not hydrogen-bond donors.

EXAMPLE 8.7

Write the IUPAC and common names for each ether:

(a) 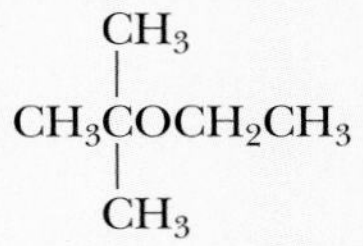

(b) 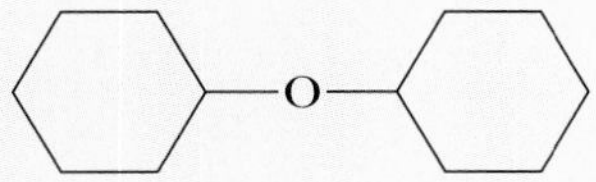

SOLUTION

(a) 2-Ethoxy-2-methylpropane. Its common name is *tert*-butyl ethyl ether.
(b) Cyclohexoxycyclohexane. Its common name is dicyclohexyl ether.

Practice Problem 8.7

Write the IUPAC and common names for each ether:

(a) 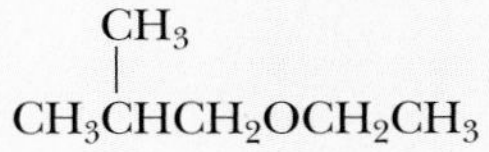

(b) 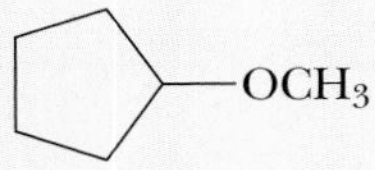

EXAMPLE 8.8

Arrange these compounds in order of increasing solubility in water:

$CH_3OCH_2CH_2OCH_3$	$CH_3CH_2OCH_2CH_3$	$CH_3CH_2CH_2CH_2CH_2CH_3$
Ethylene glycol dimethyl ether	Diethyl ether	Hexane

SOLUTION

Water is a polar solvent. Hexane, a nonpolar hydrocarbon, has the lowest solubility in water. Both diethyl ether and ethylene glycol dimethyl ether are polar compounds, due to the presence of their polar C—O—C groups, and each interacts with water as a hydrogen-bond acceptor. Because ethylene glycol dimethyl ether has more sites within its molecules for hydrogen bonding, it is more soluble in water than diethyl ether:

$CH_3CH_2CH_2CH_2CH_2CH_3$	$CH_3CH_2OCH_2CH_3$	$CH_3OCH_2CH_2OCH_3$
Insoluble	8 g/100 g water	Soluble in all proportions

Practice Problem 8.8

Arrange these compounds in order of increasing boiling point:

$$CH_3OCH_2CH_2OCH_3 \qquad HOCH_2CH_2OH \qquad CH_3OCH_2CH_2OH$$

D. Reactions of Ethers

Ethers, R—O—R, resemble hydrocarbons in their resistance to chemical reaction. They do not react with oxidizing agents, such as potassium dichromate or potassium permanganate. They are not affected by most acids or bases at moderate temperatures. Because of their good solvent properties and general inertness to chemical reaction, ethers are excellent solvents in which to carry out many organic reactions.

8.5 EPOXIDES

A. Structure and Nomenclature

Epoxide A cyclic ether in which oxygen is one atom of a three-membered ring.

An **epoxide** is a cyclic ether in which oxygen is one atom of a three-membered ring:

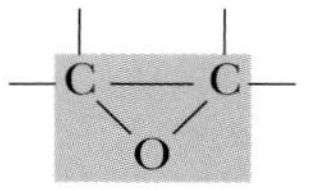

Functional group of an epoxide	Ethylene oxide

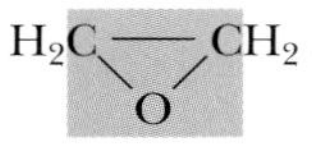

CH$_3$CH——CH$_2$ with O at bottom

Propylene oxide

Although epoxides are technically classed as ethers, we discuss them separately because of their exceptional chemical reactivity compared with other ethers.

Common names for epoxides are derived by giving the common name of the alkene from which the epoxide might have been derived, followed by the word *oxide*; an example is ethylene oxide.

B. Synthesis from Alkenes

Ethylene oxide, one of the few epoxides manufactured on an industrial scale, is prepared by passing a mixture of ethylene and air (or oxygen) over a silver catalyst:

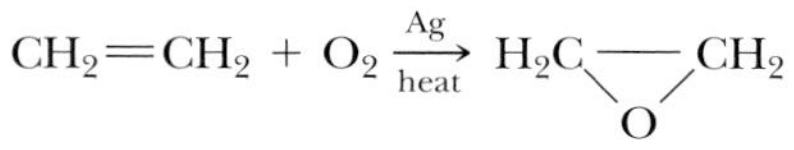

Ethylene	Ethylene oxide

In the United States, the annual production of ethylene oxide by this method is approximately 10^9 kg.

The most common laboratory method for the synthesis of epoxides from alkenes is oxidation with a peroxycarboxylic acid (a peracid), RCO_3H. One peracid used for this purpose is peroxyacetic acid:

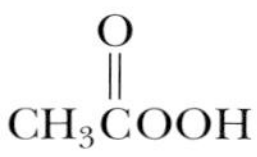

$$\overset{\displaystyle O}{\underset{\displaystyle \parallel}{}}$$
$$CH_3COOH$$

Peroxyacetic acid
(Peracetic acid)

Following is a balanced equation for the epoxidation of cyclohexene by a peroxycarboxylic acid. In the process, the peroxycarboxylic acid is reduced to a carboxylic acid:

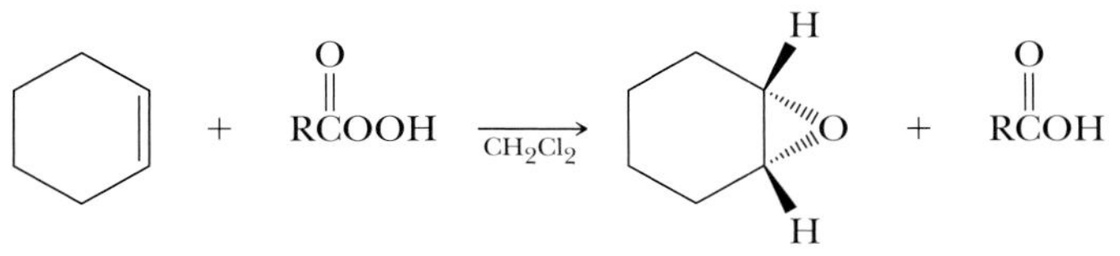

Cyclohexene   A peroxy-   1,2-Epoxycyclohexane   A carboxylic
               carboxylic acid   (Cyclohexene oxide)   acid

The epoxidation of an alkene is stereoselective. The epoxidation of *cis*-2-butene, for example, yields only *cis*-2-butene oxide:

cis-2-Butene   *cis*-2-Butene oxide

EXAMPLE 8.9

Draw a structural formula of the epoxide formed by treating *trans*-2-butene with a peroxycarboxylic acid.

SOLUTION

The oxygen of the epoxide ring is added by forming both carbon–oxygen bonds from the same side of the carbon–carbon double bond:

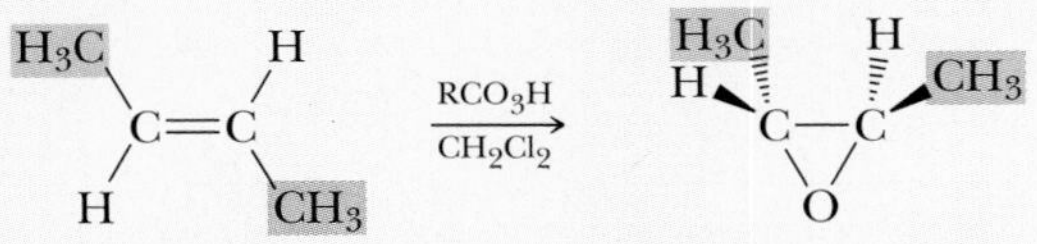

trans-2-Butene   *trans*-2-Butene oxide

Practice Problem 8.9

Draw the structural formula of the epoxide formed by treating 1,2-dimethylcyclopentene with a peroxycarboxylic acid.

C. Ring-Opening Reactions

Ethers are not normally susceptible to reaction with aqueous acid (Section 8.4D). Epoxides, however, are especially reactive because of the angle strain in the three-membered ring. The normal bond angle about an sp^3 hybridized carbon or oxygen atom is 109.5°. Because of the strain associated with the compression of bond angles in the three-membered epoxide ring from the normal 109.5° to 60°, epoxides undergo ring-opening reactions with a variety of reagents.

In the presence of an acid catalyst—most commonly, perchloric acid—epoxides are hydrolyzed to glycols. As an example, the acid-catalyzed hydrolysis of ethylene oxide gives 1,2-ethanediol:

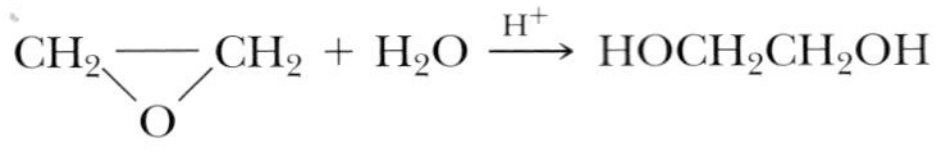
CH2 CH2 + H2O
O
H+
HOCH2CH2OH

Ethylene oxide

1,2-Ethanediol
(Ethylene glycol)

Annual production of ethylene glycol in the United States is approximately 10^{10} kg. Two of its largest uses are in automotive antifreeze and as one of the two starting materials for the production of polyethylene terephthalate (PET), which is fabricated into such consumer products as Dacron® polyester, Mylar®, and packaging films (Section 17.5B).

The acid-catalyzed ring opening of epoxides shows a stereoselectivity typical of S_N2 reactions: The nucleophile attacks anti to the leaving hydroxyl group, and the —OH groups in the glycol thus formed are anti. As a result, the hydrolysis of an epoxycycloalkane yields a *trans*-1,2-cycloalkanediol:

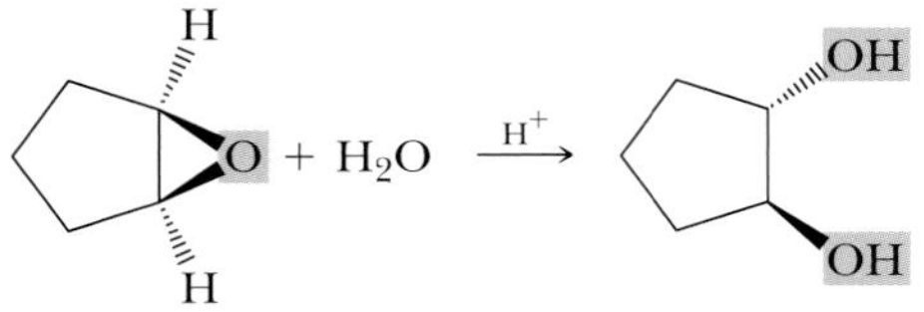
H
O + H2O
H
H+
OH
OH

1,2-Epoxycyclopentane
(Cyclopentene oxide)

trans-1,2-Cyclopentanediol

At this point, let us compare the stereochemistry of the glycol formed by the acid-catalyzed hydrolysis of an epoxide with that formed by the OsO_4 oxidation of an alkene (Section 5.4). Each reaction sequence is stereoselective, but gives a different stereoisomer. The acid-catalyzed hydrolysis of cyclopentene oxide gives *trans*-1,2-cyclopentanediol, whereas the osmium tetroxide oxidation of cyclopentene gives *cis*-1,2-cyclopentanediol. Thus, a cycloalkene can be converted to either a cis glycol or a trans glycol by the proper choice of reagents.

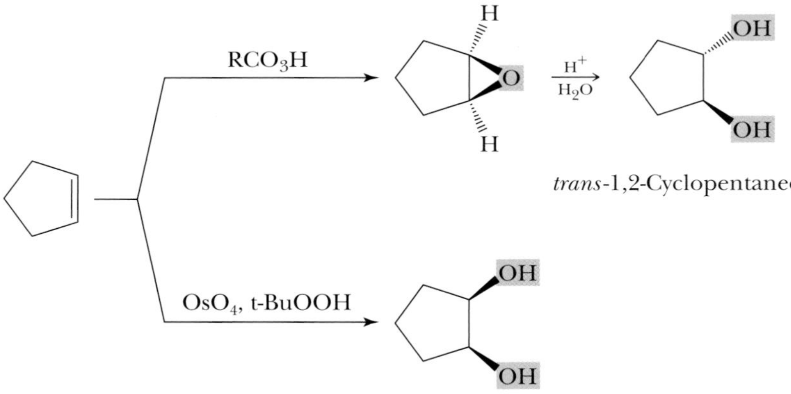
RCO3H
H
O
H
H+
H2O
OH
OH
OsO4, t-BuOOH
OH
OH

trans-1,2-Cyclopentanediol

cis-1,2-Cyclopentanediol

EXAMPLE 8.10

Draw the structural formula of the product formed by treating cyclohexene oxide with aqueous acid. Be certain to show the stereochemistry of the product.

SOLUTION

The acid-catalyzed hydrolysis of the three-membered epoxide ring gives a trans glycol:

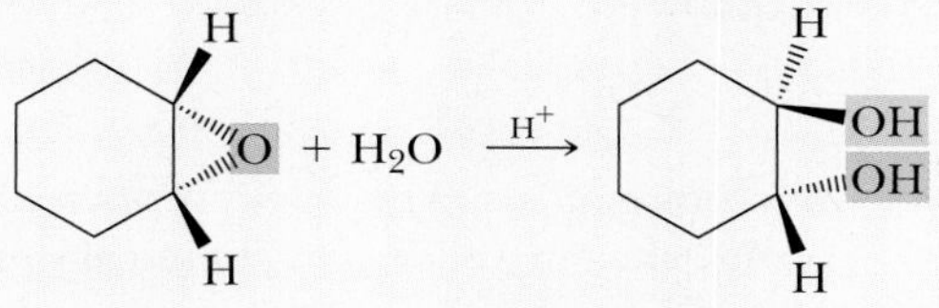

trans-1,2-cyclohexanediol

Practice Problem 8.10

Show how to convert cyclohexene to cis-1,2-cyclohexanediol.

Just as ethers are not normally susceptible to reaction with electrophiles, neither are they normally susceptible to reaction with nucleophiles. Because of the strain associated with the three-membered ring, however, epoxides undergo ring-opening reactions with good nucleophiles such as ammonia and amines (Chapter 10), alkoxide ions, and thiols and their anions (Section 8.7). Good nucleophiles attack the ring by an S_N2 mechanism and show a stereoselectivity for attack of the nucleophile at the less hindered carbon of the three-membered ring. An illustration is the reaction of cyclohexene oxide with ammonia to give trans-2-aminocyclohexanol:

Cyclohexene oxide

trans-2-Aminocyclohexanol
(major product)

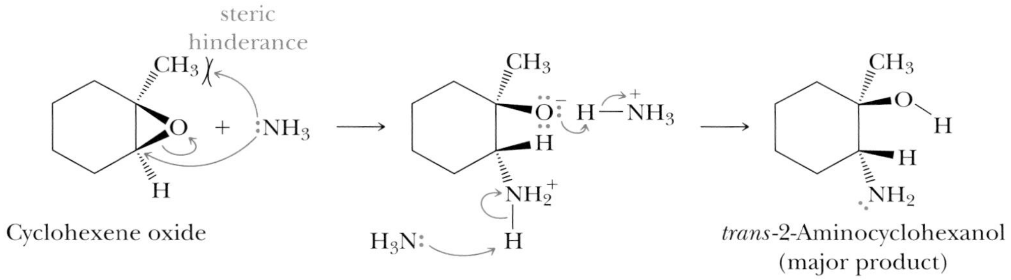

The value of epoxides lies in the number of nucleophiles that bring about ring opening and the combinations of functional groups that can be prepared from them. The following chart summarizes the three most important of these nucleophilic ring-opening reactions (the characteristic structural feature of each ring-opening product is shown in color):

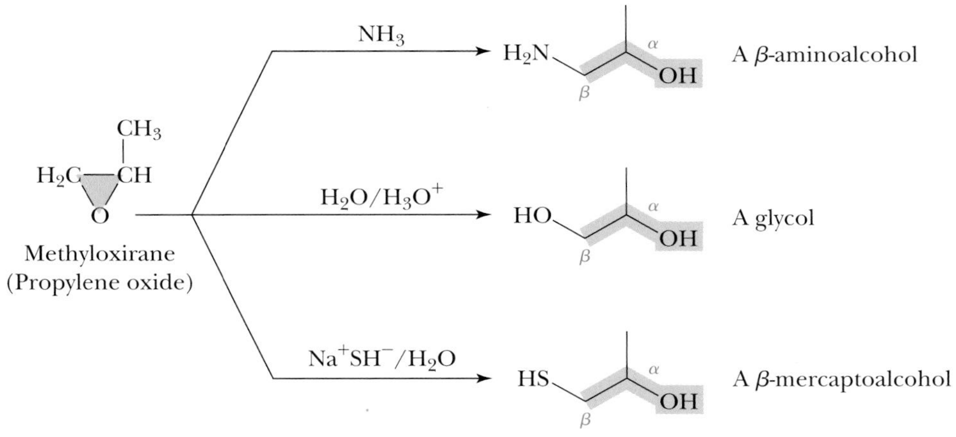

Methyloxirane
(Propylene oxide)

A β-aminoalcohol

A glycol

A β-mercaptoalcohol

CHEMICAL CONNECTIONS 8C

Ethylene Oxide: A Chemical Sterilant

Because ethylene oxide is such a highly strained molecule, it reacts with the types of nucleophilic groups present in biological materials. At sufficiently high concentrations, ethylene oxide reacts with enough molecules in cells to cause the death of microorganisms. This toxic property is the basis for using ethylene oxide as a chemical sterilant. In hospitals, surgical instruments and other items that cannot be made disposable are now sterilized by exposure to ethylene oxide.

Ethylene oxide and substituted ethylene oxides are valuable building blocks for the synthesis of larger organic molecules. Following are structural formulas for two common drugs, each synthesized in part from ethylene oxide:

Procaine
(Novocaine)

Diphenhydramine
(Benadryl)

Novocaine was the first injectable local anesthetic. Benadryl was the first synthetic antihistamine. The portion of the carbon skeleton of each that is derived from the reaction of ethylene oxide with a nitrogen–nucleophile is shown in color.

In later chapters, after we have developed the chemistry of more functional groups, we will show how to synthesize Novocaine and Benadryl from readily available starting materials. For the moment, however, it is sufficient to recognize that the unit —O—C—C—Nu can be derived by nucleophilic opening of ethylene oxide or a substituted ethylene oxide.

8.6 THIOLS

The most outstanding property of low-molecular-weight thiols is their stench. They are responsible for unpleasant odors such as those from skunks, rotten eggs, and sewage. The scent of skunks is due primarily to two thiols:

$$CH_3CH=CHCH_2SH$$
2-Butene-1-thiol

$$CH_3CHCH_2CH_2SH$$
|
$$CH_3$$
3-Methyl-1-butanethiol

The scent of skunks is a mixture of two thiols, 3-methyl-1-butanethiol and 2-butene-1-thiol. (*Stephen J. Krasemann/ Photo Researchers, Inc.*)

Thiol A compound containing an —SH (sulfhydryl) group.

A. Structure

The functional group of a **thiol** is an —SH (sulfhydryl) group. Figure 8.7 shows a Lewis structure and a ball-and-stick model of methanethiol, CH_3SH, the simplest thiol.

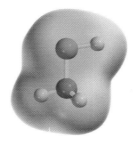

Figure 8.7
Methanethiol, CH_3SH.
(a) Lewis structure and
(b) ball-and-stick model.
The C—S—H bond angle
is 100.9°, somewhat smaller
than the tetrahedral angle
of 109.5°.

Methanethiol. The electronegativities of carbon and sulfur are virtually identical (2.5 each), while sulfur is slightly more electronegative than hydrogen (2.5 versus 2.1). The electron density model shows some slight partial positive charge on hydrogen of the S—H group and some slight partial negative charge on sulfur.

B. Nomenclature

The sulfur analog of an alcohol is called a thiol (thi- from the Greek: *theion*, sulfur) or, in the older literature, a **mercaptan**, which literally means "mercury capturing." Thiols react with Hg^{2+} in aqueous solution to give sulfide salts as insoluble precipitates. Thiophenol, C_6H_5SH, for example, gives $(C_6H_5S)_2Hg$.

In the IUPAC system, thiols are named by selecting as the parent alkane the longest chain of carbon atoms that contains the —SH group. To show that the compound is a thiol, we add *-thiol* to the name of the parent alkane and number the parent chain in the direction that gives the —SH group the lower number.

Common names for simple thiols are derived by naming the alkyl group bonded to —SH and adding the word *mercaptan*. In compounds containing other functional groups, the presence of an —SH group is indicated by the prefix **mercapto-**. According to the IUPAC system, —OH takes precedence over —SH in both numbering and naming:

Mercaptan A common name for any molecule containing an —SH group.

$$CH_3CH_2SH \qquad CH_3\overset{\overset{\displaystyle CH_3}{|}}{C}HCH_2SH \qquad HSCH_2CH_2OH$$

Ethanethiol 2-Methyl-1-propanethiol 2-Mercaptoethanol
(Ethyl mercaptan) (Isobutyl mercaptan)

Sulfur analogs of ethers are named by using the word *sulfide* to show the presence of the —S— group. Following are common names of two sulfides:

$$CH_3SCH_3 \qquad CH_3CH_2S\overset{\overset{\displaystyle CH_3}{|}}{C}HCH_3$$

Dimethyl sulfide Ethyl isopropyl sulfide

Mushrooms, onions, garlic, and coffee all contain sulfur compounds. One of these present in coffee is

(Charles D. Winters)

EXAMPLE 8.11

Write the IUPAC name for each thiol:

(a)   (b)

SOLUTION

(a) The parent alkane is pentane. We show the presence of the —SH group by adding *thiol* to the name of the parent alkane. The IUPAC name of this thiol is 1-pentanethiol. Its common name is pentyl mercaptan.

(b) The parent alkane is butane. The IUPAC name of this thiol is 2-butanethiol. Its common name is *sec*-butyl mercaptan.

Practice Problem 8.11

Write the IUPAC name for each thiol:

(a)

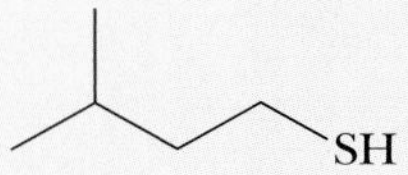

(b)

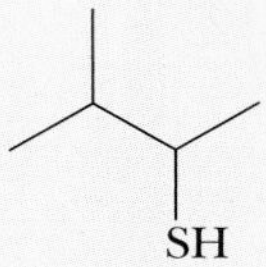

C. Physical Properties

Because of the small difference in electronegativity between sulfur and hydrogen (2.5 − 2.1 = 0.4), we classify the S—H bond as nonpolar covalent. Because of this lack of polarity, thiols show little association by hydrogen bonding. Consequently, they have lower boiling points and are less soluble in water and other polar solvents than are alcohols of similar molecular weight. Table 8.4 gives the boiling points of three low-molecular-weight thiols. For comparison, the table also gives the boiling points of alcohols with the same number of carbon atoms.

TABLE 8.4   Boiling Points of Three Thiols and Three Alcohols with the Same Number of Carbon Atoms

Thiol	Boiling Point (°C)	Alcohol	Boiling Point (°C)
methanethiol	6	methanol	65
ethanethiol	35	ethanol	78
1-butanethiol	98	1-butanol	117

Earlier, we illustrated the importance of hydrogen bonding in alcohols by comparing the boiling points of ethanol (78°C) and its constitutional isomer dimethyl ether (24°C). By comparison, the boiling point of ethanethiol is 35°C, and that of its constitutional isomer dimethyl sulfide is 37°C:

$$CH_3CH_2SH \qquad CH_3SCH_3$$

Ethanethiol   Dimethyl sulfide
bp 35°C   bp 37°C

The fact that the boiling points of these constitutional isomers are almost identical indicates that little or no association by hydrogen bonding occurs between thiol molecules.

8.7   REACTIONS OF THIOLS

In this section, we discuss the acidity of thiols and their reaction with strong bases, such as sodium hydroxide, and with molecular oxygen.

A. Acidity

Hydrogen sulfide is a stronger acid than water:

$$H_2O + H_2O \rightleftharpoons HO^- + H_3O^+ \qquad pK_a = 15.7$$

$$H_2S + H_2O \rightleftharpoons HS^- + H_3O^+ \qquad pK_a = 7.0$$

Similarly, thiols are stronger acids than alcohols. Compare, for example, the pK_a's of ethanol and ethanethiol in dilute aqueous solution:

$$CH_3CH_2OH + H_2O \rightleftharpoons CH_3CH_2O^- + H_3O^+ \qquad pK_a = 15.9$$

$$CH_3CH_2SH + H_2O \rightleftharpoons CH_3CH_2S^- + H_3O^+ \qquad pK_a = 8.5$$

Thiols are sufficiently strong acids that, when dissolved in aqueous sodium hydroxide, they are converted completely to alkylsulfide salts:

$$CH_3CH_2SH + Na^+OH^- \longrightarrow CH_3CH_2S^-Na^+ + H_2O$$

pK_a 8.5			pK_a 15.7
Stronger acid	Stronger base	Weaker base	Weaker acid

To name salts of thiols, give the name of the cation first, followed by the name of the alkyl group to which the suffix *-sulfide* is added. For example, the sodium salt derived from ethanethiol is named sodium ethylsulfide.

B. Oxidation to Disulfides

Many of the chemical properties of thiols stem from the fact that the sulfur atom of a thiol is oxidized easily to several higher oxidation states. The most common reaction of thiols in biological systems is their oxidation to disulfides, the functional group of which is a **disulfide** ($-S-S-$) bond. Thiols are readily oxidized to disulfides by molecular oxygen. In fact, they are so susceptible to oxidation that they must be protected from contact with air during storage. Disulfides, in turn, are easily reduced to thiols by several reagents. This easy interconversion between thiols and disulfides is very important in protein chemistry, as we will see in Chapter 20:

$$2\ HOCH_2CH_2SH \underset{\text{reduction}}{\overset{\text{oxidation}}{\rightleftharpoons}} HOCH_2CH_2S-SCH_2CH_2OH$$

A thiol	A disulfide

We derive common names of simple disulfides by listing the names of the groups bonded to sulfur and adding the word *disulfide*, as, for example, CH_3S-SCH_3, which is named dimethyldisulfide.

SUMMARY

The functional group of an **alcohol** (Section 8.2A) is an $-OH$ (**hydroxyl**) group bonded to an sp^3 hybridized carbon. Alcohols are classified as **1°**, **2°**, or **3°** (Section 8.2A), depending on whether the $-OH$ group is bonded to a primary, secondary, or tertiary carbon. IUPAC names of alcohols (Section 8.2B) are derived by changing the suffix of the parent alkane from *-e* to *-ol*. The chain is numbered to give the carbon bearing $-OH$ the lower number. Common names for alcohols are derived by naming the alkyl group bonded to $-OH$ and adding the word *alcohol*.

Alcohols are polar compounds (Section 8.2C) with oxygen bearing a partial negative charge and both the carbon and hydrogen bonded to it bearing partial positive charges. Because of intermolecular association by **hydrogen bonding**, the boiling points of alcohols are higher than those of hydrocarbons of comparable molecular weight. Because of increased dispersion forces, the boiling points of alcohols increase with increasing molecular weight. Alcohols interact with water by hydrogen bonding and therefore are more soluble in water than are hydrocarbons of comparable molecular weight.

The functional group of an **ether** is an atom of oxygen bonded to two carbon atoms (Section 8.4A). In the IUPAC name of an ether (Section 8.4B), the parent alkane is named, and then the $-OR$ group is named as an alkoxy substituent. Common names are derived by naming the two groups bonded to oxygen, followed by the word *ether*. Ethers are weakly polar compounds (Section 8.4C). Their boiling points are close to those of hydrocarbons of comparable molecular weight. Because ethers are hydrogen-bond acceptors, they are more soluble in water than are hydrocarbons of comparable molecular weight. An epoxide is a cyclic ether in which oxygen is one of the atoms of the three-membered ring (Section 8.5A).

A **thiol** (Section 8.6A) is the sulfur analog of an alcohol; it contains an —**SH** (**sulfhydryl**) group in place of an —OH group. Thiols are named in the same manner as alcohols, but the suffix *-e* is retained, and **-thiol** is added (Section 8.6B). Common names for thiols are derived by naming the alkyl group bonded to —SH and adding the word *mercaptan*. In compounds containing functional groups of higher precedence, the presence of —SH is indicated by the prefix **mercapto-**. For **thioethers**, name the two groups bonded to sulfur, followed by the word *sulfide*. The S—H bond is nonpolar, and the physical properties of thiols are more like those of hydrocarbons of comparable molecular weight (Section 8.6C).

KEY REACTIONS

1. Acidity of alcohols (Section 8.3A)

In dilute aqueous solution, methanol and ethanol are comparable in acidity to water. Secondary and tertiary alcohols are weaker acids than water.

$$CH_3OH + H_2O \rightleftharpoons CH_3O^- + H_3O^+ \qquad pK_a = 15.5$$

2. Reaction of alcohols with Active Metals (Section 8.3C)

Alcohols react with Li, Na, K, and other active metals to form metal alkoxides, which are somewhat stronger bases than NaOH and KOH:

$$2\ CH_3CH_2OH + 2\ Na \longrightarrow 2\ CH_3CH_2O^-Na^+ + H_2$$

3. Reaction of alcohols with HCl, HBr, and HI (Section 8.3D)

Primary alcohols react with HBr and HI by an S_N2 mechanism:

$$CH_3CH_2CH_2CH_2OH + HBr \longrightarrow CH_3CH_2CH_2CH_2Br + H_2O$$

Tertiary alcohols react with HCl, HBr, and HI by an S_N1 mechanism, with the formation of a carbocation intermediate:

$$(CH_3)_3COH + HCl \xrightarrow{25°C} (CH_3)_3CCl + H_2O$$

Secondary alcohols may react with HCl, HBr, and HI by an S_N2 or an S_N1 mechanism, depending on the alcohol and experimental conditions.

4. Reaction of alcohols with $SOCl_2$ (Section 8.3D)

This is often the method of choice for converting an alcohol to an alkyl chloride:

$$CH_3(CH_2)_5OH + SOCl_2 \longrightarrow CH_3(CH_2)_5Cl + SO_2 + HCl$$

5. Acid-catalyzed dehydration of alcohols (Section 8.3E)

When isomeric alkenes are possible, the major product is generally the more substituted alkene (Zaitsev's rule):

$$CH_3CH_2CH(OH)CH_3 \xrightarrow[heat]{H_3PO_4} \underset{\text{Major product}}{CH_3CH{=}CHCH_3} + CH_3CH_2CH{=}CH_2 + H_2O$$

6. Oxidation of a Primary Alcohol to an Aldehyde (Section 8.3F)

This oxidation is most conveniently carried out by using pyridinium chlorochromate (PCC):

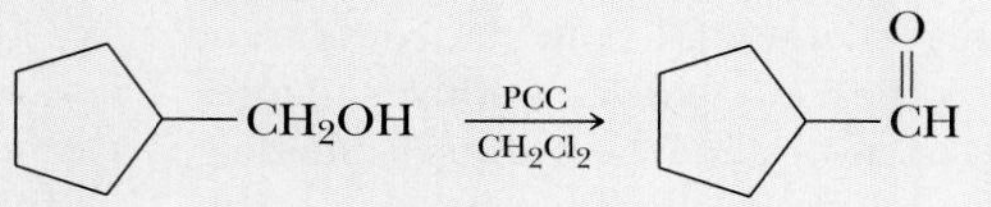

7. Oxidation of a Primary Alcohol to a Carboxylic Acid (Section 8.3F)

A primary alcohol is oxidized to a carboxylic acid by chromic acid:

$$CH_3(CH_2)_4CH_2OH + H_2CrO_4 \xrightarrow[\text{acetone}]{H_2O} CH_3(CH_2)_4\overset{\overset{O}{\|}}{C}OH + Cr^{3+}$$

8. Oxidation of a Secondary Alcohol to a Ketone (Section 8.3F)

A secondary alcohol is oxidized to a ketone by chromic acid and by PCC:

$$CH_3(CH_2)_4\overset{\overset{OH}{|}}{C}HCH_3 + H_2CrO_4 \longrightarrow CH_3(CH_2)_4\overset{\overset{O}{\|}}{C}CH_3 + Cr^{3+}$$

9. Oxidation of an alkene to an epoxide (Section 8.5B)

The most common method for the synthesis of an epoxide from an alkene is oxidation with a peroxycarboxylic acid, such as peroxyacetic acid:

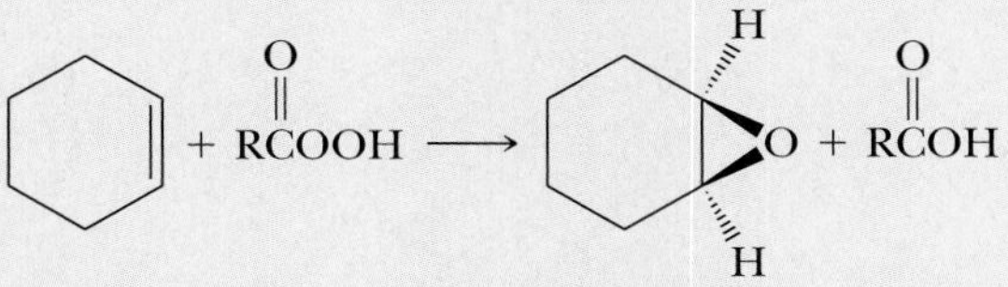

10. Acid-catalyzed hydrolysis of epoxides (Section 8.5C)

Acid-catalyzed hydrolysis of an epoxide derived from a cycloalkene gives a trans glycol (hydrolysis of cycloalkene oxide is stereoselective, giving the trans glycol):

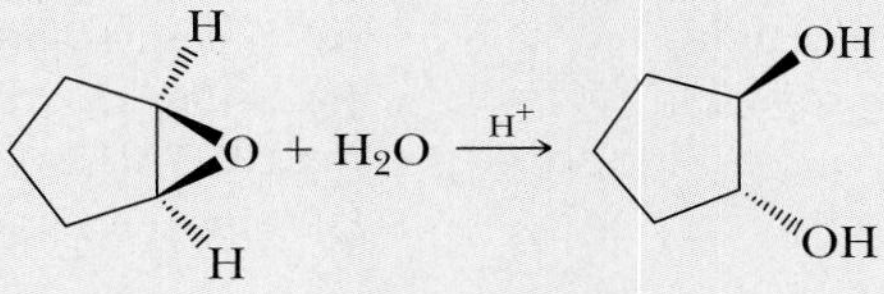

11. Nucleophilic ring opening of epoxides (Section 8.5C)

Good nucleophiles, such as ammonia and amines, open the highly strained epoxide ring by an S_N2 mechanism and show a stereoselectivity for attack of the nucleophile at the less hindered carbon of the three-membered ring:

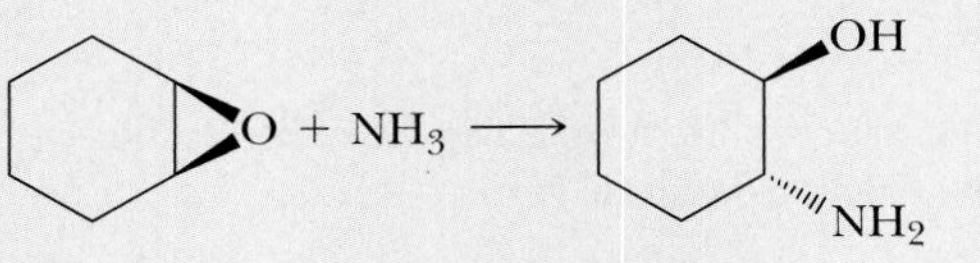

Cyclohexene oxide   *trans*-2-Aminocyclohexanol

12. Acidity of thiols (Section 8.7A)

Thiols are weak acids, pK_a 8–9, but are considerably stronger acids than alcohols, pK_a 16–18.

$$CH_3CH_2SH + H_2O \rightleftharpoons CH_3CH_2S^- + H_3O^+ \qquad pK_a = 8.5$$

13. Oxidation to Disulfides (Section 8.7B)

Oxidation of a thiol by O_2 gives a disulfide:

$$2\ RSH + \tfrac{1}{2}O_2 \longrightarrow RS\text{-}SR + H_2O$$

PROBLEMS

A problem number set in red indicates an applied "real-world" problem.

Structure and Nomenclature

8.12 Which of the following compounds are secondary alcohols?

(a) 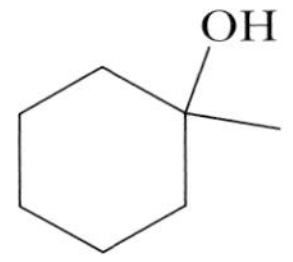 [cyclohexane with OH and CH₃]

(b) $(CH_3)_3COH$

(c) 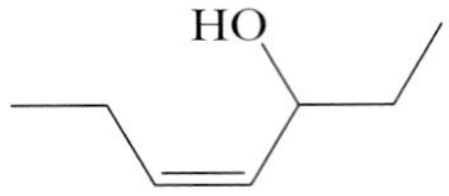 [structure with HO]

(d) 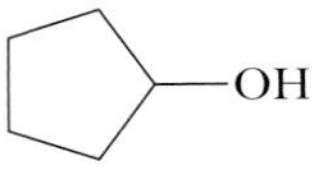 [cyclopentane—OH]

8.13 Name these compounds:

(a) 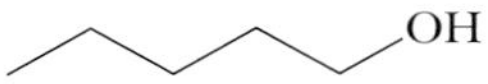 [structure with OH]   (b)  HO[structure]OH   (c) 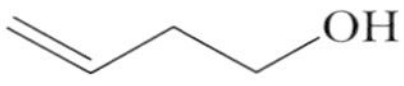 [structure with OH]

(d) 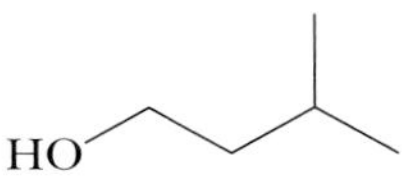 HO[structure]   (e) 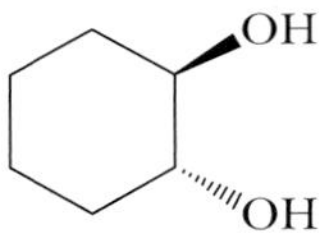 [cyclohexane with two OH groups]   (f) 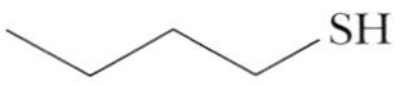 [structure with SH]

8.14 Draw a structural formula for each alcohol:
  (a) Isopropyl alcohol
  (b) Propylene glycol
  (c) (R)-5-Methyl-2-hexanol
  (d) 2-Methyl-2-propyl-1,3-propanediol
  (e) 2,2-Dimethyl-1-propanol
  (f) 2-Mercaptoethanol
  (g) 1,4-Butanediol
  (h) (Z)-5-Methyl-2-hexen-1-ol
  (i) cis-3-Pentene-1-ol
  (j) trans-1,4-Cyclohexanediol

8.15 Write names for these ethers:

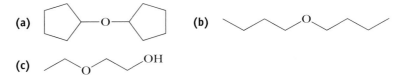

(a) [two cyclopentanes joined by O]   (b) [dipropyl ether structure]

(c) [structure with O and OH]

8.16 Name and draw structural formulas for the eight isomeric alcohols with molecular formula $C_5H_{12}O$. Which are chiral?

Physical Properties

8.17 Arrange these compounds in order of increasing boiling point (values in °C are −42, 78, 117, and 198):
  (a) $CH_3CH_2CH_2CH_2OH$
  (b) CH_3CH_2OH
  (c) $HOCH_2CH_2OH$
  (d) $CH_3CH_2CH_3$

8.18 Arrange these compounds in order of increasing boiling point (values in °C are −42, −24, 78, and 118):
  (a) CH_3CH_2OH
  (b) CH_3OCH_3
  (c) $CH_3CH_2CH_3$
  (d) CH_3COOH

8.19 Propanoic acid and methyl acetate are constitutional isomers, and both are liquids at room temperature:

One of these compounds has a boiling point of 141°C; the other has a boiling point of 57°C. Which compound has which boiling point?

8.20 Draw all possible staggered conformations of ethylene glycol ($HOCH_2CH_2OH$). Can you explain why the gauche conformation is more stable than the anti conformation by approximately 1 kcal/mol?

8.21 Following are structural formulas for 1-butanol and 1-butanethiol:

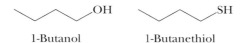

One of these compounds has a boiling point of 98.5°C; the other has a boiling point of 117°C. Which compound has which boiling point?

8.22 From each pair of compounds, select the one that is more soluble in water:

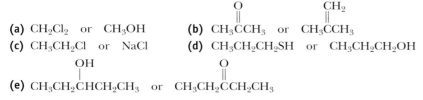

8.23 Arrange the compounds in each set in order of decreasing solubility in water:
   (a) Ethanol; butane; diethyl ether   **(b)** 1-Hexanol; 1,2-hexanediol; hexane

8.24 Each of the following compounds is a common organic solvent:
   (a) CH_2Cl_2 or CH_3CH_2OH
   (b) $CH_3CH_2OCH_2CH_3$ or CH_3CH_2OH

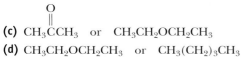

   (d) $CH_3CH_2OCH_2CH_3$ or $CH_3(CH_2)_3CH_3$

   From each pair of compounds, select the solvent with the greater solubility in water.

Synthesis of Alcohols

8.25 Give the structural formula of an alkene or alkenes from which each alcohol or glycol can be prepared:
   (a) 2-Butanol   **(b)** 1-Methylcyclohexanol
   (c) 3-Hexanol   **(d)** 2-Methyl-2-pentanol
   (e) Cyclopentanol   **(f)** 1,2-Propanediol

8.26 The addition of bromine to cyclopentene and the acid-catalyzed hydrolysis of cyclopentene oxide are both stereoselective; each gives a trans product. Compare the mechanisms of these two reactions, and show how each mechanism accounts for the formation of the trans product.

Acidity of Alcohols and Thiols

8.27 From each pair, select the stronger acid, and, for each stronger acid, write a structural formula for its conjugate base:

(a) H_2O or H_2CO_3 **(b)** CH_3OH or CH_3COOH

(c) CH_3COOH or CH_3CH_2SH

8.28 Arrange these compounds in order of increasing acidity (from weakest to strongest):

$$CH_3CH_2CH_2OH \qquad CH_3CH_2\overset{\displaystyle O}{\overset{\|}{C}}OH \qquad CH_3CH_2CH_2SH$$

8.29 From each pair, select the stronger base, and, for each stronger base, write the structural formula of its conjugate acid:

(a) OH^- or CH_3O^- **(b)** $CH_3CH_2S^-$ or $CH_3CH_2O^-$

(c) $CH_3CH_2O^-$ or NH_2^-

8.30 Label the stronger acid, stronger base, weaker acid, and weaker base in each of the following equilibria, and then predict the position of each equilibrium (for pK_a values, see Table 2.1):

(a) $CH_3CH_2O^- + HCl \rightleftharpoons CH_3CH_2OH + Cl^-$

(b) $CH_3\overset{\displaystyle O}{\overset{\|}{C}}OH + CH_3CH_2O^- \rightleftharpoons CH_3\overset{\displaystyle O}{\overset{\|}{C}}O^- + CH_3CH_2OH$

8.31 Predict the position of equilibrium for each acid–base reaction; that is, does each lie considerably to the left, does each lie considerably to the right, or are the concentrations evenly balanced?

(a) $CH_3CH_2OH + Na^+OH^- \rightleftharpoons CH_3CH_2O^-Na^+ + H_2O$

(b) $CH_3CH_2SH + Na^+OH^- \rightleftharpoons CH_3CH_2S^-Na^+ + H_2O$

(c) $CH_3CH_2OH + CH_3CH_2S^-Na^+ \rightleftharpoons CH_3CH_2O^-Na^+ + CH_3CH_2SH$

(d) $CH_3CH_2S^-Na^+ + CH_3\overset{\displaystyle O}{\overset{\|}{C}}OH \rightleftharpoons CH_3CH_2SH + CH_3\overset{\displaystyle O}{\overset{\|}{C}}O^-Na^+$

Reactions of Alcohols

8.32 Show how to distinguish between cyclohexanol and cyclohexene by a simple chemical test. (*Hint:* Treat each with Br_2 in CCl_4 and watch what happens.)

8.33 Write equations for the reaction of 1-butanol, a primary alcohol, with these reagents:

(a) Na metal   **(b)** HBr, heat

(c) $K_2Cr_2O_7$, H_2SO_4, heat   **(d)** $SOCl_2$

(e) Pyridinium chlorochromate (PCC)

8.34 Write equations for the reaction of 2-butanol, a secondary alcohol, with these reagents:

(a) Na metal   **(b)** H_2SO_4, heat

(c) HBr, heat   **(d)** $K_2Cr_2O_7$, H_2SO_2, heat

(e) $SOCl_2$ **(f)** Pyridinium chlorochromate (PCC)

8.35 When (R)-2-butanol is left standing in aqueous acid, it slowly loses its optical activity. When the organic material is recovered from the aqueous solution, only 2-butanol is found. Account for the observed loss of optical activity.

8.36 What is the most likely mechanism of the following reaction?

Draw a structural formula for the intermediate(s) formed during the reaction.

8.37 Complete the equations for these reactions:

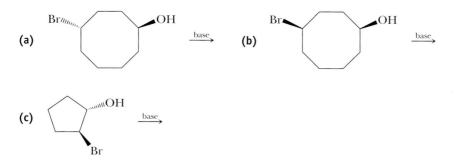

(a) 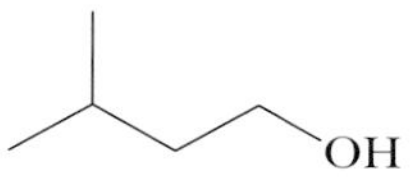 [structure] + H_2CrO_4 ⟶   **(b)** 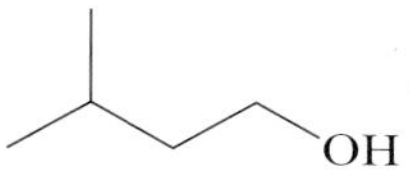 [structure] + $SOCl_2$ ⟶

(c) 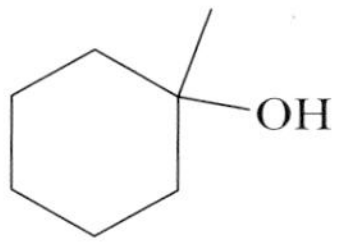 [structure] OH   HCl ⟶   **(d)** 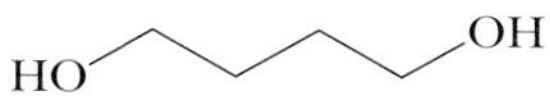 HO [structure] OH + HBr ⟶ (excess)

(e) 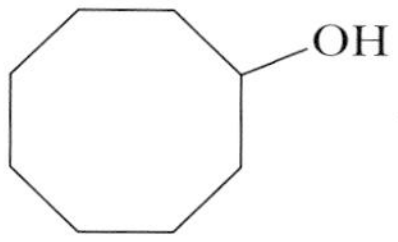 [structure] OH + H_2CrO_4 ⟶   **(f)** 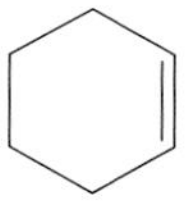 [structure] + OsO_4, H_2O_2 ⟶

8.38 In the commercial synthesis of methyl *tert*-butyl ether (MTBE), once used as an anti-knock, octane-improving gasoline additive, 2-methylpropene and methanol are passed over an acid catalyst to give the ether:

$$CH_3C{=}CH_2 + CH_3OH \xrightarrow[\text{catalyst}]{\text{acid}} CH_3\underset{\underset{CH_3}{|}}{\overset{\overset{CH_3}{|}}{C}}OCH_3$$

2-Methylpropene Methanol   2-Methoxy-2-methyl-
(Isobutylene)   propane (Methyl
                        tert-butyl ether, MTBE)

Propose a mechanism for this reaction.

8.39 Cyclic bromoalcohols, upon treatment with base, can sometimes undergo intramolecular S_N2 reactions to form bicyclic ethers. Determine whether each of the following compounds is capable of forming a bicyclic ether, and draw the product for those which can:

(a) Br⁄⁄⁄ [cyclooctane structure] OH $\xrightarrow{\text{base}}$ **(b)** Br [cyclooctane structure] OH $\xrightarrow{\text{base}}$

(c) [cyclopentane structure] OH, Br $\xrightarrow{\text{base}}$

Syntheses

8.40 Show how to convert
(a) 1-Propanol to 2-propanol in two steps.
(b) Cyclohexene to cyclohexanone in two steps.
(c) Cyclohexanol to *cis*-1,2-cyclohexanediol in two steps.
(d) Propene to propanone (acetone) in two steps.

8.41 Show how to convert cyclohexanol to these compounds:
(a) Cyclohexene   (b) Cyclohexane   (c) Cyclohexanone

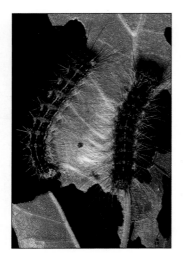

Gypsy moth caterpillars.
(William D. Griffin/Animals, Animals)

8.42 Show reagents and experimental conditions that can be used to synthesize these compounds from 1-propanol (any derivative of 1-propanol prepared in an earlier part of this problem may be used for a later synthesis):

(a) Propanal   (b) Propanoic acid
(c) Propene   (d) 2-Propanol
(e) 2-Bromopropane   (f) 1-Chloropropane
(g) Propanone   (h) 1,2-Propanediol

8.43 Show how to prepare each compound from 2-methyl-1-propanol (isobutyl alcohol):

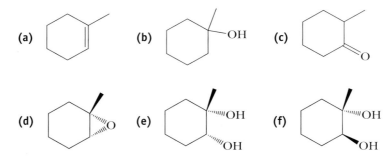

For any preparation involving more than one step, show each intermediate compound formed.

8.44 Show how to prepare each compound from 2-methylcyclohexanol:

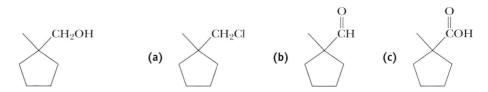

For any preparation involving more than one step, show each intermediate compound formed.

8.45 Show how to convert the alcohol on the left to compounds (a), (b), and (c).

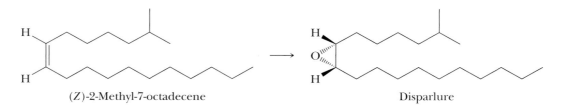

8.46 Disparlure, a sex attractant of the gypsy moth (*Porthetria dispar*), has been synthesized in the laboratory from the following (*Z*)-alkene:

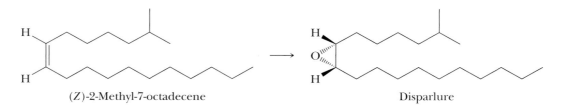

(*Z*)-2-Methyl-7-octadecene   Disparlure

(a) How might the (*Z*)-alkene be converted to disparlure?
(b) How many stereoisomers are possible for disparlure? How many are formed in the sequence you chose?

8.47 The chemical name for bombykol, the sex pheromone secreted by the female silkworm moth to attract male silkworm moths, is *trans*-10-*cis*-12-hexadecadien-1-ol. (The compound has one hydroxyl group and two carbon–carbon double bonds in a 16-carbon chain.)

    (a) Draw a structural formula for bombykol, showing the correct configuration about each carbon–carbon double bond.

    (b) How many cis–trans isomers are possible for the structural formula you drew in part (a)? All possible cis–trans isomers have been synthesized in the laboratory, but only the one named bombykol is produced by the female silkworm moth, and only it attracts male silkworm moths.

Looking Ahead

8.48 Compounds that contain an N—H group associate by hydrogen bonding.

    (a) Do you expect this association to be stronger or weaker than that between compounds containing an O—H group?

    (b) Based on your answer to part (a), which would you predict to have the higher boiling point, 1-butanol or 1-butanamine?

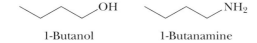

          1-Butanol   1-Butanamine

8.49 Write balanced equations for the reactions of phenol and cyclohexanol, with NaOH:

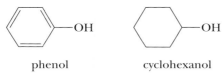

          phenol   cyclohexanol

    (a) Which compound is more acidic? (See Table 2.2.)

    (b) Which conjugate base is more nucleophilic?

8.50 Draw a resonance structure for each of the following compounds in which the heteroatom (O or S) is positively charged:

      methyl vinyl ether   methyl vinyl sulfide

    (a) Compared with ethylene, how does each resonance structure influence the reactivity of the alkene towards an electrophile?

    (b) Peracids are known to be electrophilic reagents. Based on the resonance picture and your knowledge of periodic properties of the elements, would an epoxide be more likely to form with methyl vinyl ether or methyl vinyl sulfide?

    (c) Would your answer to part (b) above be the same or different if only inductive effects were taken into consideration?

8.51 Rank the members in each set of reagents from most to least nucleophilic:

    (a)

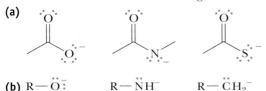

    (b) R—$\ddot{\text{O}}$:⁻   R—$\ddot{\text{N}}$H⁻   R—$\overset{..}{\text{C}}$H₂⁻

8.52 Which of the following compounds is more basic?

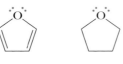

furan   tetrahydrofuran

8.53 In Chapter 15 we will see that the reactivity of the following carbonyl compounds is directly proportional to the stability of the leaving group. Rank the order of reactivity of these carbonyl compounds from most reactive to least reactive based on the stability of the leaving group.

9

Benzene and Its Derivatives

Peppers of the capsicum family. See Chemical Connections "Capsaicin - For Those Who Like it Hot". Inset: A model of capsaicin. *(Douglas Brown)*

9.1 INTRODUCTION

Benzene, a colorless liquid, was first isolated by Michael Faraday in 1825 from the oily residue that collected in the illuminating gas lines of London. Benzene's molecular formula, C_6H_6, suggests a high degree of unsaturation. For comparison, an alkane with six carbons has a molecular formula of C_6H_{14}, and a cycloalkane with six carbons has a molecular formula of C_6H_{12}. Considering benzene's high degree of unsaturation, it might be expected to show many of the reactions characteristic of alkenes. Yet, benzene is remarkably *un*reactive! It does not undergo the addition, oxidation, and reduction reactions characteristic of alkenes. For example, benzene does not react with bromine, hydrogen chloride, or other reagents that usually add to carbon–carbon double bonds. Nor is benzene oxidized by chromic acid or osmium tetroxide under conditions that readily oxidize alkenes. When benzene reacts, it does so by substitution in which a hydrogen atom is replaced by another atom or a group of atoms.

The term "aromatic" was originally used to classify benzene and its derivatives because many of them have distinctive odors. It became clear, however, that a sounder classification for these compounds would be one based on structure and chemical reactivity, not aroma. As it is now used, the term **aromatic** refers instead to the fact that benzene and its derivatives are highly unsaturated compounds which are unexpectedly stable toward reagents that react with alkenes.

We use the term **arene** to describe aromatic hydrocarbons, by analogy with alkane and alkene. Benzene is the parent arene. Just as we call a group derived by the removal of an H from an alkane an alkyl group and give it the symbol R—, we call a group derived by the removal of an H from an arene an **aryl group** and give it the symbol **Ar**—.

Aromatic compound A term used to classify benzene and its derivatives.

Arene An aromatic hydrocarbon.

Aryl group A group derived from an aromatic compound (an arene) by the removal of an H; given the symbol Ar—.

Ar— The symbol used for an aryl group, by analogy with R— for an alkyl group.

9.2 THE STRUCTURE OF BENZENE

Let us imagine ourselves in the mid-19th century and examine the evidence on which chemists attempted to build a model for the structure of benzene. First, because the molecular formula of benzene is C_6H_6, it seemed clear that the molecule must be highly unsaturated. Yet benzene does not show the chemical properties of alkenes, the only unsaturated hydrocarbons known at that time. Benzene does undergo chemical reactions, but its characteristic reaction is substitution rather than addition. When benzene is treated with bromine in the presence of ferric chloride as a catalyst, for example, only one compound with molecular formula C_6H_5Br forms:

$$C_6H_6 + Br_2 \xrightarrow{FeCl_3} C_6H_5Br + HBr$$
Benzene   Bromobenzene

Chemists concluded, therefore, that all six carbons and all six hydrogens of benzene must be equivalent. When bromobenzene is treated with bromine in the presence of ferric chloride, three isomeric dibromobenzenes are formed:

$$C_6H_5Br + Br_2 \xrightarrow{FeCl_3} C_6H_4Br_2 + HBr$$
Bromobenzene   Dibromobenzene
(formed as a mixture of
three constitutional isomers)

For chemists in the mid-19th century, the problem was to incorporate these observations, along with the accepted tetravalence of carbon, into a structural formula for benzene. Before we examine their proposals, we should note that the problem of the structure of benzene and other aromatic hydrocarbons has occupied the efforts of chemists for over a century. It was not until 1930s that chemists developed a general understanding of the unique chemical properties of benzene and its derivatives.

A. Kekulé's Model of Benzene

The first structure for benzene, proposed by August Kekulé in 1872, consisted of a six-membered ring with alternating single and double bonds and with one hydrogen bonded to each carbon. Kekulé further proposed that the ring contains three double bonds which shift back and forth so rapidly that the two forms cannot be separated. Each structure has become known as a **Kekulé structure**.

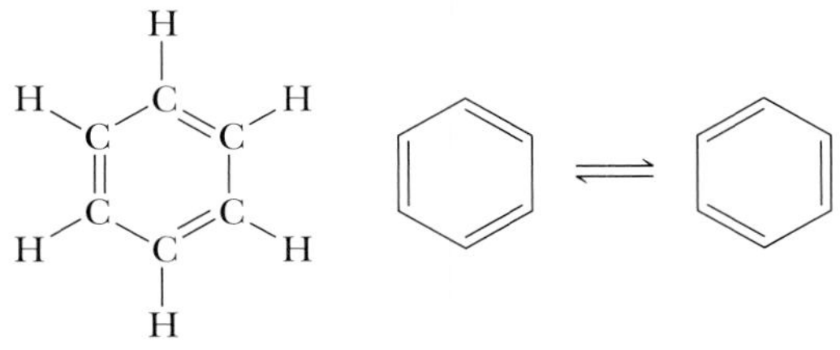

A Kekulé structure,   Kekulé structures
showing all atoms   as line-angle formulas

Because all of the carbons and hydrogens of Kekulé's structure are equivalent, substituting bromine for any one of the hydrogens gives the same compound. Thus, Kekulé's proposed structure was consistent with the fact that treating benzene with bromine in the presence of ferric chloride gives only one compound with molecular formula C_6H_5Br.

His proposal also accounted for the fact that the bromination of bromobenzene gives three (and only three) isomeric dibromobenzenes:

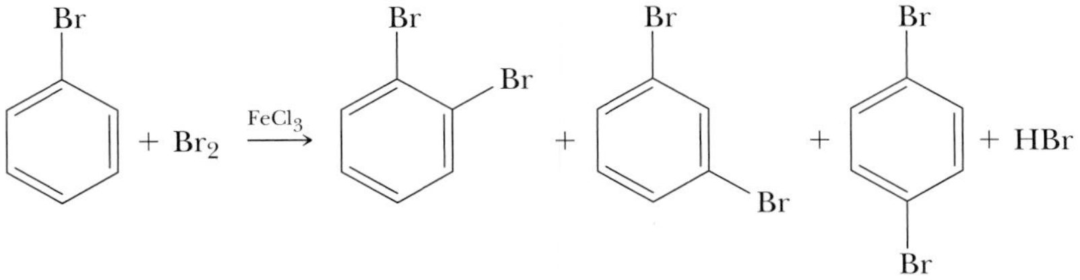

The three isomeric dibromobenzenes

Although Kekulé's proposal was consistent with many experimental observations, it was contested for years. The major objection was that it did not account for the unusual chemical behavior of benzene. If benzene contains three double bonds, why, his critics asked, doesn't it show the reactions typical of alkenes? Why doesn't it add three moles of bromine to form 1,2,3,4,5,6-hexabromocyclohexane? Why, instead, does benzene react by substitution rather than addition?

B. The Orbital Overlap Model of Benzene

The concepts of the **hybridization of atomic orbitals** and the **theory of resonance**, developed by Linus Pauling in the 1930s, provided the first adequate description of the structure of benzene. The carbon skeleton of benzene forms a regular hexagon with C—C—C and H—C—C bond angles of 120°. For this type of bonding, carbon uses sp^2 hybrid orbitals (Section 1.7E). Each carbon forms sigma bonds to two adjacent carbons by the overlap of sp^2–sp^2 hybrid orbitals and one sigma bond to hydrogen by the overlap of sp^2–$1s$ orbitals. As determined experimentally, all carbon–carbon bonds in benzene are the same length, 1.39 Å, a value almost midway between the length of a single bond between sp^3 hybridized carbons (1.54 Å) and that of a double bond between sp^2 hybridized carbons (1.33 Å):

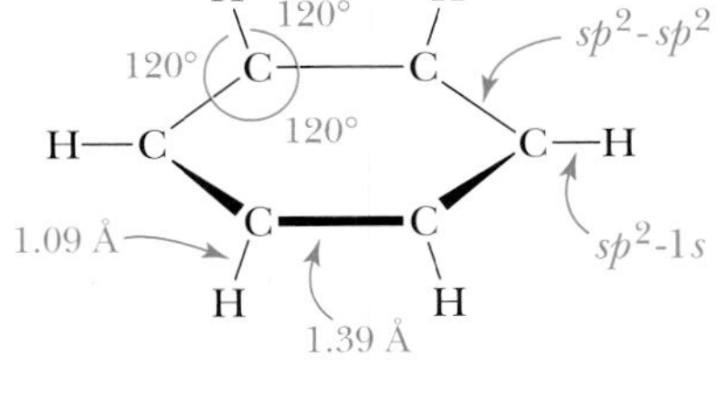

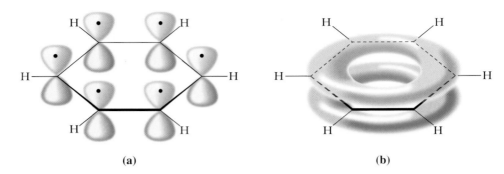

(a)   (b)

Each carbon also has a single unhybridized 2*p* orbital that contains one electron. These six 2*p* orbitals lie perpendicular to the plane of the ring and overlap to form a continuous pi cloud encompassing all six carbons. The electron density of the pi system of a benzene ring lies in one torus (a doughnut-shaped region) above the plane of the ring and a second torus below the plane (Figure 9.1).

C. The Resonance Model of Benzene

One of the postulates of resonance theory is that, if we can represent a molecule or ion by two or more contributing structures, then that molecule cannot be adequately represented by any single contributing structure. We represent benzene as a hybrid of two equivalent contributing structures, often referred to as *Kekulé structures*:

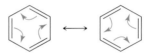

Benzene as a hybrid of two equivalent
contributing structures

Each Kekulé structure makes an equal contribution to the hybrid; thus, the C—C bonds are neither single nor double bonds, but something intermediate. We recognize that neither of these contributing structures exists (they are merely alternative ways to pair 2*p* orbitals, with no reason to prefer one over the other) and that the actual structure is a superposition of both. Nevertheless, chemists continue to use a single contributing structure to represent this molecule because it is as close as we can come to an accurate structure within the limitations of classical valence bond structures and the tetravalence of carbon.

D. The Resonance Energy of Benzene

Resonance energy The difference
in energy between a resonance
hybrid and the most stable of its
hypothetical contributing structures.

Resonance energy is the difference in energy between a resonance hybrid and its most stable hypothetical contributing structure. One way to estimate the resonance energy of benzene is to compare the heats of hydrogenation of cyclohexene and benzene. In the presence of a transition metal catalyst, hydrogen readily reduces cyclohexene to cyclohexane (Section 5.5):

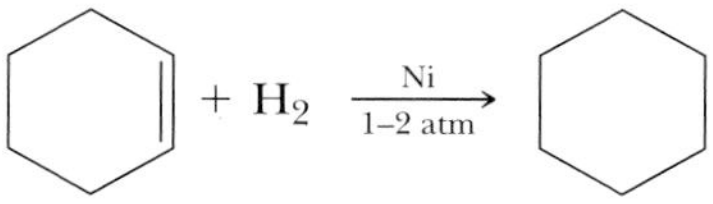

$$\bigcirc + H_2 \xrightarrow[\text{1–2 atm}]{\text{Ni}} \bigcirc \qquad \Delta H^0 = -28.6 \text{ kcal/mol}$$
$$(-120 \text{ kJ/mol})$$

By contrast, benzene is reduced only very slowly to cyclohexane under these conditions. It is reduced more rapidly when heated and under a pressure of several hundred atmospheres of hydrogen:

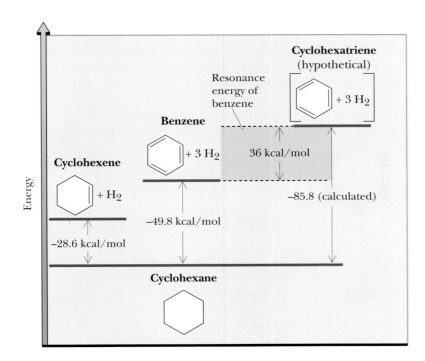

$$\Delta H^0 = -49.8 \text{ kcal/mol}$$
$$(-208 \text{ kJ/mol})$$

The catalytic reduction of an alkene is an exothermic reaction (Section 5.5B). The heat of hydrogenation per double bond varies somewhat with the degree of substitution of the double bond; for cyclohexene, $\Delta H^0 = -28.6$ kcal/mol (-120 kJ/mol). If we consider benzene to be 1,3,5-cyclohexatriene, a hypothetical compound with alternating single and double bonds, we might expect its heat of hydrogenation to be $3 \times -28.6 = -85.8$ kcal/mol (-359 kJ/mol). Instead, the heat of hydrogenation of benzene is only -49.8 kcal/mol (-208 kJ/mol). The difference of 36.0 kcal/mol (151 kJ/mol) between the expected value and the experimentally observed value is the **resonance energy of benzene**. Figure 9.2 shows these experimental results in the form of a graph.

For comparison, the strength of a carbon–carbon single bond is approximately 80–100 kcal/mol (333–418 kJ/mol), and that of hydrogen bonding in water and low-molecular-weight alcohols is approximately 2–5 kcal/mol (8.4–21 kJ/mol). Thus, although the resonance energy of benzene is less than the strength of a carbon–carbon single bond, it is considerably greater than the strength of hydrogen bonding in water and alcohols. In Section 8.2C, we saw that hydrogen bonding has a dramatic effect on the physical properties of alcohols compared with those of alkanes. In this chapter, we see that the resonance energy of benzene and other aromatic hydrocarbons has a dramatic effect on their chemical reactivity.

Figure 9.2
The resonance energy of benzene, as determined by a comparison of the heats of hydrogenation of cyclohexene, benzene, and the hypothetical 1,3,5-cyclohexatriene.

Following are resonance energies for benzene and several other aromatic hydrocarbons:

| Resonance energy [kcal/mol (kJ/mol)] | Benzene 36 (151) | Naphthalene 61 (255) | Anthracene 83 (347) | Phenanthrene 91 (381) |

9.3 THE CONCEPT OF AROMATICITY

Many other types of molecules besides benzene and its derivatives show aromatic character; that is, they contain high degrees of unsaturation, yet fail to undergo characteristic alkene addition and oxidation–reduction reactions. What chemists had long sought to understand were the principles underlying aromatic character. The German chemical physicist Eric Hückel solved this problem in the 1930s.

Hückel's criteria are summarized as follows. To be aromatic, a ring must

1. Have one $2p$ orbital on each of its atoms.
2. Be planar or nearly planar, so that there is continuous overlap or nearly continuous overlap of all $2p$ orbitals of the ring.
3. Have 2, 6, 10, 14, 18, and so forth pi electrons in the cyclic arrangement of $2p$ orbitals.

Benzene meets these criteria. It is cyclic, planar, has one $2p$ orbital on each carbon atom of the ring, and has 6 pi electrons (an aromatic sextet) in the cyclic arrangement of its $2p$ orbitals.

Heterocyclic compound An organic compound that contains one or more atoms other than carbon in its ring.

Let us apply these criteria to several **heterocyclic compounds**, all of which are aromatic. Pyridine and pyrimidine are heterocyclic analogs of benzene. In pyridine, one CH group of benzene is replaced by a nitrogen atom, and in pyrimidine, two CH groups are replaced by nitrogen atoms:

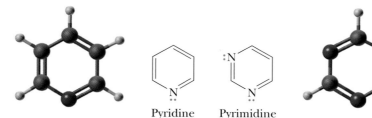

Pyridine   Pyrimidine

Each molecule meets the Hückel criteria for aromaticity: Each is cyclic and planar, has one $2p$ orbital on each atom of the ring, and has six electrons in the pi system. In pyridine, nitrogen is sp^2 hybridized, and its unshared pair of electrons occupies an sp^2 orbital perpendicular to the $2p$ orbitals of the pi system and thus is not a part of the pi system. In pyrimidine, neither unshared pair of electrons of nitrogen is part of the pi system. The resonance energy of pyridine is 32 kcal/mol (134 kJ/mol), slightly less than that of benzene. The resonance energy of pyrimidine is 26 kcal/mol (109 kJ/mol).

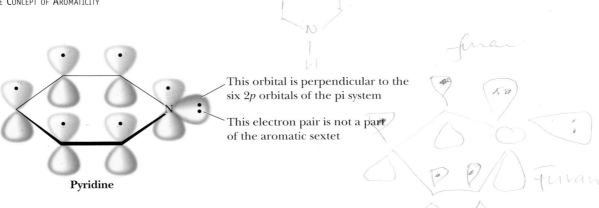

Pyridine

This orbital is perpendicular to the six 2p orbitals of the pi system

This electron pair is not a part of the aromatic sextet

The five-membered-ring compounds furan, pyrrole, and imidazole are also aromatic:

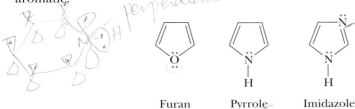

Furan   Pyrrole   Imidazole

In these planar compounds, each heteroatom is sp^2 hybridized, and its unhybridized $2p$ orbital is part of a continuous cycle of five $2p$ orbitals. In furan, one unshared pair of electrons of the heteroatom lies in the unhybridized $2p$ orbital and is a part of the pi system (Figure 9.3). The other unshared pair of electrons lies in an sp^2 hybrid orbital, perpendicular to the $2p$ orbitals, and is not a part of the pi system. In pyrrole, the unshared pair of electrons on nitrogen is part of the aromatic sextet. In imidazole, the unshared pair of electrons on one nitrogen is part of the aromatic sextet; the unshared pair on the other nitrogen is not.

This electron pair is a part of the aromatic sextet

This electron pair is not a part of the aromatic sextet

Furan

This electron pair is a part of the aromatic sextet

Pyrrole

Figure 9.3
Origin of the 6 pi electrons (the aromatic sextet) in furan and pyrrole. The resonance energy of furan is 16 kcal/mol (67 kJ/mol); that of pyrrole is 21 kcal/mol (88 kJ/mol).

Nature abounds with compounds having a heterocyclic ring fused to one or more other rings. Two such compounds especially important in the biological world are indole and purine:

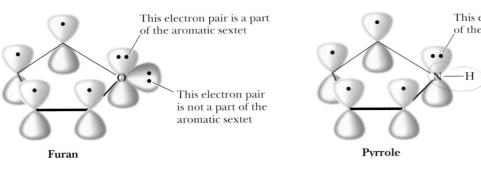

Indole   Serotonin   Purine   Adenine
              (a neurotransmitter)

Indole contains a pyrrole ring fused with a benzene ring. Compounds derived from indole include the amino acid L-tryptophan (Section 19.2A) and the neurotransmitter serotonin. Purine contains a six-membered pyrimidine ring fused with a five-membered imidazole ring. Adenine is one of the building blocks of deoxyribonucleic acids (DNA) and ribonucleic acids (RNA), as described in Chapter 20. It is also a component of the biological oxidizing agent nicotinamide adenine dinucleotide, abbreviated NAD^+ (Section 22.2B).

9.4   NOMENCLATURE

A. Monosubstituted Benzenes

Monosubstituted alkylbenzenes are named as derivatives of benzene; an example is ethylbenzene. The IUPAC system retains certain common names for several of the simpler monosubstituted alkylbenzenes. Examples are **toluene** (rather than methylbenzene) and **styrene** (rather than phenylethylene):

Benzene   Ethylbenzene   Toluene   Styrene

The common names **phenol**, **aniline**, **benzaldehyde**, **benzoic acid**, and **anisole** are also retained by the IUPAC system:

Phenol   Aniline   Benzaldehyde   Benzoic acid   Anisole

As noted in the introduction to Chapter 5, the substituent group derived by the loss of an H from benzene is a **phenyl** group (Ph); that derived by the loss of an H from the methyl group of toluene is a **benzyl group** (Bn):

> **Phenyl group** C_6H_5—, the aryl group derived by removing a hydrogen from benzene.
>
> **Benzyl group** $C_6H_5CH_2$—, the alkyl group derived by removing a hydrogen from the methyl group of toluene.

Benzene   Phenyl group (Ph)   Toluene   Benzyl group (Bn)

In molecules containing other functional groups, phenyl groups and benzyl groups are often named as substituents:

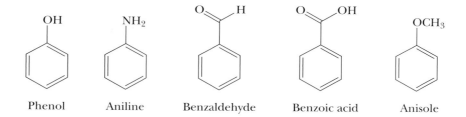

(Z)-2-Phenyl-2-butene   2-Phenylethanol   Benzyl chloride

PhCH₂CH₂OH   PhCH₂Cl

B. Disubstituted Benzenes

When two substituents occur on a benzene ring, three constitutional isomers are possible. We locate substituents either by numbering the atoms of the ring or by using the locators **ortho**, **meta**, and **para**. The numbers 1,2- are equivalent to *ortho* (Greek: straight); 1,3- to *meta* (Greek: after); and 1,4- to *para* (Greek: beyond).

When one of the two substituents on the ring imparts a special name to the compound, as, for example, toluene, phenol, and aniline, then we name the compound as a derivative of that parent molecule. In this case, the special substituent occupies ring position number 1. The IUPAC system retains the common name **xylene** for the three isomeric dimethylbenzenes. When neither group imparts a special name, we locate the two substituents and list them in alphabetical order before the ending *-benzene*. The carbon of the benzene ring with the substituent of lower alphabetical ranking is numbered C-1.

Ortho (o) Refers to groups occupying positions 1 and 2 on a benzene ring.

Meta (m) Refers to groups occupying positions 1 and 3 on a benzene ring.

Para (p) Refers to groups occupying positions 1 and 4 on a benzene ring.

4-Bromotoluene
(*p*-Bromotoluene)

3-Chloroaniline
(*m*-Chloroaniline)

1,3-Dimethylbenzene
(*m*-Xylene)

1-Chloro-4-ethylbenzene
(*p*-Chloroethylbenzene)

C. Polysubstituted Benzenes

When three or more substituents are present on a ring, we specify their locations by numbers. If one of the substituents imparts a special name, then the molecule is named as a derivative of that parent molecule. If none of the substituents imparts a special name, we number them to give the smallest set of numbers and list them in alphabetical order before the ending *-benzene*. In the following examples, the first compound is a derivative of toluene, and the second is a derivative of phenol. Because there is no special name for the third compound, we list its three substituents in alphabetical order, followed by the word *benzene*:

4-Chloro-2-nitrotoluene

2,4,6-Tribromophenol

2-Bromo-1-ethyl-4-nitrobenzene

Example 9.1

Write names for these compounds:

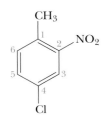

(a)   (b)   (c)   NO₂   (d)

SOLUTION

(a) 3-Iodotoluene or *m*-iodotoluene   (b) 3,5-Dibromobenzoic acid

(c) 1-Chloro-2,4-dinitrobenzene   (d) 3-Phenylpropene

Practice Problem 9.1

Write names for these compounds:

(a) 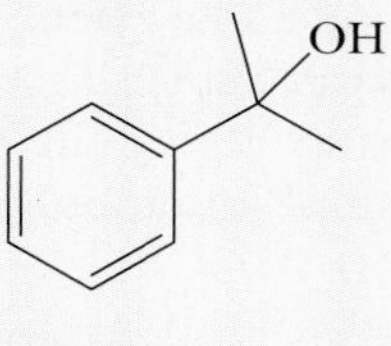   (b) 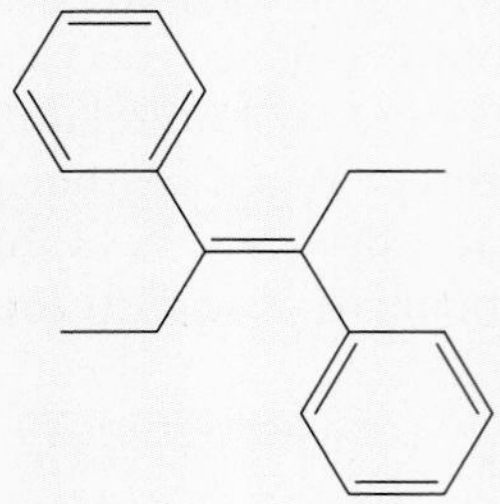   (c)

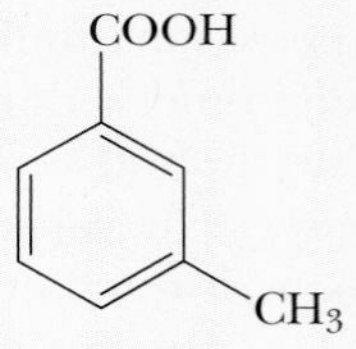

Polynuclear aromatic hydrocarbon A hydrocarbon containing two or more fused aromatic rings.

Polynuclear aromatic hydrocarbons (PAHs) contain two or more aromatic rings, each pair of which shares two ring carbon atoms. Naphthalene, anthracene, and phenanthrene, the most common PAHs, and substances derived from them are found in coal tar and high-boiling petroleum residues. At one time, naphthalene was used as a moth repellent and insecticide in preserving woolens and furs, but its use has decreased due to the introduction of chlorinated hydrocarbons such as *p*-dichlorobenzene. Also found in coal tar are lesser amounts of benzo[a]pyrene. This compound is found as well in the exhausts of gasoline-powered internal combustion engines (for example, automobile engines) and in cigarette smoke. Benzo[a]pyrene is a very potent carcinogen and mutagen.

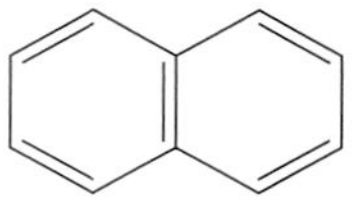

Naphthalene 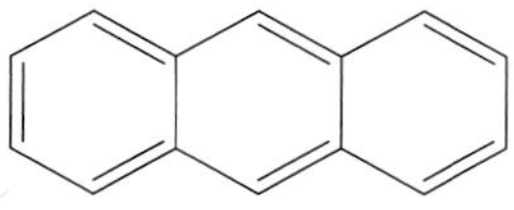   Anthracene 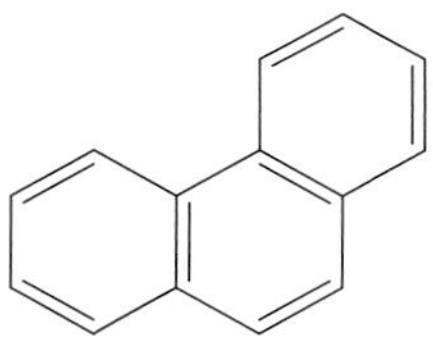   Phenanthrene 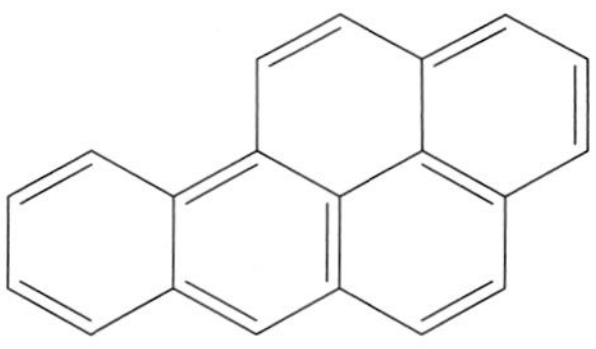   Benzo[a]pyrene

9.5 REACTIONS OF BENZENE: OXIDATION AT A BENZYLIC POSITION

As we have mentioned, benzene's aromaticity causes it to resist many of the reactions that alkenes typically undergo. However, chemists have been able to react benzene in other ways. This is fortunate, because benzene rings are abundant in many of the compounds that society depends upon, including various medications, plastics, and preservatives for food. We begin our discussion of benzene reactions with processes that take place not on the ring itself, but at the carbon immediately attached to the benzene ring. This is known as the **benzylic position**.

Benzylic carbon An *sp*³ hybridized carbon bonded to a benzene ring.

Benzene is unaffected by strong oxidizing agents, such as H_2CrO_4 and $KMnO_4$. When we treat toluene with these oxidizing agents under vigorous conditions, the side-chain methyl group is oxidized to a carboxyl group to give benzoic acid:

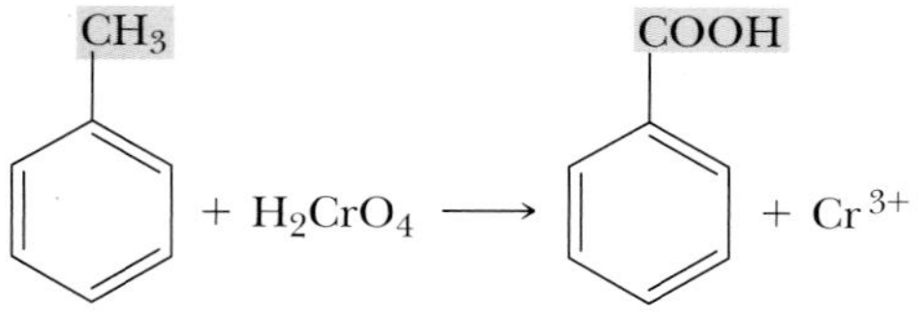

Toluene   Benzoic acid

CHEMICAL CONNECTIONS 9A

Carcinogenic Polynuclear Aromatics and Smoking

A **carcinogen** is a compound that causes cancer. The first carcinogens to be identified were a group of polynuclear aromatic hydrocarbons, all of which have at least four aromatic rings. Among them is benzo[a]pyrene, one of the most carcinogenic of the aromatic hydrocarbons. It forms whenever there is incomplete combustion of organic compounds. Benzo[a]pyrene is found, for example, in cigarette smoke, automobile exhaust, and charcoal-broiled meats.

Benzo[a]pyrene causes cancer in the following way: Once it is absorbed or ingested, the body attempts to convert it into a more soluble compound that can be excreted easily. To this end, a series of enzyme-catalyzed reactions transforms benzo[a]pyrene into a **diol epoxide**, a compound that can bind to DNA by reacting with one of its amino groups, thereby altering the structure of DNA and producing a cancer-causing mutation:

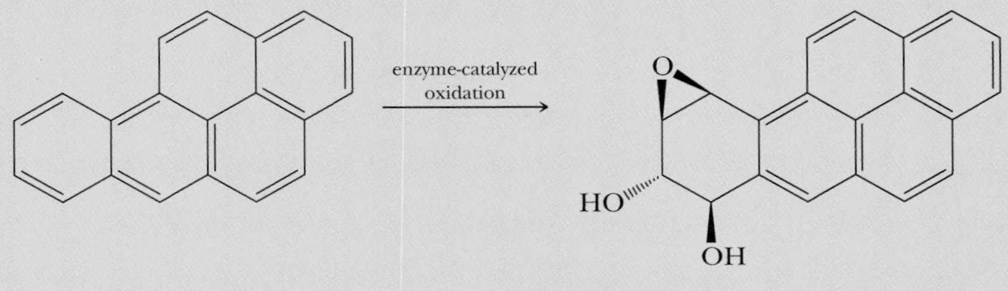

Benzo[a]pyrene   A diol epoxide

The fact that the side-chain methyl group is oxidized, but the aromatic ring is unchanged, illustrates the remarkable chemical stability of the aromatic ring. Halogen and nitro substituents on an aromatic ring are unaffected by these oxidations. For example, chromic acid oxidizes 2-chloro-4-nitrotoluene to 2-chloro-4-nitrobenzoic acid. Notice that in this oxidation, the nitro and chloro groups remain unaffected:

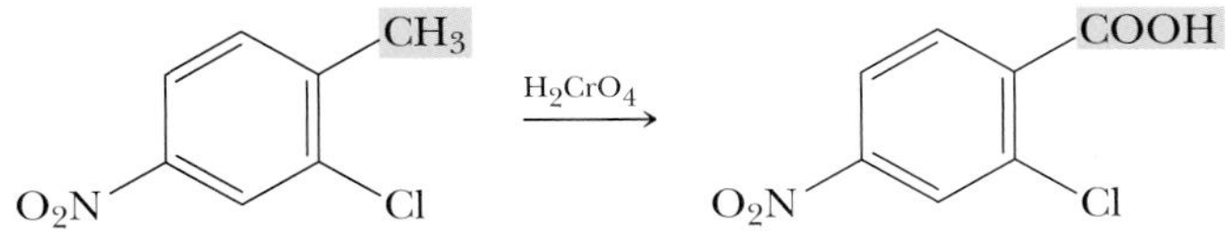

2-Chloro-4-nitrotoluene   2-Chloro-4-nitrobenzoic acid

Ethylbenzene and isopropylbenzene are also oxidized to benzoic acid under these conditions. The side chain of *tert*-butylbenzene, which has no benzylic hydrogen, is not affected by these oxidizing conditions.

From these observations, we conclude that, if a benzylic hydrogen exists, then the benzylic carbon (Section 9.4A) is oxidized to a carboxyl group and all other carbons of the side chain are removed. If no benzylic hydrogen exists, as in the case of *tert*-butylbenzene, then the side chain is not oxidized.

If more than one alkyl side chain exists, each is oxidized to —COOH. Oxidation of *m*-xylene gives 1,3-benzenedicarboxylic acid, more commonly named isophthalic acid:

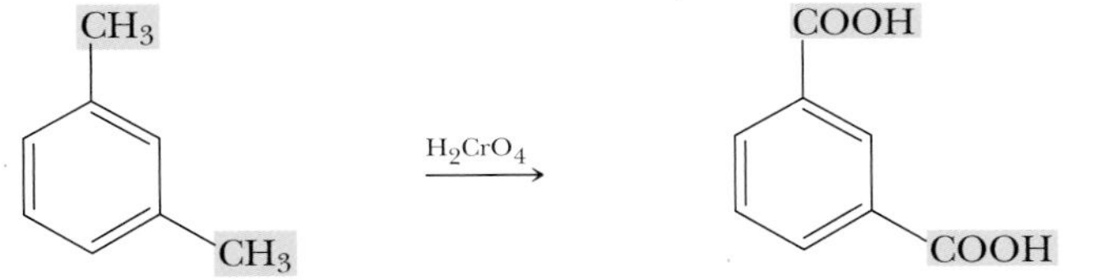

m-Xylene   1,3-Benzenedicarboxylic acid (Isophthalic acid)

EXAMPLE 9.2

Draw a structural formula for the product of vigorous oxidation of 1,4-dimethyl-benzene (p-xylene) by H_2CrO_4.

SOLUTION

Chromic acid oxidizes both alkyl groups to —COOH groups, and the product is terephthalic acid, one of two monomers required for the synthesis of Dacron® polyester and Mylar® (Section 17.5B):

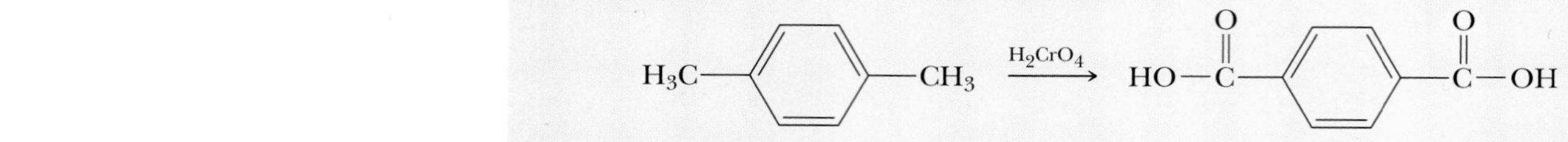

1,4-Dimethylbenzene   1,4-Benzenedicarboxylic acid
(p-Xylene)   (Terephthalic acid)

Practice Problem 9.2

Predict the products resulting from vigorous oxidation of each compound by H_2CrO_4:

(a) 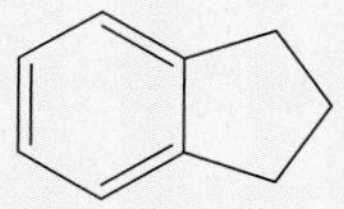   (b)

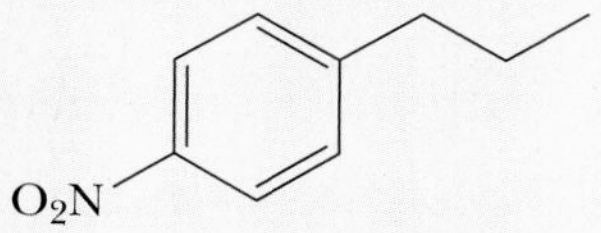

9.6   REACTIONS OF BENZENE: ELECTROPHILIC AROMATIC SUBSTITUTION

By far the most characteristic reaction of aromatic compounds is substitution at a ring carbon. Some groups that can be introduced directly onto the ring are the halogens, the nitro (—NO_2) group, the sulfonic acid (—SO_3H) group, alkyl (—R) groups, and acyl (RCO—) groups.

Halogenation:

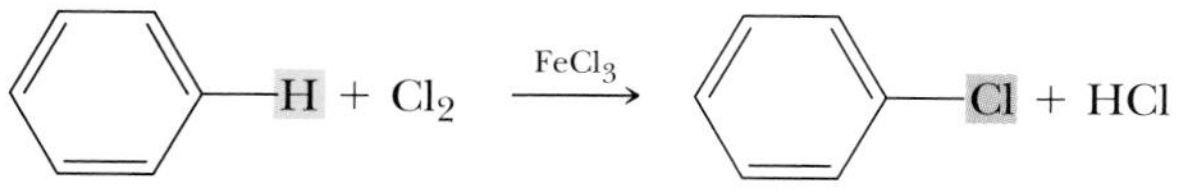

Chlorobenzene

Nitration:

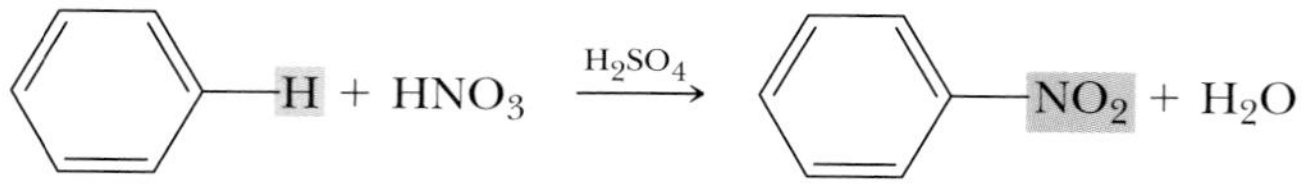

Nitrobenzene

Sulfonation:

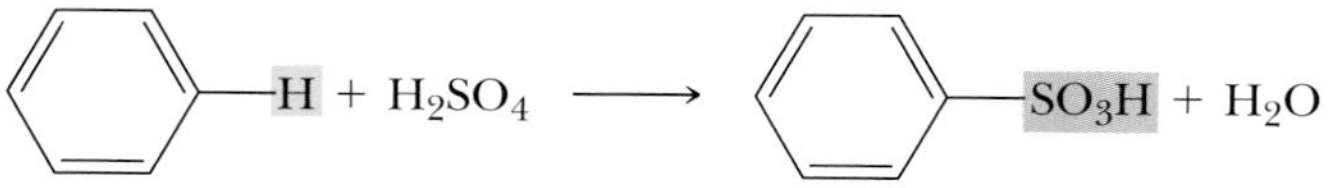

Benzenesulfonic acid

Alkylation:

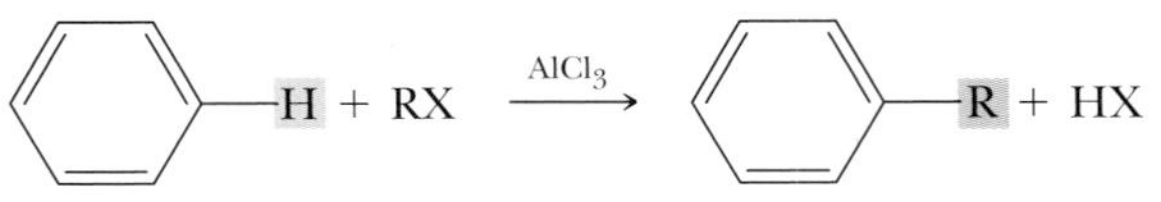

An alkylbenzene

Acylation:

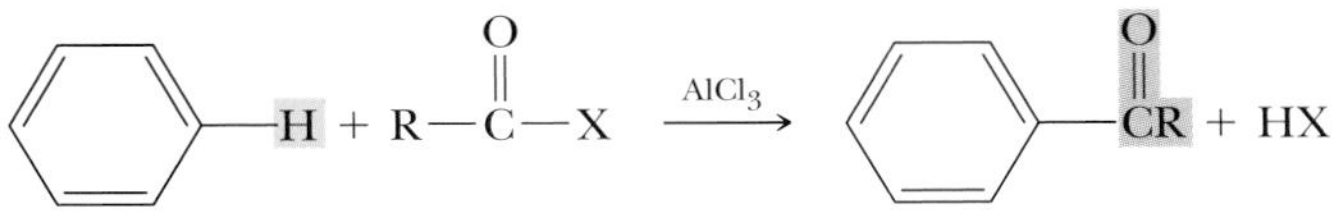

An acyl halide   An acylbenzene

9.7 MECHANISM OF ELECTROPHILIC AROMATIC SUBSTITUTION

In this section, we study several types of **electrophilic aromatic substitution** reactions—that is, reactions in which a hydrogen of an aromatic ring is replaced by an electrophile, E^+. The mechanisms of these reactions are actually very similar. In fact, they can be broken down into three common steps:

Electrophilic aromatic substitution A reaction in which an electrophile, E^+, substitutes for a hydrogen on an aromatic ring.

Step 1: Generation of the electrophile:

$$\text{Reagent(s)} \longrightarrow E^+$$

Step 2: Attack of the electrophile on the aromatic ring to give a resonance-stabilized cation intermediate:

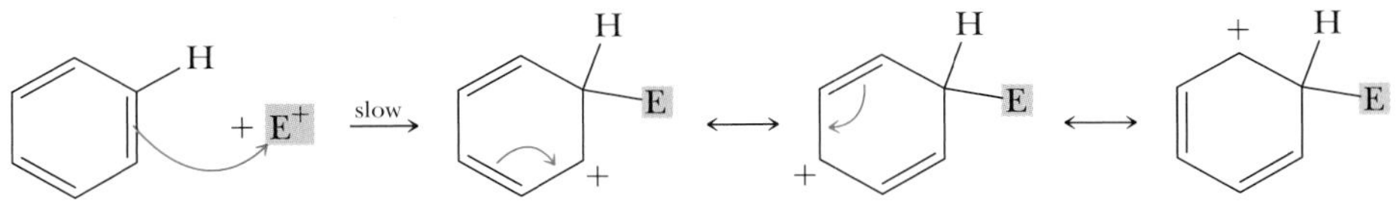

Resonance-stabilized cation intermediate

Step 3: Proton transfer to a base to regenerate the aromatic ring:

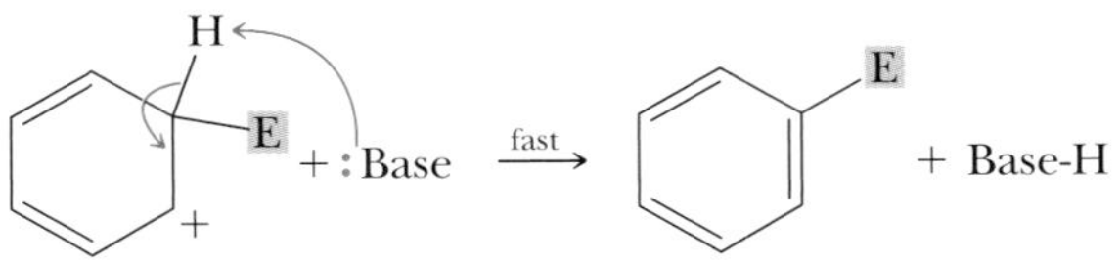

The reactions we are about to study differ only in the way the electrophile is generated and in the base that removes the proton to re-form the aromatic ring. You should keep this principle in mind as we explore the details of each reaction.

A. Chlorination and Bromination

Chlorine alone does not react with benzene, in contrast to its instantaneous addition to cyclohexene (Section 5.3C). However, in the presence of a Lewis acid catalyst, such as ferric chloride or aluminum chloride, chlorine reacts to give chlorobenzene and HCl. Chemists account for this type of electrophilic aromatic substitution by the following three-step mechanism:

Mechanism: Electrophilic Aromatic Substitution—Chlorination

Step 1: *Formation of the Electrophile:* Reaction between chlorine (a Lewis base) and FeCl$_3$ (a Lewis acid) gives an ion pair containing a chloronium ion (an electrophile):

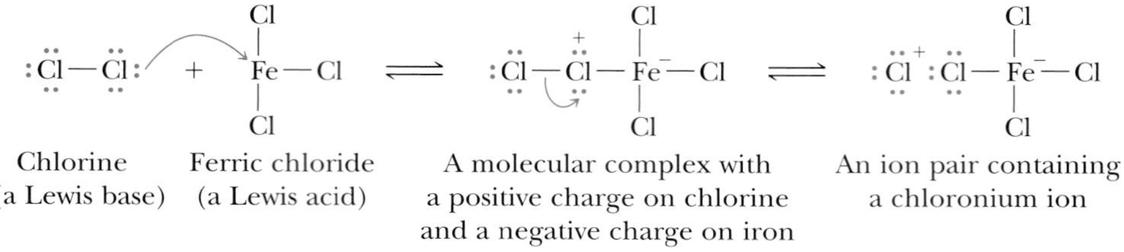

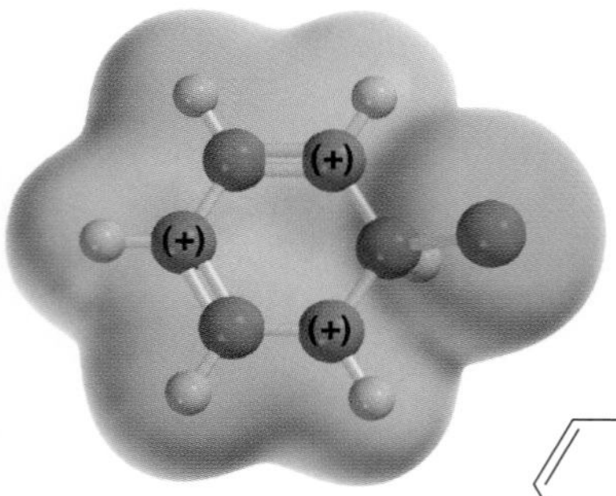

Chlorine   Ferric chloride   A molecular complex with   An ion pair containing
(a Lewis base)   (a Lewis acid)   a positive charge on chlorine   a chloronium ion
                                           and a negative charge on iron

Step 2: *Attack of the Electrophile on the Ring:* Reaction of the Cl$_2$–FeCl$_3$ ion pair with the pi electron cloud of the aromatic ring forms a resonance-stabilized cation intermediate, represented here as a hybrid of three contributing structures:

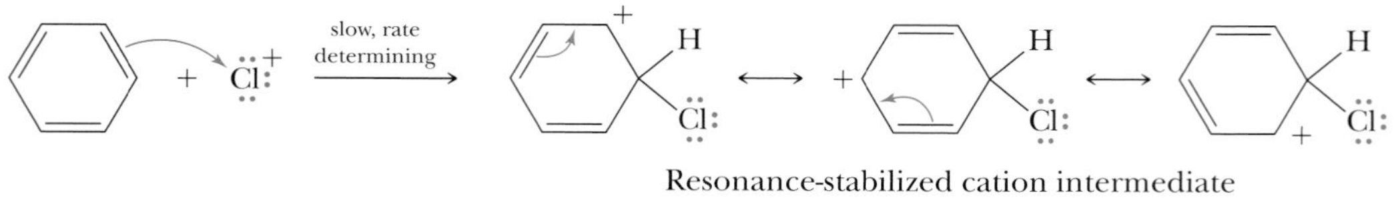

slow, rate determining

Resonance-stabilized cation intermediate

The positive charge on the resonance-stabilized intermediate is distributed approximately equally on the carbon atoms 2, 4, and 6 of the ring relative to the point of substitution.

Step 3: *Proton Transfer:* Proton transfer from the cation intermediate to FeCl$_4^-$ forms HCl, regenerates the Lewis acid catalyst, and gives chlorobenzene:

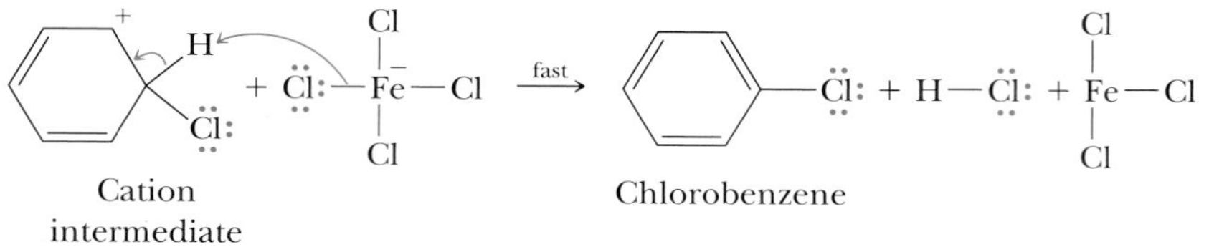

fast

Cation   Chlorobenzene
intermediate

Treatment of benzene with bromine in the presence of ferric chloride or aluminum chloride gives bromobenzene and HBr. The mechanism for this reaction is the same as that for chlorination of benzene.

The major difference between the addition of halogen to an alkene and substitution by halogen on an aromatic ring is the fate of the cation intermediate formed in the first step of each reaction. Recall from Section 5.3C that the addition of chlorine to an alkene is a two-step process, the first and slower step of which is the formation of a bridged chloronium ion intermediate. This intermediate then reacts with chloride ion to complete the addition. With aromatic compounds, the cation intermediate loses H$^+$ to regenerate the aromatic ring and regain its large resonance stabilization. There is no such resonance stabilization to be regained in the case of an alkene.

B. Nitration and Sulfonation

The sequence of steps for the nitration and sulfonation of benzene is similar to that for chlorination and bromination. For nitration, the electrophile is the **nitronium ion**, NO$_2^+$, generated by the reaction of nitric acid with sulfuric acid. In the following equations nitric acid is written HONO$_2$ to show more clearly the origin of the nitronium ion.

Mechanism: Formation of the Nitronium Ion

Step 1: Proton transfer from sulfuric acid to the OH group of nitric acid gives the conjugate acid of nitric acid:

$$H-\overset{..}{\underset{..}{O}}-NO_2 + H-\overset{..}{\underset{..}{O}}-SO_3H \rightleftharpoons H-\overset{H}{\overset{+|}{\underset{..}{O}}}-NO_2 + HSO_4^-$$

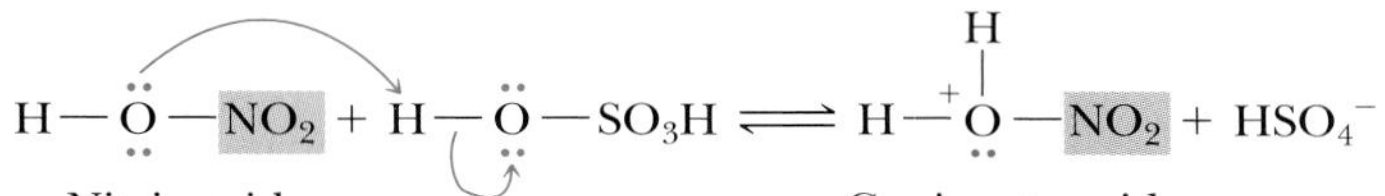

Nitric acid   Conjugate acid
                                                     of nitric acid

Step 2: Loss of water from this conjugate acid gives the nitronium ion, NO_2^+:

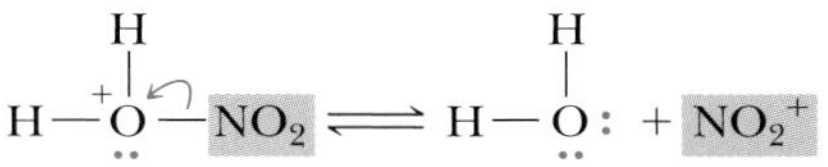

$$H-\overset{H}{\overset{+|}{\underset{..}{O}}}-NO_2 \rightleftharpoons H-\overset{H}{\underset{..}{O}}: + NO_2^+$$

The nitronium ion

The sulfonation of benzene is carried out using hot, concentrated sulfuric acid. The electrophile under these conditions is either SO_3 or HSO_3^+, depending on the experimental conditions. The HSO_3^+ electrophile is formed from sulfuric acid in the following way:

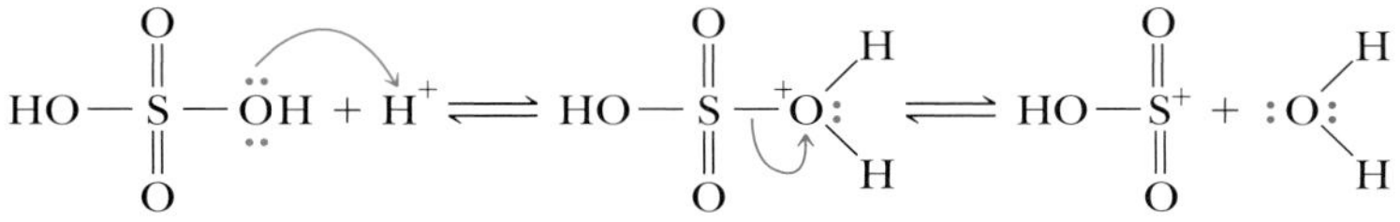

$$HO-\overset{O}{\underset{O}{\overset{||}{\underset{||}{S}}}}-\overset{..}{\underset{..}{O}}H + H^+ \rightleftharpoons HO-\overset{O}{\underset{O}{\overset{||}{\underset{||}{S}}}}-\overset{+}{\underset{H}{O}}\overset{H}{} \rightleftharpoons HO-\overset{O}{\underset{O}{\overset{||}{\underset{||}{S^+}}}} + :\overset{H}{\underset{H}{O}}:$$

Sulfuric acid   The electrophile

EXAMPLE 9.3

Write a stepwise mechanism for the nitration of benzene.

SOLUTION

Step 1: Reaction of the nitronium ion (an electrophile) with the benzene ring (a nucleophile) gives a resonance-stabilized cation intermediate.

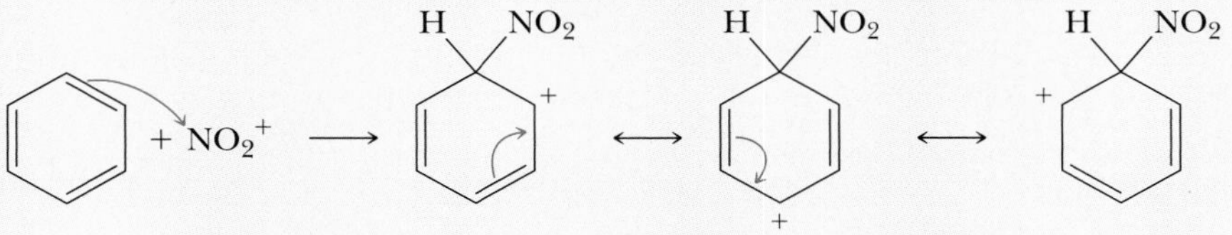

Step 2: Proton transfer from this intermediate to H_2O regenerates the aromatic ring and gives nitrobenzene:

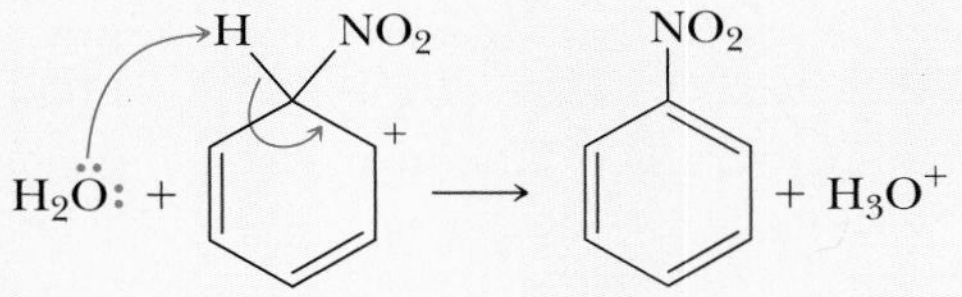

Nitrobenzene

Practice Problem 9.3

Write a stepwise mechanism for the sulfonation of benzene. Use HSO_3^+ as the electrophile.

C. Friedel–Crafts Alkylation

Alkylation of aromatic hydrocarbons was discovered in 1877 by the French chemist Charles Friedel and a visiting American chemist, James Crafts. They discovered that mixing benzene, a haloalkane, and AlCl$_3$ results in the formation of an alkylbenzene and HX. **Friedel–Crafts alkylation** forms a new carbon–carbon bond between benzene and an alkyl group, as illustrated by reaction of benzene with 2-chloropropane in the presence of aluminum chloride:

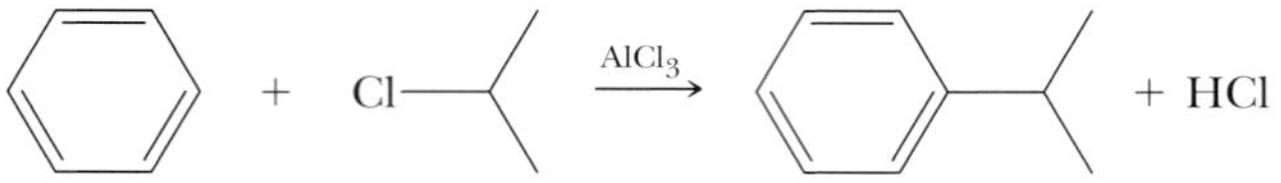

Benzene   2-Chloropropane   Isopropylbenzene   + HCl
              (Isopropyl chloride)   (Cumene)

Friedel–Crafts alkylation is among the most important methods for forming new carbon–carbon bonds to aromatic rings.

Mechanism: Friedel–Crafts Alkylation

Step 1: Reaction of a haloalkane (a Lewis base) with aluminum chloride (a Lewis acid) gives a molecular complex in which aluminum has a negative formal charge and the halogen of the haloalkane has a positive formal charge. Redistribution of electrons in this complex then gives an alkyl carbocation as part of an ion pair:

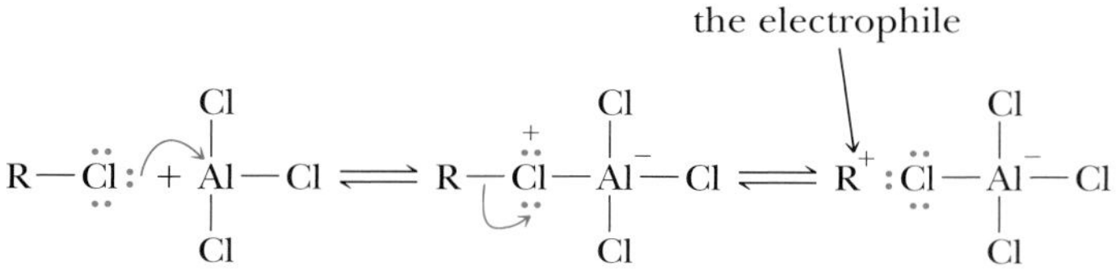

A molecular complex   An ion pair
with a positive charge on   containing
chlorine and a negative   a carbocation
charge on aluminum

Step 2: Reaction of the alkyl carbocation with the pi electrons of the aromatic ring gives a resonance-stabilized cation intermediate:

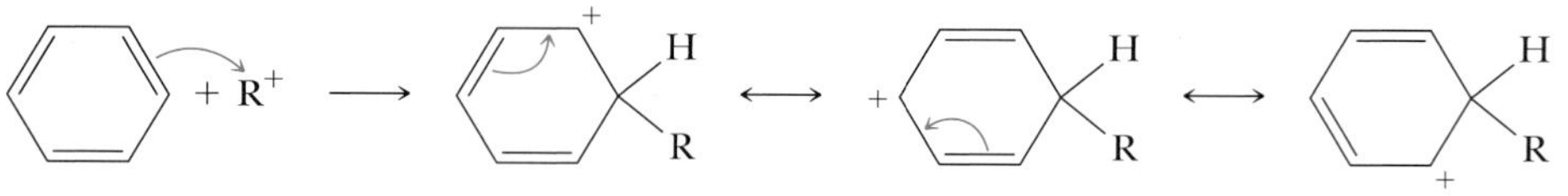

The positive charge is delocalized onto
three atoms of the ring

Step 3: Proton transfer regenerates the aromatic character of the ring and the Lewis acid catalyst:

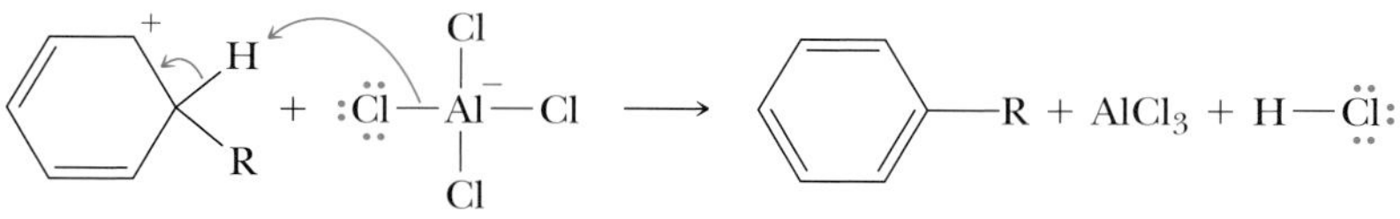

There are two major limitations on Friedel–Crafts alkylations. The first is that it is practical only with stable carbocations, such as 3° and 2° carbocations. The reasons for this limitation are beyond the scope of this text.

The second limitation on Friedel–Crafts alkylation is that it fails altogether on benzene rings bearing one or more strongly electron-withdrawing groups. The following table shows some of these groups:

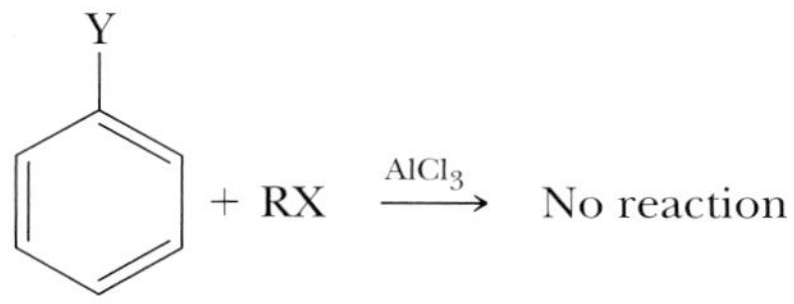

$$\text{(benzene ring with Y substituent)} + RX \xrightarrow{\text{AlCl}_3} \text{No reaction}$$

When Y Equals Any of These Groups, the Benzene Ring Does Not Undergo Friedel–Crafts Alkylation

$-\overset{\overset{\text{O}}{\|}}{\text{CH}}$	$-\overset{\overset{\text{O}}{\|}}{\text{CR}}$	$-\overset{\overset{\text{O}}{\|}}{\text{COH}}$	$-\overset{\overset{\text{O}}{\|}}{\text{COR}}$	$-\overset{\overset{\text{O}}{\|}}{\text{CNH}_2}$
$-\text{SO}_3\text{H}$	$-\text{C}\equiv\text{N}$	$-\text{NO}_2$	$-\text{NR}_3{}^+$	
$-\text{CF}_3$	$-\text{CCl}_3$			

A common characteristic of the groups listed in the preceding table is that each has either a full or partial positive charge on the atom bonded to the benzene ring. For carbonyl-containing compounds, this partial positive charge arises because of the difference in electronegativity between the carbonyl oxygen and carbon. For $-\text{CF}_3$ and $-\text{CCl}_3$ groups, the partial positive charge on carbon arises because of the difference in electronegativity between carbon and the halogens bonded to it. In both the nitro group and the trialkylamonium group, there is a positive charge on nitrogen:

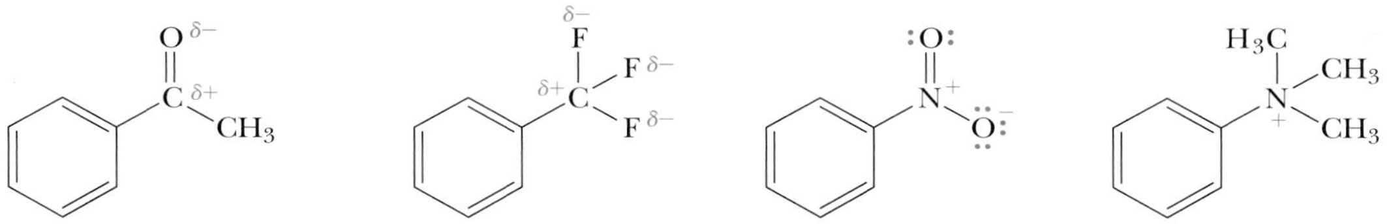

The carbonyl group of a ketone   A trifluoro-methyl group   A nitro group   A trimethyl-ammonium group

D. Friedel–Crafts Acylation

Friedel and Crafts also discovered that treating an aromatic hydrocarbon with an acyl halide (Section 15.2A) in the presence of aluminum chloride gives a ketone. An **acyl halide** is a derivative of a carboxylic acid in which the $-\text{OH}$ of the carboxyl group is replaced by a halogen, most commonly chlorine. Acyl halides are also referred to as acid halides. An $\text{RCO}-$ group is known as an acyl group; hence, the reaction of an acyl halide with an aromatic hydrocarbon is known as **Friedel–Crafts acylation**, as illustrated by the reaction of benzene and acetyl chloride in the presence of aluminum chloride to give acetophenone:

Acyl halide A derivative of a carboxylic acid in which the $-\text{OH}$ of the carboxyl group is replaced by a halogen—most commonly, chlorine.

$$\text{(benzene)} + \text{CH}_3\overset{\overset{\text{O}}{\|}}{\text{C}}\text{Cl} \xrightarrow{\text{AlCl}_3} \text{(acetophenone)} + \text{HCl}$$

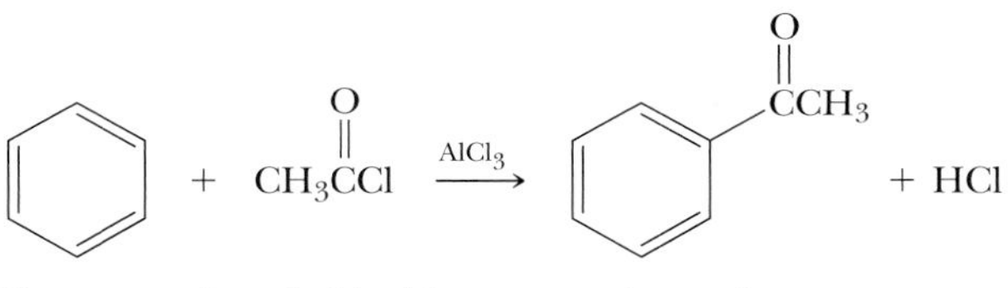

Benzene   Acetyl chloride (an acyl halide)   Acetophenone (a ketone)

In Friedel–Crafts acylations, the electrophile is an acylium ion, generated in the following way:

Mechanism: Friedel–Crafts Acylation—Generation of an Acylium Ion

Reaction between the halogen atom of the acyl chloride (a Lewis base) and aluminum chloride (a Lewis acid) gives a molecular complex. The redistribution of valence electrons in turn gives an ion pair containing an acylium ion:

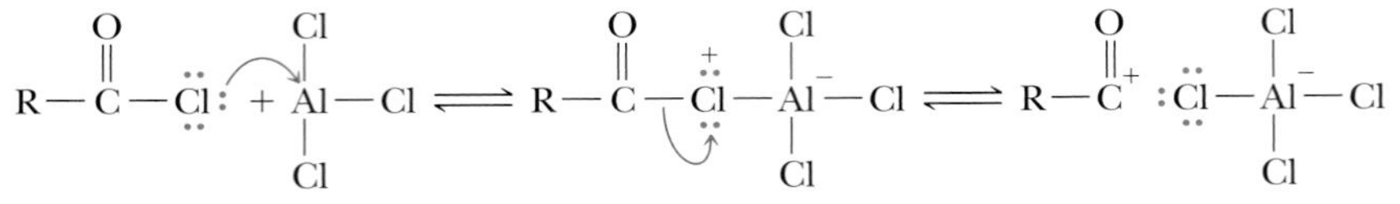

An acyl	Aluminum	A molecular complex with	An ion pair
chloride	chloride	a positive charge on	containing
(a Lewis base)	(a Lewis acid)	chlorine and a negative	an acylium ion
		charge on aluminum	

EXAMPLE 9.4

Write a structural formula for the product formed by Friedel–Crafts alkylation or acylation of benzene with

(a) $C_6H_5CH_2Cl$
    Benzyl chloride

(b)
$$\overset{O}{\overset{\|}{C_6H_5C}}Cl$$
    Benzoyl chloride

SOLUTION

(a) Treatment of benzyl chloride with aluminum chloride gives the resonance-stabilized benzyl cation. Reaction of this cation with benzene, followed by loss of H^+, gives diphenylmethane:

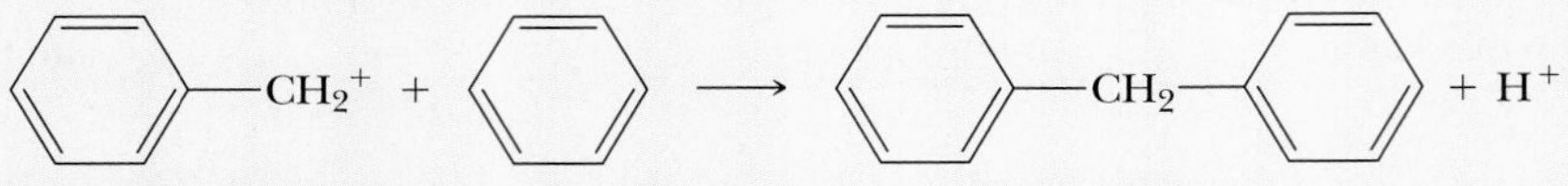

    Benzyl cation   Diphenylmethane

(b) Treatment of benzoyl chloride with aluminum chloride gives an acyl cation. Reaction of this cation with benzene, followed by loss of H^+, gives benzophenone:

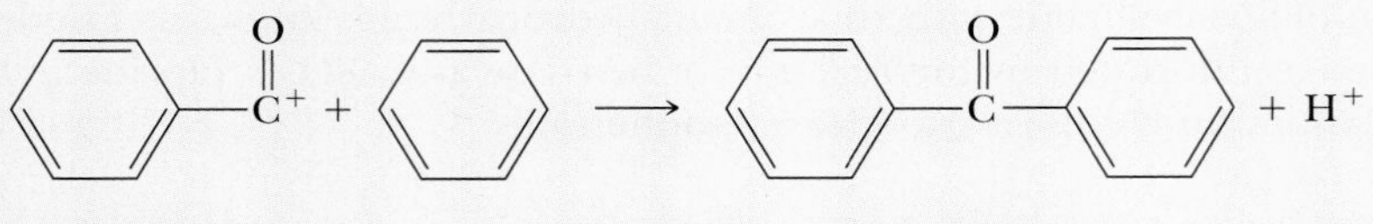

    Benzoyl   Benzophenone
    cation

Practice Problem 9.4

Write a structural formula for the product formed from Friedel–Crafts alkylation or acylation of benzene with

(a) 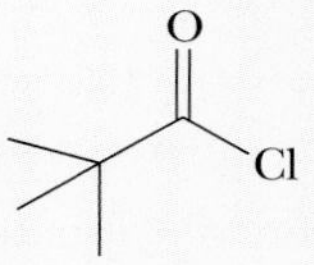 [structure: 2,2-dimethylpropanoyl chloride] (b) 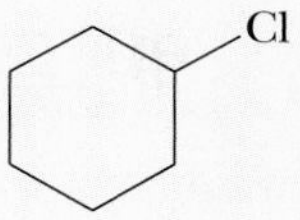 [structure: chlorocyclohexane] (c) 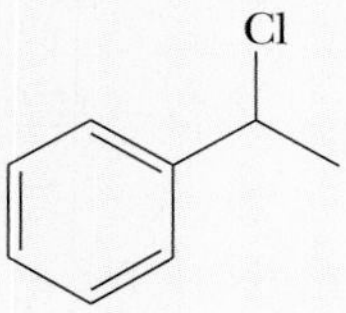 [structure: (1-chloroethyl)benzene]

E. Other Electrophilic Aromatic Alkylations

Once it was discovered that Friedel–Crafts alkylations and acylations involve cationic intermediates, chemists realized that other combinations of reagents and catalysts could give the same products. We study two of these reactions in this section: the generation of carbocations from alkenes and from alcohols.

As we saw in Section 5.3A, treatment of an alkene with a strong acid, most commonly H_2SO_4 or H_3PO_4, generates a carbocation. Isopropylbenzene is synthesized industrially by reacting benzene with propene in the presence of an acid catalyst:

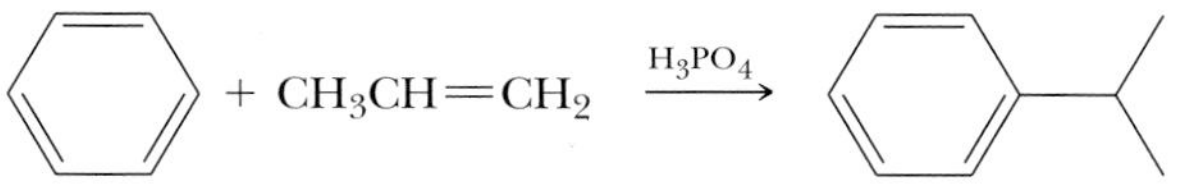

$$\text{Benzene} + CH_3CH=CH_2 \xrightarrow{H_3PO_4} \text{Isopropylbenzene (Cumene)}$$

Carbocations are also generated by treating an alcohol with H_2SO_4 or H_3PO_4 (Section 8.3E):

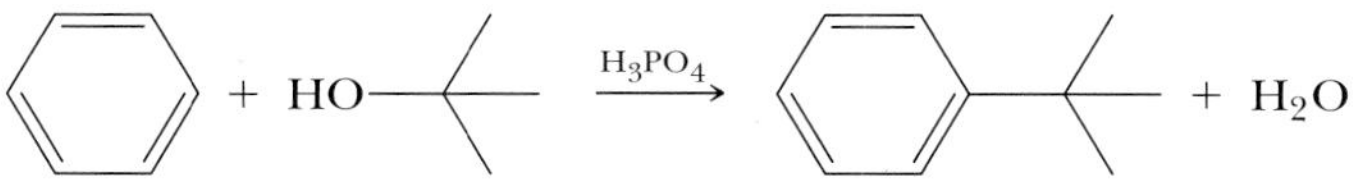

$$\text{Benzene} + HO-C(CH_3)_3 \xrightarrow{H_3PO_4} \text{2-Methyl-2-phenylpropane (tert-Butylbenzene)} + H_2O$$

EXAMPLE 9.5

Write a mechanism for the formation of isopropylbenzene from benzene and propene in the presence of phosphoric acid.

SOLUTION

Step 1: Proton transfer from phosphoric acid to propene gives the isopropyl cation:

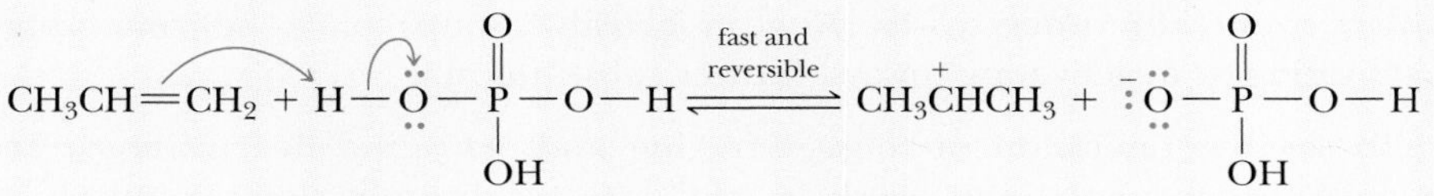

$$CH_3CH=CH_2 + H-\overset{\displaystyle O}{\underset{\displaystyle OH}{O-P-O-H}} \underset{\text{reversible}}{\overset{\text{fast and}}{\rightleftharpoons}} CH_3\overset{+}{C}HCH_3 + \overset{-}{:}\overset{\displaystyle O}{\underset{\displaystyle OH}{O-P-O-H}}$$

Step 2: Reaction of the isopropyl cation with benzene gives a resonance-stabilized carbocation intermediate:

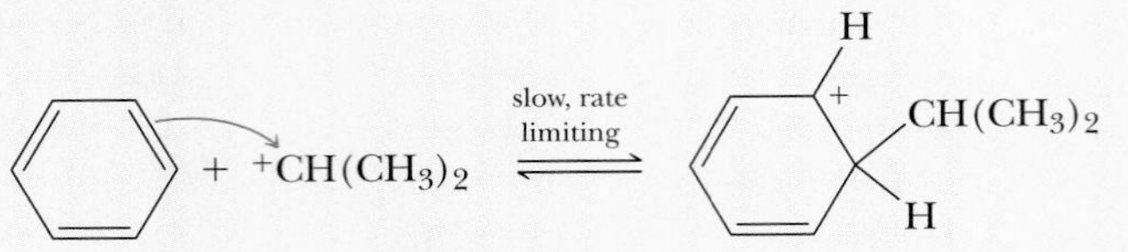

Step 3: Proton transfer from this intermediate to dihydrogen phosphate ion gives isopropylbenzene:

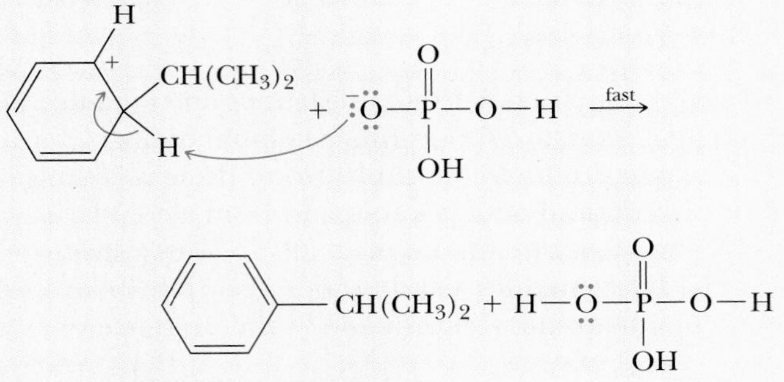

Isopropylbenzene

Practice Problem 9.5

Write a mechanism for the formation of *tert*-butylbenzene from benzene and *tert*-butyl alcohol in the presence of phosphoric acid.

9.8   DISUBSTITUTION AND POLYSUBSTITUTION

A. Effects of a Substituent Group on Further Substitution

In the electrophilic aromatic substitution of a monosubstituted benzene, three isomeric products are possible: The new group may be oriented ortho, meta, or para to the existing group. On the basis of a wealth of experimental observations, chemists have made the following generalizations about the manner in which an existing substituent influences further electrophilic aromatic substitution:

1. *Substituents affect the orientation of new groups.*   Certain substituents direct a second substituent preferentially to the ortho and para positions; other substituents direct it preferentially to a meta position. In other words, we can classify substituents on a benzene ring as **ortho–para directing** or **meta directing**.
2. *Substituents affect the rate of further substitution.*   Certain substituents cause the rate of a second substitution to be greater than that of benzene itself, whereas other substituents cause the rate of a second substitution to be lower than that of benzene. In other words, we can classify groups on a benzene ring as **activating** or **deactivating** toward further substitution.

   To see the operation of these directing and activating–deactivating effects, compare, for example, the products and rates of bromination of anisole and nitrobenzene. Bromination of anisole proceeds at a rate considerably greater than that of bromination of benzene (the methoxy group is activating), and the product is a mixture of *o*-bromoanisole and *p*-bromoanisole (the methoxy group is ortho–para directing):

Ortho–para director Any substituent on a benzene ring that directs electrophilic aromatic substitution preferentially to ortho and para positions.

Meta director Any substituent on a benzene ring that directs electrophilic aromatic substitution preferentially to a meta position.

Activating group Any substituent on a benzene ring that causes the rate of electrophilic aromatic substitution to be greater than that for benzene.

Deactivating group Any substituent on a benzene ring that causes the rate of electrophilic aromatic substitution to be lower than that for benzene.

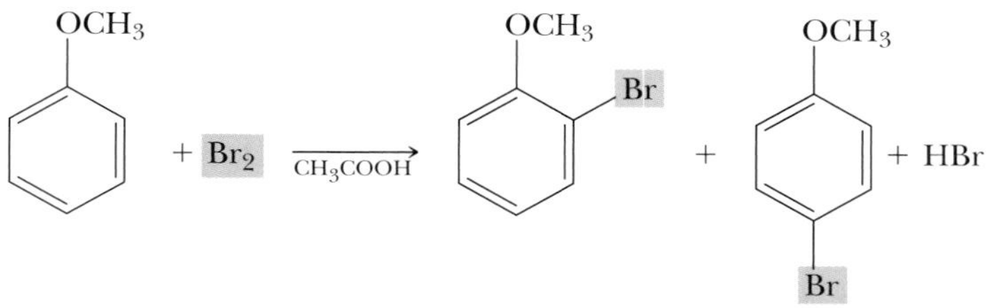

Anisole   o-Bromoanisole p-Bromoanisole
            (4%)   (96%)

We see quite another situation in the nitration of nitrobenzene, which proceeds much more slowly than the nitration of benzene itself. (A nitro group is strongly deactivating.) Also, the product consists of approximately 93% of the meta isomer and less than 7% of the ortho and para isomers combined (the nitro group is meta directing):

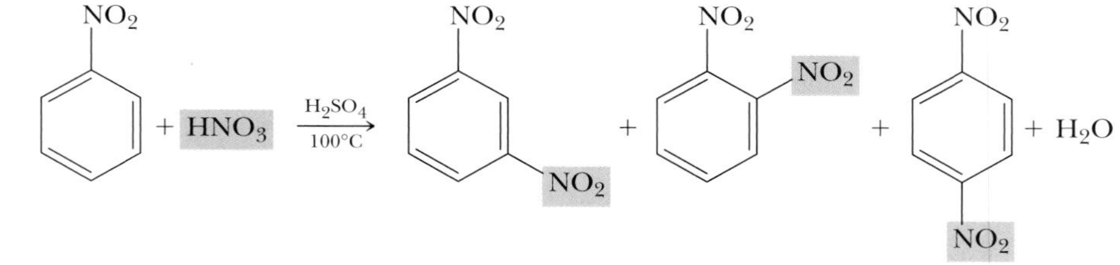

Nitrobenzene   m-Dinitro-   o-Dinitro-   p-Dinitro-
         benzene   benzene   benzene
          (93%)

Less than 7% combined

Table 9.1 lists the directing and activating–deactivating effects for the major functional groups with which we are concerned in this text.

TABLE 9.1 Effects of Substituents on Further Electrophilic Aromatic Substitution

Ortho–Para Directing	strongly activating	—NH₂	—NHR	—NR₂	—OH	—OR
	moderately activating	—NHCR (‖ O)	—NHCAr (‖ O)	—OCR (‖ O)	—OCAr (‖ O)	
	weakly activating	—R 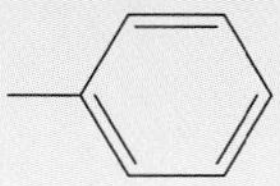	(phenyl)			
	weakly deactivating	—F:	—Cl:	—Br:	—I:	
Meta Directing	moderately deactivating	—CH (‖ O)	—CR (‖ O)	—COH (‖ O)	—COR (‖ O)	—CNH₂ (‖ O) —SOH (‖ O ‖ O)
	strongly deactivating	—NO₂	—NH₃⁺	—CF₃	—CCl₃	

Relative importance in directing further substitution

If we compare these ortho–para and meta directors for structural similarities and differences, we can make the following generalizations:

1. Alkyl groups, phenyl groups, and substituents in which the atom bonded to the ring has an unshared pair of electrons are ortho–para directing. All other substituents are meta directing.
2. Except for the halogens, all ortho–para directing groups are activating toward further substitution. The halogens are weakly deactivating.
3. All meta directing groups carry either a partial or full positive charge on the atom bonded to the ring.

We can illustrate the usefulness of these generalizations by considering the synthesis of two different disubstituted derivatives of benzene. Suppose we wish to prepare *m*-bromonitrobenzene from benzene. This conversion can be carried out in two steps: nitration and bromination. If the steps are carried out in just that order, the major product is indeed *m*-bromonitrobenzene. The nitro group is a meta director and directs bromination to a meta position:

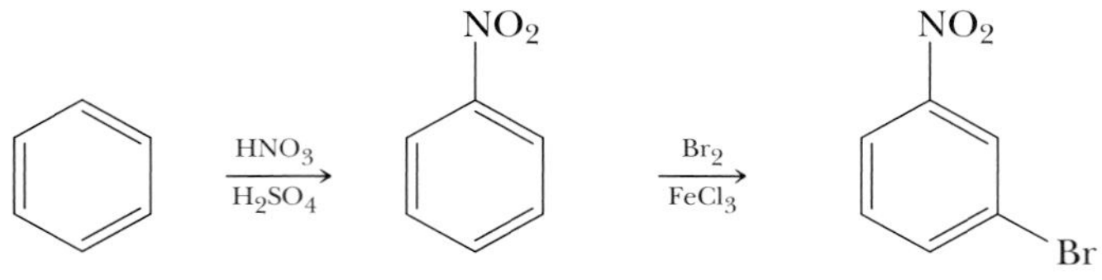

Nitrobenzene   *m*-Bromonitrobenzene

If, however, we reverse the order of the steps and first form bromobenzene, we now have an ortho–para directing group on the ring. Nitration of bromobenzene then takes place preferentially at the ortho and para positions, with the para product predominating:

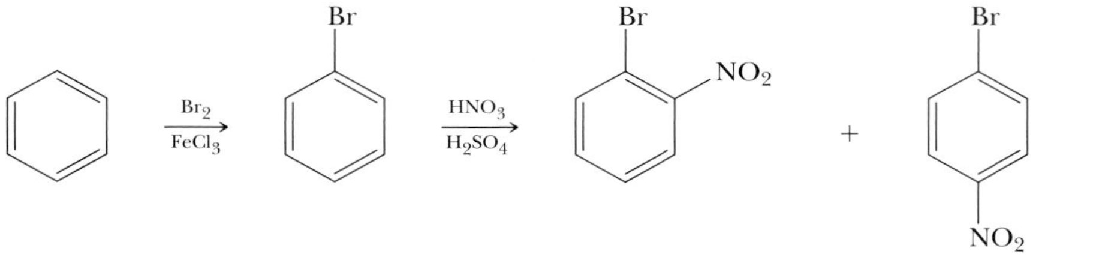

Bromobenzene   *o*-Bromonitrobenzene   *p*-Bromonitrobenzene

As another example of the importance of order in electrophilic aromatic substitutions, consider the conversion of toluene to nitrobenzoic acid. The nitro group can be introduced with a nitrating mixture of nitric and sulfuric acids. The carboxyl group can be produced by oxidation of the methyl group (Section 9.5).

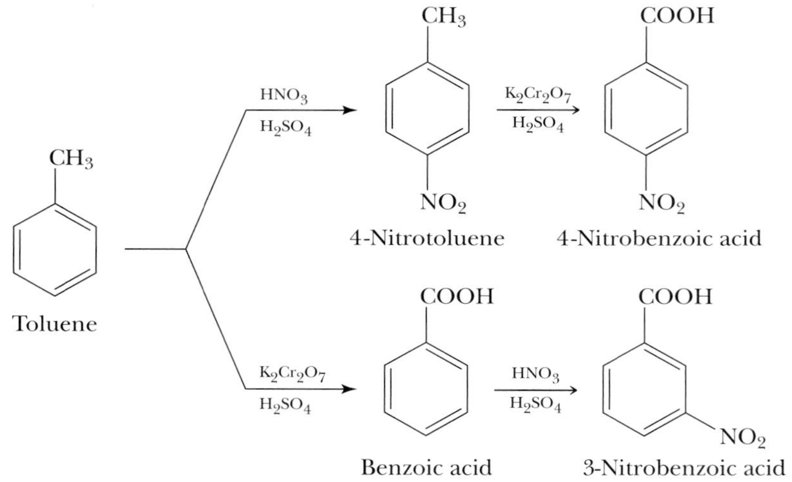

4-Nitrotoluene   4-Nitrobenzoic acid

Toluene

Benzoic acid   3-Nitrobenzoic acid

Nitration of toluene yields a product with the two substituents para to each other, whereas nitration of benzoic acid yields a product with the substituents meta to each other. Again, we see that the order in which the reactions are performed is critical.

Note that, in this last example, we show nitration of toluene producing only the para isomer. In practice because methyl is an ortho–para directing group, both ortho and para isomers are formed. In problems in which we ask you to prepare one or the other of these isomers, we assume that both form and that there are physical methods by which you can separate them and obtain the desired isomer.

EXAMPLE 9.6

Complete the following electrophilic aromatic substitution reactions. Where you predict meta substitution, show only the meta product. Where you predict ortho–para substitution, show both products:

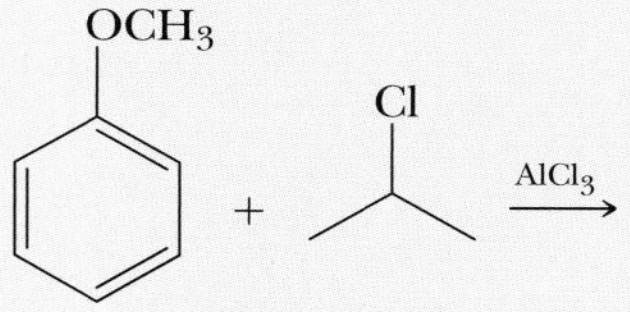

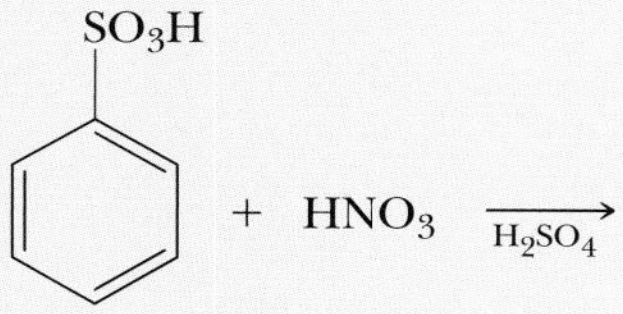

SOLUTION

The methoxyl group in (a) is ortho–para directing and strongly activating. The sulfonic acid group in (b) is meta directing and moderately deactivating:

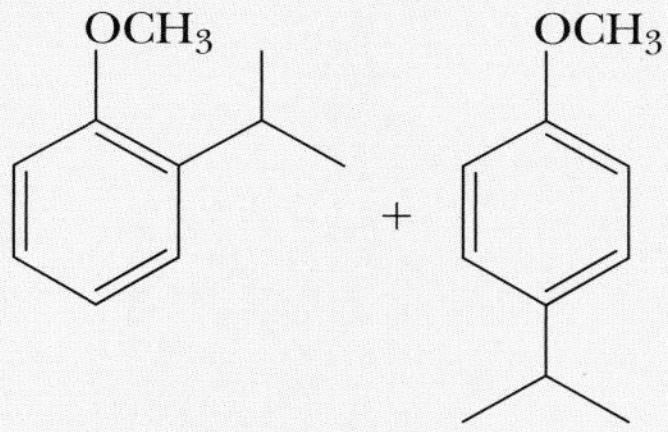

2-Isopropyl-anisole   4-Isopropyl-anisole   3-Nitrobenzene-sulfonic acid

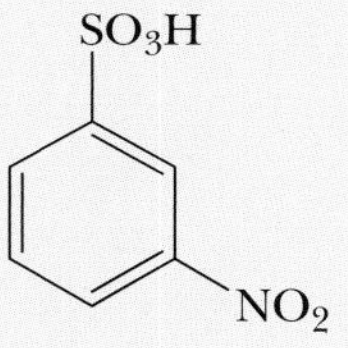

Practice Problem 9.6

Complete the following electrophilic aromatic substitution reactions. Where you predict meta substitution, show only the meta product. Where you predict ortho–para substitution, show both products:

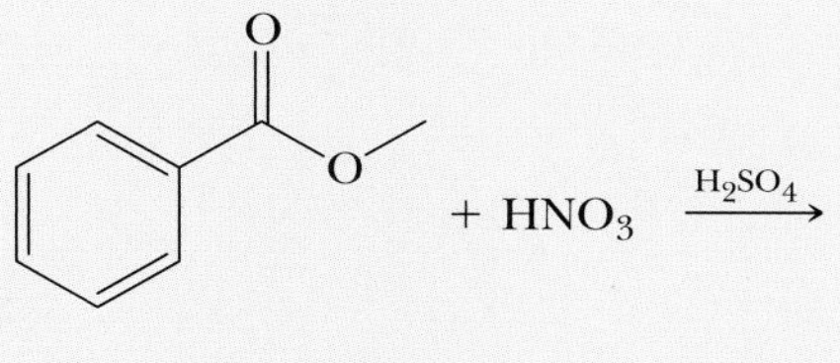

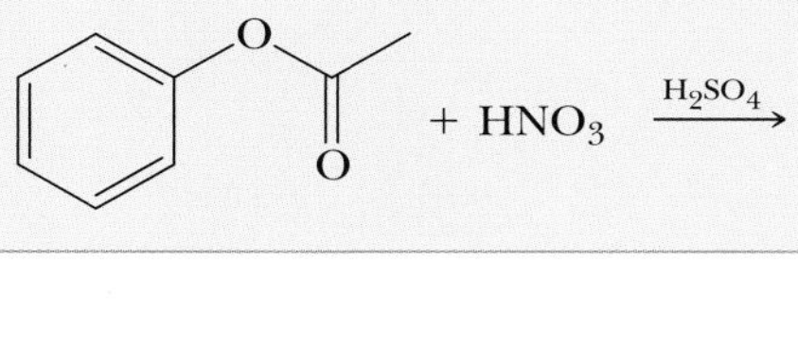

B. Theory of Directing Effects

As we have just seen, a group on an aromatic ring exerts a major effect on the patterns of further substitution. We can make these three generalizations:

1. If there is a lone pair of electrons on the atom bonded to the ring, the group is an ortho–para director.
2. If there is a full or partial positive charge on the atom bonded to the ring, the group is a meta director.
3. Alkyl groups are ortho–para directors.

We account for these patterns by means of the general mechanism for electrophilic aromatic substitution first presented in Section 9.6. Let us extend that mechanism to consider how a group already present on the ring might affect the relative stabilities of cation intermediates formed during a second substitution reaction.

We begin with the fact that the rate of electrophilic aromatic substitution is determined by the slowest step in the mechanism, which, in almost every reaction of an electrophile with the aromatic ring, is attack of the electrophile on the ring to give a resonance-stabilized cation intermediate. Thus, we must determine which of the alternative carbocation intermediates (that for ortho–para substitution or that for meta substitution) is the more stable. That is, which of the alternative cationic intermediates has the lower activation energy for its formation.

Nitration of Anisole

The rate-determining step in nitration is reaction of the nitronium ion with the aromatic ring to produce a resonance-stabilized cation intermediate. Figure 9.4 shows the cation intermediate formed by reaction meta to the methoxy group. The figure also shows the cationic intermediate formed by reaction para to the methoxy group. The intermediate formed by reaction at a meta position is a hybrid of three major contributing structures: (a), (b), and (c). These three are the only important contributing structures we can draw for reaction at a meta position.

The cationic intermediate formed by reaction at the para position is a hybrid of four major contributing structures: (d), (e), (f), and (g). What is important about structure (f) is that all atoms in it have complete octets, which means that this structure contributes more to the hybrid than structures (d), (e), or (g). Because the cation formed by reaction at an ortho or para position on anisole has a greater resonance stabilization and, hence, a lower activation energy for its formation, nitration of anisole occurs preferentially in the ortho and para positions.

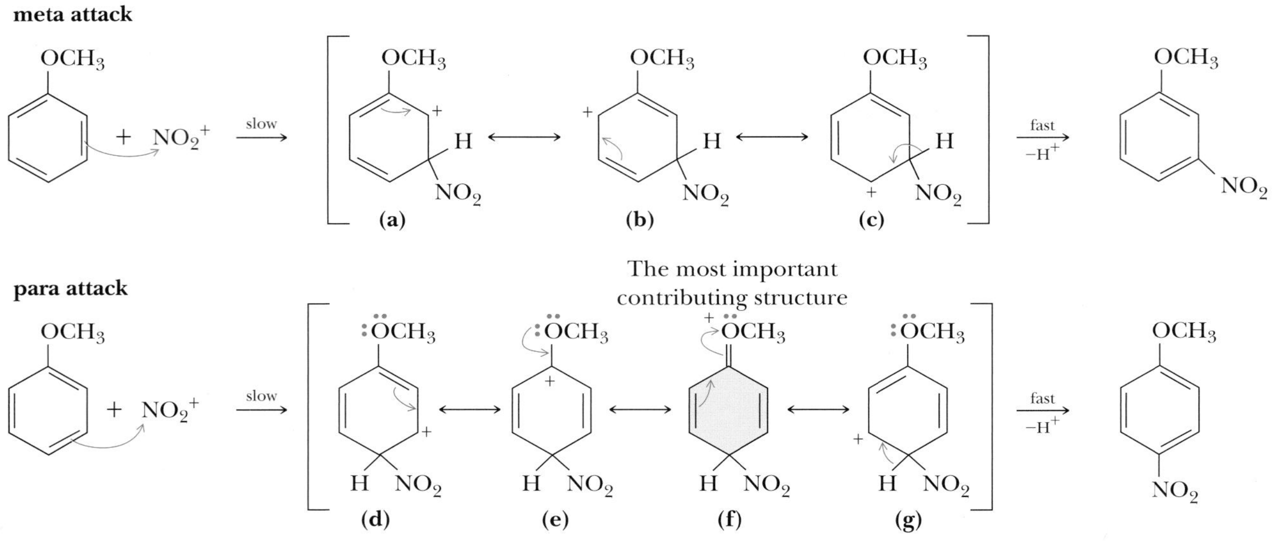

Figure 9.4
Nitration of anisole. Reaction of the electrophile meta and para to a methoxy group. Regeneration of the aromatic ring is shown from the rightmost contributing structure in each case.

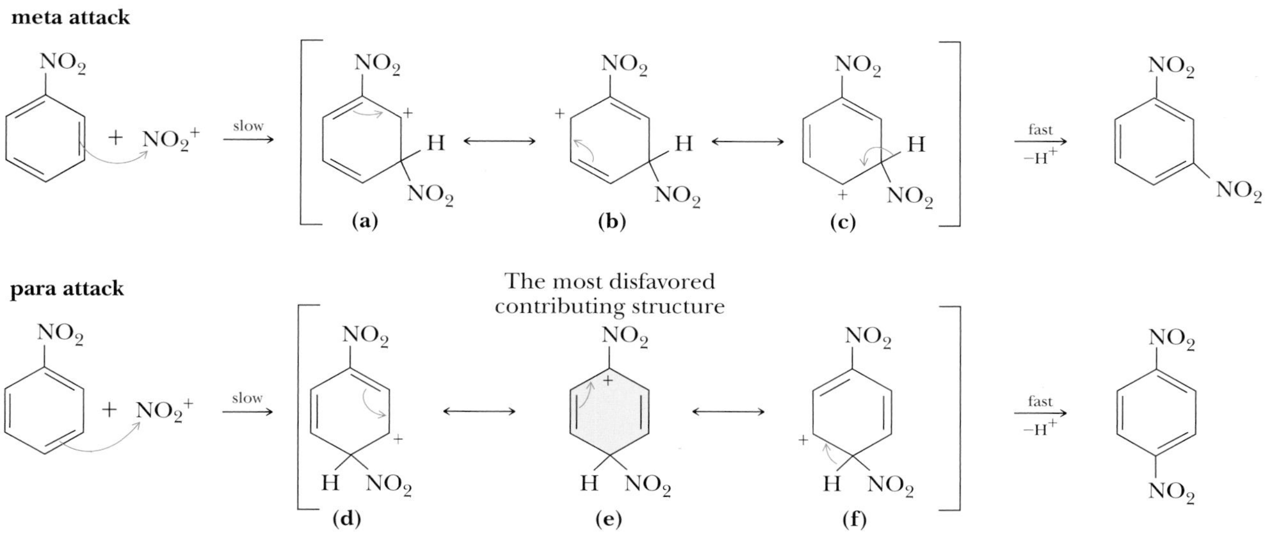

meta attack

para attack

The most disfavored
contributing structure

Figure 9.5
Nitration of nitrobenzene.
Reaction of the electrophile
meta and para to a nitro
group. Regeneration of the
aromatic ring is shown from
the rightmost contributing
structure in each case.

Nitration of Nitrobenzene

Figure 9.5 shows the resonance-stabilized cation intermediates formed by reaction of the nitronium ion meta to the nitro group and also para to it.

Each cation in the figure is a hybrid of three contributing structures; no additional ones can be drawn. Now we must compare the relative resonance stabilizations of each hybrid. If we draw a Lewis structure for the nitro group showing the positive formal charge on nitrogen, we see that contributing structure (e) places positive charges on adjacent atoms:

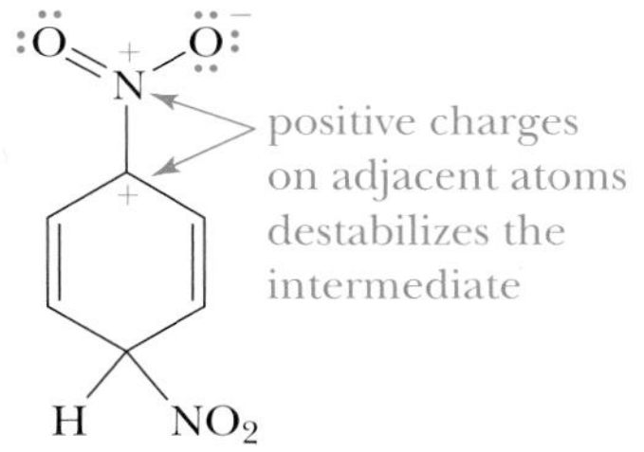

positive charges
on adjacent atoms
destabilizes the
intermediate

Because of the electrostatic repulsion thus generated, structure (e) makes only a negligible contribution to the hybrid. None of the contributing structures for reaction at a meta position places positive charges on adjacent atoms. As a consequence, resonance stabilization of the cation formed by reaction at a meta position is greater than that for the cation formed by reaction at a para (or ortho) position. Stated alternatively, the activation energy for reaction at a meta position is less than that for reaction at a para position.

A comparison of the entries in Table 9.1 shows that almost all ortho–para directing groups have an unshared pair of electrons on the atom bonded to the aromatic ring. Thus, the directing effect of most of these groups is due primarily to the ability of the atom bonded to the ring to delocalize further the positive charge on the cation intermediate.

The fact that alkyl groups are also ortho–para directing indicates that they, too, help to stabilize the cation intermediate. In Section 5.3A, we saw that alkyl groups stabilize carbocation intermediates and that the order of stability of carbocations is

$3° > 2° > 1° >$ methyl. Just as alkyl groups stabilize the cation intermediates formed in reactions of alkenes, they also stabilize the carbocation intermediates formed in electrophilic aromatic substitutions.

To summarize, any substituent on an aromatic ring that further stabilizes the cation intermediate directs ortho–para, and any group that destabilizes the cation intermediate directs meta.

EXAMPLE 9.7

Draw contributing structures formed during the para nitration of chlorobenzene, and show how chlorine participates in directing the incoming nitronium ion to ortho–para positions.

SOLUTION

Contributing structures (a), (b), and (d) place the positive charge on atoms of the ring, while contributing structure (c) places it on chlorine and thus creates additional resonance stabilization for the cation intermediate:

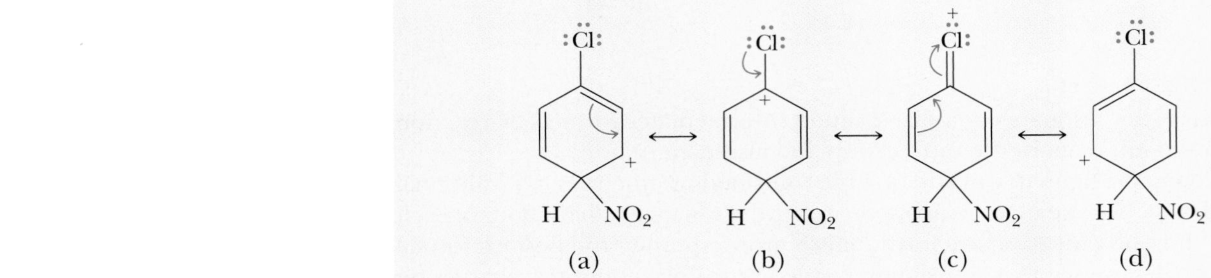

Practice Problem 9.7

Because the electronegativity of oxygen is greater than that of carbon, the carbon of a carbonyl group bears a partial positive charge, and its oxygen bears a partial negative charge. Using this information, show that a carbonyl group is meta directing:

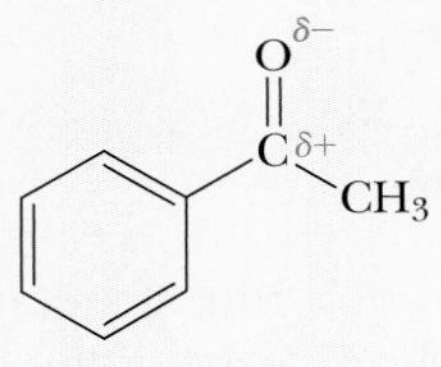

C. Theory of Activating–Deactivating Effects

We account for the activating–deactivating effects of substituent groups by a combination of resonance and inductive effects:

1. Any resonance effect, such as that of $-NH_2$, $-OH$, and $-OR$, which delocalizes the positive charge of the cation intermediate lowers the activation energy for its formation and is activating toward further electrophilic aromatic substitution. That is, these groups increase the rate of electrophilic aromatic substitution, compared with the rate at which benzene itself reacts.

2. Any resonance or inductive effect, such as that of —NO_2, —C=O, —SO_3H, —NR_3^+, —CCl_3, and —CF_3, which decreases electron density on the ring deactivates the ring to further substitution. That is, these groups decrease the rate of further electrophilic aromatic substitution, compared with the rate at which benzene itself reacts.

3. Any inductive effect (such as that of —CH_3 or another alkyl group) which releases electron density toward the ring activates the ring toward further substitution.

In the case of the halogens, the resonance and inductive effects operate in opposite directions. As Table 9.1 shows, the halogens are ortho–para directing, but, unlike other ortho–para directors listed in the table, the halogens are weakly deactivating. These observations can be accounted for in the following way.

1. *The inductive effect of halogens.* The halogens are more electronegative than carbon and have an electron-withdrawing inductive effect. Aryl halides, therefore, react more slowly in electrophilic aromatic substitution than benzene does.

2. *The resonance effect of halogens.* A halogen ortho or para to the site of electrophilic attack stabilizes the cation intermediate by delocalization of the positive charge:

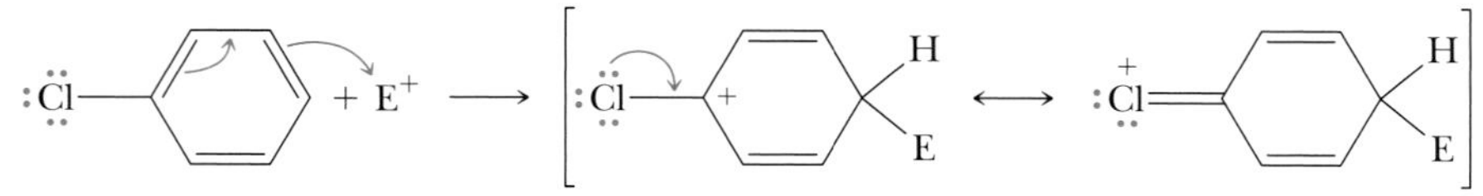

EXAMPLE 9.8

Predict the product of each electrophilic aromatic substitution:

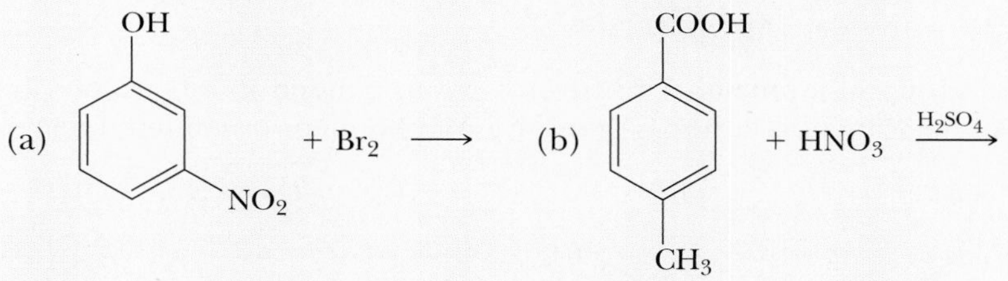

SOLUTION

The key to predicting the orientation of further substitution on a disubstituted arene is that ortho–para directing groups activate the ring toward further substitution, whereas meta directing groups deactivate it. This means that, when there is competition between ortho–para directing and meta directing groups, the ortho–para group wins.

(a) The ortho–para directing and activating —OH group determines the position of bromination. Bromination between the —OH and —NO_2 groups is only a minor product, because of steric hindrance to attack of bromine at this position:

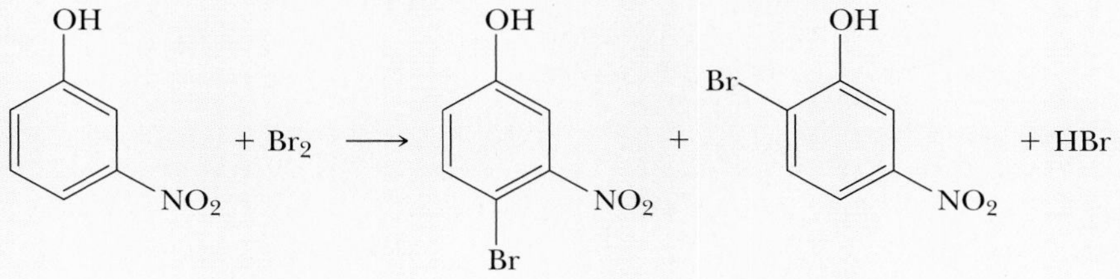

(b) The ortho–para directing and activating methyl group determines the position of nitration:

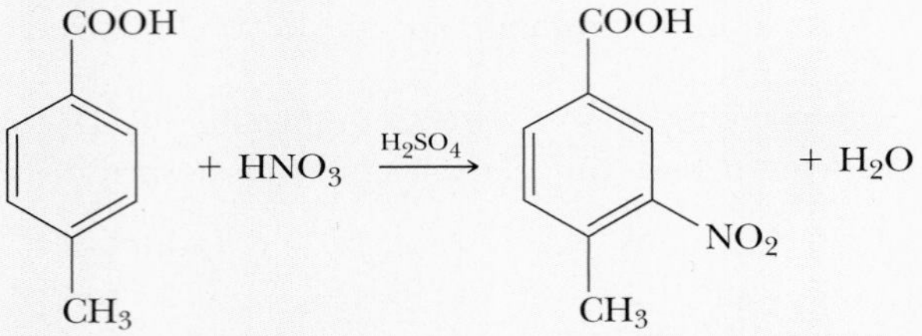

$$\text{(p-toluic acid)} + HNO_3 \xrightarrow{H_2SO_4} \text{(nitro product)} + H_2O$$

Practice Problem 9.8

Predict the product of treating each compound with HNO_3/H_2SO_4:

(a) 4-chlorotoluene

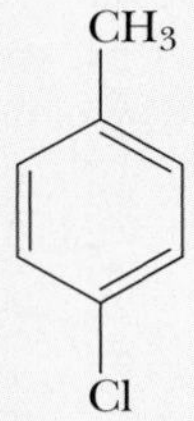

(b) 3-nitrobenzoic acid

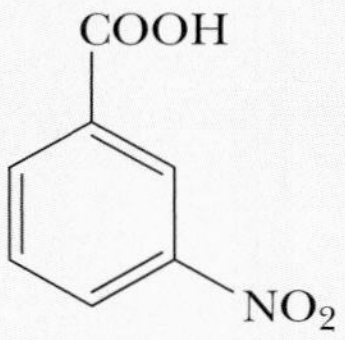

9.9 PHENOLS

A. Structure and Nomenclature

Phenol A compound that contains an —OH bonded to a benzene ring.

The functional group of a **phenol** is a hydroxyl group bonded to a benzene ring. We name substituted phenols either as derivatives of phenol or by common names:

Phenol   3-Methylphenol   1,2-Benzenediol   1,3-Benzenediol   1,4-Benzenediol
               (*m*-Cresol)   (Catechol)   (Resorcinol)   (Hydroquinone)

Phenols are widely distributed in nature. Phenol itself and the isomeric cresols (*o*-, *m*-, and *p*-cresol) are found in coal tar. Thymol and vanillin are important constituents of thyme and vanilla beans, respectively:

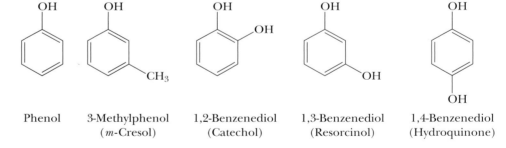

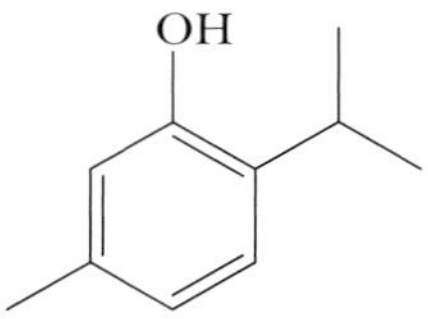

2-Isopropyl-5-methylphenol   4-Hydroxy-3-methoxy-
(Thymol)   benzaldehyde
                                    (Vanillin)

Thymol is a constituent of garden thyme, Thymus vulgaris. *(Wally Eberhart/ Visuals Unlimited)*

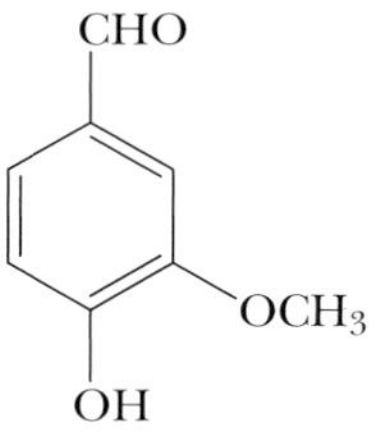

CHEMICAL CONNECTIONS 9B

Capsaicin, for Those Who Like It Hot

Capsaicin, the pungent principle from the fruit of various peppers (*Capsicum* and Solanaceae), was isolated in 1876, and its structure was determined in 1919:

Capsaicin
(from various types of peppers)

The inflammatory properties of capsaicin are well known; the human tongue can detect as little as one drop of it in 5 L of water. Many of us are familiar with the burning sensation in the mouth and sudden tear-ing in the eyes caused by a good dose of hot chili peppers. Capsaicin-containing extracts from these flaming foods are also used in sprays to ward off dogs or other animals that might nip at your heels while you are running or cycling.

Ironically, capsaicin is able to cause pain and relieve it as well. Currently, two capsaicin-containing creams, Mioton and Zostrix®, are prescribed to treat the burning pain associated with postherpetic neuralgia, a complication of shingles. They are also prescribed for diabetics, to relieve persistent foot and leg pain.

The mechanism by which capsaicin relieves pain is not fully understood. It has been suggested that, after it is applied, the nerve endings in the area responsible for the transmission of pain remain temporarily numb. Capsaicin remains bound to specific receptor sites on these pain-transmitting neurons, blocking them from further action. Eventually, capsaicin is removed from the receptor sites, but in the meantime, its presence provides needed relief from pain.

Phenol, or carbolic acid, as it was once called, is a low-melting solid that is only slightly soluble in water. In sufficiently high concentrations, it is corrosive to all kinds of cells. In dilute solutions, phenol has some antiseptic properties and was introduced into the practice of surgery by Joseph Lister, who demonstrated his technique of aseptic surgery in the surgical theater of the University of Glasgow School of Medicine in 1865. Nowadays, phenol has been replaced by antiseptics that are both more powerful and have fewer undesirable side effects. Among these is hexylresorcinol, which is widely used in nonprescription preparations as a mild antiseptic and disinfectant.

Poison Ivy. *(Charles D. Winters)*

Hexylresorcinol

Eugenol

Urushiol

Eugenol, which can be isolated from the flower buds (cloves) of *Eugenia aromatica*, is used as a dental antiseptic and analgesic. Urushiol is the main component in the irritating oil of poison ivy.

B. Acidity of Phenols

Phenols and alcohols both contain an —OH group. We group phenols as a separate class of compounds, however, because their chemical properties are quite different from those of alcohols. One of the most important of these differences is that phenols are significantly more acidic than are alcohols. Indeed, the acid ionization constant for phenol is 10^6 times larger than that of ethanol!

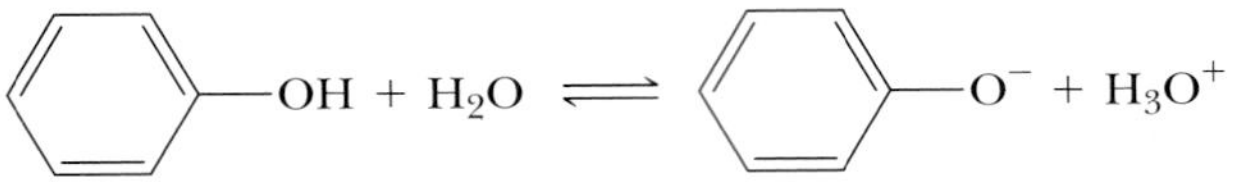

$$\text{Phenol} + H_2O \rightleftharpoons \text{Phenoxide ion} + H_3O^+ \quad K_a = 1.1 \times 10^{-10} \quad pK_a = 9.95$$

$$CH_3CH_2OH + H_2O \rightleftharpoons CH_3CH_2O^- + H_3O^+ \quad K_a = 1.3 \times 10^{-16} \quad pK_a = 15.9$$
$$\text{Ethanol} \qquad\qquad \text{Ethoxide ion}$$

Another way to compare the relative acid strengths of ethanol and phenol is to look at the hydrogen ion concentration and pH of a 0.1-M aqueous solution of each (Table 9.2). For comparison, the hydrogen ion concentration and pH of 0.1 M HCl are also included.

TABLE 9.2   Relative Acidities of 0.1-M Solutions of Ethanol, Phenol, and HCl

Acid Ionization Equation	$[H^+]$	pH
$CH_3CH_2OH + H_2O \rightleftharpoons CH_3CH_2O^- + H_3O^+$	1×10^{-7}	7.0
$C_6H_5OH + H_2O \rightleftharpoons C_6H_5O^- + H_3O^+$	3.3×10^{-6}	5.4
$HCl + H_2O \rightleftharpoons Cl^- + H_3O^+$	0.1	1.0

In aqueous solution, alcohols are neutral substances, and the hydrogen ion concentration of 0.1 M ethanol is the same as that of pure water. A 0.1-M solution of phenol is slightly acidic and has a pH of 5.4. By contrast, 0.1 M HCl, a strong acid (completely ionized in aqueous solution), has a pH of 1.0.

The greater acidity of phenols compared with alcohols results from the greater stability of the phenoxide ion compared with an alkoxide ion. The negative charge on the phenoxide ion is delocalized by resonance. The two contributing structures on the left for the phenoxide ion place the negative charge on oxygen, while the three on the right place the negative charge on the ortho and para positions of the ring. Thus, in the resonance hybrid, the negative charge of the phenoxide ion is delocalized over four atoms, which stabilizes the phenoxide ion realtive to an alkoxide ion, for which no delocalization is possible:

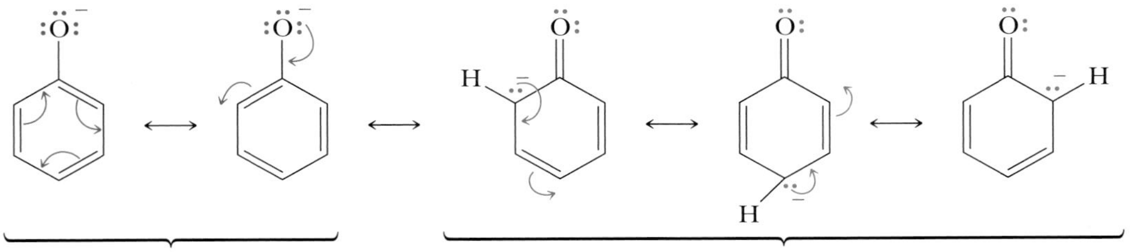

These two Kekulé structures are equivalent

These three contributing structures delocalize the negative charge onto carbon atoms of the ring

Note that, although the resonance model gives us a way of understanding why phenol is a stronger acid than ethanol, it does not provide us with any quantitative means of predicting just how much stronger an acid it might be. To find out how much stronger one acid is than another, we must determine their pK_a values experimentally and compare them.

Ring substituents, particularly halogen and nitro groups, have marked effects on the acidities of phenols through a combination of inductive and resonance effects. Because the halogens are more electronegative than carbon, they withdraw electron density from the aromatic ring, weaken the O—H bond, and stabilize the phenoxide ion. Nitro groups also withdraw electron density from the ring, weaken the O—H bond, stabilize the phenoxide ion, and thus increase acidity.

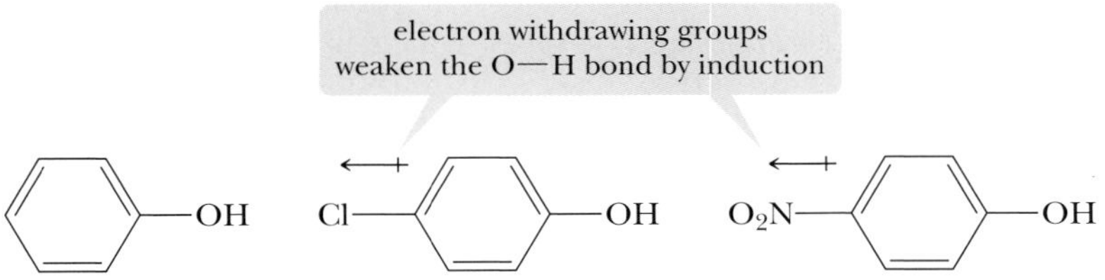

electron withdrawing groups
weaken the O—H bond by induction

Phenol
pK_a 9.95

4-Chlorophenol
pK_a 9.18

4-Nitrophenol
pK_a 7.15

Increasing acid strength

EXAMPLE 9.9

Arrange these compounds in order of increasing acidity: 2,4-dinitrophenol, phenol, and benzyl alcohol.

SOLUTION

Benzyl alcohol, a primary alcohol, has a pK_a of approximately 16–18 (Section 8.3A). The pK_a of phenol is 9.95. Nitro groups are electron withdrawing and increase the acidity of the phenolic —OH group. In order of increasing acidity, these compounds are:

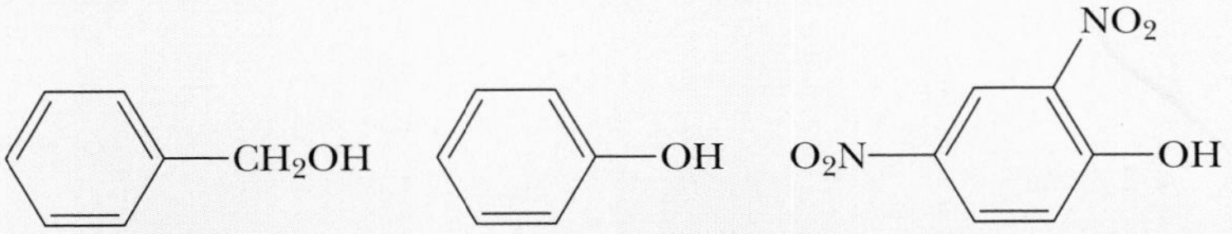

Benzyl alcohol
pK_a 16–18

Phenol
pK_a 9.95

2,4-Dinitrophenol
pK_a 3.96

Practice Problem 9.9

Arrange these compounds in order of increasing acidity: 2,4-dichlorophenol, phenol, cyclohexanol.

C. Acid–Base Reactions of Phenols

Phenols are weak acids and react with strong bases, such as NaOH, to form water-soluble salts:

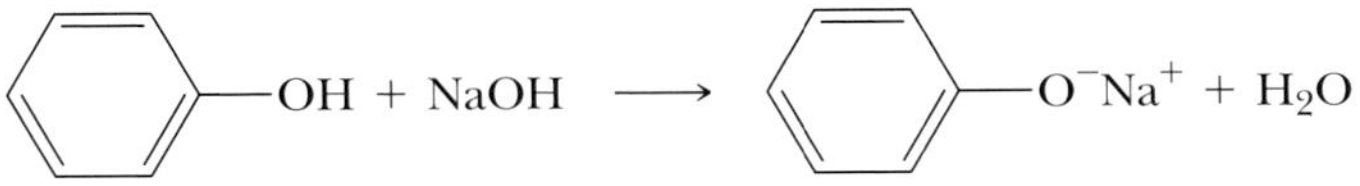

Phenol
pK_a 9.95
(stronger acid)

Sodium
hydroxide
(stronger base)

Sodium
phenoxide
(weaker base)

Water
pK_a 15.7
(weaker acid)

Most phenols do not react with weaker bases, such as sodium bicarbonate; and do not dissolve in aqueous sodium bicarbonate. Carbonic acid is a stronger acid than most phenols and, consequently, the equilibrium for their reaction with bicarbonate ion lies far to the left (see Section 2.5):

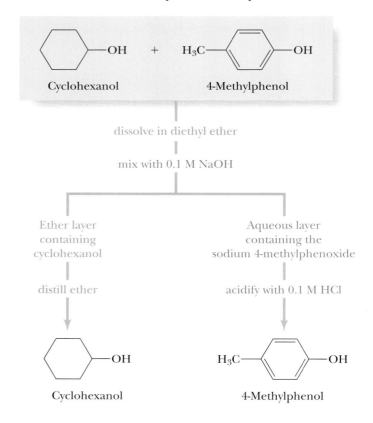

Phenol	Sodium		Sodium	Carbonic acid
pK_a 9.95	bicarbonate		phenoxide	pK_a 6.36
(weaker acid)	(weaker base)		(stronger base)	(stronger acid)

The fact that phenols are weakly acidic, whereas alcohols are neutral, provides a convenient way to separate phenols from water-insoluble alcohols. Suppose that we want to separate 4-methylphenol from cyclohexanol. Each is only slightly soluble in water; therefore, they cannot be separated on the basis of their water solubility. They can be separated, however, on the basis of their difference in acidity. First, the mixture of the two is dissolved in diethyl ether or some other water-immiscible solvent. Next, the ether solution is placed in a separatory funnel and shaken with dilute aqueous NaOH. Under these conditions, 4-methylphenol reacts with NaOH to give sodium 4-methylphenoxide, a water-soluble salt. The upper layer in the separatory funnel is now diethyl ether (density 0.74 g/cm^3), containing only dissolved cyclohexanol. The lower aqueous layer contains dissolved sodium 4-methylphenoxide. The layers are separated, and distillation of the ether (bp 35°C) leaves pure cyclohexanol (bp 161°C). Acidification of the aqueous phase with 0.1 M HCl or another strong acid converts sodium 4-methylphenoxide to 4-methylphenol, which is insoluble in water and can be extracted with ether and recovered in pure form. The following flowchart summarizes these experimental steps:

D. Phenols as Antioxidants

An important reaction in living systems, foods, and other materials that contain carbon–carbon double bonds is **autoxidation**—that is, oxidation requiring oxygen and no other reactant. If you open a bottle of cooking oil that has stood for a long time, you will notice a hiss of air entering the bottle. This sound occurs because the consumption of oxygen by autoxidation of the oil creates a negative pressure inside the bottle.

Cooking oils contain esters of polyunsaturated fatty acids. You need not worry now about what esters are; we will discuss them in Chapter 15. The important point here is that all vegetable oils contain fatty acids with long hydrocarbon chains, many of which have one or more carbon–carbon double bonds. (See Problem 4.44 for the structures of three of these fatty acids.) Autoxidation takes place at a carbon adjacent to a double bond—that is, at an **allylic carbon**.

Autoxidation is a radical chain process that converts an R—H group into an R—O—O—H group, called a *hydroperoxide*. The process begins by removing a hydrogen atom, together with one of its electrons (H·), from an allylic carbon. The carbon losing the H· now has only seven electrons in its valence shell, one of which is unpaired. An atom or molecule with an unpaired electron is called a **radical**.

Step 1: *Chain Initiation—Formation of a Radical from a Nonradical Compound* The removal of a hydrogen atom (H·) gives an allylic radical:

$$-CH_2CH=CH-\underset{\underset{\text{Section of a fatty}}{\text{acid hydrocarbon chain}}}{\overset{\overset{\text{H}}{|}}{CH}}- \xrightarrow[\text{or heat}]{\text{light}} -CH_2CH=CH-\overset{\text{Section of a fatty}}{\overset{\cdot}{C}H}-$$

Section of a fatty acid hydrocarbon chain   An allylic radical

Step 2a: *Chain Propagation—Reaction of a Radical to Form a New Radical* The allylic radical reacts with oxygen, itself a diradical, to form a hydroperoxy radical. The new covalent bond of the hydroperoxy radical forms by the combination of one electron from the allylic radical and one electron from the oxygen diradical:

$$-CH_2CH=CH-\overset{\cdot}{C}H- \ + \ \cdot O-O\cdot \longrightarrow -CH_2CH=CH-\overset{\overset{\text{O}-\text{O}\cdot}{|}}{C}H-$$

Oxygen is a diradical   A hydroperoxy radical

Step 2b: *Chain Propagation—Reaction of a Radical to Form a New Radical* The hydroperoxy radical removes an allylic hydrogen atom (H·) from a new fatty acid hydrocarbon chain to complete the formation of a hydroperoxide and, at the same time, produce a new allylic radical:

$$-CH_2CH=CH-\overset{\overset{\text{O}-\text{O}\cdot}{|}}{C}H- \ + \ -CH_2CH=CH-\overset{\overset{\text{H}}{|}}{C}H- \longrightarrow$$

Section of a new fatty acid hydrocarbon chain

$$-CH_2CH=CH-\overset{\overset{\text{O}-\text{O}-\text{H}}{|}}{C}H-CH_2- \ + \ -CH_2CH=CH-\overset{\cdot}{C}H-$$

A hydroperoxide   A new allylic radical

The most important point about the pair of chain propagation steps is that they form a continuous cycle of reactions. The new radical formed in Step 2b next reacts

Butylated hydroxytoluene (BHT) is often used as an antioxidant in baked goods to "retard spoilage." *(Charles D. Winters)*

with another molecule of O_2 in Step 2a to give a new hydroperoxy radical, which then reacts with a new hydrocarbon chain to repeat Step 2b, and so forth. This cycle of propagation steps repeats over and over in a chain reaction. Thus, once a radical is generated in Step 1, the cycle of propagation steps may repeat many thousands of times, generating thousands and thousands of hydroperoxide molecules. The number of times the cycle of chain propagation steps repeats is called the **chain length**.

Hydroperoxides themselves are unstable and, under biological conditions, degrade to short-chain aldehydes and carboxylic acids with unpleasant "rancid" smells. These odors may be familiar to you if you have ever smelled old cooking oil or aged foods that contain polyunsaturated fats or oils. A similar formation of hydroperoxides in the low-density lipoproteins deposited on the walls of arteries leads to cardiovascular disease in humans. In addition, many effects of aging are thought to be the result of the formation and subsequent degradation of hydroperoxides.

Fortunately, nature has developed a series of defenses, including the phenol vitamin E, ascorbic acid (vitamin C), and glutathione, against the formation of destructive, hydroperoxides. The compounds that defend against hydroperoxides are "natures scavengers." Vitamin E, for example, inserts itself into either Step 2a or 2b, donates an H· from its phenolic —OH group to the allylic radical, and converts the radical to its original hydrocarbon chain. Because the vitamin E radical is stable, it breaks the cycle of chain propagation steps, thereby preventing the further formation of destructive hydroperoxides. While some hydroperoxides may form, their numbers are very small and they are easily decomposed to harmless materials by one of several enzyme-catalyzed reactions.

Unfortunately, vitamin E is removed in the processing of many foods and food products. To make up for this loss, phenols such as BHT and BHA are added to foods to "retard [their] spoilage" (as they say on the packages) by autoxidation:

Vitamin E

Butylated *hydroxy-toluene* (BHT)

Butylated *hydroxy-anisole* (BHA)

Similar compounds are added to other materials, such as plastics and rubber, to protect them against autoxidation.

SUMMARY

Benzene and its alkyl derivatives are classified as **aromatic hydrocarbons**, or **arenes**. The concepts of **hybridization of atomic orbitals** and the **theory of resonance** (Section 9.2C), developed in the 1930s, provided the first adequate description of the structure of benzene. The **resonance energy** of benzene is approximately 36 kcal/mol (151 kJ/mol) (Section 9.2D).

According to the Hückel criteria for aromaticity, a five- or six-membered ring is aromatic if it (1) has one p orbital on each atom of the ring, (2) is planar, so that overlap of all p orbitals of the ring is continuous or nearly so, and (3) has 2, 6, 10, 14 and

so forth pi electrons in the overlapping system of p orbitals (Section 9.3). A **heterocyclic aromatic compound** contains one or more atoms other than carbon in an aromatic ring.

Aromatic compounds are named by the IUPAC system (Section 9.4). The common names toluene, xylene, styrene, phenol, aniline, benzaldehyde, and benzoic acid are retained. The C_6H_5— group is named **phenyl**, and the $C_6H_5CH_2$— group is named **benzyl**. To locate two substituents on a benzene ring, either number the atoms of the ring or use the locators **ortho (o)**, **meta (m)**, and **para (p)**.

Polynuclear aromatic hydrocarbons (Section 9.4C) contain two or more fused benzene rings. Particularly abundant are naphthalene, anthracene, phenanthrene, and their derivatives.

A characteristic reaction of aromatic compounds is **electrophilic aromatic substitution** (Section 9.6). Substituents on an aromatic ring influence both the site and rate of further substitution (Section 9.8). Substituent groups that direct an incoming group preferentially to the ortho and para positions are called **ortho–para directors**. Those which direct an incoming group preferentially to the meta positions are called **meta directors**. **Activating groups** cause the rate of further substitu-

tion to be faster than that for benzene; **deactivating groups** cause it to be slower than that for benzene.

A mechanistic rationale for directing effects is based on the degree of resonance stabilization of the possible cation intermediates formed upon reaction of the aromatic ring and the electrophile (Section 9.8B). Groups that stabilize the cation intermediate are ortho–para directors; groups that destabilize it are deactivators and meta directors.

The functional group of a **phenol** is an —OH group bonded to a benzene ring (Section 9.9A). Phenol and its derivatives are weak acids, with pK_a approximately 10.0, but are considerably stronger acids than alcohols, with pK_a 16–18.

KEY REACTIONS

1. Oxidation at a Benzylic Position (Section 9.5)
A benzylic carbon bonded to at least one hydrogen is oxidized to a carboxyl group:

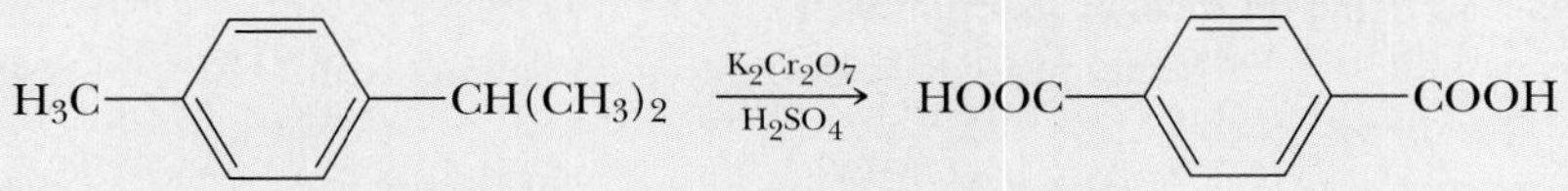

2. Chlorination and Bromination (Section 9.7A)
The electrophile is a halonium ion, Cl^+ or Br^+, formed by treating Cl_2 or Br_2 with $AlCl_3$ or $FeCl_3$:

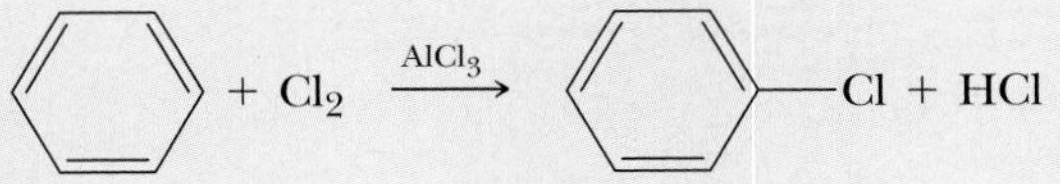

3. Nitration (Section 9.7B)
The electrophile is the nitronium ion, NO_2^+, formed by treating nitric acid with sulfuric acid:

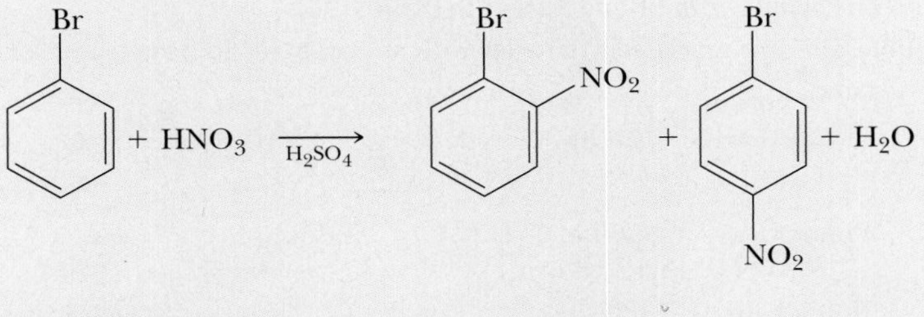

4. Sulfonation (Section 9.7B)
The electrophile is HSO_3^+:

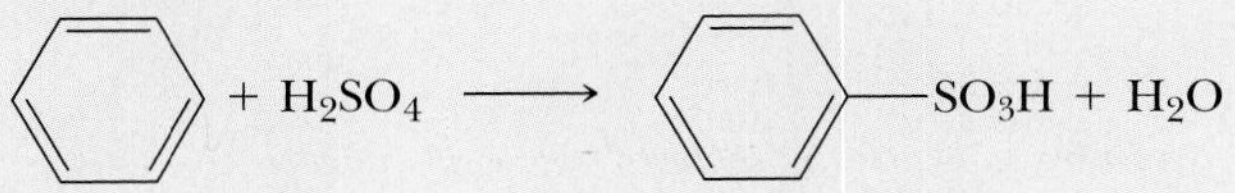

5. Friedel–Crafts Alkylation (Section 9.7C)
The electrophile is an alkyl carbocation formed by treating an alkyl halide with a Lewis acid:

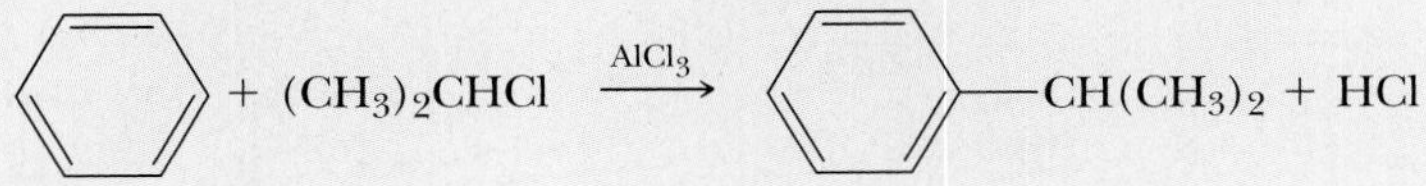

6. Friedel–Crafts Acylation (Section 9.7D)

The electrophile is an acyl cation formed by treating an acyl halide with a Lewis acid:

$$\text{(benzene)} + CH_3\overset{\displaystyle O}{\overset{\displaystyle \|}{C}}Cl \xrightarrow{\text{AlCl}_3} \text{(benzene)}-\overset{\displaystyle O}{\overset{\displaystyle \|}{C}}CH_3 + HCl$$

7. Alkylation Using an Alkene (Section 9.7E)

The electrophile is a carbocation formed by treating an alkene with H_2SO_4 or H_3PO_4:

$$\text{(4-methylphenol)} + 2\ CH_3\overset{\displaystyle CH_3}{\underset{}{C}}=CH_2 \xrightarrow{\text{H}_3\text{PO}_4} \text{(product)}$$

8. Alkylation Using an Alcohol (Section 9.7E)

The electrophile is a carbocation formed by treating an alcohol with H_2SO_4 or H_3PO_4:

$$\text{(benzene)} + (CH_3)_3COH \xrightarrow{\text{H}_3\text{PO}_4} \text{(benzene)}-C(CH_3)_3 + H_2O$$

9. Acidity of Phenols (Section 9.9B)

Phenols are weak acids:

$$\text{(benzene)}-OH + H_2O \rightleftharpoons \text{(benzene)}-O^- + H_3O^+ \qquad \begin{array}{l} K_a = 1.1 \times 10^{-10} \\ pK_a = 9.95 \end{array}$$

Phenol   Phenoxide ion

Substitution by electron-withdrawing groups, such as the halogens and the nitro group, increases the acidity of phenols.

10. Reaction of Phenols with Strong Bases (Section 9.9C)

Water-insoluble phenols react quantitatively with strong bases to form water-soluble salts:

$$\text{(benzene)}-OH + NaOH \longrightarrow \text{(benzene)}-O^-Na^+ + H_2O$$

Phenol	Sodium	Sodium	Water
pK_a 9.95	hydroxide	phenoxide	pK_a 15.7
(stronger acid)	(stronger base)	(weaker base)	(weaker acid)

PROBLEMS

A problem number set in red indicates an applied "real-world" problem.

Aromaticity

9.10 Which of the following compounds are aromatic?

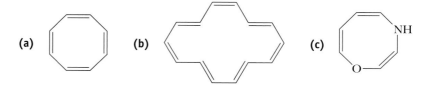

(a)   (b)   (c)

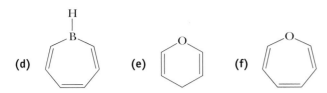

(d)   (e)   (f)

9.11 Explain why cyclopentadiene (pK_a 16) is many orders of magnitude more acidic than cyclopentane (pK_a > 50). (*Hint:* Draw the structural formula for the anion formed by removing one of the protons on the —CH$_2$— group, and then apply the Hückel criteria for aromaticity.)

Cyclopentadiene   Cyclopentane

Nomenclature and Structural Formulas

9.12 Name these compounds:

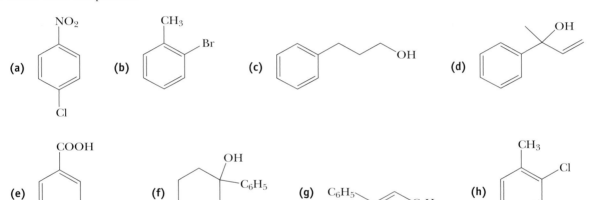

9.13 Draw structural formulas for these compounds:
- **(a)** 1-Bromo-2-chloro-4-ethylbenzene
- **(b)** 4-Iodo-1,2-dimethylbenzene
- **(c)** 2,4,6-Trinitrotoluene
- **(d)** 4-Phenyl-2-pentanol
- **(e)** *p*-Cresol
- **(f)** 2,4-Dichlorophenol
- **(g)** 1-Phenylcyclopropanol
- **(h)** Styrene (phenylethylene)
- **(i)** *m*-Bromophenol
- **(j)** 2,4-Dibromoaniline
- **(k)** Isobutylbenzene
- **(l)** *m*-Xylene

9.14 Show that pyridine can be represented as a hybrid of two equivalent contributing structures.

9.15 Show that naphthalene can be represented as a hybrid of three contributing structures. Show also, by the use of curved arrows, how one contributing structure is converted to the next.

9.16 Draw four contributing structures for anthracene.

Electrophilic Aromatic Substitution: Monosubstitution

9.17 Draw a structural formula for the compound formed by treating benzene with each of the following combinations of reagents:
- **(a)** $CH_3CH_2Cl/AlCl_3$
- **(b)** $CH_2{=}CH_2/H_2SO_4$
- **(c)** CH_3CH_2OH/H_2SO_4

9.18 Show three different combinations of reagents you might use to convert benzene to isopropylbenzene.

9.19 How many monochlorination products are possible when naphthalene is treated with $Cl_2/AlCl_3$?

9.20 Write a stepwise mechanism for the following reaction, using curved arrows to show the flow of electrons in each step:

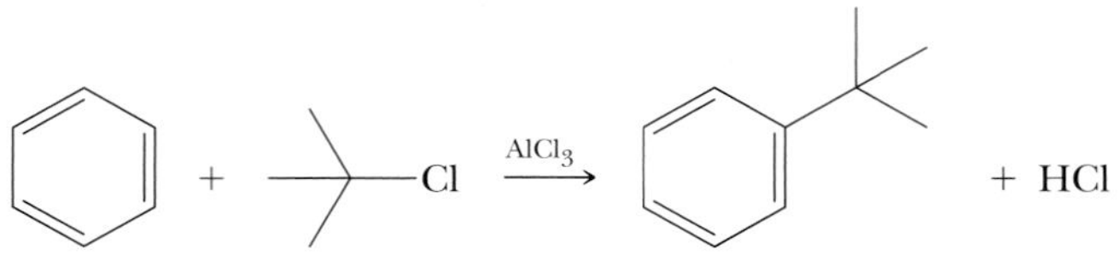

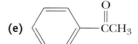

9.21 Write a stepwise mechanism for the preparation of diphenylmethane by treating benzene with dichloromethane in the presence of an aluminum chloride catalyst.

Electrophilic Aromatic Substitution: Disubstitution

9.22 When treated with $Cl_2/AlCl_3$, 1,2-dimethylbenzene (*o*-xylene) gives a mixture of two products. Draw structural formulas for these products.

9.23 How many monosubstitution products are possible when 1,4-dimethylbenzene (*p*-xylene) is treated with $Cl_2/AlCl_3$? When *m*-xylene is treated with $Cl_2/AlCl_3$?

9.24 Draw the structural formula for the major product formed upon treating each compound with $Cl_2/AlCl_3$:

(a) Toluene   (b) Nitrobenzene   (c) Chlorobenzene

(d) *tert*-Butylbenzene

(e) [benzene ring]—$\overset{\overset{\displaystyle O}{\|}}{C}CH_3$ (f) 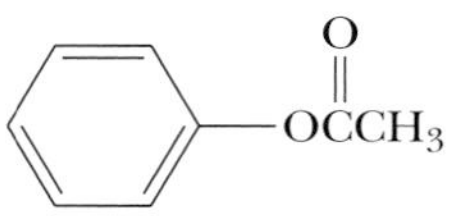 [benzene ring]—$O\overset{\overset{\displaystyle O}{\|}}{C}CH_3$

(g) 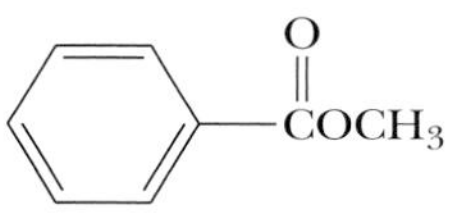 [benzene ring]—$\overset{\overset{\displaystyle O}{\|}}{C}OCH_3$

9.25 Which compound, chlorobenzene or toluene, undergoes electrophilic aromatic substitution more rapidly when treated with $Cl_2/AlCl_3$? Explain and draw structural formulas for the major product(s) from each reaction.

9.26 Arrange the compounds in each set in order of decreasing reactivity (fastest to slowest) toward electrophilic aromatic substitution:

(a)    [benzene ring]   [benzene ring]—$O\overset{\overset{\displaystyle O}{\|}}{C}CH_3$ [benzene ring]—$\overset{\overset{\displaystyle O}{\|}}{C}OCH_3$

(A) 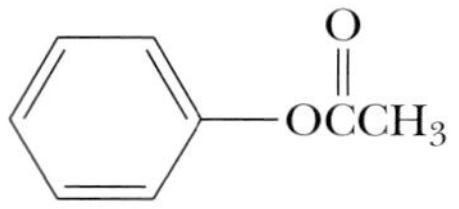   (B) 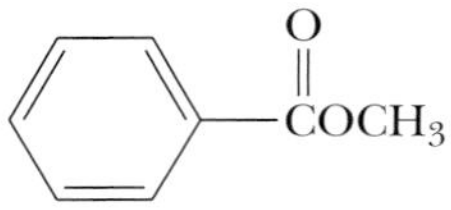   (C)

(b) 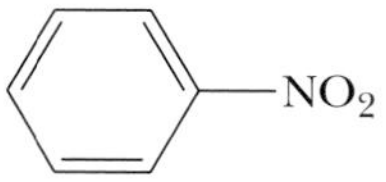   [benzene ring]—NO_2 [benzene ring]—COOH   [benzene ring]

(A) 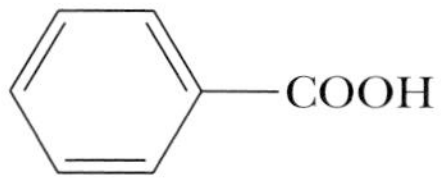   (B)    (C)

(c) 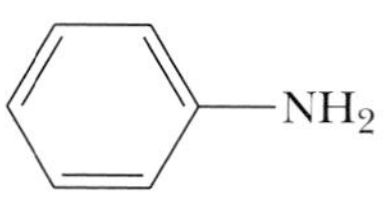   [benzene ring]—NH_2 [benzene ring]—$NH\overset{\overset{\displaystyle O}{\|}}{C}CH_3$ [benzene ring]—$\overset{\overset{\displaystyle O}{\|}}{C}NHCH_3$

(A) 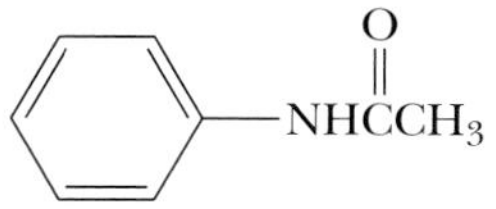   (B) 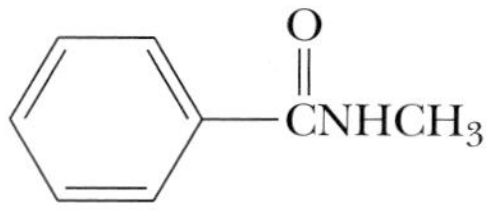   (C)

(d)    [benzene ring]   [benzene ring]—CH_3 [benzene ring]—OCH_3

(A) 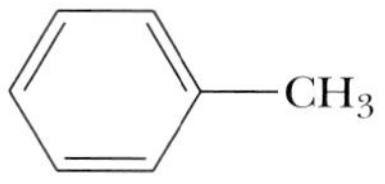   (B) 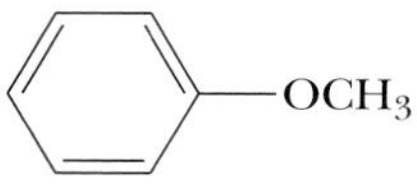   (C)

9.27 Account for the observation that the trifluoromethyl group is meta directing, as shown in the following example:

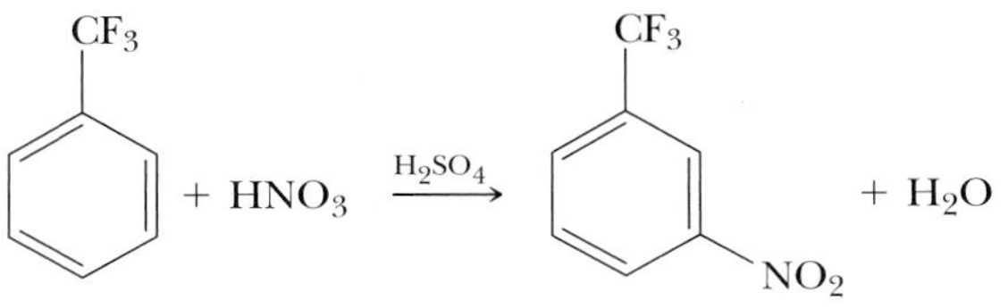

9.28 Show how to convert toluene to these carboxylic acids:
(a) 4-Chlorobenzoic acid   (b) 3-Chlorobenzoic acid

9.29 Show reagents and conditions that can be used to bring about these conversions:

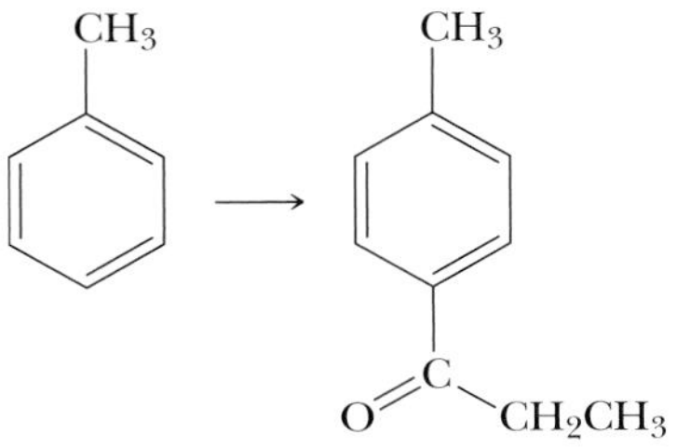

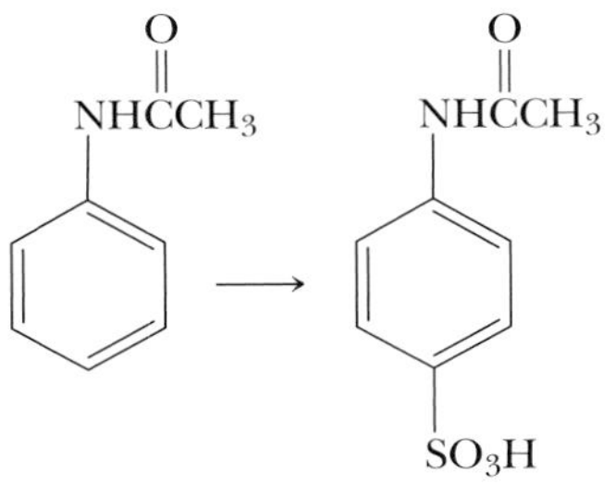

9.30 Propose a synthesis of triphenylmethane from benzene as the only source of aromatic rings. Use any other necessary reagents.

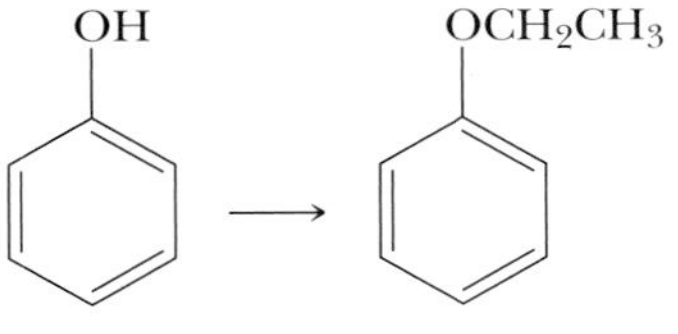

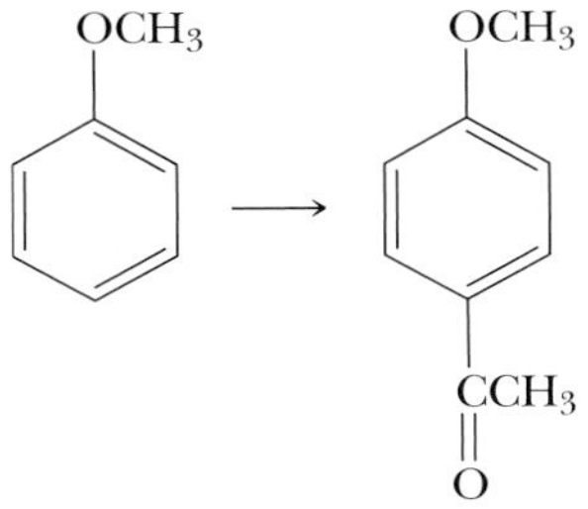

9.31 Reaction of phenol with acetone in the presence of an acid catalyst gives bisphenol A, a compound used in the production of polycarbonate and epoxy resins (Sections 17.5C and 17.5E):

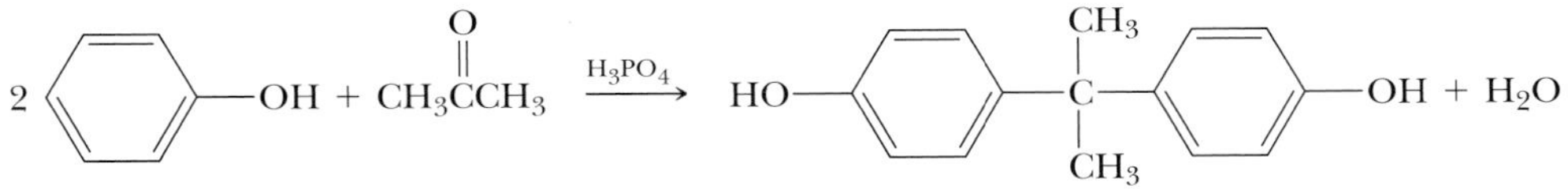

Acetone   Bisphenol A

Propose a mechanism for the formation of bisphenol A. (*Hint:* The first step is a proton transfer from phosphoric acid to the oxygen of the carbonyl group of acetone.)

9.32 2,6-Di-*tert*-butyl-4-methylphenol, more commonly known as butylated hydroxytoluene, or BHT, is used as an antioxidant in foods to "retard spoilage." BHT is synthesized industrially from 4-methylphenol (*p*-cresol) by reaction with 2-methylpropene in the presence of phosphoric acid:

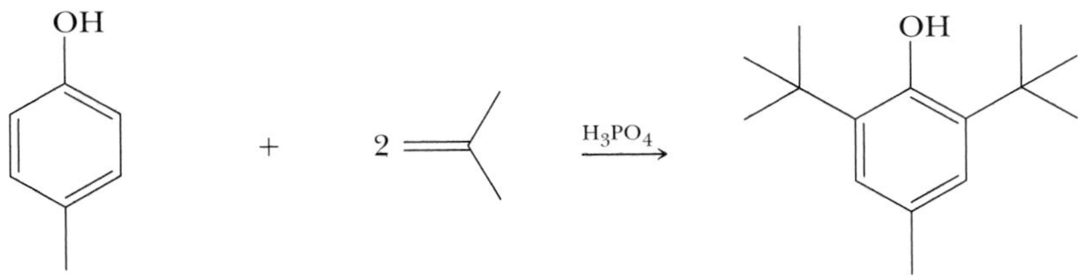

4-Methylphenol   2-Methylpropene   2,6-Di-*tert*-butyl-4-methylphenol
(Butylated hydroxytoluene, BHT)

Propose a mechanism for this reaction.

9.33 The first herbicide widely used for controlling weeds was 2,4-dichlorophenoxyacetic acid (2,4-D). Show how this compound might be synthesized from 2,4-dichlorophenol and chloroacetic acid, $ClCH_2COOH$:

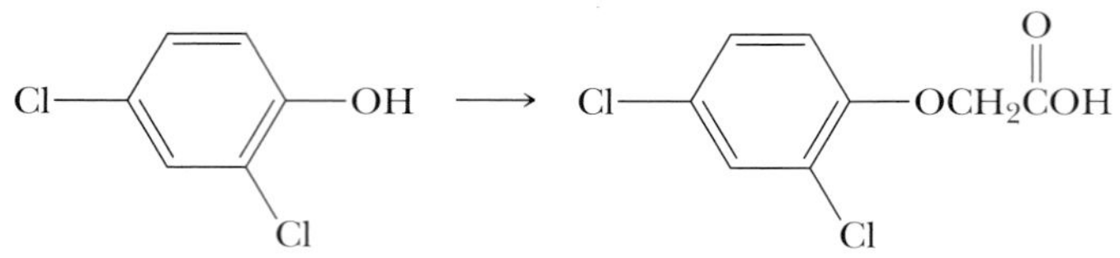

2,4-Dichlorophenol → 2,4-Dichlorophenoxyacetic acid (2,4-D)

Acidity of Phenols

9.34 Use the resonance theory to account for the fact that phenol (pK_a 9.95) is a stronger acid than cyclohexanol (pK_a 18).

9.35 Arrange the compounds in each set in order of increasing acidity (from least acidic to most acidic):

(a) 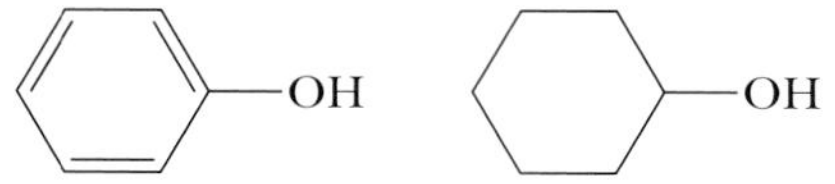 [phenol]—OH   [cyclohexanol]—OH CH_3COOH

(b) 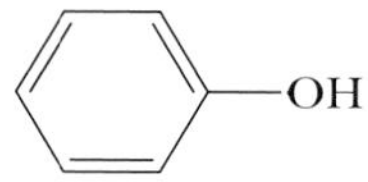 [phenol]—OH $NaHCO_3$ H_2O

(c) 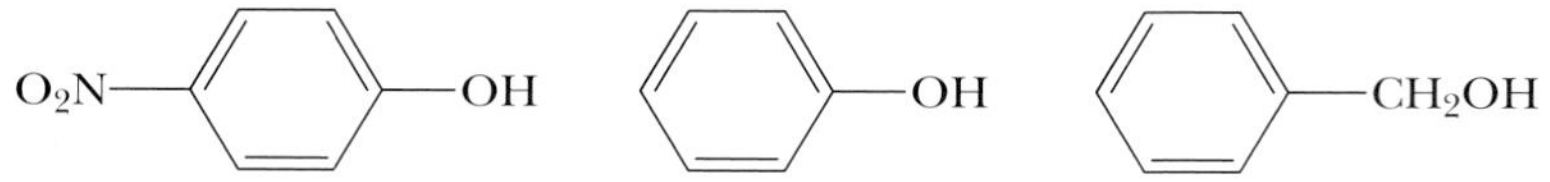 O_2N—[benzene]—OH   [benzene]—OH   [benzene]—CH_2OH

9.36 From each pair, select the stronger base:

(a) 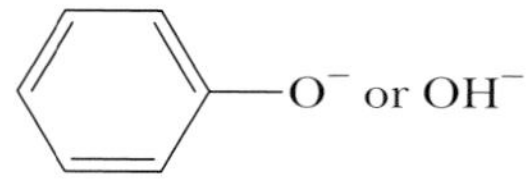 [phenyl]—O^- or OH^- (b) 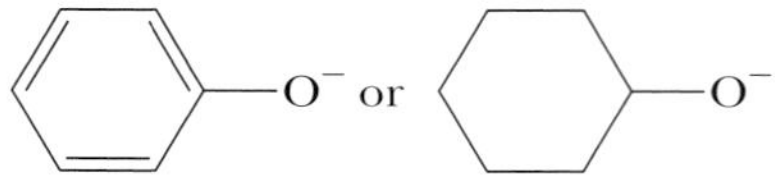 [phenyl]—O^- or [cyclohexyl]—O^-

(c) 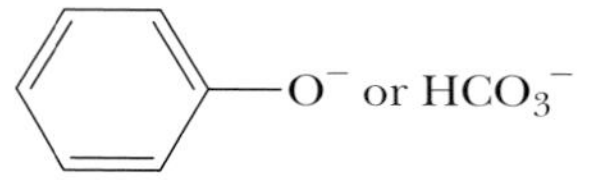 [phenyl]—O^- or HCO_3^- (d) 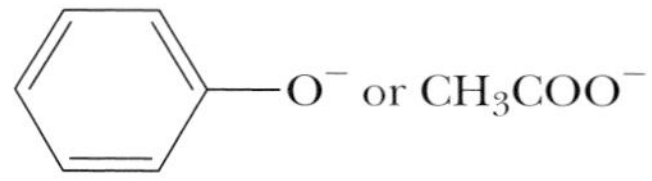 [phenyl]—O^- or CH_3COO^-

9.37 Account for the fact that water-insoluble carboxylic acids (pK_a 4–5) dissolve in 10% sodium bicarbonate with the evolution of a gas, but water-insoluble phenols (pK_a 9.5–10.5) do not show this chemical behavior.

9.38 Describe a procedure for separating a mixture of 1-hexanol and 2-methylphenol (*o*-cresol) and recovering each in pure form. Each is insoluble in water, but soluble in diethyl ether.

Syntheses

9.39 Using styrene, $C_6H_5CH=CH_2$, as the only aromatic starting material, show how to synthesize these compounds. In addition to styrene, use any other necessary organic or inorganic chemicals. Any compound synthesized in one part of this problem may be used to make any other compound in the problem:

(a) 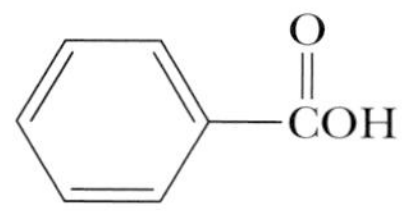 [phenyl]—COOH   (b) 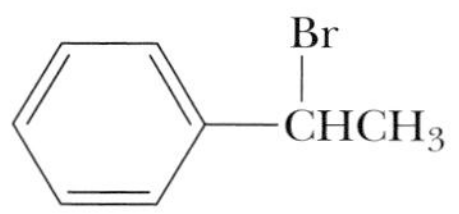 [phenyl]—CHCH$_3$ (Br)   (c) 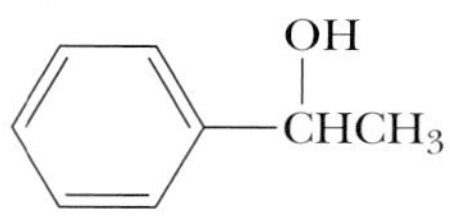 [phenyl]—CHCH$_3$ (OH)

(d) 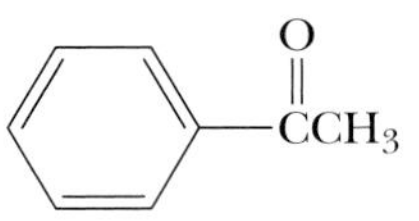 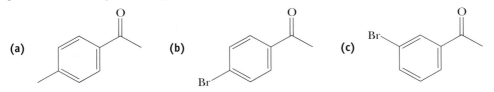 **(e)** 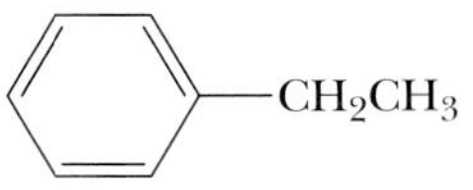 —CH$_2$CH$_3$ **(f)**

(d) structure: benzene ring with —C(=O)CH$_3$

(e) benzene ring —CH$_2$CH$_3$

(f) benzene ring with OH on —CHCH$_2$OH

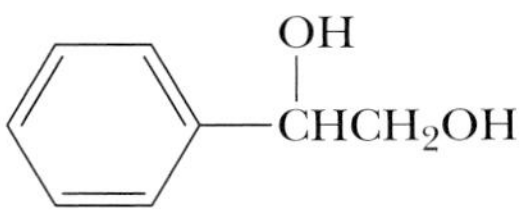

9.40 Show how to synthesize these compounds, starting with benzene, toluene, or phenol as the only sources of aromatic rings. Assume that, in all syntheses, you can separate mixtures of ortho–para products to give the desired isomer in pure form:
(a) *m*-Bromonitrobenzene
(b) 1-Bromo-4-nitrobenzene
(c) 2,4,6-Trinitrotoluene (TNT)
(d) *m*-Bromobenzoic acid
(e) *p*-Bromobenzoic acid
(f) *p*-Dichlorobenzene
(g) *m*-Nitrobenzenesulfonic acid
(h) 1-Chloro-3-nitrobenzene

9.41 Show how to synthesize these aromatic ketones, starting with benzene or toluene as the only sources of aromatic rings. Assume that, in all syntheses, mixtures of ortho–para products can be separated to give the desired isomer in pure form:

(a) **(b)** 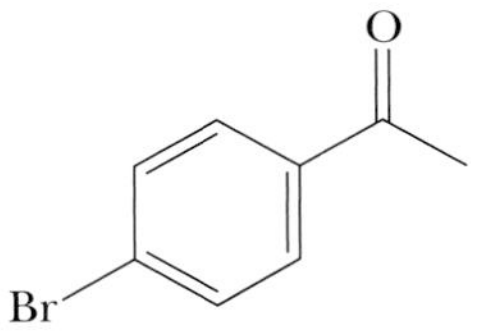 **(c)**

9.42 The following ketone, isolated from the roots of several members of the iris family, has an odor like that of violets and is used as a fragrance in perfumes. Describe the synthesis of this ketone from benzene.

4-Isopropylacetophenone

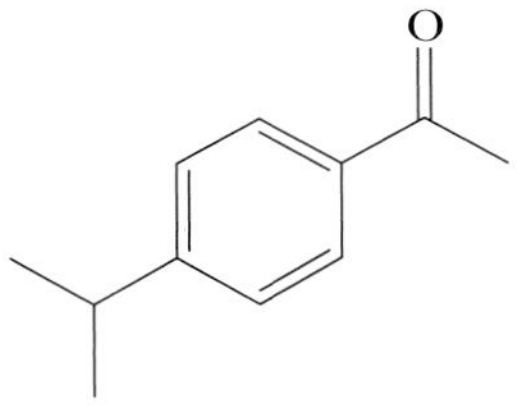

9.43 The bombardier beetle generates *p*-quinone, an irritating chemical, by the enzyme-catalyzed oxidation of hydroquinone, using hydrogen peroxide as the oxidizing agent. Heat generated in this oxidation produces superheated steam, which is ejected, along with *p*-quinone, with explosive force.

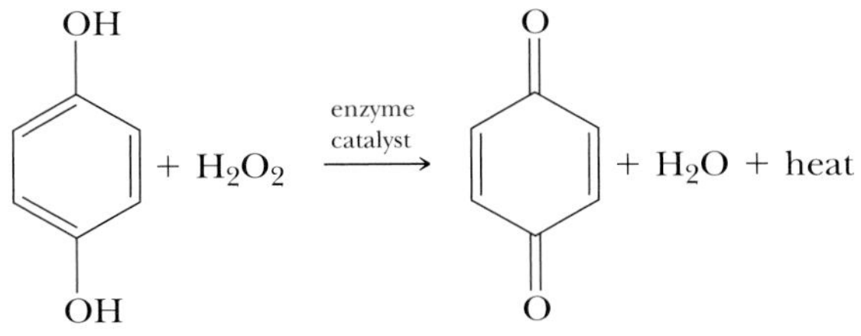

OH + H$_2$O$_2$ $\xrightarrow{\text{enzyme catalyst}}$ O + H$_2$O + heat

Hydroquinone *p*-Quinone

(a) Balance the equation.
(b) Show that this reaction of hydroquinone is an oxidation.

9.44 Following is a structural formula for musk ambrette, a synthetic musk used in perfumes to enhance and retain fragrance:

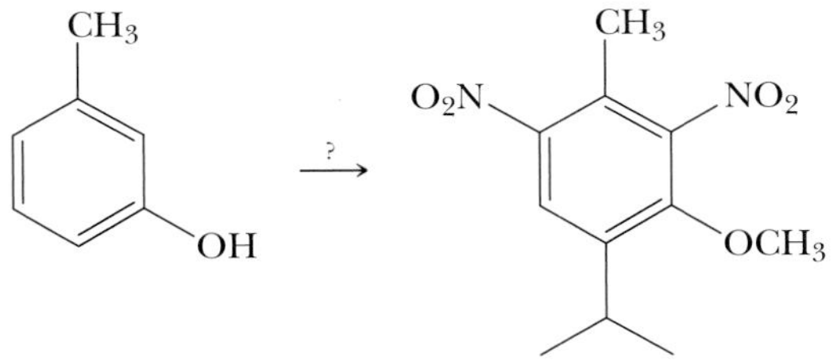

CH$_3$ ... O$_2$N ... NO$_2$ ... OCH$_3$

m-Cresol Musk ambrette

Propose a synthesis for musk ambrette from *m*-cresol.

9.45 1-(3-Chlorophenyl)propanone is a building block in the synthesis of bupropion, the hydrochloride salt of which is the antidepressant Wellbutrin. During clinical trials, researchers discovered that smokers reported a lessening in their craving for tobacco after one to two weeks on the drug. Further clinical trials confirmed this finding, and the drug is also marketed under the trade name Zyban® as an aid in smoking cessation. Propose a synthesis for this building block from benzene. (We will see in Section 13.9 how to complete the synthesis of bupropion.)

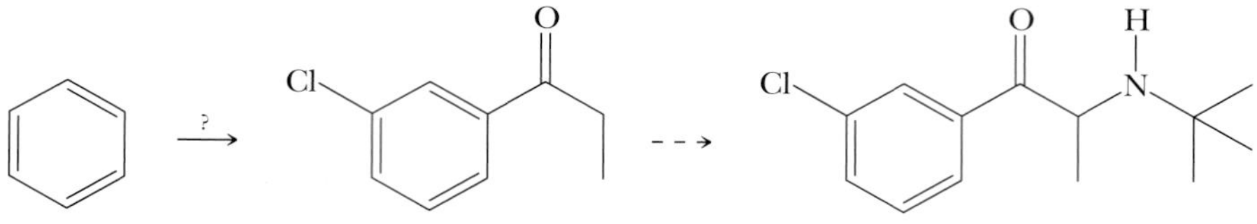

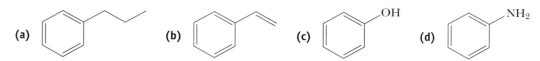

Benzene   1-(3-Chlorophenyl)-1-   Bupropion
                            propanone   (Wellbutrin, Zyban)

Looking Ahead

9.46 Which of the following compounds can be made directly by using an electrophilic aromatic substitution reaction?

(a)   (b)   (c) ⏜OH   (d) ⏜NH₂

9.47 Which compound is a better nucleophile?

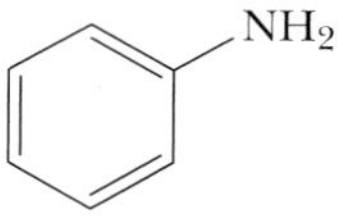

⏜NH₂   or   ⏜NH₂

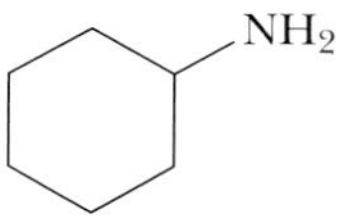

Aniline   Cyclohexanamine

9.48 Suggest a reason that the following arenes do not undergo electrophilic aromatic substitution when AlCl₃ is used in the reaction:

(a) 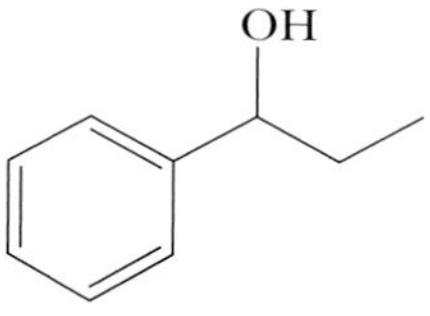 OH   (b) 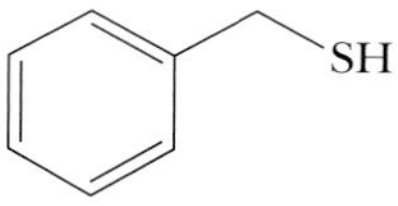 SH   (c) 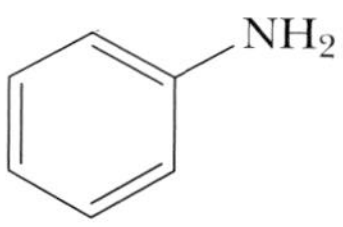 NH₂

9.49 Predict the product of the following acid–base reaction:

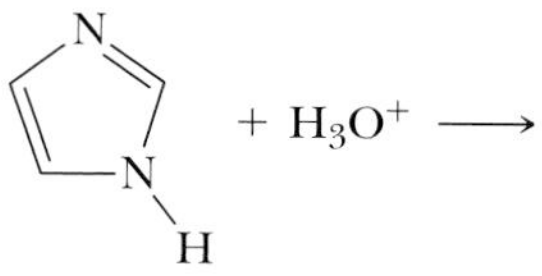

$$+ \; H_3O^+ \longrightarrow$$

9.50 Which haloalkane reacts faster in an S_N1 reaction?

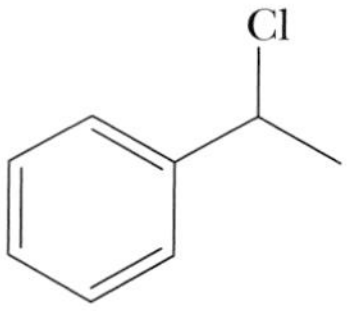

Cl   or   Cl

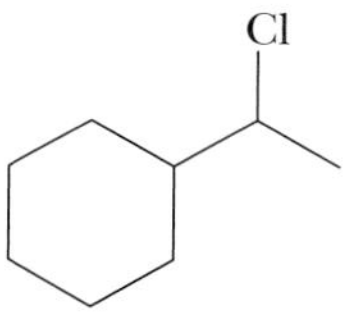

10

Amines

This inhaler delivers puffs of albuterol (Proventil), a potent synthetic bronchodilator whose structure is patterned after that of epinephrine (adrenaline). See Problem 10.13. Inset: A model of morphine. *(Mark Clarke/Photo Researchers, Inc.)*

10.1 INTRODUCTION

Carbon, hydrogen, and oxygen are the three most common elements in organic compounds. Because of the wide distribution of amines in the biological world, nitrogen is the fourth most common component of organic compounds. The most important chemical property of amines is their basicity and their nucleophilicity.

10.2 STRUCTURE AND CLASSIFICATION

Amines are derivatives of ammonia in which one or more hydrogens are replaced by alkyl or aryl groups. Amines are classified as primary (1°), secondary (2°), or

CHEMICAL CONNECTIONS 10A

Morphine as a Clue in the Design and Discovery of Drugs

The analgesic, soporific, and euphoriant properties of the dried juice obtained from unripe seed pods of the opium poppy *Papaver somniferum* have been known for centuries. By the beginning of the 19th century, the active principal, morphine, had been isolated and its structure determined:

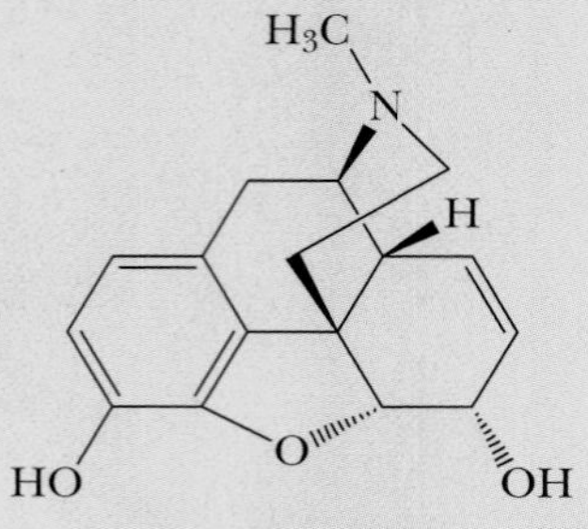

Morphine

Also occurring in the opium poppy is codeine, a monomethyl ether of morphine. Heroin is synthesized by treating morphine with two moles of acetic anhydride:

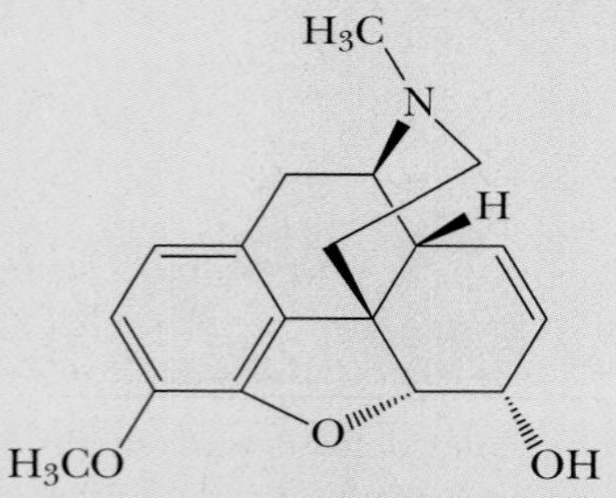

Codeine

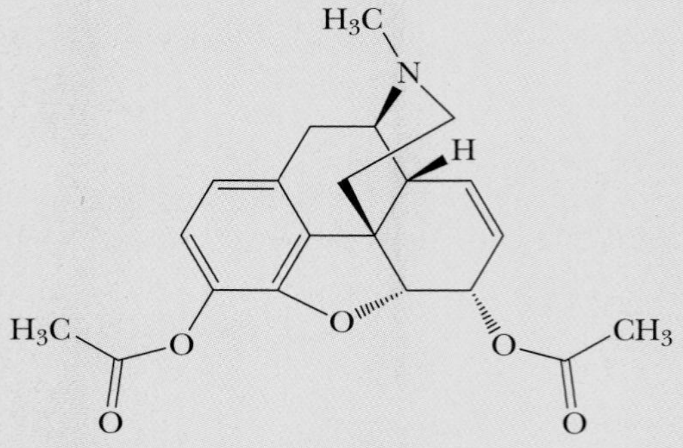

Heroin

Even though morphine is one of modern medicine's most effective painkillers, it has two serious side effects: It is addictive, and it depresses the respiratory control center of the central nervous system. Large doses of morphine (or heroin) can lead to death by respiratory failure. For these reasons, chemists have sought to produce painkillers related in structure to morphine, but without these serious side disadvantages. One strategy in this ongoing research has been to synthesize compounds related in structure to morphine, in the hope that they would be equally effective analgesics, but with diminished side effects. Following are structural formulas for two such compounds that have proven to be clinically useful:

tertiary (3°), depending on the number of hydrogen atoms of ammonia that are replaced by alkyl or aryl groups (Section 1.8B):

$:NH_3$ $CH_3—\overset{..}{N}H_2$ $CH_3—\overset{..}{N}H$ $CH_3—\overset{..}{N}—CH_3$
$\qquad\qquad\qquad\qquad\qquad\qquad\qquad\quad | \qquad\qquad\qquad |$
$\qquad\qquad\qquad\qquad\qquad\qquad\quad CH_3 \qquad\qquad CH_3$

Ammonia   Methylamine   Dimethylamine   Trimethylamine
$\qquad\qquad$(a 1° amine)   (a 2° amine)   (a 3° amine)

Aliphatic amine An amine in which nitrogen is bonded only to alkyl groups.

Amines are further divided into aliphatic amines and aromatic amines. In an **aliphatic amine**, all the carbons bonded directly to nitrogen are derived from alkyl

(−)-enantiomer = Levomethorphan
(+)-enantiomer = Dextromethorphan

Meperidine
(Demerol)

Levomethorphan is a potent analgesic. Interestingly, its dextrorotatory enantiomer, dextromethorphan, has no analgesic activity. It does, however, show approximately the same cough-suppressing activity as morphine and is used extensively in cough remedies.

It has been discovered that there can be even further simplification in the structure of morphine-like analgesics. One such simplification is represented by meperidine, the hydrochloride salt of which is the widely used analgesic Demerol®.

It was hoped that meperidine and related synthetic drugs would be free of many of the morphine-like undesirable side effects. It is now clear, however, that they are not. Meperidine, for example, is definitely addictive. In spite of much determined research, there are as yet no agents as effective as morphine for the relief of severe pain that are absolutely free of the risk of addiction.

How and in what regions of the brain does morphine act? In 1979, scientists discovered that there are specific receptor sites for morphine and other opiates and that these sites are clustered in the brain's limbic system, the area involved in emotion and the perception of pain. Scientists then asked, Why does the human brain have receptor sites specific for morphine? Could it be that the brain produces its own opiates? In 1974, scientists discovered that opiate-like compounds are indeed present in the brain; in 1975, they isolated a brain opiate that was named *enkephalin*, meaning "in the brain." Scientists have yet to understand the role of these natural brain opiates. Perhaps when we do understand their biochemistry, we will discover clues that will lead to the design and synthesis of more potent, but less addictive, analgesics.

groups; in an **aromatic amine**, one or more of the groups bonded directly to nitrogen are aryl groups:

Aniline
(a 1° aromatic amine)

N-Methylaniline
(a 2° aromatic amine)

Benzyldimethylamine
(a 3° aliphatic amine)

An amine in which the nitrogen atom is part of a ring is classified as a **heterocyclic amine**. When the nitrogen is part of an aromatic ring (Section 9.3), the

Aromatic amine An amine in which nitrogen is bonded to one or more aryl groups.

Heterocyclic amine An amine in which nitrogen is one of the atoms of a ring.

Heterocyclic aromatic amine An amine in which nitrogen is one of the atoms of an aromatic ring.

amine is classified as a **heterocyclic aromatic amine**. Following are structural formulas for two heterocyclic aliphatic amines and two heterocyclic aromatic amines:

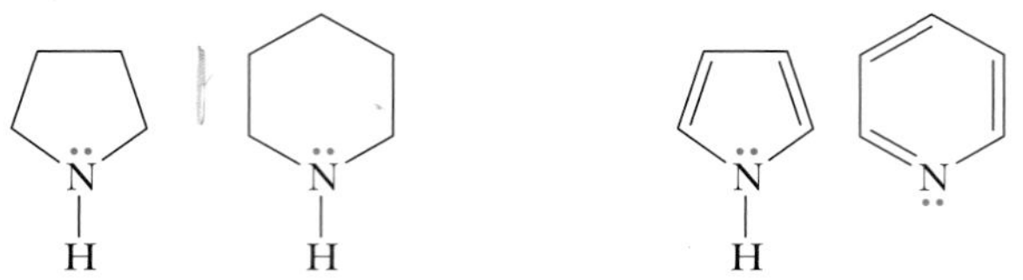

Pyrrolidine Piperidine
(heterocyclic aliphatic amines)

Pyrrole Pyridine
(heterocyclic aromatic amines)

EXAMPLE 10.1

Alkaloids are basic nitrogen-containing compounds of plant origin, many of which have physiological activity when administered to humans. The ingestion of coniine, present in water hemlock, can cause weakness, labored respiration, paralysis, and, eventually, death. Coniine was the toxic substance in "poison hemlock" that caused the death of Socrates. In small doses, nicotine is an addictive stimulant. In larger doses, it causes depression, nausea, and vomiting. In still larger doses, it is a deadly poison. Solutions of nicotine in water are used as insecticides. Cocaine is a central nervous system stimulant obtained from the leaves of the coca plant. Classify each amino group in these alkaloids according to type (that is, primary, secondary, tertiary, heterocyclic, aliphatic, or aromatic):

(a) (S)-Coniine

(b) (S)-Nicotine

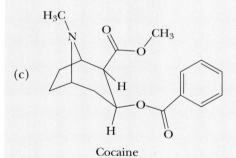

(c) Cocaine

SOLUTION

(a) A secondary heterocyclic aliphatic amine.
(b) One tertiary heterocyclic aliphatic amine and one heterocyclic aromatic amine.
(c) A tertiary heterocyclic aliphatic amine.

Practice Problem 10.1 ─────────────────────

Identify all carbon stereocenters in coniine, nicotine, and cocaine.

10.3 NOMENCLATURE

A. Systematic Names

Systematic names for aliphatic amines are derived just as they are for alcohols. The suffix -e of the parent alkane is dropped and is replaced by -amine; that is, they are named alkanamines:

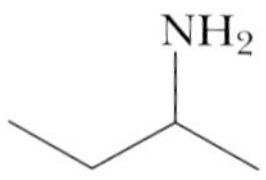

2-Butanamine 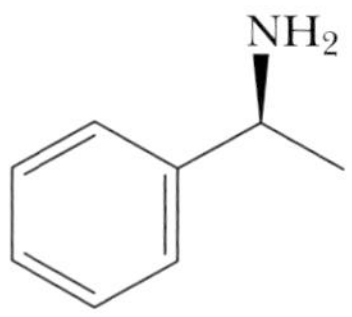   (S)-1-Phenylethanamine   1,6-Hexanediamine

---

EXAMPLE 10.2

Write the IUPAC name for each amine:

(a) 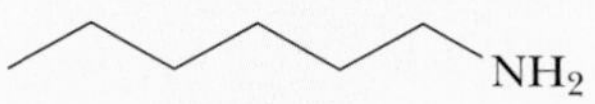   (b)

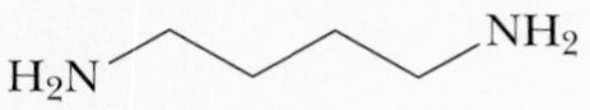

(c)

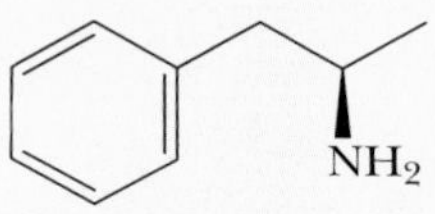

SOLUTION

(a) 1-Hexanamine
(b) 1,4-Butanediamine
(c) The systematic name of this compound is (S)-1-phenyl-2-propanamine. Its common name is amphetamine. The dextrorotatory isomer of amphetamine (shown here) is a central nervous system stimulant and is manufactured and sold under several trade names. The salt with sulfuric acid is marketed as Dexedrine® sulfate.

Practice Problem 10.2

Write a structural formula for each amine:

(a) 2-Methyl-1-propanamine   (b) Cyclohexanamine   (c) (R)-2-Butanamine

---

IUPAC nomenclature retains the common name **aniline** for $C_6H_5NH_2$, the simplest aromatic amine. Its simple derivatives are named with the prefixes o-, m-, and p-, or numbers to locate substituents. Several derivatives of aniline have common names that are still widely used. Among these are **toluidine**, for a methyl-substituted aniline, and **anisidine**, for a methoxy-substituted aniline:

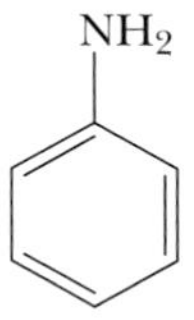

Aniline    4-Nitroaniline   4-Methylaniline   3-Methoxyaniline
                 (p-Nitroaniline)   (p-Toluidine)   (m-Anisidine)

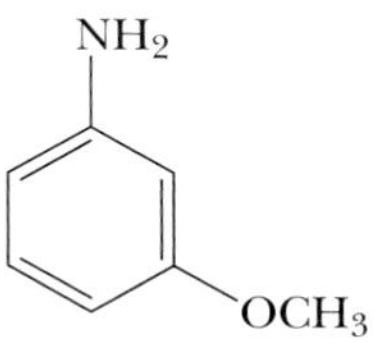

Secondary and tertiary amines are commonly named as *N*-substituted primary amines. For unsymmetrical amines, the largest group is taken as the parent amine; then the smaller group or groups bonded to nitrogen are named, and their location is indicated by the prefix *N* (indicating that they are attached to nitrogen):

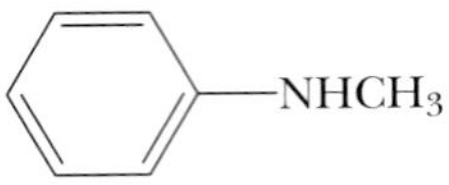

N-Methylaniline 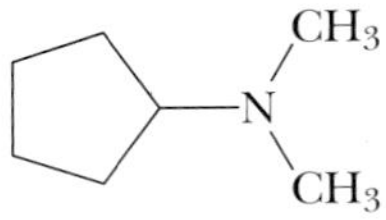   *N,N*-Dimethyl-
cyclopentanamine

Following are names and structural formulas for four heterocyclic aromatic amines, the common names of which have been retained by the IUPAC:

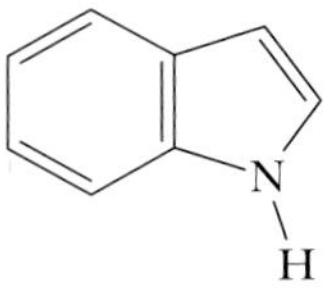

Indole 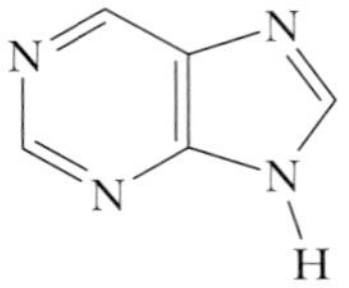   Purine 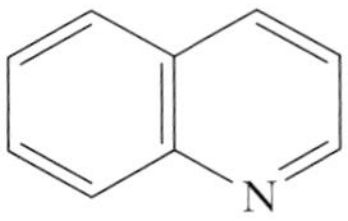   Quinoline 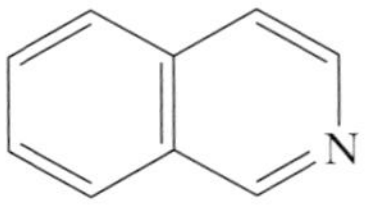   Isoquinoline

Among the various functional groups discussed in this text, the $-NH_2$ group has one of the lowest priorities. The following compounds each contain a functional group of higher precedence than the amino group, and, accordingly, the amino group is indicated by the prefix *amino-*:

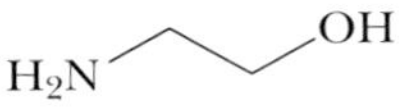

2-Aminoethanol   2-Aminobenzoic acid
(Ethanolamine)   (Anthranilic acid)

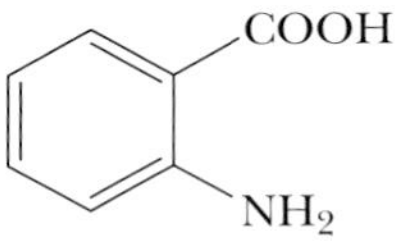

B. Common Names

Common names for most aliphatic amines are derived by listing the alkyl groups bonded to nitrogen in alphabetical order in one word ending in the suffix *-amine*; that is, they are named as **alkylamines**:

CH_3NH_2

Methylamine   *tert*-Butylamine 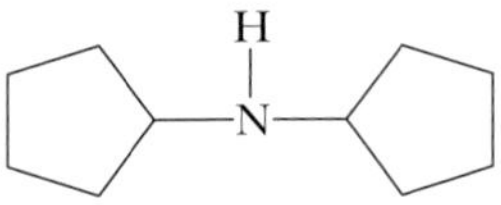   Dicyclopentylamine 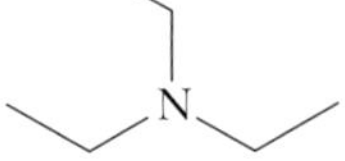   Triethylamine

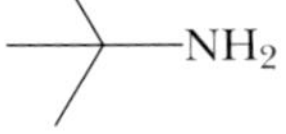

EXAMPLE 10.3

Write a structural formula for each amine:

(a) Isopropylamine   (b) Cyclohexylmethylamine   (c) Benzylamine

SOLUTION

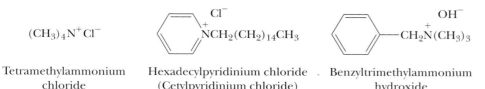

(a) $(CH_3)_2CHNH_2$ (b) ⬡—NHCH$_3$ (c) ⬡—CH$_2$NH$_2$

Practice Problem 10.3 ──────────────────────────────

Write a structural formula for each amine:

(a) Isobutylamine   (b) Triphenylamine   (c) Diisopropylamine

When four atoms or groups of atoms are bonded to a nitrogen atom, we name the compound as a salt of the corresponding amine. We replace the ending *-amine* (or aniline, pyridine, or the like) by *-ammonium* (or *anilinium, pyridinium,* or the like) and add the name of the anion (chloride, acetate, and so on). Compounds containing such ions have properties characteristic of salts. Following are three examples (cetylpyridinium chloride is used as a topical antiseptic and disinfectant):

Several over-the-counter mouthwashes contain *N*-alkylatedpyridinium chlorides as an antibacterical agent. *(Charles D. Winters)*

$(CH_3)_4N^+Cl^-$

[pyridinium ring] $\overset{+}{N}CH_2(CH_2)_{14}CH_3$ Cl$^-$

⬡—$CH_2\overset{+}{N}(CH_3)_3$ OH$^-$

Tetramethylammonium chloride

Hexadecylpyridinium chloride (Cetylpyridinium chloride)

Benzyltrimethylammonium hydroxide

10.4 PHYSICAL PROPERTIES

Amines are polar compounds, and both primary and secondary amines form intermolecular hydrogen bonds (Figure 10.1).

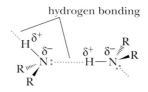

Figure 10.1
Intermolecular association of 1° and 2° amines by hydrogen bonding. Nitrogen is approximately tetrahedral in shape, with the axis of the hydrogen bond along the fourth position of the tetrahedron.

An N—H----N hydrogen bond is weaker than an O—H----O hydrogen bond, because the difference in electronegativity between nitrogen and hydrogen ($3.0 - 2.1 = 0.9$) is less than that between oxygen and hydrogen ($3.5 - 2.1 = 1.4$). We can illustrate the effect of intermolecular hydrogen bonding by comparing the boiling points of methylamine and methanol:

	CH_3NH_2	CH_3OH
molecular weight (g/mol)	31.1	32.0
boiling point (°C)	−6.3	65.0

Both compounds have polar molecules and interact in the pure liquid by hydrogen bonding. Methanol has the higher boiling point because hydrogen bonding between its molecules is stronger than that between molecules of methylamine.

CHEMICAL   CONNECTIONS   10B

The Poison Dart Frogs of South America: Lethal Amines

The Noanamá and Embrá peoples of the jungles of western Colombia have used poison blow darts for centuries, perhaps millennia. The poisons are obtained from the skin secretions of several highly colored frogs of the genus *Phyllobates* (*neará* and *kokoi* in the language of the native peoples). A single frog contains enough poison for up to 20 darts. For the most poisonous species (*Phyllobates terribilis*), just rubbing a dart over the frog's back suffices to charge the dart with poison.

Scientists at the National Institutes of Health became interested in studying these poisons when it was discovered that they act on cellular ion channels, which would make them useful tools in basic research on mechanisms of ion transport. A field station was established in western Colombia to collect the relatively common poison dart frogs. From 5,000 frogs, 11 mg of batrachotoxin and batrachotoxinin A was isolated. These names are derived from *batrachos*, the Greek word for frog.

Batrachotoxin and batrachotoxinin A are among the most lethal poisons ever discovered:

Batrachotoxin

Batrachotoxinin A

It is estimated that as little as 200 μg of batrachotoxin is sufficient to induce irreversible cardiac arrest in a human being. It has been determined that they act by causing voltage-gated Na^+ channels in nerve and muscle cells to be blocked in the open position, which leads to a huge influx of Na^+ ions into the affected cell.

The batrachotoxin story illustrates several common themes in the discovery of new drugs. First, information about the kinds of biologically active compounds and their sources are often obtained from the native peoples of a region. Second, tropical rain forests are a rich source of structurally complex, biologically active substances. Third, an entire ecosystem, not only the plants, is a potential source of fascinating organic molecules.

Poison dart frog, *Phyllobates terribilis*. (*Juan M. Renjifo/ Animals, Animals*)

All classes of amines form hydrogen bonds with water and are more soluble in water than are hydrocarbons of comparable molecular weight. Most low-molecular-weight amines are completely soluble in water (Table 10.1). Higher-molecular-weight amines are only moderately soluble or insoluble.

TABLE 10.1 Physical Properties of Selected Amines

Name	Structural Formula	Melting Point (°C)	Boiling Point (°C)	Solubility in Water
Ammonia	NH_3	−78	−33	very soluble
Primary Amines				
methylamine	CH_3NH_2	−95	−6	very soluble
ethylamine	$CH_3CH_2NH_2$	−81	17	very soluble
propylamine	$CH_3CH_2CH_2NH_2$	−83	48	very soluble
butylamine	$CH_3(CH_2)_3NH_2$	−49	78	very soluble
benzylamine	$C_6H_5CH_2NH_2$	10	185	very soluble
cyclohexylamine	$C_6H_{11}NH_2$	−17	135	slightly soluble
Secondary Amines				
dimethylamine	$(CH_3)_2NH$	−93	7	very soluble
diethylamine	$(CH_3CH_2)_2NH$	−48	56	very soluble
Tertiary Amines				
trimethylamine	$(CH_3)_3N$	−117	3	very soluble
triethylamine	$(CH_3CH_2)_3N$	−114	89	slightly soluble
Aromatic Amines				
aniline	$C_6H_5NH_2$	−6	184	slightly soluble
Heterocyclic Aromatic Amines				
pyridine	C_5H_5N	−42	116	very soluble

10.5 BASICITY OF AMINES

Like ammonia, all amines are weak bases, and aqueous solutions of amines are basic. The following acid–base reaction between an amine and water is written using curved arrows to emphasize that, in this proton-transfer reaction, the unshared pair of electrons on nitrogen forms a new covalent bond with hydrogen and displaces hydroxide ion:

$$CH_3-\underset{\underset{H}{|}}{\overset{\overset{H}{|}}{N}}: + H-\overset{..}{\underset{..}{O}}-H \rightleftharpoons CH_3-\underset{\underset{H}{|}}{\overset{\overset{H}{|}}{N}}{}^+-H \quad :\overset{..}{\underset{..}{O}}-H$$

Methylamine   Methylammonium
                       hydroxide

The equilibrium constant for the reaction of an amine with water, K_{eq}, has the following form, illustrated for the reaction of methylamine with water to give methylammonium hydroxide:

$$K_{eq} = \frac{[CH_3NH_3{}^+][OH^-]}{[CH_3NH_2][H_2O]}$$

Because the concentration of water in dilute solutions of methylamine in water is essentially a constant ($[H_2O] = 55.5$ mol/L), it is combined with K_{eq} in a new constant called a *base ionization constant*, K_b. The value of K_b for methylamine is 4.37×10^{-4} ($pK_b = 3.36$):

$$K_b = K_{eq}[H_2O] = \frac{[CH_3NH_3{}^+][OH^-]}{[CH_3NH_2]} = 4.37 \times 10^{-4}$$

It is also common to discuss the basicity of amines by referring to the acid ionization constant of the corresponding conjugate acid, as illustrated for the ionization of the methylammonium ion:

$$CH_3NH_3^+ + H_2O \rightleftharpoons CH_3NH_2 + H_3O^+$$

$$K_a = \frac{[CH_3NH_2][H_3O^+]}{[CH_3NH_3^+]} = 2.29 \times 10^{-11} \qquad pK_a = 10.64$$

Values of pK_a and pK_b for any acid–conjugate base pair are related by the equation

$$pK_a + pK_b = 14.00$$

Values of pK_a and pK_b for selected amines are given in Table 10.2.

TABLE 10.2 Base Strengths (pK_b) of Selected Amines and Acid Strengths (pK_a) of Their Conjugate Acids*

Amine	Structure	pK_b	pK_a
Ammonia	NH_3	4.74	9.26
Primary Amines			
methylamine	CH_3NH_2	3.36	10.64
ethylamine	$CH_3CH_2NH_2$	3.19	10.81
cyclohexylamine	$C_6H_{11}NH_2$	3.34	10.66
Secondary Amines			
dimethylamine	$(CH_3)_2NH$	3.27	10.73
diethylamine	$(CH_3CH_2)_2NH$	3.02	10.98
Tertiary Amines			
trimethylamine	$(CH_3)_3N$	4.19	9.81
triethylamine	$(CH_3CH_2)_3N$	3.25	10.75
Aromatic Amines			
aniline	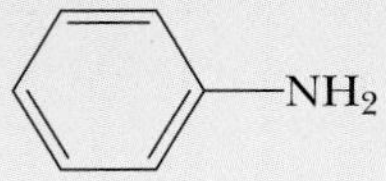 C6H5—NH2	9.37	4.63
4-methylaniline	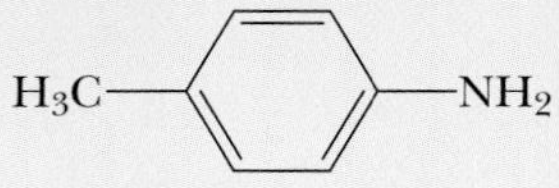 H3C—C6H4—NH2	8.92	5.08
4-chloroaniline	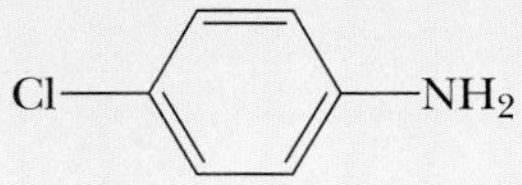 Cl—C6H4—NH2	9.85	4.15
4-nitroaniline	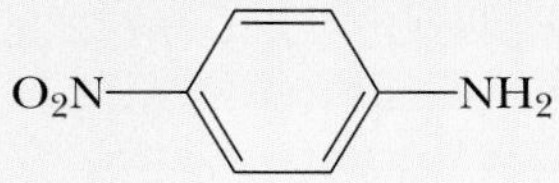 O2N—C6H4—NH2	13.0	1.0
Heterocyclic Aromatic Amines			
pyridine	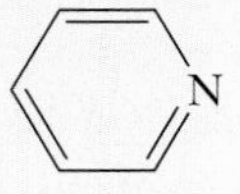 (pyridine ring, N)	8.75	5.25
imidazole	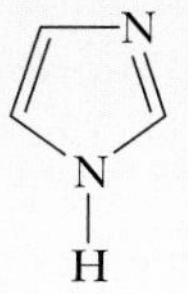 (imidazole ring, N–H)	7.05	6.95

*For each amine, $pK_a + pK_b = 14.00$.

EXAMPLE 10.4

Predict the position of equilibrium for this acid–base reaction:

$$CH_3NH_2 + CH_3COOH \rightleftharpoons CH_3NH_3^+ + CH_3COO^-$$

SOLUTION

Use the approach we developed in Section 2.5 to predict the position of equilibrium in acid–base reactions. Equilibrium favors reaction of the stronger acid and stronger base to form the weaker acid and the weaker base. Thus, in this reaction, equilibrium favors the formation of methylammonium ion and acetate ion:

$$CH_3NH_2 + CH_3COOH \rightleftharpoons CH_3NH_3^+ + CH_3COO^-$$

$$pK_a = 4.76 \qquad pK_a = 10.64$$

| Stronger base | Stronger acid | Weaker acid | Weaker base |

Practice Problem 10.4

Predict the position of equilibrium for this acid–base reaction:

$$CH_3NH_3^+ + H_2O \rightleftharpoons CH_3NH_2 + H_3O^+$$

Given information such as that in Table 10.2, we can make the following generalizations about the acid–base properties of the various classes of amines:

1. All aliphatic amines have about the same base strength, pK_b 3.0–4.0, and are slightly stronger bases than ammonia.

2. Aromatic amines and heterocyclic aromatic amines are considerably weaker bases than are aliphatic amines. Compare, for example, values of pK_b for aniline and cyclohexylamine:

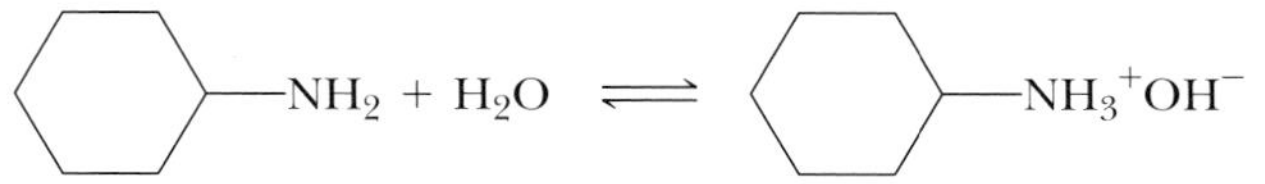

$$pK_b = 3.34$$
$$K_b = 4.5 \times 10^{-4}$$

Cyclohexylamine — Cyclohexylammonium hydroxide

$$pK_b = 9.37$$
$$K_b = 4.3 \times 10^{-10}$$

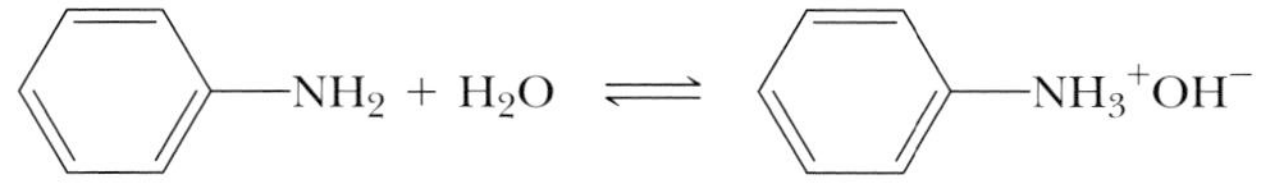

Aniline — Anilinium hydroxide

The base ionization constant for aniline is smaller (the larger the value of pK_b, the weaker is the base) than that for cyclohexylamine by a factor of 10^6.

Aromatic amines are weaker bases than are aliphatic amines because of the resonance interaction of the unshared pair on nitrogen with the pi system of the aromatic ring. Because no such resonance interaction is possible for an alkylamine, the electron pair on its nitrogen is more available for reaction with an acid:

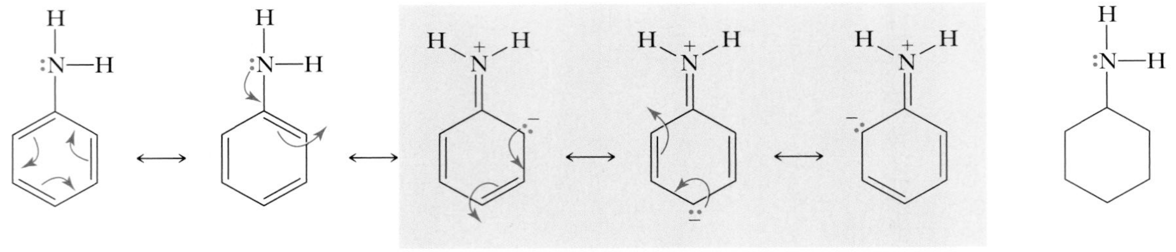

Two Kekulé structures

Interaction of the electron pair on nitrogen with the pi system of the aromatic ring

No resonance is possible with alkylamines

3. Electron-withdrawing groups such as halogen, nitro, and carbonyl decrease the basicity of substituted aromatic amines by decreasing the availability of the electron pair on nitrogen:

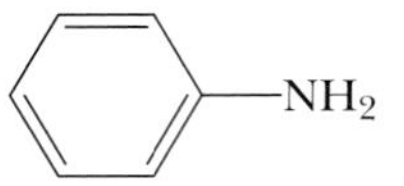

Aniline
pK_b 9.37

4-Nitroaniline
pK_b 13.0

Recall from Section 9.9B that these same substituents increase the acidity of phenols.

EXAMPLE 10.5

Select the stronger base in each pair of amines:

(a) 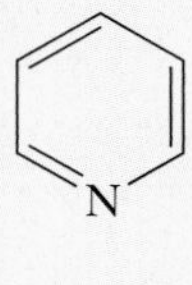   (A)   or 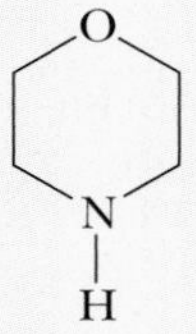   (B) 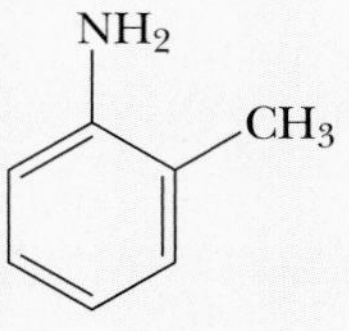   (b)   (C)   or   (D)

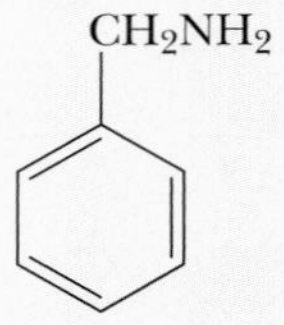

SOLUTION

(a) Morpholine (B) is the stronger base (pK_b 5.79). It has a basicity comparable to that of secondary aliphatic amines. Pyridine (A), a heterocyclic aromatic amine (pK_b 8.75), is considerably less basic than aliphatic amines.
(b) Benzylamine (D), a primary aliphatic amine, is the stronger base (pK_b 3–4). o-Toluidine (C), an aromatic amine, is the weaker base (pK_b 9–10).

Practice Problem 10.5

Select the stronger acid from each pair of ions:

(a) 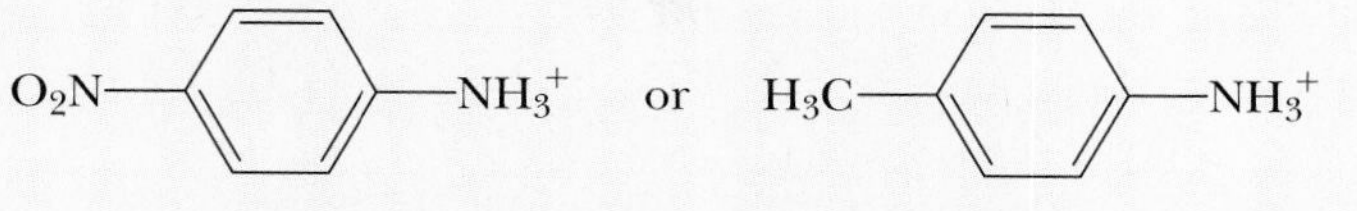 O_2N—〈benzene ring〉—NH_3^+ or H_3C—〈benzene ring〉—NH_3^+

(A)   (B)

(b) 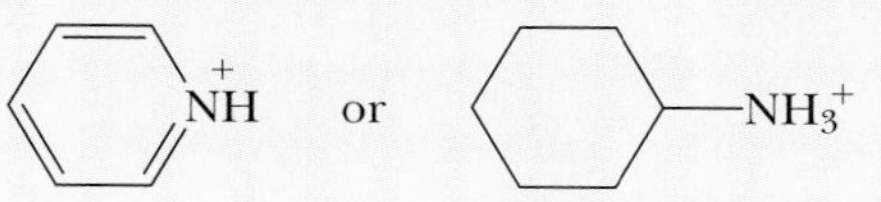 〈pyridinium ring with $\overset{+}{N}H$〉   or   〈cyclohexane ring〉—NH_3^+

(C)   (D)

Guanidine, with pK_b 0.4, is the strongest base among neutral compounds:

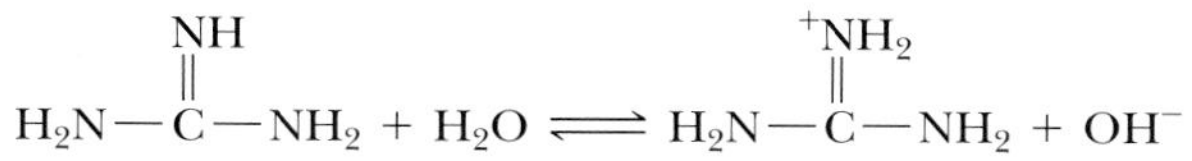

$$\underset{\text{Guanidine}}{H_2N-\overset{\overset{\displaystyle NH}{\|}}{C}-NH_2} + H_2O \rightleftharpoons \underset{\text{Guanidinium ion}}{H_2N-\overset{\overset{\displaystyle ^+NH_2}{\|}}{C}-NH_2} + OH^- \qquad pK_b = 0.4$$

The remarkable basicity of guanidine is attributed to the fact that the positive charge on the guanidinium ion is delocalized equally over the three nitrogen atoms, as shown by these three equivalent contributing structures:

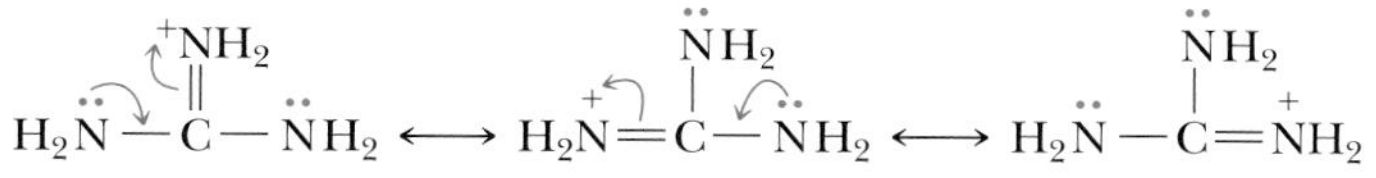

$$H_2\overset{..}{N}-\overset{\overset{\displaystyle ^+NH_2}{\|}}{C}-\overset{..}{N}H_2 \longleftrightarrow H_2\overset{+}{N}=C-\overset{..}{N}H_2 \longleftrightarrow H_2\overset{..}{N}-C=\overset{+}{N}H_2$$

Three equivalent contributing structures

Hence, the guanidinium ion is a highly stable cation. The presence of a guanidine group on the side chain of the amino acid arginine accounts for the basicity of its side chain (Section 19.2A).

10.6 REACTION WITH ACIDS

Amines, whether soluble or insoluble in water, react quantitatively with strong acids to form water-soluble salts, as illustrated by the reaction of (R)-norepinephrine (noradrenaline) with aqueous HCl to form a hydrochloride salt:

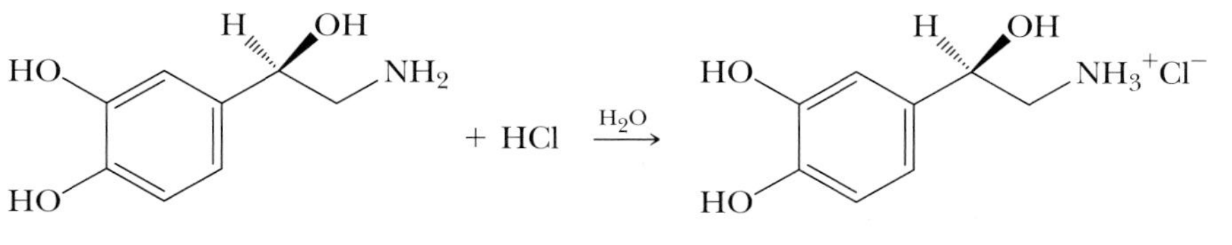

〈structure of (R)-Norepinephrine with HO, HO on benzene ring, H, OH and NH₂ side chain〉 + HCl $\xrightarrow{H_2O}$ 〈structure of (R)-Norepinephrine hydrochloride with NH₃⁺Cl⁻〉

(R)-Norepinephrine
(only slightly soluble in water)

(R)-Norepinephrine hydrochloride
(a water-soluble salt)

Norepinephrine, secreted by the medulla of the adrenal gland, is a neurotransmitter. It has been suggested that it is a neurotransmitter in those areas of the brain which mediate emotional behavior.

EXAMPLE 10.6

Complete each acid–base reaction, and name the salt formed:

(a) $(CH_3CH_2)_2NH + HCl \longrightarrow$

(b) 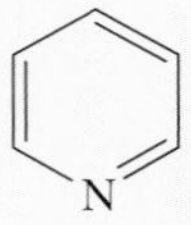 [pyridine] $+ CH_3COOH \longrightarrow$

SOLUTION

(a) $(CH_3CH_2)_2NH_2^+Cl^-$
Diethylammonium chloride

(b) 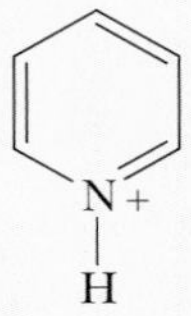 [pyridinium] CH_3COO^-
Pyridinium acetate

Practice Problem 10.6

Complete each acid–base reaction and name the salt formed:

(a) $(CH_3CH_2)_3N + HCl \longrightarrow$

(b) 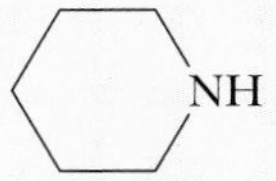 [ring]$NH + CH_3COOH \longrightarrow$

The basicity of amines and the solubility of amine salts in water can be used to separate amines from water-insoluble, nonbasic compounds. Shown in Figure 10.2 is a flowchart for the separation of aniline from anisole. Note that aniline is recovered from its salt by treatment with NaOH.

EXAMPLE 10.7

Following are two structural formulas for alanine (2-aminopropanoic acid), one of the building blocks of proteins (Chapter 19):

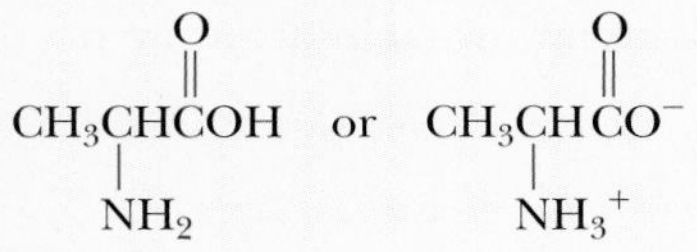

$$CH_3CHCOH \quad or \quad CH_3CHCO^-$$
$$\quad NH_2 \qquad\qquad NH_3^+$$
$$(A) \qquad\qquad (B)$$

Is alanine better represented by structural formula (A) or structural formula (B)?

SOLUTION

Structural formula (A) contains both an amino group (a base) and a carboxyl group (an acid). Proton transfer from the stronger acid (—COOH) to the stronger base (—NH$_2$) gives an internal salt; therefore, (B) is the better representation for alanine. Within the field of amino acid chemistry, the internal salt represented by (B) is called a **zwitterion** (Chapter 19).

Practice Problem 10.7

As shown in Example 10.7, alanine is better represented as an internal salt. Suppose that the internal salt is dissolved in water.

(a) In what way would you expect the structure of alanine in aqueous solution to change if concentrated HCl were added to adjust the pH of the solution to 2.0?

(b) In what way would you expect the structure of alanine in aqueous solution to change if concentrated NaOH were added to bring the pH of the solution to 12.0?

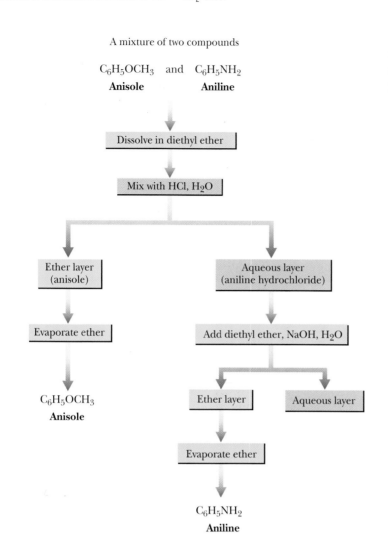

A mixture of two compounds

$C_6H_5OCH_3$ and $C_6H_5NH_2$
Anisole   **Aniline**

Dissolve in diethyl ether

Mix with HCl, H₂O

Ether layer (anisole)

Aqueous layer (aniline hydrochloride)

Evaporate ether

Add diethyl ether, NaOH, H₂O

$C_6H_5OCH_3$
Anisole

Ether layer

Aqueous layer

Evaporate ether

$C_6H_5NH_2$
Aniline

Figure 10.2
Separation and purification of an amine and a neutral compound.

10.7 SYNTHESIS OF ARYLAMINES: REDUCTION OF THE —NO₂ GROUP

As we have already seen (Section 9.7B), the nitration of an aromatic ring introduces a NO₂ group. A particular value of nitration is the fact that the resulting nitro group can be reduced to a primary amino group, —NH₂, by hydrogenation in the presence of a transition metal catalyst such as nickel, palladium, or platinum:

COOH

$+ 3H_2$ $\xrightarrow[\text{(3 atm)}]{\text{Ni}}$

COOH

$+ 2H_2O$

NO₂

NH₂

3-Nitrobenzoic
acid

3-Aminobenzoic
acid

This method has the potential disadvantage that other susceptible groups, such as a carbon–carbon double bond, and the carbonyl group of an aldehyde or ketone, may also be reduced. Note that neither the —COOH nor the aromatic ring is reduced under these conditions.

Alternatively, a nitro group can be reduced to a primary amino group by a metal in acid:

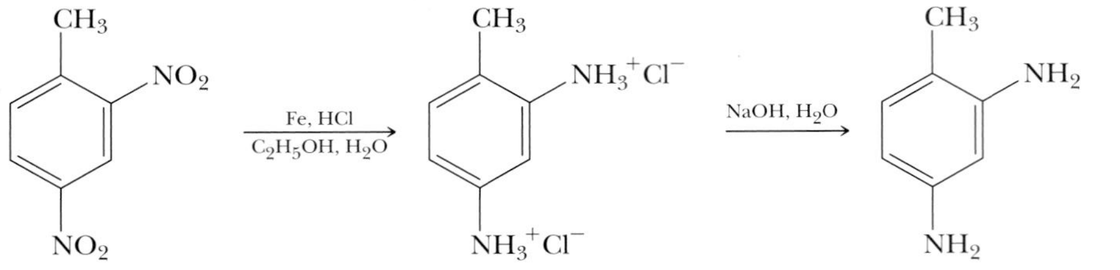

2,4-Dinitrotoluene   2,4-Diaminotoluene

The most commonly used metal-reducing agents are iron, zinc, and tin in dilute HCl. When reduced by this method, the amine is obtained as a salt, which is then treated with a strong base to liberate the free amine.

10.8   REACTION OF PRIMARY AROMATIC AMINES WITH NITROUS ACID

Nitrous acid, HNO_2, is an unstable compound that is prepared by adding sulfuric or hydrochloric acid to an aqueous solution of sodium nitrite, $NaNO_2$. Nitrous acid is a weak acid and ionizes according to the following equation:

$$HNO_2 + H_2O \rightleftharpoons H_3O^+ + NO_2^- \qquad K_a = 4.26 \times 10^{-4}$$

Nitrous $pK_a = 3.37$
acid

Nitrous acid reacts with amines in different ways, depending on whether the amine is primary, secondary, or tertiary and whether it is aliphatic or aromatic. We concentrate on the reaction of nitrous acid with primary aromatic amines, because this reaction is useful in organic synthesis.

Treatment of a primary aromatic amine—for example, aniline—with nitrous acid gives a diazonium salt:

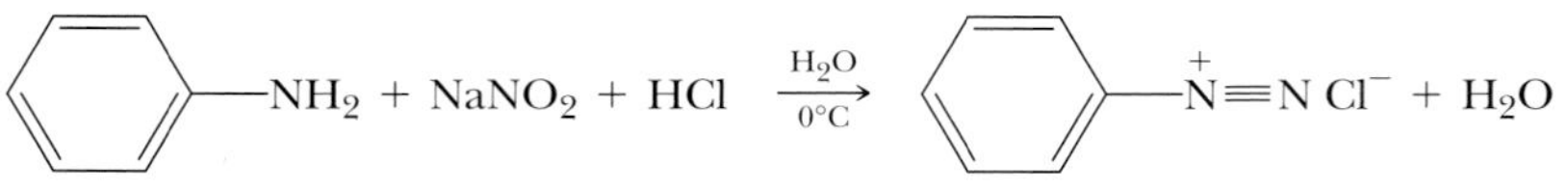

Aniline   Sodiun   Benzenediazonium
(a 1° aromatic   nitrite   chloride
amine)

We can also write the equation for this reaction in the following more abbreviated form:

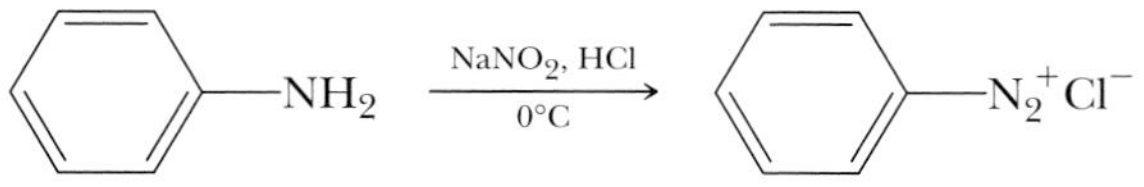

Benzenediazonium
chloride

When we warm an aqueous solution of an arenediazonium salt, the $—N_2^+$ group is replaced by an $—OH$ group. This reaction is one of the few methods we have for the synthesis of phenols. It enables us to convert an aromatic amine to a phenol by first forming the arenediazonium salt and then heating the solution. In this manner, we can convert 2-bromo-4-methylaniline to 2-bromo-4-methylphenol:

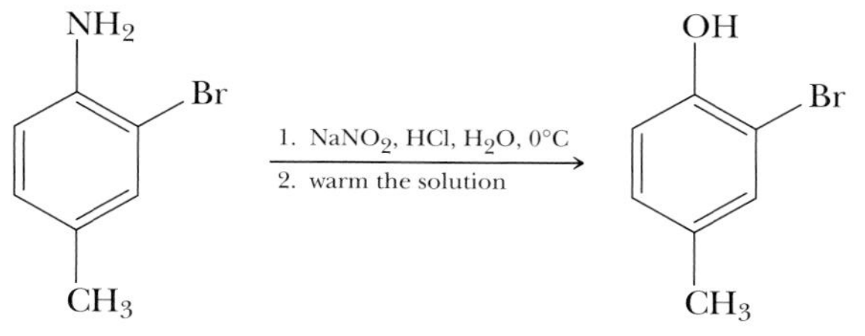

2-Bromo-4-methylaniline → 2-Bromo-4-methylphenol

1. NaNO₂, HCl, H₂O, 0°C
2. warm the solution

EXAMPLE 10.8

Show the reagents that will bring about each step in this conversion of toluene to 4-hydroxybenzoic acid:

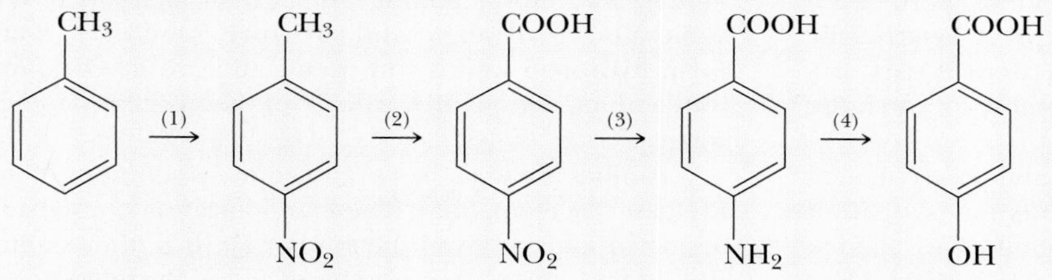

Toluene → (1) → (2) → (3) → (4) → 4-Hydroxy-benzoic acid

SOLUTION

Step 1: Nitration of toluene, using nitric acid/sulfuric acid (Section 9.7B), followed by separation of the ortho and para isomers.

Step 2: Oxidation of the benzylic carbon, using chromic acid (Section 9.5).

Step 3: Reduction of the nitro group, either using H_2 in the presence of a transition metal catalyst or using Fe, Sn, or Zn in the presence of aqueous HCl (Section 10.8).

Step 4: Treatment of the aromatic amine with $NaNO_2/HCl$ to form the diazonium ion salt and then warming the solution.

Practice Problem 10.8

Show how you can use the same set of steps in Example 10.8, but in a different order, to convert toluene to 3-hydroxybenzoic acid.

Treatment of arenediazonium salt with hypophosphorous acid, H_3PO_2, reduces the diazonium group and replaces it with $—H$, as illustrated by the conversion of aniline to 1,3,5-trichlorobenzene. Recall that the $—NH_2$ group is a powerful activating and ortho-para directing group (Section 9.8A). Treatment of aniline with chlorine requires no catalyst and gives 2,4,6-trichloroaniline (to complete the conversion, we treat the trichloroaniline with nitrous acid followed by hypophosphorous acid):

Aniline   1,3,5-Trichloro-aniline

SUMMARY

Amines are classified as **primary, secondary,** or **tertiary,** depending on the number of hydrogen atoms of ammonia replaced by alkyl or aryl groups (Section 10.2). In an **aliphatic amine,** all carbon atoms bonded to nitrogen are derived from alkyl groups. In an **aromatic amine,** one or more of the groups bonded to nitrogen are aryl groups. A **heterocyclic amine** is an amine in which the nitrogen atom is part of a ring. A **heterocyclic aromatic amine** is an amine in which the nitrogen atom is part of an aromatic ring.

In systematic nomenclature, aliphatic amines are named **alkanamines** (Section 10.3A). In the common system of nomenclature (Section 10.3B), aliphatic amines are named **alkylamines**; the alkyl groups are listed in alphabetical order in one word ending in the suffix *-amine.* An ion containing nitrogen bonded to four alkyl or aryl groups is named as a **quaternary ammonium ion.**

Amines are polar compounds, and primary and secondary amines associate by intermolecular hydrogen bonding (Section 10.4). Because an N—H----N hydrogen bond is weaker than an O—H----O hydrogen bond, amines have lower boiling points than alcohols of comparable molecular weight and structure. All classes of amines form hydrogen bonds with water and are more soluble in water than are hydrocarbons of comparable molecular weight.

Amines are weak bases, and aqueous solutions of amines are basic (Section 10.5). The base ionization constant for an amine in water is given the symbol K_b. It is also common to discuss the acid–base properties of amines by reference to the acid ionization constant, K_a, for the conjugate acid of the amine. Acid and base ionization constants for an amine in water are related by the equation $pK_a + pK_b = 14.0$.

KEY REACTIONS

1. Basicity of Aliphatic Amines (Section 10.5)
Most aliphatic amines have comparable basicities (pK_b 3.0–4.0) and are slightly stronger bases than ammonia:

$$CH_3NH_2 + H_2O \rightleftharpoons CH_3NH_3^+ + OH^- \qquad pK_b = 3.36$$

2. Basicity of Aromatic Amines (Section 10.5)
Aromatic amines (pK_b 9.0–10.0) are considerably weaker bases than are aliphatic amines. Resonance stabilization from interaction of the unshared electron pair on nitrogen with the pi system of the aromatic ring decreases the availability of that electron pair for reaction with an acid. Substitution on the ring by electron-withdrawing groups decreases the basicity of the —NH₂ group:

3. Reaction of Amines with Strong Acids (Section 10.6)

All amines react quantitatively with strong acids to form water-soluble salts:

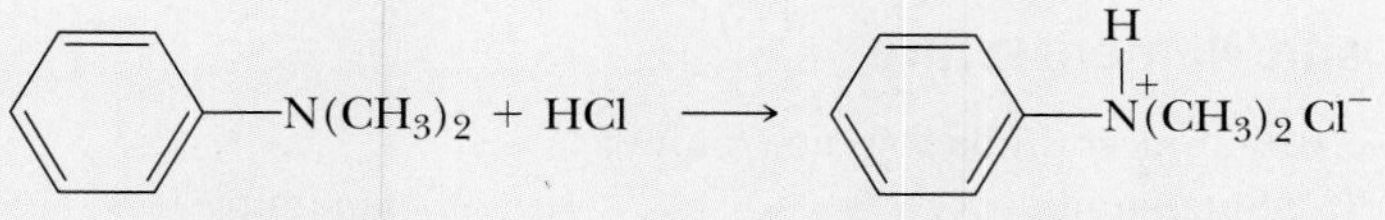

Insoluble in water   A water-soluble salt

4. Reduction of an Aromatic NO$_2$ group (Section 10.7)

An NO$_2$ group on an aromatic ring can be reduced to an amino group by catalytic hydrogenation or by treatment with a metal and hydrochloric acid, followed by a strong base to liberate the free amine:

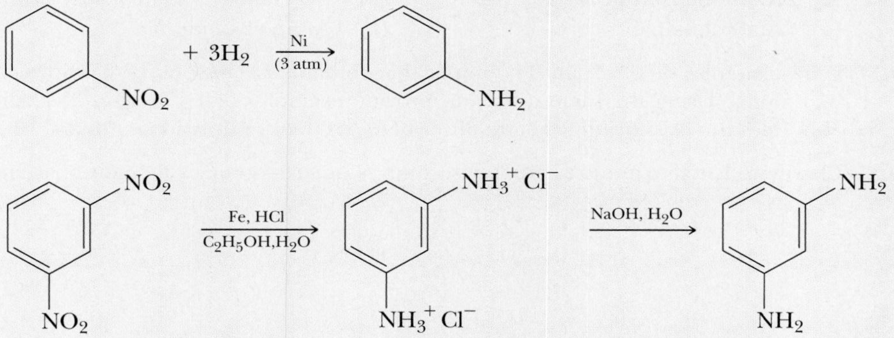

5. Conversion of a Primary Aromatic Amine to a Phenol (Section 10.8)

Treatment of a primary aromatic amine with nitrous acid gives an arenediazonium salt. Heating the aqueous solution of this salt brings about the evolution of N$_2$ and forms a phenol:

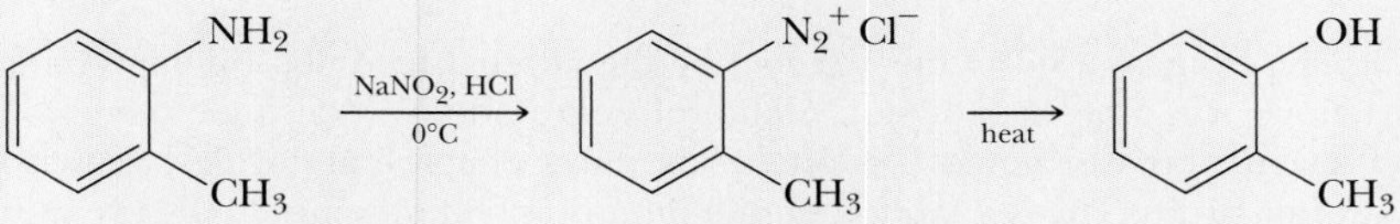

6. Reduction of an arenediazonium Salt (Section 10.8)

Treatment of an arenediazonium salt with hypophosphorous acid, H$_3$PO$_2$, results in replacement of the N$_2^+$ group by H:

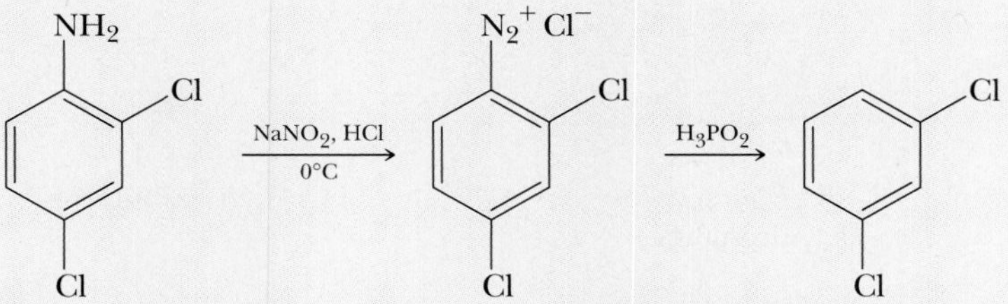

PROBLEMS

A problem number set in red indicates an applied "real-world" problem.

Structure and Nomenclature

10.9 Draw a structural formula for each amine:

(a) (*R*)-2-Butanamine
(b) 1-Octanamine
(c) 2,2-Dimethyl-1-propanamine
(d) 1,5-Pentanediamine
(e) 2-Bromoaniline
(f) Tributylamine
(g) *N,N*-Dimethylaniline
(h) Benzylamine
(i) *tert*-Butylamine
(j) *N*-Ethylcyclohexanamine
(k) Diphenylamine
(l) Isobutylamine

10.10 Draw a structural formula for each amine:

(a) 4-Aminobutanoic acid
(b) 2-Aminoethanol (ethanolamine)
(c) 2-Aminobenzoic acid
(d) (*S*)-2-Aminopropanoic acid (alanine)
(e) 4-Aminobutanal
(f) 4-Amino-2-butanone

10.11 Draw examples of 1°, 2°, and 3° amines that contain at least four sp^3 hybridized carbon atoms. Using the same criterion, provide examples of 1°, 2°, and 3° alcohols. How does the classification system differ between the two functional groups?

10.12 Classify each amino group as primary, secondary, or tertiary and as aliphatic or aromatic:

(a)

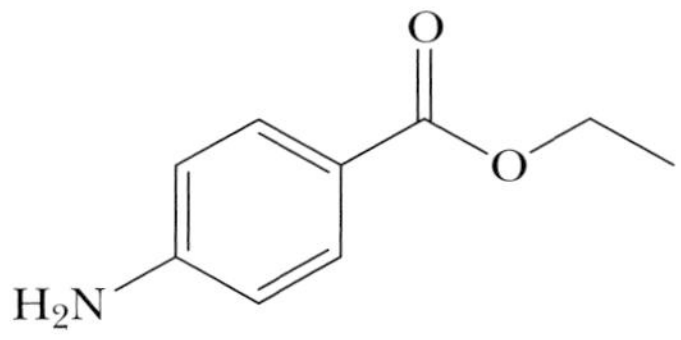

Benzocaine
(a topical anesthetic)

(b)

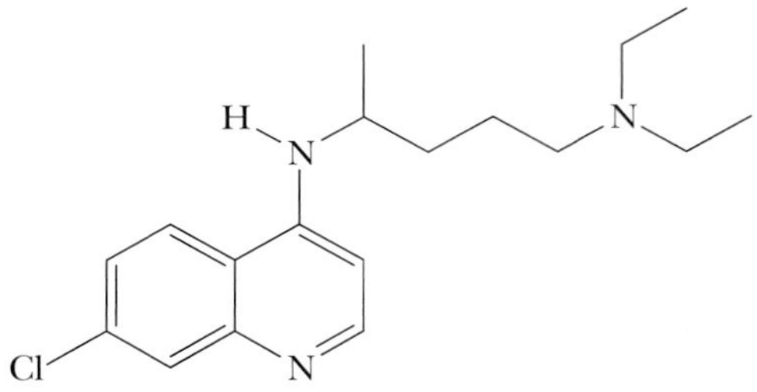

Chloroquine
(a drug for the
treatment of malaria)

10.13 Epinephrine is a hormone secreted by the adrenal medulla. Among epinephrine's actions, it is a bronchodilator. Albuterol, sold under several trade names, including Proventil® and Salbumol®, is one of the most effective and widely prescribed antiasthma drugs. The R enantiomer of albuterol is 68 times more effective in the treatment of asthma than the S enantiomer.

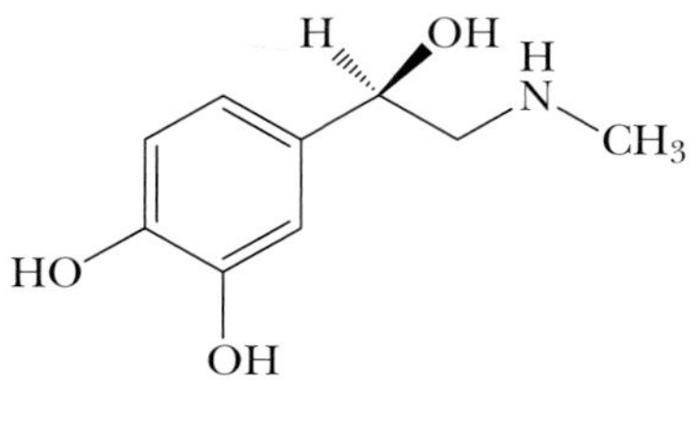

(*R*)-Epinephrine
(Adrenaline)

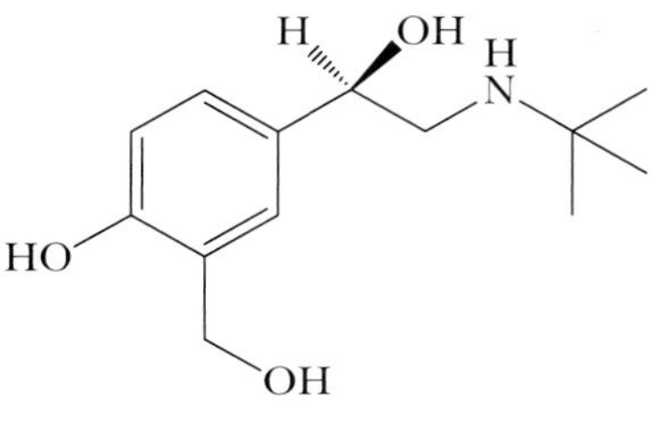

(*R*)-Albuterol

(a) Classify each amino group as primary, secondary, or tertiary.
(b) List the similarities and differences between the structural formulas of these compounds.

10.14 There are eight constitutional isomers with molecular formula $C_4H_{11}N$. Name and draw structural formulas for each. Classify each amine as primary, secondary, or tertiary.

10.15 Draw a structural formula for each compound with the given molecular formula:
- **(a)** A 2° arylamine, C_7H_9N
- **(b)** A 3° arylamine, $C_8H_{11}N$
- **(c)** A 1° aliphatic amine, C_7H_9N
- **(d)** A chiral 1° amine, $C_4H_{11}N$
- **(e)** A 3° heterocyclic amine, $C_5H_{11}N$
- **(f)** A trisubstituted 1° arylamine, $C_9H_{13}N$
- **(g)** A chiral quaternary ammonium salt, $C_9H_{22}NCl$

Physical Properties

10.16 Propylamine, ethylmethylamine, and trimethylamine are constitutional isomers with molecular formula C_3H_9N:

$CH_3CH_2CH_2NH_2$	$CH_3CH_2NHCH_3$	$(CH_3)_3N$
bp 48°C	bp 37°C	bp 3°C
Propylamine	Ethylmethylamine	Trimethylamine

Account for the fact that trimethylamine has the lowest boiling point of the three, and propylamine has the highest.

10.17 Account for the fact that 1-butanamine has a lower boiling point than 1-butanol:

bp 78°C	bp 117°C
1-Butanamine	1-Butanol

10.18 Account for the fact that putrescine, a foul-smelling compound produced by rotting flesh, ceases to smell upon treatment with two equivalents of HCl:

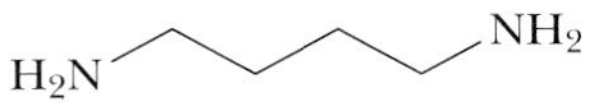

1,4-Butanediamine
(Putrescine)

Basicity of Amines

10.19 Account for the fact that amines are more basic than alcohols.

10.20 From each pair of compounds, select the stronger base:

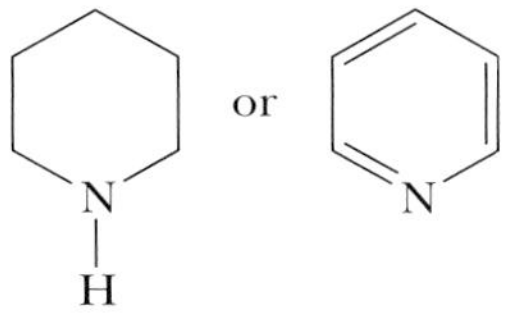

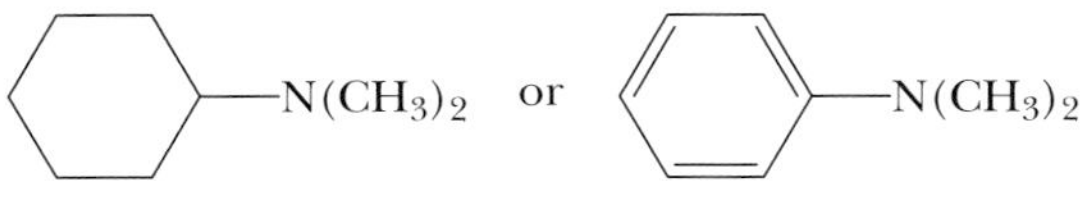

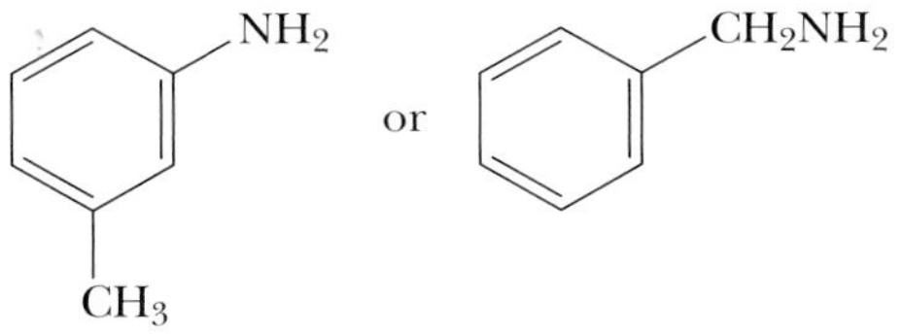

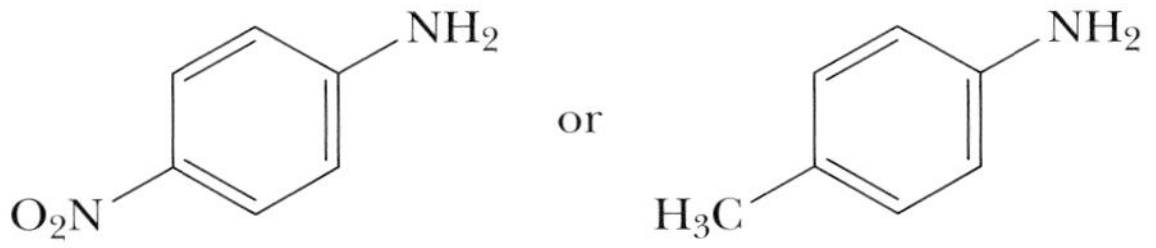

10.21 Account for the fact that substitution of a nitro group makes an aromatic amine a weaker base, but makes a phenol a stronger acid. For example, 4-nitroaniline is a weaker base than aniline, but 4-nitrophenol is a stronger acid than phenol.

10.22 Select the stronger base in this pair of compounds:

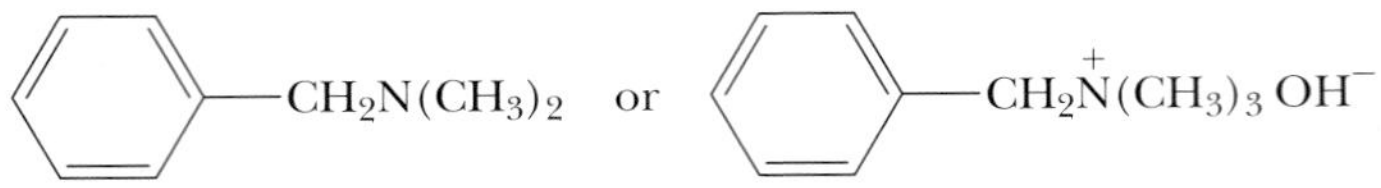

$$\text{C}_6\text{H}_5-\text{CH}_2\text{N}(\text{CH}_3)_2 \quad \text{or} \quad \text{C}_6\text{H}_5-\text{CH}_2\overset{+}{\text{N}}(\text{CH}_3)_3\,\text{OH}^-$$

10.23 Complete the following acid–base reactions and predict the position of equilibrium for each. Justify your prediction by citing values of pK_a for the stronger and weaker acid in each equilibrium. For values of acid ionization constants, consult Table 2.2 (pK_a's of some inorganic and organic acids), Table 8.2 (pK_a's of alcohols), Section 9.9B (acidity of phenols), and Table 10.2 (base strengths of amines). Where no ionization constants are given, make the best estimate from aforementioned tables and section.

(a) 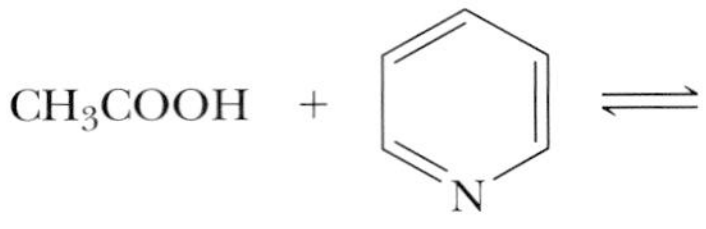 CH_3COOH + [pyridine] $\rightleftharpoons$

Acetic acid   Pyridine

(b) 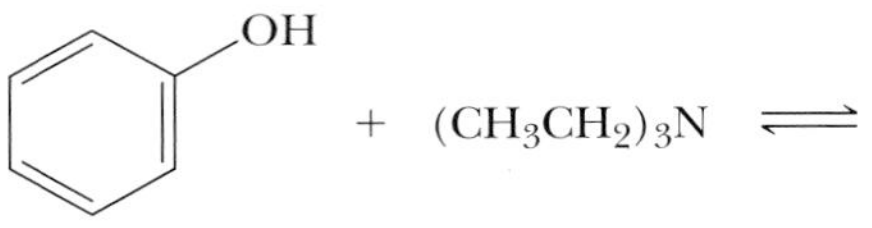   [phenol with OH]   + $(\text{CH}_3\text{CH}_2)_3\text{N}$ $\rightleftharpoons$

Phenol   Triethylamine

(c) $\text{PhCH}_2\overset{\overset{\text{CH}_3}{|}}{\text{CH}}\text{NH}_2$ + $\text{CH}_3\overset{\overset{\text{HO}}{|}}{\text{CH}}\overset{\overset{\text{O}}{\|}}{\text{C}}\text{OH}$ $\rightleftharpoons$

1-Phenyl-2-   2-Hydroxypropanoic
propanamine   acid
(Amphetamine)   (Lactic acid)

(d) $\text{PhCH}_2\overset{\overset{\text{CH}_3}{|}}{\text{CH}}\text{NHCH}_3$ + $\text{CH}_3\overset{\overset{\text{O}}{\|}}{\text{C}}\text{OH}$ $\rightleftharpoons$

Methamphetamine   Acetic acid

10.24 The pK_a of the morpholinium ion is 8.33:

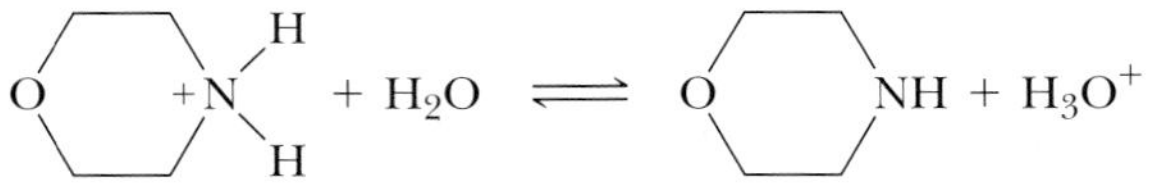

[morpholinium ion structure] $+ \text{H}_2\text{O} \rightleftharpoons$ [morpholine structure] $+ \text{H}_3\text{O}^+$ $pK_a = 8.33$

Morpholinium ion   Morpholine

(a) Calculate the ratio of morpholine to morpholinium ion in aqueous solution at pH 7.0.
(b) At what pH are the concentrations of morpholine and morpholinium ion equal?

10.25 The pK_b of amphetamine (Example 10.2) is approximately 3.2. Calculate the ratio of amphetamine to its conjugate acid at pH 7.4, the pH of blood plasma.

10.26 Calculate the ratio of amphetamine to its conjugate acid at pH 1.0, such as might be present in stomach acid.

10.27 Following is a structural formula of pyridoxamine, one form of vitamin B_6:

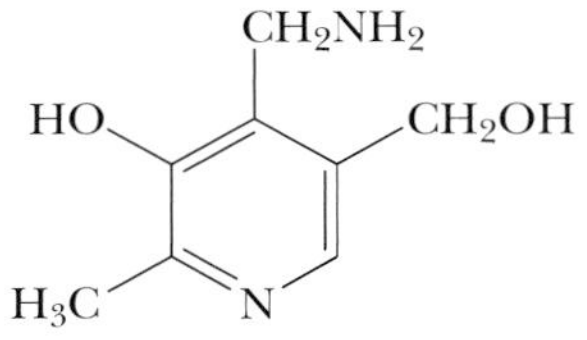

Pyridoxamine
(Vitamin B_6)

(a) Which nitrogen atom of pyridoxamine is the stronger base?
(b) Draw the structural formula of the hydrochloride salt formed when pyridoxamine is treated with one mole of HCl.

10.28 Epibatidine, a colorless oil isolated from the skin of the Ecuadorian poison frog *Epipedobates tricolor*, has several times the analgesic potency of morphine. It is the first chlorine-containing, nonopioid (nonmorphine-like in structure) analgesic ever isolated from a natural source:

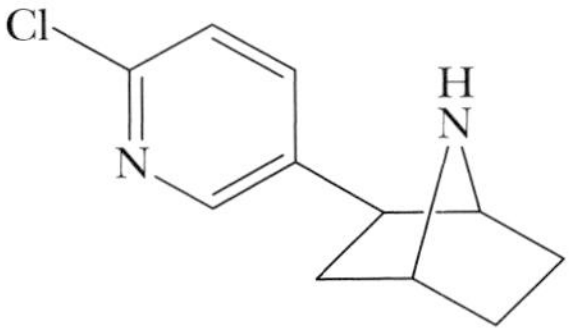

Epibatidine

(a) Which of the two nitrogen atoms of epibatidine is the more basic?
(b) Mark all stereocenters in this molecule.

Poison arrow frog.
*(Stephen J. Krasemann/
Photo Researchers, Inc.)*

10.29 Procaine was one of the first local anesthetics for infiltration and regional anesthesia:

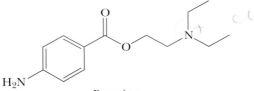

Procaine

The hydrochloride salt of procaine is marketed as Novocaine®.
(a) Which nitrogen atom of procaine is the stronger base?
(b) Draw the formula of the salt formed by treating procaine with one mole of HCl.
(c) Is procaine chiral? Would a solution of Novocaine in water be optically active or optically inactive?

10.30 Treatment of trimethylamine with 2-chloroethyl acetate gives the neurotransmitter acetylcholine as its chloride salt:

$$(CH_3)_3N + CH_3\overset{\overset{\displaystyle O}{\|}}{C}OCH_2CH_2Cl \longrightarrow C_7H_{16}ClNO_2$$

Acetylcholine chloride

Propose a structural formula for this quaternary ammonium salt and a mechanism for its formation.

10.31 Aniline is prepared by the catalytic reduction of nitrobenzene:

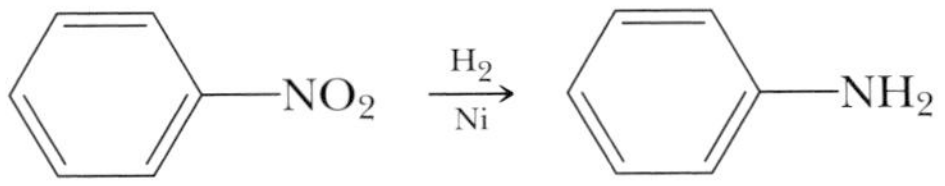

$$\text{—NO}_2 \xrightarrow[\text{Ni}]{\text{H}_2} \text{—NH}_2$$

Devise a chemical procedure based on the basicity of aniline to separate it from any unreacted nitrobenzene.

10.32 Suppose that you have a mixture of the following three compounds:

H_3C—⬡—NO_2 H_3C—⬡—NH_2 H_3C—⬡—OH

4-Nitrotoluene   4-Methylaniline   4-Methylphenol
(*p*-Nitrotoluene)   (*p*-Toluidine)   (*p*-Cresol)

Devise a chemical procedure based on their relative acidity or basicity to separate and isolate each in pure form.

10.33 Following is a structural formula for metformin, the hydrochloride salt of which is marketed as the antidiabetic Glucophage®:

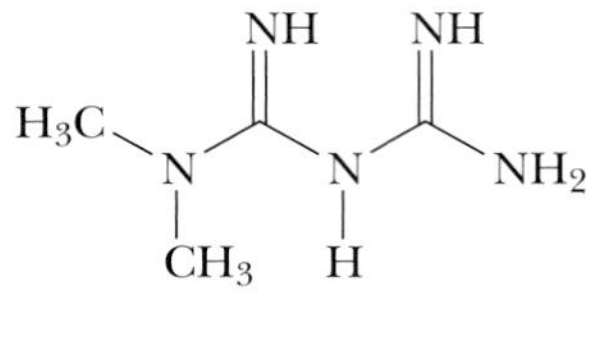

Metformin

Metformin was introduced into clinical medicine in the United States in 1995 for the treatment of type 2 diabetes. More than 25 million prescriptions for this drug were written in 2000, making it the most commonly prescribed brand-name diabetes medication in the nation.

(a) Draw the structural formula for Glucophage®.

(b) Would you predict Glucophage® to be soluble or insoluble in water? Soluble or insoluble in blood plasma? Would you predict it to be soluble or insoluble in diethyl ether? In dichloromethane? Explain your reasoning.

Synthesis

10.34 4-Aminophenol is a building block in the synthesis of the analgesic acetaminophen. Show how this building block can be synthesized in two steps from phenol (in Chapter 15, we will see how to complete the synthesis of acetaminophen):

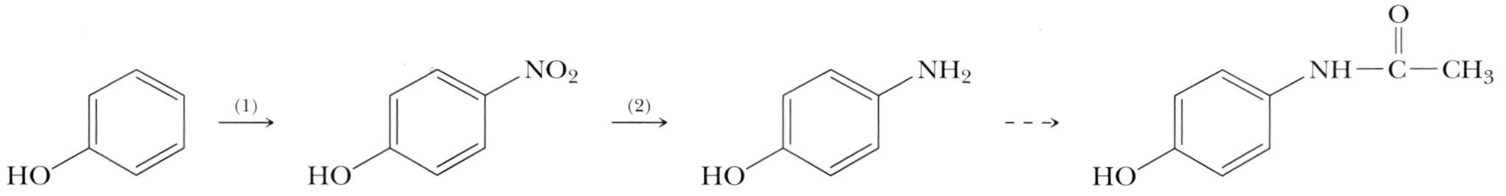

Phenol $\xrightarrow{(1)}$ 4-Nitrophenol $\xrightarrow{(2)}$ 4-Aminophenol   - - →   Acetaminophen

10.35 4-Aminobenzoic acid is a building block in the synthesis of the topical anesthetic benzocaine. Show how this building block can be synthesized in three steps from toluene (in Chapter 15, we will see how to complete the synthesis of benzocaine):

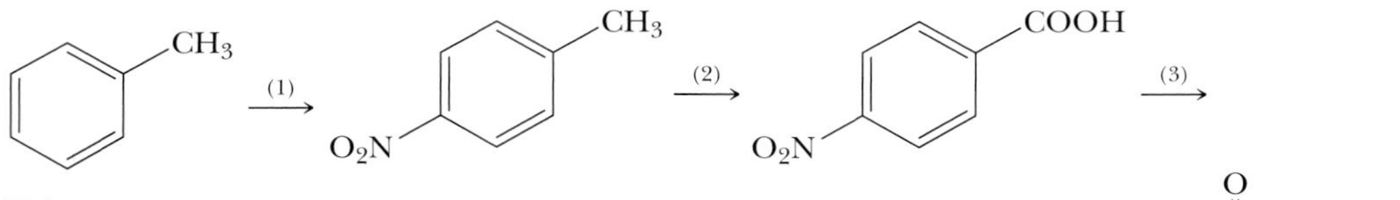

Toluene $\xrightarrow{(1)}$ ⬡ $\xrightarrow{(2)}$ ⬡ $\xrightarrow{(3)}$

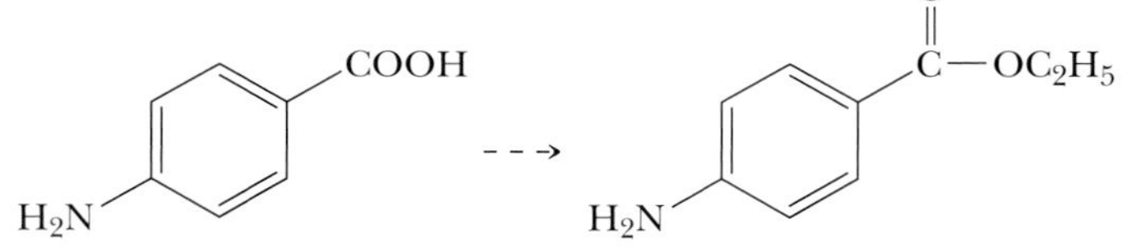

4-Aminobenzoic acid   - - →   Ethyl 4-aminobenzoate (Benzocaine)

10.36 The compound 4-aminosalicylic acid is one of the building blocks needed for the synthesis of propoxycaine, one of the family of "caine" anesthetics. Some other members of this family of local anesthetics are procaine (Novocaine®), lidocaine (Xylocaine®), and mepivicaine (Carbocaine®). 4-Aminosalicylic acid is synthesized from salicylic acid in five steps (in Chapter 15, we will see how to complete the synthesis of propoxycaine):

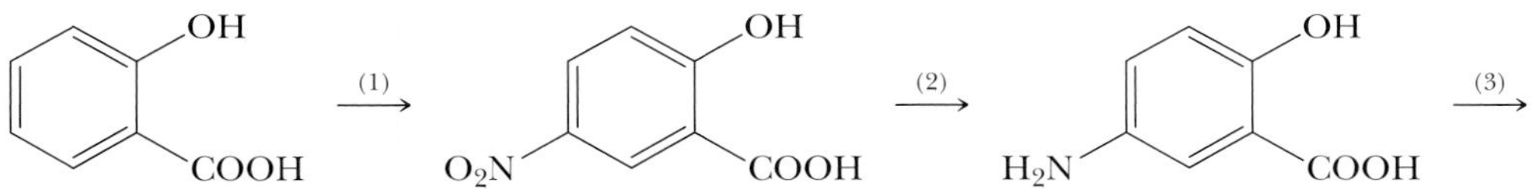

Salicylic acid

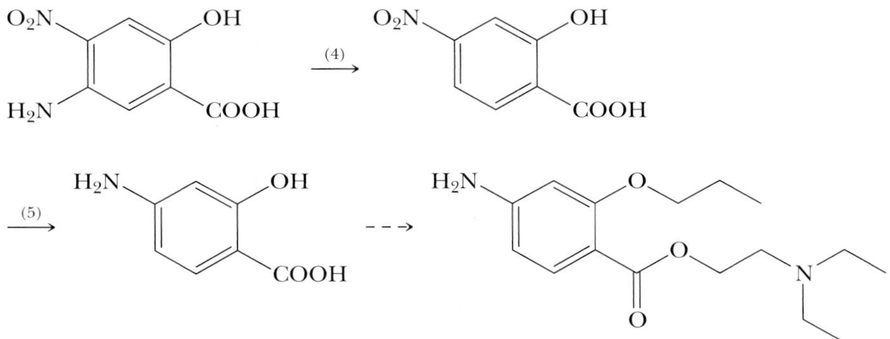

4-Aminosalicylic acid

Propoxycaine

Show reagents that will bring about the synthesis of 4-aminosalicylic acid.

10.37 A second building block for the synthesis of propoxycaine is 2-diethylaminoethanol:

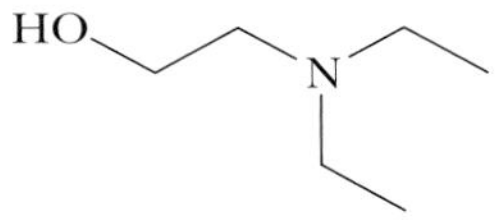

2-Diethylaminoethanol

Show how this compound can be prepared from ethylene oxide and diethylamine.

10.38 Following is a two-step synthesis of the antihypertensive drug propranolol, a so-called beta-blocker with vasodilating action:

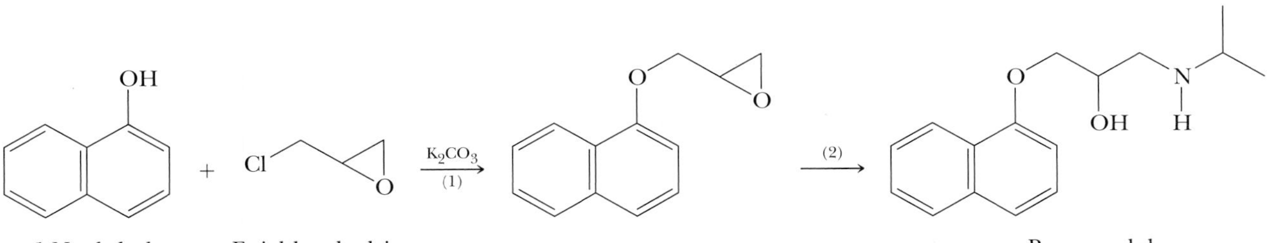

1-Naphthol   Epichlorohydrin

Propranolol
(Cardinol)

Propranolol and other beta blockers have received enormous clinical attention because of their effectiveness in treating hypertension (high blood pressure), migraine headaches, glaucoma, ischemic heart disease, and certain cardiac arrhythmias. The hydrochloride salt of propranolol has been marketed under at least 30 brand names, one of which is Cardinol®. (Note the "card-" part of the name, after *cardiac*.)

(a) What is the function of potassium carbonate, K_2CO_3, in Step 1? Propose a mechanism for the formation of the new oxygen–carbon bond in this step.

(b) Name the amine used to bring about Step 2, and propose a mechanism for this step.

(c) Is propranolol chiral? If so, how many stereoisomers are possible for it?

10.39 The compound 4-ethoxyaniline, a building block of the over-the-counter analgesic phenacetin, is synthesized in three steps from phenol:

4-Ethoxyaniline   Phenacetin

Show reagents for each step of the synthesis of 4-ethoxyaniline. (In Chapter 15, we will see how to complete this synthesis.)

10.40 Radiopaque imaging agents are substances administered either orally or intravenously that absorb X rays more strongly than body material does. One of the best known of these agents is barium sulfate, the key ingredient in the "barium cocktail" used for imaging of the gastrointestinal tract. Among other X-ray imaging agents are the so-called triiodoaromatics. You can get some idea of the kinds of imaging for which they are used from the following selection of trade names: Angiografin®, Gastrografin, Cardiografin, Cholografin, Renografin, and Urografin®. The most common of the triiodiaromatics are derivatives of these three triiodobenzenecarboxylic acids:

3-Amino-2,4,6-triiodobenzoic acid

3,5-Diamino-2,4,6-triiodobenzoic acid

5-Amino-2,4,6-triiodoisophthalic acid

3-Amino-2,4,6-triiodobenzoic acid is synthesized from benzoic acid in three steps:

3-Amino-benzoic acid

3-Amino-2,4,6-triiodobenzoic acid

(a) Show reagents for Steps (1) and (2).

(b) Iodine monochloride, ICl, a black crystalline solid with a melting point of 27.2°C and a boiling point of 97°C, is prepared by mixing equimolar amounts of I_2 and Cl_2. Propose a mechanism for the iodination of 3-aminobenzoic acid by this reagent.

(c) Show how to prepare 3,5-diamino-2,4,6-triiodobenzoic acid from benzoic acid.

(d) Show how to prepare 5-amino-2,4,6-triiodoisophthalic acid from isophthalic acid (1,3-benzenedicarboxylic acid).

10.41 The intravenous anesthetic propofol is synthesized in four steps from phenol:

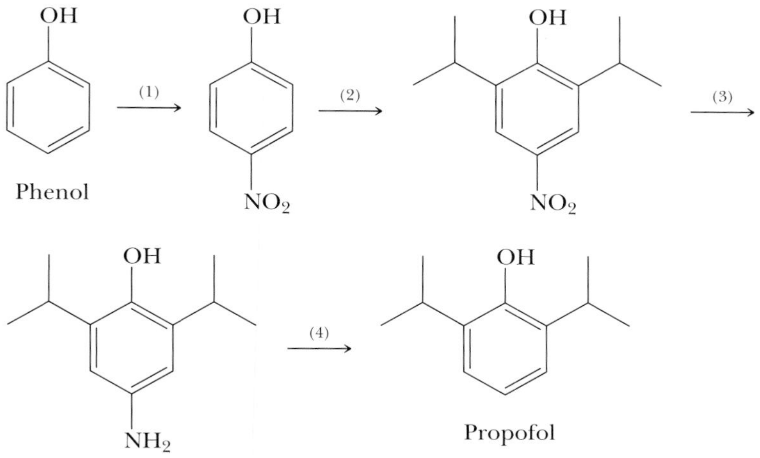

Phenol

Propofol

Show reagents to bring about each step.

Looking Ahead

10.42 State the hybridization of the nitrogen atom in each of the following compounds:

(a) 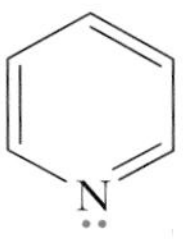

(b) 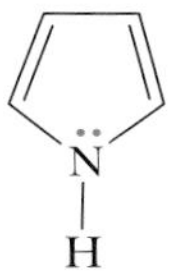

(c) 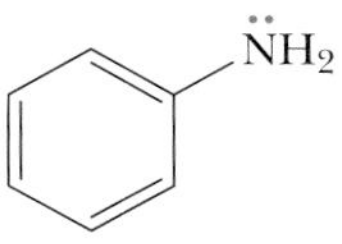

(d) 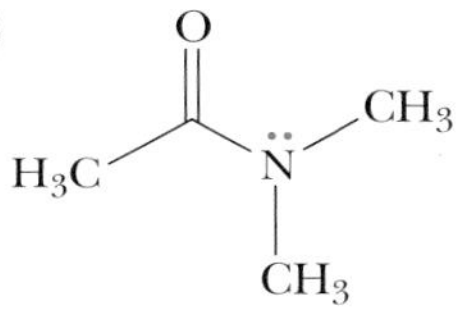

10.43 Amines can act as nucleophiles. For each of the following molecules, circle the most likely atom that would be attacked by the nitrogen of an amine:

(a) 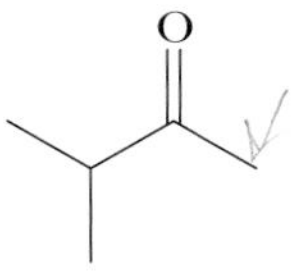

(b) 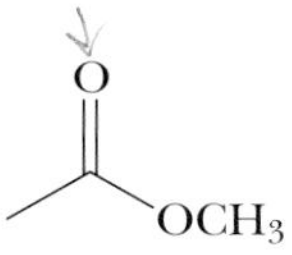

(c) 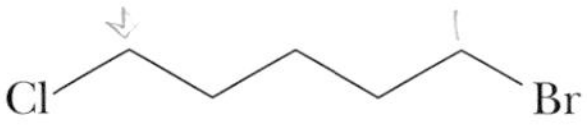

10.44 Draw a Lewis structure for a molecule with formula C_3H_7N that does not contain a ring or an alkene (a carbon–carbon double bond).

10.45 Rank the following leaving groups in order from best to worst:

$R-Cl$ $\quad$ $R-O-\overset{\overset{\displaystyle O}{\|}}{C}-R$ $\quad$ $R-OCH_3$ $\quad$ $R-N(CH_3)_2$

11

Infrared
Spectroscopy

Healthy
Heart

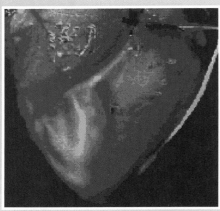

Heart
with CAD

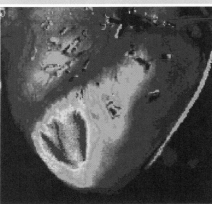

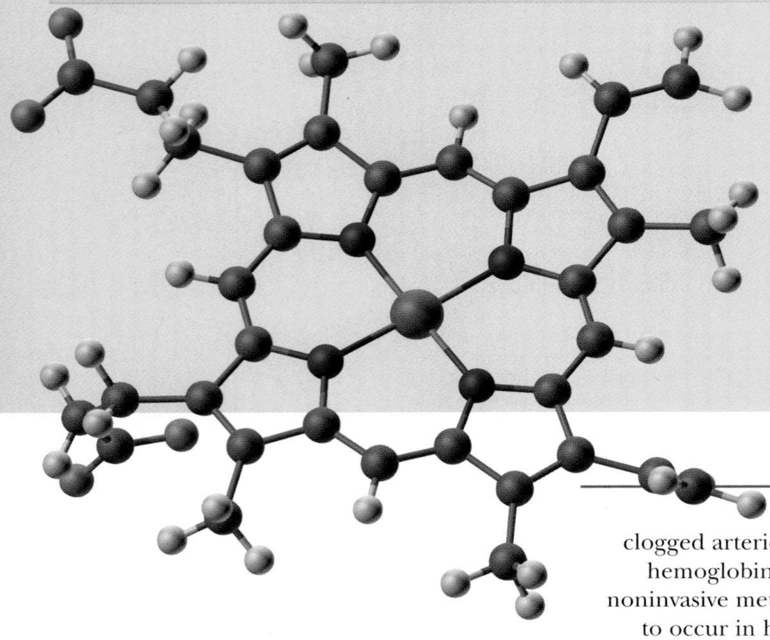

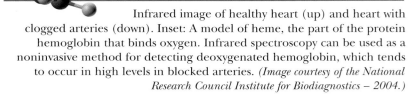

Infrared image of healthy heart (up) and heart with clogged arteries (down). Inset: A model of heme, the part of the protein hemoglobin that binds oxygen. Infrared spectroscopy can be used as a noninvasive method for detecting deoxygenated hemoglobin, which tends to occur in high levels in blocked arteries. *(Image courtesy of the National Research Council Institute for Biodiagnostics – 2004.)*

11.1 INTRODUCTION

Determining the molecular structure of a compound is a central theme in science. In medicine, for example, the structure of any drug must be known before the drug can be approved for use in patients. In the biotechnology and pharmaceutical industries, the knowledge of a compound's structure can provide new leads to promising therapeutics. In organic chemistry, knowledge of the structure of a compound is essential to its use as a reagent or a precursor to other molecules.

Chemists rely almost exclusively on instrumental methods of analysis for structure determination. We begin this chapter with a discussion of infrared (IR) spectroscopy. In the next chapter, we discuss nuclear magnetic resonance (NMR) spectroscopy. These two commonly used techniques involve the interaction of molecules with electromagnetic radiation. Thus, in order to understand the fundamentals of spectroscopy, we must first review some of the fundamentals of electromagnetic radiation.

11.2 ELECTROMAGNETIC RADIATION

Gamma rays, X rays, ultraviolet light, visible light, infrared radiation, microwaves, and radio waves are all part of the electromagnetic spectrum. Because **electromagnetic radiation** behaves as a wave traveling at the speed of light, it is described in terms of its wavelength and frequency. Table 11.1 summarizes the wavelengths, frequencies, and energies of some regions of the electromagnetic spectrum.

Electromagnetic radiation Light and other forms of radiant energy.

TABLE 11.1 Wavelength, Frequency, and Energy Relationships of Some Regions of the Electromagnetic Spectrum

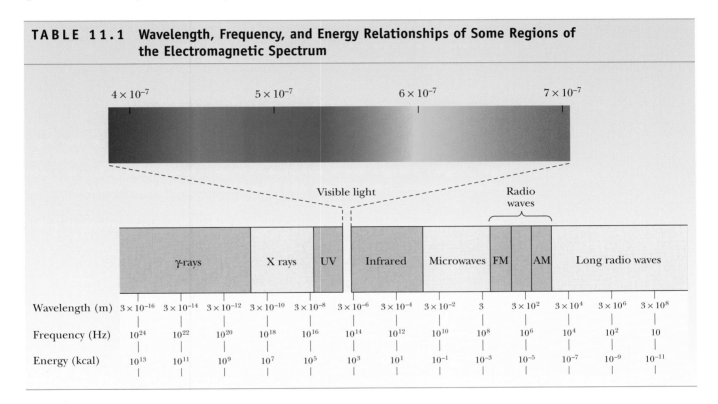

Wavelength is the distance between any two consecutive identical points on the wave. Wavelength is given the symbol λ (Greek lowercase lambda) and is usually expressed in the SI base unit of meters. Other derived units commonly used to express wavelength are given in Table 11.2.

Wavelength (λ) The distance between two consecutive identical points (or troughs) on a wave.

TABLE 11.2 Common Units Used to Express Wavelength (λ)

Unit	Relation to Meter
meter (m)	
millimeter (mm)	$1 \text{ mm} = 10^{-3} \text{ m}$
micrometer (μm)	$1 \ \mu\text{m} = 10^{-6} \text{ m}$
nanometer (nm)	$1 \text{ nm} = 10^{-9} \text{ m}$
Angstrom (Å)	$1 \text{ Å} = 10^{-10} \text{ m}$

The **frequency** of a wave is the number of full cycles of the wave that pass a given point in a second. Frequency is given the symbol **ν** (Greek nu) and is reported in **hertz** (Hz). Wavelength and frequency are inversely proportional, and we can calculate one from the other from the relationship

$$\nu\lambda = c$$

where ν is frequency in hertz, c is the velocity of light (3.00×10^8 m/s), and λ is the wavelength in meters. For example, consider infrared radiation—or heat radiation, as it is also called—with wavelength 1.5×10^{-5} m. The frequency of this radiation is

$$\nu = \frac{3.0 \times 10^8 \text{ m/s}}{1.5 \times 10^{-5} \text{ m}} = 2.0 \times 10^{13} \text{ Hz}$$

An alternative way to describe electromagnetic radiation is in terms of its properties as a stream of particles. We call these particles **photons**. The energy in a mole of photons and the frequency of radiation are related by the equation

$$E = h\nu = h\frac{c}{\lambda}$$

where E is the energy in kcal/mol and h is Planck's constant, 9.54×10^{-14} kcal·s·mol^{-1}(3.99×10^{-13} kJ·s·mol^{-1}). This equation tells us that high-energy radiation corresponds to short wavelengths, and vice versa. Thus, ultraviolet light (higher energy) has a shorter wavelength (approximately 10^{-7} m) than infrared radiation (lower energy), which has a wavelength of approximately 10^{-5} m.

EXAMPLE 11.1

Calculate the energy, in kilocalories per mole of radiation, of a wave with wavelength 2.50 μm. What type of radiant energy is this? (Refer to Table 11.1.)

SOLUTION

Use the relationship $E = hc/\lambda$. Make certain that the dimensions for distance are consistent: If the dimension of wavelength is meters, then express the velocity of light in meters per second. First convert 2.50 μm to meters, using the relationship 1 μm = 10^{-6} m (Table 11.2):

$$2.50 \; \mu\text{m} \times \frac{10^{-6} \text{ m}}{1 \; \mu\text{m}} = 2.50 \times 10^{-6} \text{ m}$$

Now substitute this value into the equation $E = hc/\lambda$:

$$E = \frac{hc}{\lambda} = 9.54 \times 10^{-14} \frac{\text{kcal·s}}{\text{mol}} \times 3.00 \times 10^8 \frac{\text{m}}{\text{s}} \times \frac{1}{2.50 \times 10^{-6} \text{ m}} = 11.4 \text{ kcal/mol}$$

$$= 47.7 \text{ kJ/mol}$$

Electromagnetic radiation with energy of 11.4 kcal/mol is radiation in the infrared region.

Practice Problem 11.1

Calculate the energy of red light (680 nm) in kilocalories per mole. Which form of radiation carries more energy, infrared radiation with wavelength 2.50 μm or red light with wavelength 680 nm?

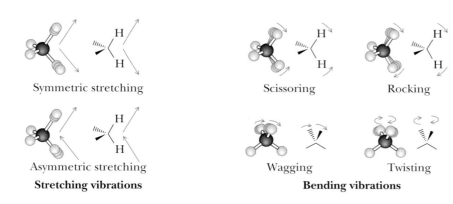

Symmetric stretching

Scissoring

Rocking

Asymmetric stretching

Wagging

Twisting

Stretching vibrations

Bending vibrations

Figure 11.3
Fundamental modes of vibration for a methylene group.

For a nonlinear molecule containing n atoms, $3n - 6$ allowed fundamental vibrations exist. For a molecule as simple as ethanol, CH_3CH_2OH, there are 21 fundamental vibrations, and for hexanoic acid, $CH_3(CH_2)_4COOH$, there are 54. Thus, even for relatively simple molecules, a large number of vibrational energy levels exists, and the patterns of energy absorption for these and larger molecules are quite complex.

The simplest vibrational motions in molecules giving rise to the absorption of infrared radiation are **stretching** and **bending** motions. Illustrated in Figure 11.3 are the fundamental stretching and bending vibrations for a methylene group.

To one skilled in the interpretation of infrared spectra, absorption patterns can yield an enormous amount of information about chemical structure. We, however, have neither the time nor the need to develop that level of competence. The value of infrared spectra for us is that we can use them to determine the presence or absence of particular functional groups. A carbonyl group, for example, typically shows strong absorption at approximately 1630–1800 cm^{-1}. The position of absorption for a particular carbonyl group depends on (1) whether it is that of an aldehyde, a ketone, a carboxylic acid, an ester, or an amide, and (2) if the carbonyl carbon is in a ring, the size of the ring.

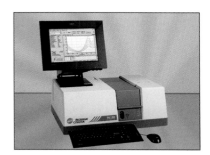

A Beckman Coulter DU 800 infrared spectrophotometer. Spectra are shown in the monitor. *(Courtesy of Beckman Coulter, Inc.)*

C. Correlation Tables

Data on absorption patterns of selected functional groups are collected in tables called **correlation tables**. Table 11.3 gives the characteristic infrared absorptions for the types of bonds and functional groups we deal with most often. Appendix 5 contains a cumulative correlation table. In these tables, we refer to the intensity of a particular absorption as **strong (s)**, **medium (m)**, or **weak (w)**.

TABLE 11.3 **Characteristic IR Absorptions of Selected Functional Groups**

Bond	Frequency (cm^{-1})	Intensity
O—H	3200–3500	strong and broad
N—H	3100–3500	medium
C—H	2850–3100	medium to strong
C≡C	2100–2260	weak
C=O	1630–1800	strong
C=C	1600–1680	weak
C—O	1050–1250	strong

In general, we will pay most attention to the region from 3650 to 1000 cm^{-1}, because the characteristic stretching vibrations for most functional groups are found in this region. Vibrations in the region from 1000 to 400 cm^{-1} can arise from phenomena such as combinations of two or more bands or harmonics of fundamental absorption bands. Such vibrations are much more complex and far more difficult to analyze. Because even slight variations in molecular structure lead to differences in absorption patterns in this region, it is often called the **fingerprint region**. If two compounds have even slightly different structures, the differences in their infrared spectra are most clearly discernible in the fingerprint region.

Fingerprint region The portion of the vibrational infrared region that extends from 1000 to 400 cm^{-1} and that is unique to every compound.

EXAMPLE 11.2

Determine the functional group that is most likely present if a compound shows IR absorption at

(a) 1705 cm^{-1} (b) 2950 cm^{-1}

SOLUTION

(a) A C=O group   (b) An aliphatic C—H group

Practice Problem 11.2

A compound shows strong, very broad IR absorption in the region from 3200 to 3500 cm^{-1} and strong absorption at 1715 cm^{-1}. What functional group accounts for both of these absorptions?

EXAMPLE 11.3

Propanone and 2-propen-1-ol are constitutional isomers. Show how to distinguish between these two isomers by IR spectroscopy.

$$CH_3-\overset{\overset{\displaystyle O}{\|}}{C}-CH_3 \qquad CH_2{=}CH-CH_2-OH$$

Propanone   2-Propen-1-ol
(Acetone)   (Allyl alcohol)

SOLUTION

Only propanone shows strong absorption in the C=O stretching region, 1630–1800 cm^{-1}. Alternatively, only 2-propen-1-ol shows strong absorption in the O—H stretching region, 3200–3500 cm^{-1}.

Practice Problem 11.3

Propanoic acid and methyl ethanoate are constitutional isomers. Show how to distinguish between these two compounds by IR spectroscopy.

$$CH_3CH_2\overset{\overset{\displaystyle O}{\|}}{C}OH \qquad CH_3\overset{\overset{\displaystyle O}{\|}}{C}OCH_3$$

Propanoic acid   Methyl ethanoate
                                        (Methyl acetate)

11.5 INTERPRETING INFRARED SPECTRA

Interpreting spectroscopic data is a skill that is easy to acquire through practice and exposure to examples. An IR spectrum will reveal not only the functional groups that are present in a sample, but also those which can be excluded from consideration. Often, we can determine the structure of a compound solely from the data in the spectrum of the compound. Other times, we may need additional information, such as the molecular formula of the compound, or knowledge of the reactions used to synthesize the molecule. In this section, we will see specific examples of IR spectra for characteristic functional groups. Familiarizing yourself with them will help you to master the technique of spectral interpretation.

A. Alkanes

Infrared spectra of alkanes are usually simple, with few peaks, the most common of which are given in Table 11.4.

Figure 11.4 shows an infrared spectrum of decane. The strong peak with multiple splittings between 2850 and 3000 cm^{-1} is characteristic of alkane C—H stretching. The C—H peak is strong in this spectrum because there are so many C—H bonds and no other functional groups. The other prominent peaks in the spectrum are a methylene bending absorption at 1465 cm^{-1} and a methyl bending absorption at 1380 cm^{-1}. Because alkane CH, CH_2, and CH_3 groups are present in many organic compounds, these peaks are among the most commonly encountered in infrared spectroscopy.

TABLE 11.4   Characteristic IR Absorptions of Alkanes, Alkenes, and Alkynes

Hydrocarbon	Vibration	Frequency (cm^{-1})	Intensity
Alkane			
C—H	stretching	2850–3000	strong
CH_2	bending	1450	medium
CH_3	bending	1375 and 1450	weak to medium
Alkene			
C—H	stretching	3000–3100	weak to medium
C=C	stretching	1600–1680	weak to medium
Alkyne			
C—H	stretching	3300	weak to medium
C≡C	stretching	2100–2260	weak to medium

B. Alkenes

An easily recognized alkene absorption is the vinylic =C—H stretching band slightly to the left of (at a greater wavenumber than) 3000 cm^{-1}. Also characteristic of alkenes is C=C stretching at 1600–1680 cm^{-1}. This vibration, however, is often weak and difficult to observe. Both vinylic =C—H stretching and C=C stretching can be seen in the infrared spectrum of cyclopentene (Figure 11.5). Also visible are the aliphatic C—H stretching near 2900 cm^{-1} and the methylene bending near 1440 cm^{-1}.

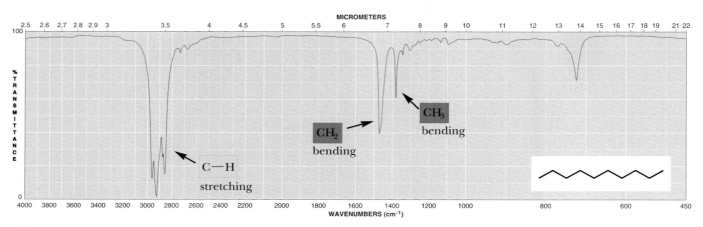

Figure 11.4
Infrared spectrum of decane.

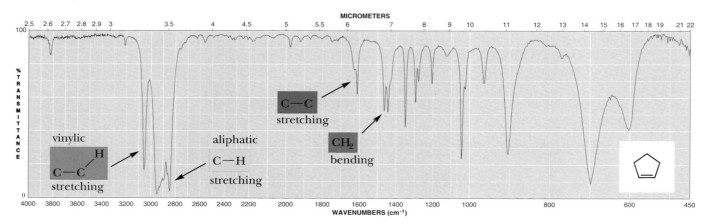

Figure 11.5
Infrared spectrum of
cyclopentene.

C. Alkynes

Terminal alkynes exhibit C≡C—H stretching at 3300 cm^{-1}. This absorption band is absent in internal alkynes, because the triple bond is not bonded to a proton. All alkynes absorb weakly between 2100–2260 cm^{-1}, due to C≡C stretching. This stretching shows clearly in the spectrum of 1-octyne (Figure 11.6).

Figure 11.6
Infrared spectrum of 1-octyne.

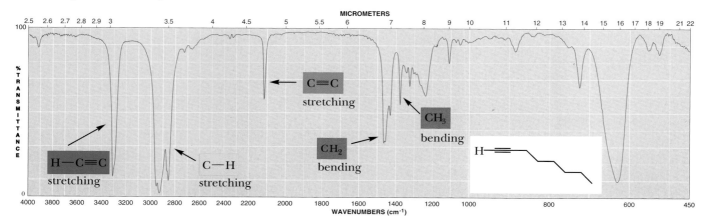

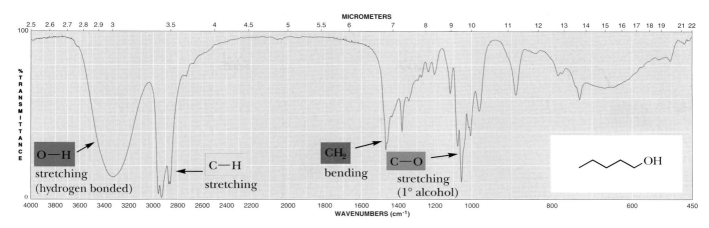

Figure 11.7
Infrared spectrum of 1-pentanol.

D. Alcohols

Alcohols are easily recognized by their characteristic O—H stretching absorption (Table 11.5). Both the position of this absorption and its intensity depend on the extent of hydrogen bonding (Section 8.2C). Under normal conditions, where there is extensive hydrogen bonding between alcohol molecules, O—H stretching occurs as a broad peak at 3200–3500 cm^{-1}. The C—O stretching vibration of alcohols appears in the range 1050–1250 cm^{-1}.

TABLE 11.5 Characteristic IR Absorptions of Alcohols

Bond	Frequency (cm^{-1})	Intensity
O—H (hydrogen bonded)	3200–3500	medium, broad
C—O	1050–1250	medium

Figure 11.7 shows an infrared spectrum of 1-pentanol. The hydrogen-bonded O—H stretching appears as a strong, broad peak centered at 3340 cm^{-1}. The C—O stretching appears near 1050 cm^{-1}, a value characteristic of primary alcohols.

E. Ethers

The C—O stretching frequencies of ethers are similar to those observed in alcohols and esters. Dialkyl ethers typically show a single absorption in this region between 1070 and 1150 cm^{-1}. The presence or absence of O—H stretching at 3200–3500 cm^{-1} for a hydrogen-bonded O—H can be used to distinguish between an ether and an alcohol. The C—O stretching vibration is also present in esters. In this case, we can use the presence or absence of C=O stretching to distinguish between an ether and an ester. Figure 11.8 shows an infrared spectrum of diethyl ether. Notice the absence of O—H stretching.

F. Benzene and Its Derivatives

Aromatic rings show a medium to weak peak in the C—H stretching region at approximately 3030 cm^{-1}, characteristic of sp^2 C—H bonds. They also show several absorptions due to C=C stretching between 1450 and 1600 cm^{-1}. In addition, aromatic

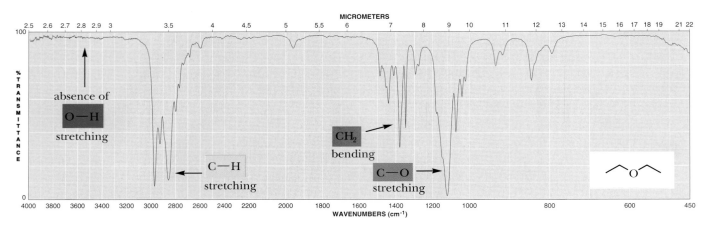

Figure 11.8
Infrared spectrum of diethyl ether.

rings show strong absorption in the region from 690 to 900 cm^{-1} due to C—H bending. Finally, the presence of weak, broad bands between 1700 and 2000 cm^{-1} are an indicator of a benzene ring (Table 11.6).

TABLE 11.6   Characteristic IR Absorptions of Aromatic Hydrocarbons

Bond	Vibration	Frequency (cm^{-1})	Intensity
C—H	stretching	3030	medium to weak
C—H	bending	690–900	strong
C=C	stretching	1475 and 1600	strong to medium

The C—H and C=C absorption patterns characteristic of aromatic rings can be seen in the infrared spectrum of toluene (Figure 11.9).

G. Amines

The most important and readily observed infrared absorptions of primary and secondary amines are due to N—H stretching vibrations and appear in the region from 3100 to 3500 cm^{-1}. Primary amines have two peaks in this region, one caused

Figure 11.9
Infrared spectrum of toluene.

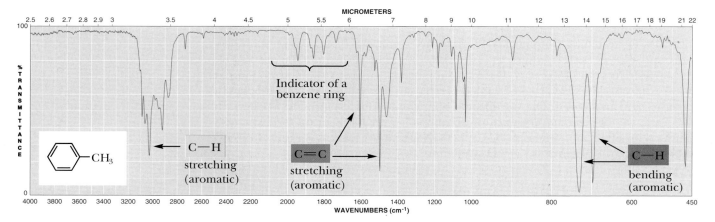

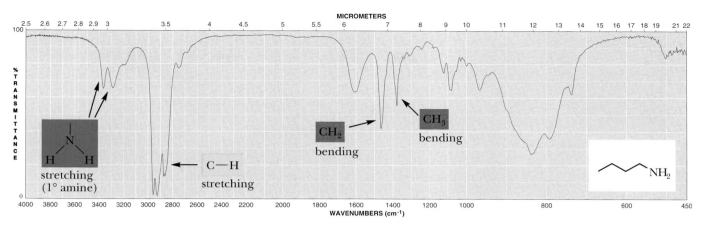

Figure 11.10
Infrared spectrum of butanamine, a primary amine.

by a symmetric stretching vibration and the other from an asymmetric stretching. The two N—H stretching absorptions characteristic of a primary amine can be seen in the IR spectrum of butanamine (Figure 11.10). Secondary amines give only one absorption in this region. Tertiary amines have no N—H and therefore are transparent in this region of the infrared spectrum.

H. Aldehydes and Ketones

Aldehydes and ketones show characteristic strong infrared absorption between 1705 and 1780 cm^{-1} associated with the stretching vibration of the carbon–oxygen double bond. The stretching vibration for the carbonyl group of menthone occurs at 1705 cm^{-1} (Figure 11.11).

Because several different functional groups contain a carbonyl group, it is often not possible to tell from absorption in this region alone whether the carbonyl-containing compound is an aldehyde, a ketone, a carboxylic acid, or an ester.

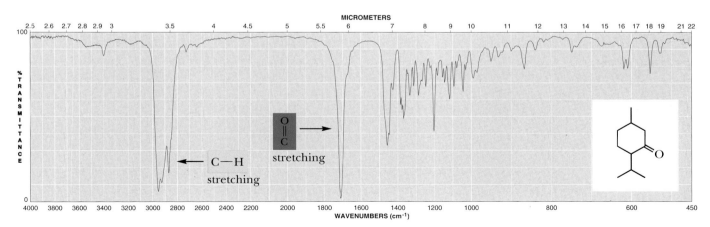

Figure 11.11
Infrared spectrum of menthone.

I. Carboxylic Acids and Their Derivatives

The most important infrared absorptions of carboxylic acids and their functional derivatives are due to the C=O stretching vibration; these absorptions are summarized in Table 11.7.

TABLE 11.7   Characteristic IR Absorptions of Carboxylic Acids, Esters, and Amides

Compound	C=O Absorption Frequency (cm^{-1})	Additional Absorptions (cm^{-1})
O‖ RCNH$_2$	1630–1680	N—H stretching at 3200 and 3400 (1° amides have two N—H peaks) (2° amides have one N—H peak)
O‖ RCOH	1700–1725	O—H stretching at 2400–3400 C—O stretching at 1210–1320
O‖ RCOR	1735–1800	C—O stretching at 1000–1100 and 1200–1250

The carboxyl group of a carboxylic acid gives rise to two characteristic absorptions in the infrared spectrum. One of these occurs in the region from 1700 to 1725 cm^{-1} and is associated with the stretching vibration of the carbonyl group. This region is essentially the same as that for the absorption of the carbonyl groups of aldehydes and ketones. The other infrared absorption characteristic of a carboxyl group is a peak between 2400 and 3400 cm^{-1} due to the stretching vibration of the O—H group. This peak, which often overlaps the C—H stretching absorptions, is generally very broad due to hydrogen bonding between molecules of the carboxylic acid. Both C=O and O—H stretchings can be seen in the infrared spectrum of butanoic acid, shown in Figure 11.12.

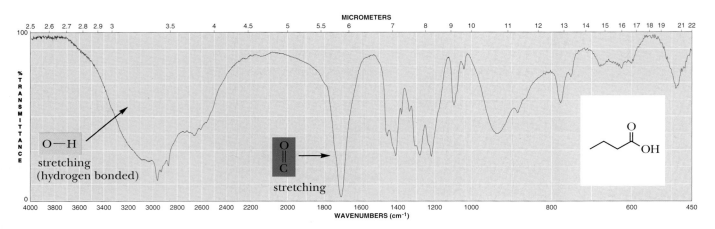

Figure 11.12
Infrared spectrum of butanoic acid.

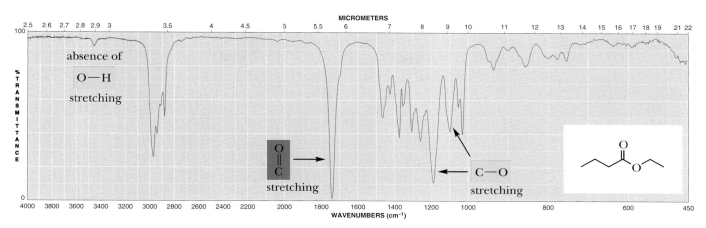

Figure 11.13
Infrared spectrum of ethyl butanoate.

Esters display strong C=O stretching absorption in the region between 1735 and 1780 cm^{-1}. In addition, they display strong C—O stretching absorption in the region from 1000 to 1250 cm^{-1} (Figure 11.13).

The carbonyl stretching of amides occurs at 1630–1680 cm^{-1}, a lower series of wavenumbers than for other carbonyl compounds. Primary and secondary amides show N—H stretching in the region from 3200 to 3400 cm^{-1}; primary amides (RCONH$_2$) show two N—H absorptions, whereas secondary amides (RCONHR) show only a single N—H absorption. Tertiary amides, of course, do not show N—H stretching absorptions. See the following three spectra (Figure 11.14).

Figure 11.14
Infrared spectra of
N,N-diethyldodecanamide
(**A**, a tertiary amide),
N-methylbenzamide
(**B**, a secondary amide),
butanamide (**C**, a primary amide).
(continues on the next page)

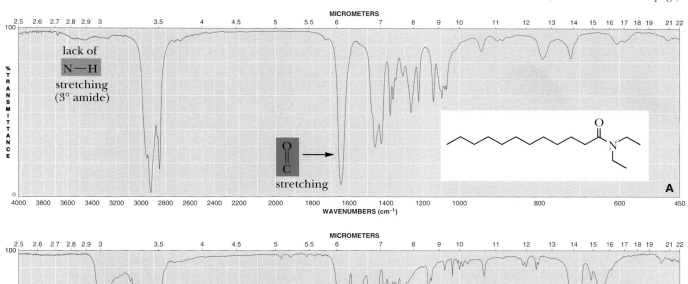

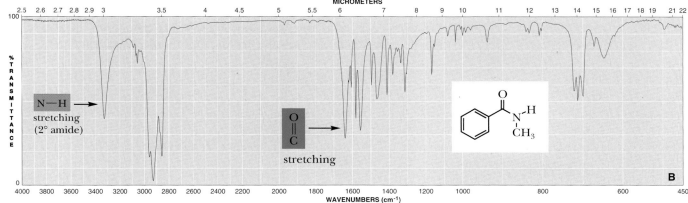

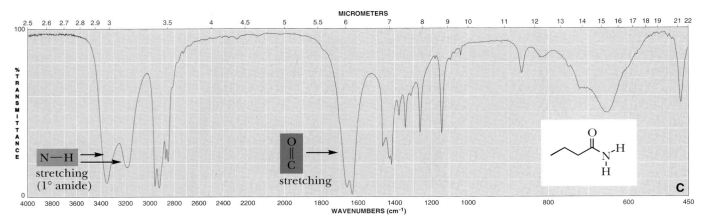

Figure 11.14 *(continued)*

EXAMPLE 11.4

An unknown compound with molecular formula $C_3H_6O_2$ yields the following IR spectrum. Draw possible structures for the unknown.

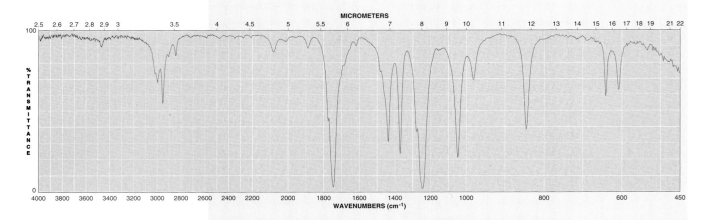

SOLUTION

The IR spectrum shows a strong absorption at approximately 1750 cm^{-1}, which is indicative of a C=O group. The spectrum also shows strong C—O absorption peaks at 1250 and 1050 cm^{-1}. Furthermore, there are no peaks above 3000 cm^{-1}, which eliminates the possibility of an O—H group. On the basis of these data, three structures are possible for the given molecular formula:

The spectrum can now be annotated as follows:

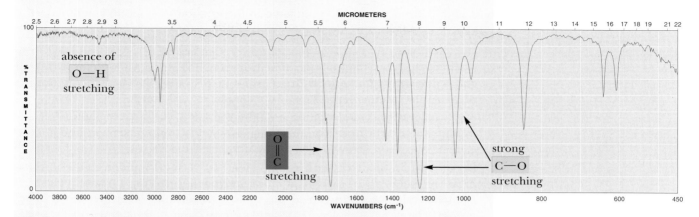

Practice Problem 11.4

What does the value of the wavenumber of the stretching frequency for a particular functional group indicate about the relative strength of the bond in that functional group?

J. Index of Hydrogen Deficiency

We can obtain valuable information about the structural formula of an unknown compound by inspecting its molecular formula. In addition to learning the number of atoms of carbon, hydrogen, oxygen, nitrogen, and so forth in a molecule of the compound, we can determine what is called its **index of hydrogen deficiency**, which is the sum of the number of rings and pi bonds in a molecule. We determine this quantity by comparing the number of hydrogens in the molecular formula of a compound of unknown structure with the number of hydrogens in a **reference compound** with the same number of carbon atoms and with no rings or pi bonds. The molecular formula of a reference hydrocarbon is C_nH_{2n+2} (Section 3.2).

Index of hydrogen deficiency The sum of the number of rings and pi bonds in a molecule.

$$\text{Index of hydrogen deficency} = \frac{(H_{reference} - H_{molecule})}{2}$$

EXAMPLE 11.5

Calculate the index of hydrogen deficiency for 1-hexene, with molecular formula C_6H_{12}, and account for this deficiency by reference to the structural formula of the compound.

SOLUTION

The molecular formula of the reference hydrocarbon with six carbon atoms is C_6H_{14}. The index of hydrogen deficiency of 1-hexene $(14 - 12)/2 = 1$ and is accounted for by the one pi bond in 1-hexene.

Practice Problem 11.5

Calculate the index of hydrogen deficiency of cyclohexene, C_6H_{10}, and account for this deficiency by reference to the structural formula of the compound.

To determine the molecular formula of a reference compound containing elements besides carbon and hydrogen, write the formula of the reference hydrocarbon, add to it other elements contained in the unknown compound, and make the following adjustments to the number of hydrogen atoms:

1. For each atom of a monovalent Group 7 element (F, Cl, Br, I) added to the reference hydrocarbon, subtract one hydrogen; halogen substitutes for hydrogen and reduces the number of hydrogens by one per halogen. The general formula of an acyclic monochloroalkane, for example, is $C_nH_{2n+1}Cl$.

2. No correction is necessary for the addition of atoms of Group 6 elements (O, S, Se) to the reference hydrocarbon. Inserting a divalent Group 6 element into a reference hydrocarbon does not change the number of hydrogens.

3. For each atom of a trivalent Group 5 element (N and P) added to the formula of the reference hydrocarbon, add one hydrogen. Inserting a trivalent Group 5 element adds one hydrogen to the molecular formula of the reference compound. The general molecular formula for an acyclic alkylamine, for example, is $C_nH_{2n+3}N$.

EXAMPLE 11.6

Isopentyl acetate, a compound with a bananalike odor, is a component of the alarm pheromone of honeybees. The molecular formula of isopentyl acetate is $C_7H_{14}O_2$. Calculate the index of hydrogen deficiency of this compound.

SOLUTION

The molecular formula of the reference hydrocarbon is C_7H_{16}. Adding oxygens to this formula does not require any correction in the number of hydrogens. The molecular formula of the reference compound is $C_7H_{16}O_2$, and the index of hydrogen deficiency is $(16 - 14)/2 = 1$, indicating either one ring or one pi bond. Following is the structural formula of isopentyl acetate, which contains one pi bond, in this case in the carbon–oxygen double bond:

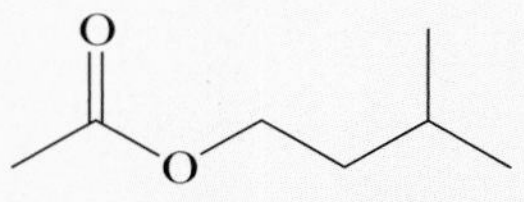

Isopentyl acetate

Practice Problem 11.6

The index of hydrogen deficiency of niacin is 5. Account for this value by reference to the structural formula of niacin.

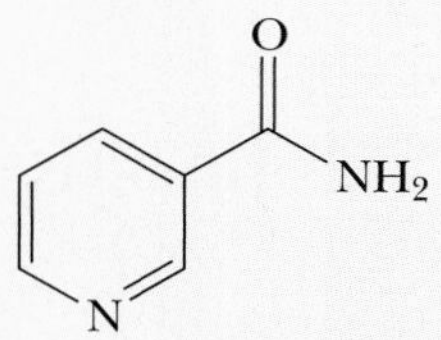

Nicotinamide
(Niacin)

EXAMPLE 11.7

Determine possible structures for a compound that yields the following IR spectrum and has a molecular formula of C_7H_8O:

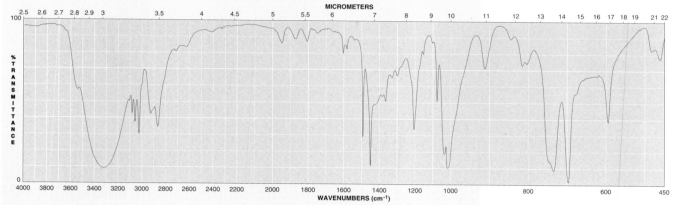

SOLUTION

The index of hydrogen deficiency for C_7H_8O is 4, based on the reference formula C_7H_{16}. While accounting for this value may seem daunting at first (consider the possible combinations of four rings or pi bonds), keep in mind that there is one common functional group in organic chemistry that has an index of hydrogen deficiency of 4, namely, a benzene ring. The characteristic aromatic $C-H$ bending bands at 690 and 740 cm^{-1} and the weak, broad bands between 1700 and 2000 cm^{-1} support the existence of a benzene ring. Also, sp^2 $C-H$ stretching is present just above the 3000-cm^{-1} mark. Aromatic $C=C$ stretching absorption bands at 1450 and 1490 cm^{-1} also indicate a benzene ring. The last piece of evidence is the strong, broad $O-H$ stretching peak at approximately 3310 cm^{-1}. Because we must have an OH group, we cannot propose any structures with an OCH_3 (ether) group. Based on this interpretation of the spectrum, the following four structures are possible:

The given spectrum can now be annotated as follows:

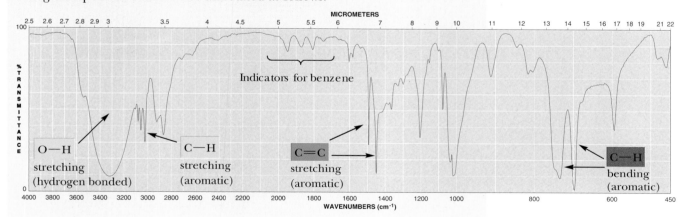

The preceding example illustrates the power and limitations of IR spectroscopy. The power lies in its ability to provide us with information regarding the functional groups in a molecule. IR spectroscopy does not, however, provide us with information on how those functional groups are connected. Fortunately, another type of spectroscopy—nuclear magnetic resonance (NMR) spectroscopy—does provide us with connectivity information. NMR spectroscopy is the topic of the next chapter.

Summary

Electromagnetic radiation (Section 11.2) can be described in terms of its **wavelength** (λ) and its **frequency** (ν). Frequency is reported in **hertz (Hz)**. An alternative way to describe electromagnetic radiation is in terms of its energy where $E = h\nu$ (Section 11.2).

Molecular spectroscopy (Section 11.3) is the experimental process of measuring which frequencies of radiation are absorbed or emitted by a substance and correlating these patterns with details of molecular structure. Interactions of molecules with **infrared radiation** excite covalent bonds to higher vibrational energy levels (Section 11.3).

The **vibrational infrared** spectrum (Section 11.4A) extends from 4000 to 400 cm^{-1}. Radiation in this region is referred to by its wavenumber $\bar{\nu}$, in reciprocal centimeters (cm^{-1}). To be **infrared active** (Section 11.4B), a bond must be polar; the more polar it is, the stronger is its absorption of IR radiation. There are $3n - 6$ allowed fundamental vibrations

for a nonlinear molecule containing n atoms. The simplest vibrations that give rise to the absorption of infrared radiation are **stretching** and **bending** vibrations. Stretching may be symmetrical or asymmetrical.

A **correlation table** is a list of the absorption patterns of functional groups. The intensity of a peak is referred to as **strong (s)**, **medium (m)**, or **weak (w)**. Stretching vibrations for most functional groups appear in the region from 3650 to 1000 cm^{-1}. The region from 1000 to 400 cm^{-1} is called the **fingerprint region** (Section 11.4C).

The **index of hydrogen deficiency** (Section 11.5J) is the sum of the number of rings and pi bonds in a molecule. It can be determined by comparing the number of hydrogens in the molecular formula of a compound of unknown structure with the number of hydrogens in a reference compound with the same number of carbon atoms and with no rings or pi bonds.

Problems

A problem number set in red indicates an applied "real-world" problem.

Index of Hydrogen Deficiency

11.7 Complete the following table:

Class of Compound	Molecular Formula	Index of Hydrogen Deficiency	Reason for Hydrogen Deficiency
alkane	C_nH_{2n+2}	0	(reference hydrocarbon)
alkene	C_nH_{2n}	1	one pi bond
alkyne	_____	_____	_____
alkadiene	_____	_____	_____
cycloalkane	_____	_____	_____
cycloalkene	_____	_____	_____

11.8 Calculate the index of hydrogen deficiency of each compound:

(a) Aspirin, $C_9H_8O_4$

(b) Ascorbic acid (vitamin C), $C_6H_8O_6$

(c) Pyridine, C_5H_5N

(d) Urea, CH_4N_2O

(e) Cholesterol, $C_{27}H_{46}O$

(f) Trichloroacetic acid, C_2HCl_3O

11.9 Compound A, with molecular formula C_6H_{10}, reacts with H_2/Ni to give compound B, with molecular formula C_6H_{12}. The IR spectrum of compound A is provided. From this information about compound A tell

(a) Its index of hydrogen deficiency.

(b) The number of rings or pi bonds (or both) in compound A.

(c) What structural feature(s) would account for compound A's index of hydrogen deficiency.

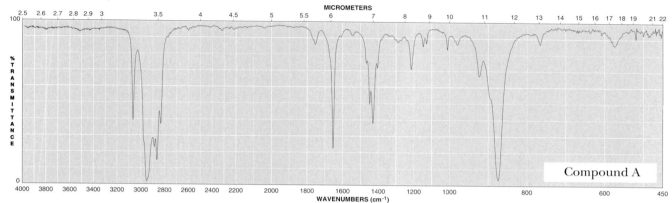

Compound A

11.10 Compound C, with molecular formula C_6H_{12}, reacts with H_2/Ni to give compound D, with molecular formula C_6H_{14}. The IR spectrum of compound C is provided. From this information about compound C, tell

(a) Its index of hydrogen deficiency.

(b) The number of rings or pi bonds (or both) in compound C.

(c) What structural feature(s) would account for compound C's index of hydrogen deficiency.

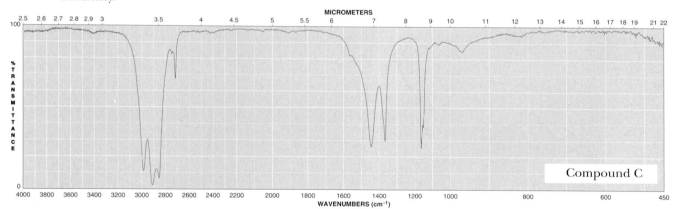

Compound C

11.11 Following are infrared spectra of compounds E and F: One spectrum is of 1-hexanol, the other of nonane. Assign each compound its correct spectrum.

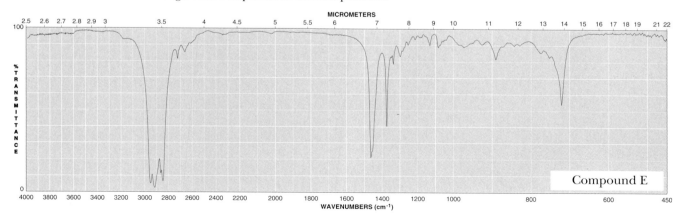

Compound E

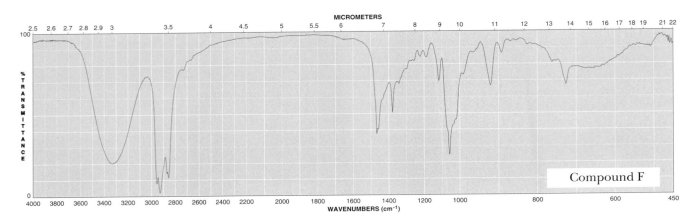

11.12 2-Methyl-1-butanol and *tert*-butyl methyl ether are constitutional isomers with molecular formula $C_5H_{12}O$. Assign each compound its correct infrared spectrum, G or H:

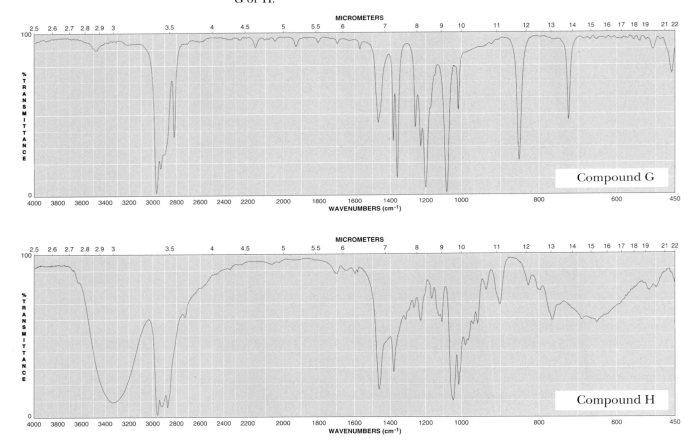

11.13 Examine the following IR spectrum and the molecular formula of compound I, $C_9H_{12}O$:

Tell

(a) Its index of hydrogen deficiency.

(b) The number of rings or pi bonds (or both) in compound I.

(c) What one structural feature would account for this index of hydrogen deficiency.

(d) What oxygen-containing functional group compound I contains.

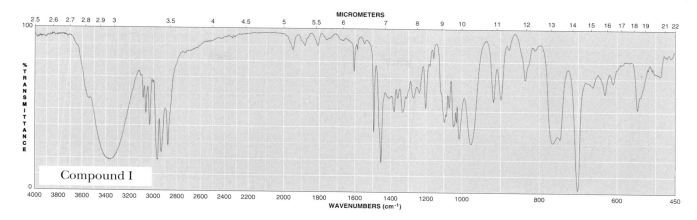

Compound I

11.14 Examine the following IR spectrum and the molecular formula of compound J, $C_5H_{13}N$:

Tell

(a) Its index of hydrogen deficiency.

(b) The number of rings or pi bonds (or both) in compound J.

(c) The nitrogen-containing functional group(s) compound J might contain.

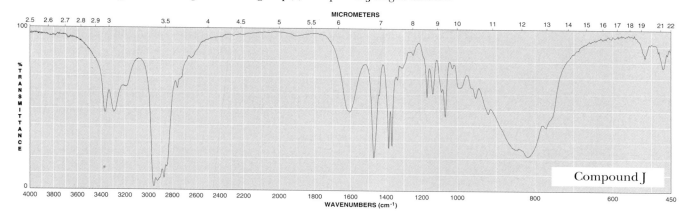

Compound J

11.15 Examine the following IR spectrum and the molecular formula of compound K, $C_6H_{12}O$:

Tell

(a) Its index of hydrogen deficiency.

(b) The number of rings or pi bonds (or both) in compound K.

(c) What structural features would account for this index of hydrogen deficiency.

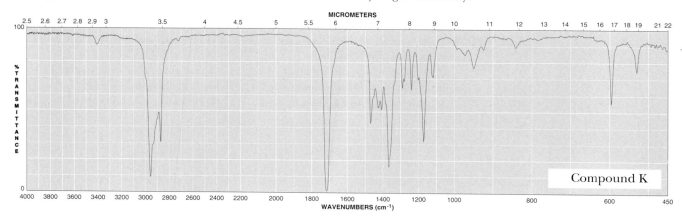

Compound K

11.16 Examine the following IR spectrum and the molecular formula of compound L, $C_6H_{12}O_2$:

Tell

(a) Its index of hydrogen deficiency.

(b) The number of rings or pi bonds (or both) in compound L.

(c) The oxygen-containing functional group(s) compound L might contain.

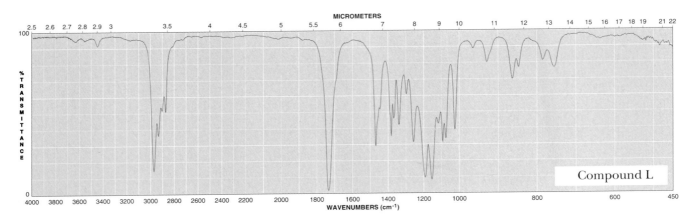

11.17 Examine the following IR spectrum and the molecular formula of compound M, C_3H_7NO:

Tell

(a) Its index of hydrogen deficiency.

(b) The number of rings or pi bonds (or both) in compound M.

(c) The oxygen- and nitrogen-containing functional group(s) in compound M.

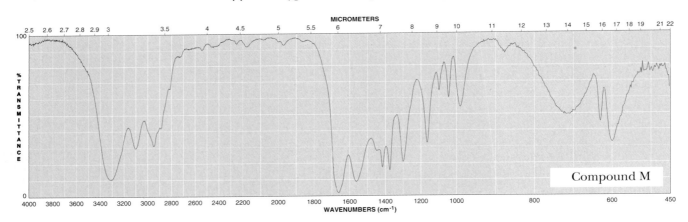

11.18 Show how IR spectroscopy can be used to distinguish between the compounds in each of the following pairs:

(a) 1-Butanol and diethyl ether   **(b)** Butanoic acid and 1-butanol

(c) Butanoic acid and 2-butanone   **(d)** Butanal and 1-butene

(e) 2-Butanone and 2-butanol   **(f)** Butane and 2-butene

11.19 For each pair of compounds that follows, list one major feature that appears in the IR spectrum of one compound, but not the other. In your answer, state what type of bond vibration is responsible for the spectral feature you list, and give its approximate position in the IR spectrum.

(a)

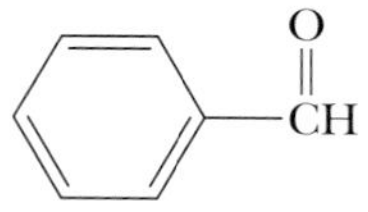

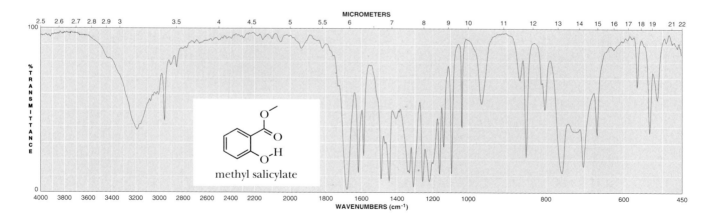

(a structures: benzaldehyde (C6H5—CH=O labeled —CH) and benzoic acid (C6H5—COOH))

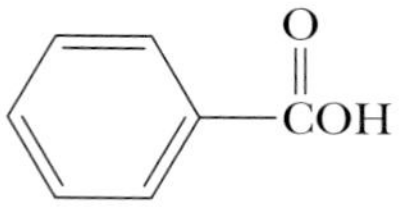

(b)

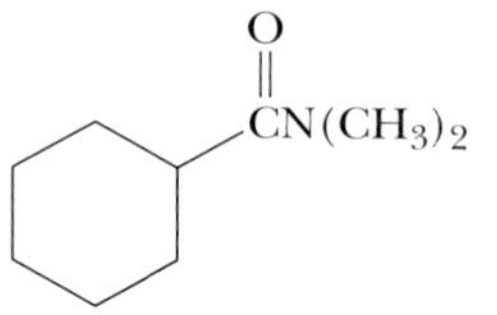

cyclohexyl—$\overset{\text{O}}{\overset{\|}{\text{C}}}$N(CH$_3$)$_2$ and   cyclohexyl—CH$_2$N(CH$_3$)$_2$

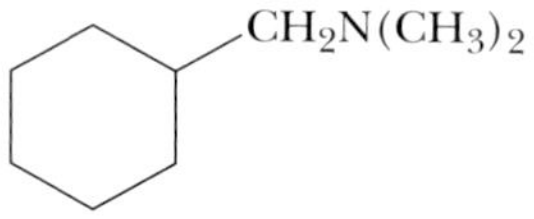

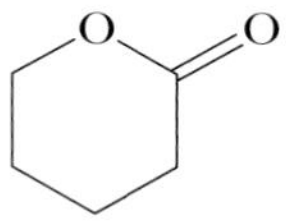

(c)

δ-valerolactone   and   HO(CH$_2$)$_4$$\overset{\text{O}}{\overset{\|}{\text{C}}}$OH

(d)

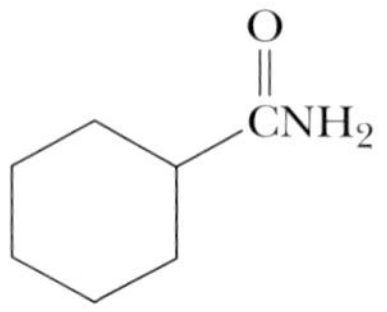

cyclohexyl—$\overset{\text{O}}{\overset{\|}{\text{C}}}NH_2$ and   cyclohexyl—$\overset{\text{O}}{\overset{\|}{\text{C}}}$N(CH$_3$)$_2$

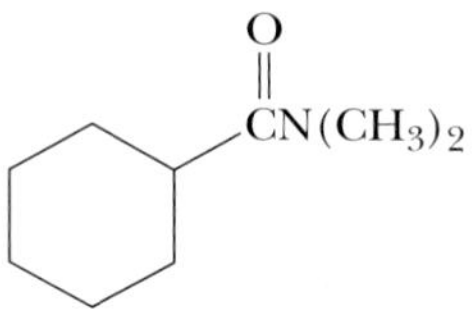

11.20 Following are an infrared spectrum and a structural formula for methyl salicylate, the fragrant component of oil of wintergreen. On this spectrum, locate the absorption peak(s) due to

(a) O—H stretching of the hydrogen-bonded —OH group (very broad and of medium intensity).

(b) C—H stretching of the aromatic ring (sharp and of weak intensity).

(c) C=O stretching of the ester group (sharp and of strong intensity).

(d) C=C stretching of the aromatic ring (sharp and of medium intensity).

methyl salicylate

Looking Ahead

11.21 In the next chapter, transitions between energy levels corresponding to frequencies on the order of 3×10^8 Hz are observed. Do these frequencies represent higher or lower energy than infrared radiation? Which region of the electromagnetic spectrum does this set of frequencies correspond to?

11.22 Predict the position of the $C{=}O$ stretching absorption in acetate ion relative to that in acetic acid:

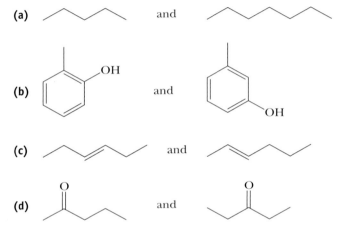

acetic acid   acetate ion

11.23 Determine whether IR spectroscopy can be used to distinguish between the following pairs of molecules (assume that you do not have the reference spectrum of either molecule):

(a)   [structure]   and   [structure]

(b)   [structure with OH]   and   [structure with OH]

(c)   [structure]   and   [structure]

(d)   [structure with O]   and   [structure with O]

11.24 Following is the IR spectrum of L-tryptophan, a naturally occurring amino acid that is abundant in foods such as turkey:

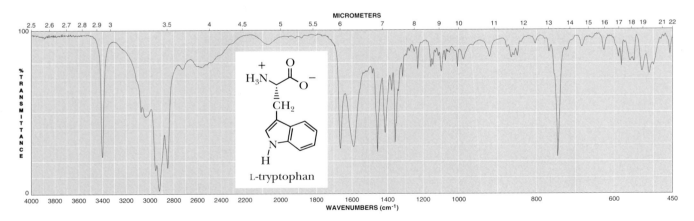

For many years, the L-tryptophan in turkey was believed to make people drowsy after Thanksgiving dinner. Scientists now know that consuming L-tryptophan makes one drowsy only if the compound is taken on an empty stomach. Therefore, it is unlikely that one's Thanksgiving day turkey is the cause of drowsiness. Notice that L-tryptophan contains one stereocenter. Its enantiomer, D-tryptophan, does not occur in nature, but can be synthesized in the laboratory. What would the IR spectrum of D-tryptophan look like?

12 Nuclear Magnetic Resonance Spectroscopy

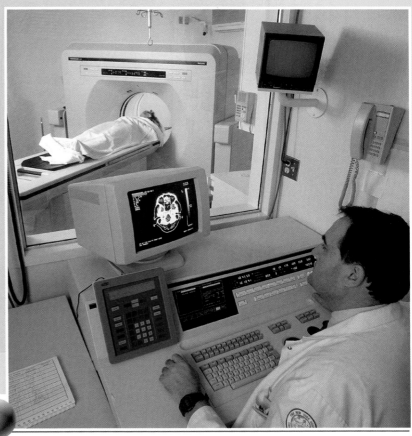

One of the most powerful diagnostic tools in modern medicine is Magnetic Resonance Imaging (MRI), a technique founded on the principles of Nuclear Magnetic Resonance (NMR) spectroscopy. Inset: A model of water. MRI differentiates the different tissues that surround water molecules in the body. Analogously, in NMR, the different environments that nuclei such as hydrogen reside in give rise to different signals. *(Larry Molvehill/Index Stock Imagery)*

12.1 INTRODUCTION

In the previous chapter, we discussed infrared spectroscopy and saw how infrared light could be used to determine the types of functional groups present in an unknown compound. In this chapter, we concentrate on the absorption of radio-frequency radiation, which causes transitions between nuclear spin energy levels; that is, we concentrate on a technique known as nuclear magnetic resonance (NMR) spectroscopy.

The phenomenon of nuclear magnetic resonance was first detected in 1946 by U.S. scientists Felix Bloch and Edward Purcell, who shared the 1952 Nobel prize for physics for their discoveries. The particular value of **nuclear magnetic resonance**

(NMR) spectroscopy is that it gives us information about the number and types of atoms in a molecule, for example, about the number and types of hydrogens using **^{1}H-NMR spectroscopy**, and about the number and types of carbons using **^{13}C-NMR spectroscopy**.

12.2   THE ORIGIN OF NUCLEAR MAGNETIC RESONANCE

From your study of general chemistry, you may already be familiar with the concept that an electron has a spin and that a spinning charge creates an associated magnetic field. In effect, an electron behaves as if it is a tiny bar magnet. An atomic nucleus that has an odd mass or an odd atomic number also has a spin and behaves as if it were a tiny bar magnet. Recall that when designating isotopes, a superscript represents the mass of the element. Thus, the nuclei of ^{1}H and ^{13}C, isotopes of the two elements most common in organic compounds, also have a spin, whereas the nuclei of ^{12}C and ^{16}O do not have a spin and do not behave as tiny bar magnets. Accordingly, in this sense, nuclei of ^{1}H and ^{13}C are quite different from nuclei of ^{12}C and ^{16}O.

EXAMPLE 12.1

Which of the following nuclei are capable of behaving like tiny bar magnets?

(a) $^{14}_{6}$C   (b) $^{14}_{7}$N

SOLUTION

(a) ^{14}C, a radioactive isotope of carbon, has neither an odd mass number nor an odd atomic number and therefore cannot behave like a tiny bar magnet.
(b) ^{14}N, the most common naturally occurring isotope of nitrogen (99.63% of all nitrogen atoms), has an odd atomic number and therefore behaves like a tiny bar magnet.

Practice Problem 12.1 ————————————————————————————

Which of the following nuclei are capable of behaving like tiny bar magnets?

(a) $^{31}_{15}$P   (b) $^{195}_{78}$Pt

Within a collection of ^{1}H and ^{13}C atoms, the spins of their tiny nuclear bar magnets are completely random in orientation. When we place them between the poles of a powerful magnet, however, interactions between their nuclear spins and the applied magnetic field are quantized, and only two orientations are allowed (Figure 12.1).

At an applied field strength of 7.05 tesla (T), which is readily available with present-day superconducting electromagnets, the difference in energy between nuclear spin states for ^{1}H is 0.0286 cal (0.120 J)/mol, which corresponds to electromagnetic radiation of approximately 300 MHz (300,000,000 Hz). At the same magnetic field strength, the difference in energy between nuclear spin states for ^{13}C is 0.0072 cal (0.035 J)/mol, which corresponds to electromagnetic radiation of 75 MHz. Thus, we can use electromagnetic radiation in the radio-frequency range to detect changes in nuclear spin states for ^{1}H and ^{13}C. In the next several sections, we describe how these measurements are made for nuclear spin states of these two isotopes and then how this information can be correlated with molecular structure.

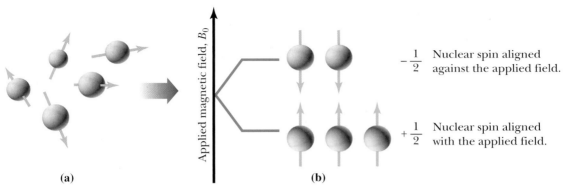

Figure 12.1
^{1}H and ^{13}C nuclei (a) in the absence of an applied magnetic field and (b) in the presence of an applied field. ^{1}H and ^{13}C nuclei with spin $+\frac{1}{2}$ are aligned with the applied magnetic field and are in the lower spin energy state; those with spin $-\frac{1}{2}$ are aligned against the applied magnetic field and are in the higher spin energy state.

12.3   NUCLEAR MAGNETIC RESONANCE

When hydrogen nuclei are placed in an applied magnetic field, a small majority of their nuclear spins align with the applied field in the lower energy state. Irradiation of the nuclei in the lower energy spin state with radio-frequency radiation of the appropriate energy causes them to absorb energy and results in their nuclear spins flipping from the lower energy state to the higher energy state, as illustrated in Figure 12.2. In this context, **resonance** is defined as the absorption of electromagnetic radiation by a spinning nucleus and the resulting flip of its nuclear spin state. The instrument we use to detect this absorption and resulting flip of nuclear spin state records it as a resonance **signal**.

Resonance The absorption of electromagnetic radiation by a spinning nucleus and the resulting "flip" of its spin from a lower energy state to a higher energy state.

Signal A recording of nuclear magnetic resonance in an NMR spectrum.

12.4   SHIELDING

If we were dealing with ^{1}H nuclei isolated from all other atoms and electrons, any combination of applied field and electromagnetic radiation that produces a resonance signal for one hydrogen nucleus would produce the same resonance signal for all other hydrogen nuclei. In other words, the same amount of energy would

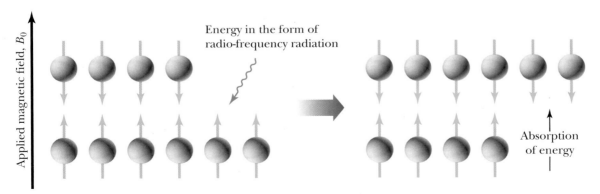

Figure 12.2
An example of resonance for nuclei of spin $\frac{1}{2}$.

cause all hydrogens to resonate, and the hydrogens would be indistinguishable one from another. NMR would then be an ineffective technique for determining the structure of a molecule, because all the hydrogens in a compound would resonate at the same frequency, giving rise to one and only one NMR signal.

Fortunately, hydrogens in most organic molecules are surrounded by electrons and by other atoms. The electrons that surround a nucleus also have spin and thereby create **local magnetic fields** that oppose the applied field. Although these local magnetic fields created by electrons are orders of magnitude weaker than the applied magnetic fields used in NMR spectroscopy, they are nonetheless significant at the molecular level. The result of these local magnetic fields is to shield hydrogens from the applied field. The greater the **shielding** of a particular hydrogen by local magnetic fields, the greater is the strength of the applied field necessary to bring that hydrogen into resonance.

As we learned in previous chapters, the electron density around a nucleus can be influenced by the atoms that surround the nucleus. For example, the electron density around the hydrogen atoms in fluoromethane is less than that around the hydrogen atoms in chloromethane, due to the greater electronegativity of fluorine relative to chlorine. Thus, we can say that the hydrogen atoms in chloromethane are *more shielded* than the hydrogen atoms in fluoromethane:

Shielding In NMR spectroscopy, electrons around a nucleus create their own local magnetic fields and thereby shield the nucleus from the applied magnetic field.

Chlorine is less electronegative than fluorine, resulting in a smaller inductive effect and thereby a greater electron density around each hydrogen. We say that the hydrogens in chloromethane are *more shielded* (by their local environment) than those in fluoromethane.

Fluorine's greater electronegativity produces a larger inductive effect and thereby reduces the electron density around each hydrogen. We say that these hydrogens are **deshielded**.

The differences in resonance frequencies among the various ^{1}H nuclei within a molecule caused by shielding are generally very small. The difference between the resonance frequencies of hydrogens in chloromethane compared with those in fluoromethane, for example, is only 360 Hz under an applied field of 7.05 tesla. Considering that the radio-frequency radiation used at this applied field is approximately 300 MHz, the difference in resonance frequencies between these two sets of hydrogens is only slightly greater than 1 part per million (1 ppm) compared with the irradiating frequency.

$$\frac{360 \text{ Hz}}{300 \times 10^6 \text{ Hz}} = \frac{1.2}{10^6} = 1.2 \text{ ppm}$$

The importance of shielding for elucidating the structure of a molecule will be discussed in Section 12.8.

12.5 AN NMR SPECTROMETER

The essential elements of an NMR spectrometer are a powerful magnet, a radio-frequency generator, a radio-frequency detector, and a sample tube (Figure 12.3).

The sample is dissolved in a solvent having no hydrogens, most commonly deuterochloroform ($CDCl_3$) or deuterium oxide (D_2O). The sample cell is a small glass tube suspended in the gap between the pole pieces of the magnet and set

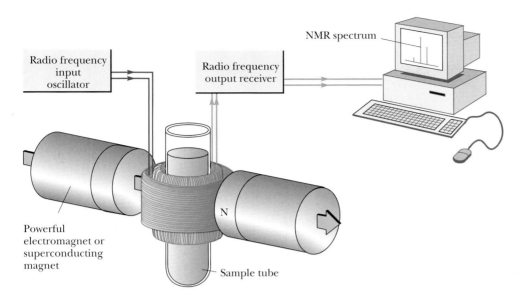

Figure 12.3
Schematic diagram of a
nuclear magnetic
resonance spectrometer.

spinning on its long axis to ensure that all parts of the sample experience a homogeneous applied field. In a typical ^{1}H-NMR spectrum, the horizontal axis represents the δ (delta) scale, with values from 0 on the right to 10 on the left. The vertical axis represents the intensity of the resonance signal.

It is customary to measure the resonance frequencies of individual nuclei relative to the resonance frequency of the same nuclei in a reference compound. The reference compound now universally accepted for ^{1}H-NMR and ^{13}C-NMR spectroscopy is **tetramethylsilane (TMS)**:

$$H_3C - \underset{\underset{CH_3}{|}}{\overset{\overset{CH_3}{|}}{Si}} - CH_3$$

Tetramethylsilane (TMS)

When we determine a ^{1}H-NMR spectrum of a compound, we report how far the resonance signals of its hydrogens are shifted from the resonance signal of the hydrogens in TMS. When we determine a ^{13}C-NMR spectrum, we report how far the resonance signals of its carbons are shifted from the resonance signal of the four carbons in TMS.

To standardize reporting of NMR data, workers have adopted a quantity called the **chemical shift(δ)**, expressed in parts per million:

$$\delta = \frac{\text{Shift in frequency of a signal from TMS (Hz)}}{\text{Operating frequency of the spectrometer (Hz)}}$$

Chemical shift, δ The position of a signal on an NMR spectrum relative to the signal of tetramethylsilane (TMS); expressed in delta (δ) units, where 1 δ equals 1 ppm.

Figure 12.4 shows a ^{1}H-NMR spectrum of methyl acetate, a compound used in the manufacture of artificial leather. The small signal at δ 0 in this spectrum represents the hydrogens of the reference compound, TMS. The remainder of the spectrum consists of two signals: one for the hydrogens of the $-OCH_3$ group and one for the hydrogens of the methyl attached to the carbonyl group. It is not our purpose at the moment to determine which hydrogens give rise to which signal but only to recognize the form in which we record an NMR spectrum and to understand the meaning of the calibration marks.

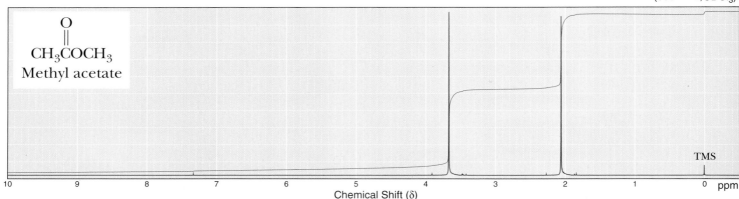

Figure 12.4
¹H-NMR spectrum of methyl acetate.

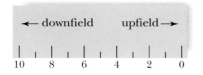

Downfield A term used to refer to the relative position of a signal on an NMR spectrum. Downfield indicates a peak to the left of the spectrum (a weaker applied field).

Upfield A term used to refer to the relative position of a signal on an NMR spectrum. Upfield indicates a peak to the right of the spectrum (a stronger applied field).

Equivalent hydrogens Hydrogens that have the same chemical environment.

A note on terminology. If a signal is shifted toward the left on the chart paper, we say that it is shifted **downfield**, meaning that nuclei giving rise to that signal are less shielded and come into resonance at a weaker applied field. Conversely, if a signal is shifted toward the right of the spectrum, we say that it is shifted **upfield**, meaning that nuclei giving rise to that signal are more shielded and come into resonance at a stronger applied field.

12.6 EQUIVALENT HYDROGENS

Given the structural formula of a compound, how do we know how many signals to expect? The answer is that **equivalent hydrogens** give the same ¹H-NMR signal; conversely, nonequivalent hydrogens give different ¹H-NMR signals. A direct way to determine which hydrogens in a molecule are equivalent is to replace each in turn by a test atom, such as a halogen atom. If replacement of two hydrogens being tested in this way gives the same compound, the two hydrogens are equivalent. If replacement gives different compounds, the two hydrogens are nonequivalent.

Using this substitution test, we can show that propane contains two sets of equivalent hydrogens: a set of six equivalent 1° hydrogens and a set of two equivalent 2° hydrogens. Thus we would expect to see two signals, one for the six equivalent —CH₃ hydrogens and one for the two equivalent —CH₂— hydrogens:

Replacement of any of the red hydrogens by chlorine gives 1-chloropropane; thus, all the red hydrogens are **equivalent**.

Replacement of either of the blue hydrogens by chlorine gives 2-chloropropane; thus, both of the blue hydrogens are **equivalent**.

EXAMPLE 12.2

State the number of sets of equivalent hydrogens in each compound and the number of hydrogens in each set:

(a)

2-Methylpropane

(b)

2-Methylbutane

SOLUTION

(a) 2-Methylpropane contains two sets of equivalent hydrogens—a set of nine equivalent 1° hydrogens and one 3° hydrogen:

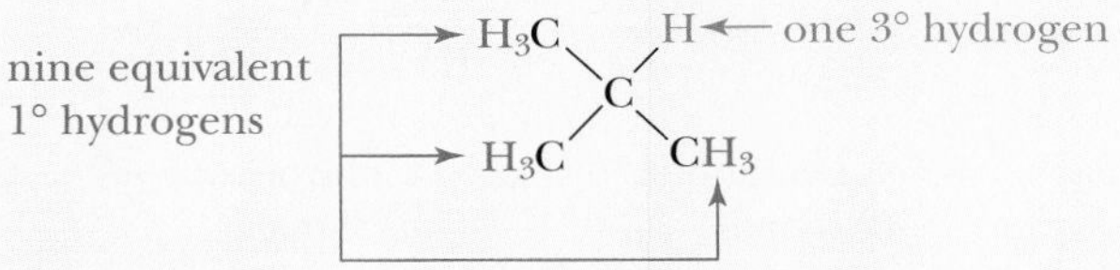

$$
\text{nine equivalent } 1° \text{ hydrogens} \quad
\begin{array}{c}
H_3C \\[2pt]
\quad \diagdown \quad \quad H \leftarrow \text{one } 3° \text{ hydrogen} \\[2pt]
\quad \quad C \\[2pt]
H_3C \diagup \quad \diagdown CH_3
\end{array}
$$

Replacing any one of the red hydrogens with a chlorine yields 1-chloro-2-methylpropane. Replacing the blue hydrogen with a chlorine yields 2-chloro-2-methylpropane.

(b) 2-Methylbutane contains four sets of equivalent hydrogens—two different sets of 1° hydrogens, one set of 2° hydrogens, and one 3° hydrogen:

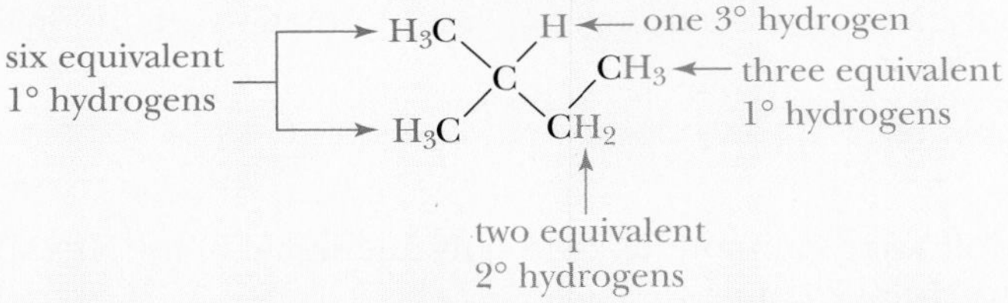

$$
\text{six equivalent } 1° \text{ hydrogens} \quad
\begin{array}{c}
H_3C \\[2pt]
\quad \diagdown \quad H \leftarrow \text{one } 3° \text{ hydrogen} \\[2pt]
\quad \quad C \quad CH_3 \leftarrow \text{three equivalent } 1° \text{ hydrogens} \\[2pt]
H_3C \diagup \quad \diagdown CH_2
\end{array}
$$

two equivalent 2° hydrogens

Replacing any one of the red hydrogens with a chlorine yields 1-chloro-2-methylbutane. Replacing the blue hydrogen with a chlorine yields 2-chloro-2-methylbutane. Replacing a purple hydrogen with a chlorine yields 2-chloro-3-methylbutane. Replacing a green hydrogen with chlorine yields 1-chloro-3-methylbutane.

Practice Problem 12.2

State the number of sets of equivalent hydrogens in each compound and the number of hydrogens in each set:

(a)

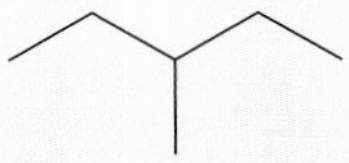

3-Methylpentane

(b)

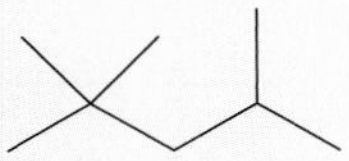

2,2,4-Trimethylpentane

Here are four organic compounds, each of which has one set of equivalent hydrogens and gives one signal in its ^{1}H-NMR spectrum:

$$
\underset{\text{Propanone}}{\underset{\text{(Acetone)}}{CH_3\overset{\overset{\displaystyle O}{\parallel}}{C}CH_3}} \qquad \underset{\text{1,2-Dichloroethane}}{ClCH_2CH_2Cl} \qquad \underset{\text{Cyclopentane}}{\bigcirc} \qquad \underset{\text{2,3-Dimethyl-2-butene}}{\overset{H_3C}{\underset{H_3C}{}}C=C\overset{CH_3}{\underset{CH_3}{}}}
$$

Molecules with two or more sets of equivalent hydrogens give rise to a different resonance signal for each set. 1,1-Dichloroethane, for example, has three equivalent 1° hydrogens (a) and one 2° hydrogen (b); there are two resonance signals in its ^{1}H-NMR spectrum.

$$
\underset{\substack{\text{1,1-Dichloroethane} \\ \text{(2 signals)}}}{\underset{(a)\quad(b)}{CH_3\overset{\overset{\displaystyle Cl}{|}}{C}HCl}} \qquad \underset{\substack{\text{Cyclopentanone} \\ \text{(2 signals)}}}{} \qquad \underset{\substack{\text{(Z)-1-Chloropropene} \\ \text{(3 signals)}}}{} \qquad \underset{\substack{\text{Cyclohexene} \\ \text{(3 signals)}}}{}
$$

You should see immediately that valuable information about a compound's molecular structure can be obtained simply by counting the number of signals in a ^{1}H-NMR spectrum of that compound. Consider, for example, the two constitutional isomers with molecular formula $C_2H_4Cl_2$. The compound 1,2-dichloroethane has four equivalent 2° hydrogens and shows one signal in its ^{1}H-NMR spectrum. Its constitutional isomer 1,1-dichloroethane has three equivalent 1° hydrogens and one 2° hydrogen; this isomer shows two signals in its ^{1}H-NMR spectrum. Thus, simply counting signals allows you to distinguish between these constitutional isomers.

EXAMPLE 12.3

Each of the following compounds gives only one signal in its ^{1}H-NMR spectrum. Propose a structural formula for each.

(a) C_2H_6O (b) $C_3H_6Cl_2$ (c) C_6H_{12}

SOLUTION

Following are structural formulas for each of the given compounds. Notice that, for each structure, the replacement of any hydrogen with a chlorine will yield the same compound regardless of the hydrogen being replaced.

(a) CH_3OCH_3 (b) $CH_3\overset{\overset{\displaystyle Cl}{|}}{\underset{\underset{\displaystyle Cl}{|}}{C}}CH_3$ (c) $\bigcirc$ or $\overset{H_3C}{\underset{H_3C}{}}C=C\overset{CH_3}{\underset{CH_3}{}}$

Practice Problem 12.3

Each of the following compounds gives only one signal in its ^{1}H-NMR spectrum. Propose a structural formula for each compound.

(a) C_3H_6O (b) C_5H_{10} (c) C_5H_{12} (d) $C_4H_6Cl_4$

12.7 SIGNAL AREAS

We have just seen that the number of signals in a ^{1}H-NMR spectrum gives us information about the number of sets of equivalent hydrogens. Signal areas in a ^{1}H-NMR spectrum can be measured by a mathematical technique called *integration*. Integration is done in the following way: As a spectrum is being recorded, the instrument's computer numerically adds together the areas under all of the signals. In the spectra shown in this text, this information is displayed in the form of a **line of integration** superposed on the original spectrum. The vertical rise of the line of integration over each signal is proportional to the area under that signal, which, in turn, is proportional to the number of hydrogens giving rise to the signal.

Figure 12.5 shows an integrated ^{1}H-NMR spectrum of the gasoline additive *tert*-butyl acetate ($C_6H_{12}O_2$). The spectrum shows signals at δ 1.44 and 1.95. The integrated height of the upfield (to the right) signal is nearly three times as tall as the height of the downfield (to the left) signal. This relationship corresponds to a ratio of 3:1. We know from the molecular formula that there is a total of 12 hydrogens in the molecule. The ratios obtained from the integration lines are consistent with the presence of one set of 9 equivalent hydrogens and one set of 3 equivalent hydrogens. Alternatively, we could use the horizontal lines on the chart as a unit of measure. (Ten chart divisions are equivalent to the distance between two consecutive solid horizontal lines.) In this fashion, we obtain an integration value of 67 for the upfield signal and 23 for the downfield signal. (Alternatively, we could measure the lines with a ruler.) These values add up to 90 chart divisions for 12 hydrogens. Dividing the total number of chart divisions by the total number of hydrogens gives $90 \div 12 = 7.5$ chart divisions per hydrogen. Thus, the signal at δ 1.44 represents $67 \div 7.5 \approx 9$ equivalent hydrogens, and the signal at δ 1.95 represents $23 \div 7.5 \approx 3$ hydrogens. We will often make use of shorthand notation in referring to an NMR spectrum of a molecule. The notation lists the chemical shift of each signal, beginning with the most deshielded signal and followed by the number of hydrogens that give rise to each signal (based on the integration). The shorthand notation describing the spectrum of *tert*-butyl acetate (Figure 12.5) would be δ 1.95 (3H) and δ 1.44 (9H).

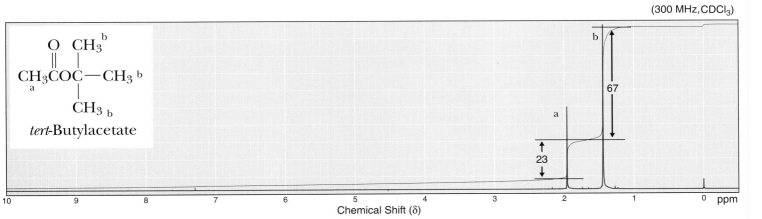

(300 MHz, CDCl$_3$)

Figure 12.5

^{1}H-NMR spectrum of *tert*-butyl acetate, $C_6H_{12}O_2$, showing a line of integration. The ratio of signal heights for the two peaks is 3:1, which, for a molecule possessing 12 hydrogens, corresponds to 9 equivalent hydrogens of one set and 3 equivalent hydrogens of another set.

EXAMPLE 12.4

Following is a ^{1}H-NMR spectrum for a compound with molecular formula $C_9H_{10}O_2$. From an analysis of the integration line, calculate the number of hydrogens giving rise to each signal.

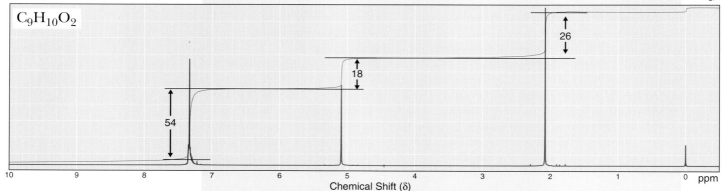

(300 MHz, CDCl$_3$)

$C_9H_{10}O_2$

SOLUTION

The ratio of the relative signal heights (obtained from the number of horizontal chart divisions) is 5:2:3 (from downfield to upfield). The molecular formula indicates that there are 10 hydrogens. Thus, the signal at δ 7.34 represents 5 hydrogens, the signal at δ 5.08 represents 2 hydrogens, and the signal at δ 2.06 represents 3 hydrogens. Consequently, the signals and the number of hydrogens each signal represents are δ 7.34 (5H), δ 5.08 (2H), and δ 2.06 (3H).

Practice Problem 12.4 ────────────────────────────

The line of integration of the two signals in the ^{1}H-NMR spectrum of a ketone with molecular formula $C_7H_{14}O$ shows a vertical rise of 62 and 10 chart divisions. Calculate the number of hydrogens giving rise to each signal, and propose a structural formula for this ketone.

12.8 CHEMICAL SHIFT

The position of a signal along the x-axis of an NMR spectrum is known as the **chemical shift** of that signal (Section 12.5). The chemical shift of a signal in a ^{1}H-NMR spectrum can give us valuable information about the type of hydrogens giving rise to that absorption. Hydrogens on methyl groups bonded to sp^3 hybridized carbons, for example, give a signal near δ 0.8–1.0 (compare Figure 12.5). Hydrogens on methyl groups bonded to a carbonyl carbon give signals near δ 2.1–2.3 (compare Figures 12.4 and 12.5), and hydrogens on methyl groups bonded to oxygen give signals near δ 3.7–3.9 (compare Figure 12.4). Table 12.1 lists the average chemical shift for most of the types of hydrogens we deal with in this text. Notice that most of the values shown fall within a rather narrow range from 0 to 13 δ units (ppm). In fact, although the table shows a variety of functional groups and hydrogens bonded to them, we can use the following rules of thumb to remember the chemical shifts of most types of hydrogen:

Chemical Shift (δ)	Type of Hydrogen
0–2	H bonded to an sp^3 carbon.
2–4.5	H bonded to an sp^3 carbon that is close to an electronegative element such as N, O, or X. The more electronegative the element, the higher is the chemical shift. Also, the closer the electronegative atom, the higher is the chemical shift.
4.6–5.7	H bonded to an sp^2 carbon in an alkene.
6.5–8.5	H bonded to an sp^2 carbon in an aromatic compound.
9.5–10.1	H bonded to a C=O (an aldehyde hydrogen).
10–13	H of a carboxyl (COOH) group.

TABLE 12.1 Average Values of Chemical Shifts of Representative Types of Hydrogens

Type of Hydrogen (R = alkyl, Ar = aryl)	Chemical Shift (δ)*	Type of Hydrogen (R = alkyl, Ar = aryl)	Chemical Shift (δ)*
$(CH_3)_4Si$	0 (by definition)	RCOCH₃ (O)	3.7–3.9
RCH_3	0.8–1.0	RCOCH₂R (O)	4.1–4.7
RCH_2R	1.2–1.4	RCH_2I	3.1–3.3
R_3CH	1.4–1.7	RCH_2Br	3.4–3.6
$R_2C=CRCHR_2$	1.6–2.6	RCH_2Cl	3.6–3.8
$RC\equiv CH$	2.0–3.0	RCH_2F	4.4–4.5
$ArCH_3$	2.2–2.5	ArOH	4.5–4.7
$ArCH_2R$	2.3–2.8	$R_2C=CH_2$	4.6–5.0
ROH	0.5–6.0	$R_2C=CHR$	5.0–5.7
RCH_2OH	3.4–4.0	ArH	6.5–8.5
RCH_2OR	3.3–4.0	RCH (O)	9.5–10.1
R_2NH	0.5–5.0	RCOH (O)	10–13
RCCH₃ (O)	2.1–2.3		
RCCH₂R (O)	2.2–2.6		

*Values are approximate. Other atoms within the molecule may cause the signal to appear outside these ranges.

EXAMPLE 12.5

Following are two constitutional isomers with molecular formula $C_6H_{12}O_2$:

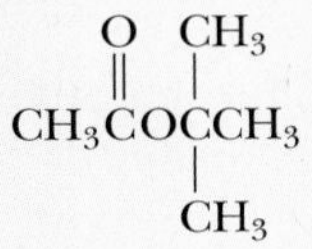

(1) CH₃COCCH₃ with O, CH₃, CH₃ groups   (2) CH₃OCCCH₃ with OCH₃, CH₃ groups

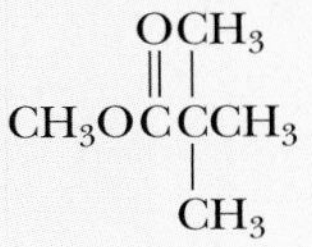

(a) Predict the number of signals in the ¹H-NMR spectrum of each isomer.
(b) Predict the ratio of areas of the signals in each spectrum.
(c) Show how to distinguish between these isomers on the basis of chemical shift.

SOLUTION

(a) Each compound contains a set of nine equivalent methyl hydrogens and a set of three equivalent methyl hydrogens.
(b) The ^{1}H-NMR spectrum of each consists of two signals in the ratio $9:3$, or $3:1$.
(c) The two constitutional isomers can be distinguished by the chemical shift of the single —CH_3 group (shown in red for each compound). Using our rules of thumb, we find that the hydrogens of CH_3O are less shielded (appear farther downfield) than the hydrogens of $CH_3C=O$. Table 12.1 gives approximate values for each chemical shift. Experimental values are as follows:

$$\delta\,1.95 \searrow\quad \underset{CH_3\overset{O}{\overset{\|}{C}}\underset{\underset{CH_3}{|}}{\overset{CH_3}{\overset{|}{O}}}CCH_3}{}\quad \delta\,1.44 \qquad \delta\,3.67 \searrow\quad \underset{CH_3O\overset{OCH_3}{\overset{\|}{C}}CCH_3}{}\quad \delta\,1.20$$

(1)   (2)

Practice Problem 12.5

Following are two constitutional isomers with molecular formula $C_4H_8O_2$:

$$CH_3CH_2O\overset{O}{\overset{\|}{C}}CH_3 \qquad CH_3CH_2\overset{O}{\overset{\|}{C}}OCH_3$$

(1)   (2)

(a) Predict the number of signals in the ^{1}H-NMR spectrum of each isomer.
(b) Predict the ratio of areas of the signals in each spectrum.
(c) Show how to distinguish between these isomers on the basis of chemical shift.

12.9 SIGNAL SPLITTING AND THE (*n* + 1) RULE

We have now seen three kinds of information that can be derived from an examination of a ^{1}H-NMR spectrum:

1. From the number of signals, we can determine the number of sets of equivalent hydrogens.
2. By integrating over signal areas, we can determine the relative numbers of hydrogens giving rise to each signal.
3. From the chemical shift of each signal, we can derive information about the types of hydrogens in each set.

We can derive a fourth kind of information from the splitting pattern of each signal. Consider, for example, the ^{1}H-NMR spectrum of 1,1,2-trichloroethane (Figure 12.6), a solvent for waxes and natural resins. This molecule contains two 2° hydrogens and one 3° hydrogen, and, according to what we have learned so far, we predict two signals with relative areas $2:1$, corresponding to the two hydrogens of the —CH_2— group and the one hydrogen of the —$CHCl_2$ group. You see from the spectrum, however, that there are in fact five **peaks**. How can this be, when we predict only two signals? The answer is that a hydrogen's resonance frequency can be affected by the tiny magnetic fields of other hydrogens close by. Those fields cause the signal to be **split** into numerous peaks. Hydrogens split each other if they are separated by no more than three bonds—for example, H—C—C—H or H—C—O—H. (There are three bonds in each case.) If there are more than three bonds, as in H—C—C—C—H, then there is normally no splitting. A signal with just one peak is called a **singlet**. A signal that is split into two peaks is called a **doublet**. Signals that are split into three and four peaks are called **triplets** and **quartets**, respectively.

The grouping of two peaks at $\delta\,3.96$ in the ^{1}H-NMR spectrum of 1,1,2-trichloroethane is the signal for the hydrogens of the —CH_2— group, and the grouping of three peaks at $\delta\,5.77$ is the signal for the single hydrogen of the —$CHCl_2$ group.

Peak (NMR) The units into which an NMR signal is split—two peaks in a doublet, three peaks in a triplet, and so on.

Singlet A signal that consists of one peak; the hydrogens that give rise to the signal have no neighboring nonequivalent hydrogens.

Doublet A signal that is split into two peaks; the hydrogens that give rise to the signal have one neighboring nonequivalent hydrogen.

Triplet A signal that is split into three peaks; the hydrogens that give rise to the signal have two neighboring nonequivalent hydrogens that are equivalent to each other.

Quartet A signal that is split into four peaks; the hydrogens that give rise to the signal have three neighboring nonequivalent hydrogens that are equivalent to each other.

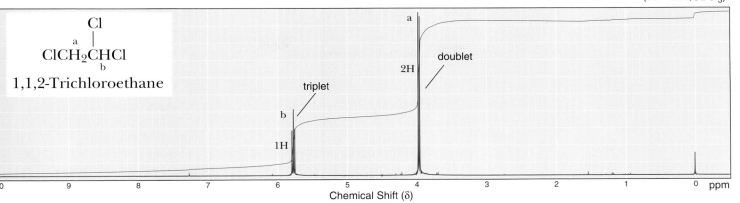

Figure 12.6
^{1}H-NMR spectrum of 1,1,2-trichloroethane.

We say that the CH_2 signal at δ 3.96 is split into a doublet and that the CH signal at δ 5.77 is split into a triplet. In this phenomenon, called **signal splitting**, the ^{1}H-NMR signal from one set of hydrogens is split by the influence of neighboring nonequivalent hydrogens.

The degree of signal splitting can be predicted on the basis of the **(*n* + 1) rule**, according to which, if a hydrogen has *n* hydrogens nonequivalent to it, but equivalent among themselves, on the same or adjacent atom(s), then the ^{1}H-NMR signal of the hydrogen is split into (*n* + 1) peaks.

Let us apply the (*n* + 1) rule to the analysis of the spectrum of 1,1,2-trichloroethane. The two hydrogens of the $-CH_2-$ group have one nonequivalent neighboring hydrogen ($n = 1$); their signal is split into a doublet ($1 + 1 = 2$). The single hydrogen of the $-CHCl_2$ group has a set of two nonequivalent neighboring hydrogens ($n = 2$); its signal is split into a triplet ($2 + 1 = 3$).

Signal splitting Splitting of an NMR signal into a set of peaks by the influence of neighboring nuclei.

(*n* + 1) Rule The ^{1}H-NMR signal of a hydrogen or set of equivalent hydrogens with *n* other hydrogens on neighboring carbons is split into (*n* + 1) peaks.

> For these hydrogens, $n = 1$;
> their signal is split into $(1 + 1)$
> or 2 peaks—a **doublet**
>
> For this hydrogen, $n = 2$;
> its signal is split into $(2 + 1)$
> or 3 peaks—a **triplet**
>
> $$Cl-CH_2-\underset{\underset{Cl}{|}}{CH}-Cl$$

The two hydrogens on carbon 2 (a CH_2 group) of 1-chloropropane are flanked on one side by a set of 2H on carbon 1, and on the other side by a set of 3H on carbon 3. Because the sets of hydrogen on carbons 1 and 3 are nonequivalent to each other and also nonequivalent to the hydrogens on carbon 2, they cause the signal for the CH_2 group on carbon 2 to be split into a complex pattern, which we will refer to simply as a multiplet.

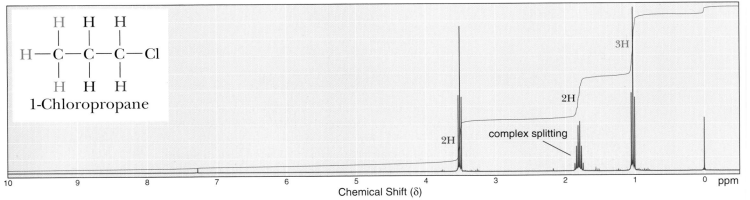

EXAMPLE 12.6

Predict the number of signals and the splitting pattern of each signal in the ^{1}H-NMR spectrum of each compound.

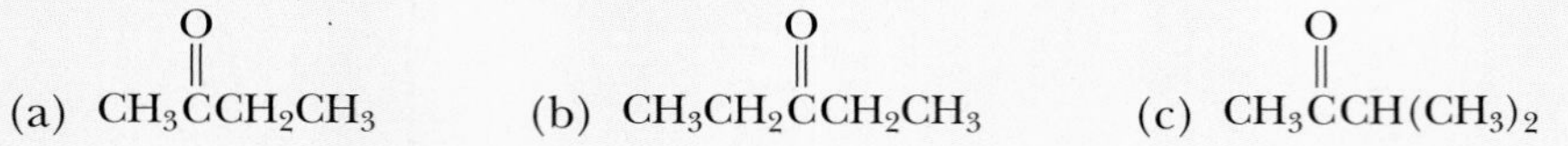

$$
\text{(a)} \quad CH_3\overset{\overset{\displaystyle O}{\|}}{C}CH_2CH_3 \qquad \text{(b)} \quad CH_3CH_2\overset{\overset{\displaystyle O}{\|}}{C}CH_2CH_3 \qquad \text{(c)} \quad CH_3\overset{\overset{\displaystyle O}{\|}}{C}CH(CH_3)_2
$$

SOLUTION

The sets of equivalent hydrogens in each molecule are color coded. In molecule (a), the signal for the red methyl group is unsplit (a singlet) because the group is too far (>3 bonds) from any other hydrogens. The blue—CH_2— group has three neighboring hydrogens ($n = 3$) and thus shows a signal split into a quartet ($3 + 1 = 4$). The green methyl group has two neighboring hydrogens ($n = 2$), and its signal is split into a triplet. The integration ratios for these signals would be $3:2:3$. Parts (b) and (c) can be analyzed in the same way. Thus, molecule (b) shows a triplet and a quartet in the ratio $3:2$. Molecule (c) shows a singlet, a septet ($6 + 1 = 7$), and a doublet in the ratio $3:1:6$.

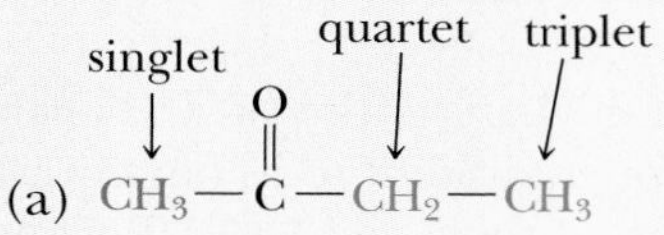

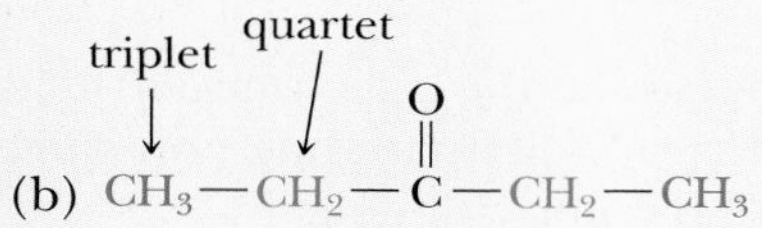

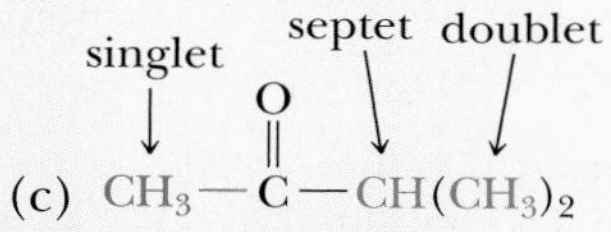

$$
\text{(a)} \quad \overset{\text{singlet}}{CH_3}-\overset{\overset{\displaystyle O}{\|}}{C}-\overset{\text{quartet}}{CH_2}-\overset{\text{triplet}}{CH_3}
$$

$$
\text{(b)} \quad \overset{\text{triplet}}{CH_3}-\overset{\text{quartet}}{CH_2}-\overset{\overset{\displaystyle O}{\|}}{C}-CH_2-CH_3
$$

$$
\text{(c)} \quad \overset{\text{singlet}}{CH_3}-\overset{\overset{\displaystyle O}{\|}}{C}-\overset{\text{septet doublet}}{CH(CH_3)_2}
$$

Practice Problem 12.6

Following are pairs of constitutional isomers. Predict the number of signals and the splitting pattern of each signal in the ^{1}H-NMR spectrum of each isomer.

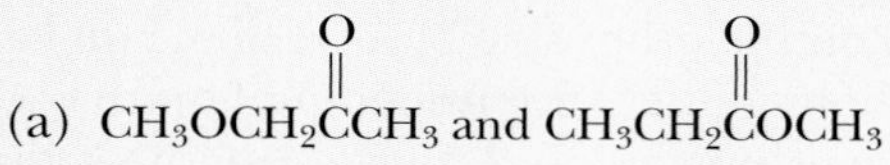

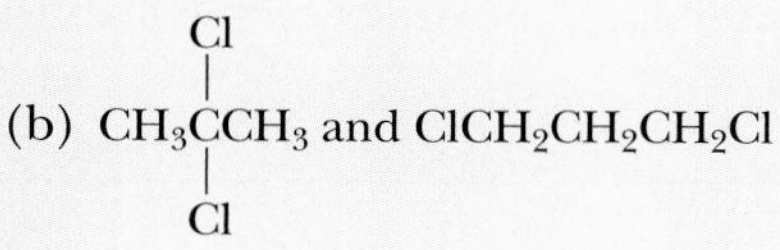

(a) $CH_3OCH_2\overset{\overset{\displaystyle O}{\|}}{C}CH_3$ and $CH_3CH_2\overset{\overset{\displaystyle O}{\|}}{C}OCH_3$

(b) $CH_3\overset{\overset{\displaystyle Cl}{|}}{\underset{\underset{\displaystyle Cl}{|}}{C}}CH_3$ and $ClCH_2CH_2CH_2Cl$

12.10 ^{13}C-NMR SPECTROSCOPY

Nuclei of carbon-12, the most abundant (98.89%) natural isotope of carbon, do not have nuclear spin and are not detected by NMR spectroscopy. Nuclei of carbon-13 (natural abundance 1.11%), however, do have nuclear spin and are detected by NMR spectroscopy in the same manner as hydrogens are detected.

CHEMICAL CONNECTIONS

Magnetic Resonance Imaging

Nuclear magnetic resonance was discovered and explained by physicists in the 1950s, and, by the 1960s, it had become an invaluable analytical tool for chemists. By the early 1970s, it was realized that the imaging of parts of the body via NMR could be a valuable addition to diagnostic medicine. Because the term "nuclear magnetic resonance" sounds to many people as if the technique might involve radioactive material, health care personnel call the technique *magnetic resonance imaging* (MRI).

The body contains several nuclei that, in principle, could be used for MRI. Of these, hydrogens, most of which come from water, triglycerides (fats), and membrane phospholipids, give the most useful signals. Phosphorus MRI is also used in diagnostic medicine.

Recall that, in NMR spectroscopy, energy in the form of radio-frequency radiation is absorbed by nuclei in the sample. The relaxation time is the characteristic time at which excited nuclei give up this energy and relax to their ground state.

In 1971, Raymond Damadian discovered that the relaxation of water in certain cancerous tumors takes much longer than the relaxation of water in normal cells. Thus, it was reasoned that if a relaxation image of the body could be obtained, it might be possible to identify tumors at an early stage. Subsequent work demonstrated that many tumors can be identified in this way.

Another important application of MRI is in the examination of the brain and spinal cord. White and gray matter, the two different layers of the brain, are easily distinguished by MRI, which is useful in the study of such diseases as multiple sclerosis. Magnetic resonance imaging and X-ray imaging are in many cases complementary: The hard, outer layer of bone is essentially invisible to MRI, but shows up extremely

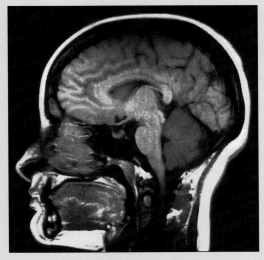

Computer-enhanced MRI scan of a normal human brain with pituitary gland highlighted (*Scott Camazine/Photo Researchers*)

well in X-ray images, whereas soft tissue is nearly transparent to X rays, but shows up in MRI.

The key to any medical imaging technique is knowing which part of the body gives rise to which signal. In MRI, the patient is placed in a magnetic field gradient that can be varied from place to place. Nuclei in the weaker magnetic field gradient absorb radiation at a lower frequency. Nuclei elsewhere, in the stronger magnetic field, absorb radiation at a higher frequency. Because a magnetic field gradient along a single axis images a plane, MRI techniques can create views of any part of the body in slicelike sections. In 2003, Paul Lauterbur and Sir Peter Mansfield were awarded the Nobel prize in physiology or medicine for their discoveries that led to the development of these imaging techniques.

Because both ^{13}C and ^{1}H have spinning nuclei and generate magnetic fields, ^{13}C couples with each ^{1}H bonded to it and gives a signal split according to the $(n + 1)$ rule. In the most common mode for recording a ^{13}C spectrum, this coupling is eliminated by instrumental techniques, so as to simplify the spectrum. In a hydrogen-decoupled spectrum, all ^{13}C signals appear as singlets. The hydrogen-decoupled ^{13}C-NMR spectrum of citric acid (Figure 12.7), a compound used to

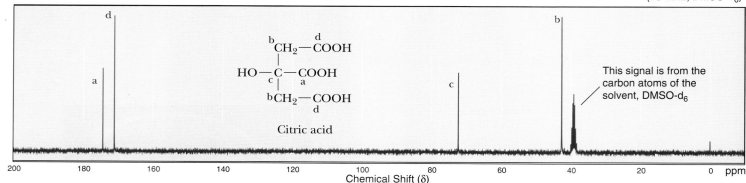

Figure 12.7
Hydrogen-decoupled ^{13}C-NMR spectrum of citric acid.

increase the solubility of many pharmaceutical drugs in water, consists of four singlets. Notice that, as in ^{1}H-NMR, equivalent carbons generate only one signal.

Table 12.2 shows approximate chemical shifts in ^{13}C-NMR spectroscopy. As with ^{1}H-NMR, we can use the following rules of thumb to remember the chemical shifts of various types of carbons:

Chemical Shift (δ)	Type of Carbon
0–50	sp^3 carbon (3° > 2° > 1°).
50–80	sp^3 carbon bonded to an electronegative element such as N, O, or X. The more electronegative the element, the larger is the chemical shift.
100–160	sp^2 carbon of an alkene or an aromatic compound.
160–180	carbonyl carbon of a carboxylic acid or carboxylic acid derivative (Chapters 14 and 15).
180–210	carbonyl carbon of a ketone or an aldehyde.

Notice how much broader the range of chemical shifts is for ^{13}C-NMR spectroscopy than for ^{1}H-NMR spectroscopy. Whereas most chemical shifts for ^{1}H-NMR spectroscopy fall within a rather narrow range of 0–13 ppm, those for ^{13}C-NMR spectroscopy cover 0–210 ppm. Because of this expanded scale, it is very unusual to find any two nonequivalent carbons in the same molecule with identical chemical shifts. Most commonly, each different type of carbon within a molecule has a distinct signal that is clearly resolved from all other signals. Notice further that the chemical shift of carbonyl carbons is quite distinct from the chemical shifts of sp^3 hybridized carbons and other types of sp^2 hybridized carbons. The presence or absence of a carbonyl carbon is quite easy to recognize in a ^{13}C-NMR spectrum.

A great advantage of ^{13}C-NMR spectroscopy is that it is generally possible to count the number of different types of carbon atoms in a molecule. There is one caution here, however: Because of the particular manner in which spin-flipped ^{13}C nuclei return to their lower energy states, integrating signal heights is often unreliable, and it is generally not possible to determine the number of carbons of each type on the basis of the signal heights.

TABLE 12.2 ^{13}C-NMR Chemical Shifts

Type of Carbon	Chemical Shift (δ)	Type of Carbon	Chemical Shift (δ)
RCH$_3$	0–40	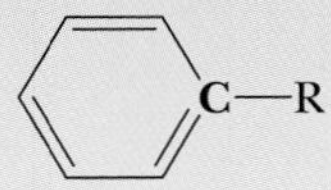	
RCH$_2$R	15–55	C—R (aromatic)	110–160
R$_3$CH	20–60		
RCH$_2$I	0–40		
RCH$_2$Br	25–65	RCOR (ester)	160–180
RCH$_2$Cl	35–80		
R$_3$COH	40–80		
R$_3$COR	40–80	RCNR$_2$ (amide)	165–180
RC≡CR	65–85		
R$_2$C=CR$_2$	100–150	RCOH (carboxylic acid)	175–185
		RCH, RCR (aldehyde, ketone)	180–210

EXAMPLE 12.7

Predict the number of signals in a proton-decoupled ^{13}C-NMR spectrum of each compound:

$$\text{(a)} \quad \underset{\parallel}{\overset{O}{CH_3CCOCH_3}} \qquad \text{(b)} \quad \underset{\parallel}{\overset{O}{CH_3CH_2CH_2CCH_3}} \qquad \text{(c)} \quad \underset{\parallel}{\overset{O}{CH_3CH_2CCH_2CH_3}}$$

SOLUTION

Here is the number of signals in each spectrum, along with the chemical shift of each, color coded to the carbon responsible for that signal. The chemical shifts of the carbonyl carbons are quite distinctive (Table 12.2) and occur at δ 171.37, 208.85, and 211.97 in these examples.

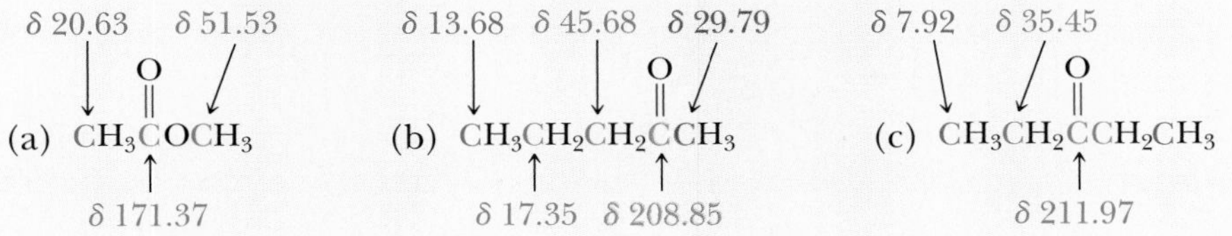

(a) δ 20.63 δ 51.53   CH$_3$COCH$_3$ δ 171.37

(b) δ 13.68 δ 45.68 δ 29.79   CH$_3$CH$_2$CH$_2$CCH$_3$ δ 17.35 δ 208.85

(c) δ 7.92 δ 35.45   CH$_3$CH$_2$CCH$_2$CH$_3$ δ 211.97

Practice Problem 12.7

Explain how to distinguish between the members of each pair of constitutional isomers, on the basis of the number of signals in the ^{13}C-NMR spectrum of each isomer:

(a) 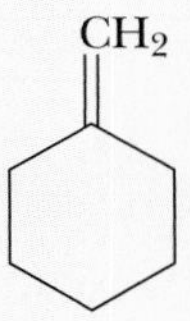 [structure: methylenecyclohexane (CH$_2$)] and [structure: 1-methylcyclohexene (CH$_3$)]

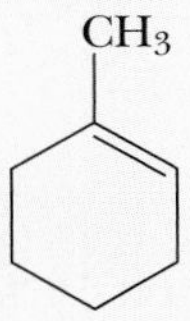

(b) 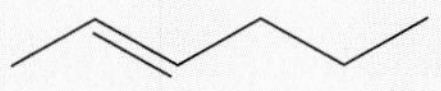 [structure: alkene] and [structure: alkene]

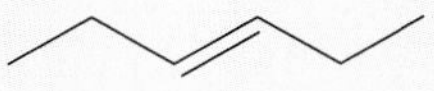

12.11 INTERPRETING NMR SPECTRA

A. Alkanes

Because all hydrogens in alkanes are in very similar chemical environments, ^{1}H-NMR chemical shifts of their hydrogens fall within a narrow range of δ 0.8–1.7. Chemical shifts for alkane carbons in ^{13}C-NMR spectroscopy fall within the considerably wider range of δ 0–60. Notice how it is relatively easy to distinguish all the signals in the ^{13}C-NMR spectrum of 2,2,4-trimethylpentane (Figure 12.8; common name isooctane), a major component in gasoline, compared with the signals in the ^{1}H-NMR spectrum of isooctane (Figure 12.9). In the ^{13}C-NMR spectrum, we can see all five signals while in the ^{1}H-NMR spectrum, we expect to see four signals, but in fact see only three. The reason is that nonequivalent hydrogens often have similar chemical shifts, which lead to an overlap of signals.

B. Alkenes

The ^{1}H-NMR chemical shifts of vinylic hydrogens (hydrogens on a carbon of a carbon-carbon double bond) are larger than those of alkane hydrogens and typically fall into the range δ 4.6–5.7. Figure 12.10 shows a ^{1}H-NMR spectrum of 1-methylcyclohexene. The signal for the one vinylic hydrogen appears at δ 5.4, split into a triplet by the two hydrogens of the neighboring —CH$_2$— group of the ring.

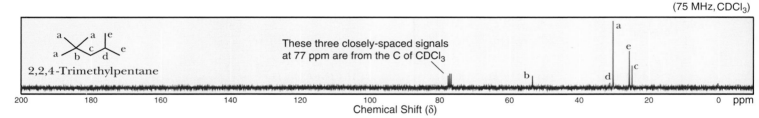

Figure 12.8
Hydrogen-decoupled ^{13}C-NMR spectrum of 2,2,4-trimethylpentane, showing all five signals.

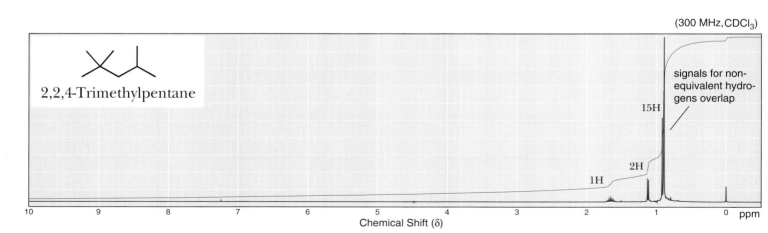

Figure 12.9
^{1}H-NMR spectrum of 2,2,4-trimethylpentane, showing only three signals. Two sets of nonequivalent hydrogens produce signals at δ 0.91, thus giving rise to an integration ratio of 1:2:15, for the peaks at δ 1.65, 1.12, and 0.91, respectively.

(300 MHz, CDCl$_3$)

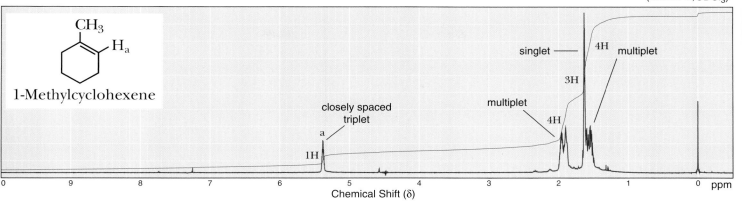

Figure 12.10
^{1}H-NMR spectrum of 1-methylcyclohexene.

The sp^2 hybridized carbons of alkenes come into resonance in ^{13}C-NMR spectroscopy in the range δ 100–150 (Table 12.2), which is considerably downfield from resonances of sp^3 hybridized carbons.

C. Alcohols

The chemical shift of a hydroxyl hydrogen in a ^{1}H-NMR spectrum is variable and depends on the purity of the sample, the solvent, and the temperature. Normally, the shift appears in the range δ 3.0–4.5, but, depending on experimental conditions, it may appear as far upfield as δ 0.5. Hydrogens on the carbon bearing the —OH group are deshielded by the electron-withdrawing inductive effect of the oxygen atom, and their absorptions typically appear in the range δ 3.4–4.0. Figure 12.11 shows a ^{1}H-NMR spectrum of 2,2-dimethyl-1-propanol. The spectrum consists of three signals. The hydroxyl hydrogen appears at δ 2.19 as a slightly broad singlet. The signal of the hydrogens on the carbon bearing the hydroxyl group in 2,2-dimethyl-1-propanol appears as a singlet at δ 3.32.

Signal splitting between the hydrogen on O—H and its neighbors on the adjacent —CH$_2$— group is not seen in the ^{1}H-NMR spectrum of 2,2-dimethyl-1-propanol. The reason is that most samples of alcohol contain traces of acid, base,

(300 MHz, CDCl$_3$)

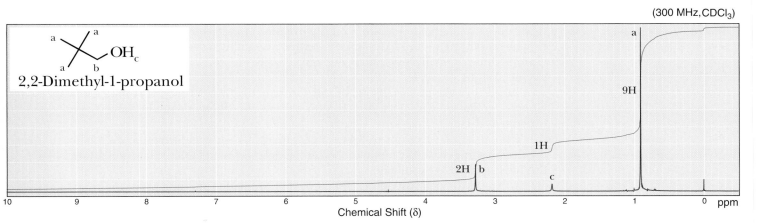

Figure 12.11
^{1}H-NMR spectrum of 2,2-dimethyl-1-propanol.

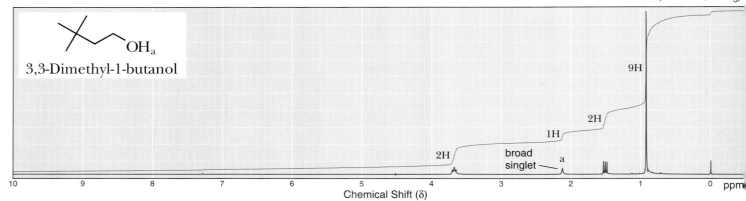

(300 MHz, CDCl$_3$)

Figure 12.12
^{1}H-NMR spectrum of 3,3-dimethyl-1-butanol, showing the decoupled hydroxyl signal at δ 2.12.

or other impurities that catalyze the transfer of the hydroxyl proton from the oxygen of one alcohol molecule to that of another alcohol molecule. This transfer, which is very fast compared with the time scale required to make an NMR measurement, decouples the hydroxyl proton from all other protons in the molecule. For this same reason, the hydroxyl proton does not usually split the signal of any α-hydrogens.

Signals from alcohol hydrogens typically appear as broad singlets, as shown in the ^{1}H-NMR spectrum of 3,3-dimethyl-1-butanol (Figure 12.12).

D. Benzene and Its Derivatives

All six hydrogens of benzene are equivalent, and their signal appears in its ^{1}H-NMR spectrum as a sharp singlet at δ 7.27. Hydrogens bonded to a substituted benzene ring appear in the region δ 6.5–8.5. Few other types of hydrogens give signals in this region; thus, aromatic hydrogens are quite easily identifiable by their distinctive chemical shifts.

Recall that vinylic hydrogens are in resonance at δ 4.6–5.7 (Section 12.11B). Hence, aromatic hydrogens absorb radiation even farther downfield than vinylic hydrogens do.

The ^{1}H-NMR spectrum of toluene (Figure 12.13) shows a singlet at δ 2.32 for the three hydrogens of the methyl group and a closely spaced multiplet at δ 7.3 for the five hydrogens of the aromatic ring.

In ^{13}C-NMR spectroscopy, carbon atoms of aromatic rings appear in the range δ 110–160. Benzene, for example, shows a single signal at δ 128. Because carbon-13 signals for alkene carbons appear in the same range, it is generally not possible to establish the presence of an aromatic ring by ^{13}C-NMR spectroscopy alone. ^{13}C-NMR

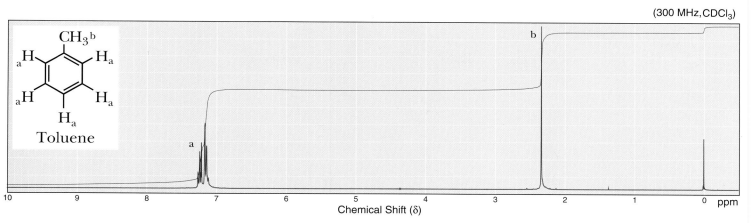

Figure 12.13
¹H-NMR spectrum of toluene.

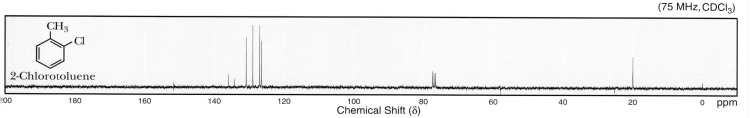

Figure 12.14
¹³C-NMR spectrum of 2-chlorotoluene.

spectroscopy is particularly useful, however, in establishing substitution patterns of aromatic rings. The spectrum ¹³C-NMR of 2-chlorotoluene (Figure 12.14) shows six signals in the aromatic region; the compound's more symmetric isomer 4-chlorotoluene (Figure 12.15) shows only four signals in the aromatic region. Thus, all one needs to do is count signals to distinguish between these constitutional isomers.

E. Amines

The chemical shifts of amine hydrogens, like those of hydroxyl hydrogens, are variable and may be found in the region δ 0.5–5.0, depending on the solvent, the concentration, and the temperature. Furthermore, the rate of intermolecular exchange

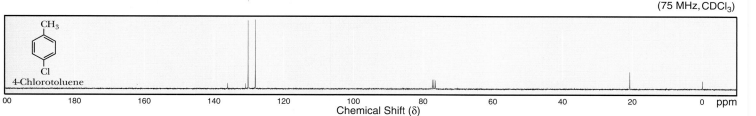

Figure 12.15
¹³C-NMR spectrum of 4-chlorotoluene.

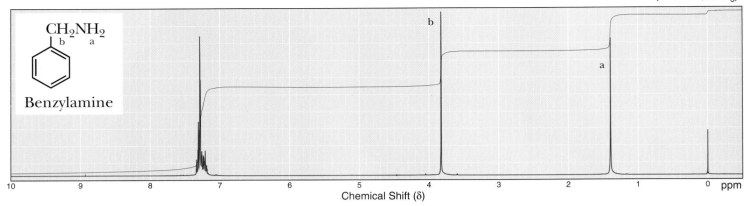

(300 MHz, CDCl₃)

Figure 12.16
^{1}H-NMR spectrum of benzylamine.

of hydrogens is sufficiently rapid, compared with the time scale of an NMR measurement, that signal splitting between amine hydrogens and hydrogens on an adjacent α-carbon is prevented.

Thus, amine hydrogens generally appear as singlets. The NH₂ hydrogens in benzylamine, $C_6H_5CH_2NH_2$, for example, appear as a singlet at δ 1.40 (Figure 12.16).

F. Aldehydes and Ketones

^{1}H-NMR spectroscopy is an important tool for identifying aldehydes and for distinguishing between aldehydes and other carbonyl-containing compounds. Just as a carbon-carbon double bond causes a downfield shift in the signal of a vinylic hydrogen (Section 12.11B), a carbon–oxygen double bond causes a downfield shift in the signal of an aldehyde hydrogen, typically to δ 9.5–10.1. Signal splitting between this hydrogen and those on the adjacent α-carbon is slight; consequently, the aldehyde hydrogen signal often appears as a closely spaced doublet or triplet. In the spectrum of butanal, for example, the triplet signal for the aldehyde hydrogen at δ 9.78 is so closely spaced that it almost looks like a singlet (Figure 12.17).

Figure 12.17
^{1}H-NMR spectrum of butanal.

(300 MHz, CDCl₃)

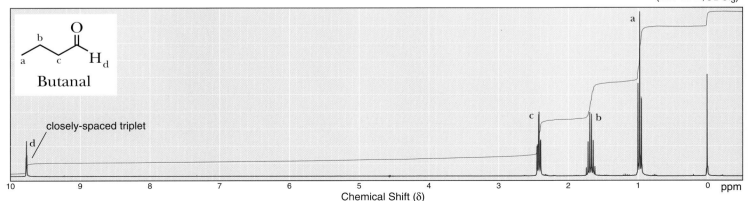

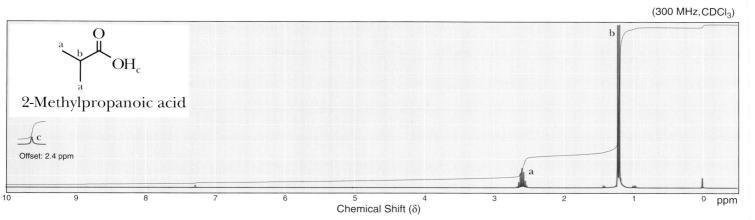

(300 MHz, CDCl₃)

Figure 12.18
¹H-NMR spectrum of 2-methylpropanoic acid (isobutyric acid).

Just as the signal for an aldehyde hydrogen is only weakly split by the adjacent nonequivalent α-hydrogens, the α-hydrogens are weakly split by the aldehyde hydrogen. Hydrogens on an α-carbon of an aldehyde or ketone typically appear around δ 2.1–2.6. The carbonyl carbons of aldehydes and ketones are readily identifiable in ¹³C-NMR spectroscopy by the position of their signal between δ 180 and 210.

G. Carboxylic Acids

The hydrogen of the carboxyl group gives a signal in the range δ 10–13. The chemical shift of a carboxyl hydrogen is so large—even larger than the chemical shift of an aldehyde hydrogen (δ 9.5–10.1)—that it serves to distinguish carboxyl hydrogens from most other types of hydrogens. The signal of the carboxyl hydrogen of 2-methylpropanoic acid is at the left of the ¹H-NMR spectrum in Figure 12.18 and has been offset by δ 2.4. (Add 2.4 to the position at which the signal appears on the spectrum.) The chemical shift of this hydrogen is δ 12.0.

H. Esters

Hydrogens on the α-carbon of the carbonyl group of an ester are slightly deshielded and give signals at δ 2.1–2.6. Hydrogens on the carbon bonded to the ester oxygen are more strongly deshielded and give signals at δ 3.7–4.7. It is thus possible to distinguish between ethyl acetate and its constitutional isomer, methyl propanoate, by the chemical shifts of either the singlet —CH₃ absorption (compare δ 2.04 with 3.68) or the quartet —CH₂— absorption (compare δ 4.11 with 2.33):

δ 2.04(s)   δ 4.11(q)   δ 2.33(q)   δ 3.68(s)

$$CH_3 - \overset{\overset{\displaystyle O}{\|}}{C} - O - CH_2 - CH_3 \qquad CH_3 - CH_2 - \overset{\overset{\displaystyle O}{\|}}{C} - O - CH_3$$

Ethyl acetate   Methyl propanoate

12.12 SOLVING NMR PROBLEMS

One of the first steps in determining the molecular structure of a compound is to establish the compound's molecular formula. In the past, this was most commonly done by elemental analysis, combustion to determine the percent composition, and so forth. More commonly today, we determine molecular weight and molecular formula by a technique known as *mass spectrometry*. (An explanation of the technique is beyond the scope of this book.) In the examples that follow, we assume that the molecular formula of any unknown compound has already been determined, and we proceed from there, using spectral analysis to determine a structural formula.

The following steps may prove helpful as a systematic approach to solving ^{1}H-NMR spectral problems:

Step 1: *Molecular formula and index of hydrogen deficiency.* Examine the molecular formula, calculate the index of hydrogen deficiency (Section 11.4J), and deduce what information you can about the presence or absence of rings or pi bonds.

Step 2: *Number of signals.* Count the number of signals, to determine the minimum number of sets of equivalent hydrogens in the compound.

Step 3: *Integration.* Use signal integration and the molecular formula to determine the number of hydrogens in each set.

Step 4: *Pattern of chemical shifts.* Examine the NMR spectrum for signals characteristic of the most common types of equivalent hydrogens. (See the general rules of thumb for ^{1}H-NMR chemical shifts in Section 12.8.) Keep in mind that the ranges are broad and that hydrogens of each type may be shifted either farther upfield or farther downfield, depending on details of the molecular structure in question.

Step 5: *Splitting patterns.* Examine splitting patterns for information about the number of nonequivalent hydrogen neighbors.

Step 6: *Structural formula.* Write a structural formula consistent with the information learned in Steps 1–5.

Spectral Problem 1   The compound is a colorless liquid with molecular formula $C_5H_{10}O$.

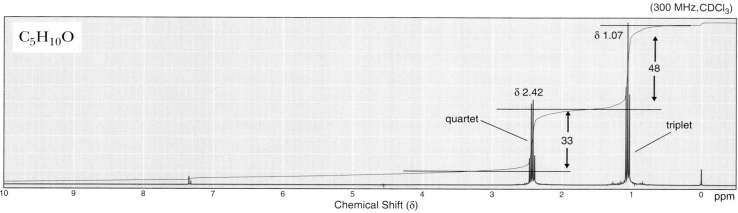

(300 MHz, CDCl$_3$)

Analysis of Spectral Problem 1

Step 1: *Molecular formula and index of hydrogen deficiency.* The reference compound is $C_5H_{12}O$; therefore, the index of hydrogen deficiency is 1. The molecule thus contains either one ring or one pi bond.

Step 2: *Number of signals.* There are two signals (a triplet and a quartet) and therefore two sets of equivalent hydrogens.

Step 3: *Integration.* By signal integration, we calculate that the number of hydrogens giving rise to each signal is in the ratio 3:2. Because there are 10 hydrogens, we conclude that the signal assignments are δ 1.07 (6H) and δ 2.42 (4H).

Step 4: *Pattern of chemical shifts.* The signal at δ 1.07 is in the alkyl region and, based on its chemical shift, most probably represents a methyl group. No signal occurs at δ 4.6 to 5.7; thus, there are no vinylic hydrogens. (If a carbon–carbon double bond is in the molecule, no hydrogens are on it; that is, it is tetrasubstituted.)

Step 5: *Splitting pattern.* The methyl signal at δ 1.07 is split into a triplet (t); hence, it must have two neighboring hydrogens, indicating —CH₂CH₃. The signal at δ 2.42 is split into a quartet (q); thus, it must have three neighboring hydrogens, which is also consistent with —CH₂CH₃. Consequently, an ethyl group accounts for these two signals. No other signals occur in the spectrum; therefore, there are no other types of hydrogens in the molecule.

Step 6: *Structural formula.* Put the information learned in the previous steps together to arrive at the following structural formula. Note that the chemical shift of the methylene group (—CH₂—) at δ 2.42 is consistent with an alkyl group adjacent to a carbonyl group.

$$\delta\ 2.42\ (q) \qquad \delta\ 1.07\ (t)$$

$$CH_3-CH_2-\overset{\displaystyle O}{\overset{\displaystyle \|}{C}}-CH_2-CH_3$$

3-Pentanone

Spectral Problem 2 The compound is a colorless liquid with molecular formula $C_7H_{14}O$.

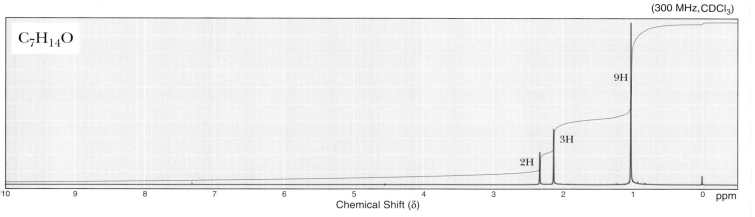

(300 MHz, CDCl₃)

$C_7H_{14}O$

9H

3H

2H

Analysis of Spectral Problem 2

Step 1: *Molecular formula and index of hydrogen deficiency.* The index of hydrogen deficiency is 1; thus, the compound contains one ring or one pi bond.

Step 2: *Number of signals.* There are three signals and therefore three sets of equivalent hydrogens.

Step 3: *Integration.* By signal integration, we calculate that the number of hydrogens giving rise to each signal is in the ratio 9 : 3 : 2, reading from left to right.

Step 4: *Pattern of chemical shifts.* The singlet at δ 1.01 is characteristic of a methyl group adjacent to an sp^3 hybridized carbon. The singlets at δ 2.11 and 2.32 are characteristic of alkyl groups adjacent to a carbonyl group.

Step 5: *Splitting pattern.* All signals are singlets (s), which means that none of the hydrogens are within three bonds of each other.

Step 6: *Structural formula.* The compound is 4,4-dimethyl-2-pentanone:

$$\delta\ 1.01\,(s) \qquad \delta\ 2.32\,(s) \qquad \delta\ 2.11\,(s)$$

$$CH_3-\overset{\displaystyle CH_3}{\underset{\displaystyle CH_3}{\overset{\displaystyle |}{\underset{\displaystyle |}{C}}}}-CH_2-\overset{\displaystyle O}{\overset{\displaystyle \|}{C}}-CH_3$$

4,4-Dimethyl-2-pentanone

SUMMARY

The interaction of molecules with **radio-frequency radiation** gives us information about nuclear spin energy levels. Nuclei of ^1H and ^{13}C, isotopes of the two elements most common to organic compounds, have a spin and behave like tiny bar magnets (Section 12.2). When placed between the poles of a powerful magnet, the nuclear spins of these elements become aligned either with the applied field or against it. Nuclear spins aligned with the applied field are in the lower energy state; those aligned against the applied field are in the higher energy state. **Resonance** is the absorption of electromagnetic radiation by a nucleus and the resulting "flip" of its nuclear spin from a lower energy spin state to a higher energy spin state. An NMR spectrometer (Section 12.5) records such a resonance as a **signal**.

The experimental conditions required to cause nuclei to resonate are affected by the local chemical and magnetic environment. Electrons around a hydrogen also have spin (Section 12.4) and create a local magnetic field that shields the hydrogen from the applied field.

In a ^1H-NMR spectrum, a resonance signal is reported by how far it is shifted from the resonance signal of the 12 equivalent hydrogens in **tetramethylsilane** (**TMS**). A resonance signal in a ^{13}C-NMR spectrum is reported by how far it is shifted from the resonance signal of the four equivalent carbons in TMS. A **chemical shift (δ)** (Section 12.5) is the frequency shift from TMS, divided by the operating frequency of the spectrometer.

Equivalent hydrogens within a molecule have identical chemical shifts (Section 12.6). The area of a ^1H-NMR signal is proportional to the number of equivalent hydrogens giving rise to that signal (Section 12.7). In **signal splitting**, the ^1H-NMR signal from one hydrogen or set of equivalent hydrogens is split by the influence of nonequivalent hydrogens on the same or adjacent carbon atoms (Section 12.9). According to the **($n + 1$) rule**, if a hydrogen has n hydrogens that are nonequivalent to it, but are equivalent among themselves, on the same or adjacent carbon atom(s), its ^1H-NMR signal is split into ($n + 1$) peaks. **Complex splitting** occurs when a hydrogen is flanked by two or more sets of hydrogens and those sets are nonequivalent. Splitting patterns are commonly referred to as singlets (s), doublets (d), triplets (t), quartets (q), quintets, and multiplets (m).

^{13}C-NMR spectra (Section 12.10) are commonly recorded in a hydrogen-decoupled instrumental mode. In this mode, all ^{13}C signals appear as singlets.

PROBLEMS

Equivalency of Hydrogens and Carbons

12.8 Determine the number of signals you would expect to see in the ^1H-NMR spectrum of each of the following compounds.

12.9 Determine the number of signals you would expect to see in the ^{13}C-NMR spectrum of each of the compounds in Problem 12.8.

Interpreting ^{1}H-NMR and ^{13}C-NMR Spectra

12.10 Following are structural formulas for the constitutional isomers of xylene and three sets of ^{13}C-NMR spectra. Assign each constitutional isomer its correct spectrum.

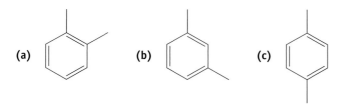

(a) (b) (c)

(75 MHz, CDCl$_3$)

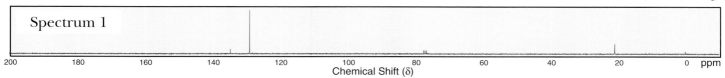

Spectrum 1

(75 MHz, CDCl$_3$)

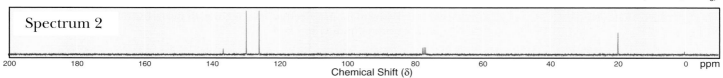

Spectrum 2

(75 MHz, CDCl$_3$)

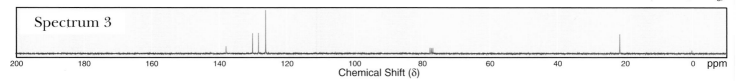

Spectrum 3

12.11 Following is a ^{1}H-NMR spectrum for compound A, with molecular formula C$_7$H$_{14}$. Compound A decolorizes a solution of bromine in carbon tetrachloride. Propose a structural formula for compound A.

(300 MHz, CDCl$_3$)

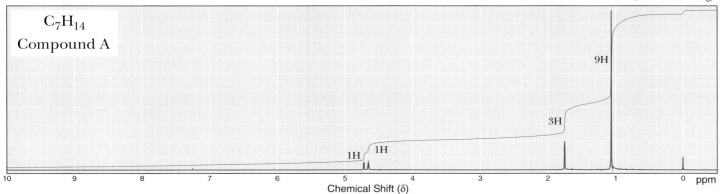

C$_7$H$_{14}$
Compound A

12.12 Following is a ^{1}H-NMR spectrum for compound B, with molecular formula C_8H_{16}. Compound B decolorizes a solution of Br_2 in CCl_4. Propose a structural formula for compound B.

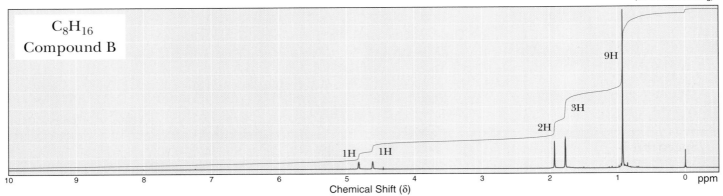

12.13 Following are the ^{1}H-NMR spectra of compounds C and D, each with molecular formula C_4H_7Cl. Each compound decolorizes a solution of Br_2 in CCl_4. Propose structural formulas for compounds C and D.

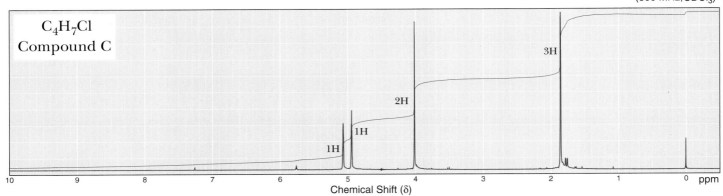

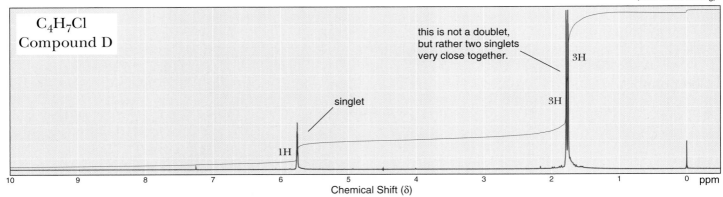

12.14 Following are the structural formulas of three alcohols with molecular formula $C_7H_{16}O$ and three sets of ^{13}C-NMR spectral data. Assign each constitutional isomer to its correct spectral data.

(a) $CH_3CH_2CH_2CH_2CH_2CH_2CH_2OH$

(b) $CH_3\overset{\displaystyle OH}{\underset{\displaystyle CH_3}{C}}CH_2CH_2CH_2CH_3$

(c) $CH_3CH_2\overset{\displaystyle OH}{\underset{\displaystyle CH_2CH_3}{C}}CH_2CH_3$

Spectrum 1	Spectrum 2	Spectrum 3
74.66	70.97	62.93
30.54	43.74	32.79
7.73	29.21	31.86
	26.60	29.14
	23.27	25.75
	14.09	22.63
		14.08

12.15 Alcohol E, with molecular formula $C_6H_{14}O$, undergoes acid-catalyzed dehydration when it is warmed with phosphoric acid, giving compound F, with molecular formula C_6H_{12}, as the major product. A ^1H-NMR spectrum of compound E shows peaks at δ 0.89 (t, 6H), 1.12 (s, 3H), 1.38 (s, 1H), and 1.48 (q, 4H). The ^{13}C-NMR spectrum of compound E shows peaks at δ 72.98, 33.72, 25.85, and 8.16. Propose structural formulas for compounds E and F.

12.16 Compound G, $C_6H_{14}O$, does not react with sodium metal and does not discharge the color of Br_2 in CCl_4. The ^1H-NMR spectrum of compound G consists of only two signals: a 12H doublet at δ 1.1 and a 2H septet at δ 3.6. Propose a structural formula for compound G.

12.17 Propose a structural formula for each haloalkane:
(a) $C_2H_4Br_2$ δ 2.5 (d, 3H) and 5.9 (q, 1H)
(b) $C_4H_8Cl_2$ δ 1.67 (d, 6H) and 2.15 (q, 4H)
(c) $C_5H_8Br_4$ δ 3.6 (s, 8H)
(d) C_4H_9Br δ 1.1 (d, 6H), 1.9 (m, 1H), and 3.4 (d, 2H)
(e) $C_5H_{11}Br$ δ 1.1 (s, 9H) and 3.2 (s, 2H)
(f) $C_7H_{15}Cl$ δ 1.1 (s, 9H) and 1.6 (s, 6H)

12.18 Following are structural formulas for esters (1), (2), and (3) and three ^1H-NMR spectra. Assign each compound its correct spectrum and assign all signals to their corresponding hydrogens.

$$\underset{(1)}{CH_3\overset{\displaystyle O}{\overset{\displaystyle \|}{C}}OCH_2CH_3} \qquad \underset{(2)}{H\overset{\displaystyle O}{\overset{\displaystyle \|}{C}}OCH_2CH_2CH_3} \qquad \underset{(3)}{CH_3O\overset{\displaystyle O}{\overset{\displaystyle \|}{C}}CH_2CH_3}$$

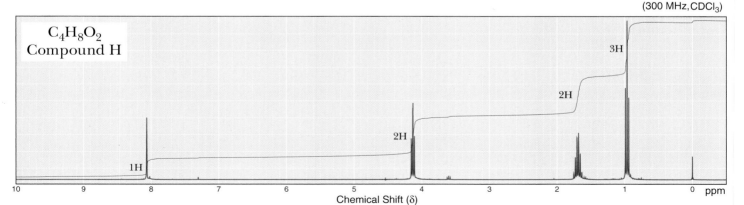

(300 MHz, CDCl₃)

$C_4H_8O_2$
Compound H

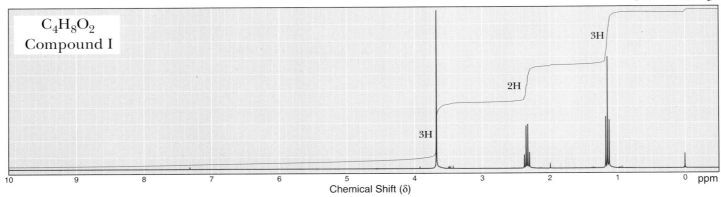

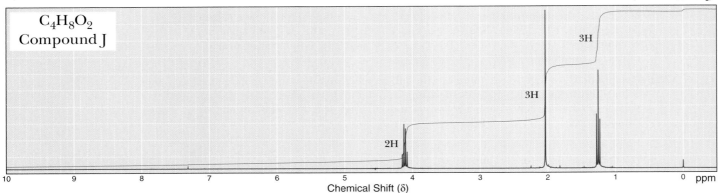

12.19 Compound K, $C_{10}H_{10}O_2$, is insoluble in water, 10% NaOH, and 10% HCl. A ^{1}H-NMR spectrum of compound K shows signals at δ 2.55 (s, 6H) and 7.97 (s, 4H). A ^{13}C-NMR spectrum of compound K shows four signals. From this information, propose a structural formula for K..

12.20 Compound L, $C_{15}H_{24}O$, is used as an antioxidant in many commercial food products, synthetic rubbers, and petroleum products. Propose a structural formula for compound L based on its ^{1}H-NMR and ^{13}C-NMR spectra.

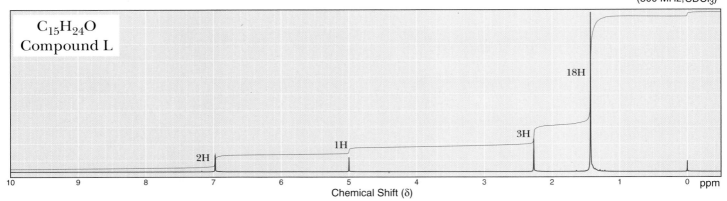

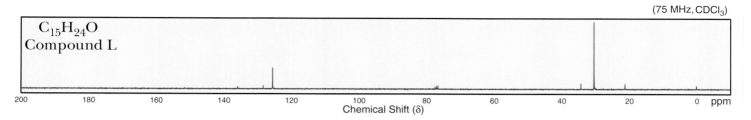

(75 MHz, CDCl₃)

$C_{15}H_{24}O$
Compound L

12.21 Propose a structural formula for these compounds, each of which contains an aromatic ring:
   (a) $C_9H_{10}O$ δ 1.2 (t, 3H), 3.0 (q, 2H), and 7.4-8.0 (m, 5H)
   (b) $C_{10}H_{12}O_2$ δ 2.2 (s, 3H), 2.9 (t, 2H), 4.3 (t, 2H), and 7.3 (s, 5H)
   (c) $C_{10}H_{14}$ δ 1.2 (d, 6H), 2.3 (s, 3H), 2.9 (septet, 1H), and 7.0 (s, 4H)
   (d) C_8H_9Br δ 1.8 (d, 3H), 5.0 (q, 1H), and 7.3 (s, 5H)

12.22 Compound M, with molecular formula $C_9H_{12}O$, readily undergoes acid-catalyzed dehydration to give compound N, with molecular formula C_9H_{10}. A ¹H-NMR spectrum of compound M shows signals at δ 0.91 (t, 3H), 1.78 (m, 2H), 2.26 (d, 1H), 4.55 (m, 1H), and 7.31 (m, 5H). From this information, propose structural formulas for compounds M and N.

12.23 Propose a structural formula for each ketone:
   (a) C_4H_8O δ 1.0 (t, 3H), 2.1 (s 3H), and 2.4 (q, 2H)
   (b) $C_7H_{14}O$ δ 0.9 (t, 6H), 1.6 (sextet, 4H), and 2.4 (t, 4H)

12.24 Propose a structural formula for compound O, a ketone with molecular formula $C_{10}H_{12}O$:

(300 MHz, CDCl₃)

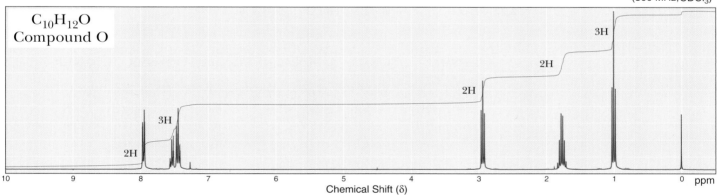

$C_{10}H_{12}O$
Compound O

12.25 Following is a ¹H-NMR spectrum for compound P, with molecular formula $C_6H_{12}O_2$. Compound P undergoes acid-catalyzed dehydration to give compound Q, $C_6H_{10}O$. Propose structural formulas for compounds P and Q.

(300 MHz, CDCl₃)

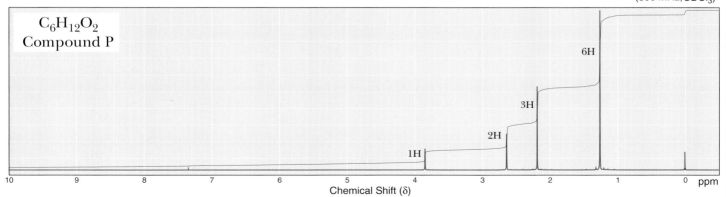

$C_6H_{12}O_2$
Compound P

12.26 Propose a structural formula for compound R, with molecular formula $C_{12}H_{16}O$. Following are its ^{1}H-NMR and ^{13}C-NMR spectra:

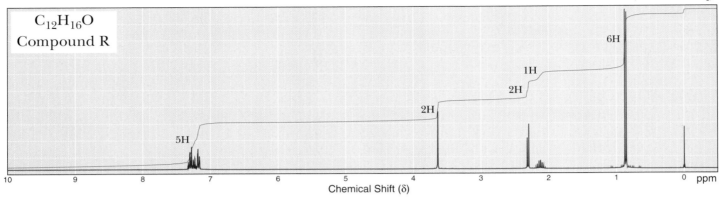

(300 MHz, CDCl$_3$)

$C_{12}H_{16}O$
Compound R

5H 2H 2H 1H 6H

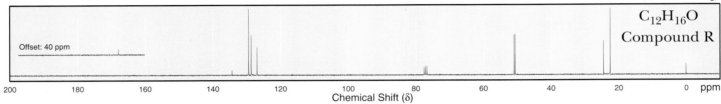

(75 MHz, CDCl$_3$)

$C_{12}H_{16}O$
Compound R

Offset: 40 ppm

12.27 Propose a structural formula for each carboxylic acid:

(a) $C_5H_{10}O_2$ (b) $C_6H_{12}O_2$ (c) $C_5H_8O_4$

^{1}H-NMR	^{13}C-NMR	^{1}H-NMR	^{13}C-NMR	^{1}H-NMR	^{13}C-NMR
0.94 (t, 3H)	180.7	1.08 (s, 9H)	179.29	0.93 (t, 3H)	170.94
1.39 (m, 2H)	33.89	2.23 (s, 2H)	46.82	1.80 (m, 2H)	53.28
1.62 (m, 2H)	26.76	12.1 (s, 1H)	30.62	3.10 (t, 1H)	21.90
2.35 (t, 2H)	22.21		29.57	12.7 (s, 2H)	11.81
12.0 (s, 1H)	13.69				

12.28 Following are ^{1}H-NMR and ^{13}C-NMR spectra of compound S, with molecular formula $C_7H_{14}O_2$. Propose a structural formula for compound S.

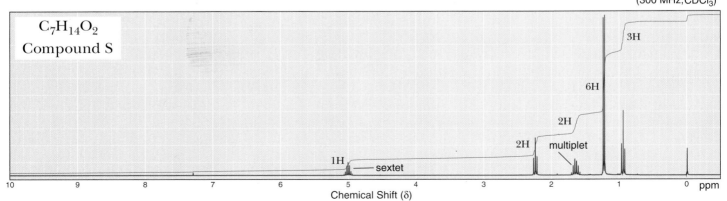

(300 MHz, CDCl$_3$)

$C_7H_{14}O_2$
Compound S

3H

6H

2H

2H multiplet

1H sextet

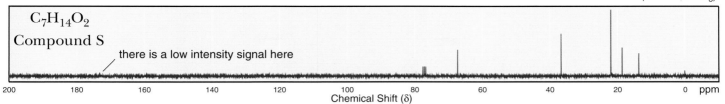

(75 MHz, CDCl₃)

$C_7H_{14}O_2$
Compound S

there is a low intensity signal here

12.29 Propose a structural formula for each ester:

(a) $C_6H_{12}O_2$ **(b)** $C_7H_{12}O_4$ **(c)** $C_7H_{14}O_2$

¹H-NMR	¹³C-NMR
1.18 (d, 6H)	177.16
1.26 (t, 3H)	60.17
2.51 (m, 1H)	34.04
4.13 (q, 2H)	19.01
	14.25

¹H-NMR	¹³C-NMR
1.28 (t, 6H)	166.52
3.36 (s, 2H)	61.43
4.21 (q, 4H)	41.69
	14.07

¹H-NMR	¹³C-NMR
0.92 (d, 6H)	171.15
1.52 (m, 2H)	63.12
1.70 (m, 1H)	37.31
2.09 (s, 3H)	25.05
4.10 (t, 2H)	22.45
	21.06

12.30 Following are ¹H-NMR and ¹³C-NMR spectra of compound T, with molecular formula $C_{10}H_{15}NO$. Propose a structural formula for this compound.

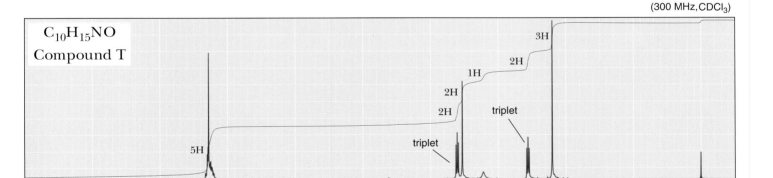

(300 MHz, CDCl₃)

$C_{10}H_{15}NO$
Compound T

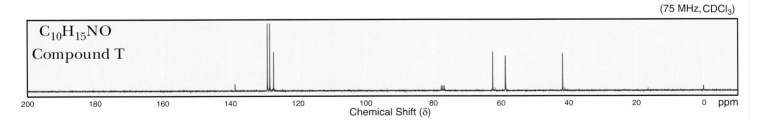

(75 MHz, CDCl₃)

$C_{10}H_{15}NO$
Compound T

12.31 Propose a structural formula for amide U, with molecular formula $C_6H_{13}NO$:

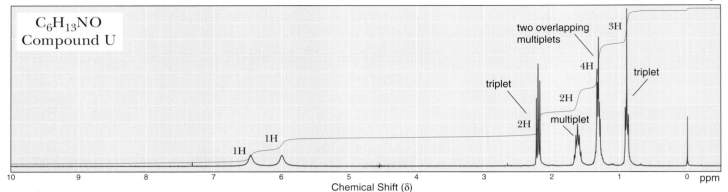

(300 MHz, CDCl₃)

12.32 Propose a structural formula for the analgesic phenacetin, with molecular formula $C_{10}H_{13}NO_2$, based on its ¹H-NMR spectrum:

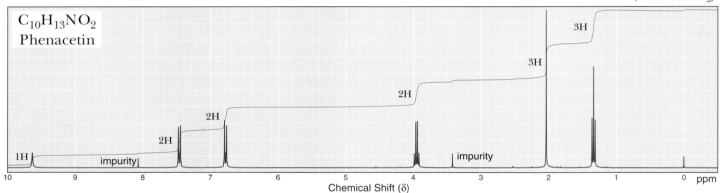

(300 MHz, CDCl₃)

12.33 Propose a structural formula for compound V, an oily liquid with molecular formula $C_8H_9NO_2$. Compound V is insoluble in water and aqueous NaOH, but dissolves in 10% HCl. When its solution in HCl is neutralized with NaOH, compound V is recovered unchanged. A ¹H-NMR spectrum of compound V shows signals at δ 3.84 (s, 3H), 4.18 (s, 2H), 7.60 (d, 2H), and 8.70 (d, 2H).

12.34 Following is a ¹H-NMR spectrum and a structural formula for anethole, $C_{10}H_{12}O$, a fragrant natural product obtained from anise. Using the line of integration, determine the number of protons giving rise to each signal. Show that this spectrum is consistent with the structure of anethole.

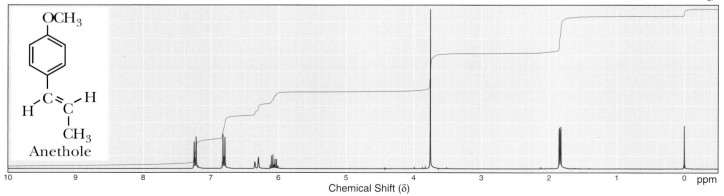

(300 MHz, CDCl₃)

12.35 Propose a structural formula for compound W, C_4H_6O, based on the following IR and ^1H-NMR spectra:

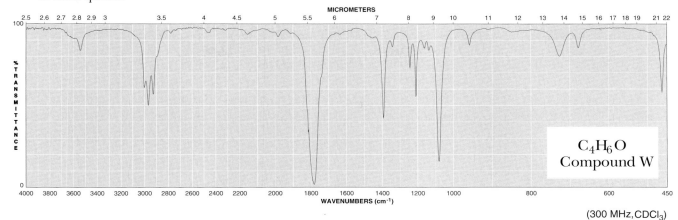

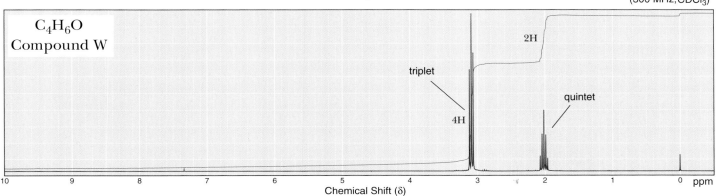

12.36 Propose a structural formula for compound X, $C_5H_{10}O_2$, based on the following IR and ^1H-NMR spectra:

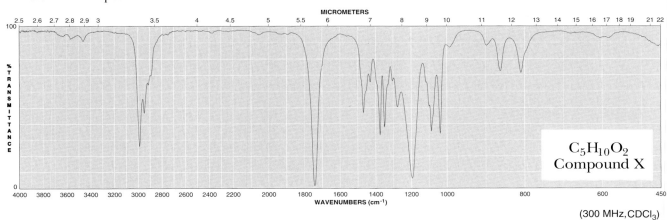

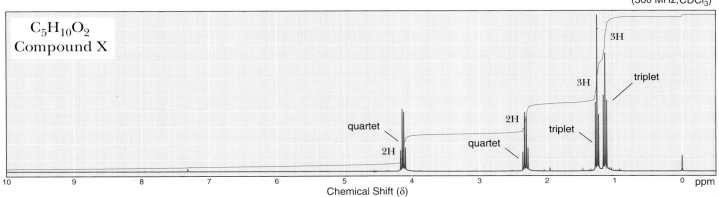

12.37 Propose a structural formula for compound Y, $C_5H_9ClO_2$, based on the following IR and ^{1}H-NMR spectra:

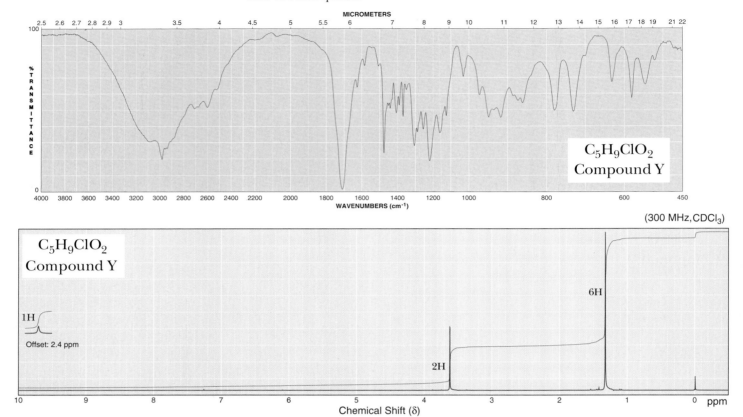

12.38 Propose a structural formula for compound Z, $C_6H_{14}O$, based on the following IR and ^{1}H-NMR spectra:

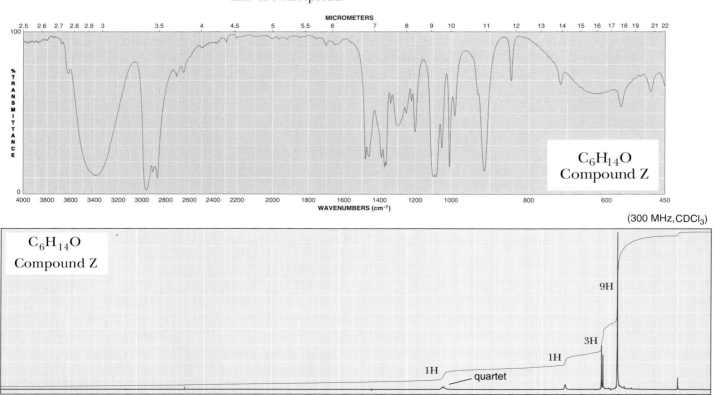

13

Aldehydes and Ketones

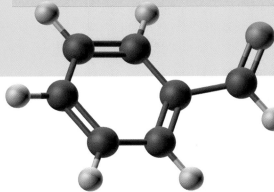

Benzaldehyde is found in the kernels of bitter almonds. Cinnamaldehyde is found in Ceylonese and Chinese cinnamon oils. Inset: A model of benzaldehyde. *(Charles D. Winters)*

13.1 INTRODUCTION

In this and several of the following chapters, we study the physical and chemical properties of compounds containing the carbonyl group, C=O. Because this group is the functional group of aldehydes, ketones, and carboxylic acids and their derivatives, it is one of the most important functional groups in organic chemistry. The chemical properties of the carbonyl group are straightforward, and an understanding of its characteristic reaction themes leads very quickly to an understanding of a wide variety of organic reactions.

13.2 STRUCTURE AND BONDING

Aldehyde A compound containing a carbonyl group bonded to hydrogen (a CHO group).

Ketone A compound containing a carbonyl group bonded to two carbons.

Dihydroxyacetone is the active ingredient in several artificial tanning preparations. *(Andy Washnik)*

The functional group of an **aldehyde** is a carbonyl group bonded to a hydrogen atom (Section 1.8C). In methanal, the simplest aldehyde, the carbonyl group is bonded to two hydrogen atoms. In other aldehydes, it is bonded to one hydrogen atom and one carbon atom. The functional group of a **ketone** is a carbonyl group bonded to two carbon atoms. Following are Lewis structures for the aldehydes methanal and ethanal and a Lewis structure for propanone, the simplest ketone (the common names of each are in parentheses underneath):

$$\overset{O}{\underset{}{\overset{\|}{HCH}}} \qquad \overset{O}{\underset{}{\overset{\|}{CH_3CH}}} \qquad \overset{O}{\underset{}{\overset{\|}{CH_3CCH_3}}}$$

Methanal   Ethanal   Propanone
(Formaldehyde)   (Acetaldehyde)   (Acetone)

A carbon–oxygen double bond consists of one sigma bond formed by the overlap of sp^2 hybrid orbitals of carbon and oxygen and one pi bond formed by the overlap of parallel $2p$ orbitals. The two nonbonding pairs of electrons on oxygen lie in the two remaining sp^2 hybrid orbitals (Figure 1.20).

13.3 NOMENCLATURE

A. IUPAC Nomenclature

The IUPAC system of nomenclature for aldehydes and ketones follows the familiar pattern of selecting the longest chain of carbon atoms that contains the functional group as the parent alkane. We show the aldehyde group by changing the suffix -*e* of the parent alkane to -*al*, as in methanal (Section 3.6). Because the carbonyl group of an aldehyde can appear only at the end of a parent chain and numbering must start with that group as carbon-1, its position is unambiguous; there is no need to use a number to locate it.

For **unsaturated aldehydes**, the presence of a carbon–carbon double bond is indicated by the infix -*en*-. As with other molecules with both an infix and a suffix, the location of the suffix determines the numbering pattern.

3-Methylbutanal   2-Propenal
                    (Acrolein)   (2*E*)-3,7-Dimethyl-2,6-octadienal
                                                    (Geranial)

For cyclic molecules in which —CHO is bonded directly to the ring, we name the molecule by adding the suffix -*carbaldehyde* to the name of the ring. We number the atom of the ring bearing the aldehyde group as number 1:

Cyclopentane-
carbaldehyde   *trans*-4-Hydroxycyclo-
                hexanecarbaldehyde

Among the aldehydes for which the IUPAC system retains common names are benzaldehyde and cinnamaldehyde:

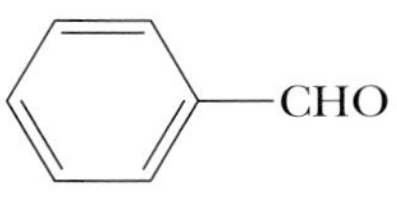

Benzaldehyde

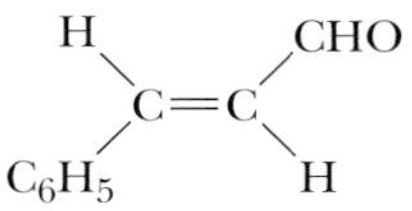

trans-3-Phenyl-2-propenal
(Cinnamaldehyde)

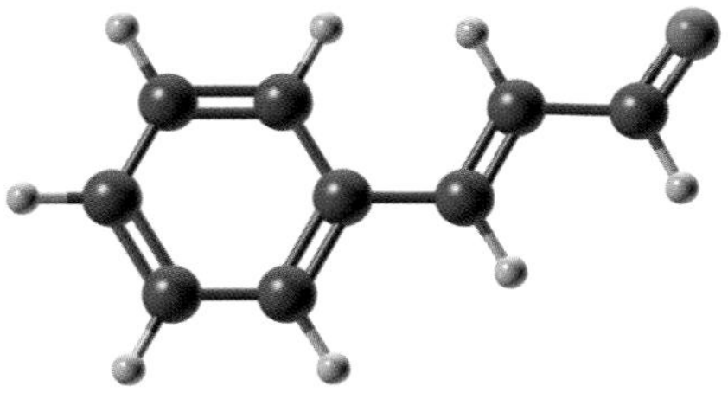

Note here the alternative ways of writing the phenyl group. In benzaldehyde, it is written as a line-angle formula and is abbreviated C_6H_5— in cinnamaldehyde. Two other aldehydes whose common names are retained in the IUPAC system are formaldehyde and acetaldehyde.

In the IUPAC system, we name ketones by selecting the longest chain that contains the carbonyl group and making that chain the parent alkane. We indicate the presence of the ketone by changing the suffix from -e to -one (Section 3.6). We number the parent chain from the direction that gives the carbonyl carbon the smaller number. The IUPAC system retains the common names acetophenone and benzophenone:

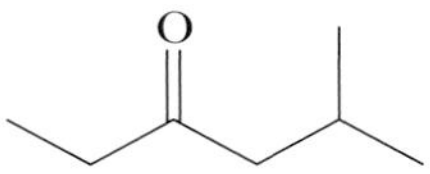

5-Methyl-3-hexanone

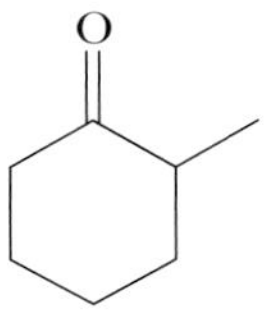

2-Methyl-
cyclohexanone

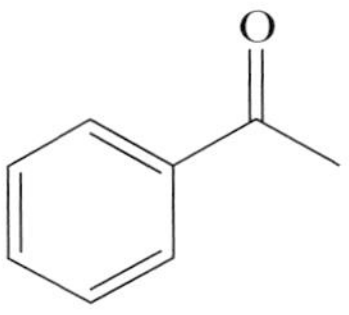

Acetophenone

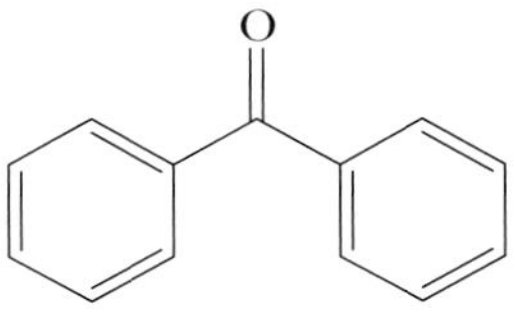

Benzophenone

EXAMPLE 13.1

Write the IUPAC name for each compound:

(a)

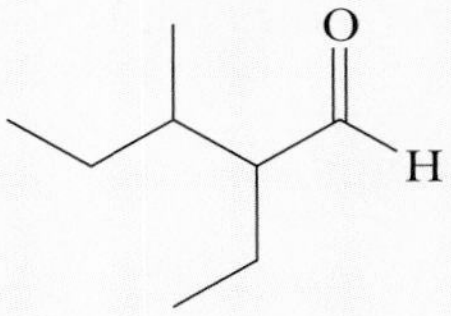

(b)

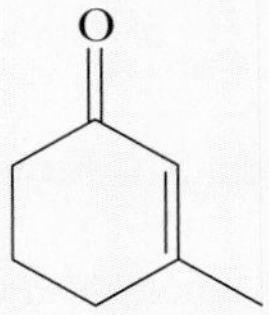

(c)

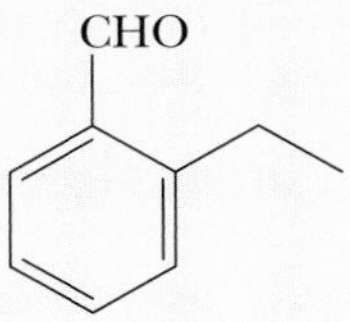

SOLUTION

(a) The longest chain has six carbons, but the longest chain that contains the carbonyl group has five carbons. The IUPAC name of this compound is 2-ethyl-3-methylpentanal.
(b) Number the six-membered ring beginning with the carbonyl carbon. The IUPAC name of this compound is 3-methyl-2-cyclohexenone.
(c) This molecule is derived from benzaldehyde. Its IUPAC name is 2-ethyl-benzaldehyde.

Practice Problem 13.1

Write the IUPAC name for each compound, and specify the configuration of (c):

(a)

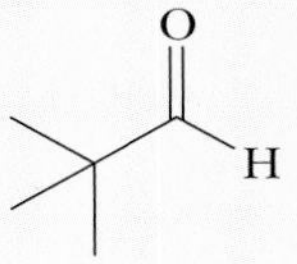

(b)

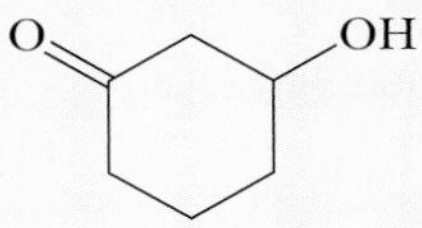

(c)

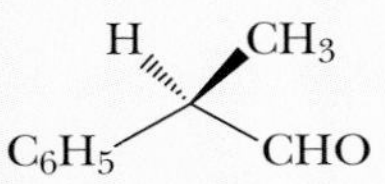

EXAMPLE 13.2

Write structural formulas for all ketones with molecular formula $C_6H_{12}O$, and give each its IUPAC name. Which of these ketones are chiral?

SOLUTION

Following are line-angle formulas and IUPAC names for the six ketones with the given molecular formula:

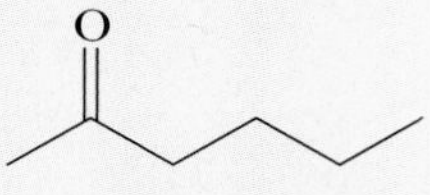

2-Hexanone 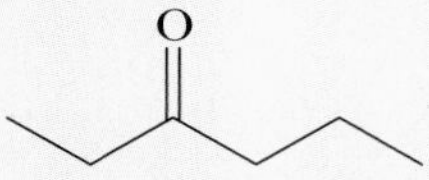   3-Hexanone 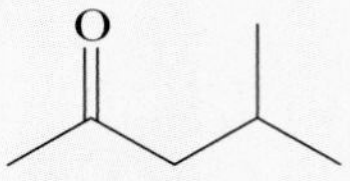   4-Methyl-2-pentanone

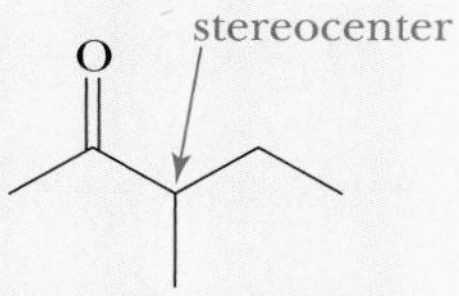

stereocenter

3-Methyl-2-pentanone 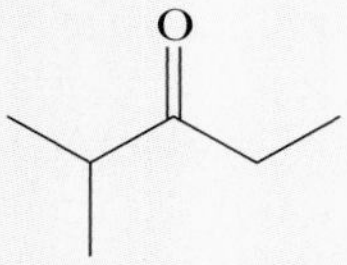   2-Methyl-3-pentanone 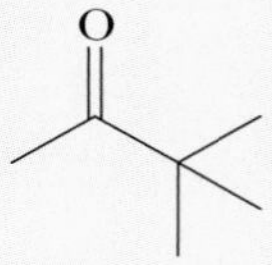   3,3-Dimethyl-2-butanone

Only 3-methyl-2-pentanone has a stereocenter and is chiral.

Practice Problem 13.2

Write structural formulas for all aldehydes with molecular formula $C_6H_{12}O$, and give each its IUPAC name. Which of these aldehydes are chiral?

Order of precedence of functional groups A system for ranking functional groups in order of priority for the purposes of IUPAC nomenclature.

B. IUPAC Names for More Complex Aldehydes and Ketones

In naming compounds that contain more than one functional group, the IUPAC has established an **order of precedence of functional groups**. Table 13.1 gives the order of precedence for the functional groups we have studied so far.

TABLE 13.1   Increasing Order of Precedence of Six Functional Groups

Functional Group	Suffix	Prefix	Example of When the Functional Group Has Lower Priority	
Carboxyl group	-oic acid	—		
Aldehyde group	-al	oxo-	3-Oxopropanoic acid	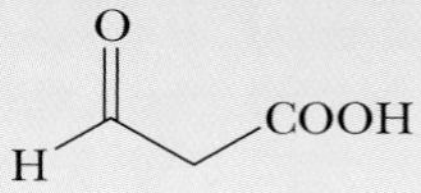
Ketone group	-one	oxo-	3-Oxobutanoic acid	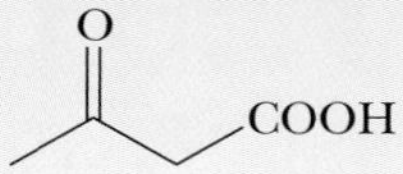
Alcohol group	-ol	hydroxy-	4-Hydroxybutanoic acid	
Amino group	-amine	amino-	3-Aminobutanoic acid	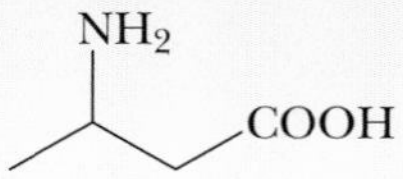
Sulfhydryl	-thiol	mercapto-	2-Mercaptoethanol	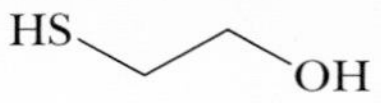

EXAMPLE 13.3

Write the IUPAC name for each compound:

(a)

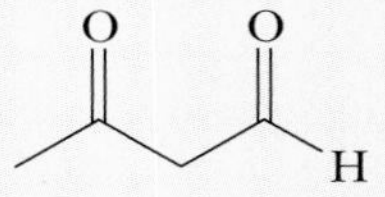

(b) 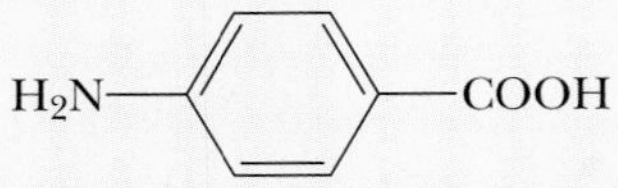 H_2N——⟨benzene ring⟩——COOH

(c)

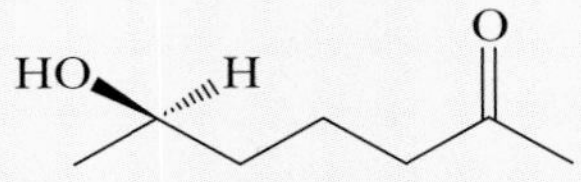

SOLUTION

(a) An aldehyde has higher precedence than a ketone, so we indicate the presence of the carbonyl group of the ketone by the prefix *oxo-*. The IUPAC name of this compound is 3-oxobutanal.

(b) The carboxyl group has higher precedence, so we indicate the presence of the amino group by the prefix *amino-*. The IUPAC name is 4-aminobenzoic acid. Alternatively, the compound may be named *p*-aminobenzoic acid, abbreviated PABA. PABA, a growth factor of microorganisms, is required for the synthesis of folic acid.

(c) The C=O group has higher precedence than the —OH group, so we indicate the —OH group by the prefix *hydroxy-*. The IUPAC name of this compound is (*R*)-6-hydroxy-2-heptanone.

Practice Problem 13.3

Write IUPAC names for these compounds, each of which is important in intermediary metabolism:

(a) CH₃CHCOOH
     |
    OH
    Lactic acid

(b) CH₃CCOOH
     ||
     O
   Pyruvic acid

(c) 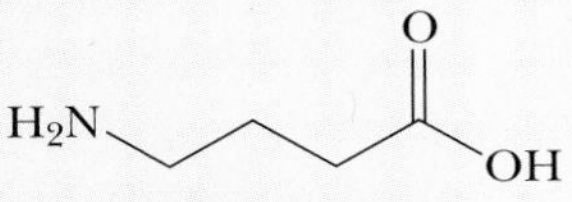 H_2N——————OH
    γ-Aminobutyric acid

The name shown is the one by which the compound is more commonly known in the biological sciences.

C. Common Names

The common name for an aldehyde is derived from the common name of the corresponding carboxylic acid by dropping the word *acid* and changing the suffix *-ic* or *-oic* to *-aldehyde*. Because we have not yet studied common names for carboxylic acids, we are not in a position to discuss common names for aldehydes. We can, however, illustrate how they are derived by reference to two common names of carboxylic acids with which you are familiar. The name formaldehyde is derived from formic acid, the name acetaldehyde from acetic acid:

HCH   HCOH   CH₃CH   CH₃COH

Formaldehyde   Formic acid   Acetaldehyde   Acetic acid

Common names for ketones are derived by naming each alkyl or aryl group bonded to the carbonyl group as a separate word, followed by the word *ketone*.

Groups are generally listed in order of increasing atomic weight. (Methyl ethyl ketone, abbreviated MEK, is a common solvent for varnishes and lacquers):

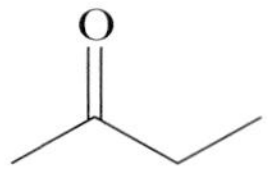

Methyl ethyl ketone (MEK) 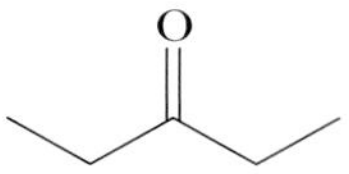   Diethyl ketone 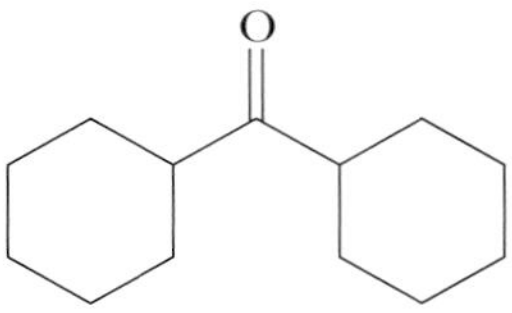   Dicyclohexyl ketone

13.4 PHYSICAL PROPERTIES

Oxygen is more electronegative than carbon (3.5 compared with 2.5; Table 1.5); therefore, a carbon–oxygen double bond is polar, with oxygen bearing a partial negative charge and carbon bearing a partial positive charge:

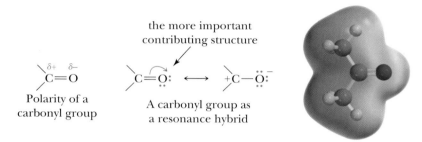

Polarity of a carbonyl group

A carbonyl group as a resonance hybrid

the more important contributing structure

The electron density model shows that the partial positive charge on an acetone molecule is distributed both on the carbonyl carbon and on the two attached methyl groups as well.

In addition, the resonance structure on the right emphasizes that, in reactions of a carbonyl group, carbon acts as an electrophile and a Lewis acid. The carbonyl oxygen, by contrast, acts as a nucleophile and a Lewis base.

Because of the polarity of the carbonyl group, aldehydes and ketones are polar compounds and interact in the liquid state by dipole–dipole interactions. As a result, aldehydes and ketones have higher boiling points than those of nonpolar compounds of comparable molecular weight. Table 13.2 lists the boiling points of six compounds of comparable molecular weight.

TABLE 13.2 Boiling Points of Six Compounds of Comparable Molecular Weight

Name	Structural Formula	Molecular Weight	Boiling Point (°C)
Diethyl ether	$CH_3CH_2OCH_2CH_3$	74	34
Pentane	$CH_3CH_2CH_2CH_2CH_3$	72	36
Butanal	$CH_3CH_2CH_2CHO$	72	76
2-Butanone	$CH_3CH_2COCH_3$	72	80
1-Butanol	$CH_3CH_2CH_2CH_2OH$	74	117
Propanoic acid	CH_3CH_2COOH	72	141

Pentane and diethyl ether have the lowest boiling points of these six compounds. Both butanal and 2-butanone are polar compounds, and because of the intermolecular attraction between carbonyl groups, their boiling points are higher than those of pentane and diethyl ether. Alcohols (Section 8.2C) and carboxylic acids (Section 14.4) are polar compounds, and their molecules associate by hydrogen bonding; their boiling points are higher than those of butanal and 2-butanone, compounds whose molecules cannot associate in that manner.

Because the carbonyl groups of aldehydes and ketones interact with water molecules by hydrogen bonding, low-molecular-weight aldehydes and ketones are more soluble in water than are nonpolar compounds of comparable molecular weight. Table 13.3 lists the boiling points and solubilities in water of several low-molecular-weight aldehydes and ketones.

TABLE 13.3 Physical Properties of Selected Aldehydes and Ketones

IUPAC Name	Common Name	Structural Formula	Boiling Point (°C)	Solubility (g/100 g water)
Methanal	Formaldehyde	$HCHO$	-21	infinite
Ethanal	Acetaldehyde	CH_3CHO	20	infinite
Propanal	Propionaldehyde	CH_3CH_2CHO	49	16
Butanal	Butyraldehyde	$CH_3CH_2CH_2CHO$	76	7
Hexanal	Caproaldehyde	$CH_3(CH_2)_4CHO$	129	slight
Propanone	Acetone	CH_3COCH_3	56	infinite
2-Butanone	Methyl ethyl ketone	$CH_3COCH_2CH_3$	80	26
3-Pentanone	Diethyl ketone	$CH_3CH_2COCH_2CH_3$	101	5

13.5 REACTIONS

The most common reaction theme of the carbonyl group is the addition of a nucleophile to form a **tetrahedral carbonyl addition intermediate**. In the following general reaction, the nucleophilic reagent is written as $Nu:^-$ to emphasize the presence of its unshared pair of electrons:

Tetrahedral carbonyl
addition intermediate

13.6 ADDITION OF GRIGNARD REAGENTS

From the perspective of the organic chemist, the addition of carbon nucleophiles is the most important type of nucleophilic addition to a carbonyl group, because these reactions form new carbon–carbon bonds. In this section, we describe the preparation and reactions of Grignard reagents and their reaction with aldehydes and ketones.

A. Formation and Structure of Organomagnesium Compounds

Alkyl, aryl, and vinylic halides react with Group I, Group II, and certain other metals to form **organometallic compounds**. Within the range of organometallic compounds, organomagnesium compounds are among the most readily available,

Organometallic compound
A compound containing a carbon–metal bond.

easily prepared, and easily handled. They are commonly named **Grignard reagents**, after Victor Grignard, who was awarded a 1912 Nobel prize in chemistry for their discovery and their application to organic synthesis.

Grignard reagents are typically prepared by the slow addition of a halide to a stirred suspension of magnesium metal in an ether solvent, most commonly diethyl ether or tetrahydrofuran (THF). Organoiodides and bromides generally react rapidly under these conditions, whereas chlorides react more slowly. Butylmagnesium bromide, for example, is prepared by adding 1-bromobutane to an ether suspension of magnesium metal. Aryl Grignards, such as phenylmagnesium bromide, are prepared in a similar manner:

$$\text{1-Bromobutane} + \text{Mg} \xrightarrow{\text{ether}} \text{Butylmagnesium bromide}$$

$$\text{Bromobenzene} + \text{Mg} \xrightarrow{\text{ether}} \text{Phenylmagnesium bromide}$$

Given that the difference in electronegativity between carbon and magnesium is 1.3 units ($2.5 - 1.2$), the carbon–magnesium bond is best described as polar covalent, with carbon bearing a partial negative charge and magnesium bearing a partial positive charge. In the structural formula on the right, the carbon–magnesium bond is shown as ionic to emphasize its nucleophilic character. Note that although we can write a Grignard reagent as a **carbanion**, a more accurate representation shows it as a polar covalent compound:

carbon is a nucleophile

$$\text{CH}_3(\text{CH}_2)_2\overset{\text{H}}{\underset{\text{H}}{\overset{|}{\underset{|}{\text{C}}}}}\overset{\delta-\quad\delta+}{-}\text{MgBr} \qquad \text{CH}_3(\text{CH}_2)_2\overset{\text{H}}{\underset{\text{H}}{\overset{|}{\underset{|}{\text{C}}}}}{:}^- \quad \overset{+}{\text{MgBr}}$$

The feature that makes Grignard reagents so valuable in organic synthesis is that the carbon bearing the halogen is now transformed into a nucleophile.

B. Reaction with Protic Acids

Grignard reagents are very strong bases and react readily with a wide variety of acids (proton donors) to form alkanes. Ethylmagnesium bromide, for example, reacts instantly with water to give ethane and magnesium salts. This reaction is an example of a stronger acid and a stronger base reacting to give a weaker acid and a weaker base (Section 2.5):

$$\overset{\delta-\quad\delta+}{\text{CH}_3\text{CH}_2-\text{MgBr}} + \text{H}-\text{OH} \longrightarrow \text{CH}_3\text{CH}_2-\text{H} + \text{Mg}^{2+} + \text{OH}^- + \text{Br}^-$$

	pK_a 15.7	pK_a 51	
Stronger	Stronger	Weaker	Weaker
base	acid	acid	base

Any compound containing an O—H, N—H, or S—H bond will react with a Grignard reagent by proton transfer. Following are examples of compounds containing those functional groups:

HOH	ROH	ArOH	RCOOH	RNH$_2$	RSH
Water	Alcohols	Phenols	Carboxylic acids	Amines	Thiols

Because Grignard reagents react so rapidly with these proton acids, Grignard reagents cannot be made from any halogen-containing compounds that also contain them.

EXAMPLE 13.4

Write an equation for the acid–base reaction between ethylmagnesium iodide and an alcohol. Use curved arrows to show the flow of electrons in this reaction. In addition, show that the reaction is an example of a stronger acid and stronger base reacting to form a weaker acid and weaker base.

SOLUTION

The alcohol is the stronger acid and ethyl carbanion is the stronger base:

$$CH_3CH_2-MgI + H-\ddot{O}R \longrightarrow CH_3CH_2-H + R\ddot{O}:^-MgI^+$$

Ethylmagnesium iodide	An alcohol pK_a 16–18	Ethane pK_a 51	A magnesium alkoxide
(stronger base)	(stronger acid)	(weaker acid)	(weaker base)

Practice Problem 13.4

Explain how these Grignard reagents react with molecules of their own kind to "self-destruct":

(a) HO—⟨benzene ring⟩—MgBr   (b)

$$\underset{HO}{\overset{O}{\parallel}}{\diagup}\diagdown\diagup\diagdown MgBr$$

C. Addition of Grignard Reagents to Aldehydes and Ketones

The special value of Grignard reagents is that they provide excellent ways to form new carbon–carbon bonds. In their reactions, Grignard reagents behave as carbanions. A carbanion is a good nucleophile and adds to the carbonyl group of an aldehyde or a ketone to form a tetrahedral carbonyl addition compound. The driving force for these reactions is the attraction of the partial negative charge on the carbon of the organometallic compound to the partial positive charge of the carbonyl carbon. In the examples that follow, the magnesium–oxygen bond is written $-O^-[MgBr]^+$ to emphasize its ionic character. The alkoxide ions formed in Grignard reactions are strong bases (Section 8.3C) and form alcohols when treated with an aqueous acid such as HCl or aqueous NH_4Cl during workup.

Addition to Formaldehyde Gives a 1° Alcohol

Treatment of a Grignard reagent with formaldehyde, followed by hydrolysis in aqueous acid, gives a primary alcohol:

$$CH_3CH_2-MgBr + H-\overset{O}{\overset{\parallel}{C}}-H \xrightarrow{\text{ether}} CH_3CH_2-\overset{:\ddot{O}:^-[MgBr]^+}{\underset{|}{C}H_2} \xrightarrow[H_2O]{HCl} CH_3CH_2-\overset{:\ddot{O}H}{\underset{|}{C}H_2} + Mg^{2+}$$

	Formaldehyde	A magnesium alkoxide	1-Propanol (a 1° alcohol)

Addition to an Aldehyde (Except Formaldehyde) Gives a 2° Alcohol

Treatment of a Grignard reagent with any aldehyde other than formaldehyde, followed by hydrolysis in aqueous acid, gives a secondary alcohol:

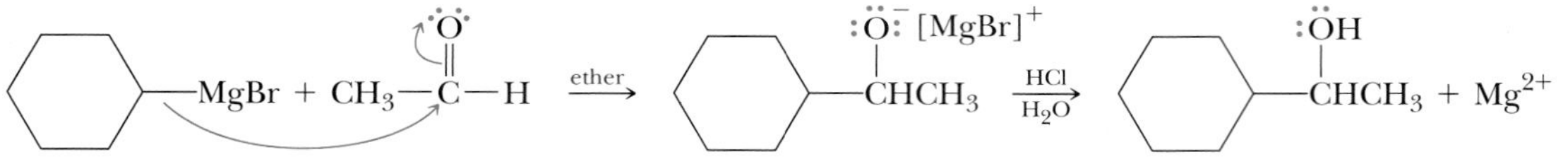

Acetaldehyde   A magnesium alkoxide   1-Cyclohexylethanol (a 2° alcohol)

Addition to a Ketone Gives a 3° Alcohol

Treatment of a Grignard reagent with a ketone, followed by hydrolysis in aqueous acid, gives a tertiary alcohol:

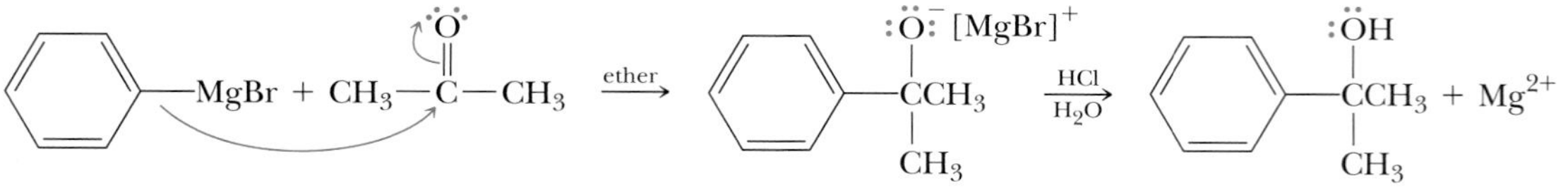

Acetone   A magnesium alkoxide   2-Phenyl-2-propanol (a 3° alcohol)

EXAMPLE 13.5

2-Phenyl-2-butanol can be synthesized by three different combinations of a Grignard reagent and a ketone. Show each combination.

SOLUTION

Curved arrows in each solution show the formation of the new carbon–carbon bond and the alkoxide ion, and labels on the final product show which set of reagents forms each bond:

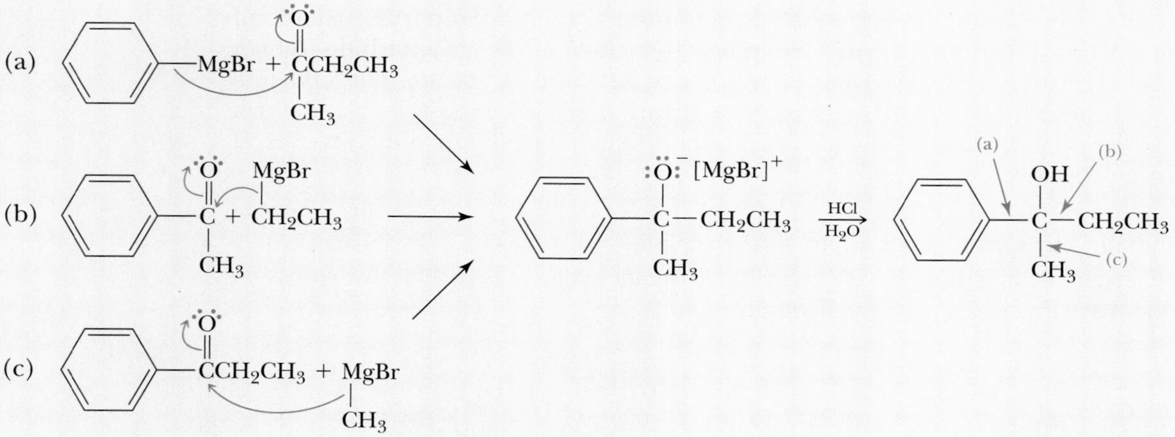

Practice Problem 13.5 ─────────────────────────────────

Show how these three compounds can be synthesized from the same Grignard reagent:

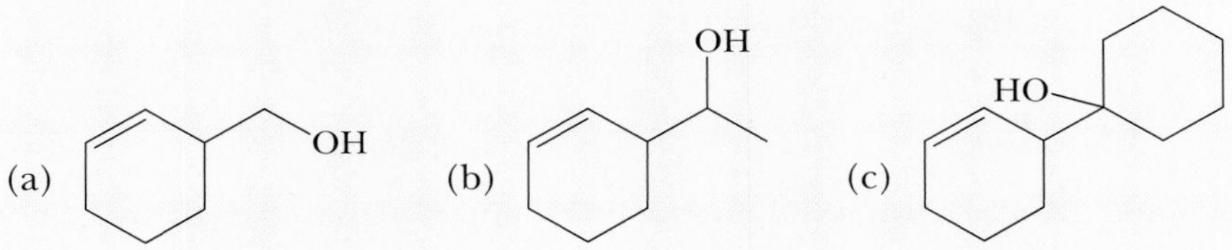

(a)

(b)

(c)

13.7 ADDITION OF ALCOHOLS

A. Formation of Acetals

The addition of a molecule of alcohol to the carbonyl group of an aldehyde or a ketone forms a **hemiacetal** (a half-acetal). This reaction is catalyzed by both acid and base: Oxygen adds to the carbonyl carbon and hydrogen adds to the carbonyl oxygen:

Hemiacetal A molecule containing an —OH and an —OR or —OAr group bonded to the same carbon.

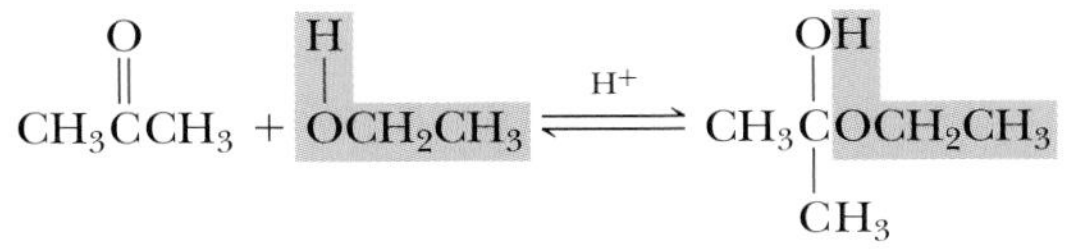

$$\underset{\text{A hemiacetal}}{CH_3\overset{\overset{\textstyle O}{\|}}{C}CH_3 + \overset{\overset{\textstyle H}{|}}{O}CH_2CH_3 \underset{}{\overset{H^+}{\rightleftharpoons} CH_3\underset{\underset{\textstyle CH_3}{|}}{\overset{\overset{\textstyle OH}{|}}{C}}OCH_2CH_3}}$$

The functional group of a hemiacetal is a carbon bonded to an —OH group and an —OR or —OAr group:

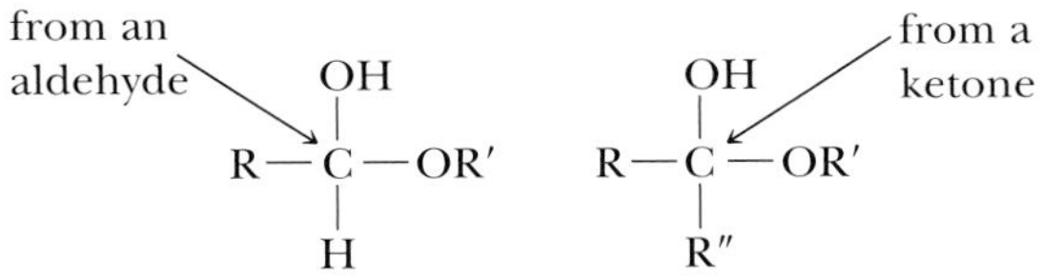

from an aldehyde

from a ketone

$$R-\overset{\overset{\textstyle OH}{|}}{\underset{\underset{\textstyle H}{|}}{C}}-OR' \qquad R-\overset{\overset{\textstyle OH}{|}}{\underset{\underset{\textstyle R''}{|}}{C}}-OR'$$

Hemiacetals

Hemiacetals are generally unstable and are only minor components of an equilibrium mixture, except in one very important type of molecule. When a hydroxyl group is part of the same molecule that contains the carbonyl group, and a five- or six-membered ring can form, the compound exists almost entirely in a cyclic hemiacetal form:

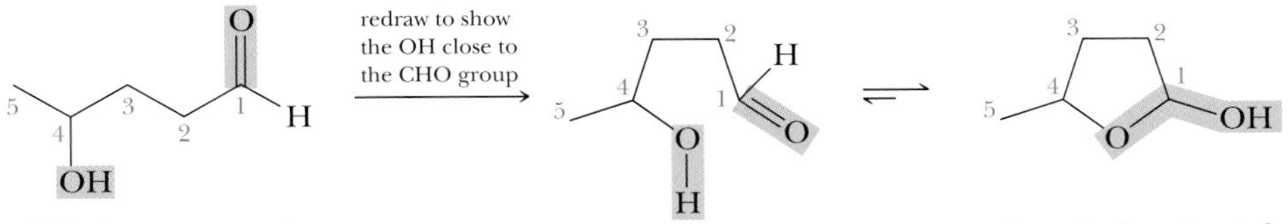

redraw to show the OH close to the CHO group

4-Hydroxypentanal

A cyclic hemiacetal (major form present at equilibrium)

We shall have much more to say about cyclic hemiacetals when we consider the chemistry of carbohydrates in Chapter 18.

Acetal A molecule containing two
—OR or —OAr groups bonded to
the same carbon.

Hemiacetals can react further with alcohols to form **acetals** plus a molecule of water. This reaction is acid catalyzed:

$$\underset{\substack{\text{A hemiacetal}}}{CH_3\overset{\displaystyle OH}{\underset{\displaystyle CH_3}{C}}OCH_2CH_3} + CH_3CH_2OH \overset{H^+}{\rightleftharpoons} \underset{\substack{\text{A diethyl acetal}}}{CH_3\overset{\displaystyle OCH_2CH_3}{\underset{\displaystyle CH_3}{C}}OCH_2CH_3} + H_2O$$

The functional group of an acetal is a carbon bonded to two —OR or —OAr groups:

from an
aldehyde

from a
ketone

$$R\overset{\displaystyle OR'}{\underset{\displaystyle H}{-C-}}OR' \qquad\qquad R\overset{\displaystyle OR'}{\underset{\displaystyle R''}{-C-}}OR'$$

Acetals

The mechanism for the acid-catalyzed conversion of a hemiacetal to an acetal can be divided into four steps. Note that acid H—A is a true catalyst in this reaction; it is used in Step 1, but a replacement H—A is generated in Step 4.

Mechanism: Acid-Catalyzed Formation of an Acetal

Step 1: Proton transfer from the acid, H—A, to the hemiacetal OH group gives an oxonium ion:

$$R\overset{\displaystyle H\ddot{O}:}{\underset{\displaystyle H}{-C-}}\ddot{O}CH_3 + H-A \rightleftharpoons R\overset{\displaystyle \overset{+}{O}\overset{H\quad H}{\diagdown\diagup}}{\underset{\displaystyle H}{-C-}}\ddot{O}CH_3 + A:^-$$

An oxonium ion

Step 2: Loss of water from the oxonium ion gives a resonance-stabilized cation:

$$R\overset{\displaystyle \overset{+}{O}\overset{H\quad H}{\diagdown\diagup}}{\underset{\displaystyle H}{-C-}}\ddot{O}CH_3 \rightleftharpoons R-\overset{+}{\underset{\displaystyle H}{C}}=\overset{+}{\ddot{O}}CH_3 \longleftrightarrow R-\overset{+}{\underset{\displaystyle H}{C}}-\ddot{O}CH_3 + H_2\ddot{O}:$$

A resonance-stabilized cation

Step 3: Reaction of the resonance-stabilized cation (an electrophile) with methanol (a nucleophile) gives the conjugate acid of the acetal:

$$CH_3-\ddot{O}:\ + R-\overset{+}{\underset{\displaystyle H}{C}}=\overset{+}{\ddot{O}}CH_3 \rightleftharpoons R\overset{\displaystyle \overset{+}{O}\overset{H\quad CH_3}{\diagdown\diagup}}{\underset{\displaystyle H}{-C-}}\ddot{O}CH_3$$

A protonated acetal

Step 4: Proton transfer from the protonated acetal to A⁻ gives the acetal and generates a new molecule of H—A, the acid catalyst:

$$A:^- + R\overset{\displaystyle \overset{+}{O}\overset{H\quad CH_3}{\diagdown\diagup}}{\underset{\displaystyle H}{-C-}}\ddot{O}CH_3 \rightleftharpoons HA + R\overset{\displaystyle \ddot{O}\overset{CH_3}{\diagup}}{\underset{\displaystyle H}{-C-}}\ddot{O}CH_3$$

A protonated acetal   An acetal

Formation of acetals is often carried out using the alcohol as a solvent and dissolving either dry HCl (hydrogen chloride) or arenesulfonic acid (Section 9.7B) in the alcohol. Because the alcohol is both a reactant and the solvent, it is present in large molar excess, which drives the reaction to the right and favors acetal formation. Alternatively, the reaction may be driven to the right by the removal of water as it is formed:

An excess of alcohol pushes the equilibrium toward acetal formation

Removal of water favors acetal formation

$$R-\overset{\overset{\displaystyle O}{\|}}{C}-R + 2CH_3CH_2OH \underset{}{\overset{H^+}{\rightleftharpoons}} R-\overset{\overset{\displaystyle OCH_2CH_3}{|}}{\underset{\underset{\displaystyle R}{|}}{C}}-OCH_2CH_3 + H_2O$$

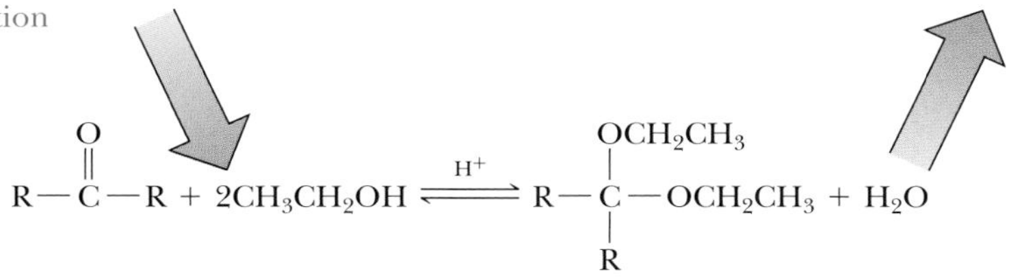

A diethyl acetal

EXAMPLE 13.6

Show the reaction of the carbonyl group of each ketone with one molecule of alcohol to form a hemiacetal and then with a second molecule of alcohol to form an acetal (note that, in part (b), ethylene glycol is a diol, and one molecule of it provides both —OH groups):

(a) 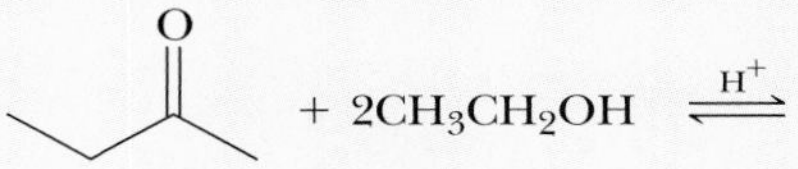 [structure] + 2CH_3CH_2OH $\overset{H^+}{\rightleftharpoons}$

(b) 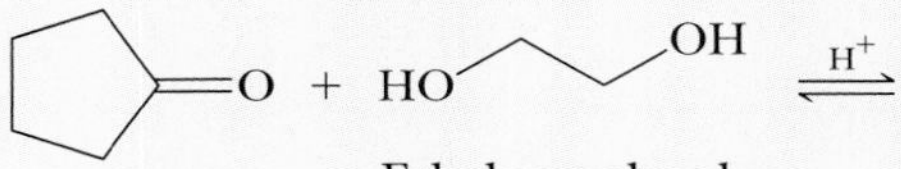 [structure] =O + HO[structure]OH $\overset{H^+}{\rightleftharpoons}$

Ethylene glycol

SOLUTION

Here are structural formulas of the hemiacetal and then the acetal:

(a) 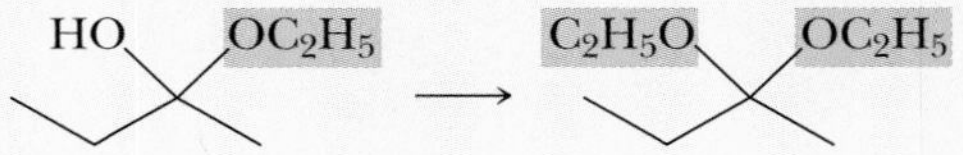 [structures: HO, OC_2H_5 → C_2H_5O, OC_2H_5]

(b) 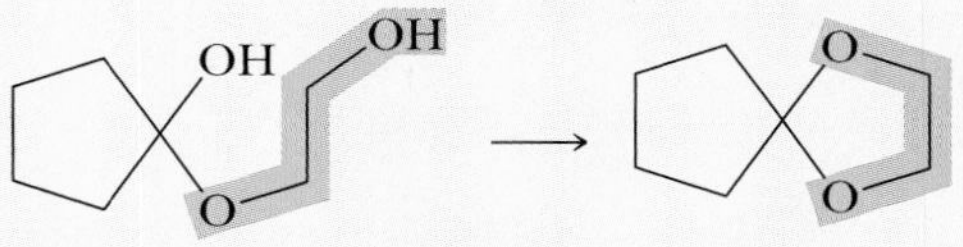 [structures]

Practice Problem 13.6

The hydrolysis of an acetal forms an aldehyde or a ketone and two molecules of alcohol. Following are structural formulas for three acetals:

(a) 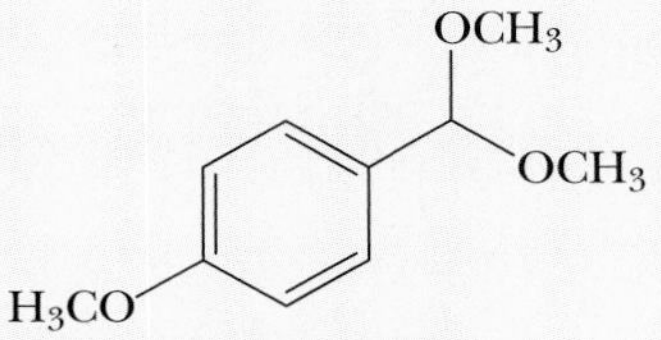 [structure with OCH_3, OCH_3, H_3CO] (b) 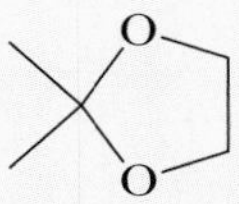 [structure] (c) 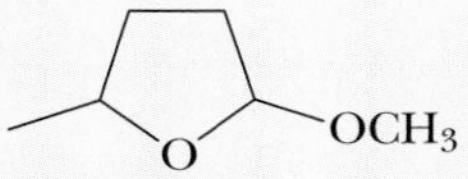 [structure with OCH_3]

Draw the structural formulas for the products of the hydrolysis of each in aqueous acid.

Like ethers, acetals are unreactive to bases, to reducing agents such as H_2/M, to Grignard reagents, and to oxidizing agents (except, of course, those which involve aqueous acid). Because of their lack of reactivity toward these reagents, acetals are often used to protect the carbonyl groups of aldehydes and ketones while reactions are carried out on functional groups in other parts of the molecule.

B. Acetals as Carbonyl-Protecting Groups

The use of acetals as carbonyl-protecting groups is illustrated by the synthesis of 5-hydroxy-5-phenylpentanal from benzaldehyde and 4-bromobutanal:

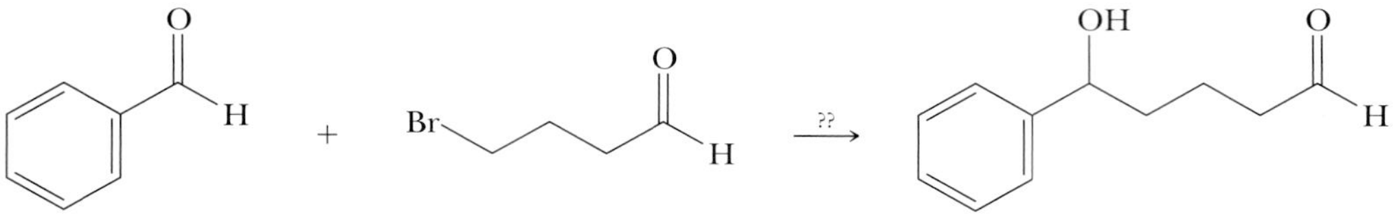

Benzaldehyde   4-Bromobutanal   5-Hydroxy-5-phenylpentanal

One obvious way to form a new carbon–carbon bond between these two molecules is to treat benzaldehyde with the Grignard reagent formed from 4-bromobutanal. This Grignard reagent, however, would react immediately with the carbonyl group of another molecule of 4-bromobutanal, causing it to self-destruct during preparation (Section 13.6B). A way to avoid this problem is to protect the carbonyl group of 4-bromobutanal by converting it to an acetal. Cyclic acetals are often used because they are particularly easy to prepare.

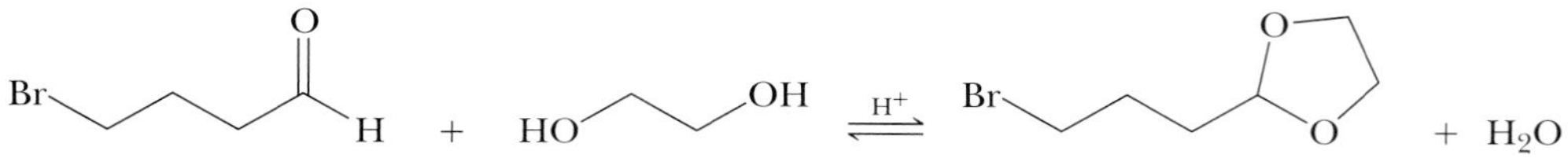

Ethylene glycol   A cyclic acetal

Treatment of the protected bromoaldehyde with magnesium in diethyl ether, followed by the addition of benzaldehyde, gives a magnesium alkoxide:

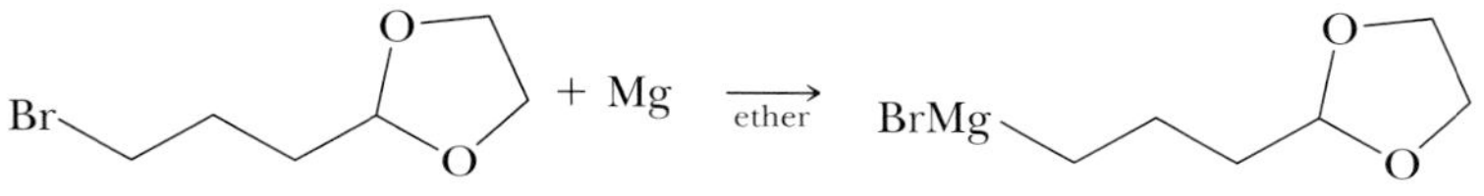

A cyclic acetal   A Grignard reagent

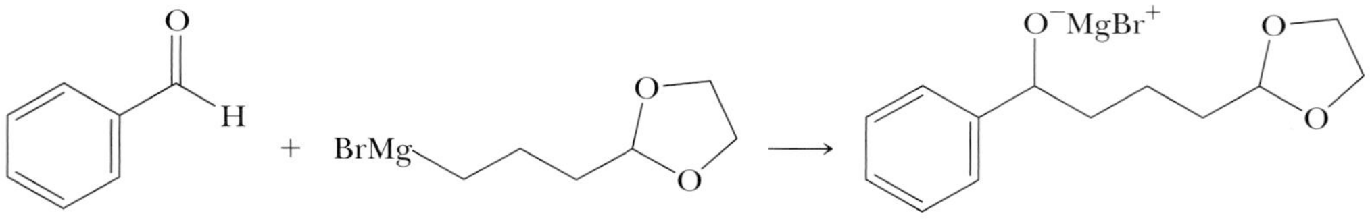

Benzaldehyde   A magnesium alkoxide

Treatment of the magnesium alkoxide with aqueous acid accomplishes two things. First, protonation of the alkoxide anion gives the desired hydroxyl group, and then, hydrolysis of the cyclic acetal regenerates the aldehyde group:

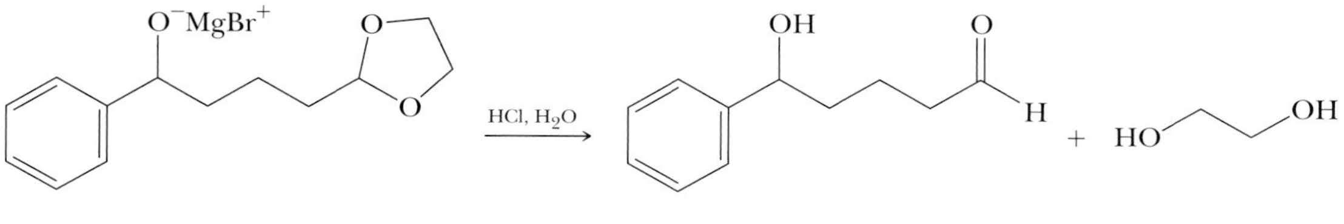

13.8 ADDITION OF AMMONIA AND AMINES

A. Formation of Imines

Ammonia, primary aliphatic amines (RNH_2), and primary aromatic amines ($ArNH_2$) react with the carbonyl group of aldehydes and ketones in the presence of an acid catalyst to give a product that contains a carbon–nitrogen double bond. A molecule containing a carbon–nitrogen double bond is called an **imine** or, alternatively, a **Schiff base**:

Imine A compound containing a carbon–nitrogen double bond; also called a Schiff base.

Schiff base An alternative name for an imine.

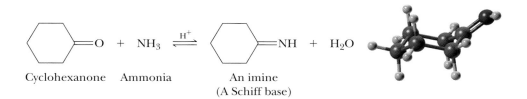

$$\underset{\text{Ethanal}}{CH_3CH\!\!=\!\!O} + \underset{\text{Aniline}}{H_2N\!\!-\!\!\bigcirc} \overset{H^+}{\rightleftharpoons} \underset{\substack{\text{An imine} \\ \text{(A Schiff base)}}}{CH_3CH\!\!=\!\!N\!\!-\!\!\bigcirc} + H_2O$$

$$\underset{\text{Cyclohexanone}}{\bigcirc\!\!=\!\!O} + \underset{\text{Ammonia}}{NH_3} \overset{H^+}{\rightleftharpoons} \underset{\substack{\text{An imine} \\ \text{(A Schiff base)}}}{\bigcirc\!\!=\!\!NH} + H_2O$$

Mechanism: Formation of an Imine from an Aldehyde or a Ketone

Step 1: Addition of the nitrogen atom of ammonia or a primary amine, both good nucleophiles, to the carbonyl carbon, followed by a proton transfer, gives a tetrahedral carbonyl addition intermediate:

A tetrahedral carbonyl addition intermediate

Step 2: Protonation of the OH group, followed by loss of water and proton transfer to solvent gives the imine. Notice that the loss of water and the proton transfer have the characteristics of an E2 reaction. Three things happen simultaneously in this dehydration: a base (in this case a water molecule) removes a proton from N, the carbon–nitrogen double bond forms, and the leaving group (in this case, a water molecule) departs:

An imine

(The flow of electrons here is similar to that in an E2 reaction.)

To give but one example of the importance of imines in biological systems, the active form of vitamin A aldehyde (retinal) is bound to the protein opsin in the human retina in the form of an imine called *rhodopsin* or *visual purple*. The amino acid lysine (Table 18.1) provides the primary amino group for this reaction:

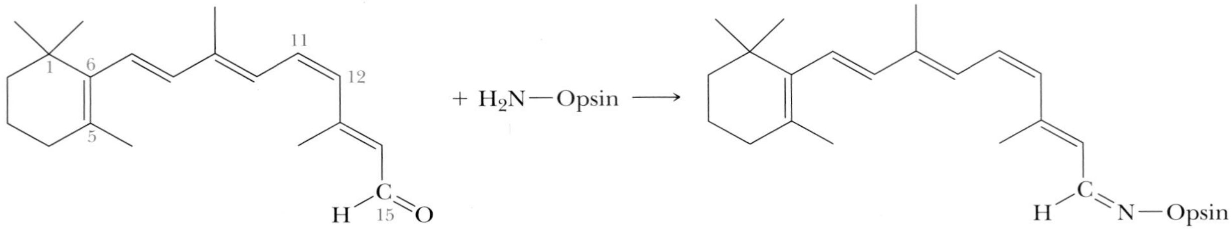

11-*cis*-Retinal

+ H₂N—Opsin ⟶

Rhodopsin
(Visual purple)

EXAMPLE 13.7

Write a structural formula for the imine formed in each reaction:

(a)

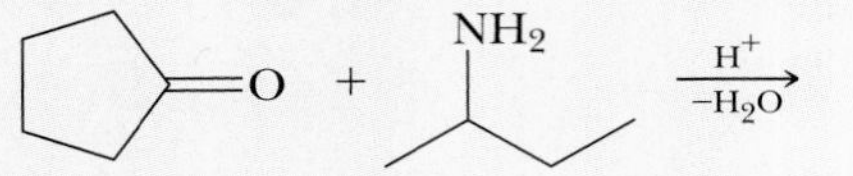

(b)

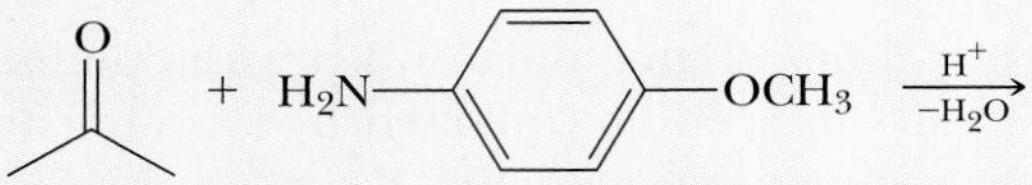

SOLUTION

Here is a structural formula for each imine:

(a)

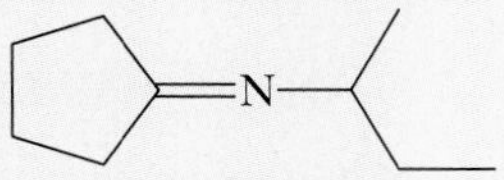

(b)

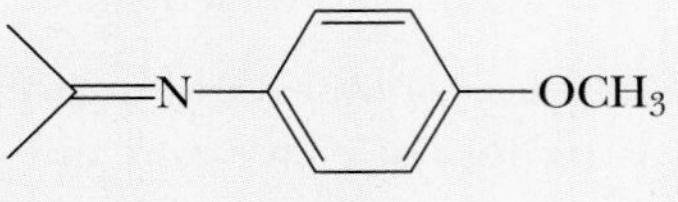

Practice Problem 13.7

Acid-catalyzed hydrolysis of an imine gives an amine and an aldehyde or a ketone. When one equivalent of acid is used, the amine is converted to its ammonium salt. For each of the following imines, write a structural formula for the products of hydrolysis, using one equivalent of HCl:

(a)   —CH=NCH₂CH₃ + H₂O $\xrightarrow{HCl}$

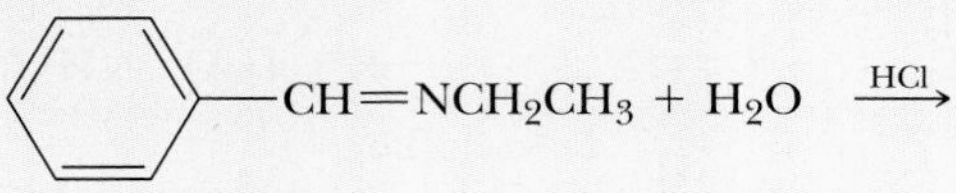

(b)   —CH₂N= + H₂O $\xrightarrow{HCl}$

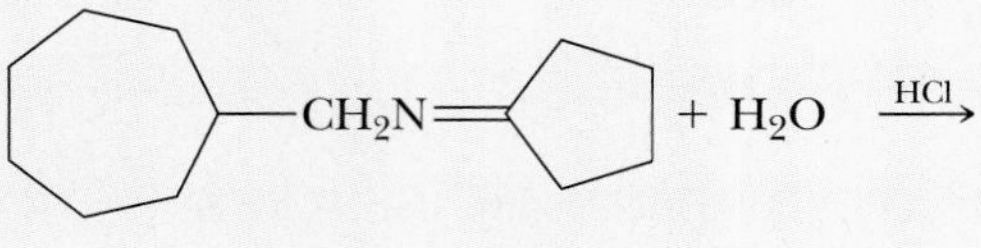

B. Reductive Amination of Aldehydes and Ketones

One of the chief values of imines is that the carbon–nitrogen double bond can be reduced to a carbon–nitrogen single bond by hydrogen in the presence of a nickel or other transition metal catalyst. By this two-step reaction, called **reductive amination**, a primary amine is converted to a secondary amine by way of an imine, as illustrated by the conversion of cyclohexylamine to dicyclohexylamine:

Reductive amination The formation of an imine from an aldehyde or a ketone, followed by the reduction of the imine to an amine.

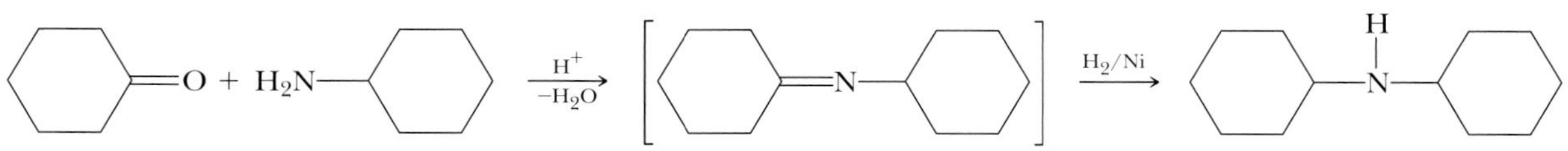

Cyclohexanone   Cyclohexyl-amine (a 1° amine)   (An imine)   Dicyclohexylamine (a 2° amine)

Conversion of an aldehyde or a ketone to an amine is generally carried out in one laboratory operation by mixing together the carbonyl-containing compound, the amine or ammonia, hydrogen, and the transition metal catalyst. The imine intermediate is not isolated.

EXAMPLE 13.8

Show how to synthesize each amine by a reductive amination:

(a) 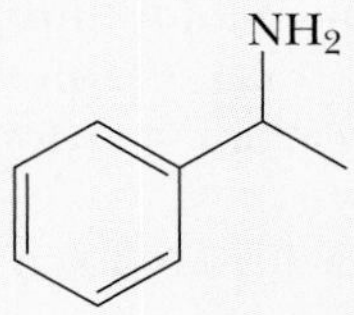   (b)

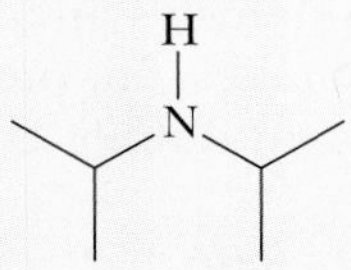

SOLUTION

Treat the appropriate compound, in each case a ketone, with ammonia or an amine in the presence of H_2/Ni:

(a) 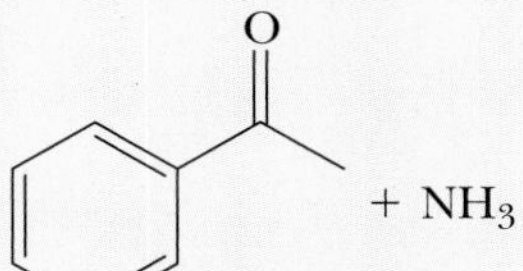 $+ NH_3$ (b) 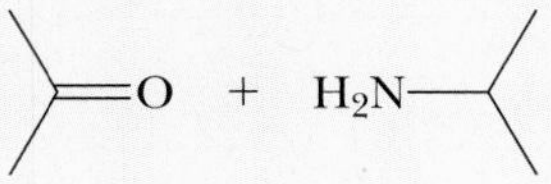 $O + H_2N-$

Practice Problem 13.8

Show how to prepare each amine by the reductive amination of an appropriate aldehyde or ketone:

(a) 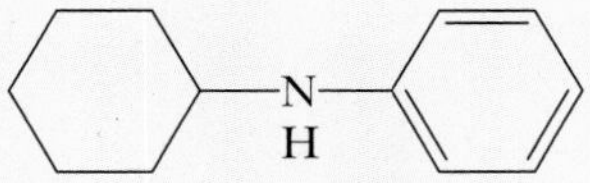   (b)

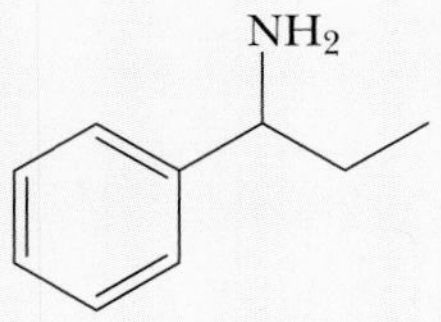

13.9   KETO–ENOL TAUTOMERISM

A. Keto and Enol Forms

α-**Carbon** A carbon atom adjacent to a carbonyl group.

α-**Hydrogen** A hydrogen on an *α*-carbon.

A carbon atom adjacent to a carbonyl group is called an **α-carbon,** and any hydrogen atoms bonded to it are called **α-hydrogens:**

$$
\overset{\alpha\text{-hydrogens}}{\underset{\alpha\text{-carbons}}{CH_3-\overset{\overset{O}{\|}}{C}-CH_2-CH_3}}
$$

Enol A molecule containing an —OH group bonded to a carbon of a carbon–carbon double bond.

An aldehyde or ketone that has at least one *α*-hydrogen is in equilibrium with a constitutional isomer called an **enol**. The name *enol* is derived from the IUPAC designation of it as both an alkene (*-en-*) and an alcohol (*-ol*):

$$
\underset{\substack{\text{Acetone}\\ \text{(keto form)}}}{CH_3-\overset{\overset{O}{\|}}{C}-CH_3} \rightleftharpoons \underset{\substack{\text{Acetone}\\ \text{(enol form)}}}{CH_3-\overset{\overset{OH}{|}}{C}=CH_2}
$$

Tautomers Constitutional isomers that differ in the location of hydrogen and a double bond relative to O, N, or S.

Keto and enol forms are examples of **tautomers**—constitutional isomers in equilibrium with each other and that differ in the location of a hydrogen atom and a double bond relative to a heteroatom, most commonly O, S, or N. This type of isomerism is called **tautomerism**.

For most simple aldehydes and ketones, the position of the equilibrium in keto–enol tautomerism lies far on the side of the keto form (Table 13.4), because a carbon–oxygen double bond is stronger than a carbon–carbon double bond.

The equilibration of keto and enol forms is catalyzed by acid, as shown in the following two-step mechanism (note that a molecule of H—A is consumed in Step 1, but another is generated in Step 2):

TABLE 13.4   The Position of Keto–Enol Equilibrium for Four Aldehydes and Ketones*

Keto form		Enol form	% Enol at Equilibrium		
$CH_3\overset{\overset{O}{\|}}{C}H$	$\rightleftharpoons$	$CH_2=\overset{\overset{OH}{|}}{C}H$	6×10^{-5}		
$CH_3\overset{\overset{O}{\|}}{C}CH_3$	$\rightleftharpoons$	$CH_3\overset{\overset{OH}{|}}{C}=CH_2$	6×10^{-7}		
cyclopentanone (=O)	$\rightleftharpoons$	cyclopentenol (—OH)	1×10^{-6}		
cyclohexanone (=O)	$\rightleftharpoons$	cyclohexenol (—OH)	4×10^{-5}		

*Data from J. March, *Advanced Organic Chemistry*, 4th ed. (New York, Wiley Interscience, 1992) p. 70.

Mechanism: Acid-Catalyzed Equilibration of Keto and Enol Tautomers

Step 1: Proton transfer from the acid catalyst, H—A, to the carbonyl oxygen forms the conjugate acid of the aldehyde or ketone:

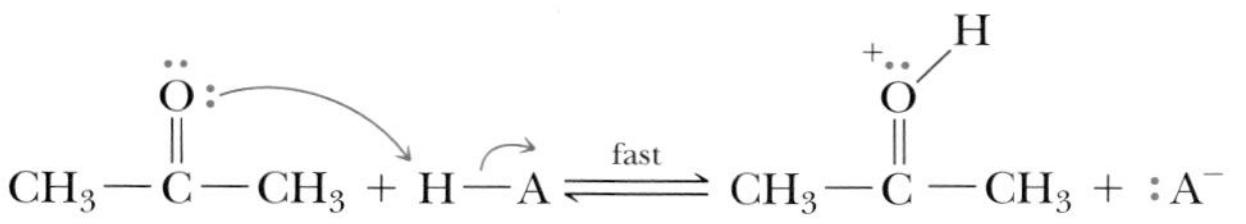

$$CH_3-\overset{\overset{\displaystyle \ddot{O}:}{||}}{C}-CH_3 + H-A \overset{fast}{\rightleftharpoons} CH_3-\overset{\overset{\displaystyle \overset{+}{O}\diagup^{H}}{||}}{C}-CH_3 + :A^-$$

Keto form   The conjugate acid
                                    of the ketone

Step 2: Proton transfer from the α-carbon to the base, A⁻, gives the enol and generates a new molecule of the acid catalyst, H—A:

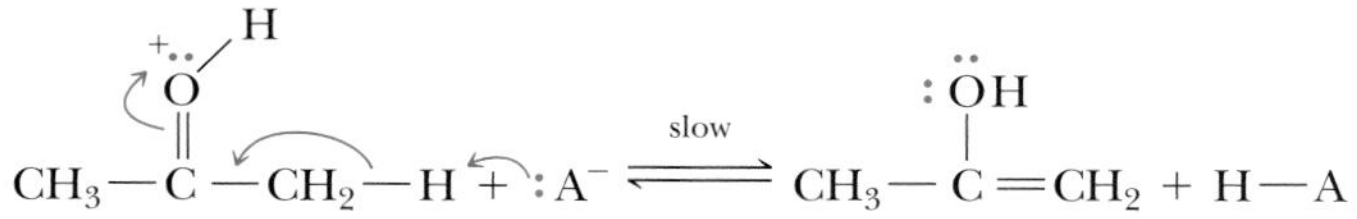

$$CH_3-\overset{\overset{\displaystyle \overset{+}{O}\diagup^{H}}{||}}{C}-CH_2-H + :A^- \overset{slow}{\rightleftharpoons} CH_3-\overset{\displaystyle :\ddot{O}H}{\underset{}{C}}=CH_2 + H-A$$

Enol form

EXAMPLE 13.9

Write two enol forms for each compound, and state which enol of each predominates at equilibrium:

(a) 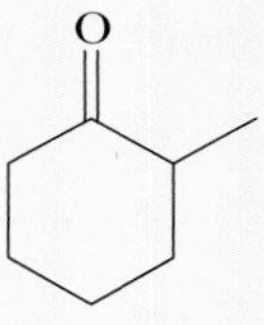   (b)

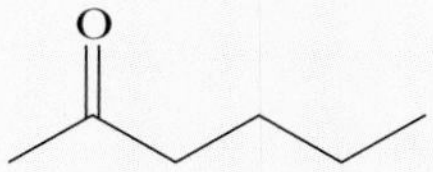

SOLUTION

In each case, the major enol form has the more substituted (the more stable) carbon–carbon double bond:

(a) 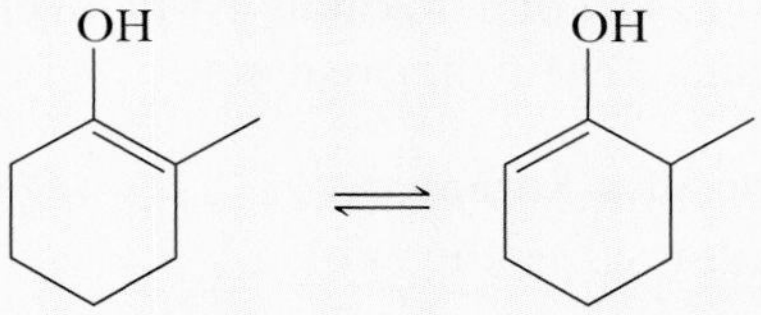   ⇌

Major enol

(b) 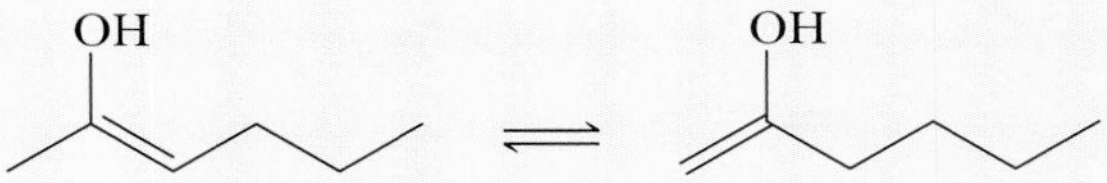   ⇌

Major enol

Practice Problem 13.9

Draw the structural formula for the keto form of each enol:

(a) 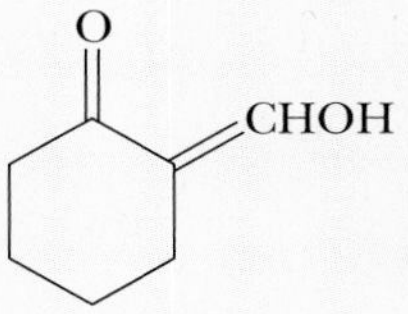   (b) 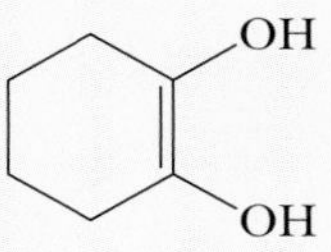   (c)

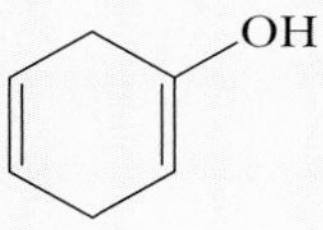

Racemization The conversion of
a pure enantiomer into a racemic
mixture.

B. Racemization at an α-Carbon

When enantiomerically pure (either R or S) 3-phenyl-2-butanone is dissolved in ethanol, no change occurs in the optical activity of the solution over time. If, however, a trace of acid (for example, HCl) is added, the optical activity of the solution begins to decrease and gradually drops to zero. When 3-phenyl-2-butanone is isolated from this solution, it is found to be a racemic mixture (Section 6.9C). This observation can be explained by the acid-catalyzed formation of an achiral enol intermediate. Tautomerism of the achiral enol to the chiral keto form generates the R and S enantiomers with equal probability:

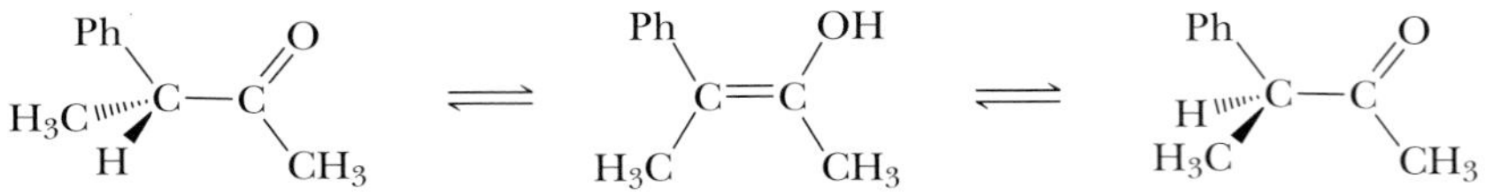

(R)-3-Phenyl-2-butanone   An achiral enol   (S)-3-Phenyl-2-butanone

Racemization by this mechanism occurs only at α-carbon stereocenters with at least one α-hydrogen.

C. α-Halogenation

Aldehydes and ketones with at least one α-hydrogen react with bromine and chlorine at the α-carbon to give an α-haloaldehyde or α-haloketone. Acetophenone, for example, reacts with bromine in acetic acid to give an α-bromoketone:

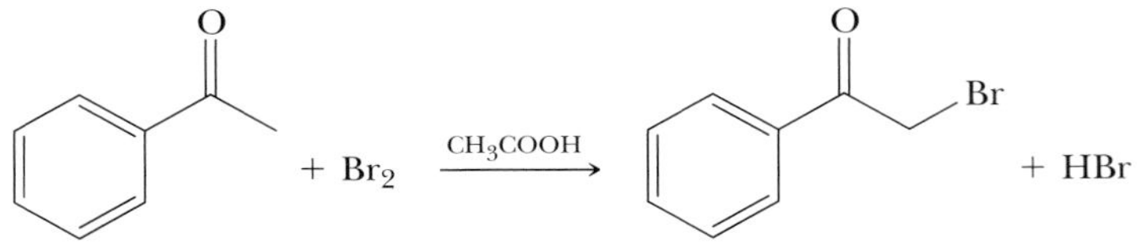

Acetophenone α-Bromoacetophenone

α-Halogenation is catalyzed by both acid and base. For acid-catalyzed halogenation, the HBr or HCl generated by the reaction catalyzes further reaction.

Mechanism: Acid-Catalyzed α-Halogenation of a Ketone

Step 1: Acid-catalyzed keto–enol tautomerism gives the enol:

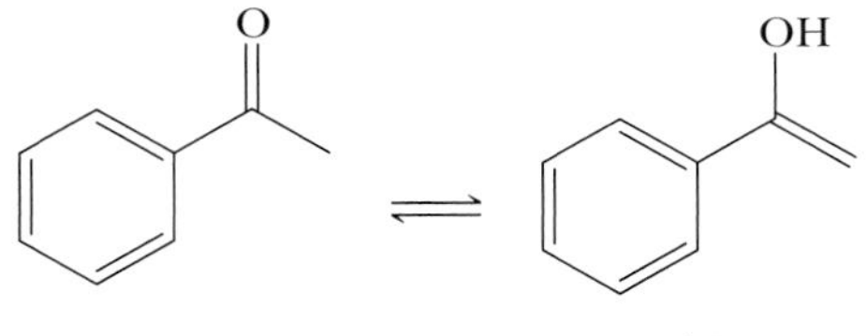

Keto form   Enol form

Step 2: Nucleophilic attack of the enol on the halogen molecule gives the α-haloketone:

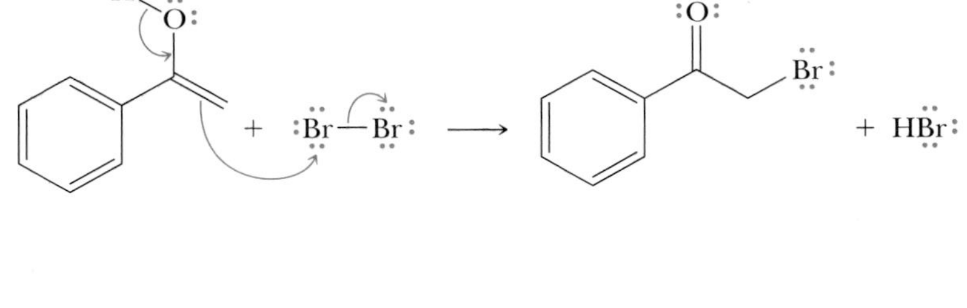

The value of α-halogenation is that it converts an α-carbon into a center that now has a good leaving group bonded to it and that is therefore susceptible to attack by a variety of good nucleophiles. In the following illustration, diethylamine (a nucleophile) reacts with the α-bromoketone to give an α-diethylaminoketone:

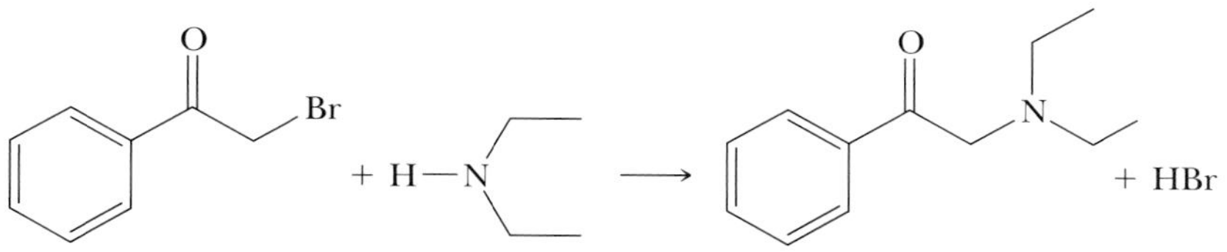

An α-bromoketone   An α-diethylaminoketone

In practice, this type of nucleophilic substitution is generally carried out in the presence of a weak base such as potassium carbonate to neutralize the HX as it is formed.

13.10 OXIDATION

A. Oxidation of Aldehydes to Carboxylic Acids

Aldehydes are oxidized to carboxylic acids by a variety of common oxidizing agents, including chromic acid and molecular oxygen. In fact, aldehydes are one of the most easily oxidized of all functional groups. Oxidation by chromic acid is illustrated by the conversion of hexanal to hexanoic acid:

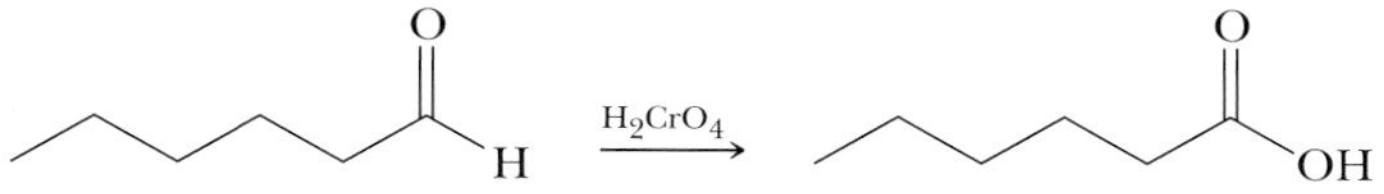

Hexanal   Hexanoic acid

Aldehydes are also oxidized to carboxylic acids by silver ion. One laboratory procedure is to shake a solution of the aldehyde dissolved in aqueous ethanol or tetrahydrofuran (THF) with a slurry of Ag_2O:

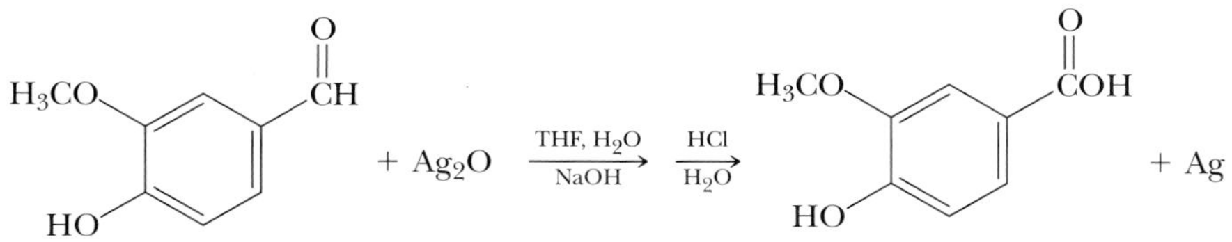

Vanillin   Vanillic acid
(from vanilla)

Tollens' reagent, another form of silver ion, is prepared by dissolving $AgNO_3$ in water, adding sodium hydroxide to precipitate silver ion as Ag_2O, and then adding aqueous ammonia to redissolve silver ion as the silver–ammonia complex ion:

$$Ag^+NO_3^- + 2NH_3 \xrightleftharpoons{NH_3, H_2O} Ag(NH_3)_2^+NO_3^-$$

When Tollens' reagent is added to an aldehyde, the aldehyde is oxidized to a carboxylic anion, and Ag^+ is reduced to metallic silver. If this reaction is carried out properly, silver precipitates as a smooth, mirrorlike deposit—hence the name **silver-mirror test**:

$$\underset{\text{RCH}}{\overset{\overset{\displaystyle O}{\|}}{}} + 2Ag(NH_3)_2^+ \xrightarrow{\text{NH}_3, \text{H}_2\text{O}} \underset{\text{RCO}^-}{\overset{\overset{\displaystyle O}{\|}}{}} + 2Ag + 4NH_3$$

Precipitates as
silver mirror

Nowadays, Ag^+ is rarely used for the oxidation of aldehydes, because of the cost of silver and because other, more convenient methods exist for this oxidation. The reaction, however, is still used for silvering mirrors. In the process, formaldehyde or glucose is used as the aldehyde to reduce Ag^+.

Aldehydes are also oxidized to carboxylic acids by molecular oxygen and by hydrogen peroxide.

A silver mirror has been deposited in the inside of this flask by the reaction of an aldehyde with Tollens' reagent. *(Charles D. Winters)*

Benzaldehyde   Benzoic acid

Molecular oxygen is the least expensive and most readily available of all oxidizing agents, and, on an industrial scale, air oxidation of organic molecules, including aldehydes, is common. Air oxidation of aldehydes can also be a problem: Aldehydes that are liquid at room temperature are so sensitive to oxidation by molecular oxygen that they must be protected from contact with air during storage. Often, this is done by sealing the aldehyde in a container under an atmosphere of nitrogen.

EXAMPLE 13.10

Draw a structural formula for the product formed by treating each compound with Tollens' reagent, followed by acidification with aqueous HCl:

(a) Pentanal   (b) Cyclopentanecarbaldehyde

SOLUTION
The aldehyde group in each compound is oxidized to a carboxyl group:

(a)   (b)

Pentanoic acid   Cyclopentanecarboxylic acid

Practice Problem 13.10

Complete these oxidations:

(a) 3-Oxobutanal + $O_2 \longrightarrow$

(b) 3-Phenylpropanal + Tollens' reagent $\longrightarrow$

B. Oxidation of Ketones to Carboxylic Acids

Ketones are much more resistant to oxidation than are aldehydes. For example, ketones are not normally oxidized by chromic acid or potassium permanganate. In fact, these reagents are used routinely to oxidize secondary alcohols to ketones in good yield (Section 8.3F).

CHEMICAL CONNECTIONS

A Green Synthesis of Adipic Acid

The current industrial production of adipic acid relies on the oxidation of a mixture of cyclohexanol and cyclohexanone by nitric acid:

Cyclohexanol

4 $+$ $6HNO_3$ $\longrightarrow$

4 COOH / COOH $+$ $3N_2O$ $+$ $3H_2O$

Hexanedioic acid   Nitrous
(Adipic acid)   oxide

A by-product of this oxidation is nitrous oxide, a gas considered to play a role in global warming and the depletion of the ozone layer in the atmosphere, as well as contributing to acid rain and acid smog. Given the fact that worldwide production of adipic acid is approximately 2.2 billion metric tons per year, the production of nitrous oxide is enormous. In spite of technological advances that allow for the recovery and recycling of nitrous oxide, it is estimated that approximately 400,000 metric tons escapes recovery and is released into the atmosphere each year.

Recently, Ryoji Noyori and coworkers at Nagoya University in Japan developed a "green" route to adipic acid, one that involves the oxidation of cyclohexene by

30% hydrogen peroxide catalyzed by sodium tungstate, Na_2WO_4:

$+$ $4H_2O_2$ $\xrightarrow[{[CH_3(C_8H_{17})_3N]HSO_4}]{Na_2WO_4}$

Cyclohexene

COOH / COOH $+$ $4H_2O$

Hexanedioic acid
(Adipic acid)

In this process, cyclohexene is mixed with aqueous 30% hydrogen peroxide, and sodium tungstate and methyltrioctylammonium hydrogen sulfate are added to the resulting two-phase system. (Cyclohexene is insoluble in water.) Under these conditions, cyclohexene is oxidized to adipic acid in approximately 90% yield.

While this route to adipic acid is environmentally friendly, it is not yet competitive with the nitric acid oxidation route because of the high cost of 30% hydrogen peroxide. What will make it competitive is either a considerable reduction in the cost of hydrogen peroxide or the institution of more stringent limitations on the emission of nitrous oxide into the atmosphere (or a combination of these).

Ketones undergo oxidative cleavage, via their enol form, by potassium dichromate and potassium permanganate at higher temperatures and by higher concentrations of nitric acid, HNO_3. The carbon–carbon double bond of the enol is cleaved to form two carboxyl or ketone groups, depending on the substitution pattern of the original ketone. An important industrial application of this reaction is the oxidation of cyclohexanone to hexanedioic acid (adipic acid), one of the two monomers required for the synthesis of the polymer nylon 66 (Section 17.5A):

Cyclohexanone   Cyclohexanone   Hexanedioic acid
(keto form)   (enol form)   (Adipic acid)

13.11   REDUCTION

Aldehydes are reduced to primary alcohols and ketones to secondary alcohols:

$$\underset{\substack{\text{An aldehyde}}}{\overset{\displaystyle O \atop \|}{R C H}} \xrightarrow{\text{reduction}} \underset{\substack{\text{A primary} \\ \text{alcohol}}}{R C H_2 O H} \qquad \underset{\substack{\text{A ketone}}}{\overset{\displaystyle O \atop \|}{R C R'}} \xrightarrow{\text{reduction}} \underset{\substack{\text{A secondary} \\ \text{alcohol}}}{\overset{\displaystyle O H \atop |}{R C H R'}}$$

A. Catalytic Reduction

The carbonyl group of an aldehyde or a ketone is reduced to a hydroxyl group by hydrogen in the presence of a transition metal catalyst, most commonly finely divided palladium, platinum, nickel, or rhodium. Reductions are generally carried out at temperatures from 25 to 100°C and at pressures of hydrogen from 1 to 5 atm. Under such conditions, cyclohexanone is reduced to cyclohexanol:

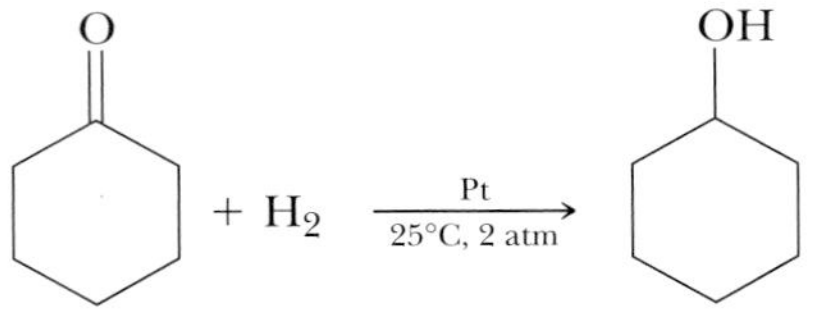

Cyclohexanone   Cyclohexanol

The catalytic reduction of aldehydes and ketones is simple to carry out, yields are generally very high, and isolation of the final product is very easy. A disadvantage is that some other functional groups (for example, carbon–carbon double bonds) are also reduced under these conditions.

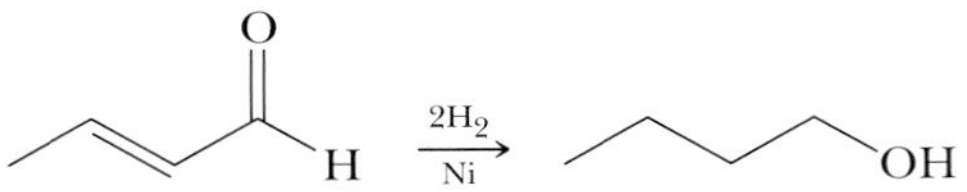

trans-2-Butenal   1-Butanol
(Crotonaldehyde)

B. Metal Hydride Reductions

By far the most common laboratory reagents used to reduce the carbonyl group of an aldehyde or a ketone to a hydroxyl group are sodium borohydride and lithium aluminum hydride. Each of these compounds behaves as a source of **hydride ion**, a very strong nucleophile. The structural formulas drawn here for these reducing agents show formal negative charges on boron and aluminum:

Hydride ion A hydrogen atom with two electrons in its valence shell; H:⁻.

$$\underset{\substack{\text{Sodium borohydride}}}{Na^+ H - \overset{\displaystyle H \atop |}{\underset{\displaystyle H \atop |}{B^-}} - H} \qquad \underset{\substack{\text{Lithium aluminum} \\ \text{hydride}}}{Li^+ H - \overset{\displaystyle H \atop |}{\underset{\displaystyle H \atop |}{Al^-}} - H} \qquad \underset{\substack{\text{Hydride ion}}}{H:^-}$$

In fact, hydrogen is more electronegative than either boron or aluminum (H = 2.1, Al = 1.5, and B = 2.0), and the formal negative charge in the two reagents resides more on hydrogen than on the metal.

Lithium aluminum hydride is a very powerful reducing agent; it rapidly reduces not only the carbonyl groups of aldehydes and ketones, but also those of carboxylic acids (Section 14.6) and their functional derivatives (Section 15.9). Sodium borohydride is a much more selective reagent, reducing only aldehydes and ketones rapidly.

Reductions using sodium borohydride are most commonly carried out in aqueous methanol, in pure methanol, or in ethanol. The initial product of reduction is a tetraalkyl borate, which is converted to an alcohol and sodium borate salts upon treatment with water. One mole of sodium borohydride reduces 4 moles of aldehyde or ketone:

$$4RCH + NaBH_4 \xrightarrow{CH_3OH} \underset{\text{A tetraalkyl borate}}{(RCH_2O)_4B^-Na^+} \xrightarrow{H_2O} 4RCH_2OH + \text{borate salts}$$

The key step in the metal hydride reduction of an aldehyde or a ketone is the transfer of a hydride ion from the reducing agent to the carbonyl carbon to form a tetrahedral carbonyl addition compound. In the reduction of an aldehyde or a ketone to an alcohol, only the hydrogen atom attached to carbon comes from the hydride-reducing agent; the hydrogen atom bonded to oxygen comes from the water added to hydrolyze the metal alkoxide salt.

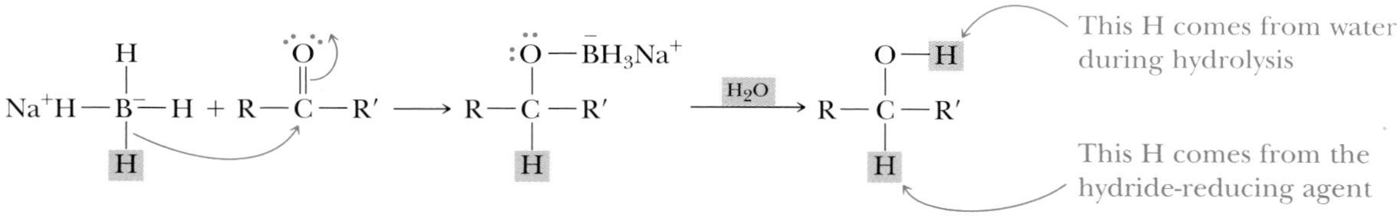

This H comes from water during hydrolysis

This H comes from the hydride-reducing agent

The next two equations illustrate the selective reduction of a carbonyl group in the presence of a carbon–carbon double bond and, alternatively, the selective reduction of a carbon–carbon double bond in the presence of a carbonyl group.

Selective reduction of a carbonyl group:

$$RCH{=}CHCR' \xrightarrow[\text{2. H}_2\text{O}]{\text{1. NaBH}_4} RCH{=}CHCHR'$$

Selective reduction of a carbon–carbon double bond:

$$RCH{=}CHCR' + H_2 \xrightarrow{Rh} RCH_2CH_2CR'$$

EXAMPLE 13.11

Complete these reductions:

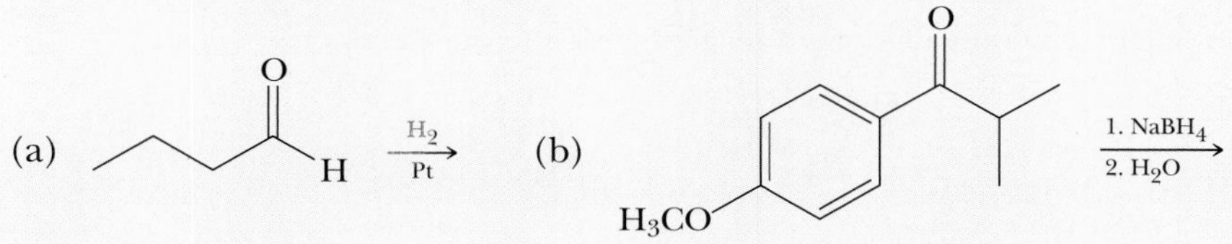

(a) [structure] $\xrightarrow[Pt]{H_2}$ (b) [structure] $\xrightarrow[\text{2. H}_2\text{O}]{\text{1. NaBH}_4}$

SOLUTION

The carbonyl group of the aldehyde in (a) is reduced to a primary alcohol, and that of the ketone in (b) is reduced to a secondary alcohol:

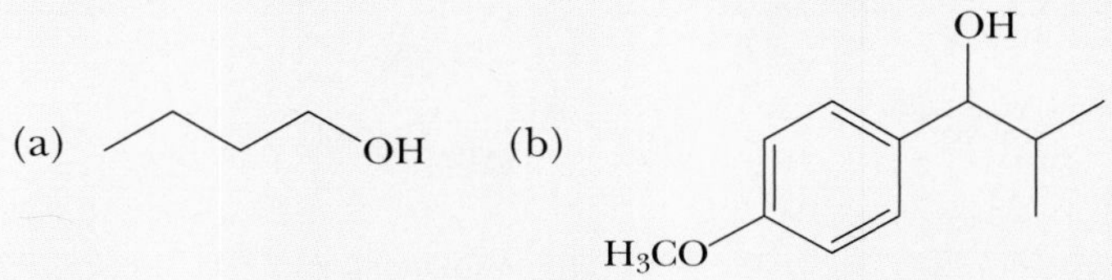

(a) [structure with OH]   (b) [structure with OH]

Practice Problem 13.11

What aldehyde or ketone gives each alcohol upon reduction by $NaBH_4$?

(a) [cyclohexane ring]—OH

(b) [benzene ring]—CH_2CH_2OH

(c) [structure with two OH groups on a carbon chain]

SUMMARY

An **aldehyde** (Section 13.2) contains a carbonyl group bonded to a hydrogen atom and a carbon atom. A **ketone** contains a carbonyl group bonded to two carbons. An aldehyde is named by changing *-e* of the parent alkane to *-al* (Section 13.3). A CHO group bonded to a ring is indicated by the suffix *-carbaldehyde*. A ketone is named by changing *-e* of the parent alkane to *-one* and using a number to locate the carbonyl group. In naming compounds that contain more than one functional group, the IUPAC system has established an **order of precedence of functional groups** (Section 13.3B). If the carbonyl group of an aldehyde or a ketone is lower in precedence than other functional groups in the molecule, it is indicated by the infix *-oxo-*.

Aldehydes and ketones are polar compounds (Section 13.4) and interact in the pure state by dipole–dipole interactions; they have higher boiling points and are more soluble in water than are nonpolar compounds of comparable molecular weight.

The carbon–metal bond in **Grignard reagents** (Section 13.5) has a high degree of partial ionic character. Grignard reagents behave as carbanions and are both strong bases and good nucleophiles.

A carbon atom adjacent to a carbonyl group is called an **α-carbon** (Section 13.9A), and a hydrogen attached to it is called an **α-hydrogen.**

KEY REACTIONS

1. Reaction with Grignard Reagents (Section 13.6C)
Treatment of formaldehyde with a Grignard reagent, followed by hydrolysis in aqueous acid, gives a primary alcohol. Similar treatment of any other aldehyde gives a secondary alcohol:

$$CH_3CH(=O) \xrightarrow[\text{2. HCl, H}_2\text{O}]{\text{1. C}_6\text{H}_5\text{MgBr}} C_6H_5CHCH_3 \text{ (OH)}$$

Treatment of a ketone with a Grignard reagent gives a tertiary alcohol:

$$CH_3CCH_3(=O) \xrightarrow[\text{2. HCl, H}_2\text{O}]{\text{1. C}_6\text{H}_5\text{MgBr}} C_6H_5C(CH_3)_2 \text{ (OH)}$$

2. Addition of Alcohols to Form Hemiacetals (Section 13.7)
Hemiacetals are only minor components of an equilibrium mixture of aldehyde or ketone and alcohol, except where the —OH and C=O groups are parts of the same molecule and a five- or six-membered ring can form:

$$CH_3CHCH_2CH_2CH(=O) \text{ (OH)} \rightleftharpoons [\text{cyclic structure: H}_3\text{C}—\text{ring with O and OH}]$$

4-Hydroxypentanal   A cyclic hemiacetal

3. Addition of Alcohols to Form Acetals (Section 13.7)

The formation of acetals is catalyzed by acid:

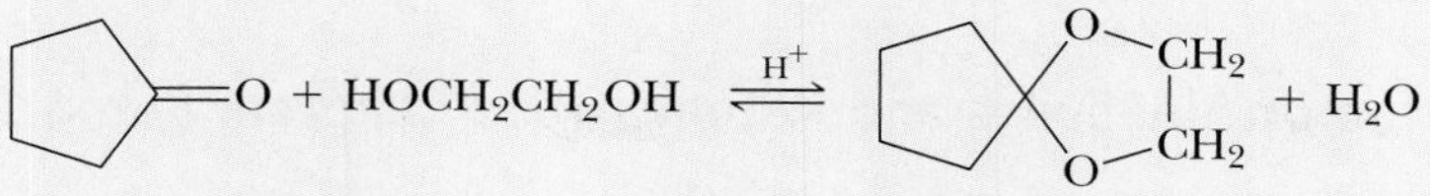

$$\text{cyclopentanone} = O + HOCH_2CH_2OH \xrightleftharpoons{H^+} \text{spiro acetal} + H_2O$$

4. Addition of Ammonia and Amines (Section 13.8)

The addition of ammonia or a primary amine to the carbonyl group of an aldehyde or a ketone forms a tetrahedral carbonyl addition intermediate. Loss of water from this intermediate gives an imine (a Schiff base):

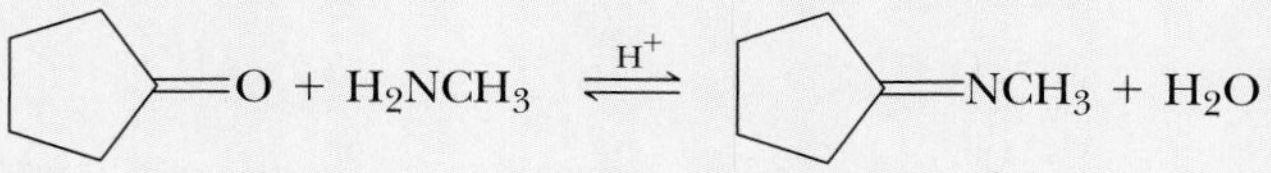

$$\text{cyclopentanone} = O + H_2NCH_3 \xrightleftharpoons{H^+} = NCH_3 + H_2O$$

5. Reductive Amination to Amines (Section 13.8B)

The carbon–nitrogen double bond of an imine can be reduced by hydrogen in the presence of a transition metal catalyst to a carbon–nitrogen single bond:

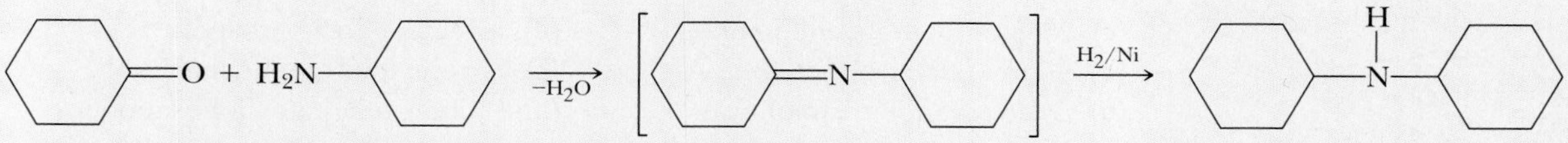

$$= O + H_2N- \xrightarrow{-H_2O} \left[= N- \right] \xrightarrow{H_2/Ni} -\overset{H}{N}-$$

6. Keto–Enol Tautomerism (Section 13.9A)

The keto form generally predominates at equilibrium:

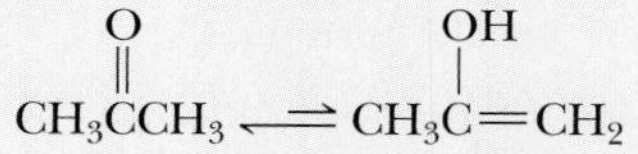

$$\underset{\substack{\text{Keto form} \\ \text{(Approx 99.9\%)}}}{CH_3\overset{O}{\overset{\|}{C}}CH_3} \xrightleftharpoons \underset{\text{Enol form}}{CH_3\overset{OH}{\overset{|}{C}}=CH_2}$$

7. Oxidation of an Aldehyde to a Carboxylic Acid (Section 13.10)

The aldehyde group is among the most easily oxidized functional groups. Oxidizing agents include H_2CrO_4, Tollens' reagent, and O_2:

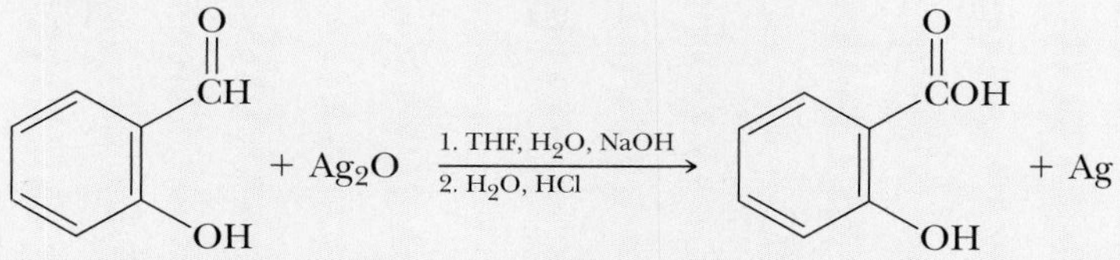

$$\text{2-hydroxybenzaldehyde} + Ag_2O \xrightarrow[\text{2. } H_2O, HCl]{\text{1. THF, } H_2O, NaOH} \text{salicylic acid} + Ag$$

8. Catalytic Reduction (Section 13.11A)

Catalytic reduction of the carbonyl group of an aldehyde or a ketone to a hydroxyl group is simple to carry out and yields of alcohols are high:

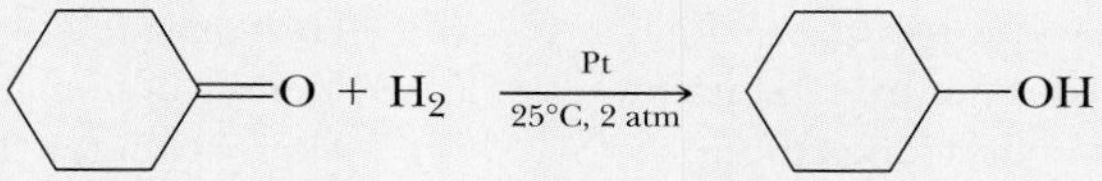

$$\text{cyclohexanone} = O + H_2 \xrightarrow[25°C, 2 \text{ atm}]{Pt} -OH$$

9. Metal Hydride Reduction (Section 13.11B)

Both $LiAlH_4$ and $NaBH_4$ reduce the carbonyl group of an aldehyde or a ketone to an hydroxyl group. They are selective in that neither reduces isolated carbon–carbon double bonds:

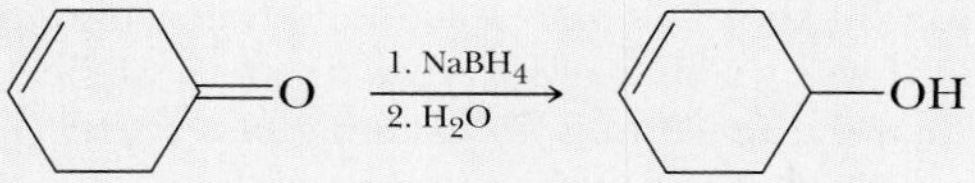

$$= O \xrightarrow[\text{2. } H_2O]{\text{1. } NaBH_4} -OH$$

PROBLEMS

A problem number set in red indicates an applied "real-world" problem.

Preparation of Aldehydes and Ketones (See Chapters 8 and 9)

13.12 Complete these reactions:

(a) [cyclooctanol with OH] $\xrightarrow[\text{H}_2\text{SO}_4]{\text{K}_2\text{Cr}_2\text{O}_7}$

(b) [cyclopentane-CH$_2$OH] $\xrightarrow[\text{CH}_2\text{Cl}_2]{\text{PCC}}$

(c) [cyclopentane-CH$_2$OH] $\xrightarrow[\text{H}_2\text{SO}_4]{\text{K}_2\text{Cr}_2\text{O}_7}$

(d) [benzene] + [acid chloride] $\xrightarrow{\text{AlCl}_3}$

13.13 Show how you would bring about these conversions:
- (a) 1-Pentanol to pentanal
- (b) 1-Pentanol to pentanoic acid
- (c) 2-Pentanol to 2-pentanone
- (d) 1-Pentene to 2-pentanone
- (e) Benzene to acetophenone
- (f) Styrene to acetophenone
- (g) Cyclohexanol to cyclohexanone
- (h) Cyclohexene to cyclohexanone

Structure and Nomenclature

13.14 Draw a structural formula for the one ketone with molecular formula C_4H_8O and for the two aldehydes with molecular formula C_4H_8O.

13.15 Draw structural formulas for the four aldehydes with molecular formula $C_5H_{10}O$. Which of these aldehydes are chiral?

13.16 Name these compounds:

(a) ... (b) ... (c) ... (d) ... (e) ... (f) ... (g) ...

13.17 Draw structural formulas for these compounds:
- (a) 1-Chloro-2-propanone
- (b) 3-Hydroxybutanal
- (c) 4-Hydroxy-4-methyl-2-pentanone
- (d) 3-Methyl-3-phenylbutanal
- (e) (S)-3-bromocyclohexanone
- (f) 3-Methyl-3-buten-2-one
- (g) 5-Oxohexanal
- (h) 2,2-Dimethylcyclohexanecarbaldehyde
- (i) 3-Oxobutanoic acid

Addition of Carbon Nucleophiles

13.18 Write an equation for the acid–base reaction between phenylmagnesium iodide and a carboxylic acid. Use curved arrows to show the flow of electrons in this reaction. In addition, show that the reaction is an example of a stronger acid and stronger base reacting to form a weaker acid and weaker base.

13.19 Diethyl ether is prepared on an industrial scale by the acid-catalyzed dehydration of ethanol:

$$2CH_3CH_2OH \xrightarrow[180°C]{H_2SO_4} CH_3CH_2OCH_2CH_3 + H_2O$$

Explain why diethyl ether used in the preparation of Grignard reagents must be carefully purified to remove all traces of ethanol and water.

13.20 Draw structural formulas for the product formed by treating each compound with propylmagnesium bromide, followed by hydrolysis in aqueous acid:

(a) CH_2O (b) 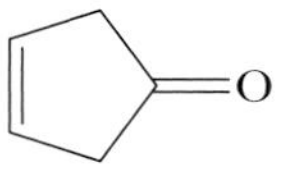 [structure] (c) 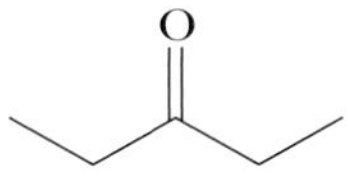 [structure] (d) 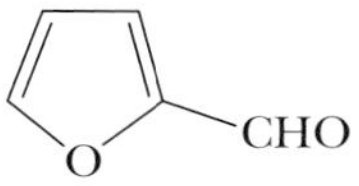 [structure] (e) 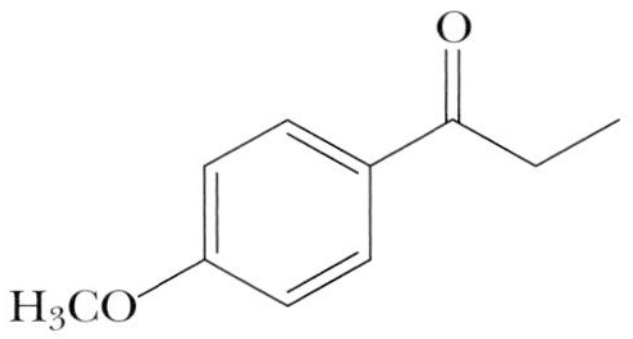 [structure]

13.21 Suggest a synthesis for each alcohol, starting from an aldehyde or a ketone and an appropriate Grignard reagent (the number of combinations of Grignard reagent and aldehyde or ketone that might be used is shown in parentheses below each target molecule):

(a) 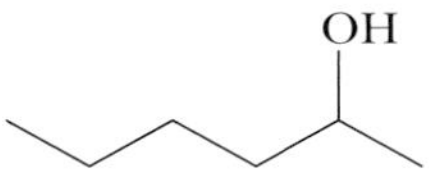 [structure]   (b) [structure]   (c) [structure]

(Two combinations)   (Two combinations)   (Three combinations)

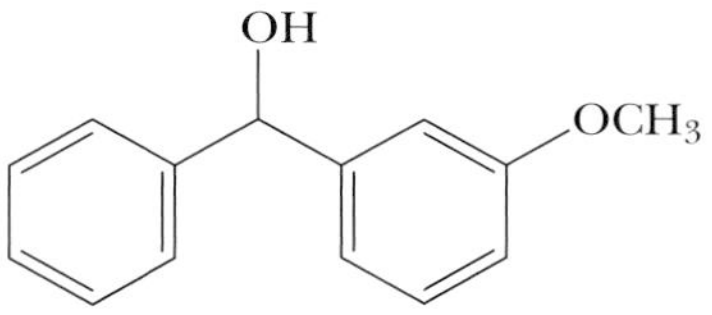

Addition of Oxygen Nucleophiles

13.22 5-Hydroxyhexanal forms a six-membered cyclic hemiacetal that predominates at equilibrium in aqueous solution:

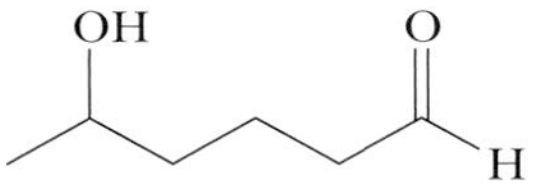

[structure] $\xrightarrow{H^+}$ a cyclic hemiacetal

5-Hydroxyhexanal

(a) Draw a structural formula for this cyclic hemiacetal.
(b) How many stereoisomers are possible for 5-hydroxyhexanal?
(c) How many stereoisomers are possible for the cyclic hemiacetal?
(d) Draw alternative chair conformations for each stereoisomer.
(e) For each stereoisomer, which alternative chair conformation is the more stable?

13.23 Draw structural formulas for the hemiacetal and then the acetal formed from each pair of reactants in the presence of an acid catalyst:

(a) 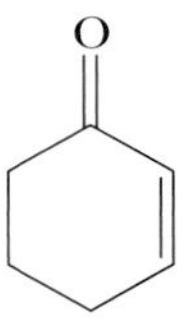 [structure] + CH_3CH_2OH (b) 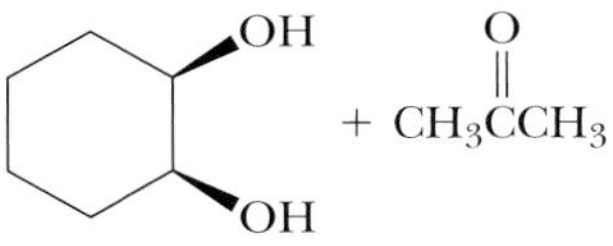 [structure] + CH_3CCH_3 (c) 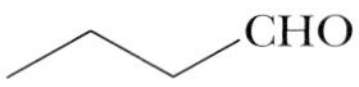 [structure] CHO + CH_3OH

13.24 Draw structural formulas for the products of hydrolysis of each acetal in aqueous acid:

(a) 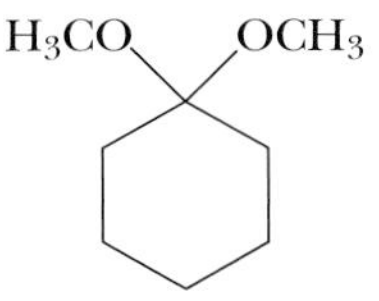 [structure]   (b) 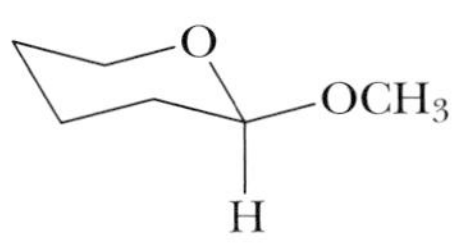 [structure]   (c) 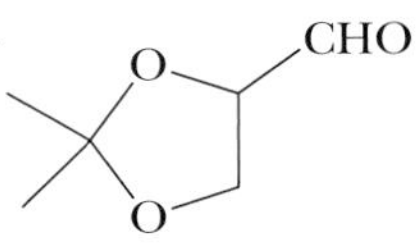 [structure]

13.25 The following compound is a component of the fragrance of jasmine: From what carbonyl-containing compound and alcohol is the compound derived?

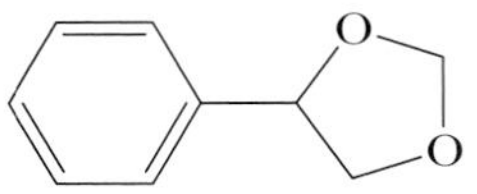

13.26 Propose a mechanism for the formation of the cyclic acetal by treating acetone with ethylene glycol in the presence of an acid catalyst. Make sure that your mechanism is consistent with the fact that the oxygen atom of the water molecule is derived from the carbonyl oxygen of acetone.

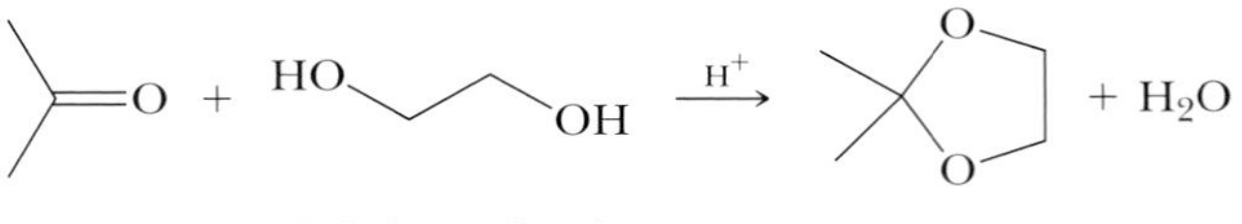

Acetone   Ethylene glycol

13.27 Propose a mechanism for the formation of a cyclic acetal from 4-hydroxypentanal and one equivalent of methanol: If the carbonyl oxygen of 4-hydroxypentanal is enriched with oxygen-18, does your mechanism predict that the oxygen label appears in the cyclic acetal or in the water? Explain.

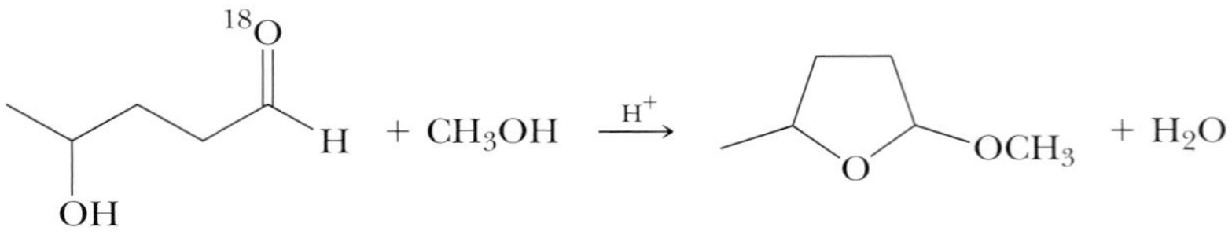

Addition of Nitrogen Nucleophiles

13.28 Show how this secondary amine can be prepared by two successive reductive aminations:

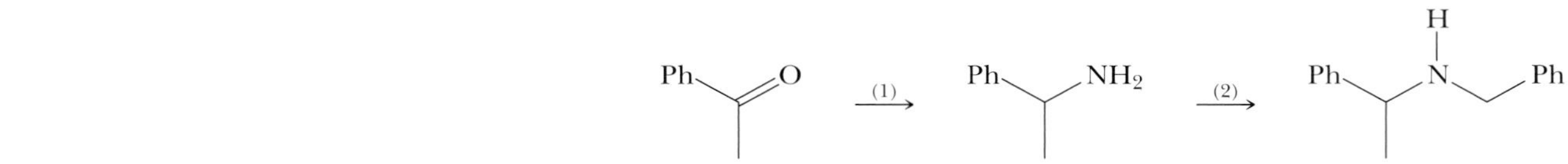

13.29 Show how to convert cyclohexanone to each of the following amines:

(a) 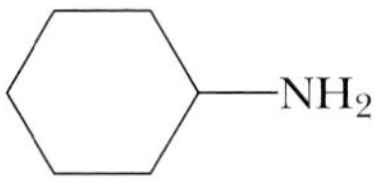 —NH$_2$ (b) 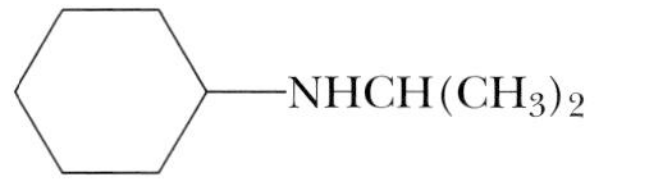 —NHCH(CH$_3$)$_2$ (c) 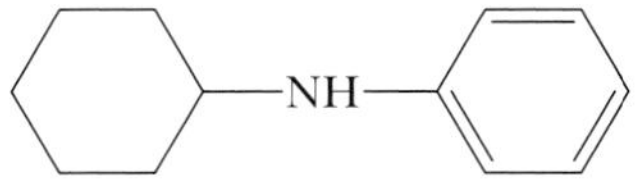 —NH—

13.30 Following are structural formulas for amphetamine and methamphetamine:

(a) 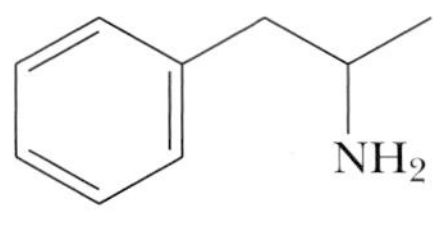   (b)

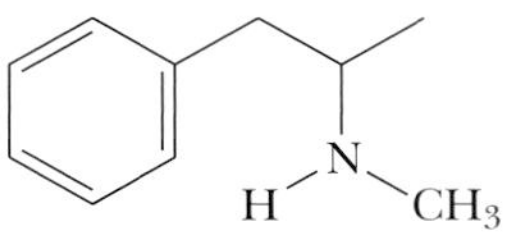

Amphetamine   Methamphetamine

The major central nervous system effects of amphetamine and amphetaminelike drugs are locomotor stimulation, euphoria and excitement, stereotyped behavior, and anorexia. Show how each drug can be synthesized by the reductive amination of an appropriate aldehyde or ketone.

13.31 Rimantadine is effective in preventing infections caused by the influenza A virus and in treating established illness. The drug is thought to exert its antiviral effect by blocking a late stage in the assembly of the virus. Following is the final step in the synthesis of rimantadine:

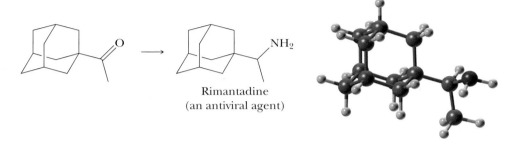

Rimantadine
(an antiviral agent)

(a) Describe experimental conditions to bring about this conversion.
(b) Is rimantadine chiral?

13.32 Methenamine, a product of the reaction of formaldehyde and ammonia, is a *prodrug*—a compound that is inactive by itself, but is converted to an active drug in the body by a biochemical transformation. The strategy behind the use of methenamine as a prodrug is that nearly all bacteria are sensitive to formaldehyde at concentrations of 20 mg/mL or higher. Formaldehyde cannot be used directly in medicine, however, because an effective concentration in plasma cannot be achieved with safe doses. Methenamine is stable at pH 7.4 (the pH of blood plasma), but undergoes acid-catalyzed hydrolysis to formaldehyde and ammonium ion under the acidic conditions of the kidneys and the urinary tract:

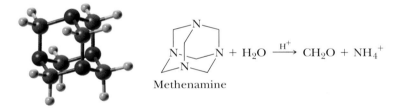

Methenamine

Thus, methenamine can be used as a site-specific drug to treat urinary infections.
(a) Balance the equation for the hydrolysis of methenamine to formaldehyde and ammonium ion.
(b) Does the pH of an aqueous solution of methenamine increase, remain the same, or decrease as a result of the hydrolysis of the compound? Explain.
(c) Explain the meaning of the following statement: The functional group in methenamine is the nitrogen analog of an acetal.
(d) Account for the observation that methenamine is stable in blood plasma, but undergoes hydrolysis in the urinary tract.

Keto–Enol Tautomerism

13.33 The following molecule belongs to a class of compounds called enediols: Each carbon of the double bond carries an —OH group:

$$\alpha\text{-hydroxyaldehyde} \rightleftharpoons \begin{array}{c} HC-OH \\ \parallel \\ C-OH \\ | \\ CH_3 \end{array} \rightleftharpoons \alpha\text{-hydroxyketone}$$

An enediol

Draw structural formulas for the α-hydroxyketone and the α-hydroxyaldehyde with which this enediol is in equilibrium.

13.34 In dilute aqueous acid, (*R*)-glyceraldehyde is converted into an equilibrium mixture of (*R,S*)-glyceraldehyde and dihydroxyacetone:

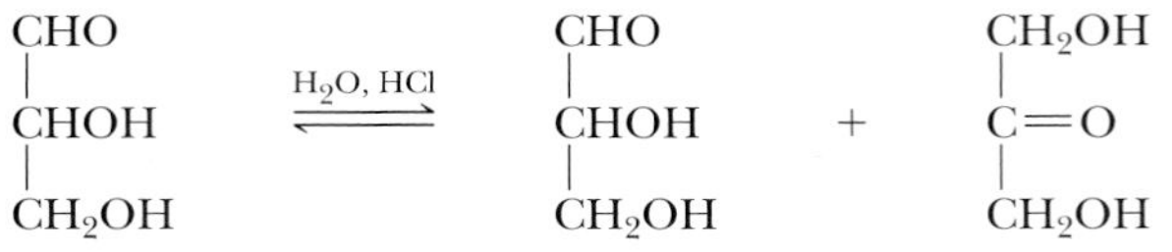

$$
\begin{array}{c}
\text{CHO} \\
| \\
\text{CHOH} \\
| \\
\text{CH}_2\text{OH}
\end{array}
\quad \underset{\text{H}_2\text{O, HCl}}{\rightleftharpoons} \quad
\begin{array}{c}
\text{CHO} \\
| \\
\text{CHOH} \\
| \\
\text{CH}_2\text{OH}
\end{array}
\quad + \quad
\begin{array}{c}
\text{CH}_2\text{OH} \\
| \\
\text{C}=\text{O} \\
| \\
\text{CH}_2\text{OH}
\end{array}
$$

(*R*)-Glyceraldehyde   (*R,S*)-Glyceraldehyde   Dihydroxyacetone

Propose a mechanism for this isomerization.

Oxidation/Reduction of Aldehydes and Ketones

13.35 Draw a structural formula for the product formed by treating butanal with each of the following sets of reagents:

(a) $LiAlH_4$ followed by H_2O (b) $NaBH_4$ in CH_3OH/H_2O
(c) H_2/Pt (d) $Ag(NH_3)_2^+$ in NH_3/H_2O and then HCl/H_2O
(e) H_2CrO_4 (f) $C_6H_5NH_2$ in the presence of H_2/Ni

13.36 Draw a structural formula for the product of the reaction of *p*-bromoacetophenone with each set of reagents in Problem 13.35.

Synthesis

13.37 Show the reagents and conditions that will bring about the conversion of cyclohexanol to cyclohexanecarbaldehyde:

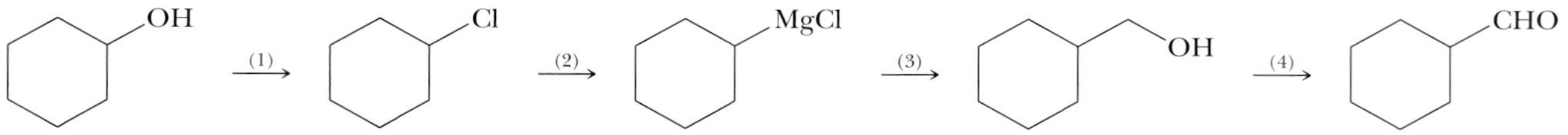

13.38 Starting with cyclohexanone, show how to prepare these compounds (in addition to the given starting material, use any other organic or inorganic reagents, as necessary):
(a) Cyclohexanol   (b) Cyclohexene
(c) *cis*-1,2-Cyclohexanediol   (d) 1-Methylcyclohexanol
(e) 1-Methylcyclohexene   (f) 1-Phenylcyclohexanol
(g) 1-Phenylcyclohexene   (h) Cyclohexene oxide
(i) *trans*-1,2-Cyclohexanediol

13.39 Show how to bring about these conversions (in addition to the given starting material, use any other organic or inorganic reagents, as necessary):

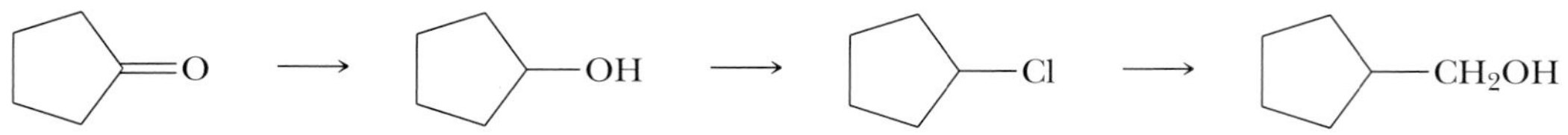

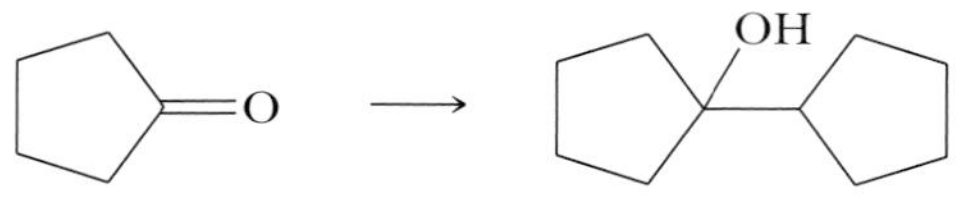

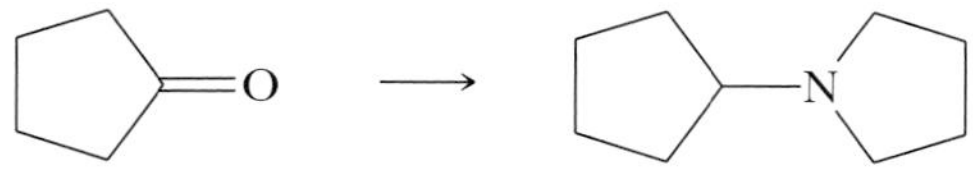

13.40 Many tumors of the breast are estrogen dependent. Drugs that interfere with estrogen binding have antitumor activity and may even help prevent the occurrence of tumors. A widely used antiestrogen drug is tamoxifen:

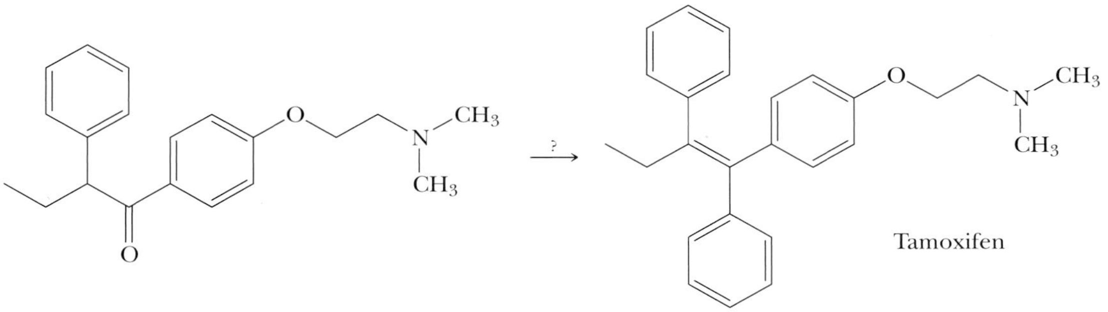

Tamoxifen

(a) How many stereoisomers are possible for tamoxifen?
(b) Specify the configuration of the stereoisomer shown here.
(c) Show how tamoxifen can be synthesized from the given ketone using a Grignard reaction, followed by dehydration.

13.41 Following is a possible synthesis of the antidepressant bupropion (Wellbutrin®):

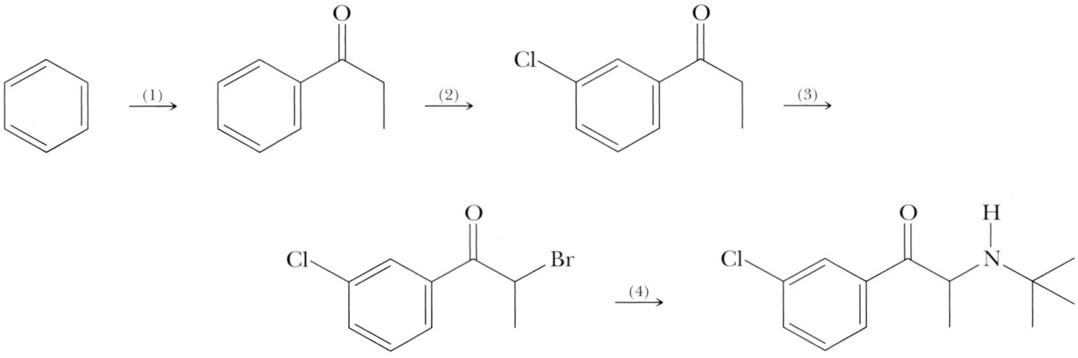

Bupropion
(Wellbutrin®)

Show the reagents that will bring about each step in this synthesis.

13.42 The synthesis of chlorpromazine in the 1950s and the discovery soon thereafter of the drug's antipsychotic activity opened the modern era of biochemical investigations into the pharmacology of the central nervous system. One of the compounds prepared in the search for more effective antipsychotics was amitriptyline.

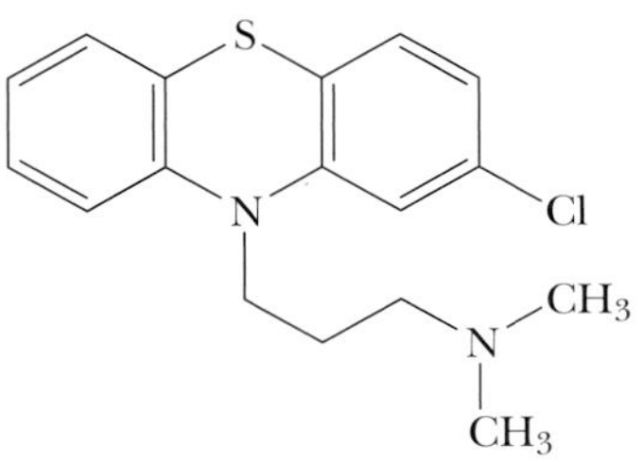

Chlorpromazine 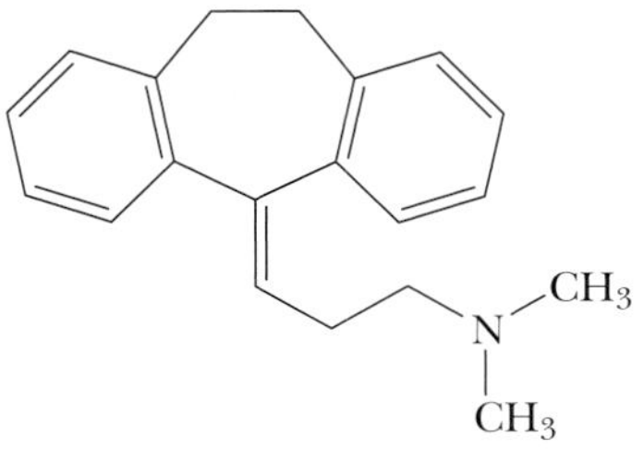   Amitriptyline

Surprisingly, amitriptyline shows antidepressant activity rather than antipsychotic activity. It is now known that amitriptyline inhibits the reuptake of norepinephrine and serotonin from the synaptic cleft. Because the reuptake of these neurotransmitters is inhibited, their effects are potentiated. That is, the two neurotransmitters remain available to interact with serotonin and norepinephrine receptor sites longer and continue to cause excitation of serotonin and norepinephrine-mediated neural pathways. The following is a synthesis for amitriptyline:

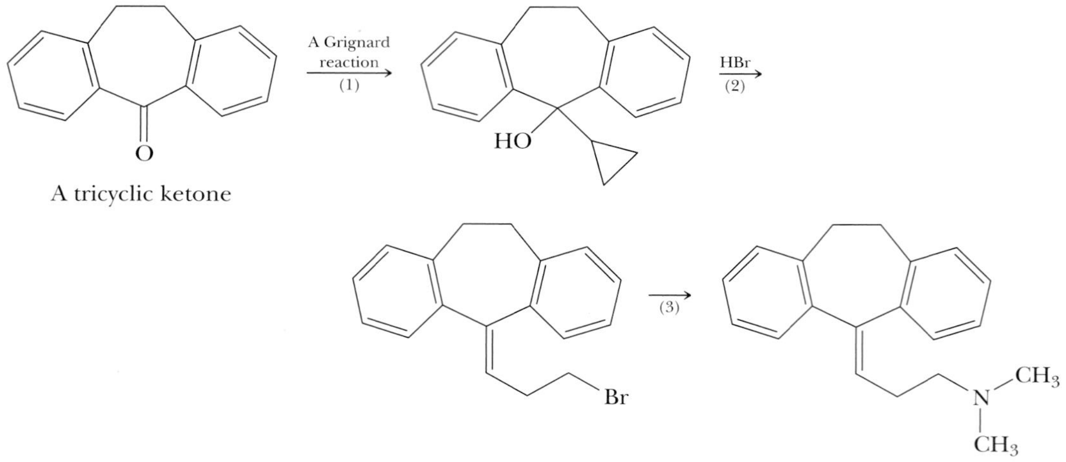

A tricyclic ketone

Amitriptyline

(a) Propose a reagent for Step 1.
(b) Propose a mechanism for Step 2. (*Note*: It is not acceptable to propose a primary carbocation as an intermediate.)
(c) Propose a reagent for Step (3).

13.43 Following is a synthesis for diphenhydramine:

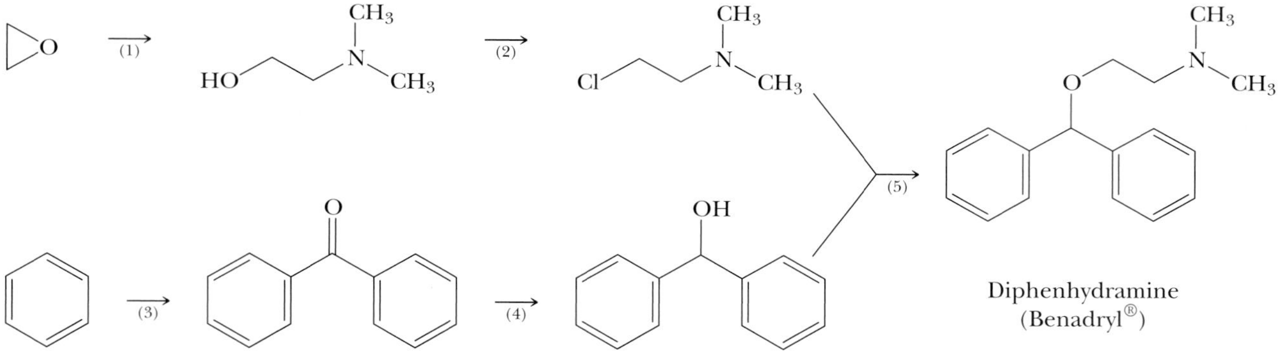

Diphenhydramine
(Benadryl®)

The hydrochloride salt of this compound, best known by its trade name, Benadryl®, is an antihistamine.
(a) Propose reagents for Steps 1 and 2.
(b) Propose reagents for Steps 3 and 4.
(c) Show that Step 5 is an example of nucleophilic aliphatic substitution. What type of mechanism—S_N1 or S_N2—is more likely for this reaction? Explain.

13.44 Following is a synthesis for the antidepressant venlafaxine:

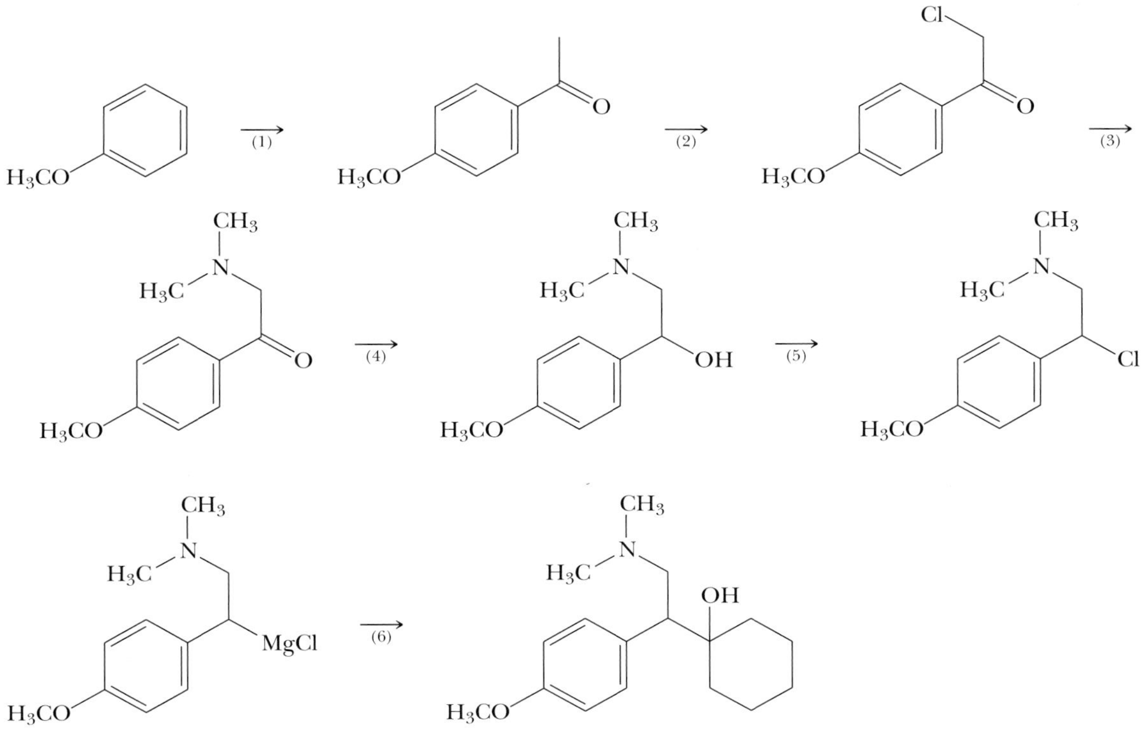

Venlafaxine

(a) Propose a reagent for Step 1, and name the type of reaction that takes place.
(b) Propose reagents for Steps 2 and 3.
(c) Propose reagents for Steps 4 and 5.
(d) Propose a reagent for Step 6, and name the type of reaction that takes place.

Spectroscopy

13.45 Compound A, $C_5H_{10}O$, is used as a flavoring agent for many foods that possess a chocolate or peach flavor. Its common name is isovaleraldehyde and it gives ^{13}C-NMR peaks at δ 202.7, 52.7, 23.6, and 22.6. Provide a structural formula for isovaleraldehyde and give its IUPAC name.

13.46 Following are ^1H-NMR and IR spectra of compound B, $C_6H_{12}O_2$:

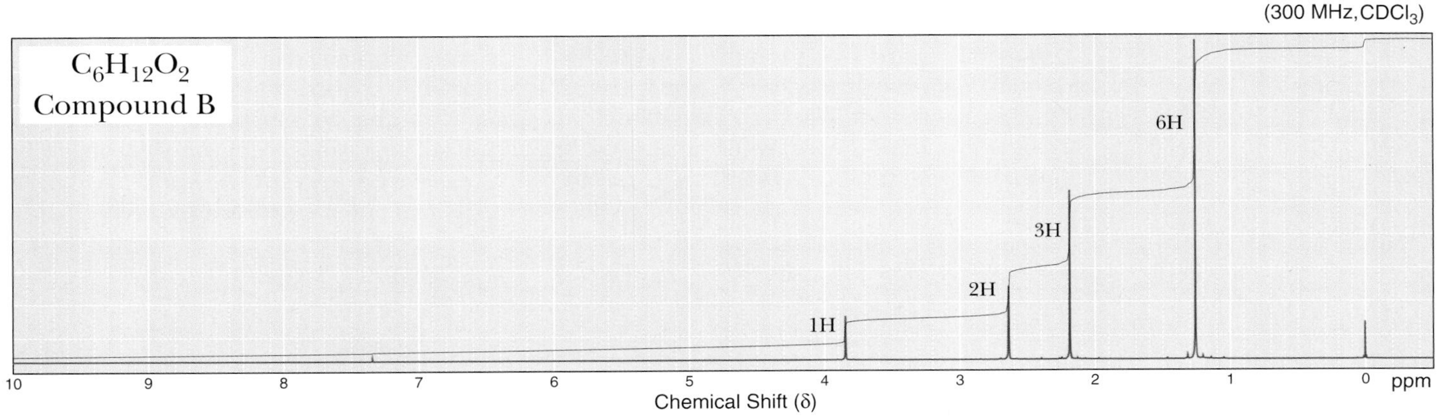

(300 MHz, CDCl₃)

$C_6H_{12}O_2$
Compound B

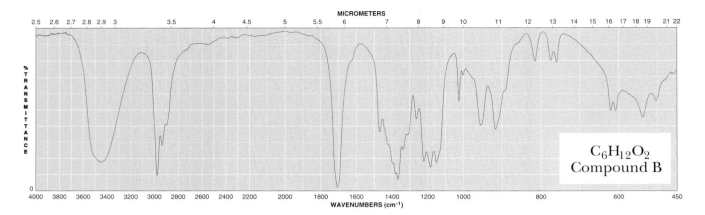

Propose a structural formula for compound B.

13.47 Compound C, $C_9H_{18}O$, is used in the automotive industry to retard the flow of solvent and thus improve the application of paints and coatings. It yields ^{13}C-NMR peaks at δ 210.5, 52.4, 24.5, and 22.6. Provide a structure and an IUPAC name for C.

Looking Ahead

13.48 Reaction of a Grignard reagent with carbon dioxide, followed by treatment with aqueous HCl, gives a carboxylic acid. Propose a structural formula for the bracketed intermediate formed by the reaction of phenylmagnesium bromide with CO_2, and propose a mechanism for the formation of this intermediate:

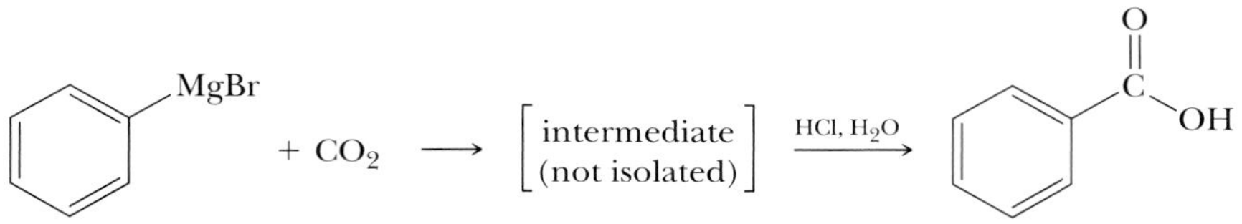

13.49 Rank the following carbonyls in order of increasing reactivity to nucleophilic attack, and explain your reasoning.

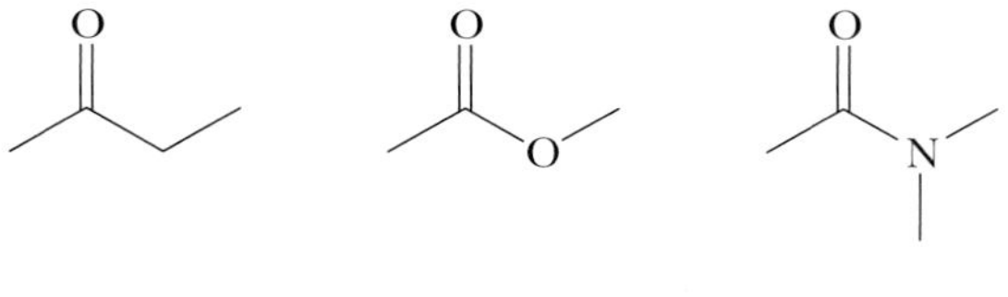

13.50 Provide the enol form of this ketone and predict the direction of equilibrium:

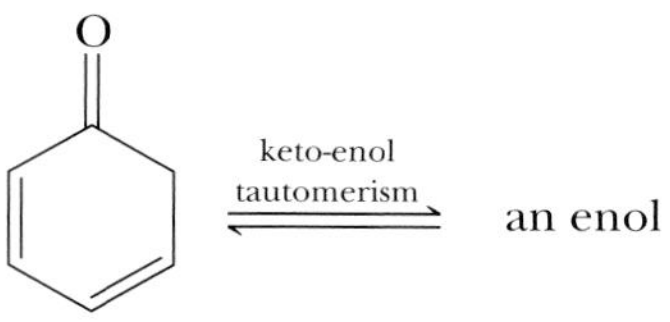

13.51 Draw the cyclic hemiacetal formed by reaction of the highlighted —OH group with the aldehyde group:

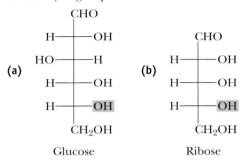

(a) Glucose

(b) Ribose

13.52 Propose a mechanism for the acid-catalyzed reaction of the following hemiacetal, with an amine acting as a nucleophile:

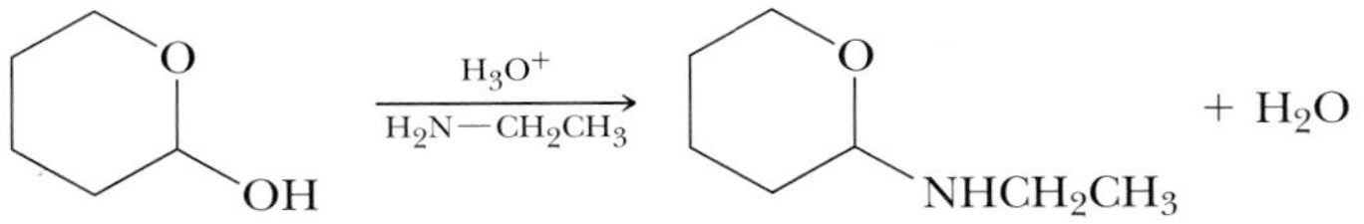

14

Carboxylic Acids

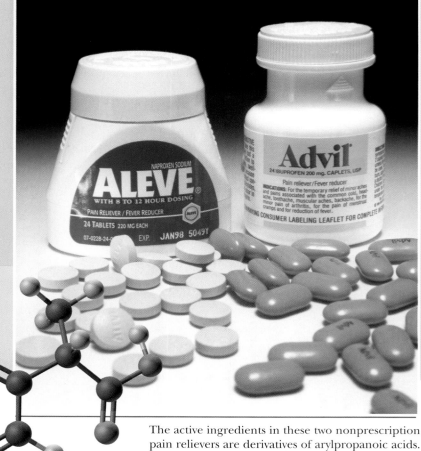

The active ingredients in these two nonprescription pain relievers are derivatives of arylpropanoic acids. See Chemical Connections "From Willow Bark to Aspirin and Beyond." Inset: A model of (S)-ibuprofen. *(Charles D. Winters)*

14.1 INTRODUCTION

The most important chemical property of carboxylic acids, another class of organic compounds containing the carbonyl group, is their acidity. Furthermore, carboxylic acids form numerous important derivatives, including esters, amides, anhydrides, and acid halides. In this chapter, we study carboxylic acids themselves; in Chapters 15 and 16, we study their derivatives.

14.2 STRUCTURE

The functional group of a carboxylic acid is a **carboxyl group**, so named because it is made up of a **carb**onyl group and a hyd**roxyl** group (Section 1.8D). Following is a Lewis structure of the carboxyl group, as well as two alternative representations of it:

Carboxyl group A —COOH group.

$$-C\underset{:\overset{..}{O}-H}{\overset{\overset{..}{O}:}{\big\|}} \qquad -COOH \qquad -CO_2H$$

The general formula of an aliphatic carboxylic acid is RCOOH; that of an aromatic carboxylic acid is ArCOOH.

14.3 NOMENCLATURE

A. IUPAC System

We derive the IUPAC name of a carboxylic acid from that of the longest carbon chain which contains the carboxyl group by dropping the final -*e* from the name of the parent alkane and adding the suffix -*oic*, followed by the word *acid* (Section 3.6). We number the chain beginning with the carbon of the carboxyl group. Because the carboxyl carbon is understood to be carbon 1, there is no need to give it a number. If the carboxylic acid contains a carbon–carbon double bond, we change the infix from -*an*- to -*en*- to indicate the presence of the double bond, and we show the location of the double bond by a number. In the following examples, the common name of each acid is given in parentheses:

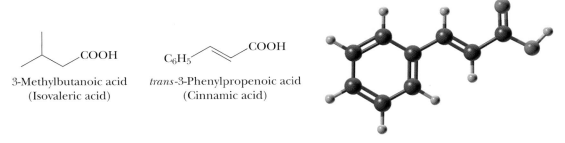

3-Methylbutanoic acid
(Isovaleric acid)

trans-3-Phenylpropenoic acid
(Cinnamic acid)

In the IUPAC system, a carboxyl group takes precedence over most other functional groups (Table 13.1), including hydroxyl and amino groups, as well as the carbonyl groups of aldehydes and ketones. As illustrated in the following examples, an —OH group of an alcohol is indicated by the prefix *hydroxy*-, an —NH$_2$ group of an amine by *amino*-, and an =O group of an aldehyde or ketone by *oxo*-:

5-Hydroxyhexanoic acid

4-Aminobutanoic acid

5-Oxohexanoic acid

Dicarboxylic acids are named by adding the suffix -*dioic*, followed by the word *acid*, to the name of the carbon chain that contains both carboxyl groups. Because the two carboxyl groups can be only at the ends of the parent chain, there is no need to number them. Following are IUPAC names and common names for several important aliphatic dicarboxylic acids:

Ethanedioic acid
(Oxalic acid)

Propanedioic acid
(Malonic acid)

Butanedioic acid
(Succinic acid)

Pentanedioic acid
(Glutaric acid)

Hexanedioic acid
(Adipic acid)

Leaves of the rhubarb plant
contain the poison oxalic acid
as its potassium and sodium
salts. *(Hans Reinhard/OKAPIA/
Photo Researchers, Inc.)*

The name *oxalic acid* is derived from one of its sources in the biological world, namely, plants of the genus *Oxalis*, one of which is rhubarb. Oxalic acid also occurs in human and animal urine, and calcium oxalate (the calcium salt of oxalic acid) is a major component of kidney stones. Adipic acid is one of the two monomers required for the synthesis of the polymer nylon 66. The U.S. chemical industry produces approximately 1.8 billion pounds of adipic acid annually, solely for the synthesis of nylon 66 (Section 17.5A).

A carboxylic acid containing a carboxyl group bonded to a cycloalkane ring is named by giving the name of the ring and adding the suffix -*carboxylic acid*. The atoms of the ring are numbered beginning with the carbon bearing the —COOH group:

2-Cyclohexenecarboxylic
acid

trans-1,3-Cyclopentane-
dicarboxylic acid

The simplest aromatic carboxylic acid is benzoic acid. Derivatives are named by using numbers and prefixes to show the presence and location of substituents relative to the carboxyl group. Certain aromatic carboxylic acids have common names by which they are more usually known. For example, 2-hydroxybenzoic acid is more often called salicylic acid, a name derived from the fact that this aromatic carboxylic acid was first obtained from the bark of the willow, a tree of the genus *Salix*. Aromatic dicarboxylic acids are named by adding the words *dicarboxylic acid* to *benzene*. Examples are 1,2-benzenedicarboxylic acid and 1,4-benzenedicarboxylic acid. Each is more usually known by its common name: phthalic acid and terephthalic acid, respectively. Terephthalic acid is one of the two organic components required for the synthesis of the textile fiber known as Dacron® polyester (Section 17.5B).

Benzoic acid

2-Hydroxybenzoic acid
(Salicylic acid)

1,2-Benzenedicarboxylic acid
(Phthalic acid)

1,4-Benzenedicarboxylic acid
(Terephthalic acid)

B. Common Names

Aliphatic carboxylic acids, many of which were known long before the development of structural theory and IUPAC nomenclature, are named according to their source or for some characteristic property. Table 14.1 lists several of the unbranched aliphatic carboxylic acids found in the biological world, along with the common name of each. Those with 16, 18, and 20 carbon atoms are particularly abundant in fats and oils (Section 21.2) and the phospholipid components of biological membranes (Section 21.4).

TABLE 14.1 Several Aliphatic Carboxylic Acids and Their Common Names

Structure	IUPAC Name	Common Name	Derivation
HCOOH	methanoic acid	formic acid	Latin: *formica*, ant
CH$_3$COOH	ethanoic acid	acetic acid	Latin: *acetum*, vinegar
CH$_3$CH$_2$COOH	propanoic acid	propionic acid	Greek: *propion*, first fat
CH$_3$(CH$_2$)$_2$COOH	butanoic acid	butyric acid	Latin: *butyrum*, butter
CH$_3$(CH$_2$)$_3$COOH	pentanoic acid	valeric acid	Latin: *valere*, to be strong
CH$_3$(CH$_2$)$_4$COOH	hexanoic acid	caproic acid	Latin: *caper*, goat
CH$_3$(CH$_2$)$_6$COOH	octanoic acid	caprylic acid	Latin: *caper*, goat
CH$_3$(CH$_2$)$_8$COOH	decanoic acid	capric acid	Latin: *caper*, goat
CH$_3$(CH$_2$)$_{10}$COOH	dodecanoic acid	lauric acid	Latin: *laurus*, laurel
CH$_3$(CH$_2$)$_{12}$COOH	tetradecanoic acid	myristic acid	Greek: *myristikos*, fragrant
CH$_3$(CH$_2$)$_{14}$COOH	hexadecanoic acid	palmitic acid	Latin: *palma*, palm tree
CH$_3$(CH$_2$)$_{16}$COOH	octadecanoic acid	stearic acid	Greek: *stear*, solid fat
CH$_3$(CH$_2$)$_{18}$COOH	eicosanoic acid	arachidic acid	Greek: *arachis*, peanut

Formic acid was first obtained in 1670 from the destructive distillation of ants, whose genus is *Formica*. It is one of the components of the venom of stinging ants. *(Ted Nelson/Dembinsky Photo Associates)*

When common names are used, the Greek letters α, β, γ, δ, and so forth are often added as a prefix to locate substituents. The α-position in a carboxylic acid is the position next to the carboxyl group; an α-substituent in a common name is equivalent to a 2-substituent in an IUPAC name. *GABA*, short for *gamma-aminobutyric acid*, is an inhibitory neurotransmitter in the central nervous system of humans:

4-Aminobutanoic acid
(γ-Aminobutyric acid, GABA)

In common nomenclature, the prefix *keto-* indicates the presence of a ketone carbonyl in a substituted carboxylic acid (as illustrated by the common name β-ketobutyric acid), and the substituent CH$_3$CO— is named an **aceto group**:

Aceto group A CH$_3$CO— group.

3-Oxobutanoic acid
(β-Ketobutyric acid;
Acetoacetic acid)

Acetyl group
(Aceto group)

An alternative common name for 3-oxobutanoic acid is acetoacetic acid. In deriving this common name, this ketoacid is regarded as a substituted acetic acid, and the CH_3CO— substituent is named an aceto group.

EXAMPLE 14.1

Write the IUPAC name for each carboxylic acid:

(a)

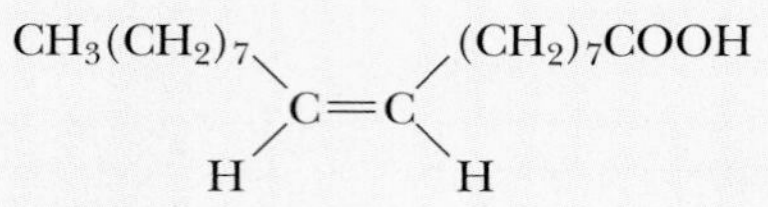

(b)

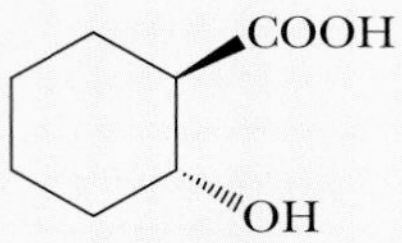

(c)

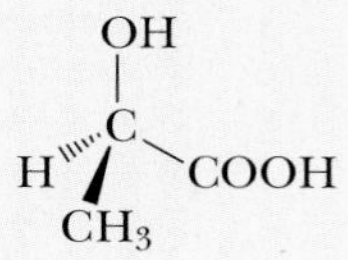

(d) $ClCH_2COOH$

SOLUTION

(a) *cis*-9-Octadecenoic acid (oleic acid)
(b) *trans*-2-Hydroxycyclohexanecarboxylic acid
(c) (R)-2-Hydroxypropanoic acid [(R)-lactic acid]
(d) Chloroethanoic acid (chloroacetic acid)

Practice Problem 14.1

Each of the following compounds has a well-recognized common name. A derivative of glyceric acid is an intermediate in glycolysis (Section 22.4). Maleic acid is an intermediate in the tricarboxylic acid (TCA) cycle. Mevalonic acid is an intermediate in the biosynthesis of steroids (Section 21.5B).

(a) 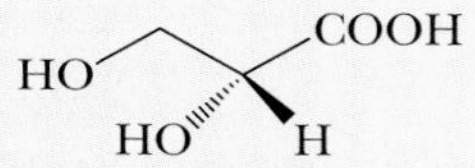 Glyceric acid   (b) 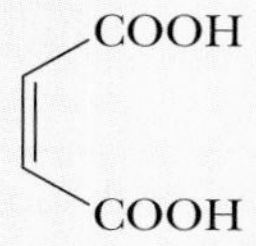 Maleic acid   (c) 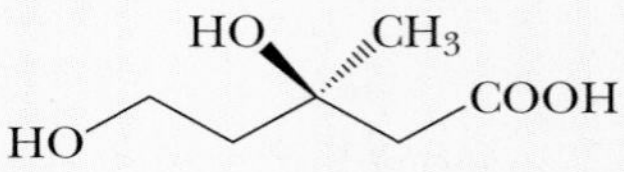 Mevalonic acid

Write the IUPAC name for each compound. Be certain to show the configuration of each.

14.4   PHYSICAL PROPERTIES

In the liquid and solid states, carboxylic acids are associated by intermolecular hydrogen bonding into dimers, as shown for acetic acid:

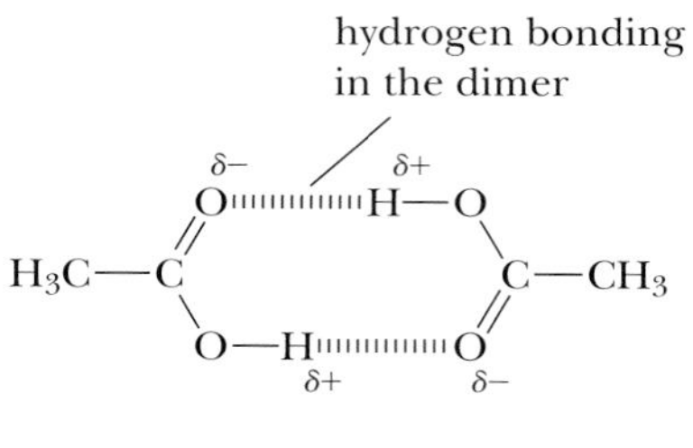

hydrogen bonding in the dimer

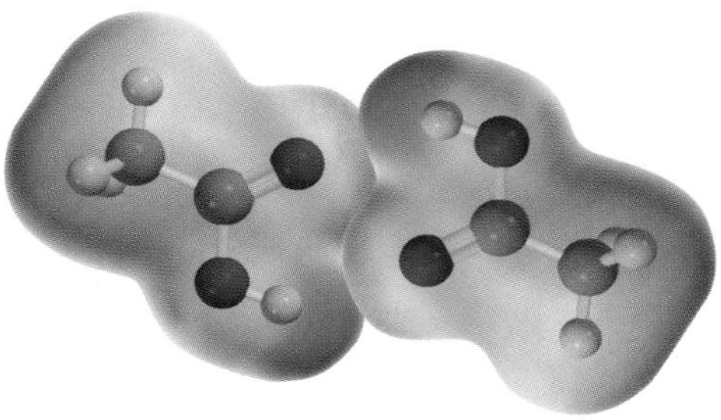

Carboxylic acids have significantly higher boiling points than other types of organic compounds of comparable molecular weight, such as alcohols, aldehydes, and ketones. For example, butanoic acid (Table 14.2) has a higher boiling point than either 1-pentanol or pentanal. The higher boiling points of carboxylic acids result from their polarity and from the fact that they form very strong intermolecular hydrogen bonds.

TABLE 14.2 Boiling Points and Solubilities in Water of Selected Carboxylic Acids, Alcohols, and Aldehydes of Comparable Molecular Weight

Structure	Name	Molecular Weight	Boiling Point (°C)	Solubility (g/100 mL H_2O)
CH_3COOH	acetic acid	60.5	118	infinite
$CH_3CH_2CH_2OH$	1-propanol	60.1	97	infinite
CH_3CH_2CHO	propanal	58.1	48	16
$CH_3(CH_2)_2COOH$	butanoic acid	88.1	163	infinite
$CH_3(CH_2)_3CH_2OH$	1-pentanol	88.1	137	2.3
$CH_3(CH_2)_3CHO$	pentanal	86.1	103	slight
$CH_3(CH_2)_4COOH$	hexanoic acid	116.2	205	1.0
$CH_3(CH_2)_5CH_2OH$	1-heptanol	116.2	176	0.2
$CH_3(CH_2)_5CHO$	heptanal	114.1	153	0.1

Carboxylic acids also interact with water molecules by hydrogen bonding through both their carbonyl and hydroxyl groups. Because of these hydrogen-bonding interactions, carboxylic acids are more soluble in water than are alcohols, ethers, aldehydes, and ketones of comparable molecular weight. The solubility of a carboxylic acid in water decreases as its molecular weight increases. We account for this trend in the following way: A carboxylic acid consists of two regions of different polarity—a polar hydrophilic carboxyl group and, except for formic acid, a nonpolar hydrophobic hydrocarbon chain. The **hydrophilic** carboxyl group increases water solubility; the **hydrophobic** hydrocarbon chain decreases water solubility.

Hydrophilic From the Greek, meaning "water loving."

Hydrophobic From the Greek, meaning "water hating."

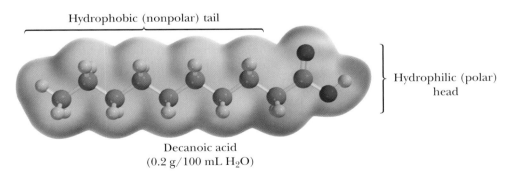

Hydrophobic (nonpolar) tail

Hydrophilic (polar) head

Decanoic acid
(0.2 g/100 mL H_2O)

The first four aliphatic carboxylic acids (formic, acetic, propanoic, and butanoic acids) are infinitely soluble in water because the hydrophilic character of the carboxyl group more than counterbalances the hydrophobic character of the hydrocarbon chain. As the size of the hydrocarbon chain increases relative to the size of the carboxyl group, water solubility decreases. The solubility of hexanoic acid in water is 1.0 g/100 g water; that of decanoic acid is only 0.2 g/100 g water.

One other physical property of carboxylic acids must be mentioned: The liquid carboxylic acids, from propanoic acid to decanoic acid, have extremely foul odors, about as bad as those of thiols, although different. Butanoic acid is found in stale perspiration and is a major component of "locker room odor." Pentanoic acid smells even worse, and goats, which secrete C_6, C_8, and C_{10} acids, are not famous for their pleasant odors.

14.5 ACIDITY

A. Acid Ionization Constants

Carboxylic acids are weak acids. Values of K_a for most unsubstituted aliphatic and aromatic carboxylic acids fall within the range from 10^{-4} to 10^{-5}. The value of K_a for acetic acid, for example, is 1.74×10^{-5}, and the pK_a of acetic acid is 4.76:

$$CH_3COOH + H_2O \rightleftharpoons CH_3COO^- + H_3O^+$$

$$K_a = \frac{[CH_3COO^-][H_3O^+]}{[CH_3COOH]} = 1.74 \times 10^{-5}$$

$$pK_a = 4.76$$

As we discussed in Section 2.6B, carboxylic acids are stronger acids (pK_a 4–5) than alcohols (pK_a 16–18) because resonance stabilizes the carboxylate anion by delocalizing its negative charge. No comparable resonance stabilization exists in alkoxide ions.

Substitution at the α-carbon of an atom or a group of atoms of higher electronegativity than carbon increase the acidity of carboxylic acids, often by several orders of magnitude (Section 2.6C). Compare, for example, the acidities of acetic acid (pK_a 4.76) and chloroacetic acid (pK_a 2.86). A single chlorine substituent on the α-carbon increases acid strength by nearly 100! Both dichloroacetic acid and trichloroacetic acid are stronger acids than phosphoric acid (pK_a 2.1):

Formula:	CH_3COOH	$ClCH_2COOH$	$Cl_2CHCOOH$	Cl_3CCOOH
Name:	Acetic acid	Chloroacetic acid	Dichloroacetic acid	Trichloroacetic acid
pK_a:	4.76	2.86	1.48	0.70

Increasing acid strength →

The acid-strengthening effect of halogen substitution falls off rather rapidly with increasing distance from the carboxyl group. Although the acid ionization constant for 2-chlorobutanoic acid (pK_a 2.83) is 100 times that for butanoic acid, the acid ionization constant for 4-chlorobutanoic acid (pK_a 4.52) is only about twice that for butanoic acid:

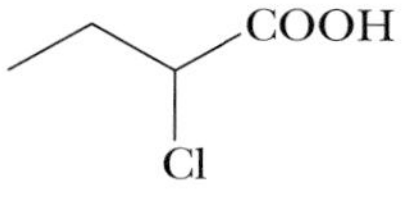

2-Chlorobutanoic acid (pK_a 2.83) 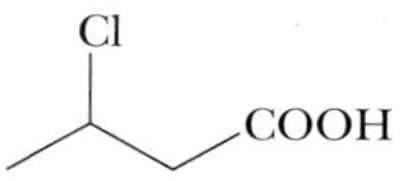   3-Chlorobutanoic acid (pK_a 3.98) 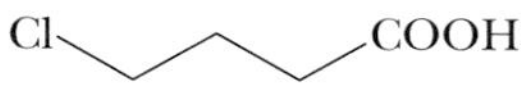   4-Chlorobutanoic acid (pK_a 4.52) 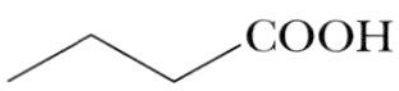   Butanoic acid (pK_a 4.82)

Decreasing acid strength →

EXAMPLE 14.2

Which acid in each set is the stronger?

(a)

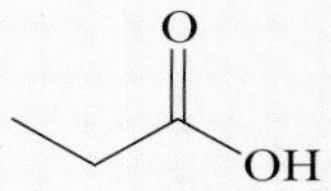

Propanoic acid

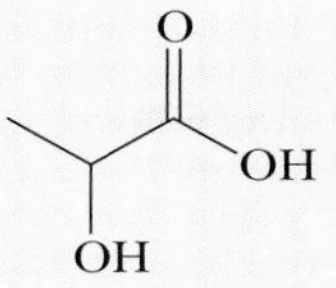

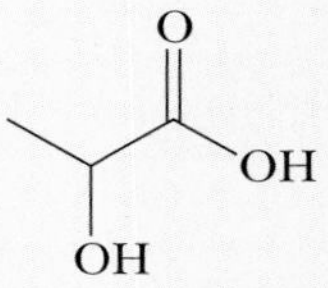

or

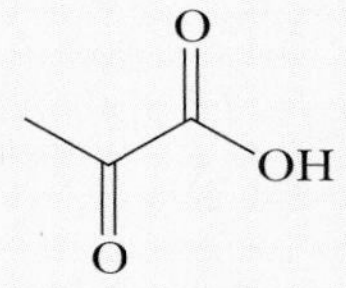

2-Hydroxy-
propanoic acid
(Lactic acid)

(b)

2-Hydroxy-
propanoic acid
(Lactic acid)

or

2-Oxopropanoic
acid
(Pyruvic acid)

SOLUTION

(a) 2-Hydroxypropanoic acid (pK_a 3.08) is a stronger acid than propanoic acid
(pK_a 4.87), because of the electron-withdrawing inductive effect of the
hydroxyl oxygen.

(b) 2-Oxopropanoic acid (pK_a 2.06) is a stronger acid than 2-hydroxypropanoic
acid (pK_a 3.08), because of the greater electron-withdrawing inductive effect of
the carbonyl oxygen compared with that of the hydroxyl oxygen.

Practice Problem 14.2

Match each compound with its appropriate pK_a value:

$$CH_3CCOOH \qquad CF_3COOH \qquad CH_3CHCOOH$$

pK_a values = 5.03, 3.08, and 0.22.

2,2-Dimethyl-
propanoic acid

Trifluoro-
acetic acid

2-Hydroxy-
propanoic acid
(Lactic acid)

B. Reaction with Bases

All carboxylic acids, whether soluble or insoluble in water, react with NaOH, KOH,
and other strong bases to form water-soluble salts:

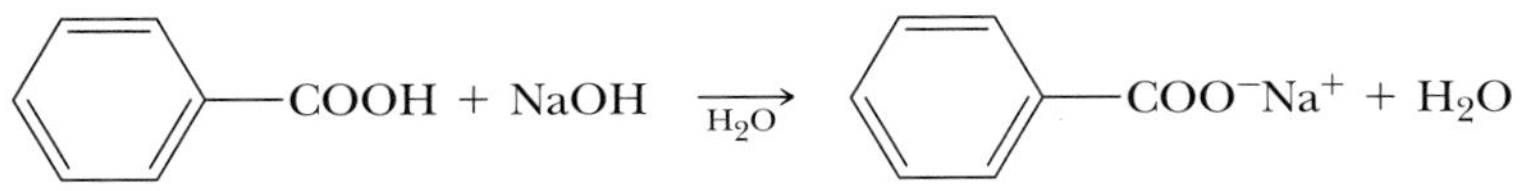

Benzoic acid
(slightly soluble
in water)

Sodium benzoate
(60 g/100 mL water)

Sodium benzoate, a fungal growth inhibitor, is often added to baked goods "to retard spoilage." Calcium propanoate is used for the same purpose.

Carboxylic acids also form water-soluble salts with ammonia and amines:

$$\text{C}_6\text{H}_5\text{—COOH} + \text{NH}_3 \xrightarrow{\text{H}_2\text{O}} \text{C}_6\text{H}_5\text{—COO}^-\text{NH}_4^+$$

Benzoic acid	Ammonium benzoate
(slightly soluble in water)	(20 g/100 mL water)

As described in Section 2.5, carboxylic acids react with sodium bicarbonate and sodium carbonate to form water-soluble sodium salts and carbonic acid (a relatively weak acid). Carbonic acid, in turn, decomposes to give water and carbon dioxide, which evolves as a gas:

$$\text{CH}_3\text{COOH} + \text{Na}^+\text{HCO}_3^- \xrightarrow{\text{H}_2\text{O}} \text{CH}_3\text{COO}^-\text{Na}^+ + \text{H}_2\text{CO}_3$$

$$\text{H}_2\text{CO}_3 \longrightarrow \text{CO}_2 + \text{H}_2\text{O}$$

$$\overline{\text{CH}_3\text{COOH} + \text{Na}^+\text{HCO}_3^- \longrightarrow \text{CH}_3\text{COO}^-\text{Na}^+ + \text{CO}_2 + \text{H}_2\text{O}}$$

Salts of carboxylic acids are named in the same manner as are salts of inorganic acids: Name the cation first and then the anion. Derive the name of the anion from the name of the carboxylic acid by dropping the suffix -ic acid and adding the suffix -ate. For example, the name of $\text{CH}_3\text{CH}_2\text{COO}^-\text{Na}^+$ is sodium propanoate, and that of $\text{CH}_3(\text{CH}_2)_{14}\text{COO}^-\text{Na}^+$ is sodium hexadecanoate (sodium palmitate).

EXAMPLE 14.3

Complete each acid–base reaction and name the salt formed:

(a) $\quad$ CH$_3$CH$_2$CH$_2$COOH + NaOH $\longrightarrow$ $\qquad$ (b) (CH$_3$)(OH)CH—COOH + NaHCO$_3$ $\longrightarrow$

SOLUTION

Each carboxylic acid is converted to its sodium salt. In (b), carbonic acid forms and decomposes to carbon dioxide and water:

(a) $\quad$ CH$_3$CH$_2$CH$_2$COOH + NaOH $\longrightarrow$ CH$_3$CH$_2$CH$_2$COO$^-$Na$^+$ + H$_2$O

Butanoic acid	Sodium butanoate

(b) $\quad$ 2-hydroxypropanoic acid + NaHCO$_3$ $\longrightarrow$ sodium 2-hydroxypropanoate + H$_2$O + CO$_2$

2-Hydroxypropanoic acid	Sodium 2-hydroxypropanoate
(Lactic acid)	(Sodium lactate)

Practice Problem 14.3

Write an equation for the reaction of each acid in Example 14.3 with ammonia, and name the salt formed.

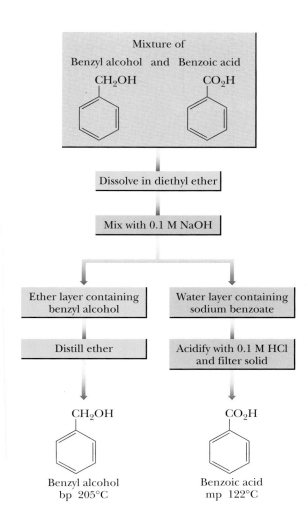

Figure 14.1
Flowchart for separation
of benzoic acid from benzyl
alcohol.

A consequence of the water solubility of carboxylic acid salts is that we can convert water-insoluble carboxylic acids to water-soluble alkali metal or ammonium salts and then extract them into aqueous solution. In turn, we can transform the salt into the free carboxylic acid by adding HCl, H_2SO_4, or some other strong acid. These reactions allow us to separate water-insoluble carboxylic acids from water-insoluble neutral compounds.

Figure 14.1 shows a flowchart for the separation of benzoic acid, a water-insoluble carboxylic acid, from benzyl alcohol, a water-insoluble nonacidic compound. First, we dissolve the mixture of benzoic acid and benzyl alcohol in diethyl ether. Next, we shake the ether solution with aqueous NaOH to convert benzoic acid to its water-soluble sodium salt. Then we separate the ether from the aqueous phase. Distillation of the ether solution yields first diethyl ether (bp 35°C) and then benzyl alcohol (bp 205°C). When we acidify the aqueous solution with HCl, benzoic acid precipitates as a water-insoluble solid (mp 122°C) and is recovered by filtration.

14.6 REDUCTION

The carboxyl group is one of the organic functional groups that is most resistant to reduction. It is not affected by catalytic reduction under conditions that easily reduce aldehydes and ketones to alcohols and that reduce alkenes to alkanes.

CHEMICAL CONNECTIONS 14A

From Willow Bark to Aspirin and Beyond

The first drug developed for widespread use was aspirin, today's most common pain reliever. Americans alone consume approximately 80 billion tablets of aspirin a year! The story of the development of this modern pain reliever goes back more than 2,000 years: In 400 B.C.E., the Greek physician Hippocrates recommended chewing bark of the willow tree to alleviate the pain of childbirth and to treat eye infections.

The active component of willow bark was found to be salicin, a compound composed of salicyl alcohol joined to a unit of β-D-glucose (Section 18.3). Hydrolysis of salicin in aqueous acid gives salicyl alcohol, which can then be oxidized to salicylic acid, an even more effective reliever of pain, fever, and inflammation than salicin and one without its extremely bitter taste:

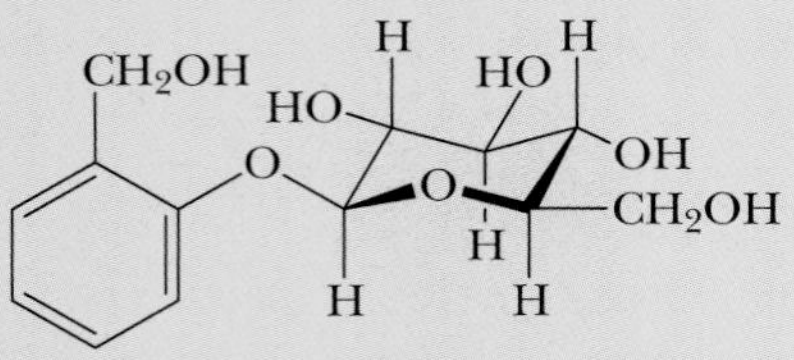

Unfortunately, patients quickly recognized salicylic acid's major side effect: It causes severe irritation of the mucous membrane lining the stomach. In the search for less irritating, but still effective, derivatives of salicylic acid, chemists at the Bayer division of I. G. Farben in Germany prepared acetylsalicylic acid in 1883 and gave it the name *aspirin*, a word derived from the German *spirsäure* (salicylic acid), with the initial *a* for the acetyl group:

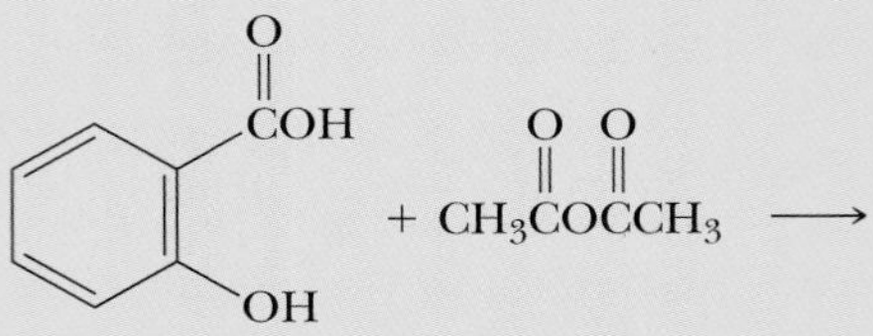

Salicylic acid   Acetic anhydride

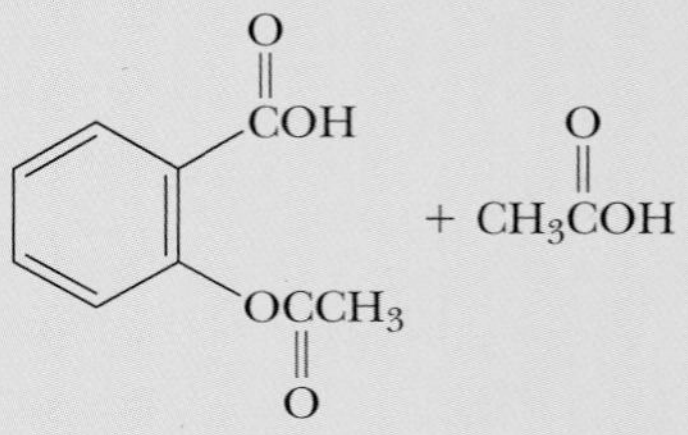

Acetyl salicylate (Aspirin)

Salicin

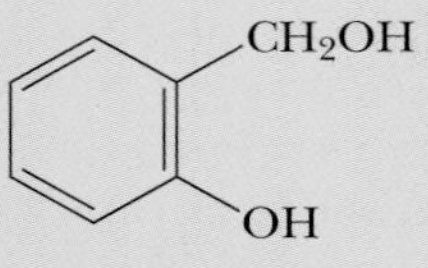

Salicyl alcohol   Salicylic acid

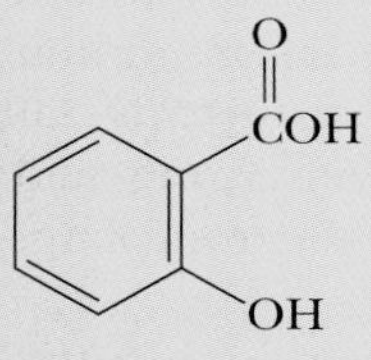

The most common reagent for the reduction of a carboxylic acid to a primary alcohol is the very powerful reducing agent lithium aluminum hydride (Section 13.11B).

A. Reduction of a Carboxyl Group

Lithium aluminum hydride, $LiAlH_4$, reduces a carboxyl group to a primary alcohol in excellent yield. Reduction is most commonly carried out in diethyl ether or tetrahydrofuran (THF). The initial product is an aluminum alkoxide, which is then treated with water to give the primary alcohol and lithium and aluminum hydroxides:

Aspirin proved to be less irritating to the stomach than salicylic acid and also more effective in relieving the pain and inflammation of rheumatoid arthritis. Bayer began large-scale production of aspirin in 1899.

In the 1960s, in a search for even more effective and less irritating analgesics and anti-inflammatory drugs, the Boots Pure Drug Company in England studied compounds related in structure to salicylic acid. They discovered an even more potent compound, which they named ibuprofen, and soon thereafter, Syntex Corporation in the United States developed naproxen and Rhone–Poulenc in France developed ketoprofen:

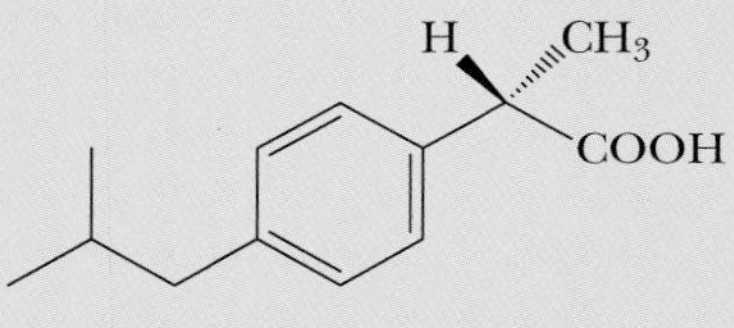

(S)-Ibuprofen

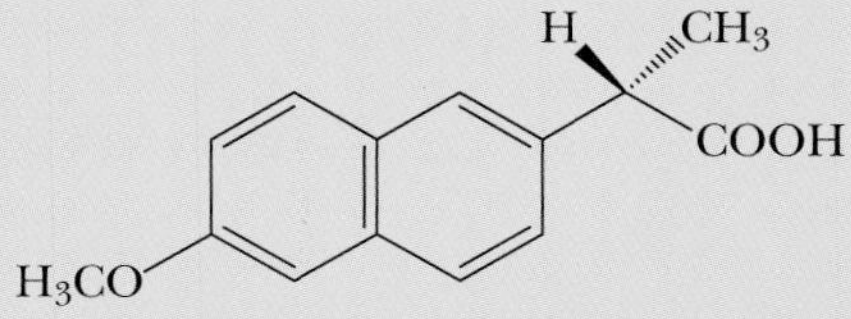

(S)-Naproxen

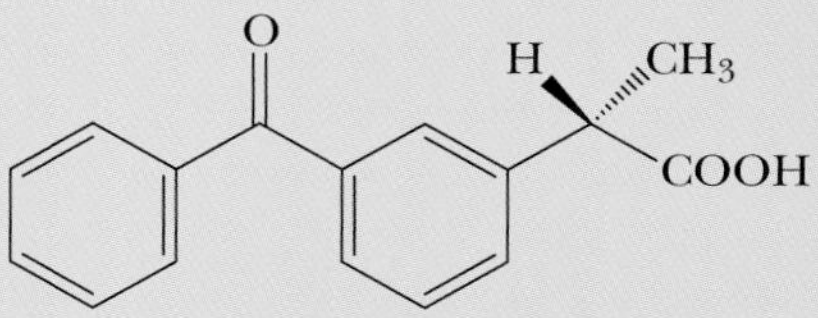

(S)-Ketoprofen

Notice that each compound has one stereocenter and can exist as a pair of enantiomers. For each drug, the physiologically active form is the S enantiomer. Even though the R enantiomer of ibuprofen has none of the analgesic or anti-inflammatory activity, it is converted in the body to the active S enantiomer.

In the 1960s, scientists discovered that aspirin acts by inhibiting cyclooxygenase (COX), a key enzyme in the conversion of arachidonic acid to prostaglandins (Section 21.6). With this discovery, it became clear why only one enantiomer of ibuprofen, naproxen, and ketoprofen is active: Only the S enantiomer of each has the correct handedness to bind to COX and inhibit its activity.

The discovery that these drugs owe their effectiveness to the inhibition of COX opened an entirely new avenue for drug research. If we know more about the structure and function of this key enzyme, might it be possible to design and discover even more effective nonsteroidal anti-inflammatory drugs for the treatment of rheumatoid arthritis and other inflammatory diseases?

And so continues the story that began with the discovery of the beneficial effects of chewing willow bark.

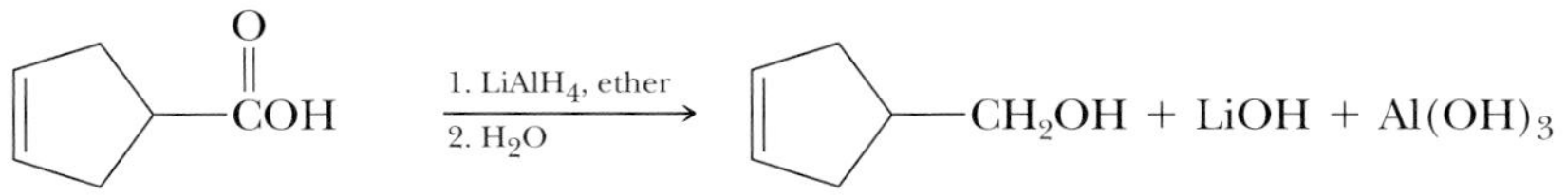

3-Cyclopentene-carboxylic acid → (1. LiAlH$_4$, ether; 2. H$_2$O) → 4-Hydroxymethyl-cyclopentene + LiOH + Al(OH)$_3$

These hydroxides are insoluble in diethyl ether or THF and are removed by filtration. Evaporation of the solvent yields the primary alcohol.

Alkenes are generally not affected by metal hydride-reducing reagents. These reagents function as hydride ion donors; that is, they function as nucleophiles, and alkenes are not normally attacked by nucleophiles.

B. Selective Reduction of Other Functional Groups

Catalytic hydrogenation does not reduce carboxyl groups, but does reduce alkenes to alkanes. Therefore, we can use H_2/M to reduce this functional group selectively in the presence of a carboxyl group:

5-Hexenoic acid   Hexanoic acid

We saw in Section 13.11B that aldehydes and ketones are reduced to alcohols by both $LiAlH_4$ and $NaBH_4$. Only $LiAlH_4$, however, reduces carboxyl groups. Thus, it is possible to reduce an aldehyde or a ketone carbonyl group selectively in the presence of a carboxyl group by using the less reactive $NaBH_4$ as the reducing agent:

5-Oxo-5-phenylpentanoic acid   5-Hydroxy-5-phenylpentanoic acid

Fischer esterification The process of forming an ester by refluxing a carboxylic acid and an alcohol in the presence of an acid catalyst, commonly sulfuric acid.

14.7   FISCHER ESTERIFICATION

Treatment of a carboxylic acid with an alcohol in the presence of an acid catalyst— most commonly, concentrated sulfuric acid—gives an ester. This method of forming an ester is given the special name **Fischer esterification** after the German chemist Emil Fischer (1852–1919). As an example of Fischer esterification, treating acetic acid with ethanol in the presence of concentrated sulfuric acid gives ethyl acetate and water:

Ethanoic acid   Ethanol   Ethyl ethanoate
(Acetic acid)   (Ethyl alcohol)   (Ethyl acetate)

These products all contain ethyl acetate as a solvent.
(Charles D. Winter)

We study the structure, nomenclature, and reactions of esters in detail in Chapter 15. In this chapter, we discuss only their preparation from carboxylic acids.

Acid-catalyzed esterification is reversible, and generally, at equilibrium, the quantities of remaining carboxylic acid and alcohol are appreciable. By controlling the experimental conditions, however, we can use Fischer esterification to prepare esters in high yields. If the alcohol is inexpensive compared with the carboxylic acid, we can use a large excess of the alcohol to drive the equilibrium to the right and achieve a high conversion of carboxylic acid to its ester.

EXAMPLE 14.4

Complete these Fischer esterification reactions:

(a) 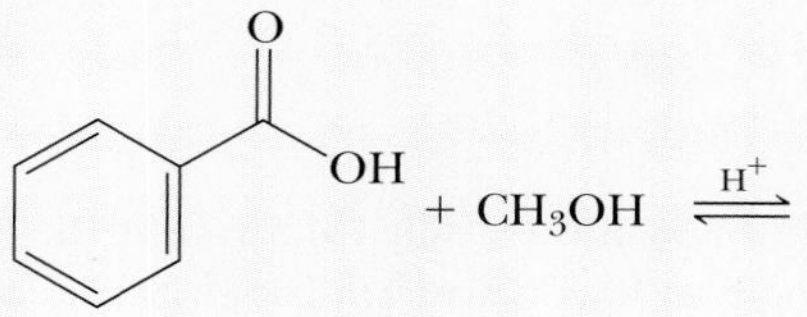 + CH$_3$OH $\overset{H^+}{\rightleftharpoons}$

(b) 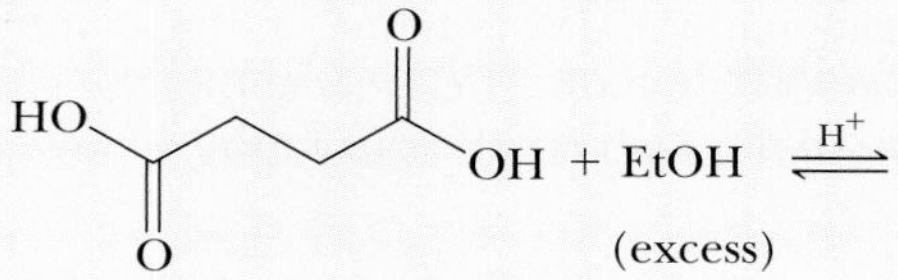 + EtOH $\overset{H^+}{\rightleftharpoons}$

(excess)

SOLUTION

Here is a structural formula for the ester produced in each reaction:

(a)

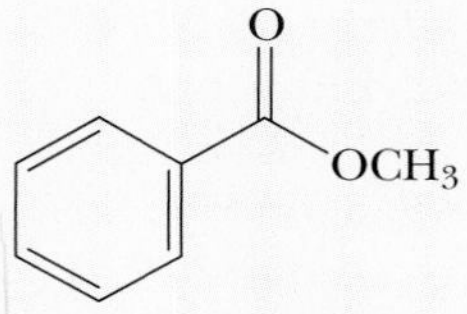

Methyl benzoate

(b)

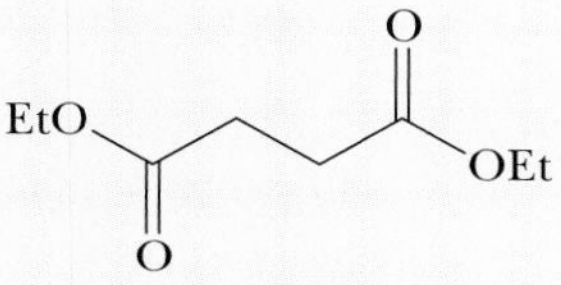

Diethyl butanedioate
(Diethyl succinate)

Practice Problem 14.4

Complete these Fischer esterification reactions:

(a) 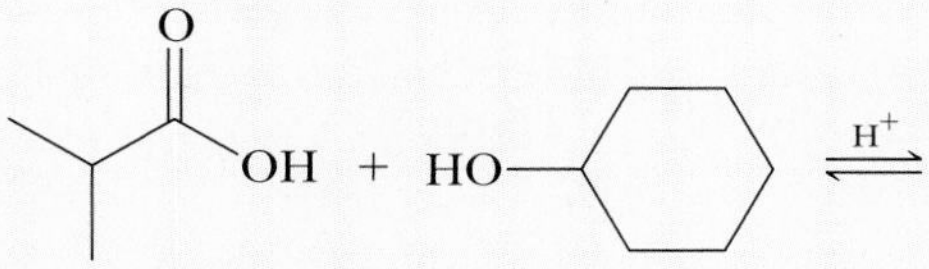 + HO— $\overset{H^+}{\rightleftharpoons}$

(b) 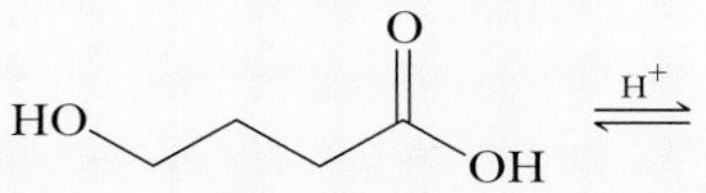 HO $\overset{H^+}{\rightleftharpoons}$ (a cyclic ester)

Following is a mechanism for Fischer esterification and we urge you to study it carefully. It is important that you understand this mechanism thoroughly, because it is a model for many of the reactions of the functional derivatives of carboxylic acids presented in Chapter 15. Note that, although we show the acid catalyst as H$_2$SO$_4$ when we write Fisher esterification reactions, the actual proton-transfer acid that initiates the reaction is the oxonium formed by the transfer of a proton

Esters as Flavoring Agents

Flavoring agents are the largest class of food additives. At present, over a thousand synthetic and natural flavors are available. The majority of these are concentrates or extracts from the material whose flavor is desired and are often complex mixtures of from tens to hundreds of compounds. A number of ester flavoring agents are synthesized industrially. Many have flavors very close to the target flavor, and adding only one or a few of them is sufficient to make ice cream, soft drinks, or candy taste natural. (Isopentane is the common name for 2-methylbutane.) The table shows the structures of a few of the esters used as flavoring agents:

Structure	Name	Flavor
	Ethyl formate	Rum
	Isopentyl acetate	Banana
	Octyl acetate	Orange
	Methyl butanoate	Apple
	Ethyl butanoate	Pineapple
	Methyl 2-aminobenzoate (Methyl anthranilate)	Grape

from H_2SO_4 (the stronger acid) to the alcohol (the stronger base) used in the esterification reaction:

Mechanism: Fischer Esterification

① Proton transfer from the acid catalyst to the carbonyl oxygen increases the electrophilicity of the carbonyl carbon . . .

② which is then attacked by the nucleophilic oxygen atom of the alcohol . . .

③ to form an oxonium ion.

④ Proton transfer from the oxonium ion to a second molecule of alcohol . . .

⑤ gives a tetrahedral carbonyl addition intermediate (TCAI).

⑥ Proton transfer to one of the —OH groups of the TCAI . . .

⑦ gives a new oxonium ion.

⑧ Loss of water from this oxonium ion . . .

⑨ gives the ester and water, and regenerates the acid catalyst.

14.8 CONVERSION TO ACID HALIDES

The functional group of an acid halide is a carbonyl group bonded to a halogen atom. Among the acid halides, acid chlorides are the most frequently used in the laboratory and in industrial organic chemistry:

Functional group of an acid halide   Acetyl chloride   Benzoyl chloride

The Pyrethrins: Natural Insecticides of Plant Origin

Pyrethrum is a natural insecticide obtained from the powdered flower heads of several species of *Chrysanthemum*, particularly *C. cinerariaefolium*. The active substances in pyrethrum—principally, pyrethrins I and II—are contact poisons for insects and cold-blooded vertebrates. Because their concentrations in the pyrethrum powder used in chrysanthemum-based insecticides are nontoxic to plants and higher animals, pyrethrum powder is used in household and livestock sprays, as well as in dusts for edible plants. Natural pyrethrins are esters of chrysanthemic acid.

While pyrethrum powders are effective insecticides, the active substances in them are destroyed rapidly in the environment. In an effort to develop synthetic compounds as effective as these natural insecticides, but with greater biostability, chemists have prepared a series of esters related in structure to chrysanthemic acid. Permethrin is one of the most commonly used synthetic pyrethrinlike compounds in household and agricultural products.

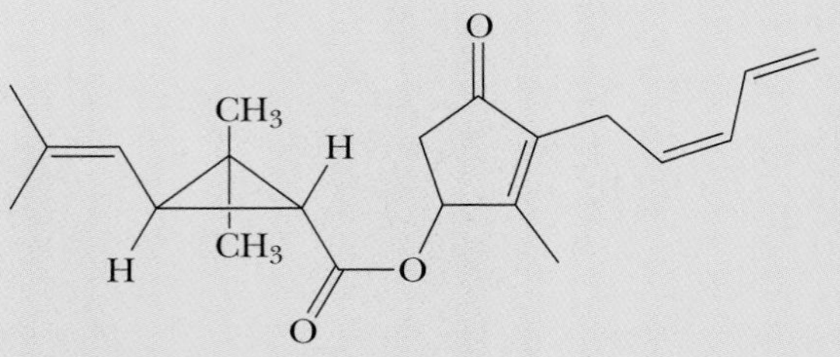

Pyrethrin I

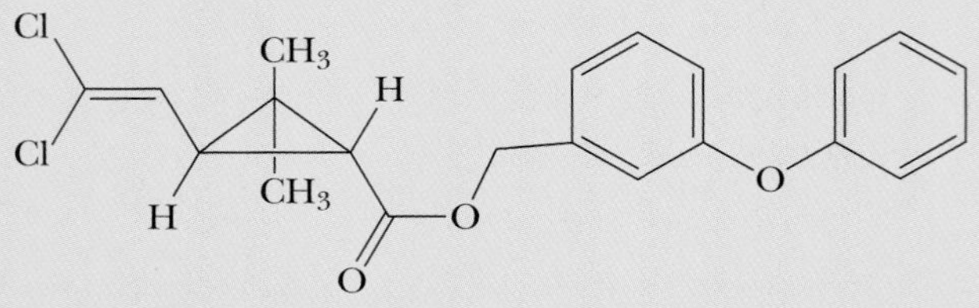

Permethrin

We study the nomenclature, structure, and characteristic reactions of acid halides in Chapter 15. In this chapter, our concern is only with their synthesis from carboxylic acids.

The most common way to prepare an acid chloride is to treat a carboxylic acid with thionyl chloride, the same reagent that converts an alcohol to a chloroalkane (Section 8.3D):

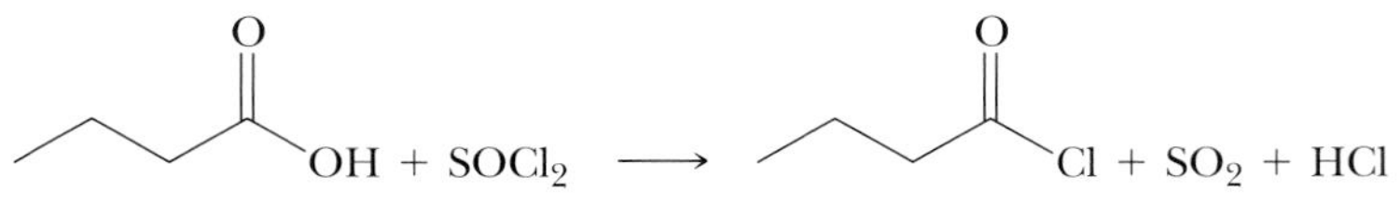

$$\text{Butanoic acid} + SOCl_2 \longrightarrow \text{Butanoyl chloride} + SO_2 + HCl$$

Butanoic acid   Thionyl chloride   Butanoyl chloride

EXAMPLE 14.5

Complete each equation:

(a) 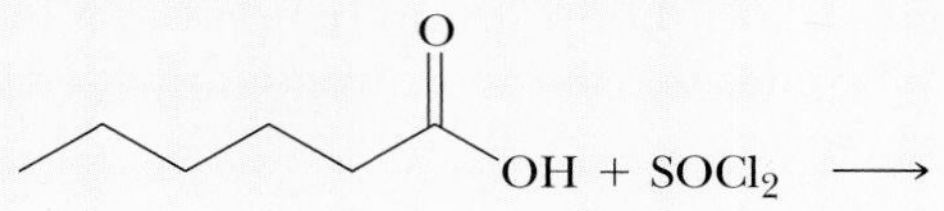 [structure] OH + SOCl₂ ⟶

(b) 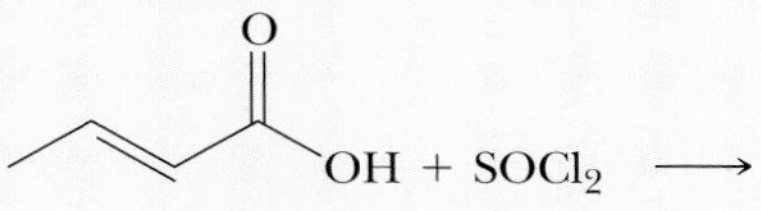 [structure] OH + SOCl₂ ⟶

SOLUTION

Following are the products for each reaction:

(a) 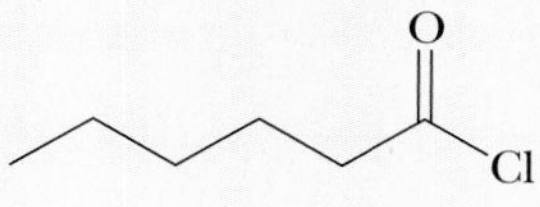 Cl + SO$_2$ + HCl   (b) 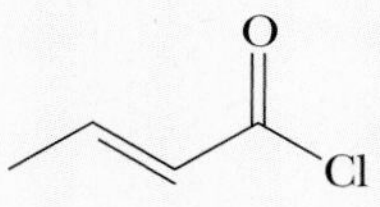 Cl + SO$_2$ + HCl

Practice Problem 14.5

Complete each equation:

(a) 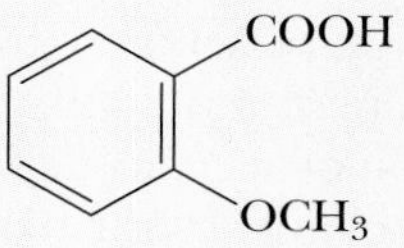 + SOCl$_2$ $\longrightarrow$ (b) 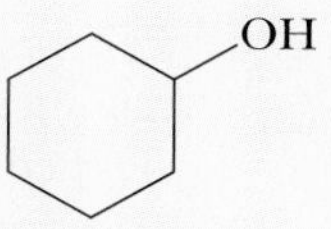 + SOCl$_2$ $\longrightarrow$

14.9 DECARBOXYLATION

A. β-Ketoacids

Decarboxylation is the loss of CO$_2$ from a carboxyl group. Almost any carboxylic acid, heated to a very high temperature, undergoes decarboxylation:

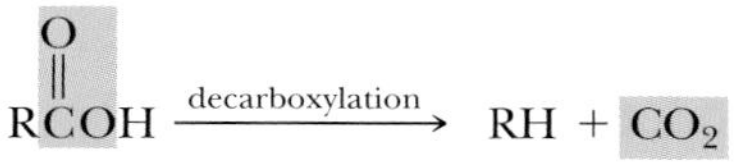

Most carboxylic acids, however, are quite resistant to moderate heat and melt or even boil without decarboxylation. Exceptions are carboxylic acids that have a carbonyl group β to the carboxyl group. This type of carboxylic acid undergoes decarboxylation quite readily on mild heating. For example, when 3-oxobutanoic acid (acetoacetic acid) is heated moderately, it undergoes decarboxylation to give acetone and carbon dioxide:

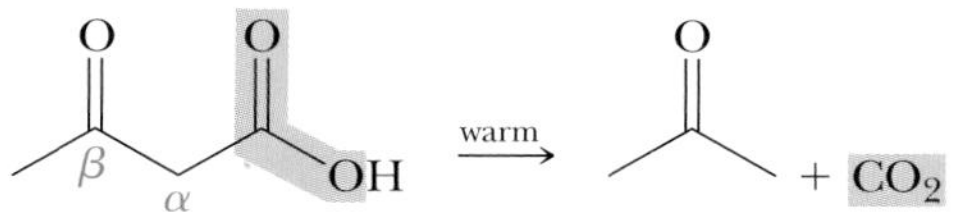

3-Oxobutanoic acid   Acetone
(Acetoacetic acid)

Decarboxylation on moderate heating is a unique property of 3-oxocarboxylic acids (β-ketoacids) and is not observed with other classes of ketoacids.

Mechanism: Decarboxylation of a β-Ketocarboxylic Acid

Step 1: Redistribution of six electrons in a cyclic six-membered transition state gives carbon dioxide and an enol:

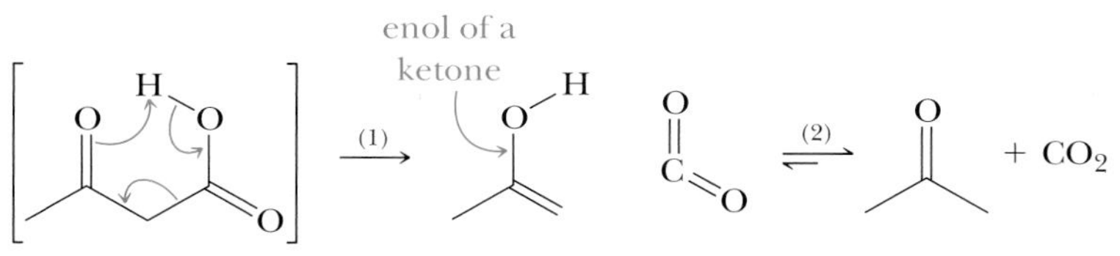

(A cyclic six-membered
transition state)

Decarboxylation Loss of CO$_2$ from a carboxyl group.

Ketone Bodies and Diabetes

3-Oxobutanoic acid (acetoacetic acid) and its reduction product, 3-hydroxybutanoic acid, are synthesized in the liver from acetyl-CoA, a product of the metabolism of fatty acids (Section 22.6C) and certain amino acids:

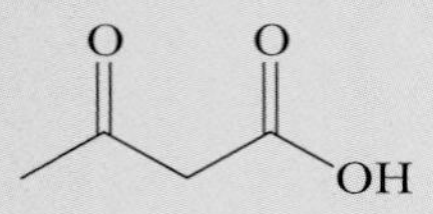

3-Oxobutanoic acid
(Acetoacetic acid)

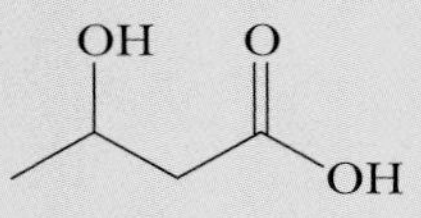

3-Hydroxybutanoic acid
(β-Hydroxybutyric acid)

3-Hydroxybutanoic acid and 3-oxobutanoic acid are known collectively as ketone bodies.

The concentration of ketone bodies in the blood of healthy, well-fed humans is approximately 0.01 mM/ L. However, in persons suffering from starvation or diabetes mellitus, the concentration of ketone bodies may increase to as much as 500 times normal. Under these conditions, the concentration of acetoacetic acid increases to the point where it undergoes spontaneous decarboxylation to form acetone and carbon dioxide. Acetone is not metabolized by humans and is excreted through the kidneys and the lungs. The odor of acetone is responsible for the characteristic "sweet smell" on the breath of severely diabetic patients.

Step 2: Keto–enol tautomerism (Section 13.9A) of the enol gives the more stable keto form of the product:

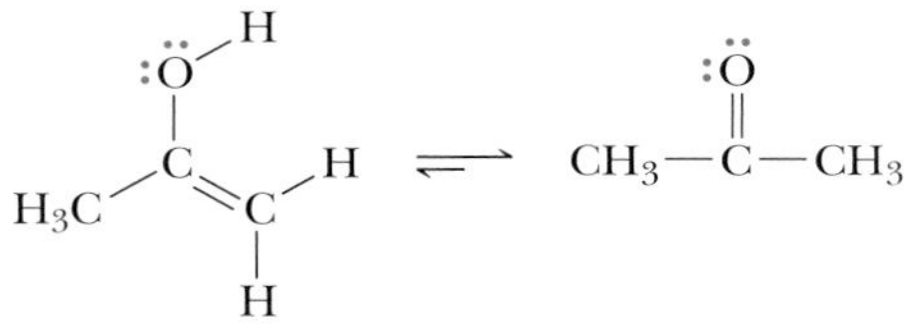

An important example of decarboxylation of a β-ketoacid in the biological world occurs during the oxidation of foodstuffs in the tricarboxylic acid (TCA) cycle. Oxalosuccinic acid, one of the intermediates in this cycle, undergoes spontaneous decarboxylation to produce α-ketoglutaric acid. Only one of the three carboxyl groups of oxalosuccinic acid has a carbonyl group in the position β to it, and it is this carboxyl group that is lost as CO_2:

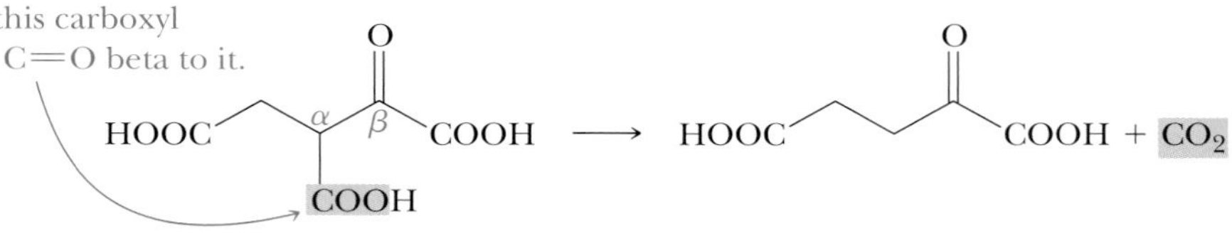

only this carboxyl
has a C=O beta to it.

Oxalosuccinic acid

α-Ketoglutaric acid

B. Malonic Acid and Substituted Malonic Acids

The presence of a ketone or an aldehyde carbonyl group on the carbon β to the carboxyl group is sufficient to facilitate decarboxylation. In the more general reaction, decarboxylation is facilitated by the presence of any carbonyl group on the β carbon, including that of a carboxyl group or ester. Malonic acid and substituted malonic acids, for example, undergo decarboxylation on heating, as illustrated by the decarboxylation of malonic acid when it is heated slightly above its melting point of 135–137°C:

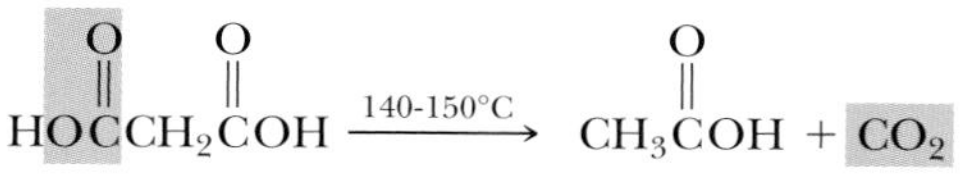

$$\underset{\substack{\text{Propanedioic acid}\\\text{(Malonic acid)}}}{HOCCH_2COH} \xrightarrow{\text{140-150°C}} CH_3COH + CO_2$$

The mechanism for decarboxylation of malonic acids is similar to what we have just studied for the decarboxylation of β-ketoacids. The formation of a cyclic, six-membered transition state involving a redistribution of three electron pairs gives the enol form of a carboxylic acid, which, in turn, isomerizes to the carboxylic acid.

Mechanism: Decarboxylation of a β-Dicarboxylic Acid

Step 1: Rearrangement of six electrons in a cyclic six-membered transition state gives carbon dioxide and the enol form of a carboxyl group.

Step 2: Keto–enol tautomerism (Section 13.9A) of the enol gives the more stable keto form of the carboxyl group.

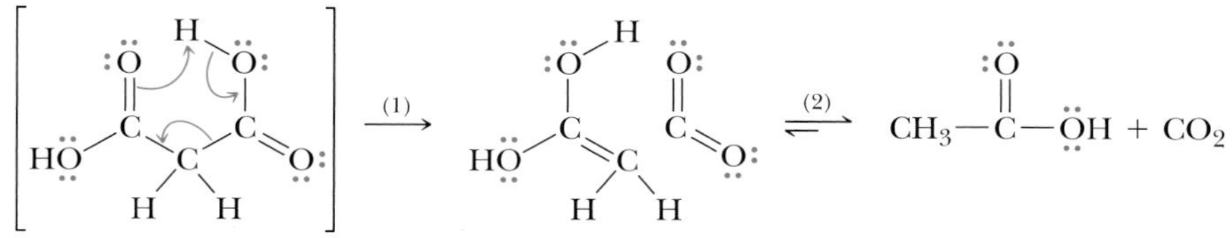

A cyclic six-membered   Enol of a
transition state   carboxyl group

EXAMPLE 14.6

Each of these carboxylic acids undergoes thermal decarboxylation:

(a) 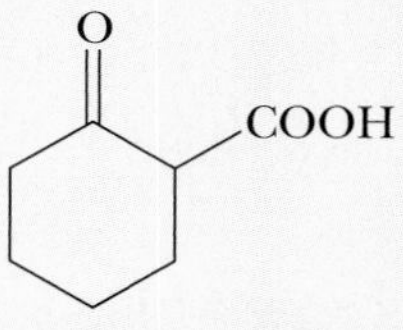   (b)

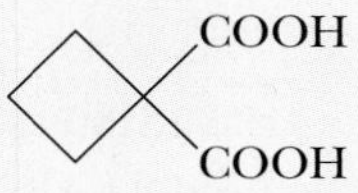

Draw a structural formula for the enol intermediate and final product formed in each reaction.

SOLUTION

(a) 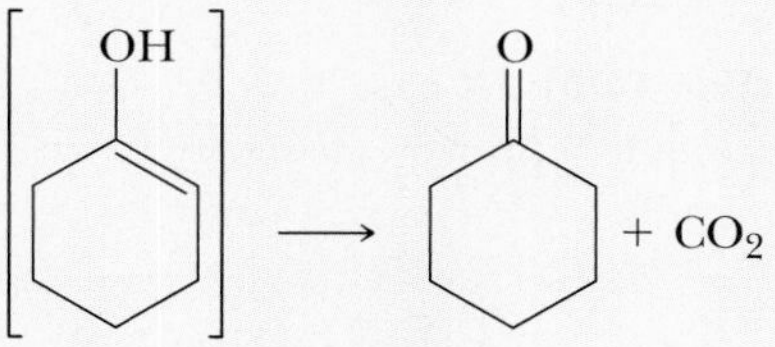   →   + CO_2

Enol
intermediate

(b) 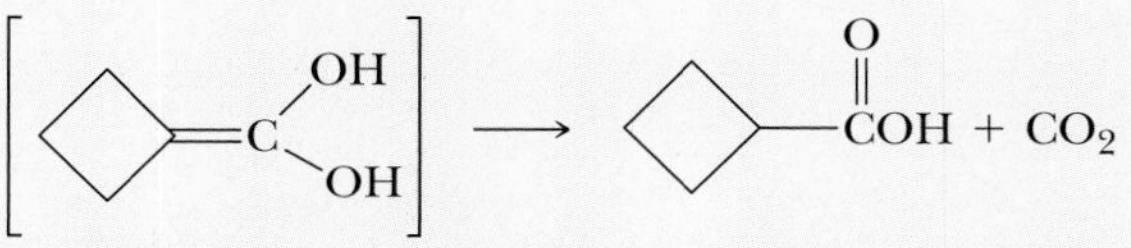   →   —COH + CO_2

Enol intermediate

Practice Problem 14.6

Draw the structural formula for the indicated β-ketoacid:

$$\beta\text{-ketoacid} \xrightarrow{\text{heat}} \qquad + CO_2$$

SUMMARY

The functional group of a **carboxylic acid** (Section 14.2) is the **carboxyl group**, —**COOH**. IUPAC names of carboxylic acids (Section 14.3) are derived from the parent alkane by dropping the suffix -e and adding -oic acid. Dicarboxylic acids are named as -dioic acids.

Carboxylic acids are polar compounds (Section 14.4) that associate by hydrogen bonding into dimers in the liquid and solid states. Carboxylic acids have higher boiling points and are more soluble in water than alcohols, aldehydes, ketones, and ethers of comparable molecular weight. A carboxylic acid consists of two regions of different polarity; a polar, **hydrophilic** car-

boxyl group, which increases solubility in water, and a nonpolar, **hydrophobic** hydrocarbon chain, which decreases solubility in water. The first four aliphatic carboxylic acids are infinitely soluble in water, because the hydrophilic carboxyl group more than counterbalances the hydrophobic hydrocarbon chain. As the size of the carbon chain increases, however, the hydrophobic group becomes dominant, and solubility in water decreases.

Values of **pK_a** for aliphatic carboxylic acids are in the range from 4.0 to 5.0 (Section 14.5A). Electron-withdrawing substituents near the carboxyl group increase acidity in both aliphatic and aromatic carboxylic acids.

KEY REACTIONS

1. Acidity of Carboxylic Acids (Section 14.5A)

Values of pK_a for most unsubstituted aliphatic and aromatic carboxylic acids are within the range from 4 to 5:

$$CH_3COH + H_2O \rightleftharpoons CH_3CO^- + H_3O^+ \quad pK_a = 4.76$$

Substitution by electron-withdrawing groups decreases pK_a (increases acidity).

2. Reaction of Carboxylic Acids with Bases (Section 14.5B)

Carboxylic acids form water-soluble salts with alkali metal hydroxides, carbonates, and bicarbonates, as well as with ammonia and amines:

$$\text{C}_6\text{H}_5{-}COOH + NaOH \xrightarrow{H_2O} \text{C}_6\text{H}_5{-}COO^-Na^+ + H_2O$$

3. Reduction by Lithium Aluminum Hydride (Section 14.6)

Lithium aluminum hydride reduces a carboxyl group to a primary alcohol:

$$\text{—COH} \xrightarrow[\text{2. H}_2\text{O}]{\text{1. LiAlH}_4} \text{—CH}_2\text{OH}$$

4. Fischer Esterification (Section 14.7)

Fischer esterification is reversible:

$$\text{—OH} + \text{HO—} \underset{\text{H}_2\text{SO}_4}{\rightleftharpoons} \text{—O—} + H_2O$$

One way to force the equilibrium to the right is to use an excess of the alcohol.

5. Conversion to Acid Halides (Section 14.8)

Acid chlorides, the most common and widely used of the acid halides, are prepared by treating carboxylic acids with thionyl chloride:

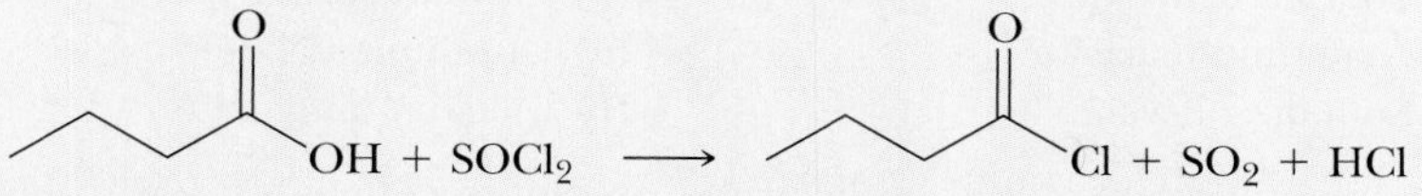

6. Decarboxylation of β-Ketoacids (Section 14.9A)

The mechanism of decarboxylation involves the redistribution of bonding electrons in a cyclic, six-membered transition state:

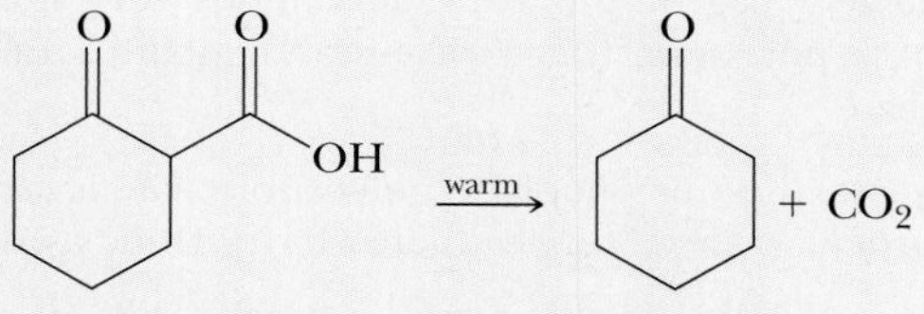

7. Decarboxylation of β-Dicarboxylic Acids (Section 14.9B)

The mechanism of decarboxylation of a β-dicarboxylic acid is similar to that of decarboxylation of a β-ketoacid:

$$HOCCH_2COH \xrightarrow{\text{heat}} CH_3COH + CO_2$$

PROBLEMS

A problem number set in red indicates an applied "real-world" problem.

Structure and Nomenclature

14.7 Name and draw structural formulas for the four carboxylic acids with molecular formula $C_5H_{10}O_2$. Which of these carboxylic acids is chiral?

14.8 Write the IUPAC name for each compound:

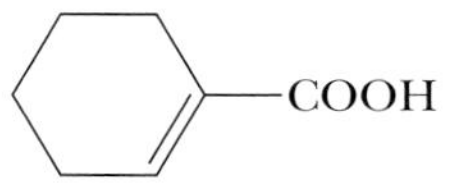

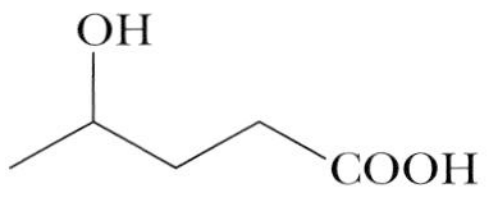

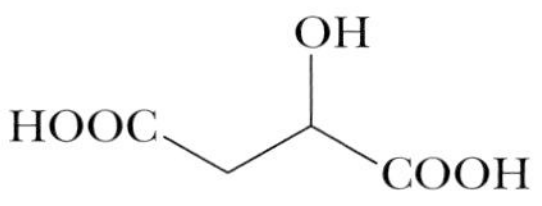

14.9 Draw a structural formula for each carboxylic acid:
(a) 4-Nitrophenylacetic acid
(b) 4-Aminopentanoic acid
(c) 3-Chloro-4-phenylbutanoic acid
(d) *cis*-3-Hexenedioic acid
(e) 2,3-Dihydroxypropanoic acid
(f) 3-Oxohexanoic acid
(g) 2-Oxocyclohexanecarboxylic acid
(h) 2,2-Dimethylpropanoic acid

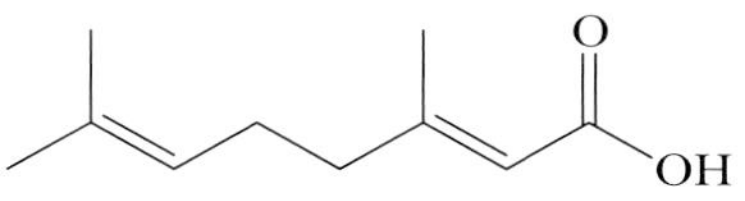

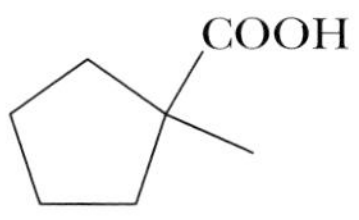

14.10 Megatomoic acid, the sex attractant of the female black carpet beetle, has the structure

$$CH_3(CH_2)_7CH=CHCH=CHCH_2COOH$$

Megatomoic acid

(a) What is the IUPAC name of megatomoic acid?
(b) State the number of stereoisomers possible for this compound.

14.11 The IUPAC name of ibuprofen is 2-(4-isobutylphenyl)propanoic acid. Draw a structural formula of ibuprofen.

14.12 Draw structural formulas for these salts:

(a) Sodium benzoate   (b) Lithium acetate
(c) Ammonium acetate   (d) Disodium adipate
(e) Sodium salicylate   (f) Calcium butanoate

14.13 The monopotassium salt of oxalic acid is present in certain leafy vegetables, including rhubarb. Both oxalic acid and its salts are poisonous in high concentrations. Draw a structural formula of monopotassium oxalate.

14.14 Potassium sorbate is added as a preservative to certain foods to prevent bacteria and molds from causing spoilage and to extend the foods' shelf life. The IUPAC name of potassium sorbate is potassium (2E,4E)-2,4-hexadienoate. Draw a structural formula of potassium sorbate.

14.15 Zinc 10-undecenoate, the zinc salt of 10-undecenoic acid, is used to treat certain fungal infections, particularly *tinea pedis* (athlete's foot). Draw a structural formula of this zinc salt.

Physical Properties

14.16 Arrange the compounds in each set in order of increasing boiling point:

(a) $CH_3(CH_2)_5COOH$ $CH_3(CH_2)_6CHO$ $CH_3(CH_2)_6CH_2OH$
(b) CH_3CH_2COOH $CH_3CH_2CH_2CH_2OH$ $CH_3CH_2OCH_2CH_3$

Preparation of Carboxylic Acids

14.17 Draw a structural formula for the product formed by treating each compound with warm chromic acid, H_2CrO_4:

(a) $CH_3(CH_2)_4CH_2OH$ (b)

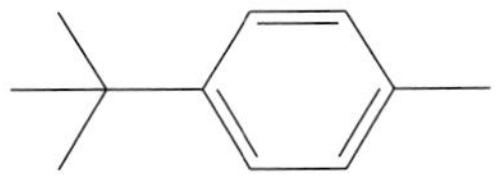

(c) 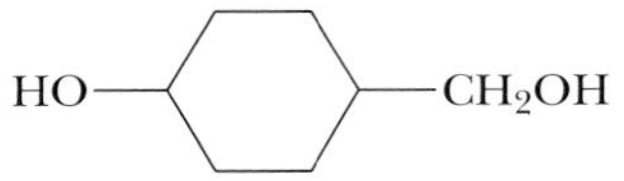 HO——CH_2OH

14.18 Draw a structural formula for a compound with the given molecular formula that, on oxidation by chromic acid, gives the carboxylic acid or dicarboxylic acid shown:

(a) $C_6H_{14}O$ $\xrightarrow{\text{oxidation}}$ COOH

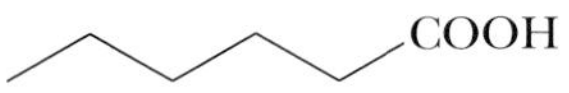

(b) $C_6H_{12}O$ $\xrightarrow{\text{oxidation}}$ COOH

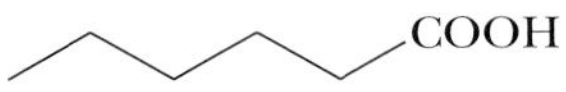

(c) $C_6H_{14}O_2$ $\xrightarrow{\text{oxidation}}$ HOOCCOOH

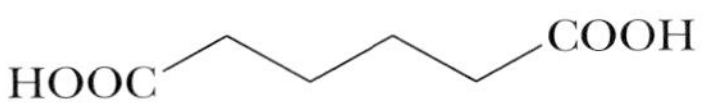

Acidity of Carboxylic Acids

14.19 Which is the stronger acid in each pair?

(a) Phenol (pK_a 9.95) or benzoic acid (pK_a 4.17)
(b) Lactic acid (K_a 8.4×10^{-4}) or ascorbic acid (K_a 7.9×10^{-5})

14.20 Arrange these compounds in order of increasing acidity: benzoic acid, benzyl alcohol, and phenol.

14.21 Assign the acid in each set its appropriate pK_a:

(a) 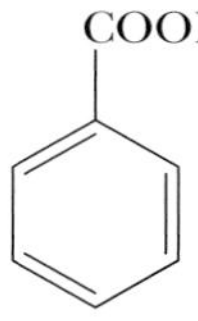 [benzoic acid structure] and [4-nitrobenzoic acid structure, with COOH and NO$_2$] (pK_a 4.19 and 3.14)

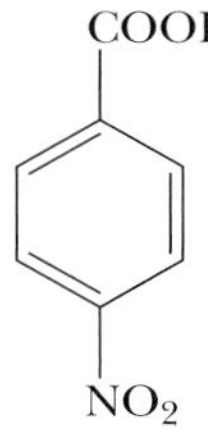

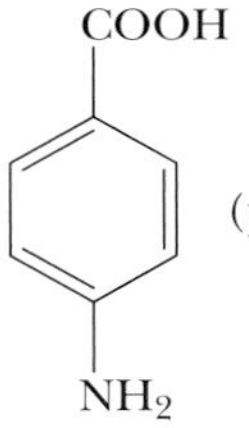

(b)  [4-nitrobenzoic acid structure, COOH and NO$_2$] and [4-aminobenzoic acid structure, COOH and NH$_2$] (pK_a 4.92 and 3.14)

(c) $CH_3\overset{\overset{\text{O}}{\|}}{C}CH_2COOH$ and $CH_3\overset{\overset{\text{O}}{\|}}{C}COOH$ (pK_a 3.58 and 2.49)

(d) $CH_3\overset{\overset{\text{OH}}{|}}{C}HCOOH$ and CH_3CH_2COOH (pK_a 4.78 and 3.08)

14.22 Complete these acid–base reactions:

(a) 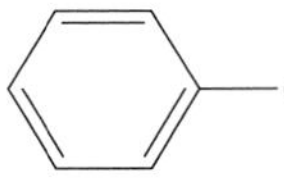 [phenyl group]—$CH_2COOH + NaOH \longrightarrow$

(b) $CH_3CH{=}CHCH_2COOH + NaHCO_3 \longrightarrow$

(c) 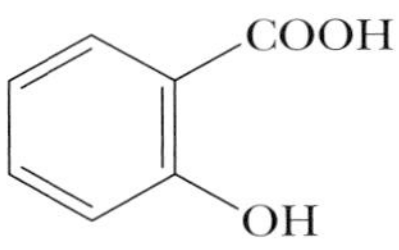 [benzene ring with COOH and OH substituents] $+ NaHCO_3 \longrightarrow$

(d) $CH_3\overset{\overset{\text{OH}}{|}}{C}HCOOH + H_2NCH_2CH_2OH \longrightarrow$

(e) $CH_3CH{=}CHCH_2COO^-Na^+ + HCl \longrightarrow$

14.23 The normal pH range for blood plasma is 7.35–7.45. Under these conditions, would you expect the carboxyl group of lactic acid (pK_a 4.07) to exist primarily as a carboxyl group or as a carboxylate anion? Explain.

14.24 The pK_a of ascorbic acid (Section 18.7) is 4.76. Would you expect ascorbic acid dissolved in blood plasma (pH 7.35–7.45) to exist primarily as ascorbic acid or as ascorbate anion? Explain.

14.25 Excess ascorbic acid is (pK_a 4.76) excreted in the urine, the pH of which is normally in the range from 4.8 to 8.4. What form of ascorbic acid, ascorbic acid itself or ascorbate anion, would you expect to be present in urine with pH 8.4?

14.26 The pH of human gastric juice is normally in the range from 1.0 to 3.0. What form of lactic acid (pK_a 4.07), lactic acid itself or its anion, would you expect to be present in the stomach?

14.27 Following are two structural formulas for the amino acid alanine (Section 19.2):

$$CH_3-\underset{\underset{NH_2}{|}}{C}H-\overset{\overset{O}{\|}}{C}-OH \qquad CH_3-\underset{\underset{NH_3^+}{|}}{C}H-\overset{\overset{O}{\|}}{C}-O^-$$

(A)   (B)

Is alanine better represented by structural formula A or B? Explain.

14.28 In Chapter 19, we discuss a class of compounds called amino acids, so named because they contain both an amino group and a carboxyl group. Following is a structural formula for the amino acid alanine in the form of an internal salt:

$$\underset{\underset{NH_3{}^+}{|}}{CH_3CHCO^-} \overset{\overset{O}{\parallel}}{} \quad \text{Alanine}$$

What would you expect to be the major form of alanine present in aqueous solution at (a) pH 2.0, (b) pH 5–6, and (c) pH 11.0? Explain.

Reactions of Carboxylic Acids

14.29 Give the expected organic products formed when phenylacetic acid, $PhCH_2COOH$, is treated with each of the following reagents:

(a) $SOCl_2$ (b) $NaHCO_3$, H_2O
(c) $NaOH$, H_2O (d) NH_3, H_2O
(e) $LiAlH_4$, followed by H_2O (f) $NaBH_4$, followed by H_2O
(g) CH_3OH + H_2SO_4 (catalyst)   (h) H_2/Ni at 25°C and 3 atm pressure

14.30 Show how to convert *trans*-3-phenyl-2-propenoic acid (cinnamic acid) to these compounds:

(a) Ph⌒⌒OH   (b) Ph⌒⌒COOH   (c) Ph⌒⌒⌒OH

14.31 Show how to convert 3-oxobutanoic acid (acetoacetic acid) to these compounds:

(a) $\underset{\underset{OH}{|}}{CH_3CHCH_2COOH}$ (b) $\underset{\underset{OH}{|}}{CH_3CHCH_2CH_2OH}$ (c) $CH_3CH{=}CHCOOH$

14.32 Complete these examples of Fischer esterification (assume an excess of the alcohol):

(a) [structure: acetic acid] + HO⌒⟨ $\overset{H^+}{\rightleftharpoons}$

(b) [benzene ring with COOH (ortho)] + CH_3OH $\overset{H^+}{\rightleftharpoons}$

(c) HO⌒⌒OH (dicarbonyl) + ⌒OH $\overset{H^+}{\rightleftharpoons}$

14.33 Formic acid is one of the components responsible for the sting of biting ants and is injected under the skin by bees and wasps. A way to relieve the pain is to rub the area of the sting with a paste of baking soda ($NaHCO_3$) and water, which neutralizes the acid. Write an equation for this reaction.

14.34 Methyl 2-hydroxybenzoate (methyl salicylate) has the odor of oil of wintergreen. This ester is prepared by the Fischer esterification of 2-hydroxybenzoic acid (salicylic acid) with methanol. Draw a structural formula of methyl 2-hydroxybenzoate.

14.35 Benzocaine, a topical anesthetic, is prepared by treating 4-aminobenzoic acid with ethanol in the presence of an acid catalyst, followed by neutralization. Draw a structural formula of benzocaine.

14.36 Examine the structural formulas of pyrethrin and permethrin. (See Chemical Connections 14C.)

(a) Locate the ester groups in each compound.
(b) Is pyrethrin chiral? How many stereoisomers are possible for it?
(c) Is permethrin chiral? How many stereoisomers are possible for it?

14.37 A commercial Clothing & Gear Insect Repellant gives the following information about permethrin, its active ingredient:

Cis/trans ratio: Minimum 35% (+/−) cis and maximum 65% (+/−) trans

(a) To what does the cis/trans ratio refer?

(b) To what does the designation "(+/−)" refer?

14.38 From what carboxylic acid and alcohol is each of the following esters derived?

(a) 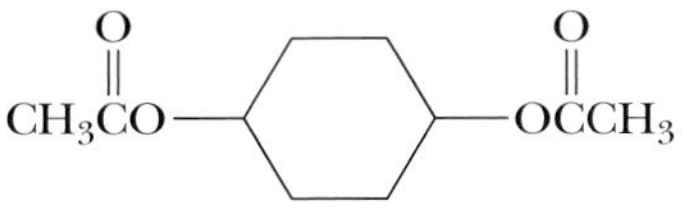 CH_3CO—⬡—$OCCH_3$ (with two C=O groups)

(b) $CH_3OCCH_2CH_2COCH_3$ (with two C=O groups)

(c) 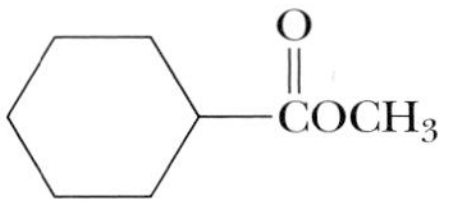 ⬡—$COCH_3$ (with C=O group)

(d) CH_3CH_2CH=$CHCOCH(CH_3)_2$ (with C=O group)

14.39 When treated with an acid catalyst, 4-hydroxybutanoic acid forms a cyclic ester (a lactone). Draw the structural formula of this lactone.

14.40 Draw a structural formula for the product formed on thermal decarboxylation of each of the following compounds:

(a) $C_6H_5CCH_2COOH$ (with C=O group)

(b) $C_6H_5CH_2CHCOOH$ (with COOH substituent)

(c) 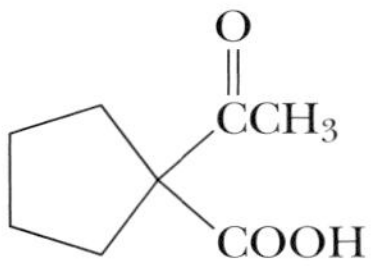 cyclopentane with CCH_3 (C=O) and $COOH$ substituents

Synthesis

14.41 Methyl 2-aminobenzoate, a flavoring agent with the taste of grapes (see "Chemical Connections 14B"), can be prepared from toluene by the following series of steps:

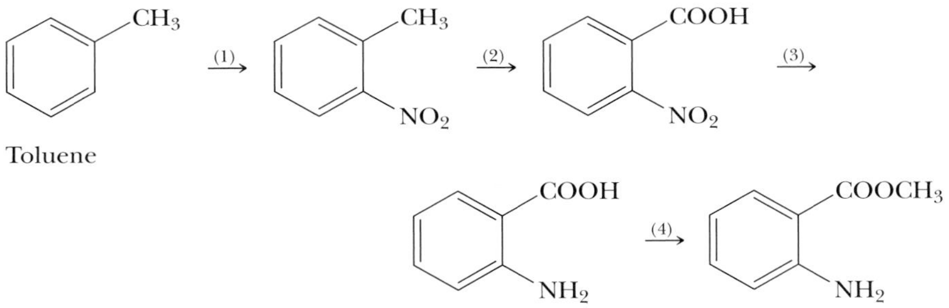

Toluene —CH_3 $\xrightarrow{(1)}$ CH_3/NO_2 benzene $\xrightarrow{(2)}$ $COOH$/NO_2 benzene $\xrightarrow{(3)}$

$COOH$/NH_2 benzene $\xrightarrow{(4)}$ $COOCH_3$/NH_2 benzene

Methyl 2-amino-
benzoate

Show how you might bring about each step in this synthesis.

14.42 Methylparaben and propylparaben are used as preservatives in foods, beverages, and cosmetics:

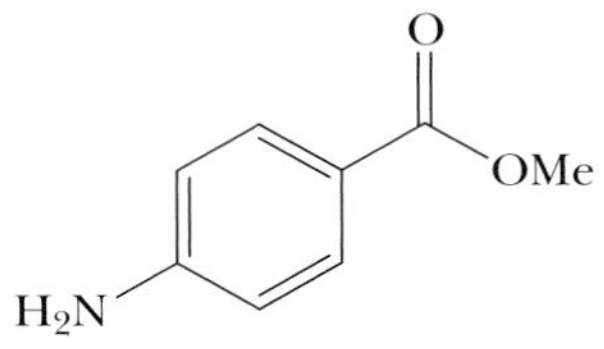

Methyl 4-aminobenzoate
(Methylparaben)

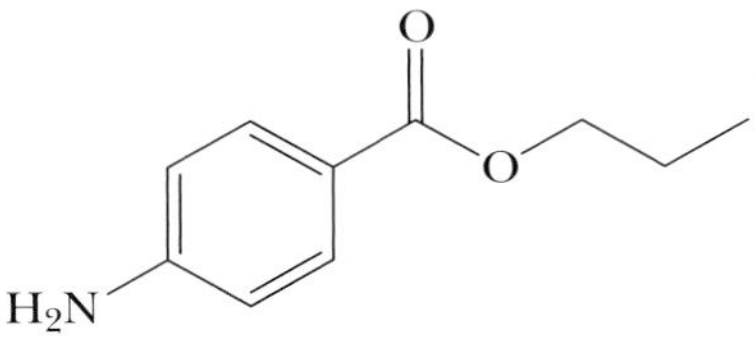

Propyl 4-aminobenzoate
(Propylparaben)

Show how the synthetic scheme in Problem 14.41 can be modified to give each of these compounds.

14.43 Procaine (its hydrochloride is marketed as Novocaine®) was one of the first local anesthetics developed for infiltration and regional anesthesia. It is synthesized by the following Fischer esterification:

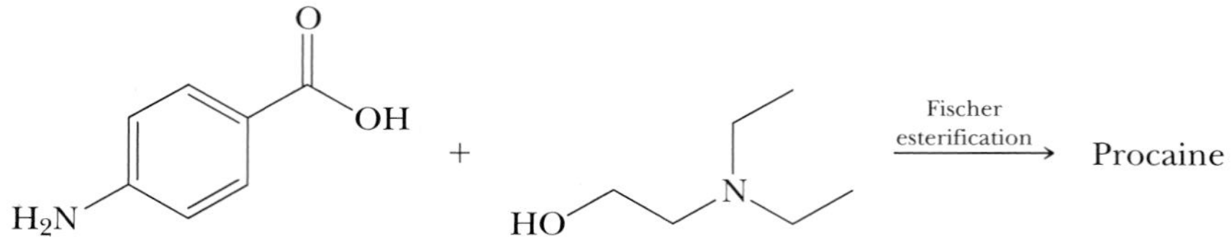

p-Aminobenzoic
acid

2-Diethylaminoethanol

Fischer
esterification
→ Procaine

Draw a structural formula for procaine.

14.44 Meclizine is an antiemetic: It helps prevent, or at least lessen, the vomiting associated with motion sickness, including seasickness. Among the names of the over-the-counter preparations of meclizine are Bonine®, Sea-Legs, Antivert®, and Navicalm®. Meclizine can be synthesized by the following series of steps:

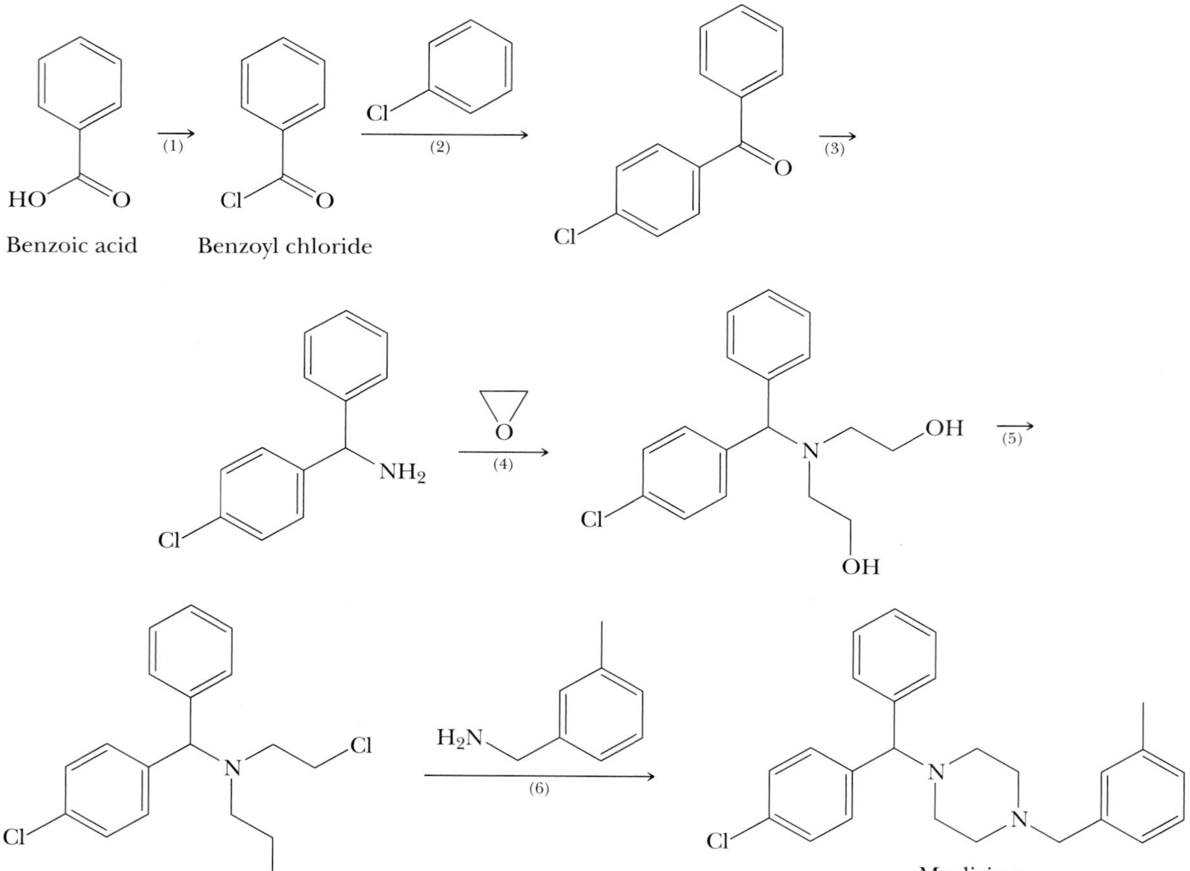

Benzoic acid   Benzoyl chloride

Meclizine

(a) Propose a reagent for Step 1.
(b) The catalyst for Step 2 is AlCl₃. Name the type of reaction that occurs in Step 2.
(c) Propose reagents for Step 3.
(d) Propose a mechanism for Step 4, and show that it is an example of nucleophilic aliphatic substitution.
(e) Propose a reagent for Step 5.
(f) Show that Step 6 is also an example of nucleophilic aliphatic substitution.

14.45 Chemists have developed several syntheses for the antiasthmatic drug albuterol (Proventil). One of these syntheses starts with salicylic acid, the same acid that is the starting material for the synthesis of aspirin:

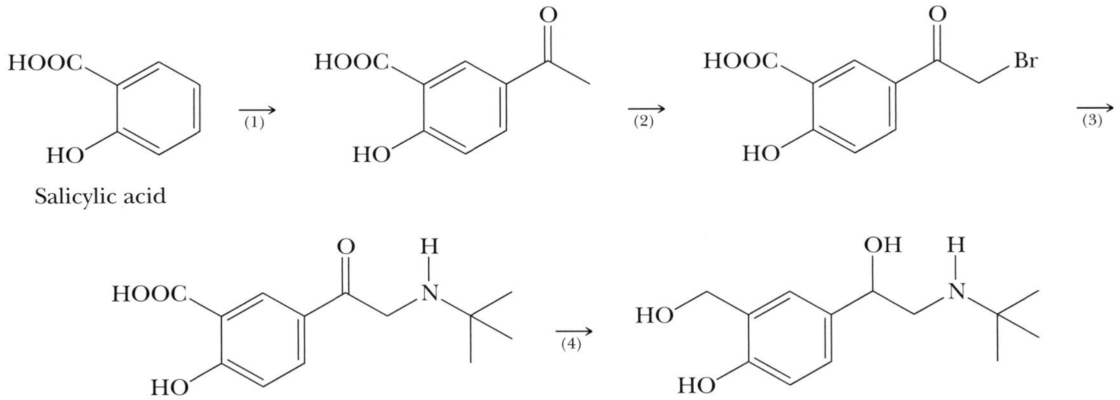

Salicylic acid

Albuterol

(a) Propose a reagent and a catalyst for Step 1. What name is given to this type of reaction?
(b) Propose a reagent for Step 2.
(c) Name the amine used to bring about Step 3.
(d) Step 4 is a reduction of two functional groups. Name the functional groups reduced and tell what reagent will accomplish the reduction.

Looking Ahead

14.46 Explain why α-amino acids, the building blocks of proteins (Chapter 20), are nearly a thousand times more acidic than aliphatic carboxylic acids:

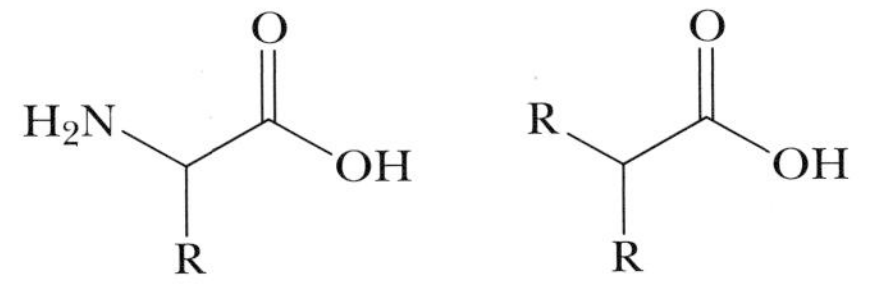

An α-amino acid   An aliphatic acid
$pK_a \approx 2$ $pK_a \approx 5$

14.47 Which is more difficult to reduce with LiAlH$_4$, a carboxylic acid or a carboxylate ion?

14.48 Show how an ester can react with H$^+$/H$_2$O to give a carboxylic acid and an alcohol (*Hint:* This is the reverse of Fischer esterification):

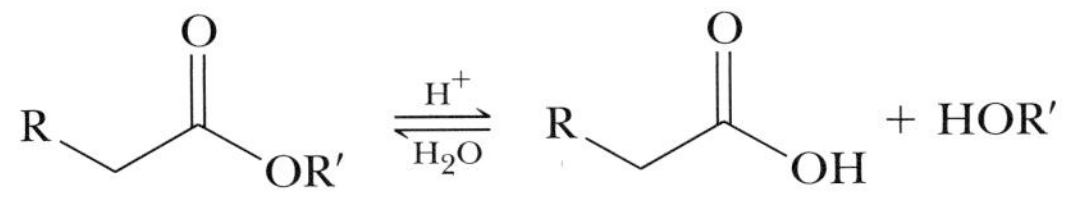

14.49 In Chapter 13, we saw how Grignard reagents readily attack the carbonyl carbon of ketones and aldehydes. Should the same process occur with Grignards and carboxylic acids? With esters?

14.50 In Section 14.7, it was suggested that the mechanism for the Fischer esterification of carboxylic acids would be a model for many of the reactions of the functional derivatives of carboxylic acids. One such reaction, the reaction of an acid halide with water, is the following:

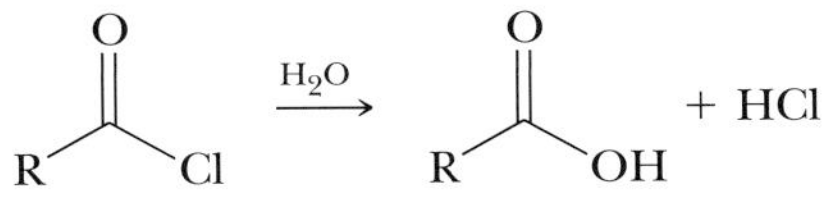

Suggest a mechanism for this reaction.

15 Functional Derivatives of Carboxylic Acids

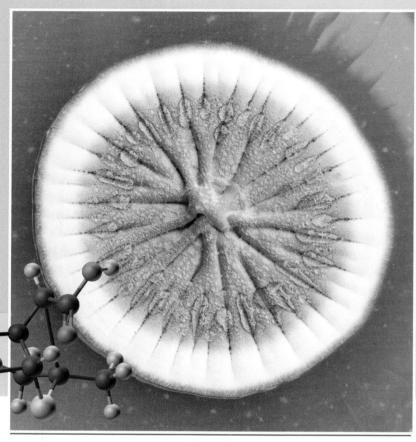

Macrophotograph of the fungus *Penicillium notatum* growing on a petri dish culture of Whickerman's agar. This fungus was used as an early source of the first penicillin antibiotic. Inset: A model of amoxicillin. *(Andrew McClenaghan/Photo Researchers, Inc.)*

15.1 INTRODUCTION

In this chapter, we study four classes of organic compounds, all derived from the carboxyl group: acid halides, acid anhydrides, esters, and amides. Under the general formula of each functional group is a drawing to help you see how the group is formally related to a carboxyl group. The loss of —OH from a carboxyl group and H— from H—Cl, for example, gives an acid chloride, and similarly, the loss of —OH from a carboxyl group and H— from ammonia gives an amide:

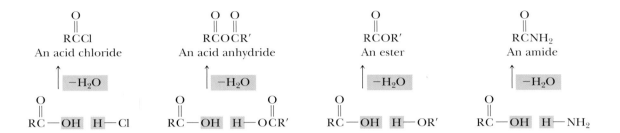

15.2 STRUCTURE AND NOMENCLATURE

A. Ethanoyl chloride

The functional group of an **acid halide** (acyl halide) is an **acyl group (RCO—)** bonded to a halogen atom (Section 14.8). The most common acid halides are acid chlorides:

Acid halide A derivative of a carboxylic acid in which the —OH of the carboxyl group is replaced by a halogen—most commonly, chlorine.

Ethanoyl chloride
(Acetyl chloride)

Benzoyl chloride

Acid halides are named by changing the suffix -*ic acid* in the name of the parent carboxylic acid to -*yl halide*.

B. Acid Anhydrides

Carboxylic Anhydrides

The functional group of a **carboxylic anhydride** (commonly referred to simply as an anhydride) is two acyl groups bonded to an oxygen atom. The anhydride may be symmetrical (having two identical acyl groups), or it may be mixed (having two different acyl groups):

Carboxylic anhydride A compound in which two acyl groups are bonded to an oxygen.

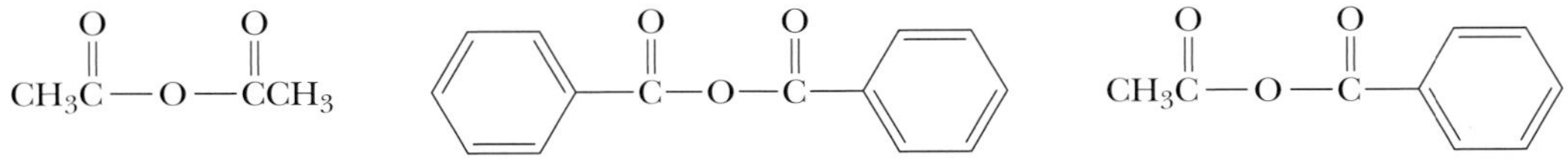

Acetic anhydride

Benzoic anhydride

Acetic benzoic anhydride
(a mixed anhydride)

Phosphoric Anhydrides

Because of the special importance of anhydrides of phosphoric acid in biochemical systems (Chapter 22), we include them here to show the similarity between them and the anhydrides of carboxylic acids. The functional group of a **phosphoric anhydride** is two phosphoryl groups bonded to an oxygen atom. Shown here are structural formulas for two anhydrides of phosphoric acid, H_3PO_4, and the ions derived by ionization of the acidic hydrogens of each:

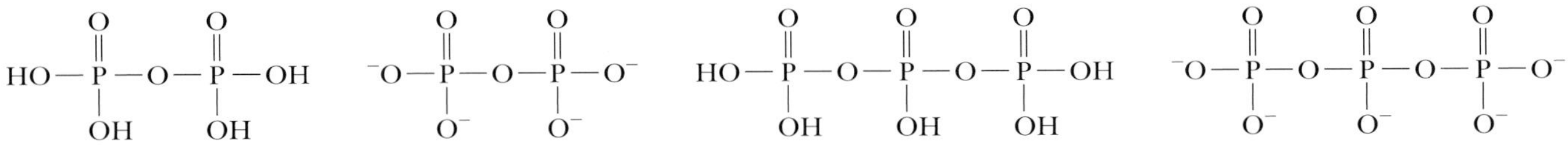

Diphosphoric acid
(Pyrophosphoric acid)

Diphosphate ion
(Pyrophosphate ion)

Triphosphoric acid

Triphosphate ion

CHEMICAL CONNECTIONS   15A

Ultraviolet Sunscreens and Sunblocks

Ultraviolet (UV) radiation penetrating the earth's ozone layer is arbitrarily divided into two regions: UVB (290–320 nm) and UVA (320–400 nm). UVB, a more energetic form of radiation than UVA, interacts directly with molecules of the skin and eyes, causing skin cancer, aging of the skin, eye damage leading to cataracts, and delayed sunburn that appears 12 to 24 hours after exposure. UVA radiation, by contrast, causes tanning. It also damages skin, albeit much less efficiently than UVB. The role of UVA in promoting skin cancer is less well understood.

Commercial sunscreen products are rated according to their sun protection factor (SPF), which is defined as the minimum effective dose of UV radiation that produces a delayed sunburn on protected skin compared with unprotected skin. Two types of active ingredients are found in commercial sunblocks and sunscreens. The most common sunblock agent is zinc oxide, ZnO, a white crystalline substance that reflects and scatters UV radiation. Sunscreens, the second type of active ingredient, absorb UV radiation and then reradiate it as heat. Sunscreens are most effective in screening out UVB radiation, but they do not screen out UVA radiation. Thus, they allow tanning, but prevent the UVB-associated damage. Given here are structural formulas for three common esters used as UVB-screening agents, along with the name by which each is most commonly listed in the "Active Ingredients" label on commercial products:

Octyl *p*-methoxycinnamate

Homosalate

Padimate A

C. Esters and Lactones

Esters of Carboxylic Acids

The functional group of a **carboxylic ester** (commonly referred to simply as an ester) is an acyl group bonded to —OR or —OAr. Both IUPAC and common names of esters are derived from the names of the parent carboxylic acids. The alkyl or aryl group bonded to oxygen is named first, followed by the name of the acid, in which the suffix -*ic acid* is replaced by the suffix -*ate*:

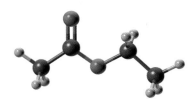

$$CH_3\overset{\text{O}}{\overset{\|}{C}}OCH_2CH_3$$

Ethyl ethanoate
(Ethyl acetate)

Diethyl butanedioate
(Diethyl succinate)

CHEMICAL CONNECTIONS 15B

From Moldy Clover to a Blood Thinner

In 1933, a disgruntled farmer delivered a pail of unclotted blood to the laboratory of Dr. Karl Link at the University of Wisconsin and told tales of cows bleeding to death from minor cuts. Over the next couple of years, Link and his collaborators discovered that when cows are fed moldy clover, their blood clotting is inhibited, and they bleed to death from minor cuts and scratches. From the moldy clover, Link isolated the anticoagulant dicoumarol, a substance that delays or prevents blood from clotting. Dicoumarol exerts its anticoagulation effect by interfering with vitamin K activity (Section 21.7D). Within a few years after its discovery, dicoumarol became widely used to treat victims of heart attack and others at risk for developing blood clots.

Dicoumarol is a derivative of coumarin, a cyclic ester that gives sweet clover its pleasant smell. Coumarin, which does not interfere with blood clotting and has been used as a flavoring agent, is converted to dicoumarol as sweet clover becomes moldy. Notice that coumarin is a lactone (cyclic ester) whereas dicoumarol is a dilactone:

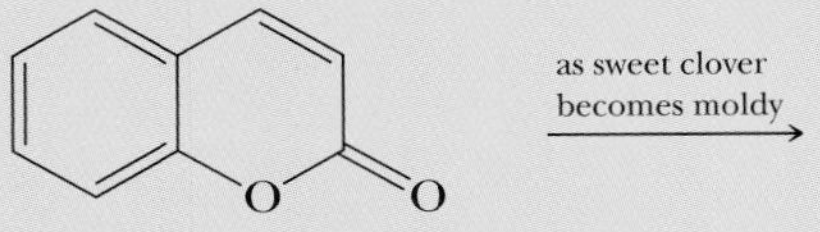

In a search for even more potent anticoagulants, Link developed warfarin (named after the Wisconsin Alumni Research Foundation), now used primarily as a rat poison: When rats consume warfarin, their blood fails to clot, and they bleed to death. Sold under the brand name Coumadin®, warfarin is also used as a blood thinner in humans. The *S* enantiomer is more active than the *R* enantiomer. The commercial product is a racemic mixture.

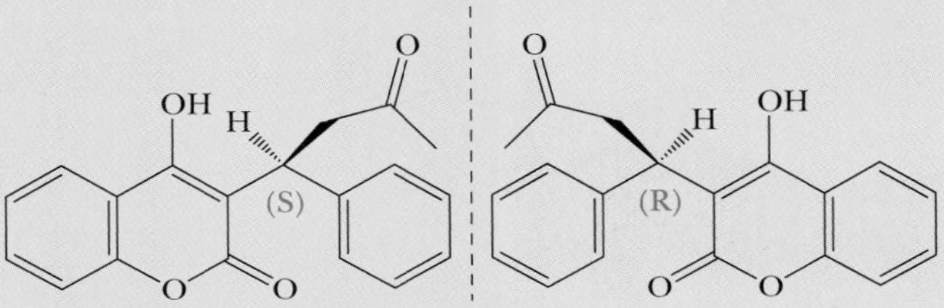

Warfarin
(a synthetic anticoagulant)

Coumarin
(from sweet clover)

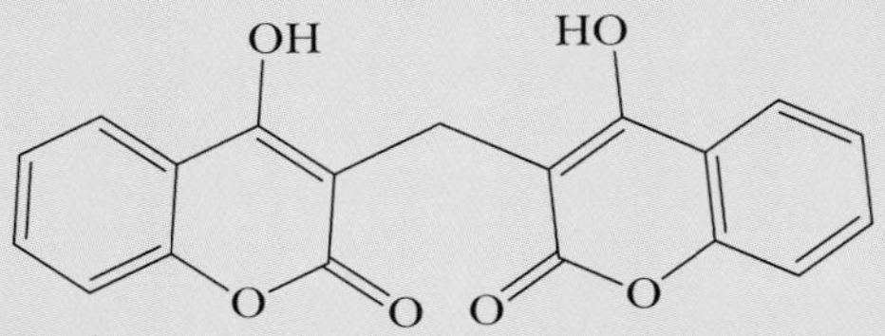

as sweet clover
becomes moldy
⟶

The powerful anticoagulant dicoumarol was first isolated from moldy clover. *(Grant Heilman/Grant Heilman Photography, Inc.)*

Dicoumarol
(an anticoagulant)

A cyclic ester is called a **lactone**. The IUPAC name of a lactone is formed by dropping the suffix *-oic acid* from the name of the parent carboxylic acid and adding the suffix *-olactone*. The common name is similarly derived. The location of

Lactone A cyclic ester.

the oxygen atom in the ring is indicated by a number if the IUPAC name of the acid is used and by a Greek letter α, β, γ, δ, ε, and so forth if the common name of the acid is used.

4-Butanolactone
(A γ-lactone)

Esters of Phosphoric Acid

Phosphoric acid has three —OH groups and forms mono-, di-, and triphosphoric esters, which are named by giving the name(s) of the alkyl or aryl group(s) bonded to oxygen, followed by the word *phosphate*—for example, dimethyl phosphate. In more complex phosphoric esters, it is common to name the organic molecule and then show the presence of the phosphoric ester by using either the word *phosphate* or the prefix *phospho-*. On the right are two phosphoric esters, each of special importance in the biological world. The first reaction in the metabolism of glucose is the formation of a phosphoric ester of D-glucose (Section 22.4), to give D-glucose 6-phosphate. Pyridoxal phosphate is one of the metabolically active forms of vitamin B_6. Each of these esters is shown as it is ionized at pH 7.4, the pH of blood plasma; the two hydrogens of each phosphate group are ionized, giving the phosphate group a charge of -2:

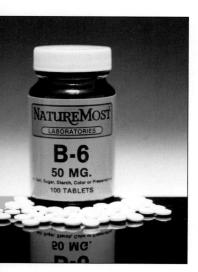

Vitamin B_6, pyridoxal.
(Charles D. Winters)

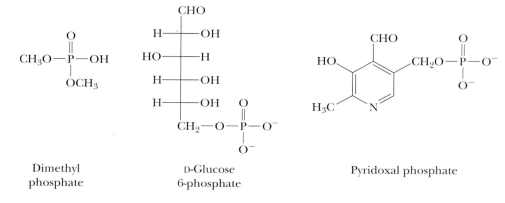

Dimethyl phosphate   D-Glucose 6-phosphate   Pyridoxal phosphate

D. Amides and Lactams

The functional group of an **amide** is an acyl group bonded to a trivalent nitrogen atom. Amides are named by dropping the suffix *-oic acid* from the IUPAC name of the parent acid, or *-ic acid* from its common name, and adding *-amide*. If the nitrogen atom of an amide is bonded to an alkyl or aryl group, the group is named and its location on nitrogen is indicated by *N-*. Two alkyl or aryl groups on nitrogen are indicated by *N,N*-di- if the groups are identical or by *N-alkyl-N-alkyl* if they are different:

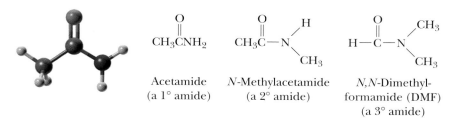

Acetamide   N-Methylacetamide   N,N-Dimethyl-
(a 1° amide)   (a 2° amide)   formamide (DMF)
                                              (a 3° amide)

Amide bonds are the key structural feature that joins amino acids together to form polypeptides and proteins (Chapter 19).

CHEMICAL CONNECTIONS 15C

The Penicillins and Cephalosporins: β-Lactam Antibiotics

The **penicillins** were discovered in 1928 by the Scottish bacteriologist Sir Alexander Fleming. As a result of the brilliant experimental work of Sir Howard Florey, an Australian pathologist, and Ernst Chain, a German chemist who fled Nazi Germany, penicillin G was introduced into the practice of medicine in 1943. For their pioneering work in developing one of the most effective antibiotics of all time, Fleming, Florey, and Chain were awarded the Nobel prize in medicine or physiology in 1945.

The mold from which Fleming discovered penicillin was *Penicillium notatum*, a strain that gives a relatively low yield of penicillin. Commercial production of the antibiotic uses *P. chrysogenum*, a strain cultured from a mold found growing on a grapefruit in a market in Peoria, Illinois. The penicillins owe their antibacterial activity to a common mechanism that inhibits the biosynthesis of a vital part of bacterial cell walls.

The structural feature common to all penicillins is a **β-lactam** ring fused to a five-membered ring containing one S atom and one N atom:

Soon after the penicillins were introduced into medical practice, penicillin-resistant strains of bacteria began to appear and have since proliferated. One approach to combating resistant strains is to synthesize newer, more effective penicillins. Among those which have been developed are ampicillin, methicillin, and amoxicillin. Another approach is to search for newer, more effective β-lactam antibiotics. The most effective of these discovered so far are the **cephalosporins**, the first of which was isolated from the fungus *Cephalosporium acremonium*. This class of β-lactam antibiotics has an even broader spectrum of antibacterial activity than the penicillins and is effective against many penicillin-resistant bacterial strains.

The cephalosporins differ in the group bonded to the carbonyl carbon...

...and the group bonded to this carbon of the six-membered ring

Keflex
(a β-lactam antibiotic)

The penicillins differ in the group bonded to the acyl carbon

β-lactam

Amoxicillin
(a β-lactam antibiotic)

Cyclic amides are given the special name **lactam**. Their common names are derived in a manner similar to those of lactones, with the difference that the suffix *-olactone* is replaced by *-olactam*:

Lactam A cyclic amide.

3-Butanolactam
(A β-lactam)

6-Hexanolactam
(An ε-lactone)

6-Hexanolactam is a key intermediate in the synthesis of nylon-6 (Section 17.5A).

EXAMPLE 15.1

Write the IUPAC name for each compound:

(a)

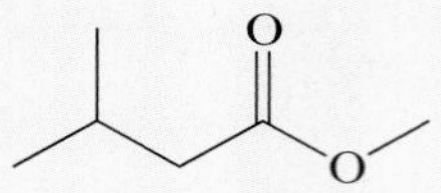

(b)

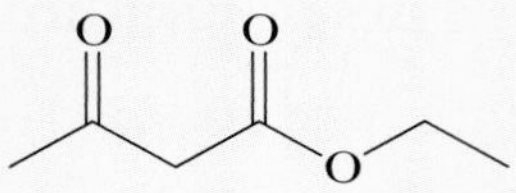

(c)

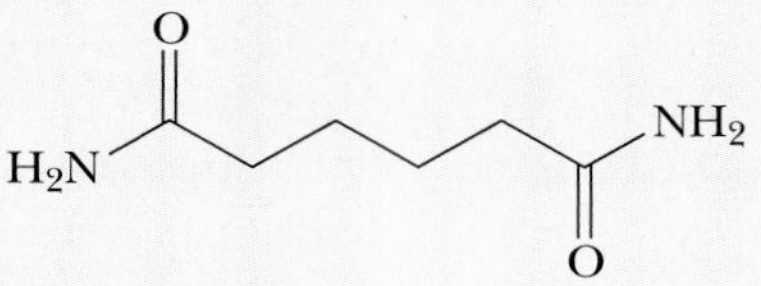

(d)

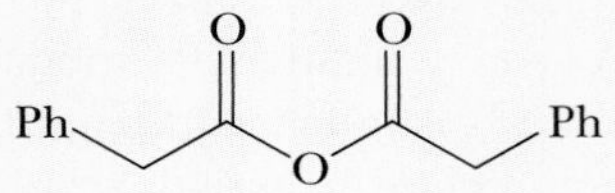

SOLUTION

Given first are IUPAC names and then, in parentheses, common names:

(a) Methyl 3-methylbutanoate (methyl isovalerate, from isovaleric acid)
(b) Ethyl 3-oxobutanoate (ethyl β-ketobutyrate, from β-ketobutyric acid)
(c) Hexanediamide (adipamide, from adipic acid)
(d) Phenylethanoic anhydride (phenylacetic anhydride, from phenylacetic acid)

Practice Problem 15.1

Draw a structural formula for each compound:

(a) *N*-Cyclohexylacetamide   (b) *sec*-Butyl acetate
(c) Cyclobutyl butanoate   (d) *N*-(2-Octyl)benzamide
(e) Diethyl adipate   (f) Propanoic anhydride

15.3 CHARACTERISTIC REACTIONS

The most common reaction theme of acid halides, anhydrides, esters, and amides is the addition of a nucleophile to the carbonyl carbon to form a tetrahedral carbonyl addition intermediate. To this extent, the reactions of these functional groups are similar to nucleophilic addition to the carbonyl groups in aldehydes and ketones (Section 13.5). The tetrahedral carbonyl addition intermediate formed from an aldehyde or a ketone then adds H^+. The result of this reaction is nucleophilic addition to a carbonyl group of an aldehyde or a ketone:

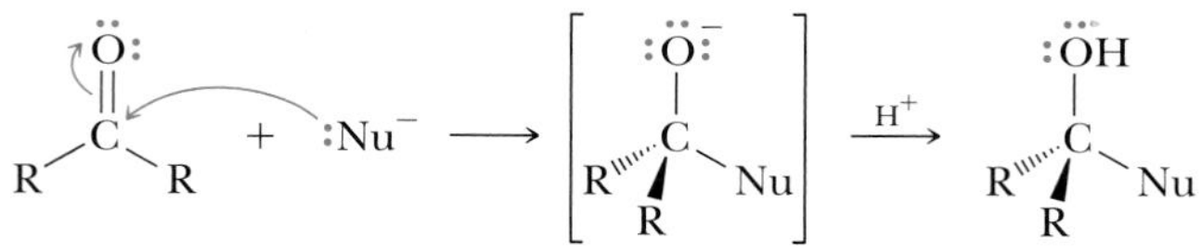

For functional derivatives of carboxylic acids, the fate of the tetrahedral carbonyl addition intermediate is quite different from that of aldehydes and ketones. This intermediate collapses to expel the leaving group and regenerate the carbonyl group. The result of this addition–elimination sequence is **nucleophilic acyl substitution**:

Nucleophilic acyl substitution
A reaction in which a nucleophile bonded to a carbonyl carbon is replaced by another nucleophile.

Nucleophilic acyl substitution:

$$
\underset{\substack{}}{R-\overset{\overset{\displaystyle \ddot{O}:}{\|}}{C}-Y} + :Nu^- \longrightarrow \left[R\overset{\overset{\displaystyle :\ddot{O}:^-}{|}}{\underset{\underset{\displaystyle Y}{|}}{C}}-Nu \right] \longrightarrow \underset{\substack{}}{R-\overset{\overset{\displaystyle \ddot{O}:}{\|}}{C}-Nu} + :Y^-
$$

Tetrahedral carbonyl   Substitution
addition intermediate   product

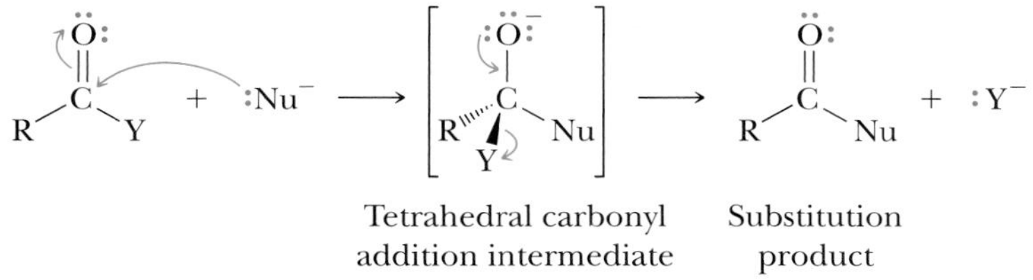

The major difference between these two types of carbonyl addition reactions is that aldehydes and ketones do not have a group, Y, that can leave as a stable anion. They undergo only nucleophilic acyl addition. The four carboxylic acid derivatives we study in this chapter do have a group, Y, that can leave as a stable anion; accordingly, they undergo nucleophilic acyl substitution.

In this general reaction, we show the nucleophile and the leaving group as anions. That need not be the case, however: Neutral molecules, such as water, alcohols, ammonia, and amines, may also serve as nucleophiles in the acid-catalyzed version of the reaction. We show the leaving groups here as anions to illustrate an important point about leaving groups, namely, that the weaker the base, the better is the leaving group (Section 7.6C):

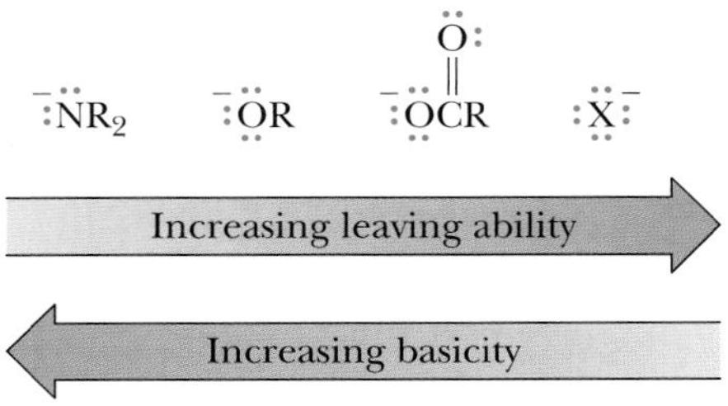

$$
:\ddot{N}R_2^- \qquad :\ddot{O}R^- \qquad :\ddot{O}\overset{\overset{\displaystyle \ddot{O}:}{\|}}{C}R^- \qquad :\ddot{X}:^-
$$

→ Increasing leaving ability →

← Increasing basicity ←

The weakest base in this series, and thus the best leaving group, is halide ion; acid halides are the most reactive toward nucleophilic acyl substitution. The strongest base, and hence the poorest leaving group, is amide ion; amides are the least reactive toward nucleophilic acyl substitution. Acid halides and acid anhydrides are so reactive that they are not found in nature. Esters and amides, however, are universally present.

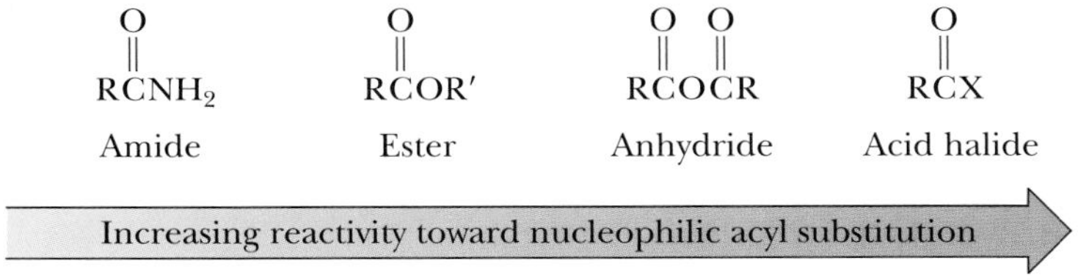

$$
\underset{\text{Amide}}{\overset{\overset{\displaystyle O}{\|}}{R C NH_2}} \qquad \underset{\text{Ester}}{\overset{\overset{\displaystyle O}{\|}}{R C O R'}} \qquad \underset{\text{Anhydride}}{\overset{\overset{\displaystyle O \quad O}{\| \quad \|}}{R C O C R}} \qquad \underset{\text{Acid halide}}{\overset{\overset{\displaystyle O}{\|}}{R C X}}
$$

→ Increasing reactivity toward nucleophilic acyl substitution →

15.4 REACTION WITH WATER: HYDROLYSIS

A. Acid Chlorides

Low-molecular-weight acid chlorides react very rapidly with water to form carboxylic acids and HCl:

$$
\underset{}{\overset{\overset{\displaystyle O}{\|}}{CH_3 C Cl}} + H_2O \longrightarrow \underset{}{\overset{\overset{\displaystyle O}{\|}}{CH_3 C OH}} + HCl
$$

Higher-molecular-weight acid chlorides are less soluble and consequently react less rapidly with water.

B. Acid Anhydrides

Acid anhydrides are generally less reactive than acid chlorides. The lower-molecular-weight anhydrides, however, react readily with water to form two molecules of carboxylic acid:

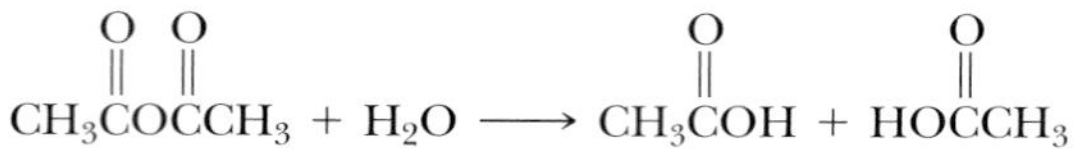

$$CH_3\overset{O}{\overset{\|}{C}}O\overset{O}{\overset{\|}{C}}CH_3 + H_2O \longrightarrow CH_3\overset{O}{\overset{\|}{C}}OH + HO\overset{O}{\overset{\|}{C}}CH_3$$

C. Esters

Esters are hydrolyzed only very slowly, even in boiling water. Hydrolysis becomes considerably more rapid, however, when esters are refluxed in aqueous acid or base. When we discussed acid-catalyzed (Fischer) esterification in Section 14.7, we pointed out that esterification is an equilibrium reaction. Hydrolysis of esters in aqueous acid is also an equilibrium reaction and proceeds by the same mechanism as esterification, except in reverse. The role of the acid catalyst is to protonate the carbonyl oxygen, thereby increasing the electrophilic character of the carbonyl carbon toward attack by water to form a tetrahedral carbonyl addition intermediate. Collapse of this intermediate gives the carboxylic acid and an alcohol. In this reaction, acid is a catalyst, it is consumed in the first step, but another is generated at the end of the reaction:

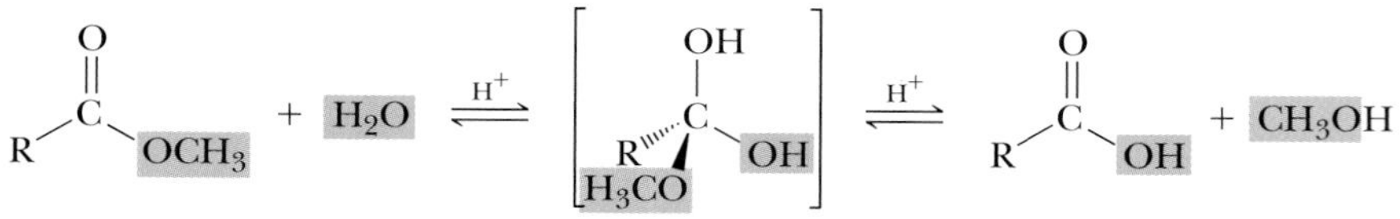

Tetrahedral carbonyl
addition intermediate

Hydrolysis of esters may also be carried out with hot aqueous base, such as aqueous NaOH. Hydrolysis of esters in aqueous base is often called **saponification**, a reference to the use of this reaction in the manufacture of soaps (Section 21.3A). Each mole of ester hydrolyzed requires 1 mole of base, as shown in the following balanced equation:

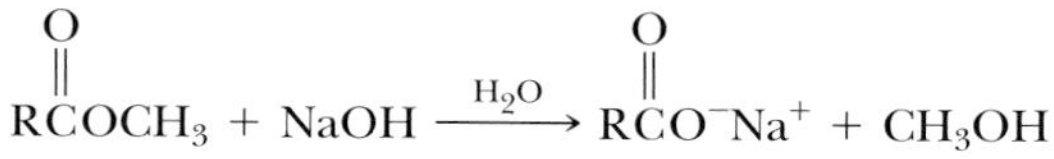

Saponification Hydrolysis of an ester in aqueous NaOH or KOH to an alcohol and the sodium or potassium salt of a carboxylic acid.

$$R\overset{O}{\overset{\|}{C}}OCH_3 + NaOH \xrightarrow{H_2O} R\overset{O}{\overset{\|}{C}}O^-Na^+ + CH_3OH$$

Mechanism: Hydrolysis of an Ester in Aqueous Base

Step 1: Addition of hydroxide ion to the carbonyl carbon of the ester gives a tetrahedral carbonyl addition intermediate:

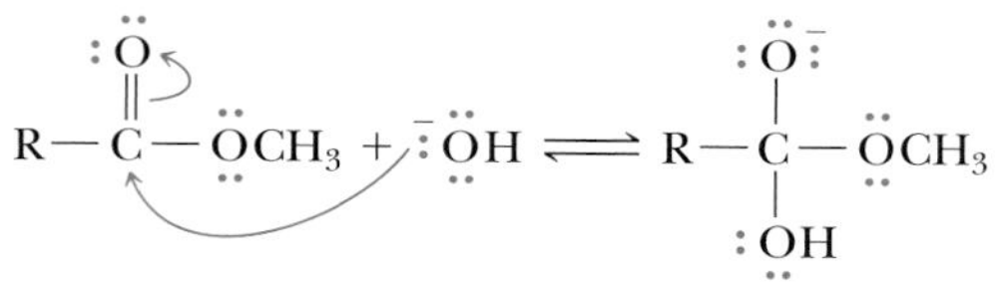

Step 2: Collapse of this intermediate gives a carboxylic acid and an alkoxide ion:

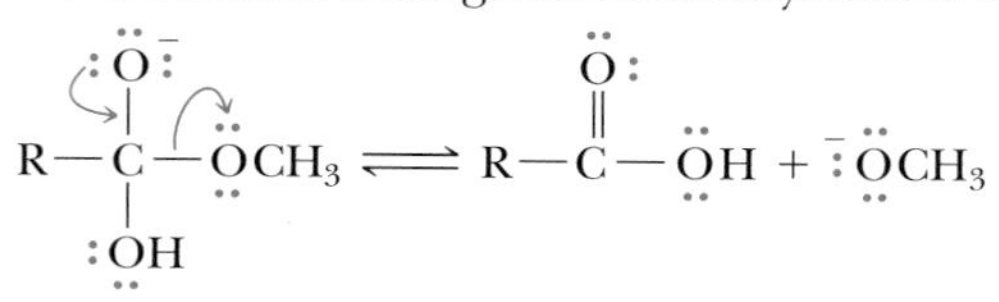

Step 3: Proton transfer from the carboxyl group (an acid) to the alkoxide ion (a base) gives the carboxylate anion. This step is irreversible because the alcohol is not a strong enough nucleophile to attack a carboxylate anion:

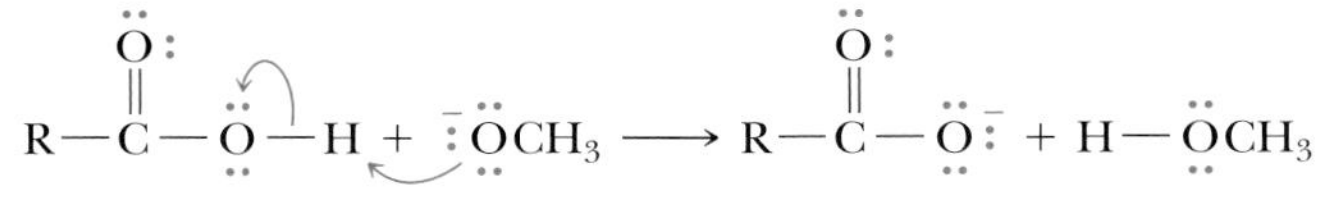

$$R-\overset{\overset{\displaystyle \ddot{O}:}{\|}}{C}-\overset{\cdot\cdot}{\underset{\cdot\cdot}{O}}-H \;+\; \overset{\cdot\cdot}{\underset{\cdot\cdot}{:}}\overset{\cdot\cdot}{O}CH_3 \longrightarrow R-\overset{\overset{\displaystyle \ddot{O}:}{\|}}{C}-\overset{\cdot\cdot}{\underset{\cdot\cdot}{O}}\overset{-}{:} \;+\; H-\overset{\cdot\cdot}{\underset{\cdot\cdot}{O}}CH_3$$

There are two major differences between the hydrolysis of esters in aqueous acid and that in aqueous base:

1. For hydrolysis in aqueous acid, acid is required in only catalytic amounts. For hydrolysis in aqueous base, base is required in equimolar amounts, because it is a reactant, not just a catalyst.
2. Hydrolysis of an ester in aqueous acid is reversible. Hydrolysis in aqueous base is irreversible, because a carboxylic acid anion is not attacked by ROH.

EXAMPLE 15.2

Complete and balance equations for the hydrolysis of each ester in aqueous sodium hydroxide, showing all products as they are ionized in aqueous NaOH:

(a)

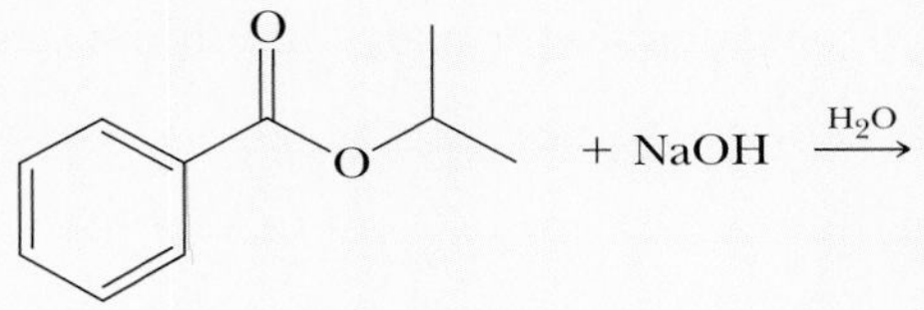

$+\ NaOH \quad \xrightarrow{\ H_2O\ }$

(b)

$+\ NaOH \quad \xrightarrow{\ H_2O\ }$

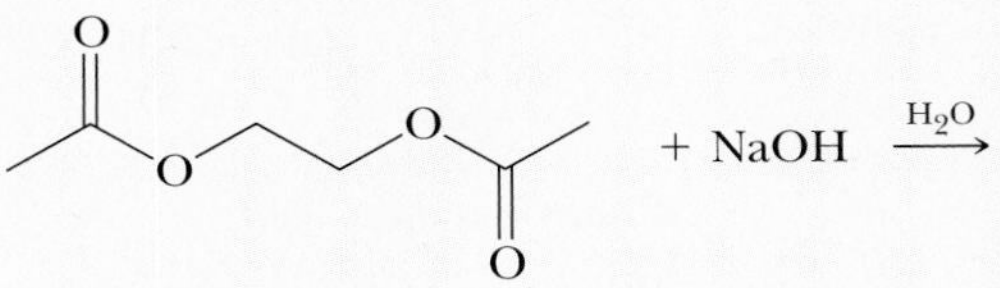

SOLUTION

The products of hydrolysis of (a) are benzoic acid and 2-propanol. In aqueous NaOH, benzoic acid is converted to its sodium salt. Therefore, 1 mole of NaOH is required for the hydrolysis of 1 mole of this ester. Compound (b) is a diester of ethylene glycol. Two moles of NaOH are required for its hydrolysis:

(a)

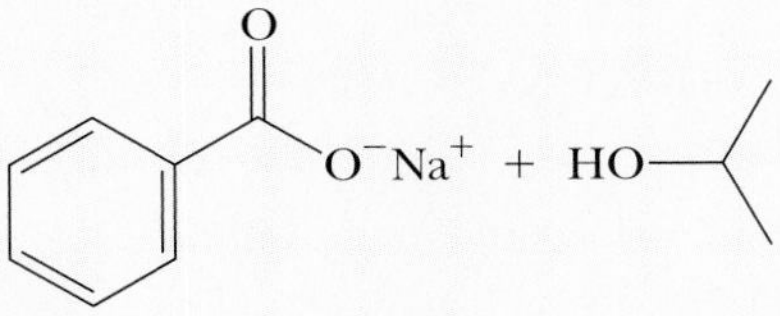

$O^-Na^+ \;+\; HO-$

    Sodium benzoate   2-Propanol
                    (Isopropyl alcohol)

(b) $2\ CH_3\overset{\overset{\displaystyle O}{\|}}{C}O^-Na^+ \quad + \quad HOCH_2CH_2OH$

    Sodium acetate   1,2-Ethanediol
                    (Ethylene glycol)

Practice Problem 15.2

Complete and balance equations for the hydrolysis of each ester in aqueous solution, showing each product as it is ionized under the given experimental conditions:

(a)

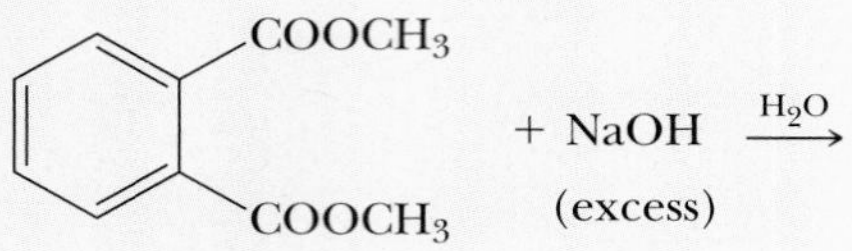

$\begin{array}{c}\text{COOCH}_3\\ \text{COOCH}_3\end{array}$ + NaOH $\xrightarrow{\text{H}_2\text{O}}$

(excess)

(b)

$+ \text{H}_2\text{O} \xrightarrow{\text{HCl}}$

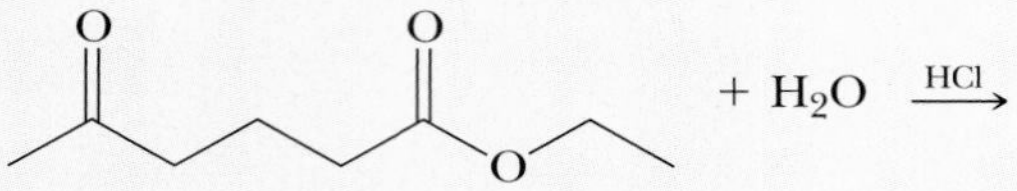

D. Amides

Amides require considerably more vigorous conditions for hydrolysis in both acid and base than do esters. Amides undergo hydrolysis in hot aqueous acid to give a carboxylic acid and ammonia. Hydrolysis is driven to completion by the acid–base reaction between ammonia or the amine and acid to form an ammonium salt. One mole of acid is required per mole of amide:

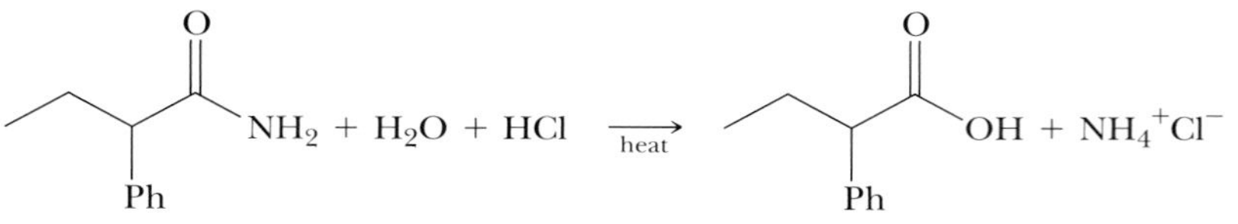

$$\underset{\substack{\text{Ph}\\ \text{2-Phenylbutanamide}}}{\text{NH}_2} + \text{H}_2\text{O} + \text{HCl} \xrightarrow{\text{heat}} \underset{\substack{\text{Ph}\\ \text{2-Phenylbutanoic acid}}}{\text{OH}} + \text{NH}_4{}^+\text{Cl}^-$$

In aqueous base, the products of amide hydrolysis are a carboxylic acid and ammonia or an amine. Base-catalyzed hydrolysis is driven to completion by the acid–base reaction between the carboxylic acid and base to form a salt. One mole of base is required per mole of amide:

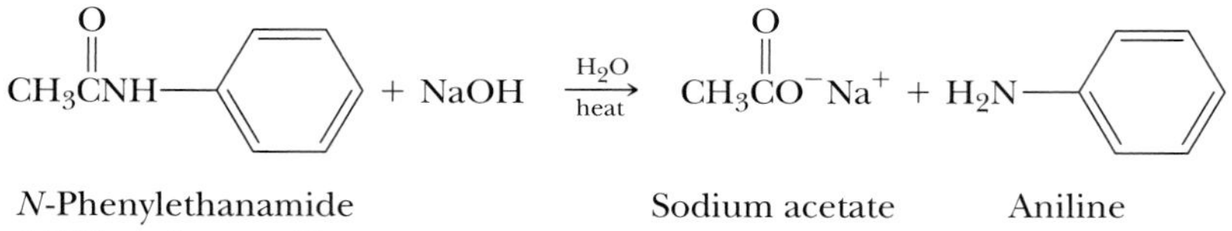

$$\underset{\substack{\text{N-Phenylethanamide}\\ \text{(N-Phenylacetamide,}\\ \text{Acetanilide)}}}{\text{CH}_3\overset{\text{O}}{\overset{\|}{\text{C}}}\text{NH}-\!\!\bigcirc} + \text{NaOH} \xrightarrow[\text{heat}]{\text{H}_2\text{O}} \underset{\text{Sodium acetate}}{\text{CH}_3\overset{\text{O}}{\overset{\|}{\text{C}}}\text{O}^-\text{Na}^+} + \underset{\text{Aniline}}{\text{H}_2\text{N}-\!\!\bigcirc}$$

The reactions of these functional groups with water are summarized in Table 15.1. Remember that, although all four functional groups react with water, there are large differences in the rates and experimental conditions under which they undergo hydrolysis.

TABLE 15.1 Summary of Reaction of Acid Chlorides, Anhydrides, Esters, and Amides with Water

$$R-\underset{\underset{O}{\parallel}}{C}-Cl + H_2O \longrightarrow R-\underset{\underset{O}{\parallel}}{C}-OH + HCl$$

$$R-\underset{\underset{O}{\parallel}}{C}-O-\underset{\underset{O}{\parallel}}{C}-R + H_2O \longrightarrow R-\underset{\underset{O}{\parallel}}{C}-OH + HO-\underset{\underset{O}{\parallel}}{C}-R$$

$$R-\underset{\underset{O}{\parallel}}{C}-OR' + H_2O \quad \begin{cases} \xrightarrow{NaOH} & R-\underset{\underset{O}{\parallel}}{C}-O^-Na^+ + R'OH \\ \xrightarrow{H_2SO_4} & R-\underset{\underset{O}{\parallel}}{C}-OH + R'OH \end{cases}$$

$$R-\underset{\underset{O}{\parallel}}{C}-NH_2 + H_2O \quad \begin{cases} \xrightarrow{NaOH} & R-\underset{\underset{O}{\parallel}}{C}-O^-Na^+ + NH_3 \\ \xrightarrow{HCl} & R-\underset{\underset{O}{\parallel}}{C}-OH + NH_4^+Cl^- \end{cases}$$

EXAMPLE 15.3

Write equations for the hydrolysis of these amides in concentrated aqueous HCl, showing all products as they exist in aqueous HCl and showing the number of moles of HCl required for the hydrolysis of each amide:

(a) $CH_3\underset{\underset{O}{\parallel}}{C}N(CH_3)_2$

(b) [structure: six-membered ring lactam with C=O and NH]

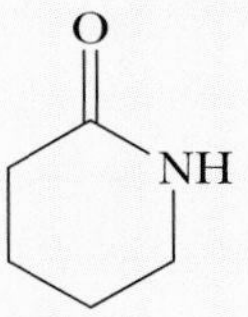

SOLUTION

(a) Hydrolysis of N,N-dimethylacetamide gives acetic acid and dimethylamine. Dimethylamine, a base, is protonated by HCl to form dimethylammonium ion and is shown in the balanced equation as dimethylammonium chloride. Complete hydrolysis of this amide requires 1 mole of HCl for each mole of the amide:

$$CH_3\underset{\underset{O}{\parallel}}{C}N(CH_3)_2 + H_2O + HCl \xrightarrow{heat} CH_3\underset{\underset{O}{\parallel}}{C}OH + (CH_3)_2NH_2^+Cl^-$$

(b) Hydrolysis of this δ-lactam gives the protonated form of 5-aminopentanoic acid. One mole of acid is required per mole of lactam:

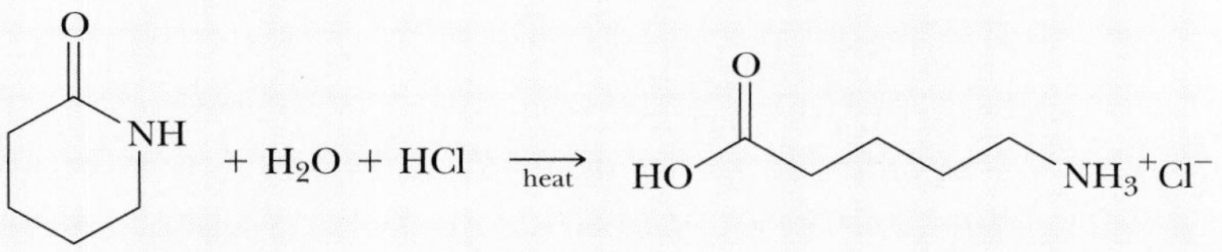

[structure: six-membered ring lactam] $+ H_2O + HCl \xrightarrow{heat}$ $HO-\underset{\underset{O}{\parallel}}{C}-CH_2CH_2CH_2CH_2-NH_3^+Cl^-$

Practice Problem 15.3

Complete equations for the hydrolysis of the amides in Example 15.3 in concentrated aqueous NaOH. Show all products as they exist in aqueous NaOH, and show the number of moles of NaOH required for the hydrolysis of each amide.

15.5 REACTION WITH ALCOHOLS

A. Acid Chlorides

Acid chlorides react with alcohols to give an ester and HCl:

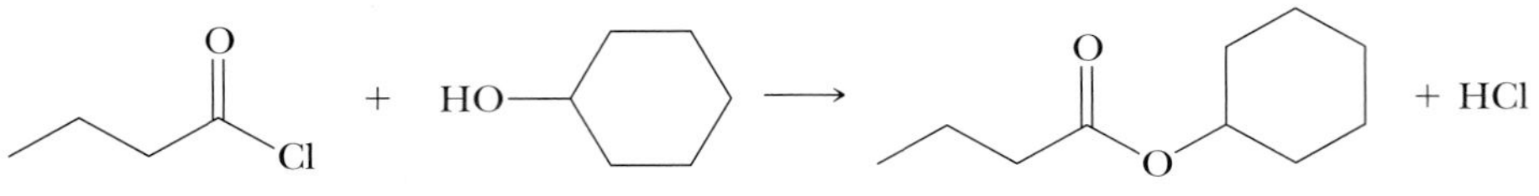

Butanoyl chloride — Cyclohexanol — Cyclohexyl butanoate

Because acid chlorides are so reactive toward even weak nucleophiles such as alcohols, no catalyst is necessary for these reactions. Phenol and substituted phenols also react with acid chlorides to give esters.

B. Acid Anhydrides

Acid anhydrides react with alcohols to give 1 mole of ester and 1 mole of a carboxylic acid.

$$CH_3\overset{\overset{O}{\|}}{C}O\overset{\overset{O}{\|}}{C}CH_3 + HOCH_2CH_3 \longrightarrow CH_3\overset{\overset{O}{\|}}{C}OCH_2CH_3 + CH_3\overset{\overset{O}{\|}}{C}OH$$

Acetic anhydride — Ethanol — Ethyl acetate — Acetic acid

Thus, the reaction of an alcohol with an anhydride is a useful method for synthesizing esters. Aspirin is synthesized on an industrial scale by reacting acetic anhydride with salicylic acid:

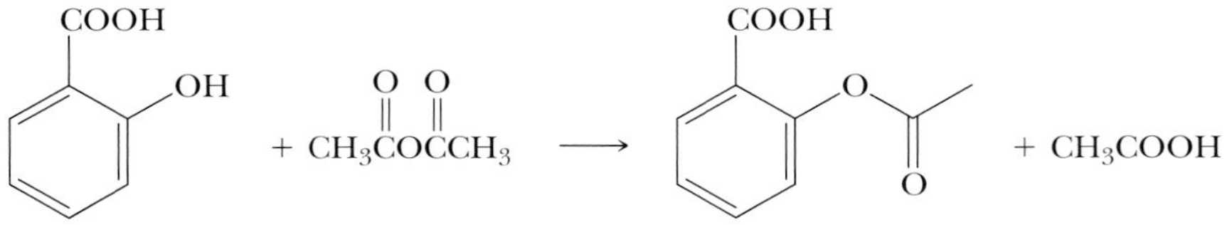

2-Hydroxybenzoic acid (Salicylic acid) — Acetic anhydride — Acetylsalicylic acid (Aspirin) — Acetic acid

C. Esters

When treated with an alcohol in the presence of an acid catalyst, esters undergo an exchange reaction called **transesterification**. In this reaction, the original —OR group of the ester is exchanged for a new —OR group. In the following example, the transesterification can be driven to completion by heating the reaction at a temperature above the boiling point of methanol (65°C) so that methanol distills from the reaction mixture:

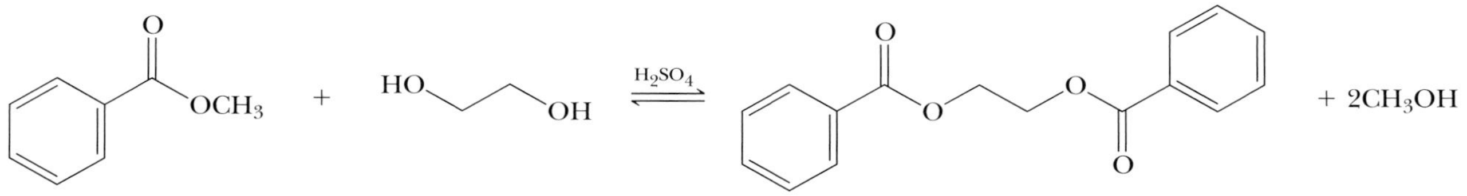

Methyl benzoate | 1,2-Ethanediol (Ethylene glycol) | (A diester of ethylene glycol)

D. Amides

Amides do not react with alcohols under any experimental conditions. Alcohols are not strong enough nucleophiles to attack the carbonyl group of an amide.

The reactions of the foregoing functional groups with alcohols are summarized in Table 15.2. As with reactions of these same functional groups with water, there are large differences in the rates and experimental conditions under which they undergo reactions with alcohols. At one extreme are acid chlorides and anhydrides, which react rapidly; at the other extreme are amides, which do not react at all.

TABLE 15.2 Summary of Reaction of Acid Chlorides, Anhydrides, Esters, and Amides with Alcohols

$$R{-}\overset{O}{\overset{\|}{C}}{-}Cl + HOR'' \longrightarrow R{-}\overset{O}{\overset{\|}{C}}{-}OR'' + HCl$$

$$R{-}\overset{O}{\overset{\|}{C}}{-}O{-}\overset{O}{\overset{\|}{C}}{-}R + R''OH \longrightarrow R{-}\overset{O}{\overset{\|}{C}}{-}OR'' + HO{-}\overset{O}{\overset{\|}{C}}{-}R$$

$$R{-}\overset{O}{\overset{\|}{C}}{-}OR' + R''OH \underset{}{\overset{H_2SO_4}{\rightleftharpoons}} R{-}\overset{O}{\overset{\|}{C}}{-}OR'' + R'OH$$

$$R{-}\overset{O}{\overset{\|}{C}}{-}NH_2 + R''OH \longrightarrow \text{No Reaction}$$

15.6 REACTIONS WITH AMMONIA AND AMINES

A. Acid Chlorides

Acid chlorides react readily with ammonia and with 1° and 2° amines to form amides. Complete conversion of an acid chloride to an amide requires 2 moles of ammonia or amine: one to form the amide and one to neutralize the hydrogen chloride formed:

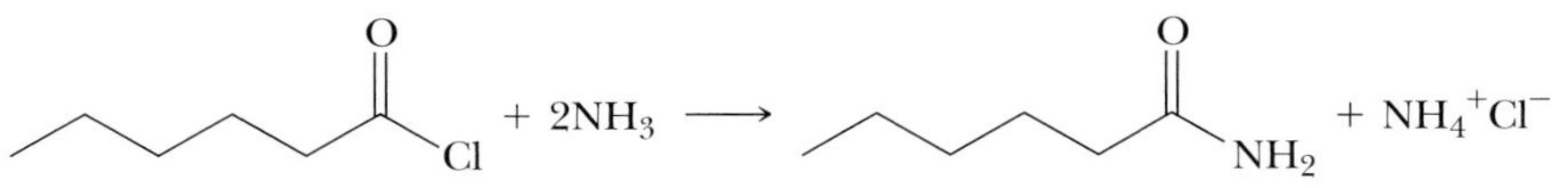

Hexanoyl chloride | Ammonia | Hexanamide | Ammonium chloride

B. Acid Anhydrides

Acid anhydrides react with ammonia and with 1° and 2° amines to form amides. As with acid chlorides, 2 moles of ammonia or amine are required—one to form the amide and one to neutralize the carboxylic acid by-product. To help you see what happens, this reaction is broken into two steps, which, when added together, give the net reaction for the reaction of an anhydride with ammonia:

$$\underset{\substack{\| \ \ \| \\ O \ \ O}}{CH_3COCCH_3} + NH_3 \longrightarrow \underset{\substack{\| \\ O}}{CH_3CNH_2} + \boxed{\underset{\substack{\| \\ O}}{CH_3COH}}$$

$$\boxed{\underset{\substack{\| \\ O}}{CH_3COH}} + NH_3 \longrightarrow \underset{\substack{\| \\ O}}{CH_3CO^-NH_4^+}$$

$$\underset{\substack{\| \ \ \| \\ O \ \ O}}{CH_3COCCH_3} + 2NH_3 \longrightarrow \underset{\substack{\| \\ O}}{CH_3CNH_2} + \underset{\substack{\| \\ O}}{CH_3CO^-NH_4{}^+}$$

C. Esters

Esters react with ammonia and with 1° and 2° amines to form amides:

$$Ph\underset{\substack{\| \\ O}}{\diagup\!\!\diagdown\!\!C}\diagdown O\diagup + NH_3 \longrightarrow Ph\underset{\substack{\| \\ O}}{\diagup\!\!\diagdown\!\!C}\diagdown NH_2 + HO\diagup\!\!\diagdown$$

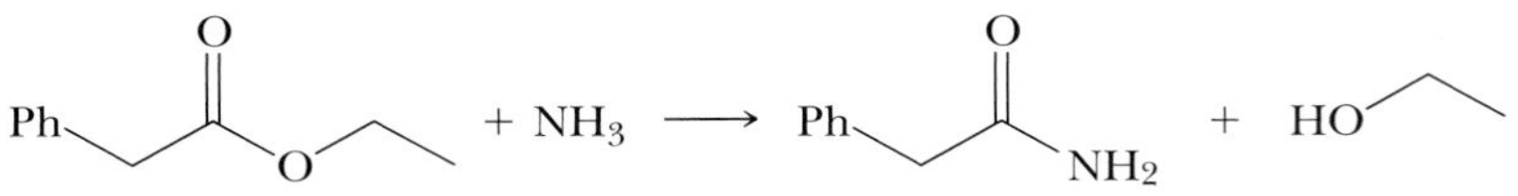

Ethyl phenylacetate   Phenylacetamide   Ethanol

Because an alkoxide anion is a poor leaving group compared with a halide or carboxylic ion, esters are less reactive toward ammonia, 1° amines, and 2° amines than are acid chlorides or acid anhydrides.

D. Amides

Amides do not react with ammonia or amines.

The reactions of the preceding four functional groups with ammonia and amines are summarized in Table 15.3.

TABLE 15.3   Summary of Reaction of Acid Chlorides, Anhydrides, Esters, and Amides with Ammonia and Amines

$$R\underset{\substack{\| \\ O}}{-C}-Cl + 2NH_3 \longrightarrow R\underset{\substack{\| \\ O}}{-C}-NH_2 + NH_4{}^+\,Cl^-$$

$$R\underset{\substack{\| \\ O}}{-C}-O\underset{\substack{\| \\ O}}{-C}-R + 2NH_3 \longrightarrow R\underset{\substack{\| \\ O}}{-C}-NH_2 + R\underset{\substack{\| \\ O}}{-C}-O^-NH_4{}^+$$

$$R\underset{\substack{\| \\ O}}{-C}-OR' + NH_3 \rightleftharpoons R\underset{\substack{\| \\ O}}{-C}-NH_2 + R'OH$$

$$R\underset{\substack{\| \\ O}}{-C}-NH_2 \text{ No reaction with ammonia or amines}$$

EXAMPLE 15.4

Complete these equations (the stoichiometry of each is given in the equation):

(a) 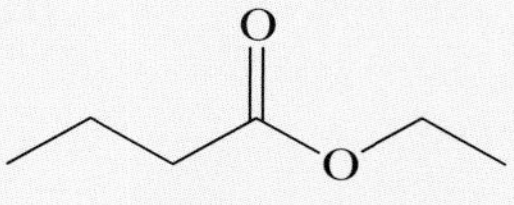 [structure of Ethyl butanoate] $+ \; NH_3 \longrightarrow$

Ethyl butanoate

(b) 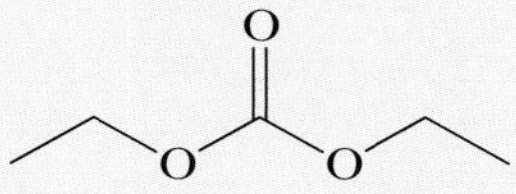 [structure of Diethyl carbonate] $+ \; 2NH_3 \longrightarrow$

Diethyl carbonate

SOLUTION

(a) 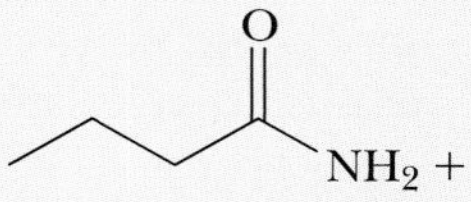 [structure] $NH_2 + CH_3CH_2OH$ (b) 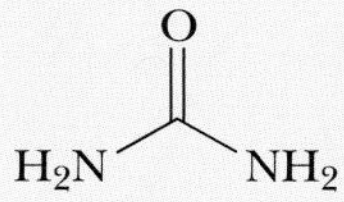 H_2N [structure] $NH_2 + 2CH_3CH_2OH$

Butanamide   Urea

Practice Problem 15.4

Complete these equations (the stoichiometry of each is given in the equation):

(a) 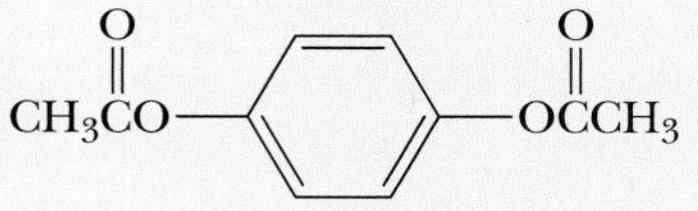 $CH_3\overset{O}{\overset{\|}{C}}O$—[benzene ring]—$O\overset{O}{\overset{\|}{C}}CH_3 + 2NH_3 \longrightarrow$ (b) 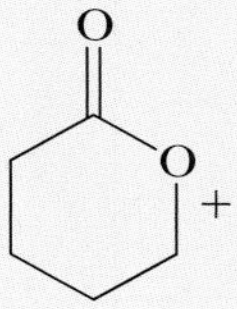 [cyclic lactone structure] $+ \; NH_3 \longrightarrow$

15.7 INTERCONVERSION OF FUNCTIONAL DERIVATIVES

In the last few sections, we have seen that acid chlorides are the most reactive carboxyl derivatives toward nucleophilic acyl substitution and that amides are the least reactive:

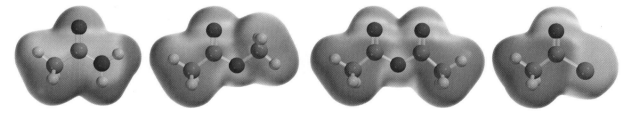

Amide < Ester < Acid anhydride < Acid halide

Increasing reactivity toward nucleophilic acyl substitution

Another useful way to think about the relative reactivities of these four functional derivatives of carboxylic acids is summarized in Figure 15.1. Any functional group in this figure can be prepared from any functional group above it by treatment with an appropriate oxygen or nitrogen nucleophile. An acid chloride, for example, can be converted to an acid anhydride, an ester, an amide, or a carboxylic acid. An acid anhydride, ester, or amide, however, does not react with chloride ion to give an acid chloride.

CHEMICAL CONNECTIONS 15D

Systematic Acquired Resistance in Plants

The use of germicides to protect plants from harmful pathogens is common in farming. Recently, plant physiologists discovered that some plant species are able to generate their own defenses against pathogens. The tobacco mosaic virus (TMV), for example, is a particularly devastating pathogen for plants such as tobacco, cucumber, and tomato.

The tobacco plant, *Nicotiana tobacum.*
(Inga Spence/Index Stock)

Scientists have found that certain strains of these plants produce large amounts of salicylic acid upon being infected with TMV. Accompanying the infection is the appearance of lesions on the leaves of the plants, which help to contain the infection to those localized areas. Furthermore, scientists have discovered that neighboring plants tend to acquire some resistance to TMV. It appears that the infected plant somehow signals neighboring plants of the impending danger by converting salicylic acid to its ester methyl salicylate:

Salicylic acid   Methyl salicylate

With a lower boiling point and higher vapor pressure than salicylic acid has, the methyl salicylate diffuses into the air from the infected plant, and the surrounding plants use it as a signal to enhance their defenses against TMV.

Notice that all carboxylic acid derivatives can be converted to carboxylic acids, which in turn can be converted to acid chlorides. Thus, any acid derivative can be used to synthesize another, either directly or via a carboxylic acid.

Figure 15.1

Relative reactivities of carboxylic acid derivatives toward nucleophilic acyl substitution. A more reactive derivative may be converted to a less reactive derivative by treatment with an appropriate reagent. Treatment of a carboxylic acid with thionyl chloride converts the carboxylic acid to the more reactive acid chloride. Carboxylic acids are about as reactive as esters under acidic conditions, but are converted to the unreactive carboxylate anions under basic conditions.

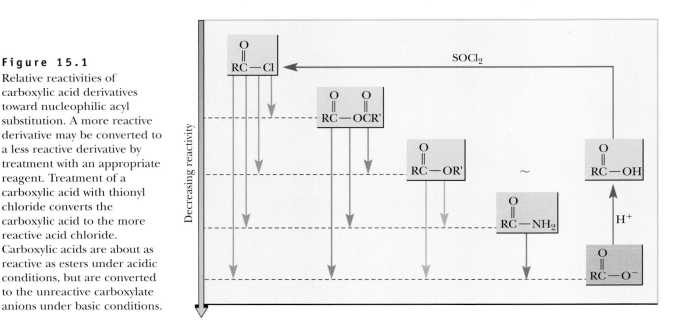

15.8 ESTERS WITH GRIGNARD REAGENTS

Treating a formic ester with 2 moles of a Grignard reagent, followed by hydrolysis of the magnesium alkoxide salt in aqueous acid, gives a 2° alcohol, whereas treating an ester other than a formate with a Grignard reagent gives a 3° alcohol in which two of the groups bonded to the carbon bearing the —OH group are the same:

$$
\underset{\substack{\text{An ester of}\\\text{formic acid}}}{\text{HCOCH}_3} + 2\text{RMgX} \longrightarrow \underset{\substack{\text{magnesium}\\\text{alkoxide}\\\text{salt}}}{} \xrightarrow{\text{H}_2\text{O, HCl}} \underset{\substack{\text{A 2° alcohol}}}{\text{HC}-\text{R}} + \text{CH}_3\text{OH}
$$

$$
\underset{\substack{\text{An ester}\\\text{other than formic acid}}}{\text{CH}_3\text{COCH}_3} + 2\text{RMgX} \longrightarrow \underset{\substack{\text{magnesium}\\\text{alkoxide}\\\text{salt}}}{} \xrightarrow{\text{H}_2\text{O, HCl}} \underset{\substack{\text{A 3° alcohol}}}{\text{CH}_3\text{C}-\text{R}} + \text{CH}_3\text{OH}
$$

Reaction of an ester with a Grignard reagent involves the formation of two successive tetrahedral carbonyl addition compounds. The first collapses to give a new carbonyl compound—an aldehyde from a formic ester, a ketone from all other esters. The second intermediate is stable and, when protonated, gives the final alcohol. It is important to realize that it is not possible to use RMgX and an ester to prepare an aldehyde or a ketone: The intermediate aldehyde or ketone is more reactive than the ester and reacts immediately with the Grignard reagent to give a tertiary alcohol.

Mechanism: Reaction of an Ester with a Grignard Reagent

Steps 1 and 2: Reaction begins in Step 1 with the addition of 1 mole of Grignard reagent to the carbonyl carbon to form a tetrahedral carbonyl addition intermediate. This intermediate then collapses in Step 2 to give a new carbonyl-containing compound and a magnesium alkoxide salt:

A magnesium salt
(a tetrahedral carbonyl
addition intermediate)

A ketone

Steps 3 and 4: The new carbonyl-containing compound reacts in Step 3 with a second mole of Grignard reagent to form a second tetrahedral carbonyl addition compound, which, after hydrolysis in aqueous acid (Step 4), gives a 3° alcohol (or a 2° alcohol if the starting ester was a formate):

A ketone   Magnesium salt   A 3° alcohol

EXAMPLE 15.5

Complete each Grignard reaction:

(a) 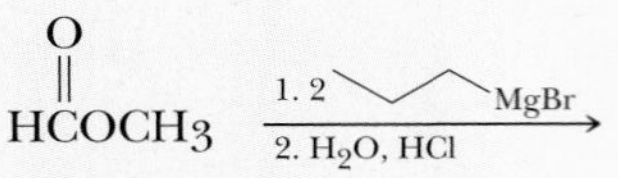 HCOCH₃ $\xrightarrow[\text{2. H}_2\text{O, HCl}]{\text{1. 2}\quad\diagdown\!\diagup\!\diagdown\text{MgBr}}$ (b) 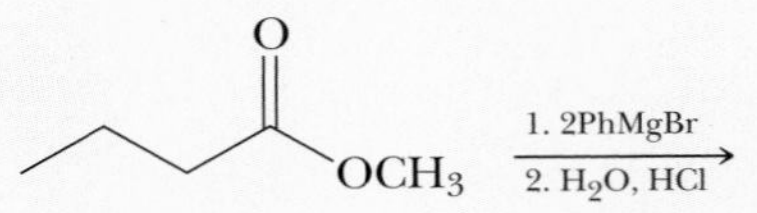 $\diagdown\!\diagup\!\diagdown\!\diagup\text{OCH}_3$ $\xrightarrow[\text{2. H}_2\text{O, HCl}]{\text{1. 2PhMgBr}}$

SOLUTION

Sequence (a) gives a 2° alcohol, and sequence (b) gives a 3° alcohol:

(a) 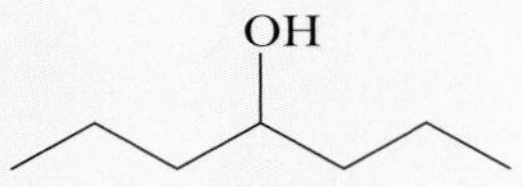 [structure with OH]   (b) 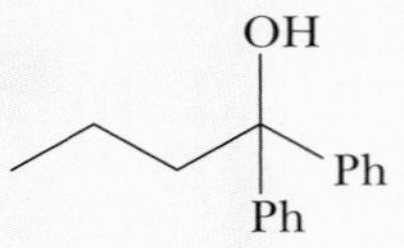 [structure with OH, Ph, Ph]

Practice Problem 15.5

Show how to prepare each alcohol by treating an ester with a Grignard reagent:

(a) 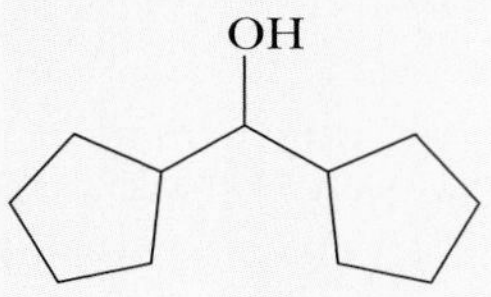 [bicyclopentyl carbinol structure with OH]   (b) 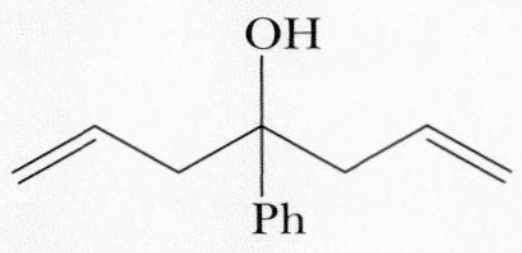 [diallyl-phenyl carbinol structure with OH, Ph]

15.9 REDUCTION

Most reductions of carbonyl compounds, including aldehydes and ketones, are now accomplished by transferring hydride ions from boron or aluminum hydrides. We have already seen the use of sodium borohydride to reduce the carbonyl groups of aldehydes and ketones to hydroxyl groups (Section 13.11B). We have also seen the use of lithium aluminum hydride to reduce not only the carbonyl groups of aldehydes and ketones, but also carboxyl groups (Section 14.6A), to hydroxyl groups.

A. Esters

An ester is reduced by lithium aluminum hydride to two alcohols. The alcohol derived from the acyl group is primary:

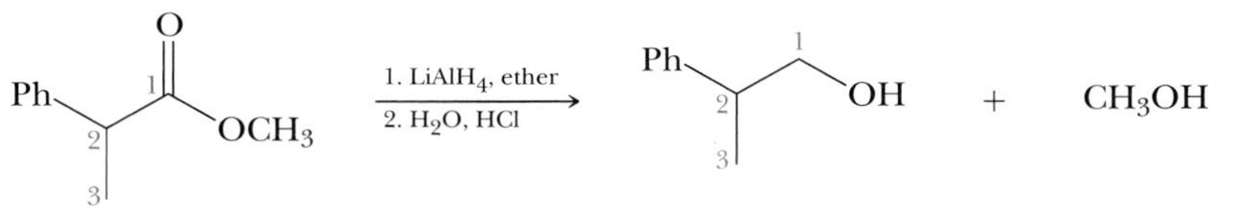

Ph $\diagup$ [C=O, OCH₃ structure] $\xrightarrow[\text{2. H}_2\text{O, HCl}]{\text{1. LiAlH}_4\text{, ether}}$ Ph $\diagup$ [OH structure] + CH₃OH

Methyl 2-phenyl-   2-Phenyl-1-propanol   Methanol
propanoate   (a 1° alcohol)

Sodium borohydride is not normally used to reduce esters, because the reaction is very slow. Because of this lower reactivity of sodium borohydride toward esters, it is possible to reduce the carbonyl group of an aldehyde or a ketone to a hydroxyl group with this reagent without reducing an ester or carboxyl group in the same molecule:

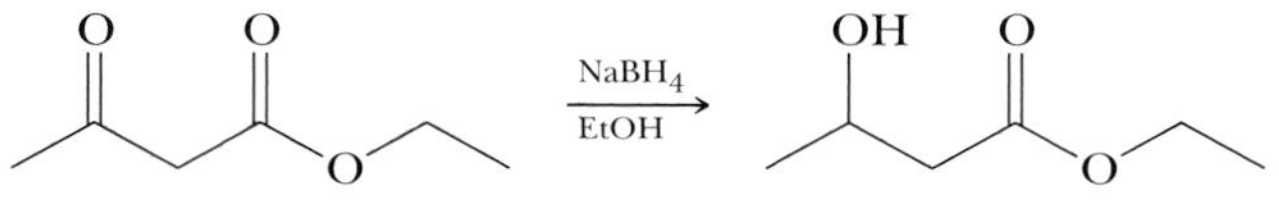

[structure with two C=O] $\xrightarrow[\text{EtOH}]{\text{NaBH}_4}$ [structure with OH and C=O]

B. Amides

Reduction of amides by lithium aluminum hydride can be used to prepare 1°, 2°, or 3° amines, depending on the degree of substitution of the amide:

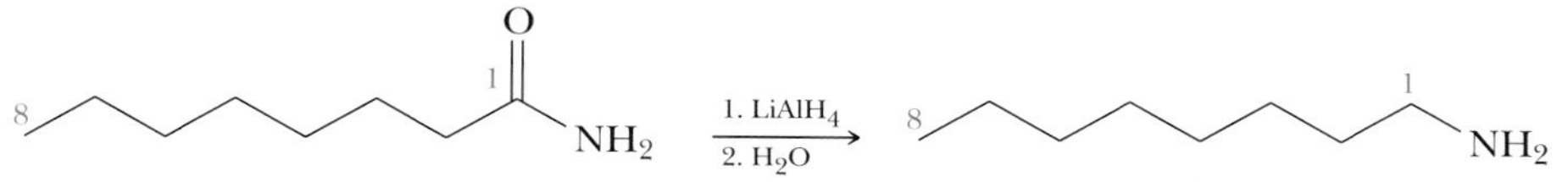

Octanamide → 1-Octanamine (a 1° amine)

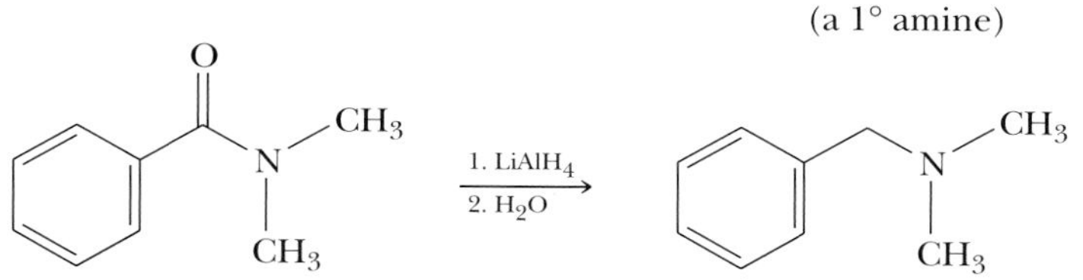

N,N-Dimethylbenzamide → *N,N*-Dimethylbenzylamine (a 3° amine)

EXAMPLE 15.6

Show how to bring about each conversion:

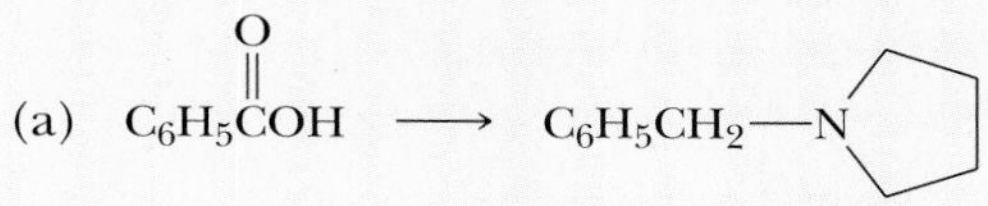

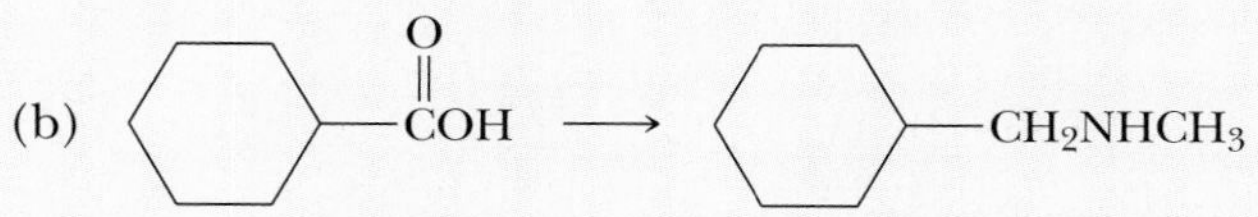

(a) $C_6H_5COH \longrightarrow C_6H_5CH_2-N$ (pyrrolidine)

(b) cyclohexyl-COH → cyclohexyl-CH₂NHCH₃

SOLUTION

The key in each part is to convert the carboxylic acid to an amide and then reduce the amide with LiAlH₄ (Section 15.9B). The amide can be prepared by treating the carboxylic acid with SOCl₂ to form the acid chloride (Section 14.8) and then treating the acid chloride with an amine (Section 15.6A). Alternatively, the carboxylic acid can be converted to an ester by Fischer esterification (Section 14.7) and the ester treated with an amine to give the amide. Solution (a) uses the acid chloride route, solution (b) the ester route:

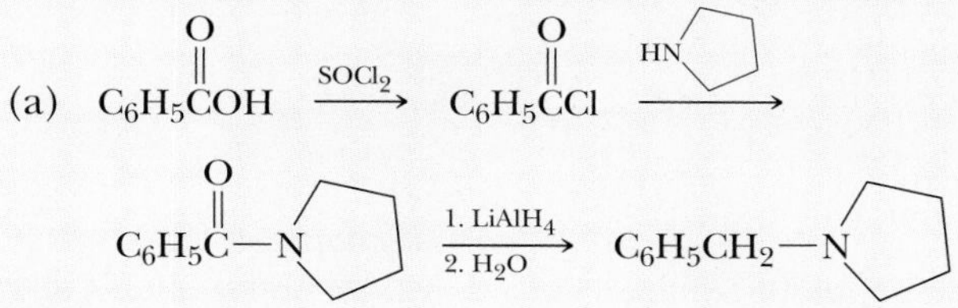

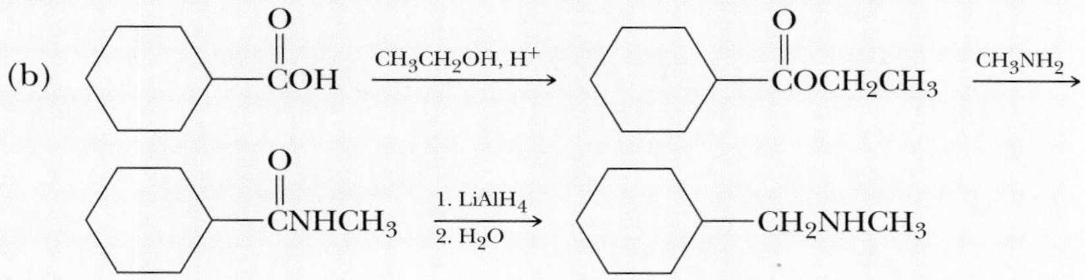

(a) $C_6H_5COH \xrightarrow{SOCl_2} C_6H_5CCl \xrightarrow{HN} C_6H_5C-N \xrightarrow[\text{2. H}_2\text{O}]{\text{1. LiAlH}_4} C_6H_5CH_2-N$

(b) cyclohexyl-COH $\xrightarrow{CH_3CH_2OH, H^+}$ cyclohexyl-COCH₂CH₃ $\xrightarrow{CH_3NH_2}$ cyclohexyl-CNHCH₃ $\xrightarrow[\text{2. H}_2\text{O}]{\text{1. LiAlH}_4}$ cyclohexyl-CH₂NHCH₃

Practice Problem 15.6

Show how to convert hexanoic acid to each amine in good yield:

(a) 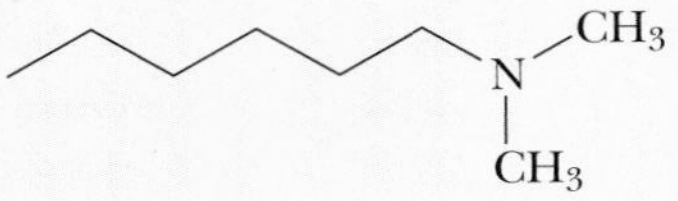 [structure: chain with N bearing two CH₃ groups]

(b) 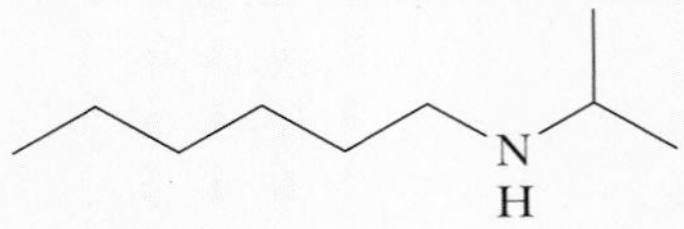 [structure: chain with N—H bearing isopropyl group]

EXAMPLE 15.7

Show how to convert phenylacetic acid to these compounds:

(a) 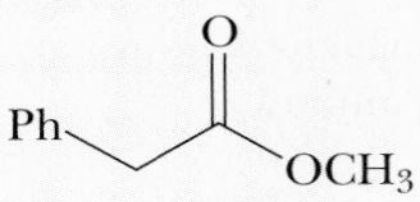 Ph—CH₂—C(=O)—OCH₃

(b) 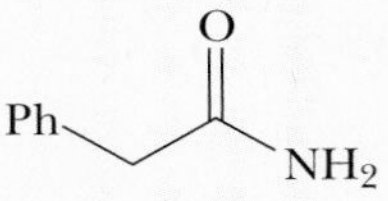 Ph—CH₂—C(=O)—NH₂

(c) Ph—CH₂—CH₂—NH₂

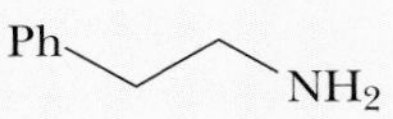

(d) 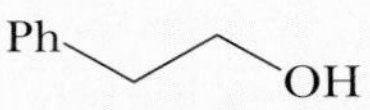 Ph—CH₂—CH₂—OH

SOLUTION

Prepare methyl ester (a) by Fischer esterification (Section 14.7) of phenylacetic acid with methanol. Then treat this ester with ammonia to prepare amide (b). Alternatively, treat phenylacetic acid with thionyl chloride (Section 14.8) to give an acid chloride, and then treat the acid chloride with two equivalents of ammonia to give amide (b). Reduction of amide (b) by LiAlH₄ gives the 1° amine (c). Similar reduction of either phenylacetic acid or ester (a) gives 1° alcohol (d):

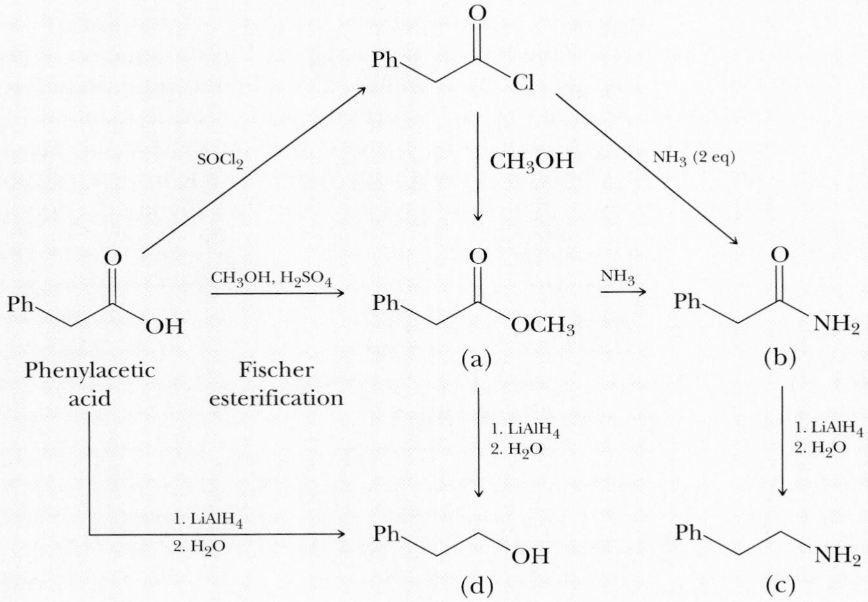

Practice Problem 15.7

Show how to convert (R)-2-phenylpropanoic acid to these compounds:

(a)

H₃C   H

Ph   OH

(R)-2-Phenyl-1-propanol

(b)

H₃C

Ph

(R)-2-Phenyl-

SUMMARY

The functional group of an **acid halide** (Section 15.2A) is an acyl group bonded to a halogen. The most common and widely used of the acid halides are the acid chlorides. The functional group of a **carboxylic anhydride** (Section 15.2B) is two acyl groups bonded to an oxygen. The functional group of a **carboxylic ester** (Section 15.2C) is an acyl group bonded to —OR or —OAr. A cyclic ester is given the name **lactone**. Phosphoric acid has three —OH groups and can form mono-, di-, and tri-esters. The functional group of an **amide** (Section 15.2D) is an acyl group bonded to a trivalent nitrogen. A cyclic amide is given the name **lactam**.

A common reaction theme of functional derivatives of carboxylic acids is **nucleophilic acyl addition** to the carbonyl carbon to form a **tetrahedral carbonyl addition intermediate**, which then collapses to regenerate the carbonyl group. The result is **nucleophilic acyl substitution** (Section 15.3). Listed in order of increasing reactivity toward nucleophilic acyl substitution, these functional derivatives are:

$$
\begin{array}{cccc}
\underset{\text{Amide}}{\overset{\displaystyle O \atop \displaystyle \|}{RCNH_2}} & \underset{\text{Ester}}{\overset{\displaystyle O \atop \displaystyle \|}{RCOR'}} & \underset{\text{Anhydride}}{\overset{\displaystyle O \quad O \atop \displaystyle \| \quad \|}{RCOCR'}} & \underset{\text{Acid chloride}}{\overset{\displaystyle O \atop \displaystyle \|}{RCCl}}
\end{array}
$$

Reactivity toward nucleophilic acyl substitution ⟶

Less reactive   More reactive

Any more reactive functional derivative can be directly converted to any less reactive functional derivative by reaction with an appropriate oxygen or nitrogen nucleophile (Section 15.7).

KEY REACTIONS

1. Hydrolysis of an Acid Chloride (Section 15.4A)

Low-molecular-weight acid chlorides react vigorously with water; higher-molecular-weight acid chlorides react less rapidly:

$$CH_3\overset{\displaystyle O \atop \displaystyle \|}{C}Cl + H_2O \longrightarrow CH_3\overset{\displaystyle O \atop \displaystyle \|}{C}OH + HCl$$

2. Hydrolysis of an Acid Anhydride (Section 15.4B)

Low-molecular-weight acid anhydrides react readily with water; higher-molecular-weight acid anhydrides react less rapidly:

$$CH_3\overset{\displaystyle O \atop \displaystyle \|}{C}O\overset{\displaystyle O \atop \displaystyle \|}{C}CH_3 + H_2O \longrightarrow CH_3\overset{\displaystyle O \atop \displaystyle \|}{C}OH + HO\overset{\displaystyle O \atop \displaystyle \|}{C}CH_3$$

3. Hydrolysis of an Ester (Section 15.4C)

Esters are hydrolyzed only in the presence of base or acid; base is required in an equimolar amount, acid is a catalyst:

$$CH_3\overset{\displaystyle O \atop \displaystyle \|}{C}O\!\!-\!\!\bigcirc + NaOH \xrightarrow{H_2O} CH_3\overset{\displaystyle O \atop \displaystyle \|}{C}O^- Na^+ + HO\!\!-\!\!\bigcirc$$

$$CH_3\overset{\displaystyle O \atop \displaystyle \|}{C}O\!\!-\!\!\bigcirc + H_2O \xrightarrow{HCl} CH_3\overset{\displaystyle O \atop \displaystyle \|}{C}OH + HO\!\!-\!\!\bigcirc$$

4. Hydrolysis of an Amide (Section 15.4D)

Either acid or base is required in an amount equivalent to that of the amide:

$$CH_3CH_2CH_2\overset{\overset{\displaystyle O}{\|}}{C}NH_2 + H_2O + HCl \xrightarrow[\text{Heat}]{H_2O} CH_3CH_2CH_2\overset{\overset{\displaystyle O}{\|}}{C}OH + NH_4^+Cl^-$$

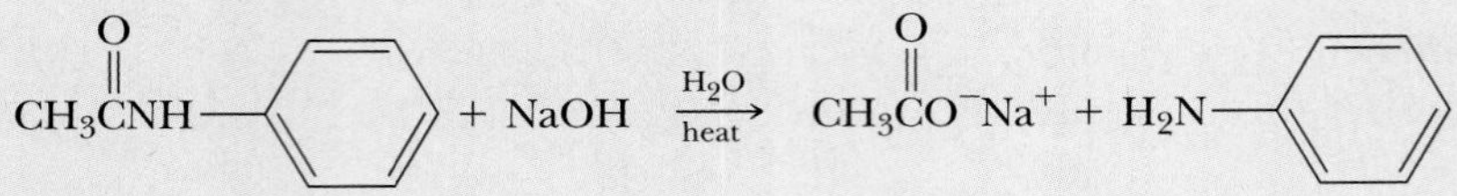

5. Reaction of an Acid Chloride with an Alcohol (Section 15.5A)

Treatment of an acid chloride with an alcohol gives an ester and HCl:

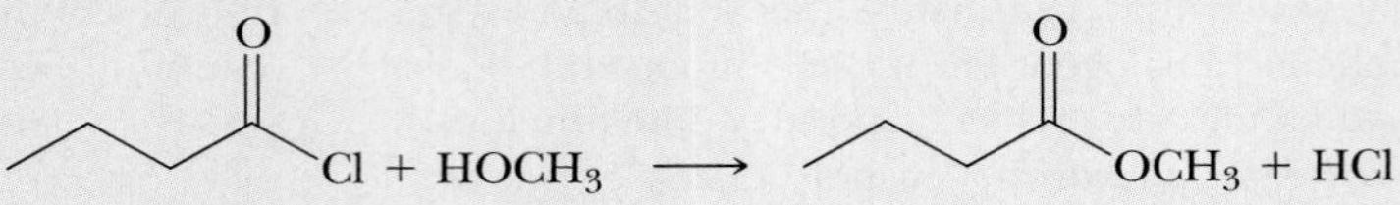

6. Reaction of an Acid Anhydride with an Alcohol (Section 15.5B)

Treatment of an acid anhydride with an alcohol gives an ester and a carboxylic acid:

$$CH_3\overset{\overset{\displaystyle O}{\|}}{C}O\overset{\overset{\displaystyle O}{\|}}{C}CH_3 + HOCH_2CH_3 \longrightarrow CH_3\overset{\overset{\displaystyle O}{\|}}{C}OCH_2CH_3 + CH_3\overset{\overset{\displaystyle O}{\|}}{C}OH$$

7. Reaction of an Ester with an Alcohol (Section 15.5C)

Treatment of an ester with an alcohol in the presence of an acid catalyst results in transesterification—that is, the replacement of one —OR group by a different —OR group:

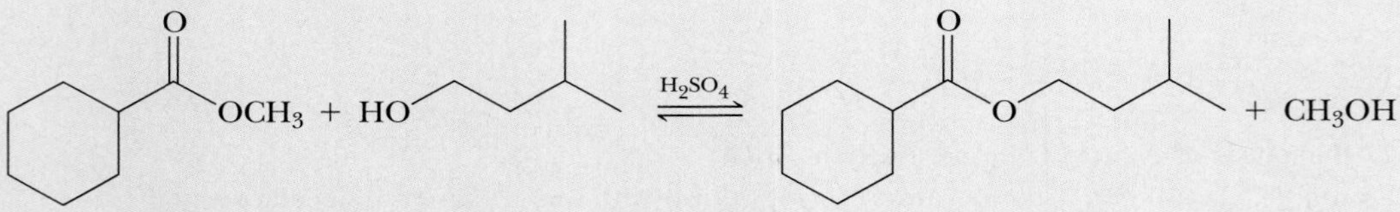

8. Reaction of an Acid Chloride with Ammonia or an Amine (Section 15.6A)

Reaction requires 2 moles of ammonia or amine—1 mole to form the amide and 1 mole to neutralize the HCl by-product:

$$CH_3\overset{\overset{\displaystyle O}{\|}}{C}Cl + 2NH_3 \longrightarrow CH_3\overset{\overset{\displaystyle O}{\|}}{C}NH_2 + NH_4^+Cl^-$$

9. Reaction of an Acid Anhydride with Ammonia or an Amine (Section 15.6B)

Reaction requires 2 moles of ammonia or amine—1 mole to form the amide and 1 mole to neutralize the carboxylic acid by-product:

$$CH_3\overset{\overset{\displaystyle O}{\|}}{C}O\overset{\overset{\displaystyle O}{\|}}{C}CH_3 + 2NH_3 \longrightarrow CH_3\overset{\overset{\displaystyle O}{\|}}{C}NH_2 + CH_3\overset{\overset{\displaystyle O}{\|}}{C}O^-NH_4^+$$

10. Reaction of an Ester with Ammonia or an Amine (Section 15.6C)

Treatment of an ester with ammonia, a 1° amine, or a 2° amine gives an amide:

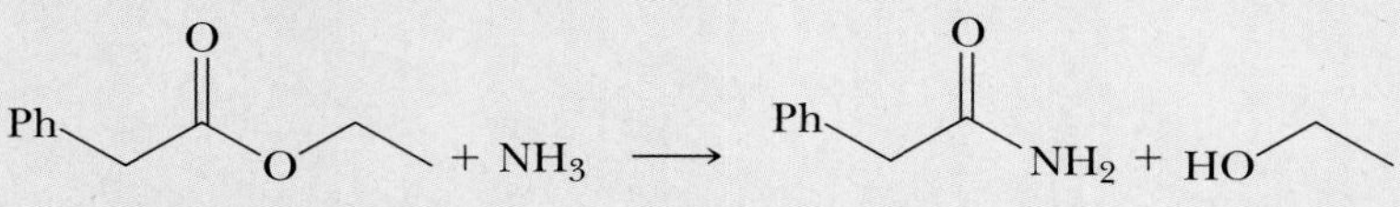

Ethyl phenylacetate   Phenylacetamide   Ethanol

11. Reaction of an Ester with a Grignard Reagent (Section 15.8)

Treating a formic ester with a Grignard reagent, followed by hydrolysis, gives a 2° alcohol, whereas treating any other ester with a Grignard reagent gives a 3° alcohol:

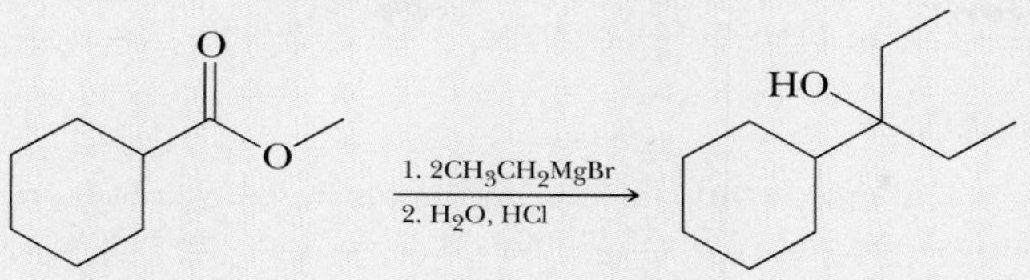
O
O
1. 2CH₃CH₂MgBr
2. H₂O, HCl
HO

12. Reduction of an Ester (Section 15.9A)

Reduction by lithium aluminum hydride gives two alcohols:

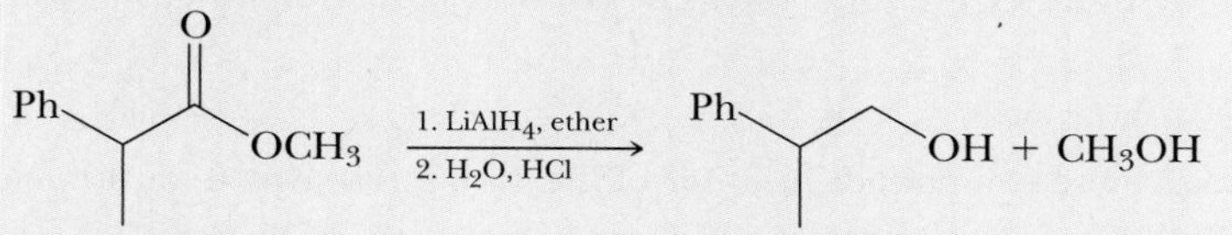
O
Ph
OCH₃
1. LiAlH₄, ether
2. H₂O, HCl
Ph
OH + CH₃OH

Methyl 2-phenylpropanoate — 2-Phenyl-1-propanol — Methanol

13. Reduction of an Amide (Section 15.9B)

Reduction by lithium aluminum hydride gives an amine:

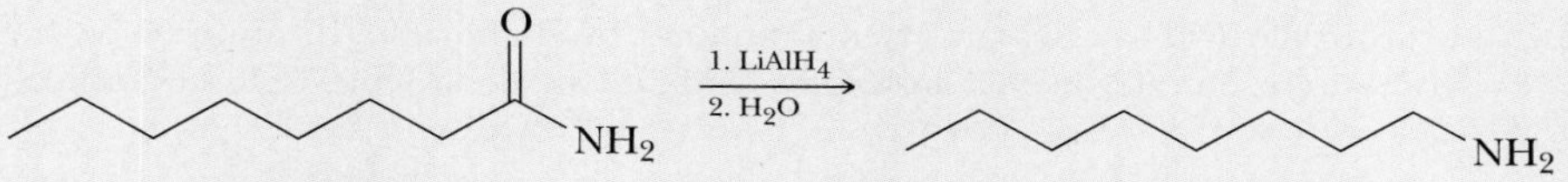
O
NH₂
1. LiAlH₄
2. H₂O
NH₂

Octanamide — 1-Octanamine

PROBLEMS

A problem number set in red indicates an applied "real-world" problem.

Structure and Nomenclature

15.8 Draw a structural formula for each compound:

(a) Dimethyl carbonate
(b) *p*-Nitrobenzamide
(c) Octanoyl chloride
(d) Diethyl oxalate
(e) Ethyl *cis*-2-pentenoate
(f) Butanoic anhydride
(g) Dodecanamide
(h) Ethyl 3-hydroxybutanoate

15.9 Write the IUPAC name for each compound:

(a) 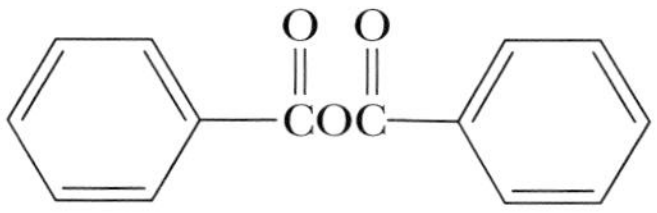
O O
COC

(b) $CH_3(CH_2)_{14}\overset{O}{\overset{\|}{C}}OCH_3$

(c) $CH_3(CH_2)_4\overset{O}{\overset{\|}{C}}NHCH_3$

(d) 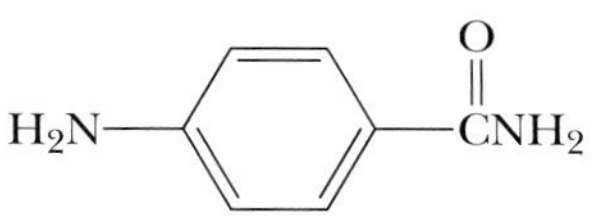
H₂N
O
CNH₂

(e) $CH_2(COOCH_2CH_3)_2$

(f) $PhCH_2\overset{O}{\overset{\|}{C}}\underset{CH_3}{\underset{|}{C}H}\overset{O}{\overset{\|}{C}}OCH_3$

15.10 When oil from the head of the sperm whale is cooled, spermaceti, a translucent wax
with a white, pearly luster, crystallizes from the mixture. Spermaceti, which makes up
11% of whale oil, is composed mainly of hexadecyl hexadecanoate (cetyl palmitate).
At one time, spermaceti was widely used in the making of cosmetics, fragrant soaps,
and candles. Draw a structural formula of cetyl palmitate.

Physical Properties

15.11 Acetic acid and methyl formate are constitutional isomers. Both are liquids at room
temperature, one with a boiling point of 32°C, the other with a boiling point of
118°C. Which of the two has the higher boiling point?

15.12 Acetic acid has a boiling point of 118°C, whereas its methyl ester has a boiling point
of 57°C. Account for the fact that the boiling point of acetic acid is higher than that
of its methyl ester, even though acetic acid has a lower molecular weight.

Reactions

15.13 Arrange these compounds in order of increasing reactivity toward nucleophilic acyl
substitution:

(1)   (2)   (3)   (4)

15.14 A carboxylic acid can be converted to an ester by Fischer esterification. Show how to
synthesize each ester from a carboxylic acid and an alcohol by Fischer esterification:

(a)   (b)

15.15 A carboxylic acid can also be converted to an ester in two reactions by first converting the carboxylic acid to its acid chloride and then treating the acid chloride with an
alcohol. Show how to prepare each ester in Problem 15.14 from a carboxylic acid and
an alcohol by this two-step scheme.

15.16 Show how to prepare these amides by reaction of an acid chloride with ammonia or
an amine:

(a)   (b)   (c)

15.17 Write a mechanism for the reaction of butanoyl chloride and ammonia to give
butanamide and ammonium chloride.

15.18 What product is formed when benzoyl chloride is treated with these reagents?
(a) C_6H_6, $AlCl_3$ (b) $CH_3CH_2CH_2CH_2OH$
(c) $CH_3CH_2CH_2CH_2SH$ (d) $CH_3CH_2CH_2CH_2NH_2$ (2 equivalents)

(e) H_2O (f) N—H (2 equivalents)

15.19 Write the product(s) of the treatment of propanoic anhydride with each reagent:

    (a) Ethanol (1 equivalent)   **(b)** Ammonia (2 equivalents)

15.20 Write the product of the treatment of benzoic anhydride with each reagent:

    (a) Ethanol (1 equivalent)   **(b)** Ammonia (2 equivalents)

15.21 The analgesic phenacetin is synthesized by treating 4-ethoxyaniline with acetic anhydride. Write an equation for the formation of phenacetin.

15.22 The analgesic acetaminophen is synthesized by treating 4-aminophenol with one equivalent of acetic anhydride. Write an equation for the formation of acetaminophen. (*Hint:* Remember from Section 7.6A that an —NH_2 group is a better nucleophile than an —OH group.)

15.23 Nicotinic acid, more commonly named niacin, is one of the B vitamins. Show how nicotinic acid can be converted to ethyl nicotinate and then to nicotinamide:

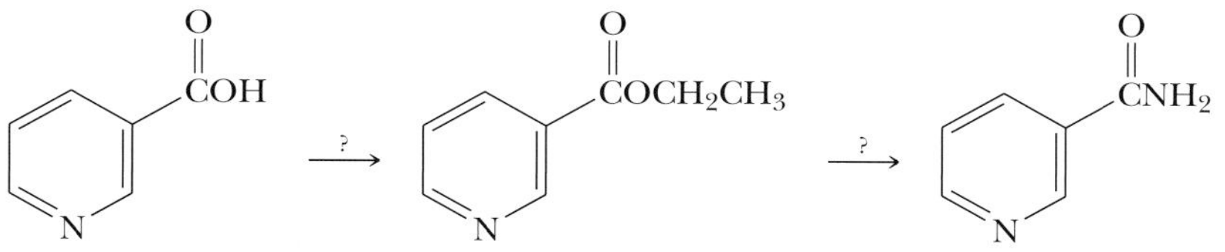

         Nicotinic acid   Ethyl nicotinate   Nicotinamide

          (Niacin)

15.24 Complete these reactions:

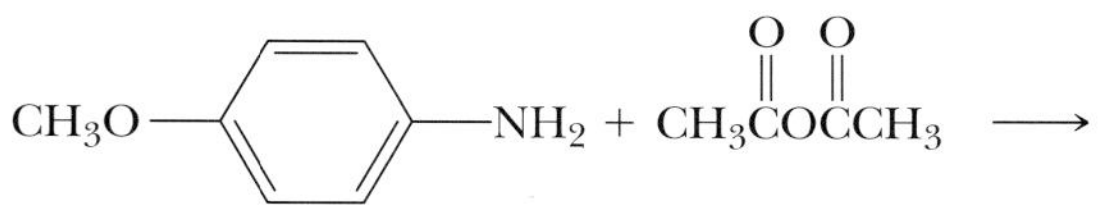

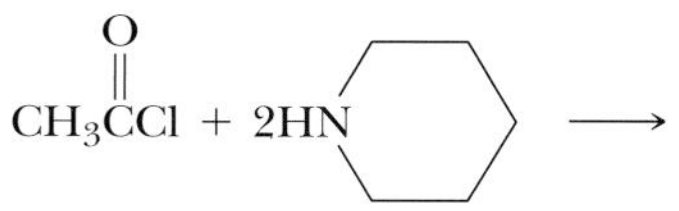

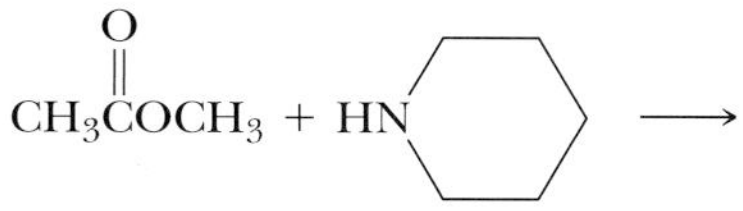

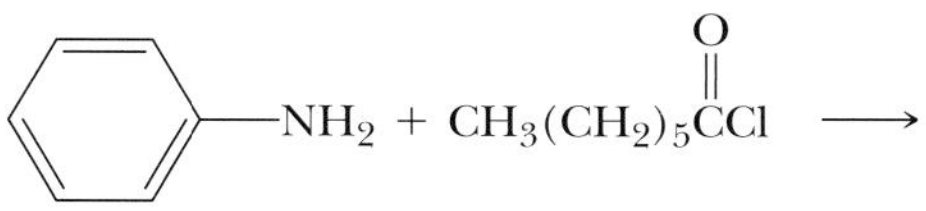

15.25 What product is formed when ethyl benzoate is treated with these reagents?

    (a) H_2O, NaOH, heat   **(b)** $LiAlH_4$, then H_2O
    (c) H_2O, H_2SO_4, heat   **(d)** $CH_3CH_2CH_2CH_2NH_2$
    (e) C_6H_5MgBr (2 moles) and then H_2O/HCl

15.26 Show how to convert 2-hydroxybenzoic acid (salicylic acid) to these compounds:

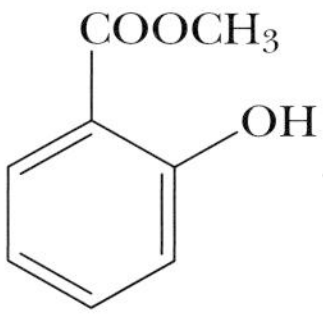

       Methyl salicylate   Acetyl salicylic acid
    (Oil of wintergreen)   (Aspirin)

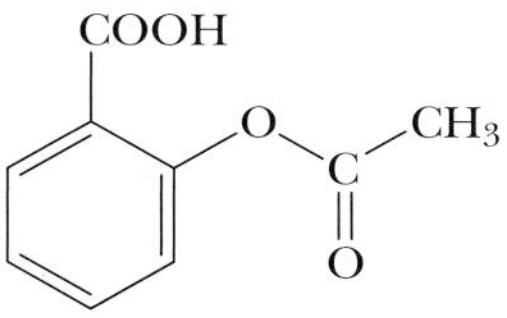

15.27 What product is formed when benzamide is treated with these reagents?

(a) H_2O, HCl, heat   (b) NaOH, H_2O, heat   (c) LiAlH$_4$, then H_2O

15.28 Treating γ-butyrolactone with two equivalents of methylmagnesium bromide, followed by hydrolysis in aqueous acid, gives a compound with the molecular formula $C_6H_{14}O_2$:

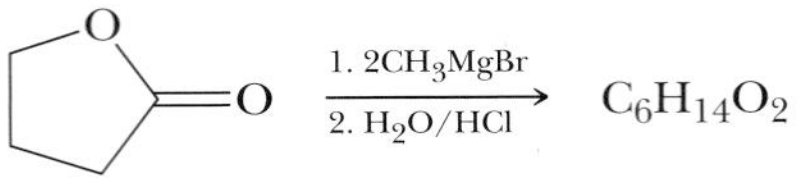

Propose a structural formula for this compound.

15.29 Show the product of treating γ-butyrolactone with each reagent:

(a) NH_3 (b) LiAlH$_4$, then H_2O (c) NaOH, H_2O, heat

15.30 Show the product of treating N-methyl-γ-butyrolactam with each reagent:

(a) H_2O, HCl, heat   (b) NaOH, H_2O, heat   (c) LiAlH$_4$, then H_2O

15.31 Complete these reactions:

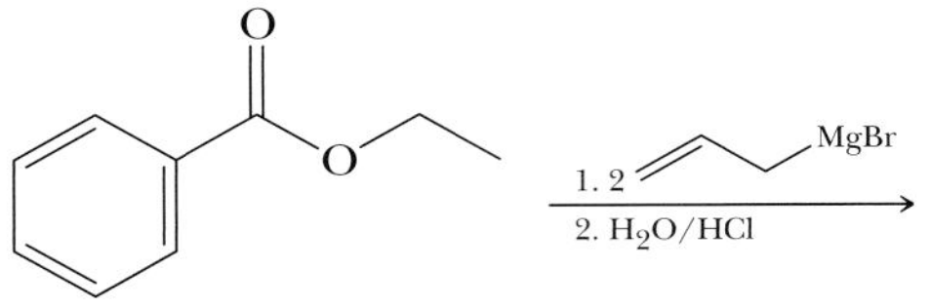

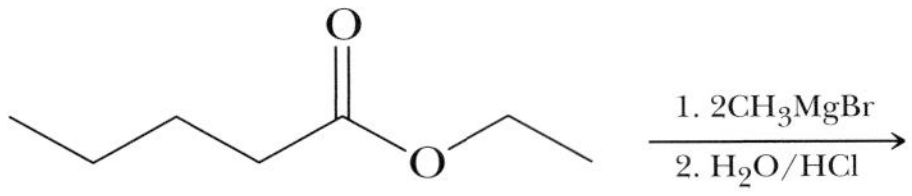

15.32 What combination of ester and Grignard reagent can be used to prepare each alcohol?

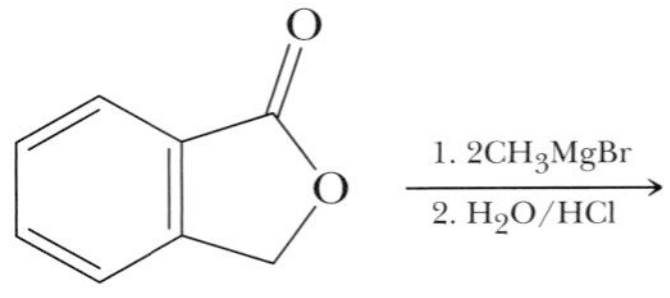

(a) 2-Methyl-2-butanol   (b) 3-Phenyl-3-pentanol   (c) 1,1-Diphenylethanol

15.33 Reaction of a 1° or 2° amine with diethyl carbonate under controlled conditions gives a carbamic ester:

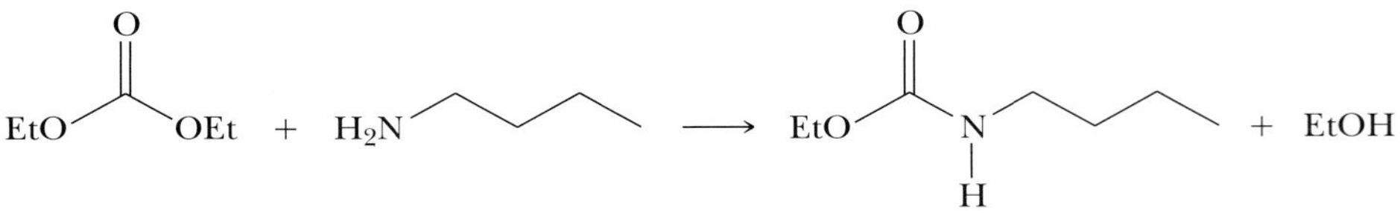

Diethyl   1-Butanamine   A carbamic ester
carbonate   (Butylamine)

Propose a mechanism for this reaction.

15.34 Barbiturates are prepared by treating diethyl malonate or a derivative of diethyl malonate with urea in the presence of sodium ethoxide as a catalyst. Following is an equation for the preparation of barbital from diethyl 2,2-diethylmalonate and urea (barbital, a long-duration hypnotic and sedative, is prescribed under a dozen or more trade names):

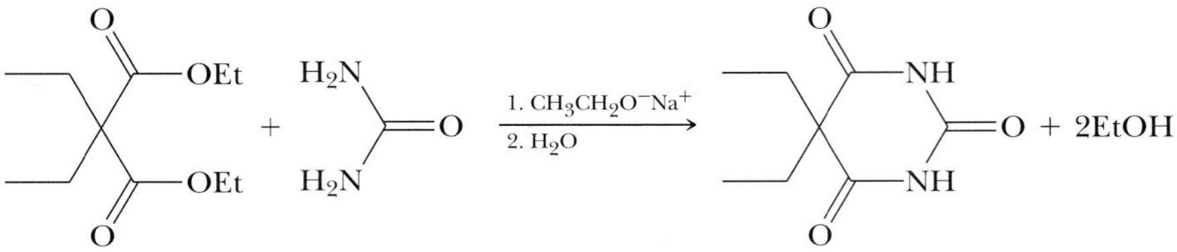

Diethyl   Urea   5,5-Diethylbarbituric acid
2,2-diethylmalonate   (Barbital)

(a) Propose a mechanism for this reaction.
(b) The pK_a of barbital is 7.4. Which is the most acidic hydrogen in this molecule, and how do you account for its acidity?

15.35 Name and draw structural formulas for the products of the complete hydrolysis of meprobamate and phenobarbital in hot aqueous acid. Meprobamate is a tranquilizer prescribed under 58 different trade names. Phenobarbital is a long-acting sedative, hypnotic, and anticonvulsant. [*Hint*: Remember that, when heated, β-dicarboxylic acids and β-ketoacids undergo decarboxylation (Section 14.9B).]

(a) 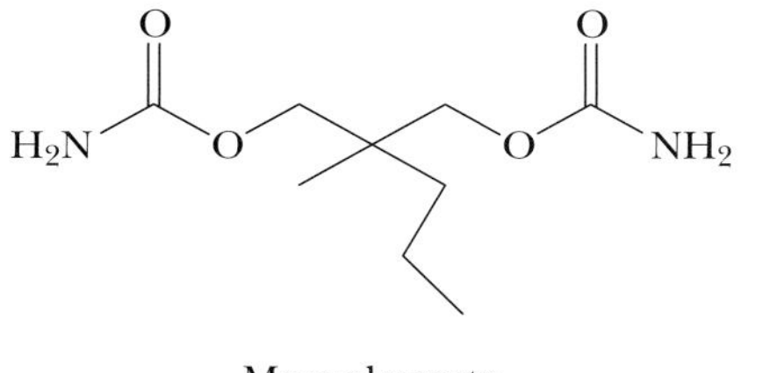   Meprobamate   (b) 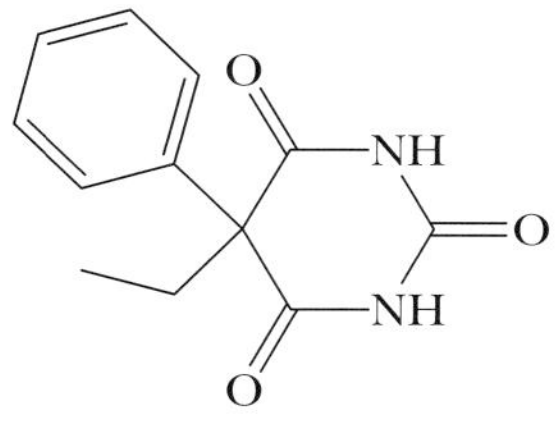   Phenobarbital

Synthesis

15.36 The active ingredient in several common insect repellents *N,N*-Diethyl-*m*-toluamide (Deet) is synthesized from 3-methylbenzoic acid (*m*-toluic acid) and diethylamine:

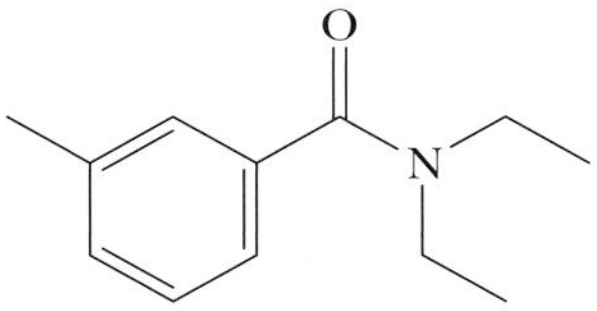

N,N-Diethyl-*m*-toluamide
(Deet)

Show how this synthesis can be accomplished.

15.37 Show how to convert ethyl 2-pentenoate into these compounds:

(a) 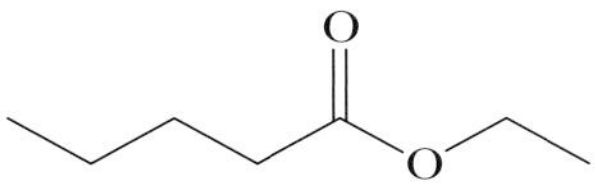   (b)

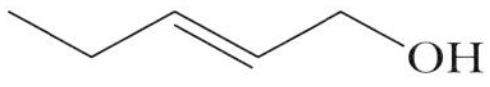

(c)

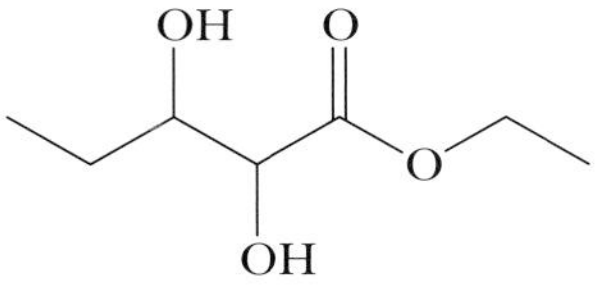

15.38 Procaine (whose hydrochloride is marketed as Novocaine®) was one of the first local anesthetics for infiltration and regional anesthesia. Show how to synthesize procaine, using the three reagents shown as the sources of carbon atoms:

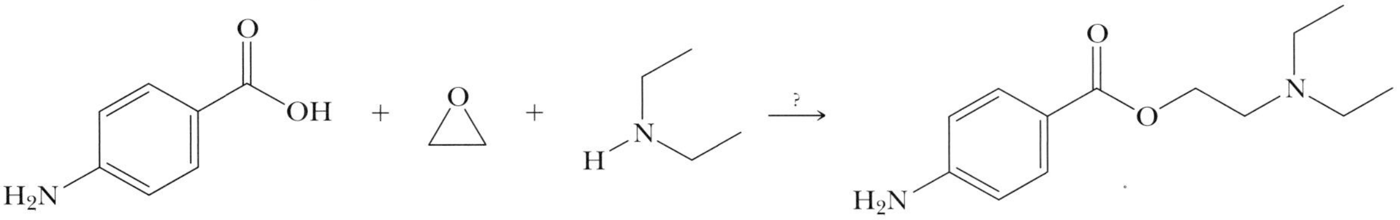

4-Aminobenzoic   Ethylene   Diethylamine   Procaine
acid   oxide

15.39 There are two nitrogen atoms in procaine. Which of the two is the stronger base? Draw the structural formula for the salt that is formed when procaine is treated with 1 mole of aqueous HCl.

15.40 Starting materials for the synthesis of the herbicide propanil, a weed killer used in rice paddies, are benzene and propanoic acid. Show reagents to bring about this synthesis:

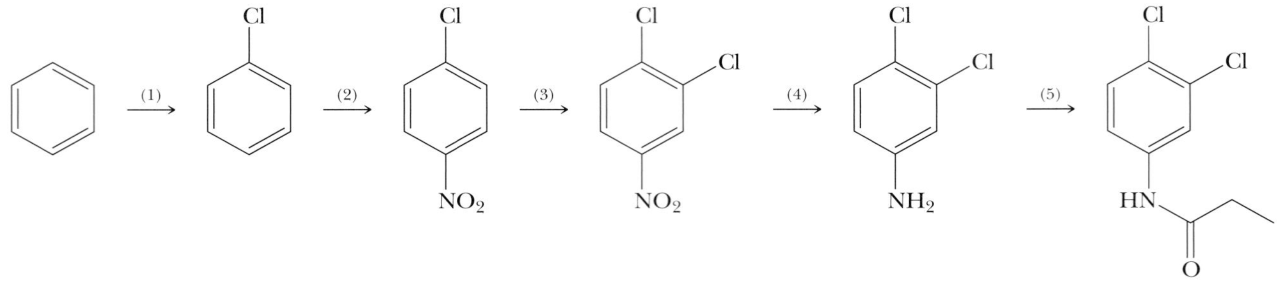

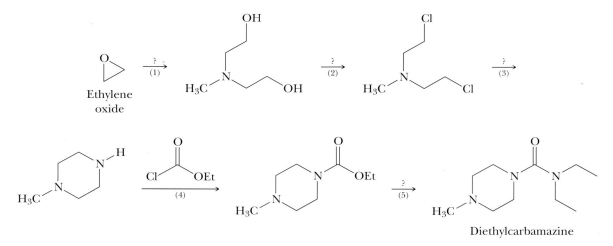

Propanil

15.41 Following are structural formulas for three local anesthetics: Lidocaine was introduced in 1948 and is now the most widely used local anesthetic for infiltration and regional anesthesia. Its hydrochloride is marketed under the name Xylocaine®. Etidocaine (its hydrochloride is marketed as Duranest®) is comparable to lidocaine in onset, but its analgesic action lasts two to three times longer. Mepivacaine (its hydrochloride is marketed as Carbocaine®) is faster and somewhat longer in duration than lidocaine.

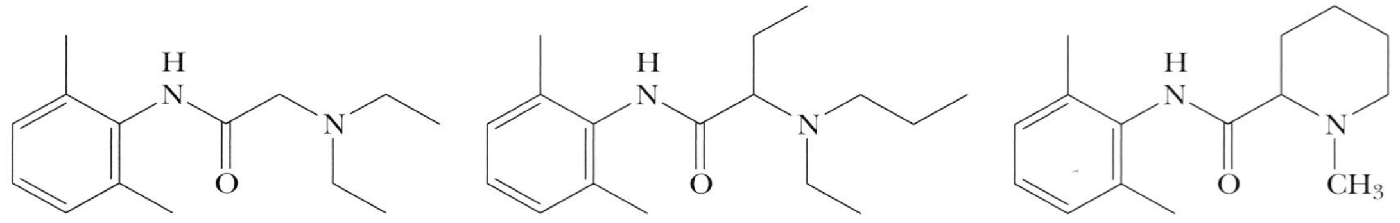

Lidocaine
(Xylocaine®)

Etidocaine
(Duranest®)

Mepivacaine
(Carbocaine®)

(a) Propose a synthesis of lidocaine from 2,6-dimethylaniline, chloroacetyl chloride (ClCH$_2$COCl), and diethylamine.

(b) Propose a synthesis of etidocaine from 2,6-dimethylaniline, 2-chlorobutanoyl chloride, and ethylpropylamine.

(c) What amine and acid chloride can be reacted to give mepivacaine?

15.42 Following is the outline of a five-step synthesis for the anthelmintic (against worms) diethylcarbamazine:

Ethylene oxide

Diethylcarbamazine

Diethylcarbamazine is used chiefly against nematodes, small cylindrical or slender threadlike worms such as the common roundworm, which are parasitic in animals and plants.

(a) Propose a reagent for Step 1. Which mechanism is more likely for this step, S_N1 or S_N2? Explain.

(b) Propose a reagent for Step 2.

(c) Propose a reagent for Step 3.

(d) Ethyl chloroformate, the reagent for Step 4, is both an acid chloride and an ester. Account for the fact that Cl, rather than OCH_2CH_3, is displaced from this reagent.

15.43 Following is an outline of a five-step synthesis for methylparaben, a compound widely used as a preservative in foods:

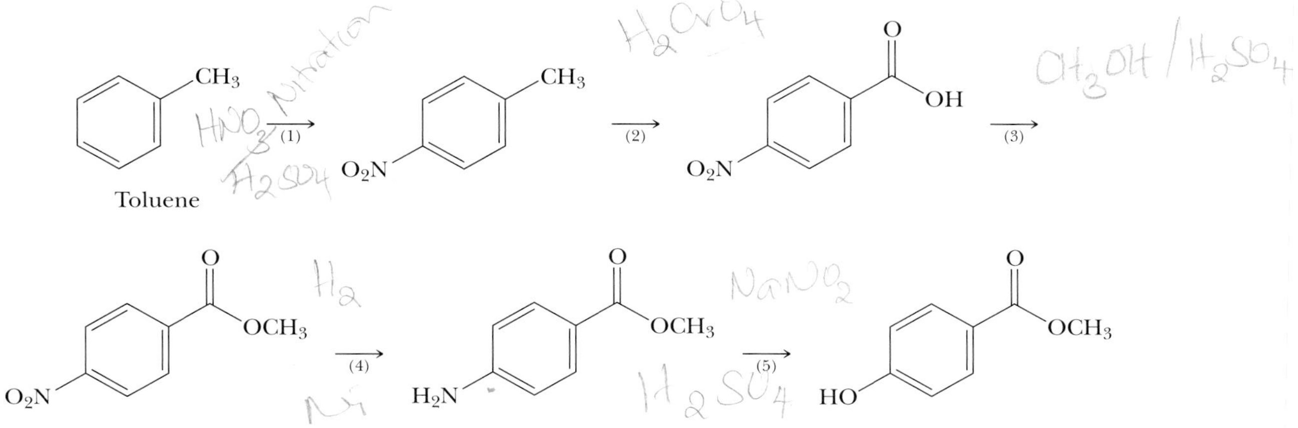

Methyl 4-hydroxybenzoate
(Methylparaben)

Propose reagents for each step.

Looking Ahead

15.44 Identify the most acidic proton in each of the following esters:

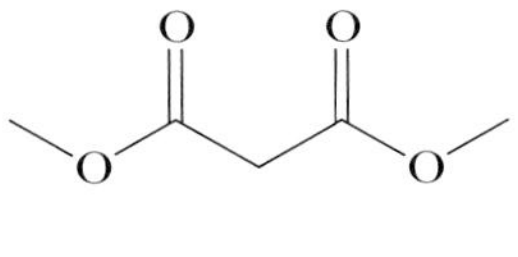

15.45 Does a nucleophilic acyl substitution occur between the ester and the nucleophile shown?

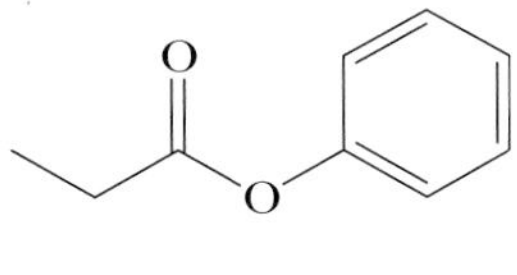

$+$ $NaOCH_3$ $\longrightarrow$

Propose an experiment that would verify your answer.

15.46 Explain why a nucleophile, Nu, attacks not only the carbonyl carbon, but also the β-carbon, as indicated in the following α,β-unsaturated ester:

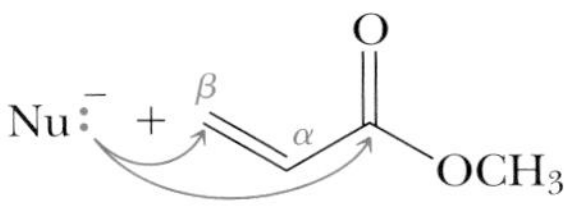

15.47 Explain why a Grignard reagent will not undergo nucleophilic acyl substitution with the following amide:

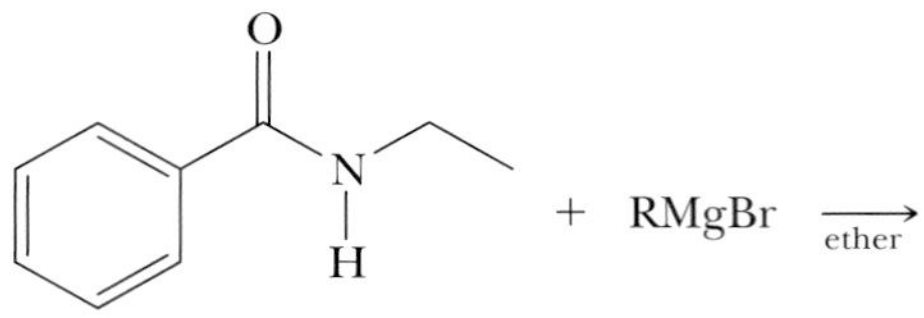

15.48 At low temperatures, the following amide exhibits *cis-trans* isomerism, while at higher temperatures it does not:

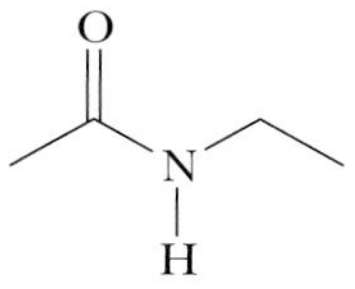

Explain how this is possible.

16

Enolate Anions

Human gallstones are almost pure cholesterol; these gallstones are about 0.5 cm in diameter. See Chemical Connections "Drugs That Lower Plasma Cholesterol Levels." Inset: A model of cholesterol. *(Carolina Biological Supply Company/Phototake, Inc.)*

16.1 INTRODUCTION

In this chapter, we continue our discussion of the chemistry of carbonyl compounds. In Chapters 13–15, we concentrated on the carbonyl group itself and on nucleophilic additions to it to form tetrahedral carbonyl addition compounds. In the current chapter, we expand on the chemistry of carbonyl-containing compounds and consider the acidity of α-hydrogens and the enolate anions formed by their removal.

16.2   FORMATION OF ENOLATE ANIONS

A. Acidity of α-Hydrogens

A carbon atom adjacent to a carbonyl group is called an **α-carbon,** and a hydrogen atom bonded to it is called an **α-hydrogen:**

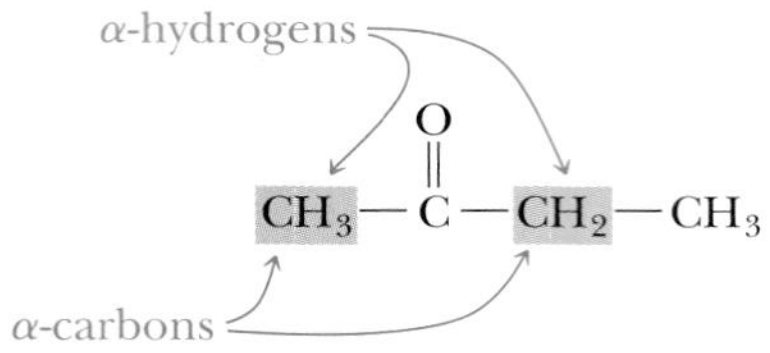

$$\alpha\text{-hydrogens} \quad \underset{\alpha\text{-carbons}}{\overset{O}{\underset{\parallel}{CH_3-C-CH_2-CH_3}}}$$

Because carbon and hydrogen have comparable electronegativities, a C—H bond normally has little polarity, and a hydrogen atom bonded to carbon shows very low acidity. The situation is different, however, for hydrogens that are alpha to a carbonyl group. As Table 16.1 shows, α-hydrogens of aldehydes, ketones, and esters are considerably more acidic than alkane and alkene hydrogens, but less acidic than the hydroxyl hydrogen of alcohols. The table also shows that hydrogens which are alpha to two carbonyl groups—for example in a β-ketoester and a β-diester—are even more acidic than alcohols.

TABLE 16.1   **The Acidity of Hydrogens Alpha to a Carbonyl Group Relative to Other Organic Hydrogens**

Class of Compound	Example	pK_a
a β-ketoester (an acetoacetic ester)	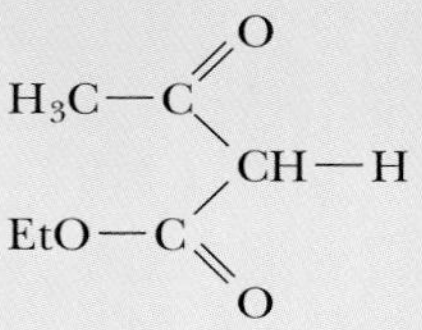 $H_3C-C{\overset{O}{\diagdown}}$ ... $CH-H$ ... $EtO-C{\underset{O}{\diagdown}}$	11
a β-diester (a malonic ester)	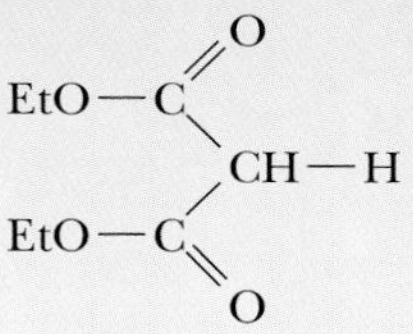 $EtO-C{\overset{O}{\diagdown}}$ ... $CH-H$ ... $EtO-C{\underset{O}{\diagdown}}$	13
an alcohol	CH_3CH_2O-H	16
an aldehyde or a ketone	$CH_3\overset{O}{\overset{\parallel}{C}}CH_2-H$	20
an ester	$EtO\overset{O}{\overset{\parallel}{C}}CH_2-H$	22
an alkene	$CH_2=CH-H$	44
an alkane	CH_3CH_2-H	51

Increasing acidity ↑

B. Enolate Anions

Carbonyl groups increase the acidity of their alpha hydrogens in two ways. First, the electron-withdrawing inductive effect of the carbonyl group weakens the bond to the alpha hydrogen and promotes its ionization. Second, the negative charge on

EXAMPLE 16.3

Draw the product of the base-catalyzed dehydration of each aldol product from Example 16.2.

SOLUTION

Loss of H_2O from aldol product (a) gives an α,β-unsaturated aldehyde, while loss of H_2O from aldol product (b) gives an α,β-unsaturated ketone:

(a)

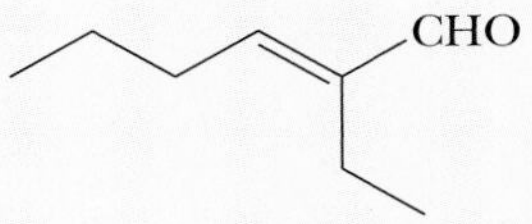

(b)

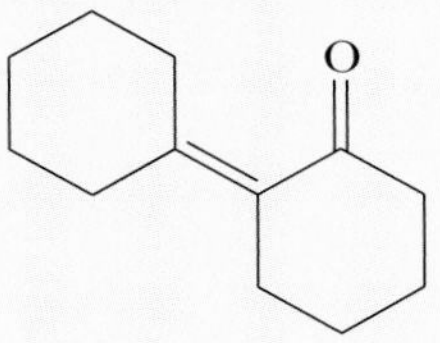

Practice Problem 16.3

Draw the product of the base-catalyzed dehydration of each aldol product from Problem 16.2.

C. Crossed Aldol Reactions

The reactants in the key step of an aldol reaction are an enolate anion and an enolate anion acceptor. In self-reactions, both roles are played by one kind of molecule. **Crossed aldol reactions** are also possible, such as the crossed aldol reaction between acetone and formaldehyde. Because it has no α-hydrogen, formaldehyde cannot form an enolate anion. It is, however, a particularly good enolate anion acceptor, because its carbonyl group is unhindered. Acetone forms an enolate anion, but its carbonyl group, which is bonded to two alkyl groups, is less reactive than that of formaldehyde. Consequently, the crossed aldol reaction between acetone and formaldehyde gives 4-hydroxy-2-butanone:

Crossed aldol reaction An aldol reaction between two different aldehydes, two different ketones, or an aldehyde and a ketone.

$$CH_3\overset{O}{\overset{\|}{C}}CH_3 + H\overset{O}{\overset{\|}{C}}H \xrightarrow{\text{NaOH}} CH_3\overset{O}{\overset{\|}{C}}CH_2CH_2OH$$

4-Hydroxy-2-butanone

As this example illustrates, for a crossed aldol reaction to be successful, one of the two reactants should have no α-hydrogen, so that an enolate anion does not form. It also helps if the compound with no α-hydrogen has the more reactive carbonyl—for example, an aldehyde. Following are examples of aldehydes that have no α-hydrogens and that can be used in crossed aldol reactions:

Formaldehyde   Benzaldehyde 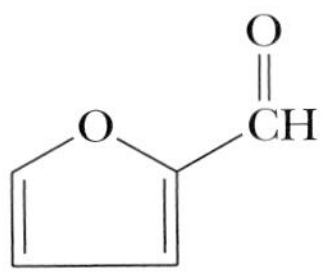   Furfural   2,2-Dimethylpropanal

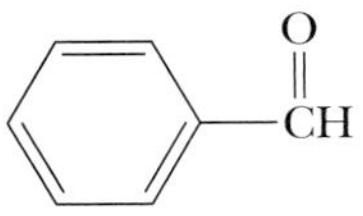

EXAMPLE 16.4

Draw the product of the crossed aldol reaction between furfural and cyclohexanone and the product formed by its base-catalyzed dehydration.

SOLUTION

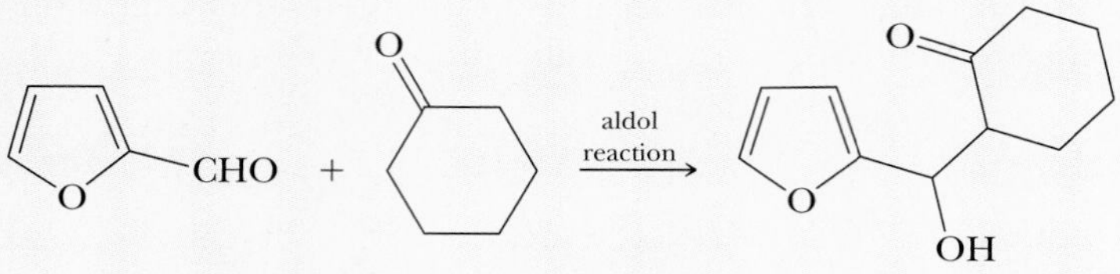

Furfural   Cyclohexanone   Aldol product

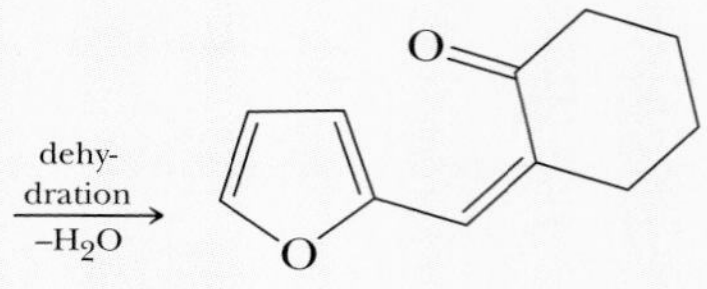

Practice Problem 16.4

Draw the product of the crossed aldol reaction between benzaldehyde and 3-pentanone and the product formed by its base-catalyzed dehydration.

D. Intramolecular Aldol Reactions

When both the enolate anion and the carbonyl group to which it adds are in the same molecule, aldol reaction results in formation of a ring. This type of **intramolecular aldol reaction** is particularly useful for the formation of five- and six-membered rings. Because they are the most stable rings, they form much more readily than four- or seven- and larger-membered rings. Intramolecular aldol reaction of 2,7-octanedione via enolate anion α_3, for example, gives a five-membered ring, whereas intramolecular aldol reaction of this same compound via enolate anion α_1 would give a seven-membered ring. In the case of 2,7-octanedione, the five-membered ring forms in preference to the seven-membered ring:

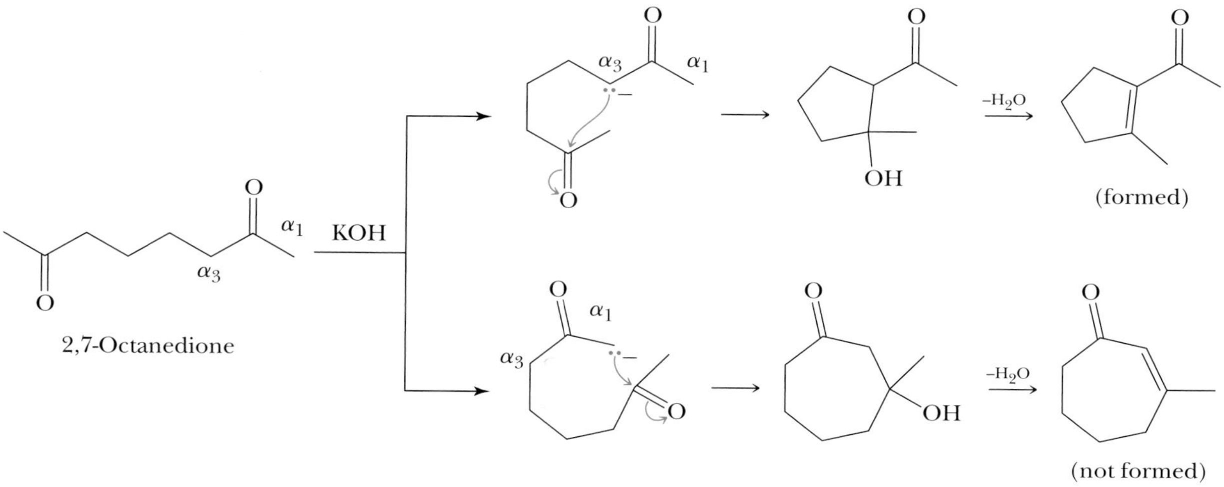

2,7-Octanedione

16.4 THE CLAISEN AND DIECKMANN CONDENSATIONS

A. Claisen Condensation

In this section, we examine the formation of an enolate anion from one ester, followed by the nucleophilic acyl substitution of the enolate anion at the carbonyl carbon of another ester. One of the first of these reactions discovered was the **Claisen condensation**, named after its discoverer, German chemist Ludwig Claisen (1851–1930). We illustrate a Claisen condensation by the reaction between two molecules of ethyl acetate in the presence of sodium ethoxide, followed by acidification, to give ethyl acetoacetate (note that, in this and many of the equations that follow, we abbreviate the ethyl group as Et):

Claisen condensation A carbonyl condensation reaction between two esters to give a β-ketoester.

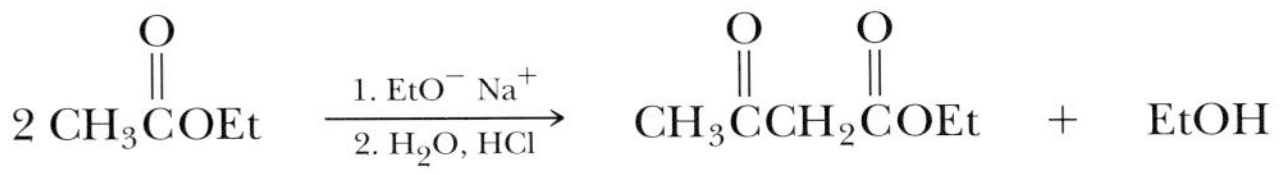

$$2\ CH_3COEt \xrightarrow[\text{2. H}_2\text{O, HCl}]{\text{1. EtO}^- \text{Na}^+} CH_3CCH_2COEt\ +\ EtOH$$

Ethyl ethanoate
(Ethyl acetate)

Ethyl 3-oxobutanoate
(Ethyl acetoacetate)

Ethanol

The functional group of the product of a Claisen condensation is a **β-ketoester**:

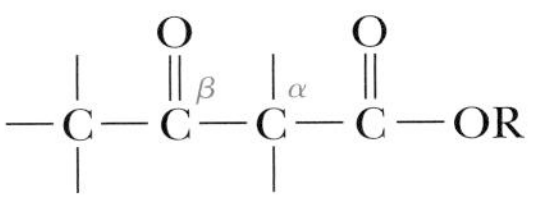

A β-ketoester

The Claisen condensation of two molecules of ethyl propanoate gives the following β-ketoester:

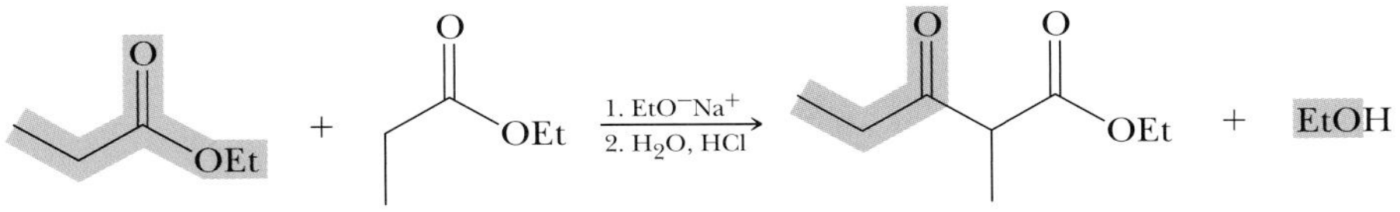

Ethyl propanoate

Ethyl propanoate

Ethyl 2-methyl-3-oxopentanoate

Claisen condensations, like the aldol reaction, require a base. Aqueous bases, such as NaOH, however, cannot be used in Claisen condensations because they would bring about hydrolysis of the ester (saponification, Section 15.4C) instead. Rather, the bases most commonly used in Claisen condensations are nonaqueous bases, such as sodium ethoxide in ethanol and sodium methoxide in methanol.

Mechanism: Claisen Condensation

As you study this mechanism, note how closely its first two steps resemble the first steps of the aldol reaction (Section 16.3). In each reaction, base removes a proton from an α-carbon in Step 1 to form a resonance-stabilized enolate anion. In Step 2, the enolate anion attacks the carbonyl carbon of another ester molecule to form a tetrahedral carbonyl addition intermediate.

Step 1: Base removes an α-hydrogen from the ester to give a resonance-stabilized enolate anion:

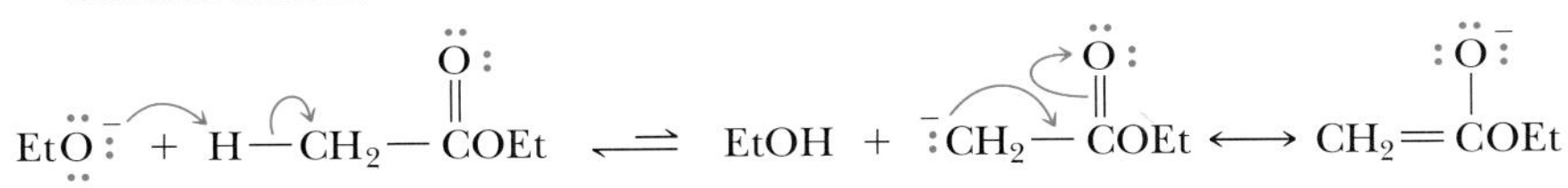

pK_a 22
(weaker acid)

(weaker base)

pK_a 15.9
(stronger acid)

Resonance-stabilized enolate anion
(stronger base)

Because the α-hydrogen of the ester is the weaker acid and ethoxide is the weaker base, the position of this equilibrium lies very much toward the left.

Step 2: Attack of the enolate anion on the carbonyl carbon of another ester molecule gives a tetrahedral carbonyl addition intermediate:

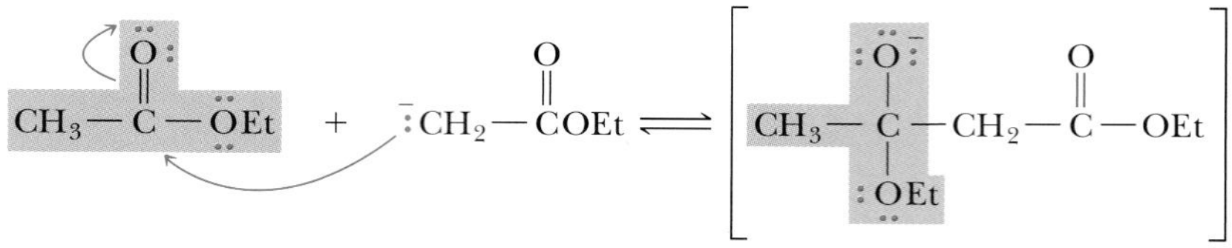

A tetrahedral cabonyl
addition intermediate

Step 3: Unlike the tetrahedral carbonyl addition intermediate in the aldol reaction, this intermediate has a leaving group (the ethoxide ion). Collapse of the tetrahedral carbonyl addition intermediate by ejection of the ethoxide ion gives a β-ketoester:

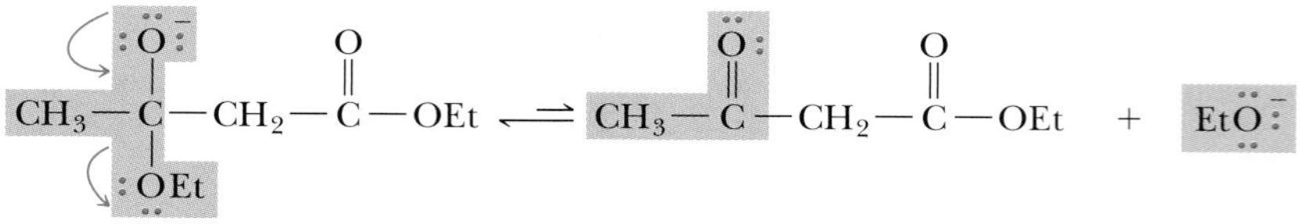

Step 4: Formation of the enolate anion of the β-ketoester drives the Claisen condensation to the right. The β-ketoester (a stronger acid) reacts with ethoxide ion (a stronger base) to give ethanol (a weaker acid) and the anion of the β-ketoester (a weaker base):

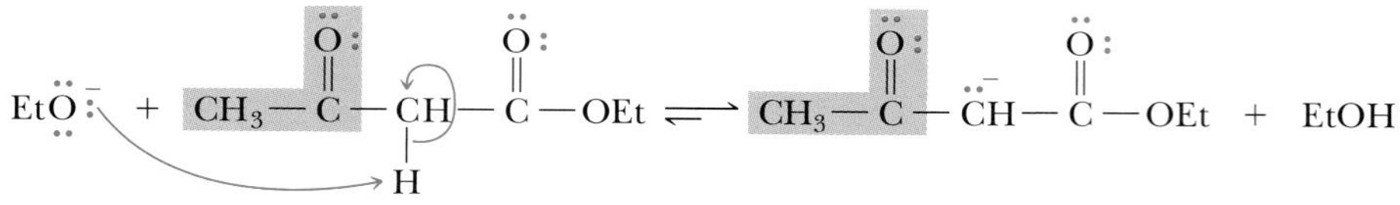

| (stronger base) | pK_a 10.7 (stronger acid) | (weaker base) | pK_a 15.9 (weaker acid) |

The position of equilibrium for this step lies very far toward the right.

Step 5: Acidification of the enolate anion gives the β-ketoester:

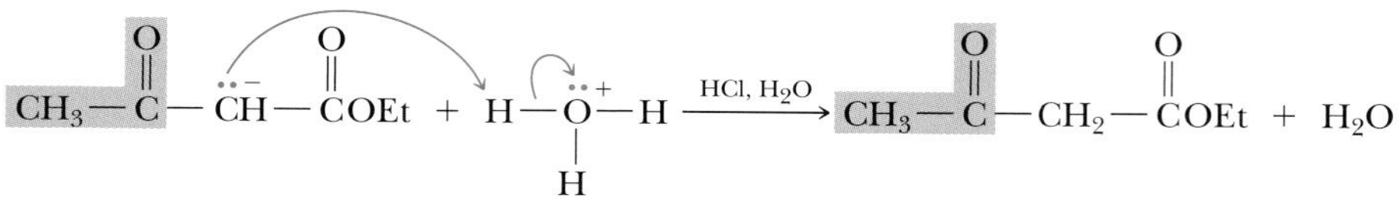

EXAMPLE 16.5

Show the product of the Claisen condensation of ethyl butanoate in the presence of sodium ethoxide followed by acidification with aqueous HCl.

SOLUTION

The new bond formed in a Claisen condensation is between the carbonyl group of one ester and the α-carbon of another:

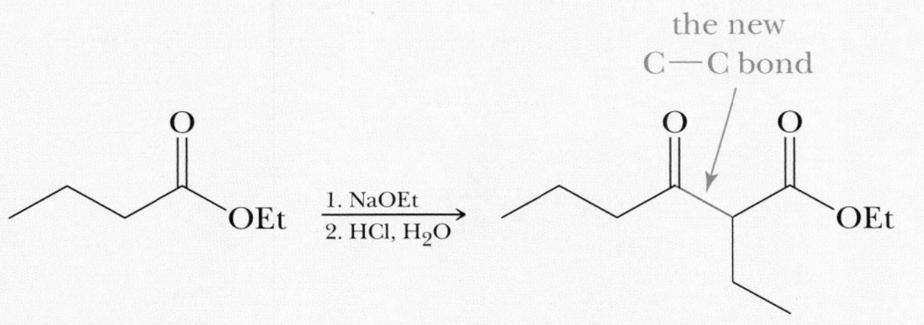

the new
C—C bond

1. NaOEt
2. HCl, H₂O

Ethyl butanoate   Ethyl 2-ethyl-3-oxohexanoate

Practice Problem 16.5

Show the product of the Claisen condensation of ethyl 3-methylbutanoate in the presence of sodium ethoxide.

B. Dieckmann Condensation

An intramolecular Claisen condensation of a dicarboxylic ester to give a five- or six-membered ring is known as a **Dieckmann condensation**. In the presence of one equivalent of sodium ethoxide, diethyl hexanedioate (diethyl adipate), for example, undergoes an intramolecular condensation to form a five-membered ring:

Dieckmann condensation An intramolecular Claisen condensation of an ester of a dicarboxylic acid to give a five- or six-membered ring.

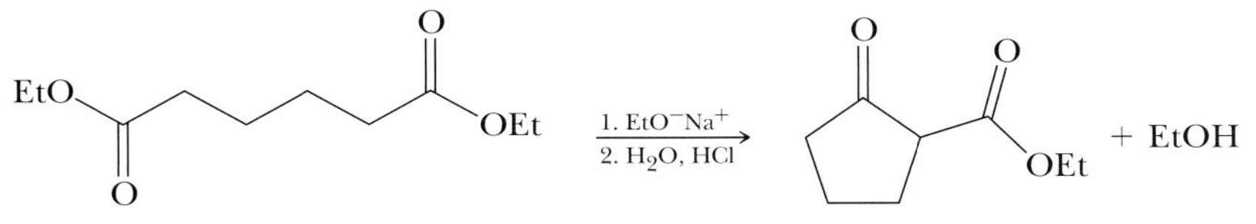

1. EtO⁻Na⁺
2. H₂O, HCl

+ EtOH

Diethyl hexanedioate   Ethyl 2-oxocyclo-
(Diethyl adipate)   pentanecarboxylate

The mechanism of a Dieckmann condensation is identical to the mechanism we described for the Claisen condensation. An anion formed at the α-carbon of one ester in Step 1 adds to the carbonyl of the other ester group in Step 2 to form a tetrahedral carbonyl addition intermediate. This intermediate ejects ethoxide ion in Step 3 to regenerate the carbonyl group. Cyclization is followed by formation of the conjugate base of the β-ketoester in Step 4, just as in the Claisen condensation. The β-ketoester is isolated after acidification with aqueous acid.

C. Crossed Claisen Condensations

In a **crossed Claisen condensation** (a Claisen condensation between two different esters, each with its own α-hydrogens), a mixture of four β-ketoesters is possible; therefore, crossed Claisen condensations of this type are generally not synthetically useful. Such condensations are useful, however, if appreciable differences in reactivity exist between the two esters, as, for example, when one of the esters has no α-hydrogens and can function only as an enolate anion acceptor. These esters have no α-hydrogens:

Crossed Claisen condensation A Claisen condensation between two different esters.

HCOEt   EtOCOEt   EtOC—COEt

Ethyl formate   Diethyl carbonate   Diethyl ethanedioate   Ethyl benzoate
                                        (Diethyl oxalate)

Crossed Claisen condensations of this type are usually carried out by using the ester with no α-hydrogens in excess. In the following illustration, methyl benzoate is used in excess:

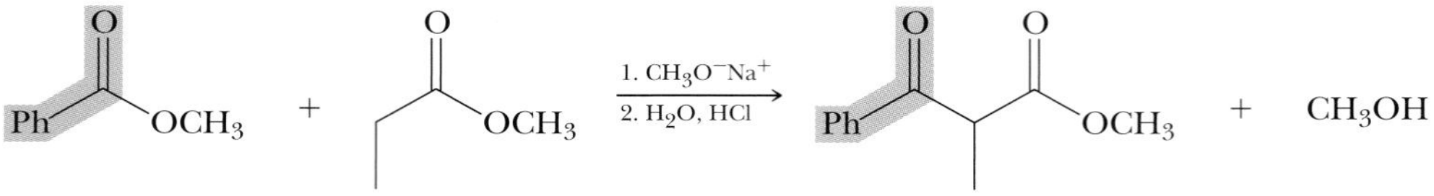

Methyl benzoate — Methyl propanoate — Methyl 2-methyl-3-oxo-3-phenylpropanoate

EXAMPLE 16.6

Complete the equation for this crossed Claisen condensation:

$$\mathrm{H\overset{O}{\overset{\|}{C}}OEt} + \mathrm{CH_3CH_2\overset{O}{\overset{\|}{C}}OEt} \xrightarrow[\text{2. } H_2O,\ HCl]{\text{1. } EtO^-Na^+}$$

SOLUTION

$$\mathrm{H\overset{O}{\overset{\|}{C}}\underset{\underset{CH_3}{|}}{C}H\overset{O}{\overset{\|}{C}}OEt} + \mathrm{EtOH}$$

Practice Problem 16.6

Complete the equation for this crossed Claisen condensation:

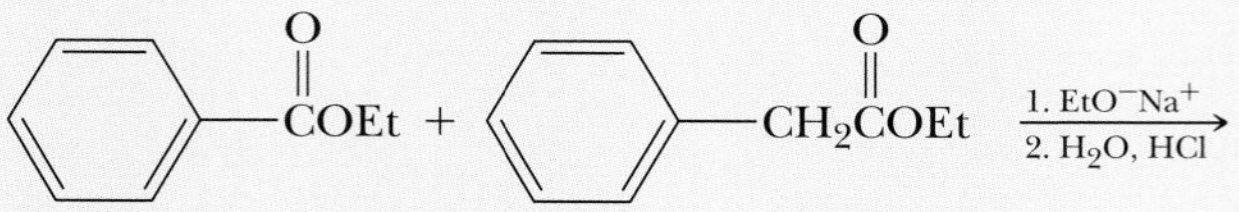

D. Hydrolysis and Decarboxylation of β-Ketoesters

Recall from Section 15.4C that the hydrolysis of an ester in aqueous sodium hydroxide (saponification), followed by acidification of the reaction mixture with HCl or other mineral acid, converts an ester to a carboxylic acid. Recall also from Section 14.9 that β-ketoacids and β-dicarboxylic acids readily undergo decarboxylation (lose CO_2) when heated. The following equations illustrate the results of a Claisen condensation, followed by saponification, acidification, and decarboxylation:

Claisen condensation:

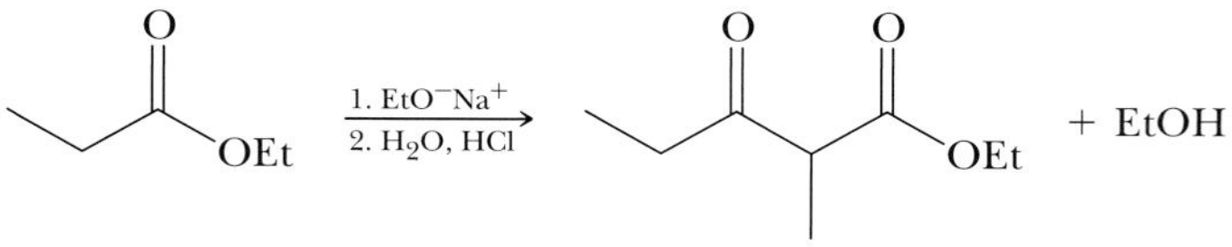

Saponification followed by acidification:

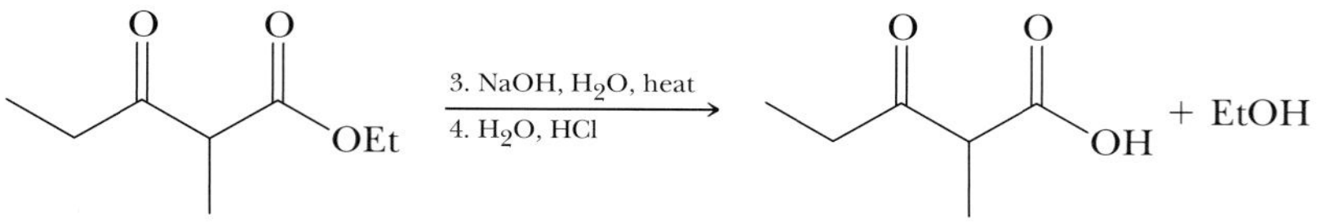

Decarboxylation:

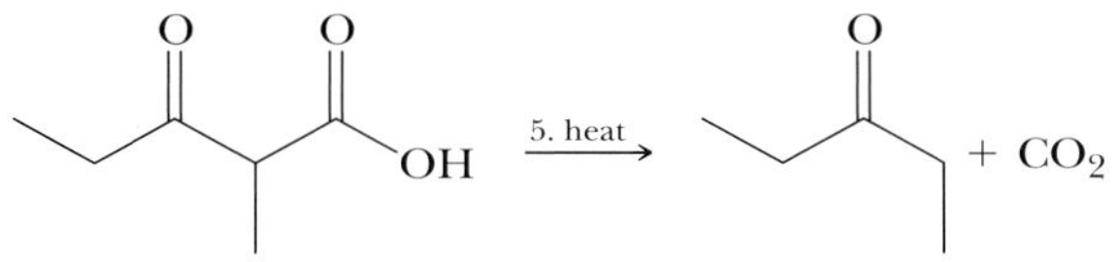

The result of these five steps is a reaction between two molecules of ester, one furnishing a carboxyl group and the other furnishing an enolate anion, to give a ketone and carbon dioxide:

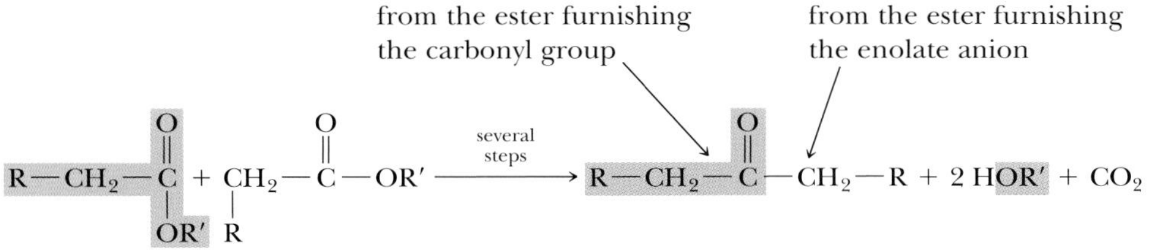

In the general reaction, both ester molecules are the same, and the product is a symmetrical ketone.

EXAMPLE 16.7

Each set of compounds undergoes (1, 2) Claisen condensation, (3) saponification followed by (4) acidification, and (5) thermal decarboxylation:

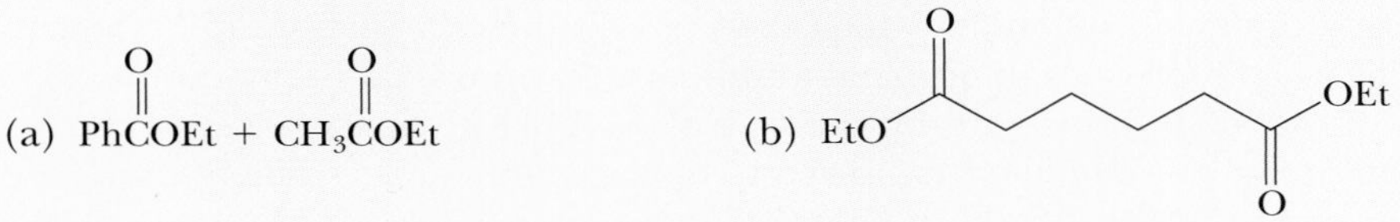

(a) PhCOEt + CH$_3$COEt

(b) EtO ... OEt

Draw a structural formula of the product after completion of this reaction sequence.

SOLUTION

Steps 1 and 2 bring about a crossed Claisen condensation in (a) and a Dieckmann condensation in (b) to form a β-ketoester. Steps 3 and 4 bring about hydrolysis of the β-ketoester to give a β-ketoacid, and Step 5 brings about decarboxylation to give a ketone:

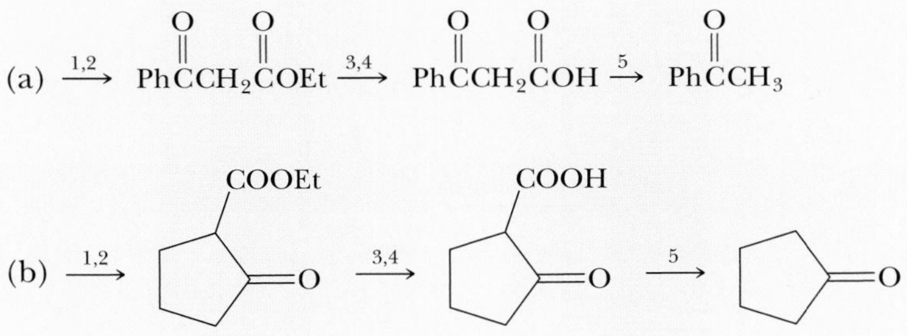

Practice Problem 16.7

Show how to convert benzoic acid to 3-methyl-1-phenyl-1-butanone by using a Claisen condensation at some stage in the synthesis:

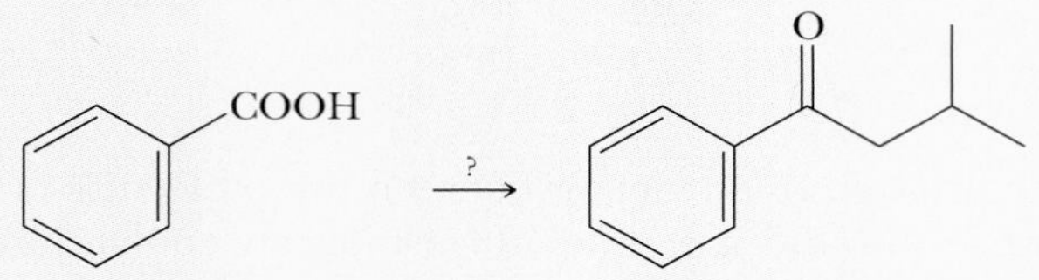

Benzoic acid   3-Methyl-1-phenyl-1-butanone

16.5 CLAISEN AND ALDOL CONDENSATIONS IN THE BIOLOGICAL WORLD

Carbonyl condensations are among the most widely used reactions in the biological world for the assembly of new carbon–carbon bonds in such important biomolecules as fatty acids, cholesterol, steroid hormones, and terpenes. One source of carbon atoms for the synthesis of these biomolecules is **acetyl-CoA**, a thioester of acetic acid and the thiol group of coenzyme A (CoA-SH, Section 22.2D). The function of the coenzyme A group of acetyl-CoA is to anchor the acetyl group on the surface of the enzyme systems that catalyze the reactions we examine in this section. In the discussions that follow, we will not be concerned with the mechanism by which each enzyme-catalyzed reaction occurs. Rather, our concern is with recognizing the type of reaction that takes place in each step.

In the Claisen condensation catalyzed by the enzyme thiolase, acetyl-CoA is converted to its enolate anion, which then attacks the carbonyl group of a second molecule of acetyl-CoA to form a tetrahedral carbonyl addition intermediate. Collapse of this intermediate by the loss of CoA-SH gives acetoacetyl-CoA. The mechanism for this condensation reaction is exactly the same as that of the Claisen condensation (Section 16.4A):

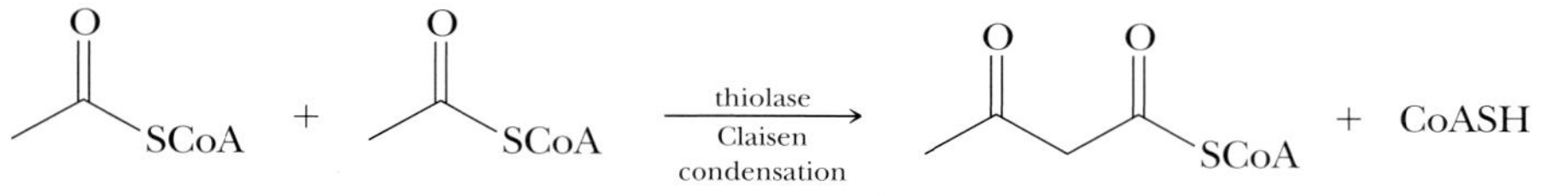

Acetyl-CoA   Acetyl-CoA   Acetoacetyl-CoA   Coenzyme A

An enzyme-catalyzed aldol reaction with a third molecule of acetyl-CoA on the ketone carbonyl of acetoacetyl-CoA gives (*S*)-3-hydroxy-3-methylglutaryl-CoA:

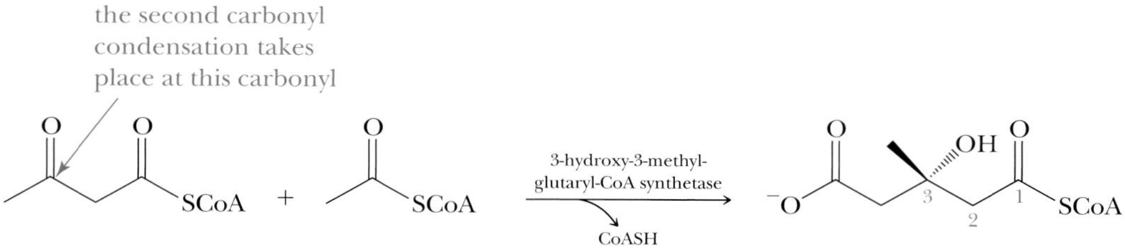

the second carbonyl condensation takes place at this carbonyl

(*S*)-3-Hydroxy-
3-methylglutaryl-CoA

Note three features of this reaction. First, the creation of the new stereocenter is stereoselective: Only the S enantiomer is formed. Although the acetyl group of each reactant is achiral, their condensation takes place in a chiral environment created by the enzyme 3-hydroxy-3-methylglutaryl-CoA synthetase. Second, hydrolysis of the thioester group of acetyl-CoA is coupled with the aldol reaction. Third, the carboxyl group is shown as it is ionized at pH 7.4, the approximate pH of blood plasma and many cellular fluids.

Enzyme-catalyzed reduction of the thioester group of 3-hydroxy-3-methylglutaryl-CoA to a primary alcohol gives mevalonic acid, shown here as its anion:

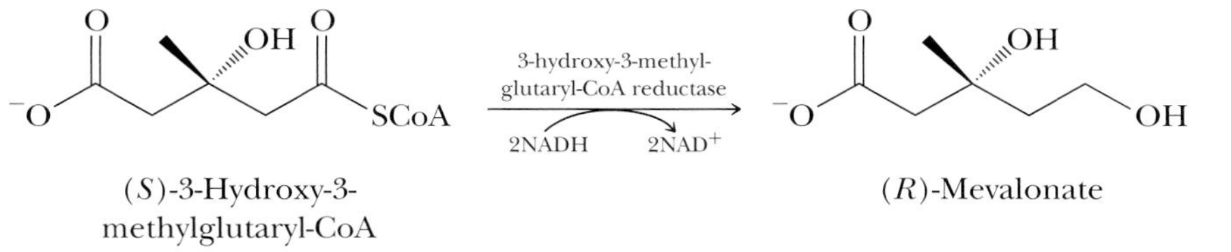

(S)-3-Hydroxy-3-methylglutaryl-CoA

(R)-Mevalonate

The reducing agent for this transformation is nicotinamide adenine dinucleotide, abbreviated NADH (Section 22.2B). This reducing agent is the biochemical equivalent of LiAlH$_4$. Each reducing agent functions by delivering a hydride ion (H:$^-$) to the carbonyl carbon of an aldehyde, a ketone, or an ester. Note that, in the reduction, a change occurs in the designation of configuration from S to R, not because of any change in configuration at the stereocenter, but rather because of a change in priority among the four groups bonded to the stereocenter.

Enzyme-catalyzed transfer of a phosphate group from adenosine triphosphate (ATP, Section 20.2) to the 3-hydroxyl group of mevalonate gives a phosphoric ester at carbon 3. Enzyme-catalyzed transfer of a pyrophosphate group (Section 15.2B) from a second molecule of ATP gives a pyrophosphoric ester at carbon 5. Enzyme-catalyzed β-elimination from this molecule results in the loss of CO_2 and PO_4^{3-}, both good leaving groups:

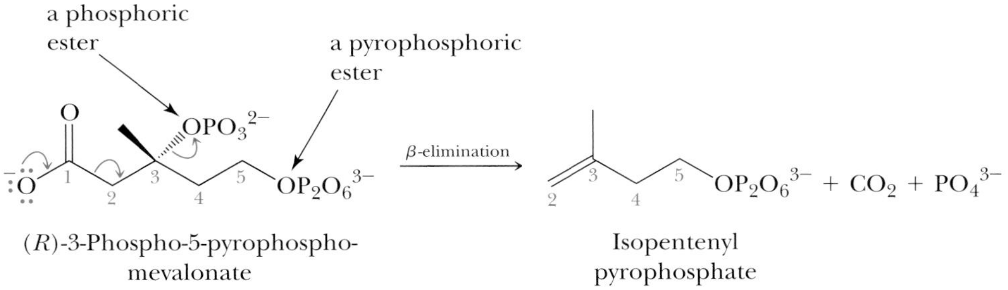

(R)-3-Phospho-5-pyrophospho-mevalonate

Isopentenyl pyrophosphate

Isopentenyl pyrophosphate has the carbon skeleton of isoprene, the unit into which the carbon skeletons of terpenes can be divided (Section 4.5). This molecule is, in fact, a key intermediate in the biosynthesis of isoprene, terpenes, cholesterol, steroid hormones, and bile acids:

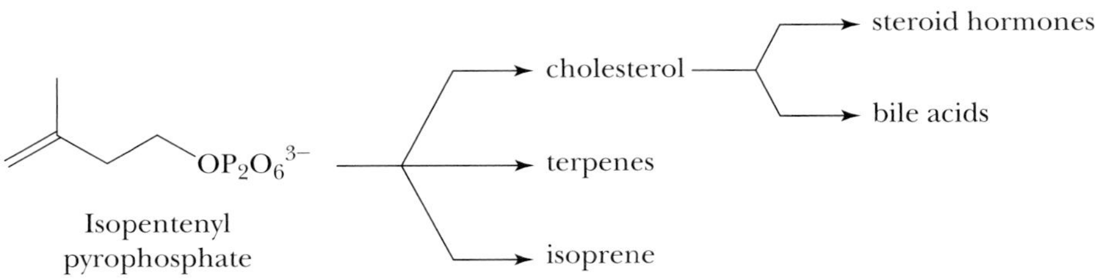

Isopentenyl pyrophosphate

CHEMICAL CONNECTIONS

Drugs That Lower Plasma Levels of Cholesterol

Coronary artery disease is the leading cause of death in the United States and other Western countries, where about one half of all deaths can be attributed to atherosclerosis. Atherosclerosis results from the buildup of fatty deposits called plaque on the inner walls of arteries. A major component of plaque is cholesterol derived from low-density-lipoproteins (LDL), which circulate in blood plasma. Because more than one half of total body cholesterol in humans is synthesized in the liver from acetyl-CoA, intensive efforts have been directed toward finding ways to inhibit this synthesis. The rate-determining step in cholesterol biosynthesis is reduction of 3-hydroxy-3-methylglutaryl-CoA (HMG-CoA) to mevalonic acid. This reduction is catalyzed by the enzyme HMG-CoA reductase and requires two moles of NADPH per mole of HMG-CoA.

Beginning in the early 1970s, researchers at the Sankyo Company in Tokyo screened more than 8000 strains of microorganisms and in 1976 announced the isolation of mevastatin, a potent inhibitor of HMG-CoA reductase, from culture broths of the fungus *Penicillium citrinum*. The same compound was isolated by researchers at Beecham Pharmaceuticals in England from cultures of *Penicillium brevicompactum*. Soon thereafter, a second, more active compound called lovastatin was isolated at the Sankyo Company from the fungus *Monascus ruber*, and at Merck Sharpe & Dohme from *Aspergillus terreus*. Both mold metabolites are extremely effective in lowering plasma concentrations of LDL. The active form of each is the 5-hydroxycarboxylate anion formed by hydrolysis of the δ-lactone.

The active form of each drug

These drugs and several synthetic modifications now available inhibit HMG-CoA reductase by forming an enzyme-inhibitor complex that prevents further catalytic action of the enzyme. It is reasoned that the 3,5-dihydroxycarboxylate anion part of the active form of each drug binds tightly to the enzyme because it mimics the hemithioacetal intermediate formed by the first reduction of HMG-CoA.

3-Hydroxy-3-methyl glutaryl-CoA (HMG-CoA)

A hemithioacetal intermediate formed by the first NADPH reduction

Mevalonate

R₁ = R₂ = H, mevastatin
R₁ = H, R₂ = CH₃, lovastatin (Mevacor)
R₁ = R₂ = CH₃, simvastatin (Zocor)

hydrolysis of the δ-lactone

Systematic studies have shown the importance of each part of the drug for effectiveness. It has been found, for example, that the carboxylate anion (—COO⁻) is essential, as are both the 3-OH and 5-OH groups. It has also been shown that almost any modification of the two fused six-membered rings and their pattern of substitution reduced potency.

16.6 THE MICHAEL REACTION: CONJUGATE ADDITION TO α,β-UNSATURATED CARBONYLS

Thus far, we have used carbon nucleophiles in two ways to form new carbon–carbon bonds:

1. Addition of organomagnesium (Grignard) reagents to the carbonyl groups of aldehydes, ketones, and esters.
2. Addition of enolate anions derived from aldehydes or ketones (aldol reactions) and esters (Claisen and Dieckmann condensations) to the carbonyl groups of other aldehydes, ketones, or esters.

Addition of an enolate anion to a carbon–carbon double bond conjugated with a carbonyl group presents an entirely new synthetic strategy. In this section, we study a type of conjugate addition involving nucleophilic addition to an electrophilic double bond.

A. Michael Addition of Enolate Anions

Nucleophilic addition of enolate anions to α,β-unsaturated carbonyl compounds was first reported in 1887 by the American chemist Arthur Michael. Following are two examples of **Michael reactions**. In the first example, the nucleophile is the enolate anion of diethyl malonate. In the second example, the nucleophile is the enolate anion of ethyl acetoacetate:

Michael reaction The conjugate addition of an enolate anion or other nucleophile to an α,β-unsaturated carbonyl compound.

EtOOC—⟍⟋—COOEt + (3-Buten-2-one) $\xrightarrow[\text{EtOH}]{\text{EtO}^-\text{Na}^+}$ EtOOC—⟍⟋—COOEt with ketone chain

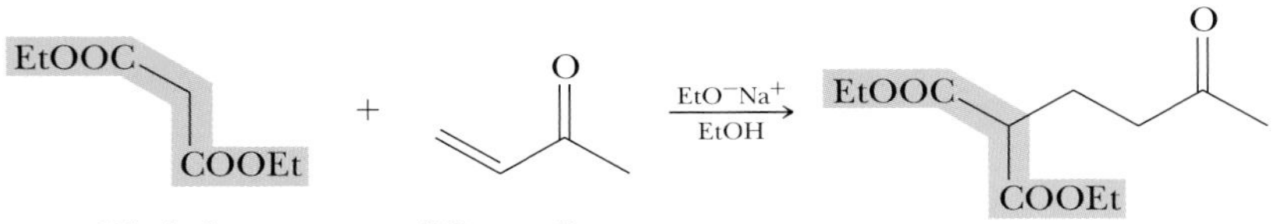

Diethyl propanedioate (Diethyl malonate) | 3-Buten-2-one (Methyl vinyl ketone)

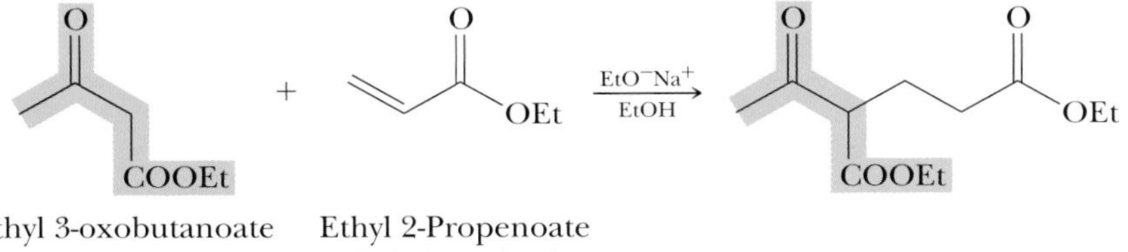

Ethyl 3-oxobutanoate (Ethyl acetoacetate) — COOEt + Ethyl 2-Propenoate —OEt $\xrightarrow[\text{EtOH}]{\text{EtO}^-\text{Na}^+}$ product —OEt with COOEt

Ethyl 3-oxobutanoate (Ethyl acetoacetate) | Ethyl 2-Propenoate (Ethyl acrylate)

Recall that nucleophiles don't ordinarily add to carbon–carbon double bonds. Rather, they add to electrophiles (Section 5.3). What activates a carbon–carbon double bond for nucleophilic attack in a Michael reaction is the presence of the adjacent carbonyl group. One important resonance structure of α,β-unsaturated carbonyl compounds puts a positive charge on the β-carbon of the double bond, making it electrophilic in its reactivity:

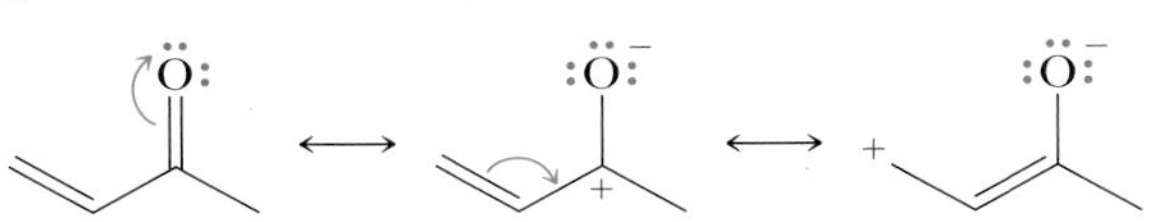

Although the major fraction of the partial positive charge of an α,β-unsaturated aldehyde or ketone is on the carbonyl carbon, there is nevertheless a significant partial positive charge on the beta carbon.

Thus, nucleophiles can add to this type of double bond, which we call "activated" for that reason.

Table 16.2 lists the most common combinations of α,β-unsaturated carbonyl compounds and nucleophiles used in Michael reactions. The most commonly used bases are metal alkoxides, pyridine, and piperidine.

TABLE 16.2   Combinations of Reagents for Effective Michael Reactions

These Types of α,β-Unsaturated Compounds Are Nucleophile Acceptors in Michael Reactions		These Types of Compounds Provide Effective Nucleophiles for Michael Reactions	
$CH_2{=}CHCH$ (O)	Aldehydes	$CH_3CCH_2CCH_3$ (O, O)	β-Diketones
$CH_2{=}CHCCH_3$ (O)	Ketones	CH_3CCH_2COEt (O, O)	β-Ketoesters
$CH_2{=}CHCOEt$ (O)	Esters	$EtOCCH_2COEt$ (O, O)	β-Diesters
		RNH_2, R_2NH	Amines

We can write the following general mechanism for a Michael reaction:

Mechanism: Michael Reaction—Conjugate Addition of Enolate Anions

Step 1: Treatment of H—Nu with base gives the nucleophile, Nu:⁻.

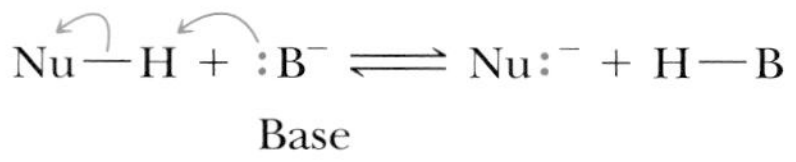

$$Nu{-}H + :B^- \rightleftharpoons Nu:^- + H{-}B$$

Base

Step 2: Nucleophilic addition of Nu:⁻ to the β-carbon of the conjugated system gives a resonance-stabilized enolate anion:

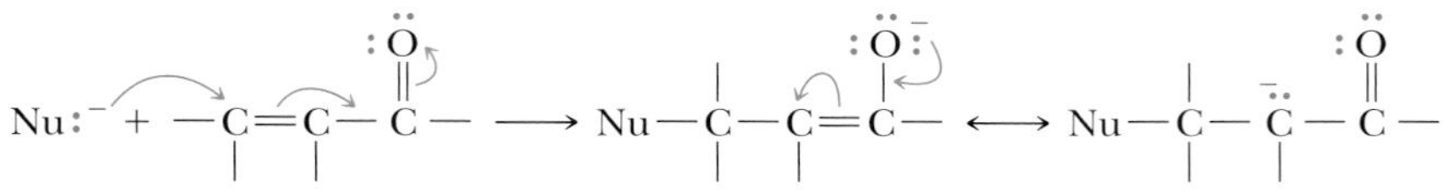

A resonance-stabilized enolate anion

Step 3: Proton transfer from H—B gives the enol and regenerates the base:

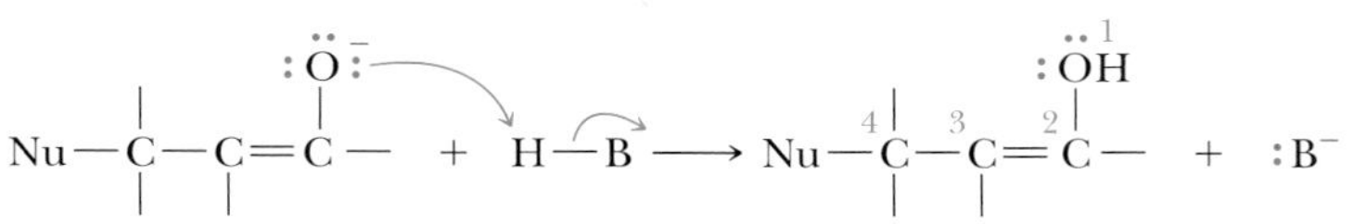

An enol
(a product of 1,4-addition)

Note that the enol formed in this step corresponds to 1,4-addition to the conjugated system of the α,β-unsaturated carbonyl compound. It is because this intermediate is formed that the Michael reaction is classified as a 1,4-, or conjugate, addition. Note also that, the base, B:⁻, is regenerated, in accordance with the experimental observation that a Michael reaction requires only a catalytic amount of base rather than a molar equivalent.

Step 4: Tautomerism (Section 13.9A) of the less stable enol form gives the more stable keto form:

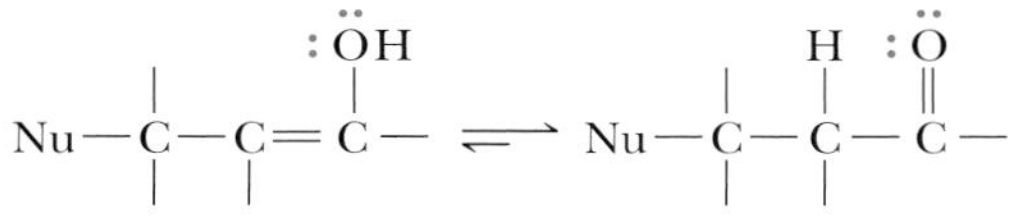

Enol form
(less stable)

Keto form
(more stable)

EXAMPLE 16.8

Draw a structural formula for the product formed by treating each set of reactants with sodium ethoxide in ethanol under conditions of the Michael reaction:

(a) 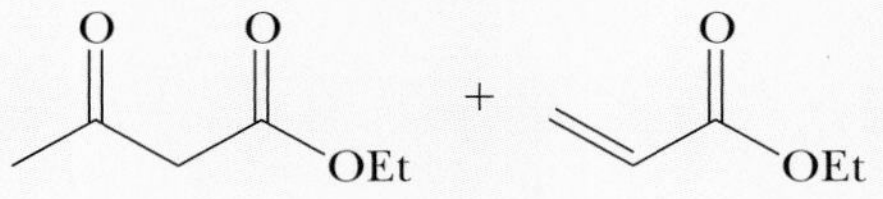 +

(b) 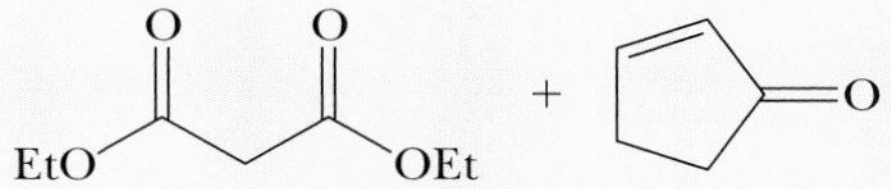 +

SOLUTION

(a)

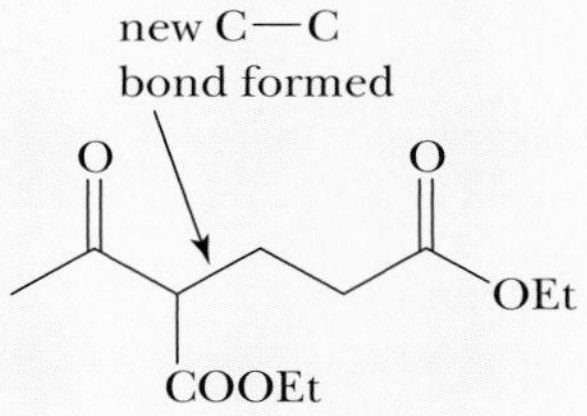

new C—C
bond formed

COOEt

(b)

new C—C
bond formed

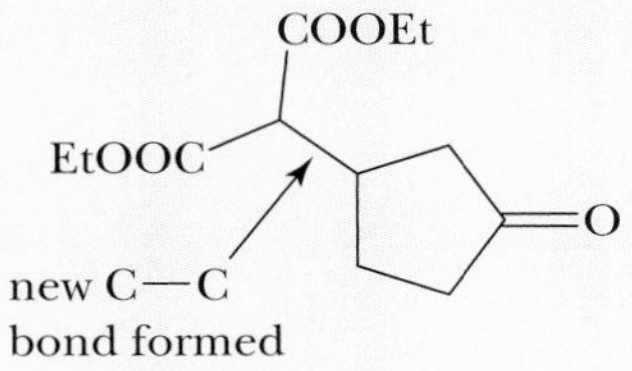

Practice Problem 16.8

Show the product formed from each Michael product in the solution to Example 16.8 after (1) hydrolysis in aqueous NaOH, (2) acidification, and (3) thermal decarboxylation of each β-ketoacid or β-dicarboxylic acid. These reactions illustrate the usefulness of the Michael reaction for the synthesis of 1,5-dicarbonyl compounds.

EXAMPLE 16.9

Show how the series of reactions in Example 16.8 and Problem 16.8 (Michael reaction, hydrolysis, acidification, and thermal decarboxylation) can be used to prepare 2,6-heptanedione.

SOLUTION

The key is to recognize that a COOH group beta to a ketone can be lost by decarboxylation. Once you find where that COOH group might have been located, you should see which carbons of the target molecule can be derived from the carbon skeleton of ethyl acetoacetate and which carbons can be derived from an α,β-unsaturated carbonyl compound. As shown here, the target molecule can be constructed from the carbon skeletons of ethyl acetoacetate and methyl vinyl ketone:

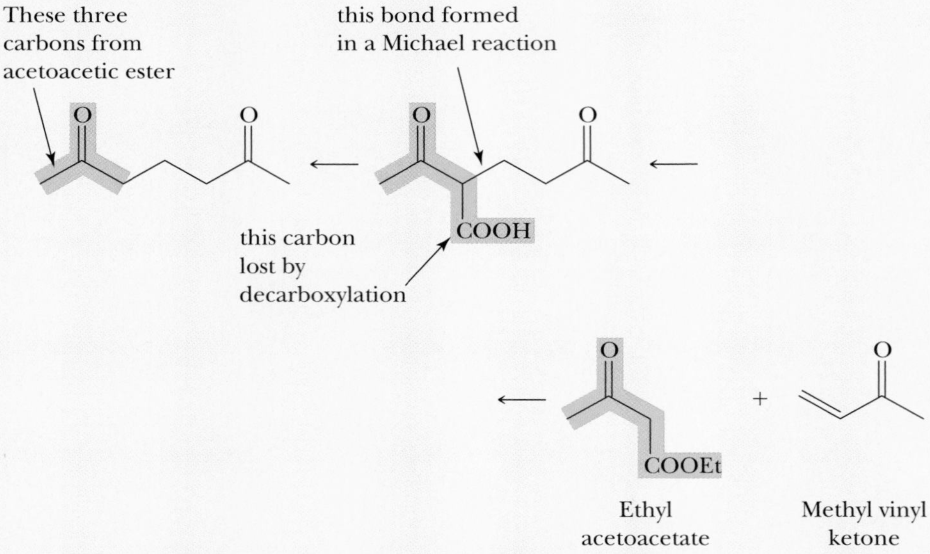

These three carbons from acetoacetic ester

this bond formed in a Michael reaction

this carbon lost by decarboxylation

Ethyl acetoacetate

Methyl vinyl ketone

Following are the steps in their conversion to 2,6-heptanedione:

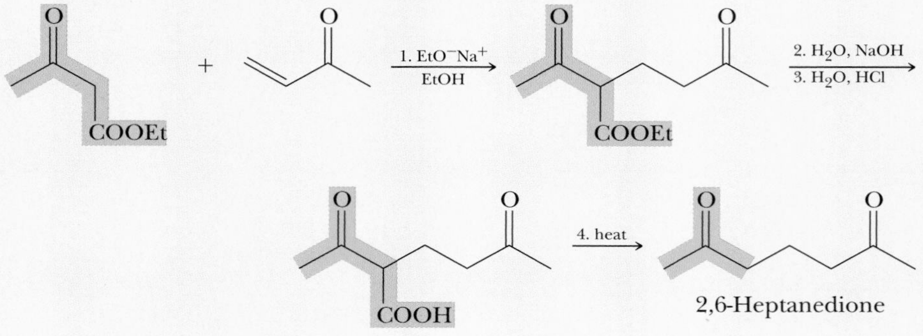

1. EtO⁻Na⁺ / EtOH

2. H₂O, NaOH
3. H₂O, HCl

4. heat

2,6-Heptanedione

Practice Problem 16.9

Show how the sequence consisting of Michael reaction, hydrolysis, acidification, and thermal decarboxylation can be used to prepare pentanedioic acid (glutaric acid).

B. Michael Addition of Amines

As Table 16.2 shows, aliphatic amines also function as nucleophiles in Michael reactions. Diethylamine, for example, adds to methyl acrylate, as shown in the following equation:

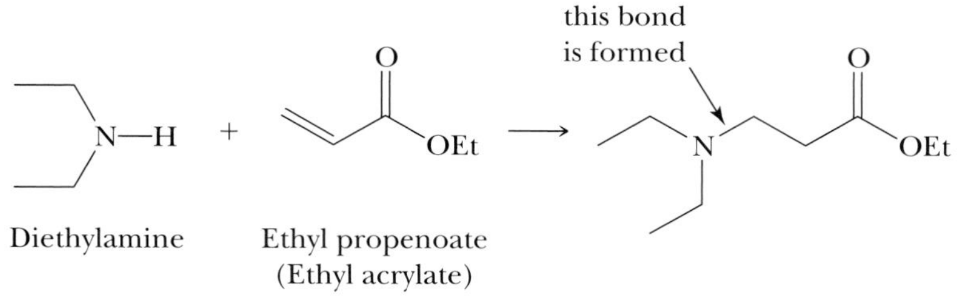

this bond is formed

Diethylamine

Ethyl propenoate (Ethyl acrylate)

EXAMPLE 16.10

Methylamine, CH_3NH_2, has two N—H bonds, and 1 mole of methylamine undergoes Michael reaction with 2 moles of ethyl acrylate. Draw a structural formula for the product of this double Michael reaction.

SOLUTION

$$H_3C-N \overset{H}{\underset{H}{\big\langle}} \quad + \ 2 \ CH_2 = CH-COOEt \ \longrightarrow \ H_3C-N \overset{CH_2-CH_2-COOEt}{\underset{CH_2-CH_2-COOEt}{\big\langle}}$$

Practice Problem 16.10

The product of the double Michael reaction in Example 16.10 is a diester that, when treated with sodium ethoxide in ethanol, undergoes a Dieckmann condensation. Draw the structural formula for the product of this Dieckmann condensation followed by acidification with aqueous HCl.

SUMMARY

An **enolate anion** is an anion formed by the removal of an α-hydrogen from a carbonyl-containing compound (Section 16.2). Aldehydes, ketones, and esters can be converted to their enolate anions (Section 16.3A) by treatment with a metal alkoxide or other strong base. An **aldol reaction** (Section 16.3B) is the addition of an enolate anion from one aldehyde or ketone to the carbonyl carbon of another aldehyde or ketone to form a **β-hydroxyaldehyde** or **β-hydroxyketone**. Dehydration of the product of an aldol reaction gives an **α,β-unsaturated aldehyde** or **ketone**. **Crossed aldol reactions** are useful only when appreciable differences in reactivity exist between the two carbonyl-containing compounds, such as when one of them has no α-hydrogens and can function only as an enolate anion acceptor. When both carbonyl groups are in the same molecule, aldol reaction results in the formation of a ring.

Intramolecular aldol reactions are particularly useful for forming five- and six-membered rings.

A key step in the **Claisen** and **Dieckmann condensations** (Section 16.4) is the addition of an enolate anion of one ester to a carbonyl group of another ester to form a tetrahedral carbonyl addition intermediate, followed by the collapse of the intermediate to give a β-ketoester.

Acetyl-CoA (Section 16.5) is the source of the carbon atoms for the synthesis of terpenes, cholesterol, steroid hormones, and fatty acids. Key intermediates in the synthesis of these biomolecules are mevalonic acid and isopentenyl pyrophosphate.

The **Michael reaction** (Section 16.6) is the addition of a nucleophile to a carbon–carbon double bond activated by an adjacent carbonyl group.

KEY REACTIONS

1. The Aldol Reaction (Section 16.3B)

The aldol reaction involves the nucleophilic addition of an enolate anion from one aldehyde or ketone to the carbonyl carbon of another aldehyde or ketone to give a β-hydroxyaldehyde or β-hydroxyketone:

2. Dehydration of the Product of an Aldol Reaction (Section 16.3)

Dehydration of the β-hydroxyaldehyde or ketone from an aldol reaction occurs readily and gives an α,β-unsaturated aldehyde or β-hydroxyketone:

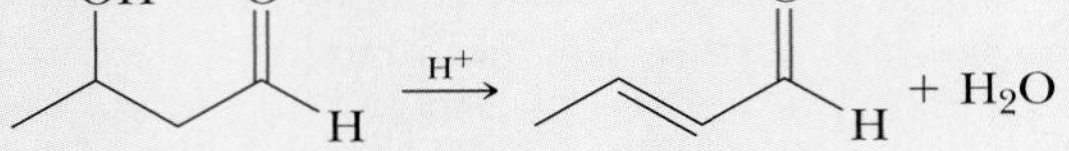

3. The Claisen Condensation (Section 16.4A)

The product of a Claisen condensation is a β-ketoester:

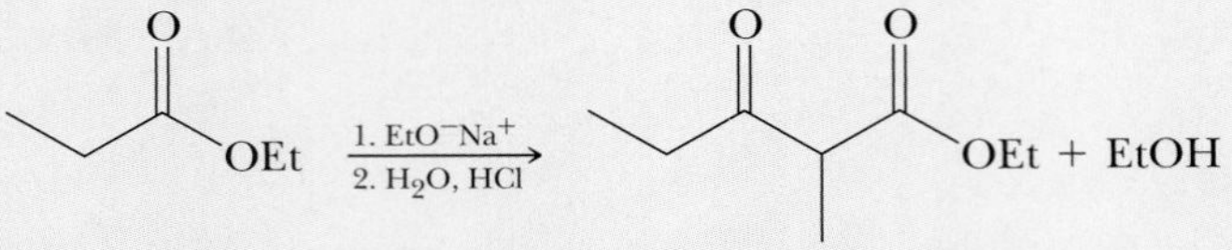

Condensation occurs by nucleophilic acyl substitution in which the attacking nucleophile is the enolate anion of an ester.

4. The Dieckmann Condensation (Section 16.4B)

An intramolecular Claisen condensation is called a Dieckmann condensation:

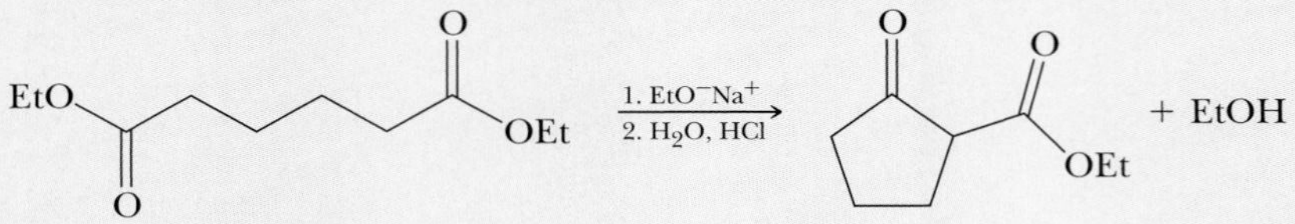

5. Crossed Claisen Condensations (Section 16.4C)

Crossed Claisen condensations are useful only when an appreciable difference exists in the reactivity between the two esters. Such is the case when an ester that has no α-hydrogens can function only as an enolate anion acceptor:

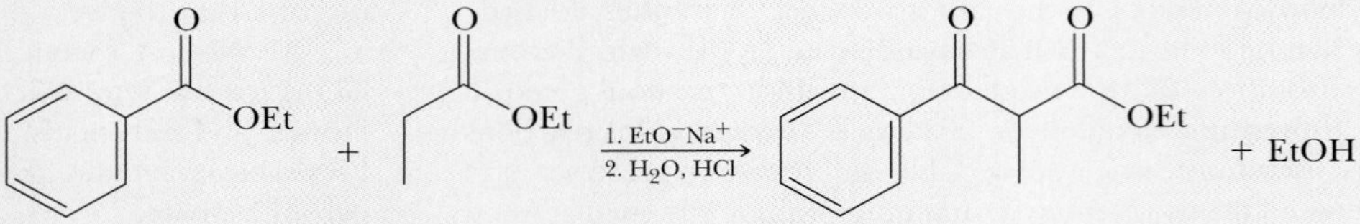

6. Hydrolysis and Decarboxylation of β-Ketoesters (Section 16.4D)

Hydrolysis of the ester, followed by decarboxylation of the resulting β-ketoacid, gives a ketone and carbon dioxide:

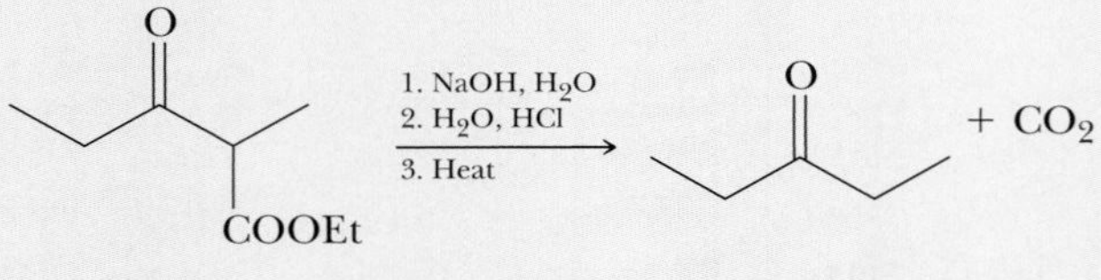

7. The Michael Reaction (Section 16.6)

Attack of a nucleophile at the β carbon of an α,β-unsaturated carbonyl compound results in conjugate addition:

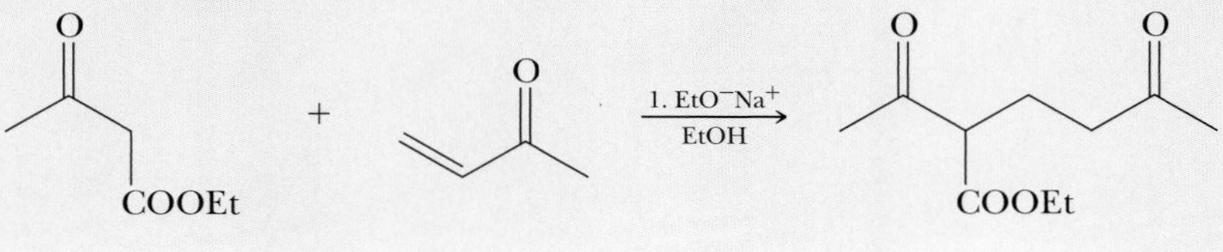

PROBLEMS

A problem number set in red indicates an applied "real-world" problem.

The Aldol Reaction

16.11 Estimate the pK_a of each compound, and then arrange them in order of increasing acidity:

(a) $CH_3\overset{\overset{\displaystyle O}{\|}}{C}CH_3$ (b) $CH_3\overset{\overset{\displaystyle OH}{|}}{C}HCH_3$ (c) $CH_3CH_2\overset{\overset{\displaystyle O}{\|}}{C}OH$

16.12 Identify the most acidic hydrogen(s) in each compound:

(a) 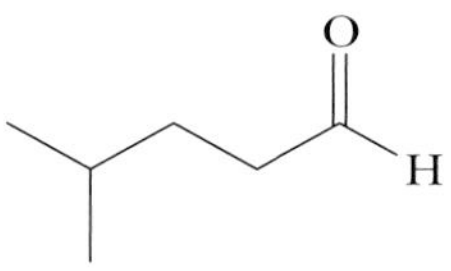   (b) 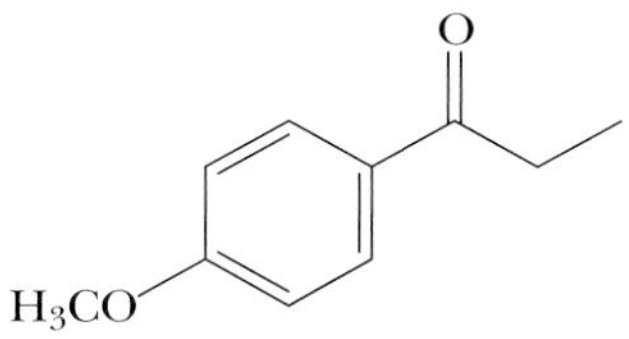   (c)

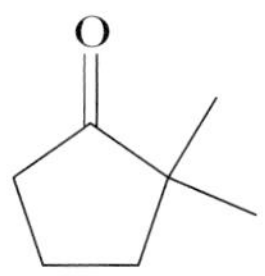

(d) 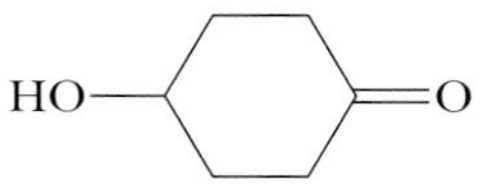   (e) 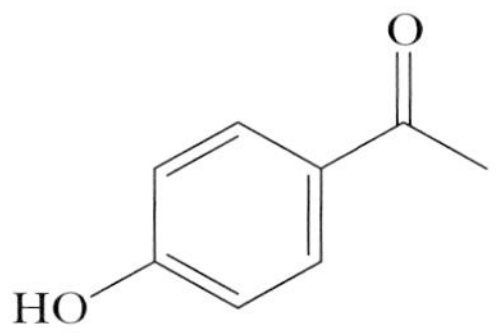   (f)

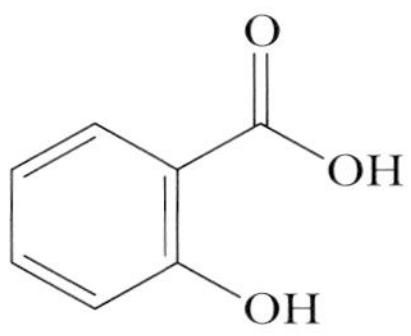

16.13 Write a second contributing structure of each anion, and use curved arrows to show the redistribution of electrons that gives your second structure:

(a) $CH_3CH_2\overset{\overset{\displaystyle :\ddot{O}:^-}{|}}{C}=CHCH_3$ (b) 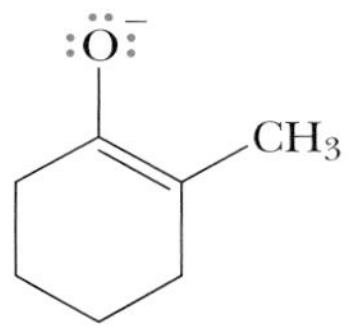   (c)

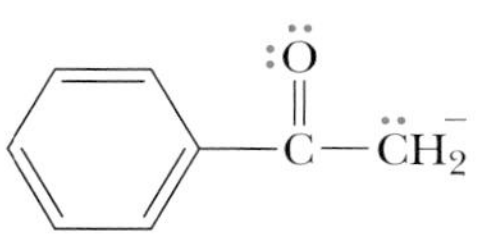

16.14 Treatment of 2-methylcyclohexanone with base gives two different enolate anions. Draw the contributing structure for each that places the negative charge on carbon.

16.15 Draw a structural formula for the product of the aldol reaction of each compound and for the α,β-unsaturated aldehyde or ketone formed by dehydration of each aldol product:

(a) 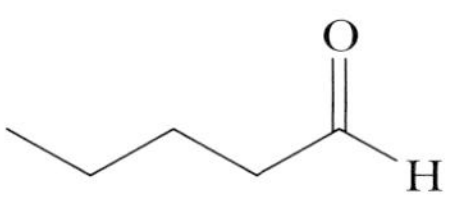   (b) 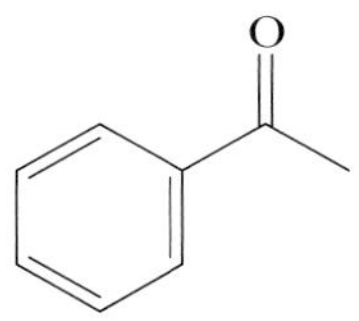   (c) 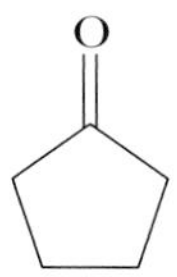   (d)

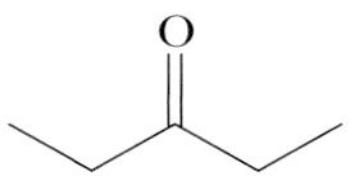

16.16 Draw a structural formula for the product of each crossed aldol reaction and for the compound formed by dehydration of each aldol product:

(a) $(CH_3)_3C\overset{\overset{\displaystyle O}{\|}}{C}H + CH_3\overset{\overset{\displaystyle O}{\|}}{C}CH_3$ (b) 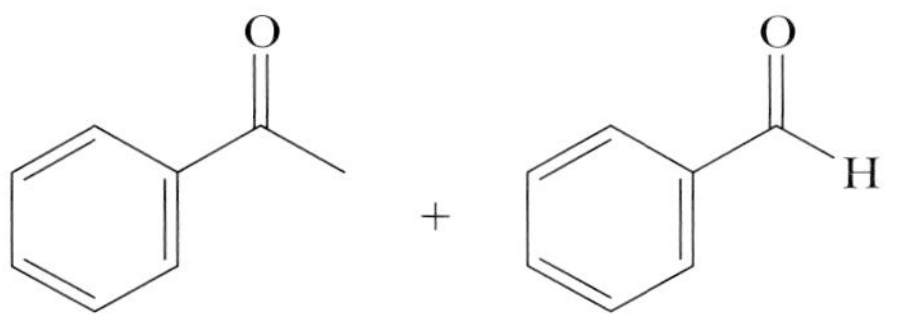 +

(c) 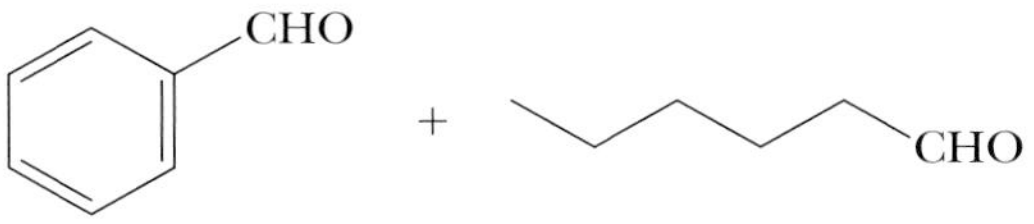 +   (d) 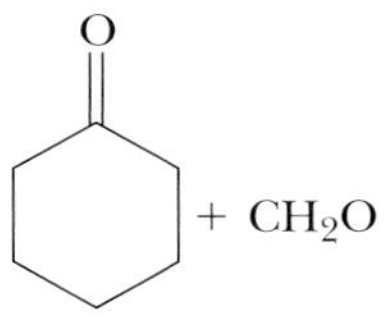 + CH_2O

16.17 When a 1:1 mixture of acetone and 2-butanone is treated with base, six aldol products are possible. Draw a structural formula for each.

16.18 Show how to prepare each α,β-unsaturated ketone by an aldol reaction followed by dehydration of the aldol product:

(a) 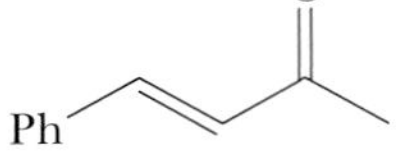   (b)

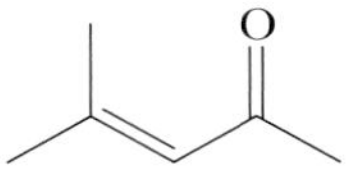

16.19 Show how to prepare each α,β-unsaturated aldehyde by an aldol reaction followed by dehydration of the aldol product:

(a) 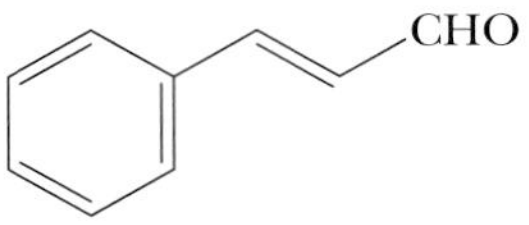   (b)

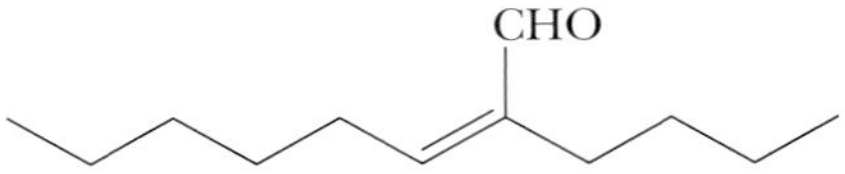

16.20 When treated with base, the following compound undergoes an intramolecular aldol reaction, followed by dehydration, to give a product containing a ring (yield 78%):

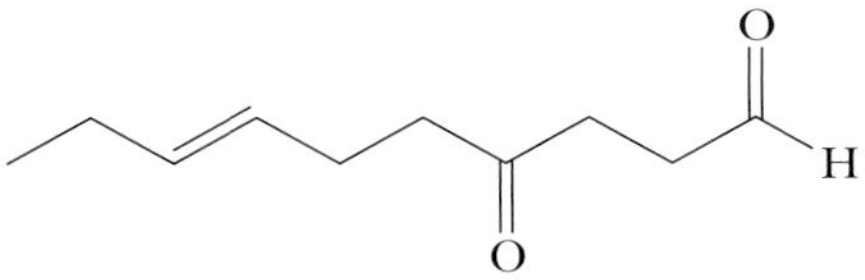

Propose a structural formula for this product.

16.21 Propose a structural formula for the compound with molecular formula $C_6H_{10}O_2$ that undergoes an aldol reaction followed by dehydration to give this α,β-unsaturated aldehyde:

$$C_6H_{10}O_2 \xrightarrow{\text{base}}$$ $+ \ H_2O$

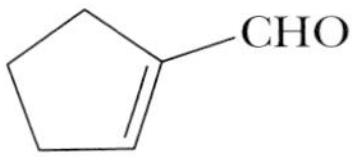

1-Cyclopentenecarbaldeyde

16.22 Show how to bring about this conversion:

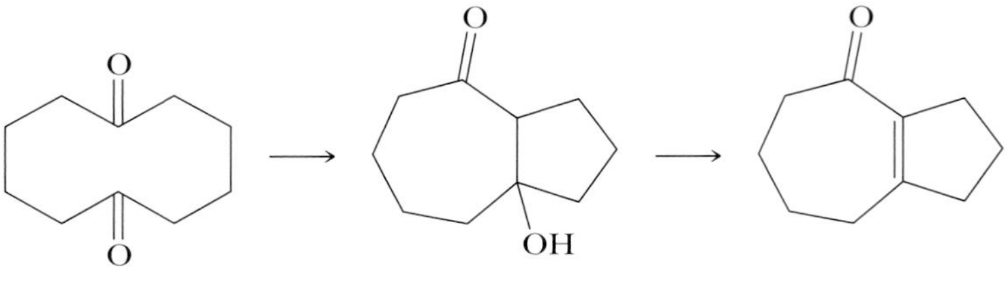

16.23 Oxanamide, a mild sedative, is synthesized from butanal in these five steps:

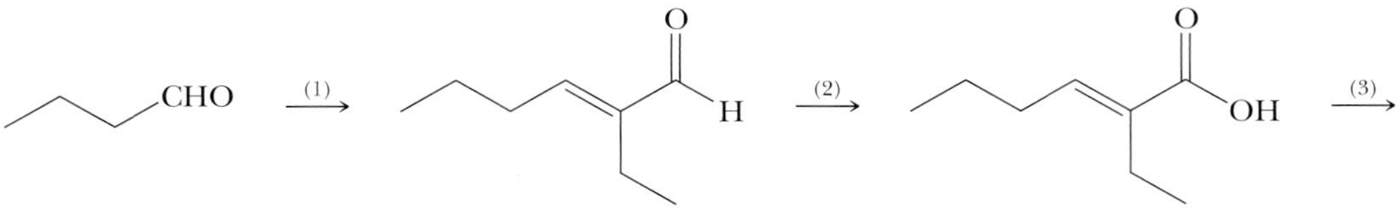

Butanal   2-Ethyl-2-hexenal   2-Ethyl-2-hexenoic acid

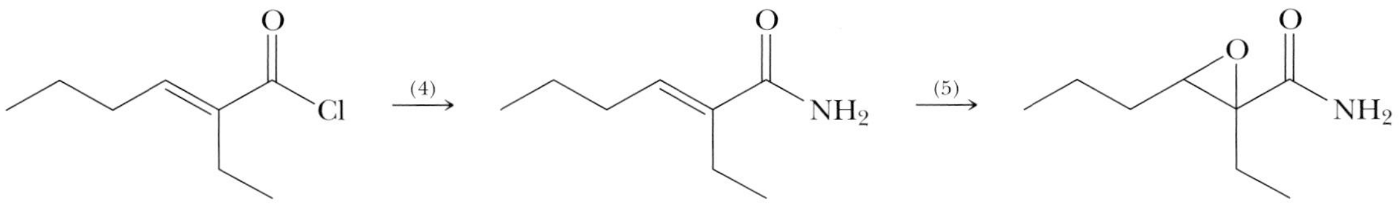

2-Ethyl-2-hexenoyl chloride   2-Ethyl-2-hexenamide   Oxanamide

(a) Show reagents and experimental conditions that might be used to bring about each step in this synthesis.

(b) How many stereocenters are in oxanamide? How many stereoisomers are possible for oxanamide?

16.24 Propose structural formulas for compounds A and B:

$$\xrightarrow[\text{H}_2\text{CrO}_4]{} \quad A(C_{11}H_{18}O_2) \xrightarrow[\text{EtOH}]{\text{EtO}^-\text{Na}^+} B(C_{11}H_{16}O)$$

The Claisen and Dieckmann Condensations

16.25 Show the product of Claisen condensation of each ester:
   (a) Ethyl phenylacetate in the presence of sodium ethoxide.
   (b) Methyl hexanoate in the presence of sodium methoxide.

16.26 Draw a structural formula for the product of saponification, acidification, and decarboxylation of each β-ketoester formed in Problem 16.25.

16.27 When a 1:1 mixture of ethyl propanoate and ethyl butanoate is treated with sodium ethoxide, four Claisen condensation products are possible. Draw a structural formula for each product.

16.28 Draw a structural formula for the β-ketoester formed in the crossed Claisen condensation of ethyl propanoate with each ester:

(a) EtOC—COEt   (b) PhCOEt   (c) HCOEt

16.29 Complete the equation for this crossed Claisen condensation:

$$+ \; CH_3COCH_2CH_3 \xrightarrow[\text{2. H}_2\text{O, HCl}]{\text{1. EtO}^-\text{Na}^+}$$

16.30 The Claisen condensation can be used as one step in the synthesis of ketones, as illustrated by this reaction sequence:

$$\xrightarrow[\text{2. HCl, H}_2\text{O}]{\text{1. EtO}^-\text{Na}^+} A \xrightarrow[\text{heat}]{\text{NaOH, H}_2\text{O}} B \xrightarrow[\text{heat}]{\text{HCl, H}_2\text{O}} C_9H_{18}O$$

Propose structural formulas for compounds A, B, and the ketone formed in the sequence.

16.31 Draw a structural formula for the product of treating each diester with sodium ethoxide followed by acidification with HCl (*Hint:* These are Dieckmann condensations):

(a)   (b)

16.32 Claisen condensation between diethyl phthalate and ethyl acetate, followed by saponification, acidification, and decarboxylation, forms a diketone, $C_9H_6O_2$. Propose structural formulas for compounds A, B, and the diketone:

$$+ \; CH_3COOEt \xrightarrow[\text{2. HCl, H}_2\text{O}]{\text{1. EtO}^-\text{Na}^+} A \xrightarrow[\text{heat}]{\text{NaOH, H}_2\text{O}} B \xrightarrow[\text{heat}]{\text{HCl, H}_2\text{O}} C_9H_6O_2$$

Diethyl phthalate   Ethyl acetate

16.33 The rodenticide and insecticide pindone is synthesized by the following sequence of reactions:

Diethyl   3,3-Dimethyl-2-   Pindone
phthalate   butanone

Propose a structural formula for pindone.

16.34 Fentanyl is a nonopoid (nonmorphinelike) analgesic used for the relief of severe pain. It is approximately 50 times more potent in humans than morphine itself. One synthesis for fentanyl begins with 2-phenylethanamine:

2-Phenylethanamine   (A)   (B)

(C)   (E)

(F)   Fentanyl

(a) Propose a reagent for Step 1. Name the type of reaction that occurs in this step.
(b) Propose a reagent to bring about Step 2. Name the type of reaction that occurs in this step.
(c) Propose a series of reagents that will bring about Step 3.
(d) Propose a reagent for Step 4. Identify the imine (Schiff base) part of Compound E.
(e) Propose a reagent to bring about Step 5.
(f) Propose two different reagents, either of which will bring about Step 6.
(g) Is fentanyl chiral? Explain.

16.35 Meclizine is an antiemetic. (It helps prevent or at least lessen the throwing up associated with motion sickness, including seasickness.) Among the names of the over-the-counter preparations of meclizine are Bonine®, Sea-Legs, Antivert®, and Navicalm®. Meclizine can be produced by the following series of reactions:

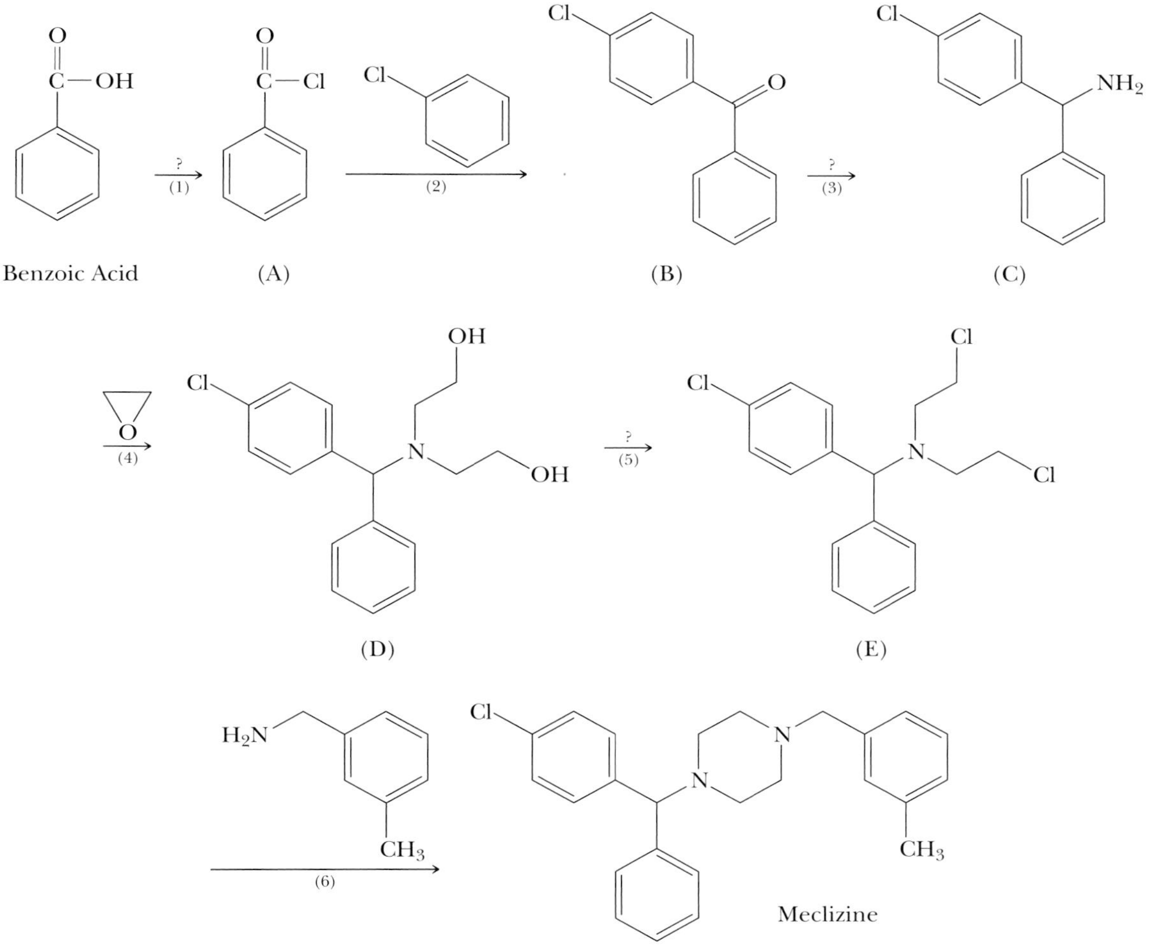

Benzoic Acid   (A)   (B)   (C)

(D)   (E)

Meclizine

(a) Name the functional group in (A). What reagent is most commonly used to convert a carboxyl group to this functional group?

(b) The catalyst for Step 2 is aluminum chloride, $AlCl_3$. Name the type of reaction that occurs in this step. The product shown here has the orientation of the new group para to the chlorine atom of chlorobenzene. Suppose you were not told the orientation of the new group. Would you have predicted it to be ortho, meta, or para to the chlorine atom? Explain.

(c) What set of reagents can be used in Step 3 to convert the C=O group to an —NH_2 group?

(d) The reagent used in Step 4 is the cyclic ether ethylene oxide. Most ethers are quite unreactive to nucleophiles such as the 1° amine in this step. Ethylene oxide, however, is an exception to this generalization. What is it about ethylene oxide that makes it so reactive toward ring-opening reactions with nucleophiles?

(e) What reagent can be used in Step 5 to convert each 1° alcohol to a 1° halide?

(f) Step 6 is a double nucleophilic displacement. Which mechanism is more likely for this reaction, S_N1 or S_N2? Explain.

16.36 2-Ethyl-1-hexanol is used for the synthesis of the sunscreen octyl *p*-methoxycinnamate. (See "Chemical Connections 15A".) This primary alcohol can be synthesized from butanal by the following series of steps:

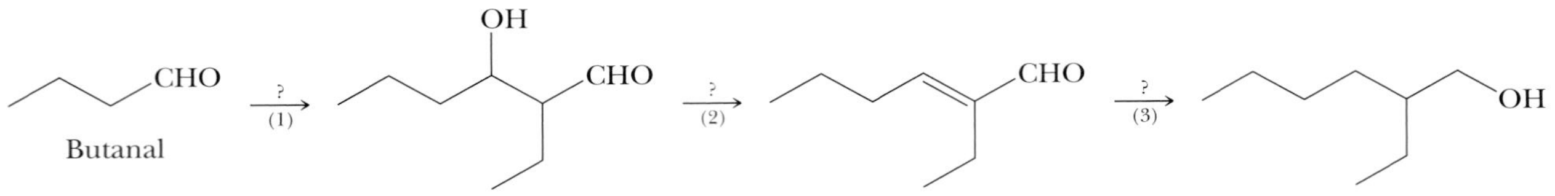

Butanal → (1) → (2) → (3) → 2-Ethyl-1-hexanol

(a) Propose a reagent to bring about Step 1. What name is given to this type of reaction?
(b) Propose a reagent for Step 2.
(c) Propose a reagent for Step 3.
(d) Following is a structural formula for the commercial sunscreening ingredient:

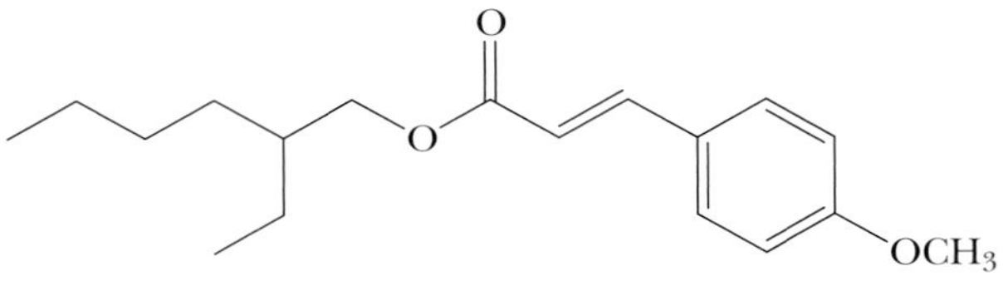

Octyl *p*-methoxycinnamate

What carboxylic acid would you use to form this ester? How would you bring about the esterification reaction?

Looking Ahead

16.37 The following reaction is one of the 10 steps in glycolysis (Section 22.4), a series of enzyme-catalyzed reactions by which glucose is oxidized to two molecules of pyruvate:

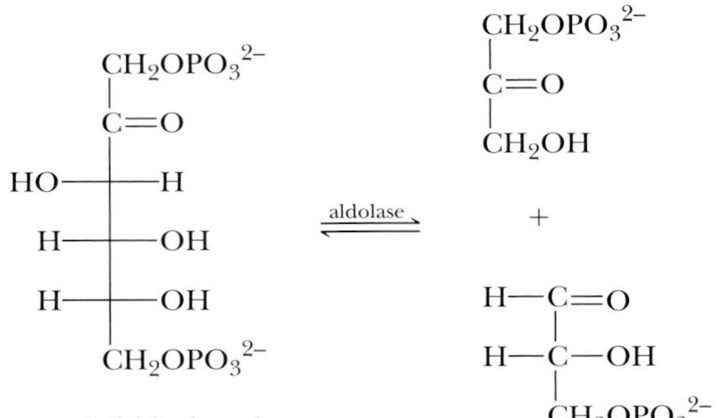

Fructose 1,6-bisphosphate $\xrightarrow{\text{aldolase}}$ Dihydroxyacetone phosphate + Glyceraldehyde 3-phosphate

Show that this step is the reverse of an aldol reaction.

16.38 The following reaction is the fourth in the set of four enzyme-catalyzed steps by which the hydrocarbon chain of a fatty acid (Section 22.6) is oxidized, two carbons at a time, to acetyl-coenzyme A:

$$R-\overset{O}{\underset{\|}{C}}-CH_2-\overset{O}{\underset{\|}{C}}SCoA + CoA-SH \longrightarrow R-\overset{O}{\underset{\|}{C}}-SCoA + CH_3\overset{O}{\underset{\|}{C}}-SCoA$$

β-Ketoacyl-CoA   Coenzyme A   An acyl-CoA   Acetyl-CoA

Show that this reaction is the reverse of a Claisen condensation.

16.39 Steroids are a major type of lipid (Section 21.5) with a characteristic tetracyclic ring system. Show how the **A** ring of the steroid testosterone can be constructed from the indicated precursors, using a Michael reaction followed by an aldol reaction (with dehydration):

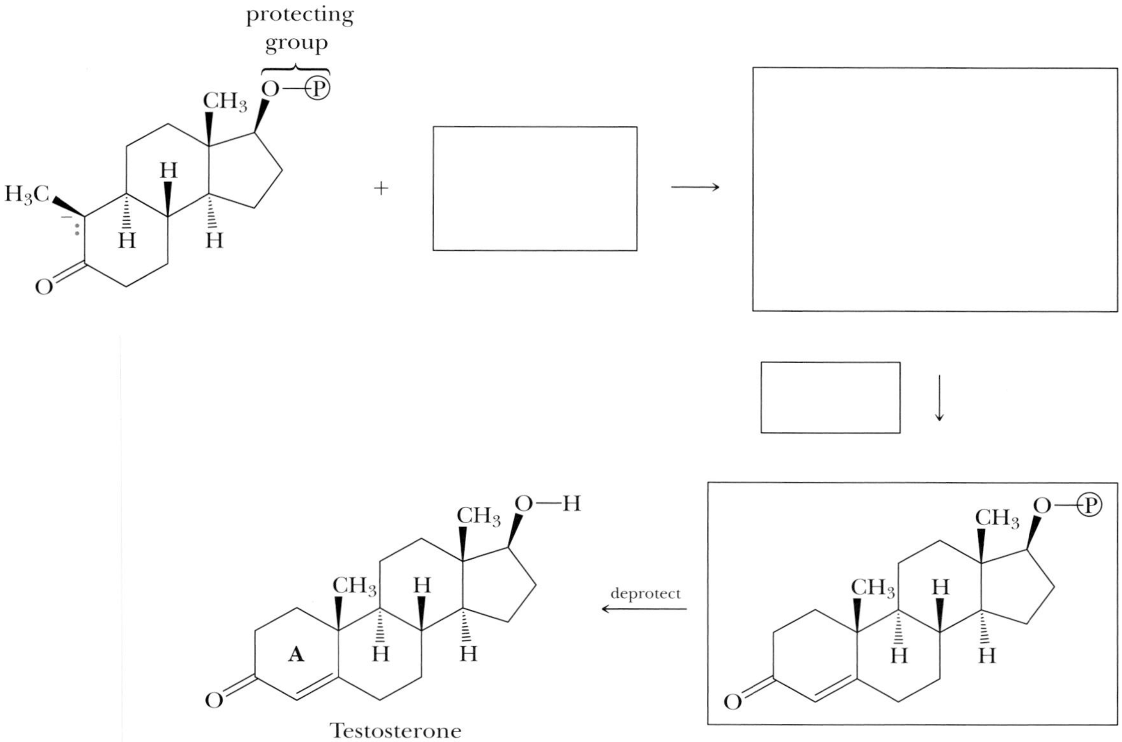

protecting group

Testosterone

deprotect

16.40 The third step of the citric acid cycle (Section 22.7) involves the protonation of one of the carboxylate groups of oxalosuccinate, a β-ketoacid, followed by decarboxylation to form α-ketoglutarate:

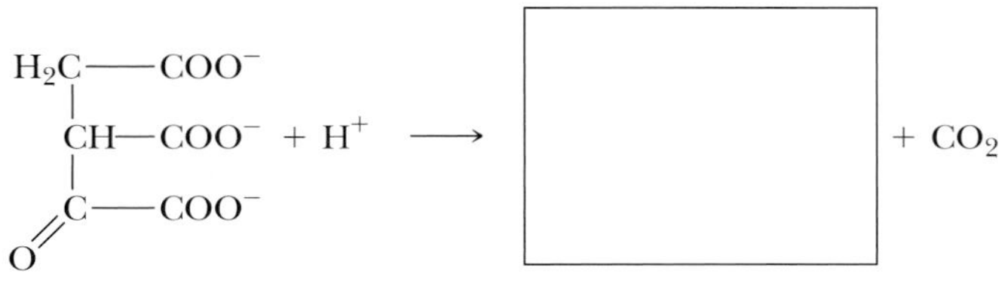

$$H_2C\!-\!COO^-$$
$$|$$
$$CH\!-\!COO^- + H^+ \longrightarrow \qquad\qquad + CO_2$$
$$|$$
$$\underset{O}{\overset{}{C}}\!-\!COO^-$$

Oxalosuccinate $\qquad\qquad\qquad$ α-Ketoglutarate

Write the structural formula of α-ketoglutarate.

17 Organic Polymer Chemistry

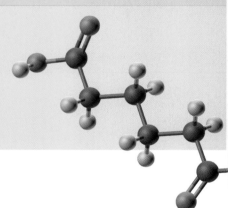

Sea of umbrellas on a rainy day in Shanghai, China.
Inset: A model of adipic acid, one of the two
monomers from which nylon 66 in made.
(Gavin Hellier/Stone/Getty Images)

17.1 INTRODUCTION

The technological advancement of any society is inextricably tied to the materials available to it. Indeed, historians have used the emergence of new materials as a way of establishing a time line to mark the development of human civilization. As part of the search to discover new materials, scientists have made increasing use of organic chemistry for the preparation of synthetic materials known as polymers. The versatility afforded by these polymers allows for the creation and fabrication of materials·with ranges of properties unattainable using such materials as wood, metals, and ceramics. Deceptively simple changes in the chemical structure of a given polymer, for example, can change its mechanical properties from those of a sandwich bag to those of a bulletproof vest. Furthermore, structural changes can introduce properties never before imagined in organic polymers. For instance, using well-defined organic reactions, chemists can turn one type of polymer into an insulator (e.g., the rubber sheath that surrounds electrical cords). Treated differently, the same type of polymer can be made into an electrical conductor with a conductivity nearly equal to that of metallic copper!

The years since the 1930s have seen extensive research and development in organic polymer chemistry, and an almost explosive growth in plastics, coatings, and rubber technology has created a worldwide multibillion-dollar industry. A few basic characteristics account for this phenomenal growth. First, the raw materials for synthetic polymers are derived mainly from petroleum. With the development of petroleum-refining processes, raw materials for the synthesis of polymers became generally cheap and plentiful. Second, within broad limits, scientists have learned how to tailor polymers to the requirements of the end use. Third, many consumer products can be fabricated more cheaply from synthetic polymers than from such competing materials as wood, ceramics, and metals. For example, polymer technology created the water-based (latex) paints that have revolutionized the coatings industry, and plastic films and foams have done the same for the packaging industry. The list could go on and on as we think of the manufactured items that are everywhere around us in our daily lives.

17.2 THE ARCHITECTURE OF POLYMERS

Polymers (Greek: *poly* + *meros*, many parts) are long-chain molecules synthesized by linking **monomers** (Greek: *mono* + *meros*, single part) through chemical reactions. The molecular weights of polymers are generally high compared with those of common organic compounds and typically range from 10,000 g/mol to more than 1,000,000 g/mol. The architectures of these macromolecules can also be quite diverse: There are polymer architectures with linear and branched chains, as well as those with comb, ladder, and star structures (Figure 17.1). Additional structural variations can be achieved by introducing covalent cross-links between individual polymer chains.

Linear and branched polymers are often soluble in solvents such as chloroform, benzene, toluene, dimethyl sulfoxide (DMSO), and tetrahydrofuran (THF). In addition, many linear and branched polymers can be melted to form highly viscous liquids. In polymer chemistry, the term **plastic** refers to any polymer that can be molded when hot and that retains its shape when cooled. **Thermoplastics** are polymers which, when melted, become sufficiently fluid that they can be molded into shapes that are retained when they are cooled. **Thermosetting plastics**, or thermosets, can be molded when they are first prepared, but once cooled, they harden irreversibly and cannot be remelted. Because of their very different physical characteristics, thermoplastics and thermosets must be processed differently and are used in very different applications.

Polymer From the Greek *poly*, many and *meros*, parts; any long-chain molecule synthesized by linking together many single parts called monomers.

Monomer From the Greek *mono*, single and *meros*, part; the simplest nonredundant unit from which a polymer is synthesized.

Plastic A polymer that can be molded when hot and retains its shape when cooled.

Thermoplastic A polymer that can be melted and molded into a shape that is retained when it is cooled.

Thermosetting plastic A polymer that can be molded when it is first prepared, but, once cooled, hardens irreversibly and cannot be remelted.

Linear   Branched   Comb   Ladder   Star   Crosslinked network   Dendritic

Figure 17.1
Various polymer architectures.

The single most important property of polymers at the molecular level is the size and shape of their chains. A good example of the importance of size is a comparison of paraffin wax, a natural polymer, and polyethylene, a synthetic polymer. These two distinct materials have identical repeat units, namely, $-CH_2-$, but differ greatly in the size of their chains. Paraffin wax has between 25 and 50 carbon atoms per chain, whereas polyethylene has between 1,000 and 3,000 carbons per chain. Paraffin wax, such as that in birthday candles, is soft and brittle, but polyethylene, from which plastic beverage bottles are fabricated, is strong, flexible, and tough. These vastly different properties arise directly from the difference in size and molecular architecture of the individual polymer chains.

17.3 POLYMER NOTATION AND NOMENCLATURE

Average degree of polymerization, n A subscript placed outside the parentheses of the simplest nonredundant unit of a polymer to indicate that the unit repeats n times in the polymer.

We typically show the structure of a polymer by placing parentheses around the **repeating unit**, which is the smallest molecular fragment that contains all the nonrepeating structural features of the chain. A subscript n placed outside the parentheses indicates that the unit repeats n times. Thus, we can reproduce the structure of an entire polymer chain by repeating the enclosed structure in both directions. An example is polypropylene, which is derived from the polymerization of propylene:

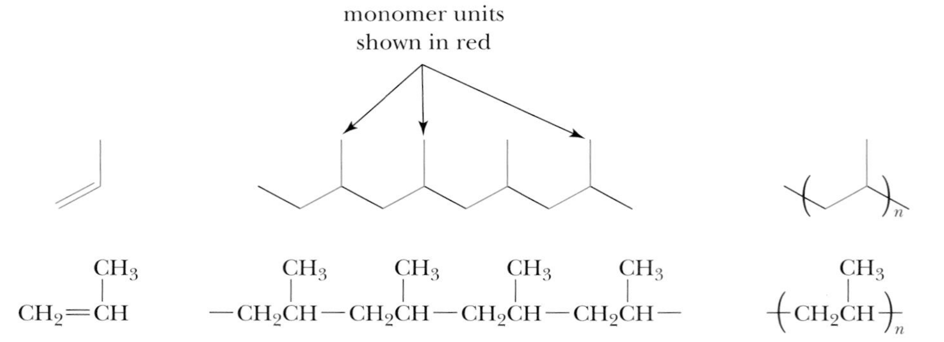

monomer units shown in red

The monomer (propylene) — Part of an extended chain of polypropylene — The repeating unit of polypropylene

The most common method of naming a polymer is to add the prefix **poly-** to the name of the monomer from which the polymer is synthesized. Examples are polyethylene and polystyrene. In the case of a more complex monomer or when the name of the monomer is more than one word (e.g., the monomer vinyl chloride), parentheses are used to enclose the name of the monomer:

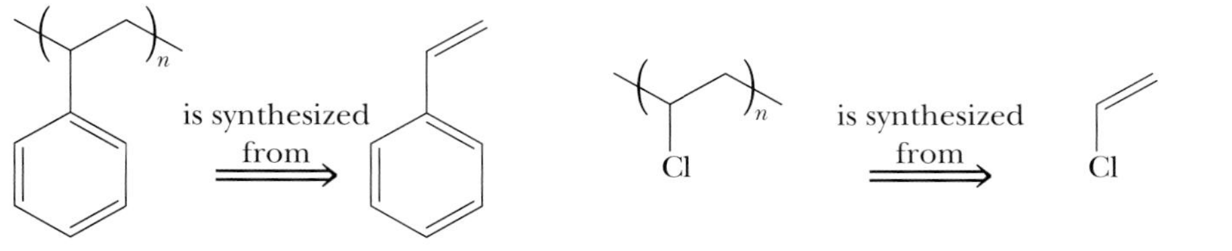

Polystyrene is synthesized from Styrene   Poly(vinyl chloride) (PVC) is synthesized from Vinyl chloride

Polyurethanes consist of flexible polyester or polyether units (blocks) alternating with rigid urethane units (blocks) derived from a diisocyanate, commonly a mixture of 2,4- and 2,6-toluene diisocyanate:

Polyurethane A polymer containing the —NHCOO— group as a repeating unit.

2,6-Toluene diisocyanate — Low-molecular-weight polyester or polyether with OH groups at each end of the chain — A polyurethane

The more flexible blocks are derived from low-molecular-weight (MW 1,000 to 4,000) polyesters or polyethers with —OH groups at each end of their chains. Polyurethane fibers are fairly soft and elastic and have found use as spandex and Lycra®, the "stretch" fabrics used in bathing suits, leotards, and undergarments.

Polyurethane foams for upholstery and insulating materials are made by adding small amounts of water during polymerization. Water reacts with isocyanate groups to form a carbamic acid that undergoes spontaneous decarboxylation to produce gaseous carbon dioxide, which then acts as the foaming agent:

$$RN{=}C{=}O + H_2O \longrightarrow \left[\begin{array}{c} O \\ \parallel \\ RNH{-}C{-}OH \end{array} \right] \longrightarrow RNH_2 + CO_2$$

An isocyanate — A carbamic acid (unstable)

E. Epoxy Resins

Epoxy resins are materials prepared by a polymerization in which one monomer contains at least two epoxy groups. Within this range, a large number of polymeric materials are possible, and epoxy resins are produced in forms ranging from low-viscosity liquids to high-melting solids. The most widely used epoxide monomer is the diepoxide prepared by treating 1 mole of bisphenol A (Problem 9.31) with 2 moles of epichlorohydrin:

Epoxy resin A material prepared by a polymerization in which one monomer contains at least two epoxy groups.

Epichloro-hydrin + Disodium salt of Bisphenol A + Epichloro-hydrin

A diepoxide   + 2NaCl

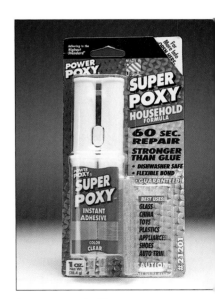

An epoxy resin kit.
(Charles D. Winters)

To prepare the following epoxy resin, the diepoxide monomer is treated with 1,2-ethanediamine (ethylene diamine):

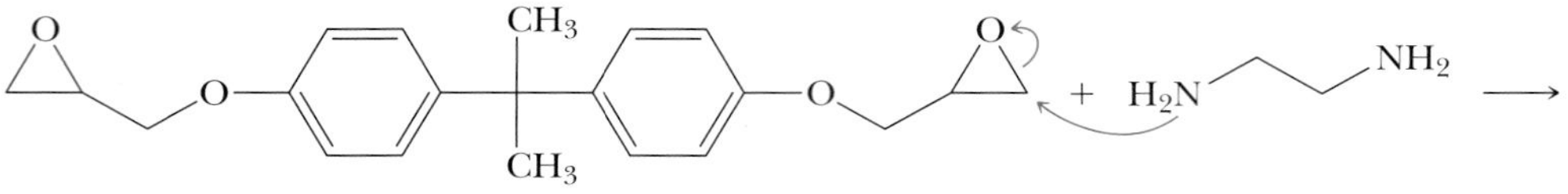

A diepoxide   A diamine

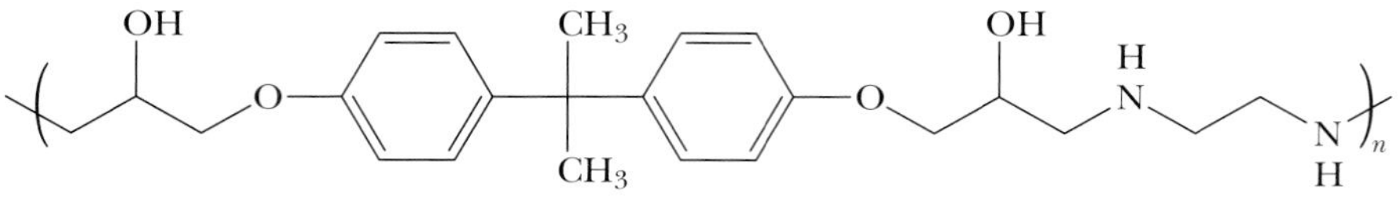

An epoxy resin

Ethylene diamine is usually called the catalyst in the two-component formulations that you buy in hardware or craft stores; it is also the component with the acrid smell. The preceding reaction corresponds to nucleophilic opening of the highly strained three-membered epoxide ring (Section 8.5C).

Epoxy resins are widely used as adhesives and insulating surface coatings. They have good electrical insulating properties, which lead to their use in encapsulating electrical components ranging from integrated circuit boards to switch coils and insulators for power transmission systems. Epoxy resins are also used as composites with other materials, such as glass fiber, paper, metal foils, and other synthetic fibers, to create structural components for jet aircraft, rocket motor casings, and so on.

EXAMPLE 17.2

By what type of mechanism does the reaction between the disodium salt of bisphenol A and epichlorohydrin take place?

SOLUTION

The mechanism is an S_N2 mechanism. The phenoxide ion of bisphenol A is a good nucleophile, and chlorine on the primary carbon of epichlorohydrin is the leaving group.

Practice Problem 17.2

Write the repeating unit of the epoxy resin formed from the following reaction:

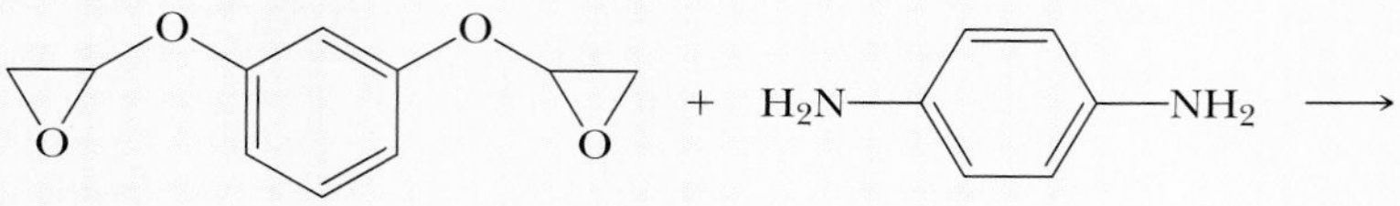

A diepoxide   A diamine

17.6   CHAIN-GROWTH POLYMERS

Chain-growth polymerization
A polymerization that involves sequential addition reactions, either to unsaturated monomers or to monomers possessing other reactive functional groups.

From the perspective of the chemical industry, the single most important reaction of alkenes is **chain-growth polymerization**, a type of polymerization in which monomer units are joined together without the loss of atoms. An example is the formation of polyethylene from ethylene:

$$n\text{CH}_2\!=\!\text{CH}_2 \xrightarrow{\text{catalyst}} -\!(\text{CH}_2\text{CH}_2)\!\!\rightarrow_n$$

Ethylene   Polyethylene

CHEMICAL CONNECTIONS 17A

Stitches That Dissolve

As the technological capabilities of medicine have grown, the demand for synthetic materials that can be used inside the body has increased as well. Polymers have many of the characteristics of an ideal biomaterial: They are lightweight and strong, are inert or biodegradable (depending on their chemical structure), and have physical properties (softness, rigidity, elasticity) that are easily tailored to match those of natural tissues. Carbon–carbon backbone polymers are resistant to degradation and are used widely in permanent organ and tissue replacements.

Even though most medical uses of polymeric materials require biostability, applications have been developed that use the biodegradable nature of some macromolecules. An example is the use of glycolic acid/lactic acid copolymers as absorbable sutures:

A copolymer of
poly(glycolic acid)–
poly(lactic acid)

Traditional suture materials such as catgut must be removed by a health-care specialist after they have served their purpose. Stitches of these hydroxyester polymers, however, are hydrolyzed slowly over a period of approximately two weeks, and by the time the torn tissues have fully healed, the stitches are fully degraded and the sutures need not be removed. Glycolic and lactic acids formed during hydrolysis of the stitches are metabolized and excreted by existing biochemical pathways.

The mechanisms of chain-growth polymerization differ greatly from the mechanism of step-growth polymerizations. In the latter, all monomers plus the polymer end groups possess equally reactive functional groups, allowing for all possible combinations of reactions to occur, including monomer with monomer, dimer with dimer, monomer with tetramer, and so forth. In contrast, chain-growth polymerizations involve end groups possessing reactive intermediates that react only with a monomer. The reactive intermediates used in chain-growth polymerizations include radicals, carbanions, carbocations, and organometallic complexes.

The number of monomers that undergo chain-growth polymerization is large and includes such compounds as alkenes, alkynes, allenes, isocyanates, and cyclic compounds such as lactones, lactams, ethers, and epoxides. We concentrate on the chain-growth polymerizations of ethylene and substituted ethylenes and show how these compounds can be polymerized by radical and organometallic-mediated mechanisms.

Table 17.1 lists several important polymers derived from ethylene and substituted ethylenes, along with their common names and most important uses.

A. Radical Chain-Growth Polymerization

The first commercial polymerizations of ethylene were initiated by radicals formed by thermal decomposition of organic peroxides, such as benzoyl peroxide. A **radical** is any molecule that contains one or more unpaired electrons. Radicals can be

Radical Any molecule that contains one or more unpaired electrons.

CHEMICAL CONNECTIONS 17B

Paper or Plastic?

Any audiophile will tell you that the quality of any sound system is highly dependent upon its speakers. Speakers create sound by moving a diaphragm in and out to displace air. Most diaphragms are in the shape of a cone, traditionally made of paper. Paper cones are inexpensive, lightweight, rigid, and nonresonant. One disadvantage is their susceptibility to damage by water and humidity. Over time and with exposure, paper cones become weakened, losing their fidelity of sound. Many of the speakers that are available today are made of polypropylene, which is also inexpensive, lightweight, rigid, and nonresonant. Furthermore, not only are polypropylene cones immune to water and humidity, but also, their performance is less influenced by heat or cold. Moreover, their added strength makes them less prone to splitting than paper. They last longer and can be displaced more frequently and for longer distances, creating deeper bass notes and higher high notes.

(Photo Courtesy of Crutchfield.com)

TABLE 17.1 Polymers Derived from Ethylene and Substituted Ethylenes

Monomer Formula	Common Name	Polymer Name(s) and Common Uses
$CH_2{=}CH_2$	ethylene	polyethylene, Polythene; break-resistant containers and packaging materials
$CH_2{=}CHCH_3$	propylene	polypropylene, Herculon; textile and carpet fibers
$CH_2{=}CHCl$	vinyl chloride	poly(vinyl chloride), PVC; construction tubing
$CH_2{=}CCl_2$	1,1-dichloroethylene	poly(1,1-dichloroethylene); Saran Wrap® is a copolymer with vinyl chloride
$CH_2{=}CHCN$	acrylonitrile	polyacrylonitrile, Orlon®; acrylics and acrylates
$CF_2{=}CF_2$	tetrafluoroethylene	polytetrafluoroethylene, PTFE; Teflon®, nonstick coatings
$CH_2{=}CHC_6H_5$	styrene	polystyrene, Styrofoam™; insulating materials
$CH_2{=}CHCOOCH_2CH_3$	ethyl acrylate	poly(ethyl acrylate); latex paints
$CH_2{=}\underset{\underset{CH_3}{\vert}}{C}COOCH_3$	methyl methacrylate	poly(methyl methacrylate), Lucite®, Plexiglas®; glass substitutes

formed by the cleavage of a bond in such a way that each atom or fragment participating in the bond retains one electron. In the following equation, **fishhook arrows** are used to show the change in position of single electrons:

Fishhook arrow A single-barbed, curved arrow used to show the change in position of a single electron.

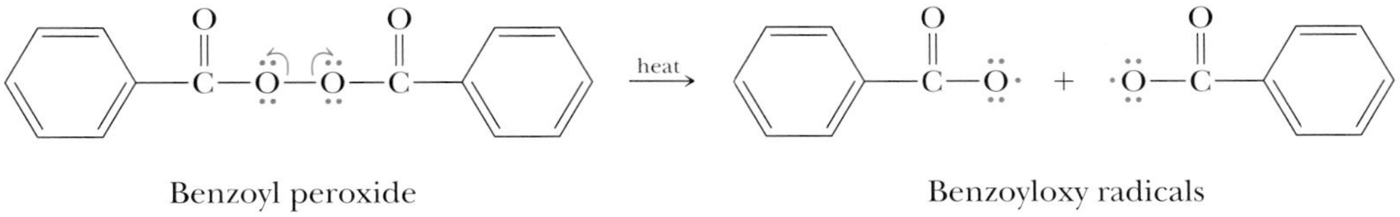

Benzoyl peroxide   Benzoyloxy radicals

Radical polymerization of ethylene and substituted ethylenes involves three steps: (1) chain initiation, (2) chain propagation, and (3) chain termination. We show these steps here and then discuss each separately in turn.

Mechanism: Radical Polymerization of Ethylene

Step 1: Chain initiation—formation of radicals from nonradical compounds:

Chain initiation In radical polymerization, the formation of radicals from molecules containing only paired electrons.

$$\text{In}\!-\!\text{In} \xrightarrow[\text{or light}]{\text{heat}} 2\text{In} \cdot$$

In this equation, In-In represents an initiator which, when heated or irradiated with radiation of a suitable wavelength, cleaves to give two radicals (In·).

Step 2: Chain propagation—reaction of a radical and a molecule to form a new radical:

Chain propagation In radical polymerization, a reaction of a radical and a molecule to give a new radical.

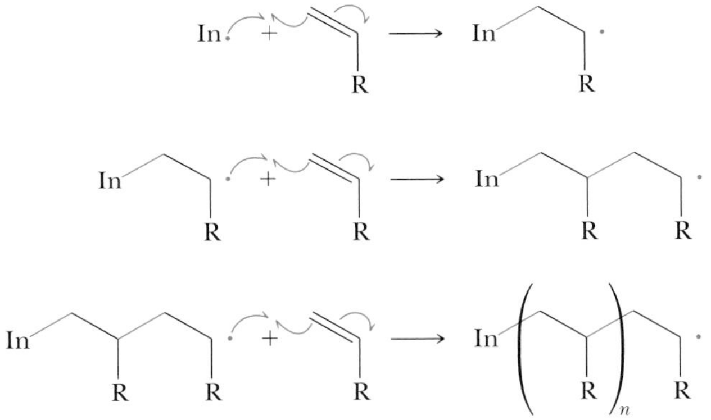

Step 3: Chain termination—destruction of radicals:

Chain termination In radical polymerization, a reaction in which two radicals combine to form a covalent bond.

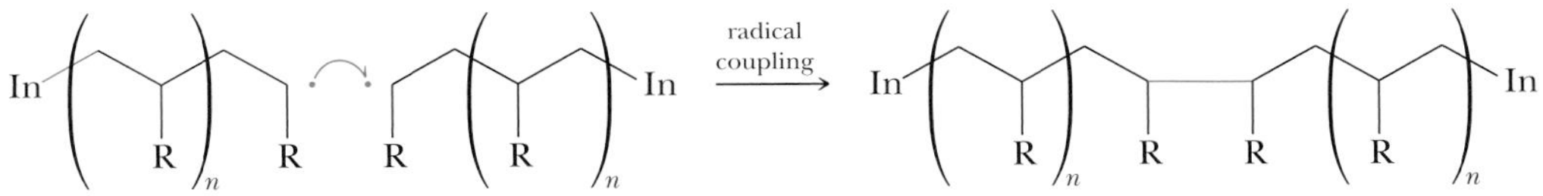

The characteristic feature of a chain-initiation step is the formation of radicals from a molecule with only paired electrons. In the case of peroxide-initiated polymerizations of alkenes, chain initiation is by (1) heat cleavage of the O—O bond of a peroxide to give two alkoxy radicals and (2) reaction of an alkoxy radical with a molecule of alkene to give an alkyl radical. In the general mechanism shown, the initiating catalyst is given the symbol In-In and its radical is given the symbol In·.

The structure and geometry of carbon radicals are similar to those of alkyl carbocations. They are planar or nearly so, with bond angles of approximately 120° about the carbon with the unpaired electron. The relative stabilities of alkyl radicals are similar to those of alkyl carbocations:

methyl < 1° < 2° < 3°

Increasing stability of alkyl radicals →

The characteristic feature of a chain-propagation step is the reaction of a radical and a molecule to give a new radical. Propagation steps repeat over and over (propagate), with the radical formed in one step reacting with a monomer to produce a new radical, and so on. The number of times a cycle of chain-propagation steps repeats is called the **chain length** and is given the symbol **n**. In the polymerization of ethylene, chain-lengthening reactions occur at a very high rate, often as fast as thousands of additions per second, depending on the experimental conditions.

Radical polymerizations of substituted ethylenes almost always give the more stable (more substituted) radical. Because additions are biased in this fashion, the polymerizations of substituted ethylene monomers tend to yield polymers with monomer units joined by the head (carbon 1) of one unit to the tail (carbon 2) of the next unit:

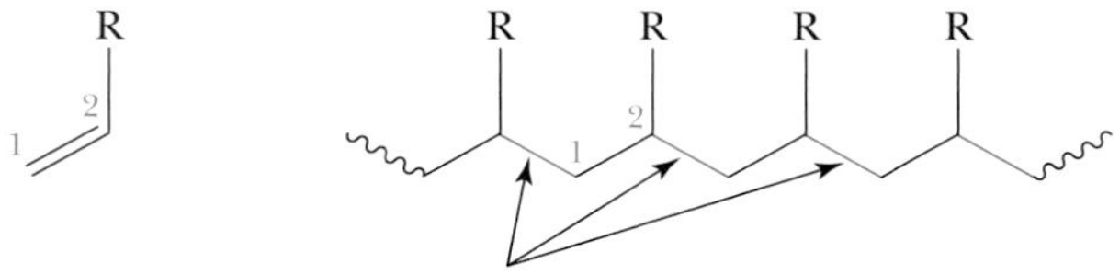

Substituted ethylene monomer   head-to-tail linkages

In principle, chain-propagation steps can continue until all starting materials are consumed. In practice, they continue only until two radicals react with each other to terminate the process. The characteristic feature of a chain-termination step is the destruction of radicals. In the mechanism shown for radical polymerization of the substituted ethylene, chain termination occurs by the coupling of two radicals to form a new carbon–carbon single bond.

The first commercial process for ethylene polymerization used peroxide catalysts at temperatures of 500°C and pressures of 1,000 atm and produced a soft, tough polymer known as low-density polyethylene (LDPE) with a density of between 0.91 and 0.94 g/cm^3 and a melt transition temperature (T_m) of about 115°C. Because LDPE's melting point is only slightly above 100°C, it cannot be used for products that will be exposed to boiling water. At the molecular level, chains of LDPE are highly branched.

The branching on chains of low-density polyethylene results from a "back-biting" reaction in which the radical end group abstracts a hydrogen from the fourth carbon back (the fifth carbon in the chain). Abstraction of this hydrogen is particularly facile because the transition state associated with the process can adopt a conformation like that of a chair cyclohexane. In addition, the less stable 1° radical is converted to a more stable 2° radical. This side reaction is called a **chain-transfer reaction**, because the activity of the end group is "transferred" from one chain to another. Continued polymerization of monomer from this new radical center leads to a branch four carbons long:

Chain-transfer reaction In radical polymerization, the transfer of reactivity of an end group from one chain to another during a polymerization.

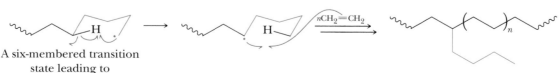

A six-membered transition state leading to 1,5-hydrogen abstraction

Approximately 65% of all LDPE is used for the manufacture of films by a blow-molding technique illustrated in Figure 17.3. LDPE film is inexpensive, which makes it ideal for packaging such consumer items as baked goods, vegetables and other produce and for trash bags.

B. Ziegler–Natta Chain-Growth Polymerization

In the 1950s, Karl Ziegler of Germany and Giulio Natta of Italy developed an alternative method for the polymerization of alkenes, work for which they shared the Nobel prize in chemistry in 1963. The early Ziegler–Natta catalysts were highly active, heterogeneous materials composed of an $MgCl_2$ support, a Group 4B transition metal halide such as $TiCl_4$, and an alkylaluminum compound—for example, diethylaluminum chloride, $Al(CH_2CH_3)_2Cl$. These catalysts bring about the polymerization of ethylene and propylene at 1–4 atm and at temperatures as low as 60°C.

The catalyst in a Ziegler–Natta polymerization is an alkyltitanium compound formed by reaction between $Al(CH_2CH_3)_2Cl$ and the titanium halide on the surface of a $MgCl_2/TiCl_4$ particle. Once formed, this alkyltitanium species repeatedly inserts ethylene units into the titanium–carbon bond to yield polyethylene.

Mechanism: Ziegler–Natta Catalysis of Ethylene Polymerization

Step 1: Formation of a titanium–ethyl bond:

$$\text{Ti—Cl} + \text{Al}(CH_2CH_3)_2Cl \longrightarrow \text{Ti—}CH_2CH_3 + \text{Al}(CH_2CH_3)Cl_2$$

Step 2: Insertion of ethylene into the titanium–carbon bond:

$$\text{Ti—}CH_2CH_3 + CH_2{=}CH_2 \longrightarrow \text{Ti—}CH_2CH_2CH_2CH_3$$

Over 60 billion pounds of polyethylene are produced worldwide every year with Ziegler–Natta catalysts. Polyethylene from Ziegler–Natta systems, termed **high-density polyethylene (HDPE)**, has a higher density (0.96 g/cm^3) and melt transition temperature (133°C) than low-density polyethylene, is 3 to 10 times stronger, and is opaque rather than transparent. The added strength and opacity are due to a much lower degree of chain branching and a resulting higher degree of crystallinity of HDPE compared with LDPE. Approximately 45% of all HDPE used in the United States is blow molded (Figure 17.4).

Even greater improvements in properties of HDPE can be realized through special processing techniques. In the melt state, HDPE chains have random coiled conformations similar to those of cooked spaghetti. Engineers have developed extrusion techniques that force the individual polymer chains of HDPE to uncoil into linear conformations. These linear chains then align with one another to form highly crystalline materials. HDPE processed in this fashion is stiffer than steel and has approximately four times its tensile strength! Because the density of polyethylene (≈ 1.0 g/cm^3) is considerably less than that of steel (8.0 g/cm^3), these comparisons of strength and stiffness are even more favorable if they are made on a weight basis.

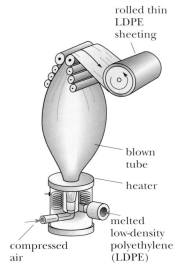

Figure 17.3
Fabrication of an LDPE film. A tube of melted LDPE along with a jet of compressed air is forced through an opening and blown into a giant, thin-walled bubble. The film is then cooled and taken up onto a roller. This double-walled film can be slit down the side to give LDPE film, or it can be sealed at points along its length to make LDPE bags.

rolled thin LDPE sheeting

blown tube

heater

melted low-density polyethylene (LDPE)

compressed air

Polyethylene films are produced by extruding the molten plastic through a ring-like gap and inflating the film into a balloon. *(Brownie Harris/Corbis)*

Figure 17.4
Blow molding of an HDPE container. (a) A short length of HDPE tubing is placed in an open die, and the die is closed, sealing the bottom of the tube. (b) Compressed air is forced into the hot polyethylene–die assembly, and the tubing is literally blown up to take the shape of the mold. (c) After cooling, the die is opened, and there is the container!

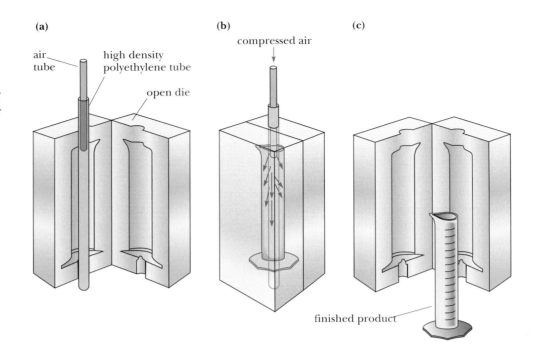

(a)

air tube

high density polyethylene tube

open die

(b)

compressed air

(c)

finished product

Some common products packaged in high-density-polyethylene containers.
(Charles D. Winters)

17.7 RECYCLING PLASTICS

Polymers in the form of plastics are materials upon which our society is incredibly dependent. Durable and lightweight, plastics are probably the most versatile synthetic materials in existence; in fact, their current production in the United States exceeds that of steel. Plastics have come under criticism, however, for their role in the current trash crisis. They make up 21% of the volume and 8% of the weight of solid waste, most of which is derived from disposable packaging and wrapping. Of the 1.5×10^8 kg of thermoplastic materials produced in the United States per year, less than 2% is recycled.

If the durability and chemical inertness of most plastics make them ideally suited for reuse, why aren't more plastics being recycled? The answer to this question has more to do with economics and consumer habits than with technological obstacles. Because curbside pickup and centralized drop-off stations for recyclables are just now becoming common, the amount of used material available for reprocessing has traditionally been small. This limitation, combined with the need for an additional sorting and separation step, rendered the use of recycled plastics in manufacturing expensive compared with virgin materials. The increase in environmental awareness over the last decade, however, has resulted in a greater demand for recycled products. As manufacturers adapt to satisfy this new market, the recycling of plastics will eventually catch up with that of other materials, such as glass and aluminum.

Six types of plastics are commonly used for packaging applications. In 1988, manufacturers adopted recycling code numbers developed by the Society of the Plastics Industry (Table 17.2). Because the plastics recycling industry still is not fully developed, only PET and HDPE are currently being recycled in large quantities. LDPE, which accounts for about 40% of plastic trash, has been slow in finding acceptance with recyclers. Facilities for the reprocessing of poly(vinyl chloride) (PVC), polypropylene (PP), and polystyrene (PS) exist, but are rare.

The process for the recycling of most plastics is simple, with separation of the desired plastics from other contaminants the most labor-intensive step. For example, PET soft-drink bottles usually have a paper label and adhesive that must be removed

TABLE 17.2 Recycling Codes for Plastics

Recycling Code	Polymer	Common Uses	Uses of Recycled Polymer
1 PET	poly(ethylene terephthalate)	soft-drink bottles, household chemical bottles, films, textile fibers	soft-drink bottles, household chemical bottles, films, textile fibers
2 HDPE	high-density polyethylene	milk and water jugs, grocery bags, bottles	bottles, molded containers
3 V	poly(vinyl chloride), PVC	shampoo bottles, pipes, shower curtains, vinyl siding, wire insulation, floor tiles, credit cards	plastic floor mats
4 LDPE	low-density polyethylene	shrink wrap, trash and grocery bags, sandwich bags, squeeze bottles	trash bags and grocery bags
5 PP	polypropylene	plastic lids, clothing fibers, bottle caps, toys, diaper linings	mixed-plastic components
6 PS	polystyrene	Styrofoam™ cups, egg cartons, disposable utensils, packaging materials, appliances	molded items such as cafeteria trays, rulers, Frisbees™, trash cans, videocasettes
7	all other plastics and mixed plastics	various	plastic lumber, playground equipment, road reflectors

before the PET can be reused. The recycling process begins with hand or machine sorting, after which the bottles are shredded into small chips. An air cyclone then removes paper and other lightweight materials. Any remaining labels and adhesives are eliminated with a detergent wash, and the PET chips are then dried. PET produced by this method is 99.9% free of contaminants and sells for about half the price of the virgin material. Unfortunately, plastics with similar densities cannot be separated with this technology, nor can plastics composed of several polymers be broken down into pure components. However, recycled mixed plastics can be molded into plastic lumber that is strong, durable, and resistant to graffiti.

An alternative to the foregoing process, which uses only physical methods of purification, is chemical recycling. Eastman Kodak salvages large amounts of its PET film scrap by a transesterification reaction. The scrap is treated with methanol in the presence of an acid catalyst to give ethylene glycol and dimethyl terephthalate, monomers that are purified by distillation or recrystallization and used as feedstocks for the production of more PET film:

Poly(ethylene terephthalate) (PET) Ethylene glycol Dimethyl terephthalate

SUMMARY

Polymerization is the process of joining together many small **monomers** into large, high-molecular-weight **polymers** (Section 17.1). The properties of polymeric materials depend on the structure of the repeat unit, as well as on the chain architecture and morphology of the material (Section 17.4).

Step-growth polymerizations involve the stepwise reaction of difunctional monomers (Section 17.5). Important commercial polymers synthesized through step-growth processes include polyamides, polyesters, polycarbonates, polyurethanes, and epoxy resins.

Chain-growth polymerization proceeds by the sequential addition of monomer units to an active chain end group (Section 17.6). **Radical chain-growth polymerization** (Section

17.6A) consists of three stages: chain initiation, chain propagation, and chain termination. In **chain initiation**, radicals are formed from nonradical molecules. In **chain propagation**, a radical and a monomer react to give a new radical. The **chain length** is the number of times a cycle of chain-propagation steps repeats. In **chain termination**, radicals are destroyed. Alkyl radicals are planar or almost so, with bond angles of 120° about the carbon with the unpaired electron. **Ziegler–Natta chain-growth polymerizations** involve the formation of an alkyl-transition metal compound and then the repeated insertion of alkene monomers into the transition metal-to-carbon bond to yield a saturated polymer chain (Section 17.6B).

KEY REACTIONS

1. Step-growth polymerization of a dicarboxylic acid and a diamine gives a polyamide (Section 17.5A)

In this equation, M and M' indicate the remainder of each monomer unit:

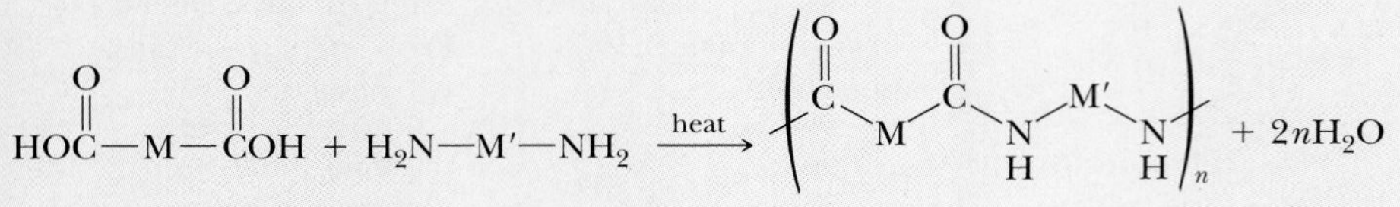

2. Step-growth polymerization of a dicarboxylic acid and a diol gives a polyester (Section 17.4B)

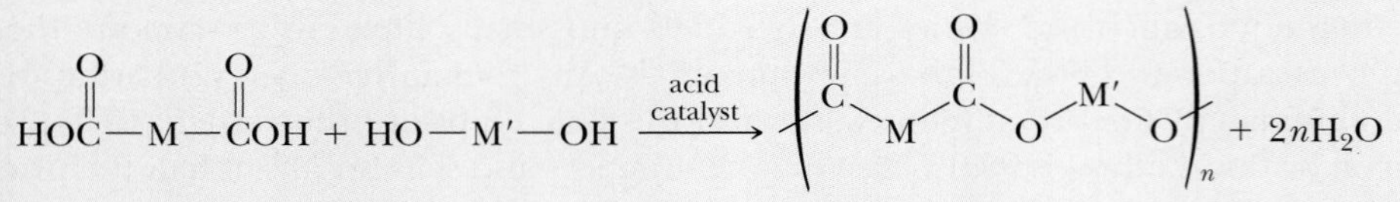

3. Step-growth polymerization of a phosgene and a diol gives a polycarbonate (Section 17.5C)

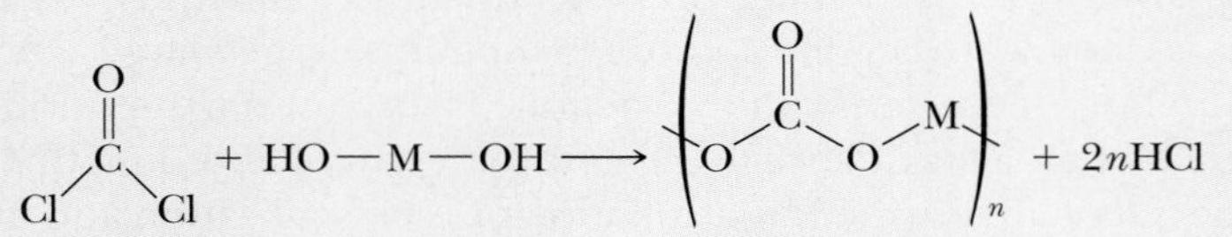

4. Step-growth polymerization of a diisocyanate and a diol gives a polyurethane (Section 17.5D)

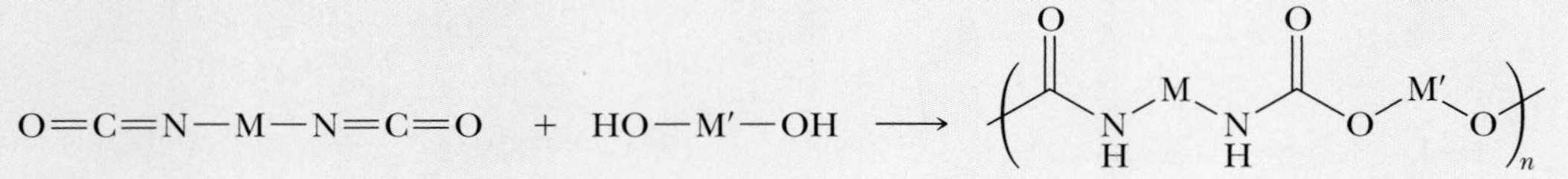

5. Step-growth polymerization of a diepoxide and a diamine gives an epoxy resin (Section 17.5E)

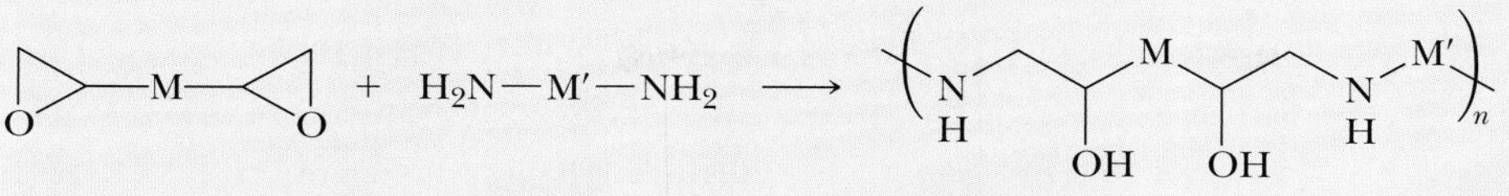

$$
\text{(diepoxide)} \quad + \quad H_2N-M'-NH_2 \quad \longrightarrow \quad \left(\begin{array}{c} N \\ H \end{array} \begin{array}{c} M \\ OH \quad OH \end{array} \begin{array}{c} N \\ H \end{array} M' \right)_n
$$

6. Radical chain-growth polymerization of ethylene and substituted ethylenes (Section 17.6A)

$$
n\text{CH}_2=\text{CHCOOCH}_3 \xrightarrow[\text{heat}]{\text{peroxide}} \left(\text{CH}_2\underset{\overset{|}{\text{COOCH}_3}}{\text{CH}}\right)_n
$$

7. Ziegler–Natta chain-growth polymerization of ethylene and substituted ethylenes (Section 17.6B)

$$
n\text{CH}_2=\text{CHCH}_3 \xrightarrow[\text{MgCl}_2]{\text{TiCl}_4/\text{Al}(\text{C}_2\text{H}_5)_2\text{Cl}} \left(\text{CH}_2\underset{\overset{|}{\text{CH}_3}}{\text{CH}}\right)_n
$$

PROBLEMS

A problem number set in red indicates an applied "real-world" problem.

Step-Growth Polymers

17.3 Identify the monomers required for the synthesis of each step-growth polymer:

(a)

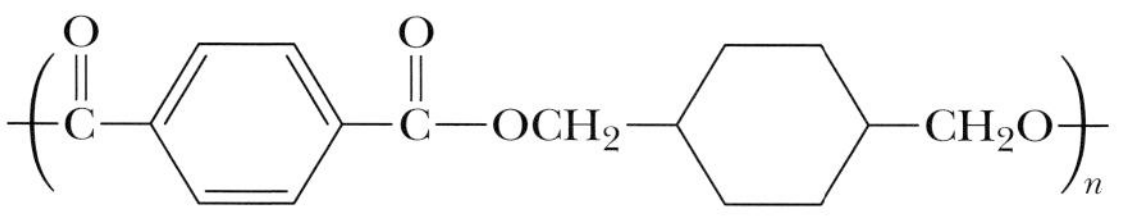

Kodel™
(a polyester)

(b)

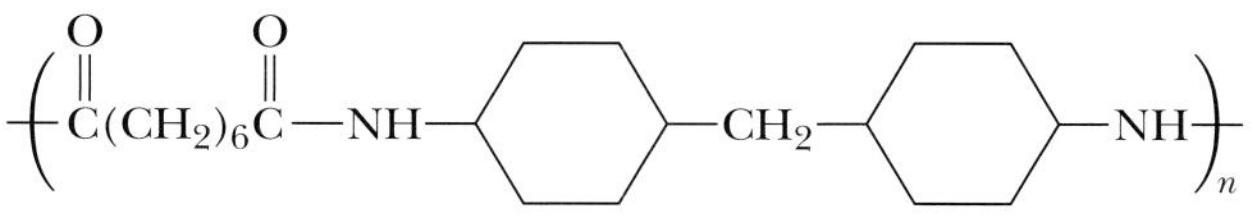

Quiana™
(a polyamide)

(c)

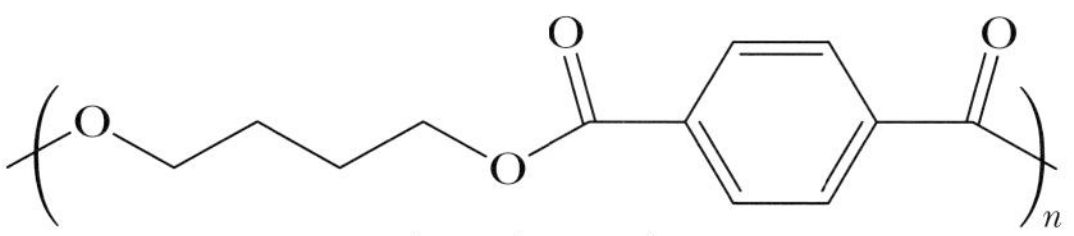

(a polyester)

(d)

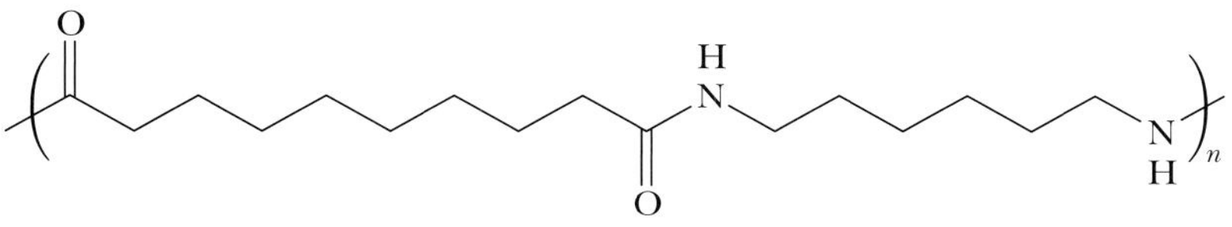

Nylon 6,10
(a polyamide)

17.4 Poly(ethylene terephthalate) (PET) can be prepared by the following reaction:

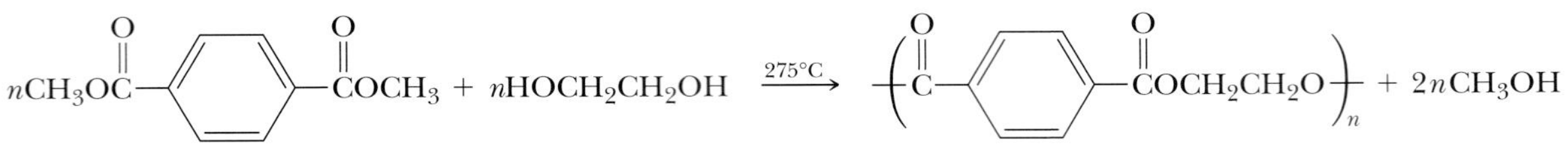

$$nCH_3OC \underset{O}{\overset{O}{\parallel}} \longrightarrow \underset{O}{\overset{O}{\parallel}}COCH_3 + nHOCH_2CH_2OH \xrightarrow{275°C} \left(C \longrightarrow COCH_2CH_2O \right)_n + 2nCH_3OH$$

| Dimethyl terephthalate | Ethylene glycol | Poly(ethylene terephthalate) | Methanol |

Propose a mechanism for the step-growth reaction in this polymerization.

17.5 Currently, about 30% of PET soft-drink bottles are being recycled. In one recycling process, scrap PET is heated with methanol in the presence of an acid catalyst. The methanol reacts with the polymer, liberating ethylene glycol and dimethyl terephthalate. These monomers are then used as feedstock for the production of new PET products. Write an equation for the reaction of PET with methanol to give ethylene glycol and dimethyl terephthalate.

17.6 Nomex® is an aromatic polyamide (aramid) prepared from the polymerization of 1,3-benzenediamine and the acid chloride of 1,3-benzenedicarboxylic acid:

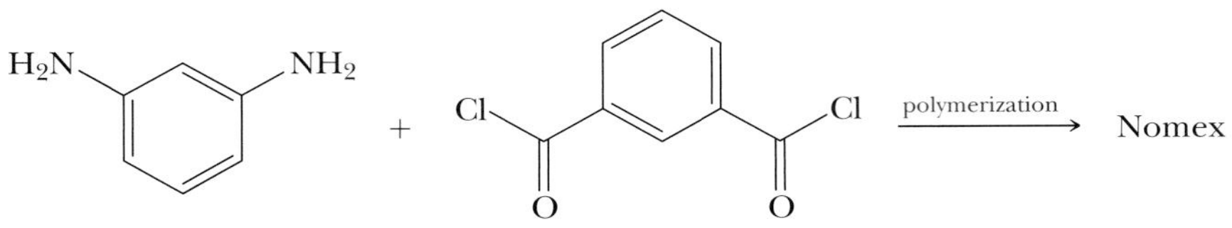

1,3-Benzenediamine   1,3-Benzene-
                                  dicarbonyl chloride

The physical properties of the polymer make it suitable for high-strength, high-temperature applications such as parachute cords and jet aircraft tires. Draw a structural formula for the repeating unit of Nomex.

17.7 Nylon 6,10 [Problem 17.3(d)] can be prepared by reacting a diamine and a diacid chloride. Draw the structural formula of each reactant.

Chain-Growth Polymerization

17.8 Following is the structural formula of a section of polypropylene derived from three units of propylene monomer:

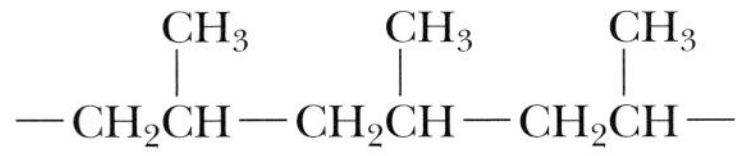

$$\begin{array}{ccc} CH_3 & CH_3 & CH_3 \\ | & | & | \\ -CH_2CH- & CH_2CH- & CH_2CH- \end{array}$$

Polypropylene

Draw a structural formula for a comparable section of

(a) Poly(vinyl chloride)   (b) Polytetrafluoroethylene (PTFE)
(c) Poly(methyl methacrylate)

17.9 Following are structural formulas for sections of two polymers:

$$\textbf{(a)} \quad \begin{array}{ccc} Cl & Cl & Cl \\ | & | & | \\ -CH_2CCH_2CCH_2C- \\ | & | & | \\ Cl & Cl & Cl \end{array} \qquad \textbf{(b)} \quad \begin{array}{ccc} F & F & F \\ | & | & | \\ -CH_2CCH_2CCH_2C- \\ | & | & | \\ F & F & F \end{array}$$

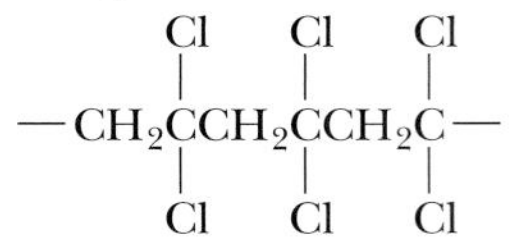

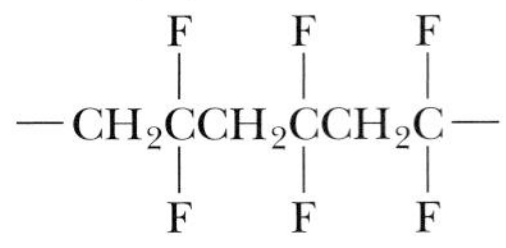

From what alkene monomer is each polymer derived?

17.10 Draw the structure of the alkene monomer used to make each chain-growth polymer:

(a) 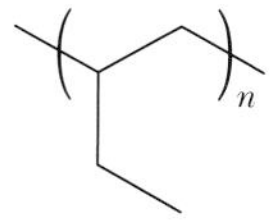 (b) 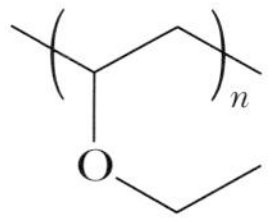 (c) 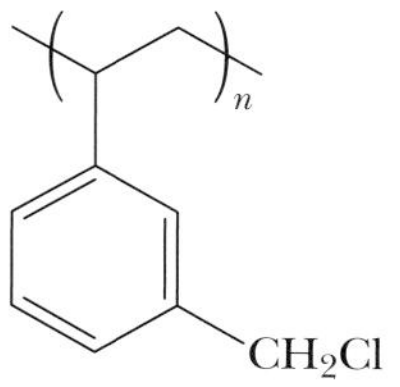 (d)

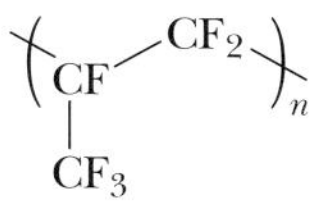

CH_2Cl

CF_2 CF CF_3

17.11 LDPE has a higher degree of chain branching than HDPE. Explain the relationship between chain branching and density.

17.12 Compare the densities of LDPE and HDPE with the densities of the liquid alkanes listed in Table 3.4. How might you account for the differences between them?

17.13 The polymerization of vinyl acetate gives poly(vinyl acetate). Hydrolysis of this polymer in aqueous sodium hydroxide gives poly(vinyl alcohol). Draw the repeat units of both poly(vinyl acetate) and poly(vinyl alcohol):

$$\text{Vinyl acetate} \quad CH_3-\overset{\overset{\displaystyle O}{\|}}{C}-O-CH=CH_2$$

17.14 As seen in the previous problem, poly(vinyl alcohol) is made by the polymerization of vinyl acetate, followed by hydrolysis in aqueous sodium hydroxide. Why is poly(vinyl alcohol) not made instead by the polymerization of vinyl alcohol, $CH_2=CHOH$?

17.15 As you know, the shape of a polymer chain affects its properties. Consider the following three polymers:

A

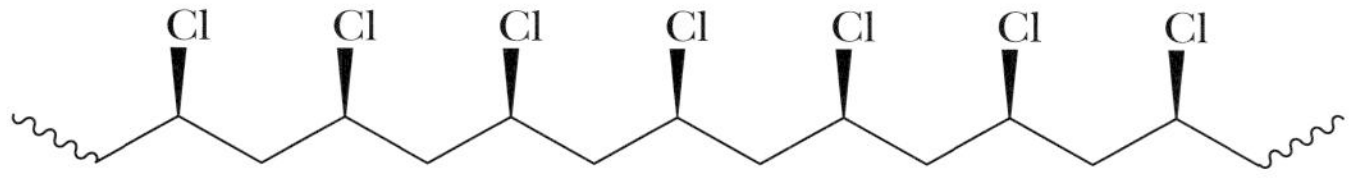

B

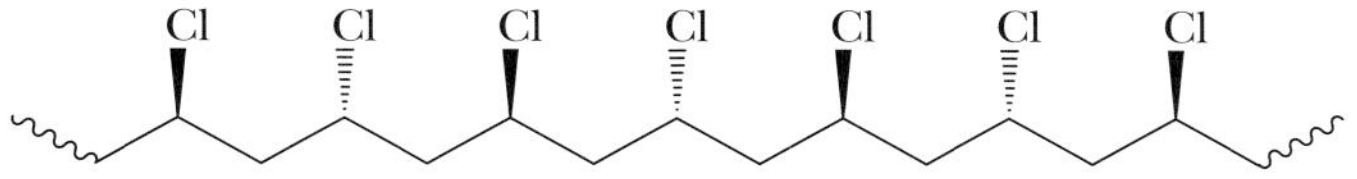

C

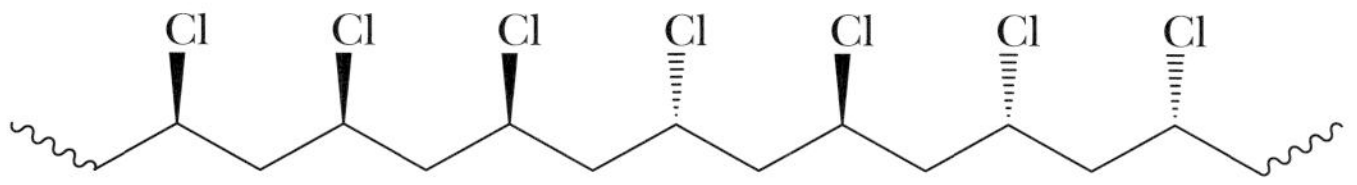

Which do you expect to be the most rigid? Which do you expect to be the most transparent? (Assume the same molecular weights.)

Looking Ahead

17.16 Cellulose, the principle component of cotton, is a polymer of D-glucose in which the monomer unit repeats at the indicated atoms:

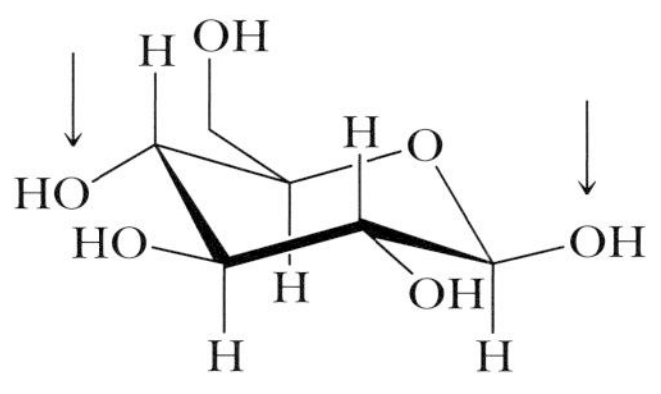

D-Glucose

Draw a three-unit section of cellulose.

17.17 Is a repeating unit a requirement for a compound to be called a polymer?

17.18 Proteins are polymers of naturally occurring monomers called amino acids:

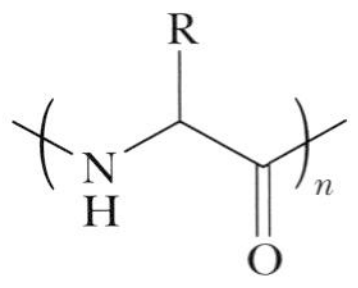

a protein

Amino acids differ in the types of R groups available in nature. Explain how the following properties of a protein might be affected upon changing the R groups from $-CH_2CH(CH_3)_2$ to $-CH_2OH$:

(a) solubility in water   (b) T_m
(c) crystallinity   (d) elasticity

18 Carbohydrates

Breads, grains, and pasta are sources of carbohydrates. Inset: A model of Glucose. *(Charles D. Winters)*

18.1 INTRODUCTION

Carbohydrates are the most abundant organic compounds in the plant world. They act as storehouses of chemical energy (glucose, starch, glycogen); are components of supportive structures in plants (cellulose), crustacean shells (chitin), and connective tissues in animals (acidic polysaccharides); and are essential components of nucleic acids (D-ribose and 2-deoxy-D-ribose). Carbohydrates account for approximately three-fourths of the dry weight of plants. Animals (including humans) get their carbohydrates by eating plants, but they do not store much of what they consume. In fact, less than 1 percent of the body weight of animals is made up of carbohydrates.

The word *carbohydrate* means "hydrate of carbon" and derives from the formula $C_n(H_2O)_m$. Two examples of carbohydrates with molecular formulas that can be written alternatively as hydrates of carbon are

- glucose (blood sugar), $C_6H_{12}O_6$, which can be written as $C_6(H_2O)_6$, and
- sucrose (table sugar), $C_{12}H_{22}O_{11}$, which can be written as $C_{12}(H_2O)_{11}$.

Not all carbohydrates, however, have this general formula. Some contain too few oxygen atoms to fit the formula, whereas some contain too many. Some also contain nitrogen. But the term "carbohydrate" has become firmly rooted in chemical nomenclature and, although not completely accurate, it persists as the name for this class of compounds.

At the molecular level, most **carbohydrates** are polyhydroxyaldehydes, polyhydroxyketones, or compounds that yield them after hydrolysis. Therefore, the chemistry of carbohydrates is essentially the chemistry of hydroxyl and carbonyl groups and of acetal bonds (Section 13.7A) formed between these two functional groups.

Carbohydrate A polyhydroxyaldehyde or polyhydroxyketone or a substance that gives these compounds on hydrolysis.

18.2   MONOSACCHARIDES

A. Structure and Nomenclature

Monosaccharides have the general formula $C_nH_{2n}O_n$, with one of the carbons being a carbonyl group of either an aldehyde or a ketone. The most common monosaccharides have from three to nine carbon atoms. The suffix *-ose* indicates that a molecule is a carbohydrate, and the prefixes *tri-*, *tetr-*, *pent-*, and so forth, indicate the number of carbon atoms in the chain. Monosaccharides containing an aldehyde group are classified as **aldoses**; those containing a ketone group are classified as **ketoses**.

There are only two trioses—glyceraldehyde, which is an aldotriose, and dihydroxyacetone, which is a ketotriose:

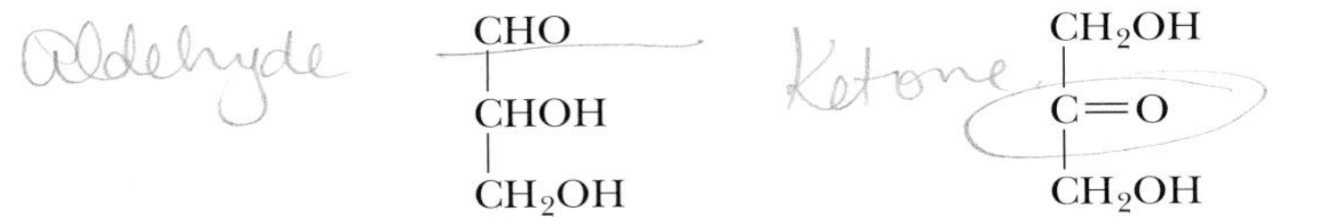

Monosaccharide A carbohydrate that cannot be hydrolyzed to a simpler compound.

Aldose A monosaccharide containing an aldehyde group.
Ketose A monosaccharide containing a ketone group.

Glyceraldehyde (an aldotriose)

Dihydroxyacetone (a ketotriose)

Often the designations *aldo-* and *keto-* are omitted, and these molecules are referred to simply as trioses, tetroses, and the like. Although these designations do not tell the nature of the carbonyl group, at least they indicate that the monosaccharide contains three and four carbon atoms, respectively.

B. Stereoisomerism

Glyceraldehyde contains one stereocenter and exists as a pair of enantiomers. The stereoisomer shown on the left has the *R* configuration and is named (*R*)-glyceraldehyde, while its enantiomer, shown on the right, is named (*S*)-glyceraldehyde:

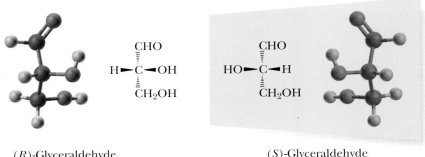

(*R*)-Glyceraldehyde   (*S*)-Glyceraldehyde

C. Fischer Projection Formulas

Chemists commonly use two-dimensional representations called **Fischer projections** to show the configuration of carbohydrates. To draw a Fischer projection, draw a three-dimensional representation with the most oxidized carbon toward the top and the molecule oriented so that the vertical bonds from the stereocenter are directed away from you and the horizontal bonds from it are directed toward you. Then write the molecule as a two-dimensional figure with the stereocenter indicated by the point at which the bonds cross. You now have a Fischer projection.

CHO   *convert to a*   CHO
H—C—OH   *Fischer projection* →   H——OH
CH₂OH   CH₂OH

(*R*)-Glyceraldehyde   (*R*)-Glyceraldehyde
(three-dimensional   (Fischer projection)
representation)

The two horizontal segments of this Fischer projection represent bonds directed toward you, and the two vertical segments represent bonds directed away from you. The only atom in the plane of the paper is the stereocenter.

D. D- and L-Monosaccharides

Even though the *R,S* system is widely accepted today as a standard for designating the configuration of stereocenters, we still commonly designate the configuration of carbohydrates by the D,L system proposed by Emil Fischer in 1891. He assigned the dextrorotatory and levorotary enantiomers of glyceraldehyde the following configurations and named them D-glyceraldehyde and L-glyceraldehyde, respectively:

CHO   CHO
H——OH   HO——H
CH₂OH   CH₂OH

D-Glyceraldehyde   L-Glyceraldehyde
$[\alpha]_D^{25} = +13.5°$ $[\alpha]_D^{25} = -13.5°$

D-glyceraldehyde and L-glyceraldehyde serve as reference points for the assignment of relative configurations to all other aldoses and ketoses. The reference point is the stereocenter farthest from the carbonyl group. Because this stereocenter is the next-to-the-last carbon on the chain, it is called the **penultimate carbon**. A **D-monosaccharide** is a monosaccharide that has the same configuration at its penultimate carbon as D-glyceraldehyde (its —OH is on the right in a Fischer projection); an **L-monosaccharide** has the same configuration at its penultimate carbon as L-glyceraldehyde (its —OH is on the left in a Fischer projection). Almost all monosaccharides in the biological world belong to the D series, and the majority of them are either hexoses or pentoses.

Table 18.1 shows the names and Fischer projection formulas for all D-aldotrioses, tetroses, pentoses, and hexoses. Each name consists of three parts. The letter D specifies the configuration at the stereocenter farthest from the carbonyl group. Prefixes, such as *rib-*, *arabin-*, and *gluc-*, specify the configurations of all other stereocenters relative to one another. The suffix *-ose* shows that the compound is a carbohydrate.

Fischer projection A two-dimensional representation showing the configuration of a stereocenter; horizontal lines represent bonds projecting forward from the stereocenter, vertical lines represent bonds projecting to the rear.

Penultimate carbon The stereocenter of a monosaccharide farthest from the carbonyl group—for example, carbon 5 of glucose.

D-Monosaccharide A monosaccharide that, when written as a Fischer projection, has the —OH on its penultimate carbon to the right.

L-Monosaccharide A monosaccharide that, when written as a Fischer projection, has the —OH on its penultimate carbon to the left.

TABLE 18.1   Configurational Relationships among the Isomeric D-Aldotetroses, D-Aldopentoses, and D-Aldohexoses

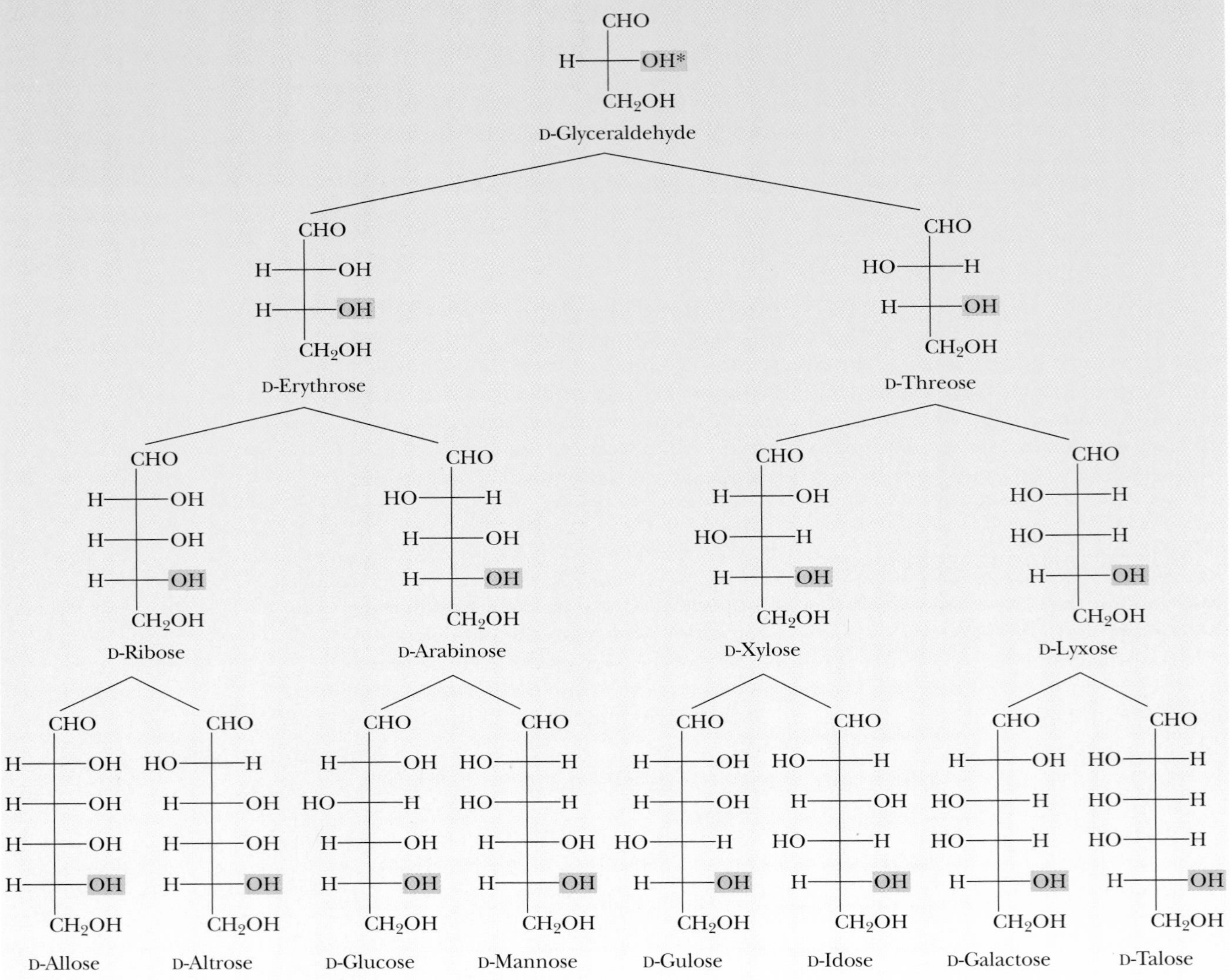

D-Glyceraldehyde

D-Erythrose   D-Threose

D-Ribose   D-Arabinose   D-Xylose   D-Lyxose

D-Allose   D-Altrose   D-Glucose   D-Mannose   D-Gulose   D-Idose   D-Galactose   D-Talose

*The configuration of the reference —OH on the penultimate carbon is shown in color.

The three most abundant hexoses in the biological world are D-glucose, D-galactose, and D-fructose. The first two are D-aldohexoses; the third, fructose, is a D-2-keto-hexose. Glucose, by far the most abundant of the three, is also known as dextrose because it is dextrorotatory. Other names for this monosaccharide include *grape sugar* and *blood sugar*. Human blood normally contains 65–110 mg of glucose/100 mL of blood. D-Fructose is one of the two monosaccharide building blocks of sucrose (table sugar, Section 18.8A):

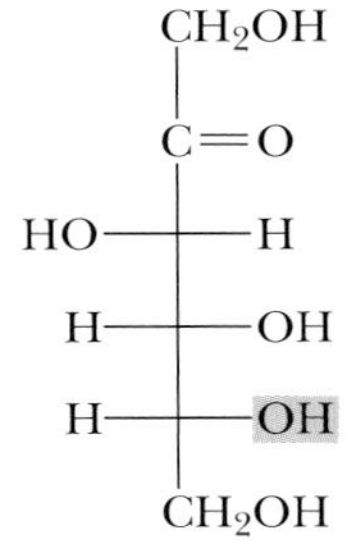

D-Fructose

EXAMPLE 18.1

(a) Draw Fischer projections for the four aldotetroses.
(b) Which of the four aldotetroses are D-monosaccharides, which are L-monosaccharides, and which are enantiomers?
(c) Refer to Table 18.1, and name each aldotetrose you have drawn.

SOLUTION

Following are Fischer projections for the four aldotetroses:

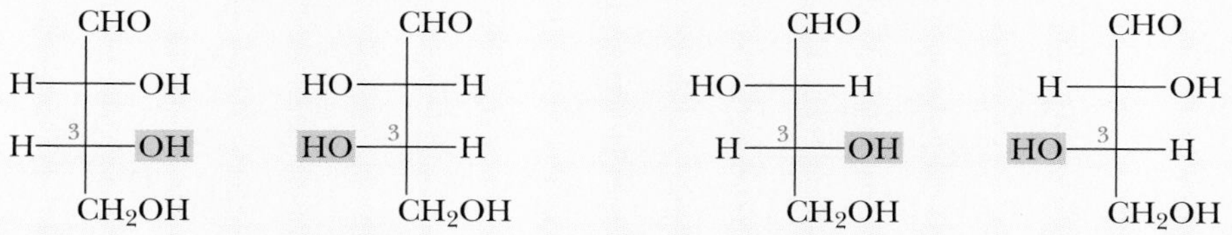 CHO	CHO	CHO	CHO
H——OH	HO——H	HO——H	H——OH
H——OH (3)	HO——H (3)	H——OH (3)	HO——H (3)
CH$_2$OH	CH$_2$OH	CH$_2$OH	CH$_2$OH
D-Erythrose	L-Erythrose	D-Threose	L-Threose
(one pair of enantiomers)		(a second pair of enantiomers)	

Note that, in the Fischer projection of a D-aldotetrose, the —OH on carbon 3 is on the right and, in an L-aldotetrose, it is on the left.

Practice Problem 18.1

(a) Draw Fischer projections for all 2-ketopentoses.
(b) Which of the 2-ketopentoses are D-ketopentoses, which are L-ketopentoses, and which are enantiomers?

E. Amino Sugars

Amino sugars contain an —NH$_2$ group in place of an —OH group. Only three amino sugars are common in nature: D-glucosamine, D-mannosamine, and D-galactosamine. N-Acetyl-D-glucosamine, a derivative of D-glucosamine, is a component of many polysaccharides, including connective tissue such as cartilage. It is also a component of chitin, the hard shell-like exoskeleton of lobsters, crabs, shrimp, and other shellfish. Several other amino sugars are components of naturally occurring antibiotics.

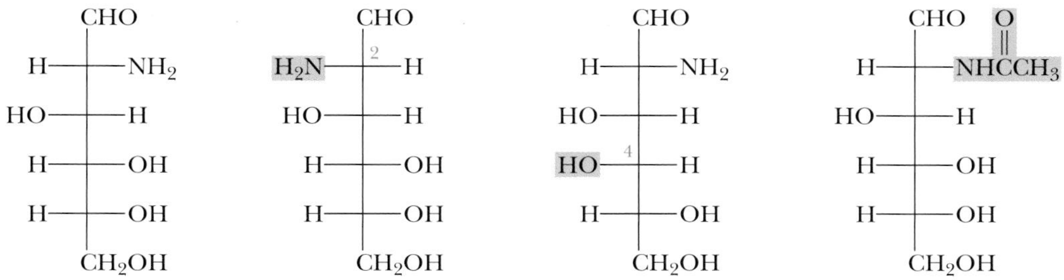

CHO	CHO	CHO	CHO O‖
H——NH$_2$	H$_2$N——H (2)	H——NH$_2$	H——NHCCH$_3$
HO——H	HO——H	HO——H	HO——H
H——OH	H——OH	HO——H (4)	H——OH
H——OH	H——OH	H——OH	H——OH
CH$_2$OH	CH$_2$OH	CH$_2$OH	CH$_2$OH
D-Glucosamine	D-Mannosamine (C-2 stereoisomer of D-glucosamine)	D-Galactosamine (C-4 stereoisomer of D-glucosamine)	N-Acetyl-D-glucosamine

F. Physical Properties

Monosaccharides are colorless, crystalline solids. Because hydrogen bonding is possible between their polar —OH groups and water, all monosaccharides are very soluble in water. They are only slightly soluble in ethanol and are insoluble in nonpolar solvents such as diethyl ether, dichloromethane, and benzene.

18.3   THE CYCLIC STRUCTURE OF MONOSACCHARIDES

In Section 13.7, we saw that aldehydes and ketones react with alcohols to form **hemi-acetals**. We also saw that cyclic hemiacetals form very readily when hydroxyl and carbonyl groups are parts of the same molecule and their interaction can form a five- or six-membered ring. For example, 4-hydroxypentanal forms a five-membered cyclic hemiacetal. Note that 4-hydroxypentanal contains one stereocenter and that a second stereocenter is generated at carbon 1 as a result of hemiacetal formation:

new stereocenter

redraw to show
—OH and —CHO
close to each other

4-Hydroxypentanal   A cyclic hemiacetal

Monosaccharides have hydroxyl and carbonyl groups in the same molecule, and they exist almost exclusively as five- and six-membered cyclic hemiacetals.

A.   Haworth Projections

Haworth projection A way of viewing the furanose and pyranose forms of monosaccharides. The ring is drawn flat and viewed through its edge, with the anomeric carbon on the right and the oxygen atom of the ring in the rear to the right.

Anomeric carbon The hemiacetal carbon of the cyclic form of a monosaccharide.

Anomers Monosaccharides that differ in configuration only at their anomeric carbons.

A common way of representing the cyclic structure of monosaccharides is the **Haworth projection**, named after the English chemist Sir Walter N. Haworth, Nobel laureate of 1937. In a Haworth projection, a five- or six-membered cyclic hemiacetal is represented as a planar pentagon or hexagon, respectively, lying roughly perpendicular to the plane of the paper. Groups bonded to the carbons of the ring then lie either above or below the plane of the ring. The new stereocenter created in forming the cyclic structure is called the **anomeric carbon**. Stereoisomers that differ in configuration only at the anomeric carbon are called **anomers**. The anomeric carbon of an aldose is carbon 1; in D-fructose, the most common ketose, it is carbon 2.

Typically, Haworth projections are written with the anomeric carbon at the right and the hemiacetal oxygen at the back right (Figure 18.1).

As you study the open-chain and cyclic hemiacetal forms of D-glucose, note that, in converting from a Fischer projection to a Haworth structure,

- groups on the right in the Fischer projection point down in the Haworth projection.
- groups on the left in the Fischer projection point up in the Haworth projection.
- for a D-monosaccharide, the terminal —CH₂OH points up in the Haworth projection.
- the configuration of the anomeric —OH group is relative to the terminal —CH₂OH group: If the anomeric —OH group is on the same side as the terminal —CH₂OH, its configuration is β; if the anomeric —OH group is on the opposite side, it is α.

Furanose A five-membered cyclic hemiacetal form of a monosaccharide.

Pyranose A six-membered cyclic hemiacetal form of a monosaccharide.

A six-membered hemiacetal ring is shown by the infix **-pyran-**, and a five-membered hemiacetal ring is shown by the infix **-furan-**. The terms **furanose** and **pyranose** are used because monosaccharide five- and six-membered rings correspond to the heterocyclic compounds pyran and furan:

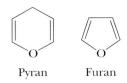

Pyran   Furan

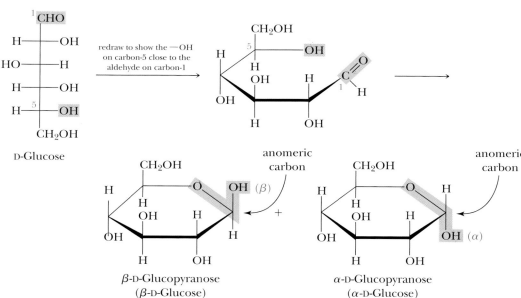

Figure 18.1
Haworth projections for
β-D-glucopyranose and
α-D-glucopyranose.

Because the α and β forms of glucose are six-membered cyclic hemiacetals, they are named α-D-glucopyranose and β-D-glucopyranose, respectively. The designations *-furan-* and *-pyran-* are not always used, however, in names of monosaccharides. Thus, the glucopyranoses are often named simply α-D-glucose and β-D-glucose.

You would do well to remember the configuration of groups on the Haworth projection of both α-D-glucopyranose and β-D-glucopyranose as reference structures. Knowing how the Fischer projection of any other monosaccharide differs from that of D-glucose, you can then construct the Haworth projection of that other monosaccharide by reference to the Haworth projection of D-glucose.

EXAMPLE 18.2

Draw Haworth projections for the α and β anomers of D-galactopyranose.

SOLUTION

One way to arrive at the structures for the α and β anomers of D-galactopyranose is to use the α and β forms of D-glucopyranose as a reference and to remember (or discover by looking at Table 18.1) that D-galactose differs from D-glucose only in the configuration at carbon 4. Thus, you can begin with the Haworth projections shown in Figure 18.1 and then invert the configuration at carbon 4:

Practice Problem 18.2

Mannose exists in aqueous solution as a mixture of α-D-mannopyranose and β-D-mannopyranose. Draw Haworth projections for these molecules.

Aldopentoses also form cyclic hemiacetals. The most prevalent forms of D-ribose and other pentoses in the biological world are furanoses. Following are Haworth projections for α-D-ribofuranose (α-D-ribose) and β-2-deoxy-D-ribofuranose (β-D-ribose):

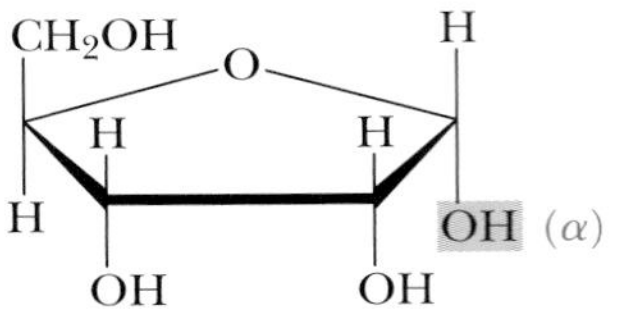

α-D-Ribofuranose
(α-D-Ribose)

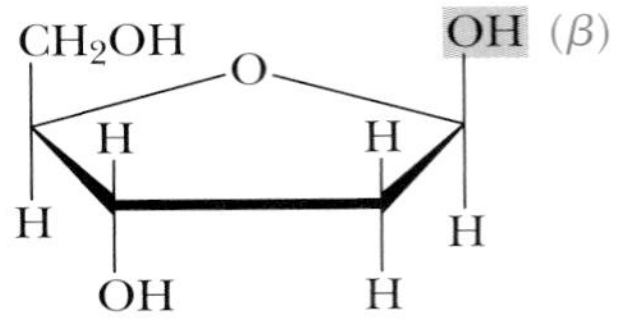

β-2-Deoxy-D-ribofuranose
(β-2-Deoxy-D-ribose)

The prefix 2-deoxy indicates the absence of oxygen at carbon 2. Units of D-ribose and 2-deoxy-D-ribose in nucleic acids and most other biological molecules are found almost exclusively in the β-configuration.

Fructose also forms five-membered cyclic hemiacetals:

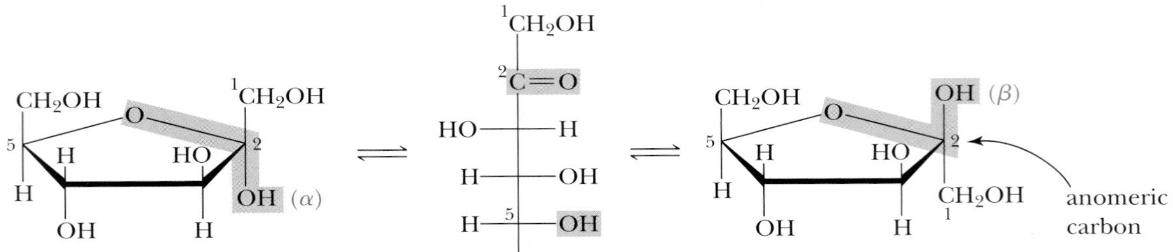

α-D-Fructofuranose
(α-D-Fructose)

D-Fructose

β-D-Fructofuranose
(β-D-Fructose)

β-D-Fructofuranose, for example, is found in the disaccharide sucrose (Section 18.8A).

B. Conformation Representations

A five-membered ring is so close to being planar that Haworth projections are adequate to represent furanoses. For pyranoses, however, the six-membered ring is more accurately represented as a **chair conformation** in which strain is a minimum (Section 3.7B). Figure 18.2 shows structural formulas for α-D-glucopyranose and β-D-glucopyranose, both drawn as chair conformations. The figure also shows the open-chain, or free, aldehyde form with which the cyclic hemiacetal forms are in equilibrium in aqueous solution. Notice that each group, including the anomeric —OH, on the chair conformation of β-D-glucopyranose is equatorial. Notice also that the —OH group on the anomeric carbon in α-D-glucopyranose is axial. Because of the equatorial orientation of the —OH on its anomeric carbon, β-D-glucopyranose is more stable and predominates in aqueous solution.

At this point, you should compare the relative orientations of groups on the D-glucopyranose ring in the Haworth projection and chair conformation:

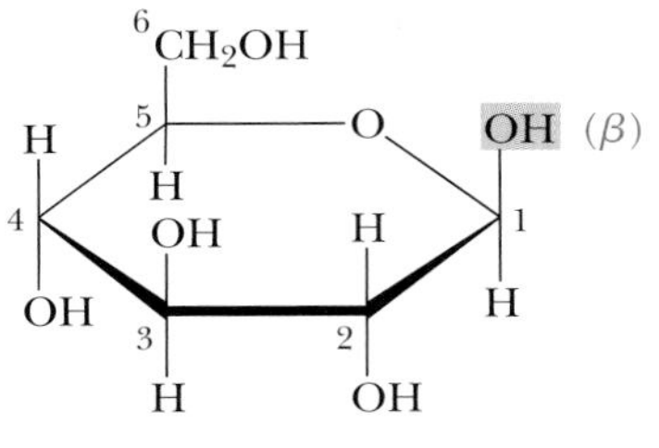

β-D-Glucopyranose
(Haworth projection)

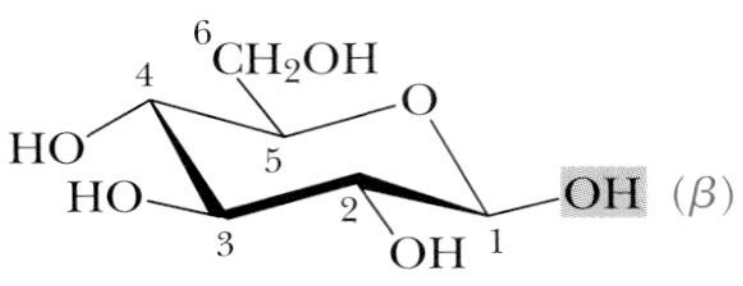

β-D-Glucopyranose
(chair conformation)

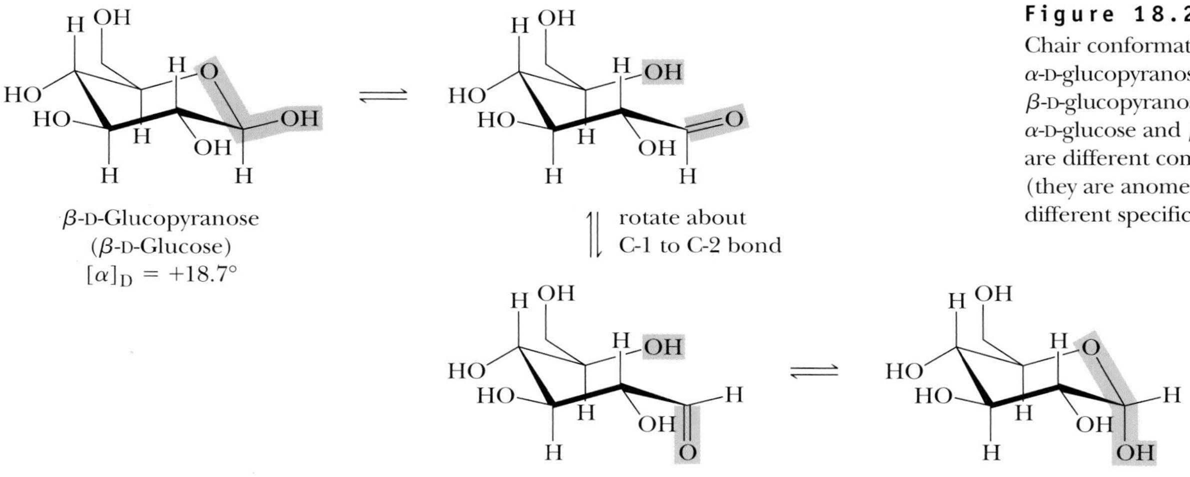

Figure 18.2
Chair conformations of
α-D-glucopyranose and
β-D-glucopyranose. Because
α-D-glucose and β-D-glucose
are different compounds
(they are anomers), they have
different specific rotations.

β-D-Glucopyranose
(β-D-Glucose)
$[\alpha]_D = +18.7°$

rotate about
C-1 to C-2 bond

α-D-Glucopyranose
(α-D-Glucose)
$[\alpha]_D = +112°$

Notice that the orientations of groups on carbons 1 through 5 in the Haworth projection of β-D-glucopyranose are up, down, up, down, and up, respectively. The same is the case in the chair conformation.

EXAMPLE 18.3

Draw chair conformations for α-D-galactopyranose and β-D-galactopyranose. Label the anomeric carbon in each cyclic hemiacetal.

SOLUTION

D-Galactose differs in configuration from D-glucose only at carbon 4. Therefore, draw the α and β forms of D-glucopyranose and then interchange the positions of the —OH and —H groups on carbon 4. Shown also are the specific rotations of each anomer.

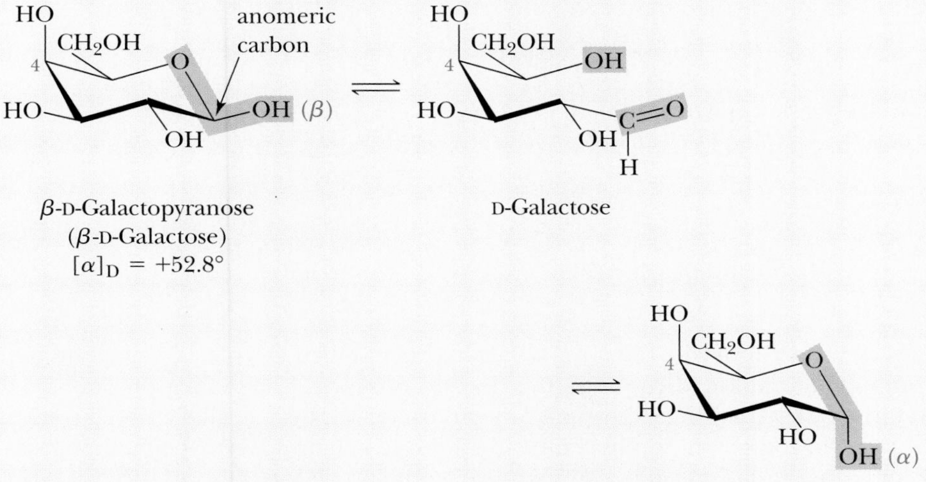

β-D-Galactopyranose
(β-D-Galactose)
$[\alpha]_D = +52.8°$

D-Galactose

α-D-Galactopyranose
(α-D-Galactose)
$[\alpha]_D = +150.7°$

Practice Problem 18.3

Practice Problem 18.3

Draw chair conformations for α-D-mannopyranose and β-D-mannopyranose. Label the anomeric carbon atom in each.

C. Mutarotation

Mutarotation is the change in specific rotation that accompanies the interconversion of α- and β-anomers in aqueous solution. As an example, a solution prepared by dissolving crystalline α-D-glucopyranose in water shows an initial rotation of $+112°$ (Figure 18.2), which gradually decreases to an equilibrium value of $+52.7°$ as α-D-glucopyranose reaches an equilibrium with β-D-glucopyranose. A solution of β-D-glucopyranose also undergoes mutarotation, during which the specific rotation changes from an initial value of $+18.7°$ to the same equilibrium value of $+52.7°$. The equilibrium mixture consists of 64% β-D-glucopyranose and 36% α-D-glucopyranose and contains only traces (0.003%) of the open-chain form. Mutarotation is common to all carbohydrates that exist in hemiacetal forms.

Mutarotation The change in optical activity that occurs when an α or β form of a carbohydrate is converted to an equilibrium mixture of the two forms.

18.4 REACTIONS OF MONOSACCHARIDES

In this section, we discuss reactions of monosaccharides with alcohols, reducing agents, and oxidizing agents. In addition, we examine how these reactions are useful in our everyday lives.

A. Formation of Glycosides (Acetals)

As we saw in Section 13.7A, treating an aldehyde or a ketone with one molecule of alcohol yields a hemiacetal, and treating the hemiacetal with a molecule of alcohol yields an acetal. Treating a monosaccharide, all forms of which exist as cyclic hemiacetals, with an alcohol gives an acetal, as illustrated by the reaction of β-D-glucopyranose (β-D-glucose) with methanol:

β-D-Glucopyranose
(β-D-Glucose)

Methyl β-D-glucopyranoside
(Methyl β-D-glucoside)

Methyl α-D-glucopyranoside
(Methyl α-D-glucoside)

Glycoside A carbohydrate in which the —OH on its anomeric carbon is replaced by —OR.

Glycosidic bond The bond from the anomeric carbon of a glycoside to an —OR group.

A cyclic acetal derived from a monosaccharide is called a **glycoside**, and the bond from the anomeric carbon to the —OR group is called a **glycosidic bond**. Mutarotation is no longer possible in a glycoside, because, unlike a hemiacetal, an acetal is no longer in equilibrium with the open-chain carbonyl-containing compound in neutral or alkaline solution. Like other acetals (Section 13.7), glycosides are stable in water and aqueous base, but undergo hydrolysis in aqueous acid to an alcohol and a monosaccharide.

We name glycosides by listing the alkyl or aryl group bonded to oxygen, followed by the name of the carbohydrate involved in which the ending -e is replaced by -ide. For example, glycosides derived from β-D-glucopyranose are named β-D-glucopyranosides; those derived from β-D-ribofuranose are named β-D-ribofuranosides.

EXAMPLE 18.4

Draw a structural formula for methyl β-D-ribofuranoside (methyl β-D-riboside). Label the anomeric carbon and the glycosidic bond.

SOLUTION

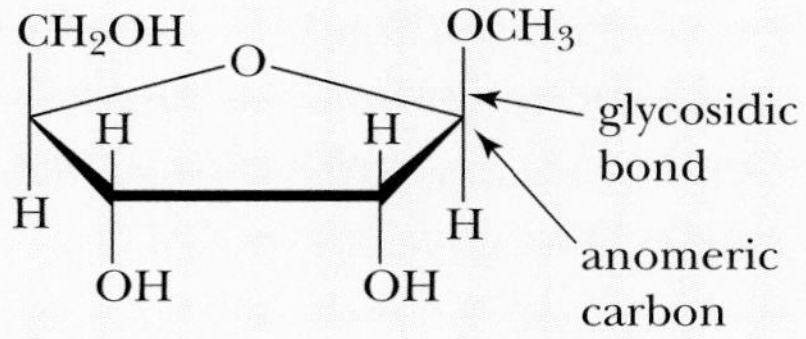

Practice Problem 18.4

Draw a structural formula for the chair conformation of methyl α-D-mannopyranoside (methyl α-D-mannoside). Label the anomeric carbon and the glycosidic bond.

Just as the anomeric carbon of a cyclic hemiacetal undergoes reaction with the —OH group of an alcohol to form a glycoside, it also undergoes reaction with the —NH group of an amine to form an N-glycoside. Especially important in the biological world are the N-glycosides formed between D-ribose and 2-deoxy-D-ribose (each as a furanose), and the heterocyclic aromatic amines uracil, cytosine, thymine, adenine, and guanine (Figure 18.3). N-Glycosides of these compounds are structural units of nucleic acids (Chapter 20).

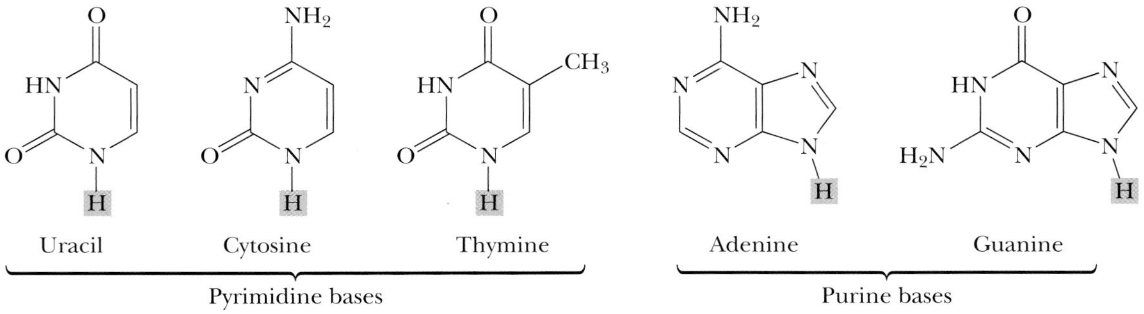

Uracil   Cytosine   Thymine   Adenine   Guanine

Pyrimidine bases   Purine bases

Figure 18.3
Structural formulas of the five most important purine and pyrimidine bases found in DNA and RNA. The hydrogen atom shown in color is lost in the formation of an N-glycoside.

EXAMPLE 18.5

Draw a structural formula for the β-N-glycoside formed between D-ribofuranose and cytosine. Label the anomeric carbon and the N-glycosidic bond.

SOLUTION

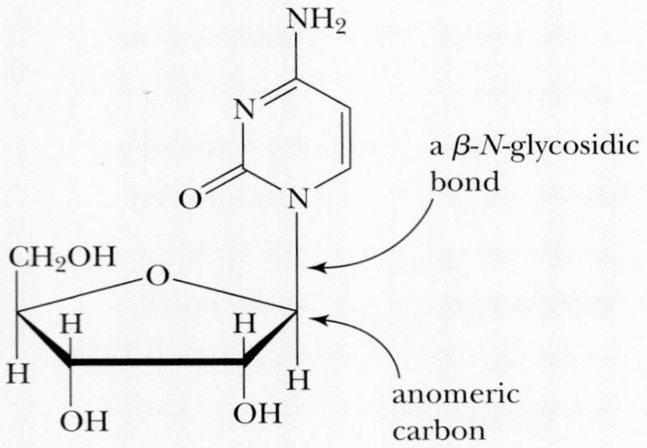

a β-N-glycosidic bond

anomeric carbon

Practice Problem 18.5

Draw a structural formula for the β-N-glycoside formed between β-D-ribofuranose and adenine.

B. Reduction to Alditols

Alditol The product formed when the C=O group of a monosaccharide is reduced to a CHOH group.

The carbonyl group of a monosaccharide can be reduced to a hydroxyl group by a variety of reducing agents, including $NaBH_4$ (Section 13.11B). The reduction products are known as **alditols**. Reduction of D-glucose gives D-glucitol, more commonly known as D-sorbitol. Here, D-glucose is shown in the open-chain form, only a small amount of which is present in solution, but, as it is reduced, the equilibrium between cyclic hemiacetal forms and the open-chain form shifts to replace the D-glucose:

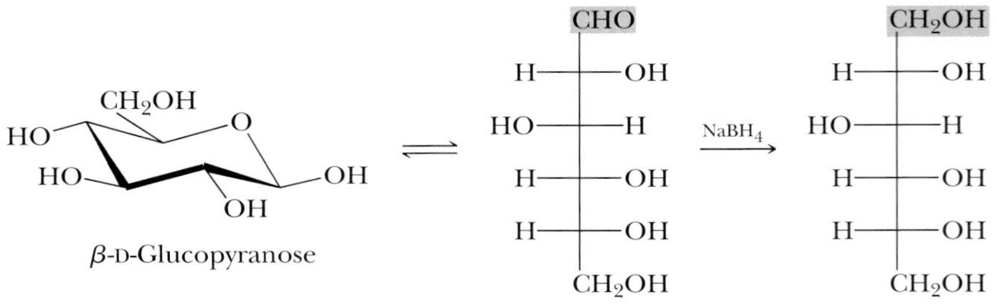

β-D-Glucopyranose   ⇌   D-Glucose   →($NaBH_4$)→   D-Glucitol (D-Sorbitol)

We name alditols by replacing the *-ose* in the name of the monosaccharide with *-itol*. Sorbitol is found in the plant world in many berries and in cherries, plums, pears, apples, seaweed, and algae. It is about 60 percent as sweet as sucrose (table sugar) and is used in the manufacture of candies and as a sugar substitute for diabetics. Among other alditols common in the biological world are erythritol, D-mannitol, and xylitol, the last of which is used as a sweetening agent in "sugarless" gum, candy, and sweet cereals:

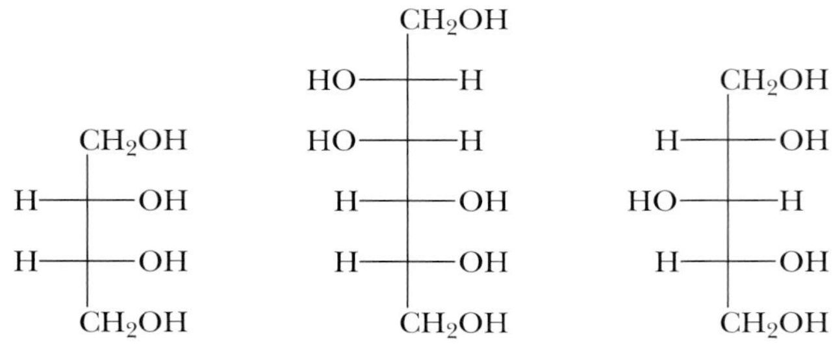

Erythritol   D-Mannitol   Xylitol

Many "sugar-free" products contain sugar alcohols, such as D-sorbitol and xylitol.
(Andy Washnik)

EXAMPLE 18.6

$NaBH_4$ reduces D-glucose to D-glucitol. Do you expect the alditol formed under these conditions to be optically active or optically inactive? Explain.

SOLUTION

D-Glucitol is chiral. Given the fact that reduction by $NaBH_4$ does not affect any of the four stereocenters in D-glucose, nor does D-glucitol have a plane of symmetry, we can predict that the product is optically active.

Practice Problem 18.6

$NaBH_4$ reduces D-erythrose to erythritol. Do you expect the alditol formed under these conditions to be optically active or optically inactive? Explain.

C. Oxidation to Aldonic Acids (Reducing Sugars)

We saw in Section 13.10A that several agents, including O_2, oxidize aldehydes (RCHO) to carboxylic acids (RCOOH). Similarly, under basic conditions, the aldehyde group of an aldose can be oxidized to a carboxylate group. Under these conditions, the cyclic form of the aldose is in equilibrium with the open-chain form, which is then oxidized by the mild oxidizing agent. D-Glucose, for example, is oxidized to D-gluconate (the anion of D-gluconic acid):

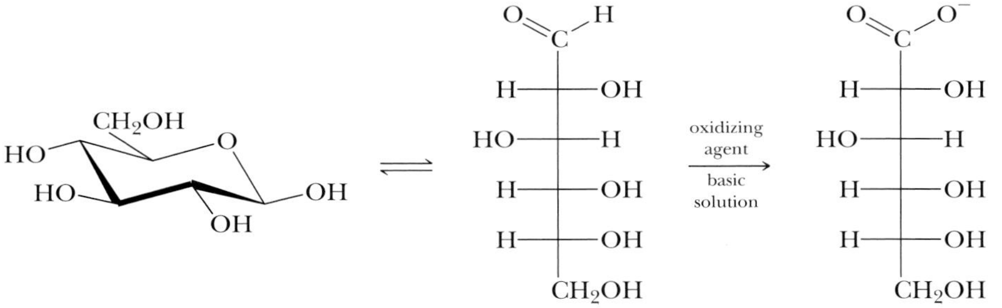

β-D-Glucopyranose
(β-D-Glucose)

D-Glucose

D-Gluconate

Any carbohydrate that reacts with an oxidizing agent to form an aldonic acid is classified as a **reducing sugar**. (It reduces the oxidizing agent.)

Reducing sugar A carbohydrate that reacts with an oxidizing agent to form an aldonic acid.

D. Oxidation to Uronic Acids

Enzyme-catalyzed oxidation of the primary alcohol at carbon 6 of a hexose yields a uronic acid. Enzyme-catalyzed oxidation of D-glucose, for example, yields D-glucuronic acid, shown here in both its open-chain and cyclic hemiacetal forms:

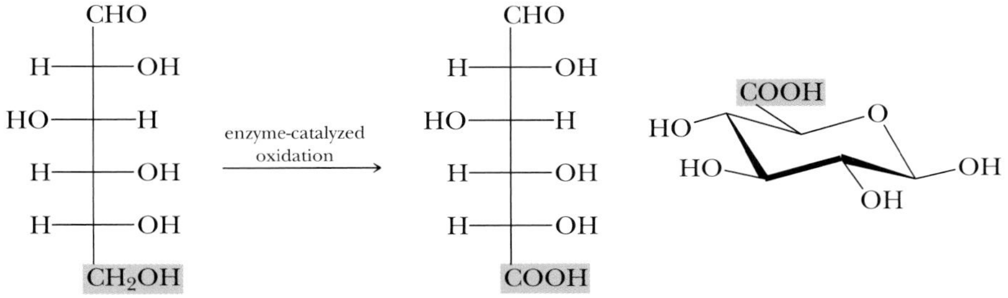

D-Glucose

Fischer projection

Chair conformation

D-Glucuronic acid
(a uronic acid)

D-Glucuronic acid is widely distributed in both the plant and animal worlds. In humans, it is an important component of the acidic polysaccharides of connective tissues. The body also uses it to detoxify foreign phenols and alcohols. In the liver, these compounds are converted to glycosides of glucuronic acid (glucuronides), to be excreted in the urine. The intravenous anesthetic propofol (Problem 10.41), for example, is converted to the following water-soluble glucuronide and then excreted in urine:

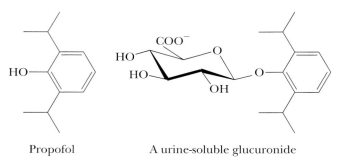

Propofol   A urine-soluble glucuronide

18.5 TESTING FOR BLOOD SUGAR (GLUCOSE)

Determining the level of glucose in blood, urine, or some other biological fluid is among those analytical procedures most often performed in a clinical chemistry laboratory. The need for a rapid and reliable test for glucose stems from the high incidence of diabetes mellitus. Approximately 18 million known diabetics live in the United States, and another 1.3 million are diagnosed each year.

In diabetes mellitus, the body has insufficient levels of the polypeptide hormone insulin (Section 19.5B). If the blood concentration of insulin is too low, muscle and liver cells do not absorb glucose, which, in turn, leads to increased levels of blood glucose (hyperglycemia), impaired metabolism of fats and proteins, ketosis, and, possibly, diabetic coma. Thus, a rapid procedure for the determination of blood glucose levels is critical for early diagnosis and effective management of this disease. In addition to being rapid, a test must be specific for D-glucose; that is, it must give a positive test for glucose, but not react with any other substance that is normally present in biological fluids.

Today, the determination of blood glucose levels is carried out by an enzyme-based procedure that uses the enzyme **glucose oxidase**. This enzyme catalyzes the oxidation of β-D-glucose to D-gluconic acid:

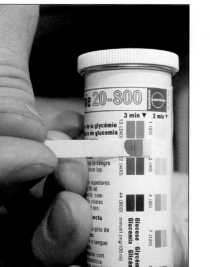

Chemstrip kit for blood glucose. *(Martin Dohrn/SPL/ Photo Researchers, Inc.)*

β-D-Glucopyranose
(β-D-Glucose)

D-Gluconic acid

Glucose oxidase is specific for β-D-glucose. Therefore, complete oxidation of any sample containing both β-D-glucose and α-D-glucose requires the conversion of the α form to the β form. Fortunately, this conversion is rapid and complete in the short time required for the test.

Molecular oxygen, O_2, is the oxidizing agent in the reaction and is reduced to hydrogen peroxide, H_2O_2, the concentration of which can be determined spectrophotometrically. In one procedure, hydrogen peroxide formed in the glucose oxidase–catalyzed reaction oxidizes colorless *o*-toluidine to a colored product in a reaction catalyzed by the enzyme peroxidase:

$$\text{2-Methylaniline} \ (o\text{-Toluidine}) + H_2O_2 \xrightarrow{\text{peroxidase}} \text{Colored product}$$

The concentration of the colored oxidation product is determined spectrophotometrically and is proportional to the concentration of glucose in the test solution. Alternatively, several commercially available test kits use the glucose oxidase reaction to render a qualitative determination of glucose in urine.

18.6 L-ASCORBIC ACID (VITAMIN C)

The structural formula of L-ascorbic acid (vitamin C) resembles that of a monosaccharide. In fact, this vitamin is synthesized both biochemically by plants and some animals and commercially from D-glucose. Humans do not have the enzyme systems required for the synthesis of L-ascorbic acid; therefore, for us, it is a vitamin. Approximately 66 million kilograms of vitamin C are synthesized every year in the United States. L-Ascorbic acid is very easily oxidized to L-dehydroascorbic acid, a diketone:

$$\text{L-Ascorbic acid (Vitamin C)} \underset{\text{reduction}}{\overset{\text{oxidation}}{\rightleftharpoons}} \text{L-Dehydroascorbic acid}$$

Both L-ascorbic acid and L-dehydroascorbic acid are physiologically active and are found together in most body fluids.

Vitamin C in an orange is identical to its synthetic tablet form. *(Andy Washnik)*

18.7 DISACCHARIDES AND OLIGOSACCHARIDES

Most carbohydrates in nature contain more than one monosaccharide unit. Those which contain two units are called **disaccharides**, those which contain three units are called **trisaccharides**, and so forth. The more general term, **oligosaccharide**, is often used for carbohydrates that contain from 6 to 10 monosaccharide units. Carbohydrates containing larger numbers of monosaccharide units are called **polysaccharides**.

In a disaccharide, two monosaccharide units are joined by a glycosidic bond between the anomeric carbon of one unit and an —OH of the other. Sucrose, lactose, and maltose are three important disaccharides.

Disaccharide A carbohydrate containing two monosaccharide units joined by a glycosidic bond.

Oligosaccharide A carbohydrate containing from 6 to 10 monosaccharide units, each joined to the next by a glycosidic bond.

Polysaccharide A carbohydrate containing a large number of monosaccharide units, each joined to the next by one or more glycosidic bonds.

A. Sucrose

Sucrose (table sugar) is the most abundant disaccharide in the biological world. It is obtained principally from the juice of sugarcane and sugar beets. In sucrose, carbon 1 of α-D-glucopyranose bonds to carbon 2 of D-fructofuranose by an α-1,2-glycosidic bond:

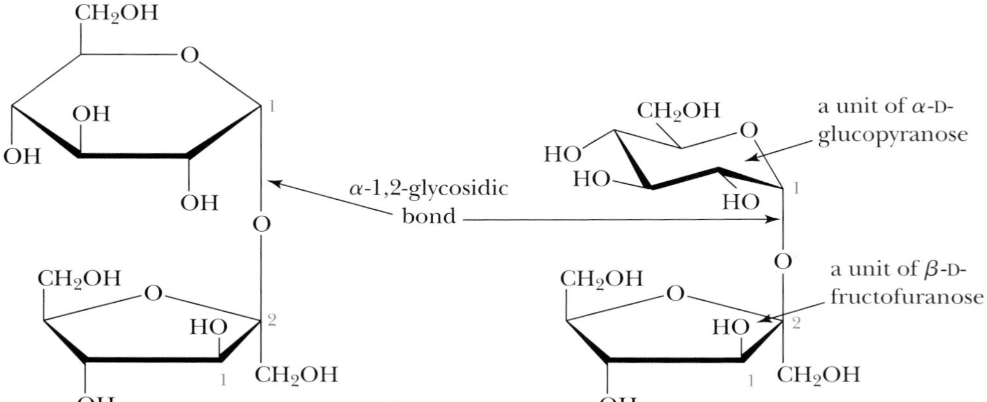

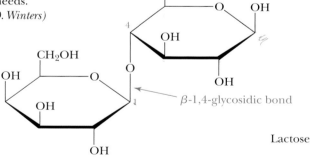

These products help individuals with lactose intolerance meet their calcium needs. *(Charles D. Winters)*

Because the anomeric carbons of both the glucopyranose and fructofuranose units are involved in formation of the glycosidic bond, neither monosaccharide unit is in equilibrium with its open-chain form. Thus, sucrose is a nonreducing sugar.

B. Lactose

Lactose, the principal sugar present in milk, accounts for 5 to 8 percent of human milk and 4 to 6 percent of cow's milk. This disaccharide consists of D-galactopyranose, bonded by a β-1,4-glycosidic bond to carbon 4 of D-glucopyranose:

Lactose is a reducing sugar, because the cyclic hemiacetal of the D-glucopyranose unit is in equilibrium with its open-chain form and can be oxidized to a carboxyl group.

CHEMICAL CONNECTIONS 18A

Relative Sweetness of Carbohydrate and Artificial Sweeteners

Among the disaccharide sweetening agents, D-fructose tastes the sweetest—even sweeter than sucrose. The sweet taste of honey is due largely to D-fructose and D-glucose. Lactose has almost no sweetness and is sometimes added to foods as a filler. Some peo-ple cannot tolerate lactose well, however, and should avoid these foods. The following table lists the sweetness of various carbohydrates and artificial sweeteners relative to that of sucrose:

Carbohydrate	Sweetness Relative to Sucrose	Artificial Sweetener	Sweetness Relative to Sucrose
fructose	1.74	saccharin	450
sucrose (table sugar)	1.00	acesulfame-K	200
honey	0.97	aspartame	180
glucose	0.74		
maltose	0.33		
galactose	0.32		
lactose (milk sugar)	0.16		

C. Maltose

Maltose derives its name from its presence in malt, the juice from sprouted barley and other cereal grains. Maltose consists of two units of D-glucopyranose, joined by a glycosidic bond between carbon 1 (the anomeric carbon) of one unit and carbon 4 of the other unit. Because the oxygen atom on the anomeric carbon of the first glucopyranose unit is alpha, the bond joining the two units is called an α-1,4-glycosidic bond. Following are a Haworth projection and a chair conformation for β-maltose, so named because the —OH group on the anomeric carbon of the glucose unit on the right are beta:

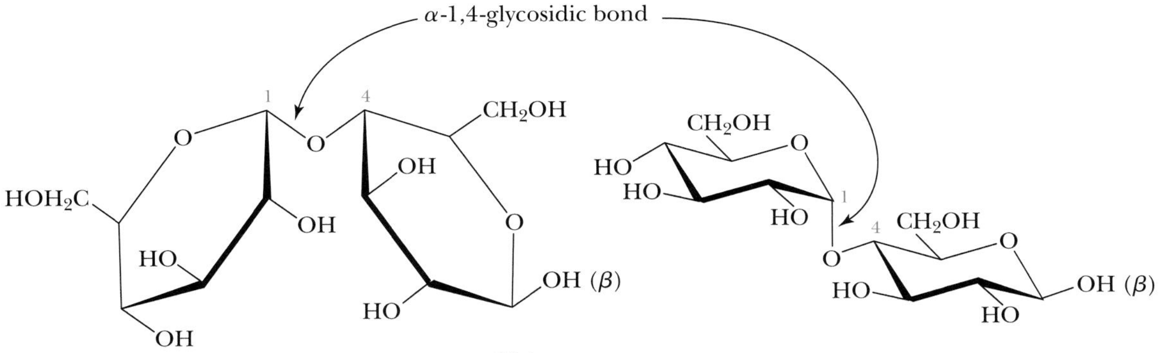

Maltose

CHEMICAL CONNECTIONS 18B

A, B, AB, and O Blood-Group Substances

Membranes of animal plasma cells have large numbers of relatively small carbohydrates bound to them. In fact, the outsides of most plasma cell membranes are literally sugarcoated. These membrane-bound carbohydrates are part of the mechanism by which the different types of cells recognize each other; in effect, the carbohydrates act as biochemical markers (antigenic determinants). Typically, the membrane-bound carbohydrates contain from 4 to 17 units consisting of just a few different monosaccharides, primarily D-galactose, D-mannose, L-fucose, N-acetyl-D-glucosamine, and N-acetyl-D-galactosamine. L-Fucose is a 6-deoxyaldohexose:

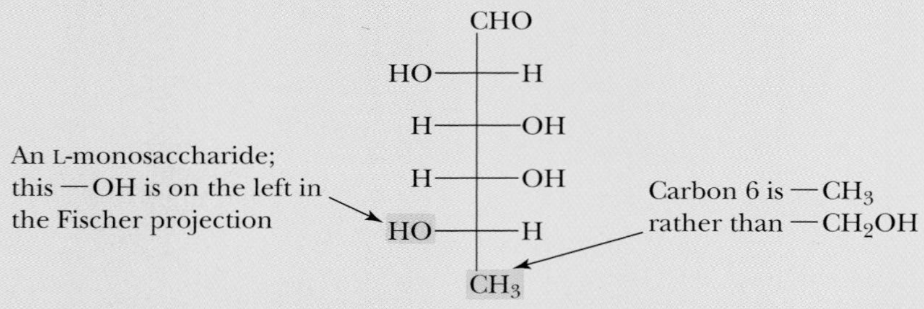

An L-monosaccharide; this —OH is on the left in the Fischer projection

Carbon 6 is —CH₃ rather than —CH₂OH

L-Fucose

EXAMPLE 18.7

Draw a chair conformation for the β anomer of a disaccharide in which two units of D-glucopyranose are joined by an α-1,6-glycosidic bond.

SOLUTION

First draw a chair conformation of α-D-glucopyranose. Then bond the anomeric carbon of this monosaccharide to carbon 6 of a second D-glucopyranose unit by an α-glycosidic bond. The resulting molecule is either α or β, depending on the orientation of the —OH group on the reducing end of the disaccharide. The disaccharide shown here is β:

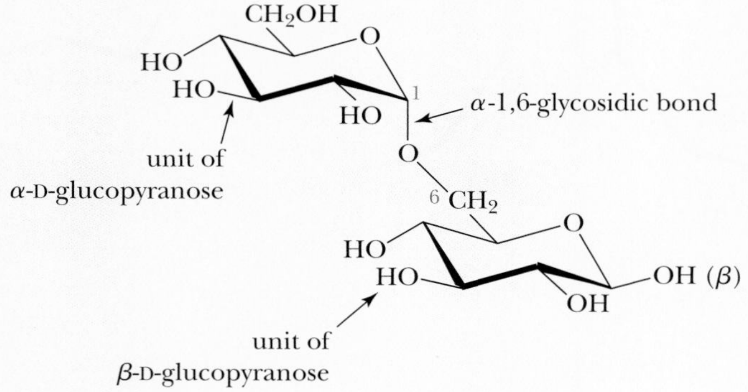

unit of
α-D-glucopyranose

α-1,6-glycosidic bond

unit of
β-D-glucopyranose

Among the first discovered and best understood of these membrane-bound carbohydrates are those of the ABO blood-group system, discovered in 1900 by Karl Landsteiner (1868–1943). Whether an individual has type A, B, AB, or O blood is genetically determined and depends on the type of trisaccharide or tetrasaccharide bound to the surface of the person's red blood cells. The monosaccharides of each blood group and the type of glycosidic bond joining them are shown in the following figure (the configurations of the glycosidic bonds are in parentheses):

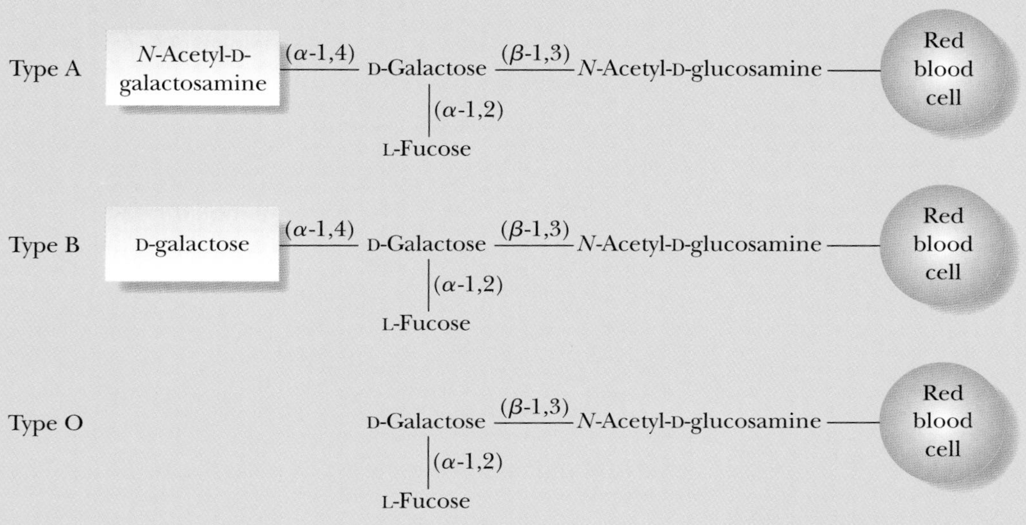

Type A N-Acetyl-D-galactosamine $\xrightarrow{(\alpha\text{-}1,4)}$ D-Galactose $\xrightarrow{(\beta\text{-}1,3)}$ N-Acetyl-D-glucosamine —— Red blood cell

$\Big|(\alpha\text{-}1,2)$

L-Fucose

Type B   D-galactose $\xrightarrow{(\alpha\text{-}1,4)}$ D-Galactose $\xrightarrow{(\beta\text{-}1,3)}$ N-Acetyl-D-glucosamine —— Red blood cell

$\Big|(\alpha\text{-}1,2)$

L-Fucose

Type O   D-Galactose $\xrightarrow{(\beta\text{-}1,3)}$ N-Acetyl-D-glucosamine —— Red blood cell

$\Big|(\alpha\text{-}1,2)$

L-Fucose

Practice Problem 18.7

Draw Haworth and chair formulas for the α form of a disaccharide in which two units of D-glucopyranose are joined by a β-1,3-glycosidic bond.

18.8 POLYSACCHARIDES

Polysaccharides consist of a large number of monosaccharide units joined together by glycosidic bonds. Three important polysaccharides, all made up of glucose units, are starch, glycogen, and cellulose.

A. Starch: Amylose and Amylopectin

Starch is found in all plant seeds and tubers and is the form in which glucose is stored for later use. Starch can be separated into two principal polysaccharides: amylose and amylopectin. Although the starch from each plant is unique, most starches contain 20 to 25 percent amylose and 75 to 80 percent amylopectin.

Complete hydrolysis of both amylose and amylopectin yields only D-glucose. Amylose is composed of continuous, unbranched chains of as many as 4,000 D-glucose units, joined by α-1,4-glycosidic bonds. Amylopectin contains chains up to 10,000 D-glucose units, also joined by α-1,4-glycosidic bonds. In addition, there is considerable branching from this linear network. At branch points, new chains of 24 to 30 units start by α-1,6-glycosidic bonds (Figure 18.4).

Figure 18.4

Amylopectin is a highly branched polymer of D-glucose. Chains consist of 24 to 30 units of D-glucose, joined by α-1,4-glycosidic bonds, and branches created by α-1,6-glycosidic bonds.

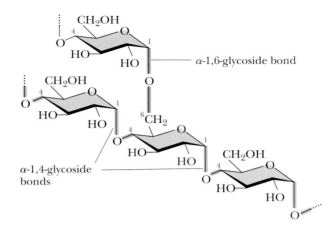

Why are carbohydrates stored in plants as polysaccharides rather than monosaccharides, a more directly usable source of energy? The answer has to do with **osmotic pressure**, which is proportional to the molar *concentration*, not the molecular weight, of a solute. If 1,000 molecules of glucose are assembled into one starch macromolecule, a solution containing 1 g of starch per 10 mL will have only 1 one-thousandth the osmotic pressure relative to a solution of 1 g of glucose in the same volume. This feat of packaging is a tremendous advantage, because it reduces the strain on various membranes enclosing solutions of such macromolecules.

B. Glycogen

Glycogen is the reserve carbohydrate for animals. Like amylopectin, glycogen is a branched polymer of D-glucose containing approximately 10^6 glucose units, joined by α-1,4- and α-1,6-glycosidic bonds. The total amount of glycogen in the body of a well-nourished adult human being is about 350 g, divided almost equally between liver and muscle.

C. Cellulose

Cellulose, the most widely distributed plant skeletal polysaccharide, constitutes almost half of the cell-wall material of wood. Cotton is almost pure cellulose.

Cellulose, a linear polymer of D-glucose units joined by β-1,4-glycosidic bonds (Figure 18.5), has an average molar mass of 400,000 g/mol, corresponding to approximately 2,800 glucose units per molecule.

Cellulose molecules act much like stiff rods, a feature that enables them to align themselves side by side into well-organized, water-insoluble fibers in which the OH groups form numerous intermolecular hydrogen bonds. This arrangement of parallel chains in bundles gives cellulose fibers their high mechanical strength and explains why cellulose is insoluble in water. When a piece of cellulose-containing material is placed in water, there are not enough —OH groups on the surface of the fiber to pull individual cellulose molecules away from the strongly hydrogen-bonded fiber.

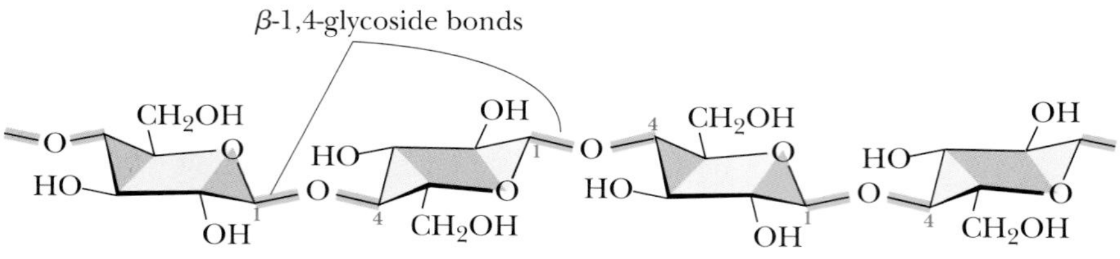

Figure 18.5

Cellulose is a linear polymer of D-glucose, joined by β-1,4-glycosidic bonds.

Humans and other animals cannot use cellulose as food, because our digestive systems do not contain β-glucosidases, enzymes that catalyze the hydrolysis of β-glucosidic bonds. Instead, we have only α-glucosidases; hence, the polysaccharides we use as sources of glucose are starch and glycogen. By contrast, many bacteria and microorganisms do contain β-glucosidases and can digest cellulose. Termites are fortunate (much to our regret) to have such bacteria in their intestines and can use wood as their principal food. Ruminants (cud-chewing animals) and horses can also digest grasses and hay, because β-glucosidase-containing microorganisms are present within their alimentary systems.

D. Textile Fibers from Cellulose

Both rayon and acetate rayon are made from chemically modified cellulose and were the first commercially important synthetic textile fibers. In the production of rayon, cellulose fibers are treated with carbon disulfide, CS_2, in aqueous sodium hydroxide. In this reaction, some of the —OH groups on a cellulose fiber are converted to the sodium salt of a xanthate ester, which causes the fibers to dissolve in alkali as a viscous colloidal dispersion:

An —OH group in
a cellulose fiber

$$\text{Cellulose—OH} \xrightarrow{\text{NaOH}} \text{Cellulose—O}^{-}\text{Na}^{+} \xrightarrow{\text{S=C=S}} \text{Cellulose—O}\overset{\overset{\displaystyle S}{\|}}{C}\text{—S}^{-}\text{Na}^{+}$$

Cellulose   Sodium salt of a xanthate ester
(insoluble in water)   (a viscous colloidal suspension)

The solution of cellulose xanthate is separated from the alkali-insoluble parts of wood and then forced through a spinneret (a metal disc with many tiny holes) into dilute sulfuric acid to hydrolyze the xanthate ester groups and precipitate regenerated cellulose. Regenerated cellulose extruded as a filament is called viscose rayon thread.

In the industrial synthesis of acetate rayon, cellulose is treated with acetic anhydride (Section 15.5B):

A glucose unit in   Acetic   A fully acetylated glucose unit
a cellulose fiber   anhydride

Acetylated cellulose is then dissolved in a suitable solvent, precipitated, and drawn into fibers known as acetate rayon. Today, acetate rayon fibers rank fourth in textile fiber production in the United States, surpassed only by Dacron® polyester, nylon, and rayon.

change in specific rotation
that accompanies the formation
equilibrium mixture

SUMMARY

Monosaccharides (Section 18.2A) are polyhydroxyaldehydes or polyhydroxyketones. The most common have the general formula $C_nH_{2n}O_n$, where n varies from 3 to 9. Their names contain the suffix *-ose*. The prefixes *tri-*, *tetr-*, *pent-*, and so on show the number of carbon atoms in the chain. The prefix *aldo-* indicates an aldehyde, the prefix *keto-* a ketone. In a **Fischer projection** (Section 18.2C) of a carbohydrate, the carbon chain is written vertically, with the most highly oxidized carbon toward the top. Horizontal lines show groups projecting above the plane of the page, vertical lines groups projecting behind the plane of the page. A monosaccharide that has the same configuration at the penultimate carbon as D-glyceraldehyde is called a **D-monosaccharide**; one that has the same configuration at the penultimate carbon as L-glyceraldehyde is called an **L-monosaccharide** (Section 18.2D). An amino sugar contains an —NH$_2$ group in place of an —OH group (Section 18.2E).

Monosaccharides exist primarily as cyclic hemiacetals (Section 18.3A). The new stereocenter resulting from hemiacetal formation is referred to as an **anomeric carbon**. The stereoisomers thus formed are called **anomers**. A six-membered cyclic hemiacetal is called a **pyranose**, a five-membered cyclic hemiacetal a **furanose**. The symbol **β-** indicates that the —OH on the anomeric carbon is on the same side of the ring as the terminal —CH$_2$OH. The symbol **α-** indicates that —OH on the anomeric carbon is on the opposite side of the ring from the terminal —CH$_2$OH. Furanoses and pyranoses can be drawn as **Haworth projections** (Section 18.3A). Pyranoses can also be shown as strain-free **chair conformations** (Section 18.3B).

Mutarotation (Section 18.3C) is the change in specific rotation that accompanies the formation of an equilibrium mixture of α- and β-anomers in aqueous solution.

A **glycoside** (Section 18.4A) is an acetal derived from a monosaccharide. The name of the glycoside is composed of the name of the alkyl or aryl group bonded to the acetal oxygen atom, followed by the name of the monosaccharide in which the terminal *-e* has been replaced by *-ide*.

An **alditol** (Section 18.4B) is a polyhydroxy compound formed by reduction of the carbonyl group of a monosaccharide to a hydroxyl group. An **aldonic acid** (Section 18.4C) is a carboxylic acid formed by oxidation of the aldehyde group of an aldose. The aldose reduces the oxidizing agent and is therefore called a **reducing sugar**. Enzyme-catalyzed oxidation of the terminal —CH$_2$OH to a —COOH gives a uronic acid (Section 18.4D).

L-**Ascorbic acid** (Section 18.6) is synthesized in nature from D-glucose by a series of enzyme-catalyzed steps.

A **disaccharide** (Section 18.7) contains two monosaccharide units joined by a glycosidic bond. Terms applied to carbohydrates containing larger numbers of monosaccharides are **trisaccharide**, **tetrasaccharide**, **oligosaccharide**, and **polysaccharide**. **Sucrose** (Section 18.7A) is a disaccharide consisting of D-glucose joined to D-fructose by an α-1,2-glycosidic bond. **Lactose** (Section 18.7B) is a disaccharide consisting of D-galactose joined to D-glucose by a β-1,4-glycosidic bond. **Maltose** (Section 18.7C) is a disaccharide of two molecules of D-glucose joined by an α-1,4-glycosidic bond.

Starch (Section 18.8A) can be separated into two fractions given the names amylose and amylopectin. **Amylose** is a linear polymer of up to 4,000 units of D-glucopyranose joined by α-1,4-glycosidic bonds. **Amylopectin** is a highly branched polymer of D-glucose joined by α-1,4-glycosidic bonds and, at branch points, by α-1,6-glycosidic bonds. **Glycogen** (Section 18.8B), the reserve carbohydrate of animals, is a highly branched polymer of D-glucopyranose joined by α-1,4-glycosidic bonds and, at branch points, by α-1,6-glycosidic bonds. **Cellulose** (Section 18.8C), the skeletal polysaccharide of plants, is a linear polymer of D-glucopyranose joined by β-1,4-glycosidic bonds. **Rayon** (Section 18.8D) is made from chemically modified and regenerated cellulose. **Acetate rayon** is made by the acetylation of cellulose.

KEY REACTIONS

1. Formation of Cyclic Hemiacetals (Section 18.3)

A monosaccharide existing as a five-membered ring is a furanose; one existing as a six-membered ring is a pyranose. A pyranose is most commonly drawn as a Haworth projection or a chair conformation:

D-Glucose → β-D-Glucopyranose (β-D-Glucose); anomeric carbon

2. Mutarotation (Section 18.3C)

Anomeric forms of a monosaccharide are in equilibrium in aqueous solution. Mutarotation is the change in specific rotation that accompanies this equilibration:

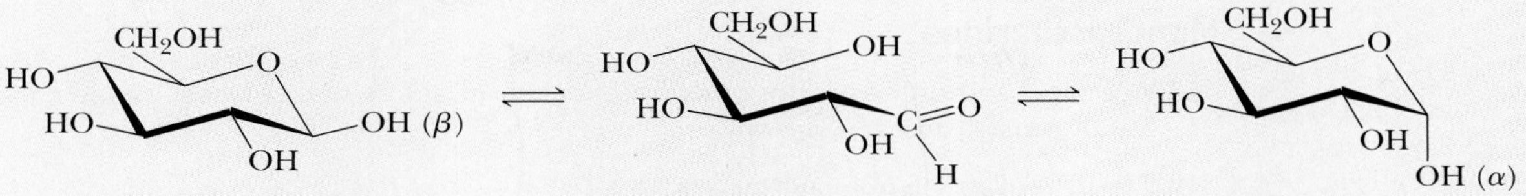

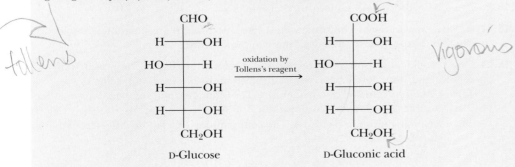

β-D-Glucopyranose
$[\alpha]_D^{25} + 18.7°$

Open-chain form

α-D-Glucopyranose
$[\alpha]_D^{25} + 112°$

3. Formation of Glycosides (Section 18.5A)

Treatment of a monosaccharide with an alcohol in the presence of an acid catalyst forms a cyclic acetal called a glycoside:

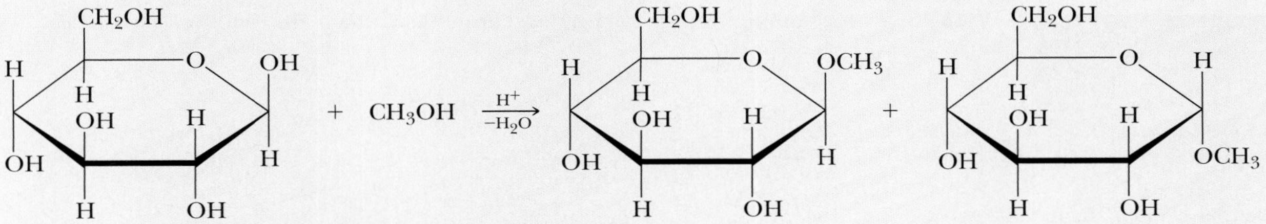

$$+ \quad CH_3OH \xrightarrow[-H_2O]{H^+}$$

The bond to the new —OR group is called a glycosidic bond.

4. Reduction to Alditols (Section 18.5B)

Reduction of the carbonyl group of an aldose or a ketose to a hydroxyl group yields a polyhydroxy compound called an alditol:

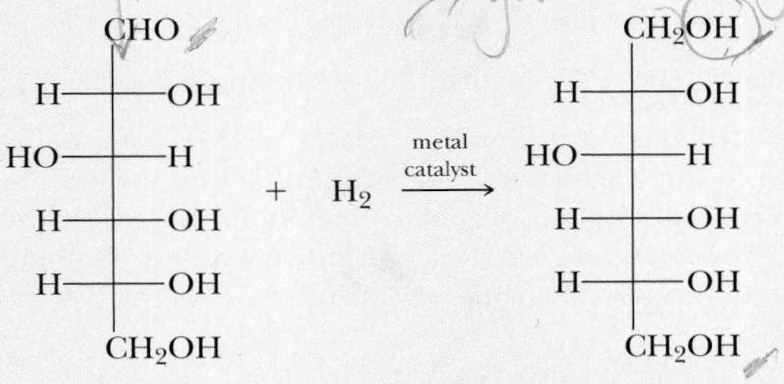

gains 2 hydrogens

$$\text{D-Glucose} \quad + \quad H_2 \xrightarrow[\text{catalyst}]{\text{metal}} \quad \text{D-Glucitol (D-Sorbitol)}$$

5. Oxidation to an Aldonic Acid (Section 18.5C)

Oxidation of the aldehyde group of an aldose to a carboxyl group by a mild oxidizing agent gives a polyhydroxycarboxylic acid called an aldonic acid:

Tollens

vigorous

$$\text{D-Glucose} \xrightarrow[\text{Tollens's reagent}]{\text{oxidation by}} \text{D-Gluconic acid}$$

PROBLEMS

A problem number set in red indicates an applied "real-world" problem.

Monosaccharides

18.8 What is the difference in structure between an aldose and a ketose? Between an aldopentose and a ketopentose?

18.9 Which hexose is also known as dextrose?

18.10 What does it mean to say that D- and L-glyceraldehyde are enantiomers?

18.11 Explain the meaning of the designations D and L as used to specify the configuration of carbohydrates.

18.12 How many stereocenters are present in D-glucose? In D-ribose? How many stereoisomers are possible for each monosaccharide?

18.13 Which compounds are D-monosaccharides and which are L-monosaccharides?

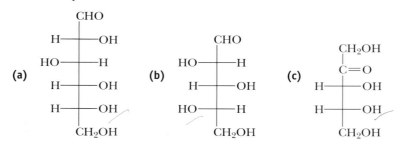

18.14 Draw Fischer projections for L-ribose and L-arabinose.

18.15 Explain why all mono- and disaccharides are soluble in water.

18.16 What is an amino sugar? Name the three amino sugars most commonly found in nature.

18.17 2,6-Dideoxy-D-altrose, known alternatively as D-digitoxose, is a monosaccharide obtained from the hydrolysis of digitoxin, a natural product extracted from purple foxglove *(Digitalis purpurea)*. Digitoxin has found wide use in cardiology because it reduces the pulse rate, regularizes heart rhythm, and strengthens the heartbeat. Draw the structural formula of 2,6-dideoxy-D-altrose.

The Cyclic Structure of Monosaccharides

18.18 Define the term *anomeric carbon*.

18.19 Explain the conventions for using α and β to designate the configurations of cyclic forms of monosaccharides.

18.20 Are α-D-glucose and β-D-glucose anomers? Explain. Are they enantiomers? Explain.

18.21 Are α-D-Gulose and α-L-Gulose anomers? Explain.

18.22 In what way are chair conformations a more accurate representation of molecular shape of hexopyranoses than are Haworth projections?

18.23 Draw α-D-glucopyranose (α-D-glucose) as a Haworth projection. Now, using only the following information, draw Haworth projections for these monosaccharides:
  (a) α-D-Mannopyranose (α-D-mannose). The configuration of D-mannose differs from that of D-glucose only at carbon 2.
  (b) α-D-Gulopyranose (α-D-gulose). The configuration of D-gulose differs from that of D-glucose at carbons 3 and 4.

The foxglove plant produces the important cardiac medication digitalis. *(Corbis Digital Stock)*

Looking Ahead

18.53 One step in glycolysis, the pathway that converts glucose to pyruvate (Section 22.4), involves an enzyme-catalyzed conversion of dihydroxyacetone phosphate to D-glyceraldehyde 3-phosphate:

Dihydroxyacetone phosphate   D-Glyceraldehyde 3-phosphate

Show that this transformation can be regarded as two enzyme-catalyzed keto-enol tautomerizations (Section 13.9).

18.54 One pathway for the metabolism of glucose 6-phosphate is its enzyme-catalyzed conversion to fructose 6-phosphate:

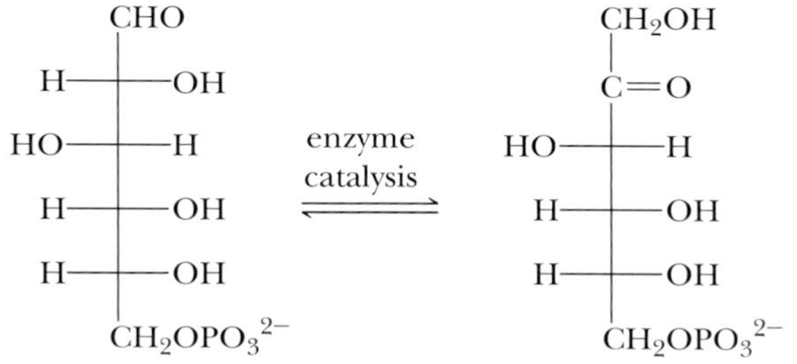

D-Glucose 6-phosphate   D-Fructose 6-phosphate

Show that this transformation can be regarded as two enzyme-catalyzed keto-enol tautomerizations.

18.55 Epimers are carbohydrates that differ in configuration at only one stereocenter.
(a) Which of the aldohexoses are epimers of each other?
(b) Are all anomer pairs also epimers of each other? Explain. Are all epimers also anomers? Explain.

18.56 Oligosaccharides are very valuable therapeutically, but are especially difficult to synthesize, even though the starting materials are readily available. Shown is the structure of globotriose, the receptor for a series of toxins synthesized by some strains of *E. coli:*

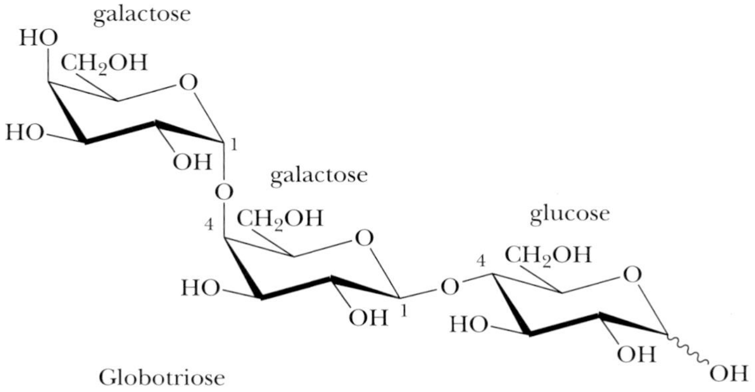

Globotriose

From left to right, globotriose consists of an α-1,4-linkage of galactose to galactose that is part of a β-1,4-linkage to glucose. The squiggly line indicates that the configuration at that carbon can be α or β. Suggest why it would be difficult to synthesize this trisaccharide, for example, by first forming the galactose–galactose glycosidic bond and then forming the glycosidic bond to glucose.

19 Amino Acids and Proteins

Spider silk is a fibrous protein that exhibits unmatched strength and toughness. Inset: Models of D-alanine and glycine, the major components of the fibrous protein of silk. *(PhotoDisc Inc./Getty Images)*

19.1 INTRODUCTION

We begin this chapter with a study of amino acids, compounds whose chemistry is built on amines (Chapter 10) and carboxylic acids (Chapter 14). We concentrate in particular on the acid–base properties of amino acids, because these properties are so important in determining many of the properties of proteins, including the catalytic functions of enzymes. With this understanding of the chemistry of amino acids, we then examine the structure of proteins themselves. Proteins are among the most important of all biological compounds. Among the functions performed by these vital molecules are the following:

- *structure*—structural proteins such as collagen and keratin are the chief constituents of skin, bones, hair, and nails.

- *catalysis*—virtually all reactions that take place in living systems are catalyzed by a special group of proteins called enzymes.
- *movement*—muscle fibers are made of proteins called myosin and actin.
- *transport*—the protein hemoglobin is responsible for the transport of oxygen from the lungs to tissues. Other proteins transport molecules across cell membranes.
- *protection*—a group of proteins called antibodies is one of the body's major defenses against disease.

Proteins have other functions as well. Even this brief list, however, should convince you of their vital role in living organisms.

19.2 AMINO ACIDS

A. Structure

An **amino acid** is a compound that contains both a carboxyl group and an amino group. Although many types of amino acids are known, the **α-amino acids** are the most significant in the biological world because they are the monomers from which proteins are constructed. A general structural formula of an α-amino acid is shown in Figure 19.1.

Although Figure 19.1(a) is a common way of writing structural formulas for amino acids, it is not accurate, because it shows an acid (—COOH) and a base (—NH$_2$) within the same molecule. These acidic and basic groups react with each other to form an internal salt (a dipolar ion) [Figure 19.1(b)]. An internal salt is given the special name **zwitterion**. Note that a zwitterion has no net charge; it contains one positive charge and one negative charge.

Because they exist as zwitterions, amino acids have many of the properties associated with salts. They are crystalline solids with high melting points and are fairly soluble in water, but insoluble in nonpolar organic solvents such as ether and hydrocarbon solvents.

B. Chirality

With the exception of glycine, H$_2$NCH$_2$COOH, all protein-derived amino acids have at least one stereocenter and therefore are chiral. Figure 19.2 shows Fischer projection formulas for the enantiomers of alanine. The vast majority of carbohydrates in the biological world are of the D-series (Section 18.2), whereas the vast majority of α-amino acids in the biological world are of the L-series.

C. Protein-Derived Amino Acids

Table 19.1 gives common names, structural formulas, and standard three-letter and one-letter abbreviations for the 20 common L-amino acids found in proteins. The amino acids shown are divided into four categories: those with nonpolar side chains; those with polar, but un-ionized, side chains; those with acidic side chains;

Amino acid A compound that contains both an amino group and a carboxyl group.

α-Amino acid An amino acid in which the amino group is on the carbon adjacent to the carboxyl group.

Zwitterion An internal salt of an amino acid.

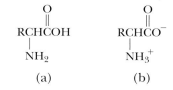

(a)   (b)

Figure 19.1
An α-amino acid.
(a) Un-ionized form and
(b) internal salt
(zwitterion) form.

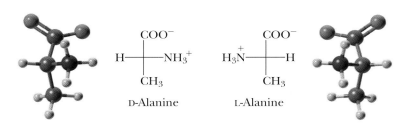

D-Alanine   L-Alanine

Figure 19.2
The enantiomers of alanine. The vast majority of α-amino acids in the biological world have the L-configuration at the α-carbon.

TABLE 19.1 The 20 Common Amino Acids Found in Proteins

Nonpolar Side Chains

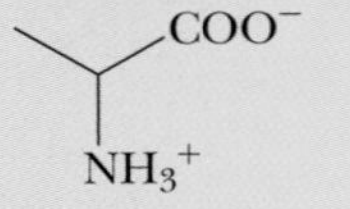

Alanine (Ala, A)

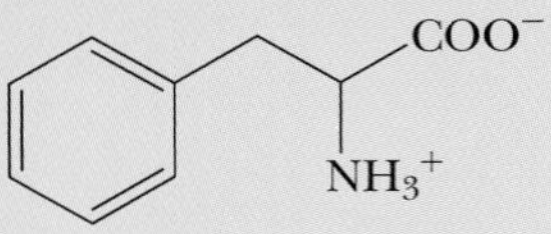

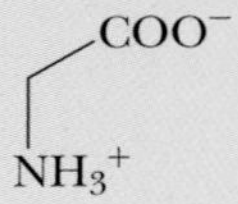

Glycine (Gly, G)

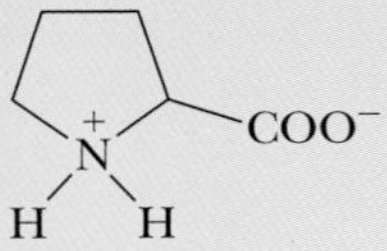

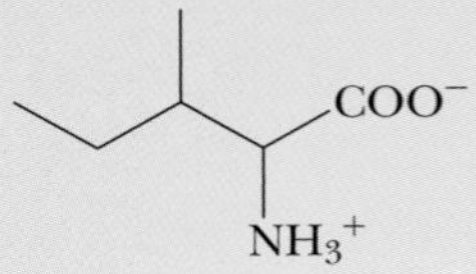

Isoleucine (Ile, I)

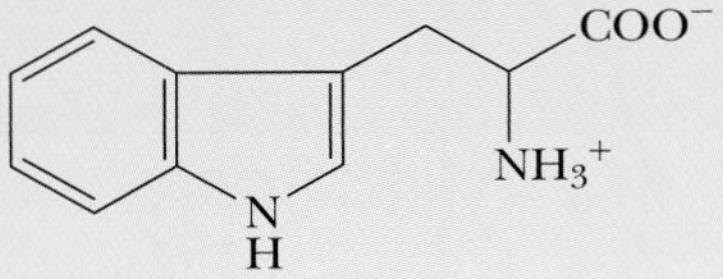

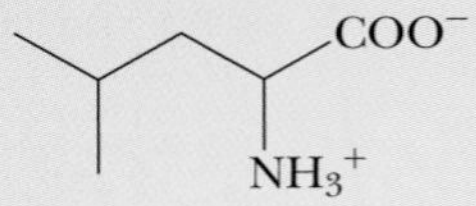

Leucine (Leu, L)

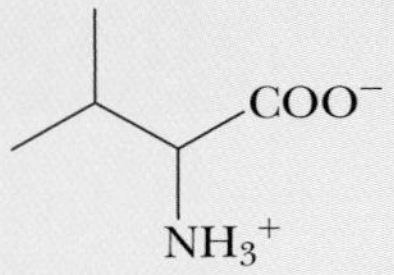

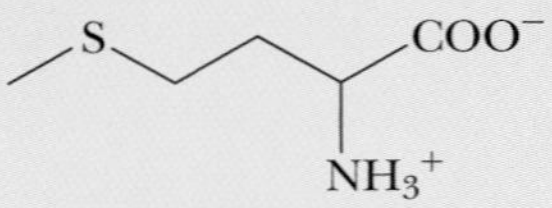

Methionine (Met, M)

Phenylalanine (Phe, F)

Proline (Pro, P)

Tryptophan (Trp, W)

Valine (Val, V)

Polar Side Chains

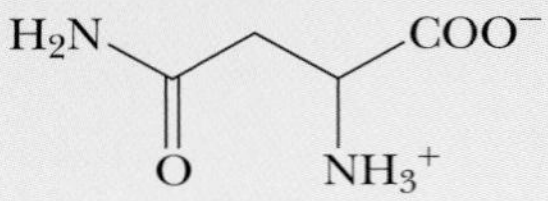

Asparagine (Asn, N)

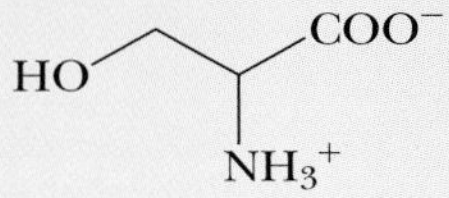

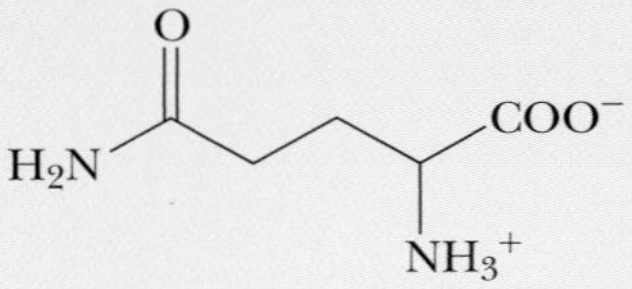

Glutamine (Gln, Q)

Serine (Ser, S)

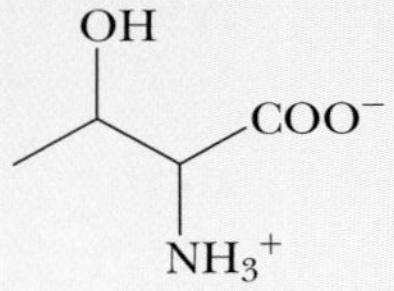

Threonine (Thr, T)

Acidic Side Chains

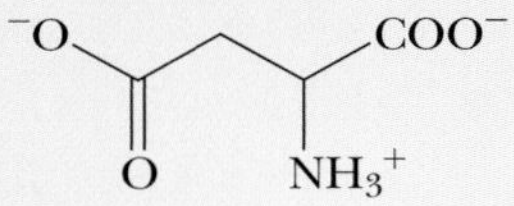

Aspartic acid (Asp, D)

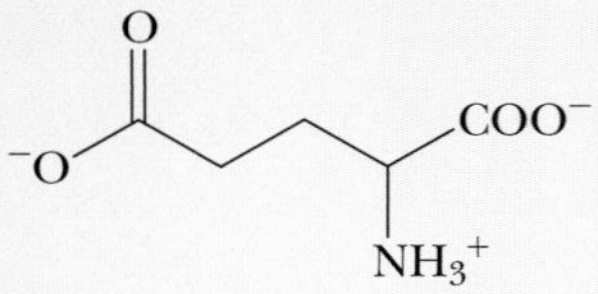

Glutamic acid (Glu, E)

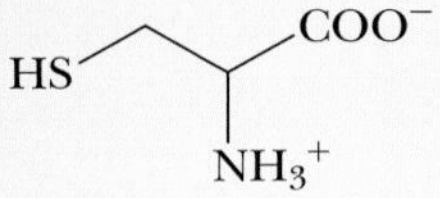

Cysteine (Cys, C)

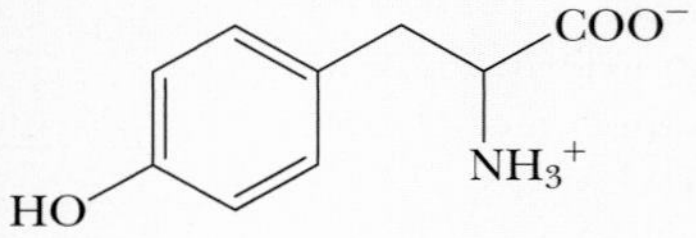

Tyrosine (Tyr, Y)

Basic Side Chains

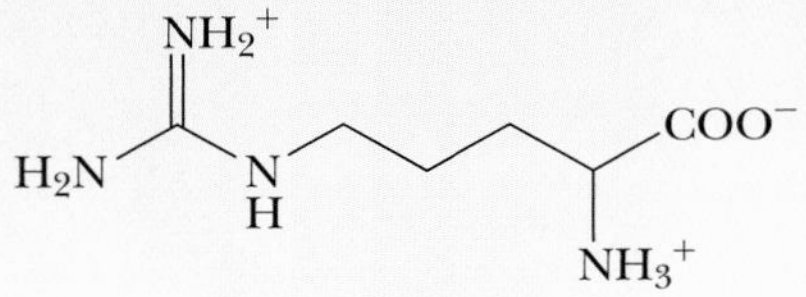

Arginine (Arg, R)

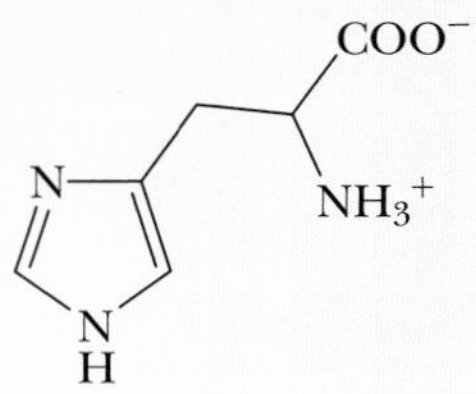

Histidine (His, H)

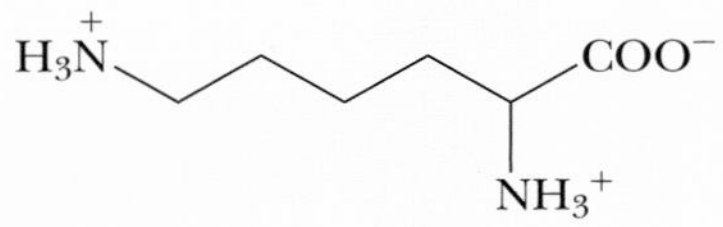

Lysine (Lys, K)

Note: Each ionizable group is shown in the form present in highest concentration in aqueous solution at pH 7.0.

and those with basic side chains. As you study the information in this table, note the following points:

1. All 20 of these protein-derived amino acids are α-amino acids, meaning that the amino group is located on the carbon alpha to the carboxyl group.
2. For 19 of the 20 amino acids, the α-amino group is primary. Proline is different: Its α-amino group is secondary.
3. With the exception of glycine, the α-carbon of each amino acid is a stereocenter. Although not shown in the table, all 19 chiral amino acids have the same relative configuration at the α-carbon. In the D,L convention, all are L-amino acids.
4. Isoleucine and threonine contain a second stereocenter. Four stereoisomers are possible for each amino acid, but only one of the four is found in proteins.
5. The sulfhydryl group of cysteine, the imidazole group of histidine, and the phenolic hydroxyl of tyrosine are partially ionized at pH 7.0, but the ionic form is not the major form present at that pH.

EXAMPLE 19.1

Of the 20 protein-derived amino acids shown in Table 19.1, how many contain (a) aromatic rings, (b) side-chain hydroxyl groups, (c) phenolic —OH groups, and (d) sulfur?

SOLUTION

(a) Phenylalanine, tryptophan, tyrosine, and histidine contain aromatic rings.
(b) Serine and threonine contain side-chain hydroxyl groups.
(c) Tyrosine contains a phenolic —OH group.
(d) Methionine and cysteine contain sulfur.

Practice Problem 19.1

Of the 20 protein-derived amino acids shown in Table 19.1, (a) which contain no stereocenter, and (b) which contain two stereocenters?

D. Some Other Common L-Amino Acids

Although the vast majority of plant and animal proteins are constructed from just these 20 α-amino acids, many other amino acids are also found in nature. Ornithine and citrulline, for example, are found predominantly in the liver and are an integral part of the urea cycle, the metabolic pathway that converts ammonia to urea:

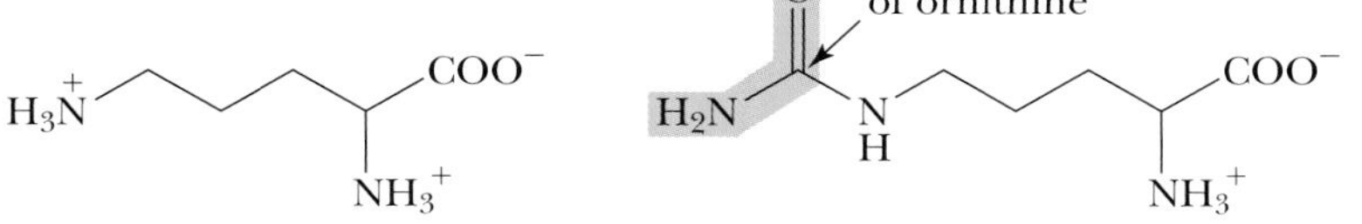

Ornithine

Citrulline

Thyroxine and triiodothyronine, two of several hormones derived from the amino acid tyrosine, are found in thyroid tissue:

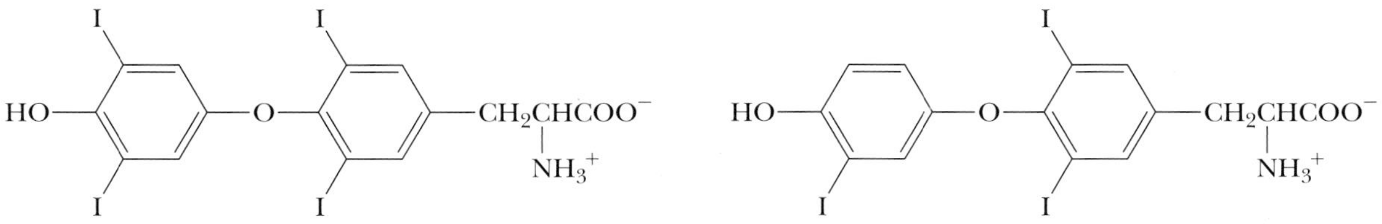

Thyroxine, T_4

Triiodothyronine, T_3

The principal function of these two hormones is to stimulate metabolism in other cells and tissues.

4-Aminobutanoic acid (γ-aminobutyric acid, or GABA) is found in high concentration (0.8 mM) in the brain, but in no significant amounts in any other mammalian tissue. GABA is synthesized in neural tissue by decarboxylation of the α-carboxyl group of glutamic acid and is a neurotransmitter in the central nervous system of invertebrates and possibly in humans as well:

Glutamic acid

4-Aminobutanoic acid
(γ-Aminobutyric acid, GABA)

Only L-amino acids are found in proteins, and only rarely are D-amino acids a part of the metabolism of higher organisms. Several D-amino acids, however, along with their L-enantiomers, are found in lower forms of life. D-Alanine and D-glutamic acid, for example, are structural components of the cell walls of certain bacteria. Several D-amino acids are also found in peptide antibiotics.

19.3 ACID–BASE PROPERTIES OF AMINO ACIDS

A. Acidic and Basic Groups of Amino Acids

Among the most important chemical properties of amino acids are their acid–base properties; all are weak polyprotic acids because of their —COOH and —NH_3^+ groups. Given in Table 19.2 are pK_a values for each ionizable group of the 20 protein-derived amino acids.

Acidity of α-Carboxyl Groups

The average value of pK_a for an α-carboxyl group of a protonated amino acid is 2.19. Thus, the α-carboxyl group is a considerably stronger acid than acetic acid (pK_a 4.76) and other low-molecular-weight aliphatic carboxylic acids. This greater acidity is accounted for by the electron-withdrawing inductive effect of the adjacent —NH_3^+ group (recall that we used similar reasoning in Section 14.5A to account for the relative acidities of acetic acid and its mono-, di-, and trichloroderivatives):

The ammonium group
has an electron-withdrawing
inductive effect

$$\text{RCHCOOH} + H_2O \rightleftharpoons \text{RCHCOO}^- + H_3O^+ \qquad pK_a = 2.19$$
$$\quad | \qquad\qquad\qquad\qquad | $$
$$\text{NH}_3^+ \qquad\qquad\qquad \text{NH}_3^+$$

Acidity of Side-Chain Carboxyl Groups

Due to the electron-withdrawing inductive effect of the α-NH_3^+ group, the side-chain carboxyl groups of protonated aspartic acid and glutamic acid are also stronger acids than acetic acid (pK_a 4.76). Notice that this acid-strengthening

TABLE 19.2 pK_a Values for Ionizable Groups of Amino Acids

Amino Acid	pK_a of α-COOH	pK_a of α-NH$_3^+$	pK_a of of Side Chain	Isoelectric Point (pI)
alanine	2.35	9.87	—	6.11
arginine	2.01	9.04	12.48	10.76
asparagine	2.02	8.80	—	5.41
aspartic acid	2.10	9.82	3.86	2.98
cysteine	2.05	10.25	8.00	5.02
glutamic acid	2.10	9.47	4.07	3.08
glutamine	2.17	9.13	—	5.65
glycine	2.35	9.78	—	6.06
histidine	1.77	9.18	6.10	7.64
isoleucine	2.32	9.76	—	6.04
leucine	2.33	9.74	—	6.04
lysine	2.18	8.95	10.53	9.74
methionine	2.28	9.21	—	5.74
phenylalanine	2.58	9.24	—	5.91
proline	2.00	10.60	—	6.30
serine	2.21	9.15	—	5.68
threonine	2.09	9.10	—	5.60
tryptophan	2.38	9.39	—	5.88
tyrosine	2.20	9.11	10.07	5.63
valine	2.29	9.72	—	6.00

Note: Dash indicates no ionizable side chain.

inductive effect decreases with increasing distance of the —COOH from the α-NH$_3^+$. Compare the acidities of the α-COOH of alanine (pK_a 2.35), the γ-COOH of aspartic acid (pK_a 3.86), and the δ-COOH of glutamic acid (pK_a 4.07).

Acidity of α-Ammonium Groups

The average value of pK_a for an α-ammonium group, α-NH$_3^+$, is 9.47, compared with an average value of 10.76 for primary aliphatic ammonium ions (Section 10.5):

$$\underset{\underset{NH_3^+}{|}}{RCHCOO^-} + H_2O \rightleftharpoons \underset{\underset{NH_2}{|}}{RCHCOO^-} + H_3O^+ \qquad pK_a = 9.47$$

$$\underset{\underset{NH_3^+}{|}}{CH_3CHCH_3} + H_2O \rightleftharpoons \underset{\underset{NH_2}{|}}{CH_3CHCH_3} + H_3O^+ \qquad pK_a = 10.60$$

Thus, the α-ammonium group of an amino acid is a slightly stronger acid than a primary aliphatic ammonium ion is, and conversely, an α-amino group is a slightly weaker base than a primary aliphatic amine is.

Basicity of the Guanidine Group of Arginine

The side-chain guanidine group of arginine is a considerably stronger base than an aliphatic amine is. As we saw in Section 10.5, guanidine (pK_b 0.4) is the strongest base of any neutral compound. The remarkable basicity of the guanidine group of arginine is attributed to the large resonance stabilization of the protonated form relative to the neutral form:

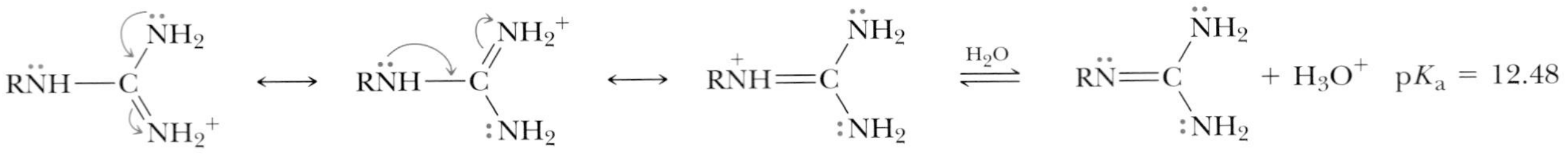

The protonated form of the guanidium ion side chain of arginine is a hybrid of three contributing structures

The deprotonated form shows no resonance stabilization without charge separation

Basicity of the Imidazole Group of Histidine

Because the imidazole group on the side chain of histidine contains six π electrons in a planar, fully conjugated ring, imidazole is classified as a heterocyclic aromatic amine (Section 9.3). The unshared pair of electrons on one nitrogen is a part of the aromatic sextet, whereas that on the other nitrogen is not. It is the pair of electrons that is not part of the aromatic sextet which is responsible for the basic properties of the imidazole ring. Protonation of this nitrogen produces a resonance-stabilized cation:

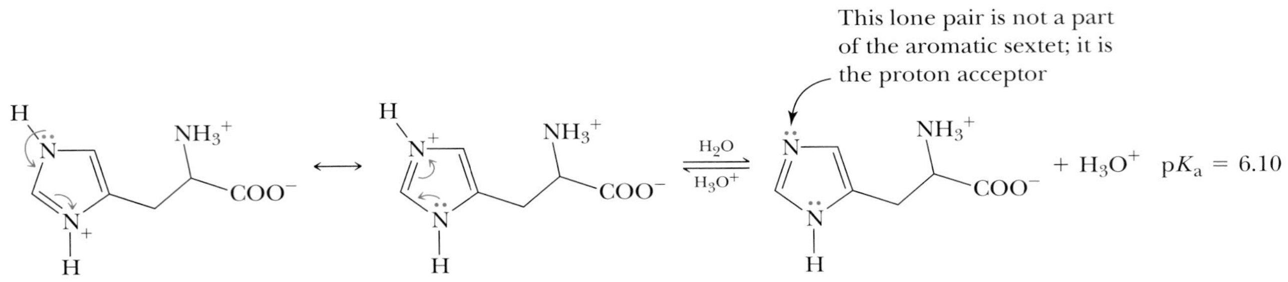

This lone pair is not a part of the aromatic sextet; it is the proton acceptor

Resonance-stabilized imidazolium cation

B. Titration of Amino Acids

Values of pK_a for the ionizable groups of amino acids are most commonly obtained by acid–base titration and by measuring the pH of the solution as a function of added base (or added acid, depending on how the titration is done). To illustrate this experimental procedure, consider a solution containing 1.00 mole of glycine to which has been added enough strong acid so that both the amino and carboxyl groups are fully protonated. Next, the solution is titrated with 1.00 M NaOH; the volume of base added and the pH of the resulting solution are recorded and then plotted as shown in Figure 19.3.

The most acidic group, and the one to react first with added sodium hydroxide, is the carboxyl group. When exactly 0.50 mole of NaOH has been added, the carboxyl group is half neutralized. At this point, the concentration of the zwitterion equals that of the positively charged ion, and the pH of 2.35 equals the pK_a of the carboxyl group (pK_{a1}):

$$\text{At pH} = pK_{a1} \qquad [\overset{+}{\text{H}}_3\text{NCH}_2\text{COOH}] = [\overset{+}{\text{H}}_3\text{NCH}_2\text{COO}^-]$$

Positive ion Zwitterion

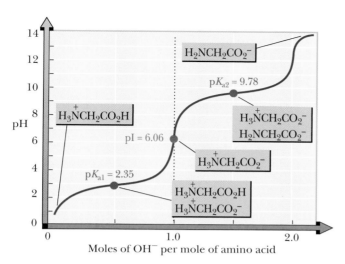

Figure 19.3
Titration of glycine with
sodium hydroxide.

The end point of the first part of the titration is reached when 1.00 mole of NaOH has been added. At this point, the predominant species present is the zwitterion, and the observed pH of the solution is 6.06.

The next section of the curve represents titration of the $-NH_3^+$ group. When another 0.50 mole of NaOH has been added (bringing the total to 1.50 moles), half of the $-NH_3^+$ groups are neutralized and converted to $-NH_2$. At this point, the concentrations of the zwitterion and negatively charged ion are equal, and the observed pH is 9.78, the pK_a of the amino group of glycine (pK_{a2}):

$$\text{At pH} = pK_{a2} \qquad [\overset{+}{H_3}NCH_2COO^-] = [H_2NCH_2COO^-]$$
$$\qquad\qquad\qquad\quad \text{Zwitterion} \qquad \text{Negative ion}$$

The second end point of the titration is reached when a total of 2.00 moles of NaOH have been added and glycine is converted entirely to an anion.

C. Isoelectric Point

Titration curves such as that for glycine permit us to determine pK_a values for the ionizable groups of an amino acid. They also permit us to determine another important property: the **isoelectric point, pI**—the pH at which most of the molecules of the amino acid in solution have a net charge of zero. (They are zwitterions.) By examining the titration curve, you can see that the isoelectric point for glycine falls half way between the pK_a values for the carboxyl and amino groups:

Isoelectric point (pI) The pH at which an amino acid, a polypeptide, or a protein has no net charge.

$$pI = \frac{1}{2}(pK_a\ \alpha\text{-COOH} + pK_a\ \alpha\text{-NH}_3^+)$$

$$= \frac{1}{2}(2.35 + 9.78) = 6.06$$

At pH 6.06, the predominant form of glycine molecules is the dipolar ion; furthermore, at this pH, the concentration of positively charged glycine molecules equals the concentration of negatively charged glycine molecules.

Given a value for the isoelectric point of an amino acid, it is possible to estimate the charge on that amino acid at any pH. For example, the charge on tyrosine at pH 5.63, the isoelectric point of tyrosine, is zero. A small fraction of tyrosine molecules is positively charged at pH 5.00 (0.63 unit less than its pI), and virtually all are positively charged at pH 3.63 (2.00 units less than its pI). As another example, the net charge on lysine is zero at pH 9.74. At pH values smaller than 9.74, an increasing fraction of lysine molecules is positively charged.

Figure 19.4
Electrophoresis of a mixture of amino acids. Those with a negative charge move toward the positive electrode; those with a positive charge move toward the negative electrode; those with no charge remain at the origin.

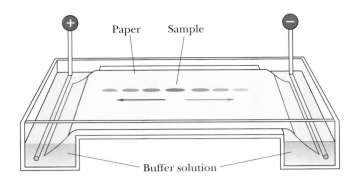

D. Electrophoresis

Electrophoresis The process of separating compounds on the basis of their electric charge.

Electrophoresis, a process of separating compounds on the basis of their electric charges, is used to separate and identify mixtures of amino acids and proteins. Electrophoretic separations can be carried out with paper, starch, agar, certain plastics, and cellulose acetate used as solid supports. In paper electrophoresis, a paper strip saturated with an aqueous buffer of predetermined pH serves as a bridge between two electrode vessels (Figure 19.4). Next, a sample of amino acids is applied as a colorless spot on the paper strip. (The amino acid mixture is colorless.) When an electrical potential is then applied to the electrode vessels, amino acids migrate toward the electrode carrying the charge opposite to their own. Molecules having a high charge density move more rapidly than do those with a lower charge density. Any molecule already at its isoelectric point remains at the origin. After the separation is complete, the paper strip is sprayed with a dye that transforms each amino acid into a colored compound, making the separated components visible.

A dye commonly used to detect amino acids is ninhydrin (1,2,3-indanetrione monohydrate). Ninhydrin reacts with α-amino acids to produce an aldehyde, carbon dioxide, and a purple-colored anion:

$$
\underset{\substack{\text{An } \alpha\text{-amino} \\ \text{acid}}}{\text{RCHCO}^- \atop \underset{\text{NH}_3^+}{|}} \;+\; 2 \;\; \underset{\text{Ninhydrin}}{\text{(structure)}} \;\longrightarrow\; \underset{\text{Purple-colored anion}}{\text{(structure)}} \;+\; \text{RCH} \;+\; \text{CO}_2 \;+\; \text{H}_3\text{O}^+
$$

This reaction is commonly used in both qualitative and quantitative analysis of amino acids. Nineteen of the 20 protein-derived α-amino acids have primary amino groups and give the same purple-colored ninhydrin-derived anion. Proline, a secondary amine, gives a different, orange-colored compound.

EXAMPLE 19.2

The isoelectric point of tyrosine is 5.63. Toward which electrode does tyrosine migrate during paper electrophoresis at pH 7.0?

SOLUTION

During paper electrophoresis at pH 7.0 (more basic than its isoelectric point), tyrosine has a net negative charge and migrates toward the positive electrode.

Practice Problem 19.2 ————————————————————————————

The isoelectric point of histidine is 7.64. Toward which electrode does histidine migrate during paper electrophoresis at pH 7.0?

EXAMPLE 19.3

The electrophoresis of a mixture of lysine, histidine, and cysteine is carried out at pH 7.64. Describe the behavior of each amino acid under these conditions.

SOLUTION

The isoelectric point of histidine is 7.64. At this pH, histidine has a net charge of zero and does not move from the origin. The pI of cysteine is 5.02; at pH 7.64 (more basic than its isoelectric point), cysteine has a net negative charge and moves toward the positive electrode. The pI of lysine is 9.74; at pH 7.64 (more acidic than its isoelectric point), lysine has a net positive charge and moves toward the negative electrode.

Practice Problem 19.3 ————————————————————————————

Describe the behavior of a mixture of glutamic acid, arginine, and valine during paper electrophoresis at pH 6.0.

19.4 POLYPEPTIDES AND PROTEINS

In 1902, Emil Fischer proposed that proteins were long chains of amino acids joined together by amide bonds between the α-carboxyl group of one amino acid and the α-amino group of another. For these amide bonds, Fischer proposed the special name **peptide bond**. Figure 19.5 shows the peptide bond formed between serine and alanine in the dipeptide serylalanine.

Peptide bond The special name given to the amide bond formed between the α-amino group of one amino acid and the α-carboxyl group of another amino acid.

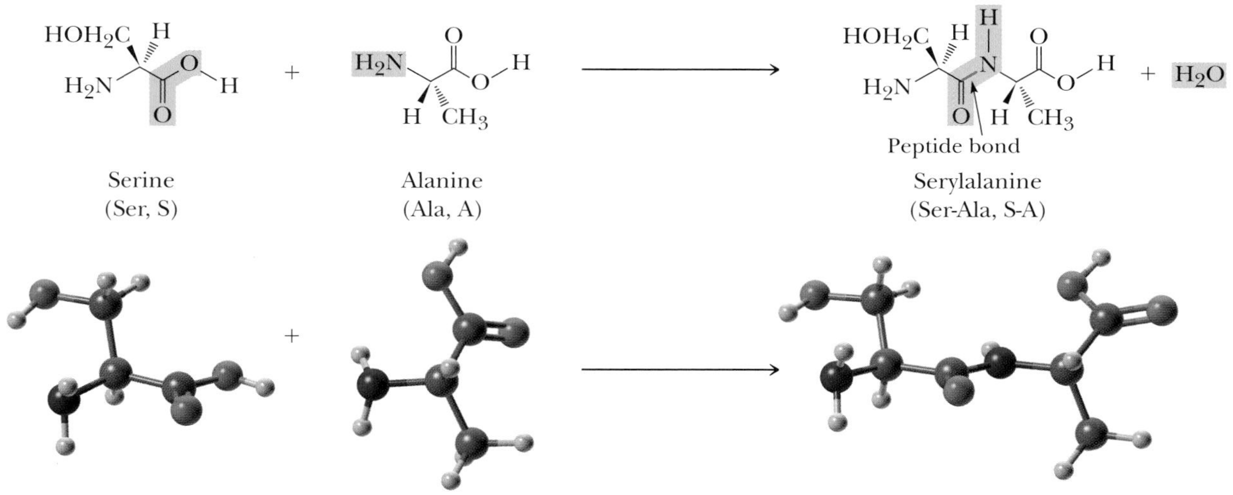

Figure 19.5
The peptide bond in serylalanine.

Peptide is the name given to a short polymer of amino acids. We classify peptides by the number of amino acid units in their chains. A molecule containing 2 amino acids joined by an amide bond is called a **dipeptide**. Those containing 3 to 10 amino acids are called **tripeptides, tetrapeptides, pentapeptides**, and so on. Molecules containing more than 10, but fewer than 20, amino acids are called **oligopeptides**. Those containing several dozen or more amino acids are called **polypeptides. Proteins** are biological macromolecules with molecular weight 5,000 or greater and consisting of one or more polypeptide chains. The distinctions in this terminology are not at all precise.

By convention, polypeptides are written from left to right, beginning with the amino acid having the free $—NH_3^+$ group and proceeding toward the amino acid with the free $—COO^-$ group. The amino acid with the free $—NH_3^+$ group is called the **N-terminal amino acid**, and that with the free $—COO^-$ group is called the **C-terminal amino acid**:

Dipeptide A molecule containing two amino acid units joined by a peptide bond.

Tripeptide A molecule containing three amino acid units, each joined to the next by a peptide bond.

Polypeptide A macromolecule containing 20 or more amino acid units, each joined to the next by a peptide bond.

N-Terminal amino acid The amino acid at the end of a polypeptide chain having the free $—NH_3^+$ group.

C-Terminal amino acid The amino acid at the end of a polypeptide chain having the free $—COO^-$ group.

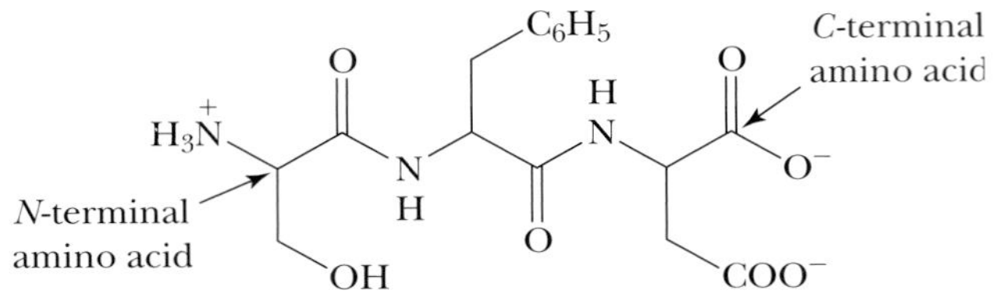

Ser-Phe-Asp

EXAMPLE 19.4

Draw a structural formula for Cys-Arg-Met-Asn. Label the *N*-terminal amino acid and the *C*-terminal amino acid. What is the net charge on this tetrapeptide at pH 6.0?

SOLUTION

The backbone of Cys-Arg-Met-Asn, a tetrapeptide, is a repeating sequence of nitrogen-α-carbon-carbonyl. The net charge on this tetrapeptide at pH 6.0 is +1. The following is a structural formula for Cys-Arg-Met-Asn:

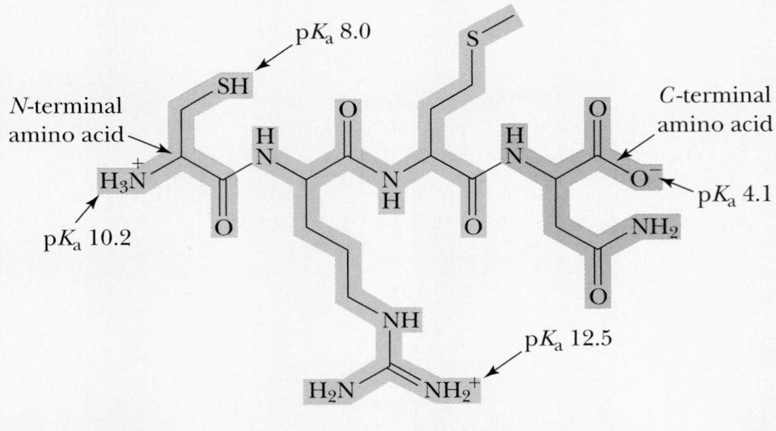

Practice Problem 19.4 ————————————————————————

Draw a structural formula for Lys-Phe-Ala. Label the *N*-terminal amino acid and the *C*-terminal amino acid. What is the net charge on this tripeptide at pH 6.0?

19.5 PRIMARY STRUCTURE OF POLYPEPTIDES AND PROTEINS

The **primary (1°) structure** of a polypeptide or protein is the sequence of amino acids in its polypeptide chain. In this sense, the primary structure is a complete description of all covalent bonding in a polypeptide or protein.

In 1953, Frederick Sanger of Cambridge University, England, reported the primary structure of the two polypeptide chains of the hormone insulin. Not only was this a remarkable achievement in analytical chemistry, but also, it clearly established that the molecules of a given protein all have the same amino acid composition and the same amino acid sequence. Today, the amino acid sequences of over 20,000 different proteins are known, and the number is growing rapidly.

Primary (1°) structure of proteins The sequence of amino acids in the polypeptide chain; read from the *N*-terminal amino acid to the *C*-terminal amino acid.

A. Amino Acid Analysis

The first step in determining the primary structure of a polypeptide is hydrolysis and quantitative analysis of its amino acid composition. Recall from Section 15.4D that amide bonds are highly resistant to hydrolysis. Typically, samples of protein are hydrolyzed in 6 M HCl in sealed glass vials at 110°C for 24 to 72 hours. (This hydrolysis can be done in a microwave oven in a shorter time.) After the polypeptide is hydrolyzed, the resulting mixture of amino acids is analyzed by ion-exchange chromatography. In this process, the mixture of amino acids is passed through a specially packed column. Each of the 20 amino acids requires a different time to pass through the column. Amino acids are detected by reaction with ninhydrin as they emerge from the column (Section 19.3D), followed by absorption spectroscopy. Current procedures for the hydrolysis of polypeptides and the analysis of amino acid mixtures have been refined to the point where it is possible to determine the amino acid composition from as little as 50 nanomoles (50×10^{-9} mole) of a polypeptide. Figure 19.6 shows the analysis of a polypeptide hydrolysate by ion-exchange chromatography. Note that, during hydrolysis, the side-chain amide groups of asparagine and glutamine are hydrolyzed, and these amino acids are detected as aspartic acid and glutamic acid. For each glutamine or asparagine hydrolyzed, an equivalent amount of ammonium chloride is formed.

B. Sequence Analysis

Once the amino acid composition of a polypeptide has been determined, the next step is to determine the order in which the amino acids are joined in the polypeptide chain. The most common sequencing strategy is to cleave the polypeptide at specific peptide bonds (by using, for example, cyanogen bromide or certain proteolytic enzymes), determine the sequence of each fragment (by using, for example, the Edman degradation), and then match overlapping fragments to arrive at the sequence of the polypeptide.

Cyanogen Bromide

Cyanogen bromide (BrCN) is specific for the cleavage of peptide bonds formed by the carboxyl group of methionine (Figure 19.7). The products of this cleavage are substituted γ-lactones (Section 15.2C), derived from the *N*-terminal portion of the polypeptide, and a second fragment containing the *C*-terminal portion of the polypeptide.

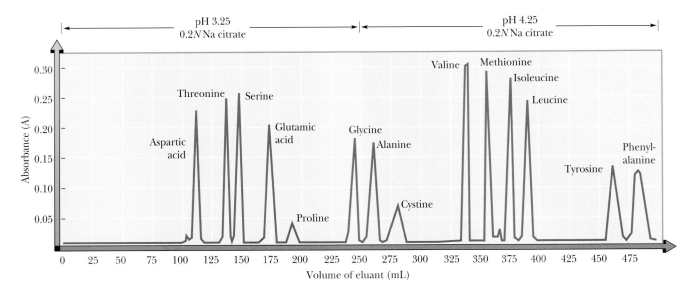

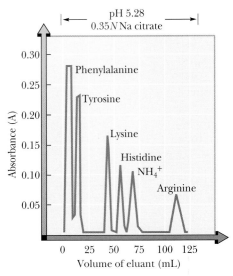

Figure 19.6

Analysis of a mixture of amino acids by ion-exchange chromatography using Amberlite IR-120, a sulfonated polystyrene resin. The resin contains phenyl-SO_3^- Na^+ groups. The amino acid mixture is applied to the column at low pH (3.25), under which conditions the acidic amino acids (Asp, Glu) are weakly bound to the resin and the basic amino acids (Lys, His, Arg) are tightly bound. Sodium citrate buffers of two different concentrations, and three different values of pH are used to elute the amino acids from the column. Cysteine is determined as cystine, Cys-S-S-Cys, the disulfide of cysteine.

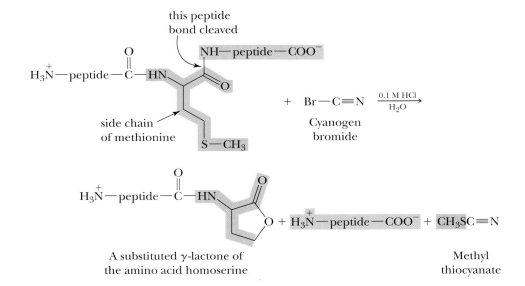

Figure 19.7

Cleavage by cyanogen bromide, BrCN, of a peptide bond formed by the carboxyl group of methionine.

Enzyme-Catalyzed Hydrolysis of Peptide Bonds

A group of proteolytic enzymes, including trypsin and chymotrypsin, can be used to catalyze the hydrolysis of specific peptide bonds. Trypsin catalyzes the hydrolysis of peptide bonds formed by the carboxyl groups of arginine and lysine; chymotrypsin catalyzes the hydrolysis of peptide bonds formed by the carboxyl groups of phenylalanine, tyrosine, and tryptophan.

EXAMPLE 19.5

Which of these tripeptides are hydrolyzed by trypsin? By chymotrypsin?

(a) Arg-Glu-Ser   (b) Phe-Gly-Lys

SOLUTION

(a) Trypsin catalyzes the hydrolysis of peptide bonds formed by the carboxyl groups of lysine and arginine. Therefore, the peptide bond between arginine and glutamic acid is hydrolyzed in the presence of trypsin:

$$\text{Arg-Glu-Ser} + H_2O \xrightarrow{\text{trypsin}} \text{Arg} + \text{Glu-Ser}$$

Chymotrypsin catalyzes the hydrolysis of peptide bonds formed by the carboxyl groups of phenylalanine, tyrosine, and tryptophan. Because none of these three aromatic amino acids is present, tripeptide (a) is not affected by chymotrypsin.

(b) Tripeptide (b) is not affected by trypsin. Although lysine is present, its carboxyl group is at the C-terminal end and therefore is not involved in peptide bond formation. Tripeptide (b) is hydrolyzed in the presence of chymotrypsin:

$$\text{Phe-Gly-Lys} + H_2O \xrightarrow{\text{chymotrypsin}} \text{Phe} + \text{Gly-Lys}$$

Practice Problem 19.5

Which of these tripeptides are hydrolyzed by trypsin? By chymotrypsin?

(a) Tyr-Gln-Val   (b) Thr-Phe-Ser   (c) Thr-Ser-Phe

Edman Degradation

Of the various chemical methods developed for determining the amino acid sequence of a polypeptide, the one most widely used today is the **Edman degradation**, introduced in 1950 by Pehr Edman of the University of Lund, Sweden. In this procedure, a polypeptide is treated with phenyl isothiocyanate, $C_6H_5N\!=\!C\!=\!S$, and then with acid. The effect of Edman degradation is to remove the N-terminal amino acid selectively as a substituted phenylthiohydantoin (Figure 19.8), which is then separated and identified.

Edman degradation A method for selectively cleaving and identifying the N-terminal amino acid of a polypeptide chain.

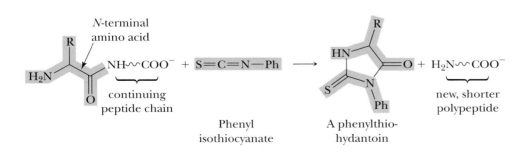

Figure 19.8
Edman degradation. Treatment of a polypeptide with phenyl isothiocyanate followed by acid selectively cleaves the N-terminal amino acid as a substituted phenylthiohydantoin.

The special value of the Edman degradation is that it cleaves the *N*-terminal amino acid from a polypeptide without affecting any other bonds in the chain. Furthermore, Edman degradation can be repeated on the shortened polypeptide, causing the next amino acid in the sequence to be cleaved and identified. In practice, it is now possible to sequence as many as the first 20 to 30 amino acids in a polypeptide by this method, using as little as a few milligrams of material.

Most polypeptides in nature are longer than 20 to 30 amino acids, the practical limit on the number of amino acids that can be sequenced by repetitive Edman degradation. The special value of cleavage with cyanogen bromide, trypsin, and chymotrypsin is that, at specific peptide bonds, a long polypeptide chain can be cleaved into smaller polypeptide fragments, and each fragment can then be sequenced separately.

EXAMPLE 19.6

Deduce the amino acid sequence of a pentapeptide from the following experimental results (note that, under the column "Amino Acids Determined from Procedure," the amino acids are listed in alphabetical order; in no way does this listing give any information about primary structure):

Experimental Procedure	Amino Acids Determined from Procedure
Amino Acid Analysis of Pentapeptide	Arg, Glu, His, Phe, Ser
Edman degradation	Glu
Hydrolysis Catalyzed by Chymotrypsin	
Fragment A	Glu, His, Phe
Fragment B	Arg, Ser
Hydrolysis Catalyzed by Trypsin	
Fragment C	Arg, Glu, His, Phe
Fragment D	Ser

SOLUTION

Edman degradation cleaves Glu from the pentapeptide; therefore, glutamic acid must be the *N*-terminal amino acid, and we have

Glu- (Arg, His, Phe, Ser)

Fragment A from chymotrypsin-catalyzed hydrolysis contains Phe. Because of the specificity of chymotrypsin, Phe must be the *C*-terminal amino acid of fragment A. Fragment A also contains Glu, which we already know is the *N*-terminal amino acid. From these observations, we conclude that the first three amino acids in the chain must be Glu-His-Phe, and we now write the following partial sequence:

Glu-His-Phe-(Arg, Ser)

The fact that trypsin cleaves the pentapeptide means that Arg must be within the pentapeptide chain; it cannot be the *C*-terminal amino acid. Therefore, the complete sequence must be

Glu-His-Phe-Arg-Ser

Practice Problem 19.6 ———————————————————

Deduce the amino acid sequence of an undecapeptide (11 amino acids) from the experimental results shown in the following table:

Experimental Procedure	Amino Acids Determined from Procedure
Amino Acid Analysis of Undecapeptide	Ala, Arg, Glu, Lys$_2$, Met, Phe, Ser, Thr, Trp, Val
Edman degradation	Ala
Trypsin-Catalyzed Hydrolysis	
Fragment E	Ala, Glu, Arg
Fragment F	Thr, Phe, Lys
Fragment G	Lys
Fragment H	Met, Ser, Trp, Val
Chymotrypsin-Catalyzed Hydrolysis	
Fragment I	Ala, Arg, Glu, Phe, Thr
Fragment J	Lys$_2$, Met, Ser, Trp, Val
Treatment with Cyanogen Bromide	
Fragment K	Ala, Arg, Glu, Lys$_2$, Met, Phe, Thr, Val
Fragment L	Trp, Ser

19.6 THREE-DIMENSIONAL SHAPES OF POLYPEPTIDES AND PROTEINS

A. Geometry of a Peptide Bond

In the late 1930s, Linus Pauling began a series of studies aimed at determining the geometry of a peptide bond. One of his first discoveries was that a peptide bond is planar. As shown in Figure 19.9, the four atoms of a peptide bond and the two α-carbons joined to it all lie in the same plane.

Had you been asked in Chapter 1 to describe the geometry of a peptide bond, you probably would have predicted bond angles of 120° about the carbonyl carbon and 109.5° about the amide nitrogen. This prediction agrees with the observed bond angles of approximately 120° about the carbonyl carbon. It does not agree, however, with bond angles of 120° about the amide nitrogen. To account for the observed geometry, Pauling proposed that a peptide bond is more accurately represented as a resonance hybrid of these two contributing structures:

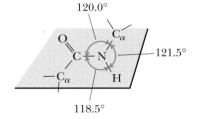

Figure 19.9
Planarity of a peptide bond. Bond angles about the carbonyl carbon and the amide nitrogen are approximately 120°.

(1)   (2)

Contributing structure (1) shows a carbon–oxygen double bond, and structure (2) shows a carbon–nitrogen double bond. The hybrid, of course, is neither of these; in the real structure, the carbon–nitrogen bond has considerable double-bond character. Accordingly, in the hybrid, the six-atom group is planar.

Figure 19.10
Hydrogen bonding between amide groups.

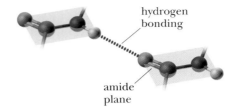

Two configurations are possible for the atoms of a planar peptide bond. In one, the two α-carbons are cis to each other; in the other, they are trans to each other. The trans configuration is more favorable, because the α-carbons with the bulky groups bonded to them are farther from each other than they are in the cis configuration. Virtually all peptide bonds in naturally occurring proteins studied to date have the trans configuration.

trans configuration   cis configuration

B. Secondary Structure

Secondary (2°) structure of proteins The ordered arrangements (conformations) of amino acids in localized regions of a polypeptide or protein.

Secondary (2°) structure is the ordered arrangement (conformation) of amino acids in localized regions of a polypeptide or protein molecule. The first studies of polypeptide conformations were carried out by Linus Pauling and Robert Corey, beginning in 1939. They assumed that, in conformations of greatest stability, all atoms in a peptide bond lie in the same plane and there is hydrogen bonding between the N—H of one peptide bond and the C=O of another, as shown in Figure 19.10.

On the basis of model building, Pauling proposed that two types of secondary structure should be particularly stable: the α-helix and the antiparallel β-pleated sheet.

The α-Helix

α-Helix A type of secondary structure in which a section of polypeptide chain coils into a spiral, most commonly a right-handed spiral.

In an **α-helix** pattern, shown in Figure 19.11, a polypeptide chain is coiled in a spiral. As you study this section of the α-helix, note the following:

1. The helix is coiled in a clockwise, or right-handed, manner. *Right-handed* means that if you turn the helix clockwise, it twists away from you. In this sense, a right-handed helix is analogous to the right-handed thread of a common wood or machine screw.

hydrogen bonding

Figure 19.11
An α-helix. The polypeptide chain is repeating units of L-alanine.

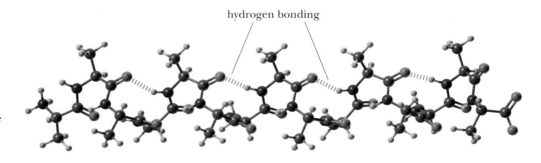

2. There are 3.6 amino acids per turn of the helix.

3. Each peptide bond is trans and planar.

4. The N—H group of each peptide bond points roughly downward, parallel to the axis of the helix, and the C=O of each peptide bond points roughly upward, also parallel to the axis of the helix.

5. The carbonyl group of each peptide bond is hydrogen bonded to the N—H group of the peptide bond four amino acid units away from it. Hydrogen bonds are shown as dotted lines.

6. All R— groups point outward from the helix.

Almost immediately after Pauling proposed the α-helix conformation, other researchers proved the presence of α-helix conformations in keratin, the protein of hair and wool. It soon became obvious that the α-helix is one of the fundamental folding patterns of polypeptide chains.

The β-Pleated Sheet

An antiparallel **β-pleated sheet** consists of an extended polypeptide chain with neighboring sections of the chain running in opposite (antiparallel) directions. In a parallel β-pleated sheet, the neighboring sections run in the same direction. Unlike the α-helix arrangement, N—H and C=O groups lie in the plane of the sheet and are roughly perpendicular to the long axis of the sheet. The C=O group of each peptide bond is hydrogen bonded to the N—H group of a peptide bond of a neighboring section of the chain (Figure 19.12).

As you study this section of β-pleated sheet, note the following:

1. The three sections of the polypeptide chain lie adjacent to each other and run in opposite (antiparallel) directions.

2. Each peptide bond is planar, and the α-carbons are trans to each other.

3. The C=O and N—H groups of peptide bonds from adjacent sections point at each other and are in the same plane, so that hydrogen bonding is possible between adjacent sections.

4. The R— groups on any one chain alternate, first above, then below, the plane of the sheet, and so on.

The β-pleated sheet conformation is stabilized by hydrogen bonding between N—H groups of one section of the chain and C=O groups of an adjacent section. By comparison, the α-helix is stabilized by hydrogen bonding between N—H and C=O groups within the same polypeptide chain.

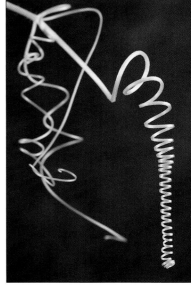

The star cucumber, *Sicyos angulatus*, uses left-handed helical tendrils to attach itself to climbing vines. Its helical pattern is analogous, but in reverse, to the right-handed α-helix of polypeptides. *(Runk Schoenberger/Grant Heilman)*

β-Pleated sheet A type of secondary structure in which two sections of polypeptide chain are aligned parallel or antiparallel to one another.

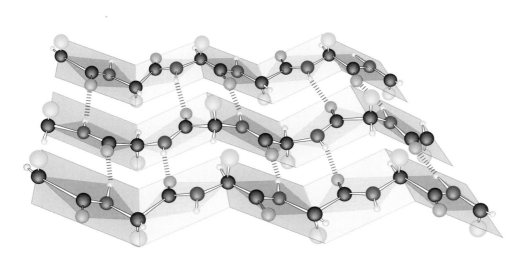

Figure 19.12
β-Pleated sheet conformation with three polypeptide chains running in opposite (antiparallel) directions. Hydrogen bonding between chains is indicated by dashed lines.

C. Tertiary Structure

Tertiary (3°) structure of proteins
The three-dimensional arrangement in space of all atoms in a single polypeptide chain.

Disulfide bond A covalent bond between two sulfur atoms; an —S—S— bond.

Tertiary (3°) structure is the overall folding pattern and arrangement in space of all atoms in a single polypeptide chain. No sharp dividing line exists between secondary and tertiary structures. Secondary structure refers to the spatial arrangement of amino acids *close to one another* on a polypeptide chain, whereas tertiary structure refers to the three-dimensional arrangement of *all* atoms in a polypeptide chain. Among the most important factors in maintaining 3° structure are disulfide bonds, hydrophobic interactions, hydrogen bonding, and salt linkages.

Disulfide bonds play an important role in maintaining tertiary structure. Disulfide bonds are formed between side chains of two cysteine units by oxidation of their thiol groups (—SH) to form a disulfide bond (Section 8.7B). Treatment of a disulfide bond with a reducing agent regenerates the thiol groups:

Figure 19.13 shows the amino acid sequence of human insulin. This protein consists of two polypeptide chains: an A chain of 21 amino acids and a B chain of 30 amino acids. The A chain is bonded to the B chain by two interchain disulfide bonds. An intrachain disulfide bond also connects the cysteine units at positions 6 and 11 of the A chain.

As an example of 2° and 3° structure, let us look at the three-dimensional structure of myoglobin, a protein found in skeletal muscle and particularly abundant in diving mammals, such as seals, whales, and porpoises. Myoglobin and its structural relative, hemoglobin, are the oxygen transport and storage molecules of vertebrates. Hemoglobin binds molecular oxygen in the lungs and transports it to myoglobin in muscles. Myoglobin stores molecular oxygen until it is required for metabolic oxidation.

Myoglobin consists of a single polypeptide chain of 153 amino acids. Myoglobin also contains a single heme unit. Heme consists of one Fe^{2+} ion, coordinated in a square planar array with the four nitrogen atoms of a molecule of porphyrin (Figure 19.14).

Figure 19.13
Human insulin. The A chain of 21 amino acids and B chain of 30 amino acids are connected by interchain disulfide bonds between A7 and B7 and between A20 and B19. In addition, a single intrachain disulfide bond occurs between A6 and A11.

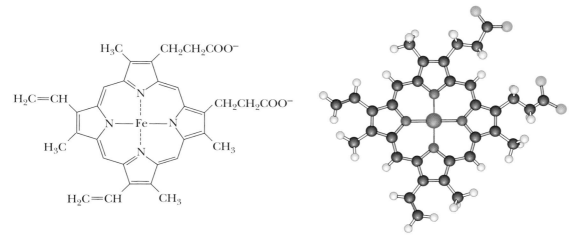

Figure 19.14
The structure of heme, found in myoglobin and hemoglobin.

Determination of the three-dimensional structure of myoglobin represented a milestone in the study of molecular architecture. For their contribution to this research, John C. Kendrew and Max F. Perutz, both of Britain, shared the 1962 Nobel prize for chemistry. The secondary and tertiary structures of myoglobin are shown in Figure 19.15. The single polypeptide chain is folded into a complex, almost boxlike shape.

Important structural features of the three-dimensional shape of myoglobin are as follows:

1. The backbone consists of eight relatively straight sections of α-helix, each separated by a bend in the polypeptide chain. The longest section of α-helix has 24 amino acids, the shortest has 7. Some 75% of the amino acids are found in these eight regions of α-helix.
2. Hydrophobic side chains of phenylalanine, alanine, valine, leucine, isoleucine, and methionine are clustered in the interior of the molecule, where they are shielded from contact with water. **Hydrophobic interactions** are a major factor in directing the folding of the polypeptide chain of myoglobin into this compact, three-dimensional shape.

The humpback whale relies on myoglobin as a storage form of oxygen. (*Stuart Westmorland/Stone/Getty Images*)

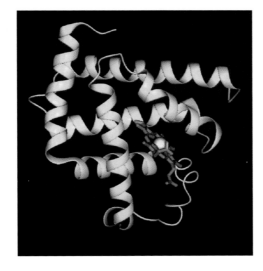

Figure 19.15
Ribbon model of myoglobin. The polypeptide chain is shown in yellow, the heme ligand in red, and the Fe atom as a white sphere.

3. The outer surface of myoglobin is coated with hydrophilic side chains, such as those of lysine, arginine, serine, glutamic acid, histidine, and glutamine, which interact with the aqueous environment by **hydrogen bonding**. The only polar side chains that point to the interior of the myoglobin molecule are those of two histidine units, which point inward toward the heme group.

4. Oppositely charged amino acid side chains close to each other in the three-dimensional structure interact by electrostatic attractions called **salt linkages**. An example of a salt linkage is the attraction of the side chains of lysine ($-NH_3^+$) and glutamic acid ($-COO^-$).

The tertiary structures of hundreds of proteins have also been determined. It is clear that proteins contain α-helix and β-pleated sheet structures, but that wide variations exist in the relative amounts of each. Lysozyme, with 129 amino acids in a single polypeptide chain, has only 25% of its amino acids in α-helix regions. Cytochrome, with 104 amino acids in a single polypeptide chain, has no α-helix structure, but does contain several regions of β-pleated sheet. Yet, whatever the proportions of α-helix, β-pleated sheet, or other periodic structure, virtually all nonpolar side chains of water-soluble proteins are directed toward the interior of the molecule, whereas polar side chains are on the surface of the molecule, in contact with the aqueous environment.

EXAMPLE 19.7

With which of the following amino acid side chains can the side chain of threonine form hydrogen bonds?

(a) Valine   (b) Asparagine   (c) Phenylalanine
(d) Histidine   (e) Tyrosine   (f) Alanine

SOLUTION

The side chain of threonine contains a hydroxyl group that can participate in hydrogen bonding in two ways: (1) Its oxygen has a partial negative charge and can function as a hydrogen bond acceptor; (2) its hydrogen has a partial positive charge and can function as a hydrogen bond donor. Therefore, the side chain of threonine can form hydrogen bonds with the side chains of tyrosine, asparagine, and histidine.

Practice Problem 19.7

At pH 7.4, with what amino acid side chains can the side chain of lysine form salt linkages?

D. Quaternary Structure

Quaternary (4°) structure of proteins The arrangement of polypeptide monomers into a noncovalently bonded aggregation.

Hydrophobic effect The tendency of nonpolar groups to cluster in such a way as to be shielded from contact with an aqueous environment.

Most proteins with molecular weight greater than 50,000 consist of two or more non-covalently linked polypeptide chains. The arrangement of protein monomers into an aggregation is known as **quaternary (4°) structure**. A good example is hemoglobin (Figure 19.16), a protein that consists of four separate polypeptide chains: two α-chains of 141 amino acids each and two β-chains of 146 amino acids each.

The major factor stabilizing the aggregation of protein subunits is the **hydrophobic effect**. When separate polypeptide chains fold into compact three-dimensional shapes to expose polar side chains to the aqueous environment and

Figure 19.16
Ribbon model of hemoglobin.
The α-chains are shown in
purple, the β-chains in yellow,
the heme ligands in red, and
the Fe atoms as white spheres.

shield nonpolar side chains from water, hydrophobic "patches" still appear on the surface, in contact with water. These patches can be shielded from water if two or more monomers assemble so that their hydrophobic patches are in contact. The numbers of subunits of several proteins of known quaternary structure are shown in Table 19.3.

TABLE 19.3 Quaternary Structure of Selected Proteins

Protein	Number of Subunits
alcohol dehydrogenase	2
aldolase	4
hemoglobin	4
lactate dehydrogenase	4
insulin	6
glutamine synthetase	12
tobacco mosaic virus protein disc	17

SUMMARY

α-Amino acids are compounds that contain an amino group alpha to a carboxyl group (Section 19.2A). Amino acids exist as zwitterions, or internal salts, at physiological pH. With the exception of glycine, all protein-derived amino acids are chiral (Section 19.2B). In the D,L convention, all are L-amino acids. Isoleucine and threonine contain a second stereocenter. The 20 protein-derived amino acids are commonly divided into four categories (Section 19.2C): nine with nonpolar side chains; four with polar, but un-ionized, side chains; four with acidic side chains; and three with basic side chains.

The **isoelectric point, pI,** of an amino acid, polypeptide, or protein is the pH at which it has no net charge (Section 19.3C). **Electrophoresis** is the process of separating

compounds on the basis of their electric charge (Section 19.3D). Compounds with a high charge density move more rapidly than those with a lower charge density. Any amino acid or protein in a solution with a pH that equals the pI of the compound remains at the origin.

A **peptide bond** is the special name given to the amide bond formed between α-amino acids (Section 19.4). A **polypeptide** is a biological macromolecule containing 20 or more amino acids joined by peptide bonds. By convention, the sequence of amino acids in a polypeptide is written beginning with the **N-terminal amino acid** toward the **C-terminal amino acid**. The **primary (1°) structure** of a polypeptide is the sequence of amino acids in its polypeptide chain (Section 19.5B).

A **peptide bond** is planar (Section 19.6A); that is, the four atoms of the amide bond and the two α-carbons of a peptide bond lie in the same plane. Bond angles about the amide nitrogen and the carbonyl carbon are approximately 120°. **Secondary (2°) structure** (Section 19.6B) is the ordered arrangement (conformation) of amino acids in localized regions of a polypeptide or protein. Two types of secondary structure are the α-helix and the β-pleated sheet. **Tertiary (3°) structure** (Section 19.6C) refers to the overall folding pattern and arrangement in space of all atoms in a single polypeptide chain. **Quaternary (4°) structure** (Section 19.6D) is the arrangement of individual polypeptide chains into a noncovalently bonded aggregate.

KEY REACTIONS

1. Acidity of an α-Carboxyl Group (Section 19.3A)

An α-COOH (pK_a approximately 2.19) of a protonated amino acid is a considerably stronger acid than acetic acid (pK_a 4.76) or other low-molecular-weight aliphatic carboxylic acid, due to the electron-withdrawing inductive effect of the α-NH$_3^+$ group:

$$\underset{\underset{NH_3^+}{|}}{RCHCOOH} + H_2O \rightleftharpoons \underset{\underset{NH_3^+}{|}}{RCHCOO^-} + H_3O^+ \qquad pK_a = 2.19$$

2. Acidity of an α-Ammonium Group (Section 19.3A)

An α-NH$_3^+$ group (pK_a approximately 9.47) is a slightly stronger acid than a primary aliphatic ammonium ion (pK_a approximately 10.76):

$$\underset{\underset{NH_3^+}{|}}{RCHCOO^-} + H_2O \rightleftharpoons \underset{\underset{NH_2}{|}}{RCHCOO^-} + H_3O^+ \qquad pK_a = 9.47$$

3. Reaction of an α-Amino Acid with Ninhydrin (Section 19.3D)

Treating an α-amino acid with ninhydrin gives a purple-colored solution:

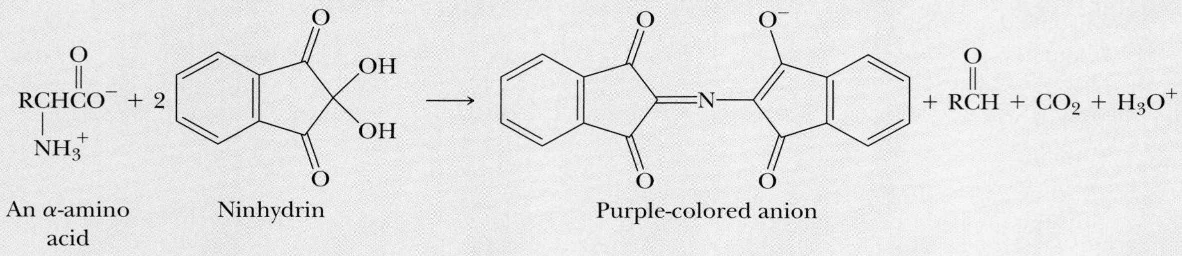

An α-amino acid   Ninhydrin   Purple-colored anion

Treating proline with ninhydrin gives an orange-colored solution.

4. Cleavage of a Peptide Bond by Cyanogen Bromide (Section 19.5B)

Cleavage is regioselective for a peptide bond formed by the carboxyl group of methionine:

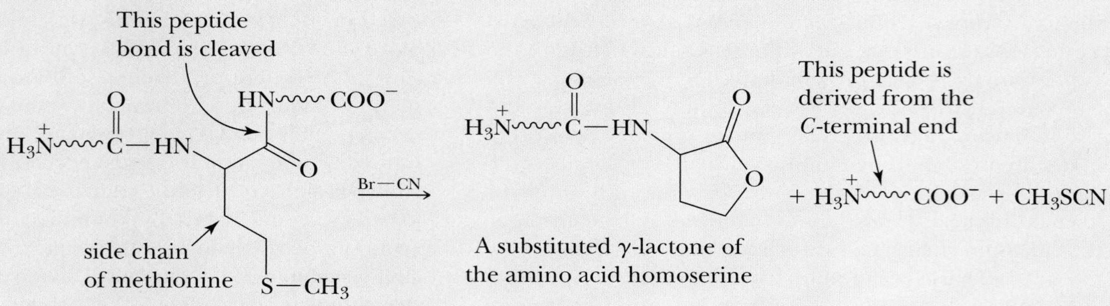

5. Edman Degradation (Section 19.5B)

Treatment with phenyl isothiocyanate followed by acid removes the *N*-terminal amino acid as a substituted phenylthiohydantoin, which is then separated and identified:

$$\underset{\substack{| \quad \| \\ \text{H}_2\text{NCHCNH-peptide}}}{\overset{\text{R O}}{}} + \text{Ph}-\text{N}=\text{C}=\text{S} \longrightarrow \text{(a phenylthiohydantoin)} + \text{H}_2\text{N-peptide}$$

Phenyl isothiocyanate

A phenylthiohydantoin

This peptide is derived from the *N*-terminal end

PROBLEMS

A problem number set in red indicates an applied "real-world" problem.

Amino Acids

19.8 What amino acid does each abbreviation stand for?
   (a) Phe   **(b)** Ser   **(c)** Asp   **(d)** Gln
   (e) His   **(f)** Gly   **(g)** Tyr

19.9 The configuration of the stereocenter in α-amino acids is most commonly specified using the D,L convention. The configuration can also be identified using the R,S convention (Section 6.4). Does the stereocenter in L-serine have the R or the S configuration?

19.10 Assign an R or S configuration to the stereocenter in each amino acid:
   (a) L-Phenylalanine   **(b)** L-Glutamic acid   **(c)** L-Methionine

19.11 The amino acid threonine has two stereocenters. The stereoisomer found in proteins has the configuration 2S,3R about the two stereocenters. Draw a Fischer projection of this stereoisomer and also a three-dimensional representation.

19.12 Define the term *zwitterion*.

19.13 Draw zwitterion forms of these amino acids:
   (a) Valine   **(b)** Phenylalanine   **(c)** Glutamine

19.14 Why are Glu and Asp often referred to as acidic amino acids?

19.15 Why is Arg often referred to as a basic amino acid? Which two other amino acids are also basic amino acids?

19.16 What is the meaning of the alpha as it is used in α-amino acid?

19.17 Several β-amino acids exist. A unit of β-alanine, for example, is contained within the structure of coenzyme A (Section 22.2D). Write the structural formula of β-alanine.

19.18 Although only L-amino acids occur in proteins, D-amino acids are often a part of the metabolism of lower organisms. The antibiotic actinomycin D, for example, contains a unit of D-valine, and the antibiotic bacitracin A contains units of D-asparagine and D-glutamic acid. Draw Fischer projections and three-dimensional representations for these three D-amino acids.

19.19 Histamine is synthesized from one of the 20 protein-derived amino acids. Suggest which amino acid is the biochemical precursor of histamine, and name the type of organic reaction(s) (e.g., oxidation, reduction, decarboxylation, nucleophilic substitution) involved in its conversion to histamine.

19.20 Both norepinephrine and epinephrine are synthesized from the same protein-derived amino acid:

(a)

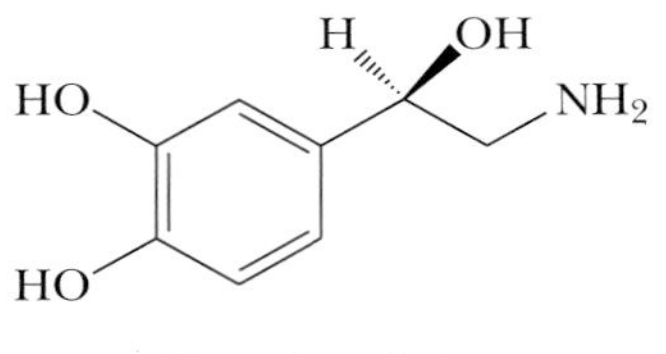

Norepinephrine

(b)

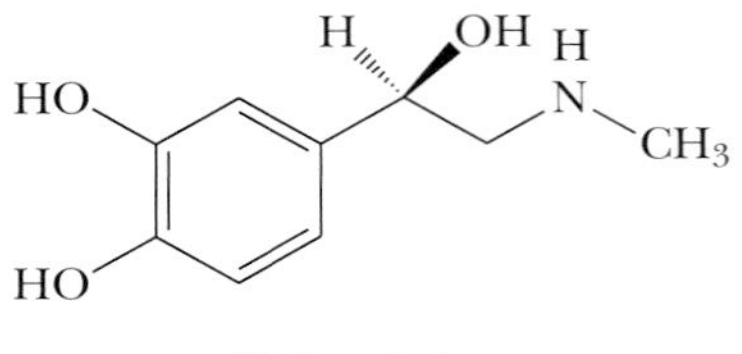

Epinephrine
(Adrenaline)

From which amino acid are the two compounds synthesized, and what types of reactions are involved in their biosynthesis?

19.21 From which amino acid are serotonin and melatonin synthesized and what types of reactions are involved in their biosynthesis?

(a)

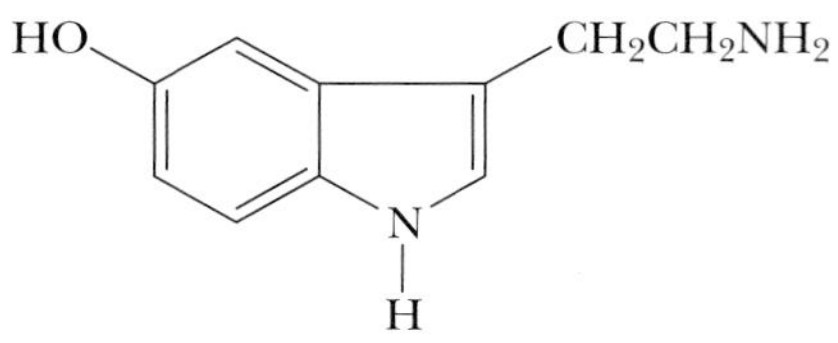

Serotonin

(b)

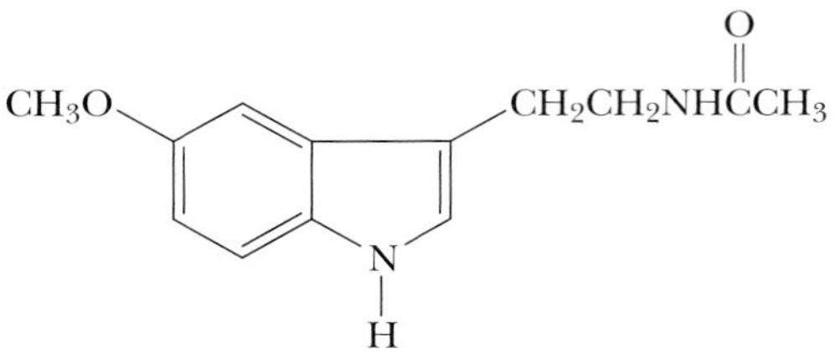

Melatonin

Acid–Base Behavior of Amino Acids

19.22 Draw a structural formula for the form of each amino acid most prevalent at pH 1.0:
    (a) Threonine   **(b)** Arginine   **(c)** Methionine   **(d)** Tyrosine

19.23 Draw a structural formula for the form of each amino acid most prevalent at pH 10.0:
    (a) Leucine   **(b)** Valine   **(c)** Proline   **(d)** Aspartic acid

19.24 Write the zwitterion form of alanine and show its reaction with
    (a) 1.0 mol NaOH   **(b)** 1.0 mol HCl

19.25 Write the form of lysine most prevalent at pH 1.0, and then show its reaction with each of the following (consult Table 19.2 for pK_a values of the ionizable groups in lysine):
    (a) 1.0 mol NaOH   **(b)** 2.0 mol NaOH   **(c)** 3.0 mol NaOH

19.26 Write the form of aspartic acid most prevalent at pH 1.0, and then show its reaction with the following (consult Table 19.2 for pK_a values of the ionizable groups in aspartic acid):
    (a) 1.0 mol NaOH   **(b)** 2.0 mol NaOH   **(c)** 3.0 mol NaOH

19.27 Given pK_a values for ionizable groups from Table 19.2, sketch curves for the titration of (a) glutamic acid with NaOH and (b) histidine with NaOH.

19.28 Draw a structural formula for the product formed when alanine is treated with each of the following reagents:
    (a) Aqueous NaOH   **(b)** Aqueous HCl
    (c) CH_3CH_2OH, H_2SO_4 **(d)** $(CH_3CO)_2O$, $CH_3COO^-\ Na^+$

19.29 Account for the fact that the isoelectric point of glutamine (pI 5.65) is higher than the isoelectric point of glutamic acid (pI 3.08).

Thus, DNA is the repository of genetic information in cells, whereas RNA serves in the transcription and translation of this information, which is then expressed through the synthesis of proteins.

In this chapter, we examine the structure of nucleosides and nucleotides and the manner in which these monomers are covalently bonded to form **nucleic acids**. Then we explore the manner in which genetic information is encoded on molecules of DNA, the function of the three types of ribonucleic acids, and, finally, how the primary structure of a DNA molecule is determined.

Nucleic acid A biopolymer containing three types of monomer units: heterocyclic aromatic amine bases derived from purine and pyrimidine, the monosaccharides D-ribose or 2-deoxy-D-ribose, and phosphate.

20.2 NUCLEOSIDES AND NUCLEOTIDES

Controlled hydrolysis of nucleic acids yields three components: heterocyclic aromatic amine bases, the monosaccharide D-ribose or 2-deoxy-D-ribose (Section 18.2), and phosphate ions. Figure 20.1 shows the five heterocyclic aromatic amine bases most common to nucleic acids. Uracil, cytosine, and thymine are referred to as pyrimidine bases after the name of the parent base; adenine and guanine are referred to as purine bases.

A **nucleoside** is a compound containing D-ribose or 2-deoxy-D-ribose bonded to a heterocyclic aromatic amine base by a β-*N*-glycosidic bond (Section 18.5A). The monosaccharide component of DNA is 2-deoxy-D-ribose (the "2-deoxy" refers to the absence of a hydroxyl group at the 2′ position), whereas that of RNA is D-ribose. The glycosidic bond is between C-1′ (the anomeric carbon) of ribose or 2-deoxyribose and *N*-1 of a pyrimidine base or N-9 of a purine base. Figure 20.2 shows the structural formula for uridine, a nucleoside derived from ribose and uracil.

Nucleoside A building block of nucleic acids, consisting of D-ribose or 2-deoxy-D-ribose bonded to a heterocyclic aromatic amine base by a β-*N*-glycosidic bond.

Pyrimidine   Uracil (U)   Cytosine (C)   Thymine (T)

Purine   Adenine (A)   Guanine (G)

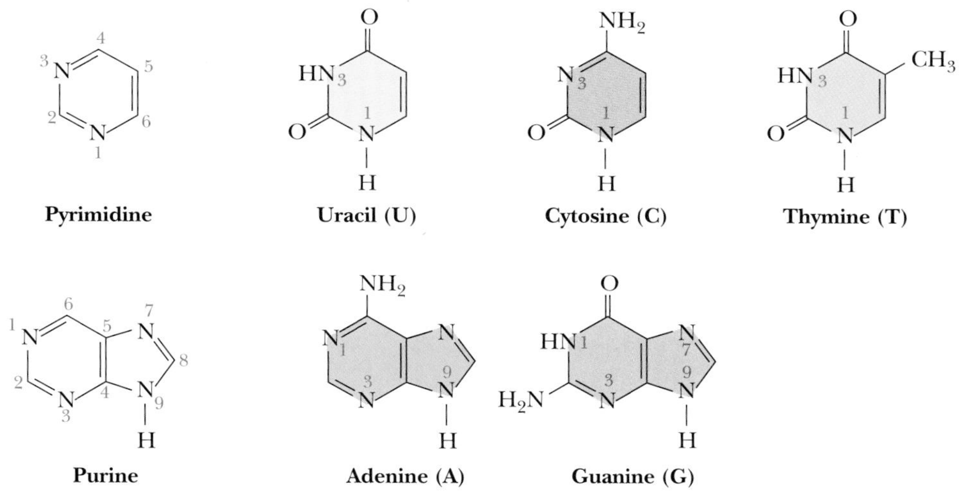

Figure 20.1
Names and one-letter abbreviations for the heterocyclic aromatic amine bases most common to DNA and RNA. Bases are numbered according to the patterns of pyrimidine and purine, the parent compounds.

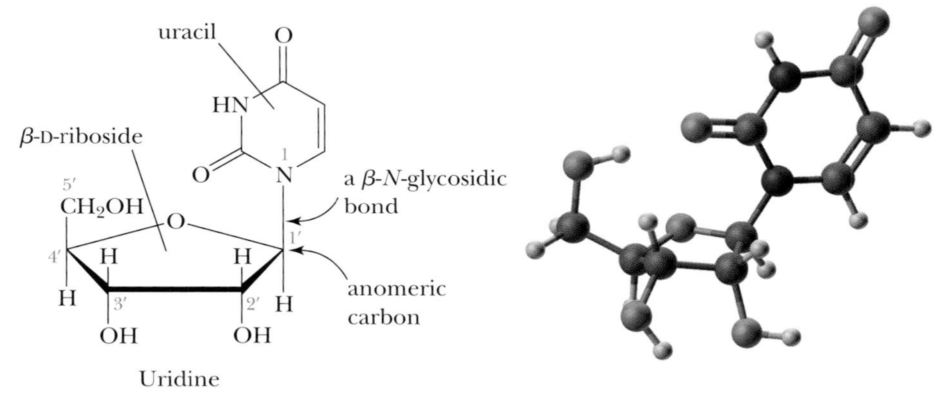

Uridine

Figure 20.2
Uridine, a nucleoside. Atom numbers on the monosaccharide rings are primed to distinguish them from atom numbers on the heterocyclic aromatic amine bases.

CHEMICAL CONNECTIONS 20A

The Search for Antiviral Drugs

The search for antiviral drugs has been more difficult than the search for antibacterial drugs primarily because viral replication depends on the metabolic processes of the invaded cell. Thus, antiviral drugs are also likely to cause harm to the cells that harbor the virus. The challenge in developing antiviral drugs is to understand the biochemistry of viruses and to develop drugs that target processes specific to them. Compared with the large number of antibacterial drugs available, there are only a handful of antiviral drugs, and they have nowhere near the effectiveness that antibiotics have on bacterial infections.

Acyclovir was one of the first of a new family of drugs for the treatment of infectious diseases caused by DNA viruses called herpesvirus. Herpes infections in humans are of two kinds: herpes simplex type 1, which gives rise to mouth and eye sores, and herpes simplex type 2, which gives rise to serious genital infections.

Acyclovir is highly effective against herpesvirus-caused genital infections. The structural formula of acyclovir is drawn in the accompanying figure to show its structural relationship to 2-deoxyguanosine. The drug is activated in vivo by the conversion of the primary —OH (which corresponds to the 5'-OH of a riboside or a deoxyriboside) to a triphosphate. Because of its close resemblance to deoxyguanosine triphosphate, an essential precursor of DNA synthesis, acyclovir triphosphate is taken up by viral DNA polymerase to form an enzyme–substrate complex on which no 3'-OH exists for replication to continue. Thus, the enzyme–substrate complex is no longer active (it is a dead-end complex), viral replication is disrupted, and the virus is destroyed.

Perhaps the best known of the HIV-fighting viral antimetabolites is zidovudine (AZT), an analog of deoxythymidine in which the 3'-OH has been replaced by an azido group, N_3. AZT is effective

---

Nucleotide A nucleoside in which a molecule of phosphoric acid is esterified with an —OH of the monosaccharide, most commonly either the 3'-OH or the 5'-OH.

A **nucleotide** is a nucleoside in which a molecule of phosphoric acid is esterified with a free hydroxyl of the monosaccharide, most commonly either the 3'-hydroxyl or the 5'-hydroxyl. A nucleotide is named by giving the name of the parent nucleoside, followed by the word "monophosphate." The position of the phosphoric ester is specified by the number of the carbon to which it is bonded. Figure 20.3 shows a structural formula and a ball-and-stick model of 5'-adenosine monophosphate. Monophosphoric esters are diprotic acids with pK_a values of approximately 1 and 6.

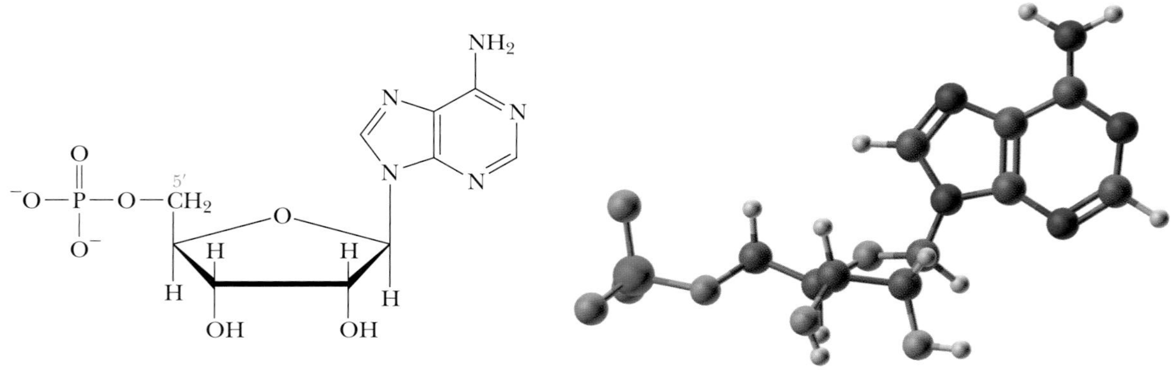

Figure 20.3
Adenosine 5'-monophosphate, a nucleotide. The phosphate group is fully ionized at pH 7.0, giving this nucleotide a charge of −2.

against HIV-1, a retrovirus that is the causative agent of AIDS. AZT is converted in vivo by cellular enzymes to the 5′-triphosphate, is recognized as deoxythymidine 5′-triphosphate by viral RNA-dependent DNA polymerase (reverse transcriptase), and is added to a growing DNA chain. There, it stops chain elongation, because no 3′-OH exists on which to add the next deoxynucleotide. AZT owes its effectiveness to the fact that it binds more strongly to viral reverse transcriptase than it does to human DNA polymerase.

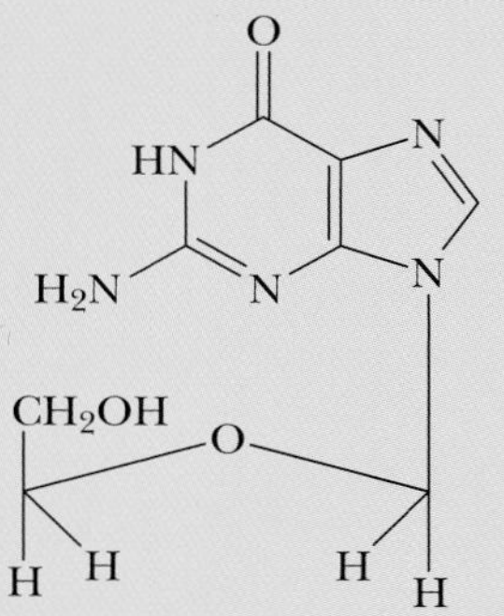

Acyclovir
(drawn to show its structural relationship to 2-deoxyguanosine)

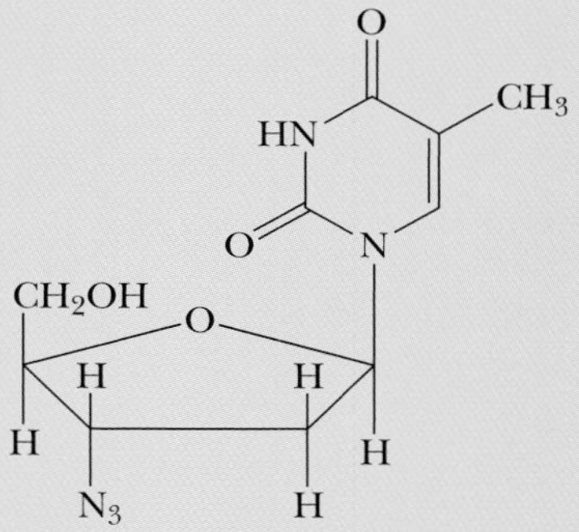

Zidovudine
(Azidothymidine; AZT)

Therefore, at a pH of 7, the two hydrogens of a phosphoric monoester are fully ionized, giving a nucleotide a charge of -2.

Nucleoside monophosphates can be further phosphorylated to form nucleoside diphosphates and nucleoside triphosphates. Shown in Figure 20.4 is a structural formula for adenosine 5′-triphosphate (ATP).

Nucleoside diphosphates and triphosphates are also polyprotic acids and are extensively ionized at pH 7.0. pK_a values of the first three ionization steps for adenosine triphosphate are less than 5.0. The value of pK_{a4} is approximately 7.0. Therefore, at pH 7.0, approximately 50% of adenosine triphosphate is present as ATP^{4-} and 50% is present as ATP^{3-}.

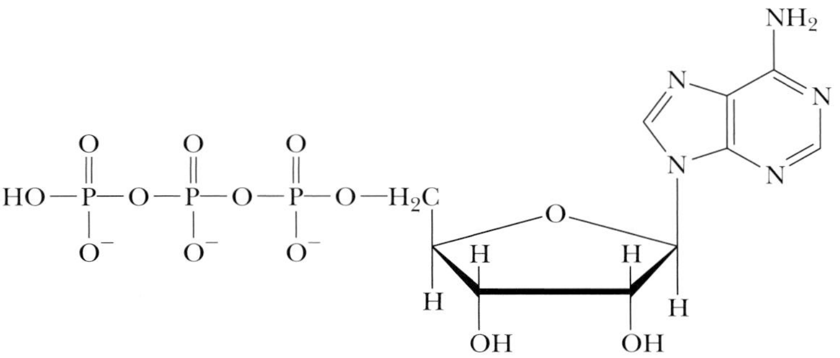

Figure 20.4
Adenosine triphosphate, ATP.

EXAMPLE 20.1

Draw a structural formula for 2′-deoxycytidine 5′-diphosphate.

SOLUTION

Cytosine is joined by a β-N-glycosidic bond between N-1 of cytosine and C-1′ of the cyclic hemiacetal form of 2-deoxy-D-ribose. The 5′-hydroxyl of the pentose is bonded to a phosphate group by an ester bond, and this phosphate is in turn bonded to a second phosphate group by an anhydride bond:

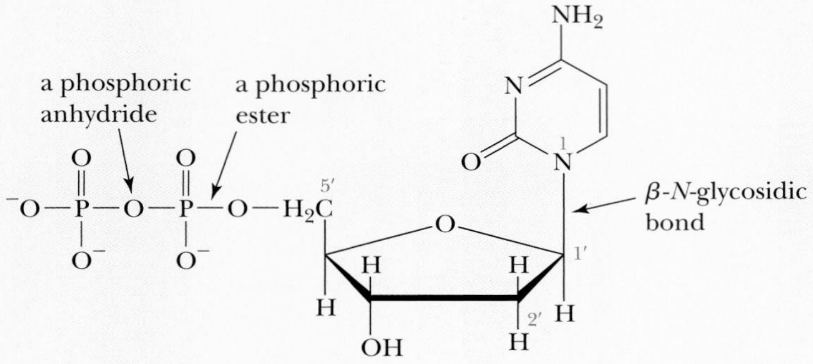

Practice Problem 20.1

Draw a structural formula for 2′-deoxythymidine 3′-monophosphate.

20.3 THE STRUCTURE OF DNA

In Chapter 19, we saw that the four levels of structural complexity in polypeptides and proteins are primary, secondary, tertiary, and quaternary. There are three levels of structural complexity in nucleic acids, and although these levels are somewhat comparable to those in polypeptides and proteins, they also differ in significant ways.

A. Primary Structure: The Covalent Backbone

Primary structure of nucleic acids The sequence of bases along the pentose–phosphodiester backbone of a DNA or RNA molecule, read from the 5′ end to the 3′ end.

5′ End The end of a polynucleotide at which the 5′-OH of the terminal pentose unit is free.

3′ End The end of a polynucleotide at which the 3′-OH of the terminal pentose unit is free.

Deoxyribonucleic acids consist of a backbone of alternating units of deoxyribose and phosphate in which the 3′-hydroxyl of one deoxyribose unit is joined by a phosphodiester bond to the 5′-hydroxyl of another deoxyribose unit (Figure 20.5). This pentose–phosphodiester backbone is constant throughout an entire DNA molecule. A heterocyclic aromatic amine base—adenine, guanine, thymine, or cytosine—is bonded to each deoxyribose unit by a β-N-glycosidic bond. The **primary structure** of a DNA molecule is the order of heterocyclic bases along the pentose–phosphodiester backbone. The sequence of bases is read from the **5′ end** to the **3′ end**.

EXAMPLE 20.2

Draw a structural formula for the DNA dinucleotide TG that is phosphorylated at the 5′ end only.

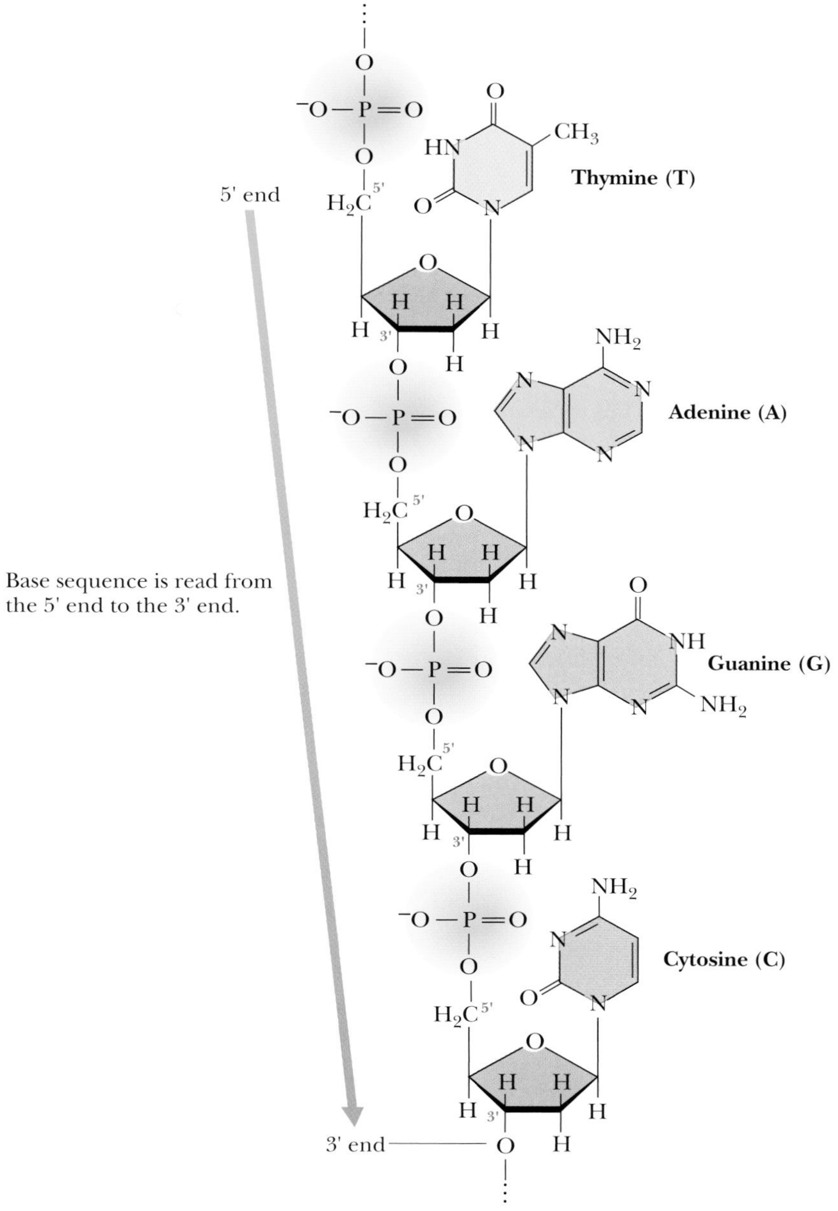

Figure 20.5
A tetranucleotide section of a single-stranded DNA.

5' end

Base sequence is read from the 5' end to the 3' end.

Thymine (T)

Adenine (A)

Guanine (G)

Cytosine (C)

3' end

SOLUTION

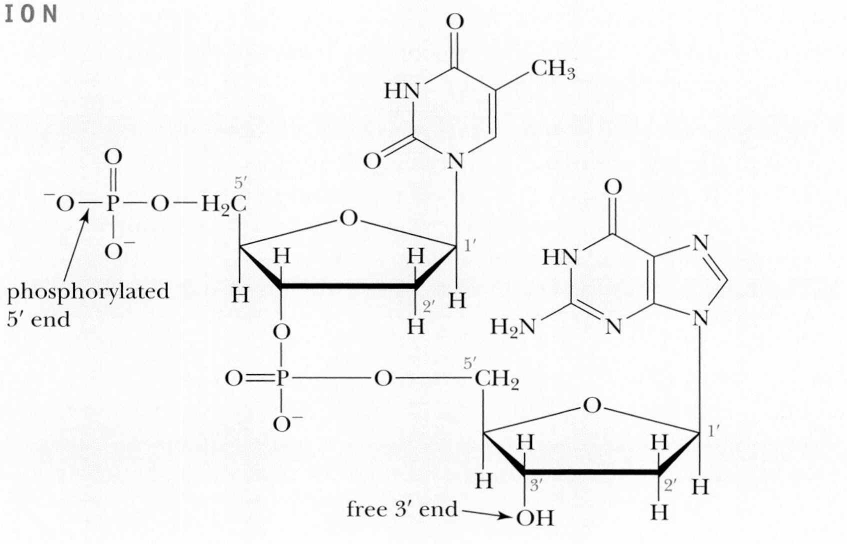

phosphorylated 5' end

free 3' end

Draw a structural formula for the section of DNA that contains the base sequence CTG and is phosphorylated at the 3′ end only.

B. Secondary Structure: The Double Helix

By the early 1950s, it was clear that DNA molecules consist of chains of alternating units of deoxyribose and phosphate joined by 3′,5′-phosphodiester bonds, with a base attached to each deoxyribose unit by a *β-N*-glycosidic bond. In 1953, the American biologist James D. Watson and the British physicist Francis H. C. Crick proposed a double-helix model for the **secondary structure of DNA**. Watson, Crick, and Maurice Wilkins shared the 1962 Nobel prize in physiology or medicine for "their discoveries concerning the molecular structure of nucleic acids, and its significance for information transfer in living material." Although Rosalind Franklin also took part in this research, her name was omitted from the Nobel list because of her death in 1958 at age 37. The Nobel foundation does not make awards posthumously.

The Watson–Crick model was based on molecular modeling and two lines of experimental observations: chemical analyses of DNA base compositions and mathematical analyses of X-ray diffraction patterns of crystals of DNA.

Base Composition

At one time, it was thought that, in all species, the four principal bases occurred in the same ratios and perhaps repeated in a regular pattern along the pentose–phosphodiester backbone of DNA. However, more precise determinations of their composition by Erwin Chargaff revealed that bases do not occur in the same ratios (Table 20.1).

Secondary structure of nucleic acids The ordered arrangement of strands of nucleic acid.

Rosalind Franklin (1920–1958). In 1951, she joined the Biophysical Laboratory at King's College, London, where she began her studies on the application of x-ray diffraction methods to the study of DNA. She is credited with discoveries that established the density of DNA, its helical conformation, and other significant aspects. Her work was, thus, important to the model of DNA developed by Watson and Crick. She died in 1958 at the age of 37 and, because a Nobel prize is never awarded posthumously, she did not share in the 1962 Nobel prize of physiology and medicine with Watson, Crick, and Wilkins. Although the relation between Watson, Crick, and Franklin was initially strained, Watson later said that "we later came to appreciate . . . the struggles the intelligent woman faces to be accepted by the scientific world which often regards women as mere diversions from serious thinking." *(Photo Researchers, Inc.)*

TABLE 20.1   Comparison of DNA from Several Organisms, Base Composition, in Mole Percent

Organism	Purines		Pyrimidines		A/T	G/C	Purines/ Pyrimidines
	A	G	C	T			
human	30.4	19.9	19.9	30.1	1.01	1.00	1.01
sheep	29.3	21.4	21.0	28.3	1.04	1.02	1.03
yeast	31.7	18.3	17.4	32.6	0.97	1.05	1.00
E. coli	26.0	24.9	25.2	23.9	1.09	0.99	1.04

Researchers drew the following conclusions from the data in the table and related data: To within experimental error,

1. The mole-percent base composition of DNA in any organism is the same in all cells of the organism and is characteristic of the organism.
2. The mole percentages of adenine (a purine base) and thymine (a pyrimidine base) are equal. The mole percentages of guanine (a purine base) and cytosine (a pyrimidine base) are also equal.
3. The mole percentages of purine bases (A + G) and pyrimidine bases (C + T) are equal.

Analyses of X-Ray Diffraction Patterns

Additional information about the structure of DNA emerged when X-ray diffraction photographs taken by Rosalind Franklin and Maurice Wilkins were analyzed. The diffraction patterns revealed that, even though the base composition of DNA isolated from different organisms varies, DNA molecules themselves are remarkably uniform in thickness. They are long and fairly straight, with an outside diameter of approximately 20 Å, and not more than a dozen atoms thick. Furthermore, the crystallographic pattern repeats every 34 Å. Herein lay one of the chief problems to be solved: How could the molecular dimensions of DNA be so regular, even though the relative percentages of the various bases differ so widely? With this accumulated information, the stage was set for the development of a hypothesis about DNA structure.

The Watson–Crick Double Helix

The heart of the Watson–Crick model is the postulate that a molecule of DNA is a complementary **double helix** consisting of two antiparallel polynucleotide strands coiled in a right-handed manner about the same axis. As illustrated in the ribbon models in Figure 20.6, chirality is associated with a double helix: Like enantiomers, left-handed and right-handed double helices are related by reflection.

To account for the observed base ratios and uniform thickness of DNA, Watson and Crick postulated that purine and pyrimidine bases project inward toward the axis of the helix and always pair in a specific manner. According to scale models, the dimensions of an adenine–thymine base pair are almost identical to the dimensions of a guanine–cytosine base pair, and the length of each pair is consistent with the core thickness of a DNA strand (Figure 20.7). Thus, if the purine base in one strand is adenine, then its complement in the antiparallel strand must be thymine. Similarly, if the purine in one strand is guanine, its complement in the antiparallel strand must be cytosine.

A significant feature of Watson and Crick's model is that no other base pairing is consistent with the observed thickness of a DNA molecule. A pair of pyrimidine bases is too small to account for the observed thickness, whereas a pair of purine bases is too large. Thus, according to the Watson–Crick model, the repeating units in a double-stranded DNA molecule are not single bases of differing dimensions, but rather base pairs of almost identical dimensions.

To account for the periodicity observed from X-ray data, Watson and Crick postulated that base pairs are stacked one on top of the other, with a distance of 3.4 Å between base pairs and with 10 base pairs in one complete turn of the helix. Thus, there is one complete turn of the helix every 34 Å. Figure 20.8 shows a ribbon model of double-stranded **B-DNA**, the predominant form of DNA in dilute aqueous solution and thought to be the most common form in nature.

Double helix A type of secondary structure of DNA molecules in which two antiparallel polynucleotide strands are coiled in a right-handed manner about the same axis.

Figure 20.6
A DNA double helix has a chirality associated with it. Right-handed and left-handed double helices of otherwise identical DNA chains are nonsuperposable mirror images.

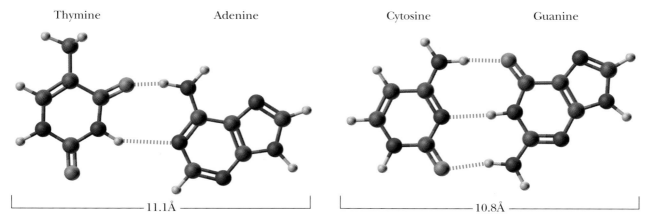

| Thymine | Adenine | Cytosine | Guanine |

Figure 20.7
Base pairing between adenine and thymine (A-T) and between guanine and cytosine (G-C). An A-T base pair is held by two hydrogen bonds, whereas a G-C base pair is held by three hydrogen bonds.

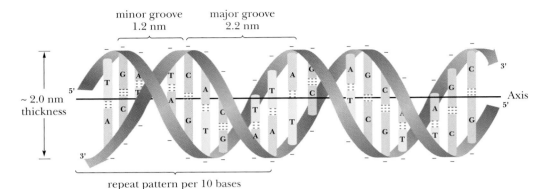

minor groove   major groove
1.2 nm   2.2 nm

~ 2.0 nm thickness

Axis

repeat pattern per 10 bases

Figure 20.8
Ribbon model of double-stranded B-DNA. Each ribbon shows the pentose–phosphodiester backbone of a single-stranded DNA molecule. The strands are antiparallel, one running to the left from the 5′ end to the 3′ end, the other running to the right from the 5′ end to the 3′ end. Hydrogen bonds are shown by three dotted lines between each G-C base pair and two dotted lines between each A-T base pair.

In the double helix, the bases in each base pair are not directly opposite from one another across the diameter of the helix, but rather are slightly displaced. This displacement and the relative orientation of the glycosidic bonds linking each base to the sugar–phosphate backbone lead to two differently sized grooves: a major groove and a minor groove (Figure 20.8). Each groove runs along the length of the cylindrical column of the double helix. The major groove is approximately 22 Å wide, the minor groove approximately 12 Å wide.

Figure 20.9 shows more detail of an idealized B-DNA double helix. The major and minor grooves are clearly recognizable in this model.

Other forms of secondary structure are known that differ in the distance between stacked base pairs and in the number of base pairs per turn of the helix. One of the most common of these, **A-DNA**, also a right-handed helix, is thicker than B-DNA, and has a repeat distance of only 29 Å. There are 10 base pairs per turn of the helix, with a spacing of 2.9 Å between base pairs.

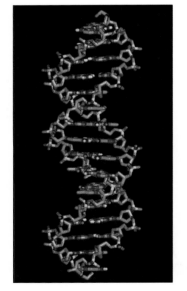

Figure 20.9
B-DNA. An idealized model of B-DNA.

EXAMPLE 20.3

One strand of a DNA molecule contains the base sequence 5′-ACTTGCCA-3′. Write its complementary base sequence.

SOLUTION

Remember that a base sequence is always written from the 5′ end of the strand to the 3′ end, that A pairs with T, and that G pairs with C. In double-stranded DNA, the two strands run in opposite (antiparallel) directions, so that the 5′ end of one strand is associated with the 3′ end of the other strand:

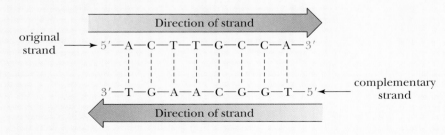

Direction of strand

original strand → 5′—A—C—T—T—G—C—C—A—3′

                   3′—T—G—A—A—C—G—G—T—5′ ← complementary strand

Direction of strand

Written from the 5′ end, the complementary strand is 5′-TGGCAAGT-3′.

Practice Problem 20.3

Write the complementary DNA base sequence for 5′-CCGTACGA-3′.

C. Tertiary Structure: Supercoiled DNA

The length of a DNA molecule is considerably greater than its diameter, and the extended molecule is quite flexible. A DNA molecule is said to be relaxed if it has no twists other than those imposed by its secondary structure. Put another way, relaxed DNA does not have a clearly defined tertiary structure. We consider two types of **tertiary structure**, one induced by perturbations in circular DNA, the other introduced by the coordination of DNA with nuclear proteins called histones. Tertiary structure, whatever the type, is referred to as **supercoiling**.

Supercoiling of Circular DNA

Circular DNA is a type of double-stranded DNA in which the two ends of each strand are joined by phosphodiester bonds [Figure 20.10(a)]. This type of DNA, the most prominent form in bacteria and viruses, is also referred to as *circular duplex* (because it is double-stranded) DNA. One strand of circular DNA may be opened, partially unwound, and then rejoined. The unwound section introduces a strain into the molecule because the nonhelical gap is less stable than hydrogen-bonded, base-paired helical sections. The strain can be localized in the nonhelical gap. Alternatively, it may be spread uniformly over the entire circular DNA by the introduction of **superhelical twists**, one twist for each turn of a helix unwound. The circular DNA shown in Figure 20.10(b) has been unwound by four complete turns of the helix. The strain introduced by this unwinding is spread uniformly over the entire molecule by the introduction of four superhelical twists [Figure 20.10(c)]. Interconversion of relaxed and supercoiled DNA is catalyzed by groups of enzymes called topoisomerases and gyrases.

Supercoiling of Linear DNA

Supercoiling of linear DNA in plants and animals takes another form and is driven by the interaction between negatively charged DNA molecules and a group of positively charged proteins called **histones**. Histones are particularly rich in lysine and arginine and, at the pH of most body fluids, have an abundance of positively charged sites along their length. The complex between negatively charged DNA and positively charged histones is called **chromatin**. Histones associate to form core particles about which double-stranded DNA then wraps. Further coiling of DNA produces the chromatin found in cell nuclei.

Tertiary structure of nucleic acids The three-dimensional arrangement of all atoms of a nucleic acid, commonly referred to as supercoiling.

Circular DNA A type of double-stranded DNA in which the 5′ and 3′ ends of each strand are joined by phosphodiester groups.

Histone A protein, particularly rich in the basic amino acids lysine and arginine, that is found associated with DNA molecules.

Chromatin A complex formed between negatively charged DNA molecules and positively charged histones.

(a) Relaxed: circular (duplex) DNA

(b) Slightly strained: circular (duplex) DNA with four twists of the helix unwound

(c) Strained: supercoiled circular DNA

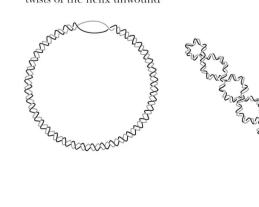

Figure 20.10
Relaxed and supercoiled DNA. (a) Circular DNA is relaxed. (b) One strand is broken, unwound by four turns, and the ends then rejoined. The strain of unwinding is localized in the nonhelical gap. (c) Supercoiling by four twists distributes the strain of unwinding uniformly over the entire molecule of circular DNA.

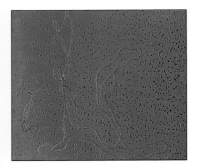

Supercoiled DNA from a mitochondrion. *(Fran Heyl Associates)*

20.4 RIBONUCLEIC ACIDS (RNA)

Ribonucleic acids (RNA) are similar to deoxyribonucleic acids (DNA) in that they, too, consist of long, unbranched chains of nucleotides joined by phosphodiester groups between the 3′-hydroxyl of one pentose and the 5′-hydroxyl of the next. There are, however, three major differences in structure between RNA and DNA:

1. The pentose unit in RNA is β-D-ribose rather than β-2-deoxy-D-ribose.
2. The pyrimidine bases in RNA are uracil and cytosine rather than thymine and cytosine.
3. RNA is single stranded rather than double stranded.

Following are structural formulas for the furanose form of D-ribose and for uracil:

β-D-Ribofuranose
(β-D-Ribose)

Uracil (U)

Cells contain up to eight times as much RNA as DNA, and, in contrast to DNA, RNA occurs in different forms and in multiple copies of each form. RNA molecules are classified, according to their structure and function, into three major types: ribosomal RNA, transfer RNA, and messenger RNA. Table 20.2 summarizes the molecular weight, number of nucleotides, and percentage of cellular abundance of the three types of RNA in cells of *E. coli*, one of the best-studied bacteria and a workhorse for cellular study.

TABLE 20.2   Types of RNA Found in Cells of *E. coli*

Type	Molecular Weight Range (amu)	Number of Nucleotides	Percentage of Cell RNA
mRNA	25,000–1,000,000	75–3,000	2
tRNA	23,000–30,000	73–94	16
rRNA	35,000–1,100,000	120–2,904	82

A. Ribosomal RNA

Ribosomal RNA (rRNA)
A ribonucleic acid found in ribosomes, the sites of protein synthesis.

The bulk of **ribosomal RNA (rRNA)** is found in the cytoplasm in subcellular particles called ribosomes, which contain about 60% RNA and 40% protein. Ribosomes are the sites in cells at which protein synthesis takes place.

B. Transfer RNA

Transfer RNA (tRNA)
A ribonucleic acid that carries a specific amino acid to the site of protein synthesis on ribosomes.

Transfer RNA (tRNA) molecules have the lowest molecular weight of all nucleic acids. They consist of 73 to 94 nucleotides in a single chain. The function of tRNA is to carry amino acids to the sites of protein synthesis on the ribosomes. Each amino acid has at least one tRNA dedicated specifically to this purpose. Several amino acids have more than one. In the transfer process, the amino acid is joined to its specific tRNA by an ester bond between the α-carboxyl group of the amino acid and the 3′ hydroxyl group of the ribose unit at the 3′ end of the tRNA:

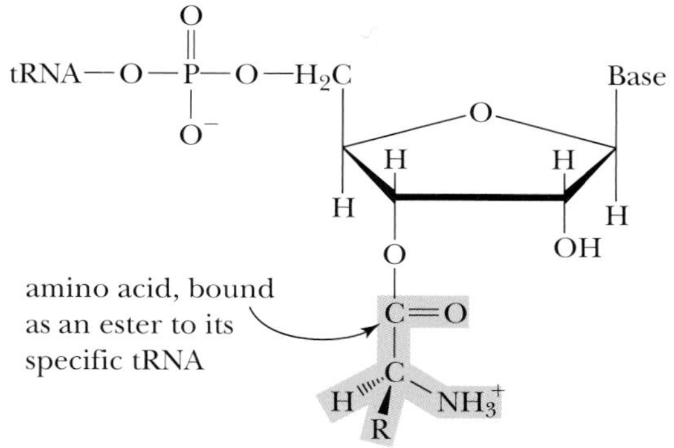

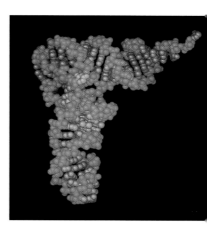

A space-filling model of yeast phenylalanine tRNA. *(After S. H. Kim, in P. Schimmel, D. Söll, and J. N. Abelson, eds., Transfer RNA: Structure, Properties, and Recognition, New York: Cold Spring Harbor Laboratories, 1979)*

C. Messenger RNA

Messenger RNA (mRNA) is present in cells in relatively small amounts and is very short-lived. Messenger RNA molecules are single stranded, and their synthesis is directed by information encoded on DNA molecules. Double-stranded DNA is unwound, and a complementary strand of mRNA is synthesized along one strand of the DNA template, beginning from the 3′ end. The synthesis of mRNA from a DNA template is called *transcription*, because genetic information contained in a sequence of bases of DNA is transcribed into a complementary sequence of bases on mRNA. The name "messenger" is derived from the function of this type of RNA, which is to carry coded genetic information from DNA to the ribosomes for the synthesis of proteins.

Messenger RNA (mRNA)
A ribonucleic acid that carries coded genetic information from DNA to ribosomes for the synthesis of proteins.

EXAMPLE 20.4

Following is a base sequence from a portion of DNA:

3′-A-G-C-C-A-T-G-T-G-A-C-C-5′

Write the sequence of bases of the mRNA synthesized with this section of DNA as a template.

SOLUTION

RNA synthesis begins at the 3′ end of the DNA template and proceeds toward the 5′ end. The complementary mRNA strand is formed using the bases C, G, A, and U. Uracil (U) is the complement of adenine (A) on the DNA template:

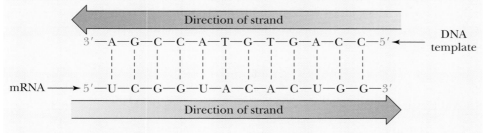

Reading from the 5′ end, we see that the sequence of mRNA is 5′-UCGGUA-CACUGG-3′.

Practice Problem 20.4

Here is a portion of the nucleotide sequence in phenylalanine tRNA:

3′-ACCACCUGCUCAGGCCUU-5′

Write the nucleotide sequence of the DNA complement of this sequence.

20.5 THE GENETIC CODE

A. Triplet Nature of the Code

It was clear by the early 1950s that the sequence of bases in DNA molecules constitutes the store of genetic information and directs the synthesis of messenger RNA, which, in turn, directs the synthesis of proteins. However, the idea that the sequence of bases in DNA directs the synthesis of proteins presents the following problem: How can a molecule containing only four variable units (adenine, cytosine, guanine, and thymine) direct the synthesis of molecules containing up to 20 variable units (the protein-derived amino acids)? How can an alphabet of only four letters code for the order of letters in the 20-letter alphabet that occurs in proteins?

An obvious answer is that there is not one base, but rather a combination of bases coding for each amino acid. If the code consists of nucleotide pairs, there are $4^2 = 16$ combinations; this is a more extensive code, but it is still not extensive enough to code for 20 amino acids. If the code consists of nucleotides in groups of three, there are $4^3 = 64$ combinations; more than enough to code for the primary structure of a protein. This appears to be a very simple solution to the problem for a system that must have taken eons of evolutionary trial and error to develop. Yet proof now exists, from comparisons of gene (nucleic acid) and protein (amino acid) sequences, that nature does indeed use a simple three-letter or triplet code to store genetic information. A triplet of nucleotides is called a **codon.**

Codon A triplet of nucleotides on mRNA that directs the incorporation of a specific amino acid into a polypeptide sequence.

B. Deciphering the Genetic Code

The next question is, Which of the 64 triplets codes for which amino acid? In 1961, Marshall Nirenberg provided a simple experimental approach to the problem, based on the observation that synthetic polynucleotides direct polypeptide synthesis in much the same manner as do natural mRNAs. Nirenberg found that when ribosomes, amino acids, tRNAs, and appropriate protein-synthesizing enzymes were incubated in vitro, no polypeptide synthesis occurred. However, when he added synthetic polyuridylic acid (poly U), a polypeptide of high molecular weight was synthesized. What was more important, the synthetic polypeptide contained only phenylalanine. With this discovery, the first element of the genetic code was deciphered: The triplet UUU codes for phenylalanine.

Similar experiments were carried out with different synthetic polyribonucleotides. It was found, for example, that polyadenylic acid (poly A) leads to the synthesis of polylysine, and that polycytidylic acid (poly C) leads to the synthesis of polyproline. By 1964, all 64 codons had been deciphered (Table 20.3). Nirenberg, R. W. Holley, and H. G. Khorana shared the 1968 Nobel prize in physiology or medicine for their seminal work.

C. Properties of the Genetic Code

Several features of the genetic code are evident from a study of Table 20.3:

1. Only 61 triplets code for amino acids. The remaining three (UAA, UAG, and UGA) are signals for chain termination; they signal to the protein-synthesizing machinery of the cell that the primary sequence of the protein is complete. The three chain termination triplets are indicated in the table by "Stop."
2. The code is degenerate, which means that several amino acids are coded for by more than one triplet. Only methionine and tryptophan are coded for by just one triplet. Leucine, serine, and arginine are coded for by six triplets, and the remaining amino acids are coded for by two, three, or four triplets.

TABLE 20.3 The Genetic Code: mRNA Codons and the Amino Acid Each Codon Directs

First Position (5′-end)	Second Position								Third Position (3′-end)
	U		**C**		**A**		**G**		
U	UUU	Phe	UCU	Ser	UAU	Tyr	UGU	Cys	U
	UUC	Phe	UCC	Ser	UAC	Tyr	UGC	Cys	C
	UUA	Leu	UCA	Ser	UAA	Stop	UGA	Stop	A
	UUG	Leu	UCG	Ser	UAG	Stop	UGG	Trp	G
C	CUU	Leu	CCU	Pro	CAU	His	CGU	Arg	U
	CUC	Leu	CCC	Pro	CAC	His	CGC	Arg	C
	CUA	Leu	CCA	Pro	CAA	Gln	CGA	Arg	A
	CUG	Leu	CCG	Pro	CAG	Gln	CGG	Arg	G
A	AUU	Ile	ACU	Thr	AAU	Asn	AGU	Ser	U
	AUC	Ile	ACC	Thr	AAC	Asn	AGC	Ser	C
	AUA	Ile	ACA	Thr	AAA	Lys	AGA	Arg	A
	AUG*	Met	ACG	Thr	AAG	Lys	AGG	Arg	G
G	GUU	Val	GCU	Ala	GAU	Asp	GGU	Gly	U
	GUC	Val	GCC	Ala	GAC	Asp	GGC	Gly	C
	GUA	Val	GCA	Ala	GAA	Glu	GGA	Gly	A
	GUG	Val	GCG	Ala	GAG	Glu	GGG	Gly	G

*AUG also serves as the principal initiation codon.

3. For the 15 amino acids coded for by two, three, or four triplets, it is only the third letter of the code that varies. For example, glycine is coded for by the triplets GGA, GGG, GGC, and GGU.
4. There is no ambiguity in the code, meaning that each triplet codes for one amino acid.

Finally, we must ask one last question about the genetic code: Is the code universal? That is, is it the same for all organisms? Every bit of experimental evidence available today from the study of viruses, bacteria, and higher animals, including humans, indicates that the code is universal. Furthermore, the fact that it is the same for all these organisms means that it has been the same over millions of years of evolution.

EXAMPLE 20.5

During transcription, a portion of mRNA is synthesized with the following base sequence:

5′-AUG-GUA-CCA-CAU-UUG-UGA-3′

(a) Write the nucleotide sequence of the DNA from which this portion of mRNA was synthesized.
(b) Write the primary structure of the polypeptide coded for by the given section of mRNA.

SOLUTION

(a) During transcription, mRNA is synthesized from a DNA strand, beginning from the 3' end of the DNA template. The DNA strand must be the complement of the newly synthesized mRNA strand:

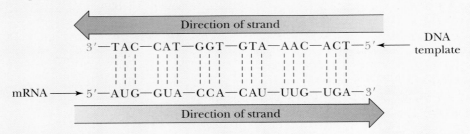

Note that the codon UGA codes for termination of the growing polypeptide chain; therefore, the sequence given in this problem codes for a pentapeptide only.

(b) The sequence of amino acids is shown in the following mRNA strand:

$$5'\text{-AUG-GUA-CCA-CAU-UUG-UGA-}3'$$

$$\text{met---val---pro---his---leu---stop}$$

Practice Problem 20.5

The following section of DNA codes for oxytocin, a polypeptide hormone:

$$3'\text{-ACG-ATA-TAA-GTT-TTA-ACG-GGA-GAA-CCA-ACT-}5$$

(a) Write the base sequence of the mRNA synthesized from this section of DNA.

(b) Given the sequence of bases in part (a), write the primary structure of oxytocin.

20.6   SEQUENCING NUCLEIC ACIDS

As recently as 1975, the task of determining the primary structure of a nucleic acid was thought to be far more difficult than determining the primary structure of a protein. Nucleic acids, it was reasoned, contain only 4 different units, whereas proteins contain 20 different units. With only 4 different units, there are fewer specific sites for selective cleavage, distinctive sequences are more difficult to recognize, and there is greater chance of ambiguity in the assignment of sequences. Two breakthroughs reversed this situation. First was the development of a type of electrophoresis called **polyacrylamide gel electrophoresis**, a technique so sensitive that it is possible to separate nucleic acid fragments that differ from one another in only a single nucleotide. The second breakthrough was the discovery of a class of enzymes called **restriction endonucleases**, isolated chiefly from bacteria.

Restriction endonuclease An enzyme that catalyzes the hydrolysis of a particular phosphodiester bond within a DNA strand.

A. Restriction Endonucleases

A restriction endonuclease recognizes a set pattern of four to eight nucleotides and cleaves a DNA strand by hydrolyzing the linking phosphodiester bonds at any site which contains that particular sequence. Molecular biologists have now isolated close to 1,000 restriction endonucleases and characterized their specificities; each cleaves DNA at a different site and produces a different set of restriction fragments.

E. coli, for example, has a restriction endonuclease EcoRI (pronounced eeko-are-one) that recognizes the hexanucleotide sequence GAATTC and cleaves it between G and A:

— cleavage here

5′---G-A-A-T-T-C-----3′ $\xrightarrow{\text{EcoRI}}$ 5′---G + 5′-A-A-T-T-C-----3′

Note that the action of restriction endonucleases is analogous to the action of trypsin (Section 19.5B), which catalyzes the hydrolysis of amide bonds formed by the carboxyl groups of Lys and Arg and the action of chymotrypsin, which catalyzes the cleavage of amide bonds formed by the carboxyl groups of Phe, Tyr, and Trp.

EXAMPLE 20.6

The following is a section of the gene coding for bovine rhodopsin, along with a table listing several restriction endonucleases, their recognition sequences, and their hydrolysis sites:

5′GTCTACAACCCGGTCATCTACTATCATGATCAACAAGCAGTTCCGGAACT-3′

Enzyme	Recognition Sequence	Enzyme	Recognition Sequence
AluI	AG↓CT	HpaII	C↓CGG
BalI	TGG↓CCA	MboI	↓GATC
FnuDII	CG↓CG	NotI	GC↓GGCCGC
HeaIII	GG↓CC	SacI	GAGCT↓C

Which endonucleases will catalyze the cleavage of the given section of DNA?

SOLUTION

Only restriction endonucleases HpaII and MboI catalyze the cleavage of this polynucleotide: HpaII does so at two sites, MboI at one:

HpaII   MboI   HpaII
↓   ↓   ↓

5′-GTCTACAACC-CGGTCATCTACTATCAT-GATCAACAAGCAGTTC-CGGAACT-3′

Practice Problem 20.6

The following is another section of the bovine rhodopsin gene:

5′-ACGTCGGGTCGTCGTCCTCTCGCGGTGGTGAGTCTTCCGGCTCTTCT-3′

Which of the endonucleases given in Example 20.6 will catalyze the cleavage of this section?

B. Methods for Sequencing Nucleic Acids

The sequencing of DNA begins with the site-specific cleavage of double-stranded DNA by one or more restriction endonucleases into smaller fragments called **restriction fragments**. Each restriction fragment is then sequenced separately, overlapping base sequences are identified, and the entire sequence of bases is subsequently deduced.

Two methods for sequencing restriction fragments have been devised. The first, developed by Allan Maxam and Walter Gilbert and known as the **Maxam–Gilbert method**, depends on base-specific chemical cleavage. The second method, developed by Frederick Sanger and known as the **chain termination** or **dideoxy method**, depends on the interruption of DNA-polymerase-catalyzed synthesis. Sanger and Gilbert shared the 1980 Nobel prize in chemistry for their "development of chemical and biochemical analysis of DNA structure." Sanger's dideoxy method is currently more widely used, so we concentrate on it.

Sanger dideoxy method A method, developed by Frederick Sanger, for sequencing DNA molecules.

C. DNA Replication in Vitro

To appreciate the rationale for the dideoxy method, we must first understand certain aspects of the biochemistry of DNA replication. First, DNA replication takes place when cells divide. During replication, the sequence of nucleotides in one strand is copied as a complementary strand to form the second strand of a double-stranded DNA molecule. Synthesis of the complementary strand is catalyzed by the enzyme DNA polymerase. As shown in the following equation, the DNA chain grows by adding each new unit to the free 3'-OH group of the chain:

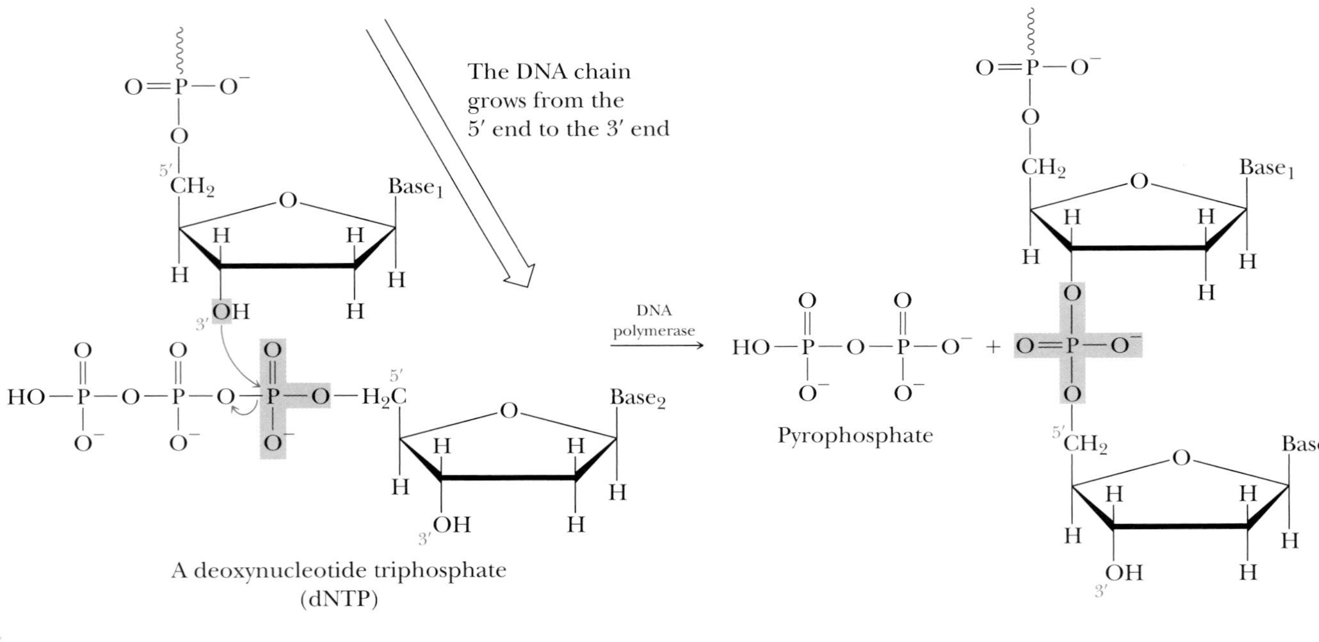

A deoxynucleotide triphosphate
(dNTP)

The DNA chain grows from the 5' end to the 3' end

Pyrophosphate

DNA polymerase will also carry out this synthesis in vitro with single-stranded DNA as a template, provided that both the four deoxynucleotide triphosphate (dNTP) monomers and a primer are present. A **primer** is an oligonucleotide that is capable of forming a short section of double-stranded DNA (dsDNA) by base pairing with its complement on a single-stranded DNA (ssDNA). Because a new DNA strand grows from its 5' to 3' end, the primer must have a free 3'-OH group to which the first nucleotide of the growing chain is added (Figure 20.11).

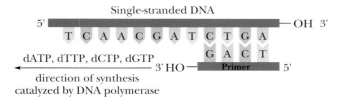

Figure 20.11

DNA polymerase catalyzes the synthesis of the complementary strand of DNA in vitro, using single-stranded DNA as a template, provided that both the four deoxynucleotide triphosphate (dNTP) monomers and a primer are present. The primer provides a short stretch of double-stranded DNA by base pairing with its complement on the single-stranded DNA.

D. The Chain Termination, or Dideoxy, Method

The key to the chain termination method is the addition of a 2′,3′-dideoxy-nucleoside triphosphate (ddNTP) to the synthesizing medium:

A 2′,3′-dideoxynucleoside triphosphate
(ddNTP)

Because a ddNTP has no —OH group at the 3′ position, it cannot serve as an acceptor for the next nucleotide to be added to the growing polynucleotide chain. Thus, chain synthesis is terminated at any point where a ddNTP becomes incorporated and hence the designation *chain termination method*.

In the chain termination method, a single-stranded DNA of unknown sequence is mixed with primer and divided into four separate reaction mixtures. To each reaction mixture are added all four deoxynucleoside triphosphates (dNTPs), one of which is labeled in the 5′ phosphoryl group with phosphorus-32 (^{32}P) so that the newly synthesized fragments can be visualized by autoradiography.

$$^{32}_{15}P \longrightarrow ^{32}_{16}S + \text{Beta particle} + \text{Gamma rays}$$

Also added to each reaction mixture are DNA polymerase and one of the four ddNTPs. The ratio of dNTPs to ddNTP in each reaction mixture is adjusted so that a ddNTP is incorporated only infrequently. In each reaction mixture, DNA synthesis takes place, but in a given population of molecules, synthesis is interrupted at every possible site (Figure 20.12).

When gel electrophoresis of each reaction mixture is completed, a piece of X-ray film is placed over the gel, and gamma rays released by the radioactive decay of ^{32}P darken the film and create a pattern on it that is an image of the resolved oligonucleotides. The base sequence of the complement of the original single-stranded template is then read directly from bottom to top of the developed film.

A variation on this method is to use a single reaction mixture, with each of the four ddNTPs labeled with a different fluorescent indicator. Each label is then detected by its characteristic spectrum. Automated DNA-sequencing machines using this variation are capable of sequencing up to 10,000 base pairs per day.

Figure 20.12

The chain termination, or dideoxy, method of DNA sequencing. The primer–DNA template is divided into four separate reaction mixtures, to each of which are added the four dNTPs, DNA polymerase, and one of the four ddNTPs. Synthesis is interrupted at every possible site. The mixtures of oligonucleotides are separated by polyacrylamide gel electrophoresis. The base sequence of the DNA complement is read from the bottom to the top (from the 5′ end to the 3′ end) of the developed gel.

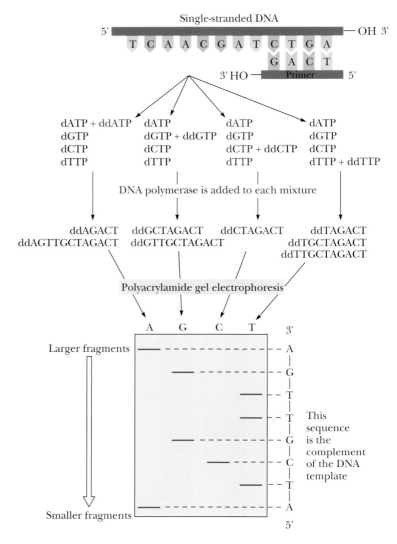

If the complement of the DNA template is 5′ A–T–C–G–T–T–G–A–3′, then the original DNA template must be 5′–T–C–A–A–C–G–A–T–3′

E. Sequencing the Human Genome

As nearly everyone knows, the sequencing of the human genome was announced in the spring of 2000 by two competing groups: the so-called Human Genome Project, a loosely linked consortium of publicly funded groups, and a private company called Celera. Actually, this milestone represents, not a complete sequence, but a "rough draft," constituting about 85% of the entire genome. The methodology for sequencing the human genome was based on a refinement of the techniques described earlier that use massively parallel separations of fragments by electrophoresis in capillary tubes. The Celera approach utilized some 300 of the fastest sequencing machines in parallel, each operating on many parallel DNA fragments. Supercomputers were employed to help assemble and compare millions of overlapping sequences.

This achievement represents the beginning of a new era of molecular medicine. From now on, specific genetic deficiencies leading to inherited diseases will be understood on a molecular basis, and new therapies targeted at shutting down undesired genes or turning on desired ones will be developed.

CHEMICAL CONNECTIONS 20B

DNA Fingerprinting

Each human being has a genetic makeup consisting of approximately 3 billion pairs of nucleotides, and, except for identical twins, the base sequence of DNA in one individual is different from that of every other individual. As a result, each person has a unique DNA "fingerprint." To determine a DNA fingerprint, a sample of DNA from a trace of blood, skin, or other tissue is treated with a set of restriction endonucleases, and the 5′ end of each restriction fragment is labeled with phosphorus-32. The resulting ^{32}P-labeled restriction fragments are then separated by polyacrylamide gel electrophoresis and visualized by placing a photographic plate over the developed gel.

In the DNA fingerprint patterns shown in the accompanying figure, lanes 1, 5, and 9 represent internal standards, or control lanes. The lanes contain the DNA fingerprint pattern of a standard virus treated with a standard set of restriction endonucleases. Lanes 2, 3, and 4 were used in a paternity suit. The DNA fingerprint of the mother in lane 4 contains five bands, which match with five of the six bands in the DNA fingerprint of the child in lane 3. The DNA fingerprint of the alleged father in lane 2 contains six bands, three of which match with bands in the DNA fingerprint of the child. Because the child inherits only half of its genes from the father, only half of the child's and father's DNA fingerprints are expected to match. In this instance, the paternity suit was won on the basis of the DNA fingerprint matching.

Lanes 6, 7, and 8 contain DNA fingerprint patterns used as evidence in a rape case. Lanes 6 and 7

are DNA fingerprints of semen obtained from the rape victim. Lane 8 is the DNA fingerprint pattern of the alleged rapist. The DNA fingerprint patterns of the semen do not match that of the alleged rapist and excluded the suspect from the case.

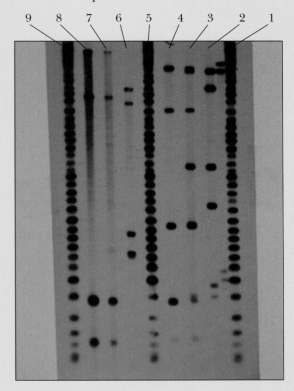

DNA fingerprint. *(Courtesy of Dr. Lawrence Koblinsky)*

SUMMARY

Nucleic acids are composed of three types of monomer units: heterocyclic aromatic amine bases derived from purine and pyrimidine, the monosaccharide D-ribose or 2-deoxy-D-ribose, and phosphate ions (Section 20.2). A **nucleoside** is a compound containing D-ribose or 2-deoxy-D-ribose, bonded to a heterocyclic aromatic amine base by a β-N-glycosidic bond. A **nucleotide** is a nucleoside in which a molecule of phosphoric acid is esterified with an —OH of the monosaccharide, most commonly either the 3′-OH or the 5′-OH. Nucleoside mono-, di- and triphosphates are strong polyprotic acids and are

extensively ionized at pH 7.0. At that pH, adenosine triphosphate, for example, is a 50:50 mixture of ATP^{3-} and ATP^{4-}.

The **primary structure** of **deoxyribonucleic acids (DNA)** consists of units of 2-deoxyribose bonded by 3′,5′-phosphodiester bonds (Section 20.3A). A heterocyclic aromatic amine base is bonded to each deoxyribose unit by a β-N-glycosidic bond. The sequence of bases is read from the 5′ end of the polynucleotide strand to the 3′ end.

The heart of the **Watson–Crick model** of the structure of DNA is the postulate that a molecule of DNA consists of two antiparallel polynucleotide strands coiled in a right-handed manner about the same axis to form a **double helix** (Section 20.3B). Purine and pyrimidine bases point inward toward the axis of the helix and are always paired G-C and A-T. In **B-DNA**, base pairs are stacked one on top of another, with a spacing of 3.4 Å and 10 base pairs per 34-Å helical repeat. In **A-DNA**, bases are stacked with a spacing of 2.9 Å between base pairs and 10 base pairs per 29-Å helical repeat.

The tertiary structure of DNA is commonly referred to as **supercoiling** (Section 20.3C). **Circular DNA** is a type of double-stranded DNA in which the ends of each strand are joined by phosphodiester groups. The opening of one strand followed by the partial unwinding and rejoining of the ends introduces strain in the nonhelical gap. The strain can be spread over the entire molecule of circular DNA by the introduction of **superhelical twists**. **Histones** are particularly rich in lysine and arginine and, therefore, have an abundance of positive charges. The association of DNA with histones produces a pigment called **chromatin.**

There are two important differences between the primary structure of **ribonucleic acids (RNA)** and DNA (Section 20.4): (1) The monosaccharide unit in RNA is D-ribose. (2) Both RNA and DNA contain the purine bases adenine (A) and guanine (G) and the pyrimidine base cytosine (C). As the fourth base, however, RNA contains uracil (U), whereas DNA contains thymine (T).

The **genetic code** (Section 20.5) consists of nucleosides in groups of three; that is, it is a triplet code. Only 61 triplets code for amino acids; the remaining 3 code for the termination of polypeptide synthesis.

Restriction endonucleases recognize a set pattern of four to eight nucleotides and cleave a DNA strand by hydrolyzing the linking phosphodiester bonds at any site which contains that particular sequence (Section 20.6A). In the **chain termination**, or **dideoxy**, **method** of DNA sequencing developed by Frederick Sanger (Section 20.6D), a primer–DNA template is divided into four separate reaction mixtures. To each is added the four dNTPs, one of which is labeled with phosphorus-32. Also added are DNA polymerase and one of the four ddNTPs. Synthesis is interrupted at every possible site. The mixtures of newly synthesized oligonucleotides are separated by polyacrylamide gel electrophoresis and visualized by autoradiography. The base sequence of the DNA complement of the original DNA template is read from the bottom to the top (from the 5′ end to the 3′ end) of the developed photographic plate.

PROBLEMS

A problem number set in red indicates an applied "real-world" problem.

Nucleosides and Nucleotides

20.7 Two drugs used in the treatment of acute leukemia are 6-mercaptopurine and 6-thioguanine:

6-Mercaptopurine   6-Thioguanine

Note that, in each of these drugs, the oxygen at carbon 6 of the parent molecule is replaced by divalent sulfur. Draw structural formulas for the enethiol (the sulfur equivalent of an enol) forms of 6-mercaptopurine and 6-thioguanine.

20.8 Following are structural formulas for cytosine and thymine:

Cytosine (C)   Thymine (T)

Draw two additional tautomeric forms for cytosine and three for thymine.

20.9 Draw a structural formula for a nucleoside composed of
    (a) β-D-Ribose and adenine   **(b)** β-2-Deoxy-D-ribose and cytosine

20.10 Nucleosides are stable in water and in dilute base. In dilute acid, however, the glycosidic bond of a nucleoside undergoes hydrolysis to give a pentose and a heterocyclic aromatic amine base. Propose a mechanism for this acid-catalyzed hydrolysis.

20.11 Explain the difference in structure between a nucleoside and a nucleotide.

20.12 Draw a structural formula for each nucleotide and estimate its net charge at pH 7.4, the pH of blood plasma:
    (a) 2'-Deoxyadenosine 5'-triphosphate (dATP)
    (b) Guanosine 3'-monophosphate (GMP)
    (c) 2'-Deoxyguanosine 5'-diphosphate (dGDP)

20.13 Cyclic-AMP, first isolated in 1959, is involved in many diverse biological processes as a regulator of metabolic and physiological activity. In this compound, a single phosphate group is esterified with both the 3' and 5' hydroxyls of adenosine. Draw a structural formula of cyclic-AMP.

The Structure of DNA

20.14 Why are deoxyribonucleic acids called acids? What are the acidic groups in their structure?

20.15 Human DNA is approximately 30.4% A. Estimate the percentages of G, C, and T and compare them with the values presented in Table 20.1.

20.16 Draw a structural formula for the DNA tetranucleotide 5'-A-G-C-T-3'. Estimate the net charge on this tetranucleotide at pH 7.0. What is the complementary tetranucleotide of this sequence?

20.17 List the postulates of the Watson–Crick model of DNA secondary structure.

20.18 The Watson–Crick model is based on certain experimental observations of base composition and molecular dimensions. Describe these observations and show how the model accounts for each.

20.19 Compare the α-helix of proteins and the double helix of DNA in terms of
    (a) The units that repeat in the backbone of the polymer chain.
    (b) The projection in space of substituents along the backbone (the R groups in the case of amino acids, purine and pyrimidine bases in the case of double-stranded DNA) relative to the axis of the helix.

20.20 Discuss the role of hydrophobic interactions in stabilizing double-stranded DNA.

20.21 Name the type of covalent bond(s) joining monomers in these biopolymers:
    (a) Polysaccharides   **(b)** Polypeptides   **(c)** Nucleic acids

20.22 In terms of hydrogen bonding, which is more stable, an A-T base pair or a G-C base pair?

20.23 At elevated temperatures, nucleic acids become denatured; that is, they unwind into single-stranded DNA. Account for the observation that the higher the G-C content of a nucleic acid, the higher is the temperature required for its thermal denaturation.

20.24 Write the DNA complement of 5'-ACCGTTAAT-3'. Be certain to label which is the 5' end and which is the 3' end of the complement strand.

20.25 Write the DNA complement of 5'-TCAACGAT-3'.

Ribonucleic Acids

20.26 Compare the degree of hydrogen bonding in the base pair A-T found in DNA with that in the base pair A-U found in RNA.

20.27 Compare DNA and RNA is these ways:

    (a) Monosaccharide units   **(b)** Principal purine and pyrimidine bases

    (c) Primary structure   **(d)** Location in the cell

    (e) Function in the cell

20.28 What type of RNA has the shortest lifetime in cells?

20.29 Write the mRNA complement of 5′-ACCGTTAAT-3′. Be certain to label which is the 5′ end and which is the 3′ end of the mRNA strand.

20.30 Write the mRNA complement of 5′-TCAACGAT-3′.

The Genetic Code

20.31 What does it mean to say that the genetic code is degenerate?

20.32 Aspartic acid and glutamic acid have carboxyl groups on their side chains and are called acidic amino acids. Compare the codons for these two amino acids.

20.33 Compare the structural formulas of the aromatic amino acids phenylalanine and tyrosine. Compare the codons for these two amino acids as well.

20.34 Glycine, alanine, and valine are classified as nonpolar amino acids. Compare their codons. What similarities do you find? What differences?

20.35 Codons in the set CUU, CUC, CUA, and CUG all code for the amino acid leucine. In this set, the first and second bases are identical, and the identity of the third base is irrelevant. For what other sets of codons is the third base also irrelevant, and for what amino acid(s) does each set code?

20.36 Compare the amino acids coded for by the codons with a pyrimidine, either U or C, as the second base. Do the majority of the amino acids specified by these codons have hydrophobic or hydrophilic side chains?

20.37 Compare the amino acids coded for by the codons with a purine, either A or G, as the second base. Do the majority of the amino acids specified by these codons have hydrophilic or hydrophobic side chains?

20.38 What polypeptide is coded for by this mRNA sequence?

        5′-GCU-GAA-GUC-GAG-GUG-UGG-3′

20.39 The alpha chain of human hemoglobin has 141 amino acids in a single polypeptide chain. Calculate the minimum number of bases on DNA necessary to code for the alpha chain. Include in your calculation the bases necessary for specifying the termination of polypeptide synthesis.

20.40 In HbS, the human hemoglobin found in individuals with sickle-cell anemia, glutamic acid at position 6 in the beta chain is replaced by valine.

    (a) List the two codons for glutamic acid and the four codons for valine.

    (b) Show that one of the glutamic acid codons can be converted to a valine codon by a single substitution mutation—that is, by changing one letter in one codon.

Looking Ahead

20.41 The loss of three consecutive units of Ts from the gene that codes for CFTR, a transmembrane conductance regulator protein, results in the disease known as cystic fibrosis. Which amino acid is missing from CFTR to cause this disease?

20.42 The following compounds have been researched as potential antiviral agents:

(a) Cordycepin (3′-deoxyadenosine)

(b)

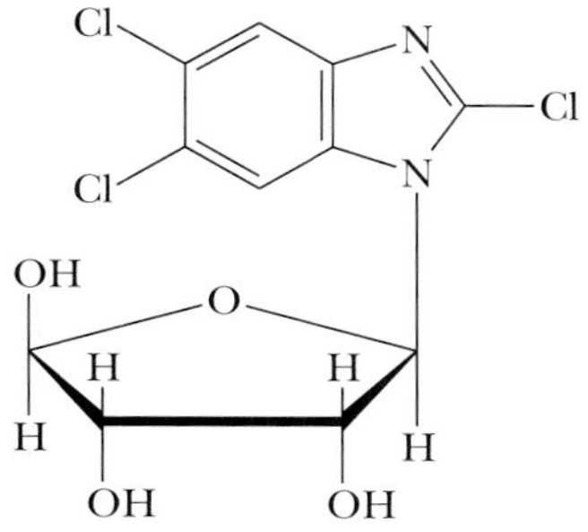

2,5,6-Trichloro-1-(β-D-ribofuranosyl) benzimidazole

(c)

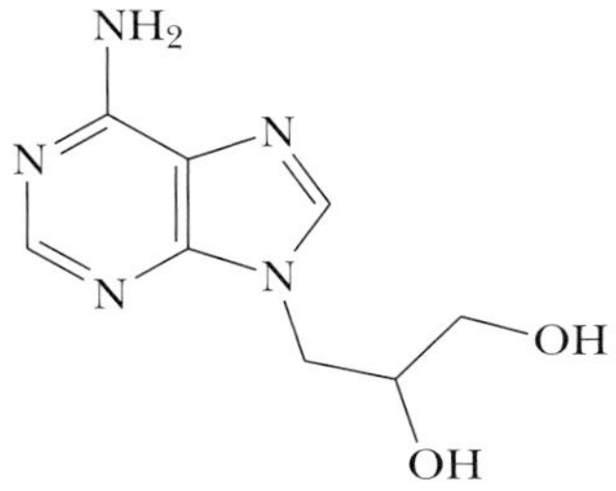

9-(2,3-Dihydroxypropyl) adenine

Suggest how each of these compounds might block the synthesis of RNA or DNA.

20.43 The ends of chromosomes, called telomeres, can form unique and nonstandard structures. One example is the presence of base pairs between units of guanosine. Show how two guanine bases can pair with each other by hydrogen bonding:

Guanine

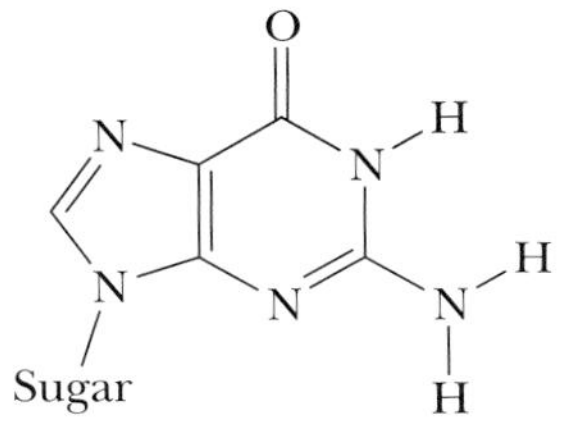

20.44 One synthesis of zidovudine (AZT) involves the following reaction (DMF is the solvent *N,N*-dimethylformamide):

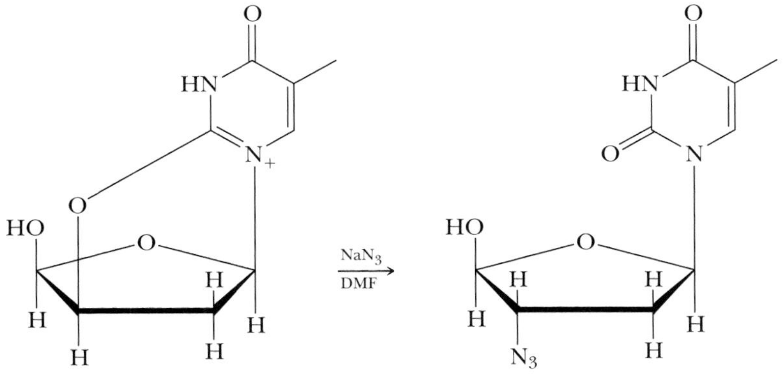

What type of reaction is this?

21

Lipids

A polar bear in snow-covered landscape, Canada. Polar bears eat only during a few weeks out of the year and then fast for periods of eight months or more, consuming no food or water during that period. Eating mainly in the winter, an adult polar bear feeds almost exclusively on seal blubber (composed of triglycerides), thus building up its own triglyceride reserves. Through the Arctic summer, the polar bear maintains normal physiological activity, roaming over long distances, but relies entirely on its body fat for sustenance, burning as much as 1.0-1.5 kg of fat per day. Inset: A model of oleic acid. (*Daniel J. Cox/Stone/Getty Images*)

21.1 INTRODUCTION

Lipid A class of biomolecules isolated from plant or animal sources by extraction with nonpolar organic solvents, such as diethyl ether and acetone.

Lipids are a heterogeneous group of naturally occurring organic compounds, classified together on the basis of their common solubility properties. Lipids are insoluble in water, but soluble in relatively nonpolar aprotic organic solvents, including diethyl ether, dichloromethane, and acetone.

Lipids play three major roles in human biology. First, they are storage depots of chemical energy in the form of triglycerides (fats). While plants store energy in the form of carbohydrates (e.g., starch), humans store it in the form of fat globules in adipose tissue. Second, lipids, in the form of phospholipids, are the water-insoluble components from which biological membranes are constructed. Third, lipids, in the form of steroid hormones, prostaglandins, thromboxanes, and leukotrienes, are chemical messengers. In this chapter, we describe the structures and biological functions of each group of lipids.

21.2 TRIGLYCERIDES

Animal fats and vegetable oils, the most abundant naturally occurring lipids, are triesters of glycerol and long-chain carboxylic acids. Fats and oils are also referred to as **triglycerides** or **triacylglycerols**. Hydrolysis of a triglyceride in aqueous base, followed by acidification, gives glycerol and three fatty acids:

Triglyceride (triacylglycerol) An ester of glycerol with three fatty acids.

$$
\begin{array}{c}
\text{O} \quad \text{CH}_2\text{OCR} \\
\text{R'COCH} \quad \text{O} \\
\text{CH}_2\text{OCR''}
\end{array}
\xrightarrow[\text{2. HCl, H}_2\text{O}]{\text{1. NaOH, H}_2\text{O}}
\begin{array}{c}
\text{CH}_2\text{OH} \\
\text{HOCH} \\
\text{CH}_2\text{OH}
\end{array}
+
\begin{array}{c}
\text{RCOOH} \\
\text{R'COOH} \\
\text{R''COOH}
\end{array}
$$

A triglyceride   1,2,3-Propanetriol   Fatty acids
          (Glycerol, glycerin)

A. Fatty Acids

More than 500 different **fatty acids** have been isolated from various cells and tissues. Given in Table 21.1 are common names and structural formulas for the most abundant of them. The number of carbons in a fatty acid and the number of carbon–carbon double bonds in its hydrocarbon chain are shown by two numbers separated by a colon. In this notation, linoleic acid, for example, is designated as an 18:2 fatty acid; its 18-carbon chain contains two carbon–carbon double bonds.

Fatty acid A long, unbranched-chain carboxylic acid, most commonly of 12 to 20 carbons, derived from the hydrolysis of animal fats, vegetable oils, or the phospholipids of biological membranes.

TABLE 21.1 The Most Abundant Fatty Acids in Animal Fats, Vegetable Oils, and Biological Membranes

Carbon Atoms/ Double Bonds*	Structure	Common Name	Melting Point(°C)
Saturated Fatty Acids			
12:0	$CH_3(CH_2)_{10}COOH$	lauric acid	44
14:0	$CH_3(CH_2)_{12}COOH$	myristic acid	58
16:0	$CH_3(CH_2)_{14}COOH$	palmitic acid	63
18:0	$CH_3(CH_2)_{16}COOH$	stearic acid	70
20:0	$CH_3(CH_2)_{18}COOH$	arachidic acid	77
Unsaturated Fatty Acids			
16:1	$CH_3(CH_2)_5CH{=}CH(CH_2)_7COOH$	palmitoleic acid	1
18:1	$CH_3(CH_2)_7CH{=}CH(CH_2)_7COOH$	oleic acid	16
18:2	$CH_3(CH_2)_4(CH{=}CHCH_2)_2(CH_2)_6COOH$	linoleic acid	−5
18:3	$CH_3CH_2(CH{=}CHCH_2)_3(CH_2)_6COOH$	linolenic acid	−11
20:4	$CH_3(CH_2)_4(CH{=}CHCH_2)_4(CH_2)_2COOH$	arachidonic acid	−49

*The first number is the number of carbons in the fatty acid; the second is the number of carbon–carbon double bonds in its hydrocarbon chain.

Some vegetable oils.
(*Charles D. Winters*)

Following are several characteristics of the most abundant fatty acids in higher plants and animals:

1. Nearly all fatty acids have an even number of carbon atoms, most between 12 and 20, in an unbranched chain.
2. The three most abundant fatty acids in nature are palmitic acid (16:0), stearic acid (18:0), and oleic acid (18:1).
3. In most unsaturated fatty acids, the cis isomer predominates; the trans isomer is rare.
4. Unsaturated fatty acids have lower melting points than their saturated counterparts. The greater the degree of unsaturation, the lower is the melting point. Compare, for example, the melting points of these four 18-carbon fatty acids:

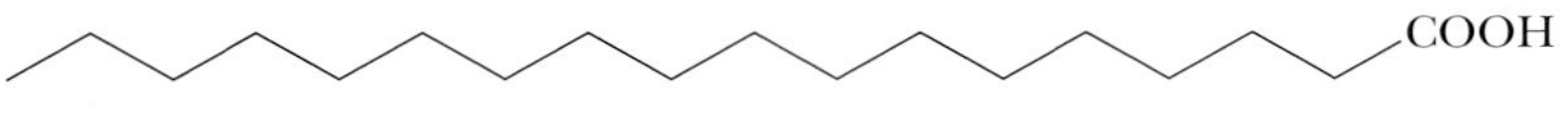

Stearic acid (18:0)
(mp 70°C)

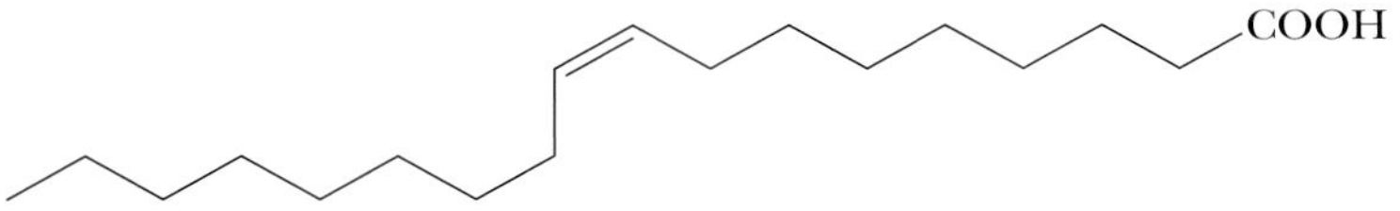

Oleic acid (18:1)
(mp 16°C)

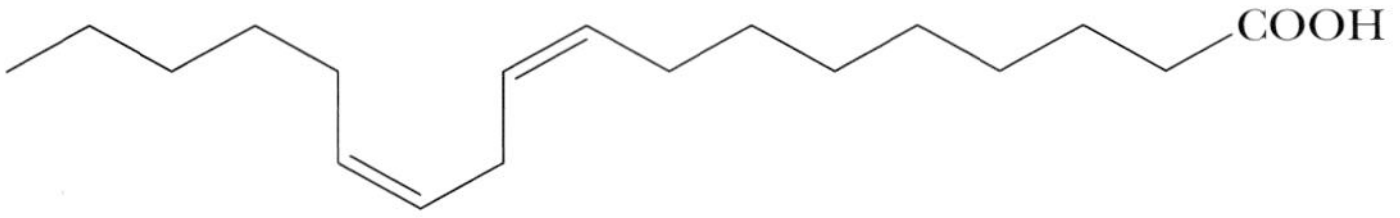

Linoleic acid (18:2)
(mp −5°C)

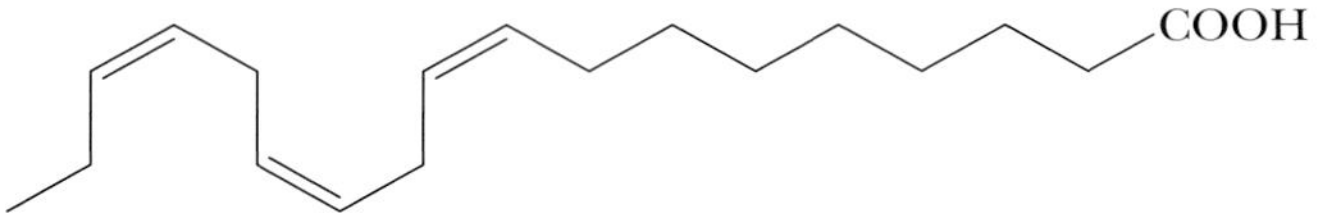

Linolenic acid (18:3)
(mp −11°C)

EXAMPLE 21.1

Draw the structural formula of a triglyceride derived from one molecule each of palmitic acid, oleic acid, and stearic acid, the three most abundant fatty acids in the biological world.

SOLUTION

In this structure, palmitic acid is esterified at carbon 1 of glycerol, oleic acid at carbon 2, and stearic acid at carbon 3:

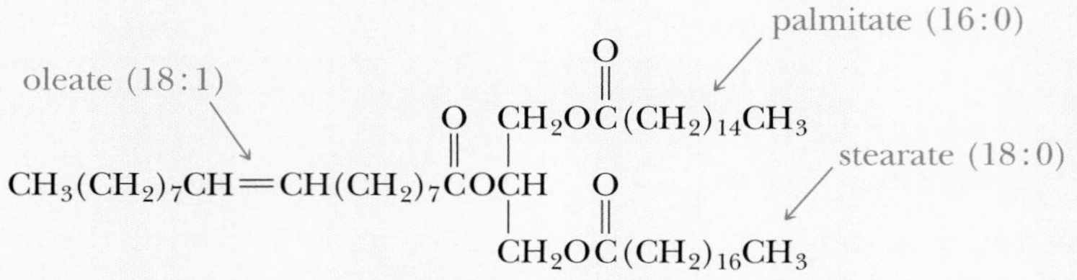

A triglyceride

Practice Problem 21.1

(a) How many constitutional isomers are possible for a triglyceride containing one molecule each of palmitic acid, oleic acid, and stearic acid?
(b) Which of the constitutional isomers that you found in Part (a) are chiral?

B. Physical Properties

The physical properties of a triglyceride depend on its fatty-acid components. In general, the melting point of a triglyceride increases as the number of carbons in its hydrocarbon chains increases and as the number of carbon–carbon double bonds decreases. Triglycerides rich in oleic acid, linoleic acid, and other unsaturated fatty acids are generally liquids at room temperature and are called **oils** (e.g., corn oil and olive oil). Triglycerides rich in palmitic, stearic, and other saturated fatty acids are generally semisolids or solids at room temperature and are called **fats** (e.g., human fat and butter fat). Fats of land animals typically contain approximately 40% to 50% saturated fatty acids by weight (Table 21.2). Most plant oils, on the other hand, contain 20% or less saturated fatty acids and 80% or more unsaturated fatty acids. The notable exception to this generalization about plant oils are the **tropical oils** (e.g., coconut and palm oils), which are considerably richer in low-molecular-weight saturated fatty acids.

Oil A triglyceride that is liquid at room temperature.

Fat A triglyceride that is semisolid or solid at room temperature.

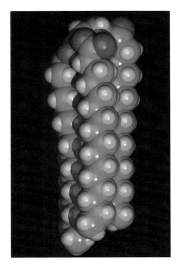

Figure 21.1
Tripalmitin, a saturated triglyceride. (*Brent Iverson, Univeristy of Texas*)

TABLE 21.2 Grams of Fatty Acid per 100 g of Triglyceride of Several Fats and Oils*

	Saturated Fatty Acids			Unsaturated Fatty Acids	
Fat or Oil	Lauric (12:0)	Palmitic (16:0)	Stearic (18:0)	Oleic (18:1)	Linoleic (18:2)
human fat	—	24.0	8.4	46.9	10.2
beef fat	—	27.4	14.1	49.6	2.5
butter fat	2.5	29.0	9.2	26.7	3.6
coconut oil	45.4	10.5	2.3	7.5	trace
corn oil	—	10.2	3.0	49.6	34.3
olive oil	—	6.9	2.3	84.4	4.6
palm oil	—	40.1	5.5	42.7	10.3
peanut oil	—	8.3	3.1	56.0	26.0
soybean oil	0.2	9.8	2.4	28.9	50.7

*Only the most abundant fatty acids are given; other fatty acids are present in lesser amounts.

The lower melting points of triglycerides that are rich in unsaturated fatty acids are related to differences in three-dimensional shape between the hydrocarbon chains of their unsaturated and saturated fatty-acid components. Figure 21.1 shows a space-filling model of tripalmitin, a saturated triglyceride. In the model, the hydrocarbon chains lie parallel to each other, giving the molecule an ordered, compact shape. Because of this compact three-dimensional shape and the resulting strength of the dispersion forces (Section 3.9B) between hydrocarbon chains of adjacent molecules, triglycerides that are rich in saturated fatty acids have melting points above room temperature.

The three-dimensional shape of an unsaturated fatty acid is quite different from that of a saturated fatty acid. Recall from Section 21.1A that unsaturated fatty acids of higher organisms are predominantly of the cis configuration; trans configurations are rare. Figure 21.2 shows a space-filling model of a **polyunsaturated triglyceride** derived from one molecule each of stearic acid, oleic acid, and linoleic acid. Each double bond in this polyunsaturated triglyceride has the cis configuration.

Polyunsaturated triglycerides have a less ordered structure and do not pack together so closely or so compactly as saturated triglycerides. As a consequence, intramolecular and intermolecular dispersion forces are weaker, with the result that polyunsaturated triglycerides have lower melting points than their saturated counterparts.

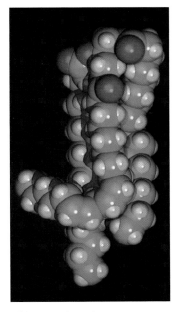

Figure 21.2
A polyunsaturated triglyceride. (*Brent Iverson, Univeristy of Texas*)

Polyunsaturated triglyceride A triglyceride having several carbon–carbon double bonds in the hydrocarbon chains of its three fatty acids.

C. Reduction of Fatty-Acid Chains

For a variety of reasons, in part convenience and in part dietary preference, the conversion of oils to fats has become a major industry. The process is called **hardening** of oils and involves the catalytic reduction (Section 5.5A) of some or all of an oil's carbon–carbon double bonds. In practice, the degree of hardening is carefully controlled to produce fats of a desired consistency. The resulting fats are sold for kitchen use (as Crisco®, Spry®, and others). Margarine and other butter substitutes are produced by partial hydrogenation of polyunsaturated oils derived from corn, cottonseed, peanut, and soybean oils. To the hardened oil are added β-carotene (to give it a yellow color and make it look like butter), salt, and about 15% milk by volume to form the final emulsion. Vitamins A and D may be added as well. Finally, because the product to this stage is tasteless, acetoin and diacetyl, two compounds that mimic the characteristic flavor of butter, are often added:

Some common products containing hydrogenated vegetable oils.
(Charles D. Winters)

$$CH_3-\underset{\underset{\displaystyle OH}{|}}{CH}-\underset{\underset{\displaystyle O}{\|}}{C}-CH_3 \qquad CH_3-\underset{\underset{\displaystyle O}{\|}}{C}-\underset{\underset{\displaystyle O}{\|}}{C}-CH_3$$

3-Hydroxy-2-butanone   2,3-Butanedione
(Acetoin)   (Diacetyl)

21.3 SOAPS AND DETERGENTS

A. Structure and Preparation of Soaps

Soap A sodium or potassium salt of a fatty acid.

Natural **soaps** are prepared most commonly from a blend of tallow and coconut oils. In the preparation of tallow, the solid fats of cattle are melted with steam, and the tallow layer that forms on the top is removed. The preparation of soaps begins by boiling these triglycerides with sodium hydroxide. The reaction that takes place is called *saponification* (Latin: *saponem*, soap):

$$\underset{\text{A triglyceride}}{RCOCH} \qquad +\quad 3NaOH \quad \xrightarrow{\text{saponification}} \quad \underset{\substack{\text{1,2,3-Propanetriol} \\ \text{(Glycerol; glycerin)}}}{CHOH} \quad + \quad \underset{\text{Sodium soaps}}{3RCO^-Na^+}$$

At the molecular level, saponification corresponds to base-promoted hydrolysis of the ester groups in triglycerides (Section 15.4C). The resulting soaps contain mainly the sodium salts of palmitic, stearic, and oleic acids from tallow and the sodium salts of lauric and myristic acids from coconut oil.

After hydrolysis is complete, sodium chloride is added to precipitate the soap as thick curds. The water layer is then drawn off, and glycerol is recovered by vacuum distillation. The crude soap contains sodium chloride, sodium hydroxide, and other impurities that are removed by boiling the curd in water and reprecipitating with more sodium chloride. After several purifications, the soap can be used as an inexpensive industrial soap without further processing. Other treatments transform the crude soap into pH-controlled cosmetic soaps, medicated soaps, and the like.

B. How Soap Cleans

Soap owes its remarkable cleansing properties to its ability to act as an emulsifying agent. Because the long hydrocarbon chains of natural soaps are insoluble in water,

(a) A soap

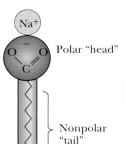

Na⁺

Polar "head"

Nonpolar "tail"

(b) Cross section of a soap micelle in water

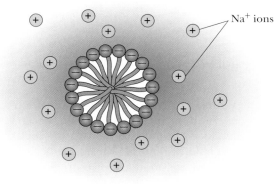

Na⁺ ions

Figure 21.3
Soap micelles. Nonpolar (hydrophobic) hydrocarbon chains are clustered in the interior of the micelle, and polar (hydrophilic) carboxylate groups are on the surface of the micelle. Soap micelles repel each other because of their negative surface charges.

they tend to cluster in such a way as to minimize their contact with surrounding water molecules. The polar carboxylate groups, by contrast, tend to remain in contact with the surrounding water molecules. Thus, in water, soap molecules spontaneously cluster into **micelles** (Figure 21.3).

Most of the things we commonly think of as dirt (such as grease, oil, and fat stains) are nonpolar and insoluble in water. When soap and this type of dirt are mixed together, as in a washing machine, the nonpolar hydrocarbon inner parts of the soap micelles "dissolve" the nonpolar dirt molecules. In effect, new soap micelles are formed, this time with nonpolar dirt molecules in the center (Figure 21.4). In this way, nonpolar organic grease, oil, and fat are "dissolved" and washed away in the polar wash water.

Soaps, however, have their disadvantages, foremost among which is the fact that they form water-insoluble salts when used in water containing Ca(II), Mg(II), or Fe(III) ions (hard water):

$$2CH_3(CH_2)_{14}COO^-Na^+ + Ca^{2+} \longrightarrow [CH_3(CH_2)_{14}COO^-]_2Ca^{2+} + 2Na^+$$

A sodium soap
(soluble in water as micelles)

Calcium salt of a fatty acid
(insoluble in water)

These calcium, magnesium, and iron salts of fatty acids create problems, including rings around the bathtub, films that spoil the luster of hair, and grayness and roughness that build up on textiles after repeated washings.

C. Synthetic Detergents

After the cleansing action of soaps was understood, chemists were in a position to design a synthetic detergent. Molecules of a good detergent, they reasoned, must have a long hydrocarbon chain—preferably, 12 to 20 carbon atoms long—and a polar group at one end of the molecule that does not form insoluble salts with the Ca(II), Mg(II), or Fe(III) ions that are present in hard water. These essential characteristics of a soap, they recognized, could be produced in a molecule containing a sulfonate (—SO₃⁻) group instead of a carboxylate (—COO⁻) group. Calcium, magnesium, and iron salts of monoalkylsulfuric and sulfonic acids are much more soluble in water than comparable salts of fatty acids.

The most widely used synthetic detergents today are the linear alkylbenzenesulfonates (LAS). One of the most common of these is sodium 4-dodecylbenzenesulfonate. To prepare this type of detergent, a linear alkylbenzene is treated with sulfuric

Micelle A spherical arrangement of organic molecules in water solution clustered so that their hydrophobic parts are buried inside the sphere and their hydrophilic parts are on the surface of the sphere and in contact with water.

Soap micelle with "dissolved" grease

Grease

Soap

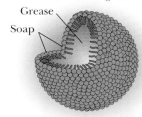

Figure 21.4
A soap micelle with a "dissolved" oil or grease droplet.

Effects of optical bleaches: *(above)* ordinary light; *(below)* black light. *(Charles D. Winters)*

Phospholipid A lipid containing glycerol esterified with two molecules of fatty acid and one molecule of phosphoric acid.

All of these products contain lecithin. *(Charles D. Winters)*

acid to form an alkylbenzenesulfonic acid (Section 9.7B), followed by neutralization of sulfonic acid with NaOH:

$$CH_3(CH_2)_{10}CH_2 \overset{1.\ H_2SO_4}{\underset{2.\ NaOH}{\longrightarrow}} CH_3(CH_2)_{10}CH_2 \longrightarrow SO_3^- Na^+$$

Dodecylbenzene   Sodium 4-dodecylbenzenesulfonate
                              (an anionic detergent)

The product is mixed with builders and spray dried to give a smooth, flowing powder. The most common builder is sodium silicate. Alkylbenzenesulfonate detergents were introduced in the late 1950s, and today they command close to 90% of the market once held by natural soaps.

Among the most common additives to detergent preparations are foam stabilizers, bleaches, and optical brighteners. A common foam stabilizer added to liquid soaps, but not laundry detergents (for obvious reasons: think of a top-loading washing machine with foam spewing out of the lid!), is the amide prepared from dodecanoic acid (lauric acid) and 2-aminoethanol (ethanolamine). The most common bleach is sodium perborate tetrahydrate, which decomposes at temperatures above 50°C to give hydrogen peroxide, the actual bleaching agent:

$$CH_3(CH_2)_{10}\overset{\overset{\displaystyle O}{\|}}{C}NHCH_2CH_2OH \qquad O{=}B{-}O{-}O^-Na^+ \cdot 4H_2O$$

N-(2-Hydroxyethyl)dodecanamide   Sodium perborate tetrahydrate
      (a foam stabilizer)   (a bleach)

Also added to laundry detergents are optical brighteners (optical bleaches). These substances are absorbed into fabrics and, after absorbing ambient light, fluoresce with a blue color, offsetting the yellow color caused by fabric as it ages. Optical brighteners produce a "whiter-than-white" appearance. You most certainly have observed their effects if you have seen the glow of white T-shirts or blouses when they are exposed to black light (UV radiation).

21.4 PHOSPHOLIPIDS

A. Structure

Phospholipids, or phosphoacylglycerols, as they are more properly named, are the second most abundant group of naturally occurring lipids. They are found almost exclusively in plant and animal membranes, which typically consist of about 40% to 50% phospholipids and 50% to 60% proteins. The most abundant phospholipids are derived from a phosphatidic acid (Figure 21.5).

The fatty acids that are most common in phosphatidic acids are palmitic and stearic acids (both fully saturated) and oleic acid (with one double bond in the hydrocarbon chain). Further esterification of a phosphatidic acid with a low-molecular-weight alcohol gives a phospholipid. Several of the most common alcohols found in phospholipids are given in Table 21.3.

B. Lipid Bilayers

Figure 21.6 shows a space-filling model of a lecithin (a phosphatidycholine). It and other phospholipids are elongated, almost rodlike molecules, with the nonpolar (hydrophobic) hydrocarbon chains lying roughly parallel to one another and the polar (hydrophilic) phosphoric ester group pointing in the opposite direction.

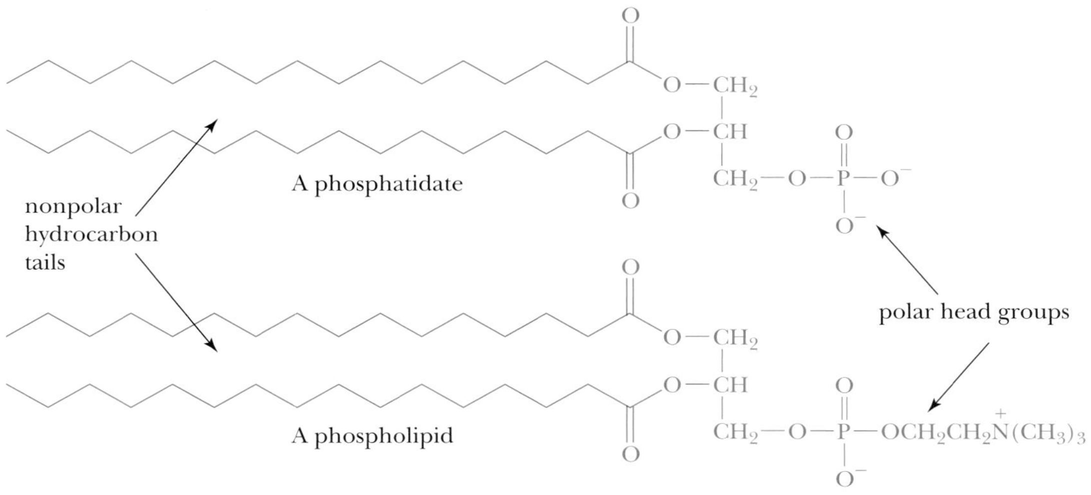

A phosphatidate

nonpolar hydrocarbon tails

polar head groups

A phospholipid

Figure 21.5
A phosphatidic acid and a phospholipid. In a phosphatidic acid, glycerol is esterified with two molecules of fatty acid and one molecule of phosphoric acid. Further esterification of the phosphoric acid group with a low-molecular-weight alcohol gives a phospholipid. Each structural formula shows all functional groups as they are ionized at pH 7.4, the approximate pH of blood plasma and of many biological fluids. Under these conditions, each phosphate group bears a negative charge and each amino group bears a positive charge.

TABLE 21.3 Low-Molecular-Weight Alcohols Most Common to Phospholipids

Alcohols Found in Phospholipids		
Structural Formula	Name	Name of Phospholipid
$HOCH_2CH_2NH_2$	ethanolamine	phosphatidylethanolamine (cephalin)
$HOCH_2CH_2\overset{+}{N}(CH_3)_3$	choline	phosphatidylcholine (lecithin)
$HOCH_2CHCOO^-$ $\quad\quad NH_3^+$	serine	phosphatidylserine
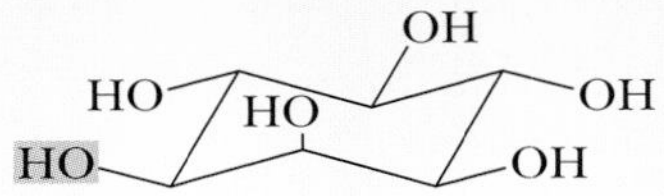 inositol structure	inositol	phosphatidylinositol

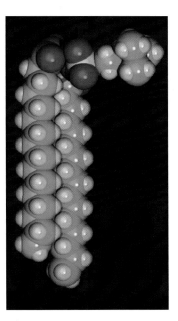

Figure 21.6
Space-filling model of a lecithin. (*Brent Iverson, Univeristy of Texas*)

Lipid bilayer A back-to-back arrangement of phospholipid monolayers.

Placed in aqueous solution, phospholipids spontaneously form a **lipid bilayer** (Figure 21.7) in which polar head groups lie on the surface, giving the bilayer an ionic coating. Nonpolar hydrocarbon chains of fatty acids lie buried within the bilayer. This self-assembly of phospholipids into a bilayer is a spontaneous process, driven by two types of noncovalent forces:

1. hydrophobic effects, which result when nonpolar hydrocarbon chains cluster together and exclude water molecules, and
2. electrostatic interactions, which result when polar head groups interact with water and other polar molecules in the aqueous environment.

CHEMICAL CONNECTIONS 21A

Snake Venom Phospholipases

The venoms of certain snakes contain enzymes called phospholipases that catalyze the hydrolysis of carboxylic ester bonds of phospholipids. The venom of the eastern diamondback rattlesnake (*Crotalus adamanteus*) and that of the Indian cobra (*Naja naja*) both contain phospholipase PLA$_2$, which catalyzes the hydrolysis of esters at carbon 2 of phospholipids. The breakdown product of this hydrolysis, a lysolecithin, acts as a detergent and dissolves the membranes of red blood cells, causing them to rupture:

$$CH_2CH_2NH_3^+$$
$$O$$
$$O=P-O^-$$
$$O$$
$$CH_2-CH-CH_2 \quad + H_2O \xrightarrow{PLA_2}$$
$$O \quad O$$
$$O=C \quad O=C$$
$$R_1 \quad R_2$$

PLA$_2$ catalyzes hydrolysis of this ester bond

A phospholipid

$$CH_2CH_2NH_3^+$$
$$O$$
$$O=P-O^-$$
$$O$$
$$CH_2-CH-CH_2 \quad + \quad R_2-C-O^- + H^+$$
$$O \quad OH \qquad\qquad O$$
$$O=C$$
$$R_1$$

A lysolecithin

Indian cobras kill several thousand people each year.

Milking an Indian Cobra for its venom. (*Dan McCoy/Rainbow*)

Recall from Section 21.3B that the formation of soap micelles is driven by these same noncovalent forces. The polar (hydrophilic) carboxylate groups of soap molecules lie on the surface of the micelle and associate with water molecules, and the nonpolar (hydrophobic) hydrocarbon chains cluster within the micelle and thus are removed from contact with water.

The arrangement of hydrocarbon chains in the interior of a phospholipid bilayer varies from rigid to fluid, depending on the degree of unsaturation of the chains themselves. Saturated hydrocarbon chains tend to lie parallel and closely packed, making the bilayer rigid. Unsaturated hydrocarbon chains, by contrast, have one or more cis double bonds, which cause "kinks" in the chains, so they pack neither as closely nor as orderly as saturated chains. The disordered packing of unsaturated hydrocarbon chains leads to fluidity of the bilayer.

Biological membranes are made of lipid bilayers. The most satisfactory model for the arrangement of phospholipids, proteins, and cholesterol in plant and animal membranes is the **fluid-mosaic model** proposed in 1972 by S. J. Singer and G. Nicolson. The term "mosaic" signifies that the various components in the membrane

Fluid-mosaic model A model of a biological membrane consisting of a phospholipid bilayer, with proteins, carbohydrates, and other lipids embedded in, and on the surface of, the bilayer.

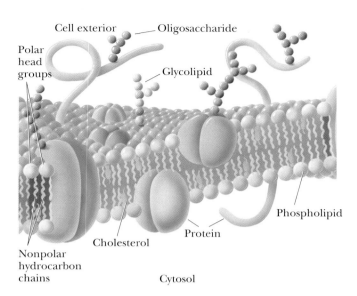

Figure 21.7
Fluid-mosaic model of a biological membrane, showing the lipid bilayer with membrane proteins oriented on the inner and outer surfaces of the membrane, and penetrating the entire thickness of the membrane.

coexist side by side, as discrete units, rather than combining to form new molecules or ions. "Fluid" signifies that the same sort of fluidity exists in membranes that we have already seen in lipid bilayers. Furthermore, the protein components of membranes "float" in the bilayer and can move laterally along the plane of the membrane.

21.5 STEROIDS

Steroids are a group of plant and animal lipids that have the tetracyclic ring system shown in Figure 21.8.

The features common to the tetracyclic ring system of most naturally occurring steroids are illustrated in Figure 21.9.

1. The fusion of the rings is trans, and each atom or group at a ring junction is axial. (Compare, for example, the orientations of —H at carbon 5 and —CH_3 at carbon 10.)

2. The pattern of atoms or groups along the points of ring fusion (carbons 5 to 10 to 9 to 8 to 14 to 13) is nearly always trans–anti–trans–anti–trans.

3. Because of the trans–anti–trans–anti–trans arrangement of atoms or groups along the points of ring fusion, the tetracyclic steroid ring system is nearly flat and quite rigid.

4. Many steroids have axial methyl groups at carbon 10 and carbon 13 of the tetracyclic ring system.

Steroid A plant or animal lipid having the characteristic tetracyclic ring structure of the steroid nucleus, namely, three six-membered rings and one five-membered ring.

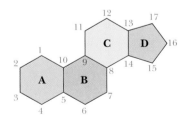

Figure 21.8
The tetracyclic ring system characteristic of steroids.

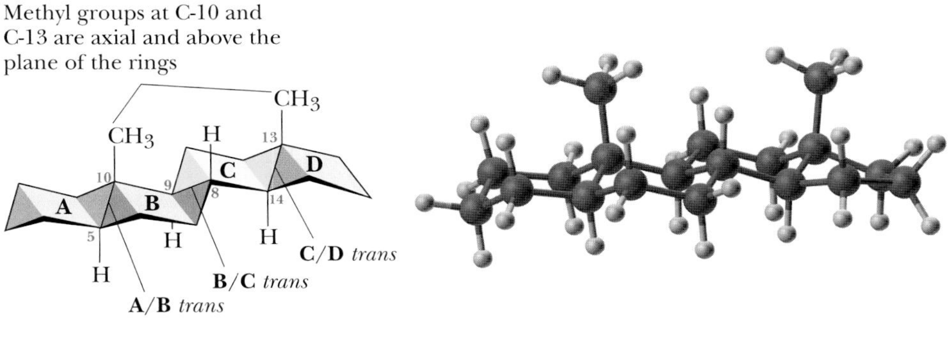

Methyl groups at C-10 and C-13 are axial and above the plane of the rings

Figure 21.9
Features common to the tetracyclic ring system of many steroids.

A. Structure of the Major Classes of Steroids

Cholesterol

Cholesterol, a white, water-insoluble, waxy solid found in blood plasma and in all animal tissues, is an integral part of human metabolism in two ways:

1. It is an essential component of biological membranes. The body of a healthy adult contains approximately 140 g of cholesterol, about 120 g of which is present in membranes. Membranes of the central and peripheral nervous systems, for example, contain about 10% cholesterol by weight.

2. It is the compound from which sex hormones, adrenocorticoid hormones, bile acids, and vitamin D are synthesized. Thus, cholesterol is, in a sense, the parent steroid.

Cholesterol has eight stereocenters and a molecule with this structural feature can exist as 2^8, or 256, stereoisomers (128 pairs of enantiomers). Only the stereoisomer shown here on the right is present in human metabolism:

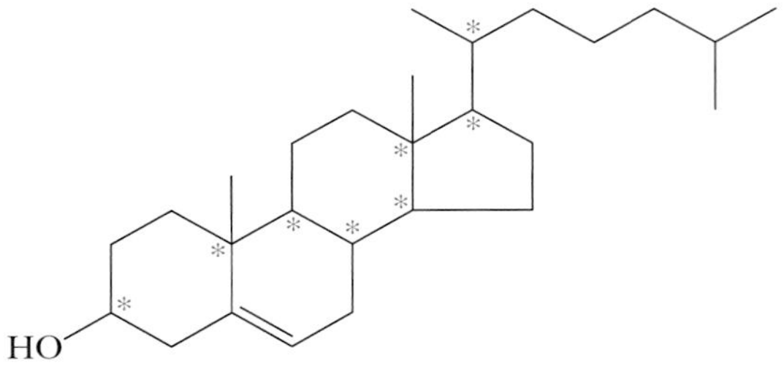

Cholesterol has eight stereocenters; 256 stereoisomers are possible

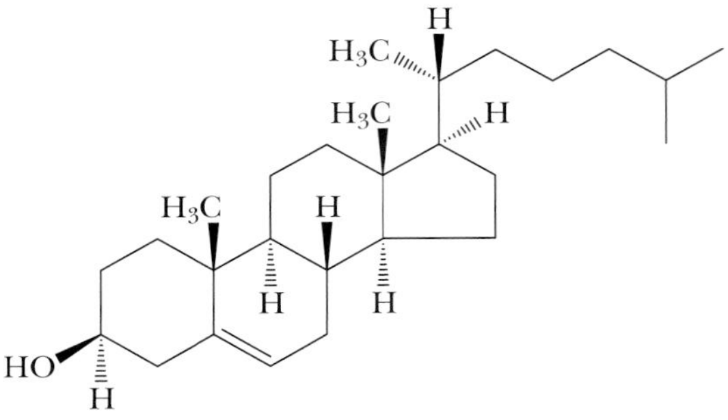

This is the stereoisomer found in human metabolism

Low-density lipoprotein (LDL) Plasma particles, of density 1.02–1.06 g/mL, consisting of approximately 25% proteins, 50% cholesterol, 21% phospholipids, and 4% triglycerides.

High-density lipoprotein (HDL) Plasma particles, of density 1.06–1.21 g/mL, consisting of approximately 33% proteins, 30% cholesterol, 29% phospholipids, and 8% triglycerides.

Estrogen A steroid hormone, such as estradiol, that mediates the development and sexual characteristics of females.

Androgen A steroid hormone, such as testosterone, that mediates the development and sexual characteristics of males.

Cholesterol is insoluble in blood plasma, but can be transported as a plasma-soluble complex formed by cholesterol with proteins called lipoproteins. **Low-density lipoproteins** (**LDL**) transport cholesterol from the site of its synthesis in the liver to the various tissues and cells of the body where it is to be used. It is primarily cholesterol associated with LDLs that builds up in atherosclerotic deposits in blood vessels. **High-density lipoproteins** (**HDL**) transport excess and unused cholesterol from cells back to the liver to be degraded to bile acids and eventual excretion in the feces. It is thought that HDLs retard or reduce atherosclerotic deposits.

Steroid Hormones

Shown in Table 21.4 are representatives of each major class of steroid hormones, along with their principal functions. Female steroid sex hormones are called **estrogens**, and male steroid sex hormones are called **androgens**.

After scientists understood the role of progesterone in inhibiting ovulation, they realized its potential as a contraceptive. Progesterone itself is relatively ineffective when taken orally. As a result of a massive research program in both industrial and academic laboratories, many synthetic progesterone-mimicking steroids became available in the 1960s. Taken regularly, these drugs prevent ovulation yet allow women to maintain a normal menstrual cycle. Some of the most effective preparations contain a progesterone analog, such as norethindrone, combined with

TABLE 21.4 Selected Steroid Hormones

Structure	Source and Major Effects

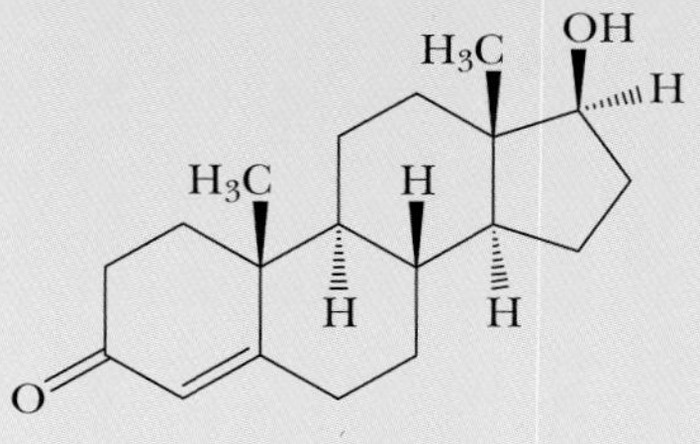

Testosterone

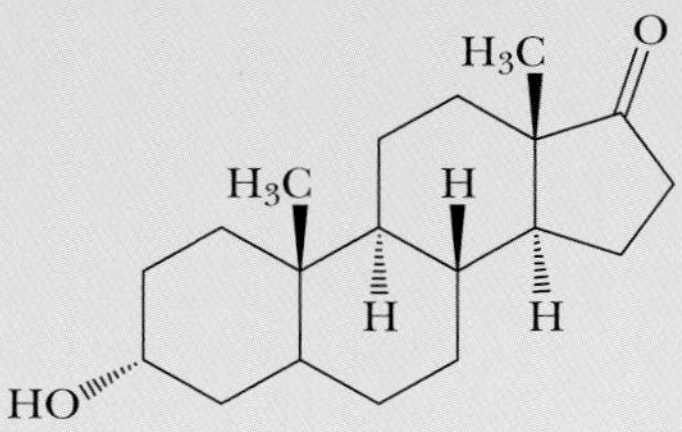

Androsterone

androgens (male sex hormones): synthesized in the testes; responsible for development of male secondary sex characteristics

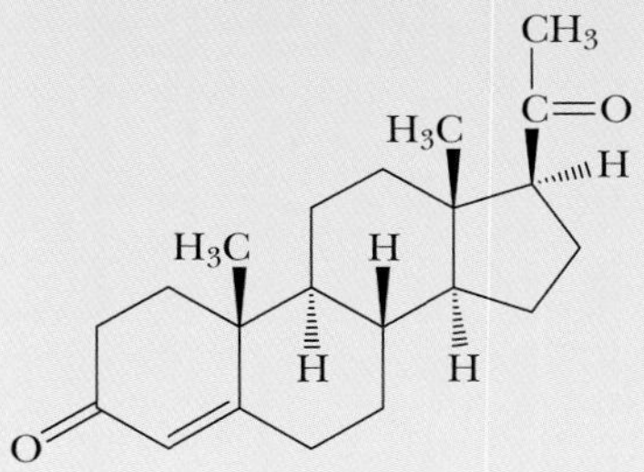

Progesterone

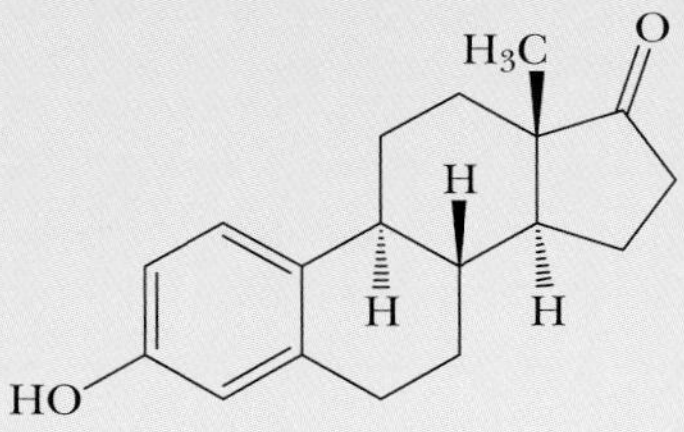

Estrone

estrogens (female sex hormones): synthesized in the ovaries; responsible for development of female secondary sex characteristics and control of the menstrual cycle

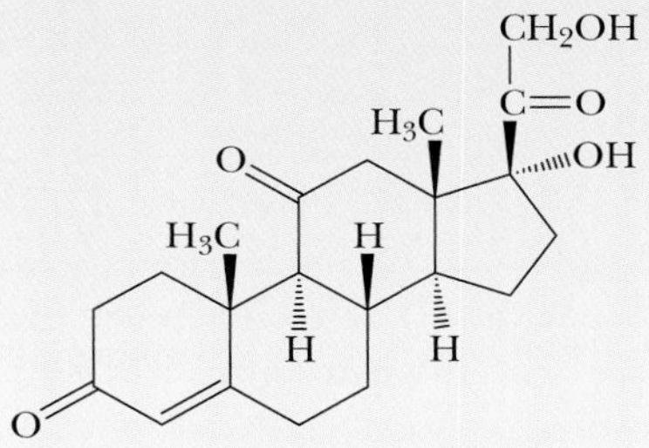

Cortisone

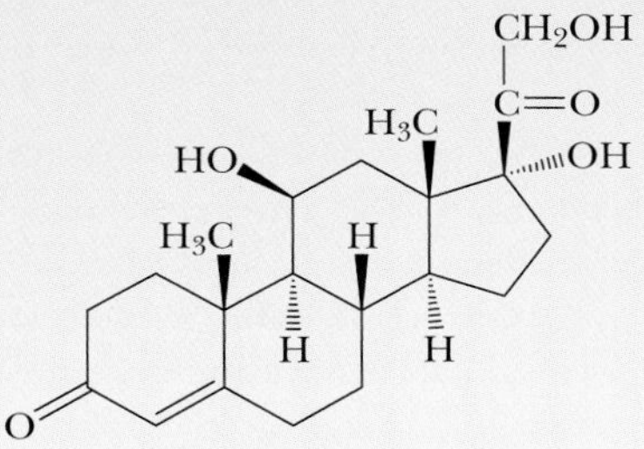

Cortisol

glucocorticoid hormones: synthesized in the adrenal cortex; regulate metabolism of carbohydrates, decrease inflammation, and involved in the reaction to stress

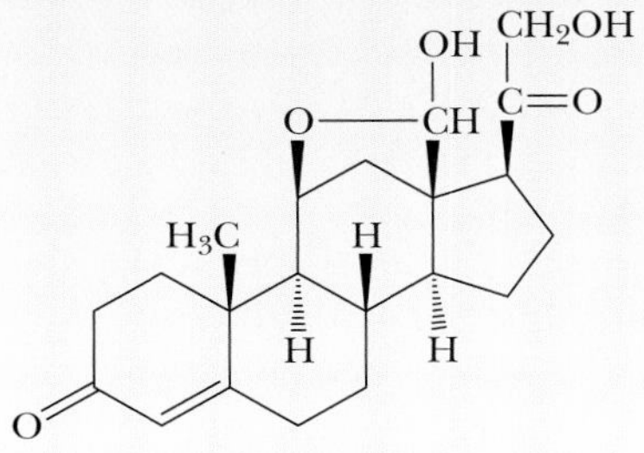

Aldosterone

a mineralocorticoid hormone: synthesized in the adrenal cortex; regulates blood pressure and volume by stimulating the kidneys to absorb Na^+, Cl^-, and HCO_3^-

a smaller amount of an estrogenlike material to help prevent irregular menstrual flow during the prolonged use of contraceptive pills.

"Nor" refers to the absence of a methyl group here. The methyl group is present in ethindrone.

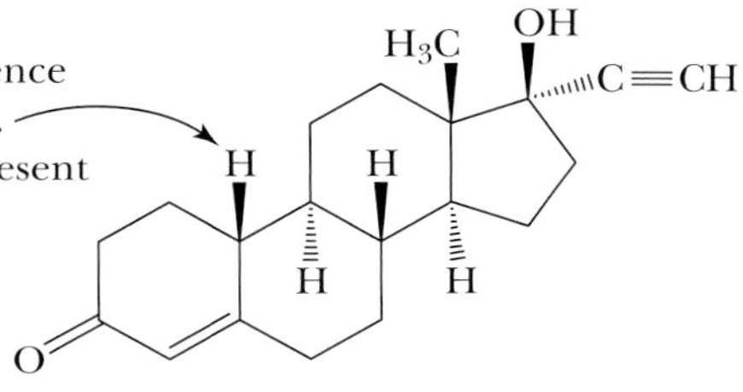

Norethindrone
(a synthetic progesterone analog)

The chief function of testosterone and other androgens is to promote the normal growth of the male reproductive organs (the primary sex characteristics) and the development of the male's characteristic deep voice, pattern of body and facial hair, and musculature (secondary sex characteristics). Although testosterone produces these effects, it is not active when taken orally because it is metabolized in the liver to an inactive steroid. A number of oral **anabolic steroids** have been developed for use in rehabilitation medicine, particularly when muscles atrophy during one's recovery from an injury. Examples include the following compounds:

Anabolic steroid A steroid hormone, such as testosterone, that promotes tissue and muscle growth and development.

Methandrostenolone   Nandrolone   Methandriol

Among certain athletes, the misuse of anabolic steroids to build muscle mass and strength, particularly for sports that require explosive action, is common. The risks associated with abusing anabolic steroids for this purpose are enormous: heightened aggressiveness, sterility, impotence, and premature death from complications of diabetes, coronary artery disease, and liver cancer.

Bile Acids

Figure 21.10 shows a structural formula for cholic acid, a constituent of human bile. The molecule is shown as an anion, as it is ionized in bile and intestinal fluids. **Bile acids**, or, more properly, bile salts, are synthesized in the liver, stored in the gallbladder, and secreted into the intestine, where their function is to emulsify dietary fats and thereby aid in their absorption and digestion. Furthermore, bile salts are the end products of the metabolism of cholesterol and, thus, are a principal pathway for the elimination of that substance from the body. A characteristic structural feature of bile salts is a cis fusion of rings A/B.

Bile acid A cholesterol-derived detergent molecule, such as cholic acid, that is secreted by the gallbladder into the intestine to assist in the absorption of dietary lipids.

B. Biosynthesis of Cholesterol

The biosynthesis of cholesterol illustrates a point we first made in our introduction to the structure of terpenes (Section 4.5): In building large molecules, one of the common patterns in the biological world is to begin with one or more smaller subunits, which are then joined by an iterative process and chemically modified by oxidation, reduction, cross-linking, addition, elimination, or related processes to give a biomolecule with a unique identity.

The building block from which all carbon atoms of steroids are derived is the two-carbon acetyl group of acetyl-CoA. Konrad Bloch of the United States and

Figure 21.10
Cholic acid, an important constituent of human bile.

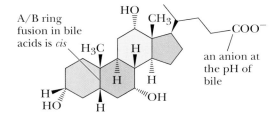

A/B ring fusion in bile acids is *cis*

an anion at the pH of bile

The eicosanoids are extremely widespread, and members of this family of compounds have been isolated from almost every tissue and body fluid.

Leukotrienes are derived from arachidonic acid and are found primarily in leukocytes (white blood cells). Leukotriene C_4 (LTC$_4$), a typical member of the family, has three conjugated double bonds (hence the suffix -*triene*) and contains the amino acids L-cysteine, glycine, and L-glutamic acid (Section 19.2). An important physiological action of LTC$_4$ is the constriction of smooth muscles, especially those of the lungs. The synthesis and release of LTC$_4$ is prompted by allergic reactions. Drugs that inhibit the synthesis of LTC$_4$ show promise for the treatment of the allergic reactions associated with asthma.

Thromboxane A_2, a potent vasoconstrictor, is also synthesized in the body from arachidonic acid. Its release triggers the irreversible phase of platelet aggregation and the constriction of injured blood vessels. It is thought that aspirin and aspirin-like drugs act as mild anticoagulants because they inhibit cyclooxygenase, the enzyme that initiates the synthesis of thromboxane A_2 from arachidonic acid.

21.7 FAT-SOLUBLE VITAMINS

Vitamins are divided into two broad classes on the basis of their solubility: those which are fat soluble (and hence classed as lipids) and those which are water soluble. The fat-soluble vitamins are A, D, E, and K.

A. Vitamin A

Vitamin A, or retinol, occurs only in the animal world, where the best sources are cod-liver oil and other fish-liver oils, animal liver, and dairy products. Vitamin A in the form of a precursor, or provitamin, is found in the plant world in a group of tetraterpene (C_{40}) pigments called carotenes. The most common of these is β-carotene, abundant in carrots, but also found in some other vegetables, particularly yellow and green ones. β-Carotene has no vitamin A activity; however, after ingestion, it is cleaved at the central carbon–carbon double bond to give retinol (vitamin A):

cleavage of this
C=C gives vitamin A

β-Carotene

enzyme-catalyzed
cleavage in the liver

Retinol
(Vitamin A)

Probably the best understood role of vitamin A is its participation in the visual cycle in rod cells. In a series of enzyme-catalyzed reactions retinol undergoes (1) a two-electron oxidation to all-*trans*-retinal and (2) isomerization about the carbon 11

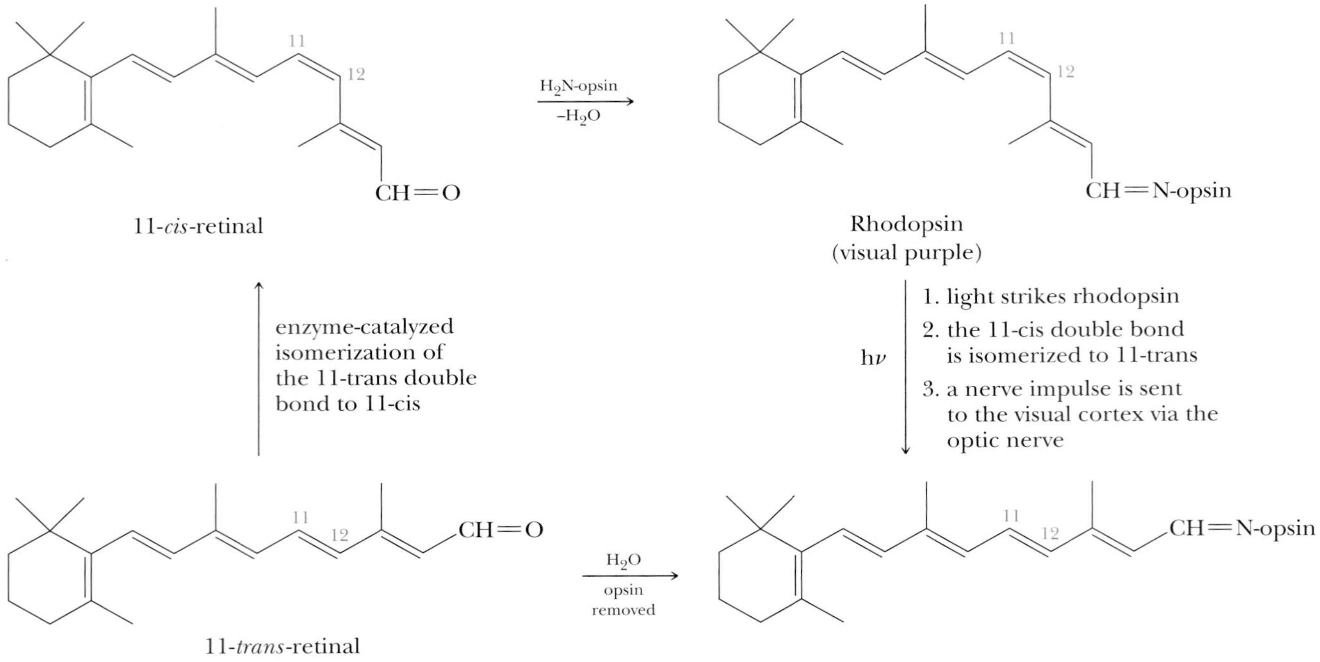

Figure 21.13
The primary chemical reaction of vision in rod cells is the absorption of light by rhodopsin, followed by the isomerization of a carbon–carbon double bond from a cis configuration to a trans configuration.

to carbon 12 double bond to give 11-*cis*-retinal, (3) thereby forming an imine (Section 13.8A) with the —NH_2 from a lysine unit of the protein opsin. The product of these reactions is rhodopsin, a highly conjugated pigment that shows intense absorption in the blue-green region of the visual spectrum.

The primary event in vision is the absorption of light by rhodopsin in rod cells of the retina of the eye to produce an electronically excited molecule. Within several picoseconds (1 picosec = 10^{-12} sec), the excess electronic energy is converted to vibrational and rotational energy, and the 11-cis double bond is isomerized to the more stable 11-trans double bond. The isomerization triggers a conformational change in the protein opsin that causes neurons in the optic nerve to fire, producing a visual image. Coupled with this light-induced change is the hydrolysis of rhodopsin to give 11-*trans*-retinal and free opsin. At this point, the visual pigment is bleached and in a refractory period. Rhodopsin is regenerated by a series of enzyme-catalyzed reactions that converts 11-*trans*-retinal to 11-*cis*-retinal and then to rhodopsin. The visual cycle is shown in abbreviated form in Figure 21.13.

B. Vitamin D

Vitamin D is the name for a group of structurally related compounds that play a major role in regulating the metabolism of calcium and phosphorus. A deficiency of vitamin D in childhood is associated with rickets, a mineral-metabolism disease that leads to bowlegs, knock-knees, and enlarged joints. Vitamin D_3, the most abundant form of the vitamin in the circulatory system, is produced in the skin of mammals by the action of ultraviolet radiation on 7-dehydrocholesterol (cholesterol with a double bond between carbons 7 and 8). In the liver, vitamin D_3 undergoes an enzyme-catalyzed, two-electron oxidation at carbon 25 of the side chain to form 25-hydroxyvitamin D_3; the oxidizing agent is molecular oxygen, O_2. 25-Hydroxyvitamin D_3 undergoes further oxidation in the kidneys, also by O_2, to form 1,25-dihydroxyvitamin D_3, the hormonally active form of the vitamin:

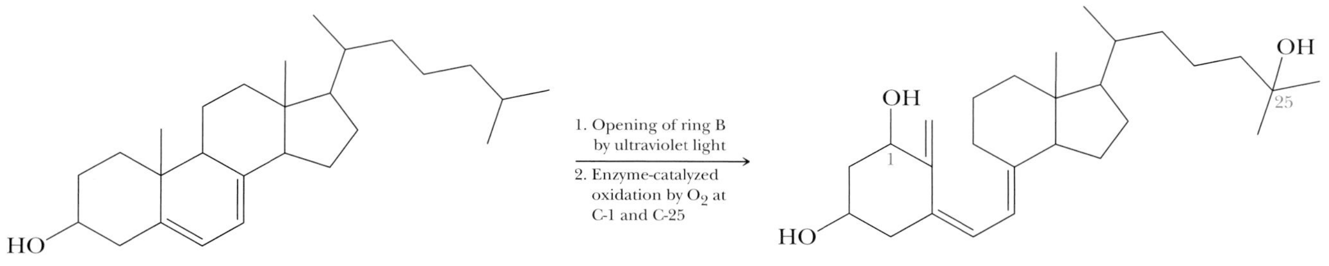

7-Dehydrocholesterol → [1. Opening of ring B by ultraviolet light; 2. Enzyme-catalyzed oxidation by O_2 at C-1 and C-25] → 1,25-Dihydroxyvitamin D_3

C. Vitamin E

Vitamin E was first recognized in 1922 as a dietary factor essential for normal reproduction in rats—hence its name *tocopherol*, from the Greek *tocos*, birth, and *pherein*, to bring about. Vitamin E is a group of compounds of similar structure, the most active of which is α-tocopherol:

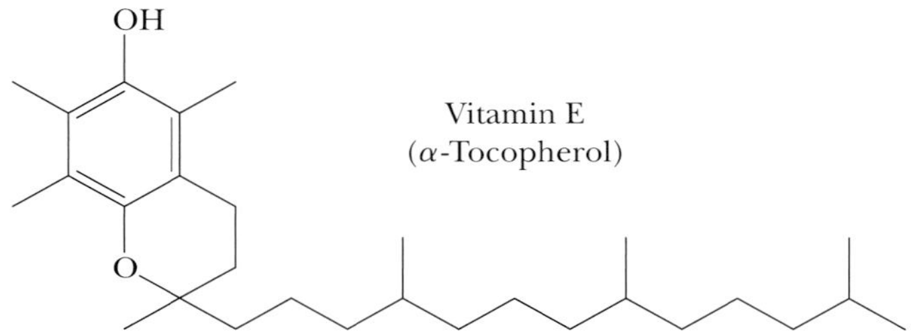

Vitamin E
(α-Tocopherol)

Vitamin E occurs in fish oil; in other oils, such as cottonseed and peanut oil; and in leafy, green vegetables. The richest source is wheat germ oil.

In the body, vitamin E functions as an antioxidant, trapping peroxy radicals of the type HOO• and ROO• formed as a result of enzyme-catalyzed oxidation by molecular oxygen of the unsaturated hydrocarbon chains in membrane phospholipids. There is speculation that peroxy radicals play a role in the aging process and that vitamin E and other antioxidants may retard that process. Vitamin E is also necessary for the proper development and function of the membranes of red blood cells.

D. Vitamin K

The name of this vitamin comes from the German word *koagulation*, signifying its important role in the blood-clotting process. A deficiency of vitamin K results in slowed blood clotting. Natural vitamins of the K family have, for the most part, been replaced in vitamin supplements by synthetic preparations. Menadione, one such synthetic material that exhibits vitamin K activity, has a hydrogen in place of the alkyl chain:

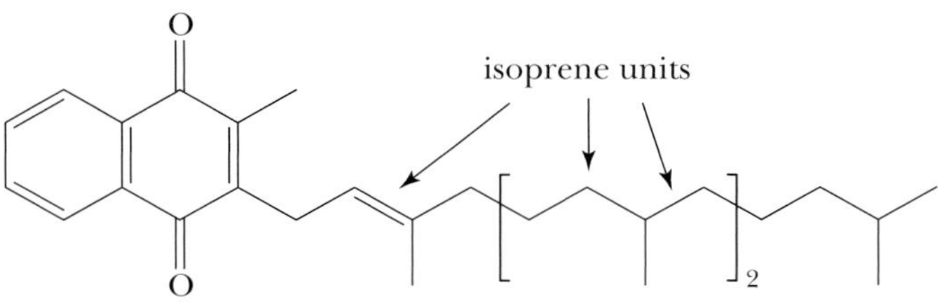

isoprene units

Vitamin K_1

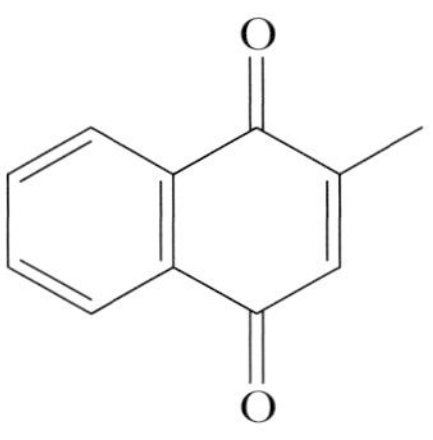

Menadione
(a synthetic vitamin K analog)

Summary

Lipids are a heterogeneous class of compounds grouped together on the basis of their solubility properties; they are insoluble in water and soluble in diethyl ether, acetone, and dichloromethane (Section 21.1). Carbohydrates, amino acids, and proteins are largely insoluble in these organic solvents.

Triglycerides (**triacylglycerols**), the most abundant lipids, are triesters of glycerol and fatty acids (Section 21.2). **Fatty acids** (Section 21.1A) are long-chain carboxylic acids derived from the hydrolysis of fats, oils, and the phospholipids of biological membranes. The melting point of a triglyceride increases as (1) the length of its hydrocarbon chains increases and (2) its degree of saturation increases. Triglycerides rich in saturated fatty acids are generally solids at room temperature; those rich in unsaturated fatty acids are generally oils at room temperature.

Soaps are sodium or potassium salts of fatty acids (Section 21.3). In water, soaps form **micelles**, which "dissolve" nonpolar organic grease and oil. Natural soaps precipitate as water-insoluble salts with Mg^{2+}, Ca^{2+}, and Fe^{3+} ions in hard water. The most common and most widely used **synthetic detergents** are linear alkylbenzenesulfonates.

Phospholipids (Section 21.4), the second most abundant group of naturally occurring lipids, are derived from phosphatidic acids—compounds containing glycerol esterified with two molecules of fatty acid and a molecule of phosphoric acid. Further esterification of the phosphoric acid part with a low-molecular-weight alcohol—most commonly, ethanolamine, choline, serine, or inositol—gives a phospholipid. Placed in aqueous solution, phospholipids spontaneously form **lipid bilayers** (Section 21.4B). According to the **fluid-mosaic model**, membrane phospholipids form lipid bilayers, with membrane proteins associated with the bilayer as both peripheral and integral proteins.

Steroids are a group of plant and animal lipids that have a characteristic tetracyclic structure of three six-membered rings and one five-membered ring (Section 21.5). **Cholesterol**, an integral part of animal membranes, is the compound from which human sex hormones, adrenocorticoid hormones, bile acids, and vitamin D are synthesized. **Low-density lipoproteins** (**LDLs**) transport cholesterol from the site of its synthesis in the liver to tissues and cells where it is to be used. **High-density lipoproteins** (**HDLs**) transport cholesterol from cells back to the liver for its degradation to bile acids and eventual excretion in the feces.

Oral contraceptive pills contain a synthetic progestin (e.g., norethindrone) that prevents ovulation, yet allows women to maintain an otherwise normal menstrual cycle. A variety of synthetic **anabolic steroids** are available for use in rehabilitation medicine to treat muscle tissue that has weakened or deteriorated due to injury. **Bile acids** differ from most other steroids in that they have a cis configuration at the junction of rings A and B.

The carbon skeleton of cholesterol and those of all biomolecules derived from it originate with the acetyl group (a C_2 unit) of **acetyl-CoA** (Section 21.4B).

Prostaglandins are a group of compounds having the 20-carbon skeleton of prostanoic acid (Section 21.6). Prostaglandins are synthesized from phospholipid-bound arachidonic acid (20:4) and other 20-carbon fatty acids in response to physiological triggers.

Vitamin A (Section 21.7A) occurs only in the animal world. The carotenes of the plant world are tetraterpenes (C_{40}) and are cleaved into vitamin A after ingestion. The best-understood role of vitamin A is its participation in the **visual cycle. Vitamin D** (Section 21.7B) is synthesized in the skin of mammals by the action of ultraviolet radiation on 7-dehydrocholesterol. This vitamin plays a major role in the regulation of calcium and phosphorus metabolism. **Vitamin E** (Section 21.7C) is a group of compounds of similar structure, the most active of which is α-tocopherol. In the body, vitamin E functions as an antioxidant. **Vitamin K** (Section 21.7D) is required for the clotting of blood.

Problems

A problem number set in red indicates an applied "real-world" problem.

Fatty Acids and Triglycerides

21.2 Define the term *hydrophobic*.

21.3 Identify the hydrophobic and hydrophilic region(s) of a triglyceride.

21.4 Explain why the melting points of unsaturated fatty acids are lower than those of saturated fatty acids.

21.5 Which would you expect to have the higher melting point, glyceryl trioleate or glyceryl trilinoleate?

21.6 Draw a structural formula for methyl linoleate. Be certain to show the correct configuration of groups about each carbon–carbon double bond.

21.7 Explain why coconut oil is a liquid triglyceride, even though most of its fatty acid components are saturated.

21.8 It is common now to see "contains no tropical oils" on cooking-oil labels, meaning that the oil contains no palm or coconut oil. What is the difference between the composition of tropical oils and that of vegetable oils, such as corn oil, soybean oil, and peanut oil?

21.9 What is meant by the term *hardening* as applied to vegetable oils?

21.10 How many moles of H_2 are used in the catalytic hydrogenation of 1 mole of a triglyceride derived from glycerol, stearic acid, linoleic acid, and arachidonic acid?

21.11 Characterize the structural features necessary to make a good synthetic detergent.

21.12 Following are structural formulas for a cationic detergent and a neutral detergent:

$$CH_3(CH_2)_6CH_2\overset{\overset{\displaystyle CH_3}{|+}}{\underset{\underset{\displaystyle CH_2C_6H_5}{|}}{N}}CH_3 \quad Cl^-$$

$$HOCH_2\overset{\overset{\displaystyle HOCH_2}{|}}{\underset{\underset{\displaystyle HOCH_2}{|}}{C}}CH_2O\overset{\displaystyle O}{\overset{\|}{C}}(CH_2)_{14}CH_3$$

Benzyldimethyloctylammonium chloride   Pentaerythrityl palmitate
(a cationic detergent)   (a neutral detergent)

Account for the detergent properties of each.

21.13 Identify some of the detergents used in shampoos and dishwashing liquids. Are they primarily anionic, neutral, or cationic detergents?

21.14 Show how to convert palmitic acid (hexadecanoic acid) into the following:
   (a) Ethyl palmitate   **(b)** Palmitoyl chloride
   (c) 1-Hexadecanol (cetyl alcohol)   **(d)** 1-Hexadecanamine
   (e) *N*,*N*-Dimethylhexadecanamide

21.15 Palmitic acid (hexadecanoic acid, 16:0) is the source of the hexadecyl (cetyl) group in the following compounds:

Cetylpyridinium chloride   Benzylcetyldimethylammonium chloride

Each compound is a mild surface-acting germicide and fungicide and is used as a topical antiseptic and disinfectant.
   (a) Cetylpyridinium chloride is prepared by treating pyridine with 1-chlorohexadecane (cetyl chloride). Show how to convert palmitic acid to cetyl chloride.
   (b) Benzylcetyldimethylammonium chloride is prepared by treating benzyl chloride with *N*,*N*-dimethyl-1-hexadecanamine. Show how this tertiary amine can be prepared from palmitic acid.

Phospholipids

21.16 Draw the structural formula of a lecithin containing one molecule each of palmitic acid and linoleic acid.

21.17 Identify the hydrophobic and hydrophilic region(s) of a phospholipid.

21.18 The hydrophobic effect is one of the most important noncovalent forces directing the self-assembly of biomolecules in aqueous solution. The hydrophobic effect arises from tendencies of biomolecules (1) to arrange polar groups so that they interact with the aqueous environment by hydrogen bonding and (2) to arrange nonpolar groups so that they are shielded from the aqueous environment. Show how the hydrophobic effect is involved in directing

   (a) the formation of micelles by soaps and detergents.

   (b) the formation of lipid bilayers by phospholipids.

   (c) the formation of the DNA double helix.

21.19 How does the presence of unsaturated fatty acids contribute to the fluidity of biological membranes?

21.20 Lecithins can act as emulsifying agents. The lecithin of egg yolk, for example, is used to make mayonnaise. Identify the hydrophobic part(s) and the hydrophilic part(s) of a lecithin. Which parts interact with the oils used in making mayonnaise? Which parts interact with the water?

Steroids

21.21 Draw the structural formula for the product formed by treating cholesterol (a) with H_2/Pd; (b) with Br_2.

21.22 List several ways in which cholesterol is essential for human life. Why do many people find it necessary to restrict their dietary intake of cholesterol?

21.23 Both low-density lipoproteins (LDL) and high-density lipoproteins (HDL) consist of a core of triacylglycerols and cholesterol esters surrounded by a single phospholipid layer. Draw the structural formula of cholesteryl linoleate, one of the cholesterol esters found in this core.

21.24 Examine the structural formulas of testosterone (a male sex hormone) and progesterone (a female sex hormone). What are the similarities in structure between the two? What are the differences?

21.25 Examine the structural formula of cholic acid, and account for the ability of this bile salt and others to emulsify fats and oils and thus aid in their digestion.

21.26 Following is a structural formula for cortisol (hydrocortisone):

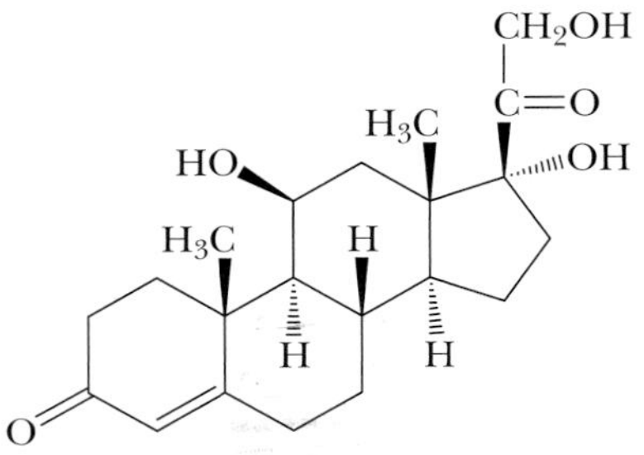

Cortisol
(Hydrocortisone)

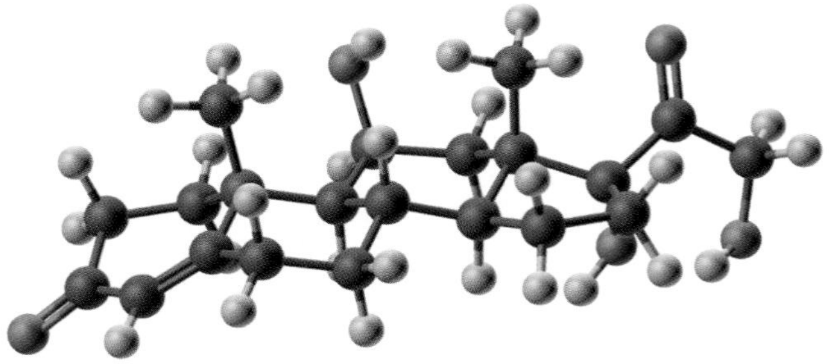

Draw a stereorepresentation of this molecule, showing the conformations of the five- and six-membered rings.

21.27 Because some types of tumors need estrogen to survive, compounds that compete with the estrogen receptor on tumor cells are useful anticancer drugs. The compound tamoxifen is one such drug. To what part of the estrone molecule is the shape of tamoxifen similar?

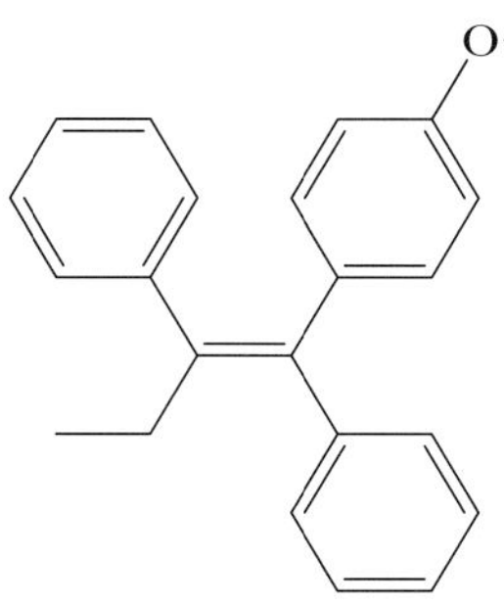

Tamoxifen 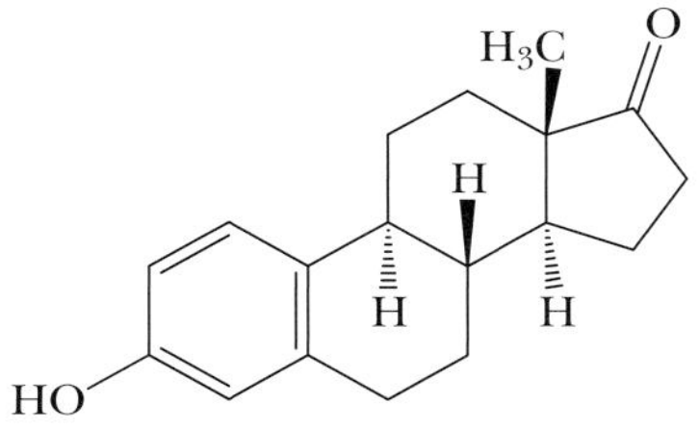   Estrone

Prostaglandins

21.28 Examine the structure of $PGF_{2\alpha}$, and
   (a) identify all stereocenters.
   (b) identify all double bonds about which cis,trans isomerism is possible.
   (c) state the number of stereoisomers possible for a molecule of this structure.

21.29 Following is the structure of unoprostone, a compound patterned after the natural prostaglandins (Section 21.6):

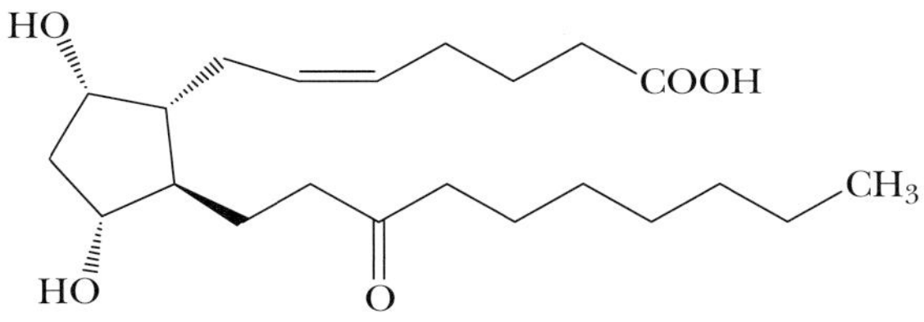

Unoprostone
(antiglaucoma)

Rescula, the isopropyl ester of unoprostone, is an antiglaucoma drug used to treat ocular hypertension. Compare the structural formula of this synthetic prostaglandin with that of $PGF_{2\alpha}$.

Fat-Soluble Vitamins

21.30 Examine the structural formula of vitamin A, and state the number of cis,trans isomers that are possible for this molecule.

21.31 The form of vitamin A present in many food supplements is vitamin A palmitate. Draw the structural formula of this molecule.

21.32 Examine the structural formulas of vitamin A, 1,25-dihydroxy-D_3, vitamin E, and vitamin K_1. Do you expect these vitamins to be more soluble in water or in dichloromethane? Do you expect them to be soluble in blood plasma?

Looking Ahead

21.33 Shown is a glycolipid, a class of lipid that contains a sugar residue:

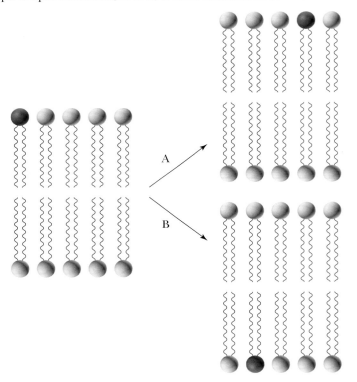

A Glycolipid

Glycolipids are found in cell membranes. (a) Which part of the molecule would you expect to reside on the extracellular side of the membrane? (b) What monosaccharide is bonded to the lipid in this glycolipid?

21.34 How would you expect temperature to affect fluidity in a cell membrane?

21.35 Which type of lipid movement, A or B, is more favorable in cell membranes?

Explain.

21.36 Aspirin works by transferring an acetyl group to the side chain of the 530th amino acid in the protein prostaglandin H_2 synthase-1. Draw the product of this reaction:

22 The Organic Chemistry of Metabolism

These U. S. runners have just won the gold medal in the 4 × 400 m relay race at the 1996 Olympic games. The buildup of lactic acid can cause severe muscle pain. Inset: A model of lactic acid. (*J. O. Atlanta 96/Gamma*)

22.1 INTRODUCTION

We have now studied the structures and typical reactions of the major types of organic functional groups. Further, we have examined the structures of carbohydrates, amino acids and proteins, nucleic acids, and lipids. Now let us apply this background to the study of the organic chemistry of metabolism. In this chapter, we study three key metabolic pathways: glycolysis, the citric acid cycle, and the β-oxidation of fatty acids. The first is a pathway by which glucose is converted to pyruvate and then to acetyl coenzyme. The second is a pathway by which the hydrocarbon chains of fatty acids are degraded, two carbons at a time, to acetyl coenzyme A. The third is the pathway by which the carbon skeletons of carbohydrates, fatty acids, and proteins are oxidized to carbon dioxide.

Those of you who go on to courses in biochemistry will undoubtedly study these metabolic pathways in considerable detail, including their role in energy production and conservation, their regulation, and the diseases associated with errors in particular metabolic steps. Our concern in this chapter is more limited. We will see that reactions of these pathways are biochemical equivalents of organic functional group reactions we have already studied in detail. For instance, we find examples of keto–enol tautomerism; oxidation of an aldehyde to a carboxylic acid; oxidation of a secondary alcohol to a ketone; an aldol reaction and a reverse aldol reaction; a reverse Claisen condensation; and the formation and hydrolysis of esters, imines, thioesters, and mixed anhydrides. In addition, we will use the mechanisms we have studied earlier to give us insights into the mechanisms of these reactions, all of which are enzyme catalyzed.

22.2 KEY PARTICIPANTS IN GLYCOLYSIS, β-OXIDATION, AND THE CITRIC ACID CYCLE

To understand the reactions that play a role in the β-oxidation of fatty acids, the tricarboxylic acid cycle, and glycolysis, we first need to introduce the principal compounds participating in these and a great many other metabolic pathways. Three of these compounds (ATP, ADP, and AMP) are central to the storage and transfer of phosphate groups. Two (NAD^+/ NADH and FAD/FADH$_2$) are **coenzymes** involved in the oxidation–reduction of metabolic intermediates. The final compound (coenzyme A) is an agent for the storage and transfer of acetyl groups.

Coenzyme A low-molecular-weight, nonprotein molecule or ion that binds reversibly to an enzyme, functions as a second substrate for the enzyme, and is regenerated by further reaction.

A. ATP, ADP, and AMP: Agents for the Storage and Transfer of Phosphate Groups

Following is a structural formula of adenosine triphosphate (Section 20.2), a compound involved in the storage and transport of phosphate groups:

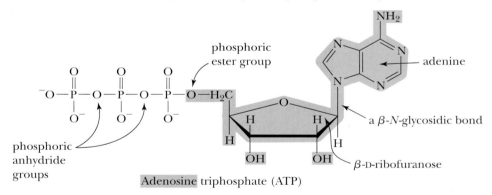

Adenosine triphosphate (ATP)

A building block for ATP, as well as for the other five key participants, is adenosine, which consists of a unit of adenine bonded to a unit of D-ribofuranose by a β-N-glycosidic bond. Three phosphate groups are bonded to the terminal —CH$_2$OH of ribose: one by a phosphoric ester bond, the remaining two by phosphoric anhydride bonds. Hydrolysis of the terminal phosphate group of ATP gives ADP. In the following abbreviated structural formulas, adenosine and its single phosphoric ester group are represented by the symbol AMP (adenosine monophosphate):

$$\text{$^-$O}\overset{\overset{\displaystyle O}{\|}}{\underset{\underset{\displaystyle O^-}{|}}{P}}\text{—O—}\overset{\overset{\displaystyle O}{\|}}{\underset{\underset{\displaystyle O^-}{|}}{P}}\text{—O—AMP} \quad + \quad H_2O \quad \longrightarrow \quad \text{$^-$O}\overset{\overset{\displaystyle O}{\|}}{\underset{\underset{\displaystyle O^-}{|}}{P}}\text{—O—AMP} \quad + \quad H_2PO_4^-$$

| Adenosine triphosphate (ATP) | Water (a phosphate acceptor) | Adenosine diphosphate (ADP) |

The reaction shown is the hydrolysis of a phosphoric anhydride; the phosphate acceptor is water. In the first two reactions of glycolysis, the phosphate acceptors are —OH groups of glucose and fructose, respectively, and are used to form phosphoric esters of these molecules. In two reactions of glycolysis, ADP is the phosphate acceptor and is converted to ATP.

B. NAD⁺/NADH: Agents for Electron Transfer in Biological Oxidation-Reductions

Nicotinamide adenine dinucleotide (NAD⁺) is one of the central agents for the transfer of electrons in metabolic oxidations and reductions. NAD⁺ is constructed of a unit of ADP, joined by a phosphoric ester bond to the terminal —CH_2OH of β-D-ribofuranose, which is in turn joined to the pyridine ring of nicotinamide by a β-N-glycosidic bond:

Nicotinamide adenine dinucleotide (NAD⁺) A biological oxidizing agent. When acting as an oxidizing agent, NAD⁺ is reduced to NADH.

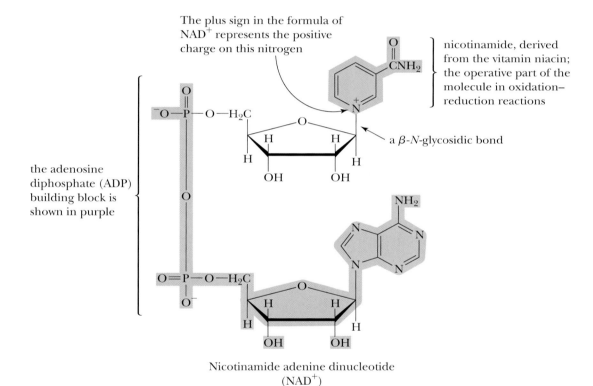

Nicotinamide adenine dinucleotide
(NAD⁺)

When NAD⁺ acts as an oxidizing agent, it is reduced to NADH, which, in turn, is a reducing agent and is oxidized to NAD⁺. In these abbreviated structural formulas, the adenine dinucleotide part of each molecule is represented by the symbol Ad:

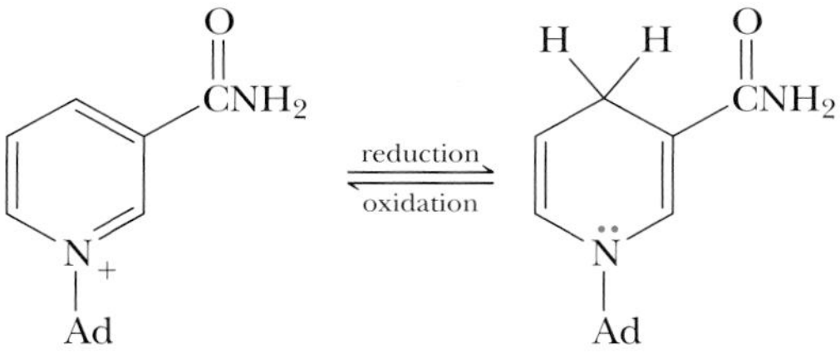

NAD⁺
(oxidized form)

NADH
(reduced form)

NAD$^+$ is involved in a variety of enzyme-catalyzed oxidation–reduction reactions. The three types of oxidations we deal with in this chapter are the following:

- oxidation of a secondary alcohol to a ketone:

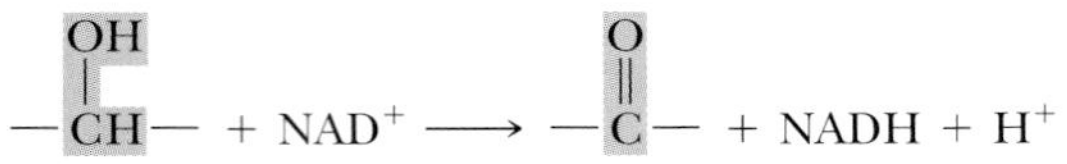

A 2° alcohol   A ketone

- oxidation of an aldehyde to a carboxylic acid:

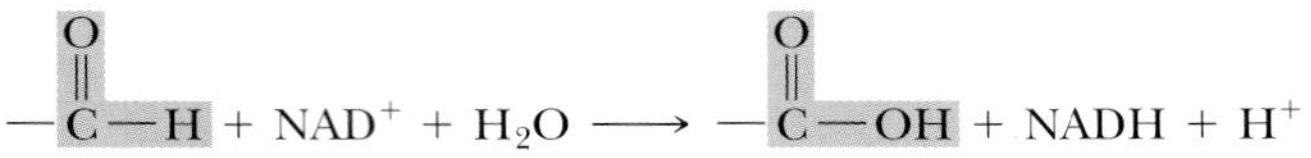

An aldehyde   A carboxylic
                               acid

- oxidation of an α-ketoacid to a carboxylic acid and carbon dioxide:

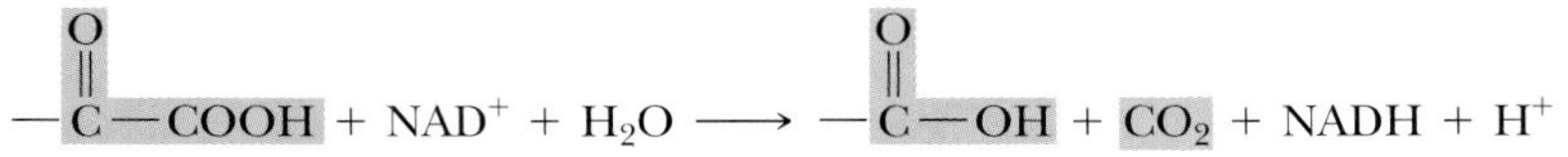

An α-ketoacid   A carboxylic
                               acid

As the mechanism that follows shows, the oxidation of each functional group involves the transfer of a hydride ion to NAD$^+$.

Mechanism: Oxidation of an Alcohol by NAD$^+$

Step 1: A basic group, B$^-$, on the surface of the enzyme removes H$^+$ from the —OH group.

Step 2: Electrons of the H—O sigma bond become the pi electrons of the C=O bond.

Step 3: A hydride ion is transferred from carbon to NAD$^+$ to create a new C—H bond.

Step 4: Electrons within the ring flow to the positively charged nitrogen.

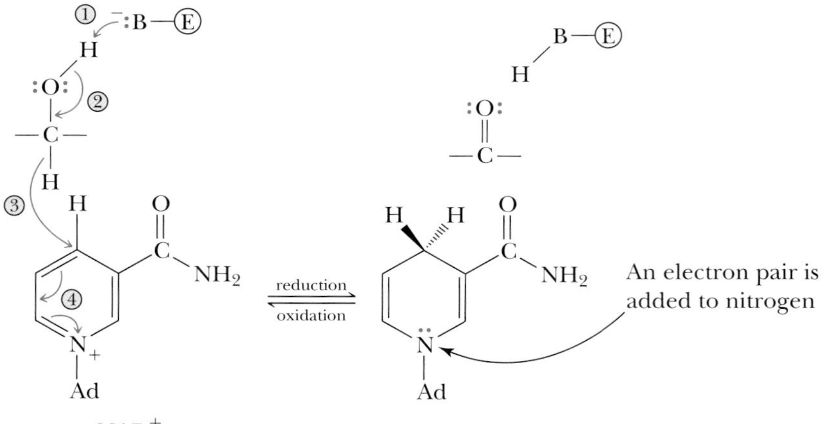

NAD$^+$ NADH

The hydride ion, H:$^-$, which is transferred from the secondary alcohol to NAD$^+$, contains two electrons; thus, NAD$^+$ and NADH function exclusively in two-electron oxidations and two-electron reductions.

C. FAD/FADH₂: Agents for Electron Transfer in Biological Oxidation-Reductions

Flavin adenine dinucleotide (FAD) is also a central component in the transfer of electrons in metabolic oxidations and reductions. In FAD, flavin is bonded to the five-carbon monosaccharide ribitol, which is, in turn, bonded to the terminal phosphate group of ADP:

Flavin adenine dinucleotide (FAD) A biological oxidizing agent. When acting as an oxidizing agent, FAD is reduced to FADH₂.

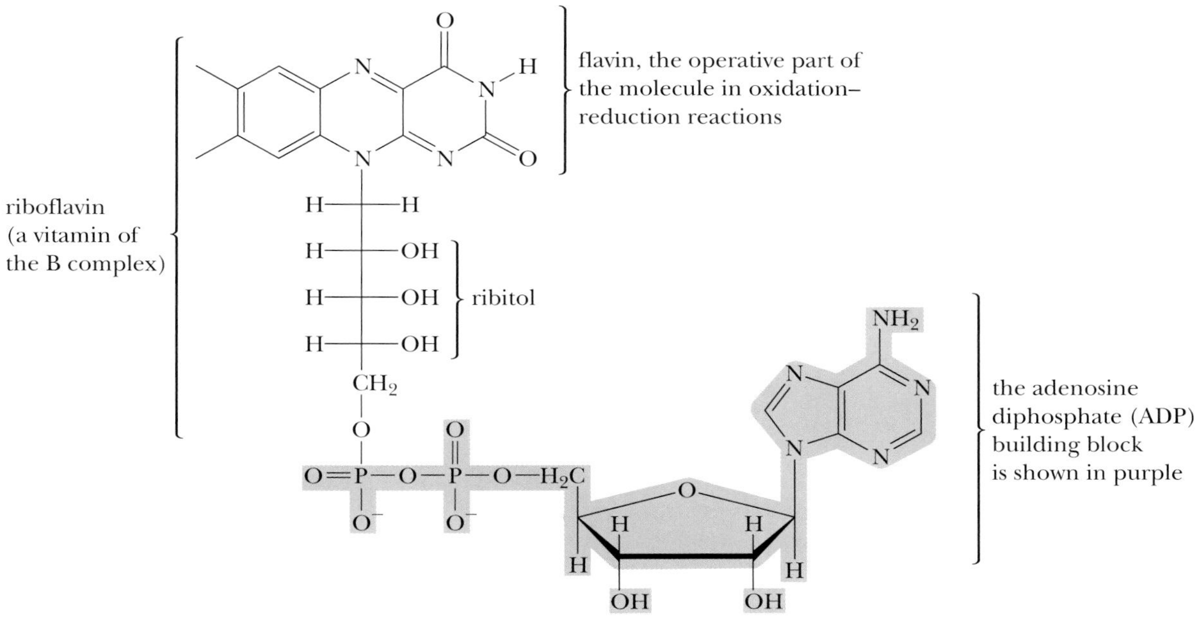

Flavin adenine dinucleotide
(FAD)

FAD participates in several types of enzyme-catalyzed oxidation–reduction reactions. Our concern in this chapter is its role in the oxidation of a carbon–carbon single bond in the hydrocarbon chain of a fatty acid to a carbon–carbon double bond, in the process of which FAD is reduced to FADH₂:

$$-CH_2-CH_2- + FAD \longrightarrow -CH=CH- + FADH_2$$

a portion of the
hydrocarbon
chain of a fatty acid

The mechanism by which FAD oxidizes $-CH_2-CH_2-$ to $-CH=CH-$ involves the transfer of a hydride ion from the hydrocarbon chain of the fatty acid to FAD.

Mechanism: Oxidation of a Fatty Acid $-CH_2-CH_2-$ to $-CH=CH-$ by FAD
The individual curved arrows in this mechanism are numbered 1–6 to help you follow the flow of electrons in the transformation.

Step 1: A basic group, B:⁻, on the surface of the enzyme removes a hydrogen from the carbon adjacent to the carboxyl group.

Step 2: Electrons from this C–H sigma bond become the pi electrons of the new C–C double bond.

Step 3: A hydride ion transfers from the carbon beta, to the carboxyl group, to a nitrogen atom of flavin.

Step 4: The pi electrons within flavin become redistributed.

Step 5: Electrons of the C=N bond remove a hydrogen from the enzyme.

Step 6: A new basic group forms on the surface of the enzyme.

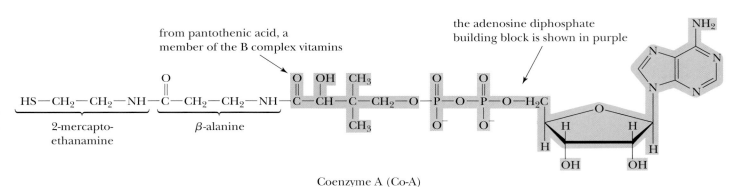

FAD   FADH$_2$

Note that, of the two hydrogen atoms added to FAD to produce FADH$_2$, one comes from the hydrocarbon chain and the other comes from an acidic group on the surface of the enzyme catalyzing this oxidation. Note also that one group on the enzyme functions as a proton acceptor and that another functions as a proton donor.

D. Coenzyme A: The Carrier of Acetyl Groups

Coenzyme A is derived from four subunits. On the left is a two-carbon unit derived from 2-mercaptoethanamine. This unit is in turn joined by an amide bond to the carboxyl group of 3-aminopropanoic acid (β-alanine). The amino group of β-alanine is joined by an amide bond to the carboxyl group of pantothenic acid, a vitamin of the B complex. Finally, the —OH group of pantothenic acid is joined by a phosphoric ester bond to the terminal phosphate group of ADP:

Coenzyme A (Co-A)

A key feature of the structure of coenzyme A is the terminal sulfhydryl (—SH) group. During the degradation of foodstuffs for the production of energy, the carbon skeletons of glucose, fructose, and galactose, along with those of fatty acids, glycerol, and several amino acids, are converted to acetate in the form of a thioester named acetyl coenzyme A, or, more commonly, acetyl-CoA:

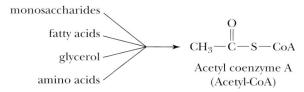

In Section 22.7, we will see how the two-carbon acetyl group is oxidized to carbon dioxide and water by the reactions of the citric acid cycle.

22.3 GLYCOLYSIS

Nearly every living cell carries out glycolysis. Living things first appeared in an environment lacking O_2, and glycolysis was an early and important pathway for extracting energy from nutrient molecules because it takes place in the absence of oxygen. Glycolysis played a central role in anaerobic metabolic processes for the first billion or so years of biological evolution on earth. Modern organisms still employ it to provide precursor molecules for aerobic pathways, such as the citric acid cycle, and as a short-term energy source when the supply of oxygen is limited.

Glycolysis is a series of 10 enzyme-catalyzed reactions that brings about the oxidation of glucose to two molecules of pyruvate. The oxidizing agent is NAD^+. Furthermore, two molecules of ATP are produced for each molecule of glucose oxidized to pyruvate. Following is the net reaction of glycolysis:

Glycolysis From the Greek *glyko*, sweet, and *lysis*, splitting; a series of 10 enzyme-catalyzed reactions by which glucose is oxidized to two molecules of pyruvate.

$$C_6H_{12}O_6 + 2NAD^+ + 2HPO_4{}^{2-} + 2ADP \xrightarrow[\substack{10 \text{ enzyme-} \\ \text{catalyzed steps}}]{\text{glycolysis}} 2CH_3\overset{\overset{\text{O}}{\|}}{C}COO^- + 2H^+ + 2NADH + 2ATP$$

Glucose   Pyruvate

EXAMPLE 22.1

Show that the conversion of glucose to two molecules of pyruvate is an oxidation. (*Hint:* That it is an oxidation is easiest to see if you take the product to be pyruvic acid; recognize, of course, that, under the pH conditions at which this reaction takes place in cells, pyruvic acid is ionized to pyruvate.)

SOLUTION

Glucose is $C_6H_{12}O_6$. Two molecules of pyruvic acid are $2(C_3H_4O_3) = C_6H_8O_6$. The number of O atoms remains the same in this conversion, but four H are lost. Therefore, the conversion of glucose to pyruvate is an oxidation.

Practice Problem 22.1

Under anaerobic (without oxygen) conditions, glucose is converted to lactate by a metabolic pathway called anaerobic glycolysis or, alternatively, lactate fermentation:

$$C_6H_{12}O_6 \xrightarrow[\text{glycolysis}]{\text{anaerobic}} 2CH_3\overset{\overset{\text{OH}}{|}}{C}HCOO^- + 2H^+$$

Glucose   Lactate

Is anaerobic glycolysis a net oxidation, a net reduction, or neither?

22.4 THE 10 REACTIONS OF GLYCOLYSIS

Although writing the net reaction of glycolysis is simple, it took several decades of patient, intensive work by scores of scientists to discover the separate reactions by which glucose is converted to pyruvate. Glycolysis is frequently called the *Embden–Meyerhof pathway*, in honor of the two German biochemists, Gustav Embden and Otto Meyerhof, who contributed so greatly to our present knowledge of it. Figure 22.1 shows the 10 reactions of glycolysis.

Reaction 1: Phosphorylation of α-D-Glucose

The transfer of a phosphate group from ATP to glucose gives α-D-glucose 6-phosphate. This conversion is an example of the reaction of an anhydride with an alcohol to form an ester (Section 15.5B); in this case, a phosphoric anhydride reacts with the primary

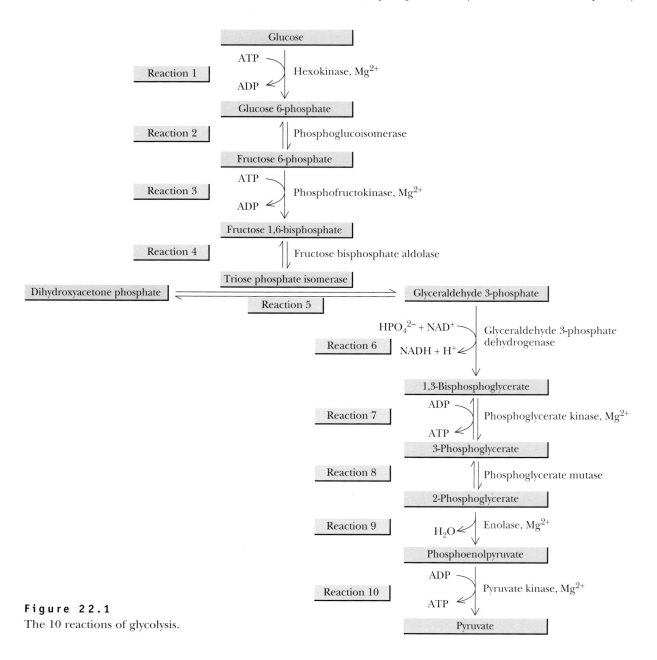

Figure 22.1
The 10 reactions of glycolysis.

alcohol group of glucose to form a phosphoric ester. In Section 15.7, we saw that a more reactive carboxyl functional group can be transformed into a less reactive carboxyl functional group. The same principle applies to functional derivatives of phosphoric acid. In Reaction 1 of glycolysis, a phosphoric anhydride, a more reactive functional group, is converted to a phosphoric ester, a less reactive functional group:

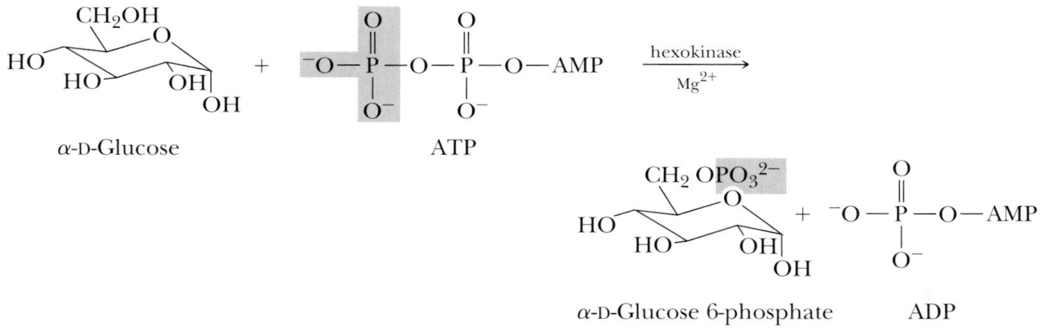

Hexokinase, the enzyme catalyzing this reaction, requires divalent magnesium ion, Mg^{2+}, whose function is to coordinate with two negatively charged oxygens of the terminal phosphate group of ATP and to facilitate attack by the $-OH$ group of glucose on the phosphorus atom of the $P{=}O$ group.

Reaction 2: Isomerization of Glucose 6-Phosphate to Fructose 6-Phosphate

In this reaction, glucose 6-phosphate, an aldohexose, is converted to fructose 6-phosphate, a 2-ketohexose:

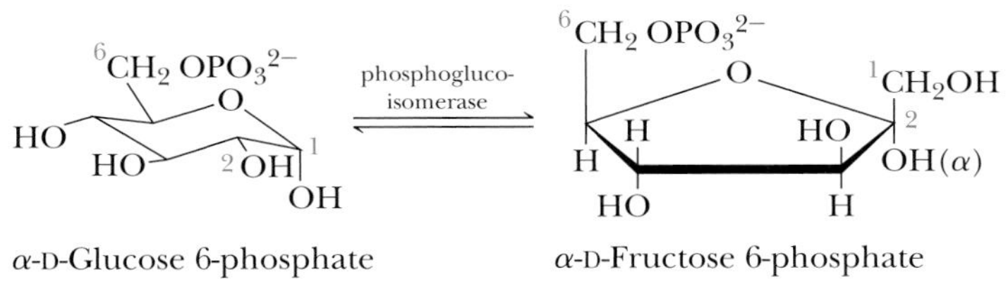

The chemistry involved in this isomerization is easiest to see by considering the open-chain (Fisher projection) forms of these two monosaccharides:

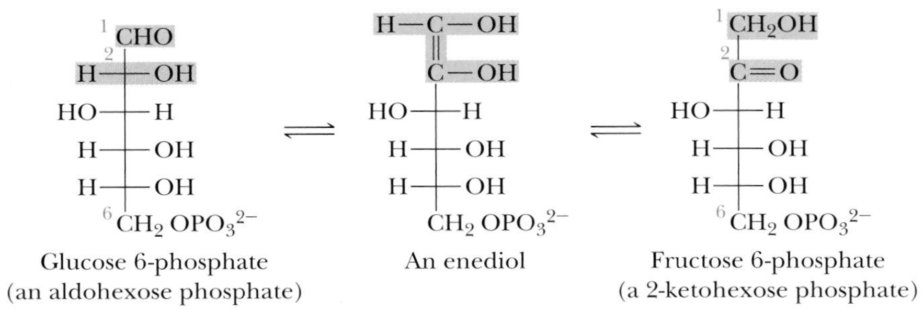

One keto–enol tautomerism forms an enediol; a second then forms the ketone carbonyl group in fructose 6-phosphate. (See Section 13.9A and Problems 13.33 and 13.34.)

Reaction 3: Phosphorylation of Fructose 6-Phosphate

In the third reaction, a second mole of ATP converts fructose 6-phosphate to fructose 1,6-bisphosphate:

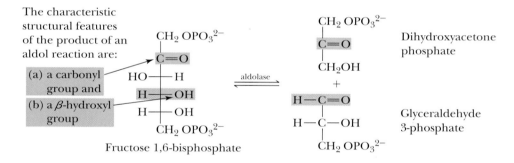

$$\underset{\text{Fructose 6-phosphate}}{\begin{array}{c}CH_2OH\\|\\C=O\\HO-H\\H-OH\\H-OH\\CH_2\,OPO_3^{2-}\end{array}} + ATP \xrightarrow[\text{Mg}^{2+}]{\text{phospho-}\atop\text{fructokinase}} \underset{\text{Fructose 1,6-bisphosphate}}{\begin{array}{c}CH_2\,OPO_3^{2-}\\|\\C=O\\HO-H\\H-OH\\H-OH\\CH_2\,OPO_3^{2-}\end{array}} + ADP$$

Reaction 4: Cleavage of Fructose 1,6-Bisphosphate to Two Triose Phosphates

In the fourth reaction, fructose 1,6-bisphosphate is cleaved to dihydroxyacetone phosphate and glyceraldehyde 3-phosphate by a reaction that is the reverse of an aldol reaction. Recall from Section 16.3 that an aldol reaction takes place between the α-carbon of one carbonyl-containing compound and the carbonyl carbon of another and that the functional group of the product of an aldol reaction is a β-hydroxyaldehyde or ketone:

The characteristic structural features of the product of an aldol reaction are:

(a) a carbonyl group and
(b) a β-hydroxyl group

$$\underset{\text{Fructose 1,6-bisphosphate}}{\begin{array}{c}CH_2\,OPO_3^{2-}\\|\\C=O\\HO-H\\H-OH\\H-OH\\CH_2\,OPO_3^{2-}\end{array}} \underset{\text{aldolase}}{\rightleftharpoons} \begin{array}{c}\underset{\text{Dihydroxyacetone phosphate}}{\begin{array}{c}CH_2\,OPO_3^{2-}\\|\\C=O\\|\\CH_2OH\end{array}}\\+\\\underset{\text{Glyceraldehyde 3-phosphate}}{\begin{array}{c}H-C=O\\|\\H-C-OH\\|\\CH_2\,OPO_3^{2-}\end{array}}\end{array}$$

Reaction 5: Isomerization of Dihydroxyacetone Phosphate to Glyceraldehyde 3-Phosphate

This interconversion of triose phosphates occurs by the same type of keto–enol tautomerism (Section 13.9A) and enediol intermediate we have already seen in the isomerization of glucose 6-phosphate to fructose 6-phosphate:

$$\underset{\substack{\text{Dihydroxyacetone}\\\text{phosphate}}}{\begin{array}{c}CH_2OH\\|\\C=O\\|\\CH_2\,OPO_3^{2-}\end{array}} \rightleftharpoons \underset{\substack{\text{An enediol}\\\text{intermediate}}}{\begin{array}{c}CHOH\\||\\C-OH\\|\\CH_2\,OPO_3^{2-}\end{array}} \rightleftharpoons \underset{\substack{\text{Glyceraldehyde}\\\text{3-phosphate}}}{\begin{array}{c}CHO\\|\\H-OH\\|\\CH_2\,OPO_3^{2-}\end{array}}$$

Reaction 6: Oxidation of the Aldehyde Group of Glyceraldehyde 3-Phosphate

To simplify structural formulas in Reaction 6, glyceraldehyde 3-phosphate is abbreviated G—CHO. Two changes occur in this molecule. First, the aldehyde group is oxidized to a carboxyl group, which is, in turn, converted to a mixed anhydride, and second, the oxidizing agent is NAD^+, which is reduced to NADH:

$$G-\overset{\overset{\textstyle O}{\|}}{C}-H + H_2O + NAD^+ \longrightarrow G-\overset{\overset{\textstyle O}{\|}}{C}-OH + H^+ + NADH$$

Glyceraldehyde   Glyceric acid
3-phosphate   3-phosphate

The reaction is considerably more complicated than might appear from the balanced equation. As the mechanism indicates, it involves (1) the formation of a thiohemiacetal, (2) hydride ion transfer to form a thioester, and (3) the conversion of a thioester to a mixed anhydride.

Mechanism: Oxidation of Glyceraldehyde 3-Phosphate to 1,3-Bisphosphoglycerate

Step 1: Reaction between glyceraldehyde 3-phosphate and a sulfhydryl group of the enzyme gives a thiohemiacetal (Section 13.7):

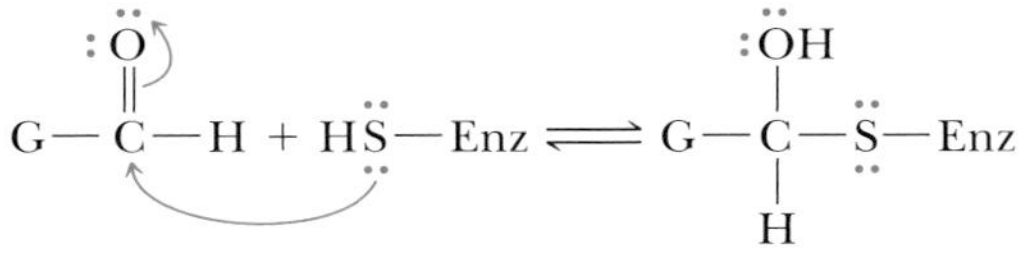

Glyceraldehyde   A thiohemiacetal
3-phosphate

Step 2: Oxidation occurs by the transfer of a hydride ion from the thiohemiacetal to NAD^+:

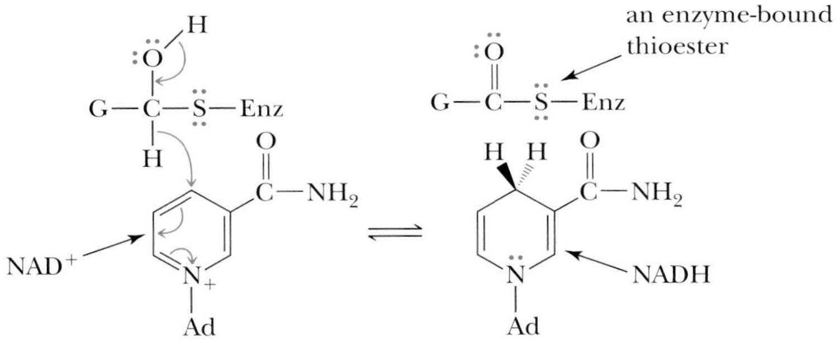

Step 3: Reaction of the thioester with phosphate ion gives a tetrahedral carbonyl addition intermediate, which then collapses to regenerate the enzyme and give a mixed anhydride of phosphoric acid and glyceric acid:

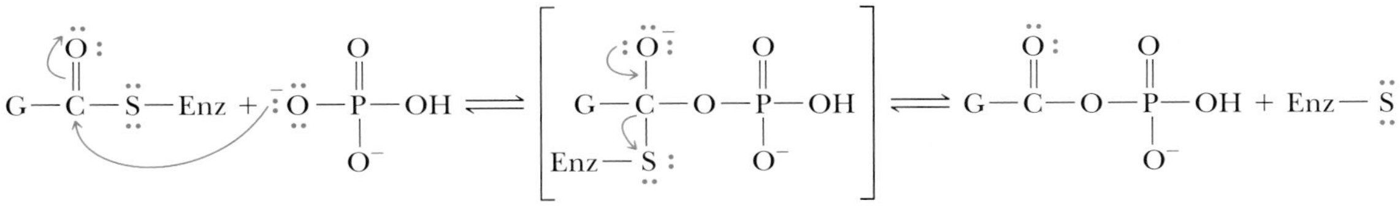

A tetrahedral   1,3-Bisphosphoglycerate
carbonyl addition   (a mixed anhydride)
intermediate

Reaction 7: Transfer of a Phosphate Group from 1,3-Bisphosphoglycerate to ADP

The transfer of a phosphate group in this reaction involves the exchange of one anhydride group for another, namely, the mixed anhydride of 1,3-bisphosphoglycerate for the new phosphoric anhydride in ATP:

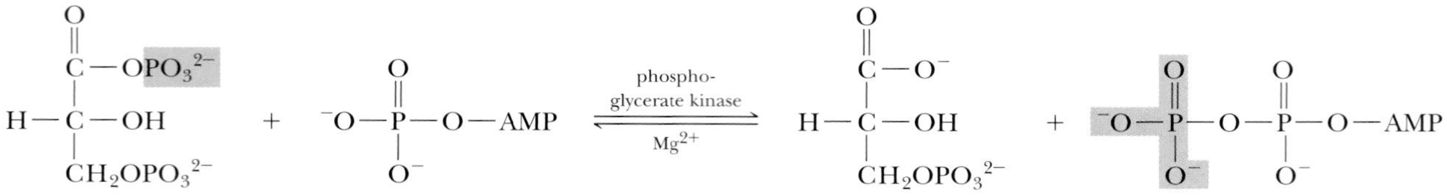

1,3-Bisphospho-
glycerate   ADP   3-Phosphoglycerate   ATP

Reaction 8: Isomerization of 3-Phosphoglycerate to 2-Phosphoglycerate

A phosphate group is transferred from the primary —OH group on carbon 3 to the secondary —OH group on carbon 2:

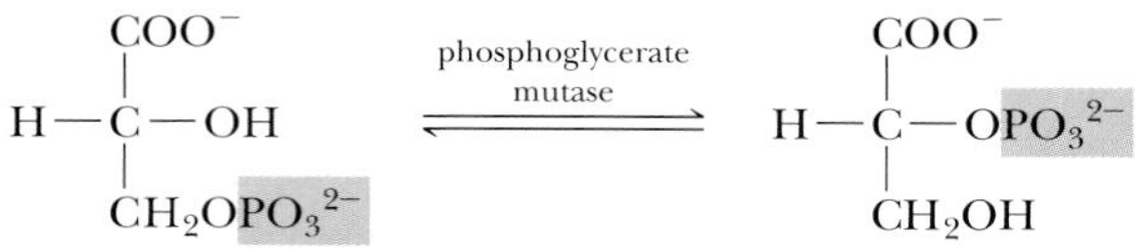

3-Phosphoglycerate   2-Phosphoglycerate

Reaction 9: Dehydration of 2-Phosphoglycerate

Dehydration of the primary alcohol (Section 8.3E) gives phosphoenolpyruvate, which is the ester of phosphoric acid and the enol form of pyruvic acid:

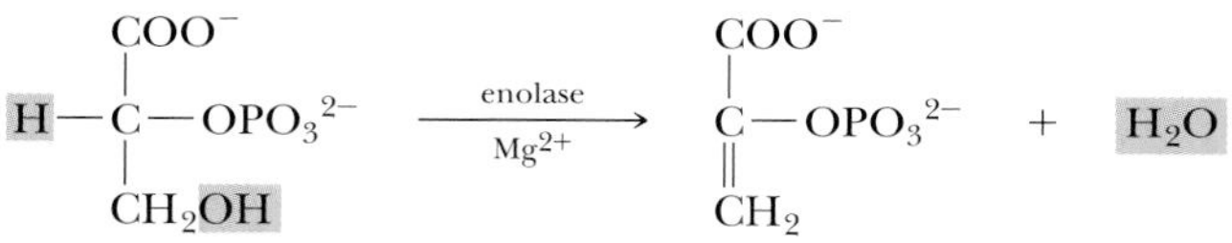

2-Phosphoglycerate   Phosphoenolpyruvate

Reaction 10: Transfer of a Phosphate Group from Phosphoenolpyruvate to ADP

Reaction 10 is divided into two steps: the transfer of a phosphate group to ADP to produce ATP and the conversion of the enol form of pyruvate to its keto form by keto–enol tautomerism (Section 13.9A):

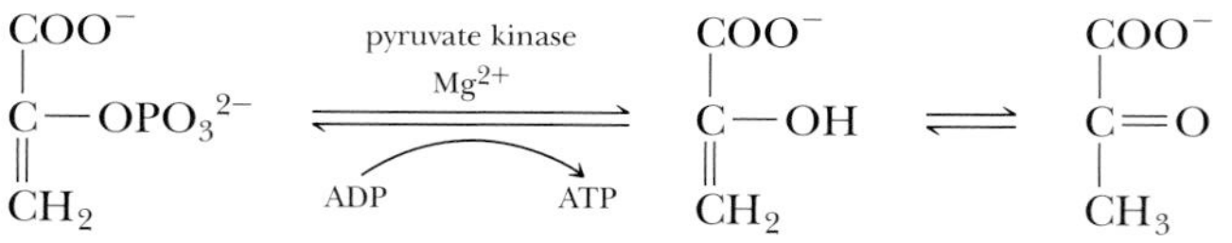

Phosphoenol-
pyruvate   Enol of
pyruvate   Pyruvate

Summing these 10 reactions gives a balanced equation for the net reaction of glycolysis:

$$C_6H_{12}O_6 + 2NAD^+ + 2HPO_4^{2-} + 2ADP \xrightarrow[\text{10 enzyme-catalyzed steps}]{\text{glycolysis}} 2CH_3\overset{\displaystyle O}{\overset{\displaystyle \|}{C}}COO^- + 2H^+ + 2NADH + 2ATP$$

Glucose   Pyruvate

22.5   THE FATES OF PYRUVATE

Pyruvate does not accumulate in cells, but rather undergoes one of three enzyme-catalyzed reactions, depending on the state of oxygenation and the type of cell in which it is produced. A key to understanding the biochemical logic responsible for two of these possible fates of pyruvate is to recognize that this compound is produced by the oxidation of glucose through the reactions of glycolysis. NAD^+ is the oxidizing agent and is reduced to NADH. For glycolysis to continue, there must be a continuing supply of NAD^+; therefore, under anaerobic conditions (in which there is no oxygen present for the reoxidation of NADH), two of the metabolic pathways we describe use pyruvate in ways that regenerate NAD^+.

A. Reduction to Lactate - Lactate Fermentation

In vertebrates, the most important pathway for the regeneration of NAD^+ under anaerobic conditions is the reduction of pyruvate to lactate, catalyzed by the enzyme lactate dehydrogenase:

$$\underset{\text{Pyruvate}}{CH_3\overset{\displaystyle O}{\overset{\|}{C}}COO^- + NADH + H_3O^+} \underset{\text{dehydrogenase}}{\overset{\text{lactate}}{\rightleftharpoons}} \underset{\text{Lactate}}{CH_3\overset{\displaystyle OH}{\overset{|}{C}HCOO^- + NAD^+ + H_2O}}$$

Even though **lactate fermentation** allows glycolysis to continue in the absence of oxygen, it also brings about an increase in the concentration of lactate, and, perhaps more importantly, it increases the concentration of hydronium ion, H_3O^+, in muscle tissue and in the bloodstream. This buildup of lactate and H_3O^+ is associated with muscle fatigue. When blood lactate reaches a concentration of about 0.4 mg/100 mL, muscle tissue becomes almost completely exhausted.

Lactate fermentation A metabolic pathway that converts glucose to two molecules of pyruvate.

EXAMPLE 22.2

Show that glycolysis, followed by the reduction of pyruvate to lactate (lactate fermentation), leads to an increase in the hydrogen ion concentration in the bloodstream.

SOLUTION

Lactate fermentation produces lactic acid, which is completely ionized at pH 7.4, the normal pH of blood plasma. Therefore, the hydronium ion concentration increases:

$$\underset{\text{Glucose}}{C_6H_{12}O_6 + 2H_2O} \xrightarrow[\text{fermentation}]{\text{lactate}} \underset{\text{Lactate}}{2CH_3\overset{\displaystyle OH}{\overset{|}{C}HCOO^- + 2H_3O^+}}$$

Practice Problem 22.2 ─────────────────────────────

Does lactate fermentation result in an increase or decrease in blood pH?

B. Reduction to Ethanol - Alcoholic Fermentation

Yeast and several other organisms have an alternative pathway to regenerate NAD^+ under anaerobic conditions. In the first step of this pathway, pyruvate undergoes enzyme-catalyzed decarboxylation to give acetaldehyde:

$$CH_3\overset{\displaystyle O}{\overset{\|}{C}}COO^- + H_3O^+ \xrightarrow[\text{decarboxylase}]{\text{pyruvate}} CH_3\overset{\displaystyle O}{\overset{\|}{C}}H + CO_2 + H_2O$$

Pyruvate   Acetaldehyde

The carbon dioxide produced in this reaction is responsible for the foam on beer and the carbonation of naturally fermented wines and champagnes. In the second step, acetaldehyde is reduced by NADH to ethanol:

$$CH_3\overset{\displaystyle O}{\overset{\|}{C}}H + NADH + H_3O^+ \xrightarrow[\text{dehydrogenase}]{\text{alcoholic}} CH_3CH_2OH + NAD^+ + H_2O$$

Acetaldehyde   Ethanol

Adding the reactions for the decarboxylation of pyruvate and the reduction of acetaldehyde to the net reaction of glycolysis gives the overall reaction of **alcoholic fermentation**:

Alcoholic fermentation
A metabolic pathway that converts glucose to two molecules of ethanol and two molecules of CO_2.

$$C_6H_{12}O_6 + 2HPO_4{}^{2-} + 2ADP + 2H^+ \xrightarrow[\text{fermentation}]{\text{alcoholic}} 2CH_3CH_2OH + 2CO_2 + 2ATP$$

Glucose   Ethanol

C. Oxidation and Decarboxylation to Acetyl-CoA

Under aerobic conditions, pyruvate undergoes oxidative decarboxylation in which the carboxylate group is converted to carbon dioxide and the remaining two carbons are converted to the acetyl group of acetyl-CoA:

$$CH_3\overset{\displaystyle O}{\overset{\|}{C}}COO^- + NAD^+ + CoASH \xrightarrow[\text{decarboxylation}]{\text{oxidative}} CH_3\overset{\displaystyle O}{\overset{\|}{C}}SCoA + CO_2 + NADH$$

Pyruvate   Acetyl-CoA

The oxidative decarboxylation of pyruvate is considerably more complex than is suggested by the preceding equation. In addition to utilizing NAD^+ and coenzyme A, this transformation also requires FAD, thiamine pyrophosphate (which is derived from thiamine, vitamin B1), and lipoic acid:

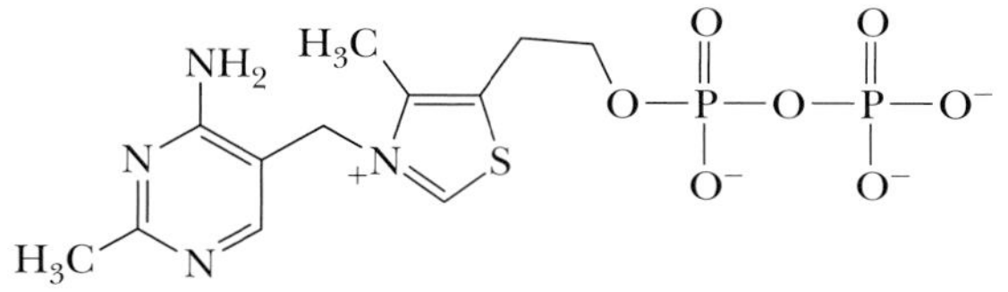

Thiamine pyrophosphate 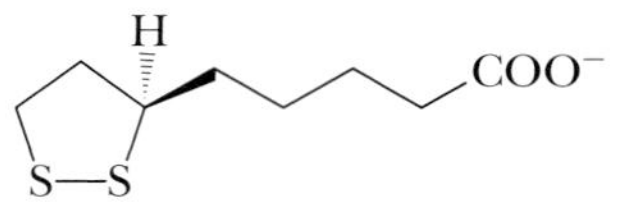   Lipoic acid
(as the carboxylate anion)

Acetyl coenzyme A then becomes a fuel for the citric acid cycle, which results in oxidation of the two-carbon chain of the acetyl group to CO_2, with the production of NADH and $FADH_2$. These reduced coenzymes are, in turn, oxidized to NAD^+ and FAD during respiration, with O_2 as the oxidizing agent.

22.6 β-OXIDATION OF FATTY ACIDS

The first phase in the catabolism of fatty acids involves their release from triglycerides, either those stored in adipose tissue or those ingested from the diet. The hydrolysis of triglycerides is catalyzed by a group of enzymes called lipases.

$$
\underset{\text{A triglyceride}}{\begin{matrix} O & CH_2OCR \\ \| & | \\ RCOCH & O \\ | & \| \\ CH_2OCR \end{matrix}} + 3H_2O \xrightarrow{\text{lipase}} \underset{\substack{\text{1,2,3-Propanetriol} \\ \text{(Glycerol; glycerin)}}}{\begin{matrix} CH_2OH \\ | \\ CHOH \\ | \\ CH_2OH \end{matrix}} + \underset{\text{Fatty acids}}{3RCOOH}
$$

The free fatty acids then pass into the bloodstream and on to cells for oxidation. There are two major stages in the **β-oxidation of fatty acids**: (1) activation of a free fatty acid in the cytoplasm and its transport across the inner mitochondrial membrane, followed by (2) β-oxidation, a repeated sequence of four reactions.

β-Oxidation of fatty acids A series of four enzyme-catalyzed reactions that cleaves carbon atoms, two at a time, from the carboxyl end of a fatty acid.

A. Activation of Fatty Acids: Formation of a Thioester with Coenzyme A

The process of β-oxidation begins in the cytoplasm with the formation of a **thioester** between the carboxyl group of a fatty acid and the sulfhydryl group of coenzyme A. The formation of this acyl-CoA derivative is coupled with the hydrolysis of ATP to AMP and pyrophosphate ion. It is common in writing biochemical reactions to show some reactants and products by a curved arrow set over or under the main reaction arrow. We use this convention here to show ATP as a reactant and AMP and pyrophosphate ion as products:

Thioester An ester in which the oxygen atom of the —OR group is replaced by an atom of sulfur.

$$
\underset{\substack{\text{Fatty acid} \\ \text{(as anion)}}}{\begin{matrix} O \\ \| \\ R-C-O^- \end{matrix}} + \underset{\text{Coenzyme A}}{HS-CoA} \xrightarrow{\overset{ATP \quad AMP+P_2O_7^{4-}}{\curvearrowright}} \underset{\substack{\text{An acyl-CoA} \\ \text{derivative}}}{\begin{matrix} O \\ \| \\ R-C-S-CoA \end{matrix}} + OH^-
$$

The mechanism of this reaction involves attack by the fatty acid carboxylate anion on P=O of a phosphoric anhydride group of ATP to form an intermediate analogous to the tetrahedral carbonyl addition intermediate formed in C=O chemistry. In the intermediate formed in the fatty acid–ATP reaction, the phosphorus attacked by the carboxylate anion becomes bonded to five groups. The collapse of this intermediate gives an acyl-AMP, which is a highly reactive mixed anhydride of the carboxyl group of the fatty acid and the phosphate group of AMP:

intermediate with one phosphorus bonded to five groups

An acyl-AMP (a mixed anhydride)

Pyrophosphate ion

634   Chapter 22 The Organic Chemistry of Metabolism

This mixed anhydride then undergoes a carbonyl addition reaction with the sulfhydryl group of coenzyme A to form a tetrahedral carbonyl addition intermediate, which collapses to give AMP and an acyl-CoA (a fatty acid thioester of coenzyme A):

Coenzyme A   An acyl-AMP

An acyl-CoA   AMP

At this point, the activated fatty acid is transported into the mitochondrion, where its carbon chain is degraded by the reactions of β-oxidation.

B. The Four Reactions of β-Oxidation

Reaction 1: Oxidation of the Hydrocarbon Chain

The first reaction of β-oxidation is oxidation of the carbon chain between the alpha- and beta-carbons of the fatty acid chain. The oxidizing agent is FAD, which is reduced to $FADH_2$. This reaction is stereoselective: only the trans alkene isomer is formed,

An acyl-CoA   A *trans*-enoyl-CoA

Reaction 2: Hydration of the Carbon–Carbon Double Bond

Enzyme-catalyzed hydration of the carbon–carbon double bond gives a β-hydroxyacyl-CoA:

A *trans*-enoyl-CoA   (R)-β-Hydroxyacyl-CoA

Note that the hydration is regioselective: The —OH is added to carbon 3 of the chain. It is also stereoselective: Only the R enantiomer is formed.

Reaction 3: Oxidation of the β-Hydroxyl Group

In the second oxidation step of β-oxidation, the secondary alcohol is oxidized to a ketone. The oxidizing agent is NAD^+, which is reduced to NADH:

(R)-β-Hydroxyacyl-CoA   β-Ketoacyl-CoA

Reaction 4: Cleavage of the Carbon Chain

The final step of β-oxidation is cleavage of the carbon chain to give a molecule of acetyl coenzyme A and a new acyl-CoA, the hydrocarbon chain of which is shortened by two carbon atoms:

$$R-\overset{\overset{\displaystyle O}{\|}}{C}-CH_2-\overset{\overset{\displaystyle O}{\|}}{C}-SCoA \ +\ HS-CoA \ \xrightarrow{\text{thiolase}}\ R-\overset{\overset{\displaystyle O}{\|}}{C}-SCoA \ +\ CH_3\overset{\overset{\displaystyle O}{\|}}{C}-SCoA$$

β-Ketoacyl-CoA   Coenzyme A   An acyl-CoA   Acetyl-CoA

Mechanism: A Reverse Claisen Condensation in β-Oxidation of Fatty Acids

Step 1: A sulfhydryl group of the enzyme thiolase attacks the carbonyl carbon of the ketone to form a tetrahedral carbonyl addition intermediate.

Step 2: The addition intermediate collapses to give the enolate anion of acetyl-CoA and an enzyme-bound thioester, which is now shortened by two carbons.

Step 3: The enolate anion reacts with a proton donor to give acetyl-CoA.

Step 4: The enzyme–thioester intermediate undergoes reaction with a molecule of coenzyme A to regenerate a sulfhydryl group on the surface of the enzyme and liberate the fatty acyl-CoA, now shortened by two carbon atoms.

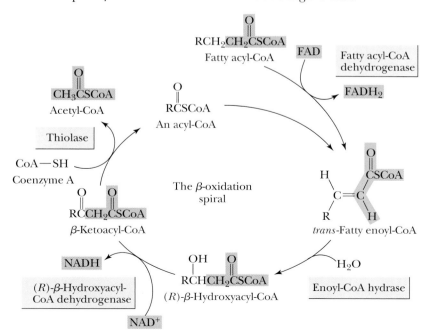

If Steps 1–3 of this mechanism are read in reverse, it is seen as an example of a **Claisen condensation** (Section 16.4A)—the attack by the enolate anion of acetyl-CoA on the carbonyl group of a thioester to form a tetrahedral carbonyl addition intermediate, followed by its collapse to give a β-ketothioester.

The four steps in β-oxidation are summarized in Figure 22.2.

Figure 22.2

The four reactions of β-oxidation. The steps of β-oxidation are called a spiral because, after each series of four reactions, the carbon chain is shortened by two carbon atoms.

C. Repetition of the β-Oxidation Spiral Yields Additional Acetate Units

The series of four reactions of β-oxidation then repeats on the shortened fatty acyl-CoA chain and continues until the entire fatty acid chain is degraded to acetyl-CoA. Seven cycles of β-oxidation of palmitic acid, for example, give eight molecules of acetyl-CoA and involve seven oxidations by FAD and seven oxidations by NAD$^+$:

$$CH_3(CH_2)_{14}\overset{\overset{O}{\|}}{C}OH + 8CoA{-}SH + 7NAD^+ + 7FAD \xrightarrow[\text{AMP+P}_2\text{O}_7^{4-}]{\text{ATP}} 8CH_3\overset{\overset{O}{\|}}{C}SCoA + 7NADH + 7FADH_2$$

Hexadecanoic acid (Palmitic acid)   Acetyl coenzyme A

22.7 THE CITRIC ACID CYCLE

Under aerobic conditions, the central metabolic pathway for the oxidation of the carbon skeletons not only of carbohydrates, but also of fatty acids and amino acids, to carbon dioxide is the citric acid cycle, also known as the *tricarboxylic acid* (*TCA*) cycle and Krebs cycle. The last-mentioned name is in honor of Sir Adolph Krebs, the biochemist who first proposed the cyclic nature of this pathway in 1937.

A. Overview of the Cycle

Through the reactions of the citric acid cycle, the carbon atoms of the acetyl group of acetyl-CoA are oxidized to carbon dioxide. There are four separate oxidations in the cycle, three involving NAD$^+$ and one involving FAD. Figure 22.3 gives an overview of the cycle, showing the four steps.

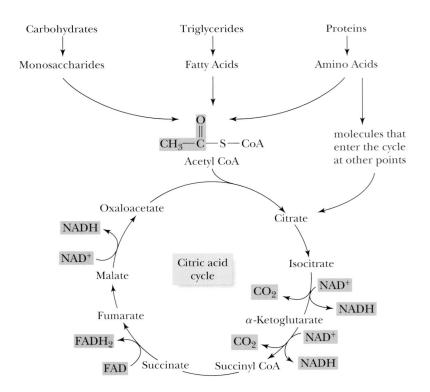

Figure 22.3
The citric acid cycle. Fuel for the cycle is derived from the catabolism (breakdown) of monosaccharides, fatty acids, and amino acids.

B. Reactions of the Citric Acid Cycle

1. Formation of Citrate

The two-carbon acetyl group of acetyl coenzyme A enters the cycle by an enzyme-catalyzed aldol reaction (Section 16.3) between the alpha carbon of acetyl-CoA and the ketone group of oxaloacetate. The product of this reaction is citrate, the tricarboxylic acid from which the cycle derives its name. In the reaction, the carbonyl condensation is coupled with the hydrolysis of the thioester to give free coenzyme A:

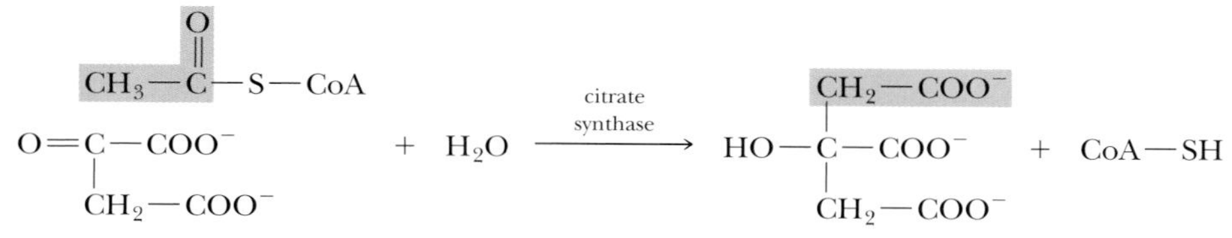

Oxaloacetate — Citrate

2. Isomerization of Citrate to Isocitrate

In the second step of the cycle, citrate is converted to the constitutional isomer isocitrate. This isomerization occurs in two steps, both catalyzed by aconitase. First, in a reaction analogous to the acid-catalyzed dehydration of an alcohol (Section 8.3E), citrate undergoes enzyme-catalyzed dehydration of aconitate. Then, in a reaction analogous to acid-catalyzed hydration of an alkene (Section 5.3B), aconitate undergoes enzyme-catalyzed hydration to give isocitrate.

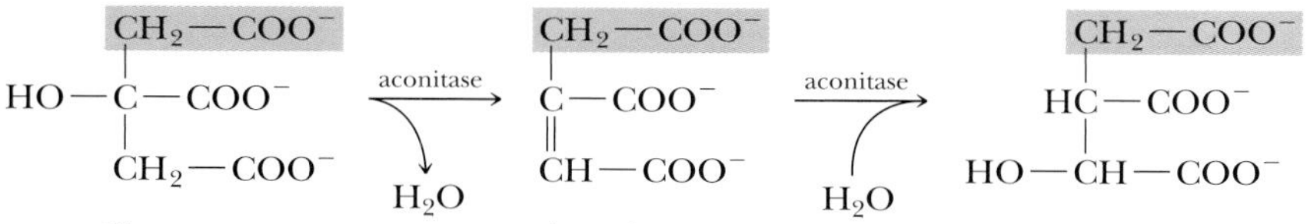

Citrate — Aconitate — Isocitrate

There are several important features to note about this transformation:

- The dehydration of citrate is completely regioselective: Dehydration is in the direction of the $-CH_2-$ group from the original molecule of oxaloacetate.
- The dehydration of citrate is completely stereoselective: It gives only the cis isomer of aconitate.
- The hydration of aconitate is completely regioselective: It gives only isocitrate.
- The hydration of aconitate is completely stereoselective: Isocitrate has two stereocenters, and four stereoisomers (two pair of enantiomers) are possible. Only one of the four stereoisomers is produced in this enzyme-catalyzed hydration.

3. Oxidation and Decarboxylation of Isocitrate

In step 3, the secondary alcohol of isocitrate is oxidized to a ketone by NAD^+ in a reaction catalyzed by the enzyme isocitrate dehydrogenase. The product, oxalosuccinate, is a β-ketoacid and undergoes decarboxylation (Section 14.9) to produce α-ketoglutarate:

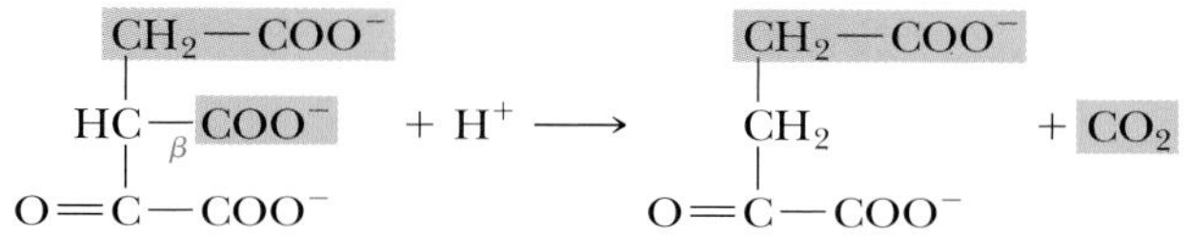

Oxalosuccinate (a β-ketoacid) — α-Ketoglutarate

Note that only one of the three carboxyl groups in oxalosuccinate is beta to the ketone carbonyl; it is this carboxyl group that undergoes the decarboxylation.

4. Oxidation and Decarboxylation of α-Ketoglutarate

The second molecule of carbon dioxide is generated in the cycle by the same type of oxidative decarboxylation as that for the conversion of pyruvate (also an α-ketoacid) to acetyl-CoA and carbon dioxide (Section 22.5C). In the oxidative decarboxylation of α-ketoglutarate, the carboxyl group is converted to carbon dioxide and the adjacent ketone is oxidized to a carboxyl group in the form of a thioester with coenzyme A:

$$
\begin{array}{ccc}
\underset{\text{α-Ketoglutarate}}{
\begin{array}{l}
CH_2{-}COO^- \\
| \\
CH_2 \\
| \\
O{=}C{-}COO^-
\end{array}}
& + \ NAD^+ + CoA{-}SH \longrightarrow &
\underset{\text{Succinyl CoA}}{
\begin{array}{l}
CH_2{-}COO^- \\
| \\
CH_2 \\
| \\
O{=}C{-}S{-}CoA
\end{array}}
& + \ CO_2 + NADH
\end{array}
$$

Note that, of the two molecules of carbon dioxide given off in this turn of the citric acid cycle, both carbons are from the carbon skeleton of oxaloacetate; neither is from the acetyl group of acetyl coenzyme A.

5. Conversion of Succinyl CoA to Succinate

Next, in coupled reactions catalyzed by succinyl CoA synthetase, succinyl CoA, HPO_4^{2-}, and guanosine diphosphate (GDP) react to form succinate, guanosine triphosphate (GTP), and coenzyme A:

$$
\begin{array}{ccc}
\underset{\text{Succinyl CoA}}{
\begin{array}{l}
CH_2{-}COO^- \\
| \\
CH_2 \\
| \\
O{=}C{-}S{-}CoA
\end{array}}
& + \ GDP + HPO_4^{2-} \longrightarrow &
\underset{\text{Succinate}}{
\begin{array}{l}
COO^- \\
| \\
CH_2 \\
| \\
CH_2 \\
| \\
COO^-
\end{array}}
& + \ GTP + CoA{-}SH
\end{array}
$$

Observe that to this point in the cycle, the two carbons of the original acetyl group of acetyl-CoA have remained differentiated from the carbon atoms of oxaloacetate. With the production of succinate, however, the two $-CH_2$ groups, as well as the two $-COO^-$ groups, are now indistinguishable.

6. Oxidation of Succinate

In the third oxidation of the cycle, succinate is oxidized to fumarate. The oxidizing agent is FAD, which is reduced to $FADH_2$:

$$
\underset{\text{Succinate}}{
\begin{array}{l}
COO^- \\
| \\
CH_2 \\
| \\
CH_2 \\
| \\
COO^-
\end{array}}
+ \ FAD \ \xrightarrow{\substack{\text{succinate} \\ \text{dehydrogenase}}} \
\underset{\text{Fumarate}}{
\begin{array}{c}
H \quad COO^- \\
\diagdown \ / \\
C \\
\| \\
C \\
/ \ \diagdown \\
{}^-OOC \quad H
\end{array}}
+ \ FADH_2
$$

This oxidation is completely stereoselective: Only the trans isomer is formed.

7. Hydration of Fumarate

In the second hydration step of the cycle, fumarate is converted to malate:

$$
\underset{\text{Fumarate}}{
\begin{array}{c}
H \quad COO^- \\
\diagdown \ / \\
C \\
\| \\
C \\
/ \ \diagdown \\
{}^-OOC \quad H
\end{array}}
+ \ H_2O \ \xrightarrow{\text{fumarase}} \
\underset{\text{Malate}}{
\begin{array}{l}
HO{-}CH{-}COO^- \\
| \\
CH_2{-}COO^-
\end{array}}
$$

Fumarase, the enzyme catalyzing this hydration, recognizes only fumarate (and not its cis isomer) and gives malate as a single enantiomer.

8. Oxidation of Malate

In the fourth oxidation of the cycle, the secondary alcohol of malate is oxidized to a ketone by NAD^+:

$$HO-\underset{\underset{\text{Malate}}{CH_2-COO^-}}{CH-COO^-} + NAD^+ \xrightarrow{\underset{\text{dehydrogenase}}{\text{malate}}} O=\underset{\underset{\text{Oxaloacetate}}{CH_2-COO^-}}{C-COO^-} + NADH + H^+$$

With the production of oxaloacetate, the reactions of the citric acid cycle are complete. Continued operation of the cycle requires two things: (1) a supply of carbon atoms in the form of acetyl groups from acetyl-CoA and (2) a supply of oxidizing agents in the form of NAD^+ and FAD. For a continuing supply of these two oxidizing agents, the operation of the cycle depends on the reactions of respiration and electron transport, a series of reactions in which the reduced coenzymes NADH and $FADH_2$ are reoxidized by molecular oxygen, O_2.

Another important feature of the cycle is best seen by examining the balanced equation for the cycle:

$$\underset{}{CH_3\overset{\overset{O}{\parallel}}{C}SCoA} + 3NAD^+ + FAD + HPO_4^{2-} + ADP \xrightarrow{\underset{\text{acid cycle}}{\text{citric}}}$$

$$2CO_2 + 3NADH + FADH_2 + ATP + CoA-SH$$

The cycle is truly catalytic: Its intermediates do not enter into the balanced equation for this pathway; they are neither destroyed nor synthesized in the net reaction. The only function of the cycle is to accept acetyl groups from acetyl-CoA, oxidize them to carbon dioxide, and at the same time produce a supply of reduced coenzymes as fuel for electron transport and oxidative phosphorylation. In fact, if any of the intermediates of the cycle are removed, the cycle ceases, because there is no way to regenerate oxaloacetate. Fortunately, the cycle is connected to other metabolic pathways through several of its intermediates. In practice, certain intermediates can be used to synthesize other biomolecules, provided that another intermediate is supplied, which in turn can be converted to oxaloacetate, thus making up for the intermediate withdrawn.

SUMMARY

ATP, ADP, and AMP (Section 22.2A) are agents for the storage and transport of phosphate groups. Nicotinamide adenine dinucleotide (NAD$^+$) (Section 22.2B) and flavin adenine dinucleotide (FAD) (Section 22.2C) are agents for the storage and transport of electrons in metabolic oxidations and reductions. NAD^+ is a two-electron oxidizing agent and is reduced to NADH, which, in turn, is a two-electron reducing agent and is oxidized to NAD^+. In the reactions of FAD that are involved in the β-oxidation of fatty acids, FAD is a two-electron oxidizing agent and is reduced to $FADH_2$. Coenzyme A (Section 22.2D) is a carrier of acetyl groups.

Glycolysis is a series of 10 enzyme-catalyzed reactions that oxidizes glucose to two molecules of pyruvate. The 10 reactions of glycolysis (Section 22.4) can be grouped in the following way:

- Transfer of a phosphate group from ATP to an —OH group of a monosaccharide to form a phosphoric ester (Reactions 1 and 3).

- Interconversion of constitutional isomers by keto–enol tautomerism (Reactions 2 and 5).
- Reverse aldol reaction (Reaction 4).
- Oxidation of an aldehyde group to the mixed anhydride of a carboxylic acid and phosphoric acid (Reaction 6).
- Transfer of a phosphate group from a monosaccharide intermediate to ADP to form ATP (Reactions 7 and 10).
- Transfer of a phosphate group from a 1° alcohol to a 2° alcohol (Reaction 8).
- Dehydration of a 1° alcohol to form a carbon–carbon double bond (Reaction 9).

Pyruvate, the product of anaerobic glycolysis, does not accumulate in cells, but rather undergoes one of three possible enzyme-catalyzed reactions, depending on the state of oxygenation and the type of cell in which the pyruvate is produced (Section 22.5). In lactate fermentation, pyruvate is reduced to lactate by NADH. In alcoholic fermentation, pyruvate is converted to acetaldehyde, which is reduced to ethanol by NADH. Under aerobic conditions, pyruvate is oxidized to acetyl coenzyme A by NAD^+.

There are two major stages in the metabolism of fatty acids (Section 22.6): (1) activation of free fatty acids in the cytoplasm through the formation of thioesters with **coenzyme A** and transport of the activated fatty acids across the inner mitochondrial membrane, followed by (2) β-oxidation. **β-Oxidation of fatty acids** (Section 22.6B) is a series of four enzyme-catalyzed reactions by which a fatty acid is degraded to acetyl-CoA.

The **citric acid cycle** (Section 22.7) accepts the two-carbon acetyl group from acetyl-CoA and oxidizes it to two molecules of carbon dioxide. Oxidizing agents are NAD^+ and FAD.

KEY REACTIONS

1. Glycolysis (Section 22.3)

Glycolysis is a series of 10 enzyme-catalyzed reactions that converts glucose to pyruvate:

$$C_6H_{12}O_6 + 2NAD^+ + 2HPO_4{}^{2-} + 2ADP \xrightarrow{\text{glycolysis}} 2CH_3\overset{\overset{\text{O}}{\|}}{C}COO^- + 2NADH + 2ATP + 2H_3O^+$$

Glucose $\qquad\qquad\qquad\qquad\qquad\qquad\qquad$ Pyruvate

2. Reduction of Pyruvate to Lactate: Lactate Fermentation (Section 22.5A)

$$CH_3\overset{\overset{\text{O}}{\|}}{C}COO^- + NADH + H^+ \underset{\text{lactate dehydrogenase}}{\rightleftharpoons} CH_3\overset{\overset{\text{OH}}{|}}{C}HCOO^- + NAD^+$$

Pyruvate $\qquad\qquad\qquad\qquad\qquad\qquad$ Lactate

3. Reduction of Pyruvate to Ethanol: Alcohol Fermentation (Section 22.5B)

The carbon dioxide formed in this reaction is responsible for the foam on beer and the carbonation of naturally fermented wines and champagnes:

$$CH_3\overset{\overset{\text{O}}{\|}}{C}COO^- + 2H^+ + NADH \xrightarrow{\text{alcoholic fermentation}} CH_3CH_2OH + CO_2 + NAD^+$$

Pyruvate $\qquad\qquad\qquad\qquad\qquad\qquad\qquad$ Ethanol

4. Oxidative Decarboxylation of Pyruvate to Acetyl-CoA (Section 22.5C)

$$CH_3\overset{\overset{\text{O}}{\|}}{C}COO^- + NAD^+ + CoA{-}SH \xrightarrow{\text{oxidative decarboxylation}} CH_3\overset{\overset{\text{O}}{\|}}{C}SCoA + CO_2 + NADH$$

Pyruvate $\qquad\qquad\qquad\qquad\qquad\qquad\qquad\qquad$ Acetyl-CoA

5. β-Oxidation of Fatty Acids (Section 22.6)

In this series of four enzyme-catalyzed reactions, the carbon chain of a fatty acid is shortened by two carbon atoms at a time:

$$CH_3(CH_2)_{14}\overset{\overset{\text{O}}{\|}}{C}OH + 8CoA{-}SH + 7NAD^+ + 7FAD \xrightarrow[\text{ATP} \quad \text{AMP}+P_2O_7{}^{4-}]{} 8CH_3\overset{\overset{\text{O}}{\|}}{C}SCoA + 7NADH + 7FADH_2$$

Hexadecanoic acid $\qquad\qquad\qquad\qquad\qquad\qquad\qquad\qquad$ Acetyl coenzyme A
(Palmitic acid)

6. Citric Acid Cycle (Section 22.7)

Through the reactions of the citric acid cycle, the carbon atoms of the acetyl group of acetyl-CoA are oxidized to carbon dioxide:

$$CH_3\overset{\overset{\text{O}}{\|}}{C}SCoA + 3NAD^+ + FAD + HPO_4{}^{2-} + ADP \xrightarrow{\text{citric acid cycle}} 2CO_2 + 3NADH + FADH_2 + ATP + CoA{-}SH$$

There are four separate oxidation steps in the cycle, three involving NAD^+ and one involving FAD.

PROBLEMS

A problem number set in red indicates an applied "real-world" problem.

Glycolysis

22.3 Name one coenzyme required for glycolysis. From what vitamin is the coenzyme derived?

22.4 Number the carbons of glucose 1 through 6. Which carbons of glucose become the carboxyl groups of the two pyruvates?

22.5 How many moles of lactate are produced from 3 moles of glucose?

22.6 Although glucose is the principal source of carbohydrates for glycolysis, fructose and galactose are also metabolized for energy.

    (a) What is the main dietary source of fructose? Of galactose?

    (b) Propose a series of reactions by which fructose might enter glycolysis.

    (c) Propose a series of reactions by which galactose might enter glycolysis.

22.7 How many moles of ethanol are produced per mole of sucrose through the reactions of glycolysis and alcoholic fermentation? How many moles of CO_2 are produced?

22.8 Glycerol that is derived from the hydrolysis of triglycerides and phospholipids is also metabolized for energy. Propose a series of reactions by which the carbon skeleton of glycerol might enter glycolysis and be oxidized to pyruvate.

22.9 Write a mechanism to show the role of NADH in the reduction of acetaldehyde to ethanol.

22.10 Ethanol is oxidized in the liver to acetate ion by NAD^+.

    (a) Write a balanced equation for this oxidation.

    (b) Do you expect the pH of blood plasma to increase, decrease, or remain the same as a result of the metabolism of a significant amount of ethanol?

22.11 When pyruvate is reduced to lactate by NADH, two hydrogens are added to pyruvate: one to the carbonyl carbon, the other to the carbonyl oxygen. Which of these hydrogens is derived from NADH?

22.12 Why is glycolysis called an anaerobic pathway?

22.13 Which carbons of glucose end up in CO_2 as a result of alcoholic fermentation?

22.14 Which steps in glycolysis require ATP? Which steps produce ATP?

β-Oxidation

22.15 Write structural formulas for palmitic, oleic, and stearic acids, the three most abundant fatty acids.

22.16 A fatty acid must be activated before it can be metabolized in cells. Write a balanced equation for the activation of palmitic acid.

22.17 Name three coenzymes that are necessary for the β-oxidation of fatty acids. From what vitamin is each derived?

22.18 We have examined β-oxidation of saturated fatty acids, such as palmitic acid and stearic acid. Oleic acid, an unsaturated fatty acid, is also a common component of dietary fats and oils. This unsaturated fatty acid is degraded by β-oxidation, but, at one stage in its degradation, requires an additional enzyme named enoyl-CoA isomerase. Why is this enzyme necessary, and what isomerization does it catalyze? (*Hint:* Consider both the configuration of the carbon–carbon double bond in oleic acid and its position in the carbon chain.)

Citric Acid Cycle

22.19 What is the main function of the citric acid cycle?

22.20 Which steps in the citric acid cycle involve

    (a) the formation of new carbon–carbon bonds

    (b) the breaking of carbon–carbon bonds

    (c) oxidation by NAD^+

    (d) oxidation by FAD

    (e) decarboxylation

    (f) the creation of new stereocenters

22.21 What does it mean to say that the citric acid cycle is catalytic—that is, that it does not produce any new compounds?

Additional Problems

22.22 Review the oxidation reactions of glycolysis, β-oxidation, and the citric acid cycle, and compare the types of functional groups oxidized by NAD^+ with those oxidized by FAD.

22.23 The *respiratory quotient* (RQ), used in studies of energy metabolism and exercise physiology, is defined as the ratio of the volume of carbon dioxide produced to the volume of oxygen used:

$$RQ = \frac{\text{Volume } CO_2}{\text{Volume } O_2}$$

(a) Show that RQ for glucose is 1.00. (*Hint:* Look at the balanced equation for the complete oxidation of glucose to carbon dioxide and water.)

(b) Calculate RQ for triolein, a triglyceride with molecular formula $C_{57}H_{104}O_6$.

(c) For an individual on a normal diet, RQ is approximately 0.85. Would this value increase or decrease if ethanol were to supply an appreciable portion of the person's caloric needs?

22.24 Acetoacetate, β-hydroxybutyrate, and acetone are commonly referred to within the health sciences as "ketone bodies," in spite of the fact that one of them is not a ketone at all. All are products of human metabolism and are always present in blood plasma. Most tissues (with the notable exception of the brain) have the enzyme systems necessary to use ketone bodies as energy sources. Ketone bodies are synthesized by the following enzyme-catalyzed reactions:

Describe the type of reaction involved in each step.

22.25 A connecting point between anaerobic glycolysis and β-oxidation is the formation of acetyl-CoA. Which carbon atoms of glucose appear as methyl groups of acetyl-CoA? Which carbon atoms of palmitic acid appear as methyl groups of acetyl-CoA?

22.26 Which of the steps in the following biochemical pathways use molecular oxygen as the oxidizing agent?

(a) Glycolysis   (b) β-Oxidation   (c) The citric acid cycle

Looking Back

22.27 Compare biological (enzyme-catalyzed) reactions with laboratory reactions in terms of
(a) efficiency of yields
(b) regiochemical outcome of products
(c) stereochemical outcome of products

22.28 Comment on the importance of stereochemistry in the synthesis of new drugs.

22.29 Of the functional groups that we have studied, which are affected by the acidity of biological environments (biological pH)?

22.30 Can you think of any aspect of your day-to-day life that does not involve or is not affected by organic chemistry? Explain.

Appendix 1

Acid Ionization Constants for the Major Classes of Organic Acids

Class and Example	Typical pK_a	Class and Example	Typical pK_a
Sulfonic acid	0–1	Alkylammonium ion	10–12
		$(CH_3CH_2)_3\overset{+}{N}-H$	
Carboxylic acid	3–5	β-ketoester	11
CH_3CO-H		$CH_3-C-CH-COCH_2CH_3$	
Arylammonium ion	4–5	Water	15.7
		$HO-H$	
		Alcohol	15–19
		CH_3CH_2O-H	
Thiol	8–12	α-Hydrogen of an aldehyde or ketone	18–20
CH_3CH_2S-H		CH_3CCH_2-H	
Phenol	9–10	α-Hydrogen of an ester	23–25
		$CH_3CH_2OCCH_2-H$	
β-Diketone	10		
$CH_3-C-CH-CCH_3$			

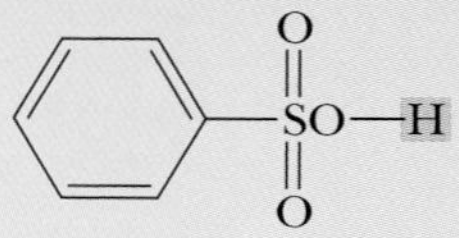

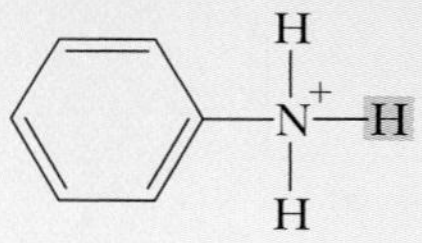

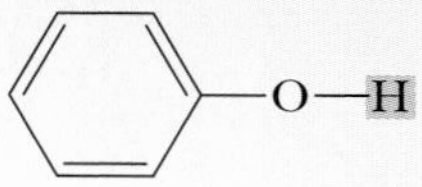

Appendix 2

Characteristic ^{1}H-NMR Chemical Shifts

Type of Hydrogen (R = alkyl, Ar = aryl)	Chemical Shift (δ)*	Type of Hydrogen (R = alkyl, Ar = aryl)	Chemical Shift (δ)*
$(CH_3)_4Si$	0 (by definition)	$\underset{\displaystyle \overset{O}{\parallel}}{RC}CH_2R$	2.2–2.6
RCH_3	0.8–1.0		
RCH_2R	1.2–1.4	$\underset{\displaystyle \overset{O}{\parallel}}{RC}OCH_3$	3.7–3.9
R_3CH	1.4–1.7		
$\overset{\diagdown}{\underset{\diagup}{C}}=\overset{\diagup}{\underset{\diagdown}{C}}\overset{\displaystyle \mid}{CH}$	1.6–2.6	$\underset{\displaystyle \overset{O}{\parallel}}{RC}OCH_2R$	4.1–4.7
		RCH_2I	3.1–3.3
$RC{\equiv}CH$	2.0–3.0	RCH_2Br	3.4–3.6
$ArCH_3$	2.2–2.5	RCH_2Cl	3.6–3.8
$ArCH_2R$	2.3–2.8	RCH_2F	4.4–4.5
ROH	0.5–6.0	$ArOH$	4.5–4.7
RCH_2OH	3.4–4.0	$R_2C{=}CH_2$	4.6–5.0
RCH_2OR	3.3–4.0	$R_2C{=}CHR$	5.0–5.7
R_2NH	0.5–5.0	ArH	6.5–8.5
$\underset{\displaystyle \overset{O}{\parallel}}{RC}CH_3$	2.1–2.3	$\underset{\displaystyle \overset{O}{\parallel}}{RC}H$	9.5–10.1
		$\underset{\displaystyle \overset{O}{\parallel}}{RC}OH$	10–13

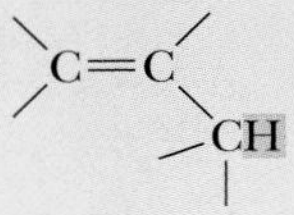

*Values are relative to tetramethylsilane. Other atoms within the molecule may cause the signal to appear outside these ranges.

Appendix 3

Characteristic ^{13}C-NMR Chemical Shifts

Type of Carbon	Chemical Shift (δ)	Type of Carbon	Chemical Shift (δ)
RCH$_3$	0–40		
RCH$_2$R	15–55		110–160
R$_3$CH	20–60		
RCH$_2$I	0–40		
RCH$_2$Br	25–65	RCOR	160–180
RCH$_2$Cl	35–80		
R$_3$COH	40–80	RCNR$_2$	165–180
R$_3$COR	40–80		
RC≡CR	65–85	RCOH	175–185
R$_2$C=CR$_2$	100–150		
		RCH, RCR	180–210

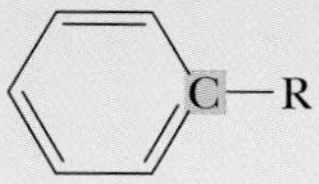

Appendix 4

Characteristic Infrared Absorption Frequencies

Bonding		Frequency (cm^{-1})	Intensity*
C—H	alkane	2850–3000	w–m
	—CH$_3$	1375 and 1450	w–m
	—CH$_2$—	1450	m
	alkene	3000–3100	w–m
		650–1000	s
	alkyne	3300	w–m
		1600–1680	w–m
	aromatic	3000–3100	s
		690–900	s
	aldehyde	2700–2800	w
		2800–2900	w
C=C	alkene	1600–1680	w–m
	aromatic	1450 and 1600	w–m
C—O	alcohol, ether,	1050–1100 (sp^3 C—O)	s
	ester, carboxylic		s
	acid, anhydride	1200–1250 (sp^2 C—O)	s
C=O	amide	1630–1680	s
	carboxylic acid	1700–1725	s
	ketone	1705–1780	s
	aldehyde	1705–1740	s
	ester	1735–1800	s
	anhydride	1760 and 1800	s
O—H	alcohol, phenol		
	free	3600–3650	m
	H-bonded	3200–3500	m
	carboxylic acid	2400–3400	m
N—H	amine and amide	3100–3500	m–s

* m = medium, s = strong, w = weak

Glossary

Acetal (Section 13.7)   A molecule containing two —OR or —OAr groups bonded to the same carbon.

Aceto group (Section 14.3B)   A CH_3CO— group.

Achiral (Section 6.3A)   An object that lacks chirality; an object that is superposable on its mirror image.

Acid halide (Section 15.2A)   A derivative of a carboxylic acid in which the —OH of the carboxyl group is replaced by a halogen—most commonly chlorine.

Activating group (Section 9.8A)   Any substituent on a benzene ring that causes the rate of electrophilic aromatic substitution to be greater than that for benzene.

Activation energy (Section 5.2A)   The difference in energy between reactants and the transition state.

Acyl halide (Section 9.7D)   A derivative of a carboxylic acid in which the —OH of the carboxyl group is replaced by a halogen—most commonly chlorine.

Alcohol (Section 8.2A)   A compound containing an —OH (hydroxyl) group bonded to an sp^3 hybridized carbon.

Alcoholic fermentation (Section 22.5B)   A metabolic pathway that converts glucose to ethanol and CO_2.

Alditol (Section 18.5B)   The product formed when the $C=O$ group of a monosaccharide is reduced to a CHOH group.

Aldol reaction (Section 16.3B)   A carbonyl condensation reaction between two aldehydes or ketones to give a β-hydroxyaldehyde or β-hydroxyketone.

Aldose (Section 18.2A)   A monosaccharide containing an aldehyde group.

Aliphatic amine (Section 10.2)   An amine in which nitrogen is bonded only to alkyl groups.

Aliphatic hydrocarbon (Section 3.1)   An alternative term used to describe an alkane.

Alkaloid (Section 10.2)   A basic nitrogen-containing compound of plant origin; many have physiological activity when administered to humans.

Alkane (Section 3.1)   A saturated hydrocarbon.

Alkene (Section 4.1)   An unsaturated hydrocarbon that contains a carbon–carbon double bond.

Alkoxy group (Section 8.3C)   An —OR group, where R is an alkyl group.

Alkyl group (Section 3.3A)   A group derived by removing a hydrogen from an alkane; given the symbol R—.

Alkyl halide (Section 7.2)   A compound containing a halogen atom covalently bonded to an alkyl group; given the symbol RX.

Alkyne (Section 4.1)   An unsaturated hydrocarbon that contains a carbon–carbon triple bond.

Amino acid (Section 19.2A)   A compound that contains both an amino group and a carboxyl group.

α-Amino acid (Section 19.2A)   An amino acid in which the amino group is on the carbon adjacent to the carboxyl group.

Amino group (Section 1.8B)   An sp^3 hybridized nitrogen atom bonded to one, two, or three carbon groups.

Amorphous domains (Section 17.4)   Disordered, noncrystalline regions in the solid state of a polymer.

Anabolic steroid (Section 21.5A)   A steroid hormone, such as testosterone, that promotes tissue and muscle growth and development.

Androgen (Section 21.5A)   A steroid hormone, such as testosterone, that mediates the development and sexual characteristics of males.

Angle strain (Section 3.6A)   The strain that arises when a bond angle is either compressed or expanded compared with its optimal value.

Anion (Section 1.3B)   An atom or a group of atoms bearing a negative charge.

Anomeric carbon (Section 18.3A)   The hemiacetal carbon of the cyclic form of a monosaccharide.

Anomers (Section 18.3A)   Monosaccharides that differ in configuration only at their anomeric carbons.

Anti addition (Section 5.3C)   Addition of atoms or groups of atoms from opposite sides or faces of a carbon–carbon double bond.

Aprotic solvent (Section 7.6D)   A solvent that cannot serve as a hydrogen bond donor—for example, acetone, diethyl ether, and dichloromethane.

Ar— (Section 9.1)   The symbol used for an aryl group, by analogy with R— for an alkyl group.

Aramid (Section 17.5A)   A poly*aromatic amide*; a polymer in which the monomer units are an aromatic diamine and an aromatic dicarboxylic acid.

Arene (Section 4.1)   A compound containing one or more benzene rings.

Arenediazonium salt (Section 10.8)   A salt of the form ArN_2^+ Cl^-, where Ar represents an aryl group.

Aromatic amine (Section 10.2)   An amine in which nitrogen is bonded to one or more aryl groups.

Aromatic compound (Section 9.1)   A term used to classify benzene and its derivatives.

Aryl group (Section 9.1)   A group derived from an aromatic compound (an arene) by the removal of an H; given the symbol Ar—.

Average degree of polymerization, n (Section 17.2)   A subscript placed outside the parentheses of the simplest nonredundant unit of a polymer in order to indicate that the unit repeats n times in the polymer.

Axial position (Section 3.6B)   A position on a chair conformation of a cyclohexane ring that extends from the ring, parallel to the imaginary axis of the ring.

Benzyl group (Section 9.4A)   The $C_6H_5CH_2$— group.

Benzylic carbon (Section 9.5)   An sp^3 hybridized carbon bonded to a benzene ring.

Bile acid (Section 21.5A)   A cholesterol-derived detergent molecule, such as cholic acid, that is secreted by the gallbladder into the intestine to assist in the absorption of dietary lipids.

Bimolecular reaction (Section 7.5A)   A reaction in which two species are involved in the reaction leading to the transition state of the rate-determining step.

Boat conformation (Section 3.6B)   A puckered conformation of a cyclohexane ring in which carbons 1 and 4 of the ring are bent toward each other.

Bond length (Section 1.3C)   The distance between atoms in a covalent bond.

Bonding electrons (Section 1.3D)   Valence electrons involved in forming a covalent bond; that is, shared electrons.

Brønsted–Lowry acid (Section 2.3)   A proton donor.

Brønsted–Lowry base (Section 2.3)   A proton acceptor.

Carbanion (Section 13.6A)   An anion in which carbon has an unshared pair of electrons and bears a negative charge.

Carbocation (Section 5.3A)   A species containing a carbon atom with only three bonds to it and bearing a positive charge.

Carbohydrate (Section 18.1)   A polyhydroxyaldehyde, a polyhydroxyketone, or a substance that gives either of these compounds on hydrolysis.

α-Carbon (Section 13.9A)   A carbon atom adjacent to a carbonyl group.

Carbonyl group (Section 1.8C)   A C=O group.

Carboxyl group (Section 1.8D)   A —COOH group.

Carboxylic anhydride (Section 15.2B)   A compound in which two acyl groups are bonded to an oxygen.

Cation (Section 1.3B)   An atom or a group of atoms bearing a positive charge.

Chain-growth polymerization (Section 17.6)   A polymerization that involves sequential addition reactions, either to unsaturated monomers or to monomers possessing other reactive functional groups.

Chain initiation (Section 17.6A)   In radical polymerization, the formation of radicals from molecules containing only paired electrons.

Chain propagation (Section 17.6A)   In radical polymerization, the reaction of a radical and a molecule to give a new radical.

Chain termination (Section 17.6A)   In radical polymerization, a reaction in which two radicals combine to form a covalent bond.

Chain-transfer reaction (Section 17.6A)   In radical polymerization, the transfer of reactivity of an end group from one chain to another during a polymerization.

Chair conformation (Section 3.6B)   The most stable puckered conformation of a cyclohexane ring.

Chemical shift (Section 12.8)   The quantity used in NMR spectroscopy to identify the positions of signals produced by the nuclei of a sample. The unit of chemical shift (δ) is expressed in parts per million (ppm).

Chiral (Section 6.3A)   From the Greek *cheir*, meaning hand; said of objects that are not superposable on their mirror images.

Chromatin (Section 20.3C)   A complex formed between negatively charged DNA molecules and positively charged histones.

Circular DNA (Section 20.3C)   A type of double-stranded DNA in which the 5′ and 3′ ends of each strand are joined by phosphodiester groups.

Cis (Section 3.7)   A prefix meaning "on the same side."

Cis–trans isomerism   Isomers that have the same order of attachment of their atoms, but a different arrangement of their atoms in space due to the presence of either a ring (Section 3.7) or a carbon–carbon double bond (Section 4.1).

Claisen condensation (Section 16.4A)   A carbonyl condensation reaction between two esters to give a β-ketoester.

Codon (Section 20.5A)   A triplet of nucleotides on mRNA that directs the incorporation of a specific amino acid into a polypeptide sequence.

Coenzyme (Section 22.2)   A low-molecular-weight nonprotein molecule or ion that binds reversibly to an enzyme, functions as a second substrate for the enzyme, and is regenerated by further reaction.

Condensation polymerization (Section 17.5)   A polymerization in which chain growth occurs in a stepwise manner between difunctional monomers. Also called *step-growth polymerization*.

Conformation (Section 3.6A)   Any three-dimensional arrangement of atoms in a molecule that results by rotation about a single bond.

Conjugate acid (Section 2.3)   The species formed when a base accepts a proton.

Conjugate base (Section 2.3)   The species formed when an acid donates a proton.

Constitutional isomers (Section 3.2)   Compounds with the same molecular formula, but a different order of attachment of their atoms.

Correlation table (Section 11.4C)   Table of data on absorption patterns of selected functional groups.

Covalent bond (Section 1.3B)   A chemical bond resulting from the sharing of one or more pairs of electrons.

Crossed aldol reaction (Section 16.3C)   An aldol reaction between two different aldehydes, two different ketones, or an aldehyde and a ketone.

Crossed Claisen condensation (Section 16.4C)   A Claisen condensation between two different esters.

Crystalline domains (Section 17.4)   Ordered crystalline regions in the solid state of a polymer; also called *crystallites*.

Curved arrow (Section 1.6B)   A symbol used to show the redistribution of valence electrons.

Cyclic ether (Section 8.2B)   An ether in which the oxygen is one of the atoms of a ring.

Cycloalkane (Section 3.4)   A saturated hydrocarbon that contains carbon atoms joined together to form a ring.

Deactivating group (Section 9.8A)   Any substituent on a benzene ring that causes the rate of electrophilic aromatic substitution to be lower than that for benzene.

Decarboxylation (Section 14.9)   Loss of CO_2 from an organic molecule.

Dehydration (Section 8.3E)   Elimination of a molecule of water from a compound.

Dehydrohalogenation (Section 7.8)   Removal of —H and —X from adjacent carbons; a type of β-elimination.

Deshielded (Section 12.4)   The phenomenon in NMR spectroscopy in which resonance or inductive effects reduce the electron density around a nucleus, thus increasing the ability of an applied magnetic field to bring the nucleus into resonance.

Dextrorotatory (Section 6.9B)   The clockwise rotation of the plane of polarized light in a polarimeter.

Diastereomers (Section 6.5A)   Stereoisomers that are not mirror images of each other, refers to relationships among objects.

Diaxial interactions (Section 3.6B)   Interactions between groups in axial positions on the same side of a chair conformation of a cyclohexane ring.

Dieckmann condensation (Section 16.4B)   An intramolecular Claisen condensation of an ester of a dicarboxylic acid to give a five- or six-membered ring.

Dipeptide (Section 19.4)   A molecule containing two amino acid units joined by a peptide bond.

Disaccharide (Section 18.8)   A carbohydrate containing two monosaccharide units joined by a glycosidic bond.

Dispersion forces (Section 3.8)   Very weak intermolecular forces of attraction resulting from the interaction between temporary induced dipoles.

Disulfide bond (Section 19.6C)   A covalent bond between two sulfur atoms; an —S—S— bond.

Double helix (Section 20.3B)   A type of secondary structure of DNA molecules in which two antiparallel polynucleotide strands are coiled in a right-handed manner about the same axis.

Double-headed arrow (Section 1.6A)   A symbol used to connect contributing structures.

Downfield (Section 12.5)   A term used in NMR spectroscopy to denote that a signal is toward the left of the spectrum or of another signal.

E (Section 4.2C)   From the German *entgegen*, opposite; specifies that groups of higher priority on the carbons of a double bond are on opposite sides.

E,Z system (Section 4.2C)   A system used to specify the configuration of groups about a carbon–carbon double bond.

Eclipsed conformation (Section 3.6A)   A conformation about a carbon–carbon single bond in which the atoms on one carbon are as close as possible to the atoms on the adjacent carbon.

Edman degradation (Section 19.5B)   A method for selectively cleaving and identifying the N-terminal amino acid of a polypeptide chain.

Elastomer (Section 17.4)   A material that, when stretched or otherwise distorted, returns to its original shape when the distorting force is released.

Electromagnetic radiation (Section 11.2)   Energy propagated through space in the form of oscillating electric and magnetic fields.

Electromagnetic spectrum (Section 11.2)   The full range of frequencies that represent energy corresponding to gamma rays, X rays, ultraviolet light, visible light, infrared radiation, microwaves, and radio waves.

Electronegativity (Section 1.3C)   A measure of the force of an atom's attraction for electrons it shares in a chemical bond with another atom.

Electrophile (Section 5.3A)   Any molecule or ion that can accept a pair of electrons to form a new covalent bond; a Lewis acid.

Electrophilic aromatic substitution (Section 9.6)   A reaction in which an electrophile E^+ substitutes for a hydrogen on an aromatic ring.

Electrophoresis (Section 19.3D)   The process of separating compounds on the basis of their electric charge.

β-Elimination reaction (Section 7.8)   The removal of atoms or groups of atoms from two adjacent carbon atoms; for example, the removal of H and OH from an alcohol to form a carbon–carbon double bond.

Enantiomer (Section 6.3A)   Stereoisomers that are nonsuperposable mirror images; refers to a relationship between pairs of objects.

Endothermic reaction (Section 5.2A)   A reaction in which the energy of the products is higher than the energy of the reactants; a reaction in which heat is absorbed.

Energy diagram (Section 5.2A)   A graph showing the changes in energy that occur during a chemical reaction; energy is plotted on the y-axis, and the progress of the reaction is plotted on the x-axis.

Enol (Section 13.9A)   A molecule containing an —OH group bonded to a carbon of a carbon–carbon double bond.

Enolate anion (Section 16.1A)   An anion formed by the removal of an α-hydrogen from a carbonyl-containing compound.

Epoxide (Section 8.5A)   A cyclic ether in which oxygen is one atom of a three-membered ring.

Epoxy resin (Section 17.5E)   A material prepared by a polymerization in which one monomer contains at least two epoxy groups.

Equatorial position (Section 3.6B)   A position on a chair conformation of a cyclohexane ring that extends from the ring, roughly perpendicular to the imaginary axis of the ring.

Equivalent nuclei (Section 12.6)   A term used in NMR spectroscopy to describe nuclei that give the same signal in terms of chemical shift.

Estrogen (Section 21.5A)   A steroid hormone, such as estradiol, that mediates development and sexual characteristics in females.

Ether (Section 8.4A)   A compound containing an oxygen atom bonded to two carbon atoms.

Exothermic reaction (Section 5.2A)   A reaction in which the energy of the products is lower than the energy of the reactants; a reaction in which heat is liberated.

Fat (Section 21.2B)   A triglyceride that is semisolid or solid at room temperature.

Fatty acid (Section 21.2A)   A long, unbranched carboxylic acid, most commonly of 12 to 20 carbons, derived from the hydrolysis of animal fats, vegetable oils, or the phospholipids of biological membranes.

Fingerprint region (Section 11.4C)   In infrared spectroscopy, the region of spectrum from 1000 to 400 cm^{-1}.

Fischer esterification (Section 14.7)   The process of forming an ester by refluxing a carboxylic acid and an alcohol in the presence of an acid catalyst, commonly sulfuric acid.

Fischer projection (Section 18.2C)   A two-dimensional representation showing the configuration of a stereocenter; horizontal lines represent bonds projecting forward from the stereocenter, vertical lines represent bonds projecting to the rear.

Fishhook arrow (Section 17.6A)   A single-barbed, curved arrow used to show the change in position of a single electron.

Fluid-mosaic model (Section 21.4B)   Model according to which a biological membrane consists of a phospholipid bilayer with proteins, carbohydrates, and other lipids embedded in and on the surface of the bilayer.

Formal charge (Section 1.3E)   The charge on an atom in a molecule or polyatomic ion.

Frequency (Section 11.2)   The number of full cycles of a wave that pass a given point in a second.

Functional group (Section 1.8)   An atom or a group of atoms within a molecule that shows a characteristic set of physical and chemical properties.

Furanose (Section 18.3A)   A five-membered cyclic hemiacetal form of a monosaccharide.

Glass transition temperature T_g (Section 17.4)   The temperature at which a polymer undergoes a transition from a hard glass to a rubbery state.

Glycol (Section 5.4B)   A diol in which the —OH groups are on adjacent carbons.

Glycolysis (Section 22.3)   A series of 10 enzyme-catalyzed reactions by which glucose is oxidized to two molecules of pyruvate.

Glycoside (Section 18.5A)   A carbohydrate in which the —OH on its anomeric carbon is replaced by —OR.

Glycosidic bond (Section 18.5A)   The bond from the anomeric carbon of a glycoside to an —OR group.

Grignard reagent (Section 13.6A)   An organomagnesium compound of the type RMgX or ArMgX.

Ground-state electron configuration (Section 1.2A)   The electron configuration of lowest energy for an atom, a molecule, or an ion.

Halonium ion (Section 5.3C)   An ion in which a halogen atom bears a positive charge.

Haworth projection (Section 18.3A)   A way of viewing furanose and pyranose forms of monosaccharides. The ring is drawn flat and viewed through its edge, with the anomeric carbon on the right and the oxygen atom of the ring in the rear to the right.

Heat of reaction (Section 5.2A)   The difference in energy between reactants and products.

α-Helix (Section 19.6B)   A type of secondary structure in which a section of polypeptide chain coils into a spiral—most commonly a right-handed spiral.

Hemiacetal (Section 13.7)   A molecule containing an —OH and an —OR or —OAr group bonded to the same carbon.

Hertz (Section 11.2)   The SI unit of frequency. One hertz equals one wave cycle per second (s^{-1}).

Heterocyclic amine (Section 10.2)   An amine in which nitrogen is one of the atoms of a ring.

Heterocyclic aromatic amine (Section 10.2)   An amine in which nitrogen is one of the atoms of an aromatic ring.

Heterocyclic compound (Section 9.3)   An organic compound that contains one or more atoms other than carbon in its ring.

High-density lipoprotein (HDL) (Section 21.5A)   Plasma particles with density 1.06–1.21 g/mL and consisting of approximately 33% proteins, 30% cholesterol, 29% phospholipids, and 8% triglycerides.

Histone (Section 20.3C)   A protein that is particularly rich in the basic amino acids lysine and arginine and that is found associated with DNA molecules.

Hybrid orbital (Section 1.7C)   An orbital produced from the combination of two or more atomic orbitals.

sp Hybrid orbital (Section 1.7F)   An orbital produced by the combination of one s atomic orbital and one p atomic orbital.

sp^2 Hybrid orbital (Section 1.7E)   An orbital produced from the combination of one s atomic orbital and two p atomic orbitals.

sp^3 Hybrid orbital (Section 1.7D)   An orbital produced from the combination of one s atomic orbital and three p atomic orbitals.

Hydration (Section 5.3B)   The addition of water.

Hydride ion (Section 13.11B)   A hydrogen atom with two electrons in its valence shell; H:$^-$.

Hydrocarbon (Section 3.1)   A compound that contains only carbon atoms and hydrogen atoms.

α-Hydrogen (Section 13.9A)   A hydrogen on an α-carbon.

Hydrogen bonding (Section 8.2C)   The attractive force between a partial positive charge on hydrogen and a partial negative charge on a nearby oxygen, nitrogen, or fluorine atom.

Hydrophilic (Section 14.4)   From the Greek, meaning "water loving."

Hydrophobic (Section 14.4)   From the Greek, meaning "water hating."

Hydrophobic effect (Section 19.6D)   The tendency of nonpolar groups to cluster in such a way as to be shielded from contact with an aqueous environment.

Hydroxyl group (Section 1.8A)   An —OH group.

Imine (Section 13.8)   A compound containing a carbon–nitrogen double bond; also called a Schiff base.

Index of hydrogen deficiency (Section 11.4J)   The sum of the number of rings and pi bonds in a molecule.

Inductive effect (Section 2.6C)   The polarization of electron density transmitted through covalent bonds caused by a nearby atom of higher electronegativity.

Infrared (IR) spectroscopy (Section 11.3)   The measurement and study of light absorption from the infrared region of the electromagnetic spectrum.

Infrared active (Section 11.4B)   Any vibration that results in the absorption of infrared irradiation. In order for a molecule to absorb infrared radiation, the bond undergoing vibration must be polar and its vibration must cause a periodic change in the bond dipole; the greater the polarity of the bond, the more intense is the absorption.

Integration (Section 12.7)   The process of determining the area under a signal in an NMR spectrum.

Ionic bond (Section 1.3B)   A chemical bond resulting from the electrostatic attraction of an anion and a cation.

Isoelectric point (pI) (Section 19.3C)   The pH at which an amino acid, a polypeptide, or a protein has no net charge.

Ketose (Section 18.2A)   A monosaccharide containing a ketone group.

Lactam (Section 15.2D)   A cyclic amide.

Lactate fermentation (Section 22.5A)   A metabolic pathway that converts glucose to two molecules of pyruvate.

Lactone (Section 15.2C)   A cyclic ester.

Levorotatory (Section 6.9B)   The counterclockwise rotation of the plane of polarized light in a polarimeter.

Lewis acid (Section 2.7)   Any molecule or ion that can form a new covalent bond by accepting a pair of electrons.

Lewis base (Section 2.7)   Any molecule or ion that can form a new covalent bond by donating a pair of electrons.

Lewis structure of an atom (Section 1.2B)   The symbol of an element, surrounded by a number of dots equal to the number of electrons in the valence shell of the atom.

Line-angle formula (Section 3.4)   An abbreviated way to draw structural formulas in which each vertex and line ending represents a carbon.

Lipid (Section 21.1)   A class of biomolecules isolated from plant or animal sources by extraction with nonpolar organic solvents, such as diethyl ether and acetone.

Lipid bilayer (Section 21.4B)   A back-to-back arrangement of phospholipid monolayers.

Local magnetic field (Section 12.4)   The magnetic field generated by electrons surrounding a nucleus.

Low-density lipoprotein (LDL) (Section 21.5A)   Plasma particles with density $1.02–1.06$ g/mL and consisting of approximately 25% proteins, 50% cholesterol, 21% phospholipids, and 4% triglycerides.

Markovnikov's rule (Section 5.3A)   In the addition of HX or H_2O to an alkene, hydrogen adds to the carbon of the double bond having the greater number of hydrogens.

Melt transition temperature, T_m (Section 17.4)   The temperature at which crystalline regions of a polymer melt.

Mercaptan (Section 8.6A)   A common name for any molecule containing an —SH group.

Meso compound (Section 6.5B)   An achiral compound possessing two or more stereocenters.

Messenger RNA (mRNA) (Section 20.4C)   A ribonucleic acid that carries coded genetic information from DNA to ribosomes for the synthesis of proteins.

Meta (m) (Section 9.4B)   Refers to groups occupying positions 1 and 3 on a benzene ring.

Meta director (Section 9.8A)   Any substituent on a benzene ring that directs electrophilic aromatic substitution preferentially to a meta position.

Micelle (Section 21.2B)   A spherical arrangement of organic molecules in water solution, clustered so that their hydrophobic parts are buried inside the sphere and their hydrophilic parts are on the surface of the sphere and in contact with water.

Michael reaction (Section 16.6)   The conjugate addition of an enolate anion or some other nucleophile to an α,β-unsaturated carbonyl compound.

Mirror image (Section 6.3A)   The reflection of an object in a mirror.

Molecular spectroscopy (Section 11.3)   The experimental process of measuring which frequencies of radiation a substance absorbs or emits and then correlating those frequencies with specific types of molecular structures.

Monomer (Section 17.2)   From the Greek *mono*, "single," and *meros*, "part"; the simplest nonredundant unit from which a polymer is synthesized.

Monosaccharide (Section 18.2A)   A carbohydrate that cannot be hydrolyzed to a simpler compound.

D-Monosaccharide (Section 18.2D)   A monosaccharide that, when written as a Fischer projection, has the —OH on its penultimate carbon to the right.

L-Monosaccharide (Section 18.2D)   A monosaccharide that, when written as a Fischer projection, has the —OH on its penultimate carbon to the left.

Mutarotation (Section 18.3C)   The change in optical activity that occurs when an α or β form of a carbohydrate is converted to an equilibrium mixture of the two forms.

(n + 1) rule (Section 12.9)   A rule for determining splitting patterns in NMR spectroscopy. According to the rule, if a hydrogen has n hydrogens nonequivalent to it, but equivalent among themselves, on the same or adjacent atom(s), then the ^{1}H-NMR signal of the hydrogen is split into $(n + 1)$ peaks.

Newman projection (Section 3.6A)   A way to view a molecule by looking along a carbon–carbon bond.

Nonbonded interaction strain (Section 3.6A)   The strain that arises when atoms that are not bonded to each other are forced abnormally close to one another.

Nonbonding electrons (Section 1.3D)   Valence electrons that are not involved in forming covalent bonds; unshared electrons.

Nonpolar covalent bond (Section 1.3C)   A covalent bond between atoms whose difference in electronegativity is less than approximately 0.5.

Nuclear magnetic resonance (NMR) spectroscopy (Section 12.2)   The measurement and study of the interaction of atomic nuclei with radio-frequency radiation under an applied magnetic field.

Nucleic acid (Section 20.2)   A biopolymer containing three types of monomer units: heterocyclic aromatic amine bases derived from purine and pyrimidine, the monosaccharide D-ribose or 2-deoxy-D-ribose, and phosphate.

Nucleophile (Section 7.3)   An atom or a group of atoms that donates a pair of electrons to another atom or group of atoms to form a new covalent bond.

Nucleophilic acyl substitution (Section 15.3)   A reaction in which a nucleophile bonded to a carbonyl carbon is replaced by another nucleophile.

Nucleophilic substitution (Section 7.3)   A reaction in which one nucleophile is substituted for another.

Nucleoside (Section 20.2)   A building block of nucleic acids, consisting of D-ribose or 2-deoxy-D-ribose bonded to a heterocyclic aromatic amine base by a β-N-glycosidic bond.

Nucleotide (Section 20.2)   A nucleoside in which a molecule of phosphoric acid is esterified with an —OH of the monosaccharide, most commonly either the $3'$—OH or the $5'$—OH.

Observed rotation (Section 6.9B)   The number of degrees through which a compound rotates the plane of polarized light.

Octane rating (Section 3.10B)   The percentage of isooctane in a mixture of isooctane and heptane that has knock properties equivalent to the gasoline being tested.

Octet rule (Section 1.3A)   The tendency among atoms of Group 1A–7A elements to react in ways that achieve an outer shell of eight valence electrons.

Oil (Section 21.2B)   A triglyceride that is liquid at room temperature.

Oligosaccharide (Section 18.8)   A carbohydrate containing from 4 to 10 monosaccharide units, each joined to the next by a glycosidic bond.

Optically active (Section 6.9)   Showing that a compound rotates the plane of polarized light.

Orbital (Section 1.2)   A region of space where an electron or a pair of electrons spends 90 to 95% of its time.

Order of precedence of functional groups (Section 13.3B)   A system for ranking functional groups in order of priority for the purposes of IUPAC nomenclature.

Organometallic compound (Section 13.6A)   A compound containing a carbon–metal bond.

Ortho (o) (Section 9.4B)   Refers to groups occupying positions 1 and 2 on a benzene ring.

Ortho–para director (Section 9.8A)   Any substituent on a benzene ring that directs electrophilic aromatic substitution preferentially to ortho and para positions.

Oxidation (Section 5.4)   The addition of O to, or the removal of H from, a carbon atom.

β-Oxidation (Section 22.6A)   A series of four enzyme-catalyzed reactions that cleaves carbon atoms, two at a time, from the carboxyl end of a fatty acid.

Oxonium ion (Section 5.3B)   An ion in which oxygen is bonded to three other atoms and bears a positive charge.

Para (p) (Section 9.4B)   Refers to groups occupying positions 1 and 4 on a benzene ring.

Penultimate carbon (Section 18.2D)   The stereocenter of a monosaccharide farthest from the carbonyl group—for example, carbon 5 of glucose.

Peptide bond (Section 19.4)   The special name given to the amide bond formed between the α-amino group of one amino acid and the α-carboxyl group of another amino acid.

Phenol (Section 9.9A)   A compound that contains an —OH bonded to a benzene ring.

Phenyl group (Section 9.4A)   The C_6H_5— group.

Phospholipid (Section 21.4A)   A lipid containing glycerol esterified with two molecules of fatty acid and one molecule of phosphoric acid.

Photon (Section 11.2)   A particle of electromagnetic radiation.

Pi (π) bond (Section 1.7E)   A covalent bond formed from the overlap of parallel p orbitals.

Plane of symmetry (Section 6.3A)   An imaginary plane passing through an object and dividing it such that one half is the mirror image of the other half.

Plane polarized light (Section 6.9A)   Light vibrating in only parallel planes.

Plastic (Section 17.2)   A polymer that can be molded when hot and that retains its shape when cooled.

β-Pleated sheet (Section 19.6B)   A type of secondary structure in which two sections of polypeptide chain are aligned parallel or antiparallel to one another.

Polar covalent bond (Section 1.3C)   A covalent bond between atoms whose difference in electronegativity is between approximately 0.5 and 1.9.

Polarimeter (Section 6.9B)   An instrument for measuring the ability of a compound to rotate the plane of polarized light.

Polyamide (Section 17.5A)   A polymer in which each monomer unit is joined to the next by an amide bond—for example, nylon 66.

Polycarbonate G17 (Section 17.5C)   A polyester in which the carboxyl groups are derived from carbonic acid.

Polyester (Section 17.5B)   A polymer in which each monomer unit is joined to the next by an ester bond—for example, poly(ethylene terephthalate).

Polymer (Section 17.2)   From the Greek *poly*, "many," and *meros*, "part"; any long-chained molecule synthesized by linking together many single parts called monomers.

Polynuclear aromatic hydrocarbon (Section 9.4C)   A hydrocarbon containing two or more fused aromatic rings.

Polypeptide (Section 19.4)   A macromolecule containing 10 or more amino acid units, each joined to the next by a peptide bond.

Polysaccharide (Section 18.9)   A carbohydrate containing a large number of monosaccharide units, each joined to the next by one or more glycosidic bonds.

Polyunsaturated fatty acid (Section 21.2A)   A fatty acid with two or more carbon–carbon double bonds in its hydrocarbon chain.

Polyunsaturated triglyceride (Section 21.2B)   A triglyceride having several carbon–carbon double bonds in the hydrocarbon chains of its three fatty acids.

Polyurethane (Section 17.5D)   A polymer containing the —NHCOO— group as a repeating unit.

Primary (1°) amine (Section 1.8B)   An amine in which one hydrogen of ammonia has been replaced by an alkyl or aryl group.

Primary (1°) carbon (Section 3.3C)   A carbon bonded to one other carbon atom.

Primary (1°) structure of proteins (Section 19.5A)   The sequence of amino acids in the polypeptide chain read from the *N*-terminal amino acid to the *C*-terminal amino acid.

Primary (1°) structure of nucleic acids (Section 20.3A)   The sequence of bases along the pentose–phosphodiester backbone of a DNA or RNA molecule read from the 5′ end to the 3′ end.

Prostaglandin (Section 21.6)   A member of the family of compounds having the 20-carbon skeleton of prostanoic acid.

Protic solvent (Section 7.6D)   A hydrogen-bond-donor solvent—for example, water, ethanol, and acetic acid.

Pyranose (Section 18.3A)   A six-membered cyclic hemiacetal form of a monosaccharide.

Quaternary (4°) carbon (Section 3.3C)   A carbon bonded to four other carbon atoms.

Quaternary (4°) structure of proteins (Section 19.6D)   The arrangement of polypeptide monomers into a noncovalently bonded aggregation.

R— (Section 3.3A)   A symbol used to represent an alkyl group.

R (Section 6.4)   From the Latin *rectus*, meaning "right"; used in the R,S system to show that the order of priority of groups on a stereocenter is clockwise.

R,S system (Section 6.4)   A set of rules for specifying the configuration about a stereocenter.

Racemic mixture (Section 6.9C)   A mixture of equal amounts of two enantiomers.

Racemization (Section 13.9B)   The conversion of a pure enantiomer into a racemic mixture.

Radical (Section 17.6A)   Any molecule that contains one or more unpaired electrons.

Rate-determining step (Section 5.2A)   The step in a reaction sequence that crosses the highest energy barrier; the slowest step in a multistep reaction.

Reaction coordinate (Section 5.2A)   A measure of the progress of a reaction, plotted on the *x*-axis in an energy diagram.

Reaction intermediate (Section 5.2A)   An unstable species that lies in an energy minimum between two transition states.

Reducing sugar (Section 18.5B)   A carbohydrate that reacts with an oxidizing agent to form an aldonic acid.

Reduction (Section 5.4)   The removal of O from, or the addition of H to, a carbon atom.

Reductive amination (Section 13.8B)   The formation of an imine from an aldehyde or a ketone, followed by its reduction to an amine.

Regioselective reaction (Section 5.3A)   A reaction in which one direction of bond forming or bond breaking occurs in preference to all other directions.

Relative nucleophilicity (Section 7.6A)   The relative rate at which a nucleophile reacts in a reference nucleophilic substitution reaction.

Resolution (Section 6.9C)   The separation of a racemic mixture into its enantiomers.

Resonance (Section 12.3)   The absorption of electromagnetic radiation by a spinning nucleus and the resulting flip of its nuclear spin state.

Resonance-contributing structures (Section 1.6A)   Representations of a molecule or an ion that differ only in the distribution of valence electrons.

Resonance energy (Section 9.2D)   The difference in energy between a resonance hybrid and the most stable of its hypothetical contributing structures.

Resonance hybrid (Section 1.6A)   A molecule or an ion that is best described as a composite of a number of contributing structures.

Restriction endonuclease (Section 20.A)   An enzyme that catalyzes the hydrolysis of a particular phosphodiester bond within a DNA strand.

Ribosomal RNA (rRNA) (Section 20.4A)   A ribonucleic acid found in ribosomes, the sites of protein synthesis.

S (Section 6.4)   From the Latin *sinister*, meaning "left"; used in the R,S system to show that the order of priority of groups on a stereocenter is counterclockwise.

Sanger dideoxy method (Section 20.6D)   A method, developed by Frederick Sanger, for sequencing DNA molecules.

Saponification (Section 15.4C)   Hydrolysis of an ester in aqueous NaOH or KOH to an alcohol and the sodium or potassium salt of a carboxylic acid.

Saturated hydrocarbon (Section 3.1)   A hydrocarbon containing only carbon–carbon single bonds.

Secondary (2°) amine (Section 1.8B)   An amine in which two hydrogens of ammonia have been replaced by alkyl or aryl groups.

Secondary (2°) carbon (Section 3.3C)   A carbon bonded to two other carbon atoms.

Secondary (2°) structure of proteins (Section 19.6B)   The ordered arrangement (conformation) of amino acids in localized regions of a polypeptide or protein.

Secondary (2°) structure of nucleic acids (Section 20.3B)   The ordered arrangement of nucleic acid strands.

Shell (Section 1.2)   A region of space around a nucleus where electrons are found.

Shielding (Section 12.4)   The phenomenon in NMR spectroscopy in which local magnetic fields from electrons surrounding a nucleus decrease the ability of an applied magnetic field to bring the nucleus into resonance.

Sigma (σ) bond (Section 1.7B)   A covalent bond in which the overlap of atomic orbitals is concentrated along the bond axis.

Signal splitting (Section 12.9)   A phenomenon in NMR spectroscopy in which the ^{1}H-NMR signal from one set of hydrogens is split by the influence of neighboring nonequivalent hydrogens.

Soap (Section 21.2A)   A sodium or potassium salt of a fatty acid.

Solvolysis (Section 7.6A)   A nucleophilic substitution reaction in which the solvent is the nucleophile.

Specific rotation (Section 6.9B)   The observed rotation of the plane of polarized light when a sample is placed in a tube 1.0 dm long and at a concentration of 1.0 g/100 mL; if a pure sample is used, its concentration is given in g/mL (i.e., its density).

Staggered conformation (Section 3.6A)   A conformation about a carbon–carbon single bond in which the atoms on one carbon are as far apart as possible from the atoms on the adjacent carbon.

Step-growth polymerization (Section 17.5)   A polymerization in which chain growth occurs in a stepwise manner between difunctional monomers; e. g., between adipic acid and hexamethylenediamine to form nylon 66.

Stereocenter (Section 6.3A)   A tetrahedral atom, most commonly carbon, that has four different groups bonded to it.

Stereoisomers (Section 6.2)   Isomers that have the same molecular formula and the same connectivity, but different orientations of their atoms in space.

Stereoselective reaction (Section 5.3C)   A reaction in which one stereoisomer is formed or destroyed in preference to all others that might be formed or destroyed.

Steric hindrance (Section 7.6B)   The ability of groups, because of their size, to hinder access to a reaction site within a molecule.

Steroid (Section 21.5A)   A plant or an animal lipid having the characteristic tetracyclic ring structure of the steroid nucleus, namely, three six-membered rings and one five-membered ring.

Strong acid (Section 2.4)   An acid that is completely ionized in aqueous solution.

Strong base (Section 2.4)   A base that is completely ionized in aqueous solution.

Syn addition (Section 5.4A)   Addition of atoms or groups of atoms from the same side or face of a carbon–carbon double bond.

Tautomers (Section 13.9A)   Constitutional isomers that differ in the location of hydrogen and a double bond relative to O, N, or S.

***C*-Terminal amino acid** (Section 19.5)   The amino acid at the end of a polypeptide chain having the free —COOH group.

***N*-Terminal amino acid** (Section 19.5)   The amino acid at the end of a polypeptide chain having the free —NH$_2$ group.

Terpene (Section 4.4)   A compound whose carbon skeleton can be divided into two or more units identical to the carbon skeleton of isoprene.

Tertiary (3°) amine (Section 1.8B)   An amine in which three hydrogens of ammonia have been replaced by alkyl or aryl groups.

Tertiary (3°) carbon (Section 3.3C)   A carbon bonded to three other carbon atoms.

Tertiary (3°) structure of proteins (Section 19.6C)   The three-dimensional arrangement in space of all atoms in a single polypeptide chain.

Tertiary (3°) structure of nucleic acids (Section 20.3C)   The three-dimensional arrangement of all atoms of a nucleic acid, commonly referred to as *supercoiling*.

Tesla (Section 12.2)   The SI unit for the strength and amount of magnetism of a magnetic field.

Thermoplastic (Section 17.2)   A polymer that can be melted and molded into a shape that is retained when it is cooled.

Thermosetting plastic (Section 17.2)   A polymer that can be molded when it is first prepared, but, once cooled, hardens irreversibly and cannot be remelted.

Thioester (Section 22.6A)   An ester in which one atom of oxygen in the carboxylate group is replaced by an atom of sulfur.

Thiol (Section 8.6A)   A compound containing an —SH (sulfhydryl) group.

Torsional strain (Section 3.6A; also called eclipsed interaction strain)   Strain that arises when atoms separated by three bonds are forced from a staggered conformation to an eclipsed conformation.

Trans (Section 3.7)   A prefix meaning "across from."

Transfer RNA (tRNA) (Section 20.4B)   A ribonucleic acid that carries a specific amino acid to the site of protein synthesis on ribosomes.

Transition state (Section 5.2A)   An unstable species of maximum energy formed during the course of a reaction; a maximum on an energy diagram.

Triglyceride (triacylglycerol) (Section 21.2)   An ester of glycerol with three fatty acids.

Tripeptide (Section 19.4)   A molecule containing three amino acid units, each joined to the next by a peptide bond.

Unimolecular reaction (Section 7.5B)   A reaction in which only one species is involved in the reaction leading to the transition state of the rate-determining step.

Upfield (Section 12.5)   A term used in NMR spectroscopy to denote that a signal is toward the right of the spectrum or of another signal.

Valence electrons (Section 1.2B)   Electrons in the valence (outermost) shell of an atom.

Valence shell (Section 1.2B)   The outermost electron shell of an atom.

Watson–Crick model (Section 20.3B)   A double-helix model for the secondary structure of a DNA molecule.

Wavelength (Section 11.2)   The distance between any two consecutive identical points on a wave.

Wavenumber (Section 11.3)   The number of waves per centimeter. The unit of wavenumber is the reciprocal centimeter (cm^{-1}).

Z (Section 4.2C)   From the German *zusammen*, "together"; specifies that groups of higher priority on the carbons of a double bond are on the same side.

Zaitsev's rule (Section 7.8)   A rule stating that the major product from a β-elimination reaction is the most stable alkene—that is, the alkene with the greatest number of substituents on the carbon–carbon double bond.

Zwitterion (Section 19.2A)   An internal salt of an amino acid.

Answers Section

Chapter 1
Covalent Bonding and Shapes of Molecules

1.1 Ground-state electron configurations are:
- **(a)** C $1s^2 2s^2 2p^2$ Si $1s^2 2s^2 2p^6 3s^2 3p^2$
  both have 4 valence electrons
- **(b)** O $1s^2 2s^2 2p^4$ S $1s^2 2s^2 2p^6 3s^2 3p^4$
  both have 6 valence electrons
- **(c)** N $1s^2 2s^2 2p^3$ P $1s^2 2s^2 2p^6 3s^2 3p^3$
  both have 5 valence electrons

1.2 The electron configuration of S(16) is $1s^2 2s^2 2p^6 3s^2 3p^4$. The electron configuration of S^{2-} is $1s^2 2s^2 2p^6 3s^2 3p^6$. In gaining two valence electrons, sulfur now has the same electron configuration as Ar, the noble gas nearest it in atomic number.

1.3 **(a)** Li, **(b)** N, and **(c)** C.

1.4 **(a)** S—H nonpolar covalent **(b)** P—H nonpolar covalent
(c) C—F polar covalent **(d)** C—Cl polar covalent

1.5 **(a)** $\overset{\delta+}{C}-\overset{\delta-}{N}$ **(b)** $\overset{\delta+}{N}-\overset{\delta-}{O}$ **(c)** $\overset{\delta+}{C}-\overset{\delta-}{Cl}$

1.6 (a, 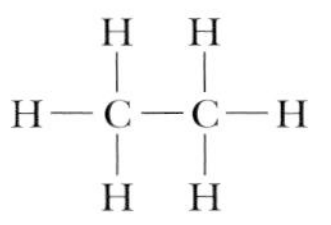 b, 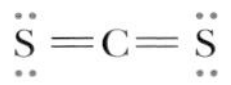 c — structural formulas)

1.7 (a, 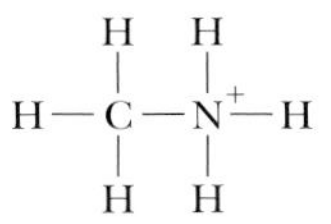 b 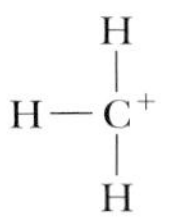 — structural formulas)

1.8 (a) 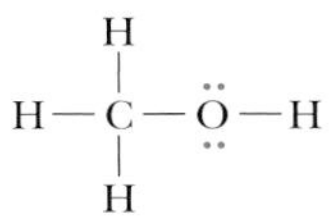 all bond angles are 109.5° (b) 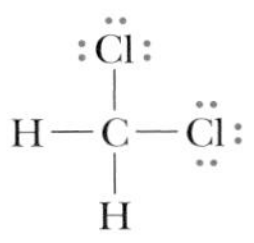 all bond angles are 109.5°
(c) 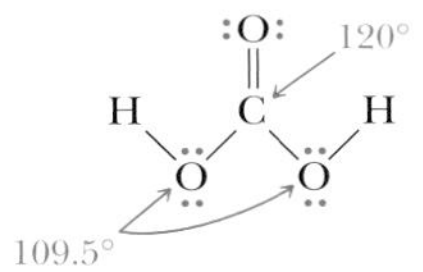 120°, 109.5°

1.9 Carbon dioxide has two polar C=O bonds but, because it is a linear molecule, it has no dipole moment. Sulfur dioxide is a bent molecule with two polar S=O bonds and, therefore, has a dipole moment.

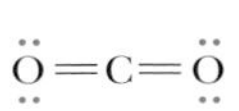

Carbon dioxide 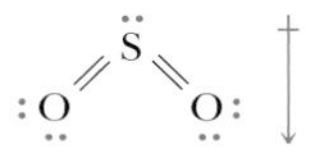   Sulfur dioxide

1.10 Only **(a)** represents a pair of resonance contributing structures.

1.11 CH_3-C ... (a) (b) (c)

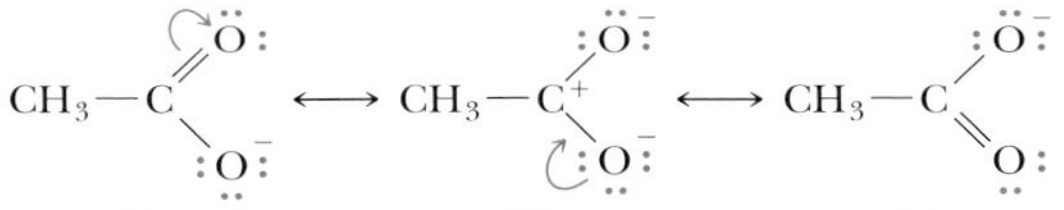

1.12 Predict bond angles of 109.5° about each sp^3 hybridized atom and bond angles of 120° about each sp^2 hybridized atom.

(a) 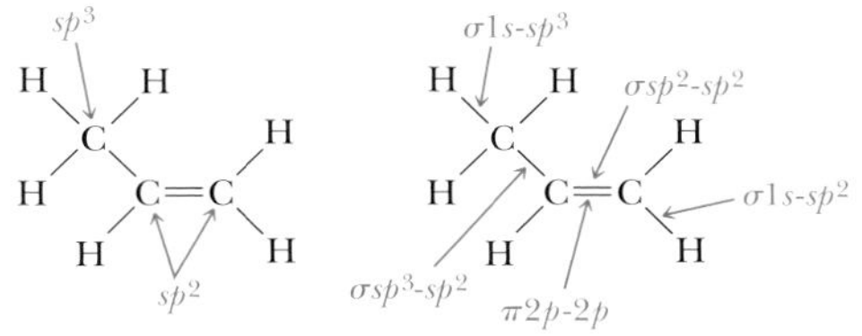 sp^3, sp^2 ... $\sigma 1s\text{-}sp^3$, $\sigma sp^2\text{-}sp^2$, $\sigma 1s\text{-}sp^2$, $\sigma sp^3\text{-}sp^2$, $\pi 2p\text{-}2p$

(b) 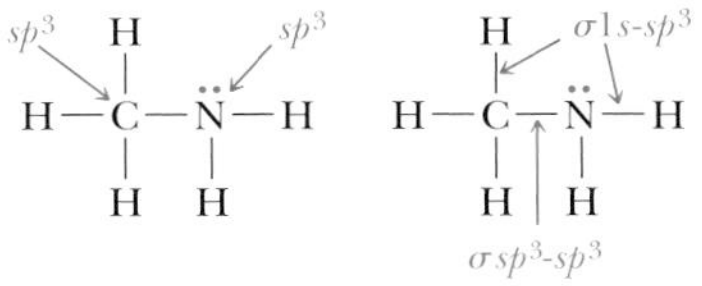 sp^3, sp^3 ... $\sigma 1s\text{-}sp^3$, $\sigma sp^3\text{-}sp^3$

1.13 $CH_3CH_2CH_2CH_2OH$ (1°) $CH_3CHCH_2CH_3$ with OH (2°)

CH_3CHCH_2OH with CH_3 (1°) CH_3COH with CH_3 and CH_3 (3°)

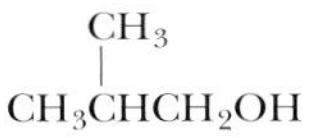

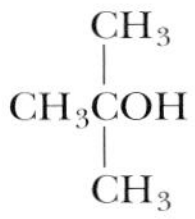

1.14 $CH_3CH_2CH_2-NH-CH_3$ $CH_3\overset{\overset{\displaystyle CH_3}{|}}{CH}-NH-CH_3$

$CH_3CH_2-NH-CH_2CH_3$

1.15 $CH_3\overset{\overset{\displaystyle O}{\|}}{C}CH_2CH_2CH_3$ $CH_3CH_2\overset{\overset{\displaystyle O}{\|}}{C}CH_2CH_3$ $CH_3\underset{\underset{\displaystyle CH_3}{|}}{CH}\overset{\overset{\displaystyle O}{\|}}{C}CH_3$

1.16 $CH_3CH_2CH_2\overset{\overset{\displaystyle O}{\|}}{C}OH$ $CH_3\underset{\underset{\displaystyle CH_3}{|}}{CH}\overset{\overset{\displaystyle O}{\|}}{C}OH$

1.17 **(a)** Na(11) $1s^2 2s^2 2p^6 3s^1$ **(b)** Mg(12) $1s^2 2s^2 2p^6 3s^2$

(c) O(8) $1s^2 2s^2 2p^4$ **(d)** N(7) $1s^2 2s^2 2p^3$

1.19 **(a)** Sulfur   **(b)** Oxygen

1.21 **(a)** Valence shell is the outermost electron shell of an atom.

(b) A valence electron is an electron in the valence (outermost) shell of an atom.

1.23 **(a)** H^+ has no electrons in its valence shell.

(b) H^- has two electrons in its valence shell.

1.25 **(a)** LiF ionic

(b) C—H nonpolar covalent, C—F polar covalent

(c) Mg—Cl polar covalent

(d) H—Cl polar covalent

1.27 Following are Lewis structures for each molecule.

(a) H—Ö—Ö—H   **(b)** H—N̈—N̈—H with H below each N

(c) H—C—Ö—H (with H above and below C)

(d) H—C—S̈—H (with H above and below C)

(e) H—C—N̈—H (with H above, below C and below N)

(f) H—C—C̈l: (with H above and below C)

(g) H—C—Ö—C—H (with H above and below each C)

(h) H—C—C—H (with H above and below each C)

(i) $H_2C=CH_2$ structure

(j) H—C≡C—H

(k) Ö=C=Ö

(l) H₂C=O structure (formaldehyde)

(m) H—C—C—C—H with :O: double bond on middle C

(n) H—Ö—C—Ö—H with :O: double bond on C

(o) H—C—C—Ö—H with :O: double bond on second C

1.29 Carbon cannot have more than four bonds. In each of these molecules, one carbon would have to have five bonds, which would put 10 electrons in its valence shell.

1.31 **(a)** H—C—C—C—H structure with :O: and O⁻

(b) H—N̈—C=C—H structure

(c) H—C—Ö⁺—H structure

(d) H—C: structure

1.33 The compound is silver oxide, Ag_2O. Given the difference in electronegativity between silver and oxygen ($3.5 - 1.9 = 1.6$) predict polar covalent bonds.

1.35 The two parameters that lead to maximum electronegativity are increasing positive charge on the nucleus and decreasing atomic radius (the distance between a nucleus and the electrons in the valance shell). Fluorine is the element for which these two parameters lead to maximum electronegativity.

1.37 **(a)** $\overset{\delta-}{O}-\overset{\delta+}{H}$ **(b)** $\overset{\delta-}{N}-\overset{\delta+}{H}$ **(c)** $\overset{\delta-}{S}-\overset{\delta+}{H}$ **(d)** $\overset{\delta+}{H}-\overset{\delta-}{F}$

1.39 **(a)** $\overset{\delta-}{C}-\overset{\delta+}{Pb}$ **(b)** $\overset{\delta+}{C}-\overset{\delta-}{Mg}-\overset{\delta-}{Cl}$ **(c)** $\overset{\delta-}{C}-\overset{\delta+}{Hg}$

polar covalent   both polar covalent   polar covalent

1.41 Unless otherwise marked, all bond angles are 109.5°.

(a) $CH_3-CH_2-CH_2-\overset{..}{\underset{..}{O}}H$ **(b)** $CH_3-CH_2-\overset{\overset{\displaystyle :O:}{\|}}{C}-H$ (120°)

(c) $CH_3-CH=CH_2$ (120°)   **(d)** $CH_3-C\equiv C-CH_3$ (180°)

(e) $CH_3-\overset{\overset{\displaystyle :O:}{\|}}{C}-\overset{..}{\underset{..}{O}}-CH_3$ (120°)   **(f)** $CH_3-\overset{..}{N}-CH_3$ with CH₃ above

1.43 All molecules, except CCl_4, have dipole moments.

(a) CH_3F **(b)** CH_2Cl_2

(c) $CHCl_3$ **(d)** CCl_4

(e) $CH_2=CCl_2$

(f) $CH_2=CHCl$ **(g)** $H_3C-C\equiv N$

(h) $(H_3C)_2C=O$ structure

1.45 CCl_3F CCl_2F_2

1.47 (a) $H-\overset{..}{\underset{..}{O}}-C\overset{\displaystyle :\overset{..}{\underset{..}{O}}:^-}{\underset{\displaystyle O:}{\Big\langle}}$ (b) $\underset{H}{\overset{H}{\diagdown}}C^+-\overset{..}{\underset{..}{O}}:^-$

(c) $CH_3-\overset{..}{\underset{..}{O}}-C^+\overset{\displaystyle :\overset{..}{\underset{..}{O}}:^-}{\underset{\displaystyle O:}{\Big\langle}}$

1.49 (a) sp^3 (b) sp^2 (c) sp (d) sp^3
(e) C is sp^2; O is sp^3 (f) sp^2

1.51 (a) structure with C, $:\overset{..}{O}:$ double bond   (b) structure with C, $:\overset{..}{O}:$, $\overset{..}{\underset{..}{O}}$, H

(c) $-\overset{..}{\underset{..}{O}}-H$ (d) $-\overset{..}{\underset{H}{N}}-H$

1.53 (a)
$$CH_3CH_2CH_2\overset{O}{\overset{\|}{C}}H \qquad \underset{\underset{CH_3}{|}}{CH_3\overset{O}{\overset{\|}{C}}HCH} \qquad CH_3CH_2\overset{O}{\overset{\|}{C}}CH_3$$

(b)
$$\underset{CH=CHCH_2CH_3}{\overset{OH}{\overset{|}{}}} \qquad \underset{CH_2=CCH_2CH_3}{\overset{OH}{\overset{|}{}}}$$

$$\underset{CH_2=CHCHCH_3}{\overset{OH}{\overset{|}{}}} \qquad \underset{CH_2=CHCH_2CH_2}{\overset{OH}{\overset{|}{}}}$$

$$\underset{CH_2CH=CHCH_3}{\overset{OH}{\overset{|}{}}} \qquad \underset{CH_3C=CHCH_3}{\overset{OH}{\overset{|}{}}}$$

$$\underset{\underset{CH_3}{|}}{\overset{OH}{\overset{|}{CH_2C=CH_2}}} \qquad \underset{\underset{CH_3}{|}}{\overset{OH}{\overset{|}{CH_3C=CH}}}$$

1.55 (a) a 2° hydroxyl group and a carboxyl group
(b) two 1° hydroxyl groups
(c) a 1° amino group and a carboxyl group
(d) one 1° and one 2° hydroxyl group, and a carbonyl (aldehyde) group
(e) a carbonyl (ketone) group and a carboxyl group
(g) two 1° amino groups

1.57 $\underset{\underset{HO}{|}\;\underset{OH}{|}}{CH_3CHCH_2}$

1.59 Carbon dioxide is a linear molecule. Ozone is a bent molecule; the O—O—O bond angle is approximately 120°.

$$\overset{sp}{\overset{\displaystyle\nwarrow}{}}\;\;\overset{..}{\underset{..}{O}}=C=\overset{..}{\underset{..}{O}} \qquad :\overset{..}{O}\;\overset{\overset{..}{O}^+}{}\;\overset{..}{\underset{..}{O}}:^- \longleftrightarrow \;^-:\overset{..}{\underset{..}{O}}\;\overset{\overset{..}{O}^+}{}\;\overset{..}{\underset{..}{O}}:$$
$$\underset{sp^2}{}$$

Ozone is best represented as a hybrid of two contributing structures; each oxygen is sp^2 hybridized

1.61 (b) Sulfur is sp^3 hybridized; it is surrounded by four regions of electron density.
(c) Predict bond angles about sulfur of 109.5°.

(d) Dimethyl sulfoxide contains a polar S=O bond and the molecular is polar.

$$H-\underset{\underset{H}{|}}{\overset{\overset{H}{|}}{C}}-\underset{..}{\overset{:O:\;\;H}{\overset{\|}{S}}}-\underset{\underset{H}{|}}{\overset{\overset{H}{|}}{C}}-H$$

1.63 (b) The positively charged carbon has six electrons in its valence shell.
(c) Predict 120° for the C—C—C bond angle.
(d) The hybridization of this carbon is sp^2.
Following is a Lewis structure for this cation.

$$\underset{\underset{H\;\;H\;\;H}{|\;\;\;\;\;|\;\;\;\;\;|}}{\overset{sp^3\;\;\overset{H\;\;\overset{sp^2}{}\;\;H}{|\;\;\;\;\;\;\;\;\;|}\;\;sp^3}{H-C-C^+-C-H}}$$

1.65 Benzene is a hybrid of two equivalent resonance contributing structures. Because the contributing structures have identical patterns of covalent bonding, the result is a resonance hybrid that has equivalent C—C bonds that have a length almost midway between a C—C bond (1.54×10^{-10}m) and C=C bond (1.33×10^{-10}m).

Chapter 2
Acids and Bases

2.1

(a) $CH_3-\overset{..}{\underset{..}{S}}-H + :\overset{..}{\underset{..}{O}}-H \longrightarrow CH_3-\overset{..}{\underset{..}{S}}:^- + H-\overset{..}{\underset{..}{O}}-H$
Acid — Base — Conjugate base of CH_3SH — Conjugate acid of OH^-

(b) $CH_3-\overset{..}{\underset{..}{O}}-H + :\underset{\underset{H}{|}}{N}-H \longrightarrow CH_3-\overset{..}{\underset{..}{O}}:^- + H-\underset{\underset{H}{|}}{N}-H$
Acid — Base — Conjugate base of CH_3OH — Conjugate acid of NH_2^-

2.2 (a) 4.76
(b) 15.7 Acetic acid (pK_a 4.76) is a stronger acid than water (pK_a 15.7).

2.3 (a) $CH_3NH_2 + CH_3COOH \rightleftharpoons CH_3NH_3^+ + CH_3COO^-$
stronger base — stronger acid — weaker acid — weaker base
pK_a 4.76 — pK_a 10.6

(b) $CH_3CH_2O^- + NH_3 \rightleftharpoons CH_3CH_2OH + NH_2^-$
weaker base — weaker acid — stronger acid — stronger base
pK_a 38 — pK_a 15.9

2.4 $CH_3-\overset{..}{\underset{..}{O}}:^- + CH_3-\overset{\overset{\overset{H}{|}}{}}{\underset{\underset{CH_3}{|}}{N^+}}-CH_3 \rightleftharpoons$

$$CH_3-\overset{..}{\underset{..}{O}}-H + CH_3-\underset{\underset{CH_3}{|}}{\overset{..}{N}}-CH_3$$

2.5 (a) $:\!\ddot{C}l\!:^- + \;Al\!-\!Cl \longrightarrow :\!Cl\!-\!\overset{-}{Al}\!-\!\ddot{C}l\!:$ (with Cl substituents above and below Al)

(b) $CH_3\!-\!\ddot{C}l\!: + \;Al\!-\!\ddot{C}l\!: \longrightarrow CH_3\!-\!\overset{+}{\ddot{C}l}\!-\!\overset{-}{Al}\!-\!\ddot{C}l\!:$

2.7 (a) $CH_3\!-\!\underset{H}{\overset{|}{N}}\!-\!H + H\!-\!\ddot{O}\!-\!H \rightleftharpoons$ pK$_a$ 15.7

$CH_3\!-\!\overset{+}{\underset{H}{\overset{H}{N}}}\!-\!H + \;^-\!\ddot{O}\!-\!H$ pK$_a$ 10.6

(b) $H\!-\!\ddot{O}\!-\!\overset{\overset{\displaystyle O}{\|}}{S}\!-\!\ddot{O}\!: + H\!-\!\ddot{O}\!-\!H \rightleftharpoons$ pK$_a$ 15.7

$H\!-\!\ddot{O}\!-\!\overset{\overset{\displaystyle O}{\|}}{\underset{\underset{\displaystyle O}{\|}}{S}}\!-\!\ddot{O}\!-\!H + \;^-\!\ddot{O}\!-\!H$ pK$_a$ −5.2

(c) $:\!\ddot{Br}\!: + H\!-\!\ddot{O}\!-\!H \rightleftharpoons H\!-\!\ddot{Br}\!: + \;^-\!\ddot{O}\!-\!H$ pK$_a$ 15.7   pK$_a$ −8

(d) $\;^-\!\ddot{O}\!-\!\overset{\overset{\displaystyle O}{\|}}{C}\!-\!\ddot{O}\!: + H\!-\!\ddot{O}\!-\!H \rightleftharpoons$ pK$_a$ 15.7

$\;^-\!\ddot{O}\!-\!\overset{\overset{\displaystyle O}{\|}}{C}\!-\!\ddot{O}\!-\!H + \;^-\!\ddot{O}\!-\!H$ pK$_a$ 6.36

2.9 Drawn first is a Lewis structure for each base and then its conjugate acid.

(a) $CH_3CH_2\!-\!\ddot{O}\!-\!H$ $CH_3CH_2\!-\!\overset{+}{\underset{H}{\overset{|}{O}}}\!-\!H$

(b) $H\!-\!\overset{\overset{\displaystyle :O:}{\|}}{C}\!-\!H$ $H\!-\!\overset{\overset{\displaystyle \overset{+}{O}:\,H}{\|}}{C}\!-\!H$

(c) $CH_3\!-\!\underset{CH_3}{\overset{|}{\ddot{N}}}\!-\!H$ $CH_3\!-\!\overset{+}{\underset{CH_3}{\overset{H}{N}}}\!-\!H$

(d) $H\!-\!\ddot{O}\!-\!\overset{\overset{\displaystyle :O:}{\|}}{C}\!-\!\ddot{O}\!:^-$ $H\!-\!\ddot{O}\!-\!\overset{\overset{\displaystyle :O:}{\|}}{C}\!-\!\ddot{O}\!-\!H$

2.11 (a) $H_3C\!-\!\overset{\overset{\displaystyle O}{\|}}{C}\!-\!\underset{\longrightarrow H}{\overset{|}{CH}}\!-\!\overset{\overset{\displaystyle O}{\|}}{C}\!-\!CH_3$

$\longrightarrow \overset{H}{\underset{\overset{\displaystyle N}{\|}}{\diagdown}}\overset{H}{\diagup}$ (with + on N)

(b) $H_2N\!-\!\overset{\overset{\displaystyle H}{\underset{+}{\|}}}{C}\!-\!NH_2$

2.13 (a) Pyruvic acid   (b) Phosphoric acid
(c) Aspirin   (d) Acetic acid

2.15 (a) $HOCO^- < NH_3 < CH_3CH_2O^-$

(b) $CH_3\overset{\overset{\displaystyle O}{\|}}{C}O^- < HO\overset{\overset{\displaystyle O}{\|}}{C}O^- < OH^-$

(c) $H_2O < CH_3\overset{\overset{\displaystyle O}{\|}}{C}O^- < NH_3$

(d) $CH_3\overset{\overset{\displaystyle O}{\|}}{C}O^- < OH^- < NH_2^-$

2.17 (a) Yes   (b) No   (c) No

2.19 In acid-base equilibria, the position of the equilibrium favors the reaction of the stronger acid and stronger base to give the weaker acid and weaker base. The acid with the higher pK$_a$ is the weaker acid, so the arrow will point toward it.

2.21 (a) $CH_3\!-\!\overset{+}{CH}\!-\!CH_3 + CH_3\!-\!\ddot{O}\!-\!H \longrightarrow CH_3\!-\!CH\!-\!CH_3$ (with $\overset{+}{\underset{H}{\overset{CH_3}{O}}}$ substituent)

(b) $CH_3\!-\!\overset{+}{CH}\!-\!CH_3 + :\!\ddot{Br}\!:^- \longrightarrow CH_3\!-\!\underset{:\ddot{Br}:}{\overset{|}{CH}}\!-\!CH_3$

(c) $CH_3\!-\!\underset{CH_3}{\overset{\overset{\displaystyle CH_3}{|}}{\overset{+}{C}}} + H\!-\!\ddot{O}\!-\!H \longrightarrow CH_3\!-\!\underset{H_3C}{\overset{\overset{\displaystyle H_3C}{|}}{C}}\!-\!\overset{+}{\underset{H}{\overset{H}{O}}}$

2.23 (a) $CH_3CH_2OH + HCO_3^- \rightleftharpoons CH_3CH_2O^- + H_2CO_3$
pK$_a$ 15.9   pK$_a$ 6.36

(b) $CH_3CH_2OH + OH^- \rightleftharpoons CH_3CH_2O^- + H_2O$
pK$_a$ 15.9   pK$_a$ 15.7

(c) $CH_3CH_2OH + NH_2^- \rightleftharpoons CH_3CH_2O^- + NH_3$
pK$_a$ 15.9   pK$_a$ 38

(d) $CH_3CH_2OH + NH_3 \rightleftharpoons CH_3CH_2O^- + NH_4^+$
pK$_a$ 15.9   pK$_a$ 9.25

2.25 (a) Yes   (b) Yes   (c) Yes

2.27 The acidity of a hydrogen on CH_3OCH_3 is determined by the stability of the anion resulting from loss of H^+ from one of the CH_3 groups. Because the electronegativity of carbon is only 2.5, the resulting anion is not a stable one, which means that a C—H bond of this compound is a very weak acid.

2.29 Alanine is better represented by (B). As drawn in (A), alanine has within its structure both an acid (the COOH group) and a base (the NH_2 group) which will undergo an internal acid-base reaction to give the internal salt (B).

Chapter 3

Alkanes and Cycloalkanes

3.1 (a) Constitutional isomers

(b) The same compound

3.2

3.3 (a) 5-Isopropyl-2-methyloctane

(b) 4-Isopropyl-4-propyloctane

3.4 (a) Isobutylcyclopentane, C_9H_{18}

(b) *sec*-Butylcycloheptane, $C_{11}H_{22}$

(c) 1-Ethyl-1-methylcyclopropane, C_6H_{12}

3.5 (a) Propanone　　**(b)** Pentanal

(c) Cyclopentanone　　**(d)** Cycloheptene

3.6

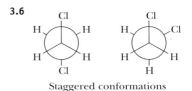

Staggered conformations

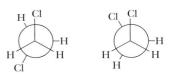

Eclipsed conformations

3.7

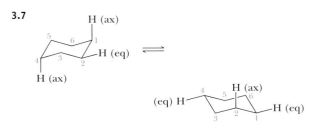

3.8 Axial methyl, ethyl, and isopropyl groups can rotate about the bond from the ring to them so that only a C—H bond faces toward the two axial hydrogens on the same side of the ring. When a *tert*-butyl group is axial, one of the three methyl groups must face toward the two axial hydrogens on the same side of the ring, creating strong axial-axial interactions, which make the equatorial conformation *tert*-butylcyclohexane considerably more stable than its axial conformation.

3.9 Both **(a)** and **(c)** show cis-trans isomerism. The isomers of each are drawn in two different stereorepresentations.

(a)

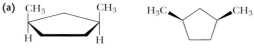

cis-1,3-Dimethylcyclopentane

trans-1,3-Dimethylcyclopentane

(c)

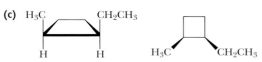

cis-1-Ethyl-2-methylcyclobutane

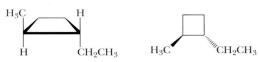

trans-1-Ethyl-2-methylcyclobutane

3.10

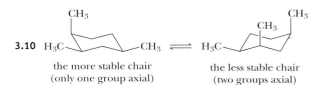

the more stable chair
(only one group axial)

the less stable chair
(two groups axial)

3.11 (a) 2,2-Dimethylpropane < 2-methylbutane < pentane

(b) 2,2,4-Trimethylhexane < 3,3-dimethylheptane < nonane

3.13 (a) $C_{10}H_{22}$, $(CH_3)_2CHCHCH_2CH_2CH_3$
$\qquad\qquad\qquad\quad |$
$\qquad\qquad\qquad CH(CH_3)_2$

(b) C_8H_{18}, $(CH_3)_3CC(CH_3)_3$

(c) $C_{11}H_{24}$, $(CH_3CH_2CH_2)_2CHC(CH_3)_3$

3.15 (a) True　　　　　　　**(b)** True

(c) False　　　　　　**(d)** False

3.17 (1) Structural formulas **(a)** and **(g)** represent the same compound, 2-butanamine. (2) Structural formulas **(a-g)**, **(c)**, **(d)**, **(e)**, and **(f)** represent constitutional isomers with molecular formula $C_4H_{11}N$. (3) Compounds **(b)** and **(h)** represent different compounds that are not constitutional isomers.

3.19 Structures **(a)** and **(d)** represent different compounds that are not constitutional isomers. Structures **(b)**, **(c)**, **(e)**, and **(f)** represent constitutional isomers.

3.21 Only **(a)**, **(b)**, **(c)**, and **(f)** represent constitutional isomers.

3.23 (a) 2-Methylpentane　　**(b)** 2,5-Dimethylhexane

(c) 3-Ethyloctane　　**(d)** 2,2,3-Trimethylbutane

(e) Isobutylcyclopentane　　**(f)** 1-*tert*-Butyl-2,4-dimethyl cyclohexane

3.25 **(a)** The longest chain is pentane. Its IUPAC name is 2-methyl-pentane.

(b) The pentane chain is numbered incorrectly. Its IUPAC name is 2-methylpentane.

(c) The longest chain is pentane. Its IUPAC name is 3-ethyl-3-methylpentane.

(d) The longest chain is hexane. Its IUPAC name is 3,4-dimethylhexane.

(e) The longest chain is heptane. Its IUPAC name is 4-methyl-heptane.

(f) The longest chain is octane. Its IUPAC name is 3-ethyl-3-methyloctane.

(g) The ring is numbered incorrectly. Its IUPAC name is 1,1-dimethylcyclopropane.

(h) The ring is numbered incorrectly. Its IUPAC name is 1-ethyl-3-methylcyclohexane.

3.27 **(a)** Propanone   **(b)** Pentanal

(c) Decanoic acid   **(d)** Cyclohexene

(e) Cyclohexanone   **(f)** Cyclobutanol

3.29 In order of most stable to least stable, they are:

Staggered   Eclipsed

Most stable ➡ Least stable

3.31 Different conformations result from the rotation about a single bond. Rotation about a C—C double bond is restricted, and therefore does not occur. *Cis*- and *trans*-3-hexene differ by the spatial orientation of the ethyl groups attached to the C—C double bond and cannot be interconverted because of the restricted rotation about a C=C bond.

3.33 **(a)** Constitutional isomers   **(b)** The same compound
(c) The same compound   **(d)** The same compound

3.35 No.

3.37 There are six cycloalkanes with molecular formula C_5H_{10}.

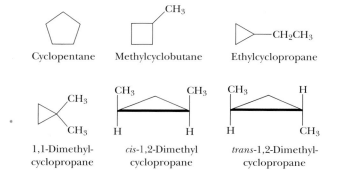

Cyclopentane   Methylcyclobutane   Ethylcyclopropane

1,1-Dimethyl-
cyclopropane

cis-1,2-Dimethyl
cyclopropane

trans-1,2-Dimethyl-
cyclopropane

3.39 In these chair conformations, axial methyl groups are shown in bold face.

1,2-Dimethyl-
cyclohexane

cis
(equal stability)

trans
(more stable)

1,3-Dimethyl-
cyclohexane

(more stable)

cis

cis
(equal stability)

1,4-Dimethyl-
cyclohexane

cis
(equal stability)

(more stable)   trans

3.41 The isopropyl group in these formulas is abbreviated iPr. The four cis-trans isomers are drawn first as planar hexagons, with the —OH group up and the methyl and isopropyl substituents either cis or trans to it. Under each is drawn its more stable chair conformation. Of these four, the chair conformation of (3) is the most stable; all three substituents on the cyclohexane ring are equatorial.

(1)   (2)

(3)   (4)

3.43 In adamantane, the cyclohexane rings all have a chair conformation.

3.45 (1) All alkanes are less dense than water; (2) As alkane molecular weight increases, density increases; (3) Constitutional isomers have similar densities.

3.47 Boiling points of unbranched alkanes are related to their surface area; the larger the surface area, the greater the strength of dispersion forces and the higher the boiling point. The relative increase in size per CH_2 group is greatest between CH_4 and CH_3CH_3, and becomes progressively smaller as molecular weight increases. Therefore, the increase in boiling point per CH_2 group is greatest between CH_4 and CH_3CH_3, and becomes progressively smaller for higher alkanes.

3.49 Water is a polar compound and cannot pass through the nonpolar hydrocarbon coating of apple skins.

3.51 On a gram-per-gram basis, methane (-13.2 kcal/g) is the better source of heat energy than propane (-12.0 kcal/g).

3.53 The cyclododecane ring (12 carbons in the ring) is flexible enough so that there is free rotation about the carbon-carbon bond between the two methyl groups.

3.55 **(a)** Rings A, B, and C are chair conformations. Ring D is an envelope conformation.

(b) The $-OH$ on ring A is equatorial; the ones on rings B and C are axial.

(c) The methyl group at the junction of rings A and B is equatorial to ring A and axial to ring B.

(d) The methyl group at the junction of rings C and D is axial to ring C.

3.57 In answering this question, assume that in each alcohol, the hydroxyl group must be bonded to an sp^3 (tetrahedral) carbon atom; similarly, in each amine, the amino group must also be bonded to an sp^3 (tetrahedral) carbon atom.

(a) $CH_3CH_2CH_2CH_3$

(b) $CH_2{=}CHCH_2CH_3$ or $CH_3{-}CH{=}CH{-}CH_3$

(c) $HC{\equiv}CCH_2CH_3$ or $CH_3C{\equiv}CCH_3$

(d) $CH_3CH_2CH_2CH_2OH$ or $CH_3CH_2\overset{OH}{\underset{|}{C}}HCH_3$

(e) $CH_2{=}CH\overset{OH}{\underset{|}{C}}HCH_3$ or $CH_2{=}CHCH_2CH_2OH$
or $CH_3CH{=}CHCH_2OH$

(f) $HC{\equiv}C\overset{OH}{\underset{|}{C}}H_2CH_2$ or $HC{\equiv}C\overset{OH}{\underset{|}{C}}HCH_3$
or $CH_3C{\equiv}CCH_2OH$

(g) $CH_3CH_2CH_2\overset{NH_2}{\underset{|}{C}}H_2$ or $CH_3CH_2\overset{NH_2}{\underset{|}{C}}HCH_3$

(h) $CH_2{=}CHCH_2\overset{NH_2}{\underset{|}{C}}H_2$ or $CH_2{=}CH\overset{NH_2}{\underset{|}{C}}HCH_3$ or
$CH_3CH{=}CHCH_2$ (with NH_2)

(i) $HC{\equiv}CCH_2\overset{NH_2}{\underset{|}{C}}H_2$ or $HC{\equiv}C\overset{NH_2}{\underset{|}{C}}HCH_3$ or
$CH_3C{\equiv}CCH_2NH_2$

(j) $CH_3CH_2CH_2\overset{O}{\overset{\|}{C}}H$

(k) $CH_2{=}CHCH_2\overset{O}{\overset{\|}{C}}H$ or $CH_3CH{=}CH\overset{O}{\overset{\|}{C}}H$

(l) $HC{\equiv}CCH_2\overset{O}{\overset{\|}{C}}H$ or $CH_3{-}C{\equiv}C{-}\overset{O}{\overset{\|}{C}}{-}H$

(m) $CH_3CH_2\overset{O}{\overset{\|}{C}}CH_3$

(n) $CH_2{=}CH\overset{O}{\overset{\|}{C}}CH_3$

(o) $HC{\equiv}C\overset{O}{\overset{\|}{C}}CH_3$

(p) $CH_3CH_2CH_2\overset{O}{\overset{\|}{C}}OH$

(q) $CH_2{=}CHCH_2\overset{O}{\overset{\|}{C}}OH$ or $CH_3CH{=}CH\overset{O}{\overset{\|}{C}}OH$

(r) $HC{\equiv}CCH_2\overset{O}{\overset{\|}{C}}OH$ or $CH_3C{\equiv}C\overset{O}{\overset{\|}{C}}OH$

Chapter 4
Alkenes and Alkynes

4.1 **(a)** 3,3-Dimethyl-1-pentene   **(b)** 2,3-Dimethyl-2-butene
(c) 3,3-Dimethyl-1-butyne

4.2 **(a)** *cis*-4-Methyl-2-pentene
(b) *trans*-2,2-Dimethyl-3-hexene

4.3 **(a)** (*E*)-1-Chloro-2,3-dimethyl-2-pentene
(b) (*Z*)-1-Bromo-1-chloropropene
(c) (*E*)-2,3,4-Trimethyl-3-heptene

4.4 **(a)** 1-Isopropyl-4-methylcyclohexene
(b) Cyclooctene
(c) 4-*tert*-Butylcyclohexene

4.5

cis,trans-2,4-Heptadiene   *cis,cis*-2,4-Heptadiene

4.6 Only the two double bonds that have two different groups bonded to each carbon of the double bond show cis-trans isomerism. Four cis-trans isomers are possible.

4.7 In ethane, each carbon is surrounded by four regions of electron density; bond angles are 109.5°. In ethylene, each carbon is surrounded by three regions of electron density; bond angles are 120°.

4.9 Labels show the hybridization of each carbon. Each C—H bond is formed by the overlap of a hybrid orbital of carbon and a 1s orbital of hydrogen. Each pi bond is formed by the overlap of parallel $2p$ orbitals.

(a)

(b)

(c)

(d)

4.11 Labels show the hybridization of each carbon. Each C—H bond is formed by the overlap of a hybrid orbital of carbon and a 1s orbital of hydrogen. Each pi bond is formed by the overlap of parallel $2p$ orbitals.

(a)

(b)

(c)

(d)

4.13 **(a)**   **(b)**

(c)   **(d)**

(e)   **(f)**

(g) Cl   **(h)**

4.15 **(a)** 2-Isobutyl-1-heptene
     (b) 1,4,4-Trimethylcyclopentene
     (c) 1,3-Cyclopentadiene   **(d)** 3,3-Dimethyl-1-butyne
     (e) 2,4-Dimethyl-2-pentene   **(f)** 1-Octyne
     (g) 2,2,5-Trimethyl-3-hexyne   **(h)** 3-Methyl-1-pentyne

4.17 **(a)** The longest chain is a butane. The correct name is 2-methyl-1-butene.
     (b) The ring is not numbered correctly. The correct name is 4-isopropylcyclohexene.
     (c) The parent chain is not numbered correctly. The correct name is 3-methyl-2-hexene.
     (d) The parent chain is a pentane. The correct name is 2-ethyl-3-methyl-1-pentene.
     (e) The ring is not numbered correctly. The correct name is 3,3-dimethylcyclohexene.
     (f) The parent chain is a heptane. The correct name is 3-methyl-3-heptene.

4.19 Only part **(b)** shows cis-trans isomerism.

        trans-2-Pentene   *cis*-2-Pentene

4.21 All three compounds have polar C—Br bonds. Because of its geometry, the isomer on the right has no dipole moment.

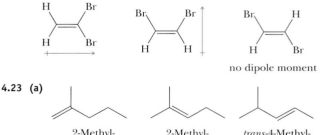

                                          no dipole moment

4.23 **(a)**

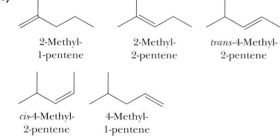

    2-Methyl-   2-Methyl-   *trans*-4-Methyl-
    1-pentene   2-pentene   2-pentene

    cis-4-Methyl-   4-Methyl-
    2-pentene   1-pentene

(b)

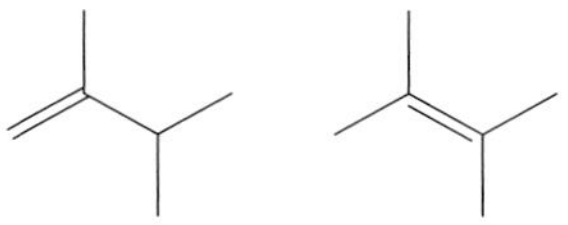

2,3-Dimethyl-
1-butene

2,3-Dimethyl-
2-butene

(c)

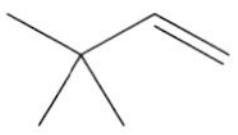

3,3-Dimethyl-
1-butene

4.25 (a) All of these compounds show cis-trans isomerism.

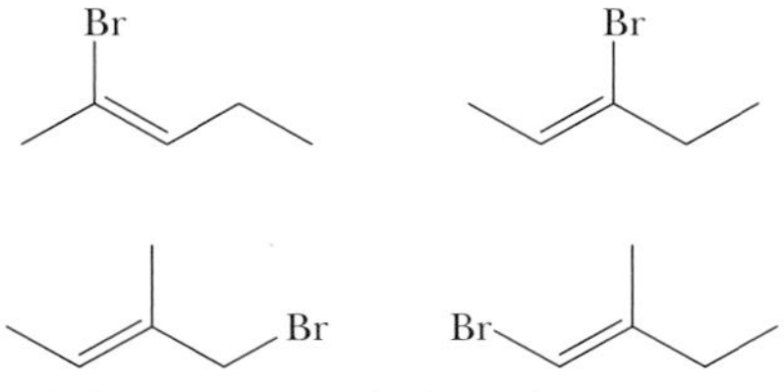

(b) None of these compounds show cis-trans isomerism.

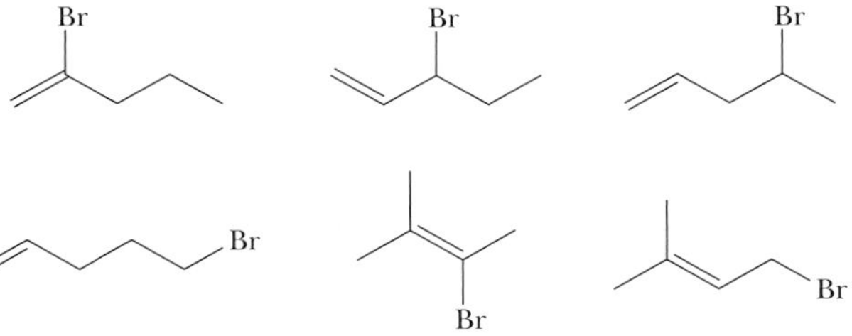

4.27 None of these compounds show E,Z or cis-trans isomerism. For the correct name, delete all designations of configuration.

4.29

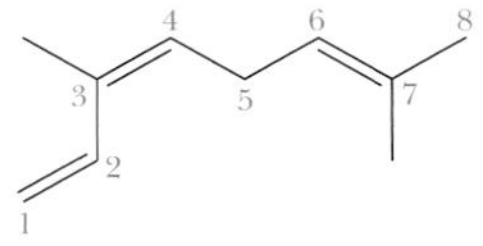

4.31 (a) Cis-trans isomers. **(b, c, d)** Different conformations of the same structure.

4.33 (a) The carbon skeleton of lycopene can be divided into 8 isoprene units, here shown in bold bonds.

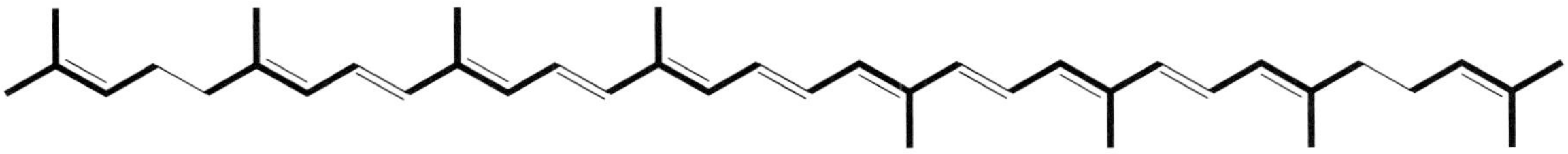

(b) Eleven of the thirteen double bonds have the possibility for cis-trans isomerism. The double bonds at either end of the molecule cannot show cis-trans isomerism.

4.35 The isoprene units are shown in bold face.

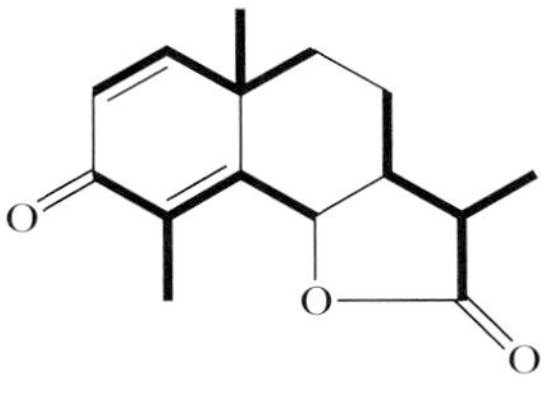

The three isoprene units are shown in bold face

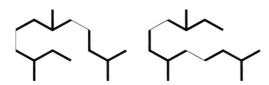

Coiling the carbon skeleton of farnesol in either of these ways gives the carbon skeleton of santonin

4.37 The six isoprene units can be connected in two ways, and are shown in bold face.

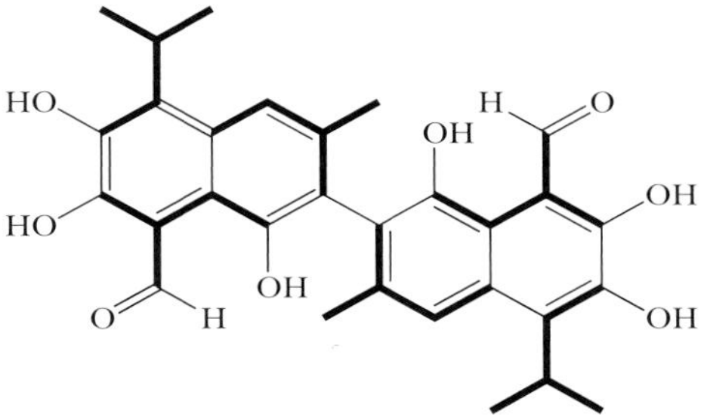

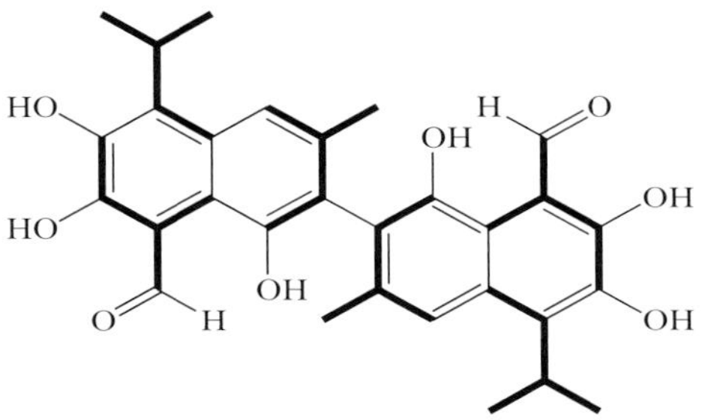

4.39 (a) There are two carbon-carbon double bonds in pyrethrin II about which cis-trans isomerism is possible, and one in pyrethrosin.

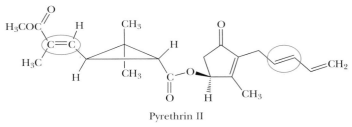

Pyrethrin II

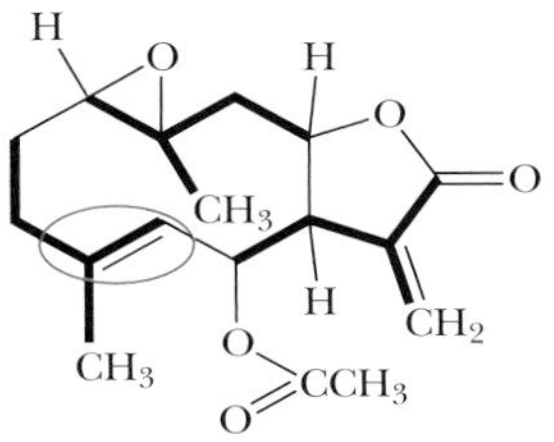

Pyrethrosin

(b) Given the geometry of a five-membered ring in pyrethrin II, substituents on the ring double bond must be cis to each other; they cannot be trans.

(c) The three isoprene units in pyrethrosin are shown in bold face.

4.41 Partial overlap of the *p* orbitals on carbons 2 and 3 of 1,3-butadiene forms a bond with some π-bond character. This delocalization of *p* electrons is known as resonance and is a powerful stabilizing process in molecules.

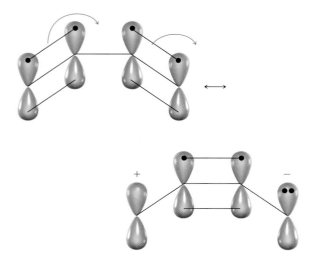

4.43 Both oxygen and nitrogen are more electronegative than carbon and, therefore, the OCH₃ group in **(a)** and the CN group in **(b)** pull electrons away from the double bond, decreasing its electron density. Silicon is less electronegative than carbon with the result that electrons from silicon are polarized toward the carbons of the double bond, thus increasing its electron density.

4.45 **(a)** Fumaric acid may be designated either *E* or trans.

(b) Aconitic acid may be designated either *Z* or trans. (Note that while the two carboxyl groups in aconitic acid are cis or *Z* to each other, the main carbon chain is trans.)

Chapter 5

Reactions of Alkenes

5.1 If the reaction were endothermic, the energy of the products would be higher than that of the reactants.

5.2 **(a)** CH₃CHCH₃
with I substituent

2-Iodopropane

(b)

1-Iodo-1-methyl-cyclohexane

5.3 In order of increasing stability, they are:

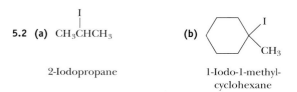

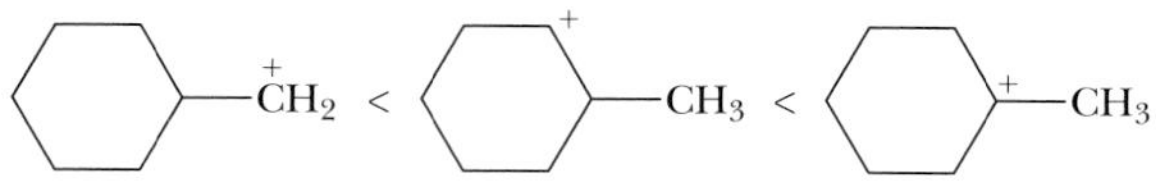

5.4 Step 1:

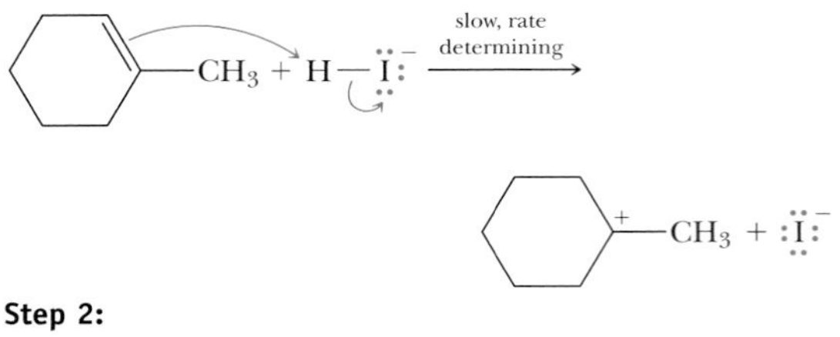

slow, rate determining

Step 2:

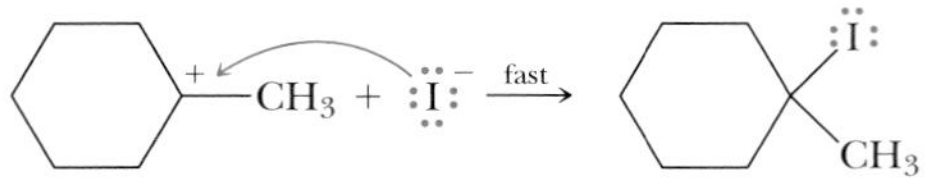

fast

5.5 **(a)**  OH   **(b)**  OH

5.6 Step 1:

slow, rate determining

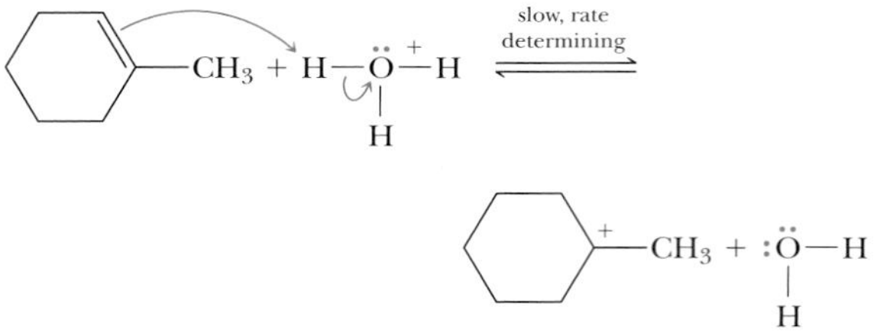

A 3° carbocation intermediate

Step 2:

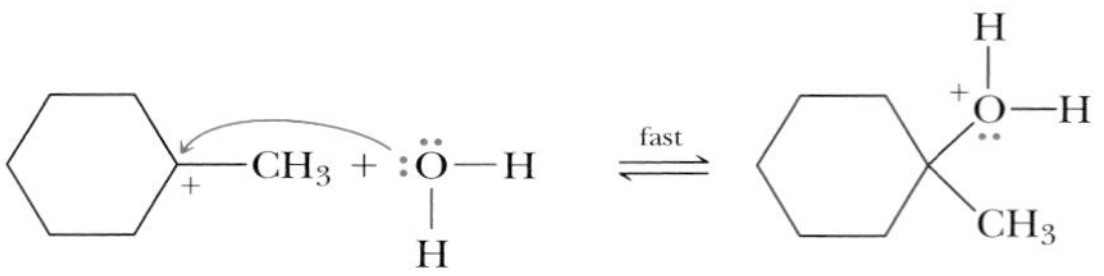

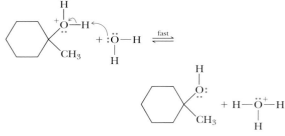

fast

Step 3:

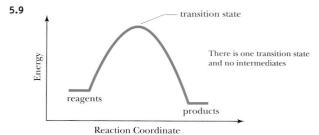

5.7 **(a)** 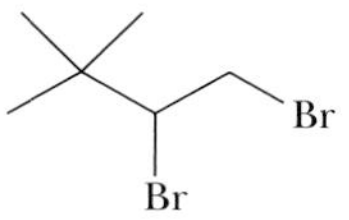 Br, Br   **(b)** 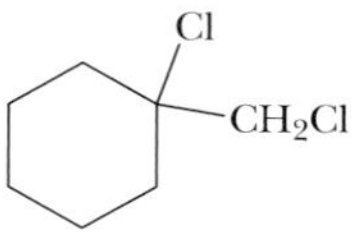 Cl, CH₂Cl

5.9

Energy vs Reaction Coordinate; transition state at peak; reagents at start, products at end. There is one transition state and no intermediates.

5.11 **(a)** True   **(b)** False   **(c)** False

5.13 **(a)**

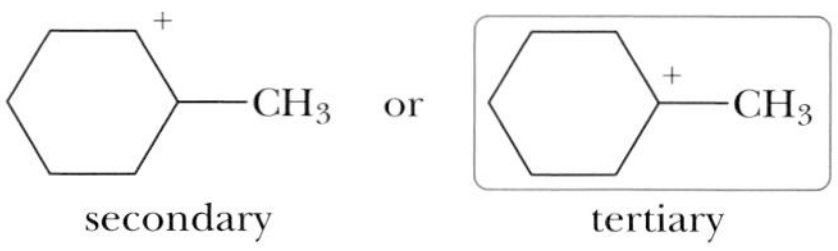

secondary   or   tertiary

(b)

secondary   or   primary

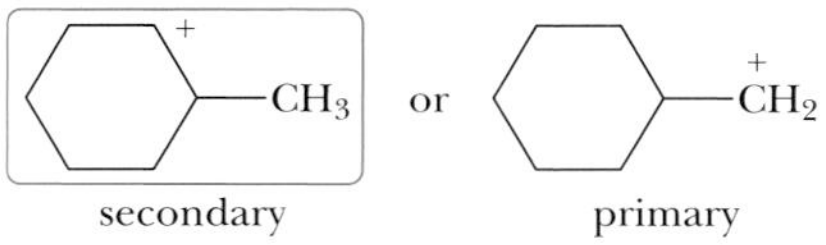

5.15 The more reactive alkene of each pair is the alkene that forms the more stable carbocation intermediate.

(a)   + HI ⟶

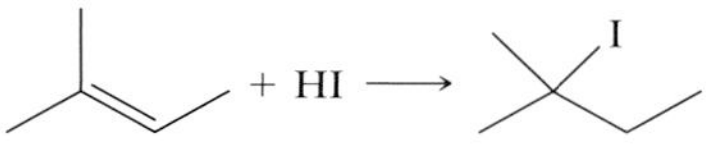

(b)   + HI ⟶

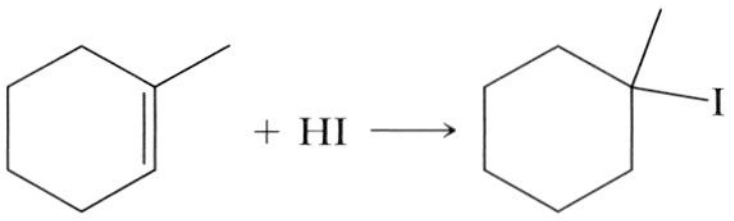

5.17 The first step in each reaction is the protonation of the alkene to generate a carbocation. The reaction path that produces the more stable carbocation occurs faster, thus producing the observed regioselectivity.

(a)   **(b)**

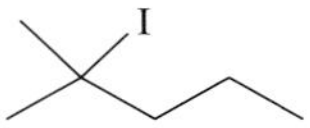

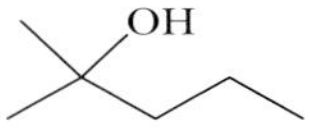

5.19 **(a)**   or

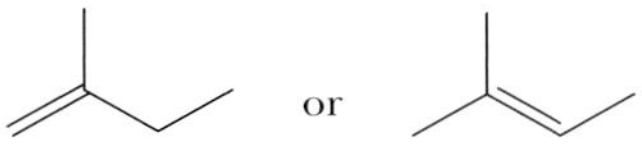

(b)

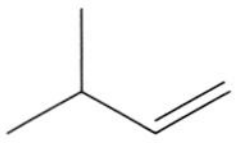

(c)   —CH₃   or   =CH₂

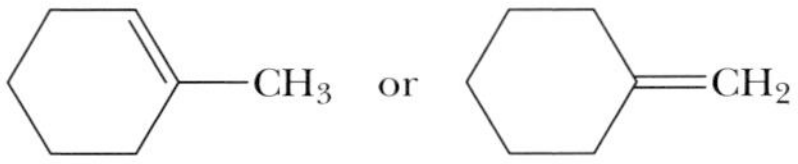

5.21 **(a)** 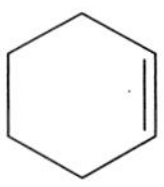   **(b)**

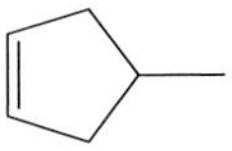

(c) 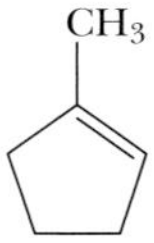   **(d)**   =CH₂

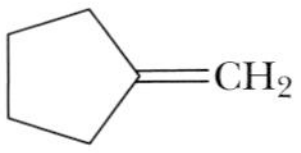

5.23 **(a)**   or

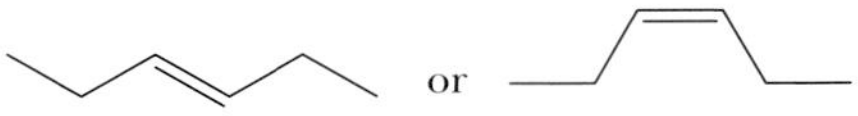

(b)   or

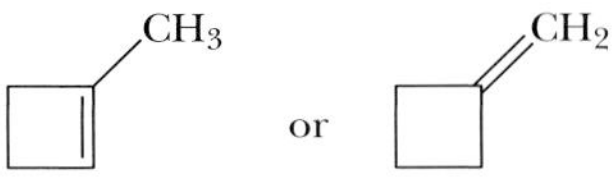

(c)   or

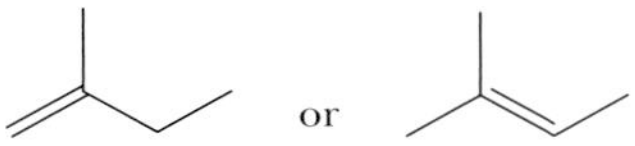

(d) CH₃CH=CH₂

5.25 **(a)** Add water to each double bond by regioselective protonation of the double bond to form 3° a carbocation. Reaction of each carbocation with water and the loss of a proton give terpin. Note that the reactions do not necessarily proceed in the order shown. Both alkenes will react at similar rates because in each case, a tertiary carbocation is formed.

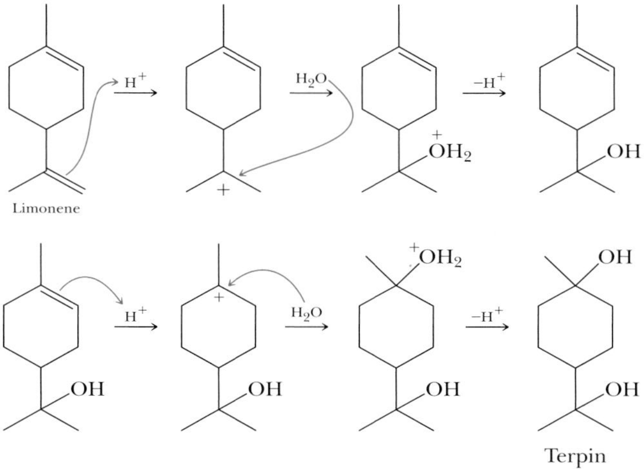

Limonene → H⁺ → → H₂O → ⁺OH₂ → –H⁺ → OH

→ H⁺ → → H₂O → ⁺OH₂ → –H⁺ → OH

Terpin

(b) Two cis-trans isomers are possible.

(c) In the more stable chair conformation, the three-carbon side chain is equatorial.

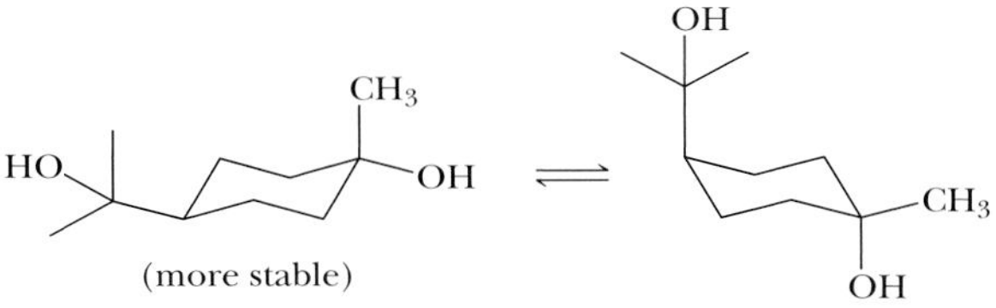

HO, CH₃, OH (more stable) ⇌ OH, CH₃, OH

5.27 Step 1:

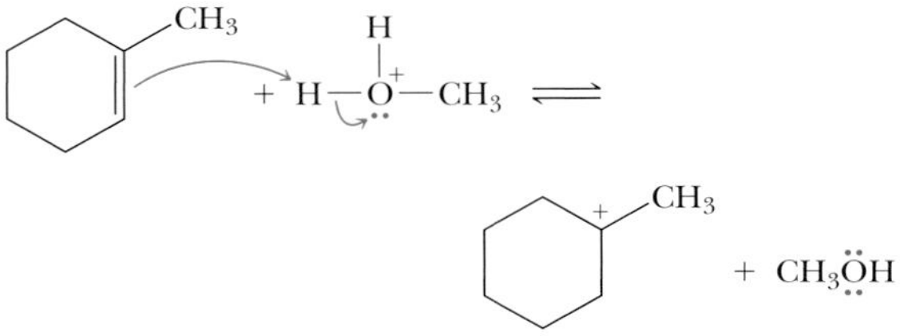

+ H—O⁺—CH₃ ⇌   + CH₃ÖH

Step 2:

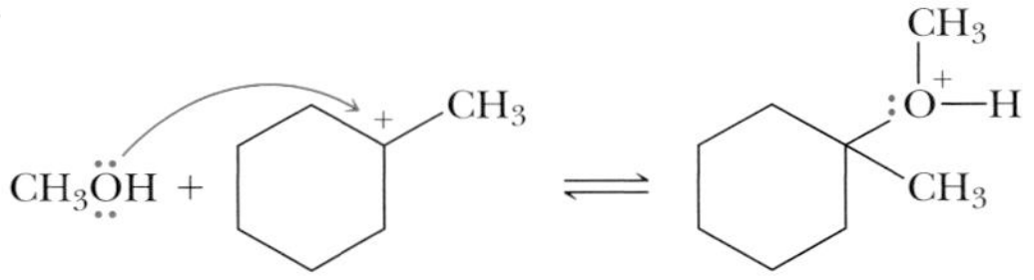

CH₃ÖH +   ⇌

Step 3:

+ :Ö—CH₃ ⇌   + H—O⁺—CH₃

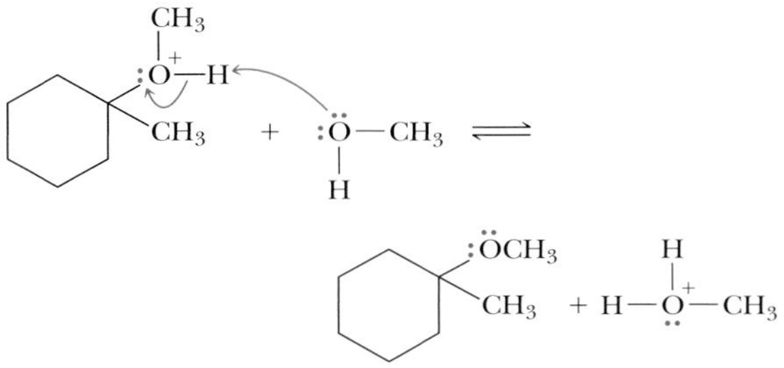

5.29 **(a)** Oxidation   **(b)** Neither   **(c)** Reduction

5.31 **(a)**   **(b)**

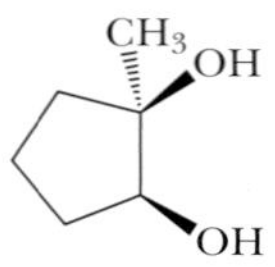

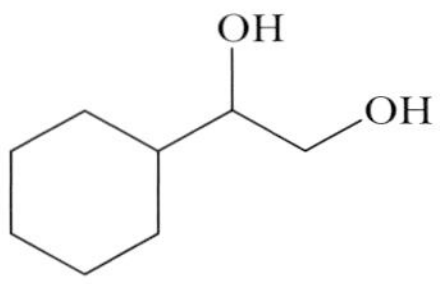

(c)

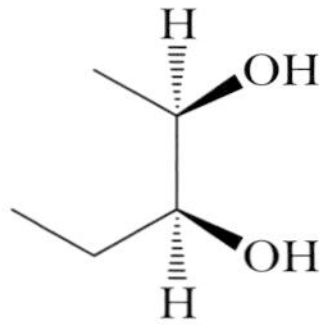

(c)
$$\xrightarrow[\text{ROOH}]{\text{OsO}_4}$$
=

(d)
$$\xrightarrow{\text{Br}_2}$$

5.33 **(a)**    **(b)**

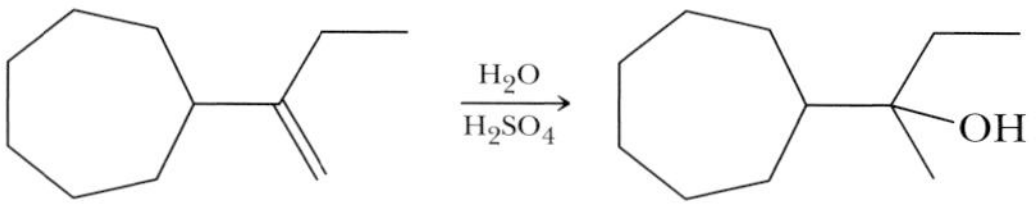

(c) 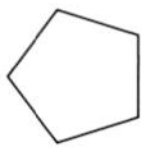 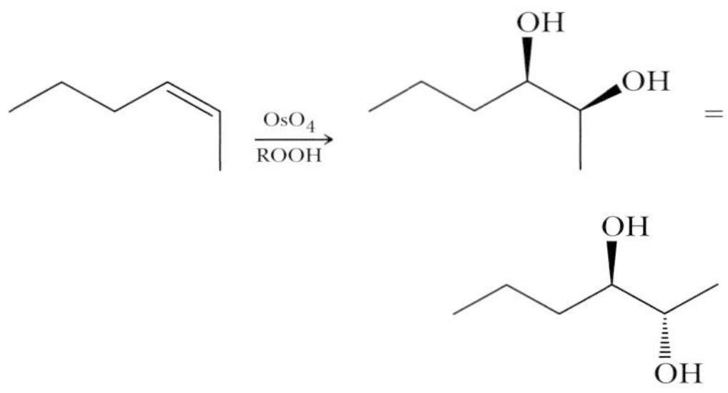   **(d)**

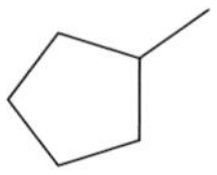

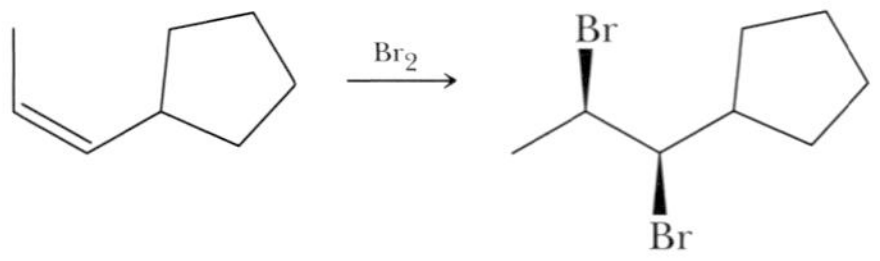

5.41 **(a)**

The same compound results from either top or bottom attack.

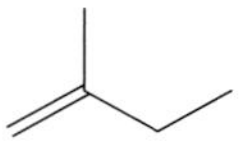

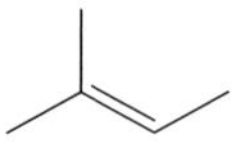

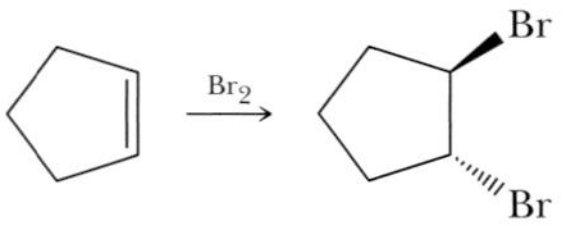

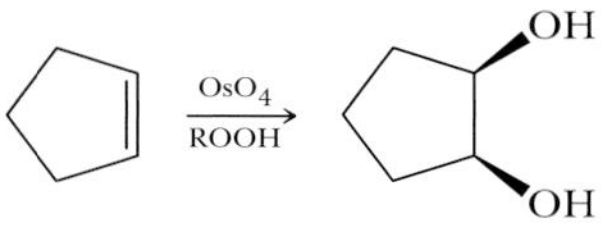

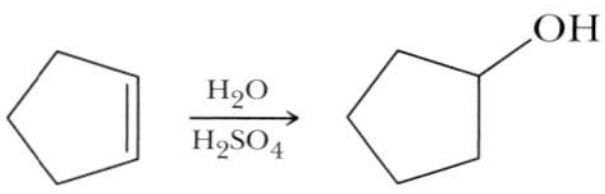

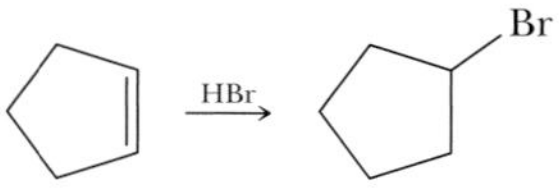

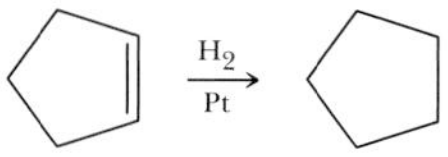

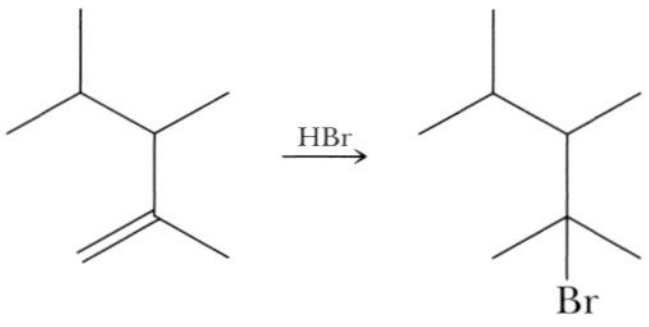

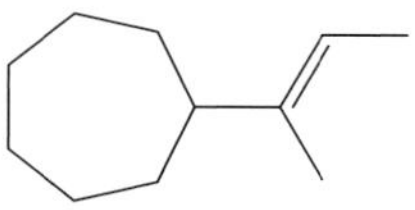

5.35 A and B are: and

(b)   +

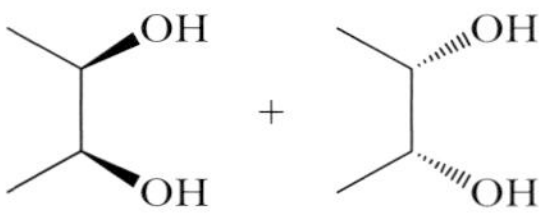

These two structures are identical; they are just oriented differently in space. If you turn the right one in space correctly, you can superpose it on the left one (Chapter 6).

5.37 **(a)**
$$\xrightarrow{\text{Br}_2}$$

(c)   +

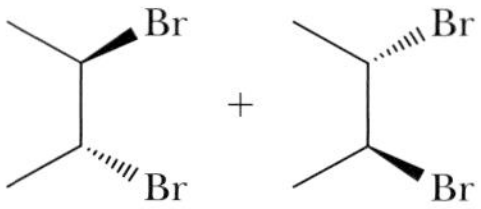

These two structures are nonsuperposable mirror images. We refer to them as enantiomers (Chapter 6).

(b)
$$\xrightarrow[\text{ROOH}]{\text{OsO}_4}$$

Chapter 6
Chirality; the Handedness of Molecules

(c)
$$\xrightarrow[\text{H}_2\text{SO}_4]{\text{H}_2\text{O}}$$

(d)
$$\xrightarrow{\text{HBr}}$$

6.1 **(a)**   S   R

(e)
$$\xrightarrow[\text{Pt}]{\text{H}_2}$$

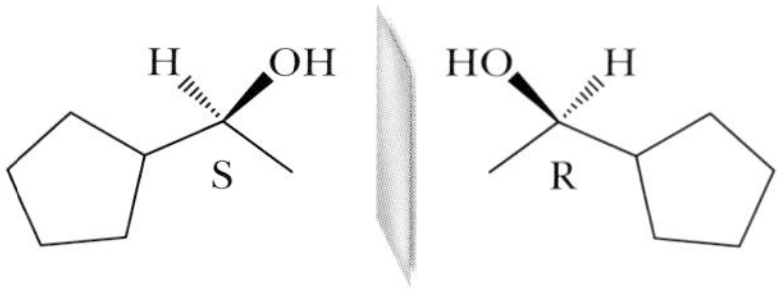

(b)   S   R

5.39 **(a)**
$$\xrightarrow{\text{HBr}}$$

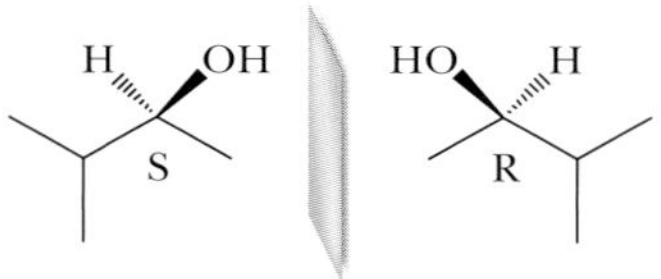

6.2 **(a)** S   **(b)** R   **(c)** R

6.3

enantiomers
(1) ⟷ (3)
(d) ↑↓   (d)   ↑↓ (d)
(2) ⟷ (4)
enantiomers

(b)   or

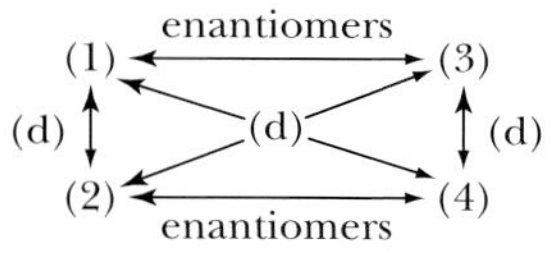

(d) shows pairs of diastereomers

$$\xrightarrow[\text{H}_2\text{SO}_4]{\text{H}_2\text{O}}$$

6.4 **(a)** Structure (2) and (3) represent the same compound (the meso compound).

(b) Structures (1) and (4) represent the enantiomers.

(c) Compound (2,3) is a meso compound.

6.5 Three stereoisomers are possible; one pair of enantiomers and one meso compound.

6.6 Two stereoisomers are possible; one pair of cis-trans isomers, each of which is achiral.

6.7 The concentration, expressed in g/100 mL of solvent is 0.040 g/100 mL. The observed rotation is 6.9°.

6.9 Constitutional isomers have the same molecular formula but a different connectivity. Stereoisomers have the same molecular formula and connectivity, but a different spatial orientation of their atoms.

6.11 A spiral with a left-handed twist when viewed from one end has a left-handed twist when viewed from the other end as well.

6.13 This question is meant to encourage you to think about chirality in everyday objects. Please share your observations with other members of the class and come to a collective conclusion.

6.15 **(a)** True **(b)** False **(c)** True **(d)** False
(e) False **(f)** True **(g)** False

6.17 **(a)**

 or

(b)

(c)

(d)

(e)

6.19 Three of the eight possible carboxylic acids are chiral.

6.21 Structures **(a)**, **(b)**, and **(c)** have stereocenters. There is no stereocenter in **(d)**.

(a)

(b)

(c)

6.23 Structures **(b)**, **(c)**, and **(f)** have stereocenters. There are no stereocenters in **(a)**, **(d)**, **(e)**, or **(g)**.

(b) $\overset{COOH}{\underset{CH_3}{H\overset{*}{C}OH}}$

(c) $CH_3\overset{}{C}H\overset{*}{C}H\underset{NH_2}{COOH}$

(f) $CH_3CH_2\overset{OH}{\underset{*}{C}}HCH{=}CH_2$

6.25 **(a)** $-H$ **(4)** $-CH_3$ **(3)** $-OH$ **(1)** $-CH_2OH$ **(2)**

(b) $-CH_2CH{=}CH_2$ **(3)** $-CH{=}CH_2$ **(1)** $-CH_3$ **(4)**
$-CH_2COOH$ **(2)**

(c) $-CH_3$ **(3)** $-H$ **(4)** $-COO^-$ **(2)** $-NH_3^+$ **(1)**

(d) $-CH_3$ **(4)** $-CH_2SH$ **(2)** $-NH_3^+$ **(1)**
$-COO^-$ **(3)**

6.27 In an achiral environment, enantiomers have the same physical and chemical properties. But, in chiral environments, enantiomers can behave very differently. The odor receptors responsible for detecting the carvone smells must be chiral themselves and, therefore, able to physiologically differentiate the enantiomers.

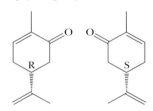

6.29

6.31 The specific rotation of the enantiomer of naturally occurring ephedrine is +41°.

6.33 Amoxicillin has four stereocenters.

6.35 **(a)**

Fluoxetine
(Prozac)
2 stereoisomers possible

(b) H, N, CH₃   **(c)**

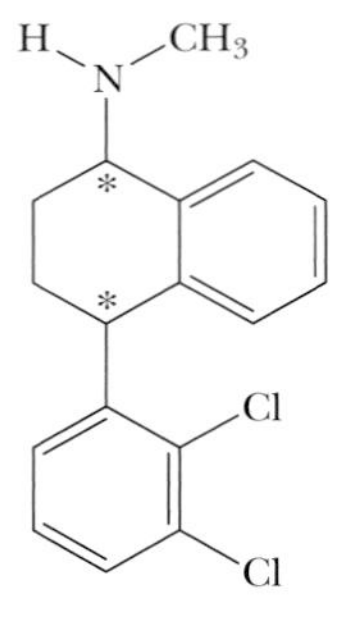

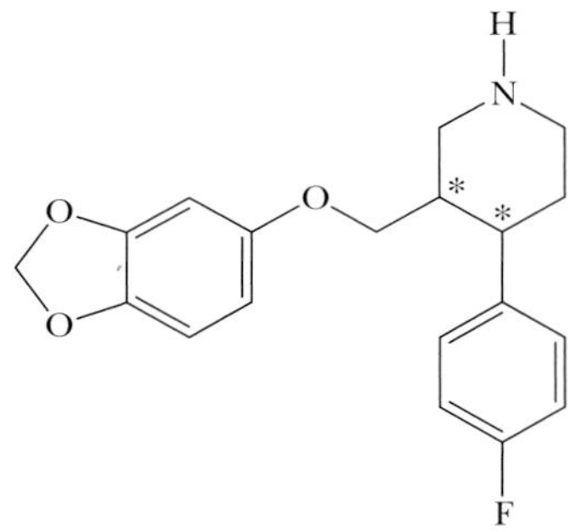

Sertraline
(Zoloft)
4 stereoisomers possible

Paroxetine
(Paxil)
4 stereosomers possible

6.37 Compounds **(a)**, **(c)**, **(d)**, and **(f)** each have at least two stereocenters and an internal mirror plane and, therefore, are meso isomers.

6.39 Three stereoisomers are possible; one pair of enantiomers and one meso compound.

6.41 A racemic mixture is a 50:50 mixture of both enantiomers of a chiral compound. Pure enantiomers rotate plane-polarized light in equal, but opposite directions. A racemic mixture is not optically active because one enantiomer cancels the optical rotation of the other.

6.43 Hydrogenation of (A) yields the same compound regardless of the side of attack of H₂ on the alkene, whereas hydrogenation of (B) gives two products, only one of which is *cis*-decalin.

(A) $\xrightarrow{H_2/Pd}$

=

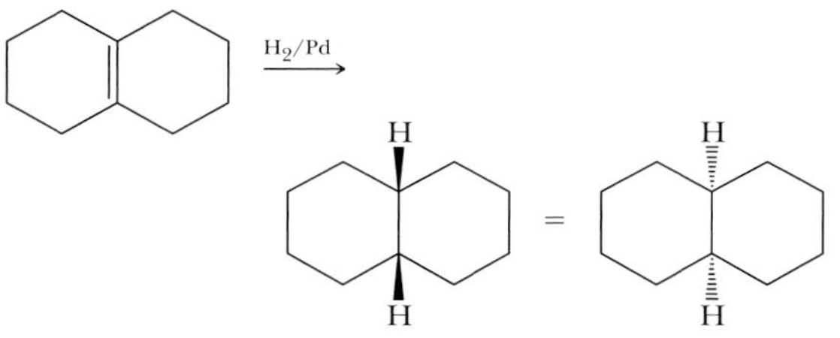

(B) $\xrightarrow{H_2/Pd}$

+

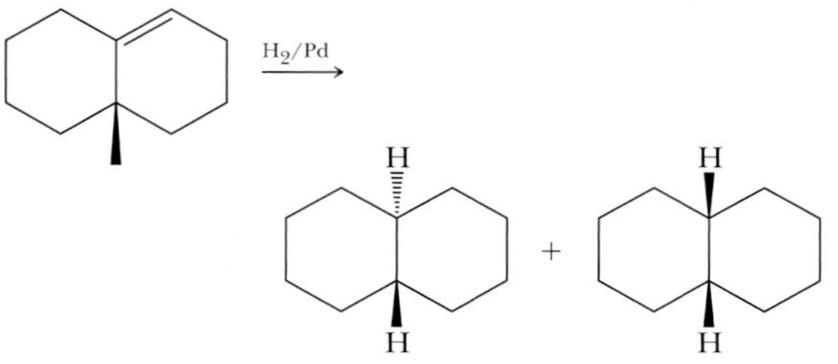

6.45 Both faces of the reactant are the same and, therefore, there is equal probability of each face reacting to form syn (same side) and anti (opposite side) addition products. The product has two stereocenters and four possible stereoisomers (two pairs of enantiomers). If stereoselectivity (the synthesis of one stereoisomer in preference to all other possibilities) is desired, this is not a useful synthetic method.

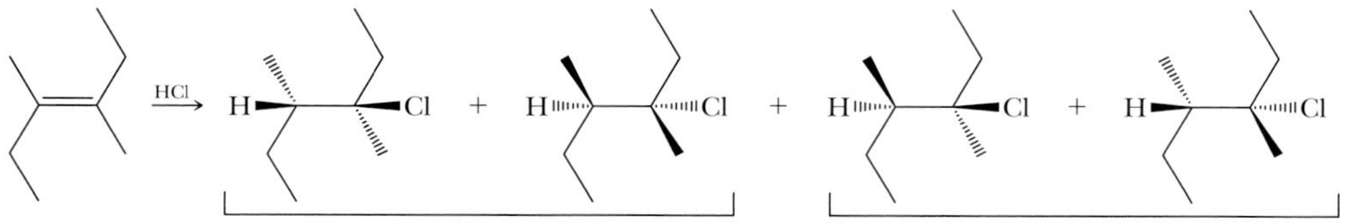

$\xrightarrow{HCl}$ H⟍C—Cl + H⟍C⟍Cl + H⟍C⟍Cl + H⟍C⟍Cl

one pair of enantiomers   a second pair of enantiomers

Chapter 7
Haloalkanes

7.1 (a) 1-Chloro-3-methyl-2-butene   **(b)** 1-Bromo-1-methylcyclohexane

(c) 1,2-Dichloropropane   **(d)** 2-Chloro-1,3-butadiene

7.2 (a) ⬠—SCH₂CH₃ + NaBr

(b) ⬠—OCCH₃ + NaBr

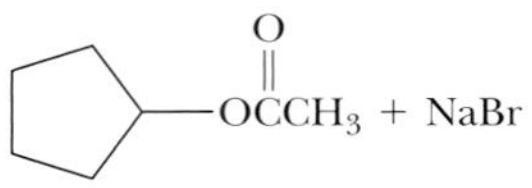

7.3 (a) ⟍⟍—SH   by S_N2

(b) CH₃CHCH₂CH₃ with OCH group   by S_N1

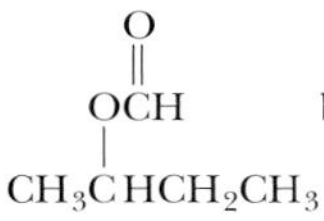

7.4 (a) [cyclohexene with CH₃] + [cyclohexane with CH₂]

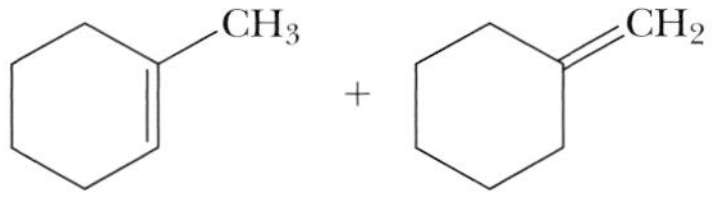

major product

(b) [cyclohexane with CH₂]

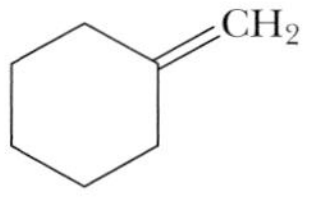

(c) [cyclohexene with CH₃] + [cyclohexene with CH₃]

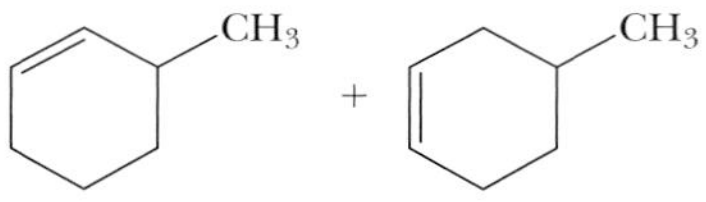

approximately equal amounts of each

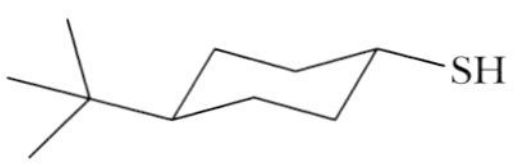

7.5 These reactions proceed by E2 mechanisms. E2 mechanisms are favored when using 2° or 3° halides with strong bases.

(a) [trans-alkene] + [cis-alkene]

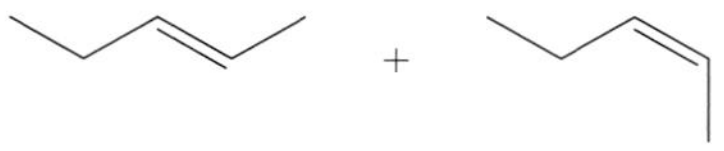

major product

(b) [cyclohexene] + [cyclohexene]

a pair of enantiomers

7.6 (a) [trans-alkene] + [cis-alkene]   by E2

major product

(b) [cyclohexane with I and CH₃]   by S_N2

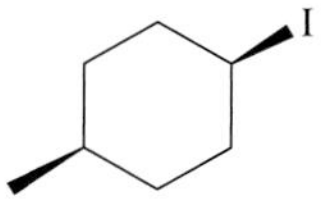

7.7 (a) 1,1-Difluoroethene   **(b)** 3-Bromocyclopentene
(c) 2-Chloro-5-methylhexane   **(d)** 1,6-Dichlorohexane
(e) Dichlorodifluoromethane   **(f)** 3-Bromo-3-ethylpentane

7.9 (a) ... (b) ... (c) ... (d) ... or ... (e) ... (f) ...

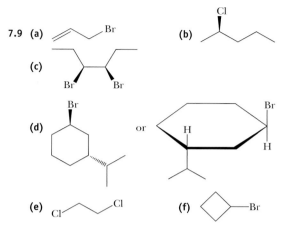

7.11 2-Iodooctane and *trans*-1-chloro-4-methylcyclohexane are 2° alkyl halides.

7.13 (a) (structure) + HCl ⟶ (structure)

(b) $CH_3CH_2CH=CH_2 + HI \longrightarrow CH_3CH_2CHCH_3$ (with I substituent)

(c) $CH_3CH=CHCH_3 + HCl \longrightarrow CH_3CHCH_2CH_3$ (with Cl substituent)

(d) (cyclopentene with CH₃) + HBr ⟶ (cyclopentane with CH₃ and Br)

7.15 In order of increasing polarity, they are:
$CH_3CH_2OH < CH_3OH < H_2O$

7.17 (a) H_2O or $\boxed{OH^-}$ **(b)** CH_3COO^- or $\boxed{OH^-}$
(c) CH_3SH or $\boxed{CH_3S^-}$

7.19 (a) $CH_3CH_2CH_2I + NaCl$ **(b)** (cyclohexyl)–$\overset{+}{N}H_3Br^-$

(c) $CH_2=CHCH_2OCH_2CH_3 + NaCl$

7.21 (a) A 2° halide, a moderate nucleophile, and a moderately ionizing solvent favor S_N2.

(b) A 2° halide, with ethyl thiolate, a good nucleophile that is a weak base, and a weakly ionizing solvent favor an S_N2 reaction.

(c) A 1° halide and a good nucleophile like I^- favor an S_N2 reaction.

(d) Methyl halides can only undergo S_N2 reactions because their extremely unstable carbocations preclude S_N1 reactions. Trimethylamine is a moderate nucleophile and acetone is a weakly ionizing solvent, further supporting an S_N2 reaction.

(e) A 1° halide and a good nucleophile like methoxide always favors S_N2.

(f) A 2° halide with methyl thiolate, which is a good nucleophile but a weak base, and a moderately ionizing solvent all favor an S_N2 reaction.

(g) Piperidine is a moderate amine nucleophile and ethanol is a moderately ionizing solvent. S_N2 reactions are favored with moderate nucleophiles reacting with 1° alkyl halides.

(h) Ammonia is a moderate nucleophile and ethanol is a moderately ionizing solvent. S_N2 is favored in this case because the alkyl halide is primary.

7.23 (a) False   **(b)** False   **(c)** True
(d) True   **(e)** False   **(f)** False

7.25 (a) Chloride is a good leaving group and the resulting 2° carbocation is a relatively stable intermediate. The most important factor influencing this reaction towards S_N1 is ethanol, which is a poor nucleophile and a moderately ionizing solvent.

(b) Chloride is a good leaving group and results in a stable 3° carbocation intermediate. An equally important factor influencing the reaction towards S_N1 is methanol, which is a poor nucleophile and a good ionizing solvent.

(c) Chloride is a good leaving group and acetic acid is a strongly ionizing solvent/poor nucleophile. 3° alkyl halides almost exclusively undergo S_N1 reactions with poor nucleophiles. All these factors strongly favor S_N1 reactions.

(d) Methanol is a good ionizing solvent and a poor nucleophile. Bromide is a good leaving group resulting in a relatively stable 2° carbocation, all favoring an S_N1 reaction.

7.27 Step 1: ionization of chlorine to give a 3° carbocation intermediate:

$$CH_3-\underset{CH_3}{\overset{CH_3}{C}}-\ddot{\overset{..}{Cl}}: \longrightarrow CH_3-\underset{CH_3}{\overset{CH_3}{C}}+ + :\ddot{\overset{..}{Cl}}:^-$$

Step 2: reaction of the carbocation (an electrophile) with either water or ethanol (both nucleophiles in this reaction) to give an oxonium ion. Transfer of a proton from the oxonium ion to solvent gives the ether and alcohol.

$$H_2\ddot{O}: \rightarrow CH_3-\underset{CH_3}{\overset{CH_3}{C}}-\overset{+}{\underset{H}{\overset{H}{O}}}: \xrightarrow{-H^+} CH_3-\underset{CH_3}{\overset{CH_3}{C}}-\ddot{O}H$$

$$CH_3-\underset{CH_3}{\overset{CH_3}{C}}+$$

$$CH_3CH_2\ddot{O}H \rightarrow CH_3-\underset{CH_3}{\overset{CH_3}{C}}-\overset{+}{\underset{CH_2CH_3}{\overset{H}{O}}}: \xrightarrow{-H^+} CH_3-\underset{CH_3}{\overset{CH_3}{C}}-\ddot{O}CH_2CH_3$$

Step 2′: Transfer of a proton from the carbocation to solvent (in this case, H_2O) gives the alkene:

$$H-\overset{..}{\underset{H}{O}}: + H-CH_2-\overset{CH_3}{\underset{CH_3}{\overset{|}{C}}}+ \longrightarrow H-\overset{..}{\underset{H}{O}}{}^+\!-H + CH_2=C\overset{CH_3}{\underset{CH_3}{\diagup}}$$

7.29 (a) [structures: pentyl chloride or isobutyl chloride]

or

(b) [structures: chloride or bromide]

or

(c) [structures: chloride or chloride]

or

(d) [structures: bromide or bromide]

or

7.31 Haloalkenes fail to undergo S_N1 reactions because the alkenyl carbocations produced through ionization of the carbon–halogen bond are too unstable. Given the geometry of an alkenyl halide, backside attack on the carbon bearing the halogen atom is impossible.

7.33

(a) [cyclohexyl]—Br + 2NH$_3$ ⟶ [cyclohexyl]—NH$_2$ + NH$_4$Br

(b) [cyclohexyl]—CH$_2$Br + 2NH$_3$ ⟶

[cyclohexyl]—CH$_2$NH$_2$ + NH$_4$Br

(c) [cyclohexyl]—Br + CH$_3$CO$^-$Na$^+$ ⟶

[cyclohexyl]—OCCH$_3$ + NaBr

(d) [propyl]—Br + [propyl]—S$^-$Na$^+$ ⟶

[propyl]—S—[propyl] + NaBr

(e) [cyclopentyl with Br] + CH$_3$CO$^-$Na$^+$ ⟶

[cyclopentyl with OCCH$_3$] + NaBr

(f) [butyl]—Br + [butyl]—O$^-$Na$^+$ ⟶

[dibutyl ether] + NaBr

7.35 Reactions **(c)** and **(d)** do not result in *cis-trans* isomers.

7.37 (a) [cyclohexylmethyl chloride]   **(b)** [isohexyl chloride]

7.39 (a)

$$HO: + \text{[trans-4-chlorocyclohexanol]} \xrightarrow{S_N2}$$

$$HO—\text{[cyclohexane]}—OH + :\overset{..}{\underset{..}{Cl}}:{}^-$$

(1)

(b)

$$HO:{}^- + \text{[cyclohexane with H, OH, Cl]} \xrightarrow{E2}$$

$$\text{[cyclohexenol]}—OH + :\overset{..}{\underset{..}{Cl}}:{}^- + H_2\overset{..}{\underset{..}{O}}:$$

(2)

(c) S_N2 reactions occur with a backside attack by the nucleophile on the carbon bearing the leaving group. In *trans-*4-chlorocyclohexanol, the nucleophile is created by deprotonation of the OH group. When the chair conformation converts to a boat conformation, the nucleophilic alkoxide ion is perfectly situated for an internal S_N2 displacement of chlorine. The nucleophilic alkoxide ion from *cis-*4-chlorocyclohexanol does not have the proper geometry for an internal S_N2 displacement of chlorine.

[mechanism structures]

7.41 Reaction **(a)** gives the better yield of the ether because it involves favorable conditions for an S_N2 reaction; a 1° halide and a good nucleophile. Reaction **(b)** occurs by an E2 reaction with a strong base deprotonating the β-proton on a sterically hindered 3° halide to give an alkene.

7.43 This transformation involves an acid-base reaction followed by an internal S_N2 reaction.

$$:\overset{..}{\underset{..}{Cl}}—CH_2CH_2—\overset{..}{\underset{..}{O}}—H + :\overset{..}{\underset{..}{O}}H{}^- \longrightarrow$$

$$:\overset{..}{\underset{..}{Cl}}—CH_2CH_2—\overset{..}{\underset{..}{O}}:{}^- \longrightarrow H_2C\overset{\overset{..}{\underset{..}{O}}:}{\diagup\diagdown}CH_2$$

7.45 Under these S_N2 conditions, the bromide nucleophile inverts the stereochemistry of the stereocenter. When 50% have been inverted, the mixture is racemic.

7.47 Although alkoxides are poor leaving groups, the release of strain in the three-membered ring is a driving force behind ring opening of epoxides by nucleophiles.

Chapter 8

Alcohols, Ethers, and Thiols

8.1 **(a)** 2-Heptanol **(b)** 2,2-Dimethyl-1-propanol
  (c) *cis*-3-Isopropylcyclohexanol

8.2 **(a)** 1° **(b)** 2° **(c)** 1° **(d)** 3°

8.3 **(a)** (*E*)-3-Penten-1-ol **(b)** 2-Cyclopentenol

8.4 It lies to the right.

8.5 **(a)** $CH_3\overset{\overset{\displaystyle CH_3}{|}}{C}=CHCH_3$ + $CH_2=\overset{\overset{\displaystyle CH_3}{|}}{C}CH_2CH_3$
      major product

  (b) [cyclopentene with CH₃] + [methylenecyclopentane with CH₂]
      major product

8.6 **(a)** [structure with O and OH]

  (b) [structure with O] **(c)** [cyclohexanone with O]

8.7 **(a)** 1-Ethoxy-2-methylpropane (ethyl isobutyl ether)
  (b) Methoxycyclopentane (cyclopentyl methyl ether)

8.8 In order of increasing boiling point, they are:

$CH_3OCH_2CH_2OCH_3 < CH_3OCH_2CH_2OH < HOCH_2CH_2OH$

8.9 [cyclopentane epoxide with two CH₃ and O]

8.10 [cyclohexene] + OsO₄, ROOH ⟶ [cyclohexane with two OH groups]

8.11 **(a)** 3-Methyl-1-butanethiol
  (b) 3-Methyl-2-butanethiol

8.13 **(a)** 1-Pentanol **(b)** 1,3-Propanediol
  (c) 3-Buten-1-ol **(d)** 3-Methyl-1-butanol
  (e) *trans*-1,2-Cyclohexanediol **(f)** 1-Butanethiol

8.15 **(a)** Dicyclopentyl ether **(b)** Dibutyl ether
  (c) 2-Ethoxyethanol

8.17 (d) < (b) < (a) < (c)

8.19 Propanoic acid has the higher boiling point (141°C) due to intermolecular hydrogen bonding. The carboxyl group can function both as a hydrogen bond donor (through the —OH group) and acceptor (through the C=O and C—O groups) and is responsible for a high degree of intermolecular association between molecules in the liquid state. Methyl acetate does not have intermolecular hydrogen bonding, thus it has the lower of the two boiling points (57°C).

8.21 The S—H bond of thiols is much less polar than the O—H bond of alcohols, therefore, thiols do not form significant hydrogen bonds. Intermolecular forces must be broken before a liquid can boil. The strong intermolecular hydrogen bonding in 1-butanol is responsible for its higher boiling point than 1-butanethiol.

8.23 **(a)** Ethanol > Diethyl ether > Butane
  (b) 1,2-Hexanediol > 1-Hexanol > Hexane

8.25 **(a)**

  (b) [cyclohexene] or [methylenecyclohexane]

  (c)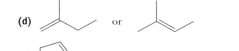

  (d) [2-methyl-1-butene] or [2-methyl-2-butene]

  (e) [cyclopentene] **(f)** [propene]

8.27 **(a)** H_2CO_3, $H-O-\overset{\overset{\displaystyle O}{||}}{C}-O^-$

  (b) and **(c)** CH_3COOH, $CH_3-\overset{\overset{\displaystyle O}{||}}{C}-O^-$

8.29 **(a)** CH_3O^-, CH_3OH **(b)** $CH_3CH_2O^-$, CH_3CH_2OH
  (c) NH_2^-, NH_3

8.31 **(a)** evenly balanced **(b)** right **(c)** left **(d)** right

8.33 **(a)** [structure] O^-Na^+ **(b)** [structure] Br
  (c) [structure] COOH **(d)** [structure] Cl
  (e) [structure] CHO

8.35 When treated with aqueous acid, 2° and 3° alcohols undergo an S_N1 reaction with water. When (*R*)-2-butanol is protonated and then loses water, an achiral planar carbocation results and stereochemistry is lost. Attack of water on the carbocation from either side results in formation of a racemic mixture.

8.37 **(a)** [structure with O, OH] **(b)** [structure with Cl]

  (c) [cyclohexane with CH₃ and Cl] **(d)** Br[structure]Br

  (e) [cyclooctanone with O] **(f)**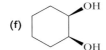

8.39 Compounds (**a**) and (**c**) undergo intramolecular cyclizations to cyclic ethers.

(**a**)   (**c**)

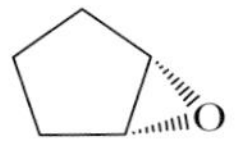

8.41

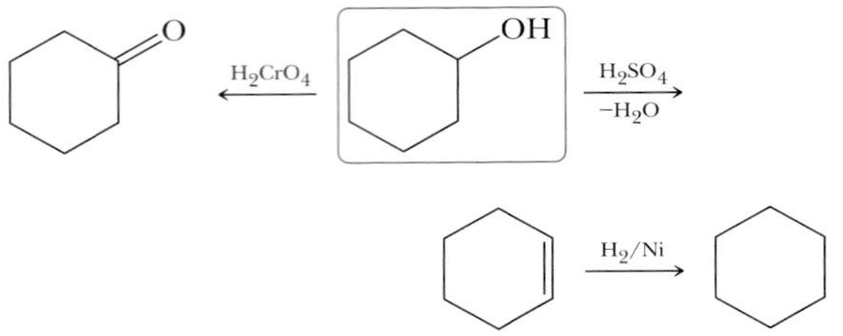

8.43

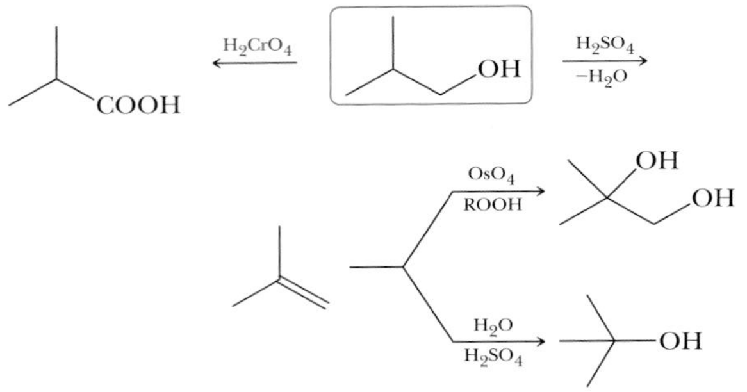

8.45

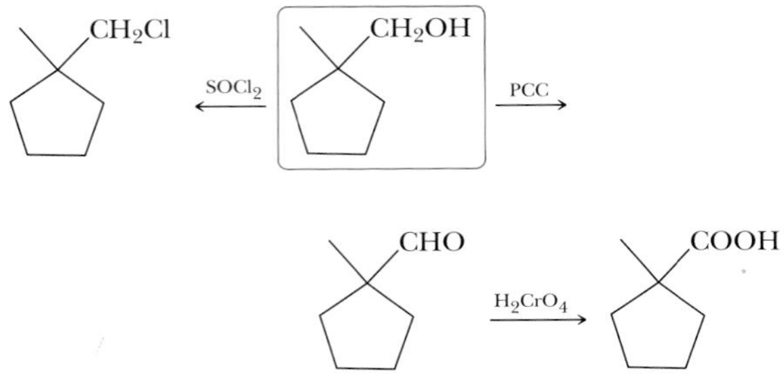

8.47 (a)

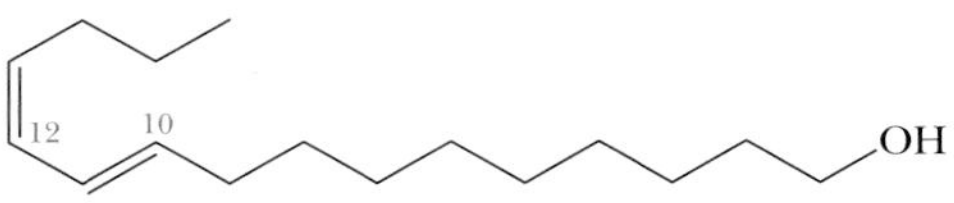

trans-10-*cis*-12-Hexadecadien-1-ol

(**b**) Four *cis-trans* isomers are possible.

8.49 (a) Phenol is the stronger acid.

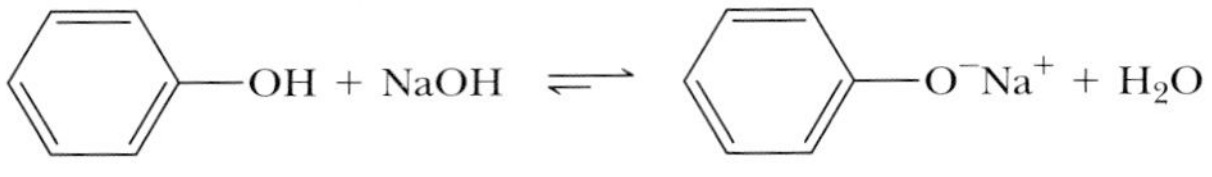

OH + NaOH ⇌ O⁻Na⁺ + H₂O

pK_a 10

OH + NaOH ⇌ O⁻Na⁺ + H₂O

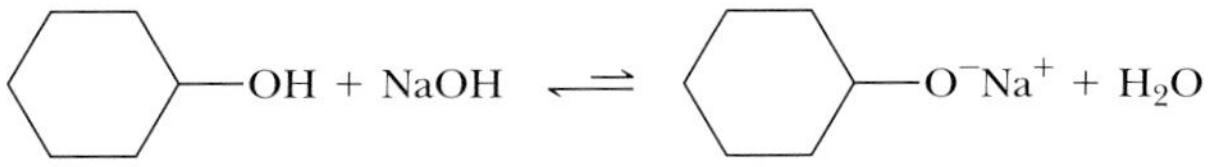

pK_a 18

(**b**) The conjugate base of cyclohexanol is a stronger nucleophile than phenoxide. When comparing the nucleophilic strength of similar atoms, nucleophilic strength increases with increased base strength.

8.51 In order of decreasing nucleophilicity, they are

(**a**)   >   >

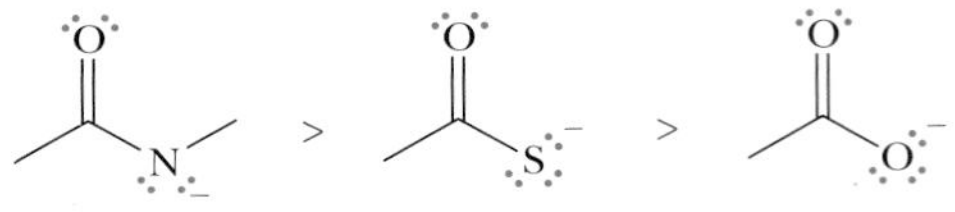

(**b**) $R-\ddot{C}H_2^- > R-\ddot{N}H^- > R-\ddot{O}\colon^-$

8.53 From most reactive to least reactive, they are

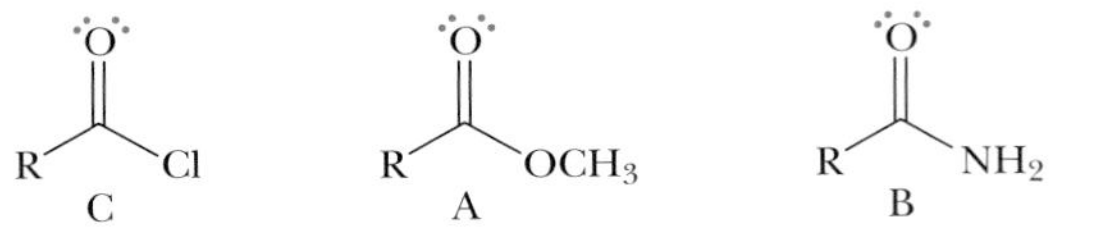

R—Cl   R—OCH₃   R—NH₂
  C   A   B

most reactive ————————→ least reactive

Chapter 9
Benzene and Its Derivatives

9.1 (a) 2-Phenyl-2-propanol
(**b**) (*E*)-3,4-Diphenyl-3-hexene
(**c**) 3-Methylbenzoic acid (*m*-methylbenzoic acid)

9.2 (a)   (**b**)

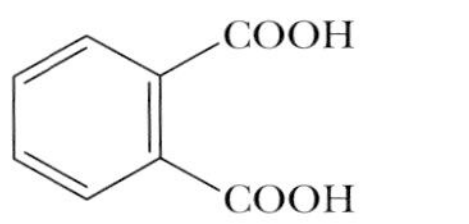

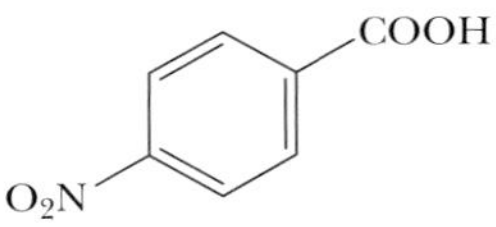

9.3 Step 1: Generation of HSO₃⁺, an electrophile:

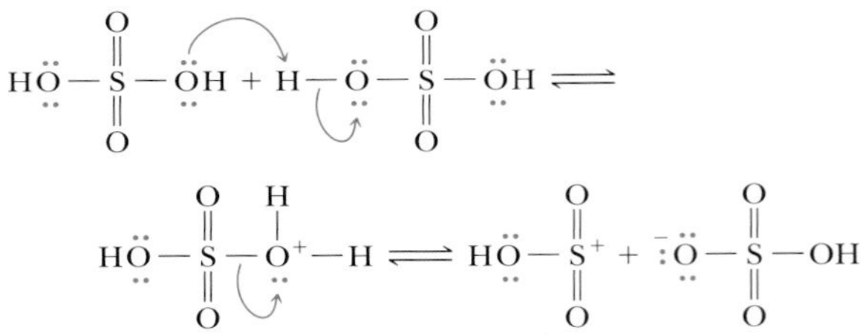

HO—S—OH + H—O—S—OH ⇌

HO—S—O⁺—H ⇌ HO—S⁺ + ⁻O—S—OH

Step 2: Nucleophilic attack of benzene on HSO₃⁺:

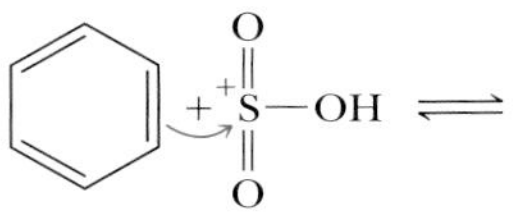

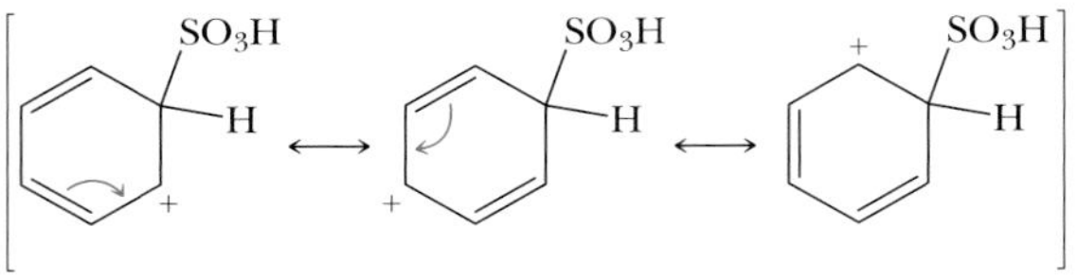

+ S—OH ⇌

resonance stabilized intermediate

Step 3: Loss of a proton to regenerate the aromatic ring:

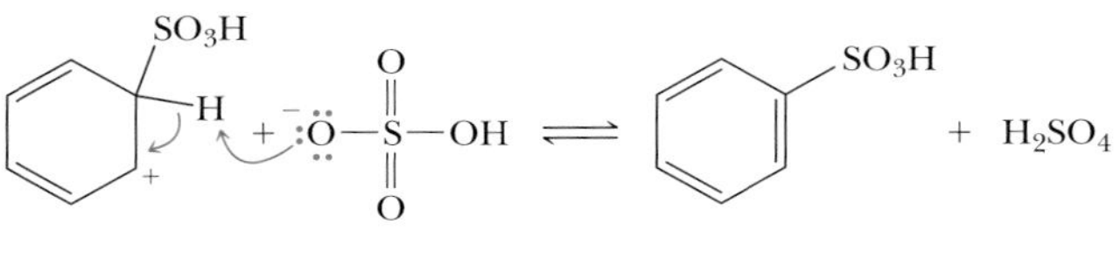

+ ⁺O—S—OH ⇌   + H₂SO₄

9.4 **(a)** 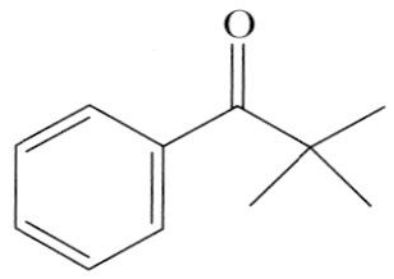 **(b)** 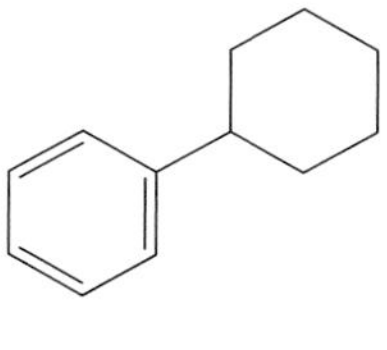

(c) 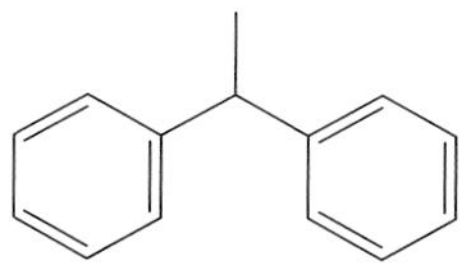

9.5 **Step 1:** Protonation of the 3° alcohol and generation of a 3° carbocation:

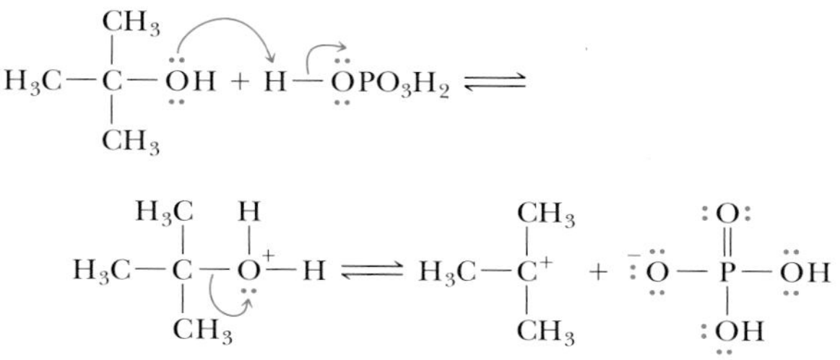

Step 2: Nucleophilic attack by benzene on the 3° carbocation:

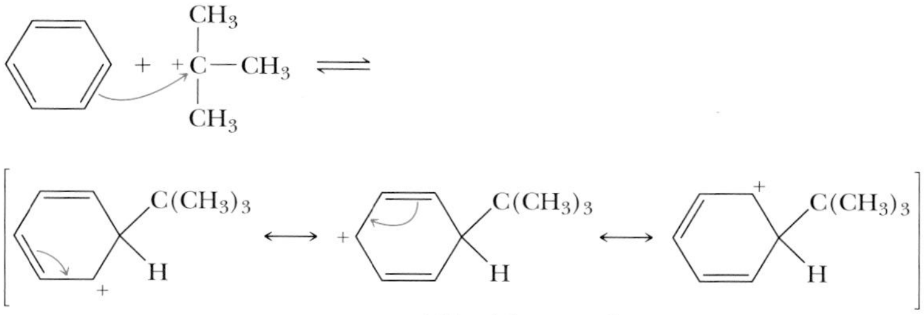

resonance stabilized intermediate

Step 3: Loss of a proton regenerates the aromatic ring:

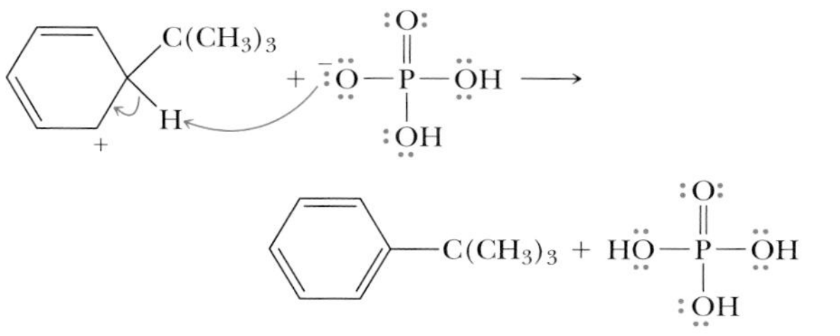

9.6 **(a)** 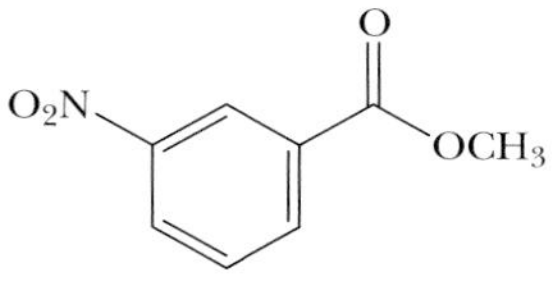

(b) 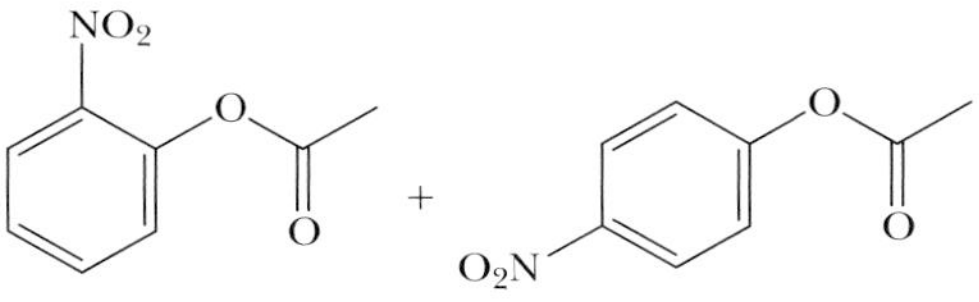

9.7 For ortho attack (and para as well), the third contributing structure places positive charges on adjacent atoms, which destabilizes the reaction intermediate and transition state for ortho, para attack.

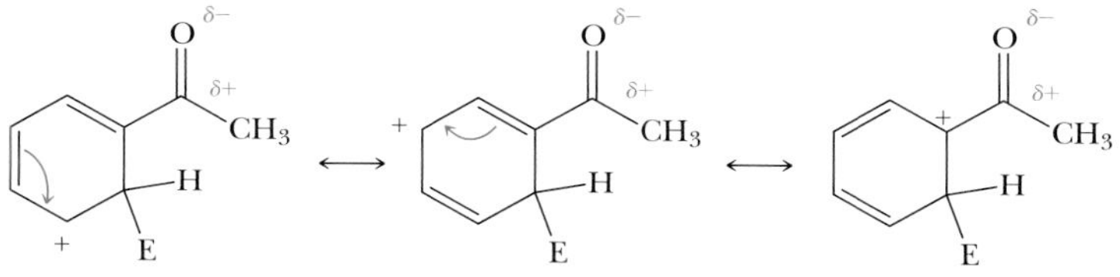

adjacent positive charges

Close inspection of the resonance hybrid for meta attack reveals no destabilizing contributing structure comparable to that for ortho, para attack. Therefore, electrophilic aromatic substitution on acetophenone takes place preferentially at the meta position.

9.8 **(a)** 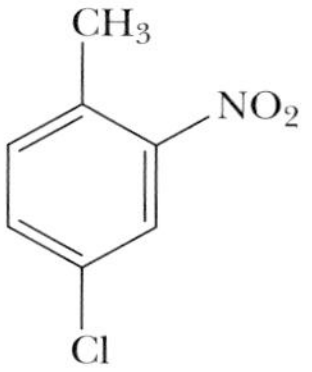 **(b)** 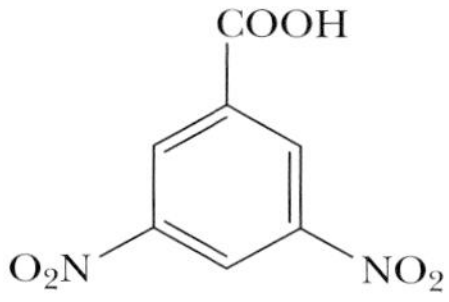

9.9 In order of increasing acidity, they are cyclohexanol < phenol < 2,4-dichlorophenol.

9.11 According to the Hückel criteria, the cyclopentadienyl anion is aromatic; it has six pi electrons in a planar, fully conjugated ring. Because of its aromatic character, this anion is considerably more stable than the cyclopentane anion, which has no resonance stabilization. Therefore, cyclopentadiene is the more acidic compound.

9.13 **(a)** 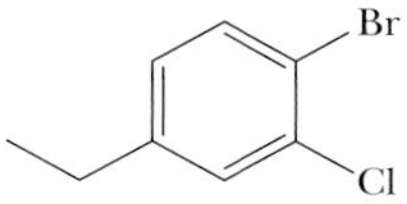 **(b)** 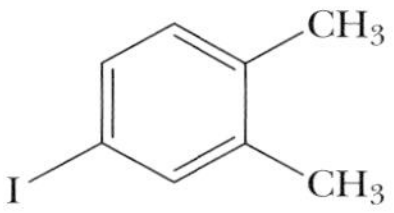

(c) 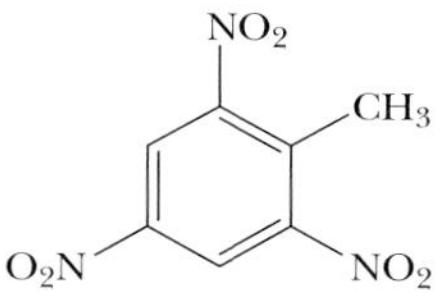 **(d)** 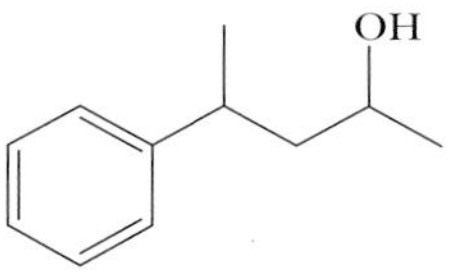

(e) 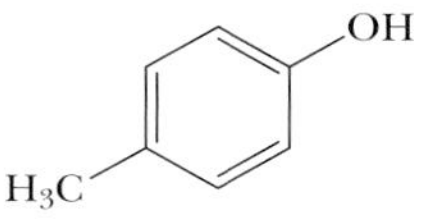 **(f)** 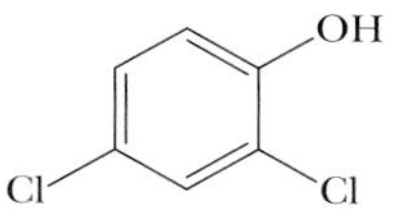

(g) 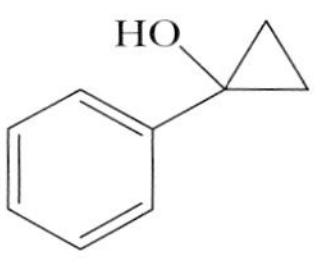 **(h)** 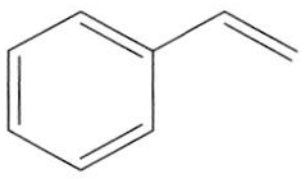

(i) 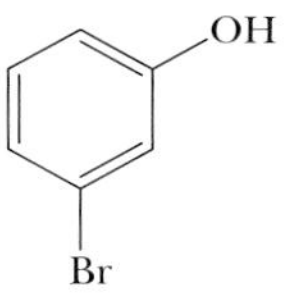 **(j)** 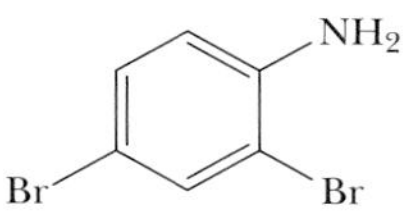

(k) 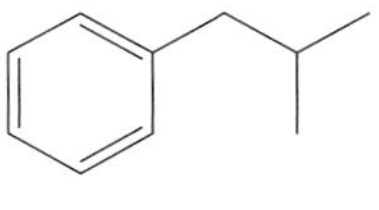 **(l)** 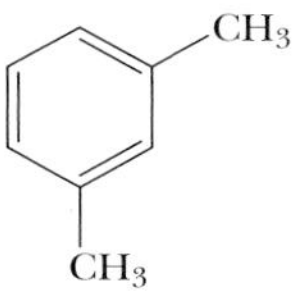

9.15

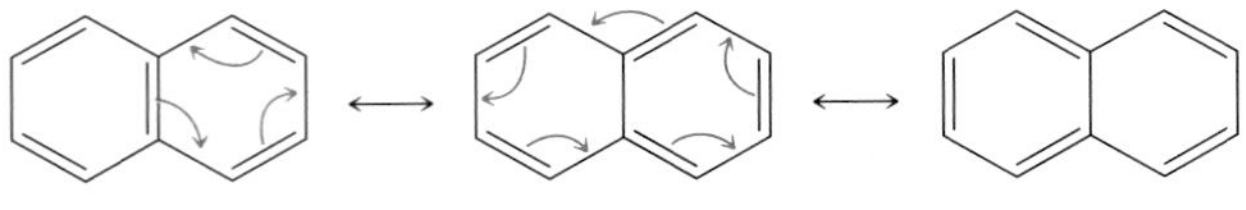

9.17 (a)

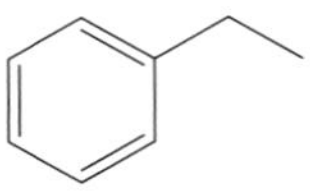

(b)

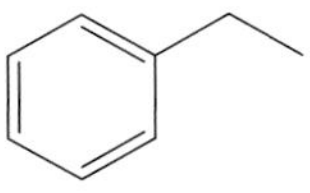

(c)

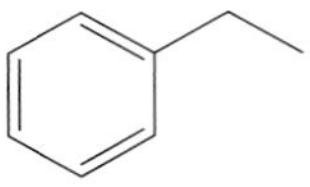

9.19 Two

9.21 Step 1: Formation of a Lewis-acid-base complex:

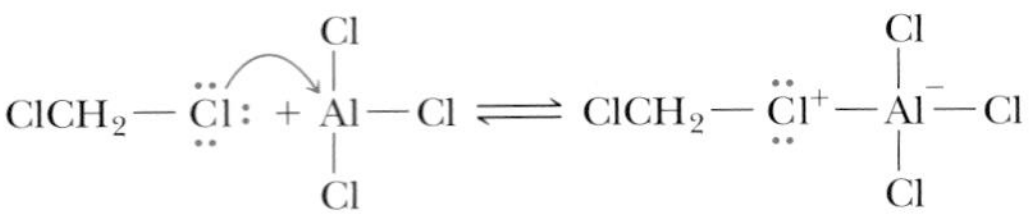

Step 2: Nucleophilic attack of benzene on the electrophilic Lewis acid-Lewis base complex gives a resonance-stabilized carbocation:

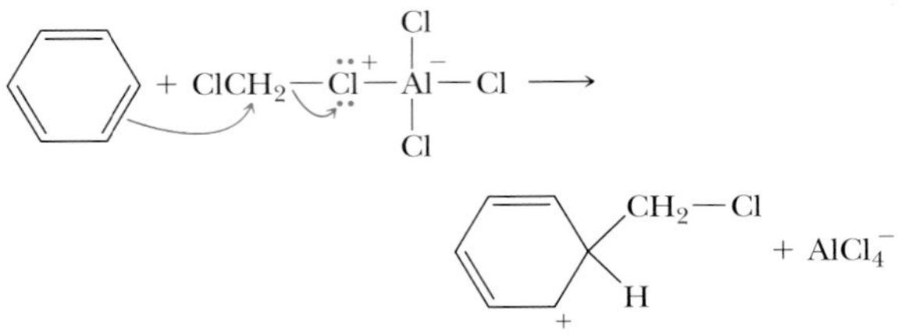

Step 3: Deprotonation of the carbocation gives benzyl chloride, HCl, and AlCl$_3$:

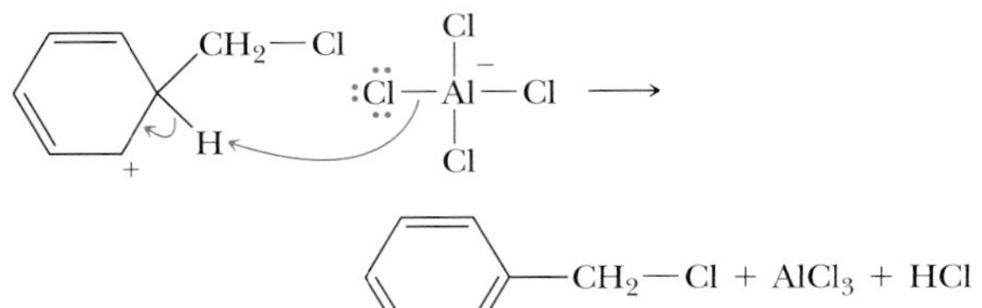

Step 4: Formation of Lewis acid-base complex between benzyl chloride and AlCl$_3$:

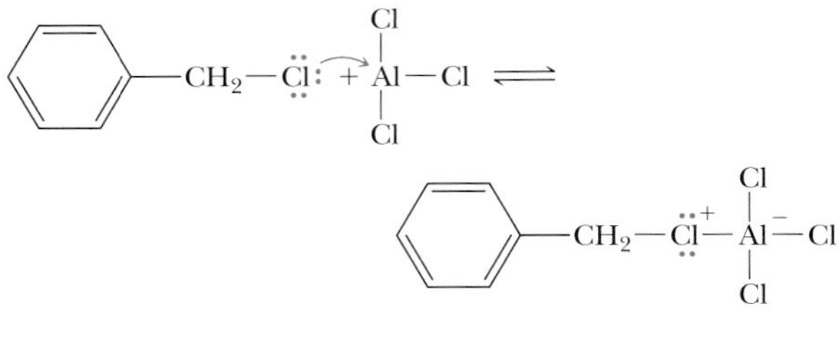

Step 5: Dissociation of the complex gives a resonance stabilized benzyl carbocation and AlCl$_4^-$:

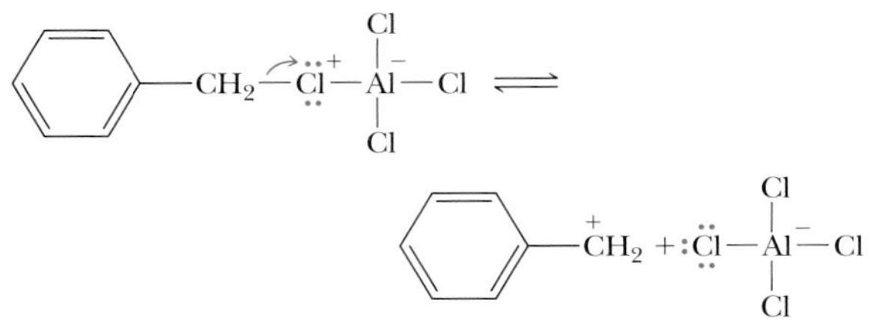

Step 6: Nucleophilic attack of the second molecule of benzene on the benzylic carbocation gives another resonance-stabilized carbocation:

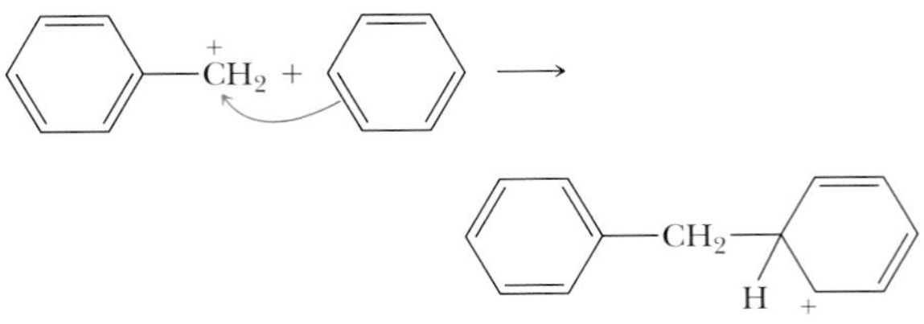

Step 7: Deprotonation of the carbocation intermediate regenerates the aromatic ring giving diphenylmethane, HCl, and AlCl$_3$.

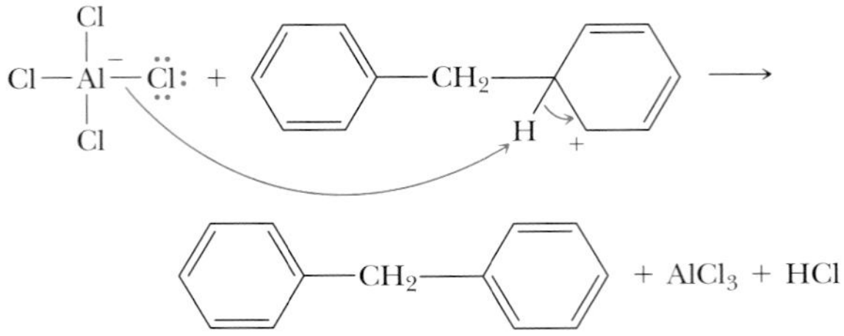

9.23 1,4-Dimethylbenzene gives one monochlorination product. 1,3-Dimethylbenzene gives two monochlorination products.

9.25 Toluene undergoes electrophilic aromatic substitution faster than chlorobenzene. Chlorine is ortho-para directing and deactivating; methyl is ortho-para directing and activating.

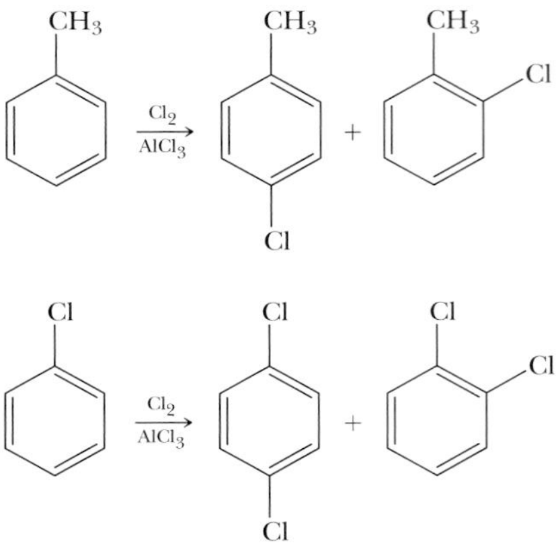

9.27 The trifluoromethyl group is highly electron withdrawing because the very electronegative fluorine atoms pull electron density away from the attached carbon atom, thereby creating a partial positive charge on the carbon bonded to the benzene ring.

9.29 (a)

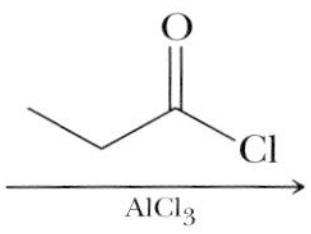

(b) 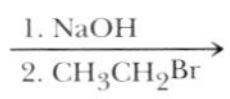 $\xrightarrow[\text{2. CH}_3\text{CH}_2\text{Br}]{\text{1. NaOH}}$

(c)

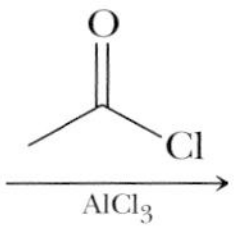

(d) 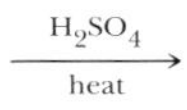 $\xrightarrow[\text{heat}]{\text{H}_2\text{SO}_4}$

9.31 Step 1: The reaction begins with protonation of acetone to form its conjugate acid, which may be written as a hybrid of two contributing structures:

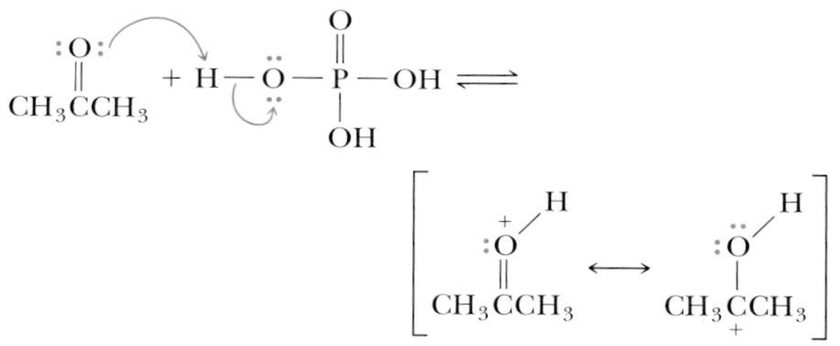

Step 2: The conjugate acid of acetone is an electrophile and reacts with phenol at the para position to give a resonance-stabilized carbocation intermediate:

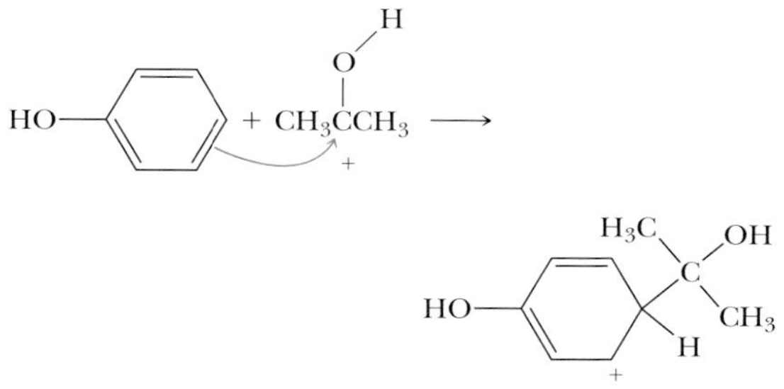

Step 3: Deprotonation of the carbocation intermediate gives 2-(4-hydroxyphenyl)-2-propanol:

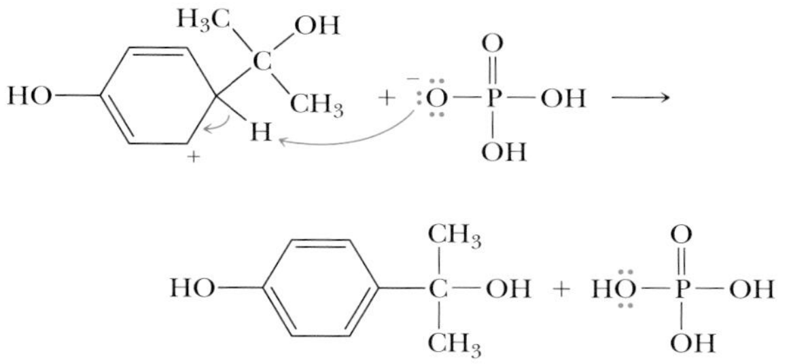

Step 4: The tertiary alcohol is protonated:

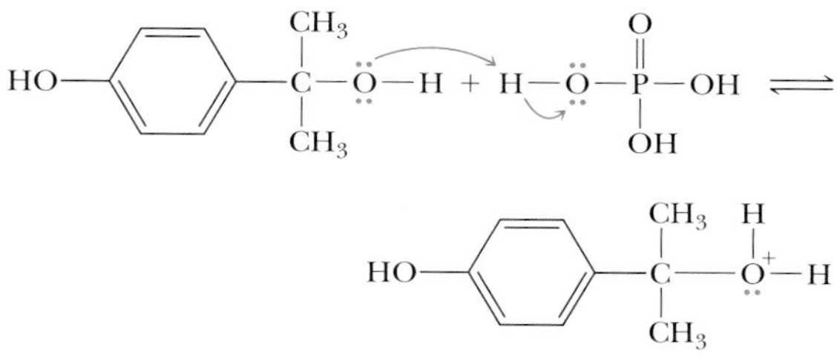

Step 5: The protonated alcohol loses water to give a resonance-stabilized carbocation intermediate:

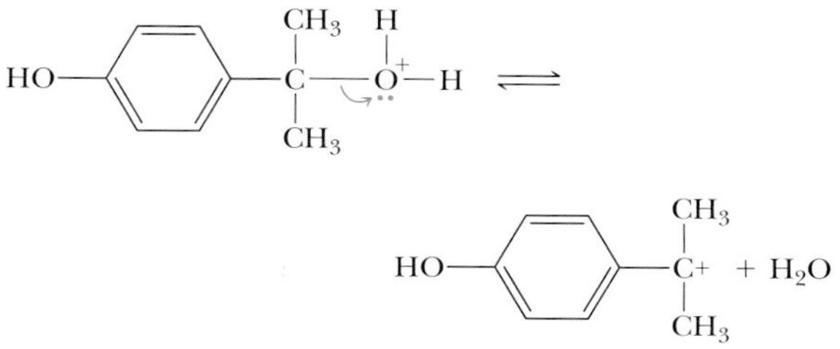

Step 6: Attack of phenol on the carbocation intermediate gives a resonance-stabilized carbocation intermediate:

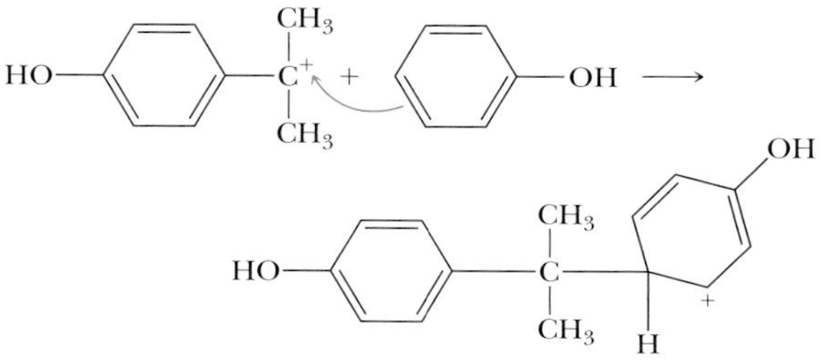

Step 7: Deprotonation of the carbocation intermediate yields bisphenol A:

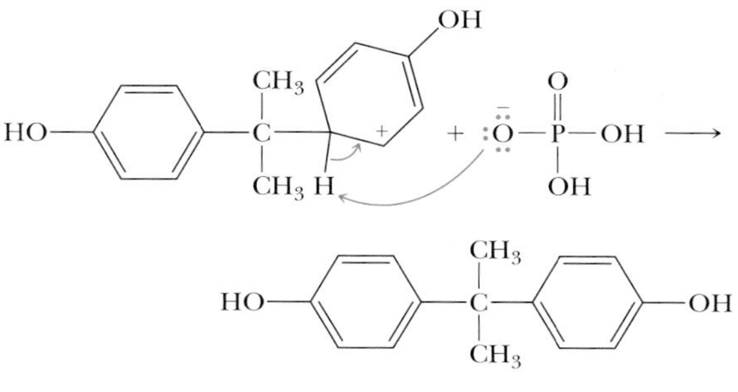

9.33

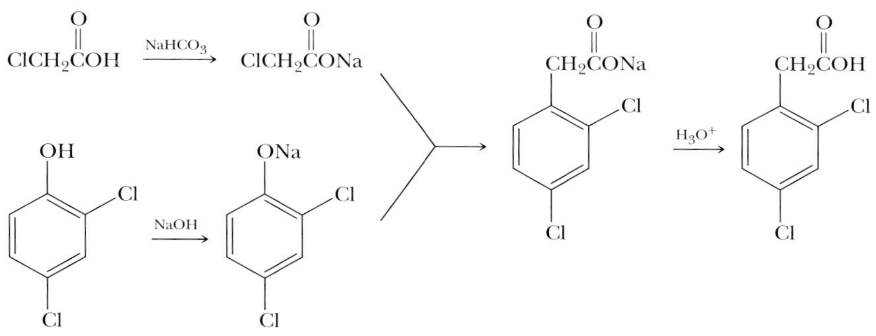

9.35 In order of increasing acidity they are:

(a) cyclohexanol < phenol < acetic acid

(b) water (pK_a 15.7) < sodium bicarbonate (pK_a 10.33) < phenol (pK_a 9.95)

(c) benzyl alcohol < phenol < 4-nitrophenol

9.37 When carbonic acid is formed, it decomposes to carbon dioxide and water according to the following equation:

$$H_2CO_3 \longrightarrow CO_2 + H_2O$$

Acid–base equilibria favor the side with the weaker acid and weaker base. Carboxylic acids (stronger acids than carbonic acid) react with bicarbonate to form carbonic acid (the weaker acid), which decomposes to carbon dioxide and water. Phenols are weaker acids than carbonic acid, so their equilibria favor the left side; phenols will not react with sodium bicarbonate to form carbonic acid.

9.39

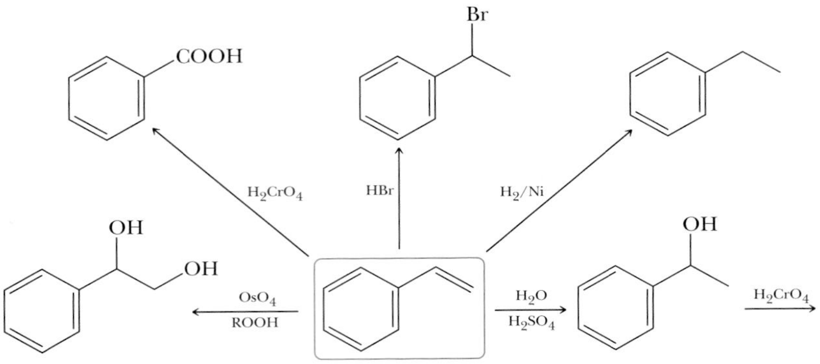

9.41 (a)

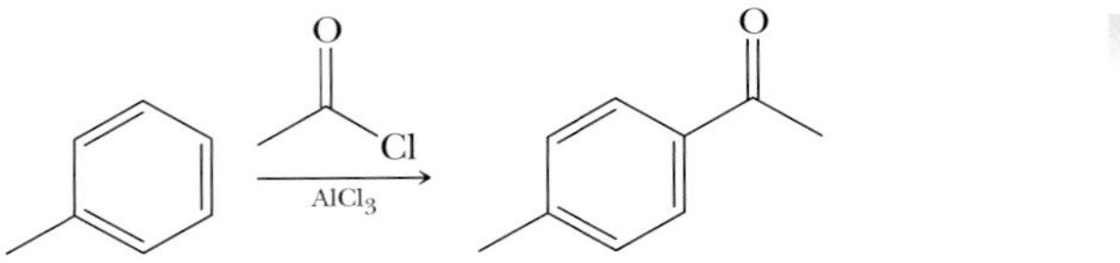

(b, c)

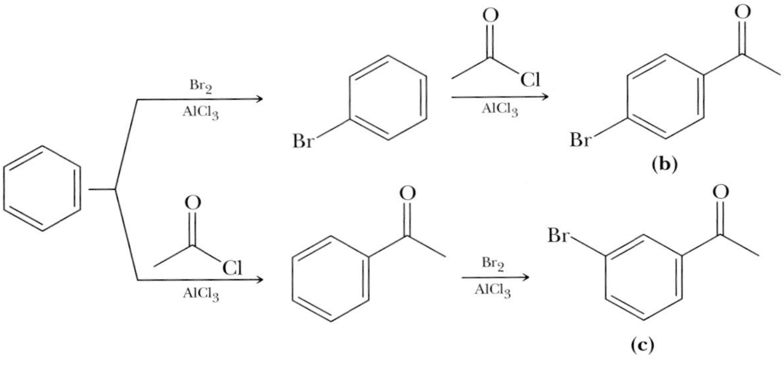

9.43 (a)

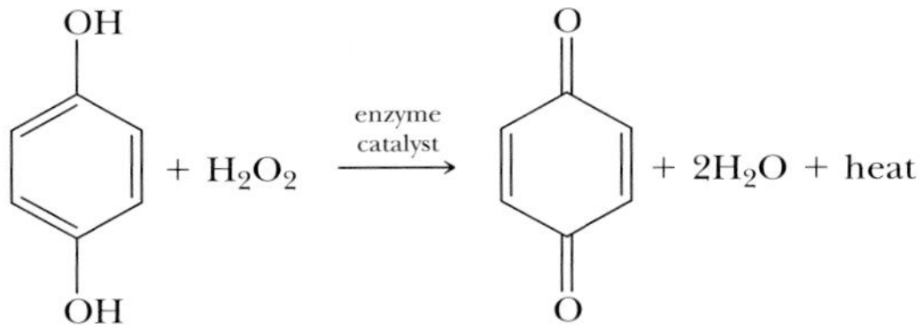

$$+ H_2O_2 \xrightarrow[\text{catalyst}]{\text{enzyme}} + 2H_2O + \text{heat}$$

(b) The starting quinone loses hydrogens; therefore, it is oxidized.

9.45

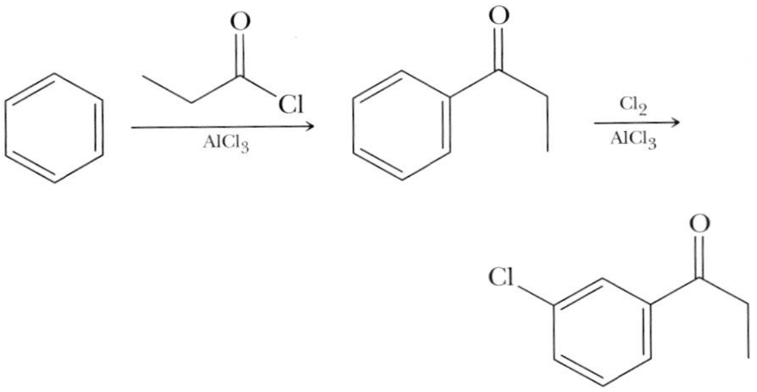

9.47 Cyclohexylamine is the better nucleophile; it is also the stronger base.

9.49

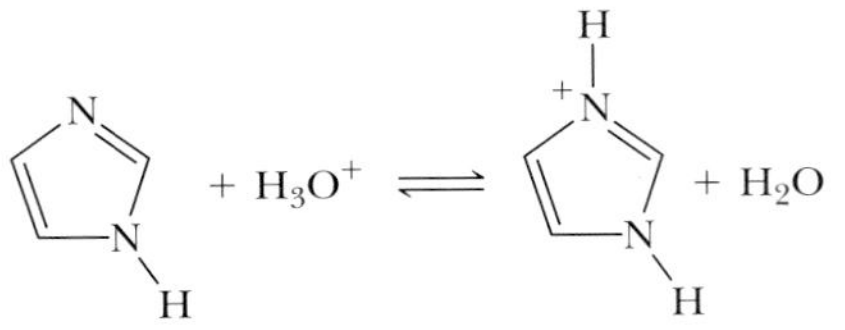

Imidazole

Chapter 10
Amines

10.1 (a)

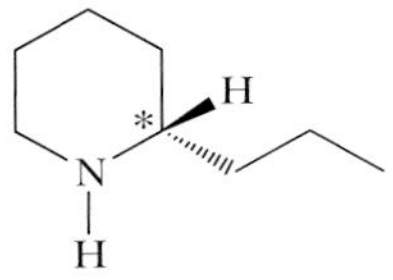

(S)-Coniine

(b)

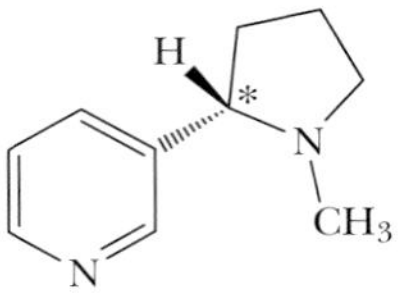

(S)-Nicotine

(c)

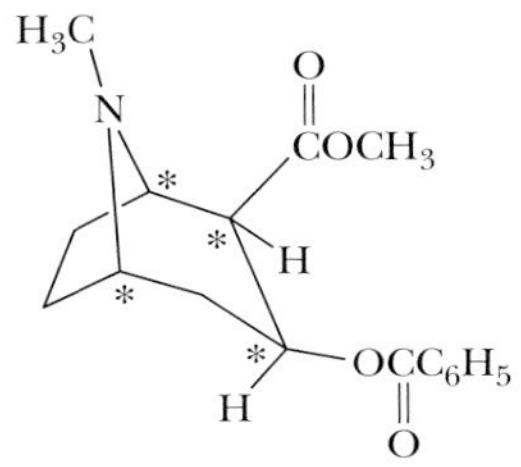

Cocaine

10.2 (a)

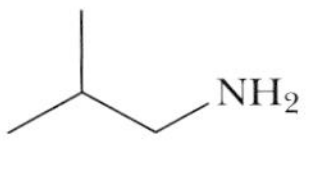

(b)

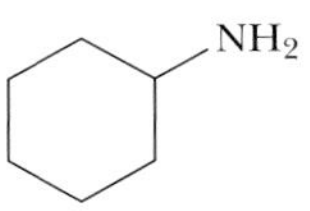

(c)

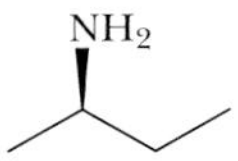

10.3 (a) [structure: isobutyl-NH₂]   **(b)** [structure: triphenylamine]

(c) [structure: diisopropylamine, N—H]

(k) [structure: diphenylamine N—H]   **(l)** [structure: isobutyl-NH₂]

10.4 To the left.

10.5 (a) A is the stronger acid.   **(b)** C is the stronger acid.

10.6 (a) $(CH_3CH_2)_3NH^+Cl^-$ **(b)** [structure: piperidine] $NH_2^+ + CH_3COO^-$

Triethylammonium
chloride

Piperidinium
acetate

10.7 (a) $CH_3\overset{O}{\overset{\|}{C}}HCOH$ **(b)** $CH_3\overset{O}{\overset{\|}{C}}HCO^-$
$\quad\quad\quad \underset{NH_3^+}{}$ $\quad\quad\quad \underset{NH_2}{}$
at pH 2.0   at pH 12.0

10.8 Reverse the order of the first and second steps:

[scheme: toluene → (1) → benzoic acid → (2) → m-nitrobenzoic acid → (3)]

(1) Oxidation   (2) Nitration   (3) Reduction

[scheme: m-aminobenzoic acid → (4) → m-hydroxybenzoic acid]

(4) HNO₂, HCl, heat

10.9 (a) [structure: sec-butylamine, NH₂]

(b) [structure: long chain amine, NH₂]

(c) [structure: neopentylamine, NH₂]

(d) H_2N [chain] NH_2

(e) [structure: 2-bromoaniline, NH₂, Br]

(f) $(CH_3CH_2CH_2CH_2)_3N$

(g) [structure: N,N-dimethylaniline, N(CH₃)₂]

(h) [structure: benzylamine, CH₂NH₂]

(i) $CH_3\underset{CH_3}{\overset{CH_3}{\underset{|}{\overset{|}{C}}}}NH_2$

(j) [structure: N-ethylcyclohexylamine, N—H]

10.11 The classification of amines as 1°, 2°, and 3° is based on how many hydrogen atoms of ammonia are replaced by alkyl or aryl groups. Classification of alcohols as 1°, 2°, and 3° is based on how many carbon groups are bonded to the carbon bearing the hydroxyl group.

$CH_3CH_2CH_2CH_2NH_2$ $CH_3CH_2NHCH_2CH_3$ $CH_3CH_2N(CH_3)_2$
a 1° amine   a 2° amine   a 3° amine

$CH_3CH_2CH_2CH_2OH$ $CH_3CH_2\overset{OH}{\overset{|}{C}}HCH_3$ $CH_3\overset{CH_3}{\underset{CH_3}{\overset{|}{\underset{|}{C}}}}OH$
a 1° alcohol   a 2° alcohol   a 3° alcohol

10.13 (a) Both amino groups are 2° aliphatic amines. **(b)** The structural similarities are indicated in bold on the structural formulas.

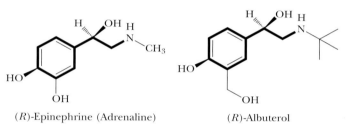

(R)-Epinephrine (Adrenaline)   (R)-Albuterol

10.15 For parts **(f)** and **(g)**, several other isomers are possible.

(a) [structure: N-methylaniline] —NHCH₃   **(b)** [structure] —N(CH₃)₂

(c) [structure] —CH₂NH₂   **(d)** $CH_3\overset{NH_2}{\overset{|}{C}}HCH_2CH_3$ (* on C)

(e) [structure: N-methylpyrrolidine] N—CH₃   **(f)** H_3C [trimethylaniline structure with CH₃, CH₃, NH₂]

(g) [structure: quaternary ammonium, H_3C, CH_3, Cl^-, N^+, ethyl, * on C]

10.17 An N—H------N hydrogen bond is not as strong as an O—H----O hydrogen bond because an N—H bond is not as polar as an O—H bond. Stronger intermolecular hydrogen bonds between molecules of 1-butanol require greater energy to separate molecules of 1-butanol and convert them to the vapor phase.

10.19 Nitrogen is less electronegative than oxygen, so the lone pair of electrons on nitrogen is held less tightly and is more available to bond with protons than are the lone pairs on oxygen; therefore, amines are more basic than alcohols.

10.21 The nitro group withdraws electron density from the aromatic ring. For 4-nitroaniline, delocalization of the lone pair of electrons on the amine nitrogen through resonance makes it less able to bond with protons, making 4-nitroaniline a weaker base than aniline. The same delocalizing effect helps to stabilize the conjugate base of 4-nitrophenol, making 4-nitrophenol a stronger acid than phenol.

10.23 Under each acid is given its estimated pK_a. The position of equilibrium is indicated by the relative lengths of the equilibrium arrows.

(a) CH_3COOH + $\rightleftharpoons$ CH_3COO^- +

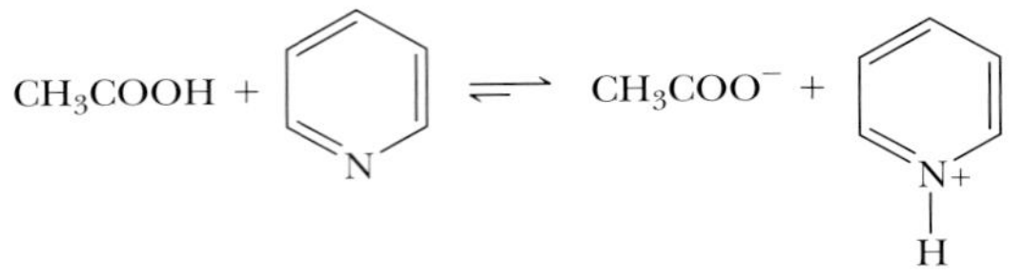

     pK_a 4.76 pK_a 5.25

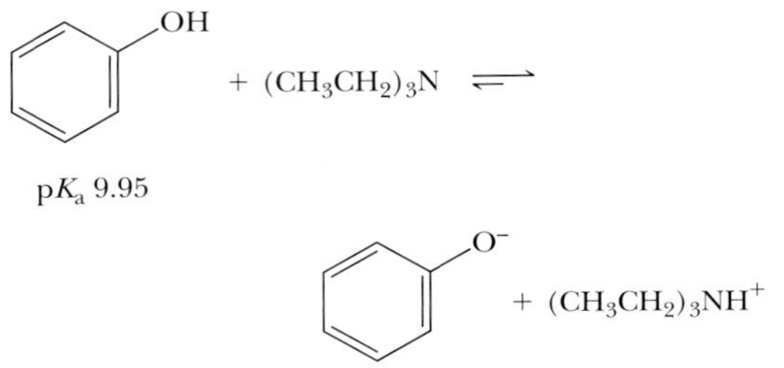

(b) + $(CH_3CH_2)_3N$ $\rightleftharpoons$

     pK_a 9.95

+ $(CH_3CH_2)_3NH^+$

                 pK_a 10.75

(c) + $\rightleftharpoons$

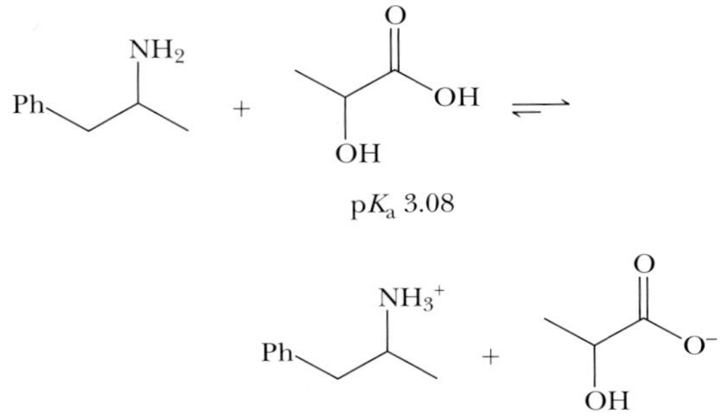

                 pK_a 3.08

+

                 pK_a ~11

(d) + $\rightleftharpoons$

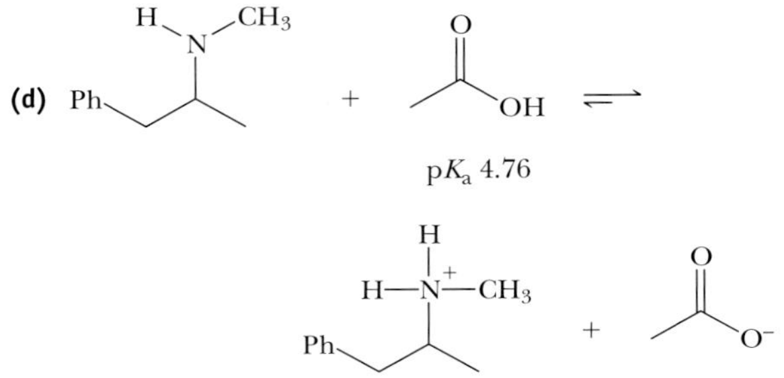

                 pK_a 4.76

+

                 pK_a ~11

10.25 At pH 7.4, amphetamine is almost entirely in its conjugate acid form; the ratio of base to conjugate acid is 1 : 2,500.

10.27 (a)

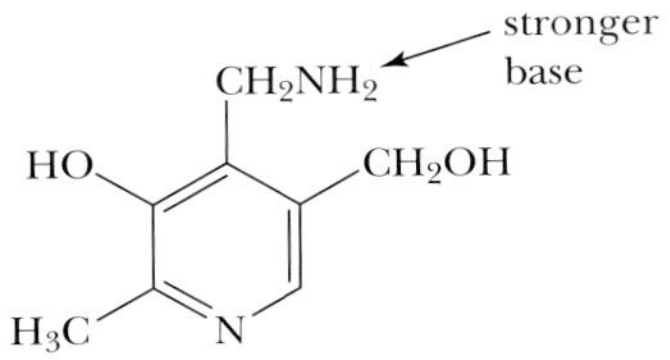

(b)

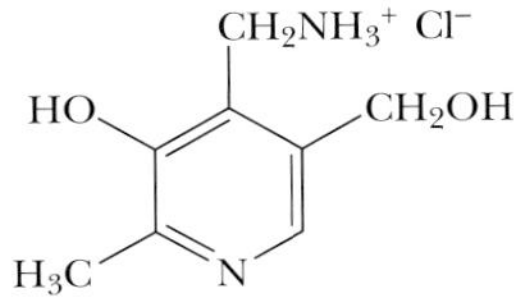

10.29 (a)

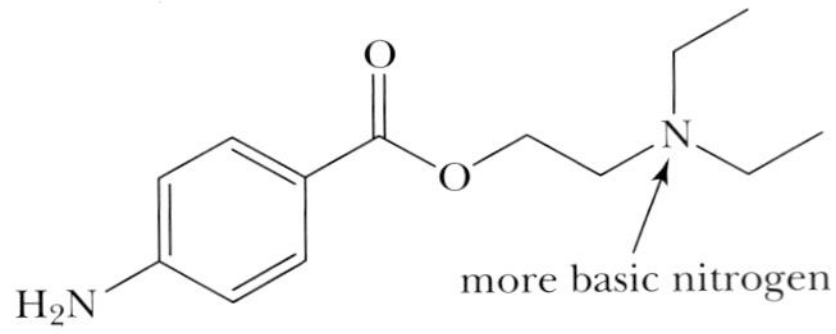

(b)

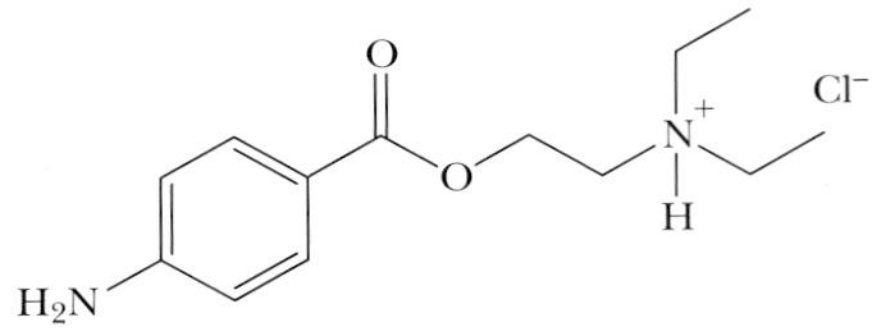

(c) Procaine is not chiral; a solution of Novocaine is not optically active.

10.31 Dissolve the mixture in an organic solvent such as diethyl ether, and then extract the ether layer with an aqueous HCl. Aniline reacts with HCl to form a water-soluble salt. Nitrobenzene is neutral and remains in the ether layer. Separate the ether and aqueous layers, and recover nitrobenzene by distillation of the ether. Recover aniline by making the aqueous solution basic using NaOH, and extracting the water-insoluble aniline with diethyl ether. Remove the ether layer and evaporate the ether to recover aniline.

10.33 Glucophage is an amine hydrochloride salt, which is water soluble and insoluble in non-polar organic solvents. Therefore, predict that Glucophage will be soluble in water and blood plasma, and insoluble in non-polar solvents such as diethyl ether and dichloromethane. The $=NH$ nitrogens are stronger bases than the other nitrogens.

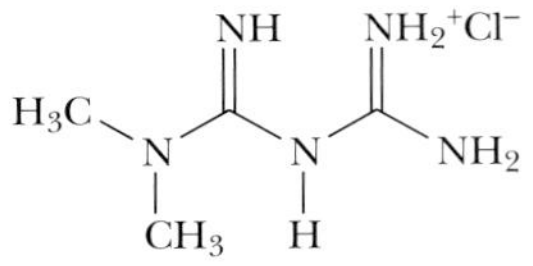

10.35 (1) HNO_3/H_2SO_4 **(2)** H_2CrO_4 **(3)** H_2/Ni

10.37 Treat ethylene oxide with diethylamine.

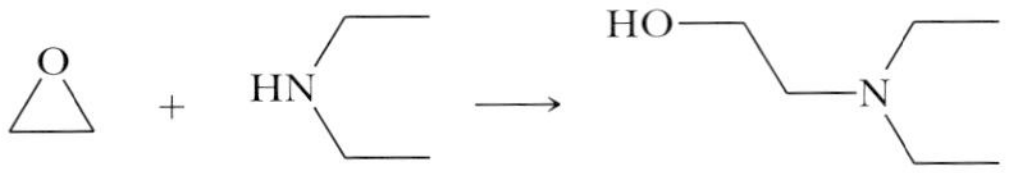

10.39 (1) 1. NaOH **2.** CH_3CH_2Br **(2)** HNO_3/H_2SO_4
     (3) H_2/Ni

10.41 **(1)** HNO_3/H_2SO_4 **(2)** $2\ CH_3CH{=}CH_2/H_3PO_4$
(3) H_2/Ni **(4)** **1.** $NaNO_2$; $HCl/0°C$ **2.** H_3PO_2

10.43 **(a)** **(b)**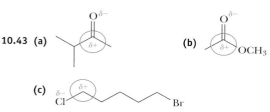

(c)

but bromine is a better
leaving group

10.45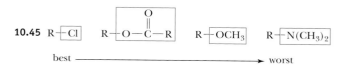

best ⟶ worst

Chapter 11

Infrared Spectroscopy

11.1 The energy of red light at 680 nm (42.1 kcal/mol) is greater than the energy of infrared radiation at 2.50 μm (11.4 kcal/mol).

11.2 Carboxylic acids have two strong absorptions. The first, a broad peak between 3200 and 3500 cm^{-1}, is due to the carboxyl OH group. The second, a strong peak around 1715 cm^{-1}, is due to the carboxyl C=O group.

11.3 Propanoic acid will have two major absorptions, a broad absorption between 2400–3400 cm^{-1} due to the carboxyl —OH group, and a strong absorption around 1700–1725 cm^{-1} due to the carboxyl C=O group. Methyl acetate will not have the hydroxyl absorption.

11.4 The higher the wavenumber, the higher is the energy of vibration. A bond that requires higher energy to stretch will be a stronger bond.

11.5 The index of hydrogen deficiency is 2. The ring accounts for one index of hydrogen deficiency; and the double bond accounts for the other.

11.6 Niacin has four double bonds (each worth one hydrogen deficiency) and a ring (worth one hydrogen deficiency) for a total index of hydrogen deficiency = 5.

11.7 The completed table is:

Class of Compound	Molecular Formula	Index of Hydrogen Deficiency	Reason for Hydrogen Deficiency
alkane	C_nH_{2n+2}	0	(reference hydrocarbon)
alkene	C_nH_{2n}	1	one pi bond
alkyne	C_nH_{2n-2}	2	two pi bonds
alkadiene	C_nH_{2n-2}	2	two pi bonds
cycloalkane	C_nH_{2n}	1	one ring
cycloalkene	C_nH_{2n-2}	2	one pi bond + one ring

11.9 **(a)** Its index of hydrogen deficiency is 2.
(b) One ring and one double bond.
(c) A cycloalkene

11.11 Compound E is nonane and compound F is 1-hexanol.

11.13 **(a)** Compound I has an index of hydrogen deficiency of 4.
(b) Compound I has 4 double bonds and/or rings.
(c) A single benzene ring can account for an index of hydrogen deficiency of 4.
(d) Compound I contains an —OH group.

11.15 **(a)** Compound K has an index of hydrogen deficiency of 1.
(b) Compound K can have one ring or one pi bond.
(c) A carbonyl group of an aldehyde or ketone.

11.17 **(a)** Compound M has an index of hydrogen deficiency of 1.
(b) Compound M must have one ring or one pi bond.
(c) The oxygen and nitrogen-containing functional group is an amide.

11.19 **(a)** Benzoic acid has a strong, broad O—H stretching absorption between 2400 and 3400 cm^{-1}; benzaldehyde does not absorb in this region.
(b) The amide has a strong C=O stretching absorption between 1630 and 1680 cm^{-1}; the amine does not absorb in this region.
(c) The carboxylic acid has a strong, broad O—H stretching absorption between 2400 and 3400 cm^{-1} from the carboxyl OH group, and a strong O—H stretching absorption between 3200 and 3400 cm^{-1} from the alcohol OH. The cyclic ester will have neither of these absorptions.
(d) The primary amide has two broad N—H stretching absorptions between 3200 and 3400 cm^{-1}; the tertiary amide has no absorptions in this region.

11.21 The frequency of 3×10^8 Hz is lower in energy than the infrared region and exists in the near microwave/radio-frequency region of the electromagnetic spectrum.

11.23 **(a)** IR spectroscopy cannot distinguish between pentane and heptane without a reference spectrum.
(b) IR spectroscopy cannot distinguish between 2-methylphenol and 3-methylphenol without a reference spectrum.
(c) The symmetric C=C stretch in *trans*-3-hexene does not result in change in molecular dipole; therefore, a C=C stretch between 1600–1680 cm^{-1} will be absent.
(d) IR spectroscopy cannot distinguish between 2-pentanone and 3-pentanone without a reference spectrum.

Chapter 12

Nuclear Magnetic Resonance Spectroscopy

12.1 **(a)** Phosphorus-31 **(b)** Platinum-195

12.2 **(a)** 3-Methylpentane has four sets of equivalent hydrogens.
(b) 2,2,4-Trimethylpentane has four sets of equivalent hydrogens.

12.3 **(a)**

$$CH_3-\overset{\overset{\displaystyle O}{\|}}{C}-CH_3$$

(b)

(c)

$$CH_3-\overset{\overset{\displaystyle CH_3}{|}}{\underset{\underset{\displaystyle CH_3}{|}}{C}}-CH_3$$

(d)

$$CH_3-\overset{\overset{\displaystyle Cl}{|}}{\underset{\underset{\displaystyle Cl}{|}}{C}}-\overset{\overset{\displaystyle Cl}{|}}{\underset{\underset{\displaystyle Cl}{|}}{C}}-CH_3$$

12.4 The ratio of signals is approximately 6:1, which corresponds to a 12:2 ratio for a total of 14 hydrogens. The larger signal represents 12 hydrogens and the smaller signal represents 2 hydrogens. The structure consistent with the data is 2,4-dimethyl-3-pentanone.

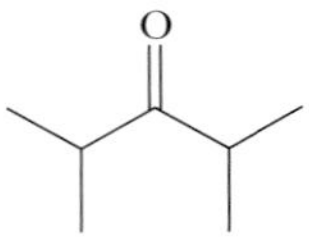

12.5 (a) Each compound will exhibit three signals: one for each of the different methyl groups and a third signal for the methylene groups.

(b) The ratio of signals will be 3:3:2.

(c) The CH_3 group of (2) will be more downfield due to the electron-withdrawing effect of oxygen.

12.6 (a) Each compound will display three signals:

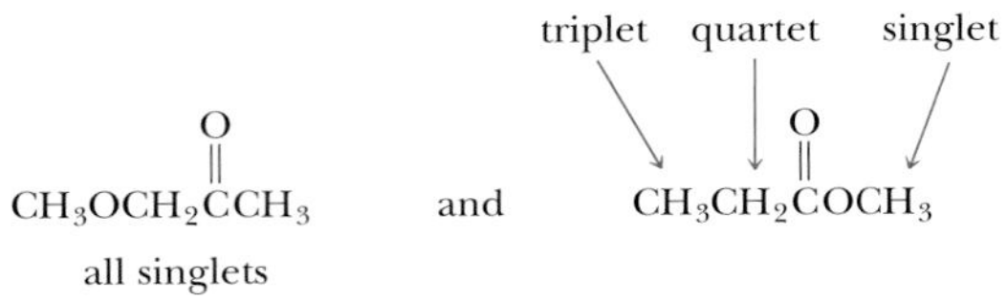

triplet quartet singlet

$CH_3OCH_2\overset{O}{\overset{\|}{C}}CH_3$ and $CH_3CH_2\overset{O}{\overset{\|}{C}}OCH_3$

all singlets

(b) The compound on the left has one signal. The one on the right has two signals.

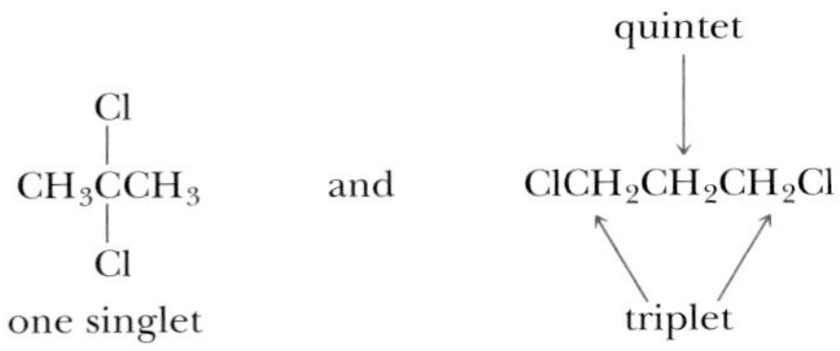

quintet

$CH_3\overset{Cl}{\underset{Cl}{C}}CH_3$ and $ClCH_2CH_2CH_2Cl$

one singlet   triplet

12.7 (a) These molecules can be distinguished by comparing the number of ^{13}C signals. The molecule on the left has 5 signals; the one on the right has 7 signals.

(b) The molecule on the left has six ^{13}C signals; the molecule on the right has three ^{13}C signals.

12.9 (a) 4 signals **(b)** 5 signals **(c)** 4 signals
(d) 2 signals **(e)** 4 signals **(f)** 3 signals
(g) 3 signals **(h)** 5 signals

12.11 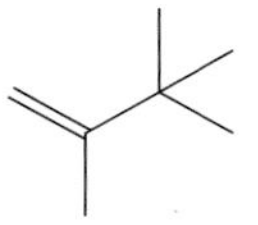 2,3,3-Trimethyl-1-butene

12.13 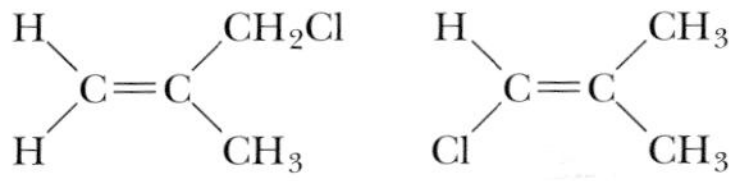
Compound C   Compound D

12.15 $CH_3CH_2\overset{OH}{\underset{CH_3}{C}}CH_2CH_3 \xrightarrow[\text{heat}]{H_3PO_4}$ $+ H_2O$

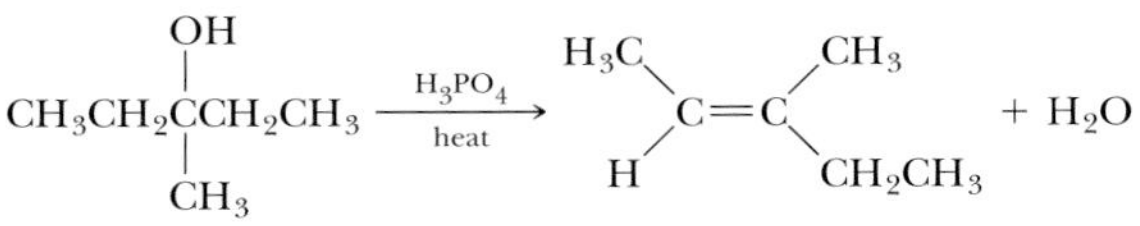

Compound E   Compound F

12.17 (a) CH_3-CHBr_2

(b) 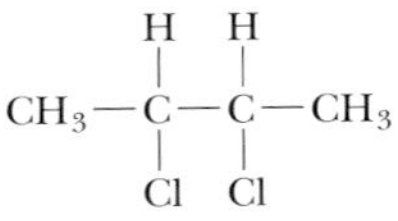 $CH_3-\overset{H}{\underset{Cl}{C}}-\overset{H}{\underset{Cl}{C}}-CH_3$

(c) $BrCH_2-\overset{CH_2Br}{\underset{CH_2Br}{C}}-CH_2Br$

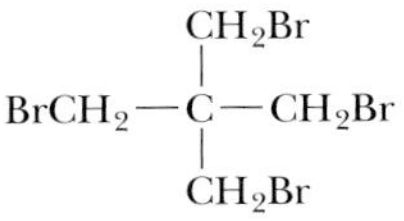

(d) 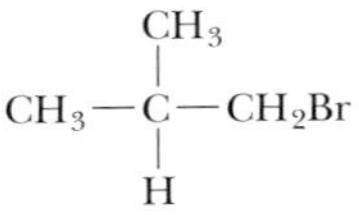 $CH_3-\overset{CH_3}{\underset{H}{C}}-CH_2Br$

(e) 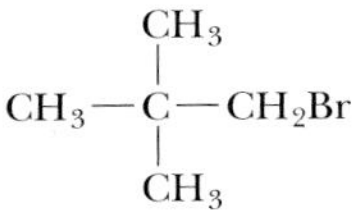 $CH_3-\overset{CH_3}{\underset{CH_3}{C}}-CH_2Br$

(f) 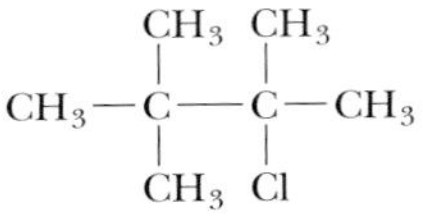 $CH_3-\overset{CH_3}{\underset{CH_3}{C}}-\overset{CH_3}{\underset{Cl}{C}}-CH_3$

12.19

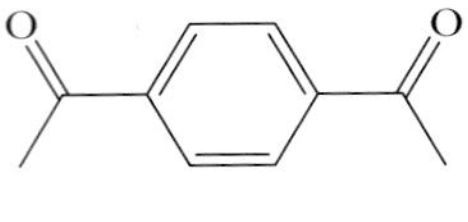

12.21 (a) 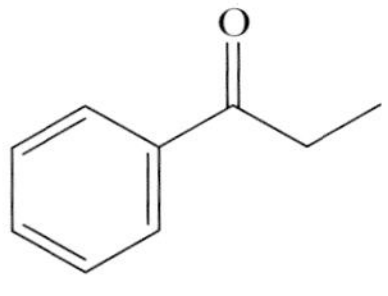 **(b)**

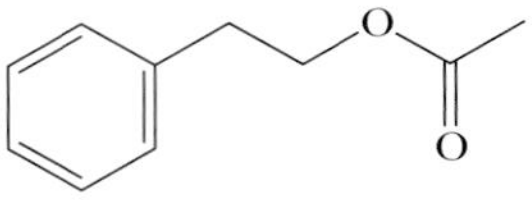

(c) 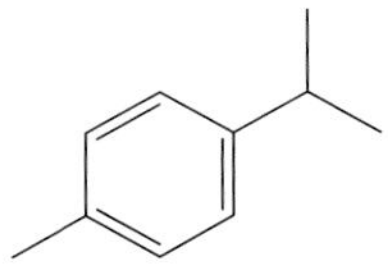 **(d)**

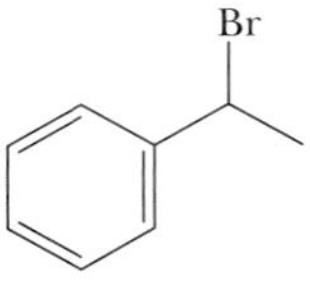

12.23 (a) 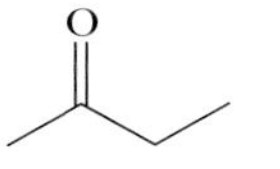 **(b)**

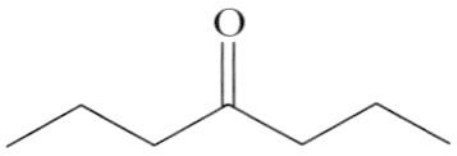

12.25 $\xrightarrow[\text{heat}]{H^+}$

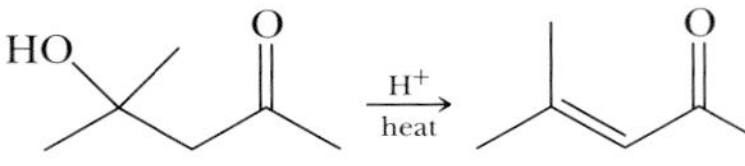

Compound P   Compound Q

12.27 (a) 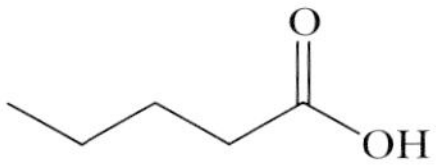 **(b)**

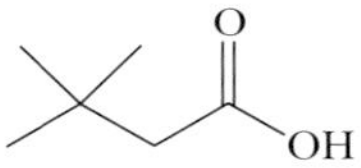

(c)

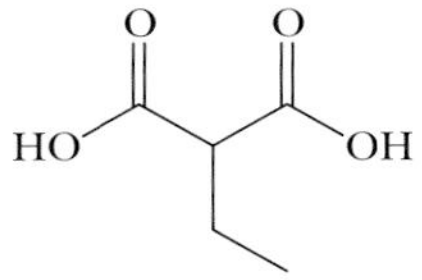

12.29 (a)

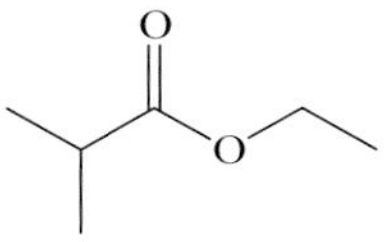

(b)

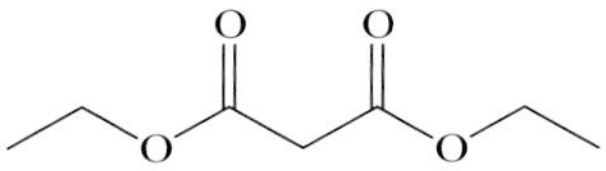

(c)

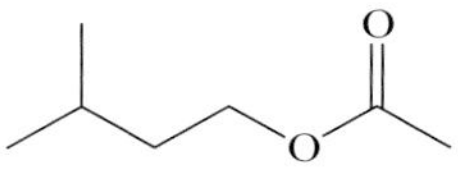

12.31

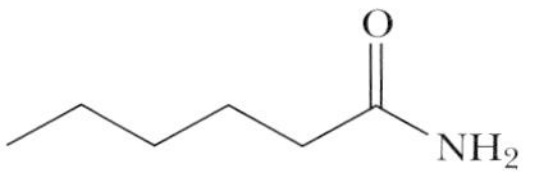

12.33

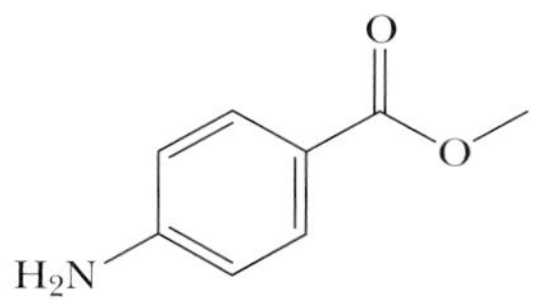

12.35

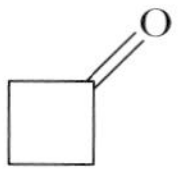

12.37

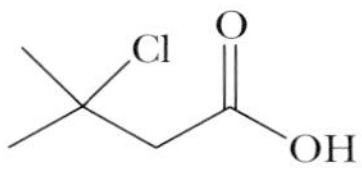

Chapter 13

Aldehydes and Ketones

13.1 (a) 2,2-Dimethylpropanal **(b)** 3-Hydroxycyclohexanone
(c) (R)-2-Phenylpropanal

13.2 Following are line-angle formulas for each aldehyde with the molecular formula $C_6H_{12}O$. The stereocenter in each chiral aldehyde is marked by an asterisk.

Hexanal

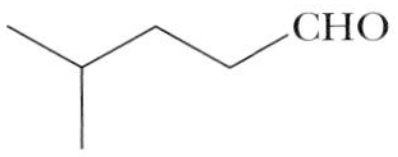

4-Methylpentanal

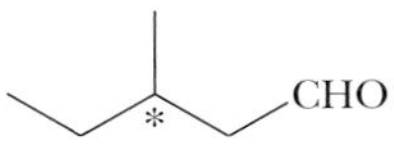

3-Methylpentanal

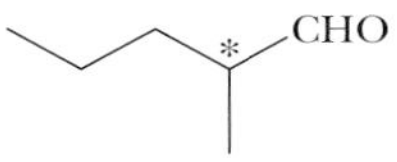

2-Methylpentanal

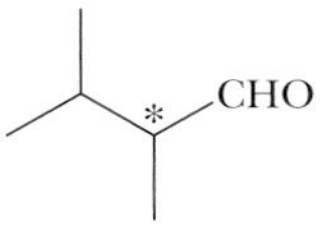

2,3-Dimethylbutanal 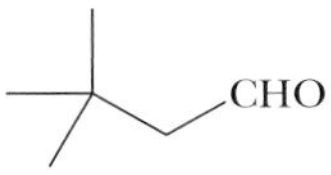   3,3-Dimethylbutanal    2,2-Dimethylbutanal

13.3 (a) 2-Hydroxypropanoic acid **(b)** 2-Oxopropanoic acid
(c) 4-Aminobutanoic acid

13.4 As soon as the Grignard reagent forms, it will react with the acidic hydrogen in the same or nearby molecule.

13.5 (a)

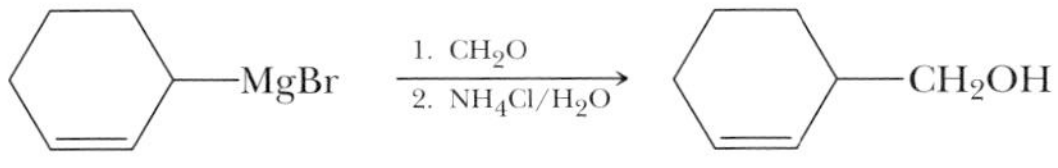

(b)

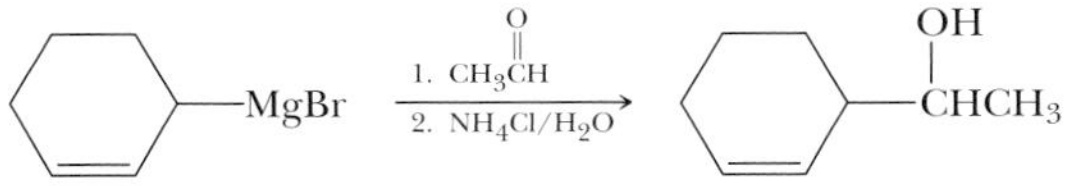

(c)

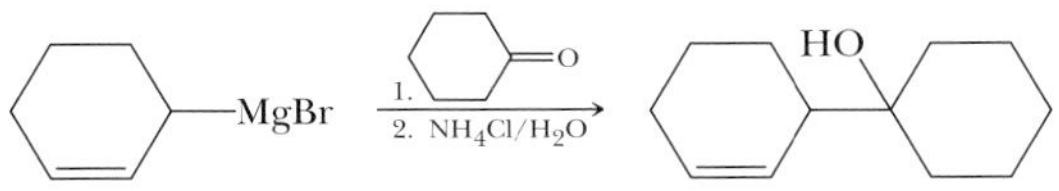

13.6 (a) 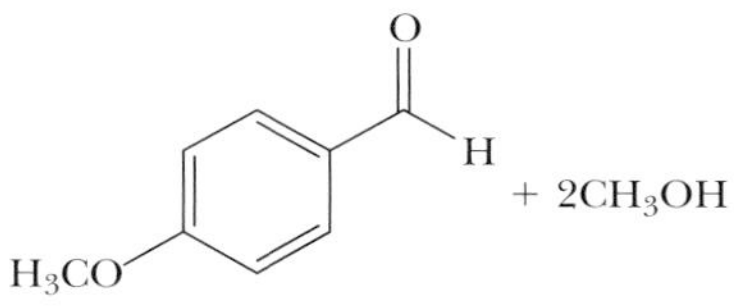 $+ 2CH_3OH$

(b)

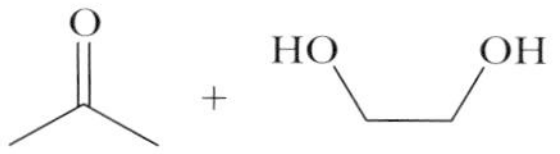

(c) 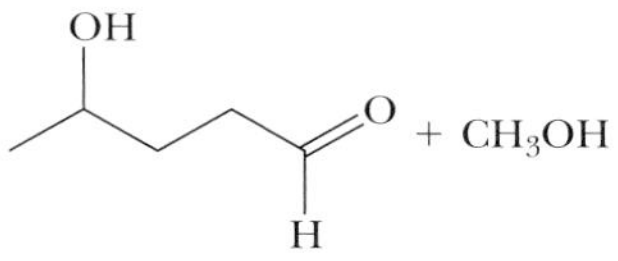 $+ CH_3OH$

13.7 (a) 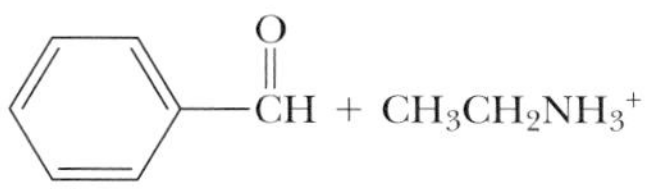 $+ CH_3CH_2NH_3^+$

(b) 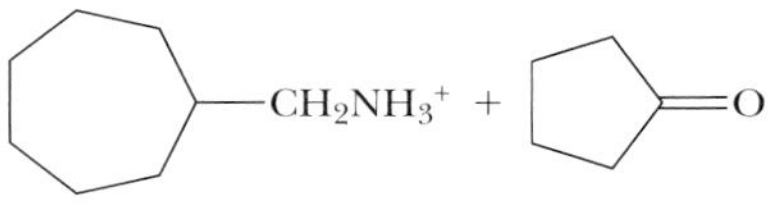 $-CH_2NH_3^+ +$

13.8 (a) 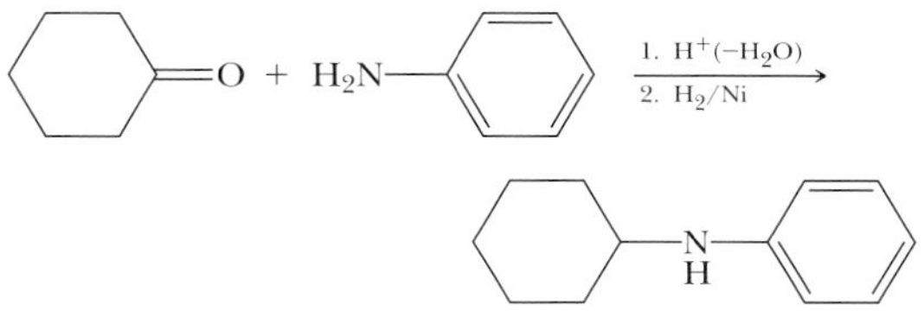 $=O + H_2N-$ $\xrightarrow[\text{2. } H_2/Ni]{\text{1. } H^+(-H_2O)}$

(b) 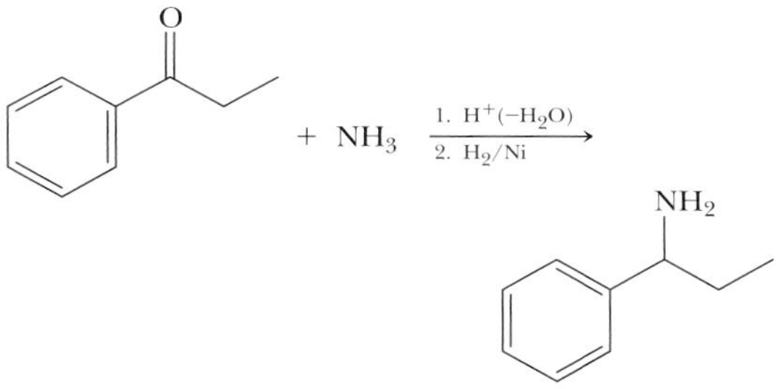 $+ NH_3$ $\xrightarrow[\text{2. } H_2/Ni]{\text{1. } H^+(-H_2O)}$

13.9 (a) 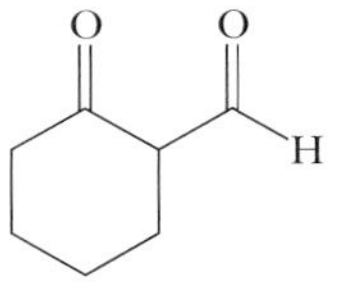 **(b)**

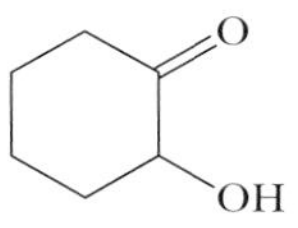

(c)

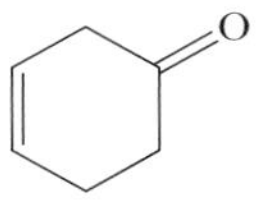

13.10 (a) 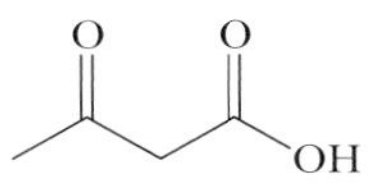 **(b)**

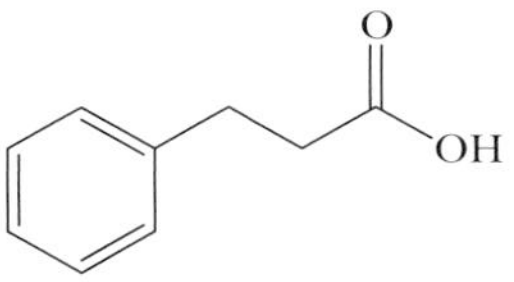

13.11 (a) 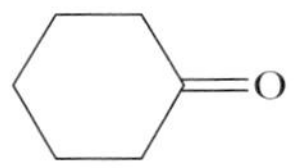 **(b)** 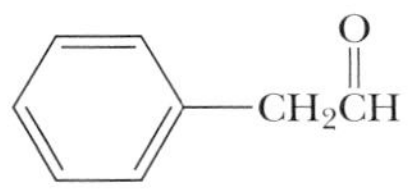 $-CH_2CH$

(c)

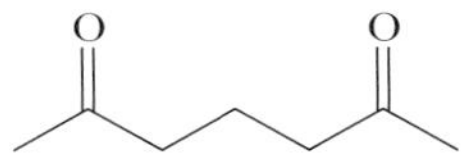

13.13 (a) $\xrightarrow[\text{CH}_2\text{Cl}_2]{\text{PCC}}$ **(b)** $\xrightarrow{\text{H}_2\text{CrO}_4}$

(c) $\xrightarrow{\text{H}_2\text{CrO}_4}$ **(d)** $\xrightarrow[\text{H}_2\text{SO}_4]{\text{H}_2\text{O}}$ $\xrightarrow{\text{H}_2\text{CrO}_4}$

(e) 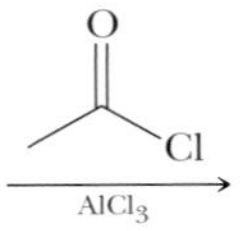 $\xrightarrow[\text{AlCl}_3]{\overset{\displaystyle \text{O}}{\overset{\|}{\text{C}}}\text{Cl}}$ **(f)** $\xrightarrow[\text{H}_2\text{SO}_4]{\text{H}_2\text{O}}$ $\xrightarrow{\text{H}_2\text{CrO}_4}$

(g) $\xrightarrow{\text{H}_2\text{CrO}_4}$ **(h)** $\xrightarrow[\text{H}_2\text{SO}_4]{\text{H}_2\text{O}}$ $\xrightarrow{\text{H}_2\text{CrO}_4}$

13.15 There are four aldehydes with this molecular formula; one is chiral.

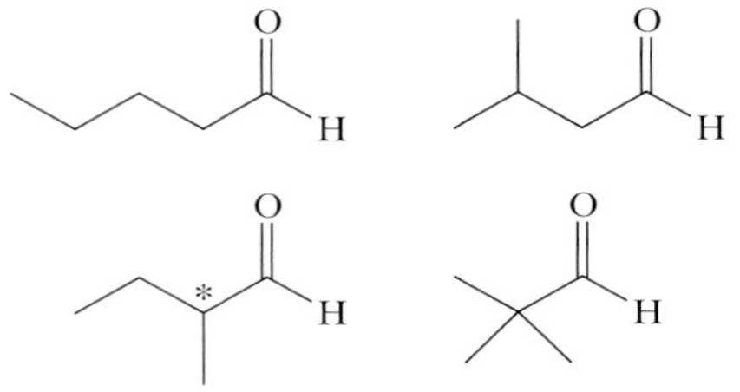

13.17 (a) 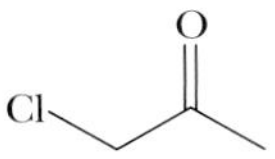 ... **(b)** 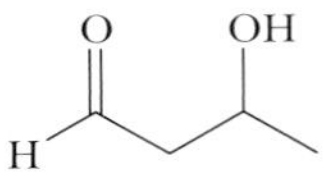 ... **(c)** 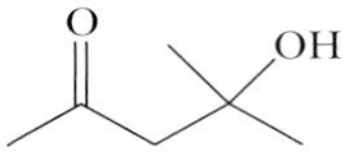 ... **(d)** 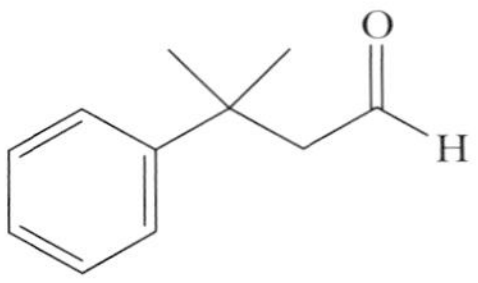 ... **(e)** 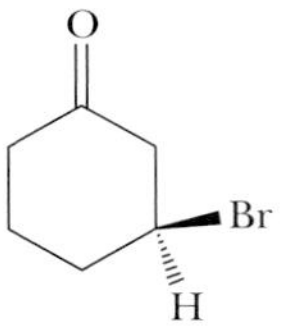 ... **(f)** 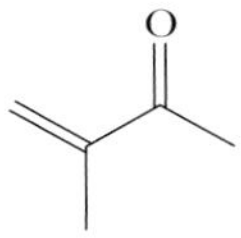 ... **(g)** 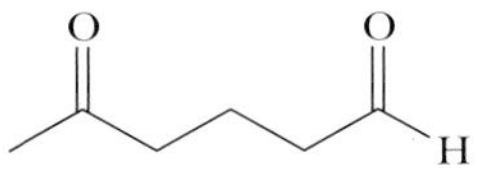 ... **(h)** 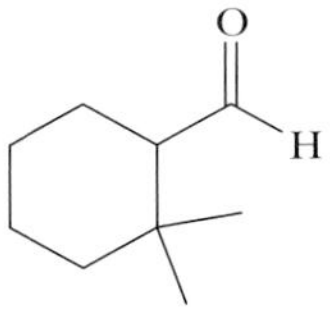 ... **(i)** 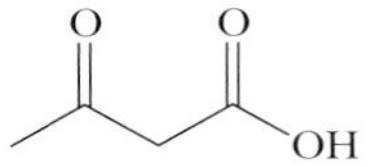 ...

13.19 Grignard reagents are very strong bases and will deprotonate ethanol and water.

13.21 Following are alternative syntheses for each target molecule.

(a)

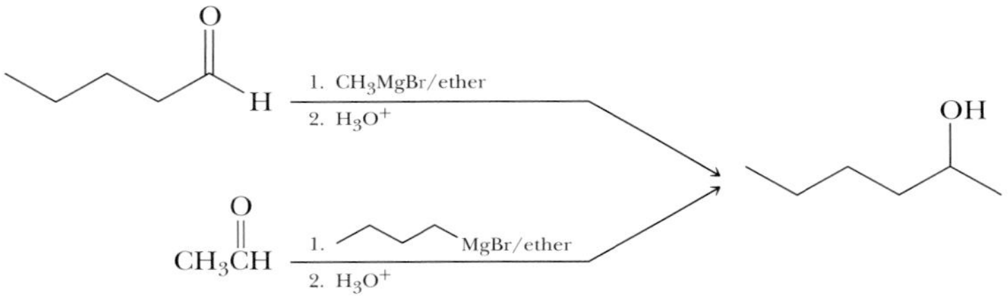

(b)

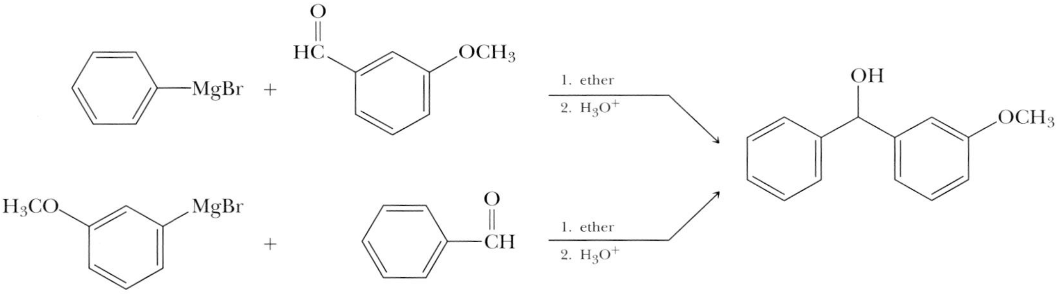

(c)

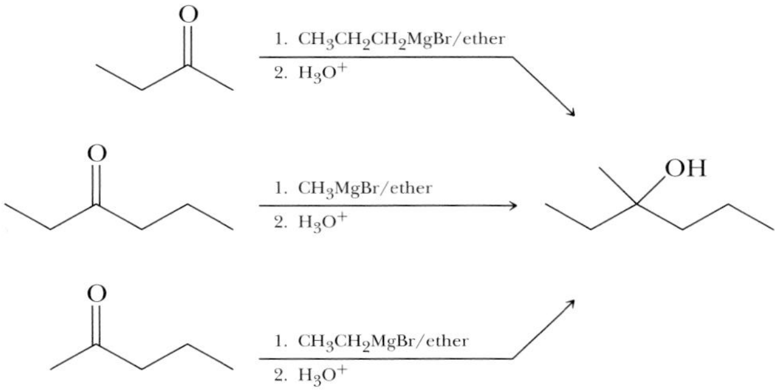

13.23 (a)

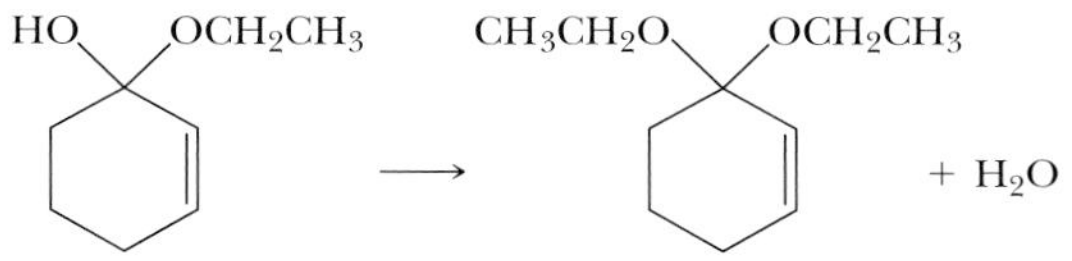

HO OCH$_2$CH$_3$ → CH$_3$CH$_2$O OCH$_2$CH$_3$ + H$_2$O

(b)

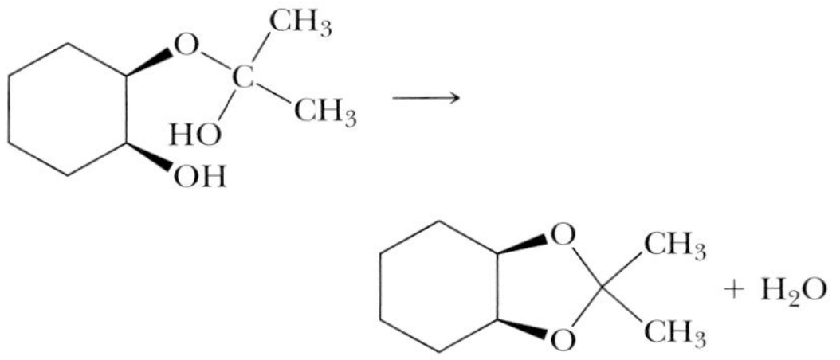

→

+ H$_2$O

(c)

OCH$_3$ ... OH → OCH$_3$ ... OCH$_3$ + H$_2$O

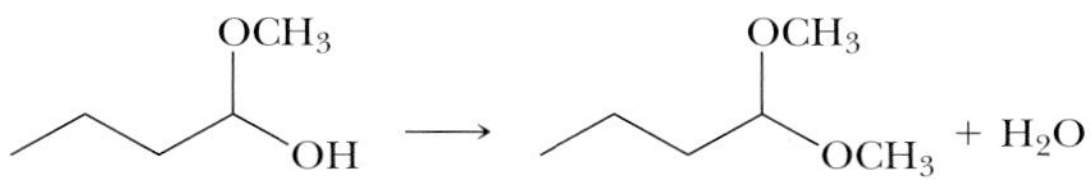

13.25

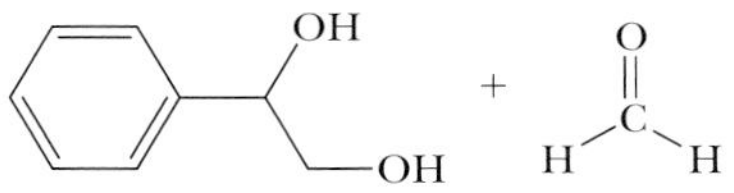

13.27 Propose a seven-step mechanism ($*$ = oxygen-18).

Step 1: Protonation of the carbonyl oxygen gives a resonance-stabilized cation:

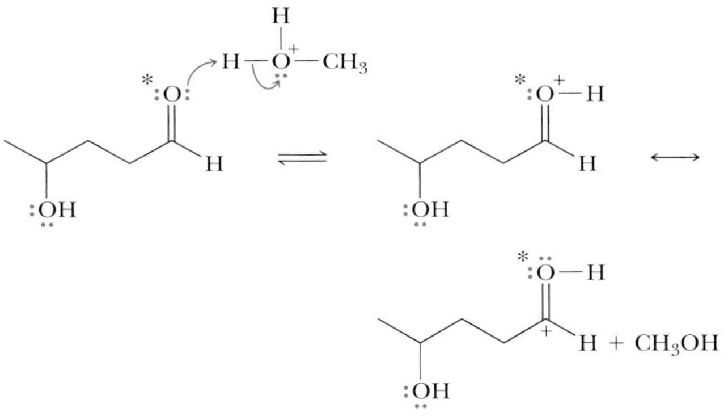

Step 2: The hydroxyl group attacks the carbon atom of the protonated carbonyl group to give a protonated cyclic hemiacetal:

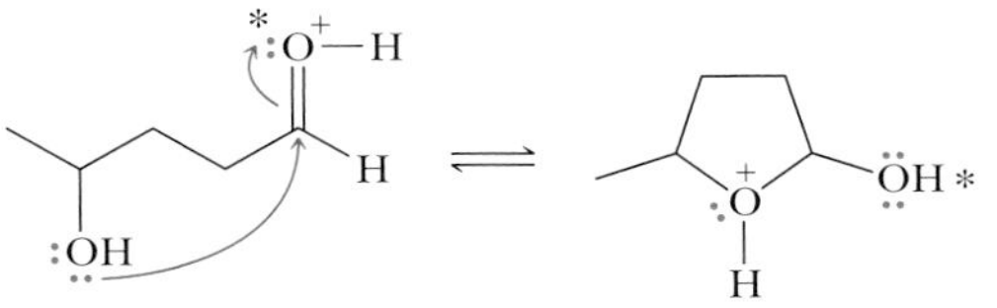

Step 3: Loss of a proton from the protonated cyclic hemiacetal gives a hemiacetal:

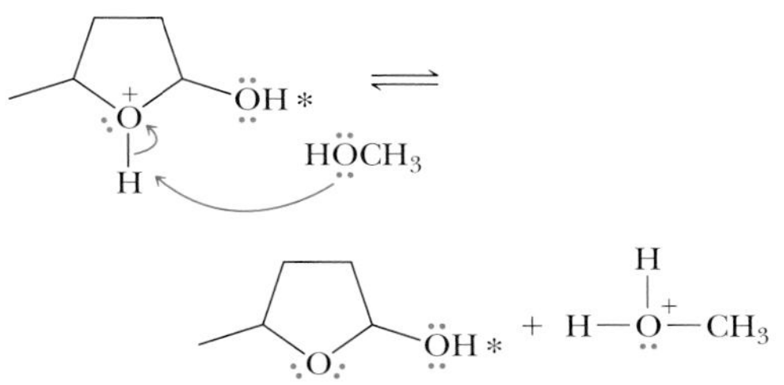

Step 4: Protonation of the hydroxyl group converts it to a better leaving group:

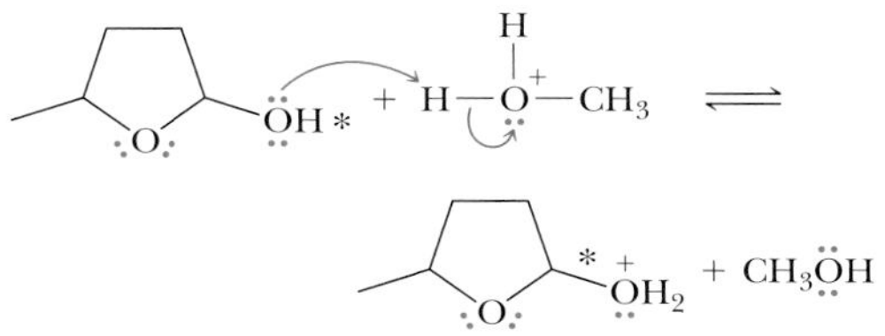

Step 5: Loss of water gives a new resonance-stabilized cation. Note that in this step the oxygen-18 label appears in the water molecule.

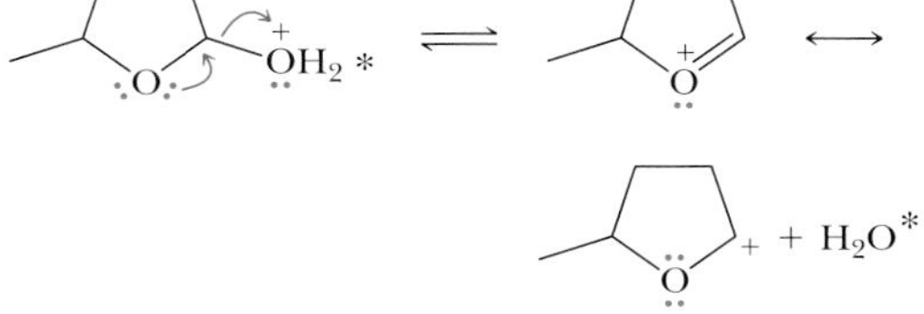

Step 6: Nucleophilic attack of methanol on the electrophilic carbon atom gives a protonated acetal:

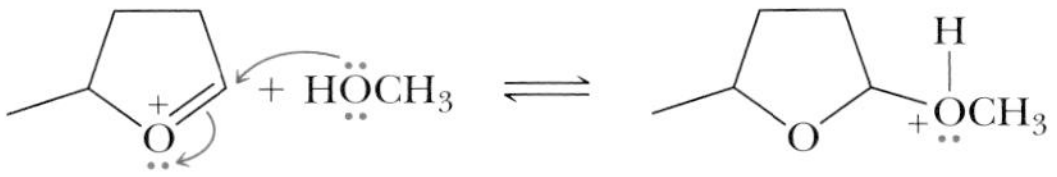

Step 7: Loss of the proton from the protonated acetal to give the acetal:

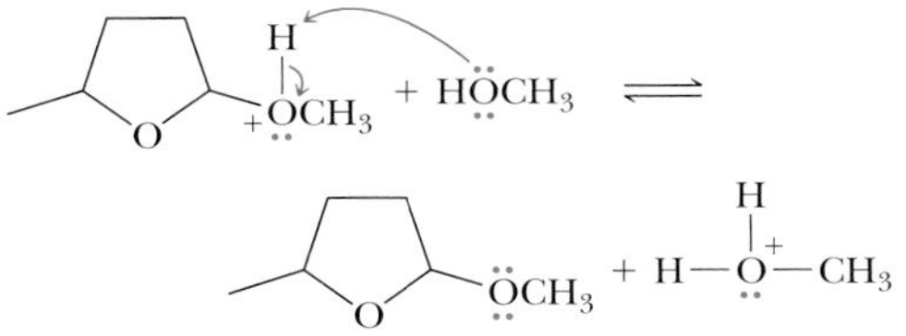

13.29

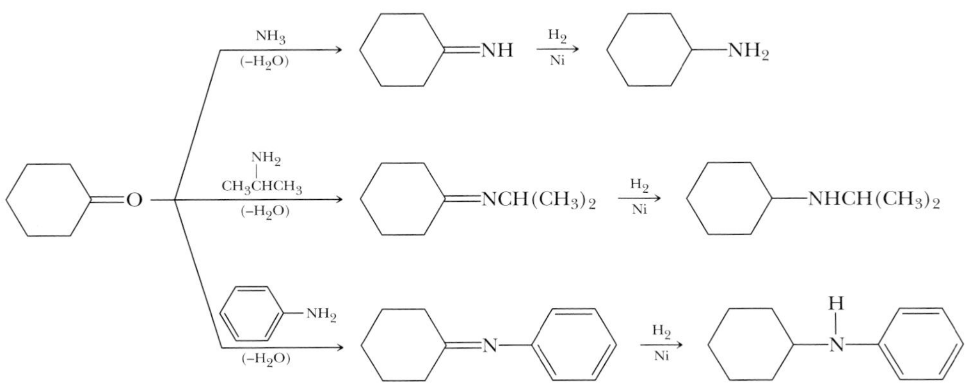

13.31

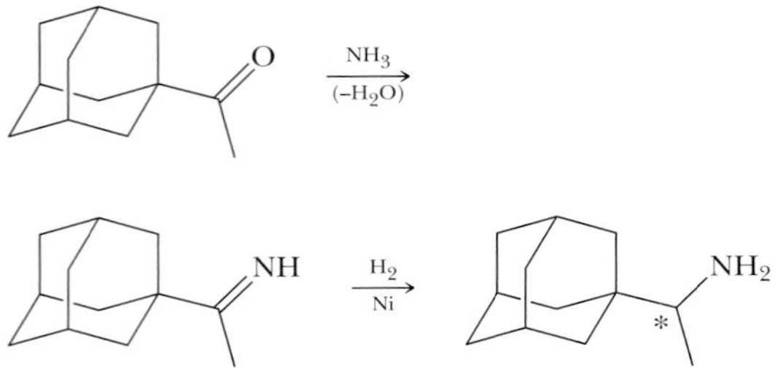

Rimantadine
is chiral

13.33

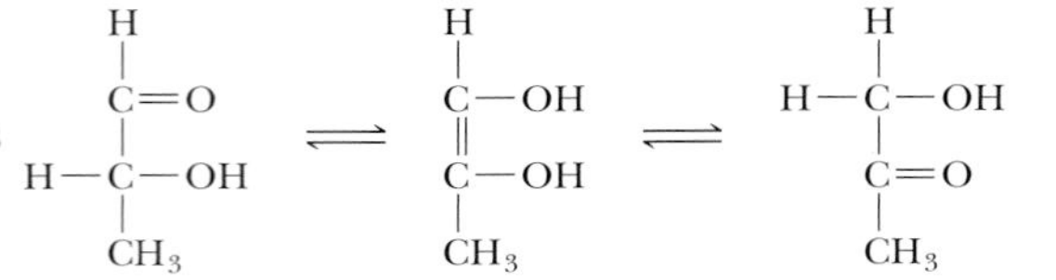

α-Hydroxyaldehyde   An enediol   α-Hydroxyketone

13.35 (a, b, c)  **(d, e)**

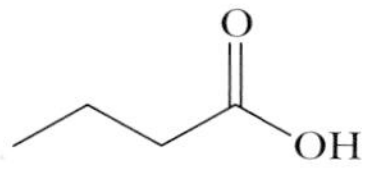

(f)

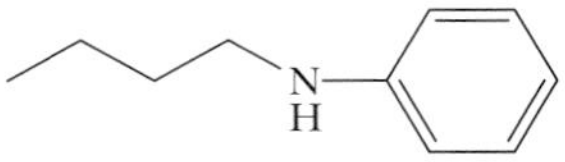

13.37 (1) SOCl$_2$ **(2)** Mg/ether
      (3) CH$_2$O, then HCl   **(4)** PCC/CH$_2$Cl$_2$

13.39 Reduction of the carbonyl group in parts **(a)**, **(b)**, and **(c)** can be accomplished using NaBH$_4$, LiAlH$_4$, or H$_2$/Ni. In these answers, we use NaBH$_4$.

(a) C$_6$H$_5$CCH$_2$CH$_3$ $\xrightarrow{\text{NaBH}_4}$ C$_6$H$_5$CHCH$_2$CH$_3$

$\xrightarrow[\text{heat}]{\text{H}_2\text{SO}_4}$ C$_6$H$_5$CH=CHCH$_3$

(b)

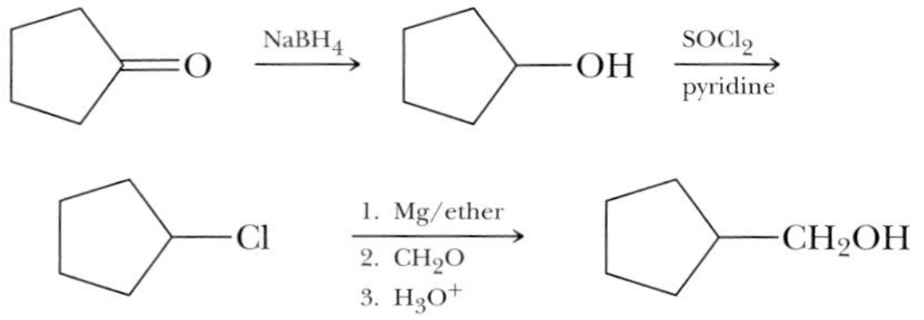

(c)

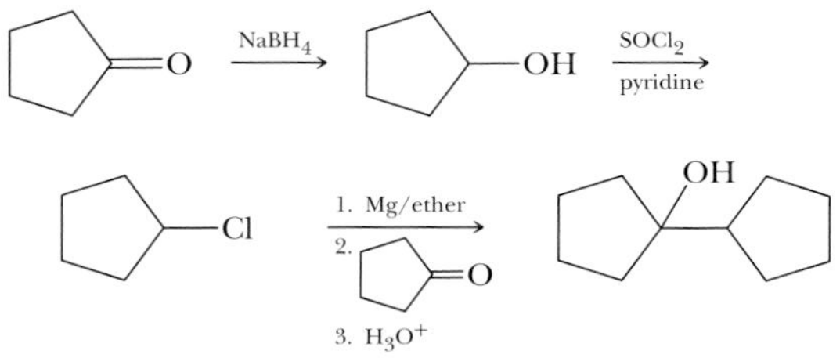

(d)

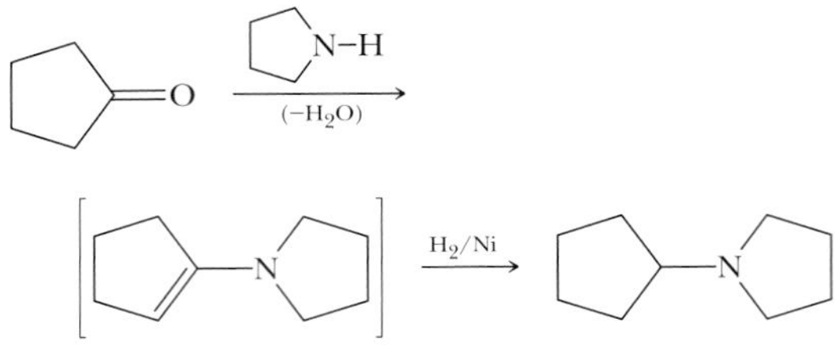

13.41 The sets of reagents for steps 1–4 are:

(1) CH$_3$CH$_2$CCl/AlCl$_3$ **(2)** Cl$_2$/AlCl$_3$

(3) Br$_2$/CH$_3$COOH   **(4)** 2(CH$_3$)$_3$CNH$_2$

13.43 (1) (CH$_3$)$_2$NH   **(2)** SOCl$_2$/pyridine

(3) C$_6$H$_5$CCl/AlCl$_3$ **(4)** NaBH$_4$

(5) The substrate is a 1° alkyl halide, which can only undergo an S$_N$2 reaction. To ensure a facile S$_N$2 reaction, the diphenylmethanol needs to be deprotonated by a strong base to yield an alkoxide ion, which is a strong nucleophile.

13.45

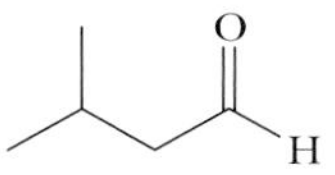

13.47

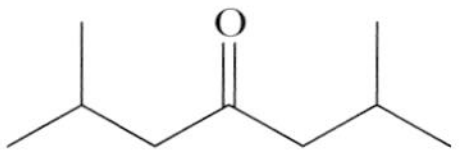

13.49

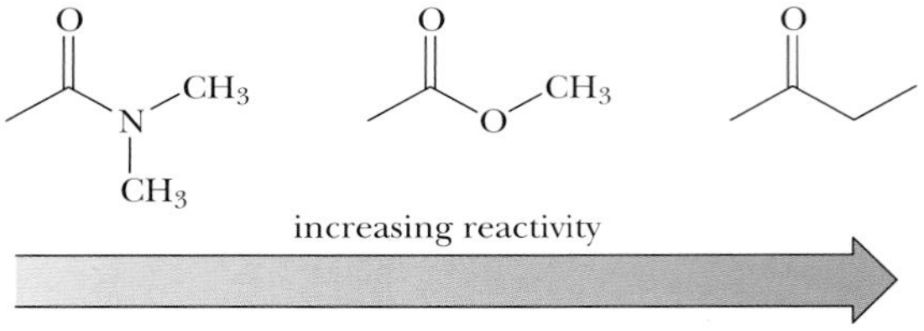

increasing reactivity

Both an amide group and an ester group are best represented as hybrids to two important contributing structures. To the extent that each functional group is stabilized by resonance, its carbonyl group is less reactive toward nucleophilic attack than the carbonyl group of a ketone. The resonance stabilization is greater for an amide (nitrogen is less electronegative than oxygen) than it is for an ester; thus an amide is less reactive toward nucleophilic attack than is an ester.

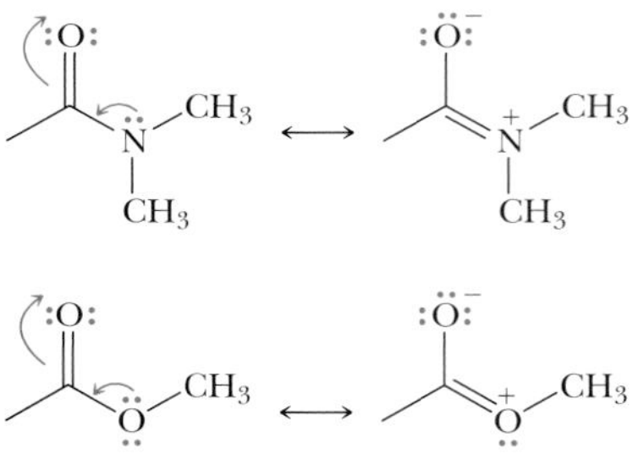

13.51 Following are the cyclic hemiacetals from each molecule. That for **(a)** is drawn as a chair conformation. For now, we only ask you to recognize that a cyclic hemiacetal forms, the size of the ring, and that a new stereocenter is created in forming the cyclic hemiacetal. We will explain more about the relative stereochemistry of each group on the ring when we discuss carbohydrates in Chapter 18.

(a) In forming this cyclic hemiacetal, the —OH group of carbon-5 of glucose bonds to the carbonyl group on carbon-1 forming a six-membered cyclic hemiacetal.

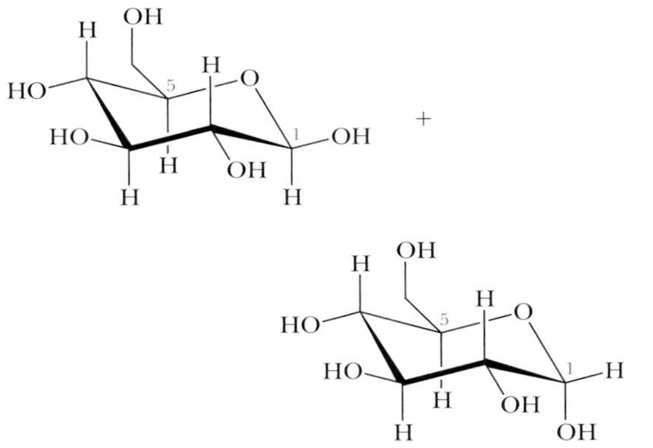

(b) In forming this cyclic hemiacetal, the —OH group on carbon-4 of ribose bonds to the carbonyl carbon on carbon-1 forming a five-membered cyclic hemiacetal.

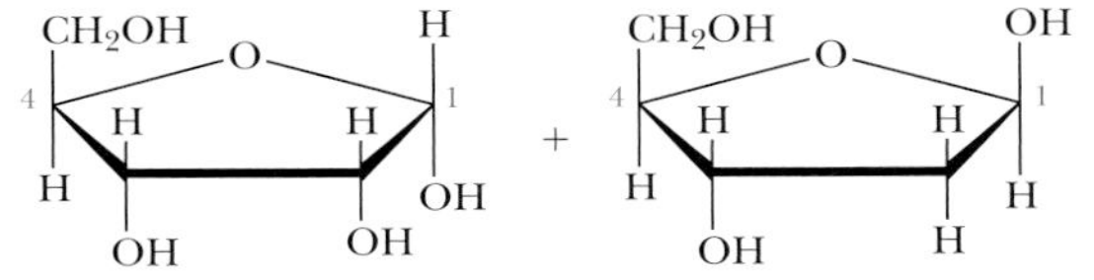

Chapter 14

Carboxylic Acids

14.1 **(a)** (*R*)-2,3-Dihydroxypropanoic acid
(b) *cis*-2-Butenedioic acid
(c) (*R*)-3,5-Dihydroxypentanoic acid

14.2 $CH_3\underset{\underset{CH_3}{|}}{\overset{\overset{CH_3}{|}}{C}}COOH$ CF_3COOH $CH_3\underset{\underset{OH}{|}}{CH}COOH$

pK_a 5.03 pK_a 0.22 pK_a 3.08

14.3 **(a)**

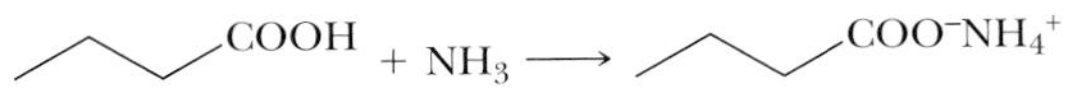

$\diagup\!\!\diagdown\!\!\diagup COOH + NH_3 \longrightarrow \diagup\!\!\diagdown\!\!\diagup COO^-NH_4^+$

Butanoic acid   Ammonium butanoate

(b)

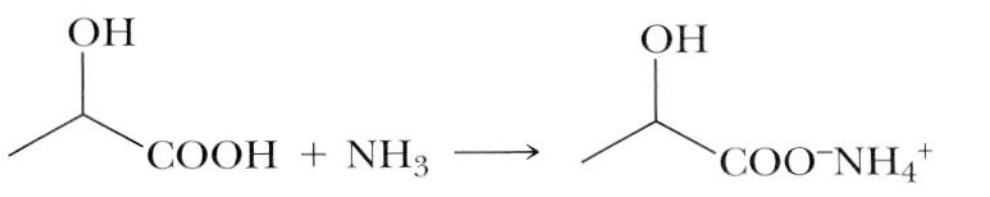

2-Hydroxypropanoic acid + $NH_3 \longrightarrow$ Ammonium 2-hydroxypropanoate
(Lactic acid)   (Ammonium lactate)

14.4 **(a)** 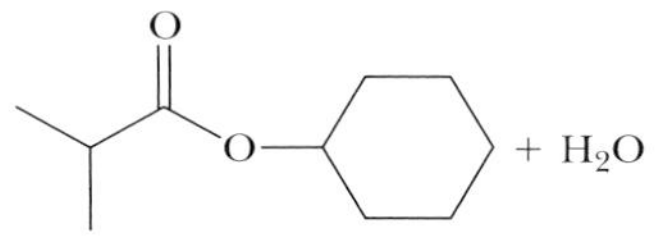 [ester structure] + H_2O

(b) 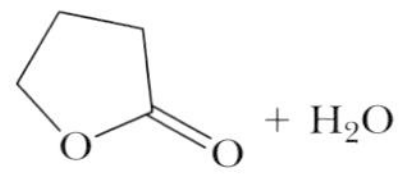 [lactone structure] + H_2O

14.5 **(a)** 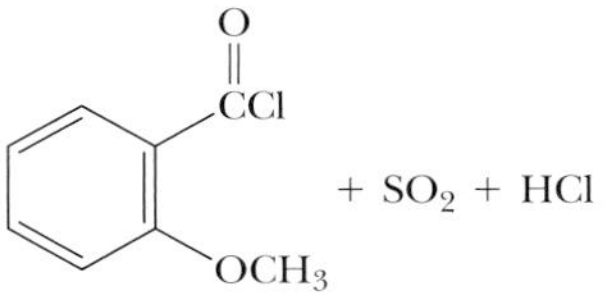 [acyl chloride with OCH₃ structure] + SO_2 + HCl

(b) 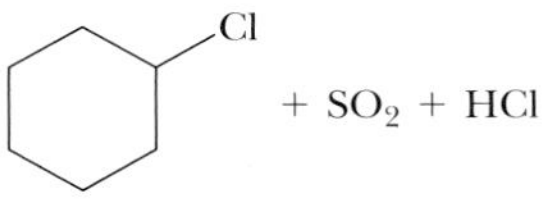 [cyclohexyl chloride structure] + SO_2 + HCl

14.6 [structure]

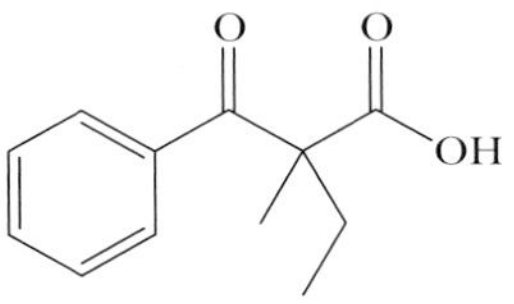

14.7 Only one of these carboxylic acids is chiral. Its stereocenter is marked by an asterisk.

Pentanoic acid 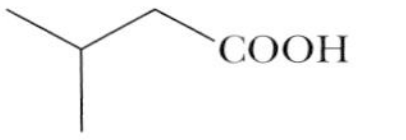   3-Methylbutanoic acid

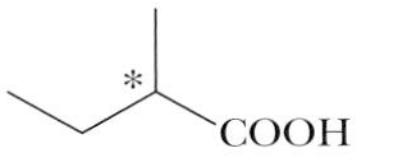

2-Methylbutanoic acid 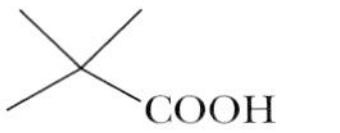   2,2-Dimethylpropanoic acid

14.9 **(a)** 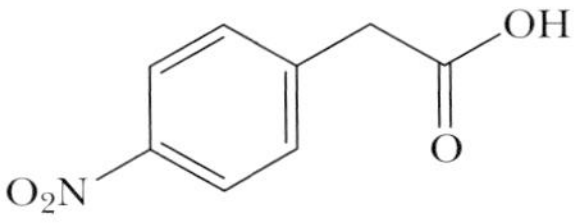 [structure with O_2N]

(b) 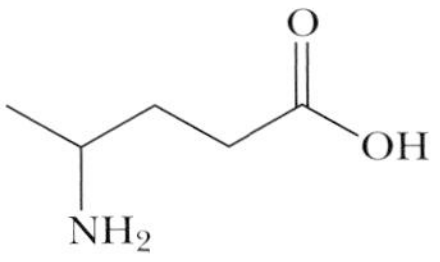 [structure with NH_2]   **(c)** 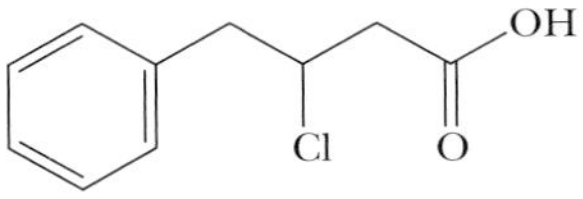 [structure with Cl]

(d) 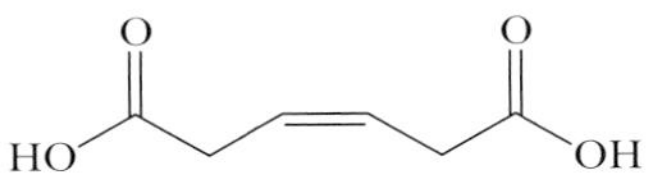 [structure with HO]

(e) 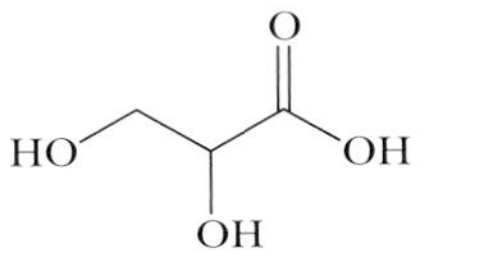 [structure with HO, OH]   **(f)** 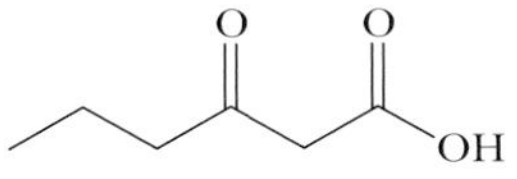 [structure]

(g) 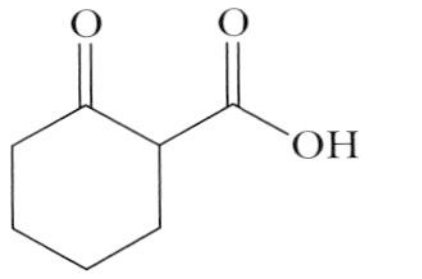 [structure]   **(h)** 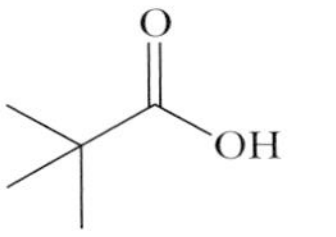 [structure]

14.11

14.13 $HO-\overset{O}{\underset{\|}{C}}-\overset{O}{\underset{\|}{C}}-O^-K^+$

14.15

14.17 (a) $CH_3(CH_2)_4\overset{O}{\underset{\|}{C}}OH$ **(b)**

(c)

14.19 (a) Benzoic acid (pK_a 4.17)

(b) lactic acid (K_a 8.4 × 10^{-4})

14.21 (a)

pK$_a$ 4.19   pK$_a$ 3.14

(b)

pK$_a$ 3.14   pK$_a$ 4.92

(c) $CH_3\overset{O}{\underset{\|}{C}}CH_2COOH$ and $CH_3\overset{O}{\underset{\|}{C}}COOH$

pK$_a$ 3.58   pK$_a$ 2.49

(d) $CH_3\overset{OH}{\underset{|}{C}}HCOOH$ and CH_3CH_2COOH

pK$_a$ 3.08   pK$_a$ 4.78

14.23 Lactic acid will exist primarily as the carboxylate anion in blood plasma. The pH of blood plasma (pH 7.35–7.45) is more than 2.0 units higher (more basic) than the pK_a of lactic acid (pK_a 4.07).

14.25 Ascorbic acid will be present in urine predominantly as ascorbate anion.

14.27 Alanine is better represented by B. Alanine has within its structure both a basic group (—NH$_2$) and an acidic group (—COOH), which react to form an internal salt.

14.29 (a) $PhCH_2\overset{O}{\underset{\|}{C}}Cl + SO_2 + HCl$

(b) $PhCH_2\overset{O}{\underset{\|}{C}}O^-Na^+ + CO_2 + H_2O$

(c) $PhCH_2\overset{O}{\underset{\|}{C}}O^-Na^+ + H_2O$ **(d)** $PhCH_2\overset{O}{\underset{\|}{C}}O^-NH_4{}^+$

(e) $PhCH_2CH_2OH$ **(f)** no reaction

(g) $PhCH_2\overset{O}{\underset{\|}{C}}OCH_3 + H_2O$ **(h)** no reaction

14.31 (a) $\xrightarrow[CH_3OH]{NaBH_4}$ **(b)** $\xrightarrow[2.\ H_2O]{1.\ LiAlH_4}$

(c) $\xrightarrow[CH_3OH]{NaBH_4}$ $\xrightarrow[heat]{H_2SO_4}$

14.33 $H-\overset{O}{\underset{\|}{C}}-OH + NaHCO_3 \longrightarrow$

$H-\overset{O}{\underset{\|}{C}}-O^-Na^+ + CO_2 + H_2O$

14.35 H_2N-

$\overset{O}{\underset{\|}{C}}-OCH_2CH_3$

14.37 (a) The cis/trans ratio refers to the percentage ratio of the cis isomer to the trans isomer in the commercial mixture.

(b) The (+) and (−) indicate the direction a chiral stereoisomer rotates plane polarized light. The (+/−) refers to a racemic mixture of enantiomers.

14.39

14.41 (1) HNO_3/H_2SO_4 **(2)** H_2CrO_4

(3) Ni/H_2 **(4)** CH_3OH/H^+

14.43

Procaine

14.45 (a) This reaction is a Friedel-Crafts acylation and it uses acetyl chloride and aluminum chloride.

(b) Bromine (Br$_2$) in acetic acid

(c) 2-Methyl-2-propanamine (*tert*-butylamine)

(d) A carboxylic acid is reduced to 1° alcohol and a ketone is reduced to a 2° alcohol. The reducing agent is LiAlH$_4$.

14.47 The carboxylate anion is more difficult to reduce than the carboxylic acid because the negative charge on the carboxylate makes the carbonyl carbon less electrophilic towards the hydride anion.

14.49 Grignard reagents do not attack the carbonyl carbon of carboxylic acids, but instead, act as strong bases and deprotonate the carboxyl group to form the carboxylate anion. Grignard reagents react with esters of carboxylic acids and attack the carbonyl carbon twice to form tertiary alcohols (except in the case of formic acid esters, which will form secondary alcohols) upon subsequent hydrolysis.

Chapter 15

Functional Derivatives of Carboxylic Acids

15.1 (a) 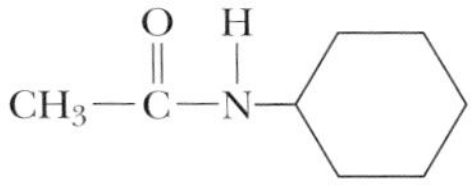 **(b)**

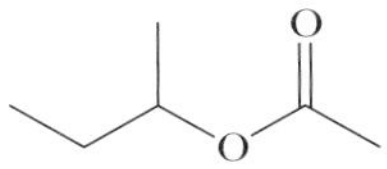

(c)

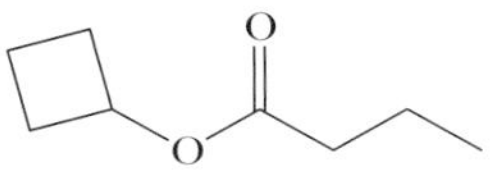

(d)

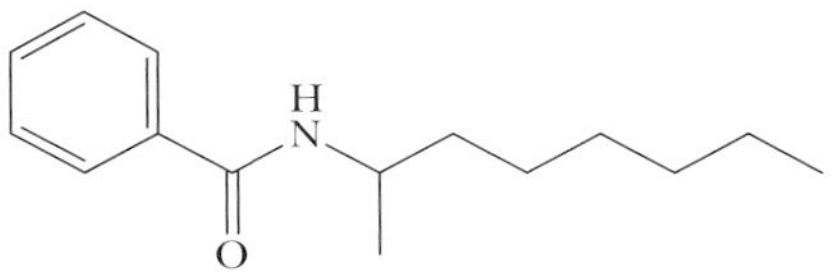

(e) $C_2H_5OC(CH_2)_4COC_2H_5$ **(f)**

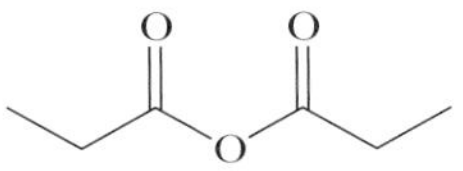

15.2 (a)

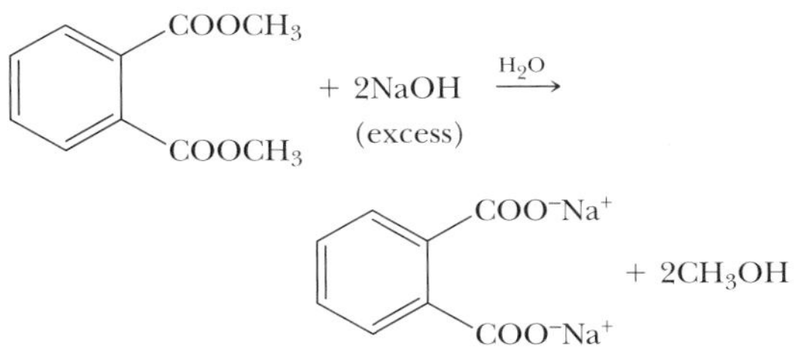

(b) 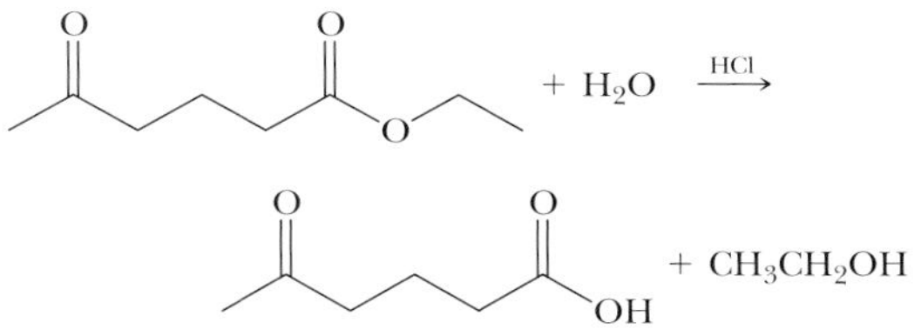 $+ H_2O \xrightarrow{HCl}$

$+ CH_3CH_2OH$

15.3 (a) $CH_3\overset{O}{\overset{\|}{C}}N(CH_3)_2 + NaOH \longrightarrow CH_3CO^-Na^+ + (CH_3)_2NH$

(b) 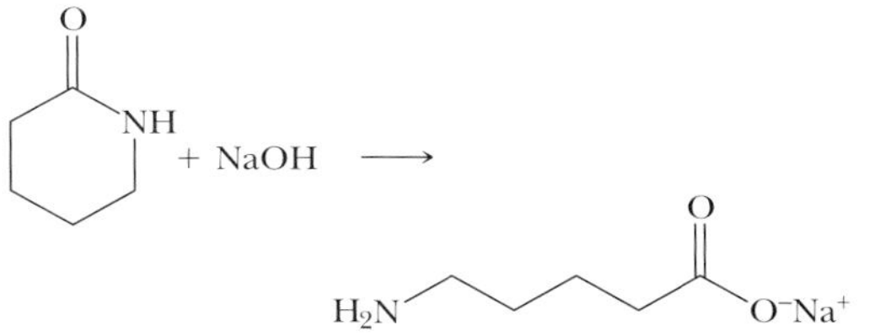 $+ NaOH \longrightarrow$

15.4 (a) 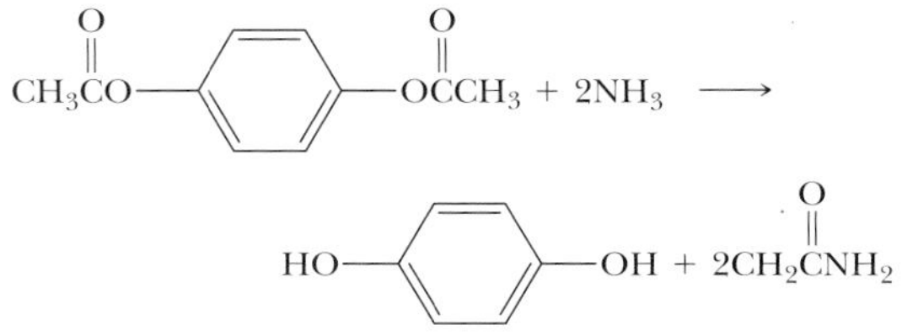 $CH_3\overset{O}{\overset{\|}{C}}O-$ $-O\overset{O}{\overset{\|}{C}}CH_3 + 2NH_3 \longrightarrow$

$HO-$ $-OH + 2CH_2\overset{O}{\overset{\|}{C}}NH_2$

(b) 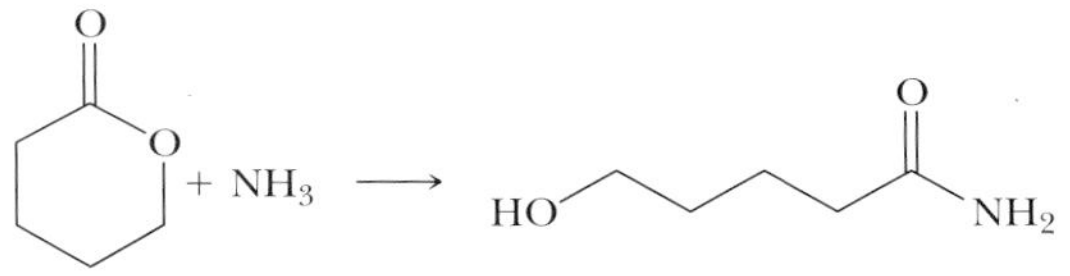 $+ NH_3 \longrightarrow$

15.5 (a) 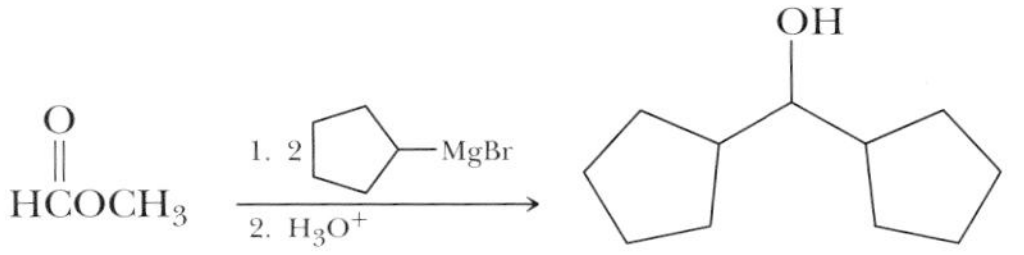 $H\overset{O}{\overset{\|}{C}}OCH_3 \xrightarrow[\text{2. } H_3O^+]{\text{1. 2 } \text{—MgBr}}$

(b) 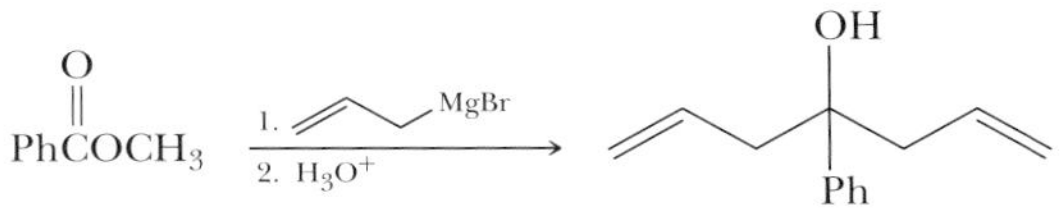 $Ph\overset{O}{\overset{\|}{C}}OCH_3 \xrightarrow[\text{2. } H_3O^+]{\text{1. } \text{MgBr}}$

15.6

(a) 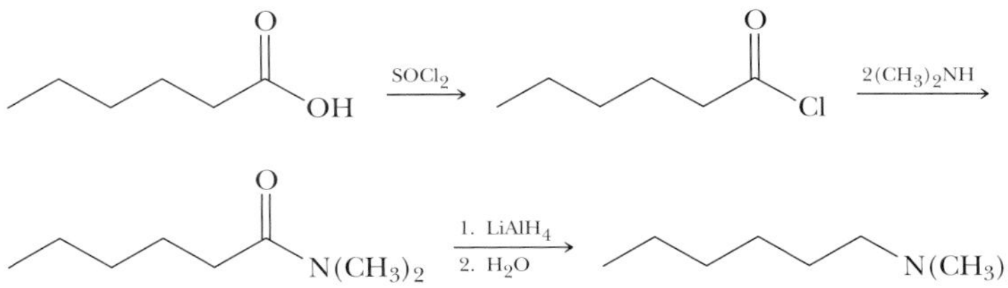 $\xrightarrow{SOCl_2}$ $\xrightarrow{2(CH_3)_2NH}$

$\xrightarrow[\text{2. } H_2O]{\text{1. } LiAlH_4}$

(b) 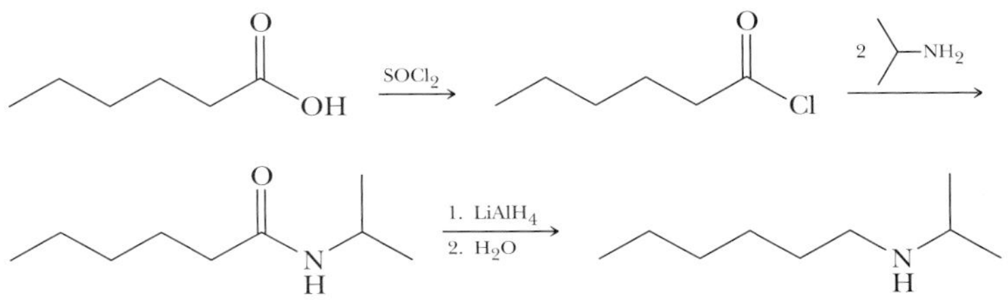 $\xrightarrow{SOCl_2}$ $\xrightarrow{2 \text{—NH}_2}$

$\xrightarrow[\text{2. } H_2O]{\text{1. } LiAlH_4}$

15.7 (a)

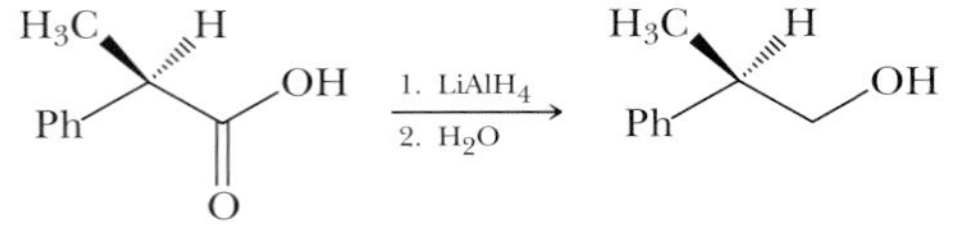

(b)

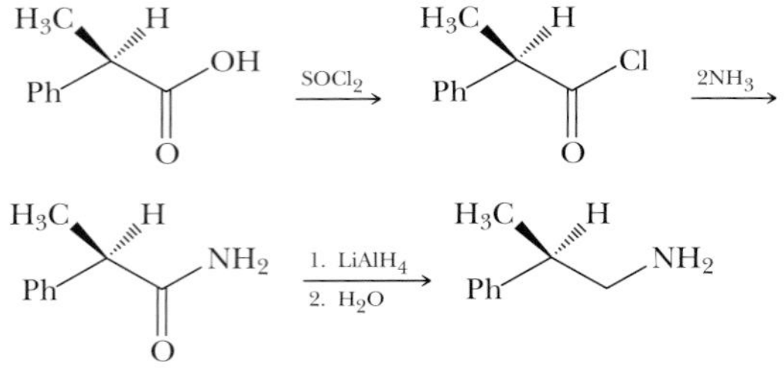

15.9 (a) Benzoic anhydride   (b) Methyl hexadecanoate
    (c) *N*-Methylhexanamide   (d) 4-Aminobenzamide
    (e) Diethyl propanedioate   (f) Methyl 2-methyl-3-oxo-
                                      4-phenylbutanoate

15.11 Acetic acid has a higher boiling point (118°C); methyl formate has the lower boiling point (32°C). Acetic acid has strong intermolecular hydrogen bonding between its molecules in the liquid state, which must be broken before molecules can escape to the vapor phase during boiling. There is no intermolecular hydrogen bonding in methyl formate, thus less energy is needed to break up intermolecular association leading to a lower boiling point relative to acetic acid.

15.13 In order of increasing reactivity, they are
(3) < (1) < (4) < (2).

15.15 (a)

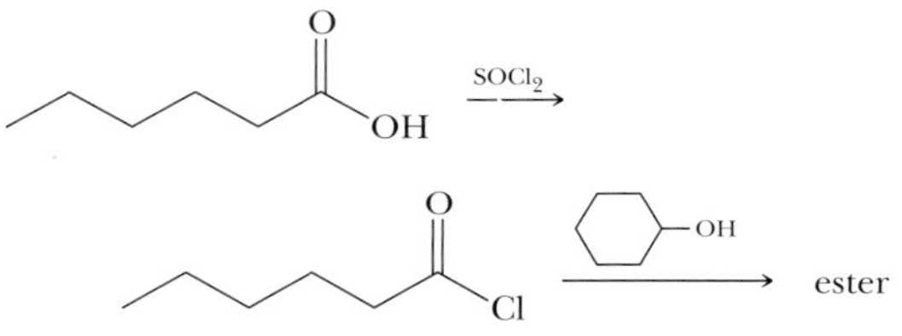

(b)

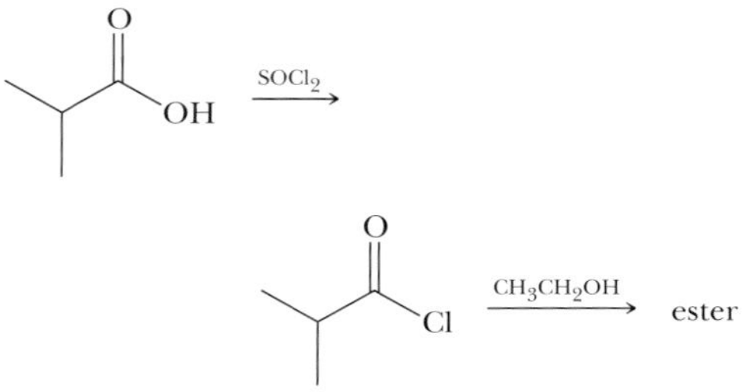

15.17 Propose a three-step mechanism:

Step 1: Nucleophilic acyl attack by ammonia on the carbonyl carbon to form a tetrahedral carbonyl addition intermediate:

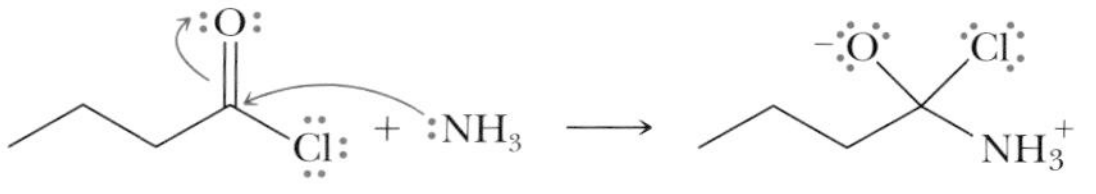

Step 2: The intermediate collapses with chloride leaving and the formation of a protonated amide:

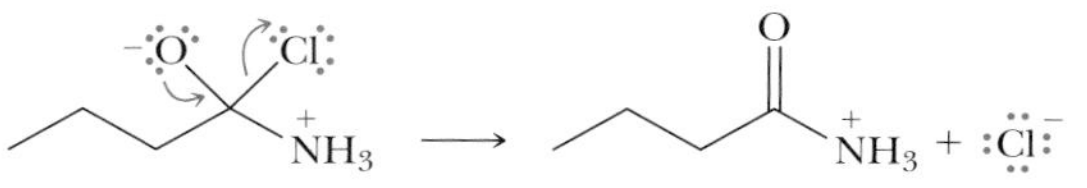

Step 3: The second molecule of ammonia removes the proton from the protonated amide producing the amide and an ammonium ion:

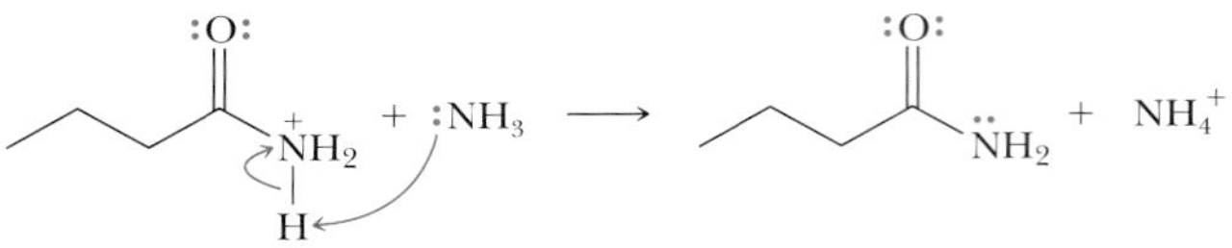

15.19 (a)

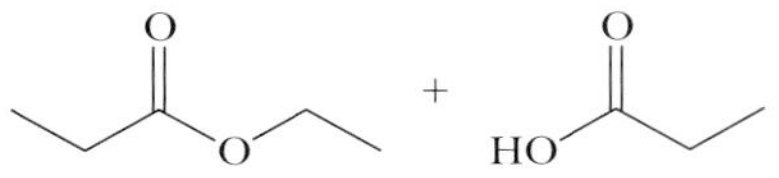

(b)

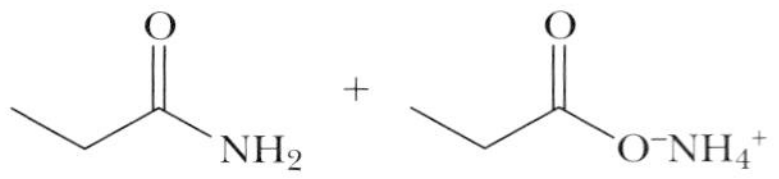

15.21

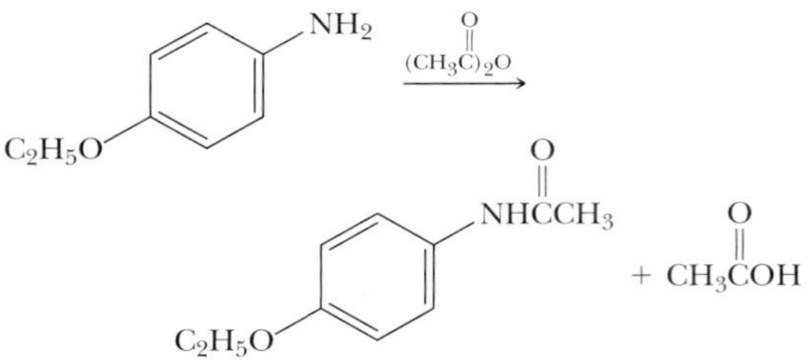

15.23

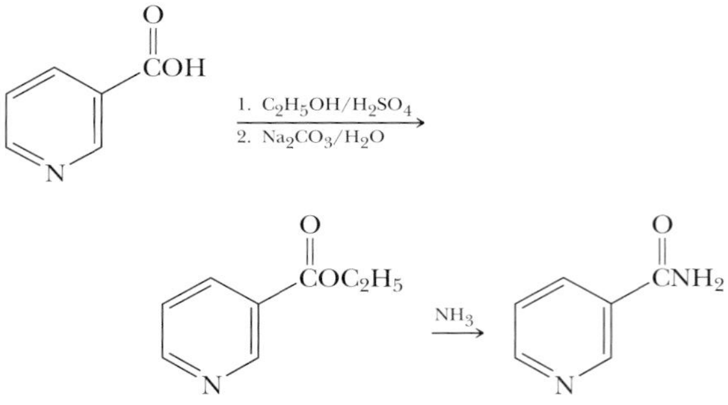

15.25 (a)

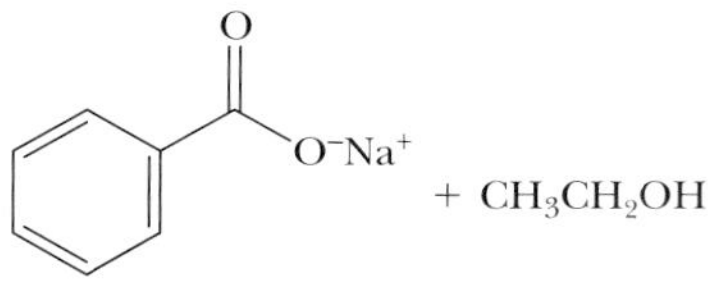

(b)

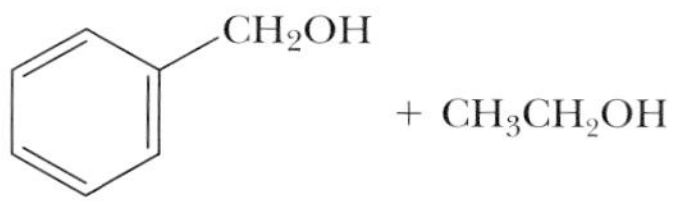

(c)

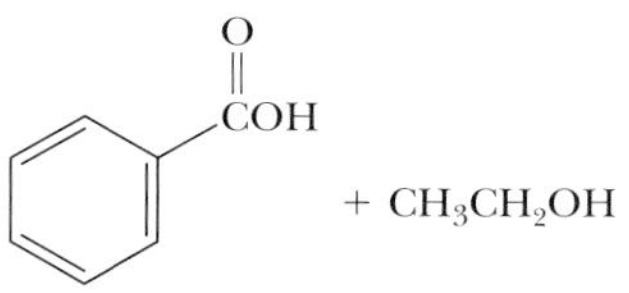

(d)

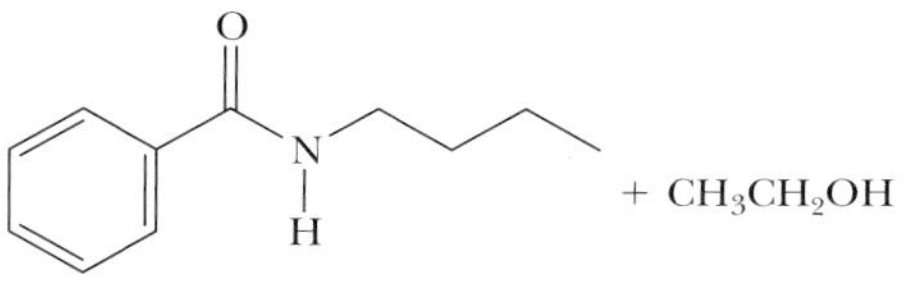

(e) 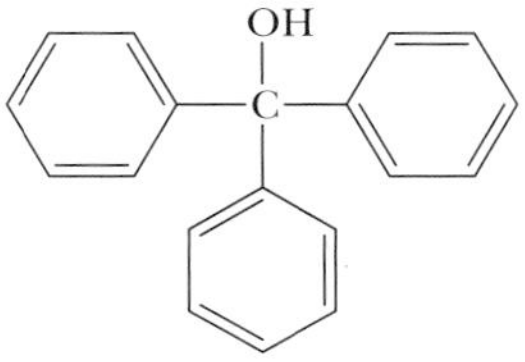 [structure: triphenylmethanol] + CH₃CH₂OH

$+ CH_3CH_2OH$

15.27 (a) 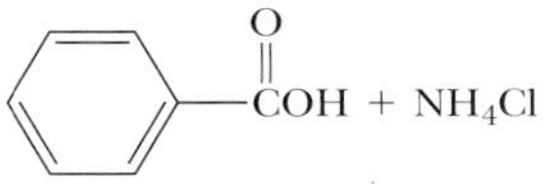 [benzoic acid structure] —COH + NH₄Cl

(b) 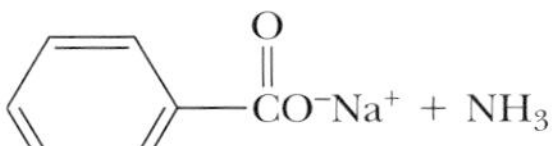 [sodium benzoate structure] —CO⁻Na⁺ + NH₃ **(c)** 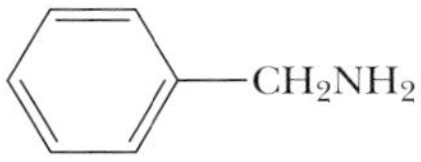 [benzylamine structure] —CH₂NH₂

15.29 (a) 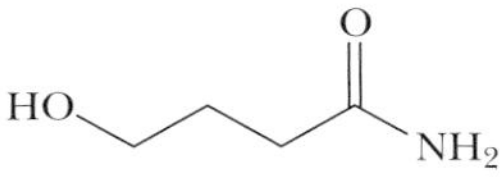 HO [chain] NH₂ **(b)** 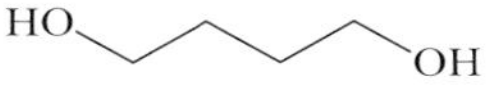 HO [chain] OH

(c) 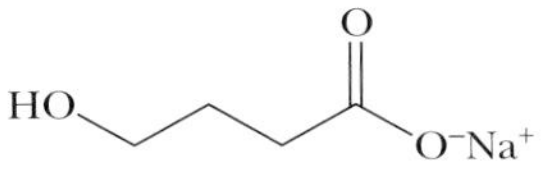 HO [chain] O⁻Na⁺

15.31 (a) 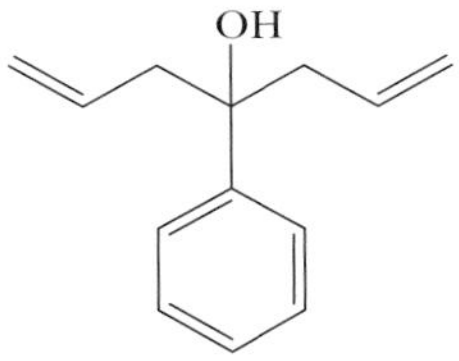 [structure with OH and phenyl, diallyl] **(b)** 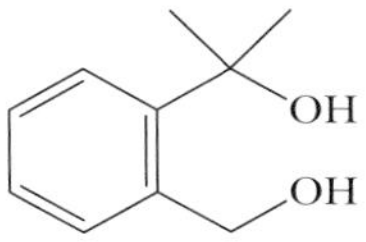 [structure: benzene with C(CH₃)₂OH and CH₂OH]

(c) [chain with OH]

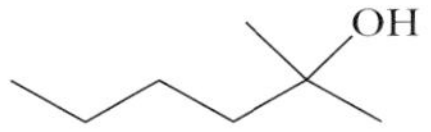

15.33 Propose a three-step mechanism. In these steps, butyl is abbreviated Bu.

Step 1: Nucleophilic acyl attack by the amine forming unstable tetrahedral intermediate:

EtO—C—OEt + :N—Bu ⟶ EtO—C—OEt
                H   H—N—H
                                            ⁺
                                            Bu

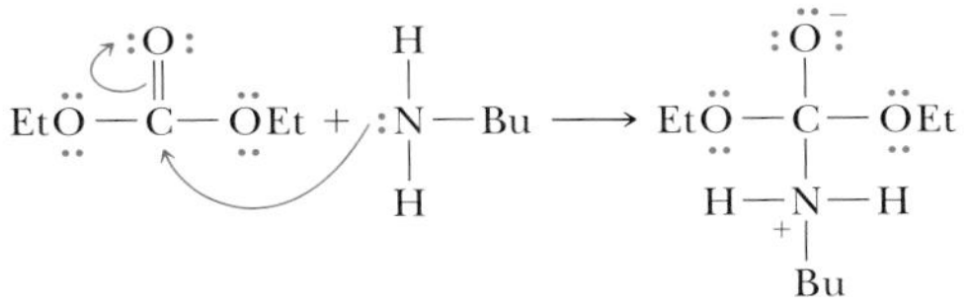

Step 2: Collapse of the tetrahedral intermediate forming a protonated carbamic ester:

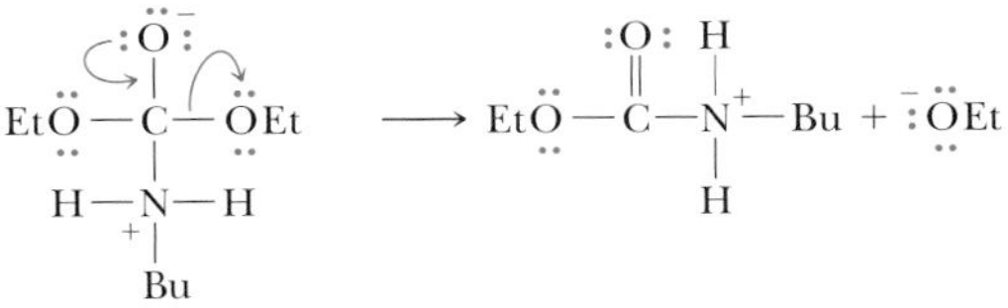

EtO—C—OEt ⟶ EtO—C—N⁺—Bu + ⁻:OEt
   H—N—H   H
      ⁺
      Bu

Step 3: Formation of a carbamic ester by deprotonation of the ammonium hydrogen:

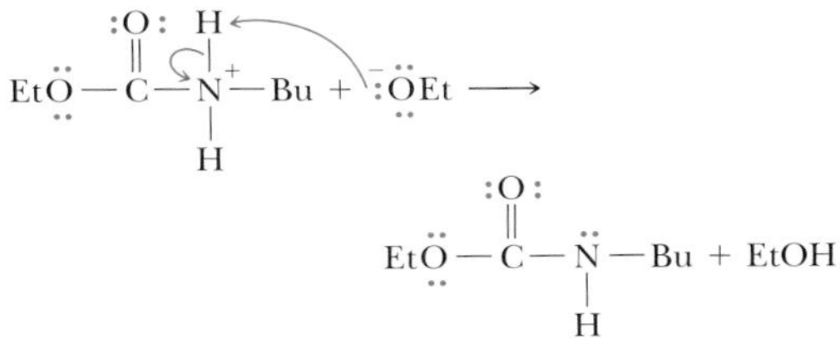

EtO—C—N⁺—Bu + :OEt ⟶
          H

EtO—C—N—Bu + EtOH
          H

15.35 (a) 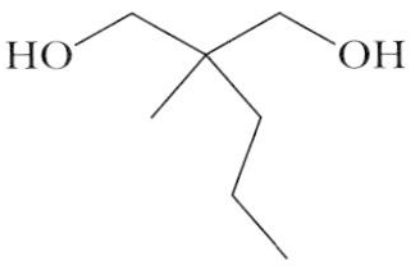 HO [structure] OH + 2CO₂ + 2NH₄Cl

$+ 2CO_2 + 2NH_4Cl$

(b) 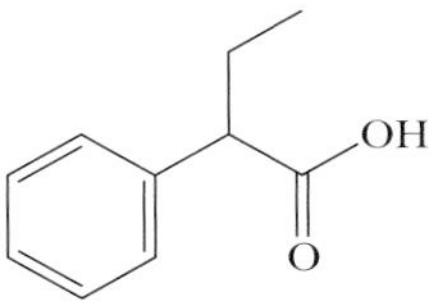 [2-phenylbutanoic acid structure] + 2CO₂ + 2NH₄Cl

$+ 2CO_2 + 2NH_4Cl$

15.37 (a) $\xrightarrow{H_2/Pd}$ **(b)** $\xrightarrow[2.\ H_3O^+]{1.\ LiAlH_4}$

(c) $\xrightarrow[H_2O_2]{OsO_4}$

15.39 The 3° aliphatic amine is the more basic of the two amines. Its HCl salt is:

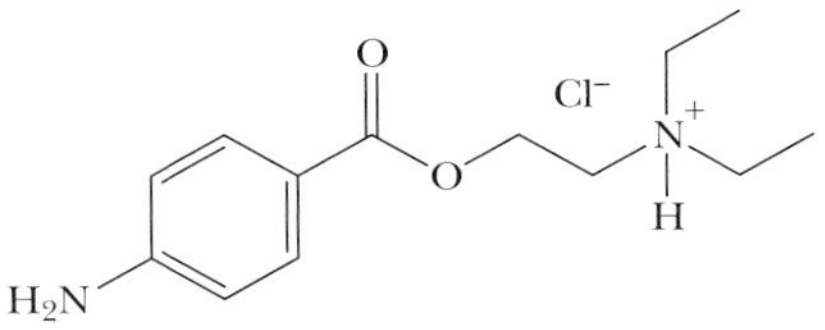

[structure: H₂N—benzene—C(=O)—O—CH₂CH₂—N⁺(Et)₂H, Cl⁻]

15.41 (a) 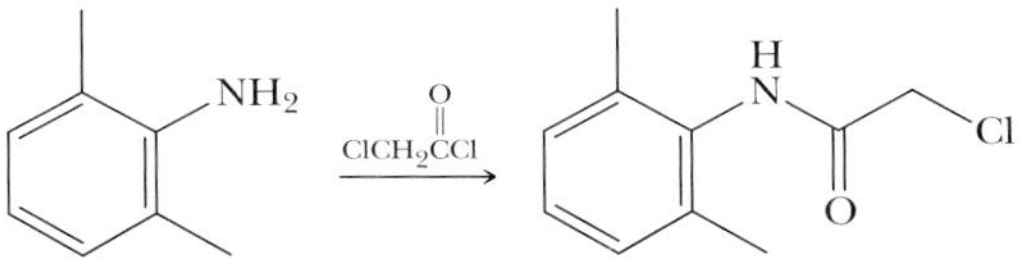 [2,6-dimethylaniline] $\xrightarrow{ClCH_2CCl}$ [amide with CH₂Cl]

[amide] $\xrightarrow{2(CH_3CH_2)_2NH}$ [amide with N(Et)₂]

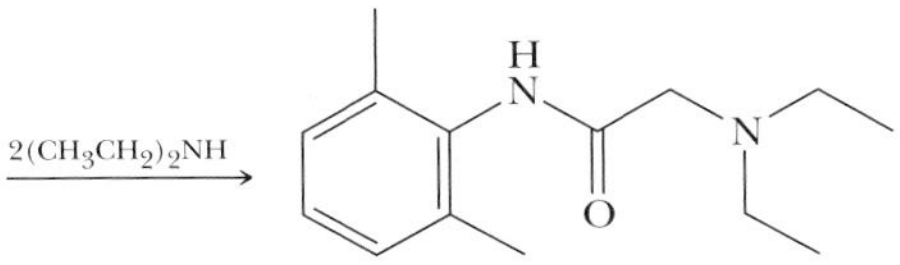

(b) 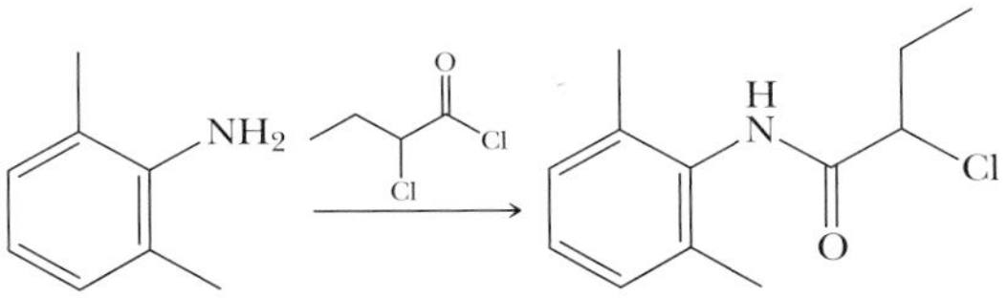 [2,6-dimethylaniline] + [acyl chloride with Cl] ⟶ [amide with Cl]

$\xrightarrow{2\ (CH_3CH_2NH)}$ [amide with N(Et)(propyl)]

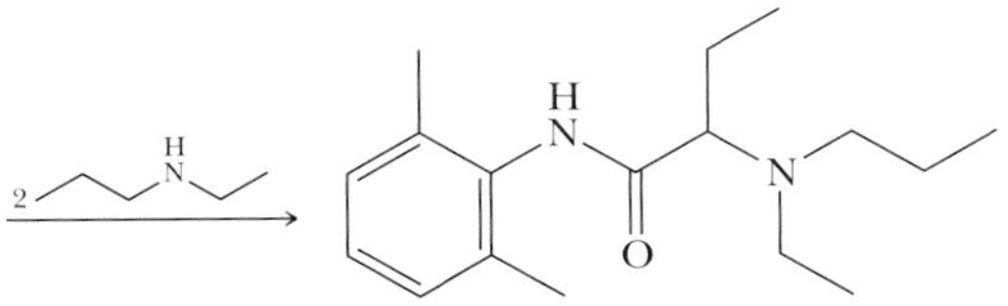

(c) 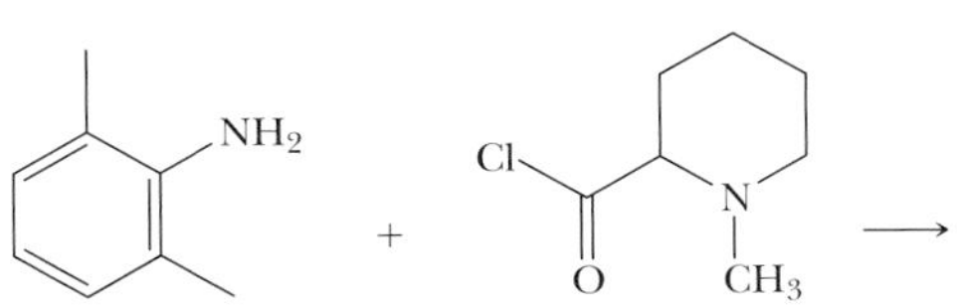 [2,6-dimethylaniline] + [N-methylpiperidine-2-carbonyl chloride] ⟶

[amide product: bupivacaine-like structure with N—CH₃ piperidine]

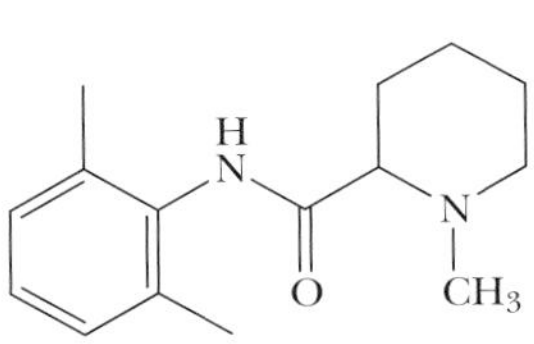

15.43 (1) HNO_3/H_2SO_4 (2) H_2CrO_4

(3) CH_3OH/H_2SO_4 (4) H_2/Ni

(5) $NaNO_2/H_2SO_4$, then heat

15.45 Label the oxygen in sodium methoxide with an isotope such as oxygen-18 (O-18). If substitution of an O-18 methoxyl group for O-16 methoxyl occurs, then acyl substitution has occurred.

15.47 The amide proton is relatively acidic to a Grignard reagent; therefore, the Grignard reagent will deprotonate the amide instead of undergoing acyl substitution.

Chapter 16

Enolate Anions

16.1 There are four acidic hydrogens in cyclohexanone and three in acetophenone.

(a) [structure]   (b) [structure]

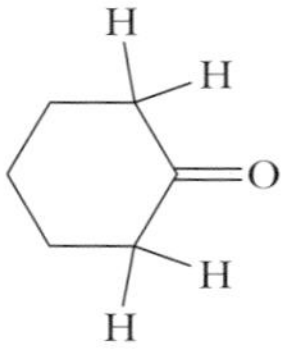

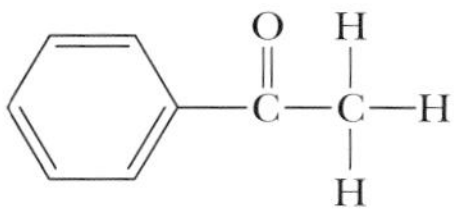

16.2 (a) [structure]

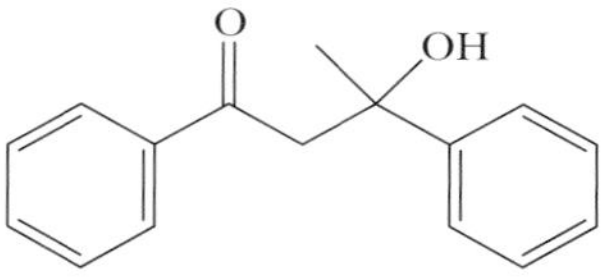

(b) [structure]

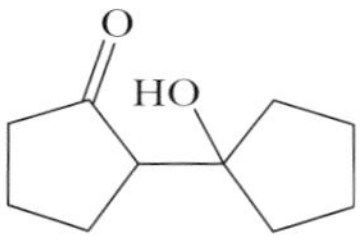

16.3 (a) [structure]

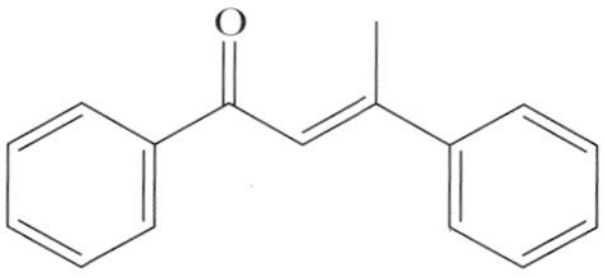

(b) [structure]

16.4 [structure] $\xrightarrow{-H_2O}$ [structure]

16.5 [structure]

16.6 [structure]

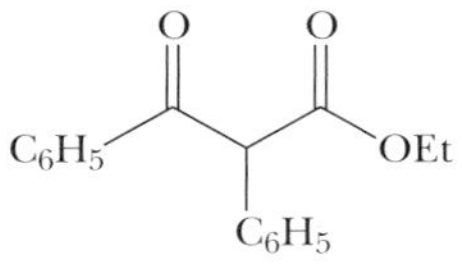

16.7 First convert benzoic acid to ethyl benzoate by Fischer esterification. Claisen condensation between ethyl benzoate and ethyl 3-methylbutanoate, followed by saponification of the ester, acidification, and heating gives the desired product.

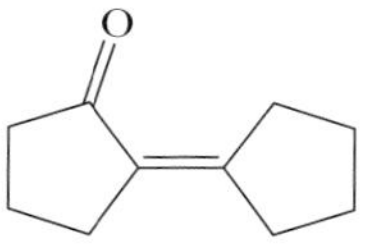

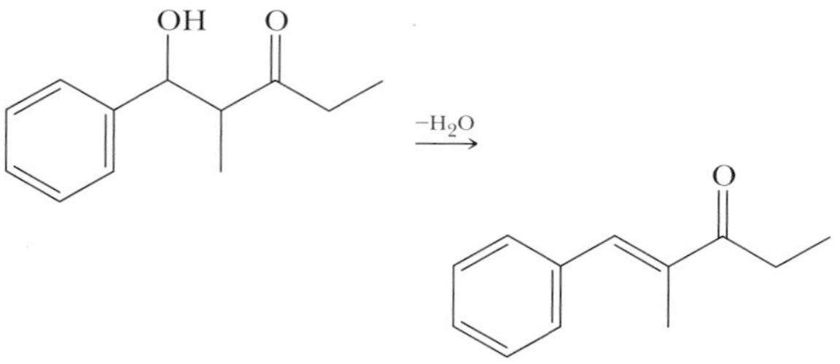

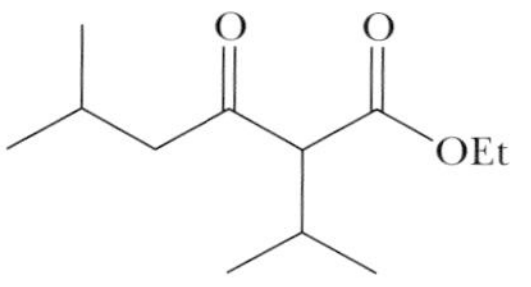

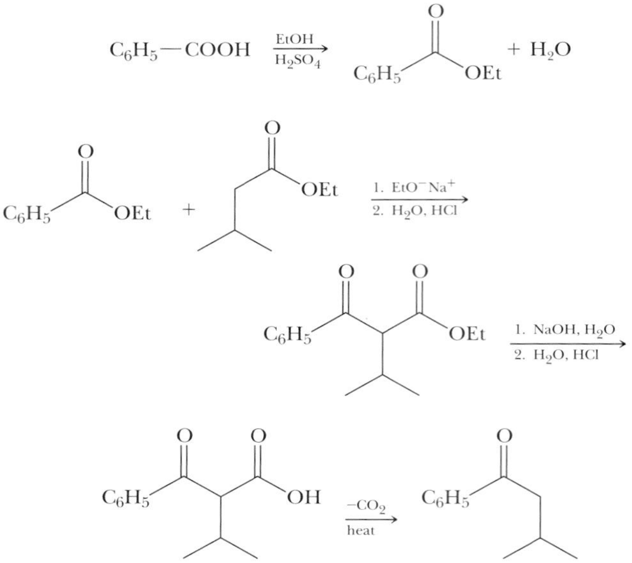

$C_6H_5-COOH \xrightarrow[H_2SO_4]{EtOH}$ [structure] $+ H_2O$

[structure] $+$ [structure] $\xrightarrow[\text{2. } H_2O, HCl]{\text{1. } EtO^-Na^+}$

[structure] $\xrightarrow[\text{2. } H_2O, HCl]{\text{1. } NaOH, H_2O}$

[structure] $\xrightarrow[\text{heat}]{-CO_2}$ [structure]

16.8 (a) [structure]

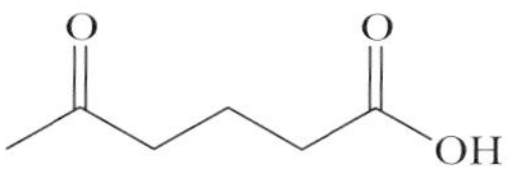

(b) [structure]

16.9 $EtOOC$ [structure] $COOEt$ $+$ [structure] $COOEt$ $\xrightarrow[EtOH]{EtO^-Na^+}$

[structure] $EtOOC$ ... $COOEt$ / $COOEt$

$\xrightarrow[\text{2. } HCl, H_2O]{\text{1. } NaOH, H_2O}$ [structure] $HOOC$ ... $COOH$ / $COOH$ $\xrightarrow{\text{heat}}$

$HOOC$ [structure] $COOH$

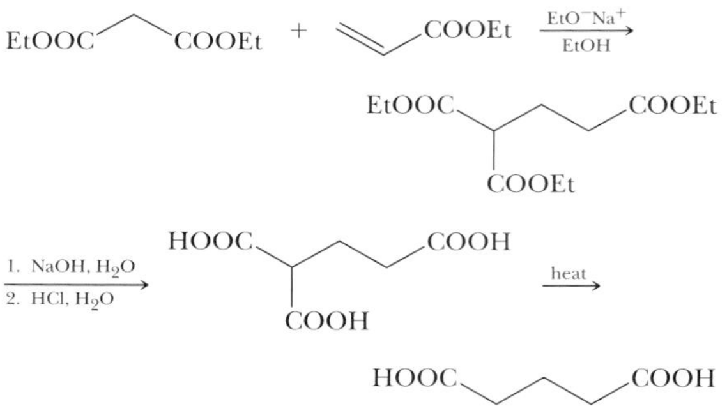

Pentanedioic acid
(Glutaric acid)

16.10 H_3C-N [structure] $COOEt$ / $COOEt$ $\xrightarrow[\text{2. } HCl, H_2O]{\text{1. } EtO^-Na^+, EtOH}$

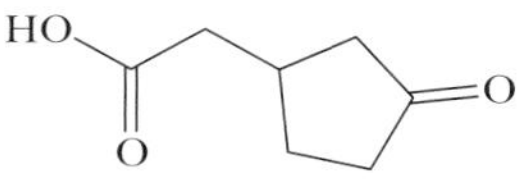

[structure] $COOEt$ H_3C-N ... O

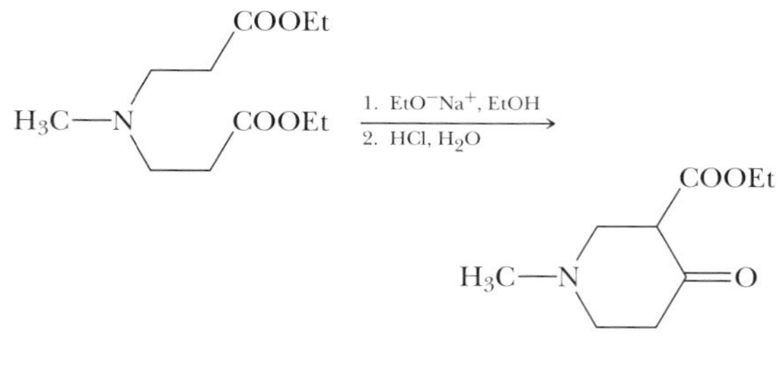

16.11

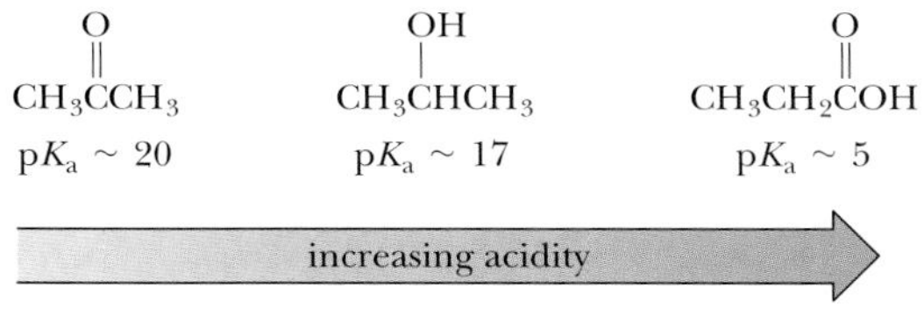

$$CH_3\overset{O}{\overset{\|}{C}}CH_3 \qquad CH_3\overset{OH}{\overset{|}{C}HCH_3} \qquad CH_3CH_2\overset{O}{\overset{\|}{C}}OH$$

pK_a ~ 20 $\qquad\qquad$ pK_a ~ 17 $\qquad\qquad$ pK_a ~ 5

increasing acidity →

16.13 (a)

(b)

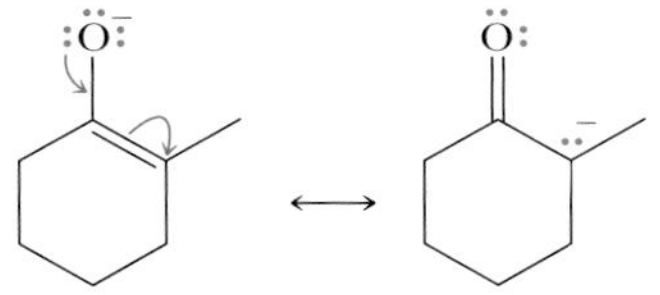

(c)

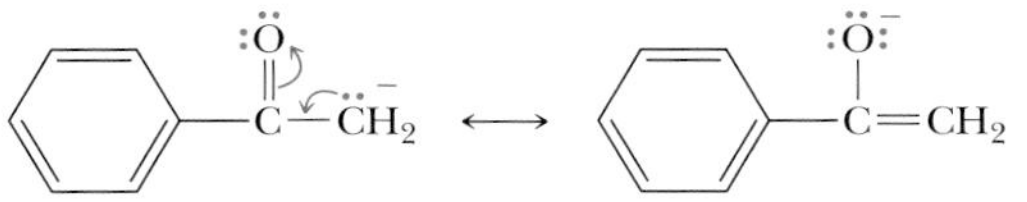

16.15 (a)

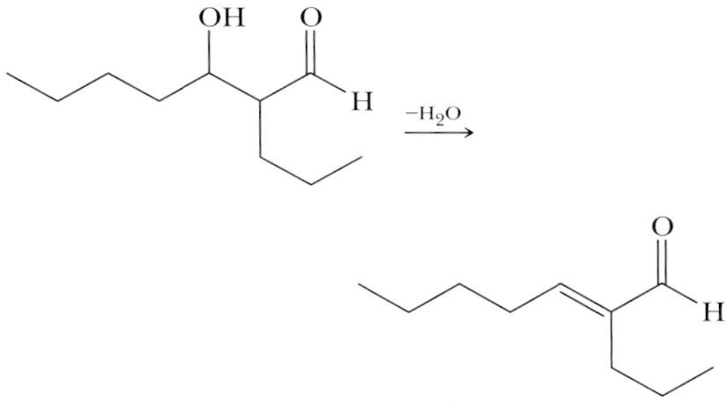

(b)

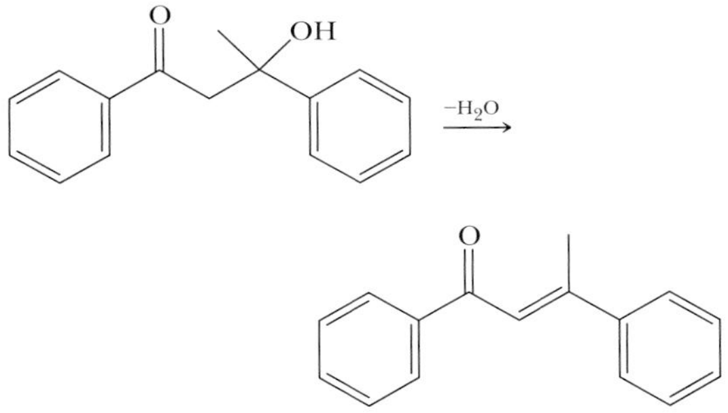

(c)

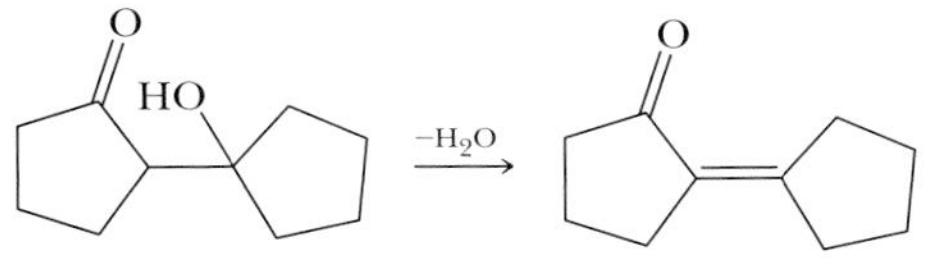

(d)

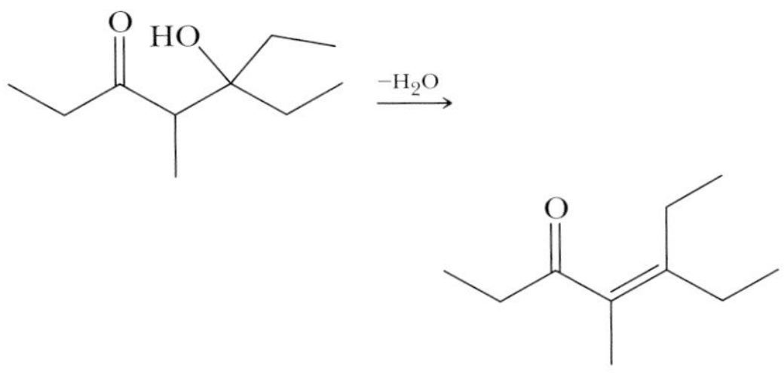

16.17 Six aldol products are possible; one for self-condensation of acetone; one for condensation of acetone with 2-butanone, two for self-condensation of 2-butanone; one for carbon-1 of 2-butanone on acetone, and one for carbon-3 of 2-butanone on acetone.

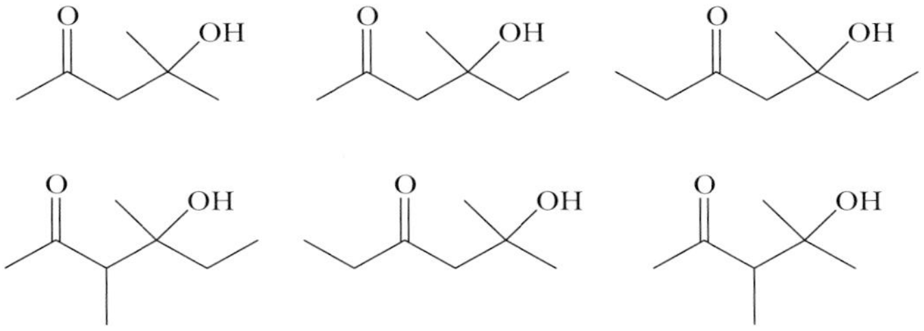

16.19 (a)

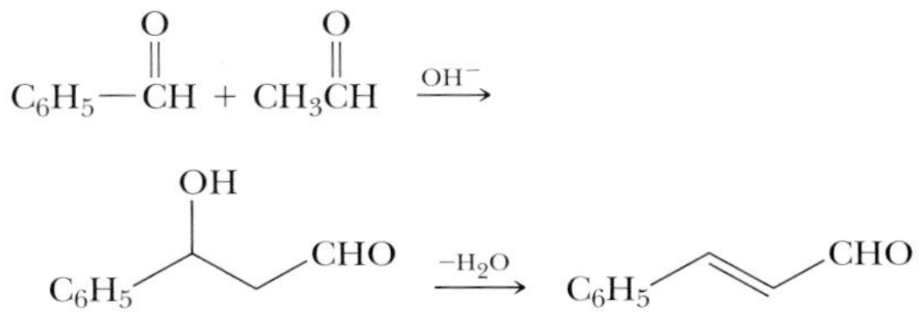

$$C_6H_5-\overset{O}{\overset{\|}{C}}H + CH_3\overset{O}{\overset{\|}{C}}H \xrightarrow{OH^-}$$

(b)

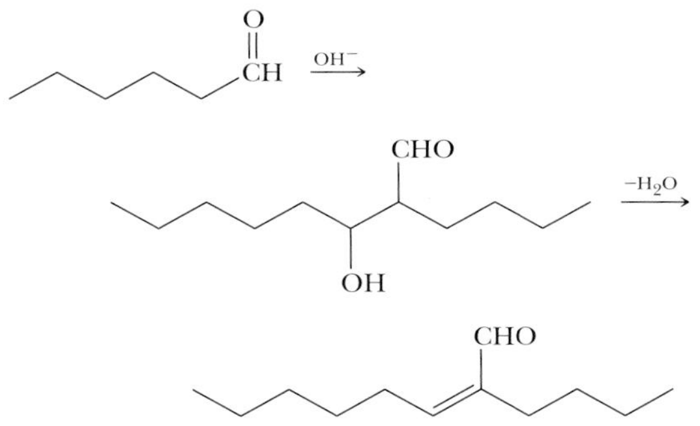

16.21

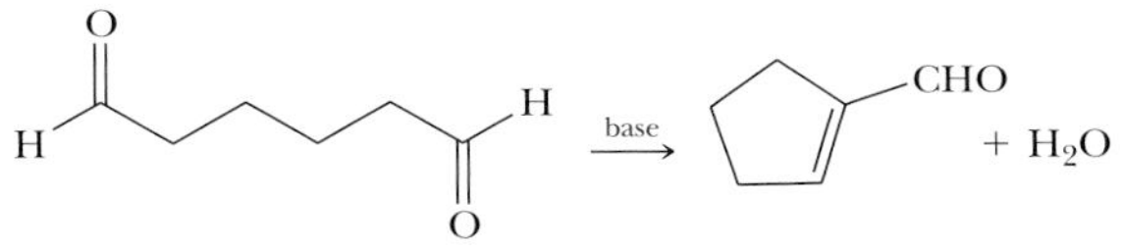

16.23 (a) (1) NaOH for an aldol reaction

(2) H_2CrO_4 to oxidize the aldehyde

(3) $SOCl_2$ to make the acid chloride

(4) $2\ NH_3$ to make the amide

(5) A peracid, RCO_3H, to make the epoxide

(b) Oxanamide has 2 stereocenters; 4 stereoisomers are possible.

16.25 (a)

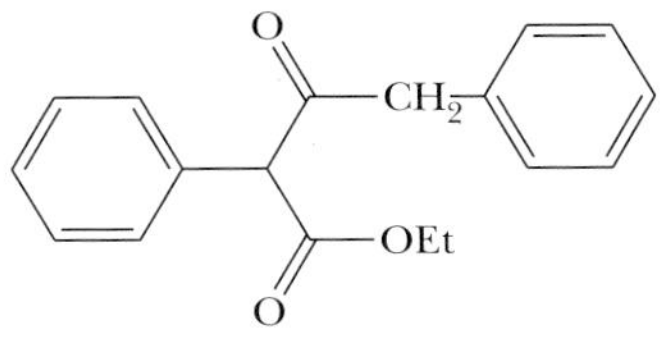

(b)

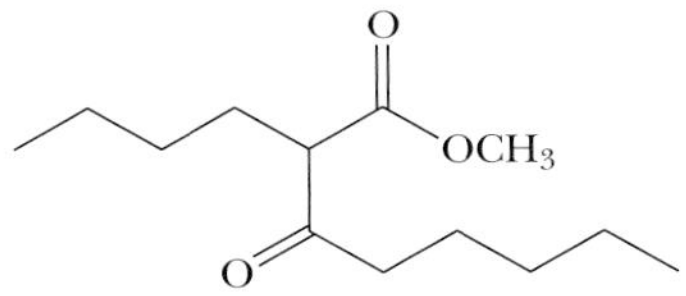

16.27

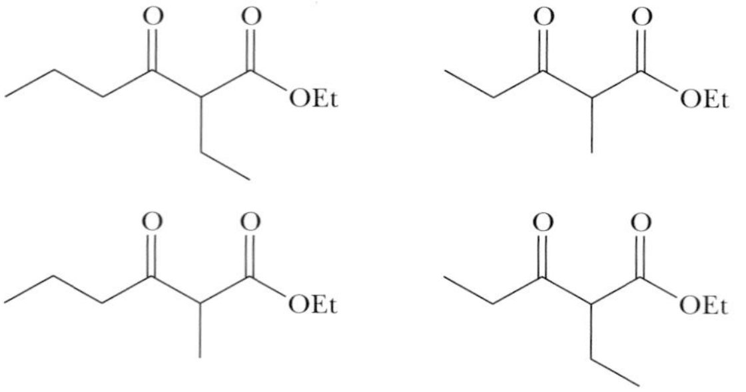

16.29

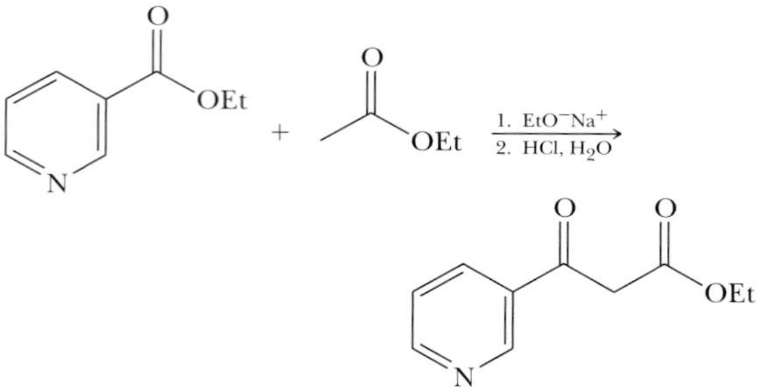

1. EtO⁻Na⁺
2. HCl, H₂O

16.31 (a) 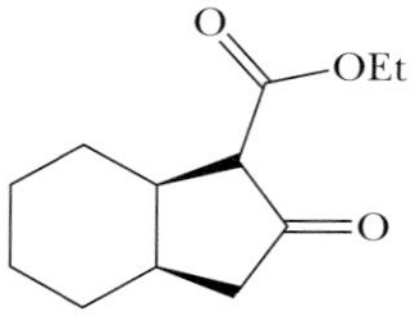 **(b)**

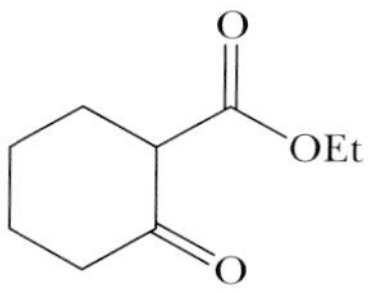

16.33

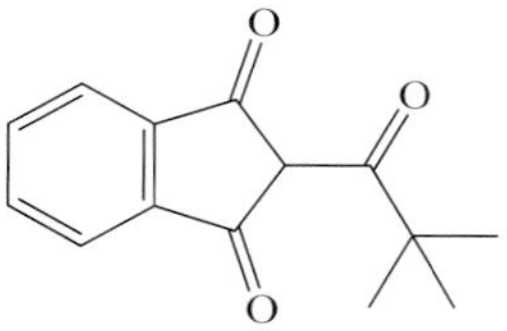

16.35 (a) The functional group is an acid chloride; the reagent is thionyl chloride, $SOCl_2$.

(b) Step 2 is a Friedel-Crafts reaction. The chlorine substituent is an ortho-para director and, therefore, ortho, para products are favored in this step.

(c) Ammonia (NH_3) and hydrogenation of the imine intermediate with H_2/Ni.

(d) The driving force behind the ring-opening reactions of epoxides with nucleophiles is the release of the angle strain in the three-membered ring.

(e) Thionyl chloride, $SOCl_2$.

(f) The most likely mechanism is an S_N2 reaction; 1° halides do not undergo S_N1 reactions.

16.37 The characteristic feature of the product of an aldol reaction is a β-hydroxyaldehyde or β-hydroxyketone. Fructose 1,6-bisphosphate contains a β-hydroxyketone. Base initiates reaction by removing a proton from the β-hydroxyl group. A reverse aldol reaction followed by protonation of the anion intermediate and keto-enol tautomerism gives the observed triose phosphates.

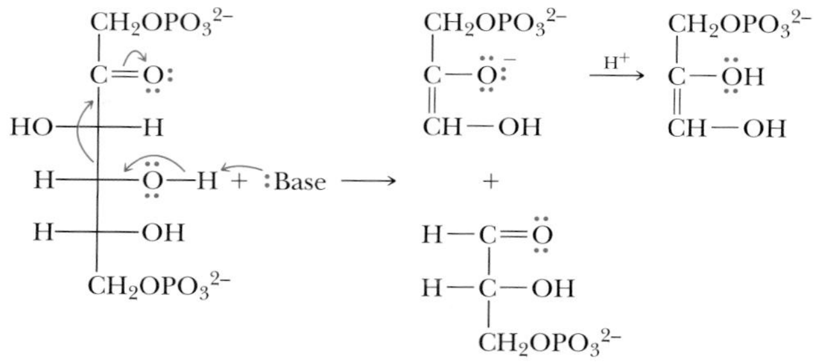

16.39 Given are the structures for the boxed intermediates/reagents.

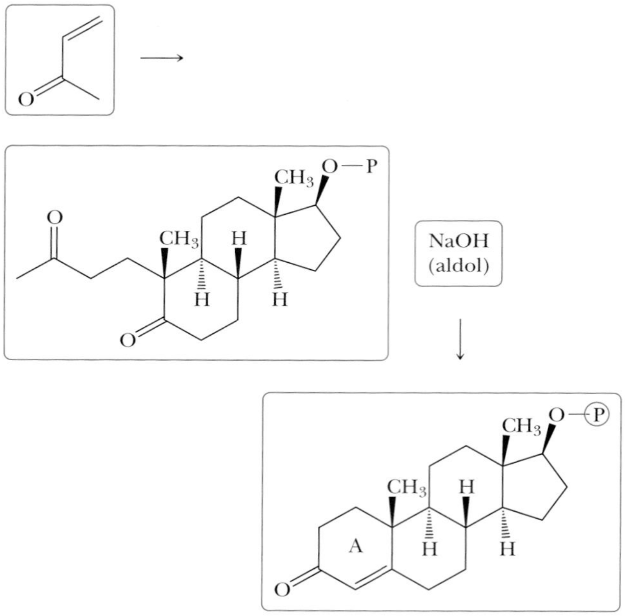

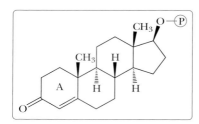

NaOH (aldol)

Chapter 17
Organic Polymer Chemistry

17.1 Cl Poly(vinyl chloride)

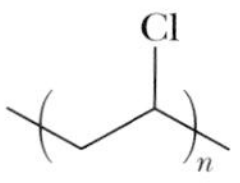

17.2

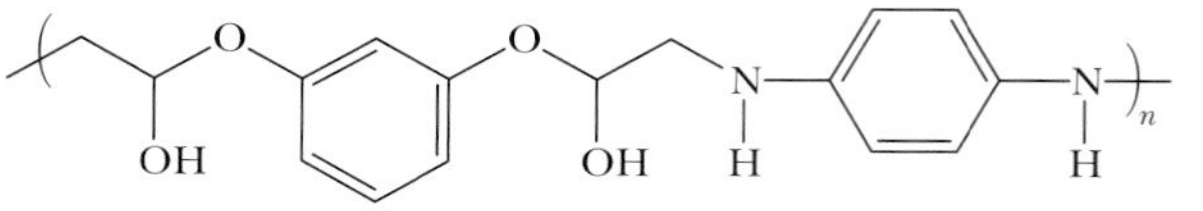

17.3

(a) HOC—❬❭—COH + HOH_2C—❬❭—CH_2OH

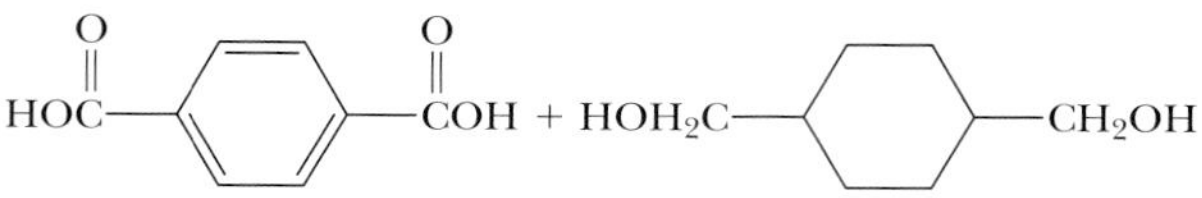

(b) $HOC(CH_2)_6COH$ + H_2N—❬❭—CH_2—❬❭—NH_2

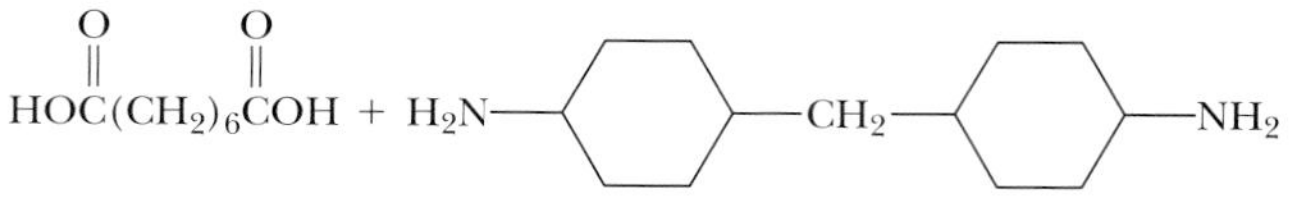

(c) HO∿OH + HOC—❬❭—COH

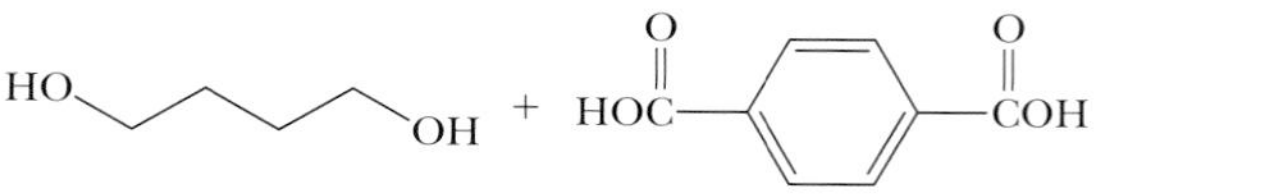

(d)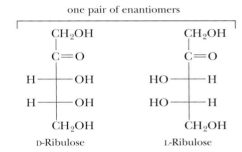

17.5 This reaction is the reverse of the reaction in Problem 15.4.

$$\xrightarrow{2n\text{CH}_3\text{OH}}$$

17.7 The 6 refers to the number of carbons in the amine and the 10 refers to the number of carbons in the carboxyl derivative.

17.9 (a) $\text{CH}_2{=}\text{CCl}_2$ (b) $\text{CH}_2{=}\text{CF}_2$

17.11 The greater the degree of branching of the main polymer chains, the less well the chains pack together in the solid state, and, therefore, the lower the density of the polymer.

17.13

Poly(vinyl acetate)   Poly(vinyl alcohol)

17.15 Predict that A and B are both rigid, opaque polymers, and that C is more flexible and transparent. Both of these physical characteristics depend on the degree of crystallinity of the polymer. These three polymer chains all have repeating stereocenters. Both A and B are termed stereoregular; that is, the configurations of the stereocenters repeat in a consistent pattern over the length of the chain. In A, all stereocenters have the same configuration, whereas in B they alternative first R, then S, and so forth. Because of this stereoregular pattern, molecules of both polymers A and B pack well in the solid state with strong intermolecular interactions between molecules. Because of this

form of packing, polymers A and B have a high degree of crystallinity and are rigid polymers. Polymer C, on the other hand, has a random orientation of stereocenters and, therefore, molecules of its chains do not pack as well in the solid state; it has a low degree of crystallinity. The lower the degree of crystallinity, the more transparent the polymer.

17.17 Yes. A polymer is constructed of monomers, which are the repeating units.

Chapter 18
Carbohydrates

18.1 There are four 2-ketopentoses; two pairs of enantiomers.

one pair of enantiomers

CH$_2$OH	CH$_2$OH
C=O	C=O
H——OH	HO——H
H——OH	HO——H
CH$_2$OH	CH$_2$OH
D-Ribulose	L-Ribulose

a second pair of enantiomers

CH$_2$OH	CH$_2$OH
C=O	C=O
HO——H	H——OH
H——OH	HO——H
CH$_2$OH	CH$_2$OH
D-Xylulose	L-Xylulose

18.2 Mannose differs from glucose in the configuration at carbon-2.

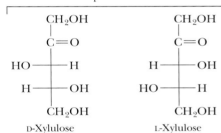

β-D-Mannopyranose   α-D-Mannopyranose
(β-D-Mannose)   (α-D-Mannose)

18.3 Mannose differs from glucose in the configuration at carbon-2.

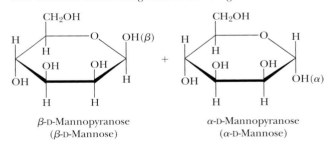

β-D-Mannopyranose   α-D-Mannopyranose

18.4

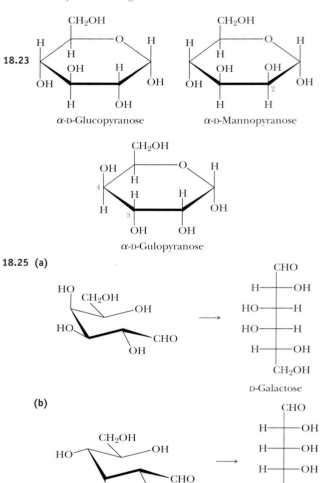

α-D-Mannopyranoside

18.5

a β-N-glycosidic bond

anomeric carbon

18.6 Optically inactive. Erythritol is a meso compound.

18.7

β-1,3-glycosidic bond

18.9 D-Glucose

18.11 The designations D and L refer to the configuration at the stereocenter farthest from the aldehyde or ketone group of a monosaccharide. When a monosaccharide is drawn as a Fischer projection, a D-monosaccharide has the —OH of this carbon on the right; an L-monosaccharide has it on the left.

18.13 Compounds (a) and (c) are D-monosaccharides; compound (b) is an L-monosaccharide.

18.15 Each carbon of a monosaccharide has an oxygen that is able to participate in hydrogen bonding with water molecules.

18.17

    CHO
H ──┬── H
H ──┬── OH   2,6-Dideoxy-D-Altrose
H ──┬── OH
H ──┬── OH
    CH₃

18.19 The designation β means that the —OH group on the anomeric carbon of a cyclic hemiacetal is on the same side of the ring as the terminal —CH₂OH group. The designation α means that it is on the opposite side from the terminal —CH₂OH group.

18.21 No, they are not anomers; they are enantiomers. Anomers differ only in the configuration at the anomeric carbon.

18.23

α-D-Glucopyranose   α-D-Mannopyranose

α-D-Gulopyranose

18.25 (a)

→ D-Galactose

    CHO
H ──┬── OH
HO ──┬── H
HO ──┬── H
H ──┬── OH
    CH₂OH

(b)

→ D-Allose

    CHO
H ──┬── OH
H ──┬── OH
H ──┬── OH
H ──┬── OH
    CH₂OH

18.27 During mutarotation, the α and β forms of a carbohydrate convert to an equilibrium mixture of the two forms. Mutarotation can be detected by observing the change in optical activity over time as the two forms equilibrate.

18.29 Yes. The specific rotation of α-L-glucose changes to −52.7°.

18.31 D-Ribose is reduced to ribitol; it is oxidized to D-ribonic acid.

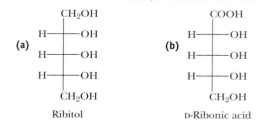

(a)

    CH₂OH
H ──┬── OH
H ──┬── OH
H ──┬── OH
    CH₂OH
Ribitol

(b)

    COOH
H ──┬── OH
H ──┬── OH
H ──┬── OH
    CH₂OH
D-Ribonic acid

18.33 D-Arabinose and D-lyxose yield optically active alditols. D-Ribose and D-xylose yield optically inactive (because they are meso compounds) alditols.

18.35 Reduction of D-allose and D-galactose give optically inactive meso alditols.

18.37 **(a)** Reactions are:

(1) Formation of a cyclic hemiacetal from a carbonyl group and a 2° alcohol.

(2) Oxidation of the 2° alcohol on carbon-5 to a ketone.

(3) Dehydration of a 1° alcohol to a carbon-carbon double bond.

(4) Reduction of a carbon-carbon double bond to a carbon-carbon single bond.

(5) Keto-enol tautomerism.

(6) Keto-enol tautomerism.

(7) Reduction of a ketone to a 2° alcohol.

(8) Opening of a cyclic hemiacetal to an aldehyde and an alcohol.

(b) Reactions (3) and (4) result in inversion of configuration at carbon-5.

18.39 **(a)**

(b) The anion derived from ionization of the —OH on carbon-3 is stabilized by resonance interaction with the carbonyl group. There is no comparable resonance stabilization for the anion derived from ionization of the —OH on carbon-3.

18.41 A glycosidic bond is a bond from the anomeric carbon of any monosaccharide to an —OR group. A glucosidic bond is a bond from the anomeric carbon of glucose to an —OR group.

18.43 Sucrose is a disaccharide composed of the monosaccharides D-glucose and D-fructose linked by a glycosidic bond. The acid catalyzes the hydrolysis of the glycosidic bond to give D-glucose and D-fructose. D-Fructose has a relative sweetness of 174 compared with 100 for sucrose. Thus, converting sucrose into fructose increases the sweetness of the mixture.

18.45 **(a)** Trehalose is not a reducing sugar, **(b)** it does not undergo mutarotation, and **(c)** it is composed of two units of D-glucose.

18.47 An oligosaccharide contains approximately 6 to 10 monosaccharide units. A polysaccharide contains more, generally many more than 10 monosaccharide units.

18.49 The difference is in the degree of chain branching. Amylose is composed of unbranched chains, whereas amylopectin is a branched network with branches starting at β-1,6 glycosidic bonds.

18.51 **(a)**

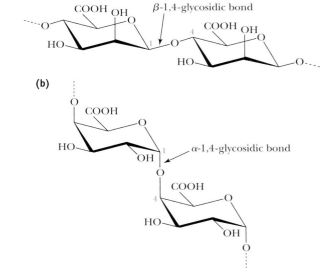

(b)

18.53 The intermediate in this conversion is an enediol; that is, a molecule that contains a carbon-carbon double bond with two OH groups on it.

Dihydroxyacetone phosphate ⇌ An enediol intermediate ⇌ D-Glyceraldehyde 3-phosphate

18.55 **(a)** There are a number of sets of epimers. One set includes D-allose, D-altrose (differs at C-2), D-glucose (differs at C-3) and D-gulose (differs at C-4).

(b) Anomers always come in pairs; anomers of D-aldoses differ in configuration only at carbon-1 in the cyclic hemiacetal form. Anomers of 2-ketoses differ in configuration only at C-2 in the cyclic hemiacetal form. Epimers differ in configuration only at a carbon other than the anomeric carbon.

Chapter 19

Amino Acids and Proteins

19.1 **(a)** Only glycine does not contain a stereocenter.

(b) Isoleucine and threonine each contain two stereocenters.

19.2 At pH 7.0, histidine (pI 7.64) has a net positive charge and will move toward the negative electrode.

19.3 At pH 6.0, glutamic acid (pI 3.08) has a net negative charge and will move toward the positive electrode. At this pH, arginine (pI 10.76) will have a net positive charge and move toward the negative electrode; valine (pI 6.00) will be neutral (have no net charge) and, therefore, remains at the origin.

19.4

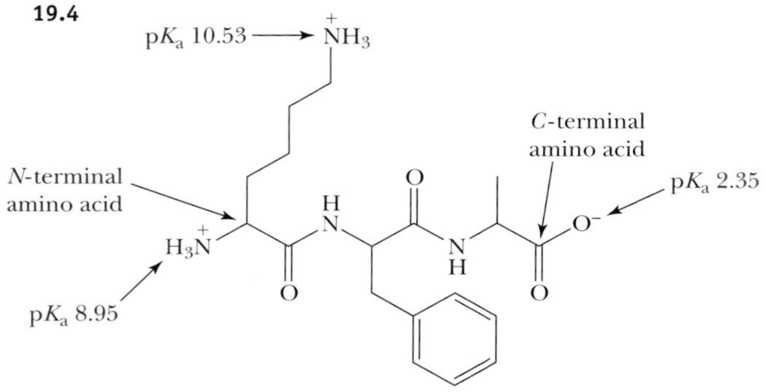

pKa 10.53 ⟶ NH₃⁺

C-terminal amino acid

N-terminal amino acid

pKa 2.35

H₃N⁺

pKa 8.95

Net charge at pH 6 is +1

19.5 None of these tripeptides contains Arg or Lys and, therefore, are not hydrolyzed by trypsin. Tripeptides **(a)** and **(b)** are hydrolyzed by chymotrypsin.

19.6 Ala-Glu-Arg-Thr-Phe-Lys-Lys-Val-Met-Ser-Trp

19.7 At pH 7.4, the side chain of lysine can form salt linkages with the side chains of Asp, Glu, and His.

19.9 The stereocenter in L-serine has the S configuration.

19.11

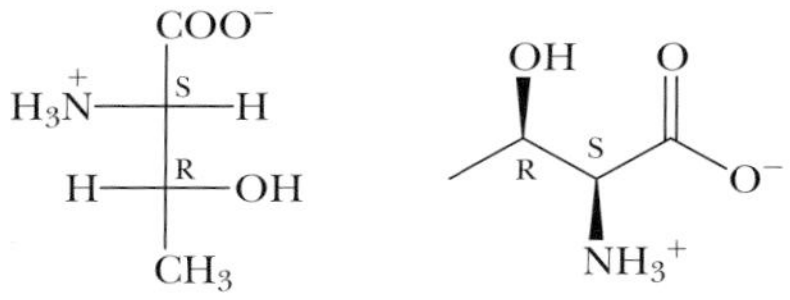

19.13 (a) 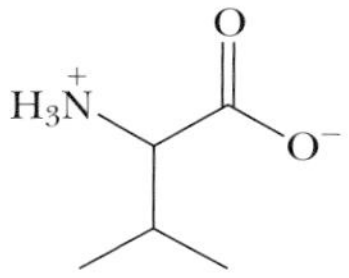 Valine　**(b)** Phenylalanine

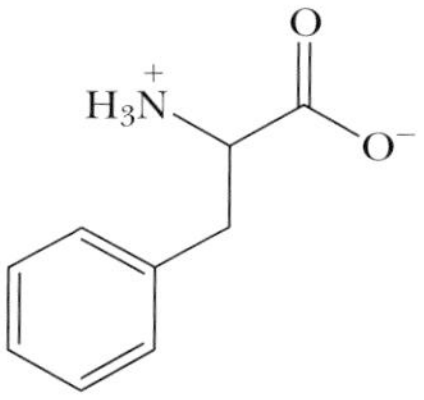

(c) Glutamine

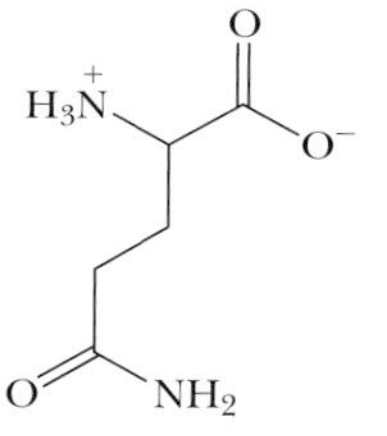

19.15 Arg is referred to as a basic amino acid because its side chain contains a basic group (an —NH₂ group). His and Lys also have basic side chains and are considered basic amino acids as well.

19.17

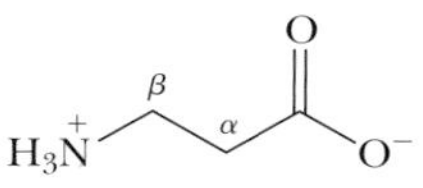

β-Alanine

19.19 Histidine is the biological precursor to histamine and decarboxylation is involved in the conversion of His to histamine.

19.21 Serotonin and melatonin are synthesized from tryptophan.
  (a) To synthesize serotonin, tryptophan undergoes a decarboxylation and the indole ring is oxidized to add the OH group.
  (b) To synthesize melatonin, the carboxyl group of tryptophan undergoes decarboxylation and acetylation of its 1° amine, and a C—H group of indole ring is oxidized to a C—OH group followed by methylation.

19.23 Following are the major forms of each amino acid present in solution at pH 10.0:

(a) H₂N　**(b)** H₂N

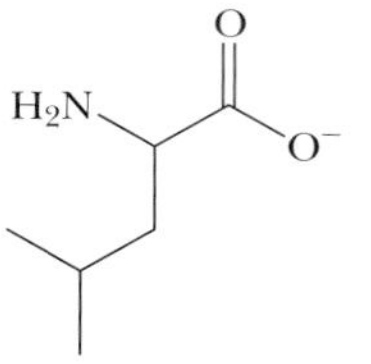

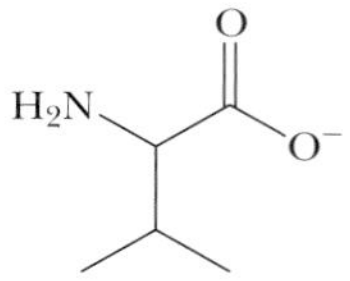

(c) 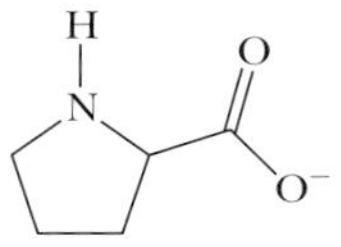　**(d)** H₂N

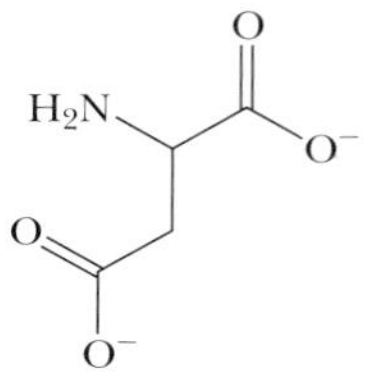

19.25

2.0 mol NaOH

1.0 mol NaOH

3.0 mol NaOH

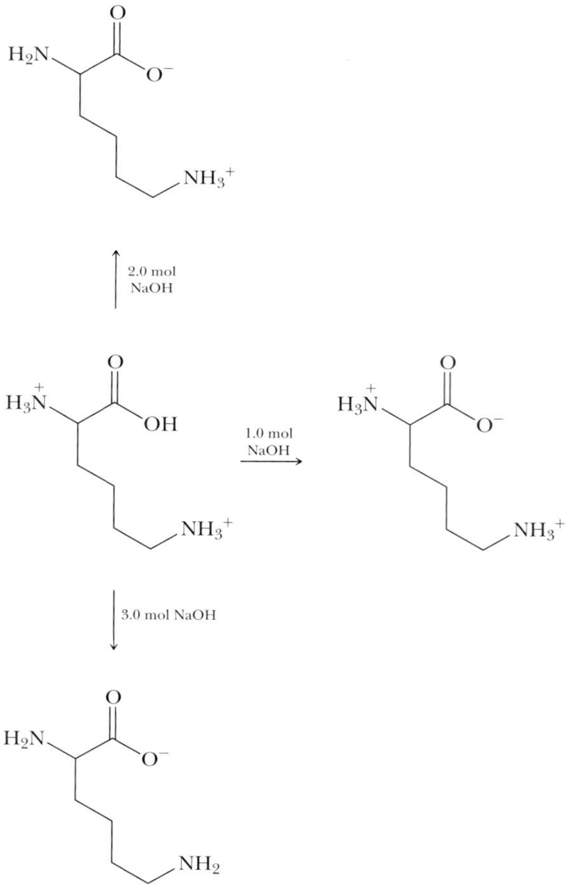

19.27 (a)

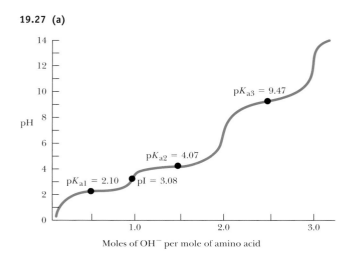

(b)

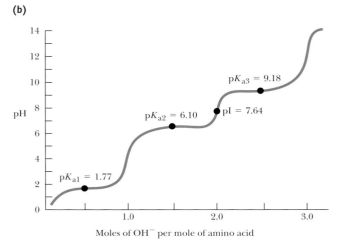

19.29 Glutamic acid possesses an acidic side chain, so in order for Glu to be neutral, the net charge on the two carboxyl groups must be -1. Thus, pI of glutamic acid falls between the pK_a values of the two carboxyl groups (pK_a 2.10 and 4.07). The pI of glutamine falls between the pK_a values for the α-COOH group (pK_a 2.17) and the α-NH_3^+ group (pK_a 9.13).

19.31 The basicity of guanidine and the guanidino group is due to the large resonance stabilization of the protonated form, as described in Section 19.3A.

19.33 The migratory properties of a compound in electrophoresis depend on the net charge of the molecule. Following is the net charge of each compound and the direction of its migration:

(a) His ($+1$, cathode)   (b) Lys ($+1$, cathode)

(c) Glu (-1, anode)   (d) Gln ($+1$, cathode)

(e) Glu-Ile-Val (0, remains at origin)

(f) Lys-Gln-Tyr ($+$, cathode)

19.35 Insulin has 4 acidic side chains and 4 basic side chains. Its isoelectric point must be at a pH in which all the acidic groups are deprotonated (pH $>$ 4) and all the basic groups are protonated (pH $<$ 6). The pI of insulin should fall around pH 5, which is similar to that of a neutral amino acid.

19.37 (a) $4 \times 3 \times 2 \times 1 = 24$ possibilities.

    (b) $20 \times 19 \times 18 \times 17 = 116{,}280$ possibilities.

19.39 Following are the peptide bonds that each reagent or enzyme hydrolyzes; amine acids reacting are by that numbers

(a) 1,2

(b) 6-7, 10-11, 13-14, 22-23, 25-26

(c) 12-13, 17-18, 18-19

(d) 27-28

19.41 (a)

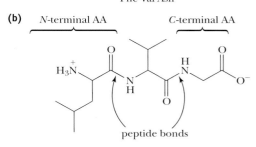

Phe-Val-Asn

(b)

Leu-Val-Gly

19.43 (a) Glu-Cys-Gly

    (b) The peptide bond between Glu and Cys is formed between the *side chain carboxyl group* of Glu and Cys. A normal peptide bond between Glu and Cys would be between the *C*-terminal carboxyl of Glu and the amino group of Cys.

    (c) Glutathione is a biological reducing agent:

$$2\ GSH \longrightarrow G\text{-}S\text{-}S\text{-}G + 2\ e^- + 2\ H^+$$

    (d) O_2 is reduced to H_2O:

$$4\ GSH + O_2 \longrightarrow 2\ G\text{-}S\text{-}S\text{-}G + 2\ H_2O$$

19.45 Amino acid side chains in α-helices are arranged on the outside of the helix.

19.47 Polar, acidic, and basic side chains prefer to be in contact with the aqueous environment to maximize hydrophilic interactions. Of the choices, these are **(b)** Arg, **(c)** Ser, and **(d)** Lys. Nonpolar side chains prefer to avoid contact with the aqueous environment and to maximize hydrophobic interactions. Of the choices, these are **(a)** Leu and **(e)** Phe.

19.49 Homoserine forms the following five-membered lactone:

from homoserine

Serine does not react in the same way because reaction would result in formation of a highly strained four-membered lactone ring.

19.51 The process shown in the second energy diagram is a lowering of the activation energy. This is associated with catalysis and a function of proteins known as enzymes.

Chapter 20
Nucleic Acids

20.1

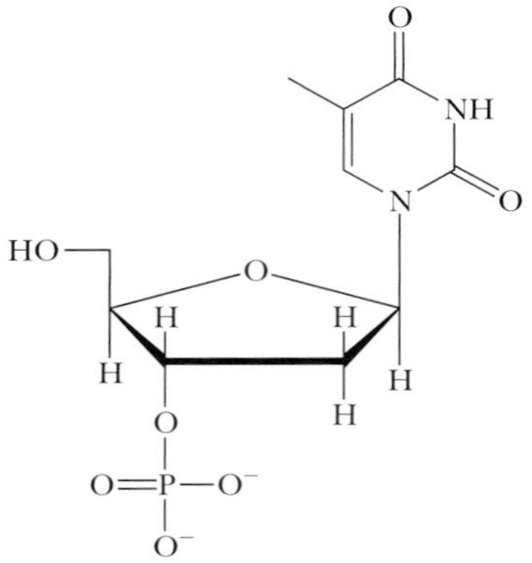

20.2

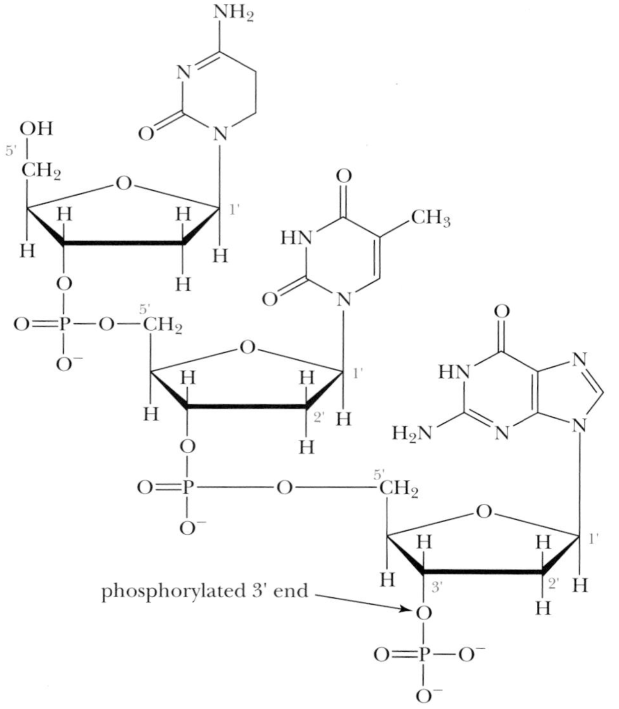

20.3 5'-TCGTACGG-3'

20.4 Remember that the base uracil (U) in RNA is the complement to adenine (A) in DNA. The complement DNA sequence is 5'-TGGTGGACGAGTCCGGAA-3'.

20.5 **(a)** 5'-UGC-UAU-AUU-CAA-AAU-UGC-CCU-
CUU-GGU-UGA-3'

  (b) The primary structure of oxytocin is Cys-Tyr-Ile-Gln-Asn-Cys-Pro-Leu-Gly.

20.6 The FnuDII and HpaII cleavage sites are:

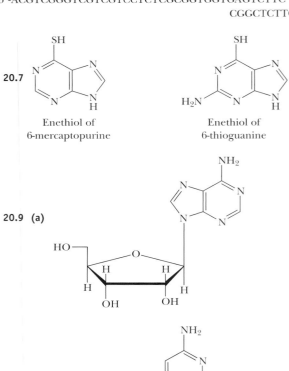

  FnuDII   HpaII
     ↓   ↓
5'-ACGTCGGGTCGTCGTCCTCTCGCGGTGGTGAGTCTTC
                                CGGCTCTTCT-3'

20.7

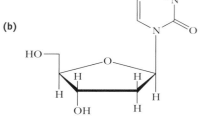

  Enethiol of   Enethiol of
6-mercaptopurine   6-thioguanine

20.9 **(a)**

(b)

20.11 Nucleotides differ from nucleosides in that nucleotides contain phosphate groups at the 5' or 3' position.

20.13

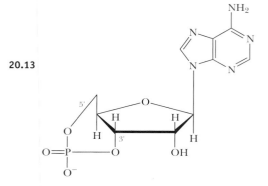

20.15 Because A pairs with T, there must also be 30.4% T in human DNA. Thus, A and T comprise 60.8% of human DNA. The remainder (39.2%) must be G and C, split equally to give 19.6% G and 19.6% C. These values agree very well with the experimental values found in Table 20.1.

20.17 **(1)** DNA consists of two antiparallel polynucleotide strands coiled in a right-handed manner about the same axis to form a double helix.

(2) Purine and pyrimidine bases project inward toward the axis of the helix and are always paired in a very specific manner, A with T and G with C.

(3) Base pairs are stacked one on top of the other with a distance of 3.4Å between base pairs and with ten base pairs in one complete turn of the helix.

(4) There is one complete turn of the helix every 34Å.

20.19 **(a)** The repeating units in the backbone of the α-helices of proteins are α-amino acids while those in DNA are units of phosphate esters of β-2′-deoxy-D-ribose.

(b) The R groups of amino acids in α-helices point outward from the helix, whereas the purine and pyrimidine bases of a DNA double helix point inward and away from the cellular environment.

20.21 **(a)** Glycosidic bonds join monomer units in polysaccharides.

(b) Peptide bonds join monomer units in polypeptides.

(c) Phosphate ester bonds join monomer units in nucleic acids.

20.23 G-C base pairs have 3 hydrogen bonds, whereas A-T base pairs have only two hydrogen bonds. The more hydrogen bonds, the more energy required to break them. Thus, nucleic acids with more G-C content will require more energy (in the form of heat) to denature.

20.25 5′-ATCGTTGA-3′

20.27 **(a)** In DNA, the monosaccharide units are β-2′-deoxy-D-ribose, whereas in RNA they are β-D-ribose.

(b) Both DNA and RNA use the same purines, G and C. They differ in that RNA uses U, while DNA uses T. Both use the pyrimidine A.

(c) The primary structure of DNA consists of pairs of A, T, G, and C, whereas that of RNA consists of single strands of A, U, G, and C.

(d) RNA is found in the cytoplasm (the space between the cell wall and the nucleus) of cells while DNA is found in the nucleus of cells.

(e) DNA's function is information storage, while RNA's is translation and transcription.

20.29 5′-AUUAACGGU-3′

20.31 Degenerate refers to the fact that several amino acids are coded for by more than one triplet.

20.33 Phe and Tyr differ in their side chains. Although both have aromatic rings in their side chains, Phe contains a benzene ring, whereas Tyr contains a phenol. Their codons differ only in the second position.

20.35 Following are the codons in which the third base is irrelevant (X = the third base) and the amino acids that they code for:

GUX–Val GCX–Ala GGX–Gly CGX–Arg
UCX–Ser CCX–Pro ACX–Thr

20.37 With the exception of Trp and Gly, all the codons with A or G as the second base code for amino acids with polar, hydrophilic side chains.

20.39 A polypeptide of 141 amino acids requires one triplet codon for each amino acid plus one stop codon. Therefore, the minimum number of DNA bases required to code for the alpha chain is (141 + 1) × 3 = 426 bases.

20.41 The loss of three consecutive units of T on a gene would cause the loss on the complementary mRNA strand of AAA. This codon triplet codes for lysine, which is the amino acid missing from CFTR in cystic fibrosis.

20.43

Chapter 21
Lipids

21.1 **(a)** There are 3 constitutional isomers as shown:

H₂C—palmitate	H₂C—oleate	H₂C—oleate
HC—oleate	HC—palmitate	HC—stearate
H₂C—stearate	H₂C—stearate	H₂C—palmitate

(b) All contain just one stereocenter and thus are chiral.

21.3 The hydrophobic regions are the three fatty acid hydrocarbon chains; the hydrophilic regions are the three ester groups.

21.5 Glyceryl trioleate would have the higher melting point because it contains fewer C=C bonds. The fewer C=C bonds a fatty acid contains, the more compact is its structure and the greater is the ability of its molecules to pack together.

21.7 Coconut oil is one of the few saturated liquid triglycerides because it contains mostly lower-molecular-weight fatty acids. The lower-molecular-weight fatty acids have fewer dispersion forces between their molecules.

21.9 Hardening of vegetable oils is the catalytic reduction, via hydrogenation, of the C=C bonds in the vegetable oils.

21.11 A good synthetic detergent should have a long hydrocarbon chain and a polar group at one end that does not form insoluble salts with Ca(II), Mg(II), or Fe(III) ions that are present in hard water.

21.13 The detergents used as shampoos and dishwashing liquids are alkylbenzenesulfonates and have the general formula:

where R represents a long hydrocarbon chain. They are primarily anionic detergents.

21.15 (a) $CH_3(CH_2)_{14}-\overset{\displaystyle O}{\overset{\|}{C}}-OH \xrightarrow[\text{2) H}^+\text{, H}_2\text{O}]{\text{1) LiAlH}_4} CH_3(CH_2)_{14}-CH_2-OH \xrightarrow{\text{SOCl}_2} \text{cetyl chloride}$

(b) $CH_3(CH_2)_{14}-\overset{\displaystyle O}{\overset{\|}{C}}-OH \xrightarrow{\text{SOCl}_2} CH_3(CH_2)_{14}-\overset{\displaystyle O}{\overset{\|}{C}}-Cl \xrightarrow{\text{(CH}_3)_2\text{NH}} CH_3(CH_2)_{14}-\overset{\displaystyle O}{\overset{\|}{C}}-\overset{\displaystyle CH_3}{\underset{\displaystyle CH_3}{N}}$

palmitic acid

$\downarrow$ 1) LiAlH$_4$
2) H$^+$, H$_2$O

N,N-dimethyl-1-hexadecanamine

21.17

hydrophobic region   hydrophilic region

21.19 The presence of unsaturated fatty acids increases the fluidity of biological membranes because their cis $C{=}C$ bonds create kinks in the chains that make them harder to pack closer together.

21.21 Each reaction is stereoselective. In **(a)**, hydrogen adds from the least hindered side of the carbon-carbon double bond, which in this case is from the side opposite the OH on ring A and the CH_3 group at the junction of rings A and B. Bromination occurs via a cyclic bromonium ion and anti addition of Br_2. In additions to cyclohexene rings, anti addition corresponds to trans diaxial addition. In **(b)** the added bromine atoms are both axial.

(a)

(b)

21.23

$CH_3(CH_2)_4(CH{=}CHCH_2)_2(CH_2)_6-\overset{\displaystyle O}{\overset{\|}{C}}-O$

21.25 Cholic acid (Figure 21.10) possesses the steroid A-D ring system with a *cis* ring fusion between rings A and B. It is able to emulsify fats and oils because its steroid skeleton acts as a hydrophobic region and its carboxylate and hydroxyl groups act as a hydrophilic region much like the analogous regions of fatty acids.

21.27 Tamoxifen has several benzene rings that may resemble the phenol ring of estrone.

21.29 The synthetic prostaglandins unoprostone and rescula have similar carbon skeletons as $PGF_{2\alpha}$. They differ in that the synthetic analogs do not have a $C{=}C$ bond at C-13, the analogs have two extra carbons (C-21, C-22), and instead of an alcohol at C-15, the synthetics possess a carbonyl group.

21.31

$CH_3(CH_2)_{14}\overset{\displaystyle O}{\overset{\|}{C}}O$

vitamin A palmitate

21.33 (a) Because sugars are water-soluble molecules, expect the sugar residue of glycolipids to lie on the extracellular side of membranes.

(b) D-Galactose

21.35 Lipid movement A is most favorable. The movement illustrated in B would require the polar head group of the lipid to pass through the nonpolar part of the membrane.

Chapter 22
The Organic Chemistry of Metabolism

22.1 The conversion of glucose to two molecules of lactate is neither oxidation nor reduction.

22.2 In lactate fermentation, glucose is converted to two molecules of lactic acid, the ionization of which decreases the pH of blood.

22.3 Nicotinamide, which is derived from the vitamin niacin, is required for the oxidation steps of glycolysis.

22.5 Three moles of glucose give six moles of lactate.

22.7 By the reactions of glycolysis and alcoholic fermentation, one mole of sucrose gives four moles of ethanol and two moles of CO_2.

22.9 In the following mechanism, the steps are numbered 1–5. The key feature of this mechanism is the transfer in Step 3 of a hydrogen with a pair of electrons (a hydride ion, $H:^-$) from NADH to the carbonyl group of acetaldehyde.

NADH   NAD$^+$

Arrows 1–2: Electrons within the ring flow from nitrogen.
Arrow 3: Transfer of a hydride ion from the CH_2 of the six-membered ring to the carbonyl carbon creates the new $C-H$ bond to the carbonyl carbon of acetaldehyde.
Arrow 4: The $C=O$ pi bond breaks as the new $C-H$ bond forms.
Arrow 5: An acidic group, $-BH$, on the surface of the enzyme transfers a proton to the newly-formed alkoxide ion to complete formation of the hydroxyl group of ethanol.

22.11 The hydrogen added to the carbonyl carbon comes from NADH.

22.13 The two molecules of CO_2 produced from glucose as a result of alcoholic fermentation are derived from carbons 3 and 4 of glucose.

22.15 Note that the carbon-carbon double bonds in oleic acid have a cis configuration.

22.17 The three coenzymes and the vitamins from which they are derived are FAD (riboflavin), NAD$^+$ (niacin), and coenzyme A (pantothenic acid).

22.19 The main function of the citric acid cycle is to produce reduced coenzymes (NADH and FADH$_2$). Their reoxidation during respiration is coupled with the production of energy in the form of ATP.

22.21 The citric acid cycle accepts acetate units in the form of acetyl coenzyme A and generates two molecules of carbon dioxide per entering acetyl group. It also accepts NAD$^+$ and FAD and generates NADH and FADH$_2$. Other than these chemical transformations, there is no other net change in any of the intermediates in the cycle.

22.23 **(a)** $C_6H_{12}O_6 + 6\,O_2 \longrightarrow 6\,CO_2 + 6\,H_2O$

$$RQ = \frac{6\,CO_2}{6\,O_2} = 1.00$$

(b) In the balanced equation for the oxidation of one mole of triolein, 80 moles of O_2 are used to produce 57 moles of CO_2. The RQ is $57/80 = 0.71$.

$$C_{57}H_{104}O_6 + 80\,O_2 \longrightarrow 57\,CO_2 + 52\,H_2O$$

(c) In the balanced equation for the oxidation of ethanol, 3 moles of O_2 are used and 2 moles of CO_2 are produced. The RQ $= 2/3 = 0.67$. Thus an individual's RQ would decrease if ethanol were to supply an appreciable portion of caloric needs.

$$C_2H_6O + 3\,O_2 \longrightarrow 2\,CO_2 + 3\,H_2O$$

22.25 Carbons 1 and 6 of glucose appear as methyl groups of acetyl-CoA. Carbon atoms 2, 4, 6, 8, 10, 12, 14, and 16 of palmitic acid appear as methyl groups of acetyl-CoA.

22.27 **(a)** Enzyme-catalyzed reactions give 100% of the desired product; laboratory reactions rarely approach this efficiency.

(b) Enzyme-catalyzed reactions are 100% regioselective. Although some laboratory reactions approach this efficiency, most do not.

(c) Enzyme-catalyzed reactions are 100% stereoselective. Although some laboratory reactions approach this efficiency, most do not.

22.29 Of the functional groups we have studied in this chapter, carboxylic acids and amines are affected by the acidity of their biological environment; their degree of protonation or deprotonation, as the case may be, depends on the pH of their environment.

Palmitic acid (C16)

Stearic acid (C18)

Oleic acid (C18)

Index

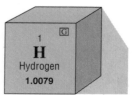

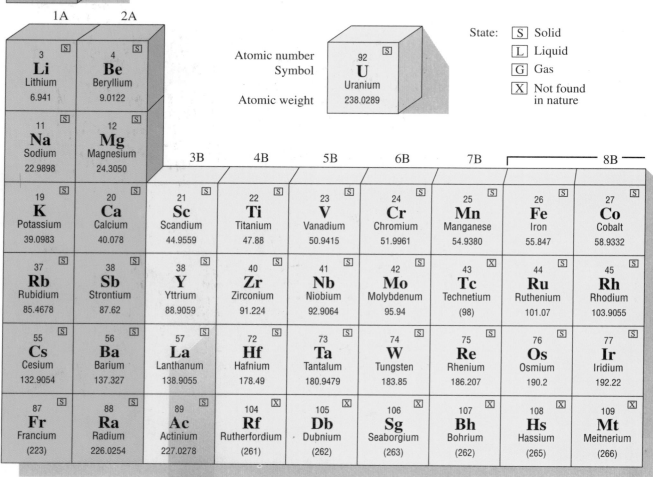

1 [G] **H** Hydrogen **1.0079**								

1A **2A**

State: [S] Solid [L] Liquid [G] Gas [X] Not found in nature

Atomic number
Symbol

92 [S] **U** Uranium 238.0289

Atomic weight

3 [S] **Li** Lithium 6.941	4 [S] **Be** Beryllium 9.0122
11 [S] **Na** Sodium 22.9898	12 [S] **Mg** Magnesium 24.3050

3B **4B** **5B** **6B** **7B** **8B**

19 [S] **K** Potassium 39.0983	20 [S] **Ca** Calcium 40.078	21 [S] **Sc** Scandium 44.9559	22 [S] **Ti** Titanium 47.88	23 [S] **V** Vanadium 50.9415	24 [S] **Cr** Chromium 51.9961	25 [S] **Mn** Manganese 54.9380	26 [S] **Fe** Iron 55.847	27 [S] **Co** Cobalt 58.9332
37 [S] **Rb** Rubidium 85.4678	38 [S] **Sb** Strontium 87.62	38 [S] **Y** Yttrium 88.9059	40 [S] **Zr** Zirconium 91.224	41 [S] **Nb** Niobium 92.9064	42 [S] **Mo** Molybdenum 95.94	43 [X] **Tc** Technetium (98)	44 [S] **Ru** Ruthenium 101.07	45 [S] **Rh** Rhodium 103.9055
55 [S] **Cs** Cesium 132.9054	56 [S] **Ba** Barium 137.327	57 [S] **La** Lanthanum 138.9055	72 [S] **Hf** Hafnium 178.49	73 [S] **Ta** Tantalum 180.9479	74 [S] **W** Tungsten 183.85	75 [S] **Re** Rhenium 186.207	76 [S] **Os** Osmium 190.2	77 [S] **Ir** Iridium 192.22
87 [S] **Fr** Francium (223)	88 [S] **Ra** Radium 226.0254	89 [S] **Ac** Actinium 227.0278	104 [X] **Rf** Rutherfordium (261)	105 [X] **Db** Dubnium (262)	106 [X] **Sg** Seaborgium (263)	107 [X] **Bh** Bohrium (262)	108 [X] **Hs** Hassium (265)	109 [X] **Mt** Meitnerium (266)

Lanthanides

58 [S] **Ce** Cerium 140.115	59 [S] **Pr** Praseodymium 140.9076	60 [S] **Nd** Neodymium 144.24	61 [X] **Pm** Promethium (145)	62 [X] **Sm** Samarium 150.36

Actinides

90 [S] **Th** Thorium 232.0381	91 [S] **Pa** Protactinium 231.0359	92 [S] **U** Uranium 238.0289	93 [X] **Np** Neptunium 237.0482	94 [S] **Pu** Plutonium (244)

* Elements 110–112 have not yet been named.